D0161861

Spectrometric Identification of Organic Compounds

Sixth Edition

Robert M. Silverstein
Francis X. Webster
State University of New York
College of Environmental Science & Forestry

John Wiley & Sons, Inc.
New York Chichester Weinheim Brisbane Singapore Toronto

Dedication

The Sixth Edition is dedicated to Dr. G. Clayton Bassler (deceased, December 27, 1996). "Clayt" was the coauthor of the First and Second Editions, and a longtime friend.

ACQUISITION EDITOR Nedah Rose
MARKETING MANAGER Karen Allman
SENIOR PRODUCTION EDITOR Elizabeth Swain
DESIGNER Ann Marie Renzi
ILLUSTRATION COORDINATOR Edward Starr
COVER DESIGN David Levy

This book was set in 10/12 Times Ten by Progressive Information Technologies and printed by Courier/Westford.

This book is printed on acid-free paper.♾

The paper in this book was manufactured by a mill whose forest management programs include sustained yield harvesting of its timberlands. Sustained yield harvesting principles ensure that the number of trees cut each year does not exceed the amount of new growth.

Library of Congress Cataloging in Publication Data:

Silverstein, Robert M. (Robert Milton), 1916–
Spectrometric identification of organic compounds.—6th ed. /
p. cm.
Includes bibliographical references and index.
ISBN 0-471-13457-0 (cloth : alk. paper)
1. Spectrum analysis. 2. Organic compounds—Spectra.
I. Webster, Francis X. II. Title.
QD272.S6S55 1997
547′.30858—dc21 97-21336
CIP

Printed in the United States of America

10 9 8 7 6 5 4 3 2

PREFACE

The authors welcome this opportunity to include new material, discard the old, and improve the presentation. Overall, the following major items are noteworthy:

- The continuing advances in NMR spectrometry are acknowledged by major revisions in three NMR chapters and by the addition of a fourth NMR chapter entitled "Spectrometry of Other Important Nuclei."
- Spectra have been upgraded throughout—the NMR spectra in particular; almost all were run at 300 or 500 MHz (75.5 and 126.0 MHz for ^{13}C), and the ^{1}H peaks have been expanded as insets.
- An overall List of Spectra has been added to the Contents. Detailed explanations have been added to the more complicated tables and charts throughout. The thorough Index provides accessibility; acronyms are included. New end-of-chapter problems have been added.
- As a consequence of the advances in NMR, UV spectrometry has been further marginalized *for our purposes,* and the UV chapter has been dropped—a difficult decision (nostalgia perhaps) because UV spectrometry is widely used for other purposes. Students should understand the relationship between absorption of visible-UV frequencies and molecular structure. But such general understanding is presented in most first-year organic texts. We cannot justify 26 pages of text and tables to describe a technique that is outmoded for structure elucidation and, in practice, virtually abandoned except for special situations.

Major changes in each chapter are summarized below.

Mass Spectrometry (Chapter 2)

Recent advances in instrumentation and ionization techniques are described briefly. Several sections have been rewritten and expanded for greater clarity. The useful Table of Formula Masses (four decimal places) has been shortened by eliminating entries that are unlikely in the present context. The Table is convenient for selecting tentative molecular formulas and fragments on the basis of unit-mass peaks. Given high-resolution peaks, specific molecular or fragment formulas can be selected.

Infrared Spectrometry (Chapter 3)

Modifications have been made in the full chart of characteristic absorptions. A simplified chart of common functional groups has also been included for rapid scanning. Students are advised to start with the simplified chart and to avoid the full chart until further leads are obtained from the mass and NMR spectra.

Proton Magnetic Resonance Spectrometry (Chapter 4)

This chapter has been extensively revised and expanded to clarify difficult concepts because the narrow discrete peaks of a high-resolution proton NMR are subject to interpretation in detail by inspection.

NMR, without question, has become the most useful tool available to the organic chemist. It is possible to interpret the 300 MHz and 500 MHz spectra (with expanded insets), presented throughout the book, in exquisite detail. But to do so requires an appreciation of the relationship between chemical structure and spectra—which means to understand how a spectrum is produced and what its promises and limitations are.

We realize that the organic chemist wants to get on with the task of identifying compounds without first mastering arcane areas of electronic engineering and quantum mechanics. But the alternative "black box" approach is not acceptable either. We believe that a pictorial, nonmathematical approach to spectra has been satisfactory for our purposes through five editions. We propose that elaboration of the same approach—together with coverage of current techniques—will be welcomed in the Sixth Edition. The undergraduate student will be engaged, and the graduate student challenged.

The following revisions are noted:

- Interpretation of NMR spectra depends on the concept of chemical-shift equivalence, an understanding of which depends on stereochemical concepts; these are reviewed with special emphasis on interchange through symmetry operations within the molecule, and through rapid structural changes.
- For pedagogical and practical reasons, this chapter treats only protons. ^{13}C NMR spectrometry is treated in Chapter 5, and other useful nuclei are treated in Chapter 7. Chapter 6 is devoted to 2-D NMR.
- Pulsed Fourier transform spectrometry follows the

brief discussion of the more intuitively obvious continuous-wave scanning NMR.

- Although "chirality" is mentioned in the stereochemistry discussion, it warrants a separate section.
- "Virtual coupling" is always a challenge for students—and for teachers and authors. We have given it appropriate emphasis.
- A separate section is devoted to the concept of magnetic equivalence.
- The utility of NOE difference spectrometry for stereochemical problems is demonstrated.
- It has recently been pointed out that the usual "stick" diagrams of a first-order pattern explain the pattern after the fact, but students are not taught to *analyze* the pattern. A brief addendum addresses this topic.
- Several tables for the prediction of chemical shifts for substituted CH_3, CH_2, and CH groups have been updated. Earlier tables for substituted CH groups in particular have not been satisfactory.
- A number of 300 MHz spectra are included in the Problems at the end of the chapter.

^{13}C NMR Spectrometry (Chapter 5)

Outdated sections have been condensed or deleted, including off-resonance decoupling and selective proton decoupling. A section on the very useful DEPT procedure has been added. Several sections have been revised, and Problems (at 75.5 MHz) have been added.

Correlation NMR Spectrometry (Chapter 6)

Chapter 6 has been almost completely rewritten. There is more emphasis on pulse sequences and on the use of inverse detection (e.g., HMQC and HMBC experiments). Some experiments from the Fifth Edition have been eliminated (e.g., *J*-Resolved), and others have been added. The chapter has been renamed "Correlation NMR Spectrometry" to better reflect the emphasis of the chapter. Because of this name change, the DEPT experiment has been moved to Chapter 5; the APT experiment has been eliminated. Gradient field NMR is presented as a recent development. Problems are assigned.

By its nature, correlation spectrometry poses a challenge even at the graduate level, as do the latter problems in Chapter 9 that depend heavily on correlated spectra.

Spectrometry of Other Important Nuclei (Chapter 7)

This is a completely new chapter. Over the past decade or so, NMR spectrometry of several nuclei besides ^{1}H and ^{13}C have played an increasing role in the identification and analysis of organic compounds. Chapter 7 introduces the student to the most important of these nuclei with spectra, discussion, and Problems. Nuclei discussed are ^{15}N, ^{19}F, ^{29}Si, and ^{31}P.

Solved Problems (Chapter 8)

Chapter 8 consists of an Introduction to Solved Problems (Chapter 8A), followed by four solved Problems (Chapter 8B).

Our suggested approaches to problem solving have been expanded and should be helpful to students. We have refrained from being overly prescriptive. Students are urged to develop their own approaches, but suggestions are offered and caveats posted.

The four Problems are arranged in increasing order of difficulty. They involve several common functional groups, a highly substituted aromatic ring, a substituted heteroatom ring with a chiral center, and a diunsaturated terpenol. Extensive use is made of 2-D spectra.

Assigned Problems (Chapter 9)

Following a brief introduction (Chapter 9A), Chapter 9B represents the culmination: spectrometric identification of organic compounds by the students. These 55 Problems are arranged in increasing order of difficulty; the early Problems are designed to build confidence, and the later Problems challenge at the graduate level.

The Problems are represented by the following spectra (^{1}H NMR spectra at 300 or 500 MHz, ^{13}C NMR at 75.5 MHz or 126.0 MHz):

	No. of Spectra
MS (EI)	55
MS (CI)	7
IR	55
^{1}H (with expanded insets)	55
^{13}C/DEPT (or APT)	50
COSY	26
HETCOR (or HMQC)	19
HMBC	5
INADEQUATE	2
Difference NOE	3

Answer Manual

An Answer Manual, available on letterhead request from the publisher, covers the Problems of Chapter 9 and those at the end of the chapters.

Acknowledgments

Spectra are the backbone of this book, and it is a pleasure to acknowledge the high-quality ^{1}H, ^{13}C, ^{15}N, ^{29}Si, and ^{31}P spectra supplied by D.J. Kiemle (SUNY-ESF, Syracuse). The ^{19}F spectra in Chapter 7 were contributed by Ron Carroll (Bristol-Myers Squibb). The FTIR spectra (Chapter 9) and the ^{1}H and ^{13}C spectra at the end of Chapter 4 were contributed by C.J. Pouchert (Aldrich Chemical Co.). Three chemical ionization spectra were contributed by J.T. Tumlinson and A.T. Proveaux (USDA, Gainesville, FL). Rong Tang (SUNY-ESF) helped us greatly with the preparation of several "synthezied" spectra in the Answer Manual.

Permission to use published material was granted by Finnigan MAT, American Society of Mass Spectrometry, John Wiley and Sons, Inc., Journal of Chemical Education, and Organic Magnetic Resonance. Processing software was furnished by Herbert Thiele (Bruker Instrument Corp.).

A generous gift of 3-propyl-1,2-dithiolane and of ipsenol was received from D. Wackerchuck (Phero Tech, Inc.). J.H. Borden and L. Chong (Simon Fraser University) contributed a sample of sulcatol.

Over the years, informal discussions with Dr. R.T. LaLonde (SUNY-ESF, Syracuse) have helped shape our presentation of several topics. The typed manuscript was made possible through the patience of Laurie DuFore and Ragan Feidt with unending revisions.

The staff at John Wiley and Sons has been highly cooperative in transforming a complex manuscript with all seams showing into a handsome Sixth Edition.

The following reviewers offered encouragement and many useful suggestions. We thank them for the considerable time expended:

Steven Bertman
Western Michigan University

Richard Bowen
University of Bradford

Albert Burgstahler
Robert Carlson
University of Kansas

Donald Dittmer
Syracuse University

Tammy Dwyer
University of San Diego

Fyaz M.D. Ismail
University of Hertfordshire

W.B. Jennings
Dan McCarthy
University of College Cork

James Leahy
University of California, Berkeley

James Louey
Sacred Heart University

John Marx
Texas Tech University

James Nowick
University of California, Irvine

Francis J. Schmitz
University of Oklahoma

Our wives (Olive and Kathryn) offered constant patience and support. There is no adequate way to express our appreciation.

Robert M. Silverstein
Francis X. Webster

PREFACE TO FIRST EDITION

During the past several years, we have been engaged in isolating small amounts of organic compounds from complex mixtures and identifying these compounds spectrometrically.

At the suggestion of Dr. A.J. Castro of San Jose State College, we developed a one unit course entitled "Spectrometric Identification of Organic Compounds," and presented it to a class of graduate students and industrial chemists during the 1962 spring semester. This book has evolved largely from the material gathered for the course and bears the same title as the course.*

We should first like to acknowledge the financial support we received from two sources: The Perkin-Elmer Corporation and Stanford Research Institute.

A large debt of gratitude is owed to our colleagues at Stanford Research Institute. We have taken advantage of the generosity of too many of them to list them individually, but we should like to thank Dr. S.A. Fuqua, in particular, for many helpful discussions of NMR spectrometry. We wish to acknowledge also the cooperation at the management level, of Dr. C.M. Himel, chairman of the Organic Research Department, and Dr. D.M. Coulson, chairman of the Analytical Research Department.

Varian Associates contributed the time and talents of its NMR Applications Laboratory. We are indebted to Mr. N.S. Bhacca, Mr. L.F. Johnson, and Dr. J.N. Shoolery for the NMR spectra and for their generous help with points of interpretation.

The invitation to teach at San Jose State College was extended by Dr. Bert M. Morris, head of the Department of Chemistry, who kindly arranged the administrative details.

The bulk of the manuscript was read by Dr. R.H. Eastman of the Stanford University whose comments were most helpful and are deeply appreciated.

Finally, we want to thank our wives. As a test of a wife's patience, there are few things to compare with an author in the throes of composition. Our wives not only endured, they also encouraged, assisted, and inspired.

R.M. Silverstein
G.C. Bassler

Menlo Park, California
April 1963

* A brief description of the methodology had been published: R.M. Silverstein and G.C. Bassler, *J. Chem. Educ.* **39,** 546 (1962).

CONTENTS

LIST OF SPECTRA

CHAPTER 1

Introduction

This book was originally written to teach the organic chemist how to identify organic compounds from the synergistic information afforded by the combination of mass (MS), infrared (IR), nuclear magnetic resonance (NMR), and ultraviolet (UV) spectra. Essentially, the molecule is perturbed by these energy probes and the molecule's responses are recorded as spectra.

In the present edition, the goal remains unchanged, but the format has evolved to respond to the remarkable evolution of instrumentation. NMR, without question, has become the most sophisticated tool available to the organic chemist, and it now requires four chapters to do it justice. In comparison, ultraviolet spectrometry has become relatively less useful for our purpose, and we have discarded it despite nostalgic ties.

We aim at a rather modest level of expertise in each area of spectrometry, recognizing that the organic chemist wants to get on with the task of identifying the compound without first mastering arcane areas of electronic engineering and quantum mechanics. But the alternative black-box approach is not acceptable either. We avoid these extremes with a pictorial, nonmathematical, vector-diagram approach to theory and instrumentation. Since NMR spectra can be interpreted in exquisite detail with some mastery of theory, we present theory in corresponding detail—but still descriptive. Since an understanding of stereochemistry is essential to the concept of "chemical-shift equivalence," we briefly review the relevant material.

Even this modest level of expertise will permit solution of a gratifying number of identification problems with *no history and no other chemical or physical data.* Of course, in practice other information is usually available: the sample source, details of isolation, a synthesis sequence, or information on analogous material. Often, complex molecules can be identified because partial structures are known, and specific questions can be formulated; the process is more confirmation than identification. In practice, however, difficulties arise in physical handling of minute amounts of compound: trapping, elution from adsorbents, solvent removal, prevention of contamination, and decomposition of unstable compounds. Water, air, stopcock greases, solvent impurities, and plasticizers have frustrated many investigations.

For pedagogical reasons, we deal only with pure organic compounds. "Pure" in this context is a relative term, and all we can say is the purer, the better. A good criterion of purity for a sufficiently volatile compound (no nonvolatile impurities present) is gas chromatographic homogeneity on both polar and nonpolar substrates in capillary columns. Various forms of liquid-phase chromatography (adsorption and liquid–liquid columns, thin layer) are applicable to less volatile compounds. The spectra presented in this book were obtained on purified samples.

In many cases, identification can be made on a fraction of a milligram, or even on several micrograms, of sample. Identification on the milligram scale is routine. Of course, not all molecules yield so easily. Chemical manipulations may be necessary but the information obtained from the spectra will permit intelligent selection of chemical treatment, and the energy probe methodology can be applied to the resulting products.

When we proposed in the first edition of this book that the synergistic combination of spectra sufficed to identify organic compounds, we did so in 177 pages after exploring the possibilities in a series of lectures at San Jose State University, CA, in 1962. The methodology thus elaborated was being rapidly adopted by practicing organic chemists, and we predicted that "in one form or another, such material would soon become part of the training of every organic chemist." Now every first-year organic textbook provides an introduction to spectrometry.

References and problems are provided at the end of each chapter.* Chapter 8 presents several solved problems, and Chapter 9 has unsolved problems.

The charts and tables throughout the text are extensive and are designed for rapid, convenient access. They—together with the numerous spectra, including those of the problem sets—should furnish useful reference material.

* Specific references are provided as footnotes. General periodical reviews in spectrometry are available in *Analytical Chemistry* and in *Annual Reports of the Royal Society.*

CHAPTER 2

Mass Spectrometry

2.1 *Introduction*

In the commonly used electron-impact (EI) mode, a mass spectrometer bombards molecules in the vapor phase with a high-energy electron beam and records the result of electron impact as a spectrum of positive ions separated on the basis of mass/charge (*m/z*); most of these ions are singly charged.* The mass spectrum of benzamide $\left(C_6H_5-\overset{\overset{\displaystyle O}{\|}}{C}-NH_2\right)$ is presented as a computer-plot bar graph of abundance (vertical peak intensity) versus *m/z* (Fig. 2.1). The positive ion peak at *m/z* 121 represents the intact molecule (M) less one electron removed by the impacting beam and is designated the molecular ion, $M^{\cdot +}$. The molecular ion in turn produces a series of fragment ions as shown for benzamide:

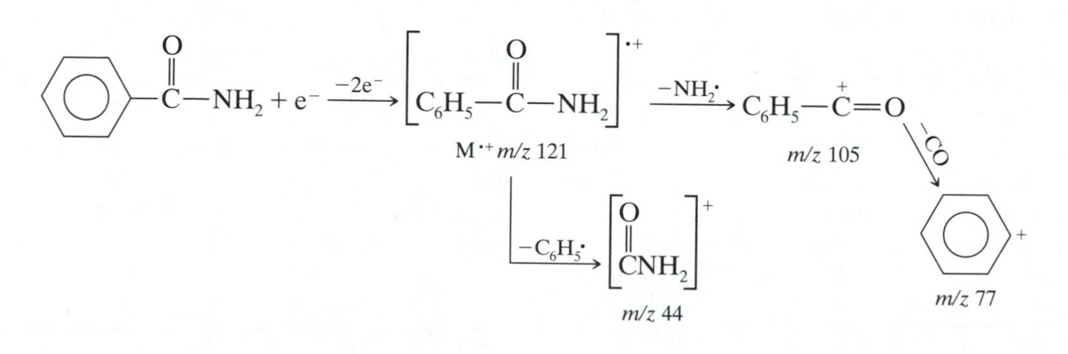

Various methods of producing molecular ions (including the EI method) are discussed in Section 2.5. Details of fragmentation patterns for representative organic compounds are described in Sections 2.7 and 2.10.

In this chapter we describe mass spectrometry (MS) in sufficient detail to appreciate its application to organic structure determination. For more details, mass spectrometry texts and spectral compilations are listed at the end of this chapter.

2.2 *Instrumentation*

The minimum instrumental requirement for the organic chemist is the ability to record the molecular weight of the compound under examination to the nearest whole number. Thus, the recording should show a peak at, say, mass 400, which is distinguishable from a peak at mass 399 or at mass 401. In order to select possible molecular formulas by measuring isotope peak intensities (see Section 2.4), adjacent peaks must be cleanly separated. Arbitrarily, the valley between two such peaks should not be more than 10% of the height of the larger peak. This degree of resolution is termed "unit" resolution and can be obtained up to a mass of approximately 2000 Da on readily available "unit-resolution" instruments.

* The unit of mass is the dalton (Da), defined as 1/12 of the mass of an atom of the isotope ^{12}C, which is arbitrarily 12.0000 . . . mass units.

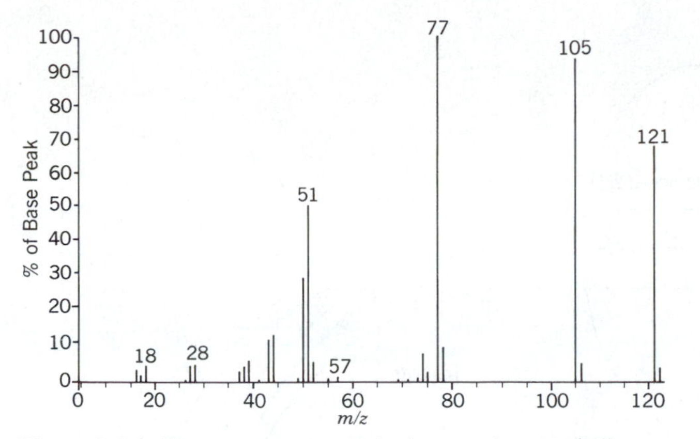

Figure 2.1. Computer-generated, electron-impact (EI) mass spectrum of benzamide ($C_6H_5—\overset{O}{\overset{\|}{C}}—NH_2$) in bar graph form. Peak abundances as percentages of base peak (100%) are reported versus mass to charge (m/z). Peaks of <0.5% of the base peak were omitted. A Hewlett-Packard HP 5995 96A GC-MS instrument was used.

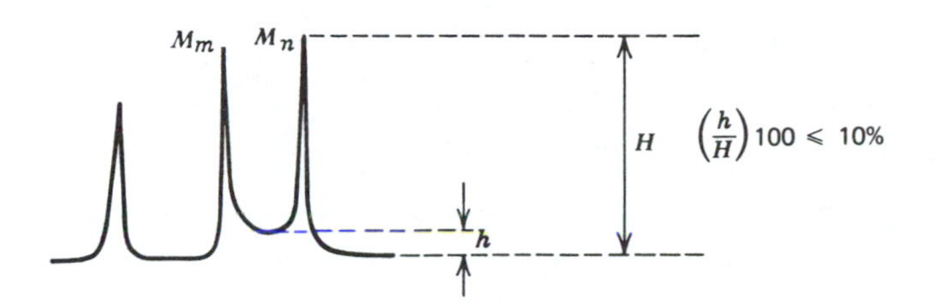

To determine the resolution of an instrument, consider two adjacent peaks of approximately equal intensity. These peaks should be chosen so that the height of the valley between the peaks is less than 10% of the intensity of the peaks. The resolution (R) is

$$R = \frac{M_n}{M_n - M_m}$$

where M_n is the higher mass number of the two adjacent peaks, and M_m is the lower mass number.

There are two important categories of magnetic-deflection mass spectrometers: low (unit) resolution and high resolution. Low-resolution instruments can be defined arbitrarily as the instruments that separate unit masses up to m/z 2000[$R = 2000/(2000 - 1999) = 2000$]. A high-resolution instrument with $R = 20{,}000$ can distinguish between $C_{16}H_{26}O_2$ and $C_{15}H_{24}NO_2$:

$$R = \frac{250.1933}{250.1933 - 250.1807} \simeq 20{,}000$$

This important class of mass spectrometers can measure the mass of an ion with sufficient accuracy to determine its atomic composition.

Mass spectrometers for structure elucidation can be classified according to the method of separating the charged particles:

A. Magnetic Field Deflection
   1. Magnetic Field Only (Unit Resolution)
   2. Double Focusing (Electrostatic Field and Magnetic Field, High Resolution)

B. Quadrupole Mass Spectrometry
   1. Quadrupole Mass Filter
   2. Quadrupole Ion Storage (Ion Trap)

C. Time of Flight

D. FT-ICR (Ion Cyclotron Resonance)

E. MS/MS (Tandem Mass Spectrometry)

2.2.1 Magnetic Field Only (A.1)

The gas stream from the inlet system (Fig. 2.2) enters the ionization chamber (operated at a pressure of about $10^{-6} - 10^{-5}$ torr) in which it is bombarded at right angles by an electron beam emitted from a hot filament. Positive ions produced by interaction with the electron beam are forced through the first accelerating slit by a weak electrostatic field. A strong electrostatic field then accelerates the ions to their final velocities. To obtain a spectrum, the applied magnetic field is increased, bringing successively heavier ions into the collector slit. A scan from mass (strictly m/z; see above) 12 – 500 may be performed in seconds.

The analyzer tube is an evacuated ($10^{-7} - 10^{-8}$ torr), curved, metal tube through which the ion beam passes from the ion source to the collector. The magnetic field is imposed perpendicular to the plane of the diagram (Fig. 2.2). The main requirement is a uniform magnetic field that can be smoothly varied in strength.

A typical ion collector consists of collimating slits that direct only one set of ions at a time into the collector, where they are detected and amplified by an electron multiplier.

Mass spectrometers provide computer output as bar graphs (Fig. 2.1) and as tabular data. Minor peaks, many of them resulting from possible impurities, occur at almost every mass unit. The minor peaks are frequently deleted in the bar graph (those <0.5% have been omitted in Fig. 2.1). A search of the computer's library and a fit to these peaks may either identify the compound or suggest "near structures." Peak heights are proportional to the number of ions of each mass.

As mentioned previously, most ions are singly charged, but double ionization does occur and this is indicated by peaks at half-mass units. These represent odd-numbered masses that carry a double charge. For example, a doubly charged ion of mass 89 gives rise to a peak at 89/2 or m/z 44.5.

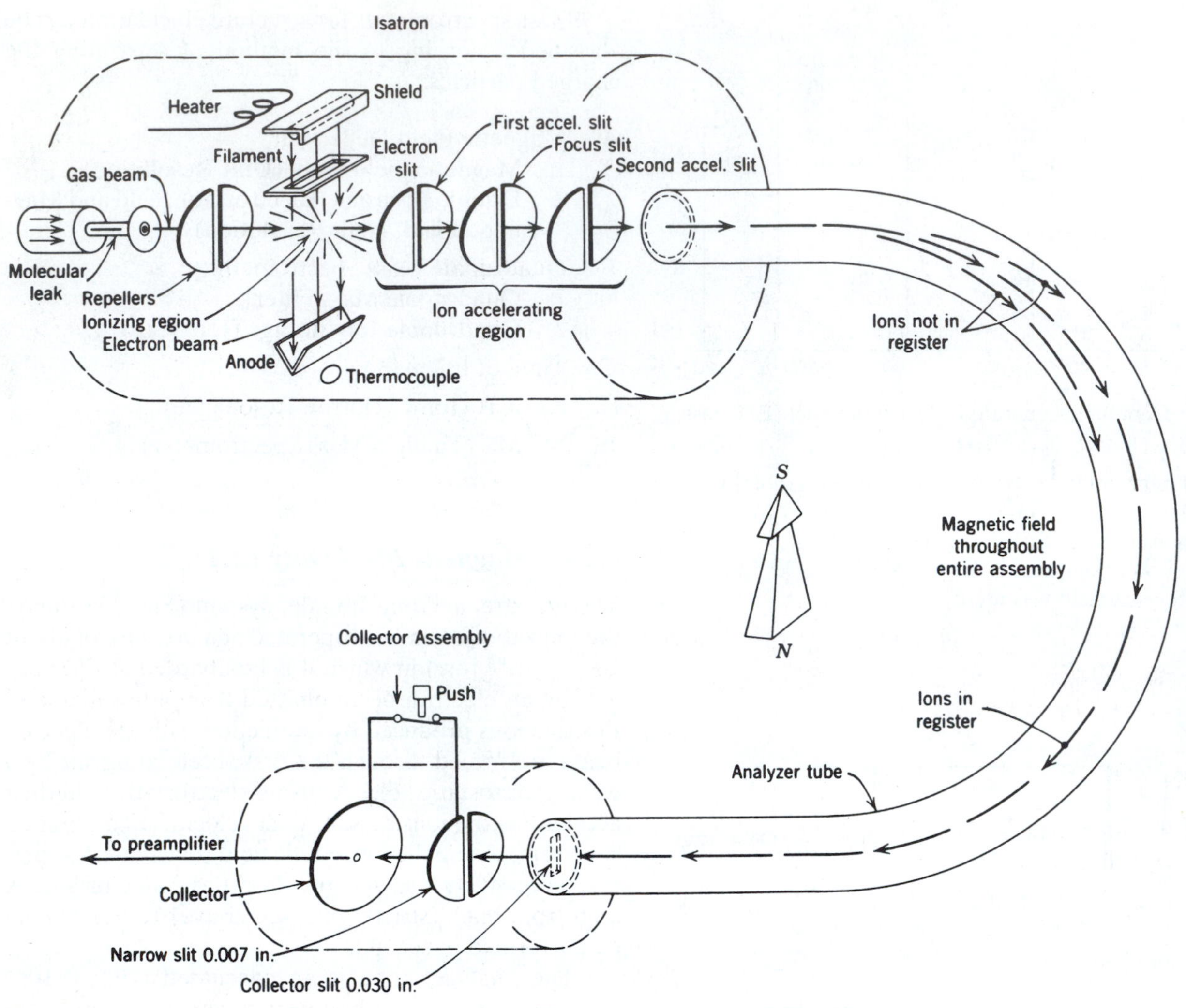

FIGURE 2.2. Schematic diagram of a single-focusing, 180° sector mass analyzer. The magnetic field is perpendicular to the page. The radius of curvature varies from one instrument to another.

2.2.2 Double Focusing (Electrostatic and Magnetic Fields) (A.2)

The introduction of an electrostatic field after (or before) the magnetic field permits high resolution so that the mass of a particle can be obtained to four decimal places. Figure 2.3 shows a double-focusing instrument. Ions generated in the source are accelerated toward the analyzer. The magnetic field provides directional focusing. The path of the positive ion is again curved by the electric field applied perpendicular to the flight path of the ions. This double focusing provides resolution as high as 60,000.

2.2.3 Quadrupole Mass Filter (B.1)

This mass filter uses four voltage-carrying rods (the "quadrupole") (Fig. 2.4*a*). Ions entering from one end travel with constant velocity in the direction parallel to the poles (z direction), but acquire complex oscillations in the x and y directions by application of both a direct current (dc) voltage (V_{dc}) and a radiofrequency (rf) voltage (V_{rf}) to the poles. There is a "stable oscillation" that allows a particular ion to pass from one end of the quad-

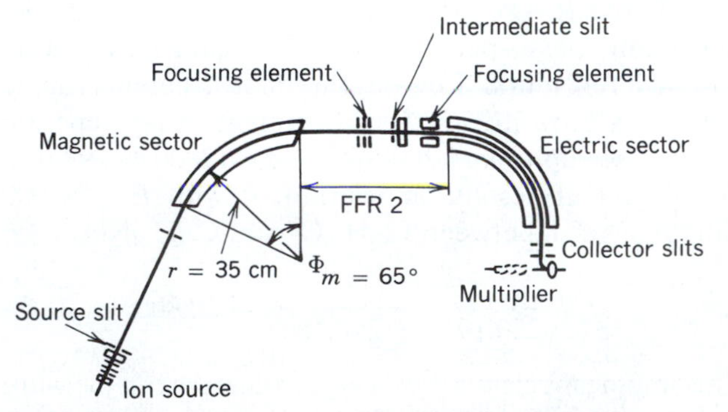

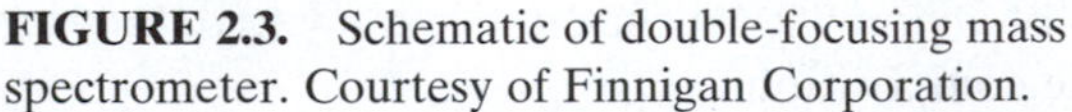

FIGURE 2.3. Schematic of double-focusing mass spectrometer. Courtesy of Finnigan Corporation.

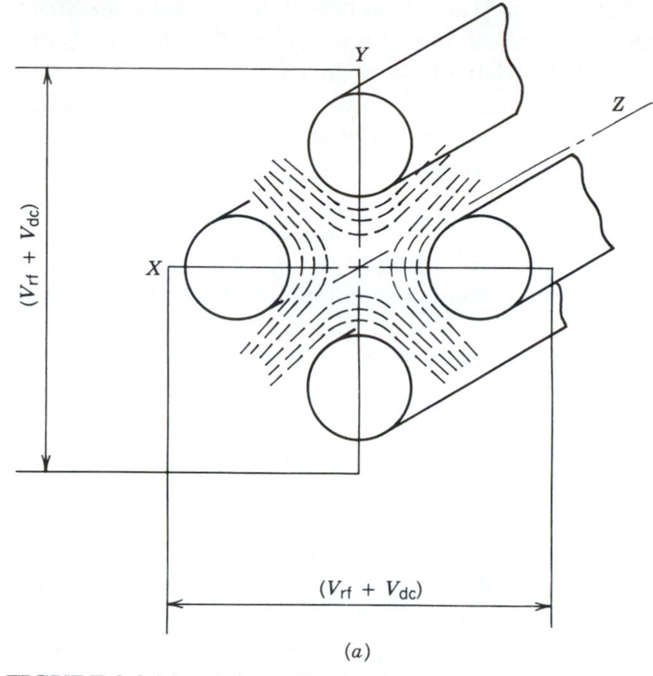

FIGURE 2.4 (*a*). Schematic of quadrupole mass filter. Courtesy of Finnigan Corporation.

rupole to the other without striking the poles; this oscillation is dependent on the *m/z* ratio of an ion. Therefore, ions of only a single *m/z* value will traverse the entire length of the filter at a given set of conditions. All other ions will have unstable oscillations and will strike the poles and be lost. Mass scanning is carried out by varying each of the rf and dc frequencies while keeping their ratios constant.

2.2.4 Quadrupole Ion Storage (Ion Trap) (B.2)

Essentially, the ion storage trap is a spherical configuration of the linear quadrupole mass filter. The operations, however, differ in that the linear filter passes the sorted ions directly through to the detector, whereas the ion trap retains the unsorted ions temporarily within the trap. They are then released to the detector sequentially by scanning the electric field. These instruments are compact (benchtop), relatively inexpensive, convenient to use, and very sensitive. They also provide an inexpensive method to carry out GC/MS/MS experiments (Section 2.2.7) (GC is gas chromatography).

In general, the quadrupole instruments do not achieve the mass range and the high resolution of sector instruments. However, the mass range and resolution are adequate for unit-resolution mass spectrometry, and the rapid scan and sensitivity make them especially suitable for use with capillary gas chromatography (Fig. 2.4*b*).

2.2.5 Time of Flight (C)

In the time-of-flight (TOF) mass spectrometers, all singly charged particles subjected to a potential difference *V* attain the same translational energy in electron volts (eV). Thus lighter particles have the shorter TOF over a given distance. The accelerated particles are passed into a field-free region where they are separated in time by their *m/z* values and collected. Since arrival times between successive ions can be less than 10^{-7} s, fast electronics are necessary for adequate resolution. Time-of-flight devices are used with sophisticated ionizing meth-

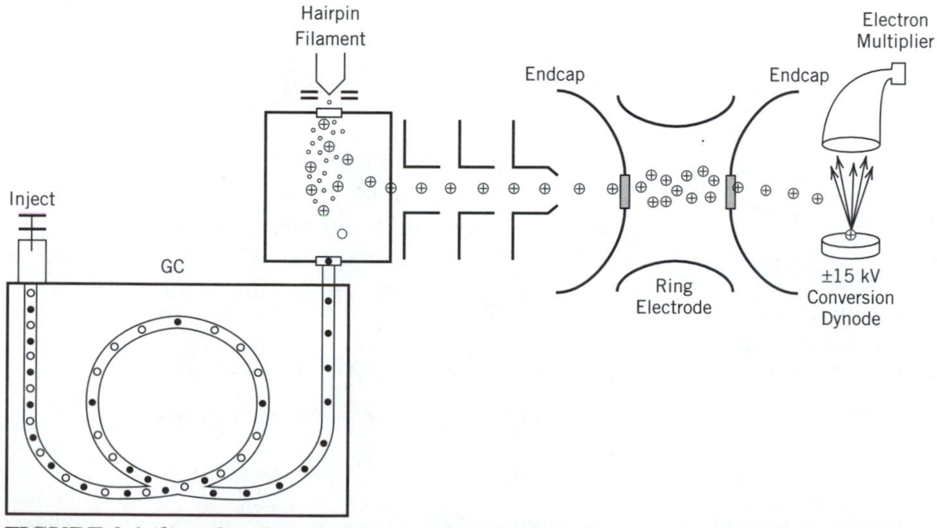

FIGURE 2.4 (*b*). Quadrupole ion storage trap with attached gas chromatograph. The ionizing unit is external to the ion trap. With permission of the American Society of Mass Spectrometry. From "What is Mass Spectrometry?"

ods (FAB, laser desorption, and plasma desorption). Resolution is modest; sensitivity is high (Section 2.5.1.3).

2.2.6 FT-ICR (Fourier Transform-Ion Cyclotron Resonance) (D) (Also termed FT-MS)

Ions generated by an electron beam from a heated filament are passed into a cubic cell where they are held by an electric trapping potential and a constant magnetic field. Each ion assumes a cycloidal orbit at its own characteristic frequency, which depends on m/z; the cell is maintained under high vacuum. Originally, these frequencies were scanned by varying the electric field until each cycloidal frequency was, in turn, in resonance with an applied constant radiofrequency. At resonance, the motion of the ions of the same frequency is coherent and a signal can be detected.

The newer instruments (Figure 2.4*c*) utilize a radiofrequency pulse in place of the scan. The pulse brings all of the cycloidal frequencies into resonance simultaneously to yield a signal as an interferogram (a time-domain spectrum). This is converted by Fourier Transform to a frequency-domain spectrum, which then yields the conventional m/z spectrum. Pulsed Fourier transform spectrometry applied to nuclear magnetic resonance spectrometry is explained in Chapters 4 and 5.

Since the resonance can be measured with high accuracy, precise m/z values can be obtained. These values can yield unambiguous molecular formulas—a most desirable goal; unfortunately, the instruments are still very expensive.

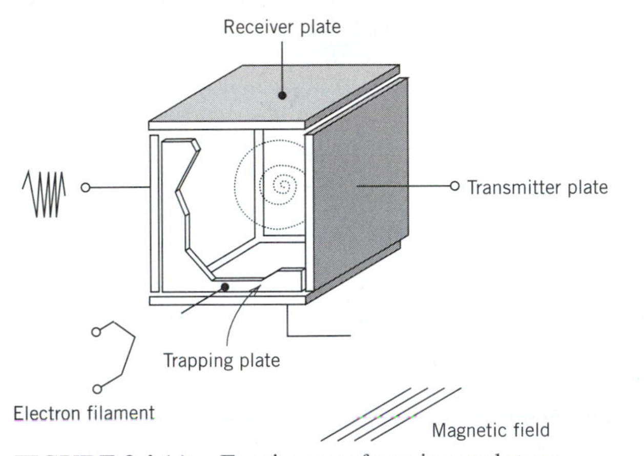

FIGURE 2.4 (*c*). Fourier transform-ion cyclotron resonance (FT-ICR). With permission from the American Society of Mass Spectrometry. From an ASMS pamphlet, "What is Mass Spectrometry?"

2.2.7 MS/MS (Tandem Mass Spectrometry) (E)

The tandem mass spectrometer consists of a mass spectrometer followed by a field-free collision chamber, followed by a second mass spectrometer. Specific ions ("parent ions") are separated in the first mass spectrometer and passed, one at a time, into the collision chamber where "daughter ions" are formed by collision with an introduced gas (helium); these ions are passed into the second mass spectrometer, where a daughter-ion spectrum is produced. Thus, we have a mass spectrum of each selected ion of the first spectrum—hence, MS/MS. Hybrid instruments are available with different types of spectrometers for the first and third stages.

Typically these instruments are used for large molecules and especially for the resolution of mixtures. In a recent development, the ion trap is used to store a selected ion from electron impact (EI) or chemical ionization (CI) by ejection of all of the other ions. This selected parent ion can then undergo collision in the ion trap with an introduced gas (helium) and the daughter ions are ejected to the detector to furnish a daughter-ion spectrum. The collisions with the introduced gas are induced by subjecting the parent ions to a high-energy waveform. Thus MS/MS spectrometry is achieved within a single ion trap, rather than in three separate compartments, at an appreciable saving in cost.

2.3 The Mass Spectrum

Mass spectra (EI) are routinely obtained at an electron beam energy of 70 eV. The simplest event that occurs is the removal of a single electron from the molecule in the gas phase by an electron of the electron beam to form the molecular ion, which is a radical cation ($M^{\cdot+}$). For example, methanol forms a molecular ion in which the single dot represents the remaining odd electron:

$$CH_3OH + e^- \longrightarrow \underset{m/z\ 32}{CH_3OH^{\cdot+}} + 2e^-$$

When the charge can be localized on one particular atom, the charge is shown on that atom:

$$CH_3\ddot{\overset{\cdot+}{O}}H$$

Many of these molecular ions disintegrate in $10^{-10} - 10^{-3}$ s to give, in the simplest case, a positively charged fragment and a radical. A number of fragment ions are thus formed, and each of these can cleave to yield smaller fragments. Again, illustrating with methanol

$$CH_3OH^{\cdot+} \longrightarrow CH_2OH^+ \ (m/z\ 31) + H^{\cdot}$$

$$CH_3OH^{\cdot+} \longrightarrow CH_3^+ \ (m/z\ 15) + \cdot OH$$

$$CH_2OH^+ \longrightarrow CHO^+ \ (m/z\ 29) + H_2$$

If some of the molecular ions remain intact long enough to reach the detector, we see a molecular ion peak. It is important to recognize the molecular ion peak because this gives the molecular weight of the compound. With unit resolution, this weight is the molecular weight to the nearest whole number.

A mass spectrum is a presentation of the masses of the positively charged fragments (including the molecular ion) versus their relative concentrations. The most intense peak in the spectrum, called the base peak, is assigned a value of 100%, and the intensities (height × sensitivity factor) of the other peaks, including the molecular ion peak, are reported as percentages of the base peak. Of course, the molecular ion peak may sometimes be the base peak. In Figure 2.1, the molecular ion peak is *m/z* 121, and the base peak is *m/z* 77.

A tabular or graphic presentation of a spectrum may be used. A graph has the advantage of presenting patterns that, with experience, can be quickly recognized. However, a graph must be drawn so that there is no difficulty in distinguishing mass units. Mistaking a peak at, say, *m/z* 79 for *m/z* 80 can result in total confusion. The molecular ion peak is usually the peak of highest mass number except for the isotope peaks.

2.4 Determination of a Molecular Formula

2.4.1 Unit-Mass Molecular Ion and Isotope Peaks

So far, we have discussed the mass spectrum in terms of unit resolutions: The unit mass of the molecular ion of C_7H_7NO (Fig. 2.1) is *m/z* 121—that is, the sum of the unit masses of the most abundant isotopes:

$$7 \times {}^{12}C = 84$$

$$7 \times {}^{1}H = 7$$

$$1 \times {}^{14}N = 14$$

$$1 \times {}^{16}O = 16$$

In addition, molecular species exist that contain the less abundant isotopes, and these give use to the "isotope peaks" at M + 1, M + 2, etc. In Figure 2.1, the M + 1 peak is approximately 8% of the intensity of the molecular ion peak, which for this purpose, is assigned an intensity of 100%. Contributing to the M + 1 peak are the isotopes, ^{13}C, ^{2}H, ^{15}N, and ^{17}O. Table 2.1 gives the abundances of these isotopes relative to those of the most abundant isotopes. The only contributor to the M + 2 peak of C_7H_7NO is ^{18}O, whose relative abundance is very low; thus the M + 2 peak is undetected. If only C, H, N, O, F, P, and I are present, the approximate expected percentage (M + 1) and percentage (M + 2) intensities can be calculated by use of the following formulas:

$$\%(M + 1) \approx 1.1 \times \text{number of C atoms} + 0.36 \times \text{number of N atoms}$$

$$\%(M + 2) \approx \frac{(1.1 \times \text{number of C atoms})^2}{200} + 0.20 \times \text{number of O atoms}$$

If these isotope peaks are intense enough to be measured accurately, the above calculations may be useful in determining the molecular formula.*

* There are limitations beyond the difficulty of measuring small peaks: The $^{13}C/^{12}C$ ratio differs with the source of the compound—synthetic compared with a natural source. A natural product from different organisms or regions may show differences. Furthermore, isotope peaks may be more intense than the calculated value because of ion–molecule interactions that vary with the sample concentration or with the class of compound involved. For example:

$$R{-}\overset{O}{\overset{\|}{C}}{-}\overset{+}{\ddot{O}}{-}CH_3 \; (M^{\cdot+}) + R\overset{O}{\overset{\|}{C}}{-}\ddot{O}{-}\underset{H}{CH_2} \longrightarrow R{-}\overset{O}{\overset{\|}{C}}{-}\ddot{O}{-}\dot{C}H_2 + R\overset{O}{\overset{\|}{C}}{-}\underset{H}{\overset{+}{\ddot{O}}}{-}CH_3 \; [M+1]^+$$

This represents transfer of a hydrogen atom from the excess of the compound to the molecular ion (see Section 2.10.7.1 and Figure 2.14).

Table 2.1 Relative Isotope Abundances of Common Elements

Elements	Isotope	Relative Abundance	Isotope	Relative Abundance	Isotope	Relative Abundance
Carbon	^{12}C	100	^{13}C	1.11		
Hydrogen	^{1}H	100	^{2}H	0.016		
Nitrogen	^{14}N	100	^{15}N	0.38		
Oxygen	^{16}O	100	^{17}O	0.04	^{18}O	0.20
Fluorine	^{19}F	100				
Silicon	^{28}Si	100	^{29}Si	5.10	^{30}Si	3.35
Phosphorus	^{31}P	100				
Sulfur	^{32}S	100	^{33}S	0.78	^{34}S	4.40
Chlorine	^{35}Cl	100			^{37}Cl	32.5
Bromine	^{79}Br	100			^{81}Br	98.0
Iodine	^{127}I	100				

If sulfur or silicon, is present, the M + 2 will be more intense. In the case of a single sulfur atom, ^{34}S contributes approximately 4.40% to the M + 2 peak; for a single silicon in the molecule, ^{30}Si contributes about 3.35% to the M + 2 peak (see Section 2.10.15). The effect of several bromine and chlorine atoms is described in Section 2.10.16. Note the appearance of additional isotope peaks in the case of multiple bromine and chlorine atoms. Obviously the mass spectrum should be routinely scanned for the relative intensities of the M + 2, M + 4, and higher isotope peaks, and the relative intensities should be carefully measured. Note that F and I are monoisotopic.

For most of the Problems in this text, the unit-resolution molecular ion, used in conjunction with IR and NMR, will suffice for determining the molecular formula by browsing in Appendix A. For several more difficult Problems, the high-resolution formula masses—for use with Appendix A (see Section 2.4.2)—have been supplied.

Table 2.1 lists the principal stable isotopes of the common elements and their relative abundance calculated on the basis of 100 molecules containing the most common isotope. Note that this presentation differs from many isotope abundance tables, in which the sum of all the isotopes of an element adds up to 100%.

Table 2.2 Exact Masses of Isotopes

Element	Atomic Weight	Nuclide	Mass
Hydrogen	1.00794	^{1}H	1.00783
		$D(^{2}H)$	2.01410
Carbon	12.01115	^{12}C	12.00000 (std)
		^{13}C	13.00336
Nitrogen	14.0067	^{14}N	14.0031
		^{15}N	15.0001
Oxygen	15.9994	^{16}O	15.9949
		^{17}O	16.9991
		^{18}O	17.9992
Fluorine	18.9984	^{19}F	18.9984
Silicon	28.0855	^{28}Si	27.9769
		^{29}Si	28.9765
		^{30}Si	29.9738
Phosphorus	30.9738	^{31}P	30.9738
Sulfur	32.066	^{32}S	31.9721
		^{33}S	32.9715
		^{34}S	33.9679
Chlorine	35.4527	^{35}Cl	34.9689
		^{37}Cl	36.9659
Bromine	79.9094	^{79}Br	78.9183
		^{81}Br	80.9163
Iodine	126.9045	^{127}I	126.9045

2.4.2 High-Resolution Molecular Ion

A unique molecular formula (or fragment formula) can often be derived from a sufficiently accurate mass measurement alone (high-resolution mass spectrometry). This is possible because the nuclide masses are not integers (see Table 2.2). For example, we can distinguish at a unit mass of 28 among CO, N_2, CH_2N, and C_2H_4.

CO	N_2	CH_2N	C_2H_4
^{12}C 12.0000	$^{14}N_2$ 28.0062	^{12}C 12.0000	$^{12}C_2$ 24.0000
^{16}O 15.9949		$^{1}H_2$ 2.0156	$^{1}H_4$ 4.0312
27.9949		^{14}N 14.0031	28.0312
		28.0187	

Thus, the mass observed for the molecular ion of CO, for example, is the sum of the exact formula masses of the most abundant isotope of carbon and of oxygen. This differs from a molecular weight of CO based on atomic weights that are the average of weights of all natural isotopes of an element (e.g., C = 12.01, O = 15.999).

Table 2.2 gives the masses to four or five decimal places for the common nuclides; it also gives the familiar atomic weights (average weights for the elements).

Appendix A lists molecular and fragment formulas in order of the unit masses. Under each unit mass, the formulas are listed in the familiar *Chemical Abstract* system. The formula mass (FM) to four decimal places is given for each formula. Appendix A is designed for browsing, on the assumption that the student has a unit molecular mass from a unit-resolution mass spectrometer and clues from other spectra. Note that the Table includes only C, H, N, and O. In Compound 8.3 (Chapter 8), for example, the molecular ion peak is 148. The relative intensity of the M + 2 peak (9.7%) suggests the presence of two sulfur atoms. Even without help from the NMR spectra, the student selects C_6H_{12} under the unit mass of 84—having allowed for two sulfur atoms (64 mass units)—and writes $C_6H_{12}S_2$.

2.5 Recognition of the Molecular Ion Peak

Quite often, under electron impact (EI), recognition of the molecular ion peak (M) poses a problem. The peak may be very weak or it may not appear at all; how can we be sure that it is the molecular ion peak and not a fragment peak or an impurity? Often the best solution is to obtain a chemical ionization spectrum (see below in this section). The usual result is an intense peak at M + 1 and little fragmentation.

Many peaks can be ruled out as possible molecular ions simply on grounds of reasonable structure requirements. The "nitrogen rule" is often helpful. It states that a molecule of even-numbered molecular weight must contain either no nitrogen or an even number of nitrogen atoms; an odd-numbered molecular weight requires an odd number of nitrogen atoms.* This rule holds for all compounds containing carbon, hydrogen, oxygen, nitrogen, sulfur, and the halogens, as well as many of the less usual atoms such as phosphorus, boron, silicon, arsenic, and the alkaline earths.

A useful corollary states that fragmentation at a single bond gives an odd-numbered ion fragment from an even-numbered molecular ion, and an even-numbered ion fragment from an odd-numbered molecular ion. For this corollary to hold, the ion fragment must contain all of the nitrogen (if any) of the molecular ion.

Consideration of the breakdown pattern coupled with other information will also assist in identifying molecular ions. It should be kept in mind that Appendix A contains fragment formulas as well as molecular formulas. Some of the formulas may be discarded as trivial in attempts to solve a particular problem.

The intensity of the molecular ion peak depends on the stability of the molecular ion. The most stable molecular ions are those of purely aromatic systems. If substituents that have favorable modes of cleavage are present, the molecular ion peak will be less intense, and the fragment peaks relatively more intense. In general, the following group of compounds will, in order of decreasing ability, give prominent molecular ion peaks: aromatic compounds > conjugated alkenes > cyclic compounds > organic sulfides > short, normal alkanes > mercaptans. Recognizable molecular ions are usually produced for these compounds in order of decreasing ability: ketones > amines > esters > ethers > carboxylic acids ~ aldehydes ~ amides ~ halides. The molecular ion is frequently not detectable in aliphatic alcohols, nitrites, nitrates, nitro compounds, nitriles, and in highly branched compounds.

The presence of an M − 15 peak (loss of CH_3), or an M − 18 peak (loss of H_2O), or an M − 31 peak (loss of OCH_3 from methyl esters), and so on, is taken as confirmation of a molecular ion peak. An M − 1 peak is common, and occasionally an M − 2 peak (loss of H_2 by either fragmentation or thermolysis), or even a rare M − 3 peak (from alcohols) is reasonable. Peaks in the range of M − 3 to M − 14, however, indicate that contaminants may be present or that the presumed molecular ion peak is actually a fragment ion peak. Losses of fragments of masses 19–25 are also unlikely (except for loss of F = 19 or HF = 20 from fluorinated compounds). Loss of 16 (O), 17 (OH), or 18 (H_2O) are likely only if an oxygen atom is in the molecule.

2.5.1 Other Useful Ionization Techniques

For organic compounds that are sufficiently volatile, introduction by vaporization or by gas chromatography through a very small orifice followed by ionization by electron impact (EI) is standard procedure. But if this procedure does not give an unambiguous molecular ion, the next step is *chemical ionization* (CI), which usually yields a prominent $[M + H]^+$ peak with little fragmentation.

Less volatile but thermally stable compounds can be thermally vaporized in the direct inlet probe (DIP) situated close to the ionizing molecular beam. This DIP is standard equipment on most instruments. An electron-impact spectrum results.

For compounds that are not thermally stable enough for the direct inlet probe, field desorption (FD) is the next resort.

In recent years, several procedures have been developed for handling high molecular weight, water-soluble biomolecules. Several of these procedures are here briefly described. [See the Chapman (1993) and Watson (1985) references for a thorough discussion of these techniques. The Harrison (1992) reference presents a thorough treatment of chemical ionizations.]

2.5.1.1 Chemical Ionization (CI) The vaporized sample is introduced into the mass spectrometer with an excess of a "reagent" gas (commonly methane) at a pressure of about 1 torr. The excess carrier gas is ionized by electron impact to the primary ions CH_4^+ and CH_3^+. These react with the excess methane to give secondary ions.

$$CH_4^{\cdot+} + CH_4 \longrightarrow CH_5^+ \text{ and } CH_3^{\cdot}$$

$$CH_3^+ + CH_4 \longrightarrow C_2H_5^+ \text{ and } H_2$$

$$CH_4 + C_2H_5^+ \longrightarrow C_3H_5^+ + 2H_2$$

The secondary ions react with the sample (M).

$$CH_5^+ + M \longrightarrow [M + H]^+ + CH_4$$

$$C_2H_5^+ + M \longrightarrow [M + H]^+ + C_2H_4$$

These CI $[M + H]^+$ ions *(quasimolecular ions)* are often prominent. Chemical ionization spectra sometimes have prominent $[M - H]^+$ ions because of hydride ion abstraction from the $M^{\cdot+}$ ion by CH_5^+. Since the $[M + H]^+$ ions are chemically produced, they do not have the great excess of energy associated with ionization by electron impact, and they undergo less fragmentation. For example, the EI spectrum of 3,4-dimethoxyacetophenone shows, in addition to the molecular ion at *m/z* 180, 49 fragment peaks in the range of *m/z* 40–167; these include the base peak at *m/z* 165 and prominent peaks at *m/z* 137 and *m/z* 77. The CH_4 induced CI spectrum shows the quasimolecular ion ($M + H^+$, *m/z* 181) as the base peak (100%), and virtually the only other peaks, each of just a few percent intensity, are the

* For the nitrogen rule to hold, only unit atomic unit masses (i.e., integers) are used in calculating the formula masses.

molecular ion peak, *m/z* 180, and *m/z* 209 and 221 peaks resulting from reaction with carbocations.

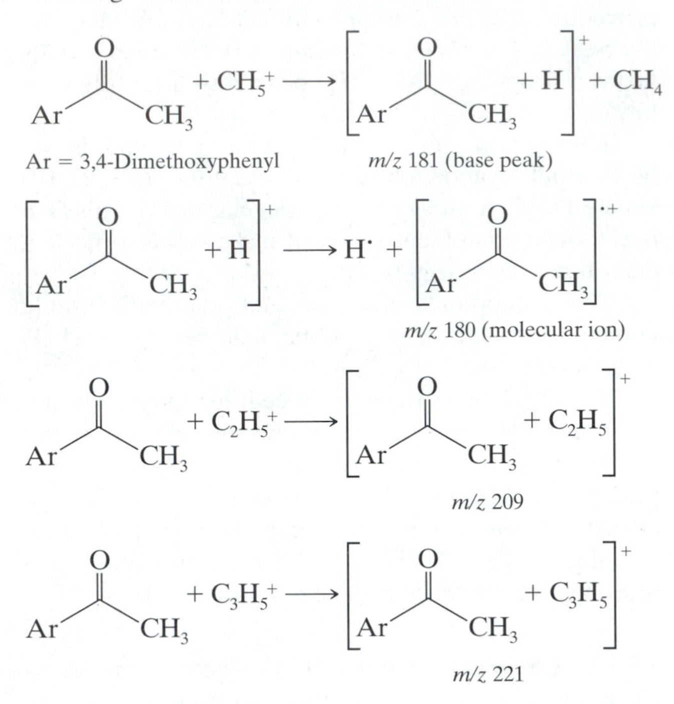

The energy content of the various secondary ions (from, respectively, CH_4, isobutane, and ammonia) decreases in this order:

$$CH_5^+ > t\text{-}C_4H_9^+ > NH_4^+$$

Thus by choice of "reagent" gas, we can control the tendency of the CI-produced $M + H^+$ ion to fragment. For example, when methane is the carrier gas, dioctyl phthalate shows its $M + H^+$ peak (*m/z* 391) as the base peak; more importantly, the fragment peaks (e.g., *m/z* 113 and 149) are 30–60% of the intensity of the base peak. When isobutane is used, the $M + H^+$ peak is large and the fragment peaks are only roughly 5% as intense as the $M + H^+$ peak.

In many laboratories, the EI spectrum and the CI spectrum (with methane or isobutane) are obtained routinely since they are complementary. The CI spectrum will frequently provide the $[M + H]^+$ peak when the EI spectrum shows only a weak or undetectable $M^{\cdot +}$ peak. The $[M - H]^+$ peak may also appear in the CI spectrum by hydride abstraction. The CI fragmentation pattern is usually difficult to predict or rationalize. Note that the "nitrogen rule" (see Section 2) does not apply to the $[M + H]^+$ or the $[M - H]^+$ peaks; neither does it apply to the CI fragmentation ions.

One general statement may be made: If a molecule MX (X is a functional group) is protonated by the reagent ion to give the quasimolecular ion MXH^+, fragmentation usually produces the neutral protonated functional group XH and the fragment ion M^+.

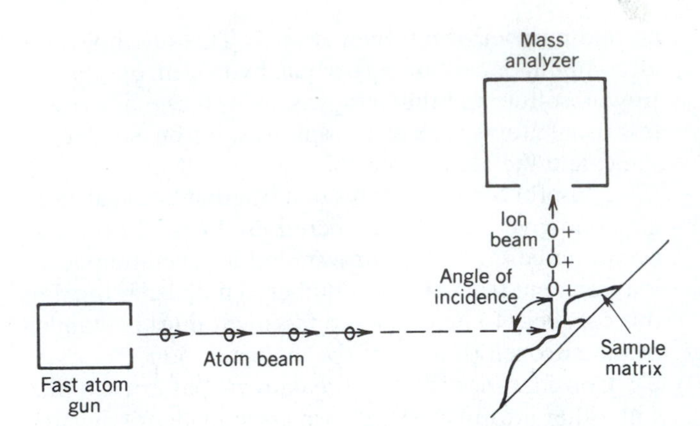

FIGURE 2.5. Schematic for FAB mass spectrometry. The intense activity at the surface of the sample produces neutrals, sample ions, ions from the matrix, etc. Only the ions are accelerated toward the analyzer.

$$MXH^+ \longrightarrow M^+ + XH$$

Thus an alcohol, ROH, is protonated to give ROH_2^+, which fragments with loss of a neutral molecule of water to give an R^+ peak.*

2.5.1.2 Field Desorption (FD) Stable molecular ions are obtained from a sample of low volatility, which is placed on the anode of a pair of electrodes, between which there is an intense electric field. Desorption occurs, and molecular and quasimolecular ions are produced with insufficient internal energy for extensive fragmentation. Usually the major peak represents the $[M + H]^+$ ion.

Synthetic polymers with molecular weights on the order of 10,000 Da have been analyzed, but there is a much lower molecular weight limit for polar biopolymers; here the FAB procedure and others (see below) are superior.

2.5.1.3 Fast Atom Bombardment (FAB) Polar molecules, such as peptides, with molecular weights up to 10,000 Da can be analyzed by a "soft" ionization technique called fast atom bombardment (FAB, Fig. 2.5). The bombarding beam consists of xenon (or argon) atoms of high translational energy ($\overrightarrow{Xe}$). This beam is produced by first ionizing xenon atoms with electrons to give xenon radical cations:

$$Xe \xrightarrow{e^-} Xe^{\cdot +} + 2e^-$$

The radical cations are accelerated to 6–10 keV to give radical cations of high translational energy $(\overrightarrow{Xe})^{\cdot +}$,

* See Harrison (1992), Chapman (1993), or Watson (1985) in the references at the end of this chapter. See also the review: Kingston, E. E., Shannon, J. S., and Lacey, M. J. *Org. Mass Spectrom.* **18,** 183–192 (1983).

which are then passed through xenon. During this passage, the charged high-energy xenon obtains electrons from the xenon atoms to become high-energy atoms ($\overrightarrow{Xe}$), and the $Xe^{\cdot+}$ ions are removed by an electric field.

$$Xe^{\cdot+} \xrightarrow{\text{accelerate}} \overrightarrow{Xe}^{\cdot+}$$

$$\overrightarrow{Xe}^{\cdot+} + Xe \longrightarrow \overrightarrow{Xe} + Xe^{\cdot+}$$

The compound of interest is dissolved in a high-boiling viscous solvent such as glycerol; a drop is placed on a thin metal sheet, and the compound is ionized by the high-energy beam of xenon atoms ($\overrightarrow{Xe}$). Ionization by translational energy minimizes the amount of vibrational excitation, and this results in less destruction of the ionized molecules. The polar solvent promotes ionization and allows diffusion of fresh sample to the surface. Thus ions are produced over a period of 20–30 min, in contrast to a few seconds for ions produced from solid samples.

The molecular ion itself is usually not seen, but adduct ions such as $[M + H]^+$ are prominent. Other adduct ions can be formed from salt impurities or upon addition of salts such as NaCl or KCl, which give $[M + Na]^+$ and $[M + K]^+$ additions. Glycerol adduct peaks are prominent and troublesome in the spectrum. Fragment ions are prominent and useful.

2.5.1.4 Electrospray Ionization (ESI) Electrospray ionization involves placing an ionizing voltage—several kilovolts—across the nebulizer needle attached to the outlet from a high-performance liquid chromatograph (HPLC).

This technique is widely used on water-soluble biomolecules—proteins, peptides, and carbohydrates in particular. The result is a spectrum whose major peaks consist of the molecular ion with a different number of charges attached. A molecular ion of, for example, about 10,000 Da with a charge (z) of 10 would behave in a mass spectrometer as though its mass were about 1000 daltons. Its mass, therefore, can be determined with a spectrometer of modest resolution—and cost.

Electrospray ionization is one of several variations of *atmospheric pressure ionization* (API) as applied to the outlet of an HPLC unit attached to the inlet of the mass spectrometer. These variations have in common the formation of a very fine spray (nebulization) from which the solvent can be quickly removed. The small particles are then ionized by a corona discharge at atmospheric pressure and swept by the continuous flow of the particles and a small electrical potential that moves the positively charged particles through a small orifice into the evacuated mass spectrometer.

2.5.1.5 Matrix Assisted Laser Desorption/Ionization (MALDI) In the MALDI procedure—used mainly for large biomolecules—the sample in a matrix is dispersed on a surface, and is desorbed and ionized by the energy of a laser beam. The matrix serves the same purpose as it does in the FAB procedure (Section 2.5.1.3.).

The MALDI procedure has been used recently in several variations to determine the molecular weight of large protein molecules—up to several hundred kDa. The combination of a pulsed laser beam and a time-of-flight mass spectrometer (Section 2.2.5.) is particularly effective.

Peptide sequencing is another application. Matrix selection is critical and depends on the wavelength of the laser beam and on the nature of the sample. Such polar compounds as carboxylic acids (e.g., nicotinic acid), urea, and glycerol have been used.

At the time of writing, ESI and MALDI are the preferred procedures for large biopolymers.

2.6 *Use of the Molecular Formula. Index of Hydrogen Deficiency*

If organic chemists had to choose a single item of information above all others that are usually available from spectra or from chemical manipulations, they would certainly choose the molecular formula.

In addition to the kinds and numbers of atoms, the molecular formula gives the index of hydrogen deficiency. The index of hydrogen deficiency is the number of *pairs* of hydrogen atoms that must be removed from the corresponding "saturated" formula to produce the molecular formula of the compound of interest. The index of hydrogen deficiency is also called the number of "sites (or degrees) of unsaturation"; this description is incomplete since hydrogen deficiency can result from cyclic structures as well as from multiple bonds. The index is thus the sum of the number of rings, the number of double bonds, and twice the number of triple bonds.

The index of hydrogen deficiency can be calculated for compounds containing carbon, hydrogen, nitrogen, halogen, oxygen, and sulfur from the formula

$$\text{Index} = \text{carbons} - \frac{\text{hydrogens}}{2} - \frac{\text{halogens}}{2} + \frac{\text{nitrogens}}{2} + 1$$

Thus, the compound C_7H_7NO has an index of 7 − 3.5 + 0.5 + 1 = 5. Note that divalent atoms (oxygen and sulfur) are not counted in the formula.

For the generalized molecular formula $\alpha_I\beta_{II}\gamma_{III}\delta_{IV}$, the index $= \text{IV} - \frac{1}{2}\text{I} + \frac{1}{2}\text{III} + 1$, where

α is H, D, or halogen (i.e., any monovalent atom)

β is O, S, or any other bivalent atom

γ is N, P, or any other trivalent atom

δ is C, Si, or any other tetravalent atom

The numerals I–IV designate the numbers of the mono-, di-, tri-, and tetravalent atoms, respectively.

For simple molecular formulas, we can arrive at the index by comparison of the formula of interest with the molecular formula of the corresponding saturated compound. Compare C_6H_6 and C_6H_{14}; the index is 4 for the former and 0 for the latter.

The index for C_7H_7NO is 5, and a possible structure is

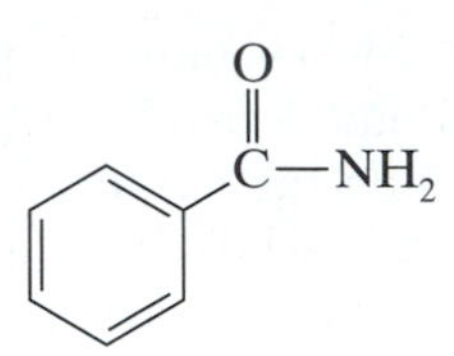

Of course, other isomers (i.e., compounds with the same molecular formula) are possible, such as

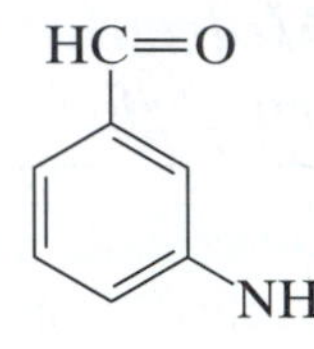

Note that the benzene ring itself accounts for four "sites of unsaturation": three for the double bonds and one for the ring.

Polar structures must be used for compounds containing an atom in a higher valence state, such as sulfur or phosphorus. Thus, if we treat sulfur in dimethyl sulfoxide (DMSO) formally as a divalent atom, the calculated index, 0, is compatible with the structure $CH_3—\overset{+}{\ddot{S}}—CH_3$ (with $:\ddot{O}:^-$ on S). We must use only formulas with filled valence shells; that is, the Lewis octet rule must be obeyed.

Similarly, if we treat the nitrogen in nitromethane as a trivalent atom, the index is 1, which is compatible with $CH_3—N^+$ (=O, —O^-). If we treat phosphorus in triphenylphosphine oxide as trivalent, the index is 12, which fits $(C_6H_5)_3P^+—O^-$. As an example, let us consider the molecular formula $C_{13}H_9N_2O_4BrS$. The index of hydrogen deficiency would be $13 - \frac{10}{2} + \frac{2}{2} + 1 = 10$ and a consistent structure would be

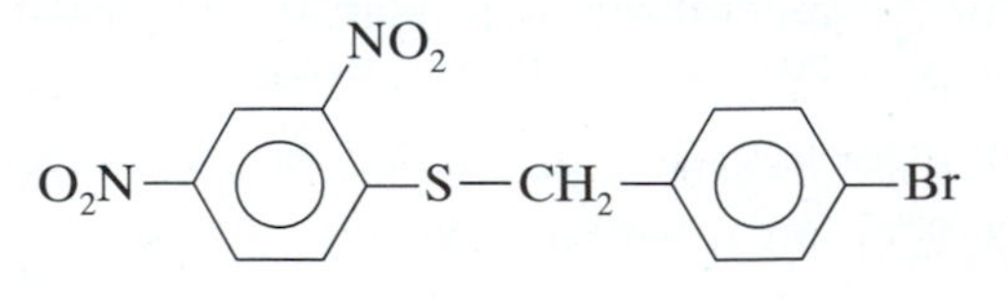

(Index of hydrogen deficiency = 4 per benzene ring and 1 per NO_2 group.)

The formula above for the index can be applied to fragment ions as well as to the molecular ion. When it is applied to even-electron (all electrons paired) ions, the result is always an odd multiple of 0.5. As an example, consider $C_7H_5O^+$ with an index of 5.5. A reasonable structure is

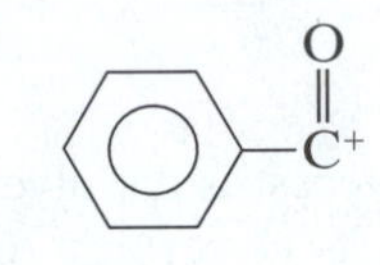

since $5\frac{1}{2}$ pairs of hydrogen atoms would be necessary to obtain the corresponding saturated formula $C_7H_{16}O$ ($C_nH_{2n+2}O$). Odd-electron fragment ions will always give integer values of the index.

Terpenes often present a choice between a double bond and a ring structure. This question can readily be resolved on a microgram scale by catalytically hydrogenating the compound and rerunning the mass spectrum. If no other easily reducible groups are present, the increase in the mass of the molecular ion peak is a measure of the number of double bonds; and other "unsaturated sites" must be rings.

Such simple considerations give the chemist very ready information about structure. As another example, a compound containing a single oxygen atom might quickly be determined to be an ether or a carbonyl compound simply by counting "unsaturated sites."

2.7 *Fragmentation*

As a first impression, fragmenting a molecule with a huge excess of energy would seem a brute-force approach to molecular structure. The rationalizations used to correlate spectral patterns with structure, however, can only be described as elegant, though sometimes arbitrary. The insight of such pioneers as McLafferty, Beynon, Stenhagen, Ryhage, and Meyerson led to a number of rational mechanisms for fragmentation. These were masterfully summarized and elaborated by Biemann (1962), Budzikiewicz (1967), and others.

Generally, the tendency is to represent the molecular ion with a delocalized charge. Djerassi's (1967) approach is to localize the positive charge on either a π bond (except in conjugated systems), or on a heteroatom. Whether or not this concept is totally rigorous, it is at the least a pedagogic *tour de force.* We shall use such locally charged molecular ions in this book.

Structures **A** and **B,** for example, represent the molecular ion of cyclohexadiene. Compound **A** is a delocalized structure with one less electron than the original uncharged diene; both the electron and the positive charge are delocalized over the π system. Since the electron removed to form the molecular ion is a π electron, other structures, such as **B** or **C** (valence bond structures) can be used. Structures such as **B** and **C** localize the electron and the positive charge and thus are useful for describing fragmentation processes.

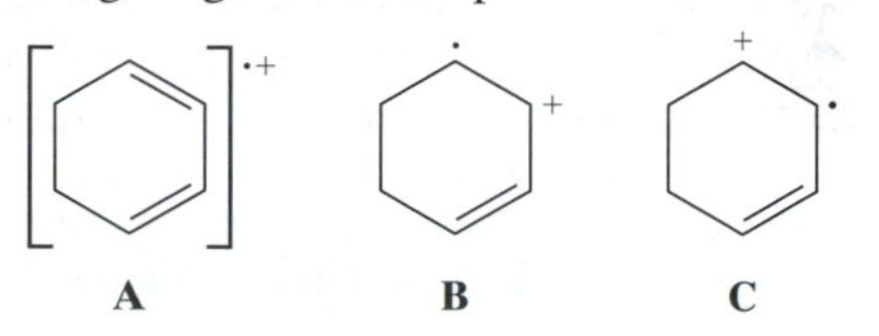

Fragmentation is initiated by electron impact. Only a small part of the driving force for fragmentation is energy transferred as the result of the impact. The major driving force is the cation–radical character that is imposed upon the structure.

Fragmentation of the odd-electron molecular ion (radical-cation, $M^{\cdot+}$) may occur by homolytic or heterolytic cleavage of a single bond. In homolytic cleavage, each electron "moves" independently as shown by a (single-barbed) fishhook; the fragments here are an even-electron cation and a free radical (odd electron).

$$CH_3{-}CH_2{-}\ddot{O}^{\cdot+}{-}R \longrightarrow CH_3\cdot + CH_2{=}\overset{+}{\ddot{O}}{-}R$$

To prevent clutter, only one of each pair of fishhooks need be shown:

$$CH_3{-}CH_2{-}\ddot{O}^{\cdot+}{-}R \longrightarrow CH_3\cdot + CH_2{=}\overset{+}{\ddot{O}}{-}R$$

In heterolytic cleavage, a pair of electrons "move" together toward the charged site as shown by the conventional curved arrow; the fragments are again an even-electron cation and a radical, but here the final charge site is on the alkyl product.

$$CH_3CH_2CH_2{-}\ddot{Br}^{\cdot+}: \longrightarrow CH_3CH_2CH_2^+ + :\dot{\ddot{Br}}:$$

In the absence of rings (whose fragmentation requires cleavage of two or more bonds), most of the prominent fragments in a mass spectrum are even-electron cations formed as above by a single cleavage. Further fragmentation of an even-electron cation usually results in another even-electron cation and an even-electron neutral molecule or fragment.

$$CH_3{-}CH_2{-}CH_2^+ \longrightarrow CH_3^+ + CH_2{=}CH_2$$

Simultaneous or consecutive cleavage of several bonds may occur when energy benefits accrue from formation of a highly stabilized cation and/or a stable radical, or a neutral molecule, often through a well-defined low-energy pathway. These are treated in Section 2.8 (rearrangements) and in Section 2.10 under individual chemical classes.

The probability of cleavage of a particular bond is related to the bond strength, to the possibility of low-energy transitions, and to the stability of the fragments, both charged and uncharged, formed in the fragmentation process. Our knowledge of pyrolytic cleavages can be used, to some extent, to predict likely modes of cleavage of the molecular ion. Because of the extremely low pressure in the mass spectrometer, there are very few fragment collisions; we are dealing largely with unimolecular decompositions. This assumption, backed by a file of reference spectra, is the basis for the vast amount of information available from the fragmentation pattern of a molecule. Whereas conventional organic chemistry deals with reactions initiated by chemical reagents, by thermal energy, or by light, mass spectrometry is concerned with the consequences suffered by an organic molecule at a vapor pressure of about 10^{-6} mm Hg struck by an ionizing electron beam.

A number of general rules for predicting prominent peaks in EI spectra can be written and rationalized by using standard concepts of physical organic chemistry.

1. The relative height of the molecular ion peak is greatest for the straight-chain compound and decreases as the degree of branching increases (see rule 3).
2. The relative height of the molecular ion peak usually decreases with increasing molecular weight in a homologous series. Fatty esters appear to be an exception.
3. Cleavage is favored at alkyl-substituted carbon atoms; the more substituted, the more likely is cleavage. This is a consequence of the increased stability of a tertiary carbocation over a secondary, which in turn is more stable than a primary.

$$\left[R{-}\overset{|}{\underset{|}{C}}{-}\right]^{\cdot+} \longrightarrow R\cdot + {}^{+}\overset{|}{\underset{|}{C}}{-}$$

Cation stability order:

$$CH_3^+ < R'CH_2^+ < R_2'CH^+ < R_3'C^+$$

Generally, the largest substituent at a branch is eliminated most readily as a radical, presumably because a long-chain radical can achieve some stability by delocalization of the lone electron.

4. Double bonds, cyclic structures, and especially aromatic (or heteroaromatic) rings stabilize the molecular ion and thus increase the probability of its appearance.
5. Double bonds favor allylic cleavage and give the resonance-stabilized allylic carbocation. This rule

does not hold for simple alkenes because of the ready migration of the double bond, but it does hold for cycloalkenes.

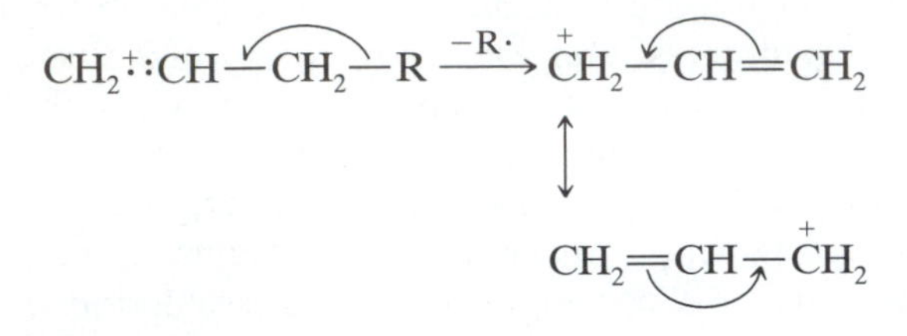

6. Saturated rings tend to lose alkyl side chains at the α bond. This is merely a special case of branching (rule 3). The positive charge tends to stay with the ring fragment.

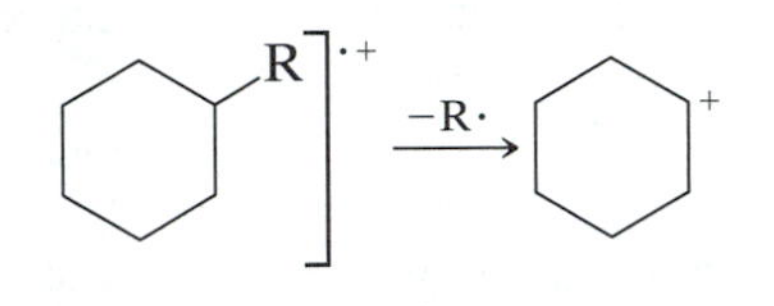

Unsaturated rings can undergo a *retro*-Diels-Alder reaction:

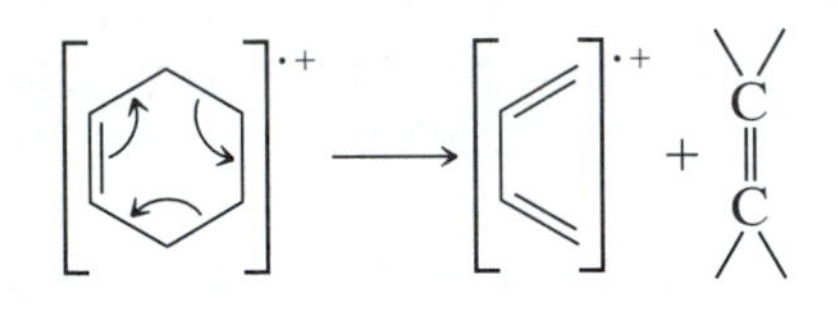

7. In alkyl-substituted aromatic compounds, cleavage is very probable at the bond β to the ring, giving the resonance-stabilized benzyl ion or, more likely, the tropylium ion:

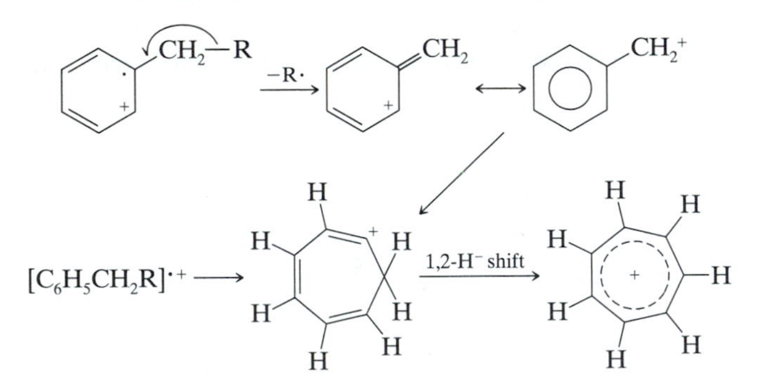

8. The C—C bonds next to a heteroatom are frequently cleaved, leaving the charge on the fragment containing the heteroatom whose nonbonding electrons provide resonance stabilization.

$$CH_3\text{—}CH_2\text{—}\overset{\cdot+}{Y}\text{—}R \xrightarrow{-CH_3^{\cdot}} CH_2\text{=}\overset{+}{Y}\text{—}R \longleftrightarrow {}^{+}CH_2\text{—}\ddot{Y}\text{—}R$$

where Y = O, NH, or S;

$$R\text{—}\underset{\underset{\cdot O:^{+}}{\|}}{C}\text{—}CH_2R' \xrightarrow{-R\cdot} \underset{\underset{O:^{+}}{\|}}{C}\text{—}CH_2R' \longleftrightarrow \underset{\underset{:O:}{\|}}{\overset{+}{C}}\text{—}CH_2R'$$

9. Cleavage is often associated with elimination of small, stable, neutral molecules, such as carbon monoxide, olefins, water, ammonia, hydrogen sulfide, hydrogen cyanide, mercaptans, ketene, or alcohols, often with rearrangement (Section 2.8).

It should be kept in mind that the fragmentation rules above apply to EI mass spectrometry. Since other ionizing (CI, etc.) techniques often produce molecular ions with much lower energy or quasimolecular ions with very different fragmentation patterns, different rules govern the fragmentation of these molecular ions.

2.8 *Rearrangements*

Rearrangement ions are fragments whose origin cannot be described by simple cleavage of bonds in the molecular ion but are a result of intramolecular atomic rearrangement during fragmentation. Rearrangements involving migration of hydrogen atoms in molecules that contain a heteroatom are especially common. One important example is the so-called McLafferty rearrangement.

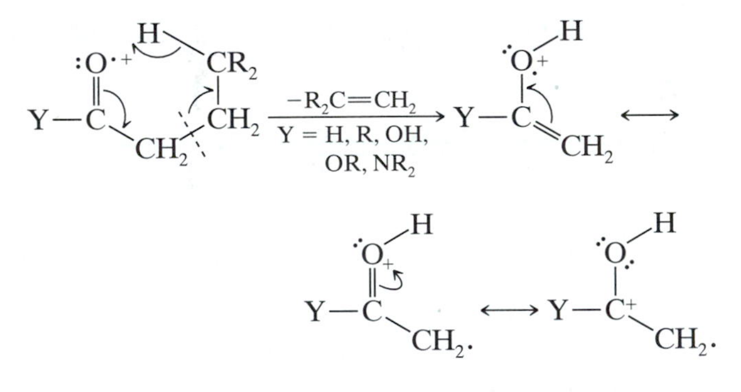

To undergo a McLafferty rearrangement, a molecule must possess an appropriately located heteroatom (e.g., O), a π system (usually a double bond), and an abstractable hydrogen atom γ to the C=O system.

Such rearrangements often account for prominent characteristic peaks and are consequently very useful for our purpose. They can frequently be rationalized on the basis of low-energy transitions and increased stability of the products. Rearrangements resulting in elimination of a stable neutral molecule are common (e.g., the alkene product in the McLafferty rearrangement) and will be encountered in the discussion of mass spectra of chemical classes.

Rearrangement peaks can be recognized by considering the mass (m/z) number for fragment ions and for their corresponding molecular ions. A simple (no rearrangement) cleavage of an even-numbered molecular ion gives an odd-numbered fragment ion and simple cleavage of an odd-numbered molecular ion gives an even-numbered fragment. Observation of a fragment ion mass different by 1 unit from that expected for a fragment resulting from simple cleavage (e.g., an even-numbered fragment mass from an even-numbered molecular ion mass) indicates rearrangement of hydrogen has accompanied fragmentation. Rearrangement peaks may be recognized by considering the corollary to the "nitrogen rule" (Section 2.5). Thus, an even-numbered peak derived from an even-numbered molecular ion is a result of two cleavages, which may involve a rearrangement.

"Random" rearrangements of hydrocarbons were noted by the early mass spectrometrists in the petroleum industry. For example,

$$\left[CH_3-C(CH_3)_2-CH_3 \right]^{\cdot+} \longrightarrow [C_2H_5]^+$$

These rearrangements defy straightforward explanations.

2.9 Derivatives

If a compound has low volatility or if the parent mass cannot be determined, it may be possible to prepare a suitable derivative. The derivative selected should provide enhanced volatility, a predictable mode of cleavage, a simplified fragmentation pattern, or an increased stability of the molecular ion.

Compounds containing several polar groups may have very low volatility (e.g., sugars, peptides, and dibasic carboxylic acids). Acetylation of hydroxyl and amino groups and methylation of free acids are obvious and effective choices to increase volatility and give characteristic peaks. Perhaps less immediately obvious is the use of trimethylsilyl derivatives of hydroxyl, amino, sulfhydryl, and carboxylic acid groups. Trimethylsilyl derivatives of sugars and of amino acids are volatile enough to pass through GC columns. The molecular ion peak of trimethylsilyl derivatives may not always be present, but the M − 15 peak resulting from cleavage of one of the $Si-CH_3$ bonds is often prominent.

Reduction of ketones to hydrocarbons has been used to elucidate the carbon skeleton of the ketone molecule. Polypeptides have been reduced with $LiAlH_4$ to give volatile polyamino alcohols with predictable fragmentation patterns. Methylation and trifluoroacetylation of tri- and tetrapeptides have lead to useful mass spectra.

2.10 Mass Spectra of Some Chemical Classes

Mass spectra of a number of chemical classes are briefly described in this section in terms of the most useful generalizations for identification. For more details, the references cited (in particular, the thorough treatment by Budzikiewicz, Djerassi, and Williams, 1967) should be consulted. Databases are available both from publishers and as part of instrument capabilities. The references are selective rather than comprehensive. A table of frequently encountered fragment ions is given in Appendix B. A table of fragments (uncharged) that are commonly eliminated and some structural inferences are presented in Appendix C. More exhaustive listings of common fragment ions have been compiled (see References). The cleavage patterns described here in Section 2.10 are for EI spectra, unless stated otherwise.

2.10.1 Hydrocarbons

2.10.1.1 Saturated Hydrocarbons Most of the work early in mass spectrometry was done on hydrocarbons of interest to the petroleum industry. Rules 1–3, (Section 2.7) apply quite generally; rearrangement peaks, though common, are not usually intense (random rearrangements), and numerous reference spectra are available.

The molecular ion peak of a straight-chain, saturated hydrocarbon is always present, though of low intensity for long-chain compounds. The fragmentation pattern is characterized by clusters of peaks, and the corresponding peaks of each cluster are 14 (CH_2) mass units apart. The largest peak in each cluster represents a C_nH_{2n+1} fragment and thus occurs at $m/z = 14n + 1$; this is accompanied by C_nH_{2n} and C_nH_{2n-1} fragments. The most abundant fragments are at C_3 and C_4, and the fragment abundances decrease in a smooth curve down to $[M - C_2H_5]^+$; the $[M - CH_3]^+$ peak is characteristically very weak or missing. Compounds containing more than eight carbon atoms show fairly similar spectra; identification then depends on the molecular ion peak.

Spectra of branched saturated hydrocarbons are grossly similar to those of straight-chain compounds, but the smooth curve of decreasing intensities is broken by preferred fragmentation at each branch. The smooth

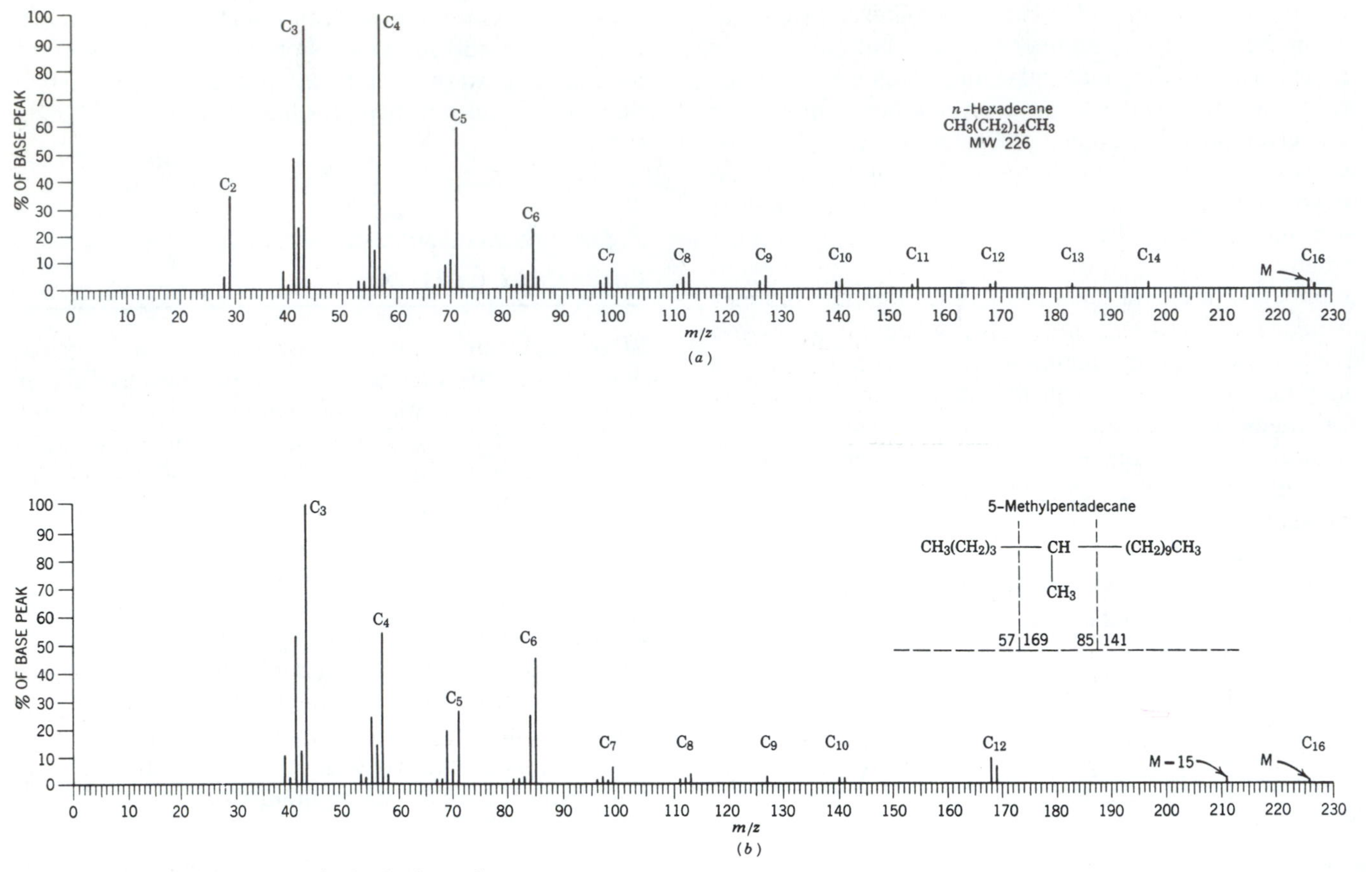

FIGURE 2.6 (*a*, *b*). Isomeric C_{16} hydrocarbons.

curve for the *n*-alkane in Figure 2.6*a* is in contrast to the discontinuity at C_{12} for the branched alkane (Fig. 2.6*b*). This discontinuity indicates that the longest branch of 5-methylpentadecane has 10 carbon atoms.

In Figure 2.6*b*, the peaks at *m/z* 169 and 85 represent cleavage on either side of the branch with charge retention on the substituted carbon atom. Subtraction of the molecular weight from the sum of these fragments accounts for the fragment $—CH—CH_3$. Again, we appreciate the absence of a C_{11} unit, which cannot form by a single cleavage. Finally, the presence of a distinct M − 15 peak also indicates a methyl branch. The fragment resulting from cleavage at a branch tends to lose a single hydrogen atom so that the resulting C_nH_{2n} peak is prominent and sometimes more intense than the corresponding C_nH_{2n+1} peak.

A saturated ring in a hydrocarbon increases the relative intensity of the molecular ion peak and favors cleavage at the bond connecting the ring to the rest of the molecule (rule 6, Section 2.7). Fragmentation of the ring is usually characterized by loss of two carbon atoms as C_2H_4 (28) and C_2H_5 (29). This tendency to lose even-numbered fragments, such as C_2H_4, gives a spectrum that contains a greater proportion of even-numbered mass ions than the spectrum of an acyclic hydrocarbon. As in branched hydrocarbons, C—C cleavage is accompanied by loss of a hydrogen atom. The characteristic peaks are therefore in the C_nH_{2n-1} and C_nH_{2n-2} series.

The mass spectrum of cyclohexane (Fig. 2.6*c*) shows a much more intense molecular ion than those of acyclic compounds, since fragmentation requires the cleavage of two carbon–carbon bonds. This spectrum has its base

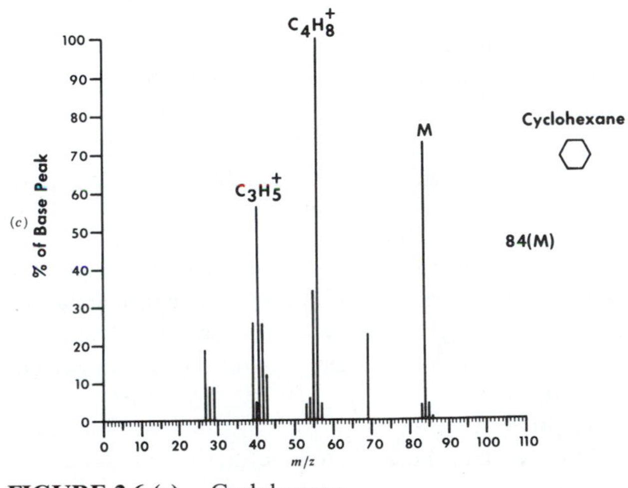

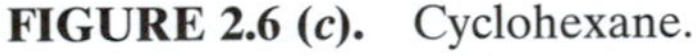

FIGURE 2.6 (*c*). Cyclohexane.

peak at m/z 56 (because of loss of C_2H_4) and a large peak at m/z 41, which is a fragment in the C_nH_{2n-1} series with $n = 3$.

2.10.1.2 Alkenes (Olefins) The molecular ion peak of alkenes, especially polyalkenes, is usually distinct. Location of the double bond in acyclic alkenes is difficult because of its facile migration in the fragments. In cyclic (especially polycyclic) alkenes, location of the double bond is frequently evident as a result of a strong tendency for allylic cleavage without much double-bond migration (rule 5, Section 2.7). Conjugation with a carbonyl group also fixes the position of the double bond. As with saturated hydrocarbons, acyclic alkenes are characterized by clusters of peaks at intervals of 14 units. In these clusters the C_nH_{2n-1} and C_nH_{2n} peaks are more intense than the C_nH_{2n+1} peaks.

The mass spectrum of β-myrcene, a terpene, is shown in Figure 2.7. The peaks at m/z 41, 55, and 69 correspond to the formula C_nH_{2n-1} with $n = 3$, 4, and 5, respectively. Formation of the m/z 41 peak must involve rearrangement. The peaks at m/z 67 and 69 are the fragments from cleavage of a bi-allylic bond.

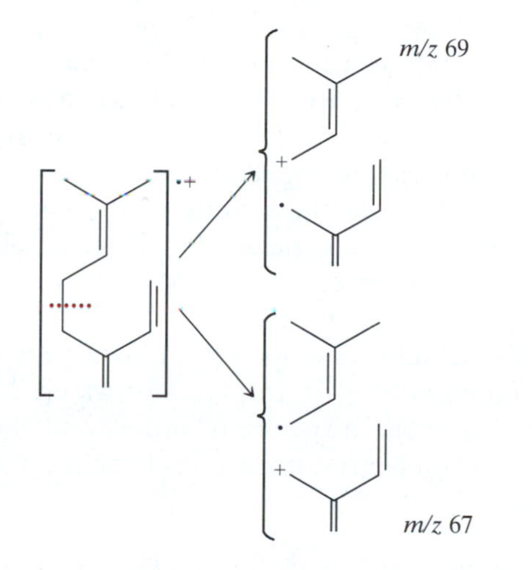

The peak at m/z 93 may be rationalized as a structure of formula $C_7H_9^+$ formed by isomerization (resulting in increased conjugation), followed by allylic cleavage.

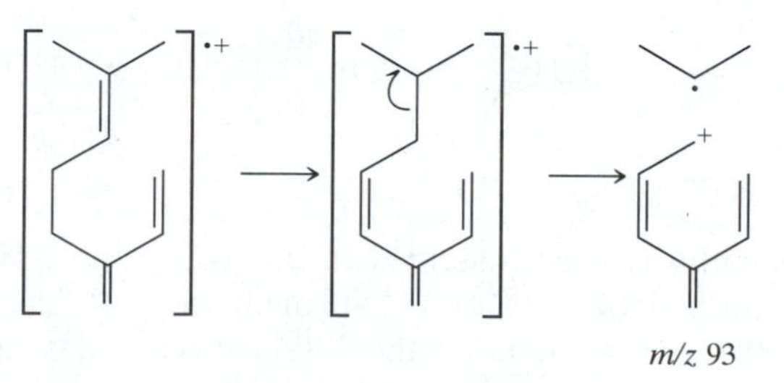

Cyclic alkenes usually show a distinct molecular ion peak. A unique mode of cleavage is the *retro*-Diels-Alder reaction shown by limonene:

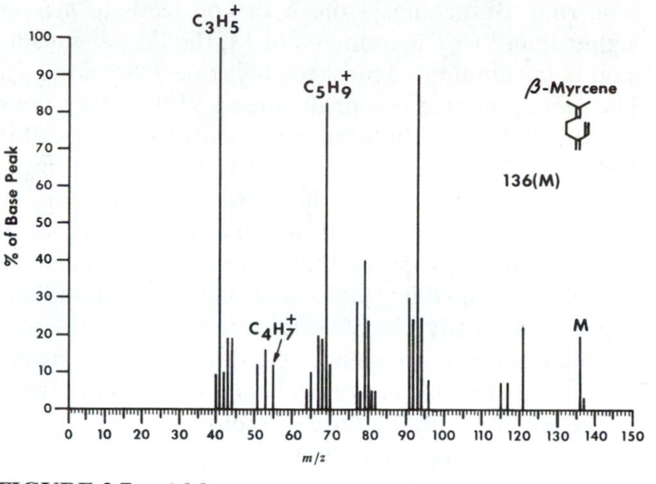

FIGURE 2.7. β-Myrcene.

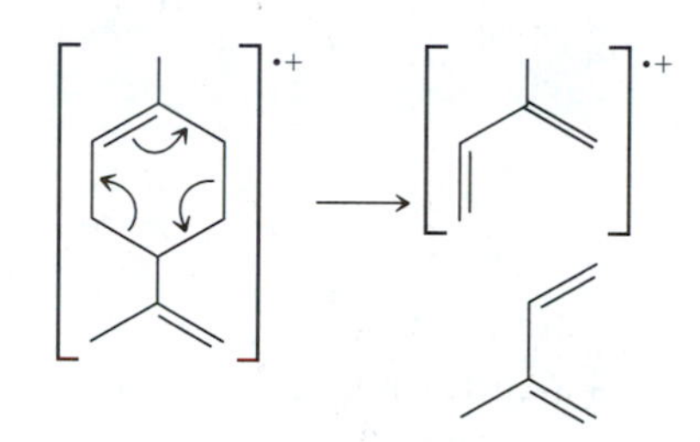

2.10.1.3 Aromatic and Aralkyl Hydrocarbons An aromatic ring in a molecule stabilizes the molecular ion peak (rule 4, Section 2.7), which is usually sufficiently large that accurate intensity measurements can be made on the M + 1 and M + 2 peaks.

Figure 2.8 is the mass spectrum of naphthalene. The molecular ion peak is also the base peak, and the largest fragment peak, m/z 51, is only 12.5% as intense as the molecular ion peak.

A prominent peak (often the base peak) at m/z 91 ($C_6H_5CH_2^+$) is indicative of an alkyl-substituted ben-

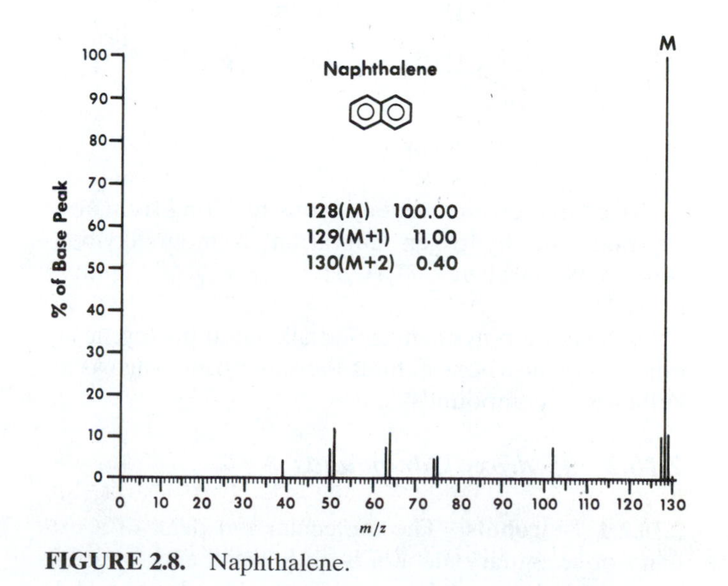

FIGURE 2.8. Naphthalene.

zene ring. Branching at the α carbon leads to masses higher than 91 by increments of 14, the largest substituent being eliminated most readily (rule 3, Section 2.7). The mere presence of a peak at mass 91, however, does not preclude branching at the α carbon because this highly stabilized fragment may result from rearrangements. A distinct and sometimes prominent M − 1 peak results from similar benzylic cleavage of a C—H bond.

It has been proposed that, in most cases, the ion of mass 91 is a tropylium rather than a benzylic cation. This explains the ready loss of a methyl group from xylenes, although toluene does not easily lose a methyl group. The incipient molecular radical ion of xylene rearranges to the methylcycloheptatriene radical ion, which then cleaves to the tropylium ion ($C_7H_7^+$).

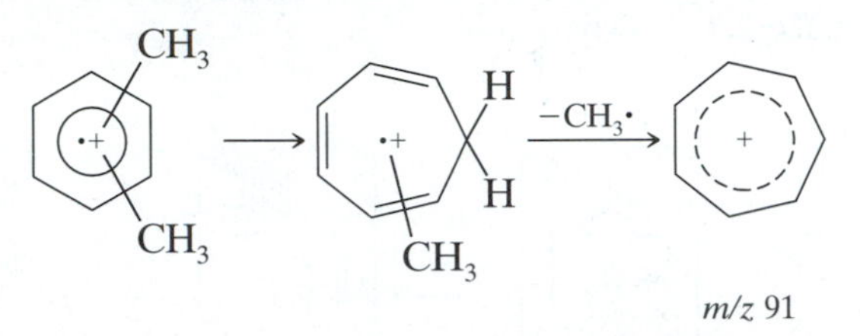

m/z 91

The frequently observed peak at *m/z* 65 results from elimination of a neutral acetylene molecule from the tropylium ion.

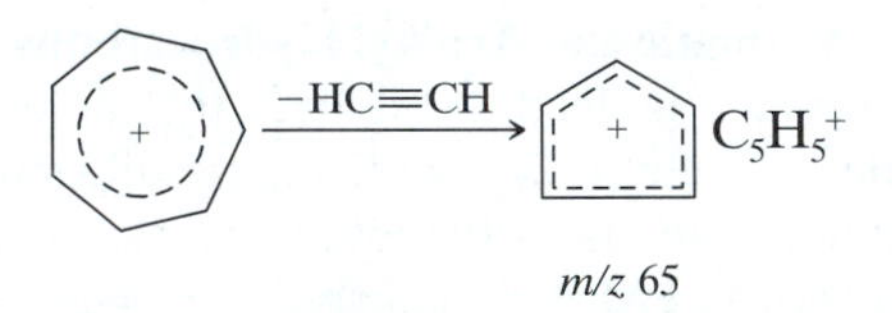

m/z 65

Hydrogen migration with elimination of a neutral alkene molecule accounts for the peak at *m/z* 92 observed when the alkyl group is longer than C_2.

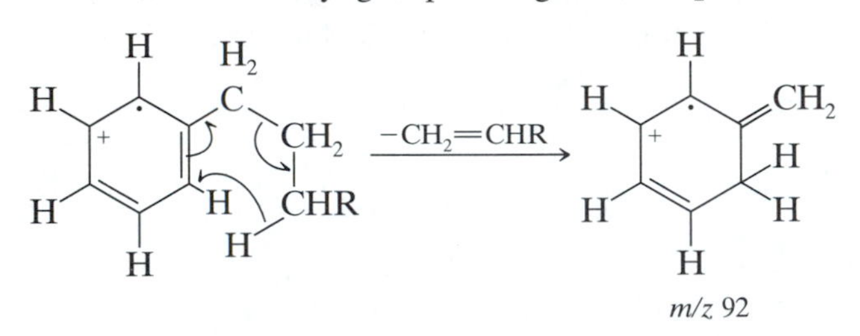

m/z 92

A characteristic cluster of ions resulting from an α cleavage and hydrogen migration in monoalkylbenzenes appears at *m/z* 77 ($C_6H_5^+$), 78 ($C_6H_6^+$), and 79 ($C_6H_7^+$).

Alkylated polyphenyls and alkylated polycyclic aromatic hydrocarbons exhibit the same β cleavage as alkylbenzene compounds.

2.10.2 *Hydroxy Compounds*

2.10.2.1 Alcohols The molecular ion peak of a primary or secondary alcohol is quite small and for a tertiary alcohol is undetectable. The molecular ion of 1-pentanol is extremely weak compared with its near homologs. Expedients such as CI, or derivatization, may be used to obtain the molecular weight.

Cleavage of the C—C bond next to the oxygen atom is of general occurrence (rule 8, Section 2.7). Thus, primary alcohols show a prominent peak resulting from $CH_2{=}\overset{+}{O}H$ (*m/z* 31). Secondary and tertiary alcohols cleave analogously to give a prominent peak resulting from R(H)C=$\overset{+}{O}$H (*m/z* 45, 59, 73, etc.), and R(R′)C=$\overset{+}{O}$H (*m/z* 59, 73, 87, etc.), respectively. The largest substituent is expelled most readily (rule 3).

R(R′)(R″)C—$\overset{\cdot +}{O}$—H $\xrightarrow{-R''\cdot}$ R(R′)C=$\overset{+}{O}$H ⟷ R(R′)$\overset{+}{C}$—OH

where R″ > R′ or R. When R and/or R′ = H, an M − 1 peak can usually be seen.

Primary alcohols, in addition to the principal C—C cleavage next to the oxygen atom, show a homologous series of peaks of progressively decreasing intensity resulting from cleavage at C—C bonds successively removed from the oxygen atom. In long-chain (>C_6) alcohols, the fragmentation becomes dominated by the hydrocarbon pattern; in fact, the spectrum resembles that of the corresponding alkene.

The spectrum in the vicinity of the very weak or missing molecular ion peak of a primary alcohol is sometimes complicated by weak M − 2 (R—CH=$\overset{+}{O}$) and M − 3 (R—C≡$\overset{+}{O}$) peaks.

A distinct and sometimes prominent peak can usually be found at M − 18 from loss of water. This peak is most noticeable in spectra of primary alcohols. This elimination by electron impact has been rationalized as follows:

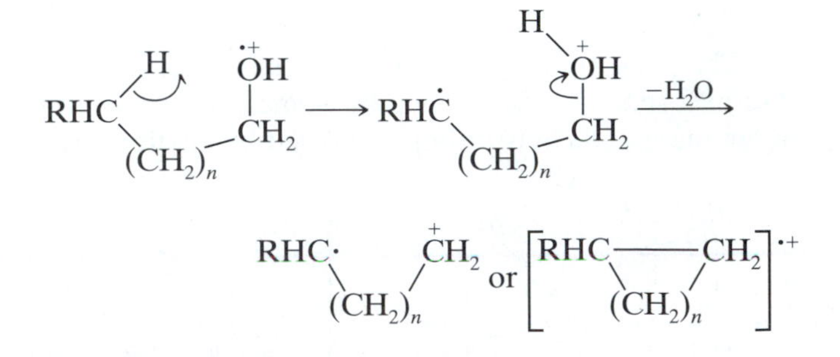

This pathway is consistent with the loss of the OH and γ hydrogen ($n = 1$) or δ hydrogen ($n = 2$); the ring structure is not proved by the observations and is merely one possible structure for the product radical cation. The M − 18 peak is frequently exaggerated by thermal decomposition of higher alcohols on hot inlet surfaces. Elimination of water, together with elimination of an alkene from primary alcohols, accounts for the presence of a peak at M − (alkene + H_2O), that is, a peak at M − 46, M − 74, M − 102, ...

$$\text{RCH(H)–}CH_2\text{–}CH_2\text{–}CH_2\text{–}\dot{O}^{+}H \xrightarrow[-H_2O]{-CH_2{=}CH_2} [CH_2{=}CHR]^{\cdot+}$$

M − (alkene + H_2O)

The alkene ion then decomposes by successive eliminations of ethylene.

Alcohols containing branched methyl groups (e.g., terpene alcohols) frequently show a fairly strong peak at M − 33 resulting from loss of CH_3 and H_2O.

Cyclic alcohols undergo fragmentation by complicated pathways; for example, cyclohexanol (M = *m/z* 100) forms $C_6H_{11}O^+$ by simple loss of the α hydrogen, loses H_2O to form $C_6H_{10}^+$ (which appears to have more than one possible bridged bicyclic structure), and forms $C_3H_5O^+$ (*m/z* 57) by a complex ring cleavage pathway.

A peak at *m/z* 31 (see above) is quite diagnostic for a primary alcohol provided it is more intense than peaks at *m/z* 45, 59, 73 . . . . However, the first-formed ion

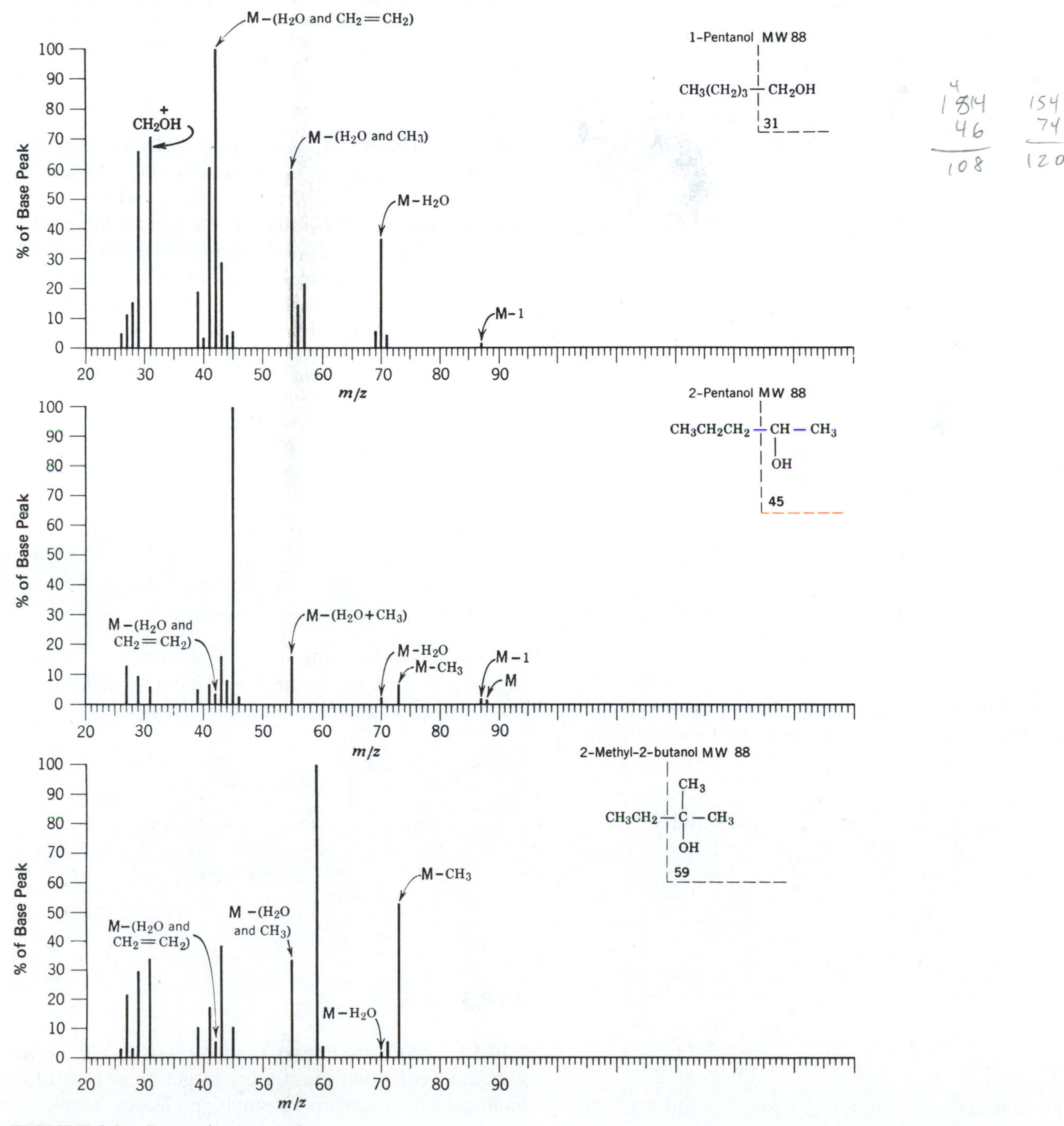

FIGURE 2.9. Isomeric pentanols.

of a secondary alcohol can decompose further to give a moderately intense m/z 31 ion.

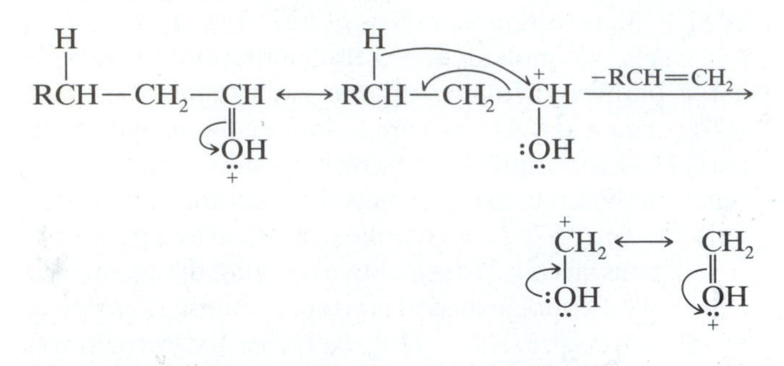

Figure 2.9 gives the characteristic spectra of isomeric primary, secondary, and tertiary C_5 alcohols.

Benzyl alcohols and their substituted homologs and analogs constitute a distinct class. Generally the parent peak is strong. A moderate benzylic peak (M − OH) may be present as expected from cleavage β to the ring. A complicated sequence leads to prominent M − 1, M − 2, and M − 3 peaks. Benzyl alcohol itself fragments to give sequentially the M − 1 ion, the $C_6H_7^+$ ion by loss of CO, and the $C_6H_5^+$ ion by loss of H_2.

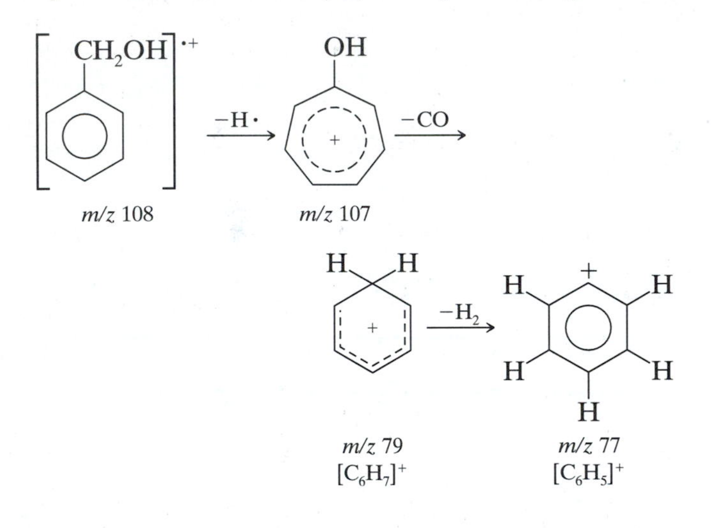

Loss of H_2O to give a distinct M − 18 peak is a common feature, especially pronounced and mechanistically straightforward in some ortho-substituted benzyl alcohols.

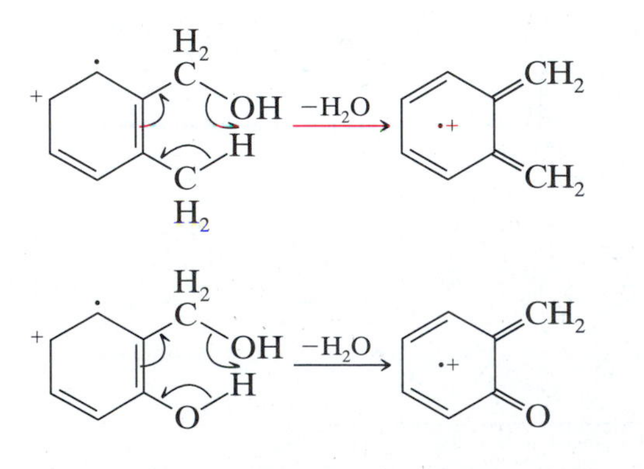

The aromatic cluster at m/z 77, 78, and 79 resulting from complex degradation is prominent here also.

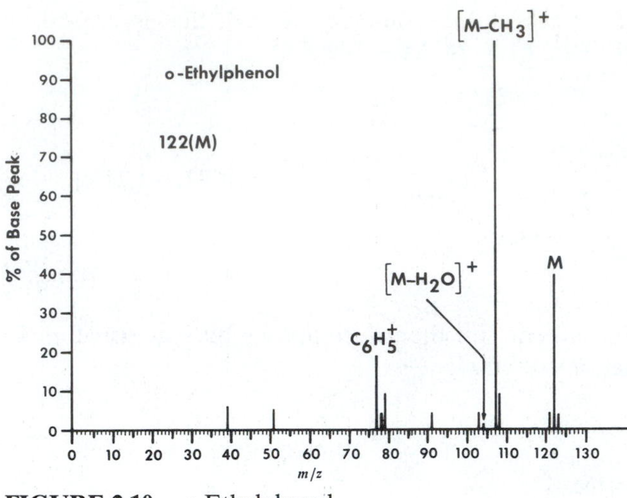

FIGURE 2.10. *o*-Ethylphenol.

2.10.2.2 Phenols A conspicuous molecular ion peak facilitates identification of phenols. In phenol itself, the molecular ion peak is the base peak, and the M − 1 peak is small. In cresols, the M − 1 peak is larger than the molecular ion as a result of a facile benzylic C—H cleavage. A rearrangement peak at m/z 77 and peaks resulting from loss of CO (M − 28) and CHO (M − 29) are usually found in phenols.

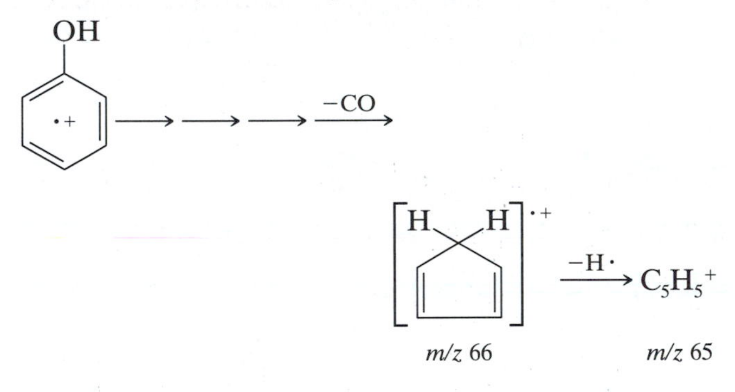

The mass spectrum of a typical phenol is shown in Figure 2.10. This spectrum shows that a methyl group is lost much more readily than an α hydrogen.

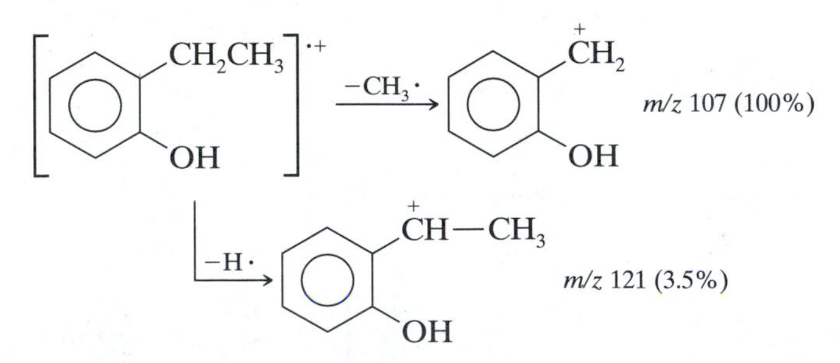

2.10.3 Ethers

2.10.3.1 Aliphatic Ethers (and Acetals) The molecular ion peak (two mass units larger than that of an analogous hydrocarbon) is small, but larger sample size usually will make the molecular ion peak or the M +

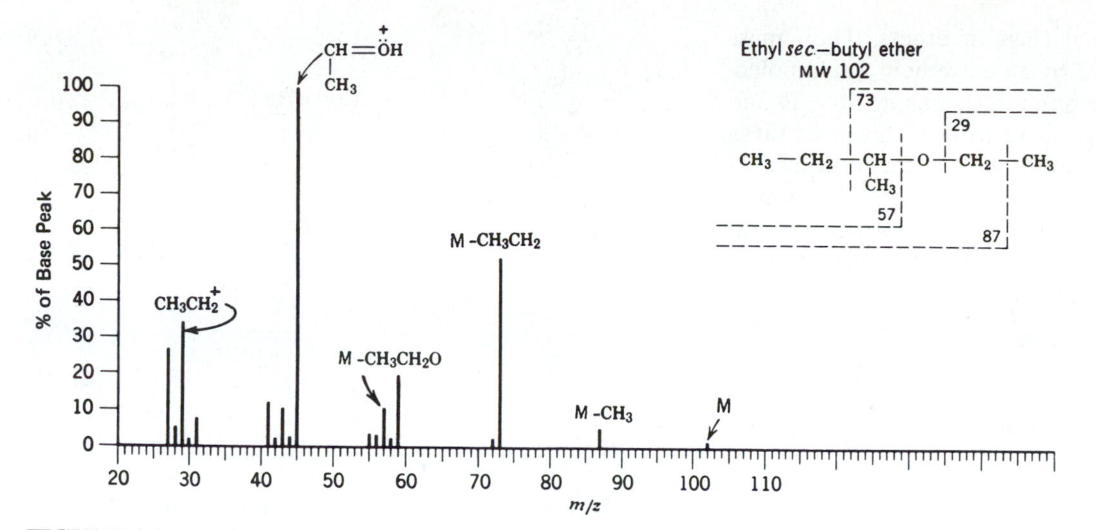

FIGURE 2.11. Ethyl *sec*-butyl ether.

1 peak obvious (H· transfer during ion-molecule collision, see Section 2.4.1).

The presence of an oxygen atom can be deduced from strong peaks at m/z 31, 45, 59, 73, ... These peaks represent the RO^+ and $ROCH_2^+$ fragments.

Fragmentation occurs in two principal ways:

1. Cleavage of the C—C bond next to the oxygen atom (α, β bond, rule 8, Section 2.7)

$$RCH_2{-}CH_2{-}\underset{\displaystyle CH_3}{\underset{|}{CH}}{-}\ddot{O}^{+\cdot}{-}CH_2{-}CH_3 \xrightarrow{-RCH_2CH_2\cdot}$$

In Figure 2.11, R = H.

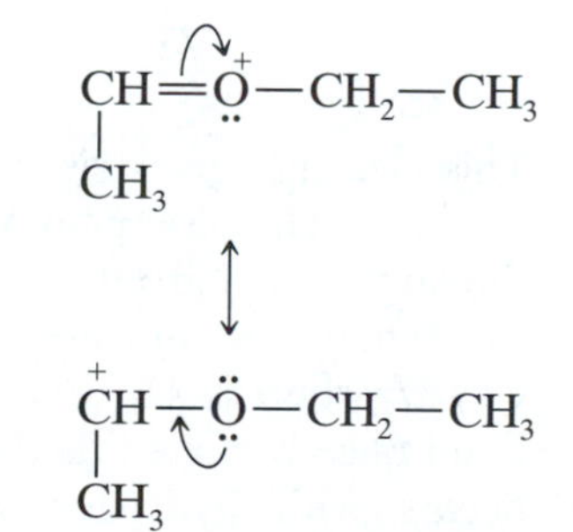

m/z 73

$$RCH_2{-}CH_2{-}\underset{\displaystyle CH_3}{\underset{|}{CH}}{-}\ddot{O}^{+\cdot}{-}CH_2{-}CH_3 \xrightarrow{-CH_3\cdot}$$

$$RCH_2{-}CH_2\underset{\displaystyle CH_3}{\underset{|}{CH}}{-}\overset{+}{O}{=}CH_2$$

In Figure 2.11, R = H, m/z 87.

$$\updownarrow$$

$$RCH_2{-}CH_2\underset{\displaystyle CH_3}{\underset{|}{CH}}{-}\ddot{O}{-}\overset{+}{C}H_2$$

One or the other of these oxygen-containing ions may account for the base peak. In the case shown, the first cleavage (i.e., at the branch positions to lose the larger fragment) is preferred. However, the first-formed fragment decomposes further by the following process, often to give the base peak (Fig. 2.11); the decomposition is important when the α carbon is substituted (See McLafferty rearrangement, Section 2.8).*

$$CH_3CH{=}\overset{+}{\ddot{O}}{-}\underset{H{-}CH_2}{\underset{|}{CH_2}} \xrightarrow{-CH_2=CH_2} \underset{\displaystyle CH_3}{\underset{|}{CH}}{=}\overset{+}{\ddot{O}}H$$

$$\updownarrow$$

$$\underset{\displaystyle CH_3}{\underset{|}{\overset{+}{C}H}}{-}\ddot{O}H$$

m/z 45

2. C—O bond cleavage with the charge remaining on the alkyl fragment.

$$R{-}\overset{+\cdot}{\ddot{O}}{-}R' \xrightarrow{-\cdot\ddot{O}R'} R^+$$

$$R{-}\overset{+\cdot}{\ddot{O}}{-}R' \xrightarrow{-R\ddot{O}\cdot} R'^+$$

The spectrum of long-chain ethers becomes dominated by the hydrocarbon pattern.

* Transfer of the hydrogen atom by a four-membered ring mechanism is an oversimplification. Deuterium labeling showed that three-, five-, and six-membered rings are also involved in longer chain compounds with relative dominance dependent on the compound. See Djerassi, C., and Fenselau, C. *J. Am. Chem. Soc.* **87,** 5747 (1965); McLafferty, F. W. and Tureček, F. *Interpretation of Mass Spectra,* 4th ed. Mill Valley, CA: University Science Books, 1993, pp. 261–262.

Acetals are a special class of ethers. Their mass spectra are characterized by an extremely weak molecular ion peak, by the prominent peaks at M − R and M − OR, and a weak peak at M − H. Each of these cleavages is mediated by an oxygen atom and thus facile. As usual, elimination of the largest group is preferred. As with aliphatic ethers, the first-formed oxygen-containing fragments can decompose further with hydrogen migration and alkene elimination.

$$\left[R{-}\overset{H}{\underset{OR}{C}}{-}OR \right]^{\cdot +} \longrightarrow \left[R{-}\underset{OR}{C}{-}OR \right]^{+} + \left[R{-}\overset{H}{C}{-}OR \right]^{+} + \left[\underset{OR}{HC}{-}OR \right]^{+}$$

Ketals behave similarly.

2.10.3.2 Aromatic Ethers The molecular ion peak of aromatic ethers is prominent. Primary cleavage occurs at the bond β to the ring, and the first-formed ion can decompose further. Thus anisole, MW 108, gives ions of *m/z* 93 and 65.

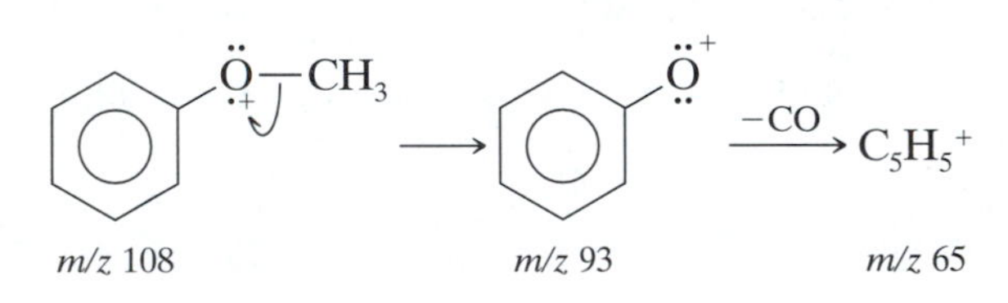

The characteristic aromatic peaks at *m/z* 78 and 77 may arise from anisole as follows:

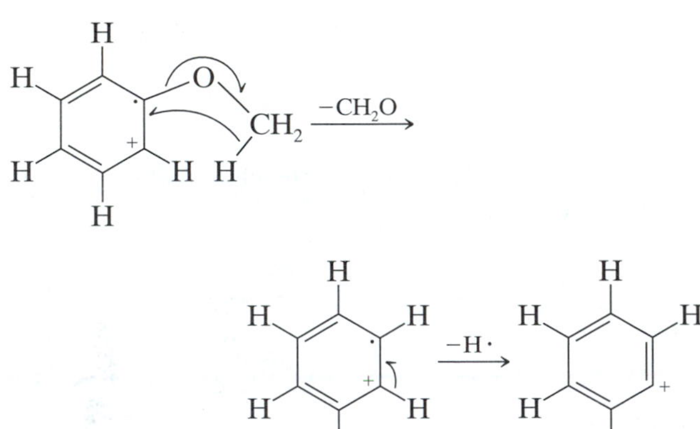

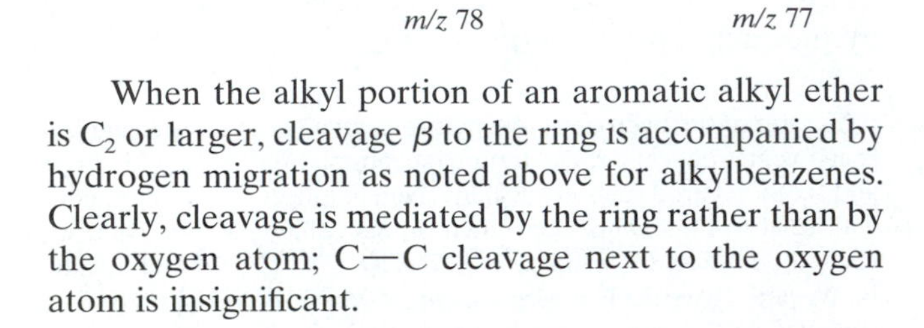

When the alkyl portion of an aromatic alkyl ether is C_2 or larger, cleavage β to the ring is accompanied by hydrogen migration as noted above for alkylbenzenes. Clearly, cleavage is mediated by the ring rather than by the oxygen atom; C—C cleavage next to the oxygen atom is insignificant.

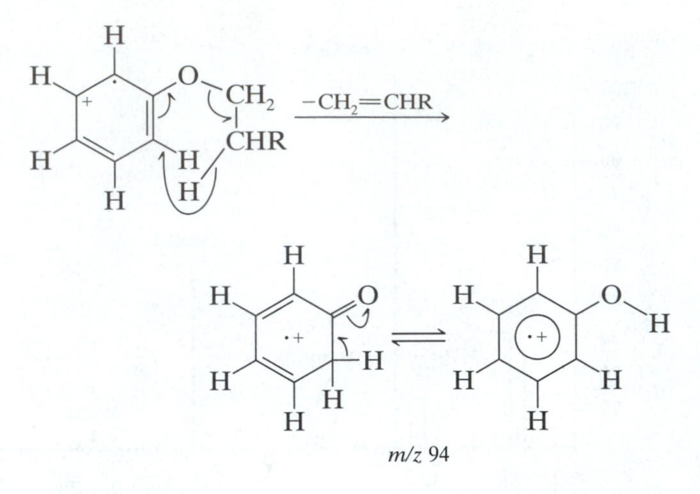

Diphenyl ethers show peaks at M − H, M − CO, and M − CHO by complex rearrangements.

2.10.4 Ketones

2.10.4.1 Aliphatic Ketones The molecular ion peak of ketones is usually quite pronounced. Major fragmentation peaks of aliphatic ketones result from cleavage at the C—C bonds adjacent to the oxygen atom, the charge remaining with the resonance-stabilized acylium ion. Thus, as with alcohols and ethers, cleavage is mediated by the oxygen atom.

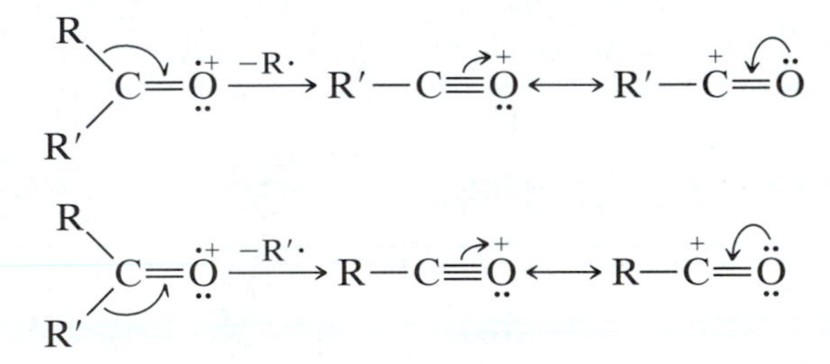

This cleavage gives rise to a peak at *m/z* 43 or 57 or 71 . . . . The base peak very often results from loss of the larger alkyl group.

When one of the alkyl chains attached to the C=O group is C_3 or longer, cleavage of the C—C bond once removed (α,β bond) from the C=O group occurs with hydrogen migration to give a major peak (McLafferty rearrangement).

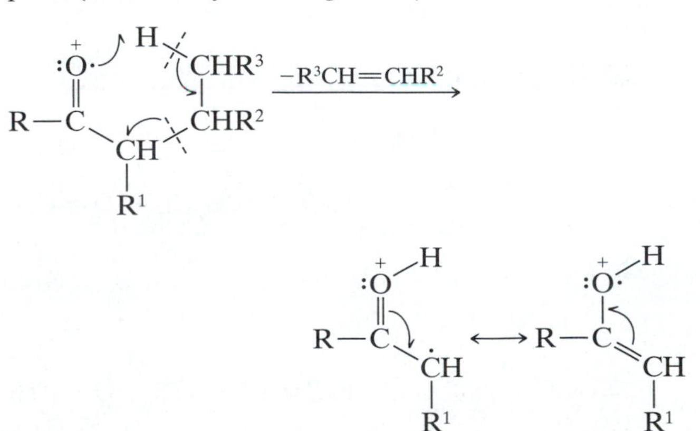

Simple cleavage of the α,β bond, which does not occur to any extent, would give an ion of low stability with two adjacent positive centers $R-\overset{\delta+}{C}(=\overset{\delta-}{O})-\overset{+}{C}H_2\cdot$ When R is C_3 or longer, the first-formed ion can cleave again with hydrogen migration:

$$\xrightarrow{-CH_2=CH_2}$$

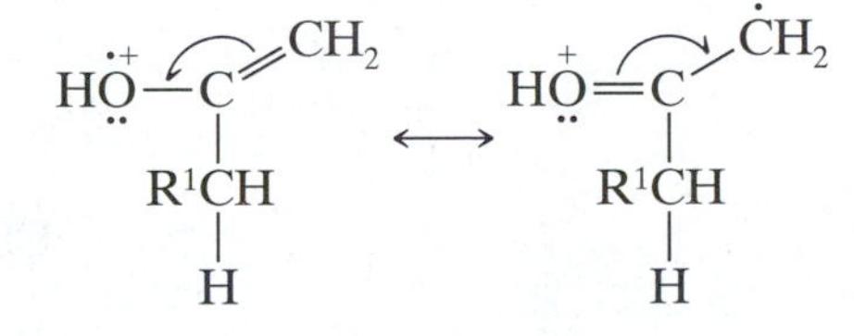

Note that in long-chain ketones the hydrocarbon peaks are indistinguishable (without the aid of high-resolution techniques) from the acyl peaks, since the mass of the C=O unit (28) is the same as two methylene units.

The multiple cleavage modes in ketones sometimes make difficult the determination of the carbon chain configuration. Reduction of the carbonyl group to a methylene group yields the corresponding hydrocarbon whose fragmentation pattern leads to the carbon skeleton.

2.10.4.2 Cyclic Ketones The molecular ion peak in cyclic ketones is prominent. As with aliphatic ketones, the primary cleavage of cyclic ketones is adjacent to the C=O group, but the ion thus formed must undergo further cleavage in order to produce a fragment. The base peak in the spectrum of cyclopentanone and of cyclohexanone is *m/z* 55. The mechanisms are similar in both cases: hydrogen shift to convert a primary radical to a conjugated secondary radical followed by formation of the resonance-stabilized ion, *m/z* 55.

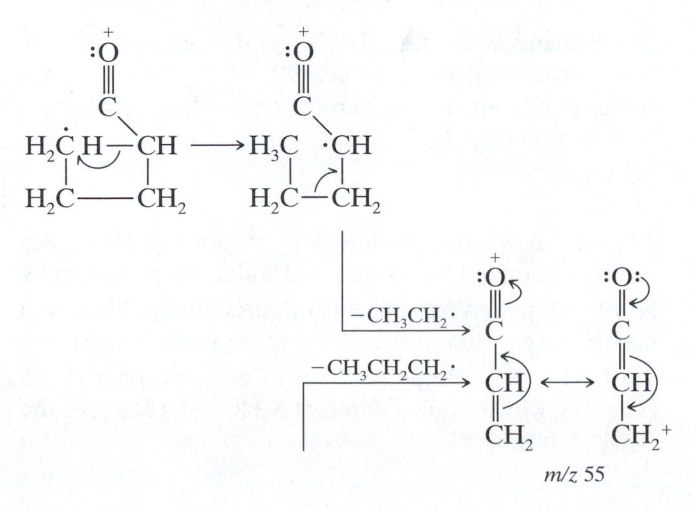

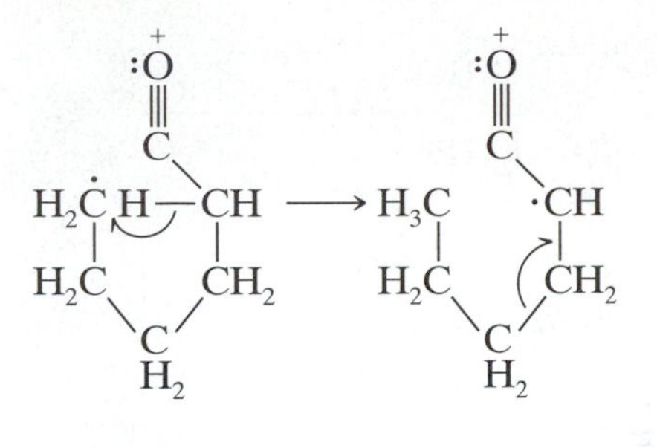

The other distinctive peaks at *m/z* 83 and 42 in the spectrum of cyclohexanone have been rationalized as follows:

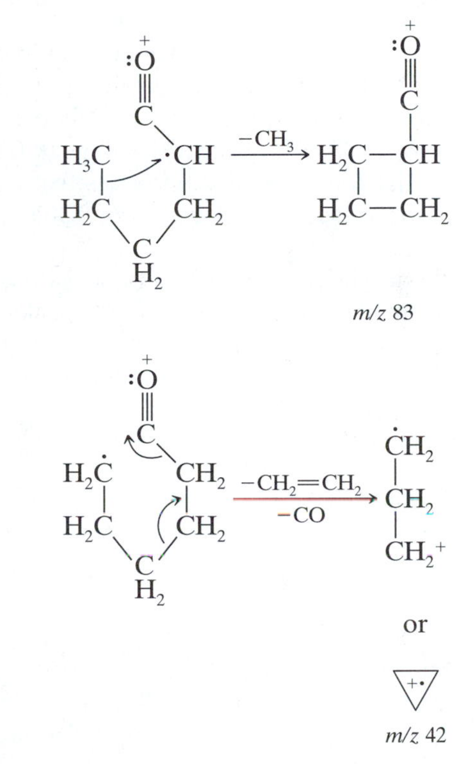

2.10.4.3 Aromatic Ketones The molecular ion peak of aromatic ketones is prominent. Cleavage of aryl alkyl ketones occurs at the bond β to the ring, leaving a characteristic $ArC\equiv\overset{+}{O}$ fragment, which usually accounts for the base peak. Loss of CO from this fragment gives the "aryl" ion (*m/z* 77 in the case of acetophenone). Cleavage of the bond adjacent to the ring to form a $RC\equiv\overset{+}{O}$ fragment is less important though somewhat enhanced by electron-withdrawing groups (and diminished by electron-donating groups) in the para position of the Ar group.

When the alkyl chain is C_3 or longer, cleavage of the C—C bond once removed from the C=O group occurs with hydrogen migration. This is the same cleavage noted for aliphatic ketones that proceeds through a cyclic transition state and results in elimination of an alkene and formation of a stable ion.

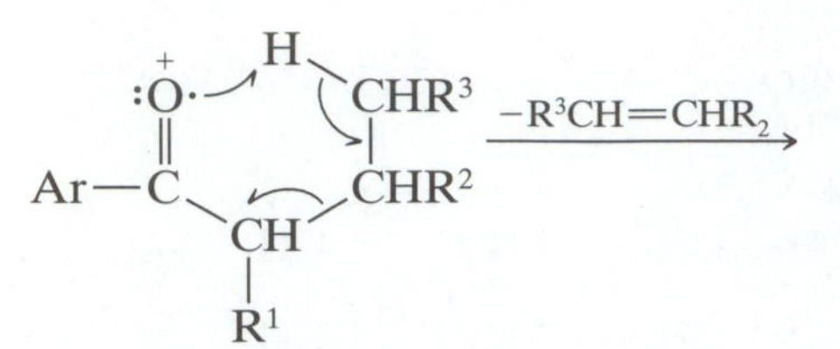

McLafferty rearrangement

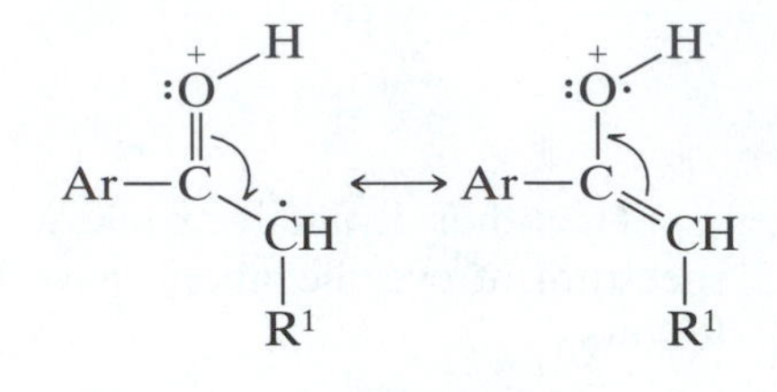

The mass spectrum of an unsymmetrical diaryl ketone, *p*-chlorobenzophenone, is displayed in Figure 2.12. The molecular ion peak (*m/z* 216) is prominent and the intensity (33.99%) of the M + 2 peak (relative to the molecular ion peak) demonstrates that chlorine is in the structure (see the discussion of Table 2.3 and Fig. 2.16 in Section 2.10.16).

Since the intensity of the *m/z* 141 peaks is about $\frac{1}{3}$ the intensity of the *m/z* 139 peak, these peaks correspond to fragments containing 1 chlorine each. The same can be said about the fragments producing the *m/z* 111 and 113 peaks.

The major peaks in Figure 2.12 arise as follows:

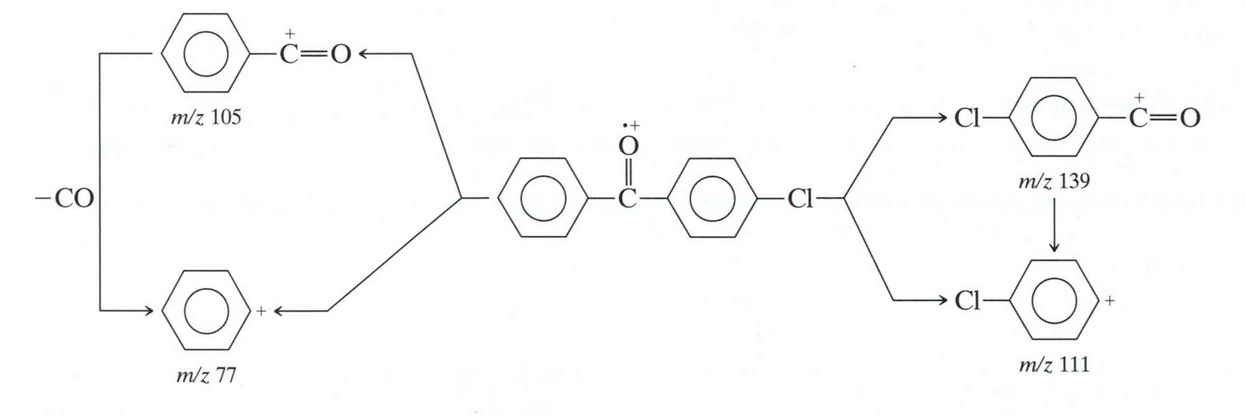

The Cl—ArCO⁺ peak is larger than the Cl—Ar⁺ peak, and the ArCO⁺ peak is larger than the Ar⁺ peak (β cleavage favored). If the fragment + 2 peaks for the Cl-substituted moieties are taken into account, there is little difference in abundance between Cl—ArCO⁺ and ArCO⁺, or between ClAr⁺ and Ar⁺; the inductive and resonance affects of the para-substituted Cl are roughly balanced out.

2.10.5 *Aldehydes*

2.10.5.1 Aliphatic Aldehydes The molecular ion peak of aliphatic aldehydes is usually discernible. Cleavage of the C—H and C—C bonds next to the oxygen atom results in an M − 1 peak and in an M − R peak (*m/z* 29, CHO^+). The M − 1 peak is a good diagnostic peak even for long-chain aldehydes, but the *m/z* 29 peak present in C_4 and higher aldehydes results from the hydrocarbon $C_2H_5^+$ ion.

In the C_4 and higher aldehydes, McLafferty cleavage of the α,β C—C bond occurs to give a major peak at *m/z* 44, 58, or 72, . . . , depending on the α substituents. This is the resonance-stabilized ion formed through the cyclic transition state as shown above for aliphatic ketones (R = H).

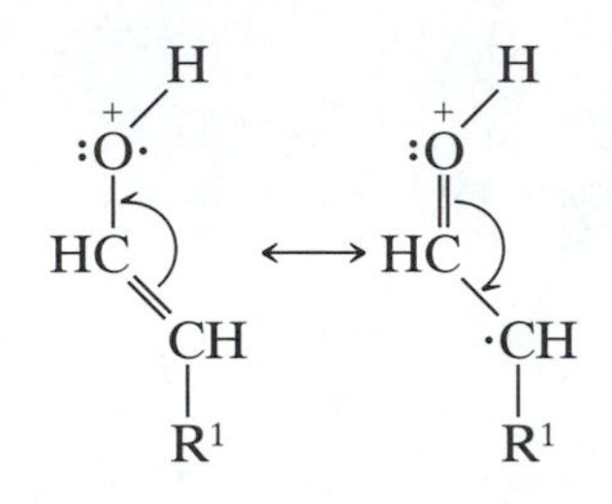

In straight-chain aldehydes, the other unique, diagnostic peaks are at M − 18 (loss of water), M − 28 (loss of ethylene), M − 43 (loss of CH_2=CH—O·), and M − 44 (loss of CH_2=CH—OH). The rearrangements leading to these peaks have been rationalized (see Budzikiewicz et al., 1967). As the chain lengthens, the hydrocarbon pattern (*m/z* 29, 43, 57, 71, . . .) becomes dominant. These features are evident in the spectrum of nonanal (Fig. 2.13).

2.10.5.2 Aromatic Aldehydes Aromatic aldehydes are characterized by a large molecular ion peak and by an M − 1 peak (Ar—C≡$\ddot{O}^+$) that is always large and may be larger than the molecular ion peak. The M − 1 ion C_6H_5C≡$\ddot{O}^+$ eliminates CO to give the phenyl ion (*m/z* 77), which in turn eliminates HC≡CH to give the $C_4H_3^+$ ion (*m/z* 51).

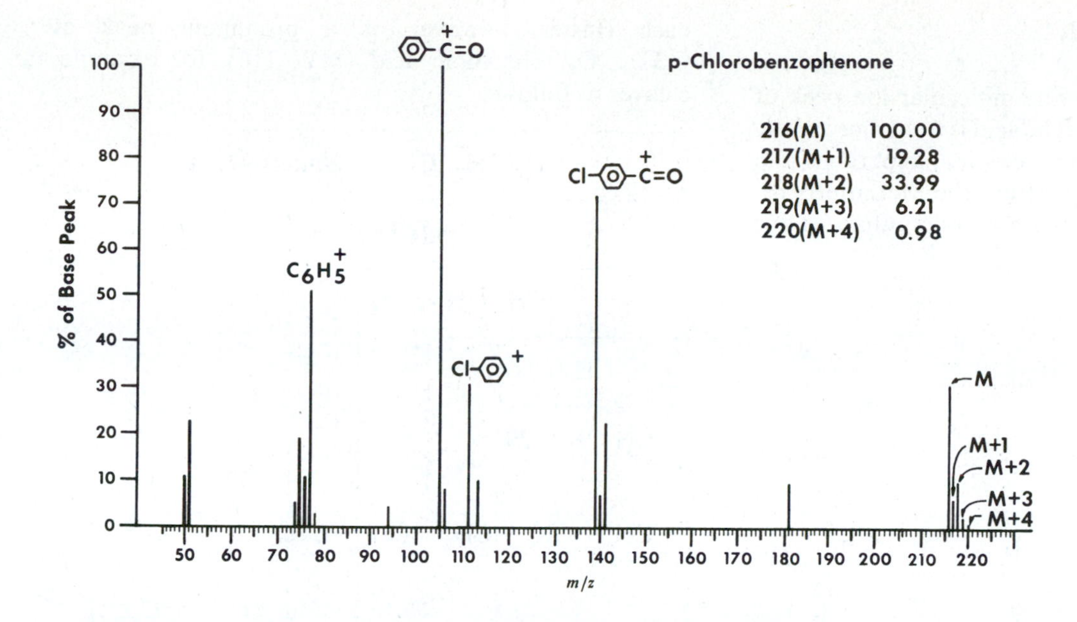

FIGURE 2.12. *p*-Chlorobenzophenone. The M peak is arbitrarily set in the table above at intensity 100% for discussion of the molecular ion cluster.

Table 2.3 Intensities of Isotope Peaks (Relative to the Molecular Ion) for Combinations of Bromine and Chlorine

Halogen Present	% M + 2	% M + 4	% M + 6	% M + 8	% M + 10	% M + 12
Br	97.9					
Br_2	195.0	95.5				
Br_3	293.0	286.0	93.4			
Cl	32.6					
Cl_2	65.3	10.6				
Cl_3	97.8	31.9	3.47			
Cl_4	131.0	63.9	14.0	1.15		
Cl_5	163.0	106.0	34.7	5.66	0.37	
Cl_6	196.0	161.0	69.4	17.0	2.23	0.11
BrCl	130.0	31.9				
Br_2Cl	228.0	159.0	31.2			
Cl_2Br	163.0	74.4	10.4			

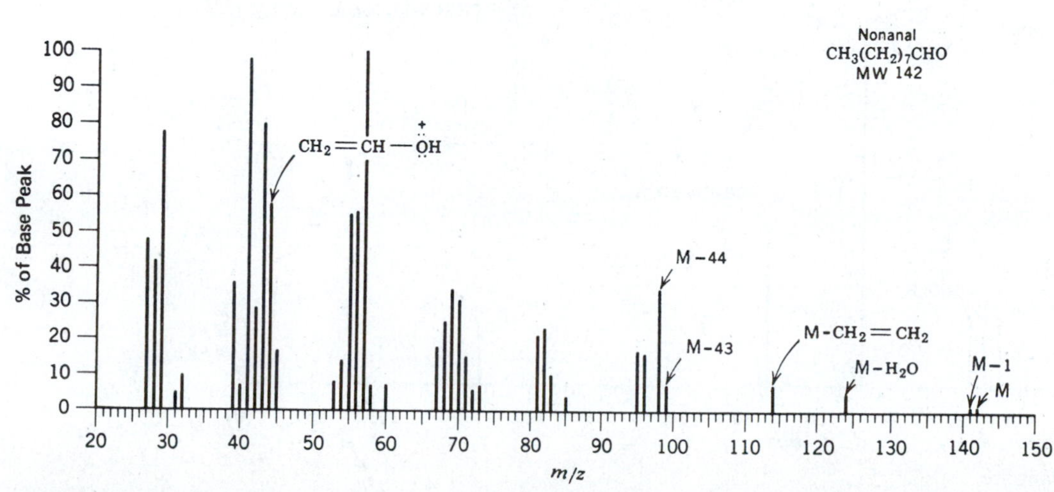

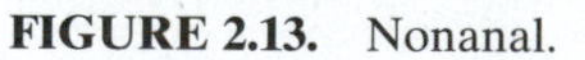

FIGURE 2.13. Nonanal.

2.10.6 Carboxylic Acids

2.10.6.1 Aliphatic Acids The molecular ion peak of a straight-chain monocarboxylic acid is weak but usually discernible. The most characteristic (sometimes the base) peak is *m/z* 60 resulting from the McLafferty rearrangement. Branching at the α carbon enhances this cleavage.

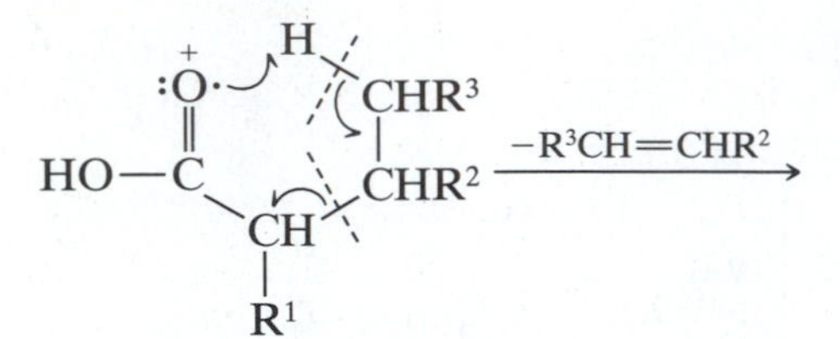

McLafferty rearrangement

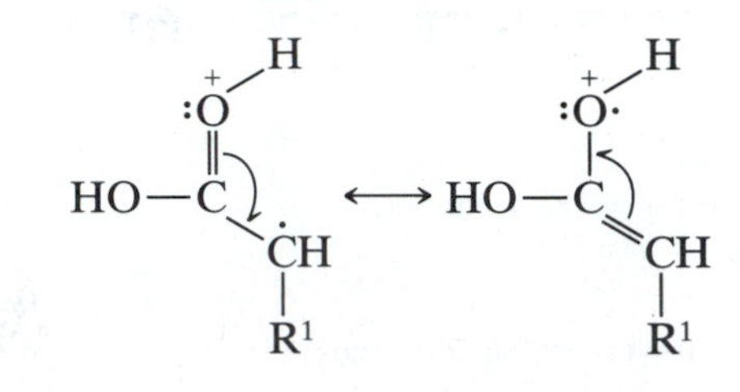

In short-chain acids, peaks at M − OH and M − CO_2H are prominent; these represent cleavage of bonds next to C=O. In long-chain acids, the spectrum consists of two series of peaks resulting from cleavage at each C—C bond with retention of charge either on the oxygen-containing fragment (*m/z* 45, 59, 73, 87, . . .) or on the alkyl fragment (*m/z* 29, 43, 57, 71, 85, . . .). As previously discussed, the hydrocarbon pattern also shows peaks at *m/z* 27, 28; 41, 42; 55, 56; 69, 70; . . . . In summary, besides the McLafferty rearrangement peak, the spectrum of a long-chain acid resembles the series of "hydrocarbon" clusters at intervals of 14 mass units. In each cluster, however, is a prominent peak at $C_nH_{2n-1}O_2$. Hexanoic acid (MW 116), for example, cleaves as follows:

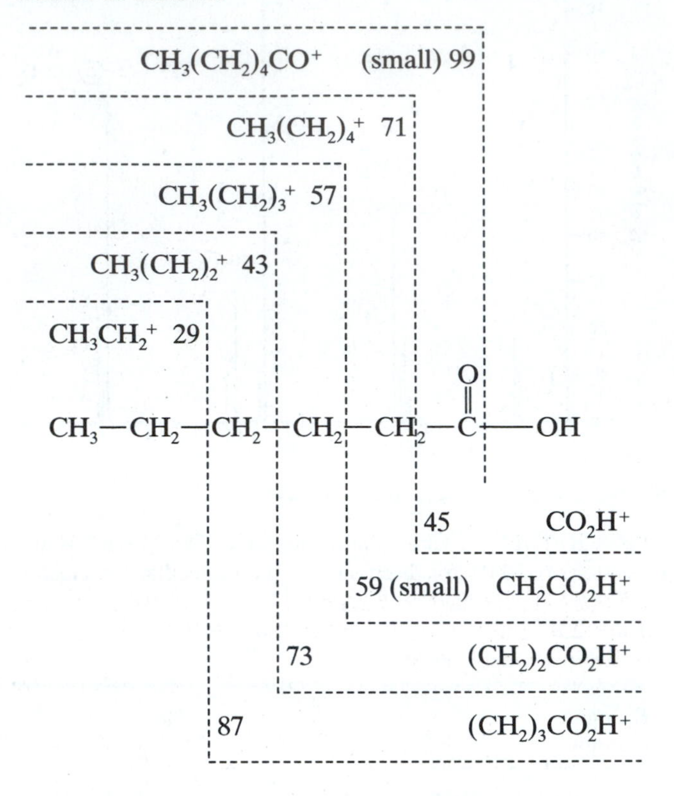

Dibasic acids are usually converted to esters to increase volatility. Trimethylsilyl esters are often successful.

2.10.6.2 Aromatic Acids The molecular ion peak of aromatic acids is large. The other prominent peaks are formed by loss of OH (M − 17) and of CO_2H (M − 45). Loss of H_2O (M − 18) is noted if a hydrogen-bear-

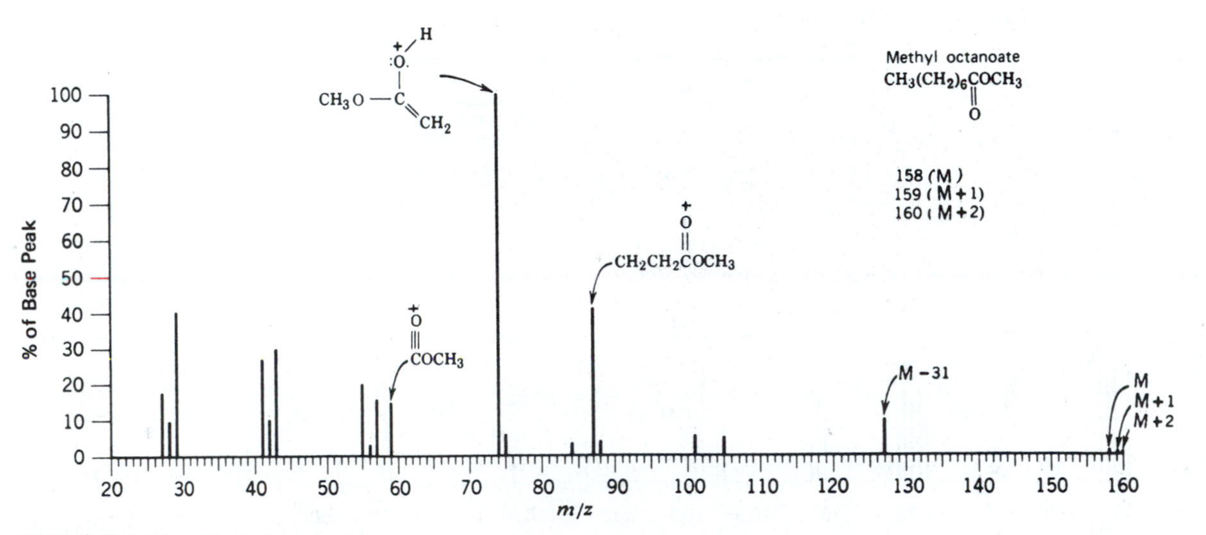

FIGURE 2.14. Methyl octanoate.

ing ortho group is available. This is one example of the general "ortho effect" noted when the substituents can be in a six-membered transition state to facilitate loss of a neutral molecule of H_2O, ROH, or NH_3.

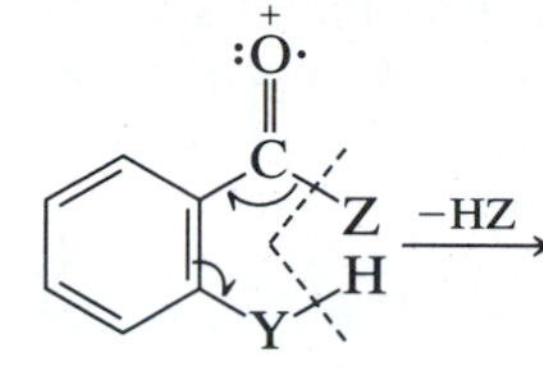

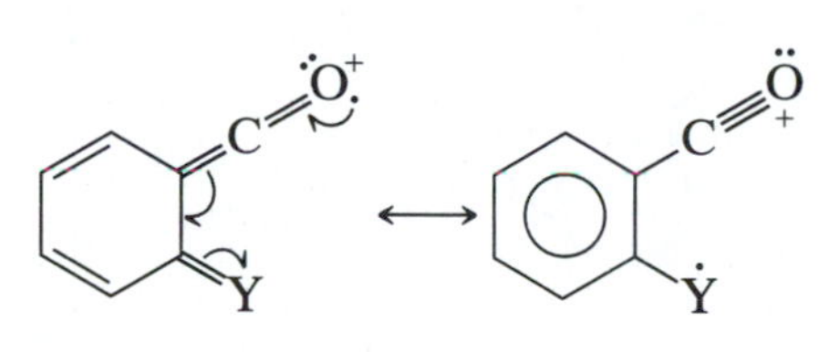

where Z = OH, OR, NH_2; Y = CH_2, O, NH.

2.10.7 *Carboxylic Esters*

2.10.7.1 Aliphatic Esters The molecular ion peak of a methyl ester of a straight-chain aliphatic acid is usually distinct. Even waxes usually show a discernible molecular ion peak. The molecular ion peak is weak in the range *m/z* 130 to ~200, but becomes somewhat more intense beyond this range. The most characteristic peak results from the familiar McLafferty rearrangement and cleavage one bond removed from the C=O group. Thus a methyl ester of an aliphatic acid unbranched at the α carbon gives a strong peak at *m/z* 74, which, in fact, is the base peak in straight-chain methyl esters from C_6 to C_{26}. The alcohol moiety and/or the α substituent can often be deduced by the location of the peak resulting from this cleavage.

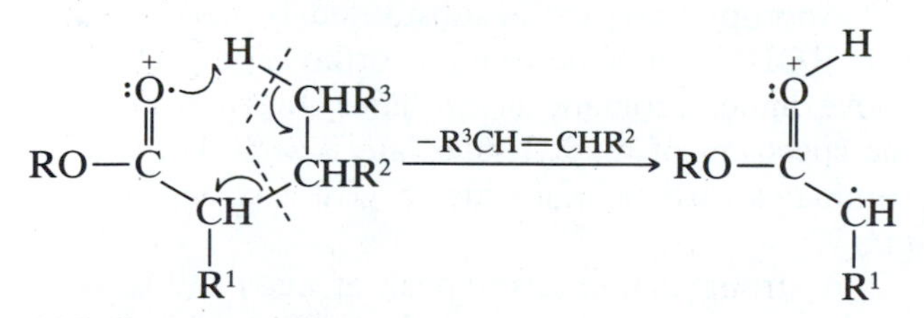

McLafferty rearrangement

Four ions can result from bond cleavage next to C=O.

$$[\text{R}+\text{C(=O)}-\text{OR}']^{\cdot+} \longrightarrow \cdot\text{R and } [\text{C(=O)}-\text{OR}']^{+}$$

$$[\text{R}-\text{C(=O)}+\text{OR}']^{\cdot+} \longrightarrow [\text{R}-\text{C(=O)}]^{+} \text{ and } \cdot\text{OR}'$$

The ion R^+ is prominent in the short-chain esters but diminishes rapidly with increasing chain length and is barely perceptible in methyl hexanoate. The ion $\text{R}-\text{C}{\equiv}\overset{+}{\text{O}}$ gives an easily recognizable peak for esters. In methyl esters it occurs at M − 31. It is the base peak in methyl acetate and is still 4% of the base peak in the C_{26} methyl ester. The ions $[\text{OR}']^+$ and $[\text{C(=O)OR}']^+$ are usually of little importance. The latter is discernible when R′ = CH_3 (see *m/z* 59 peak of Fig. 2.14).

First, consider esters in which the acid portion is the predominant portion of the molecule. The fragmentation pattern for methyl esters of straight-chain acids can be described in the same terms used for the pattern of the free acid. Cleavage at each C—C bond gives an alkyl ion (*m/z* 29, 43, 57, . . .) and an oxygen-containing ion, $C_nH_{2n-1}O_2^+$ (59, 73, 87, . . .). Thus, there are hydrocarbon clusters at intervals of 14 mass units; in each cluster is a prominent peak at $C_nH_{2n-1}O_2$. The peak (*m/z* 87) formally represented by the ion $[CH_2CH_2COOCH_3]^+$ is always more intense than its homologs, but the reason is not immediately obvious. However, it seems clear that the $C_nH_{2n-1}O_2$ ions do not at all arise from simple cleavage.

The spectrum of methyl octanoate is presented as Figure 2.14. This spectrum illustrates one difficulty in using the M + 1 peak to arrive at a molecular formula (previously mentioned, Section 2.4.1). The measured value for the M + 1 peak is 12%. The calculated value is 10.0%. The measured value is high because of an ion–molecule reaction because a relatively large sample was used to see the weak molecular ion peak.

Now let us consider esters in which the alcohol portion is the predominant portion of the molecule. Esters of fatty alcohols (except methyl esters) eliminate a molecule of acid in the same manner that alcohols eliminate water. A scheme similar to that described earlier for alcohols, involving a single hydrogen transfer to the alcohol oxygen of the ester, can be written. An alternative mechanism involves a hydride transfer to the carbonyl oxygen (McLafferty rearrangement).

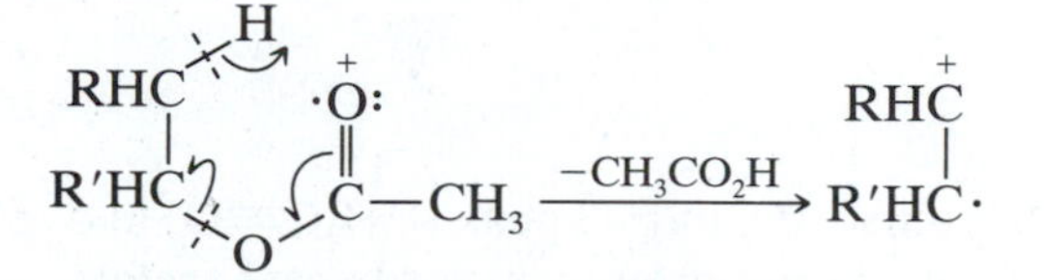

The preceding loss of acetic acid is so facile in steroidal acetates that they frequently show no detectable molecular ion peak. Steroidal systems also seem unusual in that they often display significant molecular ions as alcohols, even when the corresponding acetates do not.

Esters of long-chain alcohols show a diagnostic peak at *m/z* 61, 75, or 89, . . . from elimination of the alkyl moiety and transfer of *two* hydrogen atoms to the fragment containing the oxygen atoms.

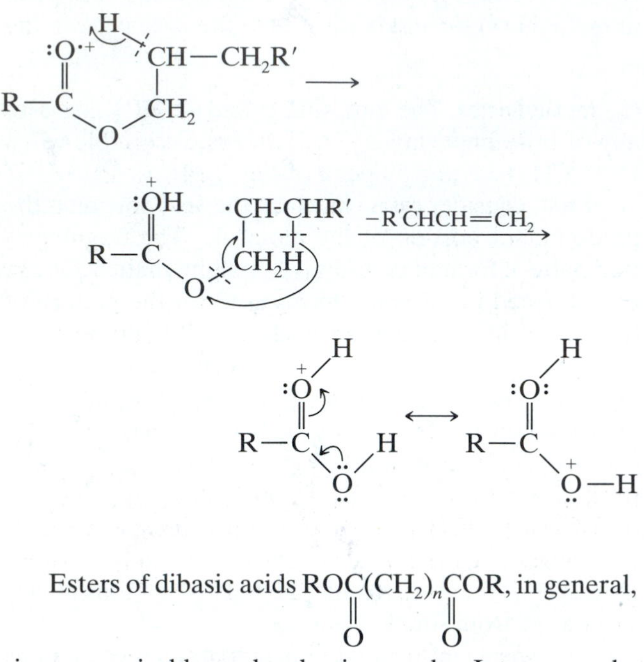

Esters of dibasic acids $\mathrm{RO\overset{\displaystyle O}{\underset{}{C}}(CH_2)_n\overset{\displaystyle O}{C}OR}$ (both C=O), in general, give recognizable molecular ion peaks. Intense peaks are found at $[\mathrm{ROC(=O)(CH_2)_nC(=O)}]^+$ and at $[\mathrm{ROC(=O)(CH_2)_n}]^+$.

2.10.7.2 Benzyl and Phenyl Esters Benzyl acetate (also furfuryl acetate and other similar acetates) and phenyl acetate eliminate the neutral molecule ketene; frequently this gives rise to the base peak.

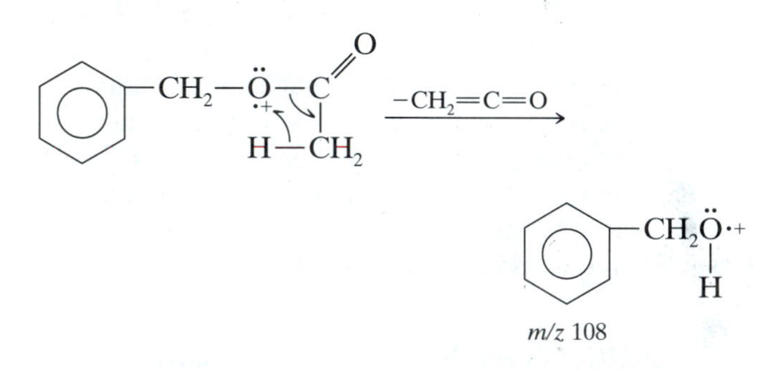

Of course, the *m/z* 43 peak ($CH_3C{\equiv}\overset{+}{O}$) and *m/z* 91 ($C_7H_7^+$) peaks are prominent for benzyl acetate.

2.10.7.3 Esters of Aromatic Acids The molecular ion peak of methyl esters of aromatic acids is prominent. As the size of the alcohol moiety increases, the intensity of the molecular ion peak decreases rapidly to practically zero at C_5. The base peak results from elimination of $\cdot$OR, and elimination of $\cdot$COOR accounts for another prominent peak. In methyl esters, these peaks are at M − 31, and M − 59, respectively.

As the alkyl moiety increases in length, three modes of cleavage become important: (1) McLafferty rearrangement, (2) rearrangement of two hydrogen atoms with elimination of an allylic radical, and (3) retention of the positive charge by the alkyl group.

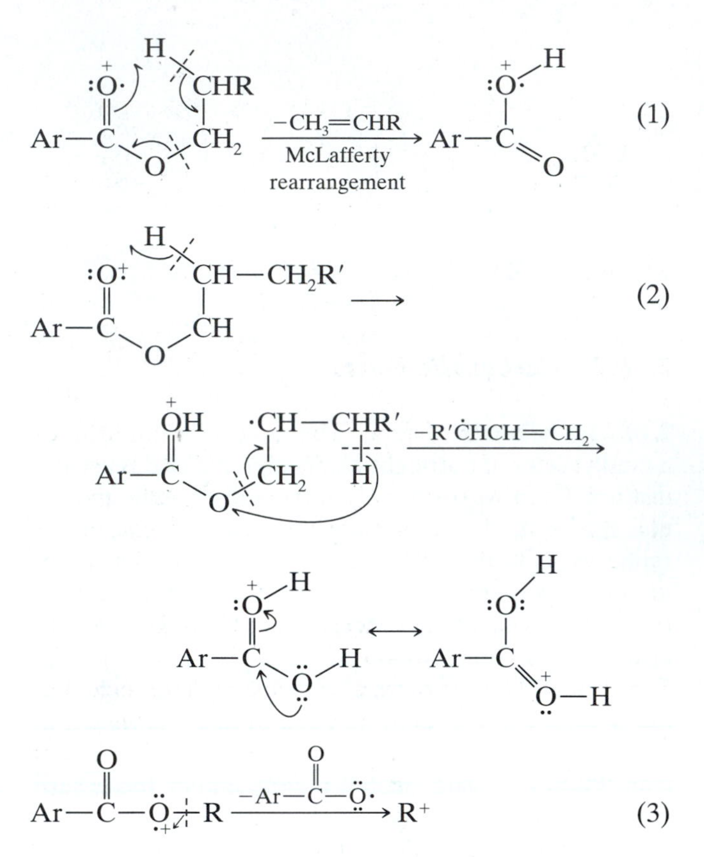

Appropriately, ortho-substituted benzoates eliminate ROH through the general "ortho" effect described above under aromatic acids. Thus, the base peak in the spectrum of methyl salicylate is *m/z* 120; this ion eliminates carbon monoxide to give a strong peak at *m/z* 92.

A strong characteristic peak at mass 149 is found in the spectra of all esters of phthalic acid, starting with the diethyl ester. This peak is not significant in the dimethyl or methyl ethyl ester of phthalic acid, nor in esters of isophthalic or terephthalic acids, all of which give the expected peaks at M − R, M − 2R, M − CO_2R, and M − $2CO_2R$. Since long-chain phthalate esters are widely used as plasticizers, a strong peak at *m/z* 149 may indicate contamination. The *m/z* 149 fragment is probably formed by two ester cleavages involving the shift of two hydrogen atoms and then another hydrogen atom, followed by elimination of H_2O.

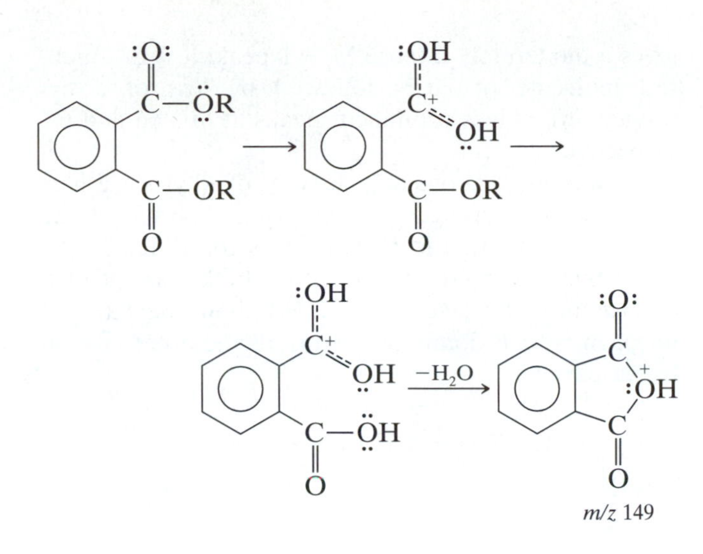

2.10.8 *Lactones*

The molecular ion peak of five-membered ring lactones is distinct but is weaker when an alkyl substitutent is present at C_4. Facile cleavage of the side chain at C_4 (rules 3 and 8, Section 2.7) gives a strong peak at M − alkyl.

The base peak (*m/z* 56) of γ-valerolactone and the same strong peak of butyrolactone probably arise as follows:

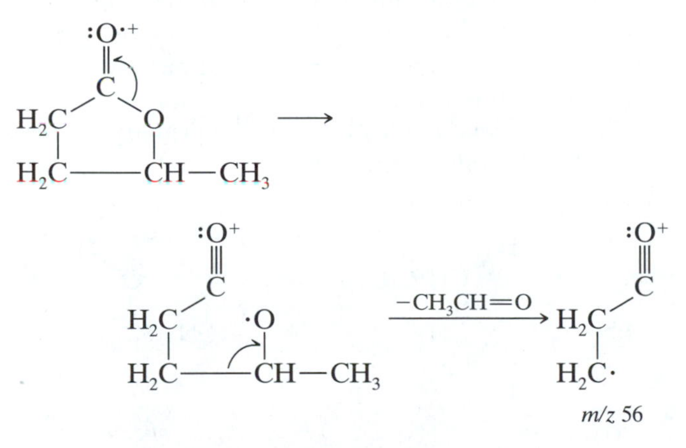

m/z 56

Labeling experiments indicate that some of the *m/z* 56 peak in γ-valerolactone arises from the $C_4H_8^+$ ion. The other intense peaks in γ-valerolactone are at *m/z* 27 ($C_2H_3^+$), 28 ($C_2H_4^+$), 29 ($C_2H_5^+$), 41 ($C_3H_5^+$), and 43 ($C_3H_7^+$), and 85 ($C_4H_5O_2^+$, loss of the methyl group). In butyrolactone, there are strong peaks at *m/z* 27, 28, 29, 41, and 42 ($C_3H_6^+$).

2.10.9 *Amines*

2.10.9.1 Aliphatic Amines The molecular ion peak of an aliphatic monoamine is an odd number, but it is usually quite weak and, in long-chain or highly branched amines, undetectable. The base peak frequently results from C—C cleavage next (α,β) to the atom (rule 8, Section 2.7); for primary amines unbranched at the α carbon, this is *m/z* 30 ($CH_2NH_2^+$). This cleavage accounts for the base peak in all primary amines and secondary and tertiary amines that are not branched at the α carbon. Loss of the largest branch from the α-C atom is preferred.

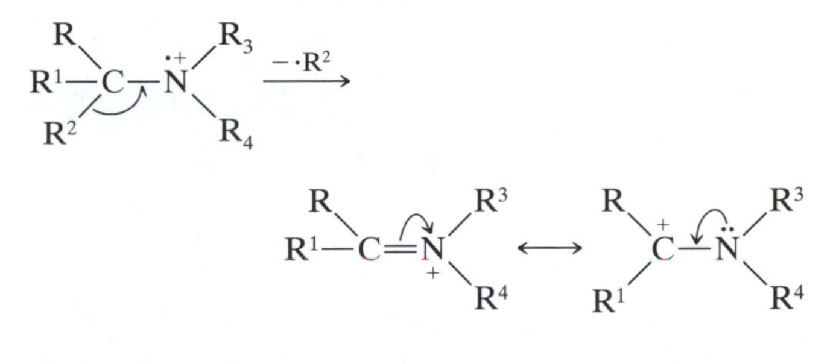

where $R^2 > R^1$ or R.

When R and/or R^1 = H, an M − 1 peak is usually visible. This is the same type of cleavage noted above for alcohols. The effect is more pronounced in amines because of the better resonance stabilization of the ion fragment by the less electronegative N atom compared with the O atom.

Primary straight-chain amines show a homologous series of peaks of progressively decreasing intensity (the cleavage at the ϵ bond is slightly more important than at the neighboring bonds) at *m/z* 30, 44, 58, . . . resulting from cleavage at C—C bonds successively removed from the nitrogen atom with retention of the charge on the N-containing fragment. These peaks are accompanied by the hydrocarbon pattern of C_nH_{2n+1}, C_nH_{2n}, and C_nH_{2n-1} ions. Thus, we note characteristic clusters at intervals of 14 mass units, each cluster containing a peak resulting from a $C_nH_{2n+2}N$ ion. Because of the very facile cleavage to form the base peak, the fragmentation pattern in the high mass region becomes extremely weak.

Cyclic fragments apparently occur during the fragmentation of longer chain amines.

$$R-CH_2-(CH_2)_n-\overset{\cdot+}{N}H_2 \longrightarrow R\cdot + CH_2-(CH_2)_n-\overset{+}{N}H_2$$

n = 3,4 *m/z* 72, 86

A peak at *m/z* 30 is good though not conclusive evidence for a straight-chain primary amine. Further decomposition of the first-formed ion from a secondary or tertiary amine leads to a peak at *m/z* 30, 44, 58, 72, ... This is a process similar to that described for aliphatic alcohols and ethers above and, similarly, is enhanced by branching at one of the α-carbon atoms*:

* See footnote in Section 2.10.3.

$$RCH_2\text{—}\underset{\underset{R''}{|}}{CH}\text{—}\dot{\overset{+}{N}}H\text{—}CH_2CH_2R' \xrightarrow{-RCH_2\cdot}$$

$$\underset{\underset{R''}{|}}{CH}{=}\overset{+}{N}H\text{—}CH_2 \quad H\text{—}CHR' \xrightarrow{-CH_2{=}CHR'} \underset{\underset{R''}{|}}{CH}{=}\overset{+}{N}H_2$$

$R'' = CH_3$, m/z 44, more intense
$R'' = H$, m/z 30, less intense

Cleavage of amino acid esters occurs at both C—C bonds (*a*, *b* below) next to the nitrogen atom, loss of the carbalkoxy group being preferred (*a*). The aliphatic amine fragment decomposes further to give a peak at m/z 30.

$$\underset{\underset{\overset{\cdot +}{N}H_2}{\|}}{CH}\text{—}COOR' \xleftarrow{b} RCH_2CH_2 \overset{b}{\vdots} \underset{\underset{\cdot\overset{+}{N}H_2}{|}}{CH} \overset{a}{\vdots} COOR' \xrightarrow{a}$$

$$RCH_2CH_2\underset{\underset{\overset{+}{N}H_2}{\|}}{CH} \xrightarrow{-RCH{=}CH_2} CH_2{=}\overset{+}{N}H_2$$

m/z 30

2.10.9.2 Cyclic Amines In contrast to acyclic amines, the molecular ion peaks of cyclic amines are usually intense unless there is substitution at the α position; for example, the molecular ion peak of pyrrolidine is strong. Primary cleavage at the bonds next to the N atom leads either to loss of an α-hydrogen atom to give a strong M − 1 peak or to opening of the ring; the latter event is followed by elimination of ethylene to give $\cdot CH_2\overset{+}{N}H{=}CH_2$ (m/z 43, base peak), hence by loss of a hydrogen atom to give $CH_2{=}\overset{+}{N}{=}CH_2$ (m/z 42). *N*-Methyl pyrrolidine also gives a $C_2H_4N^+$ (m/z 42) peak, apparently by more than one pathway.

Piperidine likewise shows a strong molecular ion and M − 1 (base) peak. Ring opening followed by several available sequences leads to characteristic peaks at m/z 70, 57, 56, 44, 43, 42, 30, 29, and 28. Substituents are cleaved from the ring (rule 6, Section 2.7).

2.10.9.3 Aromatic Amines (Anilines) The molecular ion peak (odd number) of an aromatic monoamine is intense. Loss of one of the amino H atoms of aniline gives a moderately intense M − 1 peak; loss of a neutral molecule of HCN followed by loss of a hydrogen atom gives prominent peaks at m/z 66 and 65, respectively.

It was noted above that cleavage of alkyl aryl ethers occurs with rearrangement involving cleavage of the ArO—R bond; that is, cleavage was controlled by the ring rather than by the oxygen atom. In the case of alkyl aryl amines, cleavage of the C—C bond next to the nitrogen atom is dominant; that is, the heteroatom controls cleavage.

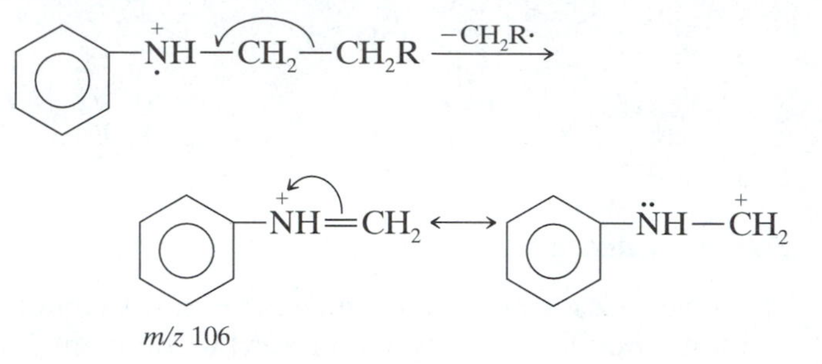

2.10.10 *Amides*

2.10.10.1 Aliphatic Amides The molecular ion peak of straight-chain monoamides is usually discernible. The dominant modes of cleavage depend on the length of the acyl moiety, and on the lengths and number of the alkyl groups attached to the nitrogen atom.

The base peak in all straight-chain primary amides higher than propionamide results from the familiar McLafferty rearrangement.

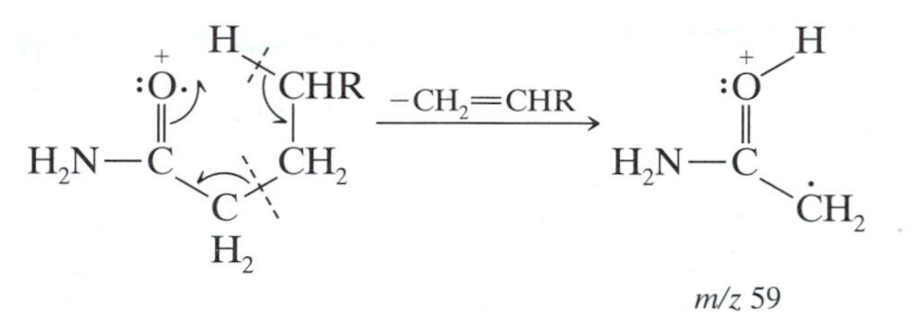

Branching at the α carbon (CH_3, etc.) gives a homologous peak at m/z 73 or 87, ...

Primary amides give a strong peak at m/z 44 from cleavage of the $R\text{—}CONH_2$ bond: ($\overset{+}{O}{\equiv}C\text{—}\ddot{N}H_2 \longleftrightarrow \ddot{O}{=}C{=}\overset{+}{N}H_2$); this is the base peak in $C_1\text{—}C_3$ primary amides and in isobutyramide. A moderate peak at m/z 86 results from γ,δ C—C cleavage, possibly accompanied by cyclization.

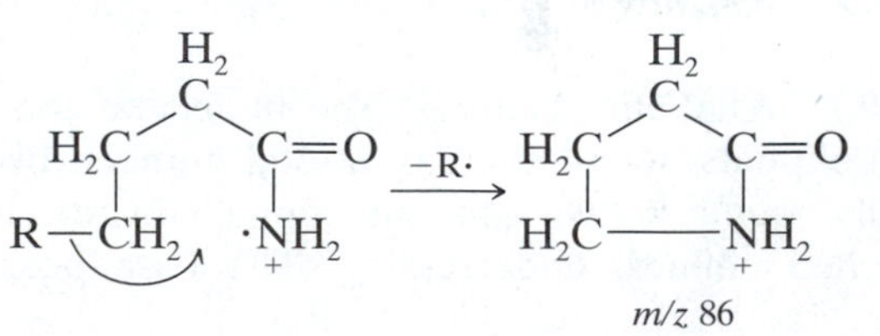

Secondary and tertiary amides with an available hydrogen on the γ-carbon of the acyl moiety and methyl groups on the N atom show the dominant peak resulting from the McLafferty rearrangement. When the *N*-alkyl groups are C_2 or longer and the acyl moiety is shorter than C_3, another mode of cleavage predominates. This is cleavage of the *N*-alkyl group β to the N atom, and cleavage of the carbonyl C—N bond with migration of an α-hydrogen atom of the acyl moiety.

$$R{-}CH_2{-}C(=O){-}\dot{N}^{+}H{-}CH_2{-}R' \xrightarrow{-R'\cdot}$$

$$R{-}CH_2{-}C(=O){-}\overset{+}{N}H{=}CH_2 \xrightarrow{-RHC{=}C{=}O} \overset{+}{N}H_2{=}CH_2 \quad m/z\ 30$$

2.10.10.2 Aromatic Amides Benzamide (Fig. 2.1) is a typical example. Loss of NH_2 from the molecular ion yields a resonance-stabilized benzoyl cation that in turn undergoes cleavage to a phenyl cation.

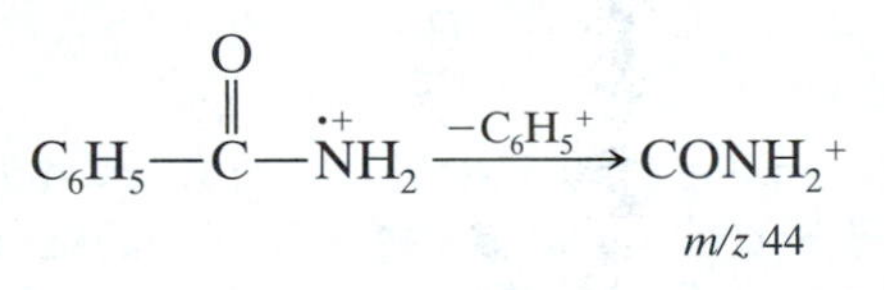

A separate fragmentation pathway gives rise to a modest m/z 44 peak.

$$C_6H_5{-}C(=O){-}\overset{\cdot +}{N}H_2 \xrightarrow{-C_6H_5^{+}} CONH_2^{+} \quad m/z\ 44$$

2.10.11 *Aliphatic Nitriles*

The molecular ion peaks of aliphatic nitriles (except for acetonitrile and propionitrile) are weak or absent, but the M + 1 peak can usually be located by its behavior on increasing inlet pressure or decreasing repeller voltage (Section 2.5). A weak but diagnostically useful M − 1 peak is formed by loss of an α hydrogen to form the stable ion: $R\dot{C}H{-}C{\equiv}N^{\cdot+} \longleftrightarrow RCH{=}C{=}\ddot{N}^{+}$.

The base peak of straight-chain nitriles between C_4 and C_9 is m/z 41. This peak is the ion resulting from hydrogen rearrangement in a six-membered transition state.

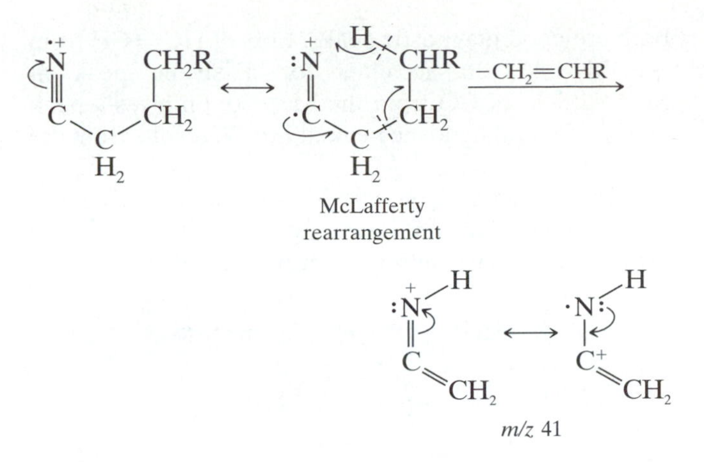

However, this peak lacks diagnostic value because of the presence of the C_3H_5 (m/z 41) for all molecules containing a hydrocarbon chain.

A peak at m/z 97 is characteristic and intense (sometimes the base peak) in straight-chain nitriles C_8 and higher. The following mechanism has been depicted:

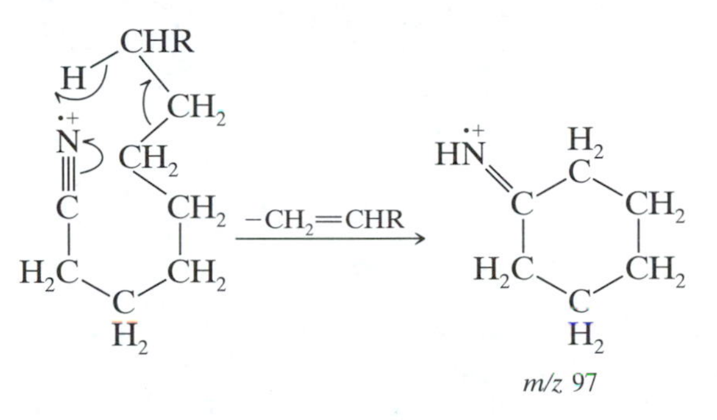

Simple cleavage at each C—C bond (except the one next to the N atom) gives a characteristic series of homologous peaks of even mass number down the entire length of the chain (m/z 40, 54, 68, 82, . . .) resulting from the $(CH_2)_nC{\equiv}N^{+}$ ions. Accompanying these peaks are the usual peaks of the hydrocarbon pattern.

2.10.12 *Nitro Compounds*

2.10.12.1 Aliphatic Nitro Compounds The molecular ion peak (odd number) of an aliphatic mononitro compound is weak or absent (except in the lower homologs). The main peaks are attributable to the hydrocarbon fragments up to M − NO_2. Presence of a nitro group is indicated by an appreciable peak at m/z 30 (NO^{+}) and a smaller peak at mass 46 (NO_2^{+}).

2.10.12.2 Aromatic Nitro Compounds The molecular ion peak of aromatic nitro compounds (odd number for one N atom) is strong. Prominent peaks result from elimination of an NO_2 radical (M − 46, the base peak in nitrobenzene), and of a neutral NO molecule with rearrangement to form the phenoxy cation (M − 30);

both are good diagnostic peaks. Loss of HC≡CH from the M − 46 ion accounts for a strong peak at M − 72; loss of CO from the M − 30 ion gives a peak at M − 58. A diagnostic peak at *m/z* 30 results from the NO^+ ion.

The isomeric *o*-, *m*-, and *p*-nitroanilines each give a strong molecular ion (even number). They all give prominent peaks resulting from two sequences.

$$m/z\ 138\ (\mathrm{M}) \xrightarrow{-\mathrm{NO_2}} m/z\ 92 \xrightarrow{-\mathrm{HCN}} m/z\ 65$$

$$m/z\ 138\ (\mathrm{M}) \xrightarrow{-\mathrm{NO}} m/z\ 108 \xrightarrow{-\mathrm{CO}} m/z\ 80$$

Aside from differences in intensities, the three isomers give very similar spectra. The meta and para compounds give a small peak at *m/z* 122 from loss of an O atom, whereas the ortho compound eliminates ·ÖH as follows to give a small peak at *m/z* 121.

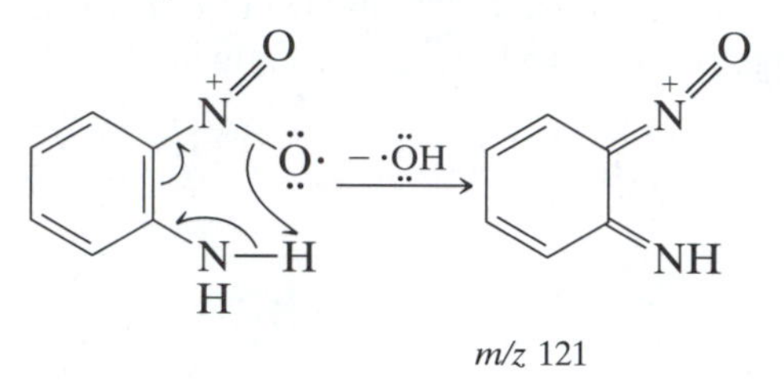

m/z 121

2.10.13 Aliphatic Nitrites

The molecular ion peak (odd number) of aliphatic nitrites (one N present) is weak or absent. The peak at *m/z* 30 (NO^+) is always large and is often the base peak. There is a large peak at *m/z* 60 ($CH_2{=}\overset{+}{O}NO$) in all nitrites unbranched at the α carbon; this represents cleavage of the C—C bond next to the ONO group. An α branch can be identified by a peak at *m/z* 74, 88, or 102, ... Absence of a large peak at *m/z* 46 permits differentiation from nitro compounds. Hydrocarbon peaks are prominent, and their distribution and intensities describe the arrangement of the carbon chain.

2.10.14 Aliphatic Nitrates

The molecular ion peak (odd number) of aliphatic nitrates (one nitrogen present) is weak or absent. A prominent (frequently the base) peak is formed by cleavage of the C—C bond next to the ONO_2 group with loss of the heaviest alkyl group attached to the α carbon.

$$\mathrm{R{-}\underset{\displaystyle R'}{CH}{-}\overset{\cdot+}{O}{-}NO_2} \xrightarrow{-\mathrm{R\cdot}} \mathrm{\underset{\displaystyle R'}{CH}{=}\overset{+}{O}{-}NO_2}$$

where R > R′. The NO_2^+ peak at *m/z* 46 is also prominent. As in the case of aliphatic nitrites, the hydrocarbon fragment ions are distinct.

2.10.15 Sulfur Compounds

The contribution (4.4%, see Table 2.2 and Fig. 2.15) of the ^{34}S isotope to the M + 2 peak, and often to a (fragment + 2) peak, affords ready recognition of sulfur-containing compounds. A homologous series of sulfur-containing fragments is four mass units higher than the hydrocarbon fragment series. The number of sulfur atoms can be determined from the size of the contribution of the ^{34}S isotope to the M + 2 peak. The mass of the sulfur atom(s) present is subtracted from the molecular weight. In diisopentyl disulfide, for example, the molecular weight is 206, and the molecule contains two sulfur atoms. The formula for the rest of the molecule is therefore found under mass 142, that is, 206 − (2 × 32).

2.10.15.1 Aliphatic Mercaptans (Thiols) The molecular ion peak of aliphatic mercaptans, except for higher tertiary mercaptans, is usually strong enough so that the M + 2 peak can be accurately measured. In general, the cleavage modes resemble those of alcohols. Cleavage of the C—C bond (α,β bond) next to the SH group gives the characteristic ion $CH_2{=}\overset{+}{S}H \longleftrightarrow \overset{+}{C}H_2{-}\ddot{S}H$ (*m/z* 47). Sulfur is poorer than nitrogen, but better than oxygen, at stabilizing such a fragment. Cleavage at the β,γ bond gives a peak at *m/z* 61 of about one-half the intensity of the *m/z* 47 peak. Cleavage at the γ,δ bond gives a small peak at *m/z* 75, and cleavage at the δ,ϵ bond gives a peak at *m/z* 89 that is more intense than the peak at *m/z* 75; presumably the *m/z* 89 ion is stabilized by cyclization:

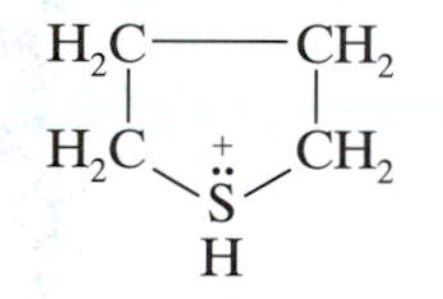

Again analogous to alcohols, primary mercaptans split out H_2S to give a strong M − 34 peak, the resulting ion then eliminating ethylene; thus the homologous series M − H_2S − $(CH_2{=}CH_2)_n$ arises.

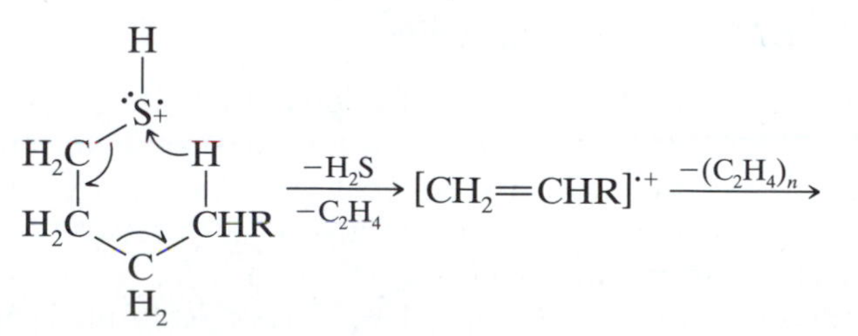

Secondary and tertiary mercaptans cleave at the α-carbon atom with loss of the largest group to give a prominent peak M − CH_3, M − C_2H_5, M − C_3H_7, ... However, a peak at *m/z* 47 may also appear as a rearrangement peak of secondary and tertiary

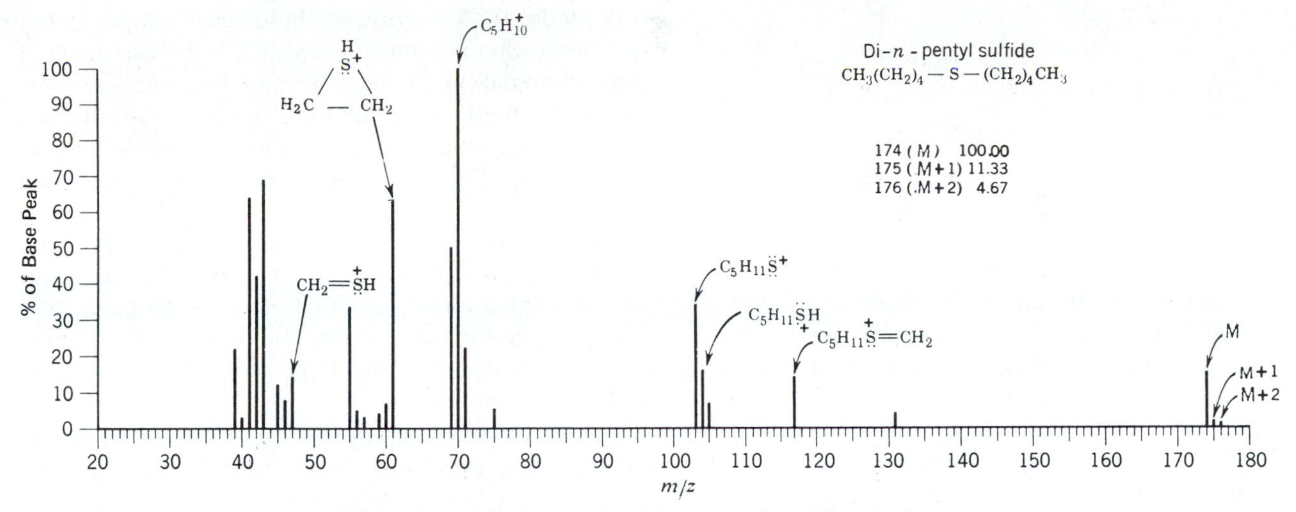

FIGURE 2.15. Di-*n*-pentyl sulfide. The table (upper right) has reset the molecular ion at an intensity of 100% for discussion of the molecular ion cluster.

mercaptans. A peak at M − 33 (loss of HS) is usually present for secondary mercaptans.

In long-chain mercaptans, the hydrocarbon pattern is superimposed on the mercaptan pattern. As for alcohols, the alkenyl peaks (i.e., *m/z* 41, 55, 69, . . .) are as large or larger than the alkyl peaks (*m/z* 43, 57, 71, . . .).

2.10.15.2 Aliphatic Sulfides The molecular ion peak of aliphatic sulfides is usually intense enough so that the M + 2 peak can be accurately measured. The cleavage modes generally resemble those of ethers. Cleavage of one or the other of the α,β C—C bonds occurs, with loss of the largest group being favored. These first-formed ions decompose further with hydrogen transfer and elimination of an alkene. The steps for aliphatic ethers also occur for sulfides; the end result is the ion $RCH{=}\overset{+}{S}H$ (see Fig. 2.15 for an example.)

$$CH_3{-}\underset{|}{\overset{CH_3}{CH}}{-}\overset{\cdot +}{S}{-}CH_2CH_3 \xrightarrow{-CH_3\cdot}$$

$$CH_3CH{=}\overset{+}{S}{-}CH_2{-}H{-}CH_2 \xrightarrow{-CH_2=CH_2} \underset{CH_3}{CH}{=}\overset{+}{S}H \longleftrightarrow \underset{CH_3}{\overset{+}{CH}}{-}SH$$

m/z 61

For a sulfide unbranched at either δ-carbon atom, this ion is $CH_2{=}\overset{+}{S}H$ (*m/z* 47), and its intensity may lead to confusion with the same ion derived from a mercaptan. However, the absence of M − H_2S or M − SH peaks in sulfide spectra makes the distinction.

A moderate to strong peak at *m/z* 61 is present (see alkyl sulfide cleavage, above) in the spectrum of all except tertiary sulfides. When an α-methyl substituent is present, *m/z* 61 is the ion $CH_3CH{=}\overset{+}{S}H$ resulting from the double cleavage described above. Methyl primary sulfides cleave at the α,β bond to give the *m/z* 61 ion, $CH_3{-}\overset{+}{S}{=}CH_2$.

However, a strong *m/z* 61 peak in the spectrum of a straight-chain sulfide calls for a different explanation. The following rationalization is offered:

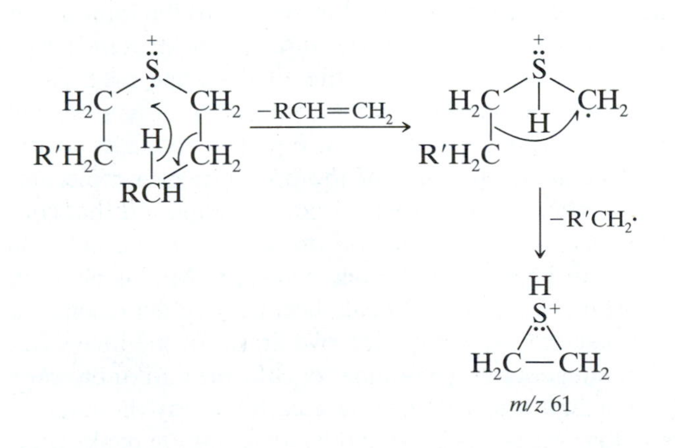

Sulfides give a characteristic ion by cleavage of the C—S bond with retention of charge on sulfur. The resulting $R\overset{+}{S}$ ion gives a peak at *m/z* 32 + CH_3, 32 + C_2H_5, 32 + C_3H_7, ... The ion of *m/z* 103 seems especially favored possibly because of formation of a rearranged cyclic ion.

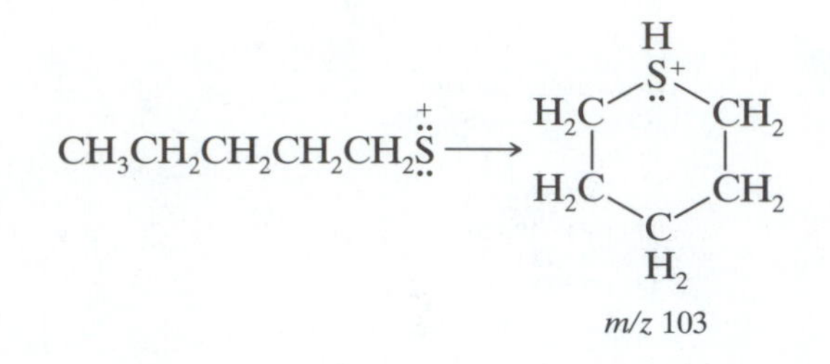

These features are illustrated by the spectrum of di-*n*-pentyl sulfide (Fig. 2.15).

As with long-chain ethers, the hydrocarbon pattern may dominate the spectrum of long-chain sulfides; the C_nH_{2n} peaks seem especially prominent. In branched-chain sulfides, cleavage at the branch may reduce the relative intensity of the characteristic sulfide peaks.

2.10.15.3 Aliphatic Disulfides The molecular ion peak, at least up to C_{10} disulfides, is strong.

A major peak results from cleavage of one of the C—S bonds with retention of the charge on the alkyl fragment. Another major peak results from the same cleavage with shift of a hydrogen atom to form the RSSH fragment, which retains the charge. Other peaks apparently result from cleavage between the sulfur atoms without rearrangement, and with migration of one or two hydrogen atoms to give, respectively, $R\overset{+}{S}$, $R\overset{+}{S} - 1$, and $R\overset{+}{S} - 2$.

2.10.16 Halogen Compounds

A compound that contains one chlorine atom will have an M + 2 peak approximately one-third the intensity of the molecular ion peak because of the presence of a molecular ion containing the ^{37}Cl isotope (see Table 2.2). A compound that contains one bromine atom will have an M + 2 peak almost equal in intensity to the molecular ion because of the presence of a molecular ion containing the ^{81}Br isotope. A compound that contains two chlorines, or two bromines, or one chlorine and one bromine will show a distinct M + 4 peak, in addition to the M + 2 peak, because of the presence of a molecular ion containing two atoms of the heavy isotope. In general, the number of chlorine and/or bromine atoms in a molecule can be ascertained by the number of alternate peaks beyond the molecular ion peak. Thus, three chlorine atoms in a molecule will give peaks at M + 2, M + 4, and M + 6; in polychloro compounds, the peak of highest mass may be so weak as to escape notice.

The relative abundances of the peaks (molecular ion, M + 2, M + 4, and so on) have been calculated by Beynon et al. (1968) for compounds containing chlorine and bromine (atoms other than chlorine and bromine were ignored). A portion of these results is presented here, somewhat modified, as Table 2.3. We can now tell what combination of chlorine and bromine atoms is present. It should be noted that Table 2.3 presents the isotope contributions in terms of percent of the molecular ion peak. Figure 2.16*a* provides the corresponding bar graphs.

As required by Table 2.3, the M + 2 peak in the spectrum of *p*-chlorobenzophenone (Fig. 2.12) is about one-third the intensity of the molecular ion peak (*m/z* 218). As mentioned earlier, the chlorine-containing fragments (*m/z* 141 and 113) show (fragment + 2) peaks of the proper intensity.

Unfortunately, the application of isotope contributions, though generally useful for aromatic halogen compounds, is limited by the weak molecular ion peak of many aliphatic halogen compounds of more than about six carbon atoms for a straight chain, or fewer for a branched chain. However, the halogen-containing fragments are recognizable by the ratio of the (fragment + 2) peaks to fragment peaks in monochlorides or monobromides. In polychloro or polybromo compounds, these (fragment + isotope) peaks form a distinctive series of multiplets (Fig. 2.16*b*). Coincidence of a fragment ion with one of the isotope fragments, with another disruption of the characteristic ratios, must always be kept in mind.

Neither fluorine nor iodine has a heavier isotope.

2.10.16.1 Aliphatic Chlorides The molecular ion peak is detectable only in the lower monochlorides. Fragmentation of the molecular ion is mediated by the chlorine atom but to a much lesser degree than is the case in oxygen-, nitrogen-, or sulfur-containing compounds. Thus, cleavage of a straight-chain monochloride at the C—C bond adjacent to the chlorine atom accounts for a small peak at *m/z* 49 (and, of course, the isotope peak at *m/z* 51).

$$R-CH_2-\overset{\cdot+}{Cl}: \xrightarrow{-R\cdot} CH_2=\overset{+}{Cl}: \longleftrightarrow \overset{+}{C}H_2-Cl:$$

Cleavage of the C—Cl bond leads to a small Cl^+ peak and to a R^+ peak, which is prominent in the lower chlorides but quite small when the chain is longer than about C_5.

Straight-chain chlorides longer than C_6 give $C_3H_6\overset{+}{Cl}$, $C_4H_8\overset{+}{Cl}$, and $C_5H_{10}\overset{+}{Cl}$ ions. Of these the $C_4H_8\overset{+}{Cl}$ ion forms the most intense (sometimes the base) peak; a five-membered cyclic structure may explain its stability.

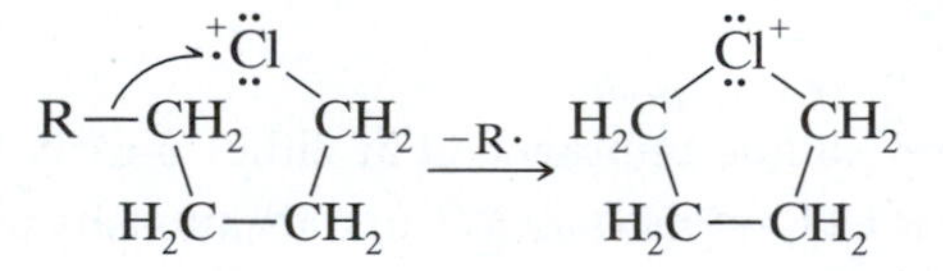

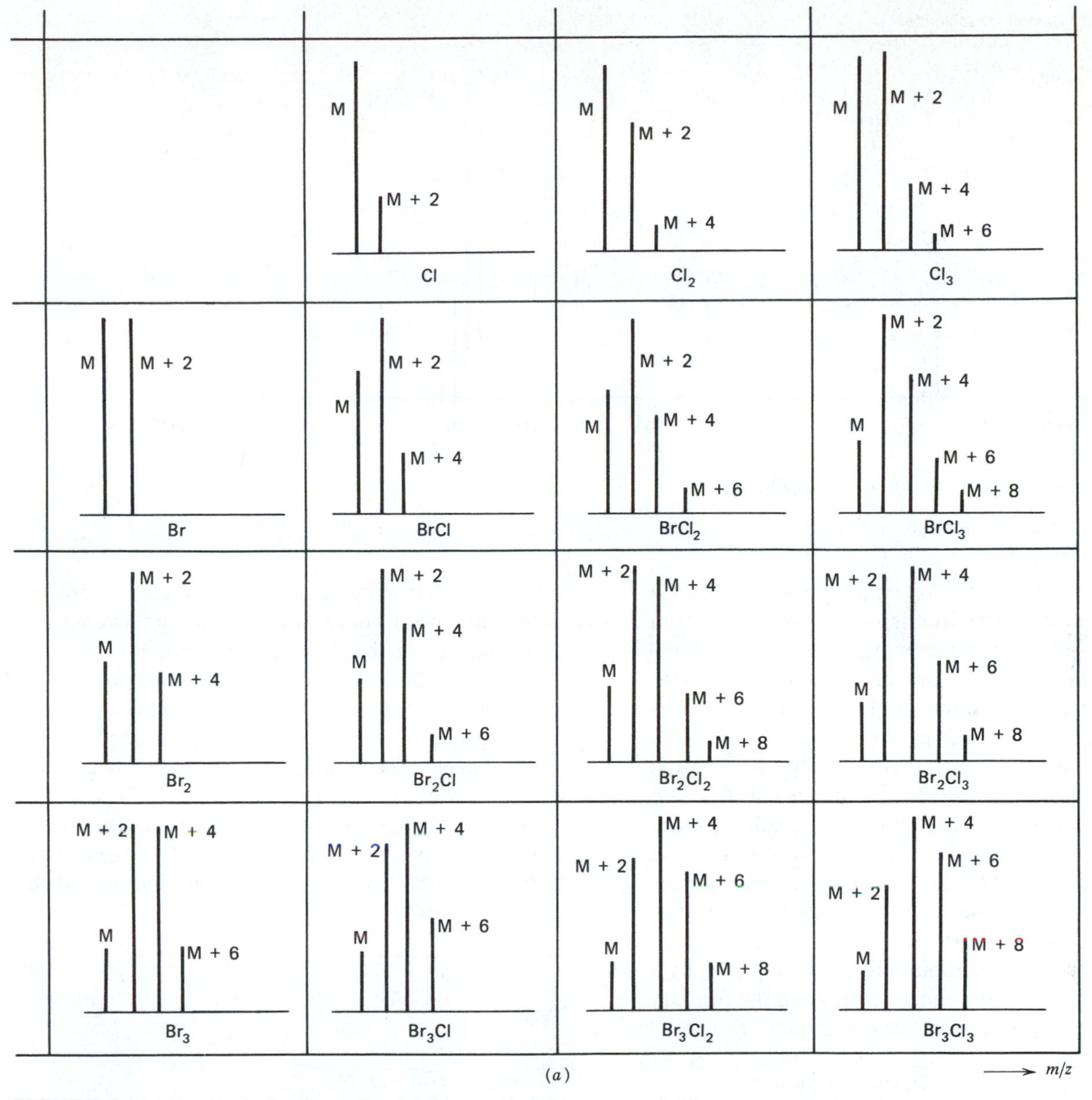

FIGURE 2.16 (*a*). Peaks in the molecular ion region of bromo and chloro compounds. Contributions due to C, H, N, and O are usually small compared to those for Br and Cl.

Loss of HCl occurs, possibly by 1,3 elimination, to give a peak (weak or moderate) at M − 36.

In general, the spectrum of an aliphatic monochloride is dominated by the hydrocarbon pattern to a greater extent than that of a corresponding alcohol, amine, or mercaptan.

2.10.16.2 Aliphatic Bromides The remarks under aliphatic chlorides apply quite generally to the corresponding bromides.

2.10.16.3 Aliphatic Iodides Aliphatic iodides give the strongest molecular ion peak of the aliphatic halides. Since iodine is monoisotopic, there is no distinctive isotope peak. The presence of an iodine atom can sometimes be deduced from isotope peaks that are suspiciously low in relation to the molecular ion peaks, and from several distinctive peaks; in polyiodo compounds, the large interval between major peaks is characteristic.

Iodides cleave much as do chlorides and bromides, but the $C_4H_8\overset{+}{I}$ ion is not as evident as the corresponding chloride and bromide ions.

2.10.16.4 Aliphatic Fluorides Aliphatic fluorides give the weakest molecular ion peak of the aliphatic halides. Fluorine is monoisotopic, and its detection in polyfluoro compounds depends on suspiciously small isotopic peaks relative to the molecular ion, on the in-

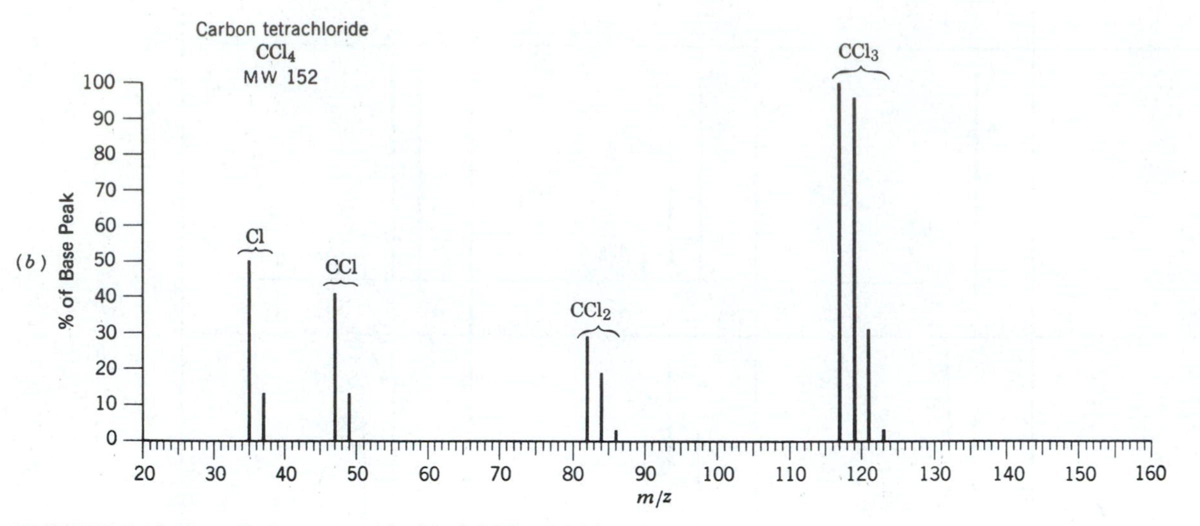

FIGURE 2.16 (*b*). Carbon tetrachloride (cf. Fig. 2.16*a*).

tervals between peaks, and on characteristic peaks. Of these, the most characteristic is *m/z* 69 resulting from the ion CF_3^+, which is the base peak in all perfluorocarbons. Prominent peaks are noted at *m/z* 119, 169, 219 . . . ; these are increments of CF_2. The stable ions $C_3F_5^+$ and $C_4F_7^+$ give large peaks at *m/z* 131 and 181. The M − F peak is frequently visible in perfluorinated compounds. In monofluorides, cleavage of the α,β C—C bond is less important than in the other monohalides, but cleavage of a C—H bond on the α-carbon atom is more important. This reversal is a consequence of the high electronegativity of the F atom and is rationalized by placing the positive charge on the α-carbon atom. The secondary carbonium ion thus depicted by a loss of a hydrogen atom is more stable than the primary carbonium ion resulting from loss of an alkyl radical.

$$[\mathrm{R{-}CH_2{-}F}]^{\cdot +} \xrightarrow{-\mathrm{H}\cdot} \mathrm{R{-}\overset{+}{C}H{-}F}$$

$$\xrightarrow{-\mathrm{R}\cdot} \mathrm{\overset{+}{C}H_2{-}F}$$

2.10.16.5 Benzyl Halides The molecular ion peak of benzyl halides is usually detectable. The benzyl (or tropylium) ion from loss of the halide (rule 8, Section 2.7) is favored even over β-bond cleavage of an alkyl substituent. A substituted phenyl ion (α-bond cleavage) is prominent when the ring is polysubstituted.

2.10.16.6 Aromatic Halides The molecular ion peak of an aryl halide is readily apparent. The M − X peak is large for all compounds in which X is attached directly to the ring.

2.10.17 Heteroaromatic Compounds

The molecular ion peak of heteroaromatics and alkylated heteroaromatics is intense. Cleavage of the bond β to the ring, as in alkylbenzenes, is the general rule; in pyridine, the position of substitution determines the ease of cleavage of the β bond (see below).

Localizing the charge of the molecular ion on the heteroatom, rather than in the ring π structure, provides a satisfactory rationale for the observed mode of cleavage. The present treatment follows that used by Djerassi (Budzikiewicz et al., 1967).

The five-membered ring heteroaromatics (furan, thiophene, and pyrrole) show very similar ring cleavage patterns. The first step in each case is cleavage of the carbon-heteroatom bond.

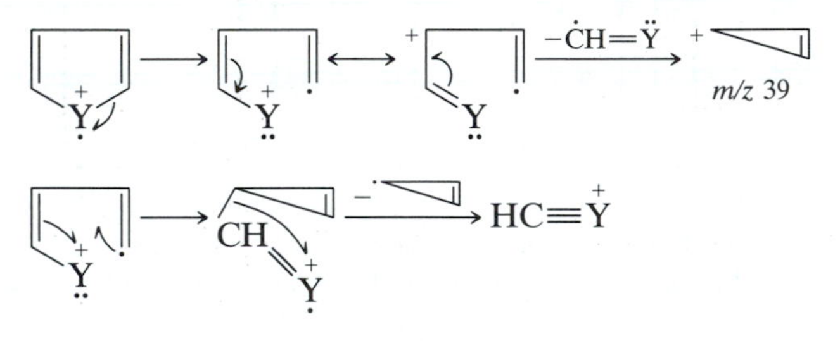

where Y = O, S, NH;

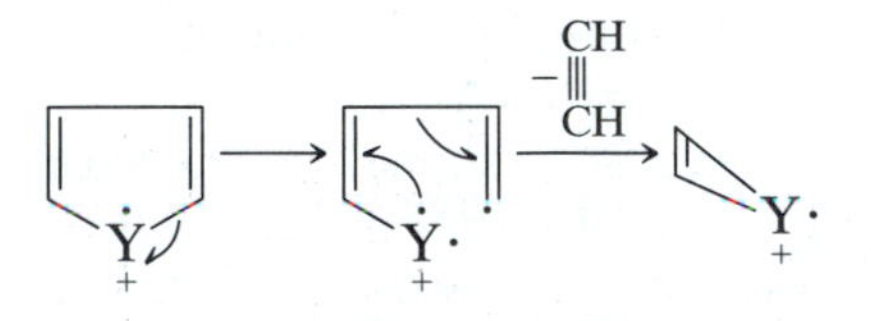

where Y = S, NH. Thus, furan exhibits two principal peaks: $C_3H_3^+$ (*m/z* 39) and $HC{\equiv}\overset{+}{O}$ (*m/z* 29). For thiophene, there are three peaks, $C_3H_3^+$ (*m/z* 39), $HC{\equiv}\overset{+}{S}$ (*m/z* 45), and $C_2H_2\overset{\cdot+}{S}$ (*m/z* 58). And for pyrrole, there are three peaks: $C_3H_3^+$ (*m/z* 39), $HC{\equiv}\overset{+}{N}H$ (*m/z* 28) and $C_2H_2\overset{\cdot+}{N}H$ (*m/z* 41). Pyrrole also eliminates a neutral molecule of HCN to give an intense peak at *m/z* 40. The base peak of 2,5-dimethylfuran is *m/z* 43 ($CH_3C{\equiv}\overset{+}{O}$).

Cleavage of the β C—C bond in alkylpyridines depends on the position of the ring substitution, being more pronounced when the alkyl group is in the 3 position. An alkyl group of more than three carbon atoms in the 2 position can undergo migration of a hydrogen atom to the ring nitrogen.

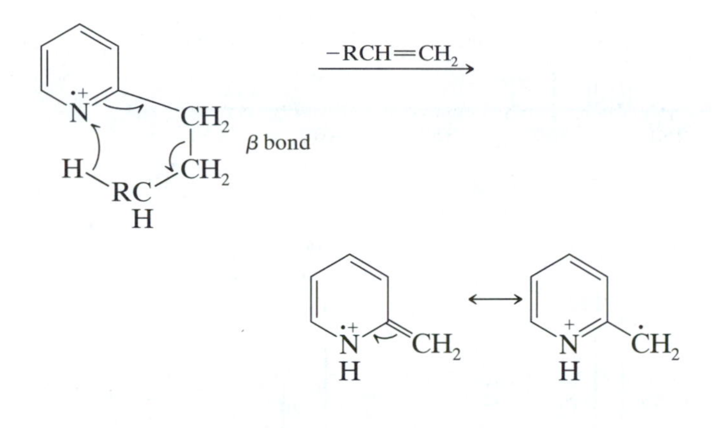

A similar cleavage is found in pyrazines since all ring substituents are necessarily ortho to one of the nitrogen atoms.

2.10.18 Natural Products

2.10.18.1 Amino Acids Detection of the molecular ion peaks of amino acids can be difficult. If we examine the mass spectra of amino acids, as well as of steroids and triglycerides, by a variety of ionization techniques, we can appreciate their relative merits.

The EI spectra of amino acids (Fig. 2.17*a*) or their esters give weak or nonexistent molecular ion peaks, but CI and FD (Fig. 2.17*b* and *c*) give either molecular or quasimolecular ion peaks. The weak molecular ions in the EI spectra arise since amino acids easily lose their carboxyl group and the esters easily lose their carboalkoxyl group upon electron impact.

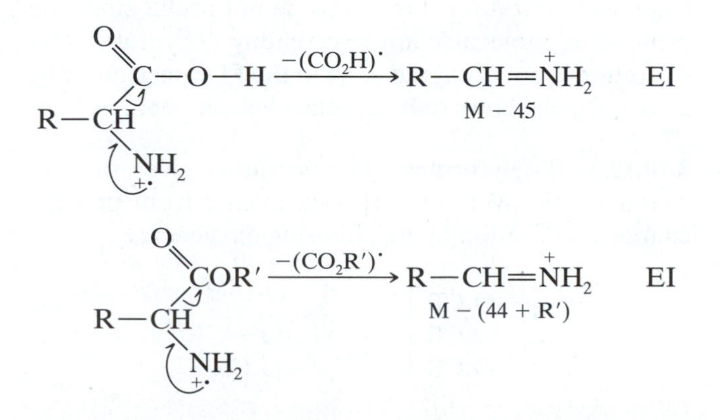

(*a*) Leucine, M=131, EI

(*b*) Leucine, M=131, CI (Isobutane, 200°C)

(*c*) Leucine, M=131, FD (12 MA)

FIGURE 2.17. Mass spectra of leucine. (*a*) Electron impact (EI). (*b*) Chemical ionization (CI). (*c*) Field desorption (FD).

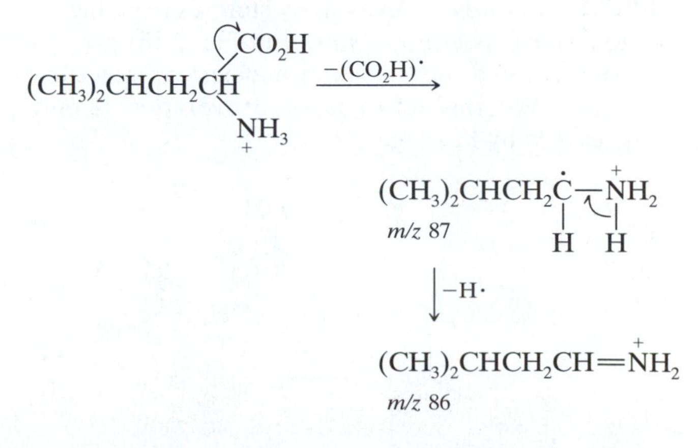

The FD fragmentation pattern for leucine shows an MH^+ (m/z 132) ion, that readily loses a carboxyl group to form the m/z 87 ion, which in turn loses a hydrogen atom to form the m/z 86 ion.

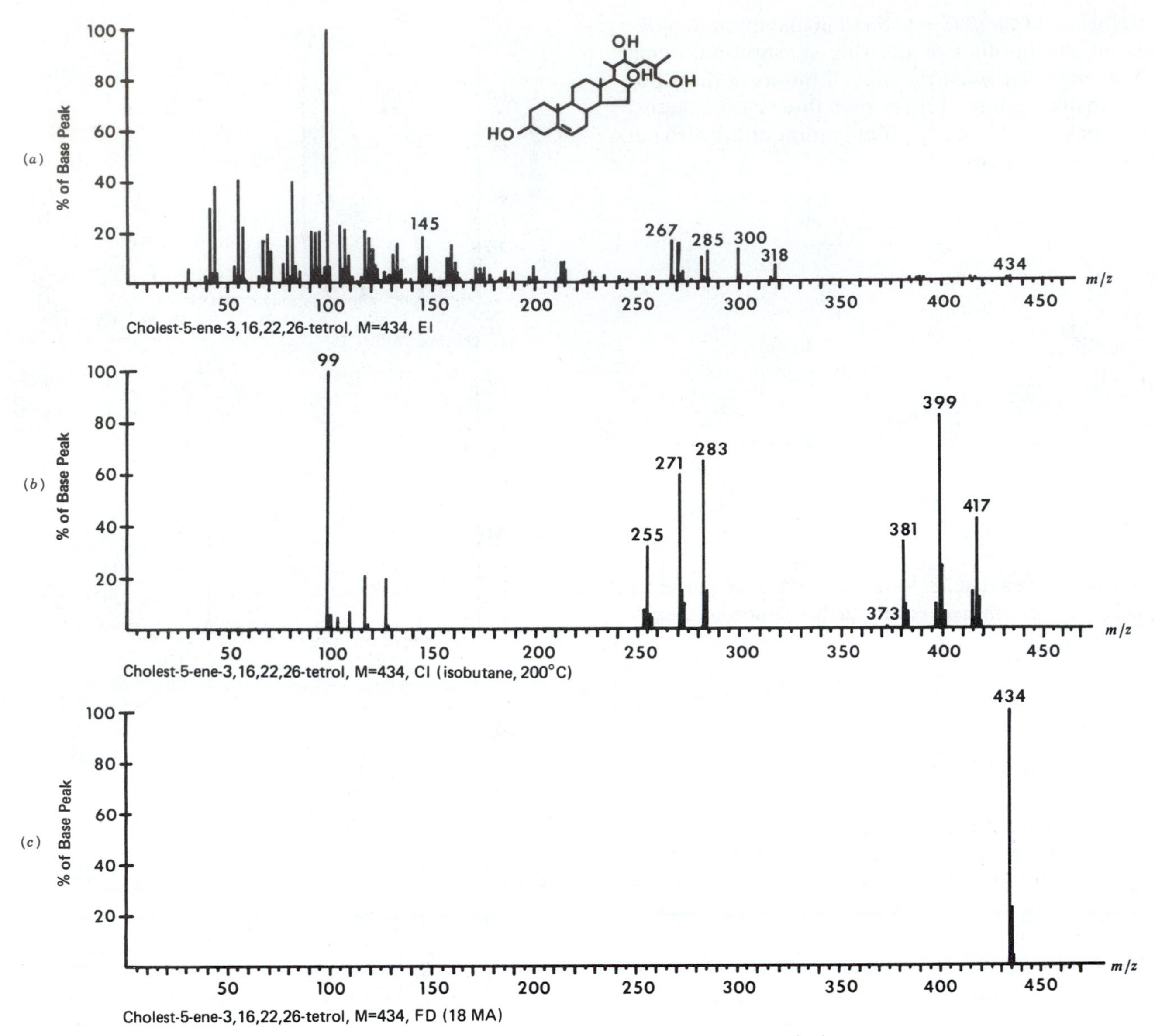

FIGURE 2.18. Mass spectra of cholest-5-ene-3,16,22,26-tetrol. (*a*) Electron impact (EI). (*b*) Chemical ionization (CI). (*c*) Field desorption (FD).

2.10.18.2 Steroids Polyhydroxy steroids (e.g., the tetrol for which spectra are shown in Fig. 2.18) give EI spectra that often show weak or nonexistent molecular ion peaks. For this tetrol facile dehydration is only partly caused by heating.

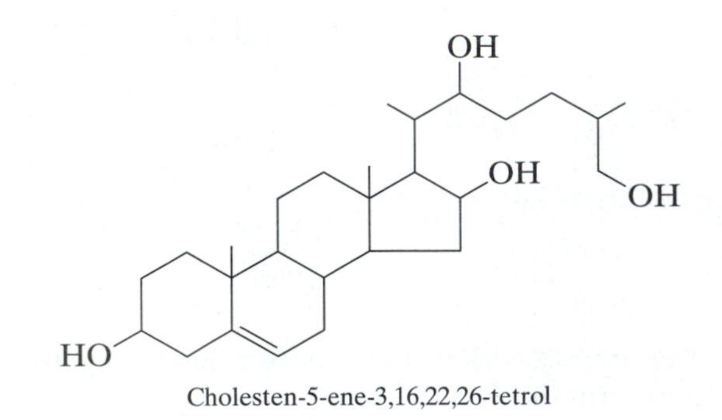

Cholesten-5-ene-3,16,22,26-tetrol

Chemical ionization (Fig. 2.18*b*) is not useful since the protonated molecular ion also readily dehydrates. Dehydration is not observed at all in the FD spectrum (Fig. 2.18*c*), which shows only a molecular ion peak.

2.10.18.3 Triglycerides Triglycerides give rise to characteristic [M − O_2CR] ions arising from positive charge stabilization by neighboring oxygen.

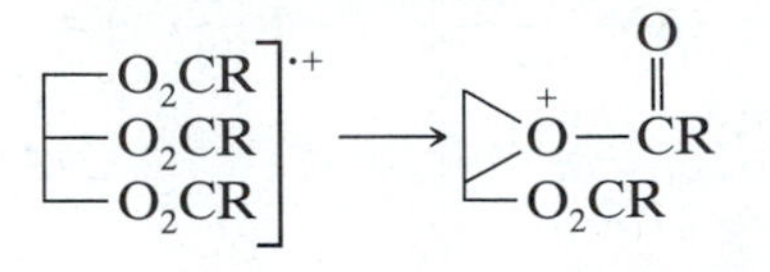

Thus the loss of the RCO_2 fragment and the appearance of $(RCO_2H + H)^+$ fragments reveal the identity of their acid components.

Naturally occurring triglycerides, which have high molecular weights and low volatility, give weak or nonexistent molecular ion (or MH^+) peaks by EI and CI techniques.

2.10.19 Miscellaneous Classes

The following classes of organic compounds are discussed in Biemann's book (B) or in Djerassi's books (D, DI, DII).*

Alkaloids	D, Chapter 5; DI; DII, Chapter 17; B, p. 305
Amino acids and peptides	DII, Chapter 26; B, Chapter 7
Antibiotics	DII, p, 172
Carbohydrates	DII, Chapter 27
Cyanides and isothiocyanates	D, Chapter 11
Estrogens	DII, Chapter 19
Glycerides	B, p. 255
Sapogenins	DII, Chapter 22
Silicones	B, p. 172
Steroids	B, Chapter 9; DII, Chapters 17, 20, 21, 22
Terpenes	B, p. 334; DII, Chapters 23, 24
Thioketals	D, Chapter 7
Tropone and tropolones	D, Chapter 18

In this chapter we have covered a large amount of information. Now let us apply some of it to the spectrum of benzamide (Fig. 2.1) as if it were an "unknown." The intense molecular ion peak (m/z 121) suggests aromaticity. The fact that 121 is an odd number indicates an odd number of nitrogen atoms in the structure. Aromatic character is supported and its identity clarified by the intense phenyl cation peak at m/z 77: We are clearly dealing with a monosubstituted benzene ring. The nitrogen rule corollary suggests that neither the m/z 77 nor 105 peak contains nitrogen. Since the IR spectrum indicates that the compound is a primary amide, we use the m/z 44 peak as confirmation of this functional group.

References

General

Beynon, J. H. (1960). *Mass Spectrometry and Its Application to Organic Chemistry.* Amsterdam: Elsevier.

Beynon, J. H., and Brenton, A. G. (1982). *Introduction to Mass Spectrometry.* Cardiff: University of Wales Publications.

Beynon, J. H., Saunders, R. A., and Williams A. E. (1968). *The Mass Spectra of Organic Molecules.* New York: Elsevier.

Biemann, K. (1962). *Mass Spectrometry, Applications to Organic Chemistry.* New York: McGraw-Hill.

Budzikiewicz, H., Djerassi, C., and Williams, D. H. (1967). *Mass Spectrometry of Organic Compounds.* San Francisco: Holden-Day.

Budzikiewicz, H., Djerassi, C., and Williams, D. H. (1964). *Structure Elucidation of Natural Products by Mass Spectrometry,* Vols. I and II. San Francisco: Holden-Day.

Chapman, J. R. (1993). *Practical Organic Mass Spectrometry,* 2nd ed. New York: Wiley.

Constatin, E., Schnell, A., and Thompson, M. (1990). *Mass Spectrometry.* Englewood Cliffs, NJ: Prentice-Hall.

Davis, R., and Frierson, M. (1987). *Mass Spectrometry.* New York: Wiley. Self-study guide.

Hamming, M., and Foster, N. (1972). *Interpretation of Mass Spectra of Organic Compounds.* New York: Academic Press.

Howe, I., Williams, D. H., and Bowen, R. D. (1981). *Mass Spectrometry-Principles and Application.* New York: McGraw-Hill.

McLafferty, F. W., and Tureček, A. (1993). *Interpretation of Mass Spectra,* 4th ed. Mill Valley, CA: University Scientific Books.

McLafferty, F. W., and Venkataraghavan, R. (1982). *Mass Spectral Correlations.* Washington, DC: American Chemical Society.

McNeal, C. J., Ed. (1986). *Mass Spectrometry in the Analysis of Large Molecules.* New York: Wiley.

Middleditch, B. S., Ed. (1979). *Practical Mass Spectrometry.* New York: Plenum Press.

Milne, G. W. A. (1971). *Mass Spectrometry: Techniques and Applications.* New York: Wiley-Interscience.

Rose, M., and Johnston, R. A. W. (1982). *Mass Spectrometry for Chemists and Biochemists.* New York: Cambridge University Press.

Shrader, S. R. (1971). *Introduction to Mass Spectrometry.* Boston: Allyn and Bacon.

Watson, J. T. (1985). *Introduction to Mass Spectrometry,* 2nd ed. New York: Raven Press.

Williams, D. H. (June 1968–June 1979). *Mass Spectrometry, A Specialist Periodical Report,* Vols. I–V. London: Chemical Society.

(1980). *Advances in Mass Spectrometry; Applications in Organic and Analytical Chemistry;* New York: Pergamon Press.

Mass Spectrometry Reviews. New York: Wiley, 1982 to date.

Mass Spectrometry Bulletin. Aldermaston, England, 1966 to date.

* B: Biemann (1962). D: Djerassi et al. (1967). DI: Djerassi et al. (1964, Vol. I). DII: Djerassi et al. (1964, Vol. II).

Data and Spectral Compilations

American Petroleum Institute Research Project 44 and Thermodynamics Research Center (formerly MCA Research Project). (1947 to date). *Catalog of Selected Mass Spectral Data.* College Station, TX; Texas A & M University, Dr. Bruno Zwolinski, Director.

ASTM (1963). "Index of Mass Spectral Data," American Society for Testing and Materials. STP-356, 244 pp.

ASTM (1969). "Index of Mass Spectral Data," American Society for Testing and Materials, AMD 11, 632 pp.

Beynon, J. H., and Williams, A. E. (1963). *Mass and Abundance Tables for Use in Mass Spectrometry.* Amsterdam: Elsevier.

Beyon, J. H., Saunders, R. A., and Williams, A. E. (1965). *Table of Meta-Stable Transitions.* New York: Elsevier.

Cornu, A., and Massot, R. (1966). *Compilation of Mass Spectral Data.* London: Heyden. Supplements issued.

(1992). *Eight Peak Index of Mass Spectra,* 4th ed. Boca Raton, FL: CRC Press.

(1974). *Handbook of Spectroscopy,* Vol. II. Cleveland: CRC Press, pp. 317–330. Electron impact data for 15 compounds in each of 16 classes of organic compounds.

Heller, S. R., and Milne, G. W. EPA/NIH Mass Spectral Search System (MSSS), A Division of CIS. Washington, DC: U.S. Government Printing Office. An interactive computer searching system containing the spectra of over 32,000 compounds. These can be searched on the basis of peak intensities as well as by Biemann and probability matching techniques.

McLafferty, F. W. (1982). *Mass Spectral Correlations,* 2nd ed. Washington, DC: American Chemical Society.

McLafferty, F. W., and Penzelik, J. (1967). *Index and Bibliography of Mass Spectrometry,* 1963–1965. New York: Wiley-Interscience.

McLafferty, F. W., and Stauffer, D. B. (1988). *The Wiley/NBS Registry of Mass Spectral Data* (7 volumes). New York: Wiley-Interscience.

McLafferty, F. W., and Stauffer, D. B. (1992). *Registry of Mass Spectral Data,* 5th ed. New York: Wiley. Magnetic disc, hard disc, or CD-ROM, 220,000 spectra.

McLafferty, F. W., and Stauffer, D. B. (1991). *The Important Peak Index of the Registry of Mass Spectral Data,* 3 Vols. New York: Wiley.

Stenhagen, E., Abrahamsson, S., and McLafferty, F., Eds. (1969). *Atlas of Mass Spectral Data.* New York: Wiley-Interscience. The three volumes have complete EI data for about 6000 compounds.

Stenhagen, E., Abrahamsson, S., and McLafferty, F., Eds. (1974). *Registry of Mass Spectral Data.* New York: Wiley-Interscience. The four volumes contain bar graphs of 18,806 compounds. Volume IV also contains the index for all four volumes.

Special Monographs

Harrison, G. (1992). *Chemical Ionization Mass Spectrometry,* 2nd ed. Boca Raton, FL: CRC Press.

Linskens, H. F., and Berlin, J. Eds. (1986). *Gas Chromatography-Mass Spectrometry.* New York: Springer-Verlag.

March, R. E., and Hughes, R. J. (1989). *Quadrupole Storage Mass Spectrometry.* New York: Wiley-Interscience.

McFadden, W. H. (1973). *Techniques of Combined Gas Chromatography/Mass Spectrometry: Applications in Organic Analysis.* New York: Wiley-Interscience.

McLafferty, F. S. (1983). *Tandem Mass Spectrometry,* 2nd ed. New York: Wiley-Interscience.

Message, G. M. (1984). *Practical Aspects of Gas Chromatography-Mass Spectrometry.* New York: Wiley.

Porter, Q. N., and Baldas, J. (1971). Mass Spectrometry of Heterocyclic Compounds, in A. Weissberger and E. C. Taylor (Eds.), *General Heterocycle Chemistry Series,* New York: Wiley-Interscience.

Safe, S., and Hutzinger, O. (1973). *Mass Spectrometry of Pesticides and Pollutants.* Cleveland: CRC Press.

Siuzdak, G. (1996). *Mass Spectrometry for Biotechnology.* New York: Academic Press.

Waller, G. R., Ed. (1972). *Biochemical Applications of Mass Spectrometry.* New York: Wiley-Interscience.

Waller, G. R., and Dermer, O. C., Eds. (1980). *Biochemical Applications of Mass Spectrometry,* First Suppl. Vol. New York: Wiley-Interscience.

Problems

All spectra except the CI spectrum of Problem 2.9 were determined by EI methods.

2.1 The exact mass of a compound determined by high-resolution mass spectrometry is 212.0833. What is the molecular formula of the compound?

2.2 The compound whose molecular formula is deduced in Problem 2.1 gives rise to the mass spectrum shown. Deduce the structure of this compound.

PROBLEM 2.2

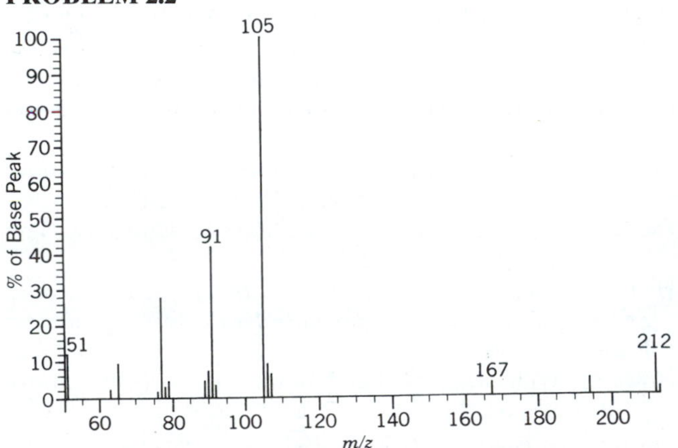

2.3 The mass spectrum of 2-butenal shows a peak at m/z 69 that is 28.9% as intense as the base peak. Propose at least one fragmentation route to account for this peak, and explain why this fragment would be reasonably stable.

2.4 The mass spectrum of 3-butyn-2-ol shows the base peak at m/z 55. Explain why the fragment giving rise to this peak would be very stable.

2.5 Consider the mass spectra below of two isomers (**A** and **B**) of molecular formula $C_{10}H_{14}$. Determine their structures and explain the major spectral features for each.

2.6 The mass spectrum of *o*-nitrotoluene shows a substantial peak at m/z 120. Similar analysis of α,α,α-tri-deutero-*o*-nitrotoluene does not give a peak at m/z 120 but rather at m/z 122. Explain.

2.7 Determine the structure for the mass spectrum shown below.

2.8 Below find mass spectra for compounds **C–F.** Compound **C** has an M + 1 peak that is 2.5% of M (where M = 100%). Compound **F** can easily be converted to compounds **D** and **E.** Compounds **C–E** each give precipitates when treated with alcoholic silver nitrate. The precipitate from **D** is white, the other two are yellow. Deduce the structures of **C–F.**

PROBLEM 2.5, Isomer A.

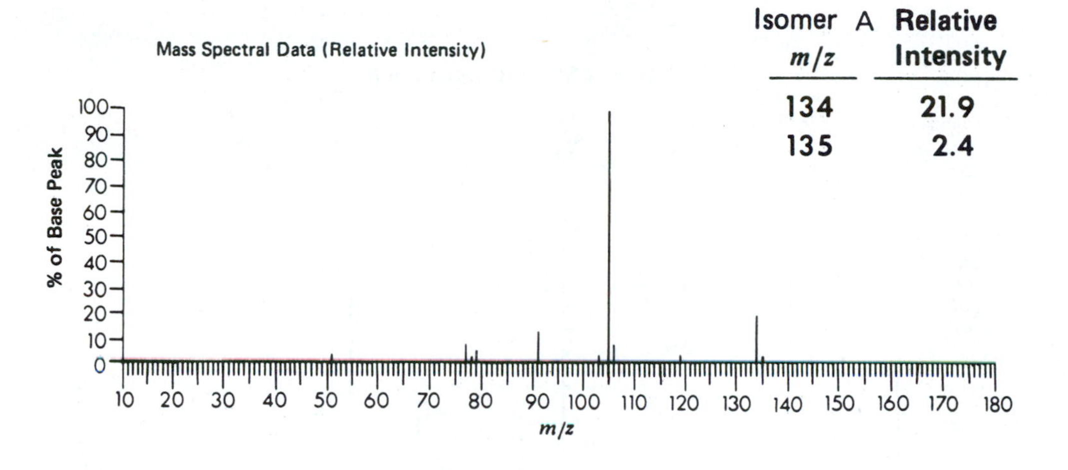

Isomer A m/z	Relative Intensity
134	21.9
135	2.4

PROBLEM 2.5, Isomer B.

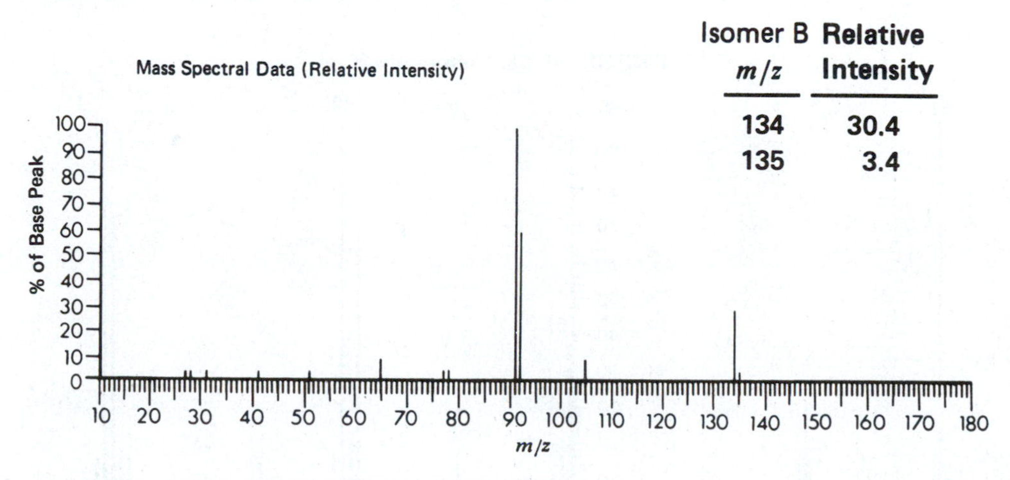

Isomer B m/z	Relative Intensity
134	30.4
135	3.4

PROBLEM 2.7.

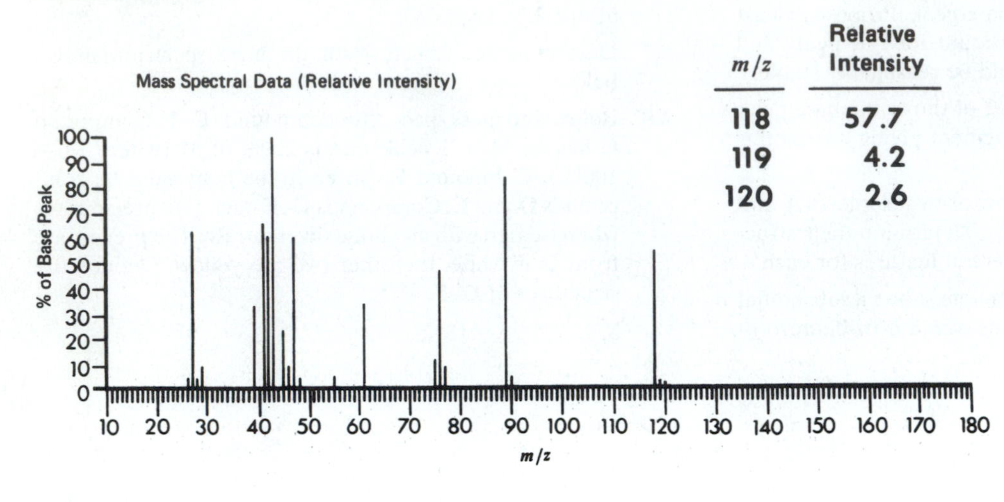

m/z	Relative Intensity
118	57.7
119	4.2
120	2.6

PROBLEM 2.8, Compound C.

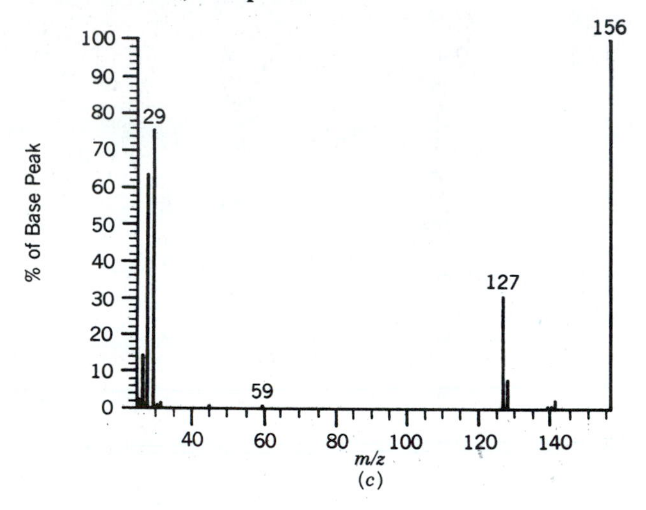

PROBLEM 2.8, Compound E.

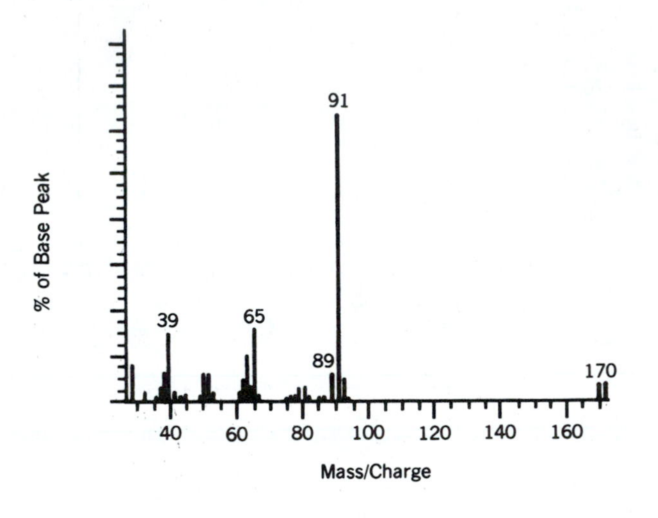

PROBLEM 2.8, Compound D.

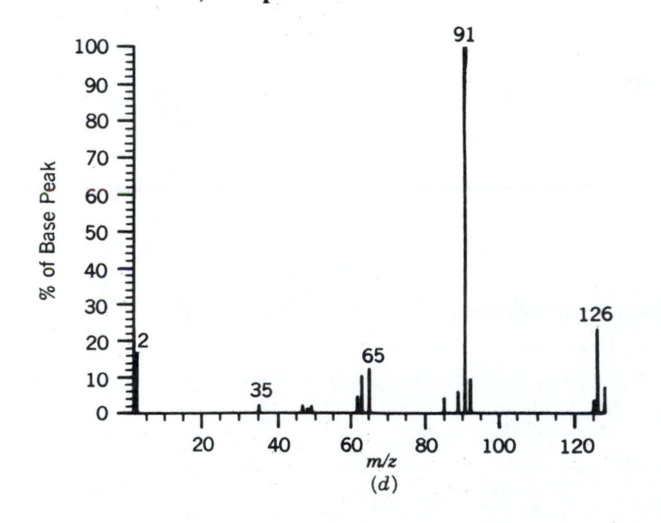

PROBLEM 2.8, Compound F.

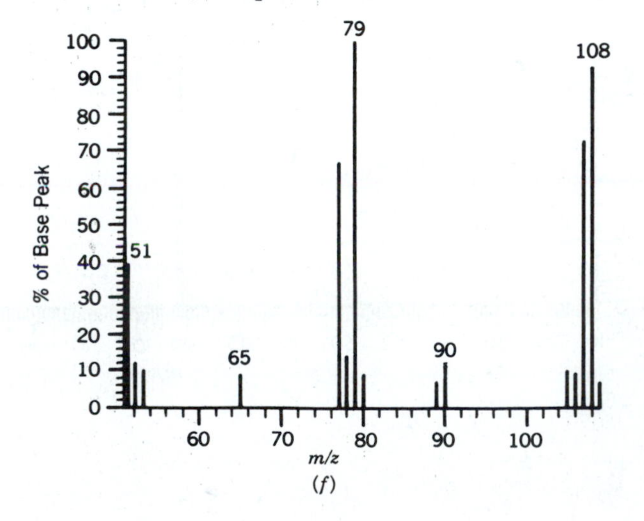

2.9 The compound represented by an electron-impact spectrum and a chemical-ionization (methane) spectrum is an ester of a long-chain, aliphatic alcohol. Interpret the spectra and identify the compound.

PROBLEM 2.9, Electron Impact.

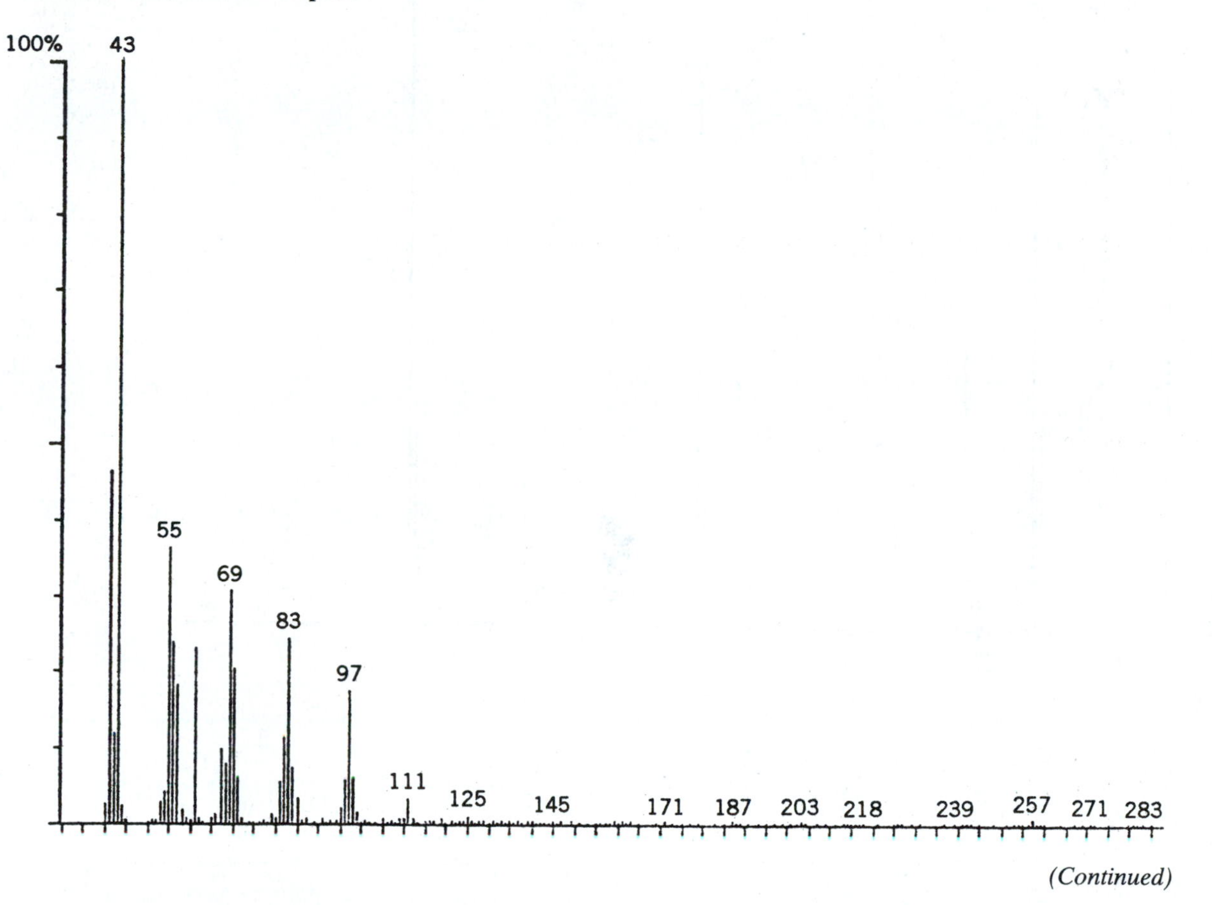

(Continued)

PROBLEM 2.9, CI Methane. *(Continued)*

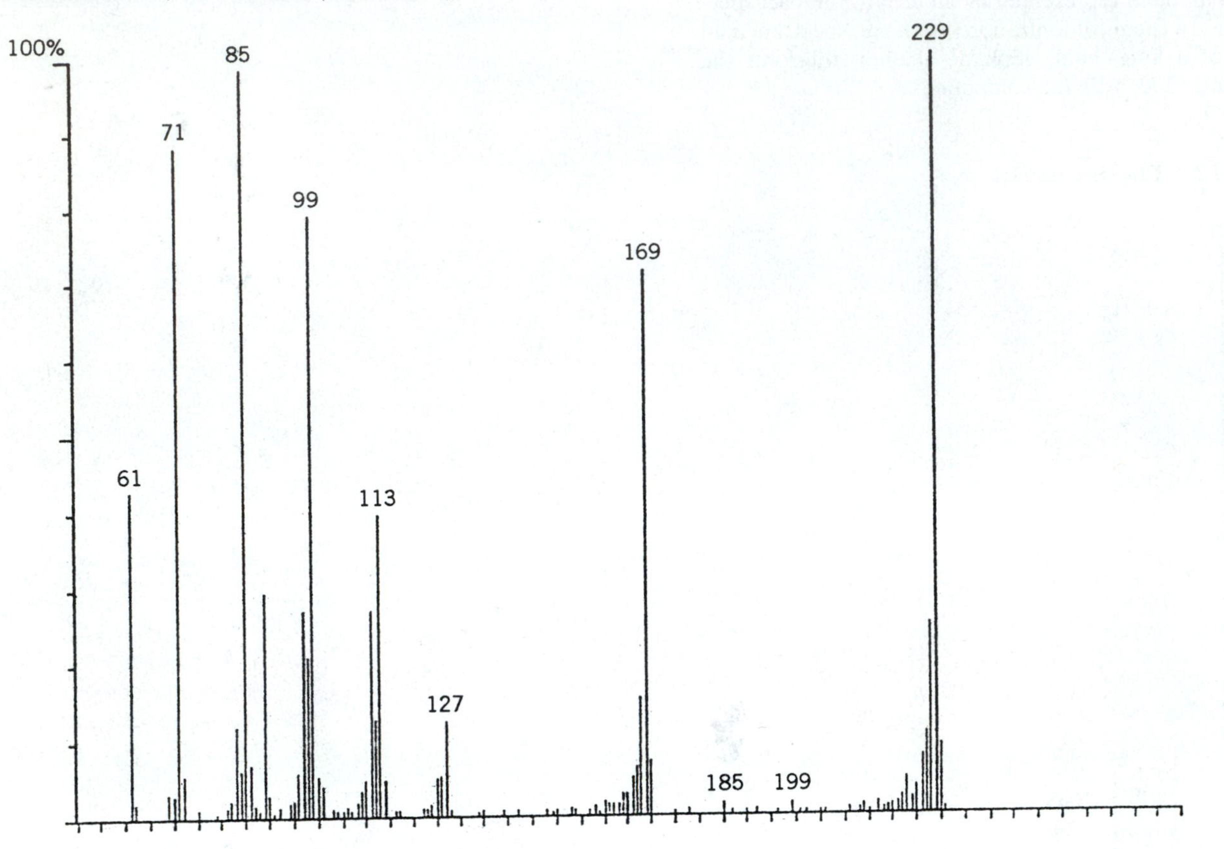

Appendix A Formula Masses (FM) for Various Combinations of Carbon, Hydrogen, Nitrogen, and Oxygen[a]

	FM		FM		FM		FM
12		H_4N_2	32.0375	C_2H_6O	46.0419	CH_3N_2O	59.0246
C	12.0000	CH_4O	32.0262	**47**		CH_5N_3	59.0484
13		**33**		HNO_2	47.0007	$C_2H_3O_2$	59.0133
CH	13.0078	HO_2	32.9976	CH_3O_2	47.0133	C_2H_5NO	59.0371
14		H_3NO	33.0215	CH_5NO	47.0371	$C_2H_7N_2$	59.0610
N	14.0031	**34**		**48**		C_3H_7O	59.0497
CH_2	14.0157	H_2O_2	34.0054	O_3	47.9847	C_3H_9N	59.0736
15		**38**		H_2NO_2	48.0085	**60**	
HN	15.0109	C_3H_2	38.0157	H_4N_2O	48.0324	CH_2NO_2	60.0085
CH_3	15.0235	**39**		CH_4O_2	48.0211	CH_4N_2O	60.0324
16		C_2HN	39.0109	**49**		CH_6N_3	60.0563
O	15.9949	C_3H_3	39.0235	H_3NO_2	49.0164	$C_2H_4O_2$	60.0211
H_2N	16.0187	**40**		**52**		C_2H_6NO	60.0450
CH_4	16.0313	C_2H_2N	40.0187	C_4H_4	52.0313	$C_2H_8N_2$	60.0688
17		C_3H_4	40.0313	**53**		C_3H_8O	60.0575
HO	17.0027	**41**		C_3H_3N	53.0266	C_5	60.0000
H_3N	17.0266	CHN_2	41.0140	C_4H_5	53.0391	**61**	
18		C_2H_3N	41.0266	**54**		CH_3NO_2	61.0164
H_2O	18.0106	C_3H_5	41.0391	$C_2H_2N_2$	54.0218	CH_5N_2O	61.0402
24		**42**		C_3H_2O	54.0106	CH_7N_3	61.0641
C_2	24.0000	N_3	42.0093	C_3H_4N	54.0344	$C_2H_5O_2$	61.0289
26		CNO	41.9980	C_4H_6	54.0470	C_2H_7NO	61.0528
CN	26.0031	CH_2N_2	42.0218	**55**		**62**	
C_2H_2	26.0157	C_2H_2O	42.0106	$C_2H_3N_2$	55.0297	CH_2O_3	62.0003
27		C_2H_4N	42.0344	C_3H_3O	55.0184	CH_4NO_2	62.0242
CHN	27.0109	C_3H_6	42.0470	C_3H_5N	55.0422	CH_6N_2O	62.0480
C_2H_3	27.0235	**43**		C_4H_7	55.0548	$C_2H_6O_2$	62.0368
28		HN_3	43.0170	**56**		**63**	
N_2	28.0062	CHNO	43.0058	C_2O_2	55.9898	HNO_3	62.9956
CO	27.9949	CH_3N_2	43.0297	C_2H_2NO	56.0136	CH_5NO_2	63.0320
CH_2N	28.0187	C_2H_3O	43.0184	$C_2H_4N_2$	56.0375	**64**	
C_2H_4	28.0313	C_2H_5N	43.0422	C_3H_4O	56.0262	C_5H_4	64.0313
29		C_3H_7	43.0548	C_3H_6N	56.0501	**65**	
HN_2	29.0140	**44**		C_4H_8	56.0626	C_4H_3N	65.0266
CHO	29.0027	N_2O	44.0011	**57**		C_5H_5	65.0391
CH_3N	29.0266	CO_2	43.9898	C_2H_3NO	57.0215	**66**	
C_2H_5	29.0391	CH_2NO	44.0136	$C_2H_5N_2$	57.0453	C_4H_4N	66.0344
30		CH_4N_2	44.0375	C_3H_5O	57.0340	C_5H_6	66.0470
NO	29.9980	C_2H_4O	44.0262	C_3H_7N	57.0579	**67**	
H_2N_2	30.0218	C_2H_6N	44.0501	C_4H_9	57.0705	$C_3H_3N_2$	67.0297
CH_2O	30.0106	C_3H_8	44.0626	**58**		C_4H_3O	67.0184
CH_4N	30.0344	**45**		CH_2N_2O	58.0167	C_4H_5N	67.0422
C_2H_6	30.0470	CH_3NO	45.0215	CH_4N_3	58.0406	C_5H_7	67.0548
31		CH_5N_2	45.0453	$C_2H_2O_2$	58.0054	**68**	
HNO	31.0058	C_2H_5O	45.0340	C_2H_4NO	58.0293	$C_3H_4N_2$	68.0375
H_3N_2	31.0297	C_2H_7N	45.0579	$C_2H_6N_2$	58.0532	C_4H_4O	68.0262
CH_3O	31.0184	**46**		C_3H_6O	58.0419	C_4H_6N	68.0501
CH_5N	31.0422	NO_2	45.9929	C_3H_8N	58.0657	C_5H_8	68.0626
32		CH_2O_2	46.0054	C_4H_{10}	58.0783	**69**	
O_2	31.9898	CH_4NO	46.0293	**59**		C_3H_3NO	69.0215
H_2NO	32.0136	CH_6N_2	46.0532	$CHNO_2$	59.0007	$C_3H_5N_2$	69.0453

[a] With permission from J.H. Beynon, *Mass Spectrometry and its Application to Organic Chemistry,* Amsterdam, 1960. The columns headed FM contain the *formula masses* based on the exact mass of the most abundant isotope of each element; these masses are based on the most abundant isotope of carbon having a mass of 12.0000. Note that the table includes only C, H, N, and O.

Appendix A (Continued)

	FM
C_4H_5O	69.0340
C_4H_7N	69.0579
C_5H_9	69.0705
70	
$C_2H_4N_3$	70.0406
$C_3H_2O_2$	70.0054
C_3H_4NO	70.0293
$C_3H_6N_2$	70.0532
C_4H_6O	70.0419
C_4H_8N	70.0657
C_5H_{10}	70.0783
71	
$C_2H_3N_2O$	71.0246
$C_2H_5N_3$	71.0484
$C_3H_3O_2$	71.0133
C_3H_5NO	71.0371
$C_3H_7N_2$	71.0610
C_4H_7O	71.0497
C_4H_9N	71.0736
C_5H_{11}	71.0861
72	
$C_2H_2NO_2$	72.0085
$C_2H_4N_2O$	72.0324
$C_2H_6N_3$	72.0563
$C_3H_4O_2$	72.0211
C_3H_6NO	72.0449
$C_3H_8N_2$	72.0688
C_4H_8O	72.0575
$C_4H_{10}N$	72.0814
C_5H_{12}	72.0939
73	
$C_2H_3NO_2$	73.0164
$C_2H_5N_2O$	73.0402
$C_2H_7N_3$	73.0641
$C_3H_5O_2$	73.0289
C_3H_7NO	73.0528
$C_3H_9N_2$	73.0767
C_4H_9O	73.0653
$C_4H_{11}N$	73.0892
74	
$C_2H_2O_3$	74.0003
$C_2H_4NO_2$	74.0242
$C_2H_6N_2O$	74.0480
$C_2H_8N_3$	74.0719
$C_3H_6O_2$	74.0368
C_3H_8NO	74.0606
$C_3H_{10}N_2$	74.0845
$C_4H_{10}O$	74.0732
75	
$C_2H_3O_3$	75.0082
$C_2H_5NO_2$	75.0320
$C_2H_7N_2O$	75.0559
$C_2H_9N_3$	75.0798
$C_3H_7O_2$	75.0446
C_3H_9NO	75.0684
76	
$C_2H_4O_3$	76.0160
$C_2H_6NO_2$	76.0399
$C_2H_8N_2O$	76.0637
$C_3H_8O_2$	76.0524
C_5H_2N	76.0187
C_6H_4	76.0313
77	
CH_3NO_3	77.0113
$C_2H_5O_3$	77.0238
$C_2H_7NO_2$	77.0477
C_6H_5	77.0391
78	
$C_2H_6O_3$	78.0317
C_5H_4N	78.0344
C_6H_6	78.0470
79	
C_5H_5N	79.0422
C_6H_7	79.0548
80	
$C_3H_2N_3$	80.0249
$C_4H_4N_2$	80.0375
C_5H_4O	80.0262
C_5H_6N	80.0501
C_6H_8	80.0626
81	
$C_3H_3N_3$	81.0328
$C_4H_5N_2$	81.0453
C_5H_5O	81.0340
C_5H_7N	81.0579
C_6H_9	81.0705
82	
$C_3H_4N_3$	82.0406
C_4H_4NO	82.0293
$C_4H_6N_2$	82.0532
C_5H_6O	82.0419
C_5H_8N	82.0657
C_6H_{10}	82.0783
83	
$C_3H_5N_3$	83.0484
$C_4H_3O_2$	83.0133
C_4H_5NO	83.0371
$C_4H_7N_2$	83.0610
C_5H_7O	83.0497
C_5H_9N	83.0736
C_6H_{11}	83.0861
84	
$C_3H_6N_3$	84.0563
$C_4H_4O_2$	84.0211
C_4H_6NO	84.0449
$C_4H_8N_2$	84.0688
C_5H_8O	84.0575
$C_5H_{10}N$	84.0814
C_6H_{12}	84.0939
85	
$C_3H_5N_2O$	85.0402
$C_3H_7N_3$	85.0641
$C_4H_5O_2$	85.0289
C_4H_7NO	85.0528
$C_4H_9N_2$	85.0767
C_5H_9O	85.0653
$C_5H_{11}N$	85.0892
C_6H_{13}	85.1018
86	
$C_2H_2N_2O_2$	86.0116
$C_2H_4N_3O$	86.0355
$C_2H_6N_4$	86.0594
$C_3H_4NO_2$	86.0242
$C_3H_6N_2O$	86.0480
$C_3H_8N_3$	86.0719
$C_4H_6O_2$	86.0368
C_4H_8NO	86.0606
$C_4H_{10}N_2$	86.0845
$C_5H_{10}O$	86.0732
$C_5H_{12}N$	86.0970
C_6H_{14}	86.1096
87	
$C_2H_7N_4$	87.0672
$C_3H_3O_3$	87.0082
$C_3H_5NO_2$	87.0320
$C_3H_7N_2O$	87.0559
$C_3H_9N_3$	87.0798
$C_4H_7O_2$	87.0446
C_4H_9NO	87.0684
$C_4H_{11}N_2$	87.0923
$C_5H_{11}O$	87.0810
$C_5H_{13}N$	87.1049
88	
$C_2H_4N_2O_2$	88.0273
$C_2H_6N_3O$	88.0511
$C_2H_8N_4$	88.0750
$C_3H_4O_3$	88.0160
$C_3H_6NO_2$	88.0399
$C_3H_8N_2O$	88.0637
$C_3H_{10}N_3$	88.0876
$C_4H_8O_2$	88.0524
$C_4H_{10}NO$	88.0763
$C_4H_{12}N_2$	88.1001
$C_5H_{12}O$	88.0888
89	
$C_2H_5N_2O_2$	89.0351
$C_2H_7N_3O$	89.0590
$C_2H_9N_4$	89.0829
$C_3H_5O_3$	89.0238
$C_3H_7NO_2$	89.0477
$C_3H_9N_2O$	89.0715
$C_3H_{11}N_3$	89.0954
$C_4H_9O_2$	89.0603
$C_4H_{11}NO$	89.0841
C_7H_5	89.0391
90	
$C_2H_4NO_3$	90.0191
$C_2H_6N_2O_2$	90.0429
$C_2H_8N_3O$	90.0668
$C_2H_{10}N_4$	90.0907
$C_3H_6O_3$	90.0317
$C_3H_8NO_2$	90.0555
$C_3H_{10}N_2O$	90.0794
$C_4H_{10}O_2$	90.0681
C_7H_6	90.0470
91	
$C_2H_3O_4$	91.0031
$C_2H_5NO_3$	91.0269
$C_2H_7N_2O_2$	91.0508
$C_2H_9N_3O$	91.0746
$C_3H_7O_3$	91.0395
$C_3H_9NO_2$	91.0634
C_6H_5N	91.0422
C_7H_7	91.0548
92	
$C_2H_4O_4$	92.0109
$C_2H_6NO_3$	92.0348
$C_2H_8N_2O_2$	92.0586
$C_3H_8O_3$	92.0473
C_6H_4O	92.0262
C_6H_6N	92.0501
C_7H_8	92.0626
93	
$C_2H_5O_4$	93.0187
$C_2H_7NO_3$	92.0426
$C_5H_5N_2$	93.0453
C_6H_5O	93.0340
C_6H_7N	93.0579
C_7H_9	93.0705
94	
$C_2H_6O_4$	94.0266
$C_4H_4N_3$	94.0406
C_5H_4NO	94.0293
$C_5H_6N_2$	94.0532
C_6H_6O	94.0419
C_6H_8N	94.0657
C_7H_{10}	94.0783
95	
$C_4H_5N_3$	95.0484
C_5H_5NO	95.0371
$C_5H_7N_2$	95.0610
C_6H_7O	95.0497
C_6H_9N	95.0736
C_7H_{11}	95.0861
96	
$C_4H_6N_3$	96.0563
$C_5H_4O_2$	96.0211
C_5H_6NO	96.0449
$C_5H_8N_2$	96.0688
C_6H_8O	96.0575
$C_6H_{10}N$	96.0814
C_7H_{12}	96.0939
97	
$C_3H_5N_4$	97.0515
$C_4H_5N_2O$	97.0402
$C_5H_5O_2$	97.0289
C_5H_7NO	97.0528
$C_5H_9N_2$	97.0767

Appendix A (*Continued*)

	FM
C_6H_9O	97.0653
$C_6H_{11}N$	97.0892
C_7H_{13}	97.1018
98	
$C_3H_4N_3O$	98.0355
$C_3H_6N_4$	98.0594
$C_4H_4NO_2$	98.0242
$C_4H_6N_2O$	98.0480
$C_4H_8N_3$	98.0719
$C_5H_6O_2$	98.0368
C_5H_8NO	98.0606
$C_5H_{10}N_2$	98.0845
$C_6H_{10}O$	98.0732
$C_6H_{12}N$	98.0970
C_7H_{14}	98.1096
99	
$C_3H_5N_3O$	99.0433
$C_3H_7N_4$	99.0672
$C_4H_3O_3$	99.0082
$C_4H_5NO_2$	99.0320
$C_4H_7N_2O$	99.0559
$C_4H_9N_3$	99.0798
$C_5H_7O_2$	99.0446
C_5H_9NO	99.0685
$C_5H_{11}N_2$	99.0923
$C_6H_{11}O$	99.0810
$C_6H_{13}N$	99.1049
C_7H_{15}	99.1174
100	
$C_2H_4N_4O$	100.0386
$C_3H_4N_2O_2$	100.0273
$C_3H_6N_3O$	100.0511
$C_3H_8N_4$	100.0750
$C_4H_4O_3$	100.0160
$C_4H_6NO_2$	100.0399
$C_4H_8N_2O$	100.0637
$C_4H_{10}N_3$	100.0876
$C_5H_8O_2$	100.0524
$C_5H_{10}NO$	100.0763
$C_5H_{12}N_2$	100.1001
$C_6H_{12}O$	100.0888
$C_6H_{14}N$	100.1127
C_7H_{16}	100.1253
101	
$C_3H_3NO_3$	101.0113
$C_3H_5N_2O_2$	101.0351
$C_3H_7N_3O$	101.0590
$C_3H_9N_4$	101.0829
$C_4H_5O_3$	101.0238
$C_4H_7NO_2$	101.0477
$C_4H_9N_2O$	101.0715
$C_4H_{11}N_3$	101.0954
$C_5H_9O_2$	101.0603
$C_5H_{11}NO$	101.0841
$C_5H_{13}N_2$	101.1080
$C_6H_{13}O$	101.0967
$C_6H_{15}N$	101.1205
102	
$C_2H_6N_4O$	102.0542
$C_3H_4NO_3$	102.0191
$C_3H_6N_2O_2$	102.0429
$C_3H_8N_3O$	102.0668
$C_3H_{10}N_4$	102.0907
$C_4H_6O_3$	102.0317
$C_4H_8NO_2$	102.0555
$C_4H_{10}N_2O$	102.0794
$C_4H_{12}N_3$	102.1032
$C_5H_{10}O_2$	102.0681
$C_5H_{12}NO$	102.0919
$C_5H_{14}N_2$	102.1158
$C_6H_{14}O$	102.1045
C_8H_6	102.0470
103	
$C_2H_5N_3O_2$	103.0382
$C_2H_7N_4O$	103.0621
$C_3H_3O_4$	103.0031
$C_3H_5NO_3$	103.0269
$C_3H_7N_2O_2$	103.0508
$C_3H_9N_3O$	103.0746
$C_3H_{11}N_4$	103.0985
$C_4H_7O_3$	103.0395
$C_4H_9NO_2$	103.0634
$C_4H_{11}N_2O$	103.0872
$C_4H_{13}N_3$	103.1111
$C_5H_{11}O_2$	103.0759
$C_5H_{13}NO$	103.0998
C_7H_5N	103.0422
C_8H_7	103.0548
104	
$C_2H_4N_2O_3$	104.0222
$C_2H_6N_3O_2$	104.0460
$C_2H_8N_4O$	104.0699
$C_3H_4O_4$	104.0109
$C_3H_6NO_3$	104.0348
$C_3H_8N_2O_2$	104.0586
$C_3H_{10}N_3O$	104.0825
$C_3H_{12}N_4$	104.1063
$C_4H_8O_3$	104.0473
$C_4H_{10}NO_2$	104.0712
$C_4H_{12}N_2O$	104.0950
$C_5H_{12}O_2$	104.0837
$C_6H_4N_2$	104.0375
C_7H_4O	104.0262
C_7H_6N	104.0501
C_8H_8	104.0626
105	
$C_2H_5N_2O_3$	105.0300
$C_2H_7N_3O_2$	105.0539
$C_2H_9N_4O$	105.0777
$C_3H_5O_4$	105.0187
$C_3H_7NO_3$	105.0426
$C_3H_9N_2O_2$	105.0664
$C_3H_{11}N_3O$	105.0903
$C_4H_9O_3$	105.0552
$C_4H_{11}NO_2$	105.0790
$C_6H_5N_2$	105.0453
C_7H_5O	105.0340
C_7H_7N	105.0579
C_8H_9	105.0705
106	
$C_2H_4NO_4$	106.0140
$C_2H_6N_2O_3$	106.0379
$C_2H_8N_3O_2$	106.0617
$C_2H_{10}N_4O$	106.0856
$C_3H_6O_4$	106.0266
$C_3H_8NO_3$	106.0504
$C_3H_{10}N_2O_2$	106.0743
$C_4H_{10}O_3$	106.0630
C_6H_4NO	106.0293
$C_6H_6N_2$	106.0532
C_7H_6O	106.0419
C_7H_8N	106.0657
C_8H_{10}	106.0783
107	
$C_2H_5NO_4$	107.0218
$C_2H_7N_2O_3$	107.0457
$C_2H_9N_3O_2$	107.0695
$C_3H_7O_4$	107.0344
$C_3H_9NO_3$	107.0583
$C_5H_5N_3$	107.0484
C_6H_5NO	107.0371
$C_6H_7N_2$	107.0610
C_7H_7O	107.0497
C_7H_9N	107.0736
C_8H_{11}	107.0861
108	
$C_2H_6NO_4$	108.0297
$C_2H_8N_2O_3$	108.0535
$C_3H_8O_4$	108.0422
$C_4H_4N_4$	108.0437
$C_5H_4N_2O$	108.0324
$C_5H_6N_3$	108.0563
$C_6H_4O_2$	108.0211
C_6H_6NO	108.0449
$C_6H_8N_2$	108.0688
C_7H_8O	108.0575
$C_7H_{10}N$	108.0814
C_8H_{12}	108.0939
109	
$C_2H_7NO_4$	109.0375
$C_4H_5N_4$	109.0515
$C_5H_5N_2O$	109.0402
$C_5H_7N_3$	109.0641
$C_6H_5O_2$	109.0289
C_6H_7NO	109.0528
$C_6H_9N_2$	109.0767
C_7H_9O	109.0653
$C_7H_{11}N$	109.0892
C_8H_{13}	109.1018
110	
$C_4H_4N_3O$	110.0355
$C_4H_6N_4$	110.0594
$C_5H_6N_2O$	110.0480
$C_5H_8N_3$	110.0719
$C_6H_6O_2$	110.0368
C_6H_8NO	110.0606
$C_6H_{10}N_2$	110.0845
$C_7H_{10}O$	110.0732
$C_7H_{12}N$	110.0970
C_8H_{14}	110.1096
111	
$C_4H_5N_3O$	111.0433
$C_4H_7N_4$	111.0672
$C_5H_5NO_2$	111.0320
$C_5H_7N_2O$	111.0559
$C_5H_9N_3$	111.0789
$C_6H_7O_2$	111.0446
C_6H_9NO	111.0684
$C_6H_{11}N_2$	111.0923
$C_7H_{11}O$	111.0810
$C_7H_{13}N$	111.1049
C_8H_{15}	111.1174
112	
$C_3H_4N_4O$	112.0386
$C_4H_4N_2O_2$	112.0273
$C_4H_6N_3O$	112.0511
$C_4H_8N_4$	112.0750
$C_5H_4O_3$	112.0160
$C_5H_6NO_2$	112.0399
$C_5H_8N_2O$	112.0637
$C_5H_{10}N_3$	112.0876
$C_6H_8O_2$	112.0524
$C_6H_{10}NO$	112.0763
$C_6H_{12}N_2$	112.1001
$C_7H_{12}O$	112.0888
$C_7H_{14}N$	112.1127
C_8H_{16}	112.1253
113	
$C_3H_5N_4O$	113.0464
$C_4H_5N_2O_2$	113.0351
$C_4H_7N_3O$	113.0590
$C_4H_9N_4$	113.0829
$C_5H_5O_3$	113.0238
$C_5H_7NO_2$	113.0477
$C_5H_9N_2O$	113.0715
$C_5H_{11}N_3$	113.0954
$C_6H_9O_2$	113.0603
$C_6H_{11}NO$	113.0841
$C_6H_{13}N_2$	113.1080
$C_7H_{13}O$	113.0967
$C_7H_{15}N$	113.1205
C_8H_{17}	113.1331
114	
$C_3H_6N_4O$	114.0542
$C_4H_4NO_3$	114.0191
$C_4H_6N_2O_2$	114.0429
$C_4H_8N_3O$	114.0668
$C_4H_{10}N_4$	114.0907

Appendix A (Continued)

	FM
$C_5H_6O_3$	114.0317
$C_5H_8NO_2$	114.0555
$C_5H_{10}N_2O$	114.0794
$C_5H_{12}N_3$	114.1032
$C_6H_{10}O_2$	114.0681
$C_6H_{12}NO$	114.0919
$C_6H_{14}N_2$	114.1158
$C_7H_{14}O$	114.1045
$C_7H_{16}N$	114.1284
C_8H_{18}	114.1409
C_9H_6	114.0470
115	
$C_3H_5N_3O_2$	115.0382
$C_3H_7N_4O$	115.0621
$C_4H_5NO_3$	115.0269
$C_4H_7N_2O_2$	115.0508
$C_4H_9N_3O$	115.0746
$C_4H_{11}N_4$	115.0985
$C_5H_7O_3$	115.0395
$C_5H_9NO_2$	115.0634
$C_5H_{11}N_2O$	115.0872
$C_5H_{13}N_3$	115.1111
$C_6H_{11}O_2$	115.0759
$C_6H_{13}NO$	115.0998
$C_6H_{15}N_2$	115.1236
$C_7H_{15}O$	115.1123
$C_7H_{17}N$	115.1362
C_9H_7	115.0548
116	
$C_2H_4N_4O_2$	116.0335
$C_3H_4N_2O_3$	116.0222
$C_3H_6N_3O_2$	116.0460
$C_3H_8N_4O$	116.0699
$C_4H_4O_4$	116.0109
$C_4H_6NO_3$	116.0348
$C_4H_8N_2O_2$	116.0586
$C_4H_{10}N_3O$	116.0825
$C_4H_{12}N_4$	116.1063
$C_5H_8O_3$	116.0473
$C_5H_{10}NO_2$	116.0712
$C_5H_{12}N_2O$	116.0950
$C_5H_{14}N_3$	116.1189
$C_6H_{12}O_2$	116.0837
$C_6H_{14}NO$	116.1076
$C_6H_{16}N_2$	116.1315
$C_7H_4N_2$	116.0375
$C_7H_{16}O$	116.1202
C_8H_6N	116.0501
C_9H_8	116.0626
117	
$C_2H_5N_4O_2$	117.0413
$C_3H_3NO_4$	117.0062
$C_3H_5N_2O_3$	117.0300
$C_3H_7N_3O_2$	117.0539
$C_3H_9N_4O$	117.0777
$C_4H_5O_4$	117.0187
$C_4H_7NO_3$	117.0426
$C_4H_9N_2O_2$	117.0664
$C_4H_{11}N_3O$	117.0903
$C_4H_{13}N_4$	117.1142
$C_5H_9O_3$	117.0552
$C_5H_{11}NO_2$	117.0790
$C_5H_{13}N_2O$	117.1029
$C_5H_{15}N_3$	117.1267
$C_6H_{13}O_2$	117.0916
$C_6H_{15}NO$	117.1154
C_8H_7N	117.0579
C_9H_9	117.0705
118	
$C_2H_4N_3O_3$	118.0253
$C_2H_6N_4O_2$	118.0491
$C_3H_4NO_4$	118.0140
$C_3H_6N_2O_3$	118.0379
$C_3H_8N_3O_2$	118.0617
$C_3H_{10}N_4O$	118.0856
$C_4H_6O_4$	118.0266
$C_4H_8NO_3$	118.0504
$C_4H_{10}N_2O_2$	118.0743
$C_4H_{12}N_3O$	118.0981
$C_4H_{14}N_4$	118.1220
$C_5H_{10}O_3$	118.0630
$C_5H_{12}NO_2$	118.0868
$C_5H_{14}N_2O$	118.1107
$C_6H_{14}O_2$	118.0994
$C_7H_6N_2$	118.0532
C_8H_6O	118.0419
C_8H_8N	118.0657
C_9H_{10}	118.0783
119	
$C_2H_5N_3O_3$	119.0331
$C_2H_7N_4O_2$	119.0570
$C_3H_5NO_4$	119.0218
$C_3H_7N_2O_3$	119.0457
$C_3H_9N_3O_2$	119.0695
$C_3H_{11}N_4O$	119.0934
$C_4H_7O_4$	119.0344
$C_4H_9NO_3$	119.0583
$C_4H_{11}N_2O_2$	119.0821
$C_4H_{13}N_3O$	119.1060
$C_5H_{11}O_3$	119.0708
$C_5H_{13}NO_2$	119.0947
$C_6H_5N_3$	119.0484
C_7H_5NO	119.0371
$C_7H_7N_2$	119.0610
C_8H_7O	119.0497
C_8H_9N	119.0736
C_9H_{11}	119.0861
120	
$C_2H_6N_3O_3$	120.0410
$C_2H_8N_4O_2$	120.0648
$C_3H_6NO_4$	120.0297
$C_3H_8N_2O_3$	120.0535
$C_3H_{10}N_3O_2$	120.0774
$C_3H_{12}N_4O$	120.1012
$C_4H_8O_4$	120.0422
$C_4H_{10}NO_3$	120.0661
$C_4H_{12}N_2O_2$	120.0899
$C_5H_4N_4$	120.0437
$C_5H_{12}O_3$	120.0786
$C_6H_4N_2O$	120.0324
$C_6H_6N_3$	120.0563
C_7H_6NO	120.0449
$C_7H_8N_2$	120.0688
C_8H_8O	120.0575
$C_8H_{10}N$	120.0814
C_9H_{12}	120.0939
121	
$C_2H_5N_2O_4$	121.0249
$C_2H_7N_3O_3$	121.0488
$C_2H_9N_4O_2$	121.0726
$C_3H_7NO_4$	121.0375
$C_3H_9N_2O_3$	121.0614
$C_3H_{11}N_3O_2$	121.0852
$C_4H_9O_4$	121.0501
$C_4H_{11}NO_3$	121.0739
$C_5H_5N_4$	121.0515
$C_6H_5N_2O$	121.0402
$C_6H_7N_3$	121.0641
$C_7H_5O_2$	121.0289
C_7H_7NO	121.0528
$C_7H_9N_2$	121.0767
C_8H_9O	121.0653
$C_8H_{11}N$	121.0892
C_9H_{13}	121.1018
122	
$C_2H_6N_2O_4$	122.0328
$C_2H_8N_3O_3$	122.0566
$C_2H_{10}N_4O_2$	122.0805
$C_3H_8NO_4$	122.0453
$C_3H_{10}N_2O_3$	122.0692
$C_4H_{10}O_4$	122.0579
$C_5H_6N_4$	122.0594
$C_6H_4NO_2$	122.0242
$C_6H_6N_2O$	122.0480
$C_6H_8N_3$	122.0719
$C_7H_6O_2$	122.0368
C_7H_8NO	122.0606
$C_7H_{10}N_2$	122.0845
$C_8H_{10}O$	122.0732
$C_8H_{12}N$	122.0970
C_9H_{14}	122.1096
123	
$C_2H_7N_2O_4$	123.0406
$C_2H_9N_3O_3$	123.0644
$C_3H_9NO_4$	123.0532
$C_5H_5N_3O$	123.0433
$C_5H_7N_4$	123.0672
$C_6H_5NO_2$	123.0320
$C_6H_7N_2O$	123.0559
$C_6H_9N_3$	123.0798
$C_7H_7O_2$	123.0446
C_7H_9NO	123.0684
$C_7H_{11}N_2$	123.0923
$C_8H_{11}O$	123.0810
$C_8H_{13}N$	123.1049
C_9H_{15}	123.1174
124	
$C_2H_8N_2O_4$	124.0484
$C_4H_4N_4O$	124.0386
$C_5H_4N_2O_2$	124.0273
$C_5H_6N_3O$	124.0511
$C_5H_8N_4$	124.0750
$C_6H_4O_3$	124.0160
$C_6H_6NO_2$	124.0399
$C_6H_8N_2O$	124.0637
$C_6H_{10}N_3$	124.0876
$C_7H_8O_2$	124.0524
$C_7H_{10}NO$	124.0763
$C_7H_{12}N_2$	124.1001
C_8N_2	124.0062
$C_8H_{12}O$	124.0888
$C_8H_{14}N$	124.1127
C_9H_{16}	124.1253
125	
$C_4H_3N_3O_2$	125.0226
$C_4H_5N_4O$	125.0464
$C_5H_5N_2O_2$	125.0351
$C_5H_7N_3O$	125.0590
$C_5H_9N_4$	125.0829
$C_6H_5O_3$	125.0238
$C_6H_7NO_2$	125.0477
$C_6H_9N_2O$	125.0715
$C_6H_{11}N_3$	125.0954
$C_7H_9O_2$	125.0603
$C_7H_{11}NO$	125.0841
$C_7H_{13}N_2$	125.1080
$C_8H_{13}O$	125.0967
$C_8H_{15}N$	125.1205
C_9H_{17}	125.1331
126	
$C_3H_2N_4O_2$	126.0178
$C_4H_4N_3O_2$	126.0304
$C_4H_6N_4O$	126.0542
$C_5H_4NO_3$	126.0191
$C_5H_6N_2O_2$	126.0429
$C_5H_8N_3O$	126.0668
$C_5H_{10}N_4$	126.0907
$C_6H_6O_3$	126.0317
$C_6H_8NO_2$	126.0555
$C_6H_{10}N_2O$	126.0794
$C_6H_{12}N_3$	126.1032
$C_7H_{10}O_2$	126.0681
$C_7H_{12}NO$	126.0919
$C_7H_{14}N_2$	126.1158
$C_8H_{14}O$	126.1045
$C_8H_{16}N$	126.1284
C_9H_{18}	126.1409
127	

Appendix A (Continued)

	FM		FM		FM		FM
$C_3H_3N_4O_2$	127.0257	$C_8H_{19}N$	129.1519	$C_4H_{10}N_3O_2$	132.0774	C_8H_8NO	134.0606
$C_4H_5N_3O_2$	127.0382	C_9H_7N	129.0579	$C_4H_{12}N_4O$	132.1012	$C_8H_{10}N_2$	134.0845
$C_4H_7N_4O$	127.0621	$C_{10}H_9$	129.0705	$C_5H_8O_4$	132.0422	$C_9H_{10}O$	134.0732
$C_5H_5NO_3$	127.0269	**130**		$C_5H_{10}NO_3$	132.0661	$C_9H_{12}N$	134.0970
$C_5H_7N_2O_2$	127.0508	$C_3H_4N_3O_3$	130.0253	$C_5H_{12}N_2O_2$	132.0899	$C_{10}H_{14}$	134.1096
$C_5H_9N_3O$	127.0746	$C_3H_6N_4O_2$	130.0491	$C_5H_{14}N_3O$	132.1138	**135**	
$C_5H_{11}N_4$	127.0985	$C_4H_4NO_4$	130.0140	$C_5H_{16}N_4$	132.1377	$C_3H_7N_2O_4$	135.0406
$C_6H_7O_3$	127.0395	$C_4H_6N_2O_3$	130.0379	$C_6H_4N_4$	132.0437	$C_3H_9N_3O_3$	135.0644
$C_6H_9NO_2$	127.0634	$C_4H_8N_3O_2$	130.0617	$C_6H_{12}O_3$	132.0786	$C_3H_{11}N_4O_2$	135.0883
$C_6H_{11}N_2O$	127.0872	$C_4H_{10}N_4O$	130.0856	$C_6H_{14}NO_2$	132.1025	$C_4H_9NO_4$	135.0532
$C_6H_{13}N_3$	127.1111	$C_5H_6O_4$	130.0266	$C_6H_{16}N_2O$	132.1264	$C_4H_{11}N_2O_3$	135.0770
$C_7H_{11}O_2$	127.0759	$C_5H_8NO_3$	130.0504	$C_7H_9N_3$	132.0563	$C_4H_{13}N_3O_2$	135.1009
$C_7H_{13}NO$	127.0998	$C_5H_{10}N_2O_2$	130.0743	$C_7H_{16}O_2$	132.1151	$C_5H_3N_4O$	135.0308
$C_7H_{15}N_2$	127.1236	$C_5H_{12}N_3O$	130.0981	C_8H_6NO	132.0449	$C_5H_{11}O_4$	135.0657
$C_8H_{15}O$	127.1123	$C_5H_{14}N_4$	130.1220	$C_8H_8N_2$	132.0688	$C_5H_{13}NO_3$	135.0896
$C_8H_{17}N$	127.1362	$C_6H_{10}O_3$	130.0630	C_9H_8O	132.0575	$C_6H_5N_3O$	135.0433
C_9H_{19}	127.1488	$C_6H_{12}NO_2$	130.0868	$C_9H_{10}N$	132.0814	$C_6H_7N_4$	135.0672
128		$C_6H_{14}N_2O$	130.1107	$C_{10}H_{12}$	132.0939	$C_7H_5NO_2$	135.0320
$C_3H_4N_4O_2$	128.0335	$C_6H_{16}N_3$	130.1346	**133**		$C_7H_7N_2O$	135.0559
$C_4H_4N_2O_3$	128.0222	$C_7H_4N_3$	130.0406	$C_3H_5N_2O_4$	133.0249	$C_7H_9N_3$	135.0798
$C_4H_6N_3O_2$	128.0460	$C_7H_{14}O_2$	130.0994	$C_3H_7N_3O_3$	133.0488	$C_8H_7O_2$	135.0446
$C_4H_8N_4O$	128.0699	$C_7H_{16}NO$	130.1233	$C_3H_9N_4O_2$	133.0726	C_8H_9NO	135.0684
$C_5H_4O_4$	128.0109	$C_7H_{18}N_2$	130.1471	$C_4H_7NO_4$	133.0375	$C_8H_{11}N_2$	135.0923
$C_5H_6NO_3$	128.0348	$C_8H_6N_2$	130.0532	$C_4H_9N_2O_3$	133.0614	$C_9H_{11}O$	135.0810
$C_5H_8N_2O_2$	128.0586	$C_8H_{18}O$	130.1358	$C_4H_{11}N_3O_2$	133.0852	$C_9H_{13}N$	135.1049
$C_5H_{10}N_3O$	128.0825	C_9H_8N	130.0657	$C_4H_{13}N_4O$	133.1091	$C_{10}H_{15}$	135.1174
$C_5H_{12}N_4$	128.1063	$C_{10}H_{10}$	130.0783	$C_5H_9O_4$	133.0501	**136**	
$C_6H_8O_3$	128.0473	**131**		$C_5H_{11}NO_3$	133.0739	$C_3H_8N_2O_4$	136.0484
$C_6H_{10}NO_2$	128.0712	$C_3H_3N_2O_4$	131.0093	$C_5H_{13}N_2O_2$	133.0978	$C_3H_{10}N_3O_3$	136.0723
$C_6H_{12}N_2O$	128.0950	$C_3H_5N_3O_3$	131.0331	$C_5H_{15}N_3O$	133.1216	$C_3H_{12}N_4O_2$	136.0961
$C_6H_{14}N_3$	128.1189	$C_3H_7N_4O_2$	131.0570	$C_6H_5N_4$	133.0515	$C_4H_{10}NO_4$	136.0610
$C_7H_{12}O_2$	128.0837	$C_4H_5NO_4$	131.0218	$C_6H_{13}O_3$	133.0865	$C_4H_{12}N_2O_3$	136.0848
$C_7H_{14}NO$	128.1076	$C_4H_7N_2O_3$	131.0457	$C_6H_{15}NO_2$	133.1103	$C_5H_2N_3O_2$	136.0147
$C_7H_{16}N_2$	128.1315	$C_4H_9N_3O_2$	131.0695	$C_7H_5N_2O$	133.0402	$C_5H_4N_4O$	136.0386
$C_8H_{16}O$	128.1202	$C_4H_{11}N_4O$	131.0934	$C_7H_7N_3$	133.0641	$C_5H_{12}O_4$	136.0735
$C_8H_{18}N$	128.1440	$C_5H_7O_4$	131.0344	C_8H_7NO	133.0528	$C_6H_4N_2O_2$	136.0273
C_9H_{20}	128.1566	$C_5H_9NO_3$	131.0583	$C_8H_9N_2$	133.0767	$C_6H_6N_3O$	136.0511
$C_{10}H_8$	128.0626	$C_5H_{11}N_2O_2$	131.0821	C_9H_9O	133.0653	$C_6H_8N_4$	136.0750
129		$C_5H_{13}N_3O$	131.1060	$C_9H_{11}N$	133.0892	$C_7H_4O_3$	136.0160
$C_3H_3N_3O_3$	129.0175	$C_5H_{15}N_4$	131.1298	$C_{10}H_{13}$	133.1018	$C_7H_6NO_2$	136.0399
$C_3H_5N_4O_2$	129.0413	$C_6H_{11}O_3$	131.0708	**134**		$C_7H_8N_2O$	136.0637
$C_4H_5N_2O_3$	129.0300	$C_6H_{13}NO_2$	131.0947	$C_3H_6N_2O_4$	134.0328	$C_7H_{10}N_3$	136.0876
$C_4H_7N_3O_2$	129.0539	$C_6H_{15}N_2O$	131.1185	$C_3H_8N_3O_3$	134.0566	$C_8H_8O_2$	136.0524
$C_4H_9N_4O$	129.0777	$C_6H_{17}N_3$	131.1424	$C_3H_{10}N_4O_2$	134.0805	$C_8H_{10}NO$	136.0763
$C_5H_5O_4$	129.0187	$C_7H_5N_3$	131.0484	$C_4H_8NO_4$	134.0453	$C_8H_{12}N_2$	136.1001
$C_5H_7NO_3$	129.0426	$C_7H_{15}O_2$	131.1072	$C_4H_{10}N_2O_3$	134.0692	$C_9H_{12}O$	136.0888
$C_5H_9N_2O_2$	129.0664	$C_7H_{17}NO$	131.1311	$C_4H_{12}N_3O_2$	134.0930	$C_9H_{14}N$	136.1127
$C_5H_{11}N_3O$	129.0903	$C_8H_7N_2$	131.0610	$C_4H_{14}N_4O$	134.1169	$C_{10}H_{16}$	136.1253
$C_5H_{13}N_4$	129.1142	C_9H_7O	131.0497	$C_5H_{10}O_4$	134.0579	**137**	
$C_6H_9O_3$	129.0552	C_9H_9N	131.0736	$C_5H_{12}NO_3$	134.0817	$C_3H_9N_2O_4$	137.0563
$C_6H_{11}NO_2$	129.0790	$C_{10}H_{11}$	131.0861	$C_5H_{14}N_2O_2$	134.1056	$C_3H_{11}N_3O_3$	137.0801
$C_6H_{13}N_2O$	129.1029	**132**		$C_6H_4N_3O$	134.0355	$C_4H_{11}NO_4$	137.0688
$C_6H_{15}N_3$	129.1267	$C_3H_4N_2O_4$	132.0171	$C_6H_6N_4$	134.0594	$C_5H_3N_3O_2$	137.0226
$C_7H_{13}O_2$	129.0916	$C_3H_6N_3O_3$	132.0410	$C_6H_{14}O_3$	134.0943	$C_5H_5N_4O$	137.0464
$C_7H_{15}NO$	129.1154	$C_3H_8N_4O_2$	132.0648	$C_7H_6N_2O$	134.0480	$C_6H_5N_2O_2$	137.0351
$C_7H_{17}N_2$	129.1393	$C_4H_6NO_4$	132.0297	$C_7H_8N_3$	134.0719	$C_6H_7N_3O$	137.0590
$C_8H_{17}O$	129.1280	$C_4H_8N_2O_3$	132.0535	$C_8H_6O_2$	134.0368	$C_6H_9N_4$	137.0829

Appendix A (Continued)

	FM
$C_7H_5O_3$	137.0238
$C_7H_7NO_2$	137.0477
$C_7H_9N_2O$	137.0715
$C_7H_{11}N_3$	137.0954
$C_8H_9O_2$	137.0603
$C_8H_{11}NO$	137.0841
$C_8H_{13}N_2$	137.1080
$C_9H_{13}O$	137.0967
$C_9H_{15}N$	137.1205
$C_{10}H_{17}$	137.1331
138	
$C_3H_{10}N_2O_4$	138.0641
$C_5H_4N_3O_2$	138.0304
$C_5H_6N_4O$	138.0542
$C_6H_4NO_3$	138.0191
$C_6H_6N_2O_2$	138.0429
$C_6H_8N_3O$	138.0668
$C_6H_{10}N_4$	138.0907
$C_7H_6O_3$	138.0317
$C_7H_8NO_2$	138.0555
$C_7H_{10}N_2O$	138.0794
$C_7H_{12}N_3$	138.1032
$C_8H_{10}O_2$	138.0681
$C_8H_{12}NO$	138.0919
$C_8H_{14}N_2$	138.1158
$C_9H_{14}O$	138.1045
$C_9H_{16}N$	138.1284
$C_{10}H_{18}$	138.1409
139	
$C_4H_3N_4O_2$	139.0257
$C_5H_3N_2O_3$	139.0144
$C_5H_5N_3O_2$	139.0382
$C_5H_7N_4O$	139.0621
$C_6H_5NO_3$	139.0269
$C_6H_7N_2O_2$	139.0508
$C_6H_9N_3O$	139.0747
$C_6H_{11}N_4$	139.0985
$C_7H_7O_3$	139.0395
$C_7H_9NO_2$	139.0634
$C_7H_{11}N_2O$	139.0872
$C_7H_{13}N_3$	139.1111
$C_8H_{11}O_2$	139.0759
$C_8H_{13}NO$	139.0998
$C_8H_{15}N_2$	139.1236
$C_9H_3N_2$	139.0297
$C_9H_{15}O$	139.1123
$C_9H_{17}N$	139.1362
$C_{10}H_{19}$	139.1488
$C_{11}H_7$	139.0548
140	
$C_4H_4N_4O_2$	140.0335
$C_5H_4N_2O_3$	140.0222
$C_5H_6N_3O_2$	140.0460
$C_5H_8N_4O$	140.0699
$C_6H_4O_4$	140.0109
$C_6H_6NO_3$	140.0348
$C_6H_8N_2O_2$	140.0586
$C_6H_{10}N_3O$	140.0825
$C_6H_{12}N_4$	140.1063
$C_7H_8O_3$	140.0473
$C_7H_{10}NO_2$	140.0712
$C_7H_{12}N_2O$	140.0950
$C_7H_{14}N_3$	140.1189
$C_8H_{12}O_2$	140.0837
$C_8H_{14}NO$	140.1076
$C_8H_{16}N_2$	140.1315
$C_9H_{16}O$	140.1202
$C_9H_{18}N$	140.1440
$C_{10}H_6N$	140.0501
$C_{10}H_{20}$	140.1566
$C_{11}H_8$	140.0626
141	
$C_4H_3N_3O_3$	141.0175
$C_4H_5N_4O_2$	141.0413
$C_5H_3NO_4$	141.0062
$C_5H_5N_2O_3$	141.0300
$C_5H_7N_3O_2$	141.0539
$C_5H_9N_4O$	141.0777
$C_6H_5O_4$	141.0187
$C_6H_7NO_3$	141.0426
$C_6H_9N_2O_2$	141.0664
$C_6H_{11}N_3O$	141.0903
$C_6H_{13}N_4$	141.1142
$C_7H_9O_3$	141.0552
$C_7H_{11}NO_2$	141.0790
$C_7H_{13}N_2O$	141.1029
$C_7H_{15}N_3$	141.1267
$C_8H_{13}O_2$	141.0916
$C_8H_{15}NO$	141.1154
$C_8H_{17}N_2$	141.1393
$C_9H_{17}O$	141.1280
$C_9H_{19}N$	141.1519
$C_{10}H_7N$	141.0579
$C_{10}H_{21}$	141.1644
$C_{11}H_9$	141.0705
142	
$C_4H_4N_3O_3$	142.0253
$C_4H_6N_4O_2$	142.0491
$C_5H_4NO_4$	142.0140
$C_5H_6N_2O_3$	142.0379
$C_5H_8N_3O_2$	142.0617
$C_5H_{10}N_4O$	142.0856
$C_6H_6O_4$	142.0266
$C_6H_8NO_3$	142.0504
$C_6H_{10}N_2O_2$	142.0743
$C_6H_{12}N_3O$	142.0981
$C_6H_{14}N_4$	142.1220
$C_7H_{10}O_3$	142.0630
$C_7H_{12}NO_2$	142.0868
$C_7H_{14}N_2O$	142.1107
$C_7H_{16}N_3$	142.1346
$C_8H_{14}O_2$	142.0994
$C_8H_{16}NO$	142.1233
$C_8H_{18}N_2$	142.1471
$C_9H_6N_2$	142.0532
$C_9H_{18}O$	142.1358
$C_9H_{20}N$	142.1597
$C_{10}H_8N$	142.0657
$C_{10}H_{22}$	142.1722
$C_{11}H_{10}$	142.0783
143	
$C_4H_3N_2O_4$	143.0093
$C_4H_5N_3O_3$	143.0331
$C_4H_7N_4O_2$	143.0570
$C_5H_5NO_4$	143.0218
$C_5H_7N_2O_3$	143.0457
$C_5H_9N_3O_2$	143.0695
$C_5H_{11}N_4O$	143.0934
$C_6H_7O_4$	143.0344
$C_6H_9NO_3$	143.0583
$C_6H_{11}N_2O_2$	143.0821
$C_6H_{13}N_3O$	143.1060
$C_6H_{15}N_4$	143.1298
$C_7H_{11}O_3$	143.0708
$C_7H_{13}NO_2$	143.0947
$C_7H_{15}N_2O$	143.1185
$C_7H_{17}N_3$	143.1424
$C_8H_{15}O_2$	143.1072
$C_8H_{17}NO$	143.1311
$C_8H_{19}N_2$	143.1549
$C_9H_7N_2$	143.0610
$C_9H_{19}O$	143.1436
$C_9H_{21}N$	143.1675
$C_{10}H_7O$	143.0497
$C_{10}H_9N$	143.0736
$C_{11}H_{11}$	143.0861
144	
$C_4H_4N_2O_4$	144.0171
$C_4H_6N_3O_3$	144.0410
$C_4H_8N_4O_2$	144.0648
$C_5H_6NO_4$	144.0297
$C_5H_8N_2O_3$	144.0535
$C_5H_{10}N_3O_2$	144.0774
$C_5H_{12}N_4O$	144.1012
$C_6H_8O_4$	144.0422
$C_6H_{10}NO_3$	144.0661
$C_6H_{12}N_2O_2$	144.0899
$C_6H_{14}N_3O$	144.1138
$C_6H_{16}N_4$	144.1377
$C_7H_{12}O_3$	144.0786
$C_7H_{14}NO_2$	144.1025
$C_7H_{16}N_2O$	144.1264
$C_7H_{18}N_3$	144.1502
$C_8H_6N_3$	144.0563
$C_8H_{16}O_2$	144.1151
$C_8H_{18}NO$	144.1389
$C_8H_{20}N_2$	144.1628
C_9H_6NO	144.0449
$C_9H_8N_2$	144.0688
$C_9H_{20}O$	144.1515
$C_{10}H_8O$	144.0575
$C_{10}H_{10}N$	144.0814
$C_{11}H_{12}$	144.0939
145	
$C_4H_5N_2O_4$	145.0249
$C_4H_7N_3O_3$	145.0488
$C_4H_9N_4O_2$	145.0726
$C_5H_7NO_4$	145.0375
$C_5H_9N_2O_3$	145.0614
$C_5H_{11}N_3O_2$	145.0852
$C_5H_{13}N_4O$	145.1091
$C_6H_9O_4$	145.0501
$C_6H_{11}NO_3$	145.0739
$C_6H_{13}N_2O_2$	145.0978
$C_6H_{15}N_3O$	145.1216
$C_6H_{17}N_4$	145.1455
$C_7H_5N_4$	145.0515
$C_7H_{13}O_3$	145.0865
$C_7H_{15}NO_2$	145.1103
$C_7H_{17}N_2O$	145.1342
$C_7H_{19}N_3$	145.1580
$C_8H_5N_2O$	145.0402
$C_8H_7N_3$	145.0641
$C_8H_{17}O_2$	145.1229
$C_8H_{19}NO$	145.1467
C_9H_7NO	145.0528
$C_9H_9N_2$	145.0767
$C_{10}H_9O$	145.0653
$C_{10}H_{11}N$	145.0892
$C_{11}H_{13}$	145.1018
146	
$C_4H_6N_2O_4$	146.0328
$C_4H_8N_3O_3$	146.0566
$C_4H_{10}N_4O_2$	146.0805
$C_5H_8NO_4$	146.0453
$C_5H_{10}N_2O_3$	146.0692
$C_5H_{12}N_3O_2$	146.0930
$C_5H_{14}N_4O$	146.1169
$C_6H_{10}O_4$	146.0579
$C_6H_{12}NO_3$	146.0817
$C_6H_{14}N_2O_2$	146.1056
$C_6H_{16}N_3O$	146.1295
$C_7H_6N_4$	146.0594
$C_7H_{14}O_3$	146.0943
$C_7H_{16}NO_2$	146.1182
$C_7H_{18}N_2O$	146.1420
$C_8H_2O_3$	146.0003
$C_8H_6N_2O$	146.0480
$C_8H_8N_3$	146.0719
$C_8H_{18}O_2$	146.1307
$C_9H_6O_2$	146.0368
C_9H_8NO	146.0606
$C_9H_{10}N_2$	146.0845
$C_{10}H_{10}O$	146.0732

Appendix A (Continued)

	FM
$C_{10}H_{12}N$	146.0970
$C_{11}H_{14}$	146.1096
147	
$C_4H_7N_2O_4$	147.0406
$C_4H_9N_3O_3$	147.0644
$C_4H_{11}N_4O_2$	147.0883
$C_5H_9NO_4$	147.0532
$C_5H_{11}N_2O_3$	147.0770
$C_5H_{13}N_3O_2$	147.1009
$C_5H_{15}N_4O$	147.1247
$C_6H_{11}O_4$	147.0657
$C_6H_{13}NO_3$	147.0896
$C_6H_{15}N_2O_2$	147.1134
$C_6H_{17}N_3O$	147.1373
$C_7H_5N_3O$	147.0433
$C_7H_7N_4$	147.0672
$C_7H_{15}O_3$	147.1021
$C_7H_{17}NO_2$	147.1260
$C_8H_5NO_2$	147.0320
$C_8H_7N_2O$	147.0559
$C_8H_9N_3$	147.0798
$C_9H_7O_2$	147.0446
C_9H_9NO	147.0684
$C_9H_{11}N_2$	147.0923
$C_{10}H_{11}O$	147.0810
$C_{10}H_{13}N$	147.1049
$C_{11}H_{15}$	147.1174
148	
$C_4H_8N_2O_4$	148.0484
$C_4H_{10}N_3O_3$	148.0723
$C_4H_{12}N_4O_2$	148.0961
$C_5H_{10}NO_4$	148.0610
$C_5H_{12}N_2O_3$	148.0849
$C_5H_{16}N_4O$	148.1325
$C_6H_4N_4O$	148.0386
$C_6H_{12}O_4$	148.0735
$C_6H_{14}NO_3$	148.0974
$C_6H_{16}N_2O_2$	148.1213
$C_7H_6N_3O$	148.0511
$C_7H_8N_4$	148.0750
$C_7H_{16}O_3$	148.1100
$C_8H_6NO_2$	148.0399
$C_8H_8N_2O$	148.0637
$C_8H_{10}N_3$	148.0876
$C_9H_8O_2$	148.0524
$C_9H_{10}NO$	148.0763
$C_9H_{12}N_2$	148.1001
$C_{10}H_{12}O$	148.0888
$C_{10}H_{14}N$	148.1127
$C_{11}H_{16}$	148.1253
149	
$C_4H_9N_2O_4$	149.0563
$C_4H_{11}N_3O_3$	149.0801
$C_4H_{13}N_4O_2$	149.1040
$C_5H_{11}NO_4$	149.0688
$C_5H_{13}N_2O_3$	149.0927
$C_5H_{15}N_3O_2$	149.1165
$C_6H_5N_4O$	149.0464
$C_6H_{13}O_4$	149.0814
$C_6H_{15}NO_3$	149.1052
$C_7H_5N_2O_2$	149.0351
$C_7H_7N_3O$	149.0590
$C_7H_9N_4$	149.0829
$C_8H_5O_3$	149.0238
$C_8H_7NO_2$	149.0477
$C_8H_9N_2O$	149.0715
$C_8H_{11}N_3$	149.0954
$C_9H_9O_2$	149.0603
$C_9H_{11}NO$	149.0841
$C_9H_{13}N_2$	149.1080
$C_{10}H_{13}O$	149.0967
$C_{10}H_{15}N$	149.1205
$C_{11}H_{17}$	149.1331
150	
$C_4H_{10}N_2O_4$	150.0641
$C_4H_{12}N_3O_3$	150.0879
$C_4H_{14}N_4O_2$	150.1118
$C_5H_{12}NO_4$	150.0766
$C_5H_{14}N_2O_3$	150.1005
$C_6H_4N_3O_2$	150.0304
$C_6H_6N_4O$	150.0542
$C_6H_{14}O_4$	150.0892
$C_7H_6N_2O_2$	150.0429
$C_7H_8N_3O$	150.0668
$C_7H_{10}N_4$	150.0907
$C_8H_6O_3$	150.0317
$C_8H_8NO_2$	150.0555
$C_8H_{10}N_2O$	150.0794
$C_8H_{12}N_3$	150.1032
$C_9H_{10}O_2$	150.0681
$C_9H_{12}NO$	150.0919
$C_9H_{14}N_2$	150.1158
$C_{10}H_{14}O$	150.1045
$C_{10}H_{16}N$	150.1284
$C_{11}H_{18}$	150.1409
151	
$C_4H_{11}N_2O_4$	151.0719
$C_4H_{13}N_3O_3$	151.0958
$C_5H_3N_4O_2$	151.0257
$C_5H_{13}NO_4$	151.0845
$C_6H_3N_2O_3$	151.0144
$C_6H_5N_3O_2$	151.0382
$C_6H_7N_4O$	151.0621
$C_7H_5NO_3$	151.0269
$C_7H_7N_2O_2$	151.0508
$C_7H_9N_3O$	151.0746
$C_7H_{11}N_4$	151.0985
$C_8H_7O_3$	151.0395
$C_8H_9NO_2$	151.0634
$C_8H_{11}N_2O$	151.0872
$C_8H_{13}N_3$	151.1111
$C_9H_{11}O_2$	151.0759
$C_9H_{13}NO$	151.0998
$C_9H_{15}N_2$	151.1236
$C_{10}H_{15}O$	151.1123
$C_{10}H_{17}N$	151.1362
$C_{11}H_{19}$	151.1488
152	
$C_4H_{12}N_2O_4$	152.0797
$C_5H_4N_4O_2$	152.0335
$C_6H_4N_2O_3$	152.0222
$C_6H_6N_3O_2$	152.0460
$C_6H_8N_4O$	152.0699
$C_7H_6NO_3$	152.0348
$C_7H_8N_2O_2$	152.0586
$C_7H_{10}N_3O$	152.0825
$C_7H_{12}N_4$	152.1063
$C_8H_8O_3$	152.0473
$C_8H_{10}NO_2$	152.0712
$C_8H_{12}N_2O$	152.0950
$C_8H_{14}N_3$	152.1189
$C_9H_{12}O_2$	152.0837
$C_9H_{14}NO$	152.1076
$C_9H_{16}N_2$	152.1315
$C_{10}H_{16}O$	152.1202
$C_{10}H_{18}N$	152.1440
$C_{11}H_6N$	152.0501
$C_{11}H_{20}$	152.1566
$C_{12}H_8$	152.0626
153	
$C_5H_3N_3O_3$	153.0175
$C_5H_5N_4O_2$	153.0413
$C_6H_5N_2O_3$	153.0300
$C_6H_7N_3O_2$	153.0539
$C_6H_9N_4O$	153.0777
$C_7H_5O_4$	153.0187
$C_7H_7NO_3$	153.0426
$C_7H_9N_2O_2$	153.0664
$C_7H_{11}N_3O$	153.0903
$C_7H_{13}N_4$	153.1142
$C_8H_9O_3$	153.0552
$C_8H_{11}NO_2$	153.0790
$C_8H_{13}N_2O$	153.1029
$C_8H_{15}N_3$	153.1267
$C_9H_{13}O_2$	153.0916
$C_9H_{15}NO$	153.1154
$C_9H_{17}N_2$	153.1393
$C_{10}H_{17}O$	153.1280
$C_{10}H_{19}N$	153.1519
$C_{11}H_7N$	153.0579
$C_{11}H_{21}$	153.1644
$C_{12}H_9$	153.0705
154	
$C_5H_4N_3O_3$	154.0253
$C_5H_6N_4O_2$	154.0491
$C_6H_4NO_4$	154.0140
$C_6H_6N_2O_3$	154.0379
$C_6H_8N_3O_2$	154.0617
$C_6H_{10}N_4O$	154.0856
$C_7H_6O_4$	154.0266
$C_7H_8NO_3$	154.0504
$C_7H_{10}N_2O_2$	154.0743
$C_7H_{12}N_3O$	154.0981
$C_7H_{14}N_4$	154.1220
$C_8H_{10}O_3$	154.0630
$C_8H_{12}NO_2$	154.0868
$C_8H_{14}N_2O$	154.1107
$C_8H_{16}N_3$	154.1346
$C_9H_{14}O_2$	154.0994
$C_9H_{16}NO$	154.1233
$C_9H_{18}N_2$	154.1471
$C_{10}H_{18}O$	154.1358
$C_{10}H_{20}N$	154.1597
$C_{11}H_8N$	154.0657
$C_{11}H_{22}$	154.1722
$C_{12}H_{10}$	154.0783
155	
$C_5H_3N_2O_4$	155.0093
$C_5H_5N_3O_3$	155.0331
$C_5H_7N_4O_2$	155.0570
$C_6H_5NO_4$	155.0218
$C_6H_7N_2O_3$	155.0457
$C_6H_9N_3O_2$	155.0695
$C_6H_{11}N_4O$	155.0934
$C_7H_7O_4$	155.0344
$C_7H_9NO_3$	155.0583
$C_7H_{11}N_2O_2$	155.0821
$C_7H_{13}N_3O$	155.1060
$C_8H_{11}O_3$	155.0708
$C_8H_{13}NO_2$	155.0947
$C_8H_{15}N_2O$	155.1185
$C_8H_{17}N_3$	155.1424
$C_9H_{15}O_2$	155.1072
$C_9H_{17}NO$	155.1311
$C_9H_{19}N_2$	155.1549
$C_{10}H_7N_2$	155.0610
$C_{10}H_{19}O$	155.1436
$C_{10}H_{21}N$	155.1675
$C_{11}H_7O$	155.0497
$C_{11}H_9N$	155.0736
$C_{11}H_{23}$	155.1801
$C_{12}H_{11}$	155.0861
156	
$C_5H_4N_2O_4$	156.0171
$C_5H_6N_3O_3$	156.0410
$C_5H_8N_4O_2$	156.0648
$C_6H_6NO_4$	156.0297
$C_6H_8N_2O_3$	156.0535
$C_6H_{10}N_3O_2$	156.0774
$C_6H_{12}N_4O$	156.1012
$C_7H_8O_4$	156.0422
$C_7H_{10}NO_3$	156.0661
$C_7H_{12}N_2O_2$	156.0899
$C_7H_{14}N_3O$	156.1138

Appendix A (Continued)

	FM
$C_7H_{16}N_4$	156.1377
$C_8H_{12}O_3$	156.0786
$C_8H_{14}NO_2$	156.1025
$C_8H_{16}N_2O$	156.1264
$C_8H_{18}N_3$	156.1502
$C_9H_6N_3$	156.0563
$C_9H_{16}O_2$	156.1151
$C_9H_{18}NO$	156.1389
$C_9H_{20}N_2$	156.1628
$C_{10}H_6NO$	156.0449
$C_{10}H_8N_2$	156.0688
$C_{10}H_{20}O$	156.1515
$C_{10}H_{22}N$	156.1753
$C_{11}H_8O$	156.0575
$C_{11}H_{10}N$	156.0814
$C_{11}H_{24}$	156.1879
$C_{12}H_{12}$	156.0939
157	
$C_5H_5N_2O_4$	157.0249
$C_5H_7N_3O_3$	157.0488
$C_5H_9N_4O_2$	157.0726
$C_6H_7NO_4$	157.0375
$C_6H_9N_2O_3$	157.0614
$C_6H_{11}N_3O_2$	157.0852
$C_6H_{13}N_4O$	157.1091
$C_7H_9O_4$	157.0501
$C_7H_{11}NO_3$	157.0739
$C_7H_{13}N_2O_2$	157.0978
$C_7H_{15}N_3O$	157.1216
$C_7H_{17}N_4$	157.1455
$C_8H_5N_4$	157.0515
$C_8H_{13}O_3$	157.0865
$C_8H_{15}NO_2$	157.1103
$C_8H_{17}N_2O$	157.1342
$C_8H_{19}N_3$	157.1580
$C_9H_5N_2O$	157.0402
$C_9H_7N_3$	157.0641
$C_9H_{17}O_2$	157.1229
$C_9H_{19}NO$	157.1467
$C_9H_{21}N_2$	157.1706
$C_{10}H_7NO$	157.0528
$C_{10}H_9N_2$	157.0767
$C_{10}H_{21}O$	157.1593
$C_{10}H_{23}N$	157.1832
$C_{11}H_9O$	157.0653
$C_{11}H_{11}N$	157.0892
$C_{12}H_{13}$	157.1018
158	
$C_5H_6N_2O_4$	158.0328
$C_5H_8N_3O_3$	158.0566
$C_5H_{10}N_4O_2$	158.0805
$C_6H_8NO_4$	158.0453
$C_6H_{10}N_2O_3$	158.0692
$C_6H_{12}N_3O_2$	158.0930
$C_6H_{14}N_4O$	158.1169
$C_7H_{10}O_4$	158.0579
$C_7H_{12}NO_3$	158.0817
$C_7H_{14}N_2O_2$	158.1056
$C_7H_{16}N_3O$	158.1295
$C_7H_{18}N_4$	158.1533
$C_8H_6N_4$	158.0594
$C_8H_{14}O_3$	158.0943
$C_8H_{16}NO_2$	158.1182
$C_8H_{18}N_2O$	158.1420
$C_8H_{20}N_3$	158.1659
$C_9H_6N_2O$	158.0480
$C_9H_8N_3$	158.0719
$C_9H_{18}O_2$	158.1307
$C_9H_{20}NO$	158.1546
$C_{10}H_6O_2$	158.0368
$C_{10}H_8NO$	158.0606
$C_{10}H_{10}N_2$	158.0845
$C_{10}H_{22}O$	158.1672
$C_{11}H_{10}O$	158.0732
$C_{11}H_{12}N$	158.0970
$C_{12}H_{14}$	158.1096
159	
$C_5H_7N_2O_4$	159.0406
$C_5H_9N_3O_3$	159.0644
$C_5H_{11}N_4O_2$	159.0883
$C_6H_9NO_4$	159.0532
$C_6H_{11}N_2O_3$	159.0770
$C_6H_{13}N_3O_2$	159.1009
$C_6H_{15}N_4O$	159.1247
$C_7H_{11}O_4$	159.0657
$C_7H_{13}NO_3$	159.0896
$C_7H_{15}N_2O_2$	159.1134
$C_7H_{17}N_3O$	159.1373
$C_8H_5N_3O$	159.0433
$C_8H_7N_4$	159.0672
$C_8H_{15}O_3$	159.1021
$C_8H_{17}NO_2$	159.1260
$C_8H_{19}N_2O$	159.1498
$C_8H_{21}N_3$	159.1737
$C_9H_5NO_2$	159.0320
$C_9H_7N_2O$	159.0559
$C_9H_9N_3$	159.0798
$C_9H_{19}O_2$	159.1385
$C_9H_{21}NO$	159.1624
$C_{10}H_7O_2$	159.0446
$C_{10}H_9NO$	159.0684
$C_{10}H_{11}N_2$	159.0923
$C_{11}H_{11}O$	159.0810
$C_{11}H_{13}N$	159.1049
$C_{12}H_{15}$	159.1174
160	
$C_5H_8N_2O_4$	160.0484
$C_5H_{10}N_3O_3$	160.0723
$C_5H_{12}N_4O_2$	160.0961
$C_6H_{10}NO_4$	160.0610
$C_6H_{12}N_2O_3$	160.0848
$C_6H_{14}N_3O_2$	160.1087
$C_6H_{16}N_4O$	160.1325
$C_7H_{12}O_4$	160.0735
$C_7H_{14}NO_3$	160.0974
$C_7H_{16}N_2O_2$	160.1213
$C_7H_{18}N_3O$	160.1451
$C_7H_{20}N_4$	160.1690
$C_8H_6N_3O$	160.0511
$C_8H_8N_4$	160.0750
$C_8H_{16}O_3$	160.1100
$C_8H_{18}NO_2$	160.1338
$C_8H_{20}N_2O$	160.1577
$C_9H_6NO_2$	160.0399
$C_9H_8N_2O$	160.0637
$C_9H_{10}N_3$	160.0876
$C_9H_{20}O_2$	160.1464
$C_{10}H_8O_2$	160.0524
$C_{10}H_{10}NO$	160.0763
$C_{10}H_{12}N_2$	160.1001
$C_{11}H_{12}O$	160.0888
$C_{11}H_{14}N$	160.1127
$C_{12}H_{16}$	160.1253
161	
$C_5H_9N_2O_4$	161.0563
$C_5H_{11}N_3O_3$	161.0801
$C_5H_{13}N_4O_2$	161.1040
$C_6H_{11}NO_4$	161.0688
$C_6H_{13}N_2O_3$	161.0927
$C_6H_{15}N_3O_2$	161.1165
$C_6H_{17}N_4O$	161.1404
$C_7H_5N_4O$	161.0464
$C_8H_5N_2O_2$	161.0351
$C_8H_7N_3O$	161.0590
$C_8H_9N_4$	161.0829
$C_8H_{17}O_3$	161.1178
$C_8H_{19}NO_2$	161.1416
$C_9H_5O_3$	161.0238
$C_9H_7NO_2$	161.0477
$C_9H_9N_2O$	161.0715
$C_9H_{11}N_3$	161.0954
$C_{10}H_9O_2$	161.0603
$C_{10}H_{11}NO$	161.0841
$C_{10}H_{13}N_2$	161.1080
$C_{11}H_{13}O$	161.0967
$C_{11}H_{15}N$	161.1205
$C_{12}H_{17}$	161.1331
162	
$C_5H_{10}N_2O_4$	162.0641
$C_5H_{12}N_3O_3$	162.0879
$C_5H_{14}N_4O_2$	162.1118
$C_6H_{12}NO_4$	162.0766
$C_6H_{14}N_2O_3$	162.1005
$C_6H_{16}N_3O_2$	162.1244
$C_6H_{18}N_4O$	162.1482
$C_7H_6N_4O$	162.0542
$C_7H_{14}O_4$	162.0892
$C_7H_{16}NO_3$	162.1131
$C_7H_{18}N_2O_2$	162.1369
$C_8H_6N_2O_2$	162.0429
$C_8H_8N_3O$	162.0668
$C_8H_{10}N_4$	162.0907
$C_8H_{18}O_3$	162.1256
$C_9H_6O_3$	162.0317
$C_9H_8NO_2$	162.0555
$C_9H_{10}N_2O$	162.0794
$C_9H_{12}N_3$	162.1032
$C_{10}H_{10}O_2$	162.0681
$C_{10}H_{12}NO$	162.0919
$C_{10}H_{14}N_2$	162.1158
$C_{11}H_{14}O$	162.1045
$C_{11}H_{16}N$	162.1284
$C_{12}H_{18}$	162.1409
163	
$C_5H_{11}N_2O_4$	163.0719
$C_5H_{13}N_3O_3$	163.0958
$C_5H_{15}N_4O_2$	163.1196
$C_6H_{13}NO_4$	163.0845
$C_6H_{15}N_2O_3$	163.1083
$C_6H_{17}N_3O_2$	163.1322
$C_7H_5N_3O_2$	163.0382
$C_7H_7N_4O$	163.0621
$C_7H_{15}O_4$	163.0970
$C_7H_{17}NO_3$	163.1209
$C_8H_5NO_3$	163.0269
$C_8H_7N_2O_2$	163.0508
$C_8H_9N_3O$	163.0746
$C_8H_{11}N_4$	163.0985
$C_9H_7O_3$	163.0395
$C_9H_9NO_2$	163.0634
$C_9H_{11}N_2O$	163.0872
$C_9H_{13}N_3$	163.1111
$C_{10}H_{11}O_2$	163.0759
$C_{10}H_{13}NO$	163.0998
$C_{10}H_{15}N_2$	163.1236
$C_{11}H_{15}O$	163.1123
$C_{11}H_{17}N$	163.1362
$C_{12}H_{19}$	163.1488
164	
$C_5H_{12}N_2O_4$	164.0797
$C_5H_{14}N_3O_3$	164.1036
$C_5H_{16}N_4O_2$	164.1275
$C_6H_4N_4O_2$	164.0335
$C_6H_{14}NO_4$	164.0923
$C_6H_{16}N_2O_3$	164.1162
$C_7H_6N_3O_2$	164.0460
$C_7H_8N_4O$	164.0699
$C_7H_{16}O_4$	164.1049
$C_8H_6NO_3$	164.0348
$C_8H_8N_2O_2$	164.0586
$C_8H_{10}N_3O$	164.0825
$C_8H_{12}N_4$	164.1063
$C_9H_8O_3$	164.0473
$C_9H_{10}NO_2$	164.0712
$C_9H_{12}N_2O$	164.0950
$C_9H_{14}N_3$	164.1189
$C_{10}H_{12}O_2$	164.0837
$C_{10}H_{14}NO$	164.1076

Appendix A (Continued)

	FM		FM		FM		FM
$C_{10}H_{16}N_2$	164.1315	$C_7H_7N_2O_3$	167.0457	$C_8H_{11}NO_3$	169.0739	$C_7H_{13}N_3O_2$	171.1009
$C_{11}H_{16}O$	164.1202	$C_7H_9N_3O_2$	167.0695	$C_8H_{13}N_2O_2$	169.0978	$C_7H_{15}N_4O$	171.1247
$C_{11}H_{18}N$	164.1440	$C_7H_{11}N_4O$	167.0934	$C_8H_{15}N_3O$	169.1216	$C_8H_{11}O_4$	171.0657
$C_{12}H_{20}$	164.1566	$C_8H_7O_4$	167.0344	$C_8H_{17}N_4$	169.1455	$C_8H_{13}NO_3$	171.0896
165		$C_8H_9NO_3$	167.0583	$C_9H_{13}O_3$	169.0865	$C_8H_{15}N_2O_2$	171.1134
$C_5H_{13}N_2O_4$	165.0876	$C_8H_{11}N_2O_2$	167.0821	$C_9H_{15}NO_2$	169.1103	$C_8H_{17}N_3O$	171.1373
$C_5H_{15}N_3O_3$	165.1114	$C_8H_{13}N_3O$	167.1060	$C_9H_{17}N_2O$	169.1342	$C_8H_{19}N_4$	171.1611
$C_6H_5N_4O_2$	165.0413	$C_8H_{15}N_4$	167.1298	$C_9H_{19}N_3$	169.1580	$C_9H_5N_3O$	171.0433
$C_6H_{15}NO_4$	165.1001	$C_9H_{11}O_3$	167.0708	$C_{10}H_7N_3$	169.0641	$C_9H_7N_4$	171.0672
$C_7H_5N_2O_3$	165.0300	$C_9H_{13}NO_2$	167.0947	$C_{10}H_{17}O_2$	169.1229	$C_9H_{15}O_3$	171.1021
$C_7H_7N_3O_2$	165.0539	$C_9H_{15}N_2O$	167.1185	$C_{10}H_{19}NO$	169.1467	$C_9H_{17}NO_2$	171.1260
$C_7H_9N_4O$	165.0777	$C_9H_{17}N_3$	167.1424	$C_{10}H_{21}N_2$	169.1706	$C_9H_{19}N_2O$	171.1498
$C_8H_5O_4$	165.0187	$C_{10}H_{15}O_2$	167.1072	$C_{11}H_7NO$	169.0528	$C_9H_{21}N_3$	171.1737
$C_8H_7NO_3$	165.0426	$C_{10}H_{17}NO$	167.1311	$C_{11}H_9N_2$	169.0767	$C_{10}H_7N_2O$	171.0559
$C_8H_9N_2O_2$	165.0664	$C_{10}H_{19}N_2$	167.1549	$C_{11}H_{21}O$	169.1593	$C_{10}H_9N_3$	171.0798
$C_8H_{11}N_3O$	165.0903	$C_{11}H_7N_2$	167.0610	$C_{11}H_{23}N$	169.1832	$C_{10}H_{19}O_2$	171.1385
$C_8H_{13}N_4$	165.1142	$C_{11}H_{19}O$	167.1436	$C_{12}H_9O$	169.0653	$C_{10}H_{21}NO$	171.1624
$C_9H_9O_3$	165.0552	$C_{11}H_{21}N$	167.1675	$C_{12}H_{11}N$	169.0892	$C_{10}H_{23}N_2$	171.1863
$C_9H_{11}NO_2$	165.0790	$C_{12}H_9N$	167.0736	$C_{12}H_{25}$	169.1957	$C_{11}H_7O_2$	171.0446
$C_9H_{13}N_2O$	165.1029	$C_{12}H_{23}$	167.1801	$C_{13}H_{13}$	169.1018	$C_{11}H_9NO$	171.0684
$C_9H_{15}N_3$	165.1267	$C_{13}H_{11}$	167.0861	**170**		$C_{11}H_{11}N_2$	171.0923
$C_{10}H_{13}O_2$	165.0916	**168**		$C_6H_6N_2O_4$	170.0328	$C_{11}H_{23}O$	171.1750
$C_{10}H_{15}NO$	165.1154	$C_6H_4N_2O_4$	168.0171	$C_6H_8N_3O_3$	170.0566	$C_{11}H_{25}N$	171.1988
$C_{10}H_{17}N_2$	165.1393	$C_6H_6N_3O_3$	168.0410	$C_6H_{10}N_4O_2$	170.0805	$C_{12}H_{11}O$	171.0810
$C_{11}H_{17}O$	165.1280	$C_6H_8N_4O_2$	168.0648	$C_7H_8NO_4$	170.0453	$C_{12}H_{13}N$	171.1049
$C_{11}H_{19}N$	165.1519	$C_7H_6NO_4$	168.0297	$C_7H_{10}N_2O_3$	170.0692	$C_{13}H_{15}$	171.1174
$C_{12}H_7N$	165.0579	$C_7H_8N_2O_3$	168.0535	$C_7H_{12}N_3O_2$	170.0930	**172**	
$C_{12}H_{21}$	165.1644	$C_7H_{10}N_3O_2$	168.0774	$C_7H_{14}N_4O$	170.1169	$C_6H_8N_2O_4$	172.0484
$C_{13}H_9$	165.0705	$C_7H_{12}N_4O$	168.1012	$C_8H_{10}O_4$	170.0579	$C_6H_{10}N_3O_3$	172.0723
166		$C_8H_8O_4$	168.0422	$C_8H_{12}NO_3$	170.0817	$C_6H_{12}N_4O_2$	172.0961
$C_5H_{14}N_2O_4$	166.0954	$C_8H_{10}NO_3$	168.0661	$C_8H_{14}N_2O_2$	170.1056	$C_7H_{10}NO_4$	172.0610
$C_6H_4N_3O_3$	166.0253	$C_8H_{12}N_2O_2$	168.0899	$C_8H_{16}N_3O$	170.1295	$C_7H_{12}N_2O_3$	172.0848
$C_6H_6N_4O_2$	166.0491	$C_8H_{14}N_3O$	168.1138	$C_8H_{18}N_4$	170.1533	$C_7H_{14}N_3O_2$	172.1087
$C_7H_6N_2O_3$	166.0379	$C_8H_{16}N_4$	168.1377	$C_9H_6N_4$	170.0594	$C_7H_{16}N_4O$	172.1325
$C_7H_8N_3O_2$	166.0617	$C_9H_{12}O_3$	168.0786	$C_9H_{14}O_3$	170.0943	$C_8H_{12}O_4$	172.0735
$C_7H_{10}N_4O$	166.0856	$C_9H_{14}NO_2$	168.1025	$C_9H_{16}NO_2$	170.1182	$C_8H_{14}NO_3$	172.0974
$C_8H_6O_4$	166.0266	$C_9H_{16}N_2O$	168.1264	$C_9H_{18}N_2O$	170.1420	$C_8H_{16}N_2O_2$	172.1213
$C_8H_8NO_3$	166.0504	$C_9H_{18}N_3$	168.1502	$C_9H_{20}N_3$	170.1659	$C_8H_{18}N_3O$	172.1451
$C_8H_{10}N_2O_2$	166.0743	$C_{10}H_{16}O_2$	168.1151	$C_{10}H_6N_2O$	170.0480	$C_8H_{20}N_4$	172.1690
$C_8H_{12}N_3O$	166.0981	$C_{10}H_{18}NO$	168.1389	$C_{10}H_8N_3$	170.0719	$C_9H_6N_3O$	172.0511
$C_8H_{14}N_4$	166.1220	$C_{10}H_{20}N_2$	168.1628	$C_{10}H_{18}O_2$	170.1307	$C_9H_8N_4$	172.0750
$C_9H_{10}O_3$	166.0630	$C_{11}H_8N_2$	168.0688	$C_{10}H_{20}NO$	170.1546	$C_9H_{16}O_3$	172.1100
$C_9H_{12}NO_2$	166.0868	$C_{11}H_{20}O$	168.1515	$C_{10}H_{22}N_2$	170.1784	$C_9H_{18}NO_2$	172.1338
$C_9H_{14}N_2O$	166.1107	$C_{11}H_{22}N$	168.1753	$C_{11}H_8NO$	170.0606	$C_9H_{20}N_2O$	172.1577
$C_9H_{16}N_3$	166.1346	$C_{12}H_8O$	168.0575	$C_{11}H_{10}N_2$	170.0845	$C_9H_{22}N_3$	172.1815
$C_{10}H_{14}O_2$	166.0994	$C_{12}H_{10}N$	168.0814	$C_{11}H_{22}O$	170.1671	$C_{10}H_6NO_2$	172.0399
$C_{10}H_{16}NO$	166.1233	$C_{12}H_{24}$	168.1879	$C_{11}H_{24}N$	170.1910	$C_{10}H_8N_2O$	172.0637
$C_{10}H_{18}N_2$	166.1471	$C_{13}H_{12}$	168.0939	$C_{12}H_{10}O$	170.0732	$C_{10}H_{10}N_3$	172.0876
$C_{11}H_{18}O$	166.1358	**169**		$C_{12}H_{12}N$	170.0970	$C_{10}H_{20}O_2$	172.1464
$C_{11}H_{20}N$	166.1597	$C_6H_5N_2O_4$	169.0249	$C_{12}H_{26}$	170.2036	$C_{10}H_{22}NO$	172.1702
$C_{12}H_8N$	166.0657	$C_6H_7N_3O_3$	169.0488	$C_{13}H_{14}$	170.1096	$C_{10}H_{24}N_2$	172.1941
$C_{12}H_{22}$	166.1722	$C_6H_9N_4O_2$	169.0726	**171**		$C_{11}H_8O_2$	172.0524
$C_{13}H_{10}$	166.0783	$C_7H_7NO_4$	169.0375	$C_6H_7N_2O_4$	171.0406	$C_{11}H_{10}NO$	172.0763
167		$C_7H_9N_2O_3$	169.0614	$C_6H_9N_3O_3$	171.0644	$C_{11}H_{12}N_2$	172.1001
$C_6H_5N_3O_3$	167.0331	$C_7H_{11}N_3O_2$	169.0852	$C_6H_{11}N_4O_2$	171.0883	$C_{11}H_{24}O$	172.1828
$C_6H_7N_4O_2$	167.0570	$C_7H_{13}N_4O$	169.1091	$C_7H_9NO_4$	171.0532	$C_{12}H_{12}O$	172.0888
$C_7H_5NO_4$	167.0218	$C_8H_9O_4$	169.0501	$C_7H_{11}N_2O_3$	171.0770	$C_{12}H_{14}N$	172.1127

Appendix A (Continued)

	FM
$C_{13}H_{16}$	172.1253
173	
$C_6H_9N_2O_4$	173.0563
$C_6H_{11}N_3O_3$	173.0801
$C_6H_{13}N_4O_2$	173.1040
$C_7H_{11}NO_4$	173.0688
$C_7H_{13}N_2O_3$	173.0927
$C_7H_{15}N_3O_2$	173.1165
$C_7H_{17}N_4O$	173.1404
$C_8H_{13}O_4$	173.0814
$C_8H_{15}NO_3$	173.1052
$C_8H_{17}N_2O_2$	173.1291
$C_8H_{19}N_3O$	173.1529
$C_8H_{21}N_4$	173.1768
$C_9H_7N_3O$	173.0590
$C_9H_9N_4$	173.0829
$C_9H_{17}O_3$	173.1178
$C_9H_{19}NO_2$	173.1416
$C_9H_{21}N_2O$	173.1655
$C_{10}H_5O_3$	173.0238
$C_{10}H_7NO_2$	173.0477
$C_{10}H_9N_2O$	173.0715
$C_{10}H_{11}N_3$	173.0954
$C_{10}H_{21}O_2$	173.1542
$C_{10}H_{23}NO$	173.1781
$C_{11}H_9O_2$	173.0603
$C_{11}H_{11}NO$	173.0841
$C_{11}H_{13}N_2$	173.1080
$C_{12}H_{13}O$	173.0967
$C_{12}H_{15}NO_2$	173.1205
$C_{13}H_{17}$	173.1331
174	
$C_6H_{10}N_2O_4$	174.0641
$C_6H_{12}N_3O_3$	174.0879
$C_6H_{14}N_4O_2$	174.1118
$C_7H_{12}NO_4$	174.0766
$C_7H_{14}N_2O_3$	174.1005
$C_7H_{16}N_3O_2$	174.1244
$C_7H_{18}N_4O$	174.1482
$C_7H_{16}N_4O$	174.1244
$C_8H_6N_4O$	174.0542
$C_8H_{14}O_4$	174.0892
$C_8H_{16}NO_3$	174.1131
$C_8H_{18}N_2O_2$	174.1369
$C_8H_{20}N_3O$	174.1608
$C_8H_{22}N_4$	174.1846
$C_9H_6N_2O_2$	174.0429
$C_9H_{10}N_4$	174.0907
$C_9H_{18}O_3$	174.1256
$C_9H_{20}NO_2$	174.1495
$C_9H_{22}N_2O$	174.1733
$C_{10}H_6O_3$	174.0317
$C_{10}H_8NO_2$	174.0555
$C_{10}H_{10}N_2O$	174.0794
$C_{10}H_{12}N_3$	174.1032
$C_{10}H_{22}O_2$	174.1620
$C_{11}H_{10}O_2$	174.0681
$C_{11}H_{12}NO$	174.0919
$C_{11}H_{14}N_2$	174.1158
$C_{12}H_{14}O$	174.1045
$C_{12}H_{16}N$	174.1284
$C_{13}H_{18}$	174.1409
175	
$C_6H_{11}N_2O_4$	175.0719
$C_6H_{13}N_3O_3$	175.0958
$C_6H_{15}N_4O_2$	175.1196
$C_7H_{13}NO_4$	175.0845
$C_7H_{15}N_2O_3$	175.1083
$C_7H_{17}N_3O_2$	175.1322
$C_7H_{19}N_4O$	175.1560
$C_8H_7N_4O$	175.0621
$C_8H_{15}O_4$	175.0970
$C_8H_{17}NO_3$	175.1209
$C_8H_{19}N_2O_2$	175.1447
$C_8H_{21}N_3O$	175.1686
$C_9H_5NO_3$	175.0269
$C_9H_7N_2O_2$	175.0508
$C_9H_9N_3O$	175.0746
$C_9H_{11}N_4$	175.0985
$C_9H_{19}O_3$	175.1334
$C_9H_{21}NO_2$	175.1573
$C_{10}H_7O_3$	175.0395
$C_{10}H_9NO_2$	175.0634
$C_{10}H_{11}N_2O$	175.0872
$C_{10}H_{13}N_3$	175.1111
$C_{11}H_{11}O_2$	175.0759
$C_{11}H_{13}NO$	175.0998
$C_{11}H_{15}N_2$	175.1236
$C_{12}H_{15}O$	175.1123
$C_{12}H_{17}N$	175.1362
$C_{13}H_3O$	175.0184
$C_{13}H_{19}$	175.1488
176	
$C_6H_{12}N_2O_4$	176.0797
$C_6H_{14}N_3O_3$	176.1036
$C_6H_{16}N_4O_2$	176.1275
$C_7H_{14}NO_4$	176.0923
$C_7H_{16}N_2O_3$	176.1162
$C_7H_{18}N_3O_2$	176.1400
$C_7H_{20}N_4O$	176.1639
$C_8H_6N_3O_2$	176.0460
$C_8H_8N_4O$	176.0699
$C_8H_{16}O_4$	176.1049
$C_8H_{18}NO_3$	176.1287
$C_8H_{20}N_2O_2$	176.1526
$C_9H_6NO_3$	176.0348
$C_9H_8N_2O_2$	176.0586
$C_9H_{10}N_3O$	176.0825
$C_9H_{12}N_4$	176.1063
$C_9H_{20}O_3$	176.1413
$C_{10}H_8O_3$	176.0473
$C_{10}H_{10}NO_2$	176.0712
$C_{10}H_{12}N_2O$	176.0950
$C_{10}H_{14}N_3$	176.1189
$C_{11}H_{12}O_2$	176.0837
$C_{11}H_{14}NO$	176.1076
$C_{11}H_{16}N_2$	176.1315
$C_{12}H_{16}O$	176.1202
$C_{12}H_{18}N$	176.1440
$C_{13}H_{20}$	176.1566
177	
$C_6H_{13}N_2O_4$	177.0876
$C_6H_{15}N_3O_3$	177.1114
$C_6H_{17}N_4O_2$	177.1353
$C_7H_5N_4O_2$	177.0413
$C_7H_{15}NO_4$	177.1001
$C_7H_{17}N_2O_3$	177.1240
$C_7H_{19}N_3O_2$	177.1478
$C_8H_5N_2O_3$	177.0300
$C_8H_7N_3O_2$	177.0539
$C_8H_9N_4O$	177.0777
$C_8H_{17}O_4$	177.1127
$C_8H_{19}NO_3$	177.1365
$C_9H_7NO_3$	177.0426
$C_9H_9N_2O_2$	177.0664
$C_9H_{11}N_3O$	177.0903
$C_9H_{13}N_4$	177.1142
$C_{10}H_9O_3$	177.0552
$C_{10}H_{11}NO_2$	177.0790
$C_{10}H_{13}N_2O$	177.1029
$C_{10}H_{15}N_3$	177.1267
$C_{11}H_{13}O_2$	177.0916
$C_{11}H_{15}NO$	177.1154
$C_{11}H_{17}N_2$	177.1393
$C_{12}H_{17}O$	177.1280
$C_{12}H_{19}N$	177.1519
$C_{13}H_{21}$	177.1644
178	
$C_6H_{14}N_2O_4$	178.0954
$C_6H_{16}N_3O_3$	178.1193
$C_6H_{18}N_4O_2$	178.1431
$C_7H_6N_4O_2$	178.0491
$C_7H_{16}NO_4$	178.1080
$C_7H_{18}N_2O_3$	178.1318
$C_8H_6N_2O_3$	178.0379
$C_8H_8N_3O_2$	178.0617
$C_8H_{10}N_4O$	178.0856
$C_8H_{18}O_4$	178.1205
$C_9H_6O_4$	178.0266
$C_9H_8NO_3$	178.0504
$C_9H_{10}N_2O_2$	178.0743
$C_9H_{12}N_3O$	178.0981
$C_9H_{14}N_4$	178.1220
$C_{10}H_{10}O_3$	178.0630
$C_{10}H_{12}NO_2$	178.0868
$C_{10}H_{14}N_2O$	178.1107
$C_{10}H_{16}N_3$	178.1346
$C_{11}H_{14}O_2$	178.0994
$C_{11}H_{16}NO$	178.1233
$C_{11}H_{18}N_2$	178.1471
$C_{12}H_{18}O$	178.1358
$C_{12}H_{20}N$	178.1597
$C_{13}H_8N$	178.0657
$C_{13}H_{22}$	178.1722
$C_{14}H_{10}$	178.0783
179	
$C_6H_{15}N_2O_4$	179.1032
$C_6H_{17}N_3O_3$	179.1271
$C_7H_5N_3O_3$	179.0331
$C_7H_7N_4O_2$	179.0570
$C_7H_{17}NO_4$	179.1158
$C_8H_5NO_4$	179.0218
$C_8H_7N_2O_3$	179.0457
$C_8H_9N_3O_2$	179.0695
$C_8H_{11}N_4O$	179.0934
$C_9H_7O_4$	179.0344
$C_9H_9NO_3$	179.0583
$C_9H_{11}N_2O_2$	179.0821
$C_9H_{13}N_3O$	179.1060
$C_9H_{15}N_4$	179.1298
$C_{10}H_{11}O_3$	179.0708
$C_{10}H_{13}NO_2$	179.0947
$C_{10}H_{15}N_2O$	179.1185
$C_{10}H_{17}N_3$	179.1424
$C_{11}H_{15}O_2$	179.1072
$C_{11}H_{17}NO$	179.1311
$C_{11}H_{19}N_2$	179.1549
$C_{12}H_{19}O$	179.1436
$C_{12}H_{21}N$	179.1675
$C_{13}H_9N$	179.0736
$C_{13}H_{23}$	179.1801
$C_{14}H_{11}$	179.0861
180	
$C_6H_{16}N_2O_4$	180.1111
$C_7H_6N_3O_3$	180.0410
$C_7H_8N_4O_2$	180.0648
$C_8H_6NO_4$	180.0297
$C_8H_8N_2O_3$	180.0535
$C_8H_{10}N_3O_2$	180.0774
$C_8H_{12}N_4O$	180.1012
$C_9H_8O_4$	180.0422
$C_9H_{10}NO_3$	180.0661
$C_9H_{12}N_2O_2$	180.0899
$C_9H_{14}N_3O$	180.1138
$C_9H_{16}N_4$	180.1377
$C_{10}H_{12}O_3$	180.0786
$C_{10}H_{14}NO_2$	180.1025
$C_{10}H_{16}N_2O$	180.1264
$C_{10}H_{18}N_3$	180.1502
$C_{11}H_{16}O_2$	180.1151
$C_{11}H_{18}NO$	180.1389
$C_{11}H_{20}N_2$	180.1628
$C_{12}H_8N_2$	180.0688
$C_{12}H_{20}O$	180.1515
$C_{12}H_{22}N$	180.1753
$C_{13}H_8O$	180.0575
$C_{13}H_{10}N$	180.0814
$C_{13}H_{24}$	180.1879

Appendix A (Continued)

	FM
$C_{14}H_{12}$	180.0939
181	
$C_7H_5N_2O_4$	181.0249
$C_7H_7N_3O_3$	181.0488
$C_7H_9N_4O_2$	181.0726
$C_8H_7NO_4$	181.0375
$C_8H_9N_2O_3$	181.0614
$C_8H_{11}N_3O_2$	181.0852
$C_8H_{13}N_4O$	181.1091
$C_9H_9O_4$	181.0501
$C_9H_{11}NO_3$	181.0739
$C_9H_{13}N_2O_2$	181.0978
$C_9H_{15}N_3O$	181.1216
$C_9H_{17}N_4$	181.1455
$C_{10}H_{13}O_3$	181.0865
$C_{10}H_{15}NO_2$	181.1103
$C_{10}H_{17}N_2O$	181.1342
$C_{10}H_{19}N_3$	181.1580
$C_{11}H_7N_3$	181.0641
$C_{11}H_{17}O_2$	181.1229
$C_{11}H_{19}NO$	181.1467
$C_{11}H_{21}N_2$	181.1706
$C_{12}H_7NO$	181.0528
$C_{12}H_9N_2$	181.0767
$C_{12}H_{21}O$	181.1593
$C_{12}H_{23}N$	181.1832
$C_{13}H_9O$	181.0653
$C_{13}H_{11}N$	181.0892
$C_{13}H_{25}$	181.1957
$C_{14}H_{13}$	181.1018
182	
$C_7H_6N_2O_4$	182.0328
$C_7H_8N_3O_3$	182.0566
$C_7H_{10}N_4O_2$	182.0805
$C_8H_8NO_4$	182.0453
$C_8H_{10}N_2O_3$	182.0692
$C_8H_{12}N_3O_2$	182.0930
$C_8H_{14}N_4O$	182.1169
$C_9H_{10}O_4$	182.0579
$C_9H_{12}NO_3$	182.0817
$C_9H_{14}N_2O_2$	182.1056
$C_9H_{16}N_3O$	182.1295
$C_9H_{18}N_4$	182.1533
$C_{10}H_6N_4$	182.0594
$C_{10}H_{14}O_3$	182.0943
$C_{10}H_{16}NO_2$	182.1182
$C_{10}H_{18}N_2O$	182.1420
$C_{10}H_{20}N_3$	182.1659
$C_{11}H_8N_3$	182.0719
$C_{11}H_{18}O_2$	182.1307
$C_{11}H_{20}NO$	182.1546
$C_{11}H_{22}N_2$	182.1784
$C_{12}H_8NO$	182.0606
$C_{12}H_{10}N_2$	182.0845
$C_{12}H_{22}O$	182.1671
$C_{12}H_{24}N$	182.1910
$C_{13}H_{10}O$	182.0732
$C_{13}H_{12}N$	182.0970
$C_{13}H_{26}$	182.2036
$C_{14}H_{14}$	182.1096
183	
$C_7H_7N_2O_4$	183.0406
$C_7H_9N_3O_3$	183.0644
$C_7H_{11}N_4O_2$	183.0883
$C_8H_9NO_4$	183.0532
$C_8H_{11}N_2O_3$	183.0770
$C_8H_{13}N_3O_2$	183.1009
$C_8H_{15}N_4O$	183.1247
$C_9H_{11}O_4$	183.0657
$C_9H_{13}NO_3$	183.0896
$C_9H_{15}N_2O_2$	183.1134
$C_9H_{17}N_3O$	183.1373
$C_9H_{19}N_4$	183.1611
$C_{10}H_7N_4$	183.0672
$C_{10}H_{15}O_3$	183.1021
$C_{10}H_{17}NO_2$	183.1260
$C_{10}H_{19}N_2O$	183.1498
$C_{10}H_{21}N_3$	183.1737
$C_{11}H_7N_2O$	183.0559
$C_{11}H_9N_3$	183.0798
$C_{11}H_{19}O_2$	183.1385
$C_{11}H_{21}NO$	183.1624
$C_{11}H_{23}N_2$	183.1863
$C_{12}H_7O_2$	183.0446
$C_{12}H_9NO$	183.0684
$C_{12}H_{11}N_2$	183.0923
$C_{12}H_{23}O$	183.1750
$C_{12}H_{25}N$	183.1988
$C_{13}H_{11}O$	183.0810
$C_{13}H_{13}N$	183.1049
$C_{13}H_{27}$	183.2114
$C_{14}H_{15}$	183.1174
184	
$C_7H_8N_2O_4$	184.0484
$C_7H_{10}N_3O_3$	184.0723
$C_7H_{12}N_4O_2$	184.0961
$C_8H_{10}NO_4$	184.0610
$C_8H_{12}N_2O_3$	184.0848
$C_8H_{14}N_3O_2$	184.1087
$C_8H_{16}N_4O$	184.1325
$C_9H_{12}O_4$	184.0735
$C_9H_{14}NO_3$	184.0974
$C_9H_{16}N_2O_2$	184.1213
$C_9H_{18}N_3O$	184.1451
$C_9H_{20}N_4$	184.1690
$C_{10}H_6N_3O$	184.0511
$C_{10}H_8N_4$	184.0750
$C_{10}H_{16}O_3$	184.1100
$C_{10}H_{18}NO_2$	184.1338
$C_{10}H_{20}N_2O$	184.1577
$C_{10}H_{22}N_3$	184.1815
$C_{11}H_8N_2O$	184.0637
$C_{11}H_{10}N_3$	184.0876
$C_{11}H_{20}O_2$	184.1464
$C_{11}H_{22}NO$	184.1702
$C_{11}H_{24}N_2$	184.1941
$C_{12}H_8O_2$	184.0524
$C_{12}H_{10}NO$	184.0763
$C_{12}H_{12}N_2$	184.1001
$C_{12}H_{24}O$	184.1828
$C_{12}H_{26}N$	184.2067
$C_{13}H_{12}O$	184.0888
$C_{13}H_{14}N$	184.1127
$C_{13}H_{28}$	184.2192
$C_{14}H_{16}$	184.1253
185	
$C_7H_9N_2O_4$	185.0563
$C_7H_{11}N_3O_3$	185.0801
$C_7H_{13}N_4O_2$	185.1040
$C_8H_{11}NO_4$	185.0688
$C_8H_{13}N_2O_3$	185.0927
$C_8H_{15}N_3O_2$	185.1165
$C_8H_{17}N_4O$	185.1404
$C_9H_{13}O_4$	185.0814
$C_9H_{15}NO_3$	185.1052
$C_9H_{17}N_2O_2$	185.1291
$C_9H_{19}N_3O$	185.1529
$C_9H_{21}N_4$	185.1768
$C_{10}H_7N_3O$	185.0590
$C_{10}H_9N_4$	185.0829
$C_{10}H_{17}O_3$	185.1178
$C_{10}H_{19}NO_2$	185.1416
$C_{10}H_{21}N_2O$	185.1655
$C_{10}H_{23}N_3$	185.1894
$C_{11}H_9N_2O$	185.0715
$C_{11}H_{11}N_3$	185.0954
$C_{11}H_{21}O_2$	185.1542
$C_{11}H_{23}NO$	185.1781
$C_{11}H_{25}N_2$	185.2019
$C_{12}H_9O_2$	185.0603
$C_{12}H_{11}NO$	185.0841
$C_{12}H_{13}N_2$	185.1080
$C_{12}H_{25}O$	185.1906
$C_{12}H_{27}N$	185.2145
$C_{13}H_{13}O$	185.0967
$C_{13}H_{15}N$	185.1205
$C_{14}H_{17}$	185.1331
186	
$C_7H_{10}N_2O_4$	186.0641
$C_7H_{12}N_3O_3$	186.0879
$C_7H_{14}N_4O_2$	186.1118
$C_8H_{12}NO_4$	186.0766
$C_8H_{14}N_2O_3$	186.1005
$C_8H_{16}N_3O_2$	186.1244
$C_8H_{18}N_4O$	186.1482
$C_9H_6N_4O$	186.0542
$C_9H_{14}O_4$	186.0892
$C_9H_{16}NO_3$	186.1131
$C_9H_{18}N_2O_2$	186.1369
$C_9H_{20}N_3O$	186.1608
$C_{10}H_6N_2O_2$	186.0429
$C_{10}H_8N_3O$	186.0668
$C_{10}H_{10}N_4$	186.0907
$C_{10}H_{18}O_3$	186.1256
$C_{10}H_{20}NO_2$	186.1495
$C_{10}H_{22}N_2O$	186.1733
$C_{10}H_{24}N_3$	186.1972
$C_{11}H_8NO_2$	186.0555
$C_{11}H_{10}N_2O$	186.0794
$C_{11}H_{12}N_3$	186.1032
$C_{11}H_{22}O_2$	186.1620
$C_{11}H_{24}NO$	186.1859
$C_{11}H_{26}N_2$	186.2098
$C_{12}H_{10}O_2$	186.0681
$C_{12}H_{12}NO$	186.0919
$C_{12}H_{14}N_2$	186.1158
$C_{12}H_{26}O$	186.1985
$C_{13}H_{14}O$	186.1045
$C_{13}H_{16}N$	186.1284
$C_{14}H_{18}$	186.1409
187	
$C_7H_{11}N_2O_4$	187.0719
$C_7H_{13}N_3O_3$	187.0958
$C_7H_{15}N_4O_2$	187.1196
$C_8H_{13}NO_4$	187.0845
$C_8H_{15}N_2O_3$	187.1083
$C_8H_{17}N_3O_2$	187.1322
$C_8H_{19}N_4O$	187.1560
$C_9H_7N_4O$	187.0621
$C_9H_{15}O_4$	187.0970
$C_9H_{17}NO_3$	187.1209
$C_9H_{19}N_2O_2$	187.1447
$C_9H_{21}N_3O$	187.1686
$C_9H_{23}N_4$	187.1925
$C_{10}H_7N_2O_2$	187.0508
$C_{10}H_9N_3O$	187.0746
$C_{10}H_{11}N_4$	187.0985
$C_{10}H_{19}O_3$	187.1334
$C_{10}H_{21}NO_2$	187.1573
$C_{10}H_{23}N_2O$	187.1811
$C_{10}H_{25}N_3$	187.2050
$C_{11}H_7O_3$	187.0395
$C_{11}H_9NO_2$	187.0634
$C_{11}H_{11}N_2O$	187.0872
$C_{11}H_{13}N_3$	187.1111
$C_{11}H_{23}O_2$	187.1699
$C_{11}H_{25}NO$	187.1937
$C_{12}H_{11}O_2$	187.0759
$C_{12}H_{13}NO$	187.0998
$C_{12}H_{15}N_2$	187.1236
$C_{13}H_{15}O$	187.1123
$C_{13}H_{17}N$	187.1362
$C_{14}H_{19}$	187.1488
188	
$C_7H_{12}N_2O_4$	188.0797
$C_7H_{14}N_3O_3$	188.1036
$C_7H_{16}N_4O_2$	188.1275
$C_8H_{14}NO_4$	188.0923

Appendix A (Continued)

	FM
$C_8H_{16}N_2O_3$	188.1162
$C_8H_{18}N_3O_2$	188.1400
$C_8H_{20}N_4O$	188.1639
$C_9H_6N_3O_2$	188.0460
$C_9H_8N_4O$	188.0699
$C_9H_{16}O_4$	188.1049
$C_9H_{18}NO_3$	188.1287
$C_9H_{20}N_2O_2$	188.1526
$C_9H_{22}N_3O$	188.1764
$C_9H_{24}N_4$	188.2003
$C_{10}H_8N_2O_2$	188.0586
$C_{10}H_{10}N_3O$	188.0825
$C_{10}H_{12}N_4$	188.1063
$C_{10}H_{20}O_3$	188.1413
$C_{10}H_{22}NO_2$	188.1651
$C_{10}H_{24}N_2O$	188.1890
$C_{11}H_8O_3$	188.0473
$C_{11}H_{10}NO_2$	188.0712
$C_{11}H_{12}N_2O$	188.0950
$C_{11}H_{14}N_3$	188.1189
$C_{11}H_{24}O_2$	188.1777
$C_{12}H_{12}O_2$	188.0837
$C_{12}H_{14}NO$	188.1076
$C_{12}H_{16}N_2$	188.1315
$C_{13}H_{16}O$	188.1202
$C_{13}H_{18}N$	188.1440
$C_{14}H_{20}$	188.1566
189	
$C_7H_{13}N_2O_4$	189.0876
$C_7H_{15}N_3O_3$	189.1114
$C_7H_{17}N_4O_2$	189.1353
$C_8H_{15}NO_4$	189.1001
$C_8H_{17}N_2O_3$	189.1240
$C_8H_{19}N_3O_2$	189.1478
$C_8H_{21}N_4O$	189.1717
$C_9H_7N_3O_2$	189.0539
$C_9H_9N_4O$	189.0777
$C_9H_{17}O_4$	189.1127
$C_9H_{19}NO_3$	189.1365
$C_9H_{21}N_2O_2$	189.1604
$C_9H_{23}N_3O$	189.1842
$C_{10}H_7NO_3$	189.0426
$C_{10}H_9N_2O_2$	189.0664
$C_{10}H_{11}N_3O$	189.0903
$C_{10}H_{13}N_4$	189.1142
$C_{10}H_{21}O_3$	189.1491
$C_{10}H_{23}NO_2$	189.1730
$C_{11}H_9O_3$	189.0552
$C_{11}H_{11}NO_2$	189.0790
$C_{11}H_{13}N_2O$	189.1029
$C_{11}H_{15}N_3$	189.1267
$C_{12}H_{13}O_2$	189.0916
$C_{12}H_{15}NO$	189.1154
$C_{12}H_{17}N_2$	189.1393
$C_{13}H_{17}O$	189.1280
$C_{13}H_{19}N$	189.1519
$C_{14}H_{21}$	189.1644
190	
$C_7H_{14}N_2O_4$	190.0954
$C_7H_{16}N_3O_3$	190.1193
$C_7H_{18}N_4O_2$	190.1431
$C_8H_6N_4O_2$	190.0491
$C_8H_{16}NO_4$	190.1080
$C_8H_{18}N_2O_3$	190.1318
$C_8H_{20}N_3O_2$	190.1557
$C_8H_{22}N_4O$	190.1795
$C_9H_8N_3O_2$	190.0617
$C_9H_{10}N_4O$	190.0856
$C_9H_{18}O_4$	190.1205
$C_9H_{20}NO_3$	190.1444
$C_9H_{22}N_2O_2$	190.1682
$C_{10}H_8NO_3$	190.0504
$C_{10}H_{10}N_2O_2$	190.0743
$C_{10}H_{12}N_3O$	190.0981
$C_{10}H_{14}N_4$	190.1220
$C_{10}H_{22}O_3$	190.1569
$C_{11}H_{10}O_3$	190.0630
$C_{11}H_{12}NO_2$	190.0868
$C_{11}H_{14}N_2O$	190.1107
$C_{11}H_{16}N_3$	190.1346
$C_{12}H_{14}O_2$	190.0994
$C_{12}H_{16}NO$	190.1233
$C_{12}H_{18}N_2$	190.1471
$C_{13}H_{18}O$	190.1358
$C_{13}H_{20}N$	190.1597
$C_{14}H_{22}$	190.1722
$C_{15}H_{10}$	190.0783
191	
$C_7H_{15}N_2O_4$	191.1032
$C_7H_{17}N_3O_3$	191.1271
$C_7H_{19}N_4O_2$	191.1509
$C_8H_7N_4O_2$	191.0570
$C_8H_{17}NO_4$	191.1158
$C_8H_{19}N_2O_3$	191.1396
$C_8H_{21}N_3O_2$	191.1635
$C_9H_7N_2O_3$	191.0457
$C_9H_9N_3O_2$	191.0695
$C_9H_{11}N_4O$	191.0934
$C_9H_{19}O_4$	191.1284
$C_9H_{21}NO_3$	191.1522
$C_{10}H_7O_4$	191.0344
$C_{10}H_9NO_3$	191.0583
$C_{10}H_{11}N_2O_2$	191.0821
$C_{10}H_{13}N_3O$	191.1060
$C_{10}H_{15}N_4$	191.1298
$C_{11}H_{11}O_3$	191.0708
$C_{11}H_{13}NO_2$	191.0947
$C_{11}H_{15}N_2O$	191.1185
$C_{11}H_{17}N_3$	191.1424
$C_{12}H_{15}O_2$	191.1072
$C_{12}H_{17}NO$	191.1311
$C_{12}H_{19}N_2$	191.1549
$C_{13}H_{19}O$	191.1436
$C_{13}H_{21}N$	191.1675
$C_{14}H_9N$	191.0736
$C_{14}H_{23}$	191.1801
$C_{15}H_{11}$	191.0861
192	
$C_7H_{16}N_2O_4$	192.1111
$C_7H_{18}N_3O_3$	192.1349
$C_7H_{20}N_4O_2$	192.1588
$C_8H_6N_3O_3$	192.0410
$C_8H_8N_4O_2$	192.0648
$C_8H_{18}NO_4$	192.1236
$C_8H_{20}N_2O_3$	192.1475
$C_9H_6NO_4$	192.0297
$C_9H_8N_2O_3$	192.0535
$C_9H_{10}N_3O_2$	192.0774
$C_9H_{12}N_4O$	192.1012
$C_9H_{20}O_4$	192.1362
$C_{10}H_8O_4$	192.0422
$C_{10}H_{10}NO_3$	192.0661
$C_{10}H_{12}N_2O_2$	192.0899
$C_{10}H_{14}N_3O$	192.1138
$C_{10}H_{16}N_4$	192.1377
$C_{11}H_{12}O_3$	192.0786
$C_{11}H_{14}NO_2$	192.1025
$C_{11}H_{16}N_2O$	192.1264
$C_{11}H_{18}N_3$	192.1502
$C_{12}H_{16}O_2$	192.1151
$C_{12}H_{18}NO$	192.1389
$C_{12}H_{20}N_2$	192.1628
$C_{13}H_8N_2$	192.0688
$C_{13}H_{20}O$	192.1515
$C_{13}H_{22}N$	192.1753
$C_{14}H_{10}N$	192.0814
$C_{14}H_{24}$	192.1879
$C_{15}H_{12}$	192.0939
193	
$C_7H_{17}N_2O_4$	193.1189
$C_7H_{19}N_3O_3$	193.1427
$C_8H_7N_3O_3$	193.0488
$C_8H_9N_4O_2$	193.0726
$C_8H_{19}NO_4$	193.1315
$C_9H_7NO_4$	193.0375
$C_9H_9N_2O_3$	193.0614
$C_9H_{11}N_3O_2$	193.0852
$C_9H_{13}N_4O$	193.1091
$C_{10}H_9O_4$	193.0501
$C_{10}H_{11}NO_3$	193.0739
$C_{10}H_{13}N_2O_2$	193.0978
$C_{10}H_{15}N_3O$	193.1216
$C_{10}H_{17}N_4$	193.1455
$C_{11}H_{13}O_3$	193.0865
$C_{11}H_{15}NO_2$	193.1103
$C_{11}H_{17}N_2O$	193.1342
$C_{11}H_{19}N_3$	193.1580
$C_{12}H_{17}O_2$	193.1229
$C_{12}H_{19}NO$	193.1467
$C_{12}H_{21}N_2$	193.1706
$C_{13}H_9N_2$	193.0767
$C_{13}H_{21}O$	193.1593
$C_{13}H_{23}N$	193.1832
$C_{14}H_9O$	193.0653
$C_{14}H_{11}N$	193.0892
$C_{14}H_{25}$	193.1957
$C_{15}H_{13}$	193.1018
194	
$C_7H_{18}N_2O_4$	194.1267
$C_8H_6N_2O_4$	194.0328
$C_8H_8N_3O_3$	194.0566
$C_8H_{10}N_4O_2$	194.0805
$C_9H_8NO_4$	194.0453
$C_9H_{10}N_2O_3$	194.0692
$C_9H_{12}N_3O_2$	194.0930
$C_9H_{14}N_4O$	194.1169
$C_{10}H_{10}O_4$	194.0579
$C_{10}H_{12}NO_3$	194.0817
$C_{10}H_{14}N_2O_2$	194.1056
$C_{10}H_{16}N_3O$	194.1295
$C_{10}H_{18}N_4$	194.1533
$C_{11}H_{14}O_3$	194.0943
$C_{11}H_{16}NO_2$	194.1182
$C_{11}H_{18}N_2O$	194.1420
$C_{11}H_{20}N_3$	194.1659
$C_{12}H_8N_3$	194.0719
$C_{12}H_{18}O_2$	194.1307
$C_{12}H_{20}NO$	194.1546
$C_{12}H_{22}N_2$	194.1784
$C_{13}H_8NO$	194.0606
$C_{13}H_{10}N_2$	194.0845
$C_{13}H_{22}O$	194.1671
$C_{13}H_{24}N$	194.1910
$C_{14}H_{10}O$	194.0732
$C_{14}H_{12}N$	194.0970
$C_{14}H_{26}$	194.2036
$C_{15}H_{14}$	194.1096
195	
$C_8H_7N_2O_4$	195.0406
$C_8H_9N_3O_3$	195.0644
$C_8H_{11}N_4O_2$	195.0883
$C_9H_9NO_4$	195.0532
$C_9H_{11}N_2O_3$	195.0770
$C_9H_{13}N_3O_2$	195.1009
$C_9H_{15}N_4O$	195.1247
$C_{10}H_{11}O_4$	195.0657
$C_{10}H_{13}NO_3$	195.0896
$C_{10}H_{15}N_2O_2$	195.1134
$C_{10}H_{17}N_3O$	195.1373
$C_{10}H_{19}N_4$	195.1611
$C_{11}H_7N_4$	195.0672
$C_{11}H_{15}O_3$	195.1021
$C_{11}H_{17}NO_2$	195.1260
$C_{11}H_{19}N_2O$	195.1498
$C_{11}H_{21}N_3$	195.1737
$C_{12}H_7N_2O$	195.0559
$C_{12}H_9N_3$	195.0798
$C_{12}H_{19}O_2$	195.1385

Appendix A (Continued)

	FM
$C_{12}H_{21}NO$	195.1624
$C_{12}H_{23}N_2$	195.1863
$C_{13}H_9NO$	195.0684
$C_{13}H_{11}N_2$	195.0923
$C_{13}H_{23}O$	195.1750
$C_{13}H_{25}N$	195.1988
$C_{14}H_{11}O$	195.0810
$C_{14}H_{13}N$	195.1049
$C_{14}H_{27}$	195.2114
$C_{15}H_{15}$	195.1174
196	
$C_8H_8N_2O_4$	196.0484
$C_8H_{10}N_3O_3$	196.0723
$C_8H_{12}N_4O_2$	196.0961
$C_9H_{10}NO_4$	196.0610
$C_9H_{12}N_2O_3$	196.0848
$C_9H_{14}N_3O_2$	196.1087
$C_9H_{16}N_4O$	196.1325
$C_{10}H_{12}O_4$	196.0735
$C_{10}H_{14}NO_3$	196.0974
$C_{10}H_{16}N_2O_2$	196.1213
$C_{10}H_{18}N_3O$	196.1451
$C_{10}H_{20}N_4$	196.1690
$C_{11}H_8N_4$	196.0750
$C_{11}H_{16}O_3$	196.1100
$C_{11}H_{18}NO_2$	196.1338
$C_{11}H_{20}N_2O$	196.1577
$C_{11}H_{22}N_3$	196.1815
$C_{12}H_8N_2O$	196.0637
$C_{12}H_{10}N_3$	196.0876
$C_{12}H_{20}O_2$	196.1464
$C_{12}H_{22}NO$	196.1702
$C_{12}H_{24}N_2$	196.1941
$C_{13}H_8O_2$	196.0524
$C_{13}H_{10}NO$	196.0763
$C_{13}H_{12}N_2$	196.1001
$C_{13}H_{24}O$	196.1828
$C_{13}H_{26}N$	196.2067
$C_{14}H_{12}O$	196.0888
$C_{14}H_{14}N$	196.1127
$C_{14}H_{28}$	196.2192
$C_{15}H_{16}$	196.1253
197	
$C_8H_9N_2O_4$	197.0563
$C_8H_{11}N_3O_3$	197.0801
$C_8H_{13}N_4O_2$	197.1040
$C_9H_{11}NO_4$	197.0688
$C_9H_{13}N_2O_3$	197.0927
$C_9H_{15}N_3O_2$	197.1165
$C_9H_{17}N_4O$	197.1404
$C_{10}H_{13}O_4$	197.0814
$C_{10}H_{15}NO_3$	197.1052
$C_{10}H_{17}N_2O_2$	197.1291
$C_{10}H_{19}N_3O$	197.1529
$C_{10}H_{21}N_4$	197.1768
$C_{11}H_7N_3O$	197.0590
$C_{11}H_9N_4$	197.0829
$C_{11}H_{17}O_3$	197.1178
$C_{11}H_{19}NO_2$	197.1416
$C_{11}H_{21}N_2O$	197.1655
$C_{11}H_{23}N_3$	197.1894
$C_{12}H_9N_2O$	197.0715
$C_{12}H_{11}N_3$	197.0954
$C_{12}H_{21}O_2$	197.1542
$C_{12}H_{23}NO$	197.1781
$C_{12}H_{25}N_2$	197.2019
$C_{13}H_9O_2$	197.0603
$C_{13}H_{11}NO$	197.0841
$C_{13}H_{13}N_2$	197.1080
$C_{13}H_{25}O$	197.1906
$C_{13}H_{27}N$	197.2145
$C_{14}H_{13}O$	197.0967
$C_{14}H_{15}N$	197.1205
$C_{14}H_{29}$	197.2270
$C_{15}H_{17}$	197.1331
198	
$C_8H_{10}N_2O_4$	198.0641
$C_8H_{12}N_3O_3$	198.0879
$C_8H_{14}N_4O_2$	198.1118
$C_9H_{12}NO_4$	198.0766
$C_9H_{14}N_2O_3$	198.1005
$C_9H_{16}N_3O_2$	198.1244
$C_9H_{18}N_4O$	198.1482
$C_{10}H_{14}O_4$	198.0892
$C_{10}H_{16}NO_3$	198.1131
$C_{10}H_{18}N_2O_2$	198.1369
$C_{10}H_{20}N_3O$	198.1608
$C_{10}H_{22}N_4$	198.1846
$C_{11}H_8N_3O$	198.0668
$C_{11}H_{10}N_4$	198.0907
$C_{11}H_{18}O_3$	198.1256
$C_{11}H_{20}NO_2$	198.1495
$C_{11}H_{22}N_2O$	198.1733
$C_{11}H_{24}N_3$	198.1972
$C_{12}H_8NO_2$	198.0555
$C_{12}H_{10}N_2O$	198.0794
$C_{12}H_{12}N_3$	198.1032
$C_{12}H_{22}O_2$	198.1620
$C_{12}H_{24}NO$	198.1859
$C_{12}H_{26}N_2$	198.2098
$C_{13}H_{10}O_2$	198.0681
$C_{13}H_{12}NO$	198.0919
$C_{13}H_{14}N_2$	198.1158
$C_{13}H_{26}O$	198.1985
$C_{13}H_{28}N$	198.2223
$C_{14}H_{14}O$	198.1045
$C_{14}H_{16}N$	198.1284
$C_{14}H_{30}$	198.2349
$C_{15}H_{18}$	198.1409
199	
$C_8H_{11}N_2O_4$	199.0719
$C_8H_{13}N_3O_3$	199.0958
$C_8H_{15}N_4O_2$	199.1196
$C_9H_{13}NO_4$	199.0845
$C_9H_{15}N_2O_3$	199.1083
$C_9H_{17}N_3O_2$	199.1322
$C_9H_{19}N_4O$	199.1560
$C_{10}H_7N_4O$	199.0621
$C_{10}H_{15}O_4$	199.0970
$C_{10}H_{17}NO_3$	199.1209
$C_{10}H_{19}N_2O_2$	199.1447
$C_{10}H_{21}N_3O$	199.1686
$C_{10}H_{23}N_4$	199.1925
$C_{11}H_7N_2O_2$	199.0508
$C_{11}H_9N_3O$	199.0746
$C_{11}H_{11}N_4$	199.0985
$C_{11}H_{19}O_3$	199.1334
$C_{11}H_{21}NO_2$	199.1573
$C_{11}H_{23}N_2O$	199.1811
$C_{11}H_{25}N_3$	199.2050
$C_{12}H_9NO_2$	199.0634
$C_{12}H_{11}N_2O$	199.0872
$C_{12}H_{13}N_3$	199.1111
$C_{12}H_{23}O_2$	199.1699
$C_{12}H_{25}NO$	199.1937
$C_{12}H_{27}N_2$	199.2176
$C_{13}H_{11}O_2$	199.0759
$C_{13}H_{13}NO$	199.0998
$C_{13}H_{15}N_2$	199.1236
$C_{13}H_{27}O$	199.2063
$C_{13}H_{29}N$	199.2301
$C_{14}H_{15}O$	199.1123
$C_{14}H_{17}N$	199.1362
$C_{15}H_{19}$	199.1488
200	
$C_8H_{12}N_2O_4$	200.0797
$C_8H_{14}N_3O_3$	200.1036
$C_8H_{16}N_4O_2$	200.1275
$C_9H_{14}NO_4$	200.0923
$C_9H_{16}N_2O_3$	200.1162
$C_9H_{18}N_3O_2$	200.1400
$C_9H_{20}N_4O$	200.1639
$C_{10}H_8N_4O$	200.0699
$C_{10}H_{16}O_4$	200.1049
$C_{10}H_{18}NO_3$	200.1287
$C_{10}H_{20}N_2O_2$	200.1526
$C_{10}H_{22}N_3O$	200.1764
$C_{10}H_{24}N_4$	200.2003
$C_{11}H_8N_2O_2$	200.0586
$C_{11}H_{10}N_3O$	200.0825
$C_{11}H_{12}N_4$	200.1063
$C_{11}H_{20}O_3$	200.1413
$C_{11}H_{22}NO_2$	200.1651
$C_{11}H_{24}N_2O$	200.1890
$C_{11}N_{26}N_3$	200.2129
$C_{12}H_8O_3$	200.0473
$C_{12}H_{10}NO_2$	200.0712
$C_{12}H_{12}N_2O$	200.0950
$C_{12}H_{14}N_3$	200.1189
$C_{12}H_{24}O_2$	200.1777
$C_{12}H_{26}NO$	200.2015
$C_{12}H_{28}N_2$	200.2254
$C_{13}H_{12}O_2$	200.0837
$C_{13}H_{14}NO$	200.1076
$C_{13}H_{16}N_2$	200.1315
$C_{13}H_{28}O$	200.2141
$C_{14}H_{16}O$	200.1202
$C_{14}H_{18}N$	200.1440
$C_{15}H_{20}$	200.1566
201	
$C_8H_{13}N_2O_4$	201.0876
$C_8H_{15}N_3O_3$	201.1114
$C_8H_{17}N_4O_2$	201.1353
$C_9H_{15}NO_4$	201.1001
$C_9H_{17}N_2O_3$	201.1240
$C_9H_{19}N_3O_2$	201.1478
$C_9H_{21}N_4O$	201.1717
$C_{10}H_7N_3O_2$	201.0539
$C_{10}H_9N_4O$	201.0777
$C_{10}H_{17}O_4$	201.1127
$C_{10}H_{19}NO_3$	201.1365
$C_{10}H_{21}N_2O_2$	201.1604
$C_{10}H_{23}N_3O$	201.1842
$C_{10}H_{25}N_4$	201.2081
$C_{11}H_7NO_3$	201.0426
$C_{11}H_9N_2O_2$	201.0664
$C_{11}H_{11}N_3O$	201.0903
$C_{11}H_{13}N_4$	201.1142
$C_{11}H_{21}O_3$	201.1491
$C_{11}H_{23}NO_2$	201.1730
$C_{11}H_{25}N_2O$	201.1968
$C_{11}H_{27}N_3$	201.2207
$C_{12}H_9O_3$	201.0552
$C_{12}H_{11}NO_2$	201.0790
$C_{12}H_{13}N_2O$	201.1029
$C_{12}H_{15}N_3$	201.1267
$C_{12}H_{25}O_2$	201.1855
$C_{12}H_{27}NO$	201.2094
$C_{13}H_{13}O_2$	201.0916
$C_{13}H_{15}NO$	201.1154
$C_{13}H_{17}N_2$	201.1393
$C_{14}H_{17}O$	201.1280
$C_{14}H_{19}N$	201.1519
$C_{15}H_{21}$	201.1644
202	
$C_8H_{14}N_2O_4$	202.0954
$C_8H_{16}N_3O_3$	202.1193
$C_8H_{18}N_4O_2$	202.1431
$C_9H_6N_4O_2$	202.0491
$C_9H_{16}NO_4$	202.1080
$C_9H_{18}N_2O_3$	202.1318
$C_9H_{20}N_3O_2$	202.1557
$C_9H_{22}N_4O$	202.1795
$C_{10}H_8N_3O_2$	202.0617
$C_{10}H_{10}N_4O$	202.0856
$C_{10}H_{18}O_4$	202.1205
$C_{10}H_{20}NO_3$	202.1444
$C_{10}H_{22}N_2O_2$	202.1682

Appendix A (Continued)

	FM
$C_{10}H_{24}N_3O$	202.1921
$C_{10}H_{26}N_4$	202.2160
$C_{11}H_8NO_3$	202.0504
$C_{11}H_{10}N_2O_2$	202.0743
$C_{11}H_{12}N_3O$	202.0981
$C_{11}H_{14}N_4$	202.1220
$C_{11}H_{22}O_3$	202.1569
$C_{11}H_{24}NO_2$	202.1808
$C_{11}H_{26}N_2O$	202.2046
$C_{12}H_{10}O_3$	202.0630
$C_{12}H_{12}NO_2$	202.0868
$C_{12}H_{14}N_2O$	202.1107
$C_{12}H_{16}N_3$	202.1346
$C_{12}H_{26}O_2$	202.1934
$C_{13}H_{14}O_2$	202.0994
$C_{13}H_{16}NO$	202.1233
$C_{13}H_{18}N_2$	202.1471
$C_{14}H_{18}O$	202.1358
$C_{14}H_{20}N$	202.1597
$C_{15}H_{22}$	202.1722
203	
$C_8H_{15}N_2O_4$	203.1032
$C_8H_{17}N_3O_3$	203.1271
$C_8H_{19}N_4O_2$	203.1509
$C_9H_7N_4O_2$	203.0570
$C_9H_{17}NO_4$	203.1158
$C_9H_{19}N_2O_3$	203.1396
$C_9H_{21}N_3O_2$	203.1635
$C_9H_{23}N_4O$	203.1873
$C_{10}H_7N_2O_3$	203.0457
$C_{10}H_9N_3O_2$	203.0695
$C_{10}H_{11}N_4O$	203.0934
$C_{10}H_{19}O_4$	203.1284
$C_{10}H_{21}NO_3$	203.1522
$C_{10}H_{23}N_2O_2$	203.1761
$C_{10}H_{25}N_3O$	203.1999
$C_{11}H_7O_4$	203.0344
$C_{11}H_9NO_3$	203.0583
$C_{11}H_{11}N_2O_2$	203.0821
$C_{11}H_{13}N_3O$	203.1060
$C_{11}H_{15}N_4$	203.1298
$C_{11}H_{23}O_3$	203.1648
$C_{11}H_{25}NO_2$	203.1886
$C_{12}H_{11}O_3$	203.0708
$C_{12}H_{13}NO_2$	203.0947
$C_{12}H_{15}N_2O$	203.1185
$C_{12}H_{17}N_3$	203.1424
$C_{13}H_{15}O_2$	203.1072
$C_{13}H_{17}NO$	203.1311
$C_{13}H_{19}N_2$	203.1549
$C_{14}H_{19}O$	203.1436
$C_{14}H_{21}N$	203.1675
$C_{15}H_9N$	203.0736
$C_{15}H_{23}$	203.1801
204	
$C_8H_{16}N_2O_4$	204.1111
$C_8H_{18}N_3O_3$	204.1349
$C_8H_{20}N_4O_2$	204.1588
$C_9H_6N_3O_3$	204.0410
$C_9H_8N_4O_2$	204.0648
$C_9H_{18}NO_4$	204.1236
$C_9H_{20}N_2O_3$	204.1475
$C_9H_{22}N_3O_2$	204.1713
$C_9H_{24}N_4O$	204.1952
$C_{10}H_8N_2O_3$	204.0535
$C_{10}H_{10}N_3O_2$	204.0774
$C_{10}H_{12}N_4O$	204.1012
$C_{10}H_{20}O_4$	204.1362
$C_{10}H_{22}NO_3$	204.1600
$C_{10}H_{24}N_2O_2$	204.1839
$C_{11}H_8O_4$	204.0422
$C_{11}H_{10}NO_3$	204.0661
$C_{11}H_{12}N_2O_2$	204.0899
$C_{11}H_{14}N_3O$	204.1138
$C_{11}H_{16}N_4$	204.1377
$C_{11}H_{24}O_3$	204.1726
$C_{12}H_{12}O_3$	204.0786
$C_{12}H_{14}NO_2$	204.1025
$C_{12}H_{16}N_2O$	204.1264
$C_{12}H_{18}N_3$	204.1502
$C_{13}H_{16}O_2$	204.1151
$C_{13}H_{18}NO$	204.1389
$C_{13}H_{20}N_2$	204.1628
$C_{14}H_{20}O$	204.1515
$C_{14}H_{22}N$	204.1753
$C_{15}H_{10}N$	204.0814
$C_{15}H_{24}$	204.1879
$C_{16}H_{12}$	204.0939
205	
$C_8H_{17}N_2O_4$	205.1189
$C_8H_{19}N_3O_3$	205.1427
$C_8H_{21}N_4O_2$	205.1666
$C_9H_7N_3O_3$	205.0488
$C_9H_9N_4O_2$	205.0726
$C_9H_{19}NO_4$	205.1315
$C_9H_{21}N_2O_3$	205.1553
$C_9H_{23}N_3O_2$	205.1791
$C_{10}H_7NO_4$	205.0375
$C_{10}H_9N_2O_3$	205.0614
$C_{10}H_{11}N_3O_2$	205.0852
$C_{10}H_{13}N_4O$	205.1091
$C_{10}H_{21}O_4$	205.1440
$C_{10}H_{23}NO_3$	205.1679
$C_{11}H_9O_4$	205.0501
$C_{11}H_{11}NO_3$	205.0739
$C_{11}H_{13}N_2O_2$	205.0978
$C_{11}H_{15}N_3O$	205.1216
$C_{11}H_{17}N_4$	205.1455
$C_{12}H_{13}O_3$	205.0865
$C_{12}H_{15}NO_2$	205.1103
$C_{12}H_{17}N_2O$	205.1342
$C_{12}H_{19}N_3$	205.1580
$C_{13}H_{17}O_2$	205.1229
$C_{13}H_{19}NO$	205.1467
$C_{13}H_{21}N_2$	205.1706
$C_{14}H_9N_2$	205.0767
$C_{14}H_{21}O$	205.1593
$C_{14}H_{23}N$	205.1832
$C_{15}H_9O$	205.0653
$C_{15}H_{11}N$	205.0892
$C_{15}H_{25}$	205.1957
$C_{16}H_{13}$	205.1018
206	
$C_8H_{18}N_2O_4$	206.1267
$C_8H_{20}N_3O_3$	206.1506
$C_8H_{22}N_4O_2$	206.1744
$C_9H_6N_2O_4$	206.0328
$C_9H_8N_3O_3$	206.0566
$C_9H_{10}N_4O_2$	206.0805
$C_9H_{20}NO_4$	206.1393
$C_9H_{22}N_2O_3$	206.1631
$C_{10}H_8NO_4$	206.0453
$C_{10}H_{10}N_2O_3$	206.0692
$C_{10}H_{12}N_3O_2$	206.0930
$C_{10}H_{14}N_4O$	206.1169
$C_{10}H_{22}O_4$	206.1518
$C_{11}H_{10}O_4$	206.0579
$C_{11}H_{12}NO_3$	206.0817
$C_{11}H_{14}N_2O_2$	206.1056
$C_{11}H_{16}N_3O$	206.1295
$C_{11}H_{18}N_4$	206.1533
$C_{12}H_{14}O_3$	206.0943
$C_{12}H_{16}NO_2$	206.1182
$C_{12}H_{18}N_2O$	206.1420
$C_{12}H_{20}N_3$	206.1659
$C_{13}H_8N_3$	206.0719
$C_{13}H_{18}O_2$	206.1307
$C_{13}H_{20}NO$	206.1546
$C_{13}H_{22}N_2$	206.1784
$C_{14}H_{10}N_2$	206.0845
$C_{14}H_{22}O$	206.1671
$C_{14}H_{24}N$	206.1910
$C_{15}H_{10}O$	206.0732
$C_{15}H_{12}N$	206.0970
$C_{15}H_{26}$	206.2036
$C_{16}H_{14}$	206.1096
207	
$C_8H_{19}N_2O_4$	207.1345
$C_8H_{21}N_3O_3$	207.1584
$C_9H_7N_2O_4$	207.0406
$C_9H_9N_3O_3$	207.0644
$C_9H_{11}N_4O_2$	207.0883
$C_9H_{21}NO_4$	207.1471
$C_{10}H_9NO_4$	207.0532
$C_{10}H_{11}N_2O_3$	207.0770
$C_{10}H_{13}N_3O_2$	207.1009
$C_{10}H_{15}N_4O$	207.1247
$C_{11}H_{11}O_4$	207.0657
$C_{11}H_{13}NO_3$	207.0896
$C_{11}H_{15}N_2O_2$	207.1134
$C_{11}H_{17}N_3O$	207.1373
$C_{11}H_{19}N_4$	207.1611
$C_{12}H_{15}O_3$	207.1021
$C_{12}H_{17}NO_2$	207.1260
$C_{12}H_{19}N_2O$	207.1498
$C_{12}H_{21}N_3$	207.1737
$C_{13}H_9N_3$	207.0798
$C_{13}H_{19}O_2$	207.1385
$C_{13}H_{21}NO$	207.1624
$C_{13}H_{23}N_2$	207.1863
$C_{14}H_9NO$	207.0684
$C_{14}H_{11}N_2$	207.0923
$C_{14}H_{23}O$	207.1750
$C_{14}H_{25}N$	207.1988
$C_{15}H_{11}O$	207.0810
$C_{15}H_{13}N$	207.1049
$C_{15}H_{27}$	207.2114
$C_{16}H_{15}$	207.1174
208	
$C_8H_{20}N_2O_4$	208.1424
$C_9H_8N_2O_4$	208.0484
$C_9H_{10}N_3O_3$	208.0723
$C_9H_{12}N_4O_2$	208.0961
$C_{10}H_{10}NO_4$	208.0610
$C_{10}H_{12}N_2O_3$	208.0848
$C_{10}H_{14}N_3O_2$	208.1087
$C_{10}H_{16}N_4O$	208.1325
$C_{11}H_{12}O_4$	208.0735
$C_{11}H_{14}NO_3$	208.0974
$C_{11}H_{16}N_2O_2$	208.1213
$C_{11}H_{18}N_3O$	208.1451
$C_{11}H_{20}N_4$	208.1690
$C_{12}H_8N_4$	208.0750
$C_{12}H_{16}O_3$	208.1100
$C_{12}H_{18}NO_2$	208.1338
$C_{12}H_{20}N_2O$	208.1577
$C_{12}H_{22}N_3$	208.1815
$C_{13}H_8N_2O$	208.0637
$C_{13}H_{10}N_3$	208.0876
$C_{13}H_{20}O_2$	208.1464
$C_{13}H_{22}NO$	208.1702
$C_{13}H_{24}N_2$	208.1941
$C_{14}H_{10}NO$	208.0763
$C_{14}H_{12}N_2$	208.1001
$C_{14}H_{24}O$	208.1828
$C_{14}H_{26}N$	208.2067
$C_{15}H_{12}O$	208.0888
$C_{15}H_{14}N$	208.1127
$C_{15}H_{28}$	208.2192
$C_{16}H_{16}$	208.1253
209	
$C_9H_9N_2O_4$	209.0563
$C_9H_{11}N_3O_3$	209.0801
$C_9H_{13}N_4O_2$	209.1040
$C_{10}H_{11}NO_4$	209.0688
$C_{10}H_{13}N_2O_3$	209.0927
$C_{10}H_{15}N_3O_2$	209.1165
$C_{10}H_{17}N_4O$	209.1404

Appendix A (Continued)

	FM
$C_{11}H_{13}O_4$	209.0814
$C_{11}H_{15}NO_3$	209.1052
$C_{11}H_{17}N_2O_2$	209.1291
$C_{11}H_{19}N_3O$	209.1529
$C_{11}H_{21}N_4$	209.1768
$C_{12}H_9N_4$	209.0829
$C_{12}H_{17}O_3$	209.1178
$C_{12}H_{19}NO_2$	209.1416
$C_{12}H_{21}N_2O$	209.1655
$C_{12}H_{23}N_3$	209.1894
$C_{13}H_9N_2O$	209.0715
$C_{13}H_{11}N_3$	209.0954
$C_{13}H_{21}O_2$	209.1542
$C_{13}H_{23}NO$	209.1781
$C_{13}H_{25}N_2$	209.2019
$C_{14}H_9O_2$	209.0603
$C_{14}H_{11}NO$	209.0841
$C_{14}H_{13}N_2$	209.1080
$C_{14}H_{25}O$	209.1906
$C_{14}H_{27}N$	209.2145
$C_{15}H_{13}O$	209.0967
$C_{15}H_{15}N$	209.1205
$C_{15}H_{29}$	209.2270
$C_{16}H_{17}$	209.1331
210	
$C_9H_{10}N_2O_4$	210.0641
$C_9H_{12}N_3O_3$	210.0879
$C_9H_{14}N_4O_2$	210.1118
$C_{10}H_{12}NO_4$	210.0766
$C_{10}H_{14}N_2O_3$	210.1005
$C_{10}H_{16}N_3O_2$	210.1244
$C_{10}H_{18}N_4O$	210.1482
$C_{11}H_{14}O_4$	210.0892
$C_{11}H_{16}NO_3$	210.1131
$C_{11}H_{18}N_2O_2$	210.1369
$C_{11}H_{20}N_3O$	210.1608
$C_{11}H_{22}N_4$	210.1846
$C_{12}H_8N_3O$	210.0668
$C_{12}H_{10}N_4$	210.0907
$C_{12}H_{18}O_3$	210.1256
$C_{12}H_{20}NO_2$	210.1495
$C_{12}H_{22}N_2O$	210.1733
$C_{12}H_{24}N_3$	210.1972
$C_{13}H_8NO_2$	210.0555
$C_{13}H_{10}N_2O$	210.0794
$C_{13}H_{12}N_3$	210.1032
$C_{13}H_{22}O_2$	210.1620
$C_{13}H_{24}NO$	210.1859
$C_{13}H_{26}N_2$	210.2098
$C_{14}H_{10}O_2$	210.0681
$C_{14}H_{12}NO$	210.0919
$C_{14}H_{14}N_2$	210.1158
$C_{14}H_{26}O$	210.1985
$C_{14}H_{28}N$	210.2223
$C_{15}H_{14}O$	210.1045
$C_{15}H_{16}N$	210.1284
$C_{15}H_{30}$	210.2349
$C_{16}H_{18}$	210.1409
211	
$C_9H_{11}N_2O_4$	211.0719
$C_9H_{13}N_3O_3$	211.0958
$C_9H_{15}N_4O_2$	211.1196
$C_{10}H_{13}NO_4$	211.0845
$C_{10}H_{15}N_2O_3$	211.1083
$C_{10}H_{17}N_3O_2$	211.1322
$C_{10}H_{19}N_4O$	211.1560
$C_{11}H_7N_4O$	211.0621
$C_{11}H_{15}O_4$	211.0970
$C_{11}H_{17}NO_3$	211.1209
$C_{11}H_{19}N_2O_2$	211.1447
$C_{11}H_{21}N_3O$	211.1686
$C_{11}H_{23}N_4$	211.1925
$C_{12}H_9N_3O$	211.0746
$C_{12}H_{11}N_4$	211.0985
$C_{12}H_{19}O_3$	211.1334
$C_{12}H_{21}NO_2$	211.1573
$C_{12}H_{23}N_2O$	211.1811
$C_{12}H_{25}N_3$	211.2050
$C_{13}H_9NO_2$	211.0634
$C_{13}H_{11}N_2O$	211.0872
$C_{13}H_{13}N_3$	211.1111
$C_{13}H_{23}O_2$	211.1699
$C_{13}H_{25}NO$	211.1937
$C_{13}H_{27}N_2$	211.2176
$C_{14}H_{11}O_2$	211.0759
$C_{14}H_{13}NO$	211.0998
$C_{14}H_{15}N_2$	211.1236
$C_{14}H_{27}O$	211.2063
$C_{14}H_{29}N$	211.2301
$C_{15}H_{15}O$	211.1123
$C_{15}H_{17}N$	211.1362
$C_{15}H_{31}$	211.2427
$C_{16}H_{19}$	211.1488
212	
$C_9H_{12}N_2O_4$	212.0797
$C_9H_{14}N_3O_3$	212.1036
$C_9H_{16}N_4O_2$	212.1275
$C_{10}H_{14}NO_4$	212.0923
$C_{10}H_{16}N_2O_3$	212.1162
$C_{10}H_{18}N_3O_2$	212.1400
$C_{10}H_{20}N_4O$	212.1639
$C_{11}H_8N_4O$	212.0699
$C_{11}H_{16}O_4$	212.1049
$C_{11}H_{18}NO_3$	212.1287
$C_{11}H_{20}N_2O_2$	212.1526
$C_{11}H_{22}N_3O$	212.1764
$C_{11}H_{24}N_4$	212.2003
$C_{12}H_8N_2O_2$	212.0586
$C_{12}H_{10}N_3O$	212.0825
$C_{12}H_{12}N_4$	212.1063
$C_{12}H_{20}O_3$	212.1413
$C_{12}H_{22}NO_2$	212.1651
$C_{12}H_{24}N_2O$	212.1890
$C_{12}H_{26}N_3$	212.2129
$C_{13}H_8O_3$	212.0473
$C_{13}H_{10}NO_2$	212.0712
$C_{13}H_{12}N_2O$	212.0950
$C_{13}H_{14}N_3$	212.1189
$C_{13}H_{24}O_2$	212.1777
$C_{13}H_{26}NO$	212.2015
$C_{13}H_{28}N_2$	212.2254
$C_{14}H_{12}O_2$	212.0837
$C_{14}H_{14}NO$	212.1076
$C_{14}H_{16}N_2$	212.1315
$C_{14}H_{28}O$	212.2141
$C_{14}H_{30}N$	212.2380
$C_{15}H_{16}O$	212.1202
$C_{15}H_{18}N$	212.1440
$C_{15}H_{32}$	212.2505
$C_{16}H_{20}$	212.1566
213	
$C_9H_{13}N_2O_4$	213.0876
$C_9H_{15}N_3O_3$	213.1114
$C_9H_{17}N_4O_2$	213.1353
$C_{10}H_{15}NO_4$	213.1001
$C_{10}H_{17}N_2O_3$	213.1240
$C_{10}H_{19}N_3O_2$	213.1478
$C_{10}H_{21}N_4O$	213.1717
$C_{11}H_7N_3O_2$	213.0539
$C_{11}H_9N_4O$	213.0777
$C_{11}H_{17}O_4$	213.1127
$C_{11}H_{19}NO_3$	213.1365
$C_{11}H_{21}N_2O_2$	213.1604
$C_{11}H_{23}N_3O$	213.1842
$C_{11}H_{25}N_4$	213.2081
$C_{12}H_9N_2O_2$	213.0664
$C_{12}H_{11}N_3O$	213.0903
$C_{12}H_{13}N_4$	213.1142
$C_{12}H_{21}O_3$	213.1491
$C_{12}H_{23}NO_2$	213.1730
$C_{12}H_{25}N_2O$	213.1968
$C_{12}H_{27}N_3$	213.2207
$C_{13}H_9O_3$	213.0552
$C_{13}H_{11}NO_2$	213.0790
$C_{13}H_{13}N_2O$	213.1029
$C_{13}H_{15}N_3$	213.1267
$C_{13}H_{25}O_2$	213.1855
$C_{13}H_{27}NO$	213.2094
$C_{13}H_{29}N_2$	213.2332
$C_{14}H_{13}O_2$	213.0916
$C_{14}H_{15}NO$	213.1154
$C_{14}H_{17}N_2$	213.1393
$C_{14}H_{29}O$	213.2219
$C_{15}H_{17}O$	213.1280
$C_{15}H_{19}N$	213.1519
$C_{16}H_{21}$	213.1644
214	
$C_9H_{14}N_2O_4$	214.0954
$C_9H_{16}N_3O_3$	214.1193
$C_9H_{18}N_4O_2$	214.1431
$C_{10}H_{16}NO_4$	214.1080
$C_{10}H_{18}N_2O_3$	214.1318
$C_{10}H_{20}N_3O_2$	214.1557
$C_{10}H_{22}N_4O$	214.1795
$C_{11}H_8N_3O_2$	214.0617
$C_{11}H_{10}N_4O$	214.0856
$C_{11}H_{18}O_4$	214.1205
$C_{11}H_{20}NO_3$	214.1444
$C_{11}H_{22}N_2O_2$	214.1682
$C_{11}H_{24}N_3O$	214.1921
$C_{11}H_{26}N_4$	214.2160
$C_{12}H_8NO_3$	214.0504
$C_{12}H_{10}N_2O_2$	214.0743
$C_{12}H_{12}N_3O$	214.0981
$C_{12}H_{14}N_4$	214.1220
$C_{12}H_{22}O_3$	214.1569
$C_{12}H_{24}NO_2$	214.1808
$C_{12}H_{26}N_2O$	214.2046
$C_{12}H_{28}N_3$	214.2285
$C_{13}H_{10}O_3$	214.0630
$C_{13}H_{12}NO_2$	214.0869
$C_{13}H_{14}N_2O$	214.1107
$C_{13}H_{16}N_3$	214.1346
$C_{13}H_{26}O_2$	214.1934
$C_{13}H_{28}NO$	214.2172
$C_{13}H_{30}N_2$	214.2411
$C_{14}H_{14}O_2$	214.0994
$C_{14}H_{16}NO$	214.1233
$C_{14}H_{18}N_2$	214.1471
$C_{15}H_{18}O$	214.1358
$C_{15}H_{20}N$	214.1597
$C_{16}H_{22}$	214.1722
215	
$C_9H_{15}N_2O_4$	215.1032
$C_9H_{17}N_3O_3$	215.1271
$C_9H_{19}N_4O_2$	215.1509
$C_{10}H_7N_4O_2$	215.0570
$C_{10}H_{17}NO_4$	215.1158
$C_{10}H_{19}N_2O_3$	215.1396
$C_{10}H_{21}N_3O_2$	215.1635
$C_{10}H_{23}N_4O$	215.1873
$C_{11}H_7N_2O_3$	215.0457
$C_{11}H_9N_3O_2$	215.0695
$C_{11}H_{11}N_4O$	215.0934
$C_{11}H_{19}O_4$	215.1284
$C_{11}H_{21}NO_3$	215.1522
$C_{11}H_{23}N_2O_2$	215.1761
$C_{11}H_{25}N_3O$	215.1999
$C_{11}H_{27}N_4$	215.2238
$C_{12}H_9NO_3$	215.0583
$C_{12}H_{11}N_2O_2$	215.0821
$C_{12}H_{13}N_3O$	215.1060
$C_{12}H_{15}N_4$	215.1298
$C_{12}H_{23}O_3$	215.1648
$C_{12}H_{25}NO_2$	215.1886
$C_{12}H_{27}N_2O$	215.2125
$C_{12}H_{29}N_3$	215.2363
$C_{13}H_{11}O_3$	215.0708

Appendix A (Continued)

	FM
$C_{13}H_{13}NO_2$	215.0947
$C_{13}H_{15}N_2O$	215.1185
$C_{13}H_{17}N_3$	215.1424
$C_{14}H_{15}O_2$	215.1072
$C_{14}H_{17}NO$	215.1311
$C_{14}H_{19}N_2$	215.1549
$C_{15}H_{19}O$	215.1436
$C_{15}H_{21}N$	215.1675
$C_{16}H_{23}$	215.1801
216	
$C_9H_{16}N_2O_4$	216.1111
$C_9H_{18}N_3O_3$	216.1349
$C_9H_{20}N_4O_2$	216.1588
$C_{10}H_8N_4O_2$	216.0648
$C_{10}H_{18}NO_4$	216.1236
$C_{10}H_{20}N_2O_3$	216.1475
$C_{10}H_{22}N_3O_2$	216.1713
$C_{10}H_{24}N_4O$	216.1952
$C_{11}H_8N_2O_3$	216.0535
$C_{11}H_{10}N_3O_2$	216.0774
$C_{11}H_{12}N_4O$	216.1012
$C_{11}H_{20}O_4$	216.1362
$C_{11}H_{22}NO_3$	216.1600
$C_{11}H_{24}N_2O_2$	216.1839
$C_{11}H_{26}N_3O$	216.2077
$C_{11}H_{28}N_4$	216.2316
$C_{12}H_8O_4$	216.0422
$C_{12}H_{10}NO_3$	216.0661
$C_{12}H_{12}N_2O_2$	216.0899
$C_{12}H_{14}N_3O$	216.1138
$C_{12}H_{16}N_4$	216.1377
$C_{12}H_{24}O_3$	216.1726
$C_{12}H_{26}NO_2$	216.1965
$C_{12}H_{28}N_2O$	216.2203
$C_{13}H_{12}O_3$	216.0786
$C_{13}H_{14}NO_2$	216.1025
$C_{13}H_{16}N_2O$	216.1264
$C_{13}H_{18}N_3$	216.1502
$C_{13}H_{28}O_2$	216.2090
$C_{14}H_{16}O_2$	216.1151
$C_{14}H_{18}NO$	216.1389
$C_{14}H_{20}N_2$	216.1628
$C_{15}H_{20}O$	216.1515
$C_{15}H_{22}N$	216.1753
$C_{16}H_{10}N$	216.0814
$C_{16}H_{24}$	216.1879
$C_{17}H_{12}$	216.0939
217	
$C_9H_{17}N_2O_4$	217.1189
$C_9H_{19}N_3O_3$	217.1427
$C_9H_{21}N_4O_2$	217.1666
$C_{10}H_7N_3O_3$	217.0488
$C_{10}H_9N_4O_2$	217.0726
$C_{10}H_{19}NO_4$	217.1315
$C_{10}H_{21}N_2O_3$	217.1553
$C_{10}H_{23}N_3O_2$	217.1791
$C_{10}H_{25}N_4O$	217.2030
$C_{11}H_7NO_4$	217.0375
$C_{11}H_9N_2O_3$	217.0614
$C_{11}H_{11}N_3O_2$	217.0852
$C_{11}H_{13}N_4O$	217.1091
$C_{11}H_{21}O_4$	217.1440
$C_{11}H_{23}NO_3$	217.1679
$C_{11}H_{25}N_2O_2$	217.1917
$C_{11}H_{27}N_3O$	217.2156
$C_{12}H_9O_4$	217.0501
$C_{12}H_{11}NO_3$	217.0739
$C_{12}H_{13}N_2O_2$	217.0978
$C_{12}H_{15}N_3O$	217.1216
$C_{12}H_{17}N_4$	217.1455
$C_{12}H_{25}O_3$	217.1804
$C_{12}H_{27}NO_2$	217.2043
$C_{13}H_{13}O_3$	217.0865
$C_{13}H_{15}NO_2$	217.1103
$C_{13}H_{17}N_2O$	217.1342
$C_{13}H_{19}N_3$	217.1580
$C_{14}H_{17}O_2$	217.1229
$C_{14}H_{19}NO$	217.1467
$C_{14}H_{21}N_2$	217.1706
$C_{15}H_9N_2$	217.0767
$C_{15}H_{21}O$	217.1593
$C_{15}H_{23}N$	217.1832
$C_{16}H_{11}N$	217.0892
$C_{16}H_{25}$	217.1957
$C_{17}H_{13}$	217.1018
218	
$C_9H_{18}N_2O_4$	218.1267
$C_9H_{20}N_3O_3$	218.1506
$C_9H_{22}N_4O_2$	218.1744
$C_{10}H_8N_3O_3$	218.0566
$C_{10}H_{10}N_4O_2$	218.0805
$C_{10}H_{20}NO_4$	218.1393
$C_{10}H_{22}N_2O_3$	218.1631
$C_{10}H_{24}N_3O_2$	218.1870
$C_{10}H_{26}N_4O$	218.2108
$C_{11}H_8NO_4$	218.0453
$C_{11}H_{10}N_2O_3$	218.0692
$C_{11}H_{12}N_3O_2$	218.0930
$C_{11}H_{14}N_4O$	218.1169
$C_{11}H_{22}O_4$	218.1518
$C_{11}H_{24}NO_3$	218.1757
$C_{11}H_{26}N_2O_2$	218.1996
$C_{12}H_{10}O_4$	218.0579
$C_{12}H_{12}NO_3$	218.0817
$C_{12}H_{14}N_2O_2$	218.1056
$C_{12}H_{16}N_3O$	218.1295
$C_{12}H_{18}N_4$	218.1533
$C_{12}H_{26}O_3$	218.1883
$C_{13}H_{14}O_3$	218.0943
$C_{13}H_{16}NO_2$	218.1182
$C_{13}H_{18}N_2O$	218.1420
$C_{13}H_{20}N_3$	218.1659
$C_{14}H_{18}O_2$	218.1307
$C_{14}H_{20}NO$	218.1546
$C_{14}H_{22}N_2$	218.1784
$C_{15}H_{10}N_2$	218.0845
$C_{15}H_{22}O$	218.1671
$C_{15}H_{24}N$	218.1910
$C_{16}H_{10}O$	218.0732
$C_{16}H_{12}N$	218.0970
$C_{16}H_{26}$	218.2036
$C_{17}H_{14}$	218.1096
219	
$C_9H_{19}N_2O_4$	219.1345
$C_9H_{21}N_3O_3$	219.1584
$C_9H_{23}N_4O_2$	219.1822
$C_{10}H_7N_2O_4$	219.0406
$C_{10}H_9N_3O_3$	219.0644
$C_{10}H_{11}N_4O_2$	219.0883
$C_{10}H_{21}NO_4$	219.1471
$C_{10}H_{23}N_2O_3$	219.1710
$C_{10}H_{25}N_3O_2$	219.1948
$C_{11}H_9NO_4$	219.0532
$C_{11}H_{11}N_2O_3$	219.0770
$C_{11}H_{13}N_3O_2$	219.1009
$C_{11}H_{15}N_4O$	219.1247
$C_{11}H_{23}O_4$	219.1597
$C_{11}H_{25}NO_3$	219.1835
$C_{12}H_{11}O_4$	219.0657
$C_{12}H_{13}NO_3$	219.0896
$C_{12}H_{15}N_2O_2$	219.1134
$C_{12}H_{17}N_3O$	219.1373
$C_{12}H_{19}N_4$	219.1611
$C_{13}H_{15}O_3$	219.1021
$C_{13}H_{17}NO_2$	219.1260
$C_{13}H_{19}N_2O$	219.1498
$C_{13}H_{21}N_3$	219.1737
$C_{14}H_9N_3$	219.0798
$C_{14}H_{19}O_2$	219.1385
$C_{14}H_{21}NO$	219.1624
$C_{14}H_{23}N_2$	219.1863
$C_{15}H_9NO$	219.0684
$C_{15}H_{11}N_2$	219.0923
$C_{15}H_{23}O$	219.1750
$C_{15}H_{25}N$	219.1988
$C_{16}H_{11}O$	219.0810
$C_{16}H_{13}N$	219.1049
$C_{16}H_{27}$	219.2114
$C_{17}H_{15}$	219.1174
220	
$C_9H_{20}N_2O_4$	220.1424
$C_9H_{22}N_3O_3$	220.1662
$C_9H_{24}N_4O_2$	220.1901
$C_{10}H_8N_2O_4$	220.0484
$C_{10}H_{10}N_3O_3$	220.0723
$C_{10}H_{12}N_4O_2$	220.0961
$C_{10}H_{22}NO_4$	220.1549
$C_{10}H_{24}N_2O_3$	220.1788
$C_{11}H_{10}NO_4$	220.0610
$C_{11}H_{12}N_2O_3$	220.0848
$C_{11}H_{14}N_3O_2$	220.1087
$C_{11}H_{16}N_4O$	220.1325
$C_{12}H_{12}O_4$	220.0735
$C_{12}H_{14}NO_3$	220.0974
$C_{12}H_{16}N_2O_2$	220.1213
$C_{12}H_{18}N_3O$	220.1451
$C_{12}H_{20}N_4$	220.1690
$C_{13}H_8N_4$	220.0750
$C_{13}H_{16}O_3$	220.1100
$C_{13}H_{18}NO_2$	220.1338
$C_{13}H_{20}N_2O$	220.1577
$C_{13}H_{22}N_3$	220.1815
$C_{14}H_{10}N_3$	220.0876
$C_{14}H_{20}O_2$	220.1464
$C_{14}H_{22}NO$	220.1702
$C_{14}H_{24}N_2$	220.1941
$C_{15}H_{10}NO$	220.0763
$C_{15}H_{12}N_2$	220.1001
$C_{15}H_{24}O$	220.1828
$C_{15}H_{26}N$	220.2067
$C_{16}H_{12}O$	220.0888
$C_{16}H_{14}N$	220.1127
$C_{16}H_{28}$	220.2192
$C_{17}H_{16}$	220.1253
221	
$C_9H_{21}N_2O_4$	221.1502
$C_9H_{23}N_3O_3$	221.1741
$C_{10}H_9N_2O_4$	221.0563
$C_{10}H_{11}N_3O_3$	221.0801
$C_{10}H_{13}N_4O_2$	221.1040
$C_{10}H_{23}NO_4$	221.1628
$C_{11}H_{11}NO_4$	221.0688
$C_{11}H_{13}N_2O_3$	221.0927
$C_{11}H_{15}N_3O_2$	221.1165
$C_{11}H_{17}N_4O$	221.1404
$C_{12}H_{13}O_4$	221.0814
$C_{12}H_{15}NO_3$	221.1052
$C_{12}H_{17}N_2O_2$	221.1291
$C_{12}H_{19}N_3O$	221.1529
$C_{12}H_{21}N_4$	221.1768
$C_{13}H_9N_4$	221.0829
$C_{13}H_{17}O_3$	221.1178
$C_{13}H_{19}NO_2$	221.1416
$C_{13}H_{21}N_2O$	221.1655
$C_{13}H_{23}N_3$	221.1894
$C_{14}H_9N_2O$	221.0715
$C_{14}H_{11}N_3$	221.0954
$C_{14}H_{21}O_2$	221.1542
$C_{14}H_{23}NO$	221.1781
$C_{14}H_{25}N_2$	221.2019
$C_{15}H_9O_2$	221.0603
$C_{15}H_{11}NO$	221.0841
$C_{15}H_{13}N_2$	221.1080
$C_{15}H_{25}O$	221.1906
$C_{15}H_{27}N$	221.2145
$C_{16}H_{13}O$	221.0967
$C_{16}H_{15}N$	221.1205
$C_{16}H_{29}$	221.2270

Appendix A (Continued)

	FM
$C_{17}H_{17}$	221.1331
222	
$C_{9}H_{22}N_{2}O_{4}$	222.1580
$C_{10}H_{10}N_{2}O_{4}$	222.0641
$C_{10}H_{12}N_{3}O_{3}$	222.0879
$C_{10}H_{14}N_{4}O_{2}$	222.1118
$C_{11}H_{12}NO_{4}$	222.0766
$C_{11}H_{14}N_{2}O_{3}$	222.1005
$C_{11}H_{16}N_{3}O_{2}$	222.1244
$C_{11}H_{18}N_{4}O$	222.1482
$C_{11}N_{3}O_{3}$	221.9940
$C_{12}H_{14}O_{4}$	222.0892
$C_{12}H_{16}NO_{3}$	222.1131
$C_{12}H_{18}N_{2}O_{2}$	222.1369
$C_{12}H_{20}N_{3}O$	222.1608
$C_{12}H_{22}N_{4}$	222.1846
$C_{13}H_{8}N_{3}O$	222.0668
$C_{13}H_{10}N_{4}$	222.0907
$C_{13}H_{18}O_{3}$	222.1256
$C_{13}H_{20}NO_{2}$	222.1495
$C_{13}H_{22}N_{2}O$	222.1733
$C_{13}H_{24}N_{3}$	222.1972
$C_{14}H_{10}N_{2}O$	222.0794
$C_{14}H_{12}N_{3}$	222.1032
$C_{14}H_{22}O_{2}$	222.1620
$C_{14}H_{24}NO$	222.1859
$C_{14}H_{26}N_{2}$	222.2098
$C_{15}H_{10}O_{2}$	222.0681
$C_{15}H_{12}NO$	222.0919
$C_{15}H_{14}N_{2}$	222.1158
$C_{15}H_{26}O$	222.1985
$C_{15}H_{28}N$	222.2223
$C_{16}H_{14}O$	222.1045
$C_{16}H_{16}N$	222.1284
$C_{16}H_{30}$	222.2349
$C_{16}NO$	221.9980
$C_{17}H_{18}$	222.1409
223	
$C_{10}H_{11}N_{2}O_{4}$	223.0719
$C_{10}H_{13}N_{3}O_{3}$	223.0958
$C_{10}H_{15}N_{4}O_{2}$	223.1196
$C_{11}H_{13}NO_{4}$	223.0845
$C_{11}H_{15}N_{2}O_{3}$	223.1083
$C_{11}H_{17}N_{3}O_{2}$	223.1322
$C_{11}H_{19}N_{4}O$	223.1560
$C_{12}H_{7}N_{4}O$	223.0621
$C_{12}H_{15}O_{4}$	223.0970
$C_{12}H_{17}NO_{3}$	223.1209
$C_{12}H_{19}N_{2}O_{2}$	223.1447
$C_{12}H_{21}N_{3}O$	223.1686
$C_{12}H_{23}N_{4}$	223.1925
$C_{13}H_{9}N_{3}O$	223.0746
$C_{13}H_{11}N_{4}$	223.0985
$C_{13}H_{19}O_{3}$	223.1334
$C_{13}H_{21}NO_{2}$	223.1573
$C_{13}H_{23}N_{2}O$	223.1811
$C_{13}H_{25}N_{3}$	223.2050
$C_{14}H_{9}NO_{2}$	223.0634
$C_{14}H_{11}N_{2}O$	223.0872
$C_{14}H_{13}N_{3}$	223.1111
$C_{14}H_{23}O_{2}$	223.1699
$C_{14}H_{25}NO$	223.1937
$C_{14}H_{27}N_{2}$	223.2176
$C_{15}H_{11}O_{2}$	223.0759
$C_{15}H_{13}NO$	223.0998
$C_{15}H_{27}O$	223.2063
$C_{15}H_{29}N$	223.2301
$C_{16}H_{15}O$	223.1123
$C_{16}H_{17}N$	223.1362
$C_{16}H_{31}$	223.2427
$C_{17}H_{19}$	223.1488
224	
$C_{10}H_{12}N_{2}O_{4}$	224.0797
$C_{10}H_{14}N_{3}O_{3}$	224.1036
$C_{10}H_{16}N_{4}O_{2}$	224.1275
$C_{11}H_{14}NO_{4}$	224.0923
$C_{11}H_{16}N_{2}O_{3}$	224.1162
$C_{11}H_{18}N_{3}O_{2}$	224.1400
$C_{11}H_{20}N_{4}O$	224.1639
$C_{12}H_{8}N_{4}O$	224.0699
$C_{12}H_{16}O_{4}$	224.1049
$C_{12}H_{18}NO_{3}$	224.1287
$C_{12}H_{20}N_{2}O_{2}$	224.1526
$C_{12}H_{22}N_{3}O$	224.1764
$C_{12}H_{24}N_{4}$	224.2003
$C_{13}H_{8}N_{2}O_{2}$	224.0586
$C_{13}H_{10}N_{3}O$	224.0825
$C_{13}H_{12}N_{4}$	224.1063
$C_{13}H_{20}O_{3}$	224.1413
$C_{13}H_{22}NO_{2}$	224.1651
$C_{13}H_{24}N_{2}O$	224.1890
$C_{13}H_{26}N_{3}$	224.2129
$C_{14}H_{10}NO_{2}$	224.0712
$C_{14}H_{12}N_{2}O$	224.0950
$C_{14}H_{14}N_{3}$	224.1189
$C_{14}H_{24}O_{2}$	224.1777
$C_{14}H_{26}NO$	224.2015
$C_{14}H_{28}N_{2}$	224.2254
$C_{15}H_{12}O_{2}$	224.0837
$C_{15}H_{14}NO$	224.1076
$C_{15}H_{16}N_{2}$	224.1315
$C_{15}H_{28}O$	224.2141
$C_{15}H_{30}N$	224.2380
$C_{16}H_{16}O$	224.1202
$C_{16}H_{18}N$	224.1440
$C_{16}H_{32}$	224.2505
$C_{17}H_{20}$	224.1566
225	
$C_{10}H_{13}N_{2}O_{4}$	225.0876
$C_{10}H_{15}N_{3}O_{3}$	225.1114
$C_{10}H_{17}N_{4}O_{2}$	225.1353
$C_{11}H_{15}NO_{4}$	225.1001
$C_{11}H_{17}N_{2}O_{3}$	225.1240
$C_{11}H_{19}N_{3}O_{2}$	225.1478
$C_{11}H_{21}N_{4}O$	225.1717
$C_{12}H_{9}N_{4}O$	225.0777
$C_{12}H_{17}O_{4}$	225.1127
$C_{12}H_{19}NO_{3}$	225.1365
$C_{12}H_{21}N_{2}O_{2}$	225.1604
$C_{12}H_{23}N_{3}O$	225.1842
$C_{12}H_{25}N_{4}$	225.2081
$C_{13}H_{9}N_{2}O_{2}$	225.0664
$C_{13}H_{11}N_{3}O$	225.0903
$C_{13}H_{13}N_{4}$	225.1142
$C_{13}H_{21}O_{3}$	225.1491
$C_{13}H_{23}NO_{2}$	225.1730
$C_{13}H_{25}N_{2}O$	225.1968
$C_{13}H_{27}N_{3}$	225.2207
$C_{14}H_{9}O_{3}$	225.0552
$C_{14}H_{11}NO_{2}$	225.0790
$C_{14}H_{13}N_{2}O$	225.1029
$C_{14}H_{15}N_{3}$	225.1267
$C_{14}H_{25}O_{2}$	225.1855
$C_{14}H_{27}NO$	225.2094
$C_{14}H_{29}N_{2}$	225.2332
$C_{15}H_{13}O_{2}$	225.0916
$C_{15}H_{15}NO$	225.1154
$C_{15}H_{17}N_{2}$	225.1393
$C_{15}H_{29}O$	225.2219
$C_{15}H_{31}N$	225.2458
$C_{16}H_{17}O$	225.1280
$C_{16}H_{19}N$	225.1519
$C_{16}H_{33}$	225.2584
$C_{17}H_{21}$	225.1644
226	
$C_{10}H_{14}N_{2}O_{4}$	226.0954
$C_{10}H_{16}N_{3}O_{3}$	226.1193
$C_{10}H_{18}N_{4}O_{2}$	226.1431
$C_{11}H_{16}NO_{4}$	226.1080
$C_{11}H_{18}N_{2}O_{3}$	226.1318
$C_{11}H_{20}N_{3}O_{2}$	226.1557
$C_{11}H_{22}N_{4}O$	226.1795
$C_{12}H_{8}N_{3}O_{2}$	226.0617
$C_{12}H_{10}N_{4}O$	226.0856
$C_{12}H_{18}O_{4}$	226.1205
$C_{12}H_{20}NO_{3}$	226.1444
$C_{12}H_{22}N_{2}O_{2}$	226.1682
$C_{12}H_{24}N_{3}O$	226.1929
$C_{12}H_{26}N_{4}$	226.2160
$C_{13}H_{10}N_{2}O_{2}$	226.0743
$C_{13}H_{12}N_{3}O$	226.0981
$C_{13}H_{14}N_{4}$	226.1220
$C_{13}H_{22}O_{3}$	226.1569
$C_{13}H_{24}NO_{2}$	226.1808
$C_{13}H_{26}N_{2}O$	226.2046
$C_{13}H_{28}N_{3}$	226.2285
$C_{14}H_{10}O_{3}$	226.0630
$C_{14}H_{12}NO_{2}$	226.0868
$C_{14}H_{14}N_{2}O$	226.1107
$C_{14}H_{16}N_{3}$	226.1346
$C_{14}H_{26}O_{2}$	226.1934
$C_{14}H_{28}NO$	226.2172
$C_{14}H_{30}N_{2}$	226.2411
$C_{15}H_{14}O_{2}$	226.0994
$C_{15}H_{16}NO$	226.1233
$C_{15}H_{18}N_{2}$	226.1471
$C_{15}H_{30}O$	226.2298
$C_{15}H_{32}N$	226.2536
$C_{16}H_{18}O$	226.1358
$C_{16}H_{20}N$	226.1597
$C_{16}H_{34}$	226.2662
$C_{17}H_{22}$	226.1722
227	
$C_{10}H_{15}N_{2}O_{4}$	227.1032
$C_{10}H_{17}N_{3}O_{3}$	227.1271
$C_{10}H_{19}N_{4}O_{2}$	227.1509
$C_{11}H_{17}NO_{4}$	227.1158
$C_{11}H_{21}N_{3}O_{2}$	227.1635
$C_{11}H_{23}N_{4}O$	227.1873
$C_{12}H_{7}N_{2}O_{3}$	227.0457
$C_{12}H_{9}N_{3}O_{2}$	227.0695
$C_{12}H_{11}N_{4}O$	227.0934
$C_{12}H_{19}O_{4}$	227.1284
$C_{12}H_{21}NO_{3}$	227.1522
$C_{12}H_{23}N_{2}O_{2}$	227.1761
$C_{12}H_{25}N_{3}O$	227.1999
$C_{12}H_{27}N_{4}$	227.2238
$C_{13}H_{9}NO_{3}$	227.0583
$C_{13}H_{11}N_{2}O_{2}$	227.0821
$C_{13}H_{13}N_{3}O$	227.1060
$C_{13}H_{15}N_{4}$	227.1298
$C_{13}H_{25}NO_{2}$	227.1886
$C_{13}H_{27}N_{2}O$	227.2125
$C_{13}H_{29}N_{3}$	227.2363
$C_{14}H_{11}O_{3}$	227.0708
$C_{14}H_{13}NO_{2}$	227.0947
$C_{14}H_{15}N_{2}O$	227.1185
$C_{14}H_{17}N_{3}$	227.1424
$C_{14}H_{27}O_{2}$	227.2012
$C_{14}H_{29}NO$	227.2250
$C_{15}H_{15}O_{2}$	227.1072
$C_{15}H_{17}NO$	227.1311
$C_{15}H_{19}N_{2}$	227.1549
$C_{15}H_{31}O$	227.2376
$C_{15}H_{33}N$	227.2615
$C_{16}H_{19}O$	227.1436
$C_{16}H_{21}N$	227.1675
$C_{17}H_{23}$	227.1801
228	
$C_{10}H_{16}N_{2}O_{2}$	228.1111
$C_{10}H_{18}N_{3}O_{3}$	228.1349
$C_{10}H_{20}N_{4}O_{2}$	228.1588
$C_{11}H_{8}N_{4}O_{2}$	228.0648
$C_{11}H_{18}NO_{4}$	228.1236
$C_{11}H_{20}N_{2}O_{3}$	228.1475
$C_{11}H_{22}N_{3}O_{2}$	228.1713
$C_{11}H_{24}N_{4}O$	228.1952
$C_{12}H_{8}N_{2}O_{3}$	228.0535

Appendix A (Continued)

	FM		FM		FM		FM
$C_{12}H_{12}N_4O$	228.1012	$C_{14}H_{15}NO_2$	229.1103	$C_{10}H_{23}N_4O_2$	231.1822	$C_{13}H_{20}N_4$	232.1690
$C_{12}H_{20}O_4$	228.1362	$C_{14}H_{17}N_2O$	229.1342	$C_{11}H_7N_2O_4$	231.0406	$C_{13}H_{28}O_3$	232.2039
$C_{12}H_{22}NO_3$	228.1600	$C_{14}H_{19}N_3$	229.1580	$C_{11}H_9N_3O_3$	231.0644	$C_{14}H_{16}O_3$	232.1100
$C_{12}H_{24}N_2O_2$	228.1839	$C_{14}H_{29}O_2$	229.2168	$C_{11}H_{11}N_4O_2$	231.0883	$C_{14}H_{18}NO_2$	232.1338
$C_{12}H_{26}N_3O$	228.2077	$C_{14}H_{31}NO$	229.2407	$C_{11}H_{21}NO_4$	231.1471	$C_{14}H_{20}N_2O$	232.1577
$C_{12}H_{28}N_4$	228.2316	$C_{15}H_{17}O_2$	229.1229	$C_{11}H_{23}N_2O_3$	231.1710	$C_{14}H_{22}N_3$	232.1815
$C_{13}H_8O_4$	228.0422	$C_{15}H_{19}NO$	229.1467	$C_{11}H_{25}N_3O_2$	231.1948	$C_{15}H_{10}N_3$	232.0876
$C_{13}H_{10}NO_3$	228.0661	$C_{15}H_{21}N_2$	229.1706	$C_{11}H_{27}N_4O$	231.2187	$C_{15}H_{20}O_2$	232.1464
$C_{13}H_{12}N_2O_2$	228.0899	$C_{16}H_{21}O$	229.1593	$C_{12}H_9NO_4$	231.0532	$C_{15}H_{22}NO$	232.1702
$C_{13}H_{14}N_3O$	228.1138	$C_{16}H_{23}N$	229.1832	$C_{12}H_{11}N_2O_3$	231.0770	$C_{15}H_{24}N_2$	232.1941
$C_{13}H_{24}O_3$	228.1726	$C_{17}H_9O$	229.0653	$C_{12}H_{13}N_3O_2$	231.1009	$C_{16}H_{10}NO$	232.0768
$C_{13}H_{26}NO_2$	228.1965	$C_{17}H_{11}N$	229.0892	$C_{12}H_{15}N_4O$	231.1247	$C_{16}H_{12}N_2$	232.1001
$C_{13}H_{28}N_2O$	228.2203	$C_{18}H_{13}$	229.1018	$C_{12}H_{23}O_4$	231.1597	$C_{16}H_{24}O$	232.1828
$C_{13}H_{30}N_3$	228.2442	**230**		$C_{12}H_{25}NO_3$	231.1835	$C_{16}H_{26}N$	232.2067
$C_{14}H_{12}O_3$	228.0786	$C_{10}H_{18}N_2O_4$	230.1267	$C_{12}H_{27}N_2O_2$	231.2074	$C_{17}H_{12}O$	232.0888
$C_{14}H_{14}NO_2$	228.1025	$C_{10}H_{20}N_3O_3$	230.1506	$C_{12}H_{29}N_3O$	231.2312	$C_{17}H_{14}N$	232.1127
$C_{14}H_{16}N_2O$	228.1264	$C_{10}H_{22}N_4O_2$	230.1744	$C_{13}H_{11}O_4$	231.0657	$C_{17}H_{28}$	232.2192
$C_{14}H_{18}N_3$	228.1502	$C_{11}H_8N_3O_3$	230.0566	$C_{13}H_{13}NO_3$	231.0896	$C_{18}H_{16}$	232.1253
$C_{14}H_{28}O_2$	228.2090	$C_{11}H_{10}N_4O_2$	230.0805	$C_{13}H_{15}N_2O_2$	231.1134	**233**	
$C_{14}H_{30}NO$	228.2329	$C_{11}H_{20}NO_4$	230.1393	$C_{13}H_{17}N_3O$	231.1373	$C_{10}H_{23}N_3O_3$	233.1741
$C_{14}H_{32}N_2$	228.2567	$C_{11}H_{22}N_2O_3$	230.1631	$C_{13}H_{19}N_4$	231.1611	$C_{10}H_{25}N_4O_2$	233.1979
$C_{15}H_{16}O_2$	228.1151	$C_{11}H_{24}N_3O_2$	230.1870	$C_{14}H_{15}O_3$	231.1021	$C_{11}H_9N_2O_4$	233.0563
$C_{15}H_{18}NO$	228.1389	$C_{11}H_{26}N_4O$	230.2108	$C_{14}H_{17}NO_2$	231.1260	$C_{11}H_{11}N_3O_3$	233.0801
$C_{15}H_{20}N_2$	228.1628	$C_{12}H_8NO_4$	230.0453	$C_{14}H_{19}N_2O$	231.1498	$C_{11}H_{23}NO_4$	233.1628
$C_{15}H_{32}O$	228.2454	$C_{12}H_{10}N_2O_3$	230.0692	$C_{14}H_{21}N_3$	231.1737	$C_{11}H_{25}N_2O_3$	233.1866
$C_{16}H_{20}O$	228.1515	$C_{12}H_{12}N_3O_2$	230.0930	$C_{15}H_9N_3$	231.0798	$C_{11}H_{27}N_3O_2$	233.2105
$C_{16}H_{22}N$	228.1753	$C_{12}H_{14}N_4O$	230.1169	$C_{15}H_{19}O_2$	231.1385	$C_{12}H_{11}NO_4$	233.0688
$C_{17}H_{10}N$	228.0814	$C_{12}H_{22}O_4$	230.1518	$C_{15}H_{21}NO$	231.1624	$C_{12}H_{13}N_2O_3$	233.0927
$C_{17}H_{24}$	228.1879	$C_{12}H_{24}NO_3$	230.1757	$C_{15}H_{23}N_2$	231.1863	$C_{12}H_{15}N_3O_2$	233.1165
$C_{18}H_{12}$	228.0939	$C_{12}H_{26}N_2O_2$	230.1996	$C_{16}H_9NO$	231.0684	$C_{12}H_{17}N_4O$	233.1404
229		$C_{12}H_{28}N_3O$	230.2234	$C_{16}H_{11}N_2$	231.0923	$C_{12}H_{25}O_4$	233.1753
$C_{10}H_{17}N_2O_4$	229.1189	$C_{12}H_{30}N_4$	230.2473	$C_{16}H_{23}O$	231.1750	$C_{12}H_{27}NO_3$	233.1992
$C_{10}H_{19}N_3O_3$	229.1427	$C_{13}H_{10}O_4$	230.0579	$C_{17}H_{11}O$	231.0810	$C_{13}H_{13}O_4$	233.0814
$C_{10}H_{21}N_4O_2$	229.1666	$C_{13}H_{12}NO_3$	230.0817	$C_{17}H_{13}N$	231.1049	$C_{13}H_{15}NO_3$	233.1052
$C_{11}H_7N_3O_3$	229.0488	$C_{13}H_{14}N_2O_2$	230.1056	$C_{17}H_{27}$	231.2114	$C_{13}H_{17}N_2O_2$	233.1291
$C_{11}H_9N_4O_2$	229.0726	$C_{13}H_{16}N_3O$	230.1295	$C_{18}H_{15}$	231.1174	$C_{13}H_{19}N_3O$	233.1529
$C_{11}H_{19}NO_4$	229.1315	$C_{13}H_{18}N_4$	230.1533	**232**		$C_{13}H_{21}N_4$	233.1768
$C_{11}H_{21}N_2O_3$	229.1553	$C_{13}H_{26}O_3$	230.1883	$C_{10}H_{20}N_2O_4$	232.1424	$C_{14}H_9N_4$	233.0829
$C_{11}H_{23}N_3O_2$	229.1791	$C_{13}H_{28}NO_2$	230.2121	$C_{10}H_{22}N_3O_3$	232.1662	$C_{14}H_{17}O_3$	233.1178
$C_{11}H_{25}N_4O$	229.2030	$C_{13}H_{30}N_2O$	230.2360	$C_{10}H_{24}N_4O_2$	232.1901	$C_{14}H_{19}NO_2$	233.1416
$C_{12}H_9N_2O_3$	229.0614	$C_{14}H_{14}O_3$	230.0943	$C_{11}H_8N_2O_4$	232.0484	$C_{14}H_{21}N_2O$	233.1655
$C_{12}H_{11}N_3O_2$	229.0852	$C_{14}H_{16}NO_2$	230.1182	$C_{11}H_{10}N_3O_3$	232.0723	$C_{15}H_9N_2O$	233.0715
$C_{12}H_{13}N_4O$	229.1091	$C_{14}H_{18}N_2O$	230.1420	$C_{11}H_{12}N_4O_2$	232.0961	$C_{15}H_{11}N_3$	233.0954
$C_{12}H_{21}O_4$	229.1440	$C_{14}H_{20}N_3$	230.1659	$C_{11}H_{22}NO_4$	232.1549	$C_{15}H_{21}O_2$	233.1542
$C_{12}H_{23}NO_3$	229.1679	$C_{14}H_{30}O_2$	230.2247	$C_{11}H_{24}N_2O_3$	232.1788	$C_{15}H_{23}NO$	233.1781
$C_{12}H_{25}N_2O_2$	229.1917	$C_{15}H_{18}O_2$	230.1307	$C_{11}H_{26}N_3O_2$	232.2026	$C_{15}H_{25}N_2$	233.2019
$C_{12}H_{27}N_3O$	229.2156	$C_{15}H_{20}NO$	230.1546	$C_{11}H_{28}N_4O$	232.2265	$C_{16}H_9O_2$	233.0603
$C_{12}H_{29}N_4$	229.2394	$C_{15}H_{22}N_2$	230.1784	$C_{12}H_{10}NO_4$	232.0610	$C_{16}H_{11}NO$	233.0841
$C_{13}H_9O_4$	229.0501	$C_{16}H_{10}N_2$	230.0845	$C_{12}H_{12}N_2O_3$	232.0848	$C_{16}H_{13}N_2$	233.1080
$C_{13}H_{11}NO_3$	229.0739	$C_{16}H_{22}O$	230.1671	$C_{12}H_{14}N_3O_2$	232.1087	$C_{16}H_{25}O$	233.1906
$C_{13}H_{13}N_2O_2$	229.0978	$C_{16}H_{24}N$	230.1910	$C_{12}H_{16}N_4O$	232.1325	$C_{16}H_{27}N$	233.2145
$C_{13}H_{15}N_3O$	229.1216	$C_{17}H_{10}O$	230.0732	$C_{12}H_{24}O_4$	232.1675	$C_{17}H_{13}O$	233.0967
$C_{13}H_{17}N_4$	229.1455	$C_{17}H_{12}N$	230.0970	$C_{12}H_{26}NO_3$	232.1914	$C_{17}H_{15}N$	233.1205
$C_{13}H_{25}O_3$	229.1804	$C_{17}H_{26}$	230.2036	$C_{12}H_{28}N_2O_2$	232.2152	$C_{17}H_{29}$	233.2270
$C_{13}H_{27}NO_2$	229.2043	$C_{18}H_{14}$	230.1096	$C_{13}H_{12}O_4$	232.0735	$C_{18}H_{17}$	233.1331
$C_{13}H_{29}N_2O$	229.2281	**231**		$C_{13}H_{14}NO_3$	232.0974	**234**	
$C_{13}H_{31}N_3$	229.2520	$C_{10}H_{19}N_2O_4$	231.1345	$C_{13}H_{16}N_2O_2$	232.1213	$C_{10}H_{22}N_2O_4$	234.1580
$C_{14}H_{13}O_3$	229.0865	$C_{10}H_{21}N_3O_3$	231.1584	$C_{13}H_{18}N_3O$	232.1451	$C_{10}H_{24}N_3O_3$	234.1819

Appendix A (Continued)

	FM
$C_{10}H_{26}N_4O_2$	234.2057
$C_{11}H_{10}N_2O_4$	234.0641
$C_{11}H_{12}N_3O_3$	234.0879
$C_{11}H_{14}N_4O_2$	234.1118
$C_{11}H_{24}NO_4$	234.1706
$C_{11}H_{26}N_2O_3$	234.1945
$C_{12}H_{12}NO_4$	234.0766
$C_{12}H_{14}N_2O_3$	234.1005
$C_{12}H_{16}N_3O_2$	234.1244
$C_{12}H_{18}N_4O$	234.1482
$C_{12}H_{26}O_4$	234.1832
$C_{13}H_{14}O_4$	234.0892
$C_{13}H_{16}NO_3$	234.1131
$C_{13}H_{18}N_2O_2$	234.1369
$C_{13}H_{20}N_3O$	234.1608
$C_{13}H_{22}N_4$	234.1846
$C_{14}H_{10}N_4$	234.0907
$C_{14}H_{18}O_3$	234.1256
$C_{14}H_{20}NO_2$	234.1495
$C_{14}H_{22}N_2O$	234.1733
$C_{14}H_{24}N_3$	234.1972
$C_{15}H_{10}N_2O$	234.0794
$C_{15}H_{12}N_3$	234.1032
$C_{15}H_{22}O_2$	234.1620
$C_{15}H_{24}NO$	234.1859
$C_{15}H_{26}N_2$	234.2098
$C_{16}H_{10}O_2$	234.0681
$C_{16}H_{12}NO$	234.0919
$C_{16}H_{14}N_2$	234.1158
$C_{16}H_{26}O$	234.1985
$C_{16}H_{28}N$	234.2223
$C_{17}H_{16}N$	234.1284
$C_{17}H_{30}$	234.2349
$C_{18}H_{18}$	234.1409
235	
$C_{10}H_{23}N_2O_4$	235.1659
$C_{10}H_{25}N_3O_3$	235.1897
$C_{11}H_{11}N_2O_4$	235.0719
$C_{11}H_{13}N_3O_3$	235.0958
$C_{11}H_{15}N_4O_2$	235.1196
$C_{11}H_{25}NO_4$	235.1784
$C_{12}H_{13}NO_4$	235.0845
$C_{12}H_{15}N_2O_3$	235.1083
$C_{12}H_{17}N_3O_2$	235.1322
$C_{12}H_{19}N_4O$	235.1560
$C_{13}H_{15}O_4$	235.0970
$C_{13}H_{17}NO_3$	235.1209
$C_{13}H_{19}N_2O_2$	235.1447
$C_{13}H_{21}N_3O$	235.1686
$C_{13}H_{23}N_4$	235.1925
$C_{14}H_9N_3O$	235.0746
$C_{14}H_{11}N_4$	235.0985
$C_{14}H_{19}O_3$	235.1334
$C_{14}H_{21}NO_2$	235.1573
$C_{14}H_{23}N_2O$	235.1811
$C_{14}H_{25}N_3$	235.2050
$C_{15}H_9NO_2$	235.0634
$C_{15}H_{11}N_2O$	235.0872
$C_{15}H_{13}N_3$	235.1111
$C_{15}H_{23}O_2$	235.1699
$C_{15}H_{25}NO$	235.1937
$C_{15}H_{27}N_2$	235.2176
$C_{16}H_{11}O_2$	235.0759
$C_{16}H_{13}NO$	235.0998
$C_{16}H_{15}N_2$	235.1236
$C_{16}H_{27}O$	235.2063
$C_{16}H_{29}N$	235.2301
$C_{17}H_{15}O$	235.1123
$C_{17}H_{17}N$	235.1362
$C_{17}H_{31}$	235.2427
$C_{18}H_{19}$	235.1488
236	
$C_{10}H_{24}N_2O_4$	236.1737
$C_{11}H_{12}N_2O_4$	236.0797
$C_{11}H_{14}N_3O_3$	236.1036
$C_{11}H_{16}N_4O_2$	236.1275
$C_{12}H_2N_3O_3$	236.0096
$C_{12}H_4N_4O_2$	236.0335
$C_{12}H_{14}NO_4$	236.0923
$C_{12}H_{16}N_2O_3$	236.1162
$C_{12}H_{18}N_3O_2$	236.1400
$C_{12}H_{20}N_4O$	236.1639
$C_{13}H_8N_4O$	236.0699
$C_{13}H_{16}O_4$	236.1049
$C_{13}H_{18}NO_3$	236.1287
$C_{13}H_{20}N_2O_2$	236.1526
$C_{13}H_{22}N_3O$	236.1764
$C_{13}H_{24}N_4$	236.2003
$C_{14}H_{10}N_3O$	236.0825
$C_{14}H_{12}N_4$	236.1063
$C_{14}H_{20}O_3$	236.1413
$C_{14}H_{22}NO_2$	236.1651
$C_{14}H_{24}N_2O$	236.1890
$C_{14}H_{26}N_3$	236.2129
$C_{15}H_{10}NO_2$	236.0712
$C_{15}H_{12}N_2O$	236.0950
$C_{15}H_{14}N_3$	236.1189
$C_{15}H_{24}O_2$	236.1777
$C_{15}H_{26}NO$	236.2015
$C_{15}H_{28}N_2$	236.2254
$C_{16}H_{12}O_2$	236.0837
$C_{16}H_{14}NO$	236.1076
$C_{16}H_{16}N_2$	236.1315
$C_{16}H_{28}O$	236.2141
$C_{16}H_{30}N$	236.2380
$C_{17}H_{16}O$	236.1202
$C_{17}H_{18}N$	236.1440
$C_{17}H_{32}$	236.2505
$C_{18}H_{20}$	236.1566
237	
$C_{11}H_{13}N_2O_4$	237.0876
$C_{11}H_{15}N_3O_3$	237.1114
$C_{11}H_{17}N_4O_2$	237.1353
$C_{12}H_{15}NO_4$	237.1001
$C_{12}H_{17}N_2O_3$	237.1240
$C_{12}H_{19}N_3O_2$	237.1478
$C_{12}H_{21}N_4O$	237.1717
$C_{13}H_9N_4O$	237.0777
$C_{13}H_{17}O_4$	237.1127
$C_{13}H_{19}NO_3$	237.1365
$C_{13}H_{21}N_2O_2$	237.1604
$C_{13}H_{23}N_3O$	237.1842
$C_{13}H_{25}N$	237.2081
$C_{14}H_9N_2O_2$	237.0664
$C_{14}H_{11}N_3O$	237.0903
$C_{14}H_{13}N_4$	237.1142
$C_{14}H_{21}O_3$	237.1491
$C_{14}H_{23}NO_2$	237.1730
$C_{14}H_{25}N_2O$	237.1968
$C_{14}H_{27}N_3$	237.2207
$C_{15}H_9O_3$	237.0552
$C_{15}H_{11}NO_2$	237.0790
$C_{15}H_{13}N_2O$	237.1029
$C_{15}H_{15}N_3$	237.1267
$C_{15}H_{25}O_2$	237.1855
$C_{15}H_{27}NO$	237.2094
$C_{15}H_{29}N_2$	237.2332
$C_{16}H_{13}O_2$	237.0916
$C_{16}H_{15}NO$	237.1154
$C_{16}H_{17}N_2$	237.1393
$C_{16}H_{29}O$	237.2219
$C_{16}H_{31}N$	237.2458
$C_{17}H_{17}O$	237.1280
$C_{17}H_{19}N$	237.1519
$C_{17}H_{33}$	237.2584
$C_{18}H_{21}$	237.1644
238	
$C_{11}H_{14}N_2O_4$	238.0954
$C_{11}H_{16}N_3O_3$	238.1193
$C_{11}H_{18}N_4O_2$	238.1431
$C_{12}H_{16}NO_4$	238.1080
$C_{12}H_{18}N_2O_3$	238.1318
$C_{12}H_{20}N_3O_2$	238.1557
$C_{12}H_{22}N_4O$	238.1795
$C_{13}H_8N_3O_2$	238.0617
$C_{13}H_{10}N_4O$	238.0856
$C_{13}H_{18}O_4$	238.1205
$C_{13}H_{20}NO_3$	238.1444
$C_{13}H_{22}N_2O_2$	238.1682
$C_{13}H_{24}N_3O$	238.1921
$C_{13}H_{26}N_4$	238.2160
$C_{14}H_{10}N_2O_2$	238.0743
$C_{14}H_{12}N_3O$	238.0981
$C_{14}H_{14}N_4$	238.1220
$C_{14}H_{22}O_3$	238.1569
$C_{14}H_{24}NO_2$	238.1808
$C_{14}H_{26}N_2O$	238.2046
$C_{14}H_{28}N_3$	238.2285
$C_{15}H_{10}O_3$	238.0630
$C_{15}H_{12}NO_2$	238.0868
$C_{15}H_{14}N_2O$	238.1107
$C_{15}H_{16}N_3$	238.1346
$C_{15}H_{26}O_2$	238.1934
$C_{15}H_{28}NO$	238.2172
$C_{15}H_{30}N_2$	238.2411
$C_{16}H_{14}O_2$	238.0994
$C_{16}H_{16}NO$	238.1233
$C_{16}H_{18}N_2$	238.1471
$C_{16}H_{30}O$	238.2298
$C_{16}H_{32}N$	238.2536
$C_{17}H_{18}O$	238.1358
$C_{17}H_{20}N$	238.1597
$C_{17}H_{34}$	238.2662
$C_{18}H_{22}$	238.1722
239	
$C_{11}H_{15}N_2O_4$	239.1032
$C_{11}H_{17}N_3O_3$	239.1271
$C_{11}H_{19}N_4O_2$	239.1509
$C_{12}H_{17}NO_4$	239.1158
$C_{12}H_{19}N_2O_3$	239.1396
$C_{12}H_{21}N_3O_2$	239.1635
$C_{12}H_{23}N_4O$	239.1873
$C_{13}H_9N_3O_2$	239.0695
$C_{13}H_{11}N_4O$	239.0934
$C_{13}H_{19}O_4$	239.1284
$C_{13}H_{21}NO_3$	239.1522
$C_{13}H_{23}N_2O_2$	239.1761
$C_{13}H_{25}N_3O$	239.1999
$C_{13}H_{27}N_4$	239.2238
$C_{14}H_9NO_3$	239.0583
$C_{14}H_{11}N_2O_2$	239.0821
$C_{14}H_{13}N_3O$	239.1060
$C_{14}H_{15}N_4$	239.1298
$C_{14}H_{23}O_3$	239.1648
$C_{14}H_{25}NO_2$	239.1886
$C_{14}H_{27}N_2O$	239.2125
$C_{14}H_{29}N_3$	239.2363
$C_{15}H_{11}O_3$	239.0708
$C_{15}H_{13}NO_2$	239.0947
$C_{15}H_{15}N_2O$	239.1185
$C_{15}H_{17}N_3$	239.1424
$C_{15}H_{27}O_2$	239.2012
$C_{15}H_{29}NO$	239.2250
$C_{15}H_{31}N_2$	239.2489
$C_{16}H_{15}O_2$	239.1072
$C_{16}H_{17}NO$	239.1311
$C_{16}H_{19}N_2$	239.1549
$C_{16}H_{31}O$	239.2376
$C_{16}H_{33}N$	239.2615
$C_{17}H_{19}O$	239.1436
$C_{17}H_{21}N$	239.1675
$C_{17}H_{35}$	239.2740
$C_{18}H_{23}$	239.1801
240	
$C_{11}H_{16}N_2O_4$	240.1111
$C_{11}H_{18}N_3O_3$	240.1349
$C_{11}H_{20}N_4O_2$	240.1588
$C_{12}H_8N_4O_2$	240.0648
$C_{12}H_{18}NO_4$	240.1236
$C_{12}H_{20}N_2O_3$	240.1475
$C_{12}H_{22}N_3O_2$	240.1713
$C_{12}H_{24}N_4O$	240.1952

Appendix A (Continued)

	FM
$C_{13}H_{8}N_{2}O_{3}$	240.0535
$C_{13}H_{10}N_{3}O_{2}$	240.0774
$C_{13}H_{12}N_{4}O$	240.1012
$C_{13}H_{20}O_{4}$	240.1362
$C_{13}H_{22}NO_{3}$	240.1600
$C_{13}H_{24}N_{2}O_{2}$	240.1839
$C_{13}H_{28}N_{4}$	240.2316
$C_{14}H_{8}O_{4}$	240.0422
$C_{14}H_{10}NO_{3}$	240.0661
$C_{14}H_{12}N_{2}O_{2}$	240.0899
$C_{14}H_{14}N_{3}O$	240.1138
$C_{14}H_{16}N_{4}$	240.1377
$C_{14}H_{24}O_{3}$	240.1726
$C_{14}H_{26}NO_{2}$	240.1965
$C_{14}H_{28}N_{2}O$	240.2203
$C_{14}H_{30}N_{3}$	240.2442
$C_{15}H_{12}O_{3}$	240.0786
$C_{15}H_{14}NO_{2}$	240.1025
$C_{15}H_{16}N_{2}O$	240.1264
$C_{15}H_{18}N_{3}$	240.1502
$C_{15}H_{28}O_{2}$	240.2090
$C_{15}H_{30}NO$	240.2329
$C_{15}H_{32}N_{2}$	240.2567
$C_{16}H_{16}O_{2}$	240.1151
$C_{16}H_{20}N_{2}$	240.1628
$C_{16}H_{18}NO$	240.1389
$C_{16}H_{32}O$	240.2454
$C_{16}H_{34}N$	240.2693
$C_{17}H_{20}O$	240.1515
$C_{17}H_{22}N$	240.1753
$C_{17}H_{36}$	240.2819
$C_{18}H_{24}$	240.1879
241	
$C_{11}H_{17}N_{2}O_{4}$	241.1189
$C_{11}H_{19}N_{3}O_{3}$	241.1427
$C_{11}H_{21}N_{4}O_{2}$	241.1666
$C_{12}H_{19}NO_{4}$	241.1315
$C_{12}H_{21}N_{2}O_{3}$	241.1553
$C_{12}H_{23}N_{3}O_{2}$	241.1791
$C_{12}H_{25}N_{4}O$	241.2030
$C_{13}H_{11}N_{3}O_{2}$	241.0852
$C_{13}H_{13}N_{4}O$	241.1091
$C_{13}H_{21}O_{4}$	241.1440
$C_{13}H_{25}N_{2}O_{2}$	241.1679
$C_{13}H_{25}N_{2}O_{2}$	241.1917
$C_{13}H_{27}N_{3}O$	241.2156
$C_{13}H_{29}N_{4}$	241.2394
$C_{14}H_{11}NO_{3}$	241.0739
$C_{14}H_{13}N_{2}O_{2}$	241.0978
$C_{14}H_{15}N_{3}O$	241.1216
$C_{14}H_{17}N_{4}$	241.1445
$C_{14}H_{25}O_{3}$	241.1804
$C_{14}H_{27}NO_{2}$	241.2043
$C_{14}H_{29}N_{2}O$	241.2281
$C_{14}H_{31}N_{3}$	241.2520
$C_{15}H_{13}O_{3}$	241.0865
$C_{15}H_{15}NO_{2}$	241.1103
$C_{15}H_{17}N_{2}O$	241.1342
$C_{15}H_{19}N_{3}$	241.1580
$C_{15}H_{29}O_{2}$	241.2168
$C_{15}H_{31}NO$	241.2407
$C_{15}H_{33}N_{2}$	241.2646
$C_{16}H_{17}O_{2}$	241.1229
$C_{16}H_{19}NO$	241.1467
$C_{16}H_{21}N_{2}$	241.1706
$C_{16}H_{33}O$	241.2533
$C_{16}H_{35}N$	241.2771
$C_{17}H_{21}O$	241.1593
$C_{17}H_{23}N$	241.1832
$C_{18}H_{25}$	241.1957
242	
$C_{11}H_{18}N_{2}O_{4}$	242.1267
$C_{11}H_{20}N_{3}O_{3}$	242.1506
$C_{11}H_{22}N_{4}O_{2}$	242.1744
$C_{12}H_{8}N_{3}O_{3}$	242.0566
$C_{12}H_{10}N_{4}O_{2}$	242.0805
$C_{12}H_{20}NO_{4}$	242.1393
$C_{12}H_{22}N_{2}O_{3}$	242.1631
$C_{12}H_{24}N_{3}O_{2}$	242.1870
$C_{12}H_{26}N_{4}O$	242.2108
$C_{13}H_{8}NO_{4}$	242.0453
$C_{13}H_{10}N_{2}O_{3}$	242.0692
$C_{13}H_{12}N_{3}O_{2}$	242.0930
$C_{13}H_{14}N_{4}O$	242.1169
$C_{13}H_{22}O_{4}$	242.1518
$C_{13}H_{24}NO_{3}$	242.1757
$C_{13}H_{26}N_{2}O_{2}$	242.1996
$C_{13}H_{28}N_{3}O$	242.2234
$C_{13}H_{30}N_{4}$	242.2473
$C_{14}H_{10}O_{4}$	242.0579
$C_{14}H_{12}NO_{3}$	242.0817
$C_{14}H_{14}N_{2}O_{2}$	242.1056
$C_{14}H_{16}N_{3}O$	242.1295
$C_{14}H_{18}N_{4}$	242.1533
$C_{14}H_{26}O_{3}$	242.1883
$C_{14}H_{28}NO_{2}$	242.2121
$C_{14}H_{30}N_{2}O$	242.2360
$C_{14}H_{32}N_{3}$	242.2598
$C_{15}H_{14}O_{3}$	242.0943
$C_{15}H_{16}NO_{2}$	242.1182
$C_{15}H_{18}N_{2}O$	242.1420
$C_{15}H_{20}N_{3}$	242.1659
$C_{15}H_{30}O_{2}$	242.2247
$C_{15}H_{32}NO$	242.2485
$C_{15}H_{34}N_{2}$	242.2724
$C_{16}H_{18}O_{2}$	242.1307
$C_{16}H_{20}NO$	242.1546
$C_{16}H_{22}N_{2}$	242.1784
$C_{16}H_{34}O$	242.2611
$C_{16}H_{18}O_{2}$	242.1307
$C_{16}H_{20}NO$	242.1546
$C_{16}H_{22}N_{2}$	242.1784
$C_{16}H_{34}O$	242.2611
$C_{17}H_{22}O$	242.1871
$C_{17}H_{24}N$	242.1910
$C_{18}H_{10}O$	242.0732
$C_{18}H_{12}N$	242.0970
$C_{18}H_{26}$	242.2036
$C_{19}H_{14}$	242.1096
243	
$C_{11}H_{19}N_{2}O_{4}$	243.1345
$C_{11}H_{21}N_{3}O_{3}$	243.1584
$C_{11}H_{23}N_{4}O_{2}$	243.1822
$C_{12}H_{7}N_{2}O_{4}$	243.0406
$C_{12}H_{9}N_{3}O_{3}$	243.0644
$C_{12}H_{11}N_{4}O_{2}$	243.0883
$C_{12}H_{21}NO_{4}$	243.1471
$C_{12}H_{23}N_{2}O_{3}$	243.1710
$C_{12}H_{25}N_{3}O_{2}$	243.1948
$C_{12}H_{27}N_{4}O$	243.2187
$C_{13}H_{9}NO_{4}$	243.0532
$C_{13}H_{11}N_{2}O_{3}$	243.0770
$C_{13}H_{13}N_{3}O_{2}$	243.1009
$C_{13}H_{15}N_{4}O$	243.1247
$C_{13}H_{23}O_{4}$	243.1597
$C_{13}H_{25}NO_{3}$	243.1835
$C_{13}H_{27}N_{2}O_{2}$	243.2074
$C_{13}H_{29}N_{3}O$	243.2312
$C_{13}H_{31}N_{4}$	243.2551
$C_{14}H_{11}O_{4}$	243.0657
$C_{14}H_{13}NO_{3}$	243.0896
$C_{14}H_{15}N_{2}O_{2}$	243.1134
$C_{14}H_{17}N_{3}O$	243.1373
$C_{14}H_{19}N_{4}$	243.1611
$C_{14}H_{27}O_{3}$	243.1961
$C_{14}H_{29}NO_{2}$	243.2199
$C_{14}H_{31}N_{2}O$	243.2438
$C_{14}H_{33}N_{3}$	243.2677
$C_{15}H_{15}O_{3}$	243.1021
$C_{15}H_{17}NO_{2}$	243.1260
$C_{15}H_{19}N_{2}O$	243.1498
$C_{15}H_{21}N_{3}$	243.1737
$C_{15}H_{31}O_{2}$	243.2325
$C_{15}H_{33}NO$	243.2564
$C_{16}H_{19}O_{2}$	243.1385
$C_{16}H_{21}NO$	243.1624
$C_{16}H_{23}N_{2}$	243.1863
$C_{17}H_{23}O$	243.1750
$C_{17}H_{25}N$	243.1988
$C_{18}H_{11}O$	243.0810
$C_{18}H_{13}N$	243.1049
$C_{18}H_{27}$	243.2114
$C_{19}H_{15}$	243.1174
244	
$C_{11}H_{20}N_{2}O_{4}$	244.1424
$C_{11}H_{22}N_{3}O_{3}$	244.1662
$C_{11}H_{24}N_{4}O_{2}$	244.1901
$C_{12}H_{8}N_{2}O_{4}$	244.0484
$C_{12}H_{10}N_{3}O_{3}$	244.0723
$C_{12}H_{12}N_{4}O_{2}$	244.0961
$C_{12}H_{22}NO_{4}$	244.1549
$C_{12}H_{24}N_{2}O_{3}$	244.1788
$C_{12}H_{26}N_{3}O_{2}$	244.2026
$C_{12}H_{28}N_{4}O$	244.2265
$C_{13}H_{10}NO_{4}$	244.0610
$C_{13}H_{12}N_{2}O_{3}$	244.0848
$C_{13}H_{14}N_{3}O_{2}$	244.1087
$C_{13}H_{16}N_{4}O$	244.1325
$C_{13}H_{24}O_{4}$	244.1675
$C_{13}H_{26}NO_{3}$	244.1914
$C_{13}H_{28}N_{2}O_{2}$	244.2152
$C_{13}H_{30}N_{3}O$	244.2391
$C_{13}H_{32}N_{4}$	244.2629
$C_{14}H_{12}O_{4}$	244.0735
$C_{14}H_{14}NO_{3}$	244.0974
$C_{14}H_{16}N_{2}O_{2}$	244.1213
$C_{14}H_{18}N_{3}O$	244.1451
$C_{14}H_{20}N_{4}$	244.1690
$C_{14}H_{28}O_{3}$	244.2039
$C_{14}H_{30}NO_{2}$	244.2278
$C_{14}H_{32}N_{2}O$	244.2516
$C_{15}H_{16}O_{3}$	244.1100
$C_{15}H_{18}NO_{2}$	244.1338
$C_{15}H_{20}N_{2}O$	244.1577
$C_{15}H_{22}N_{3}$	244.1815
$C_{15}H_{32}O_{2}$	244.2403
$C_{16}H_{10}N_{3}$	244.0876
$C_{16}H_{20}O_{2}$	244.1464
$C_{16}H_{22}NO$	244.1702
$C_{16}H_{24}N_{2}$	244.1941
$C_{17}H_{10}NO$	244.0763
$C_{17}H_{12}N_{2}$	244.1001
$C_{17}H_{24}O$	244.1828
$C_{17}H_{26}N$	244.2067
$C_{18}H_{12}O$	244.0888
$C_{18}H_{14}N$	244.1127
$C_{18}H_{28}$	244.2192
$C_{19}H_{16}$	244.1253
245	
$C_{11}H_{21}N_{2}O_{4}$	245.1502
$C_{11}H_{23}N_{3}O_{3}$	245.1741
$C_{11}H_{25}N_{4}O_{2}$	245.1979
$C_{12}H_{9}N_{2}O_{4}$	245.0563
$C_{12}H_{11}N_{3}O_{3}$	245.0801
$C_{12}H_{13}N_{4}O_{2}$	245.1040
$C_{12}H_{23}NO_{4}$	245.1628
$C_{12}H_{25}N_{2}O_{3}$	245.1866
$C_{12}H_{27}N_{3}O_{2}$	245.2105
$C_{12}H_{29}N_{4}O$	245.2343
$C_{13}H_{11}NO_{4}$	245.0688
$C_{13}H_{13}N_{2}O_{3}$	245.0927
$C_{13}H_{15}N_{3}O_{2}$	245.1165
$C_{13}H_{17}N_{4}O$	245.1404
$C_{13}H_{25}O_{4}$	245.1753
$C_{13}H_{27}NO_{3}$	245.1992
$C_{13}H_{29}N_{2}O_{2}$	245.2230
$C_{13}H_{31}N_{3}O$	245.2469
$C_{14}H_{13}O_{4}$	245.0814

Appendix A (Continued)

	FM
$C_{14}H_{15}NO_3$	245.1052
$C_{14}H_{17}N_2O_2$	245.1291
$C_{14}H_{19}N_3O$	245.1529
$C_{14}H_{21}N_4$	245.1768
$C_{14}H_{29}O_3$	245.2117
$C_{14}H_{31}NO_2$	245.2356
$C_{15}H_{17}O_3$	245.1178
$C_{15}H_{19}NO_2$	245.1416
$C_{15}H_{21}N_2O$	245.1655
$C_{15}H_{23}N_3$	245.1894
$C_{16}H_9N_2O$	245.0715
$C_{16}H_{11}N_3$	245.0954
$C_{16}H_{21}O_2$	245.1542
$C_{16}H_{23}NO$	245.1781
$C_{16}H_{25}N_2$	245.2019
$C_{17}H_{11}NO$	245.0841
$C_{17}H_{13}N_2$	245.1080
$C_{17}H_{25}O$	245.1906
$C_{17}H_{27}N$	245.2145
$C_{18}H_{13}O$	245.0967
$C_{18}H_{15}N$	245.1205
$C_{18}H_{29}$	245.2270
$C_{19}H_{17}$	245.1331
246	
$C_{11}H_{22}N_2O_4$	246.1580
$C_{11}H_{24}N_3O_3$	246.1819
$C_{11}H_{26}N_4O_2$	246.2057
$C_{12}H_{10}N_2O_4$	246.0641
$C_{12}H_{12}N_3O_3$	246.0879
$C_{12}H_{14}N_4O_2$	246.1118
$C_{12}H_{24}NO_4$	246.1706
$C_{12}H_{26}N_2O_3$	246.1945
$C_{12}H_{28}N_3O_2$	246.2183
$C_{12}H_{30}N_4O$	246.2422
$C_{13}H_{12}NO_4$	246.0766
$C_{13}H_{14}N_2O_3$	246.1005
$C_{13}H_{16}N_3O_2$	246.1244
$C_{13}H_{18}N_4O$	246.1482
$C_{13}H_{26}O_4$	246.1832
$C_{13}H_{28}NO_3$	246.2070
$C_{13}H_{30}N_2O_2$	246.2309
$C_{14}H_{14}O_4$	246.0892
$C_{14}H_{16}NO_3$	246.1131
$C_{14}H_{18}N_2O_2$	246.1369
$C_{14}H_{20}N_3O$	246.1608
$C_{14}H_{22}N_4$	246.1846
$C_{14}H_{30}O_3$	246.2196
$C_{15}H_{10}N_4$	246.0907
$C_{15}H_{18}O_3$	246.1256
$C_{15}H_{20}NO_2$	246.1495
$C_{15}H_{22}N_2O$	246.1733
$C_{15}H_{24}N_3$	246.1972
$C_{16}H_{10}N_2O$	246.0794
$C_{16}H_{12}N_3$	246.1032
$C_{16}H_{22}O_2$	246.1620
$C_{16}H_{24}NO$	246.1859
$C_{16}H_{26}N_2$	246.2098
$C_{17}H_{10}O_2$	246.0681
$C_{17}H_{12}NO$	246.0919
$C_{17}H_{14}N_2$	246.1158
$C_{17}H_{26}O$	246.1985
$C_{17}H_{28}N$	246.2223
$C_{18}H_{14}O$	246.1045
$C_{18}H_{16}N$	246.1284
$C_{18}H_{30}$	246.2349
$C_{19}H_{18}$	246.1409
247	
$C_{11}H_{23}N_2O_4$	247.1659
$C_{11}H_{25}N_3O_3$	247.1897
$C_{11}H_{27}N_4O_2$	247.2136
$C_{12}H_{11}N_2O_4$	247.0719
$C_{12}H_{13}N_3O_3$	247.0958
$C_{12}H_{15}N_4O_2$	247.1196
$C_{12}H_{25}NO_4$	247.1784
$C_{12}H_{27}N_2O_3$	247.2023
$C_{12}H_{29}N_3O_2$	247.2261
$C_{13}H_{13}NO_4$	247.0845
$C_{13}H_{15}N_2O_3$	247.1083
$C_{13}H_{17}N_3O_2$	247.1322
$C_{13}H_{19}N_4O$	247.1560
$C_{13}H_{27}O_4$	247.1910
$C_{13}H_{29}NO_3$	247.2148
$C_{14}H_{15}O_4$	247.0970
$C_{14}H_{17}NO_3$	247.1209
$C_{14}H_{19}N_2O_2$	247.1448
$C_{14}H_{21}N_3O$	247.1686
$C_{14}H_{23}N_4$	247.1925
$C_{15}H_9N_3O$	247.0746
$C_{15}H_{11}N_4$	247.0985
$C_{15}H_{19}O_3$	247.1334
$C_{15}H_{21}NO_2$	247.1573
$C_{15}H_{23}N_2O$	247.1811
$C_{15}H_{25}N_3$	247.2050
$C_{16}H_{11}N_2O$	247.0872
$C_{16}H_{13}N_3$	247.1111
$C_{16}H_{23}O_2$	247.1699
$C_{16}H_{25}NO$	247.1937
$C_{16}H_{27}N_2$	247.2176
$C_{17}H_{11}O_2$	247.0759
$C_{17}H_{13}NO$	247.0998
$C_{17}H_{15}N_2$	247.1236
$C_{17}H_{27}O$	247.2063
$C_{17}H_{29}N$	247.2301
$C_{18}H_{15}O$	247.1123
$C_{18}H_{17}N$	247.1362
$C_{18}H_{31}$	247.2427
$C_{19}H_{19}$	247.1488
248	
$C_{11}H_{24}N_2O_4$	248.1737
$C_{11}H_{26}N_3O_3$	248.1976
$C_{11}H_{28}N_4O_2$	248.2214
$C_{12}H_{12}N_2O_4$	248.0797
$C_{12}H_{14}N_3O_3$	248.1036
$C_{12}H_{16}N_4O_2$	248.1275
$C_{12}H_{26}NO_4$	248.1863
$C_{12}H_{28}N_2O_3$	248.2101
$C_{13}H_{14}NO_4$	248.0923
$C_{13}H_{16}N_2O_3$	248.1162
$C_{13}H_{18}N_3O_2$	248.1400
$C_{13}H_{20}N_4O$	248.1639
$C_{13}H_{28}O_4$	248.1988
$C_{14}H_{16}O_4$	248.1049
$C_{14}H_{20}N_2O_2$	248.1526
$C_{14}H_{22}N_3O$	248.1764
$C_{14}H_{24}N_4$	248.2003
$C_{15}H_{10}N_3O$	248.0825
$C_{15}H_{12}N_4$	248.1063
$C_{15}H_{20}O_3$	248.1413
$C_{15}H_{22}NO_2$	248.1651
$C_{15}H_{24}N_2O$	248.1890
$C_{15}H_{26}N_3$	248.2129
$C_{16}H_{10}NO_2$	248.0712
$C_{16}H_{12}N_2O$	248.0950
$C_{16}H_{14}N_3$	248.1189
$C_{16}H_{24}O_2$	248.1777
$C_{16}H_{26}NO$	248.2015
$C_{16}H_{28}N_2$	248.2254
$C_{17}H_{12}O_2$	248.0837
$C_{17}H_{14}NO$	248.1076
$C_{17}H_{16}N_2$	248.1315
$C_{17}H_{28}O$	248.2141
$C_{17}H_{30}N$	248.2380
$C_{18}H_{16}O$	248.1202
$C_{18}H_{18}N$	248.1440
$C_{18}H_{32}$	248.2505
$C_{19}H_{20}$	248.1566
249	
$C_{11}H_{25}N_2O_4$	249.1815
$C_{11}H_{27}N_3O_3$	249.2054
$C_{12}H_{13}N_2O_4$	249.0876
$C_{12}H_{15}N_3O_3$	249.1114
$C_{12}H_{17}N_4O_2$	249.1353
$C_{12}H_{27}NO_4$	249.1941
$C_{13}H_{15}NO_4$	249.1001
$C_{13}H_{17}N_2O_3$	249.1240
$C_{13}H_{19}N_3O_2$	249.1478
$C_{13}H_{21}N_4O$	249.1717
$C_{14}H_9N_4O$	249.0777
$C_{14}H_{17}O_4$	249.1127
$C_{14}H_{19}NO_3$	249.1365
$C_{14}H_{21}N_2O_2$	249.1604
$C_{14}H_{23}N_3O$	249.1842
$C_{14}H_{25}N_4$	249.2081
$C_{15}H_9N_2O_2$	249.0664
$C_{15}H_{11}N_3O$	249.0903
$C_{15}H_{13}N_4$	249.1142
$C_{15}H_{21}O_3$	249.1491
$C_{15}H_{23}NO_2$	249.1730
$C_{15}H_{25}N_2O$	249.1968
$C_{15}H_{27}N_3$	249.2207
$C_{16}H_{11}NO_2$	249.0790
$C_{16}H_{13}N_2O$	249.1029
$C_{16}H_{15}N_3$	249.1267
$C_{16}H_{25}O_2$	249.1855
$C_{16}H_{27}NO$	249.2094
$C_{16}H_{29}N_2$	249.2332
$C_{17}H_{13}O_2$	249.0916
$C_{17}H_{15}NO$	249.1154
$C_{17}H_{17}N_2$	249.1393
$C_{17}H_{29}O$	249.2219
$C_{17}H_{31}N$	249.2458
$C_{18}H_{17}O$	249.1280
$C_{18}H_{19}N$	249.1519
$C_{18}H_{33}$	249.2584
$C_{19}H_{21}$	249.1644
250	
$C_{11}H_{26}N_2O_4$	250.1894
$C_{12}H_{14}N_2O_4$	250.0954
$C_{12}H_{16}N_3O_3$	250.1193
$C_{12}H_{18}N_4O_2$	250.1431
$C_{13}H_{16}NO_4$	250.1080
$C_{13}H_{18}N_2O_3$	250.1318
$C_{13}H_{20}N_3O_2$	250.1557
$C_{13}H_{22}N_4O$	250.1795
$C_{14}H_{10}N_4O$	250.0856
$C_{14}H_{20}NO_3$	250.1444
$C_{14}H_{22}N_2O_2$	250.1682
$C_{14}H_{24}N_3O$	250.1921
$C_{14}H_{26}N_4$	250.2160
$C_{15}H_{10}N_2O_2$	250.0743
$C_{15}H_{12}N_3O$	250.0981
$C_{15}H_{14}N_4$	250.1220
$C_{15}H_{22}O_3$	250.1569
$C_{15}H_{24}NO_2$	250.1808
$C_{15}H_{26}N_2O$	250.2046
$C_{15}H_{28}N_3$	250.2285
$C_{16}H_{10}O_3$	250.0630
$C_{16}H_{12}NO_2$	250.0868
$C_{16}H_{14}N_2O$	250.1107
$C_{16}H_{16}N_3$	250.1346
$C_{16}H_{26}O_2$	250.1934
$C_{16}H_{28}NO$	250.2172
$C_{16}H_{30}N_2$	250.2411
$C_{17}H_{14}O_2$	250.0994
$C_{17}H_{16}NO$	250.1233
$C_{17}H_{18}N_2$	250.1471
$C_{17}H_{30}O$	250.2298
$C_{17}H_{32}N$	250.2536
$C_{18}H_{18}O$	250.1358
$C_{18}H_{20}N$	250.1597
$C_{18}H_{34}$	250.2662
$C_{19}H_{22}$	250.1722

Appendix B Common Fragment Ions

All fragments listed bear +1 charges. To be used in conjunction with Appendix C. Not all members of homologous and isomeric series are given. The list is meant to be suggestive rather than exhaustive. Appendix II of Hamming and Foster (1972), Table A-7 of McLafferty's (1993) interpretative book, and the high-resolution ion data of McLafferty (1982) are recommended as supplements. Structural inferences are listed in parentheses.

m/z	Ions[a] (Structural Inference)
14	CH_2
15	CH_3
16	O
17	OH
18	H_2O, NH_4
19	F, H_3O
26	$C{\equiv}N$, C_2H_2
27	C_2H_3
28	C_2H_4, CO, N_2 (air), CH=NH
29	C_2H_5, CHO
30	CH_2NH_2 (RCH_2NH_2), NO
31	CH_2OH (RCH_2OH), OCH_3
32	O_2 (air)
33	SH, CH_2F
34	H_2S
35	Cl (^{37}Cl at 37)
36	HCl ($H\,^{37}Cl$ at 38)
39	C_3H_3
40	$CH_2C{\equiv}N$, Ar(air)
41	C_3H_5, $CH_2C{\equiv}N$ + H,[a] C_2H_2NH
42	C_3H_6, C_2H_2O
43	C_3H_7, $CH_3C{=}O$, $CH_3C{=}OG$, (G = R, Ar, NH_2, OR, OH), C_2H_5N
44	$CH_2C(H){=}O$ + H (Aldehydes, McLafferty rearrangement), CH_3CHNH_2, CO_2, $NH_2C{=}O$ ($RC{=}ONH_2$), $(CH_3)_2N$
45	CH_3CHOH, CH_2CH_2OH, CH_2OCH_3 (RCH_2OCH_3), $C({=}O){-}OH$, $CH_3CH{-}O$ + H (CH_3CHOHR)
46	NO_2
47	CH_2SH (RCH_2SH), CH_3S
48	CH_3S + H
49	CH_2Cl ($CH_2\,^{37}Cl$ at 51)
51	CHF_2, C_4H_3
53	C_4H_5
54	$CH_2CH_2C{\equiv}N$
55	C_4H_7, $CH_2{=}CHC{=}O$
56	C_4H_8
57	C_4H_9, $C_2H_5C{=}O$
58	$CH_3{-}C({=}O)CH_2$ + H, $C_2H_5CHNH_2$, $(CH_3)_2NCH_2$, $C_2H_5NHCH_2$, C_2H_2S
59	$(CH_3)_2COH$, $CH_2OC_2H_5$, $C({=}O){-}OCH_3$ (RCO_2CH_3), $NH_2C{=}O$ + H, CH_3OCHCH_3, CH_3CHCH_2OH, C_2H_5CHOH (with CH_2 attached to $NH_2C{=}O$)
60	$CH_2C({=}O)OH$ + H, CH_2ONO
61	$CH_3C({=}O){-}O$ + 2H, CH_2CH_2SH, CH_2SCH_3
65	C_5H_5
66	 (cyclopentadiene) $\equiv C_5H_6$, H_2S_2 (RSSR)
67	C_5H_7
68	$CH_2CH_2CH_2C{\equiv}N$
69	C_5H_9, CF_3, $CH_3CH{=}CHC{=}O$, $CH_2{=}C(CH_3)C{=}O$
70	C_5H_{10}
71	C_5H_{11}, $C_3H_7C{=}O$
72	$C_2H_5C({=}O)CH_2$ + H, $C_3H_7CHNH_2$, $(CH_3)_2N{=}C{=}O$, $C_2H_5NHCHCH_3$ and isomers
73	Homologs of 59, $(CH_3)_3Si$
74	$CH_2{-}C({=}O){-}OCH_3$ + H
75	$C({=}O){-}OC_2H_5$ + 2H, $C_2H_5C({=}O)O$ + 2H, $CH_2SC_2H_5$, $(CH_3)_2CSH$, $(CH_3O)_2CH$, $(CH_3)_2SiOH$
76	C_6H_4 (C_6H_5X, C_6H_4XY)
77	C_6H_5 (C_6H_5X)
78	C_6H_5 + H
79	C_6H_5 + 2H, Br (^{81}Br at 81)

Appendix B Continued

m/z	Ions[a] (Structural Inference)
80	–CH_2, N–H, N–CH_2, CH_3SS + H, HBr ($H^{81}Br$ at 82)
81	–CH_2, C_6H_9
82	$CH_2CH_2CH_2CH_2C{\equiv}N$, CCl_2 ($C^{35}Cl^{37}Cl$ at 84, $C^{37}Cl_2$ at 86), C_6H_{10}
83	C_6H_{11}, $CHCl_2$ ($CH^{35}Cl^{37}Cl$ at 85, $CH^{37}Cl_2$ at 87), S
85	C_6H_{13}, $C_4H_9C{=}O$, $CClF_2$ ($C^{37}ClF_2$ at 87), O, O O
86	$C_3H_7C(=O)CH_2$ + H, $C_4H_9CHNH_2$ and isomers
87	$C_3H_7C(=O)$, homologs of 73, $CH_2CH_2C(=O)OCH_3$
88	$CH_2{-}C(=O){-}OC_2H_5$ + H
89	$C(=O){-}OC_3H_7$ + 2H, –C
90	CH_3CHONO_2, –CH
91	–CH_2 ($C_6H_5CH_2Br$), –CH + H, –C + 2H, $(CH_2)_4Cl$ [$(CH_2)_4{}^{37}Cl$, at 93], N
92	N –CH_2, –CH_2 + H,
93	CH_2Br ($CH_2{}^{81}Br$ at 95, RCH_2Br), C_7H_9, N –C=O, –O, C_7H_9 (terpenes)
94	–O + H, N–H –C=O
95	O –C=O
96	$CH_2CH_2CH_2CH_2CH_2C{\equiv}N$
97	C_7H_{13}, S –CH_2
98	O –CH_2O + H
99	C_7H_{15}, $C_6H_{11}O$, O O
100	$C_4H_9C(=O)CH_2$ + H, $C_5H_{11}CHNH_2$
101	$C(=O){-}OC_4H_9$
102	$CH_2C(=O){-}OC_3H_7$ + H
103	$C(=O){-}OC_4H_9$ + 2H, $C_5H_{11}S$, $CH(OCH_2CH_3)_2$
104	$C_2H_5CHONO_2$
105	–C=O, $C_6H_5(C{=}O)G$[G = OH, OR, OAr, halogen, NH_2], –CH_2CH_2, –$CHCH_3$
106	$NHCH_2$

Appendix B Continued

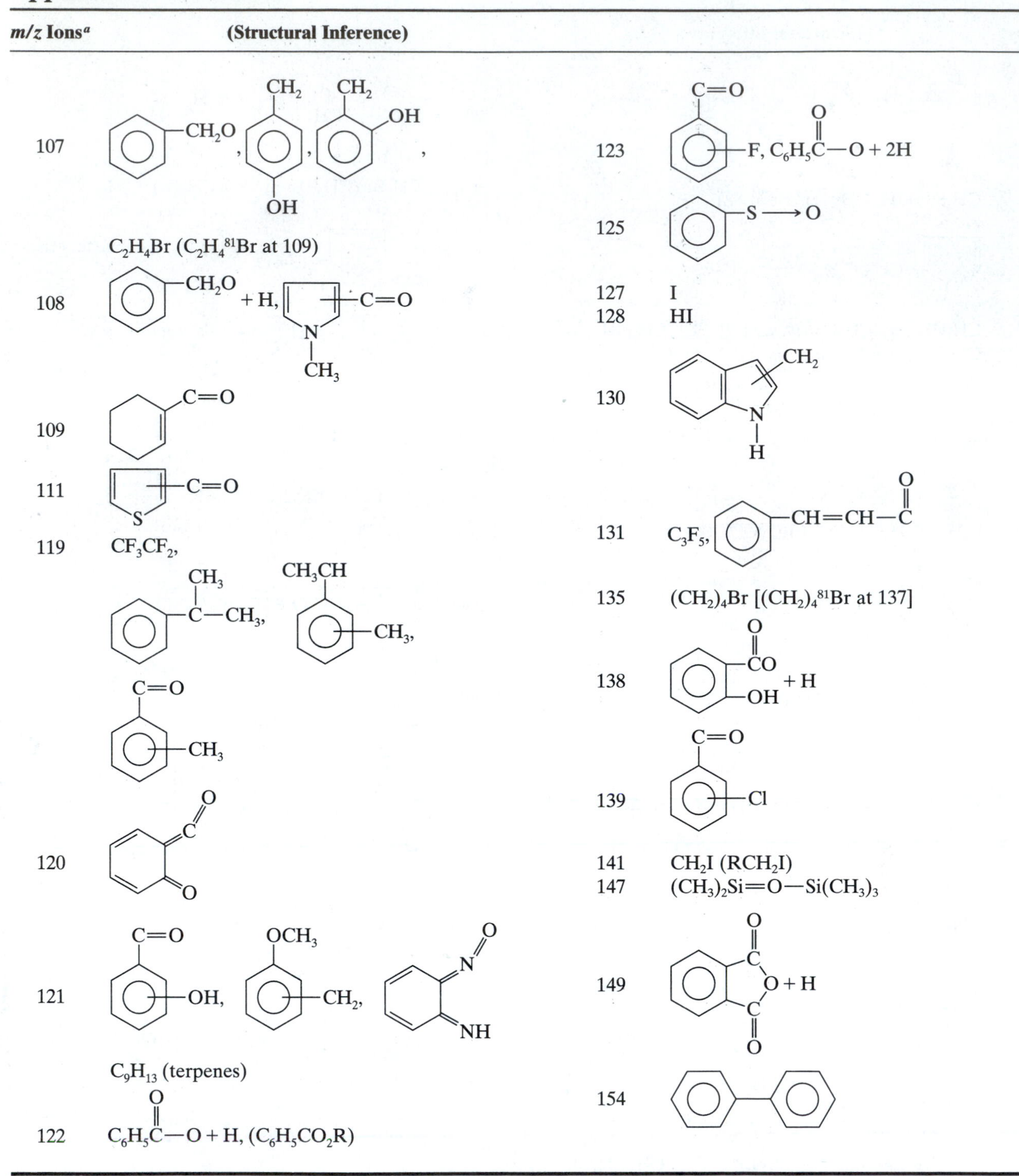

[a] Ions indicated as a fragment + nH (n = 1, 2, 3, . . .) are ions that arise via rearrangement involving hydrogen transfer.

Appendix C Common Fragments Lost

This list is suggestive rather than comprehensive. It should be used in conjunction with Appendix B. Table 5-19 of Hamming and Foster (1972) and Table A-5 of McLafferty (1993) are recommended as supplements. All of these fragments are lost as neutral species.

Molecular Ion Minus	Fragment Lost (Inference Structure)
1	$H\cdot$
2	$2H\cdot$
15	$CH_3\cdot$
16	O ($ArNO_2$, amine oxides, sulfoxides); $\cdot NH_2$ (carboxamides, sulfonamides)
17	$HO\cdot$
18	H_2O (alcohols, aldehydes, ketones)
19	$F\cdot$
20	HF
26	$CH{\equiv}CH$, $\cdot CH{\equiv}N$
27	$CH_2{=}CH\cdot$, $HC{\equiv}N$ (aromatic nitrites, nitrogen heterocycles)
28	$CH_2{=}CH_2$, CO, (quinones) (HCN + H)
29	$CH_3CH_2\cdot$, (ethyl ketones, $ArCH_2CH_2CH_3$), $\cdot CHO$
30	$NH_2CH_2\cdot$, CH_2O ($ArOCH_3$), NO ($ArNO_2$), C_2H_6
31	$\cdot OCH_3$(methyl esters), $\cdot CH_2OH$, CH_3NH_2
32	CH_3OH, S
33	$HS\cdot$ (thiols), ($\cdot CH_3$ and H_2O)
34	H_2S (thiols)
35	$Cl\cdot$
36	HCl, $2H_2O$
37	H_2Cl (or HCl + H)
38	C_3H_2, C_2N, F_2
39	C_3H_3, HC_2N
40	$CH_3C{\equiv}CH$
41	$CH_2{=}CHCH_2\cdot$
42	$CH_2{=}CHCH_3$, $CH_2{=}C{=}O$, $H_2C\overset{\overset{H_2}{C}}{\diagup\!\diagdown}CH_2$, NCO, $NCNH_2$
43	$C_3H_7\cdot$(propyl ketones, $ArCH_2{-}C_3H_7$), $CH_3\overset{\overset{O}{\|}}{C}\cdot$(methyl ketones, $CH_3\overset{\overset{O}{\|}}{C}G$, where G = various functional groups), $CH_2{=}CH{-}O\cdot$, ($CH_3\cdot$ and $CH_2{=}CH_2$), HCNO
44	$CH_2{=}CHOH$, CO_2 (esters, anhydrides), N_2O, $CONH_2$, $NHCH_2CH_3$
45	CH_3CHOH, $CH_3CH_2O\cdot$(ethyl esters), CO_2H, $CH_3CH_2NH_2$
46	(H_2O and $CH_2{=}CH_2$), CH_3CH_2OH, $\cdot NO_2$ ($ArNO_2$)
47	$CH_3S\cdot$
48	CH_3SH, SO (sulfoxides), O_3
49	$\cdot CH_2Cl$
51	$\cdot CHF_2$
52	C_4H_4, C_2N_2
53	C_4H_5
54	$CH_2{=}CH{-}CH{=}CH_2$
55	$CH_2{=}CHCHCH_3$

Appendix C Continued

Molecular Ion Minus	Fragment Lost (Inference Structure)
56	$CH_2{=}CHCH_2CH_3$, $CH_3CH{=}CHCH_3$, 2CO
57	$C_4H_9\cdot$ (butyl ketones), C_2H_5CO (ethyl ketones, EtC=OG, G = various structural units)
58	$\cdot$NCS, (NO + CO), CH_3COCH_3, C_4H_{10}
59	$CH_3OC(=O)\cdot$, $CH_3C(=O)NH_2$, (thiiranyl ring with S–H, S$\cdot$)
60	C_3H_7OH, $CH_2{=}C(OH)_2$ (acetate esters)[a]
61	$CH_3CH_2S\cdot$, (thiiranyl ring with S–H, S$\cdot$)
62	(H_2S and $CH_2{=}CH_2$)
63	$\cdot CH_2CH_2Cl$
64	C_5H_4, S_2, SO_2
68	$CH_2{=}C(CH_3){-}CH{=}CH_2$
69	$CF_3\cdot$, $C_5H_9\cdot$
71	$C_5H_{11}\cdot$
73	$CH_3CH_2OC(=O)\cdot$
74	C_4H_9OH
75	C_6H_3
76	C_6H_4, CS_2
77	C_6H_5, CS_2H
78	C_6H_6, CS_2H_2, C_5H_4N
79	$Br\cdot$, C_5H_5N
80	HBr
85	$\cdot CClF_2$
100	$CF_2{=}CF_2$
119	$CF_3{-}CF_2\cdot$
122	C_6H_5COOH
127	$I\cdot$
128	HI

[a] McLafferty rearrangement.

CHAPTER 3

Infrared Spectrometry

3.1 *Introduction*

Infrared (IR) radiation refers broadly to that part of the electromagnetic spectrum between the visible and microwave regions. Of greatest practical use to the organic chemist is the limited portion between 4000 and 400 cm^{-1}. There has been some interest in the near-IR (14,290–4000 cm^{-1}) and the far-IR regions, 700–200 cm^{-1}.

From the brief theoretical discussion that follows, it is clear that even a very simple molecule can give an extremely complex spectrum. The organic chemist takes advantage of this complexity when matching the spectrum of an unknown compound against that of an authentic sample. A peak-by-peak correlation is excellent evidence for identity. Any two compounds, except enantiomers, are unlikely to give exactly the same IR spectrum.

Although the IR spectrum is characteristic of the entire molecule, it is true that certain groups of atoms give rise to bands at or near the same frequency regardless of the structure of the rest of the molecule. It is the persistence of these characteristic bands that permits the chemist to obtain useful structural information by simple inspection and reference to generalized charts of characteristic group frequencies. We shall rely heavily on these characteristic group frequencies.

Since we are not solely dependent on IR spectra for identification, a detailed analysis of the spectrum will not be required. Following our general plan, we shall present only sufficient theory to accomplish our purpose: utilization of IR spectra in conjunction with other spectral data in order to determine molecular structure.

The importance of IR spectrometry as a tool of the practicing organic chemist is readily apparent from the number of books devoted wholly or in part to discussions of applications of IR spectrometry (see the references at the end of this chapter). There are many compilations of spectra as well as indexes to spectral collections and to the literature. Among the more commonly used compilations are those published by Sadtler (1972) and by Aldrich (1985).

3.2 *Theory*

Infrared radiation of frequencies less than about 100 cm^{-1} is absorbed and converted by an organic molecule into energy of molecular rotation. This absorption is quantized; thus a molecular rotation spectrum consists of discrete lines.

Infrared radiation in the range from about 10,000–100 cm^{-1} is absorbed and converted by an organic molecule into energy of molecular vibration. This absorption is also quantized, but vibrational spectra appear as bands rather than as lines because a single vibrational energy change is accompanied by a number of rotational energy changes. It is with these vibrational–rotational bands, particularly those occurring between 4000 and 400 cm^{-1}, that we shall be concerned. The frequency or wavelength of absorption depends on the relative masses of the atoms, the force constants of the bonds, and the geometry of the atoms.

Band positions in IR spectra are presented here as wavenumbers ($\bar{\nu}$) whose unit is the reciprocal centimeter (cm^{-1}); this unit is proportional to the energy of vibration and modern instruments are linear in reciprocal centimeters. Wavelength (λ) was used in the older literature in units of micrometers ($\mu m = 10^{-6}$ m; earlier called microns). Wavenumbers are reciprocally related to wavelength.

$$cm^{-1} = 10^4/\mu m$$

Note that wavenumbers are sometimes called "frequencies." However, this is incorrect since wavenumbers ($\bar{\nu}$ in units of cm^{-1}) are equal to $1 \times 10^4/\lambda$ in units of μm, whereas frequencies (ν in Hz) are equal to c/λ in cm, c being the speed of light (3×10^{10} cm/s). The symbol $\bar{\nu}$ is called "nu bar." Our spectra are linear in cm^{-1} except for a few linear in μm. Note also that these spectra are different from one another in appearance (see Fig. 3.6).

Band intensities can be expressed either as transmittance (T) or absorbance (A). Transmittance is the ratio of the radiant power transmitted by a sample to the radiant power incident on the sample. Absorbance is the logarithm, to the base 10, of the reciprocal of the transmittance; $A = \log_{10}(1/T)$. Organic chemists usually report intensity in semiquantitative terms (s = strong, m = medium, w = weak).

There are two types of molecular vibrations: stretching and bending. A stretching vibration is a rhythmical movement along the bond axis such that the interatomic distance is increasing or decreasing. A bending vibration may consist of a change in bond angle between bonds with a common atom or the movement of a group of atoms with respect to the remainder of the molecule without movement of the atoms in the group with respect to one another. For example, twisting, rocking, and torsional vibrations involve a change in bond angles with reference to a set of coordinates arbitrarily set up within the molecule.

Only those vibrations that result in a rhythmical change in the dipole moment of the molecule are observed in the IR. The alternating electric field, produced by the changing charge distribution accompanying a vibration, couples the molecule vibration with the oscillating electric field of the electromagnetic radiation.

A molecule has as many degrees of freedom as the total degrees of freedom of its individual atoms. Each atom has three degrees of freedom corresponding to the Cartesian coordinates (x, y, z) necessary to describe its position relative to other atoms in the molecule. A molecule of n atoms therefore has $3n$ degrees of freedom. For nonlinear molecules, three degrees of freedom describe rotation and three describe translation; the remaining $3n - 6$ degrees of freedom are vibrational degrees of freedom or fundamental vibrations. Linear molecules have $3n - 5$ vibrational degrees of freedom, for only two degrees of freedom are required to describe rotation.

Fundamental vibrations involve no change in the center of gravity of the molecule.

The three fundamental vibrations of the nonlinear, triatomic water molecule can be depicted as follows:

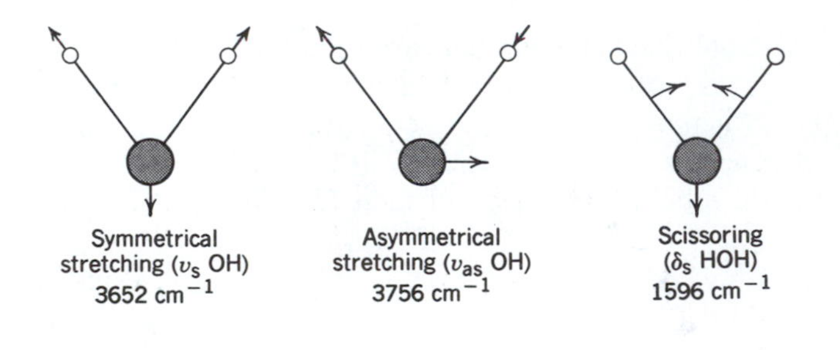

Note the very close spacing of the interacting or coupled asymmetric and symmetric stretching, above, compared with the far-removed scissoring mode.

The CO_2 molecule is linear and contains three atoms; therefore it has four fundamental vibrations [(3 × 3) − 5].

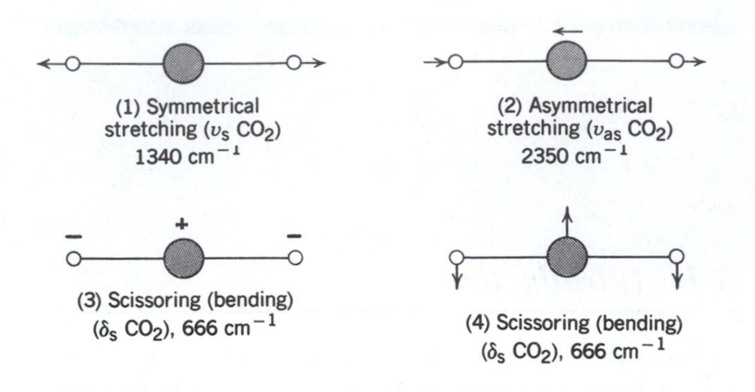

The symmetrical stretching vibration in (1) above is inactive in the IR since it produces no change in the dipole moment of the molecule. The bending vibrations in (3) and (4) above are equivalent and are the resolved components of bending motion oriented at any angle to the internuclear axis; they have the same frequency and are said to be doubly degenerate.

The various stretching and bending modes for an AX_2 group appearing as a portion of a molecule, for example, the CH_2 group in a hydrocarbon molecule, are shown in Figure 3.1. The $3n - 6$ rule does not apply since the CH_2 group represents only a portion of a molecule.

The theoretical number of fundamental vibrations (absorption frequencies) will seldom be observed because overtones (multiples of a given frequency) and combination tones (sum of two other vibrations) increase the number of bands, whereas other phenomena reduce the number of bands. The following will reduce the theoretical number of bands.

1. Fundamental frequencies that fall outside of the 4000-400 cm region^{-1}.
2. Fundamental bands that are too weak to be observed.
3. Fundamental vibrations that are so close that they coalesce.
4. The occurrence of a degenerate band from several absorptions of the same frequency in highly symmetrical molecules.
5. The failure of certain fundamental vibrations to appear in the IR because of the lack of change in molecular dipole.

Assignments for stretching frequencies can be approximated by the application of Hooke's law. In the application of the law, two atoms and their connecting

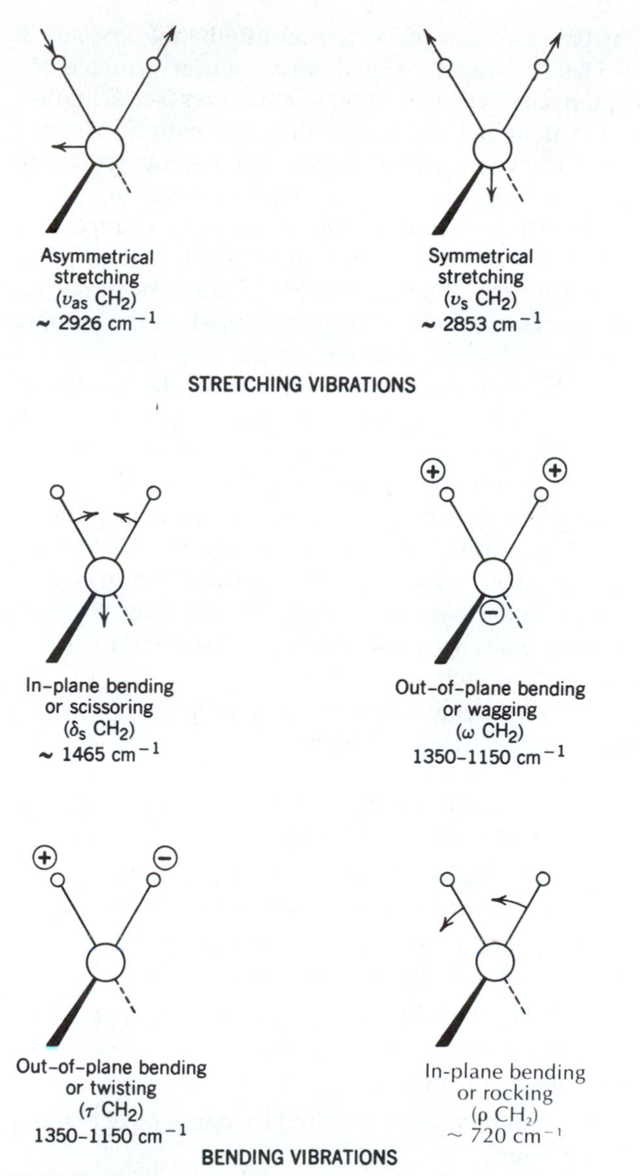

FIGURE 3.1. Vibrational modes for a CH_2 group. (+ and − indicate movement perpendicular to the plane of the page.)

bond are treated as a simple harmonic oscillator composed of two masses joined by a spring. The following equation, derived from Hooke's law, states the relationship between frequency of oscillation, atomic masses, and the force constant of the bond.

$$\bar{\nu} = \frac{1}{2\pi c}\left[\frac{f}{(MxMy)/(Mx + My)}\right]^{1/2}$$

where $\bar{\nu}$ = the vibrational frequency (cm^{-1})
c = velocity of light (cm/s)
f = force constant of bond (dyne/cm)

Mx and My = mass (g) of atom x and atom y, respectively.

The value of f is approximately 5×10^5 dyne/cm for single bonds and approximately two and three times this value for double and triple bonds, respectively.

Application of the formula to C—H stretching using 19.8×10^{-24} and 1.64×10^{-24} g as mass values for C and H, respectively, places the frequency of the C—H bond vibration at 3040 cm^{-1}. Actually, C—H stretching vibrations, associated with methyl and methylene groups, are generally observed in the region between 2960 and 2850 cm^{-1}. The calculation is not highly accurate because effects arising from the environment of the C—H within a molecule have been ignored. The frequency of IR absorption is commonly used to calculate the force constants of bonds.

The shift in absorption frequency following deuteration is often employed in the assignment of C—H stretching frequencies. The above equation can be used to estimate the change in stretching frequency as the result of deuteration. The term $MxMy/(Mx + My)$ will be equal to $M_C M_H/(M_C + M_H)$ for the C—H compound. Since $M_C \gg M_H$, this term is approximately equal to $M_C M_H/M_C$ or to M_H. Thus, for the C—D compound the term is equal to M_D, the frequency by Hooke's law application is inversely proportional to the square root of the mass of the isotope of hydrogen, and the ratio of the C—H to C—D stretching frequencies should equal $\sqrt{2}$. If the ratio of the frequencies, following deuteration, is much less than $\sqrt{2}$, we can assume that the vibration is not simply a C—H stretching vibration but instead a mixed vibration involving interaction (coupling) with another vibration.

Calculations place the stretching frequencies of the following bonds in the general absorption regions indicated:

Bond Type	Absorption Region (cm^{-1})
C—C, C—O, C—N	1300–800
C=C, C=O, C=N, N=O	1900–1500
C≡C, C≡N	2300–2000
C—H, O—H, N—H	3800–2700

To approximate the vibrational frequencies of bond stretching by Hooke's law, the relative contributions of bond strengths and atomic masses must be considered. For example, a superficial comparison of the C—H group with the F—H group, on the basis of atomic masses, might lead to the conclusion that the stretching frequency of the F—H bond should occur at a lower frequency than that for the C—H bond. However, the increase in the force constant from left to right across the first two rows of the periodic table has a greater effect than the mass increase. Thus, the F—H group absorbs at a higher frequency (4138 cm^{-1}) than the C—H group (3040 cm^{-1}).

In general, functional groups that have a strong dipole give rise to strong absorptions in the IR.

3.2.1 Coupled Interactions

When two bond oscillators share a common atom, they seldom behave as individual oscillators unless the individual oscillation frequencies are widely different. This is because there is mechanical coupling interaction between the oscillators. For example, the carbon dioxide molecule, which consists of two C═O bonds with a common carbon atom, has two fundamental stretching vibrations: an asymmetrical and a symmetrical stretching mode. The symmetrical stretching mode consists of an in-phase stretching or contracting of the C═O bonds, and absorption occurs at a wavelength longer than that observed for the carbonyl group in an aliphatic ketone. The symmetrical stretching mode produces no change in the dipole moment (μ) of the molecule and is therefore "inactive" in the IR, but it is easily observed in the Raman spectrum* near 1340 cm^{-1}. In the asymmetrical stretching mode, the two C═O bonds stretch out of phase; one C═O bond stretches as the other contracts. The asymmetrical stretching mode, since it produces a change in the dipole moment, is IR active; the absorption (2350 cm^{-1}) is at a higher frequency (shorter wavelength) than observed for a carbonyl group in aliphatic ketones.

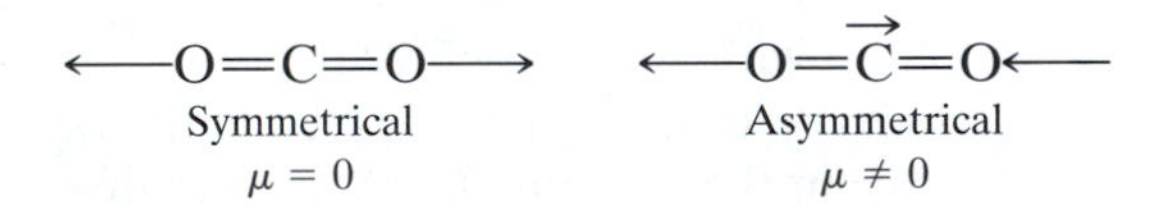

This difference in carbonyl absorption frequencies displayed by the carbon dioxide molecule results from strong mechanical coupling or interaction. In contrast, two ketonic carbonyl groups separated by one or more carbon atoms show normal carbonyl absorption near 1715 cm^{-1} because appreciable coupling is prevented by the intervening carbon atom(s).

Coupling accounts for the two N—H stretching bands in the 3497–3077 cm^{-1} region in primary amine and primary amide spectra; for the two C═O stretching bands in the 1818–1720 cm^{-1} region in carboxylic anhydride and imide spectra, and for the two C—H stretching bands in the 3000–2760 cm^{-1} region for both methylene and methyl groups.

Useful characteristic group frequency bands often involve coupled vibrations. The spectra of alcohols have a strong band in the region between 1260 and 1000 cm^{-1}, which is usually designated as the "C—O stretching band." In the spectrum of methanol this band is at 1034 cm^{-1}; in the spectrum of ethanol it occurs at 1053 cm^{-1}. Branching and unsaturation produce absorption characteristic of these structures (see alcohols). It is evident that we are dealing not with an isolated C—O stretching vibration but rather a coupled asymmetric vibration involving C—C—O stretching.

Vibrations resulting from bond angle changes frequently couple in a manner similar to stretching vibrations. Thus, the ring C—H out-of-plane bending frequencies of aromatic molecules depend on the number of adjacent hydrogen atoms on the ring; coupling between the hydrogen atoms is affected by the bending of the C—C bond in the ring to which the hydrogen atoms are attached.

Interaction arising from coupling of stretching and bending vibrations is illustrated by the absorption of secondary acyclic amides. Secondary acyclic amides, which exist predominantly in the trans conformation, show strong absorption in the 1563–1515 cm^{-1} region; this absorption involves coupling of the N—H bending and C—N stretching vibrations.

The requirements for effective coupling interaction may be summarized as follows:

1. The vibrations must be of the same symmetry species if interaction is to occur.
2. Strong coupling between stretching vibrations requires a common atom between the groups.
3. Interaction is greatest when the coupled groups absorb, individually, near the same frequency.
4. Coupling between bending and stretching vibrations can occur if the stretching bond forms one side of the changing angle.
5. A common bond is required for coupling of bending vibrations.
6. Coupling is negligible when groups are separated by one or more carbon atoms and the vibrations are mutually perpendicular.

As we have seen in our discussion of interaction, coupling of two fundamental vibrational modes will produce two new modes of vibration, with frequencies higher and lower than that observed when interaction is absent. Interaction can also occur between fundamental vibrations and overtones or combination-tone vibrations. Such interaction is known as Fermi resonance. One example of Fermi resonance is afforded by the absorption pattern of carbon dioxide. In our discussion of interaction, we indicated that the symmetrical stretching band of CO_2 appears in the Raman spectrum near 1340 cm^{-1}. Actually two bands are observed; one at 1286 cm^{-1} and one at 1388 cm^{-1}. The splitting results from coupling between the fundamental C═O stretching vibration, near 1340 cm^{-1}, and the first overtone of the bending vibration. The fundamental bending vi-

*Band intensity in Raman spectra depends on bond polarizability rather than molecular dipole changes.

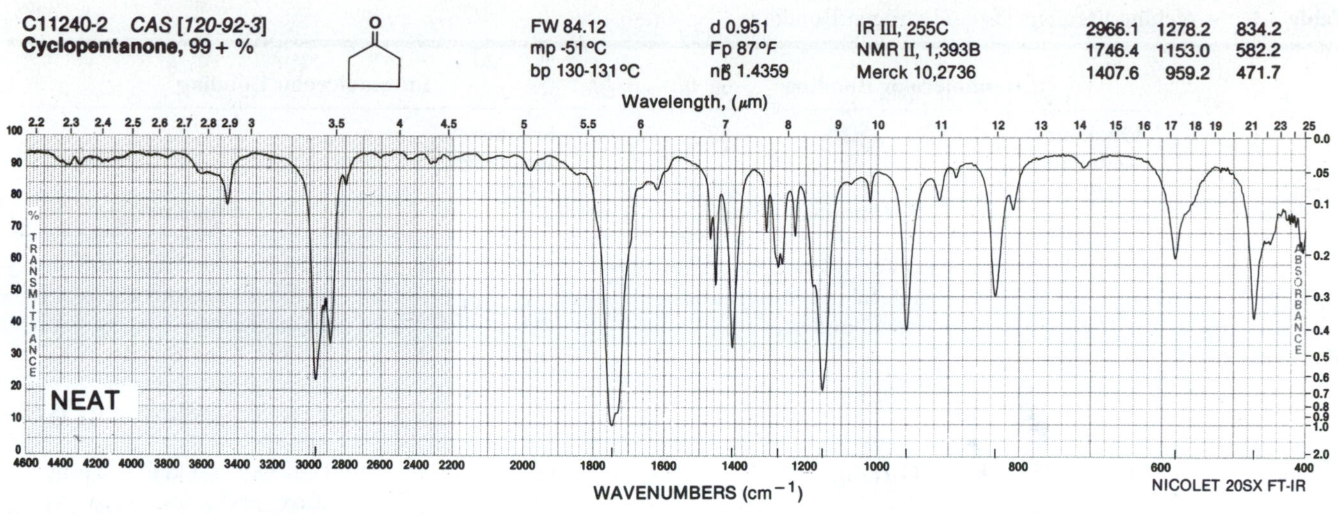

FIGURE 3.2. Cyclopentanone, thin film.

bration occurs near 666 cm^{-1}, the first overtone near 1334 cm^{-1}.

Fermi resonance is a common phenomenon in IR and Raman spectra. It requires that the vibrational levels be of the same symmetry species and that the interacting groups be located in the molecule so that mechanical coupling is appreciable.

An example of Fermi resonance in an organic structure is the "doublet" appearance of the C=O stretch of cyclopentanone under sufficient resolution conditions. Figure 3.2 shows the appearance of the spectrum of cyclopentanone under the usual conditions. With adequate resolution (Fig. 3.3), Fermi resonance with an overtone or combination band of an α-methylene group shows two absorptions in the carbonyl stretch region.

3.2.2 Hydrogen Bonding

Hydrogen bonding can occur in any system containing a proton donor group (X—H) and a proton acceptor

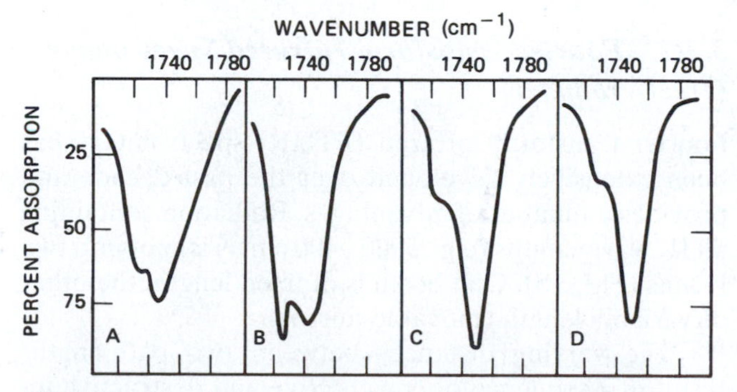

FIGURE 3.3. Infrared spectrum of cyclopentanone in various media. A. Carbon tetrachloride solution (0.15 *M*). B. Carbon disulfide solution (0.023 *M*). C. Chloroform solution (0.025 *M*). D. Liquid state (thin films). (Computed spectral slit width 2 cm^{-1}.)

($\ddot{Y}$) if the *s* orbital of the proton can effectively overlap the *p* or π orbital of the acceptor group. Atoms X and $\ddot{Y}$ are electronegative, with $\ddot{Y}$ possessing lone pair electrons. The common proton donor groups in organic molecules are carboxyl, hydroxyl, amine, or amide groups. Common proton acceptor atoms are oxygen, nitrogen, and the halogens. Unsaturated groups, such as the C=C linkage, can also act as proton acceptors.

The strength of the hydrogen bond is at a maximum when the proton donor group and the axis of the lone pair orbital are collinear. The strength of the bond decreases as the distance between X and Y increases.

Hydrogen bonding alters the force constant of both groups; thus, the frequencies of both stretching and bending vibrations are altered. The X—H stretching bands move to lower frequencies (longer wavelengths) usually with increased intensity and band widening. The stretching frequency of the acceptor group, for example, C=O, is also reduced but to a lesser degree than the proton donor group. The H—X bending vibration usually shifts to a shorter wavelength when bonding occurs; this shift is less pronounced than that of the stretching frequency.

*Inter*molecular hydrogen bonding involves association of two or more molecules of the same or different compounds. Intermolecular bonding may result in dimer molecules (as observed for carboxylic acids) or in polymeric molecular chains, which exist in neat samples or concentrated solutions of monohydroxy alcohols. *Intra*molecular hydrogen bonds are formed when the proton donor and acceptor are present in a single molecule under spatial conditions that allow the required overlap of orbitals, for example, the formation of a five- or six-membered ring. The extent of both inter- and intramolecular bonding is temperature dependent. The effect of concentration on intermolecular and intramolecular hydrogen bonding is markedly different. The

Table 3.1 Stretching Frequencies in Hydrogen Bonding

	Intermolecular Bonding			Intramolceular Bonding		
	Frequency Reduction (cm^{-1})			Frequency Reduction (cm^{-1})		
X—H ··· Y Strength	ν_{OH}	$\nu_{C=O}$	Compound Class	ν_{OH}	$\nu_{C=O}$	Compound Class
Weak	300[a]	15[b]	Alcohols, phenols, and intermolecular hydroxyl to carbonyl bonding	<100[a]	10	1,2-Diols, α- and most β-hydroxy ketones; o-chloro and o-alkoxy phenols
Medium				100–300[a]	50	1,3-Diols; some β-hydroxy ketones; β-hydroxy amino compounds; β-hydroxy nitro compounds
Strong	>500[a]	50[b]	RCO_2H dimers	>300[a]	100	o-Hydroxy aryl ketones; o-hydroxy aryl acids; o-hydroxy aryl esters; β-diketones; tropolones

[a] Frequency shift relative to "free" stretching frequencies.
[b] Carbonyl stretching only where applicable.

bands that result from intermolecular bonding generally disappear at low concentrations (less than about 0.01 *M* in nonpolar solvents). Intramolecular hydrogen bonding is an internal effect and persists at very low concentrations.

The change in frequency between "free" OH absorption and bonded OH absorption is a measure of the strength of the hydrogen bond. Ring strain, molecular geometry, and the relative acidity and basicity of the proton donor and acceptor groups affect the strength of bonding. Intramolecular bonding involving the same bonding groups is stronger when a six-membered ring is formed than when a smaller ring results from bonding. Hydrogen bonding is strongest when the bonded structure is stabilized by resonance.

The effects of hydrogen bonding on the stretching frequencies of hydroxyl and carbonyl groups are summarized in Table 3.1. Figure 3.17 (spectrum of cyclohexylcarbinol in the stretch region) clearly illustrates this effect.

An important aspect of hydrogen bonding involves interaction between functional groups of solvent and solute. If the solute is polar, then it is important to note the solvent used and the solute concentration.

3.3 *Instrumentation*

3.3.1 *Dispersion IR Spectrometer*

For many years, an infrared spectrum was obtained by passing an infrared beam though the sample and scanning the spectrum with a dispersion device (the familiar diffraction grating). The spectrum was scanned by rotating the diffraction grating; the absorption areas (peaks) were detected and plotted as frequencies versus intensities.

Figure 3.4 demonstrates a sophisticated double-beam dispersion instrument, operation of which involves splitting the beam and passing one portion through the sample cell and the other portion through the reference cell. The individual beams are then recombined into a single beam of alternating segments by means of the rotating sector mirror, *M*7, and the absorption intensities of the segments are balanced by the attenuator in the reference beam. Thus, the solvent in the reference cell and in the sample cell are balanced out, and the spectrum contains only the absorption peaks of the sample itself.

3.3.2 *Fourier Transform Infrared Spectrometer (Interferometer)*

Fourier transform infrared (FT IR) spectrometry has been extensively developed over the past decade and provides a number of advantages. Radiation containing all IR wavelengths (e.g., 5000–400 cm^{-1}) is split into two beams (Fig. 3.5). One beam is of fixed length, the other of variable length (movable mirror).

The varying distances between two pathlengths result in a sequence of constructive and destructive interferences and hence variations in intensities: an interferogram. Fourier transformation converts this interferogram from the time domain into one spectral point on the more familiar form of the frequency domain.

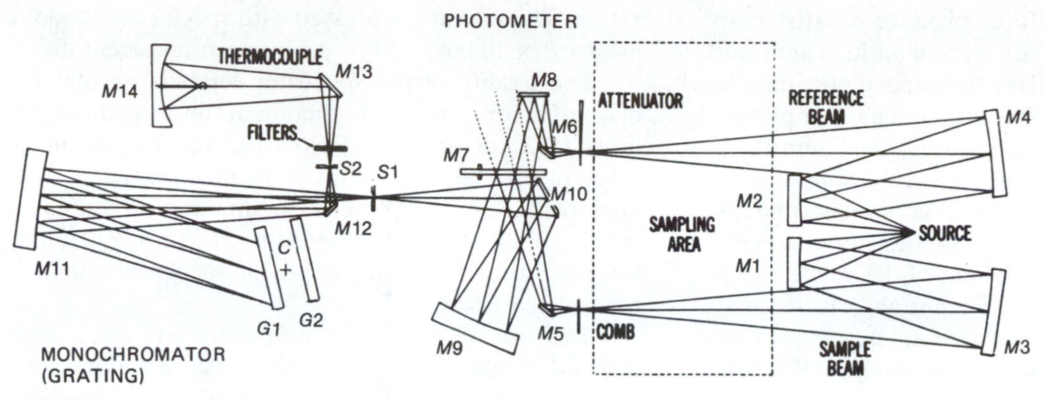

FIGURE 3.4. Optical system of double-beam IR spectrophotometer.

Smooth and continuous variation of the length of the piston adjusts the position of mirror B and varies the length of beam B; Fourier transformation at successive points throughout this variation gives rise to the complete IR spectrum. Passage of this radiation through a sample subjects the compound to a broad band of energies. In principle the analysis of one broadbanded pass of radiation through the sample will give rise to a complete IR spectrum.

There are a number of advantages to FT IR methods. Since a monochromator is not used, the entire radiation range is passed through the sample simultaneously and much time is saved (Felgett's advantage);

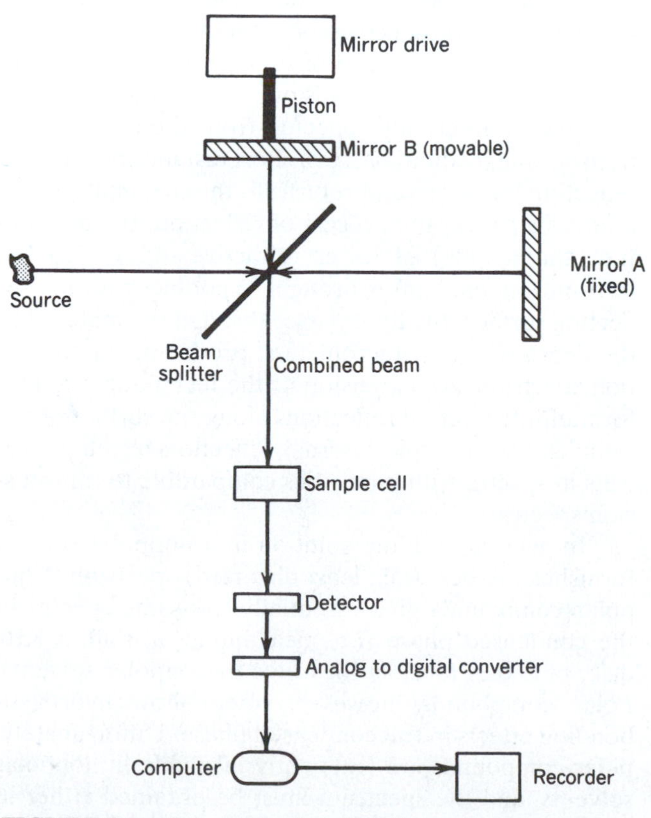

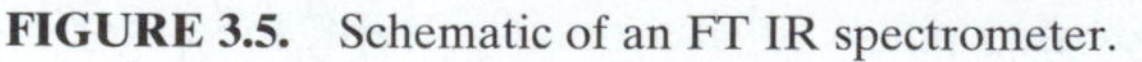

FIGURE 3.5. Schematic of an FT IR spectrometer.

FT IR instruments can have very high resolution (≤ 0.001 cm^{-1}). Moreover since the data undergo analog-to-digital conversion, IR results are easily manipulated: Results of several scans are combined to average out random absorption artifacts, and excellent spectra from very small samples can be obtained. An FT IR unit can therefore be used in conjunction with HPLC or GC. As with any computer-aided spectrometer, spectra of pure samples or solvents (stored in the computer) can be subtracted from mixtures. Flexibility in spectral printout is also available: for example, spectra linear in either wavenumber or wavelength can be obtained from the same data set.

Several manufacturers offer GC-FT IR instruments with which a vapor-phase spectrum can be obtained on nanogram amounts of a compound eluting from a capillary GC column. Vapor-phase spectra resemble those obtained at high dilution in a nonpolar solvent: Concentration-dependent peaks are shifted to higher frequency compared with those obtained from concentrated solutions, thin films, or the solid state (see Aldrich, 1985).

3.4 *Sample Handling*

Infrared spectra may be obtained for gases, liquids, or solids. The spectra of gases or low-boiling liquids may be obtained by expansion of the sample into an evacuated cell. Gas cells are available in lengths of a few centimeters to 40 m. The sampling area of a standard IR spectrophotometer will not accommodate cells much longer than 10 cm; long paths are achieved by multiple reflection optics.

Liquids may be examined neat or in solution. Neat liquids are examined between salt plates, usually without a spacer. Pressing a liquid sample between flat plates produces a film 0.01 mm or less in thickness, the plates being held together by capillary action. Samples of 1–10 mg are required. Thick samples of neat liquids usually absorb too strongly to produce a satisfactory spectrum. Volatile liquids are examined in sealed cells with very thin spacers. Silver chloride plates may be used for samples that dissolve sodium chloride plates.

Solutions are handled in cells of 0.1–1 mm thickness. Volumes of 0.1–1 mL of 0.05–10% solutions are required for readily available cells. A compensating cell, containing pure solvent, is placed in the reference beam. The spectrum thus obtained is that of the solute except in those regions in which the solvent absorbs strongly. For example, thick samples of carbon tetrachloride absorb strongly near 800 cm^{-1}; compensation for this band is ineffective since strong absorption prevents any radiation from reaching the detector.

The solvent selected must be dry and transparent in the region of interest. When the entire spectrum is of interest, several solvents must be used. A common pair of solvents is carbon tetrachloride (CCl_4) and carbon disulfide (CS_2). Carbon tetrachloride is relatively free of absorption at frequencies above 1333 cm^{-1}, whereas CS_2 shows little absorption below 1333 cm^{-1}. Solvent and solute combinations that react must be avoided. For example, CS_2 cannot be used as a solvent for primary or secondary amines. Amino alcohols react slowly with CS_2 and CCl_4.

When only very small samples are available, ultramicrocavity cells are used in conjunction with a beam condenser. A spectrum can be obtained on a few micrograms of sample in solution. When volatility permits, the solute can be recovered for examination by other spectrometric techniques. The absorption patterns of selected solvents and mulling oils are presented in Appendix A.

Solids are usually examined as a mull, as a pressed disk, or as a deposited glassy film. Mulls are prepared by thoroughly grinding 2–5 mg of a solid in a smooth agate mortar. Grinding is continued after the addition of 1 or 2 drops of the mulling oil. The suspended particles must be less than 2 μm to avoid excessive scattering of radiation. The mull is examined as a thin film between flat salt plates. Nujol® (a high-boiling petroleum oil) is commonly used as a mulling agent. When hydrocarbon bands interfere with the spectrum, Fluorolube® (a completely halogenated polymer containing F and Cl) or hexachlorobutadiene may be used. The use of both Nujol® and Fluorolube® mulls makes possible a scan, essentially free of interfering bands, over the 4000–250 cm^{-1} region.

The pellet (pressed-disk) technique depends on the fact that dry, powdered potassium bromide (or other alkali metal halides) can be compacted under pressure in vacuo to form transparent disks. The sample (0.5–1.0 mg) is intimately mixed with approximately 100 mg of dry, powdered KBr. Mixing can be effected by thorough grinding in a smooth agate mortar or, more efficiently, with a small vibrating ball mill, or by lyophilization. The mixture is pressed with special dies under a pressure of 10,000–15,000 psi into a transparent disk.

The quality of the spectrum depends on the intimacy of mixing and the reduction of the suspended particles to 2 μm or less. Microdisks, 0.5–1.5 mm in diameter, can be used with a beam condenser. The microdisk technique permits examination of samples as small as 1 μg. Bands near 3448 and 1639 cm^{-1}, resulting from moisture, frequently appear in spectra obtained by the pressed-disk technique.

The use of KBr disks or pellets has often been avoided because of the demanding task of making good pellets. Such KBr techniques can be less formidable through the Mini-Press, which affords a simple procedure; the KBr–sample mixture is placed in the nut portion of the assembly with one bolt in place. The second bolt is introduced, and pressure is applied by tightening the bolts. Removal of the bolts leaves a pellet in the nut that now serves as a cell.

Deposited films are useful only when the material can be deposited from solution or cooled from a melt as microcrystals or as a glassy film. Crystalline films generally lead to excessive light scattering. Specific crystal orientation may lead to spectra differing from those observed for randomly oriented particles such as exist in a mull or halide disk. The deposited film technique is particularly useful for obtaining spectra of resins and plastics. Care must be taken to free the sample of solvent by vacuum treatment or gentle heating.

A technique known as attenuated total reflection or internal reflection spectroscopy is now available for obtaining qualitative spectra of solids regardless of thickness. The technique depends on the fact that a beam of light that is internally reflected from the surface of a transmitting medium passes a short distance beyond the reflecting boundary and returns to the transmitting medium as a part of the process of reflection. If a material (i.e., the sample) of lower refractive index than the transmitting medium is brought in contact with the reflecting surface, the light passes through the material to the depth of a few micrometers, producing an absorption spectrum. An extension of the technique provides for multiple internal reflections along the surface of the sample. The multiple internal reflection technique results in spectra with intensities comparable to transmission spectra.

In general, a dilute solution in a nonpolar solvent furnishes the best (i.e., least distorted) spectrum. Nonpolar compounds give essentially the same spectra in the condensed phase (i.e., neat liquid, a mull, a KBr disk, or a thin film) as they give in nonpolar solvents. Polar compounds, however, often show hydrogen-bonding effects in the condensed phase. Unfortunately, polar compounds are frequently insoluble in nonpolar solvents, and the spectrum must be obtained either in a condensed phase or in a polar solvent; the latter introduces the possibility of solute-solvent hydrogen bonding.

Reasonable care must be taken in handling salt cells and plates. Moisture-free samples should be used. Fingers should not come in contact with the optical surfaces. Care should be taken to prevent contamination with silicones, which are hard to remove and have strong absorption patterns.

3.5 *Interpretation of Spectra*

There are no rigid rules for interpreting an IR spectrum. Certain requirements, however, must be met before an attempt is made to interpret a spectrum.

1. The spectrum must be adequately resolved and of adequate intensity.
2. The spectrum should be that of a reasonably pure compound.
3. The spectrophotometer should be calibrated so that the bands are observed at their proper frequencies or wavelengths. Proper calibration can be made with reliable standards, such as poly(styrene) film.
4. The method of sample handling must be specified. If a solvent is employed, the solvent, concentration, and the cell thickness should be indicated.

A precise treatment of the vibrations of a complex molecule is not feasible; thus, the IR spectrum must be interpreted from empirical comparison of spectra and extrapolation of studies of simpler molecules. Many questions arising in the interpretation of an IR spectrum can be answered by data obtained from the mass and NMR spectra.

Infrared absorption of organic molecules is summarized in the chart of characteristic group absorptions in Appendix C. Many of the group absorptions vary over a wide range because the bands arise from complex interacting vibrations within the molecule. Absorption bands may, however, represent predominantly a single vibrational mode. Certain absorption bands, for example, those arising from the C—H, O—H, and C=O stretching modes, remain within fairly narrow regions of the spectrum. Important details of structure may be revealed by the exact position of an absorption band within these narrow regions. Shifts in absorption position and changes in band contours, accompanying changes in molecular environment, may also suggest important structural details.

The two important areas for a preliminary examination of a spectrum are the regions 4000–1300 and 900–650 cm^{-1}. The high-frequency portion of the spectrum is called the functional group region. The characteristic stretching frequencies for important functional groups such as OH, NH, and C=O occur in this portion of the spectrum. The absence of absorption in the assigned ranges for the various functional groups can usually be used as evidence for the absence of such groups in the molecule. Care must be exercised, however, in such interpretations since certain structural characteristics may cause a band to become extremely broad so that it may go unnoticed. For example, intramolecular hydrogen bonding in the enolic form of acetylacetone results in a broad OH band, which may be overlooked. The absence of absorption in the 1850–1540 cm^{-1} region excludes a structure containing a carbonyl group.

Weak bands in the high-frequency region, resulting from the fundamental absorption of functional groups, such as S—H and C≡C, are extremely valuable in the determination of structure. Such weak bands would be of little value in the more complicated regions of the spectrum. Overtones and combination tones of lower frequency bands frequently appear in the high-frequency region of the spectrum. Overtone and combination-tone bands are characteristically weak except when Fermi resonance occurs. Strong skeletal bands for aromatics and heteroaromatics fall in the 1600–1300 cm^{-1} region of the spectrum.

The lack of strong absorption bands in the 900–650 cm^{-1} region generally indicates a nonaromatic structure. Aromatic and heteroaromatic compounds display strong out-of-plane C—H bending and ring bending absorption bands in this region that can frequently be correlated with the substitution pattern. Broad, moderately intense absorption in the low-frequency region suggests the presence of carboxylic acid dimers, amines, or amides, all of which show out-of-plane bending in this region. If the region is extended to 1000 cm^{-1}, absorption bands characteristic of alkene structures are included.

The intermediate portion of the spectrum, 1300–900 cm^{-1}, is usually referred to as the "fingerprint" region. The absorption pattern in this region is frequently complex, with the bands originating in interacting vibrational modes. This portion of the spectrum is extremely valuable when examined in reference to the other regions. For example, if alcoholic or phenolic O—H stretching absorption appears in the high-frequency region of the spectrum, the position of the C—C—O absorption band in the 1260–100 cm^{-1} region frequently makes it possible to assign the O—H absorption to alcohols and phenols with highly specific structures. Absorption in this intermediate region is probably unique for every molecular species.

Any conclusions reached after examination of a particular band should be confirmed where possible by examination of other portions of the spectrum. For example, the assignment of a carbonyl band to an aldehyde should be confirmed by the appearance of a band or a pair of bands in the 2900–2695 cm^{-1} region of the spectrum, arising from C—H stretching vibrations of the aldehyde group. Similarly, the assignment of a carbonyl band to an ester should be confirmed by observation of a strong band in the C—O stretching region, 1300–1100 cm^{-1}.

Similar compounds may give virtually identical spectra under normal conditions, but fingerprint differences can be detected with an expanded vertical scale or with a very large sample (major bands off scale). For example, pentane and hexane are essentially indistinguishable under normal conditions and can be differentiated only at very high recorder sensitivity.

Finally, in a "fingerprint" comparison of spectra, or any other situation in which the *shapes* of peaks are important, we should be aware of the substantial differences in the appearance of the spectrum in changing from a spectrum that is linear in wavenumber to one that is linear in wavelength (Fig. 3.6).

Admittedly, the full chart of characteristic absorption groups (Appendix C) is intimidating. The following statements and simplified chart may help (Fig. 3.7).

The first bit of advice is negative: Do not attempt a frontal, systematic assault on an infrared spectrum. Rather, look for evidence of the presence or absence of a few common functional groups with very characteristic absorptions. Start with OH, C=O, and NH groups in Figure 3.7 since a "yes/no" answer is usually available. A "yes" answer for any of these groups sharpens

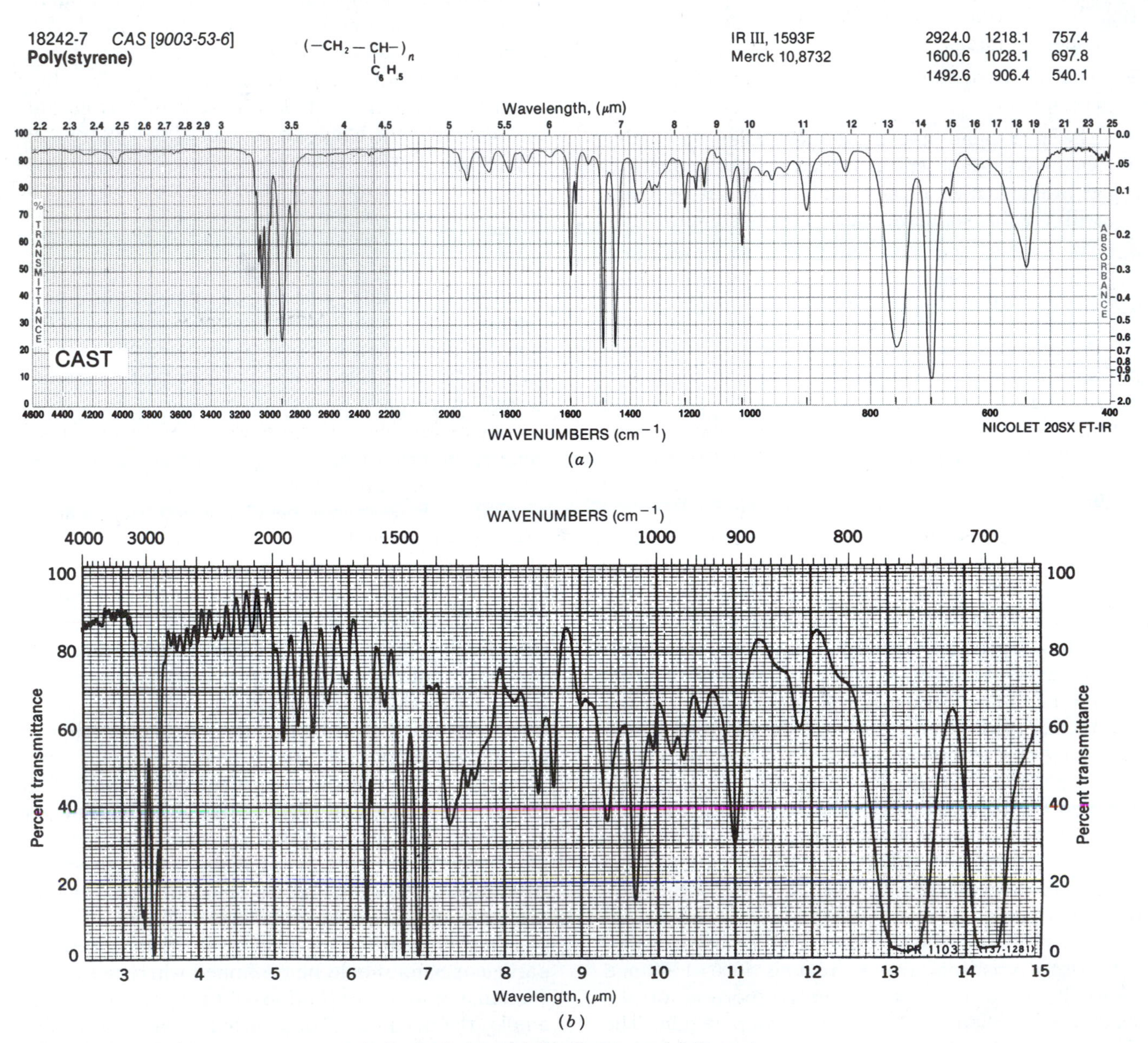

FIGURE 3.6. Polystyrene, same sample for both (*a*) and (*b*). Spectrum (*a*) linear in wavenumber (cm^{-1}); spectrum (*b*) linear in wavelength (μm).

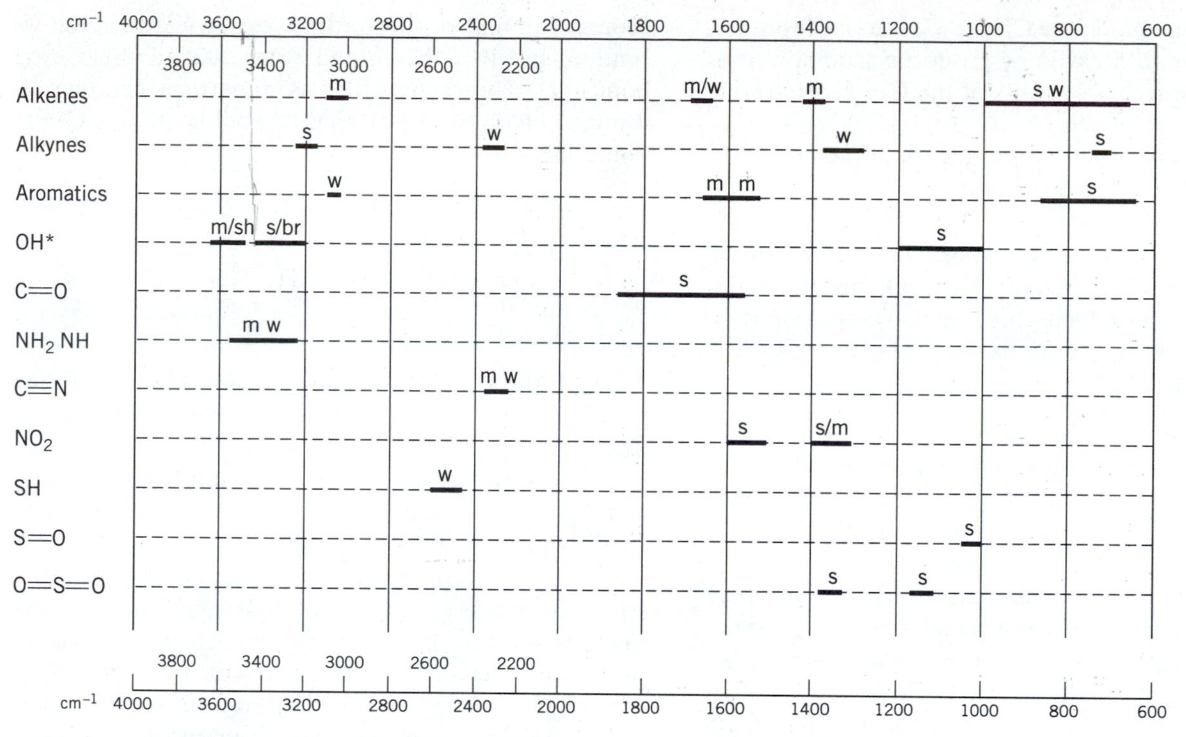

FIGURE 3.7. Simplified chart of several common functional groups with very characteristic absorptions. s = strong, m = medium, w = weak, sh = sharp, br = broad.

the focus considerably. Certainly the answer will contribute to development of a molecular formula from the mass spectrum and to an entry point for the NMR spectra. These other spectra, in turn, will suggest further leads in the IR spectrum.

Figure 3.7 lists the common groups that provide distinctive, characteristic absorptions. Section 3.6 furnishes more detailed information—including a number of caveats.

3.6 Characteristic Group Absorptions of Organic Molecules

A table of characteristic group absorptions is presented as Appendix C. The ranges presented for group absorptions have been assigned following the examination of many compounds in which the groups occur. Although the ranges are quite well defined, the precise frequency or wavelength at which a specific group absorbs is dependent on its environment within the molecule and on its physical state.

This section is concerned with a comprehensive look at these characteristic group absorptions and their relationship to molecular structure. As a major type or class of molecule or functional group is introduced in the succeeding sections, an example of an IR spectrum with the important peak assignments will be given. Spectra of common laboratory substances, representing many of the chemical classes listed below, are shown in Appendix B. Characteristic group absorptions are found in Appendix C.

3.6.1 Normal Alkanes (Paraffins)

The spectra of normal alkanes (paraffins) can be interpreted in terms of four vibrations, namely, the stretching and bending of C—H and C—C bonds. Detailed analysis of the spectra of the lower members of the alkane series has made detailed assignments of the spectral positions of specific vibrational modes possible.

Not all of the possible absorption frequencies of the paraffin molecule are of equal value in the assignment of structure. The C—C bending vibrations occur at very low frequencies (below 500 cm^{-1}) and therefore do not appear in our spectra. The bands assigned to C—C stretching vibrations are weak and appear in the broad region of 1200–800 cm^{-1}; they are generally of little value for identification.

The most characteristic vibrations are those arising from C—H stretching and bending. Of these vibrations, those arising from methylene twisting and wagging are usually of limited diagnostic value because of their weakness and instability. This instability is a result of strong coupling to the remainder of the molecule.

The vibrational modes of alkanes are common to many organic molecules. Although the positions of C—H stretching and bending frequencies of methyl and methylene groups remain nearly constant in hydrocarbons, the attachment of CH_3 or CH_2 to atoms other than carbon, or to a carbonyl group or aromatic ring, may result in aprreciable shifts of the C—H stretching and bending frequencies.

The spectrum of dodecane, Figure 3.8, is that of a typical straight-chain hydrocarbon.

3.6.1.1 C—H Stretching Vibrations Absorption arising from C—H stretching in the alkanes occurs in the general region of 3000–2840 cm^{-1}. The positions of the C—H stretching vibrations are among the most stable in the spectrum.

Methyl Groups An examination of a large number of saturated hydrocarbons containing methyl groups showed, in all cases, two distinct bands occurring at 2962 and 2872 cm^{-1}. The first of these results from the asymmetrical (as) stretching mode in which two C—H bonds of the methyl group are extending while the third one is contracting ($\nu_{as}CH_3$). The second arises from symmetrical (s) stretching (ν_sCH_3) in which all three of the C—H bonds extend and contract in phase. The presence of several methyl groups in a molecule results in strong absorption at these positions.

Methylene Groups The asymmetrical stretching ($\nu_{as}CH_2$) and symmetrical stretching (ν_sCH_2) occur, respectively, near 2926 and 2853 cm^{-1}. The positions of these bands do not vary more than ± 10 cm^{-1} in the aliphatic and nonstrained cyclic hydrocarbons. The frequency of methylene stretching is increased when the methylene group is part of a strained ring.

3.6.1.2 C—H Bending Vibrations ***Methyl Groups*** Two bending vibrations can occur within a methyl group. The first of these, the symmetrical bending vibration, involves the in-phase bending of the C—H bonds (**I**). The second, the asymmetrical bending vibration, involves out-of-phase bending of the C—H bonds (**II**).

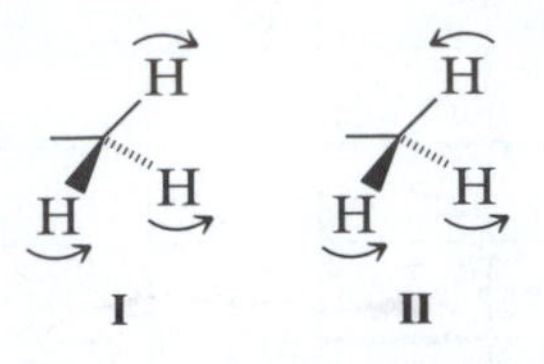

In **I**, the C—H bonds are moving like the closing petals of a flower; in **II**, one petal opens as two petals close.

The symmetrical bending vibration (δ_sCH_3) occurs near 1375 cm^{-1}, the asymmetrical bending vibration ($\delta_{as}CH_3$) near 1450 cm^{-1}.

The asymmetrical vibration generally overlaps the scissoring vibration of the methylene groups (see below). Two distinct bands are observed, however, in compounds such as diethyl ketone, in which the methylene scissoring band has been shifted to a lower frequency, 1439–1399 cm^{-1}, and increased in intensity because of its proximity to the carbonyl group.

The absorption band near 1375 cm^{-1}, arising from the symmetrical bending of the methyl C—H bonds, is very stable in position when the methyl group is attached to another carbon atom. The intensity of this band is greater for each methyl group in the compound than that for the asymmetrical methyl bending vibration or the methylene scissoring vibration.

Methylene Groups The bending vibrations of the C—H bonds in the methylene group have been shown schematically in Figure 1. The four bending vibrations

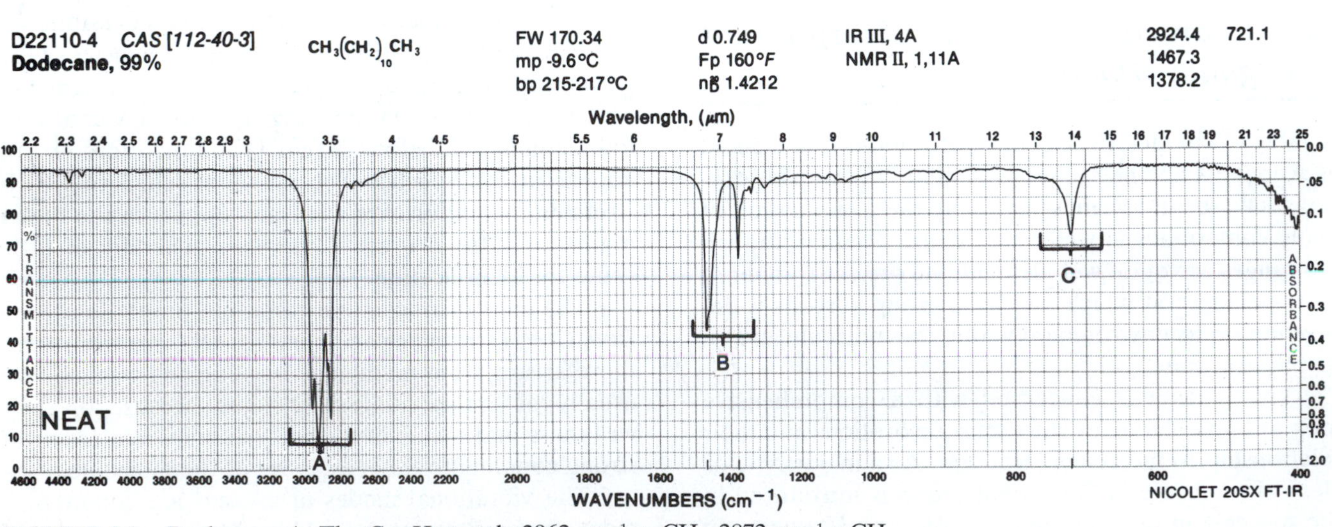

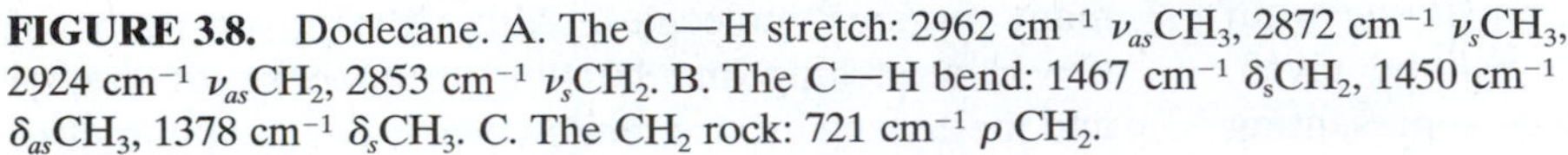

FIGURE 3.8. Dodecane. A. The C—H stretch: 2962 cm^{-1} $\nu_{as}CH_3$, 2872 cm^{-1} ν_sCH_3, 2924 cm^{-1} $\nu_{as}CH_2$, 2853 cm^{-1} ν_sCH_2. B. The C—H bend: 1467 cm^{-1} δ_sCH_2, 1450 cm^{-1} $\delta_{as}CH_3$, 1378 cm^{-1} δ_sCH_3. C. The CH_2 rock: 721 cm^{-1} ρ CH_2.

are referred to as *scissoring, rocking, wagging,* and *twisting.*

The scissoring band ($\delta_s CH_2$) in the spectra of hydrocarbons occurs at a nearly constant position near 1465 cm^{-1}.

The band resulting from the methylene rocking vibration (ρ CH_2), in which all of the methylene groups rock in phase, appears near 720 cm^{-1} for straight-chain alkanes of seven or more carbon atoms. This band may appear as a doublet in the spectra of solid samples. In the lower members of the *n*-alkane series, the band appears at somewhat higher frequencies.

Absorption of hydrocarbons, because of methylene twisting and wagging vibrations, is observed in the 1350–1150 cm^{-1} region. These bands are generally appreciably weaker than those resulting from methylene scissoring. A series of bands in this region, arising from the methylene group, is characteristic of the spectra of solid samples of long-chain acids, amides, and esters.

3.6.2 *Branched-Chain Alkanes*

In general, the changes brought about in the spectrum of a hydrocarbon by branching result from changes in skeletal stretching vibrations and methyl bending vibrations; these occur below 1500 cm^{-1}. The spectrum of Figure 3.9 is that of a typical branched alkane.

3.6.2.1 C—H Stretching Vibrations ***Tertiary*** **C—H** ***Groups*** Absorption resulting from this vibrational mode is very weak and is usually lost in other aliphatic C—H absorption. Absorption in hydrocarbons occurs near 2890 cm^{-1}.

3.6.2.2 C—H Bending Vibrations gem-*Dimethyl Groups* Configurations in which two methyl groups are attached to the same carbon atom exhibit distinctive absorption in the C—H bending region. The isopropyl group shows a strong doublet, with peaks of almost equal intensity at 1385–1380 and 1370–1365 cm^{-1}. The tertiary butyl group gives rise to two C—H bending bands, one in the 1395–1385 cm^{-1} region and one near 1370 cm^{-1}. In the *t*-butyl doublet, the long wavelength band is more intense. When the *gem*-dimethyl group occurs at an internal position, a doublet is observed in essentially the same region where absorption occurs for the isopropyl and *t*-butyl groups. Doublets are observed for *gem*-dimethyl groups because of interaction between the in-phase and out-of-phase CH_3 bending of the two methyl groups attached to a common carbon atom.

Weak bands result from methyl rocking vibrations in isopropyl and *t*-butyl groups. These vibrations are sensitive to mass and interaction with skeletal stretching modes and are generally less reliable than the C—H bending vibrations. The following assignments have been made: isopropyl group, 922–919 cm^{-1}, and *t*-butyl group, 932–926 cm^{-1}.

3.6.3 *Cyclic Alkanes*

3.6.3.1 C—H Stretching Vibrations The methylene stretching vibrations of unstrained cyclic poly(methylene) structures are much the same as those observed for acyclic paraffins. Increasing ring strain moves the C—H stretching bands progressively to high frequencies. The ring CH_2 and CH groups in a monoalkylcyclopropane ring absorb in the region of 3100–2990 cm^{-1}.

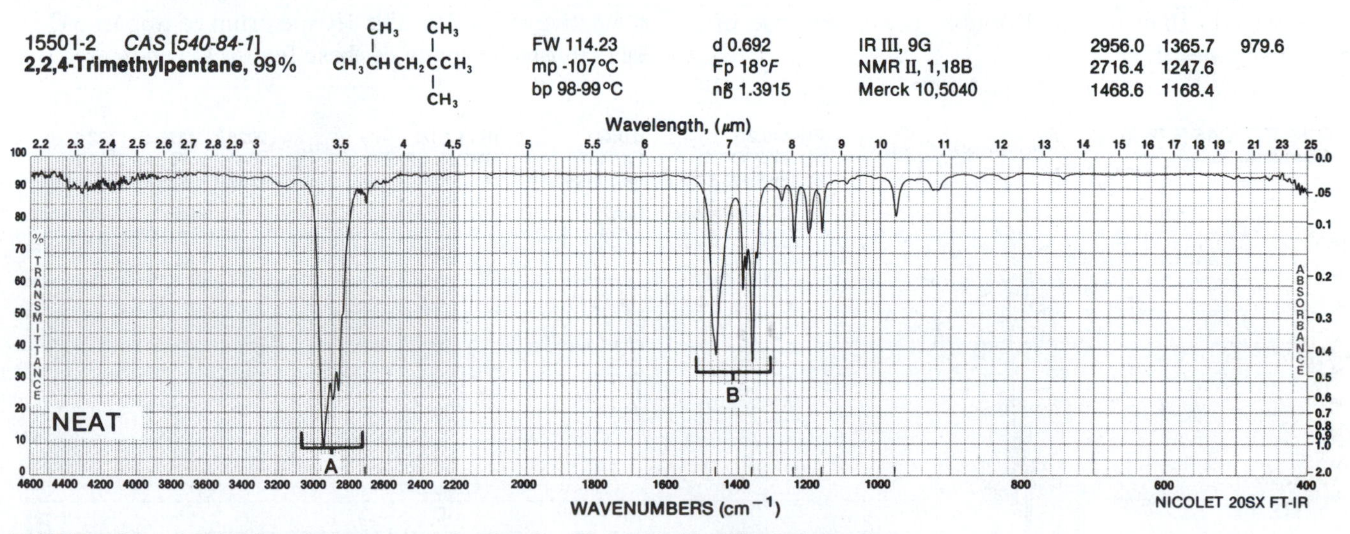

FIGURE 3.9. 2,2,4-Trimethylpentane.* A. The C—H stretch (see Fig. 3.8). B. The C—H bend (see Fig. 3.8). There are overlapping doublets for the *t*-butyl and the isopropyl groups at 1340–1400 cm^{-1}. Compare the absence of a methylene rocking band(s) 800–1000 cm^{-1} to Figure 3.8.

*Isooctane is the trivial name for 2,2,4-trimethylpentane.

3.6.3.2 C—H Bending Vibrations Cyclization decreases the frequency of the CH_2 scissoring vibration. Cyclohexane absorbs at 1452 cm^{-1}, whereas *n*-hexane absorbs at 1468 cm^{-1}. Cyclopentane absorbs at 1455 cm^{-1}, cyclopropane absorbs at 1442 cm^{-1}. This shift frequently makes it possible to observe distinct bands for methylene and methyl absorption in this region. Spectra of other saturated hydrocarbons appear in Appendix B: hexane (No. 1), Nujol® (No. 2), and cyclohexane (No. 3).

3.6.4 Alkenes

Alkene (olefinic) structures introduce several new modes of vibration into a hydrocarbon molecule: a C═C stretching vibration, C—H stretching vibrations in which the carbon atom is present in the alkene linkage, and in-plane and out-of-plane bending of the alkene C—H bond. The spectrum of Figure 3.10 is that of a typical terminal alkene.

3.6.4.1 C═C Stretching Vibrations ***Unconjugated Linear Alkenes*** The C═C stretching mode of unconjugated alkenes usually shows moderate to weak absorption at 1667–1640 cm^{-1}. Monosubstituted alkenes, that is, vinyl groups, absorb near 1640 cm^{-1}, with moderate intensity. Disubstituted *trans*-alkenes, tri- and tetraalkyl-substituted alkenes absorb at or near 1670 cm^{-1}; disubstituted *cis*-alkenes and vinylidene alkenes absorb near 1650 cm^{-1}.

The absorption of symmetrical disubstituted *trans*-alkenes or tetrasubstituted alkenes may be extremely weak or absent. The *cis*-alkenes, which lack the symmetry of the trans structure, absorb more strongly than *trans*-alkenes. Internal double bonds generally absorb more weakly than terminal double bonds because of pseudosymmetry.

Abnormally high absorption frequency is observed for —CH═CF_2 and —CF═CF_2 groups. The former absorbs near 1754 cm^{-1}, the latter near 1786 cm^{-1}. In contrast, the absorption frequency is reduced by the attachment of chlorine, bromine, or iodine.

Cycloalkenes Absorption of the internal double bond in the unstrained cyclohexene system is essentially the same as that of a cis isomer in an acyclic system. The C═C stretch vibration is coupled with the C—C stretching of the adjacent bonds. As the angle α

C
C⟷C
α

becomes smaller the interaction becomes less until it is at a minimum at 90° in cyclobutene (1566 cm^{-1}). In the cyclopropene structure, interaction again becomes appreciable, and the absorption frequency increases (1641 cm^{-1}).

The substitution of alkyl groups for a hydrogen atom in strained ring systems serves to increase the frequency of C═C absorption. Cyclobutene absorbs at 1566 cm^{-1}, 1-methylcyclobutene at 1641 cm^{-1}.

The absorption frequency of external exocyclic bonds increases with decreasing ring size. Methylenecyclohexane absorbs at 1650 cm^{-1}, methylenecyclopropane at 1781 cm^{-1}.

Conjugated Systems The alkene bond stretching vibrations in conjugated dienes without a center of symmetry interact to produce two C═C stretching bands. The spectrum of an unsymmetrical conjugated diene, such as 1,3-pentadiene, shows absorption near 1650 and 1600 cm^{-1}. The symmetrical molecule 1,3-butadiene shows only one band near 1600 cm^{-1}, resulting from asymmetric stretching; the symmetrical stretching band is inactive in the IR. The IR spectrum of isoprene (Fig. 3.11) illustrates many of these features.

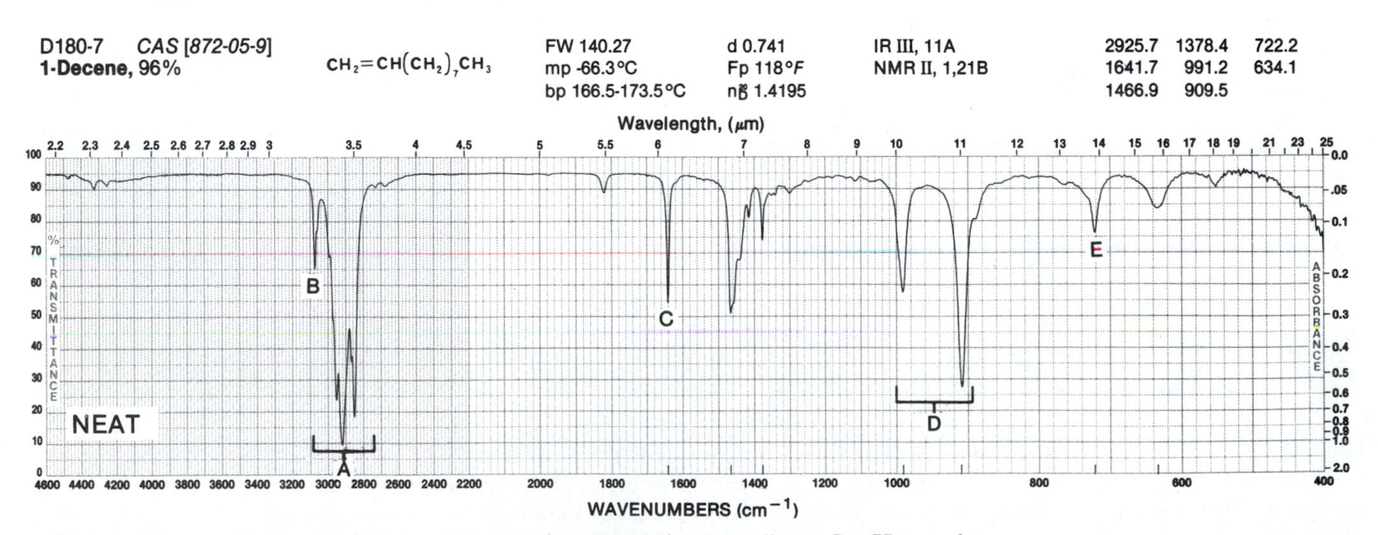

FIGURE 3.10. 1-Decene. A. The C—H stretch (see Fig. 3.8). Note alkene C—H stretch (B) at 3049 cm^{-1}. C. The C═C stretch, 1642 cm^{-1}, see Appendix Table D-1. D. Out-of-plane C—H bend: 991 cm^{-1}, (alkene) 909.5 cm^{-1}. E. Methylene rock: 722 cm^{-1}.

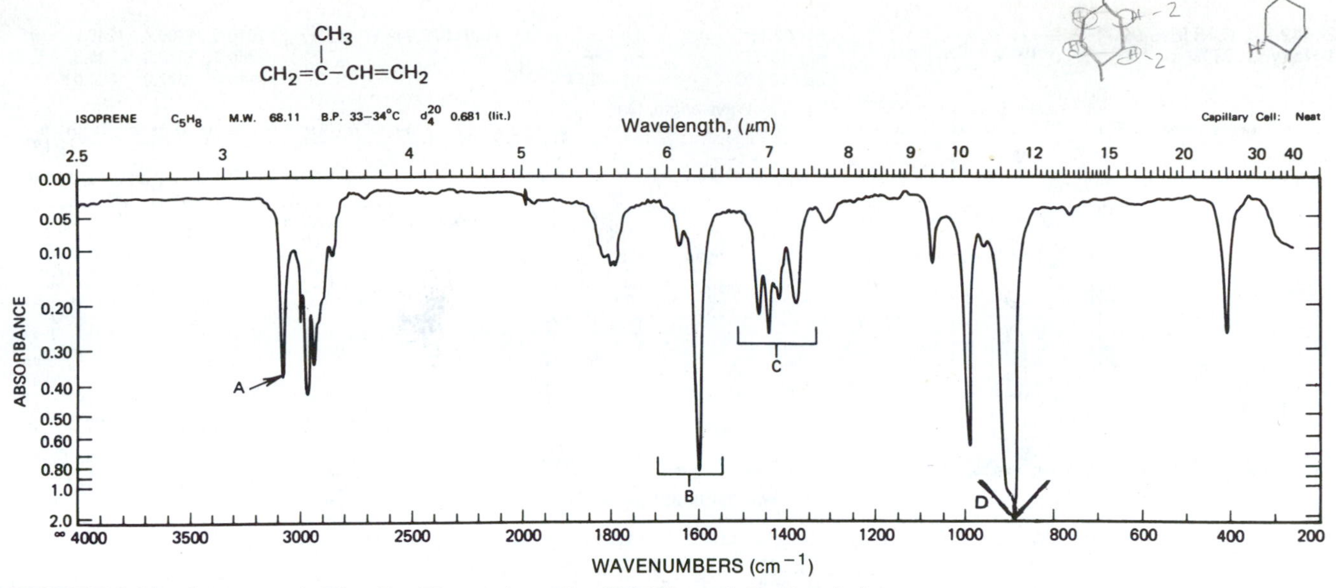

FIGURE 3.11. Isoprene. A. The C—H stretch: =C—H 3090 cm^{-1}. B. Coupled C=C—C=C stretch: symmetric 1640 cm^{-1} (weak), asymmetric 1598 cm^{-1} (strong). C. The C—H bend (saturated, alkene in-plane). D. The C—H out-of-plane bend: 990 cm^{-1}, 892 cm^{-1} (see vinyl, Appendix Table D-1.)

Conjugation of an alkene double bond with an aromatic ring produces enhanced alkene absorption near 1625 cm^{-1}.

The absorption frequency of the alkene bond in conjugation with a carbonyl group is lowered by about 30 cm^{-1}; the intensity of absorption is increased. In *s*-cis structures, the alkene absorption may be as intense as that of the carbonyl group. *s*-Trans structures absorb more weakly than *s*-cis structures.

Cumulated Alkenes A cumulated double-bond system, as occurs in the allenes ($>C=C=CH_2$), absorbs near 2000–1900 cm^{-1}. The absorption results from asymmetric C=C=C stretching. The absorption may be considered an extreme case of exocyclic C=C absorption.

3.6.4.2 Alkene C—H Stretching Vibrations In general, any C—H stretching bands above 3000 cm^{-1} result from aromatic, heteroaromatic, alkyne, or alkene C—H stretching. Also found in the same region are the C—H stretching in small rings, such as cyclopropane, and the C—H in halogenated alkyl groups. The frequency and intensity of alkene C—H stretching absorption are influenced by the pattern of substitution. With proper resolution, multiple bands are observed for structures in which stretching interaction may occur. For example, the vinyl group produces three closely spaced C—H stretching bands. Two of these result from symmetrical and asymmetrical stretching of the terminal C—H groups, and the third from the stretching of the remaining single C—H.

3.6.4.3 Alkene C—H Bending Vibrations Alkene C—H bonds can undergo bending either in the same plane as the C=C bond or perpendicular to it; the bending vibrations can be either in phase or out of phase with respect to each other.

Assignments have been made for a few of the more prominent and reliable in-plane bending vibrations. The vinyl group absorbs near 1416 cm^{-1} because of a scissoring vibration of the terminal methylene. The C—H rocking vibration of a cis-disubstituted alkene occurs in the same general region.

The most characteristic vibrational modes of alkenes are the out-of-plane C—H bending vibrations between 1000 and 650 cm^{-1}. These bands are usually the strongest in the spectra of alkenes. The most reliable bands are those of the vinyl group, the vinylidene group, and the trans-disubstituted alkene. Alkene absorption is summarized in Appendix Tables D-1 and D-2.

In allene structures, strong absorption is observed near 850 cm^{-1}, arising from $=CH_2$ wagging. The first overtone of this band may also be seen. Some spectra showing alkene features are shown in Appendix B: trichloroethylene (No. 12) and tetrachloroethylene (No. 13).

3.6.5 Alkynes

The two stretching vibrations in alkynes (acetylenes) involve C≡C and C—H stretching. Absorption due to C—H bending is characteristic of acetylene and monosubstituted alkynes. The spectrum of Figure 3.12 is that of a typical terminal alkyne.

3.6.5.1 C≡C Stretching Vibrations The weak C≡C stretching band of alkyne molecules occurs in the region

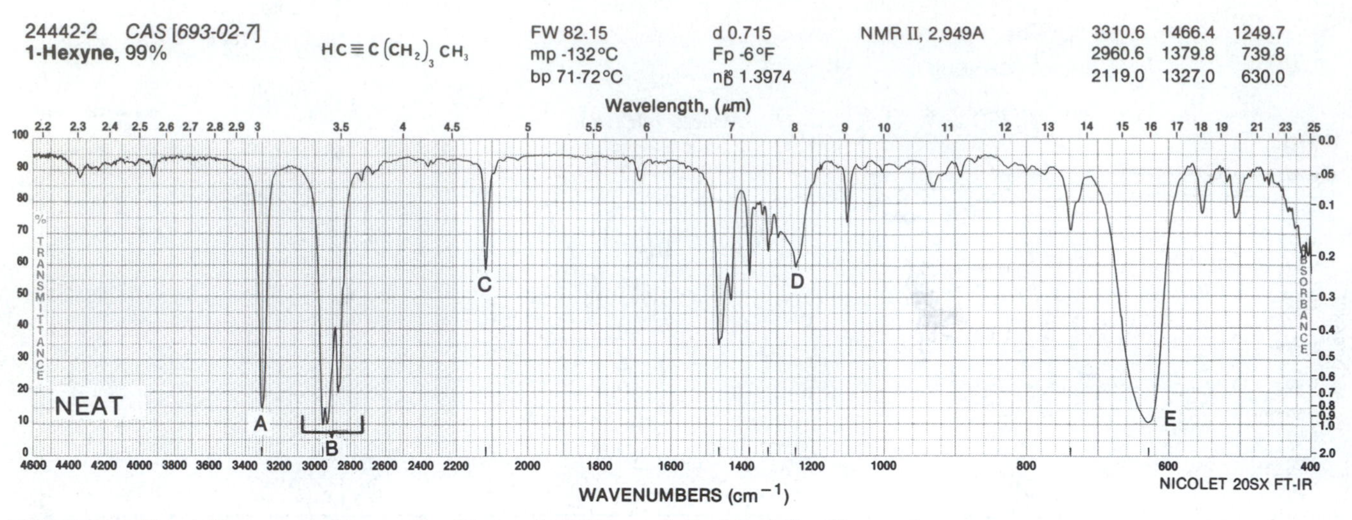

FIGURE 3.12. 1-Hexyne. A. The ≡C—H stretch, 3310 cm^{-1}. B. Alkyl C—H stretch (see Fig. 3.8), 2857–2941 cm^{-1}. C. The C≡C stretch, 2119 cm^{-1}. D. The ≡C—H bend overtone, 1250 cm^{-1}. E. The ≡C—H bend fundamental, 630 cm^{-1}.

of 2260–2100 cm^{-1}. Because of symmetry, no C≡C band is observed in the IR for symmetrically substituted alkynes. In the IR spectra of monosubstituted alkynes, the band appears at 2140–2100 cm^{-1}. Disubstituted alkynes, in which the substituents are different, absorb near 2260–2190 cm^{-1}. When the substituents are similar in mass, or produce similar inductive and resonance effects, the band may be so weak as to be unobserved in the IR spectrum. For reasons of symmetry, a terminal C≡C produces a stronger band than an internal C≡C (pseudosymmetry). The intensity of the C≡C stretching band is increased by conjugation with a carbonyl group.

3.6.5.2 C—H Stretching Vibrations The C—H stretching band of monosubstituted alkynes occurs in the general region of 3333–3267 cm^{-1}. This is a strong band and is narrower than the hydrogen-bonded OH and NH bands occurring in the same region.

3.6.5.3 C—H Bending Vibrations The C—H bending vibration of alkynes or monosubstituted alkynes leads to strong, broad absorption in the 700–610 cm^{-1} region. The first overtone of the C—H bending vibration appears as a weak, broad band in the 1370–1220 cm^{-1} region.

3.6.6 Mononuclear Aromatic Hydrocarbons

The most prominent and most informative bands in the spectra of aromatic compounds occur in the low-frequency range between 900 and 675 cm^{-1}. These strong absorption bands result from the out-of-plane ("oop") bending of the ring C—H bonds. In-plane bending bands appear in the 1300–1000 cm^{-1} region. Skeletal vibrations, involving carbon–carbon stretching within the ring, absorb in the 1600–1585 and 1500–1400 cm^{-1} regions. The skeletal bands frequently appear as doublets, depending on the nature of the ring substituents.

Aromatic C—H stretching bands occur between 3100 and 3000 cm^{-1}.

Weak combination and overtone bands appear in the 2000–1650 cm^{-1} region. The pattern of the overtone bands is not a reliable guide to the substitution pattern of the ring. Because they are weak, the overtone and combination bands are most readily observed in spectra obtained from thick samples. The spectrum of Figure 3.13 is that of a typical aromatic (benzenoid) compound.

3.6.6.1 Out-of-Plane C—H Bending Vibrations The in-phase, out-of-plane bending of a ring hydrogen atom is strongly coupled to adjacent hydrogen atoms. The position of absorption of the out-of-plane bending bands is therefore characteristic of the number of adjacent hydrogen atoms on the ring. The bands are frequently intense and appear at 900–675 cm^{-1}.

Assignments for C—H out-of-plane bending bands in the spectra of substituted benzenes appear in the chart of characteristic group absorptions (Appendix C). These assignments are usually reliable for alkyl-substituted benzenes, but caution must be observed in the interpretation of spectra when polar groups are attached directly to the ring, for example, in nitrobenzenes, aromatic acids, and esters or amides of aromatic acids.

The absorption band that frequently appears in the spectra of substituted benzenes near 710–675 cm^{-1} is attributed to out-of-plane ring bending. Some spectra showing typical aromatic absorption appear in Appendix B: benzene (No. 4), indene (No. 8), diethylphthalate (No. 21), and *m*-xylene (No. 6).

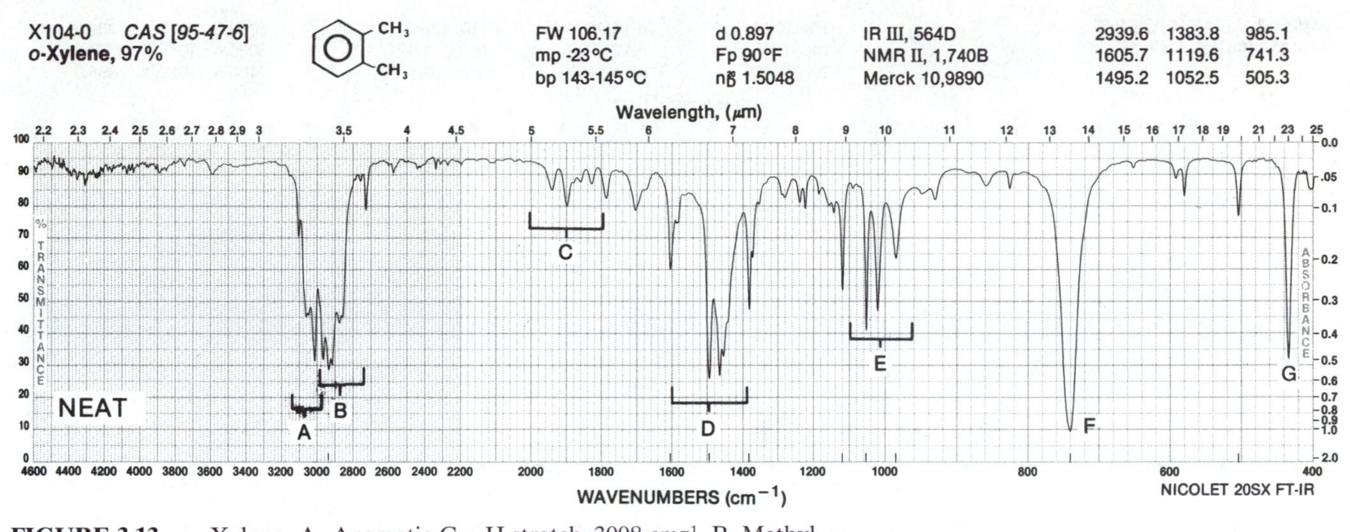

FIGURE 3.13. *o*-Xylene. A. Aromatic C—H stretch, 3008 cm^{-1}. B. Methyl C—H stretch, 2965, 2940, 2918, 2875 cm^{-1} (see Fig. 3.8). C. Overtone or combination bands, 2000–1667 cm^{-1} (see Fig. 3.14). D. The C$\overset{...}{=}$C ring stretch, 1605, 1495, 1466 cm^{-1}. E. In-plane C—H bend, 1052, 1022 cm^{-1}. F. Out-of-plane $\overset{...}{=}$C—H bend, 741 cm^{-1}. G. Out-of-plane ring C$\overset{...}{=}$C bend, 438 cm^{-1}.

3.6.7 Polynuclear Aromatic Hydrocarbons

Polynuclear aromatic compounds, like the mononuclear aromatics, show characteristic absorption in three regions of the spectrum.

The aromatic C—H stretching and the skeletal vibrations absorb in the same regions as observed for the mononuclear aromatics. The most characteristic absorption of polynuclear aromatics results from C—H out-of-plane bending in the 900–675-cm^{-1} region. These bands can be correlated with the number of adjacent hydrogen atoms on the rings. Most β-substituted naphthalenes, for example, show three absorption bands resulting from out-of-plane C—H bending; these correspond to an isolated hydrogen atom and two adjacent hydrogen atoms on one ring and four adjacent hydrogen atoms on the other ring.

In the spectra of α-substituted naphthalenes the bands for the isolated hydrogen and the two adjacent hydrogen atoms of β-naphthalenes are replaced by a band for three adjacent hydrogen atoms. This band is near 810–785 cm^{-1}.

C—H Out-of-Plane Bending Vibrations of a β-Substituted Naphthalene

Substitution Pattern	Absorption Range (cm^{-1})
Isolated hydrogen	862–835
Two adjacent hydrogen atoms	835–805
Four adjacent hydrogen atoms	760–735

Additional bands may appear because of ring bending vibrations. The position of absorption bands for more highly substituted naphthalenes and other polynuclear aromatics are summarized by Colthup et al. (1990) and by Conley (1972).

3.6.8 Alcohols and Phenols

The characteristic bands observed in the spectra of alcohols and phenols result from O—H stretching and C—O stretching. These vibrations are sensitive to hydrogen bonding. The C—O stretching and O—H bending modes are not independent vibrational modes because they couple with the vibrations of adjacent groups. Some typical spectra of alcohols and a phenol are shown in Figures 3.14–3.16.

3.6.8.1 O—H Stretching Vibrations The non–hydrogen-bonded or "free" hydroxyl group of alcohols and phenols absorbs strongly in the 3650 – 3584 cm^{-1} region. These sharp, "free" hydroxyl bands are observed in the vapor phase, in very dilute solution in nonpolar solvents or for hindered OH groups. Intermolecular hydrogen bonding increases as the concentration of the solution increases, and additional bands start to appear at lower frequencies, 3550–3200 cm^{-1}, at the expense of the "free" hydroxyl band. This effect is illustrated in Figure 3.17, in which the absorption bands in the O—H stretching region are shown for two different concentrations of cyclohexylcarbinol in carbon tetrachloride. For comparisons of this type, the path length of the cell must be altered with changing concentration, so that the same number of absorbing molecules will be present in the IR beam at each concentration. The band at 3623 cm^{-1} results from the monomer, whereas the

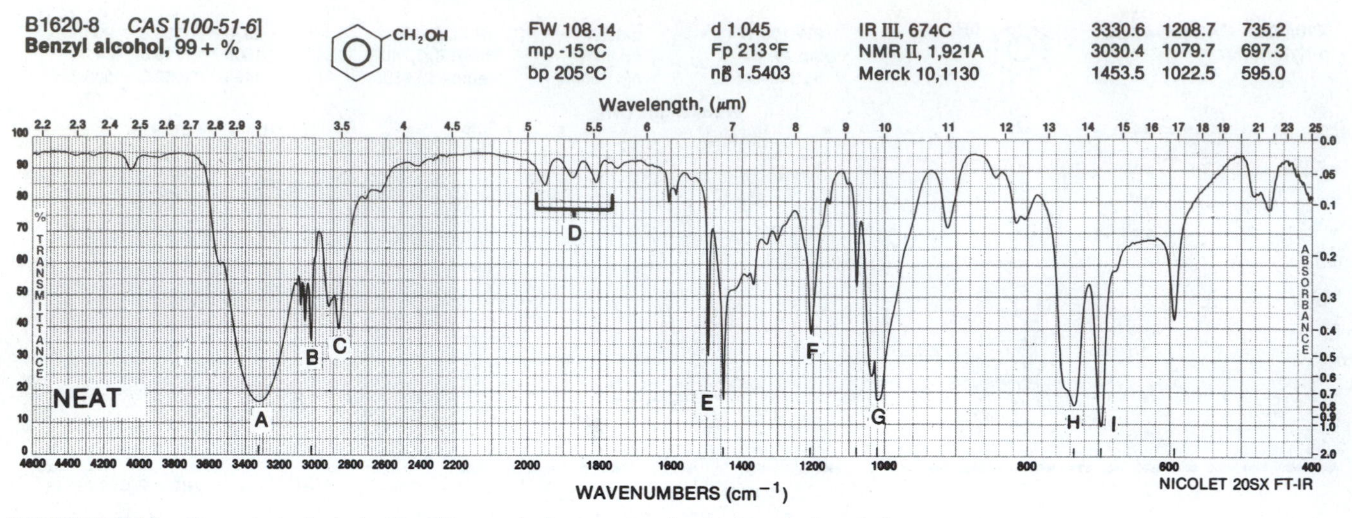

FIGURE 3.14. Benzyl alcohol. A. O—H stretch: intermolecular hydrogen bonded, 3331 cm^{-1}. B. The C—H stretch: aromatic 3100–3000 cm^{-1}. C. The C—H stretch: methylene, 2980–2840 cm^{-1}. D. Overtone or combination bands, 2000–1667 cm^{-1}. E. The C┄C ring stretch, 1497, 1454 cm^{-1}, overlapped by CH_2 scissoring, about 1471 cm^{-1}. F. The O—H bend, possibly augmented by C—H in-plane bend, 1209 cm^{-1}. G. The C—O stretch, primary alcohol (see Table 3.2) 1023 cm^{-1}. H. Out-of-plane aromatic C—H bend, 735 cm^{-1}. I. Ring C┄C bend, 697 cm^{-1}.

broad absorption near 3333 cm^{-1} arises from "polymeric" structures.

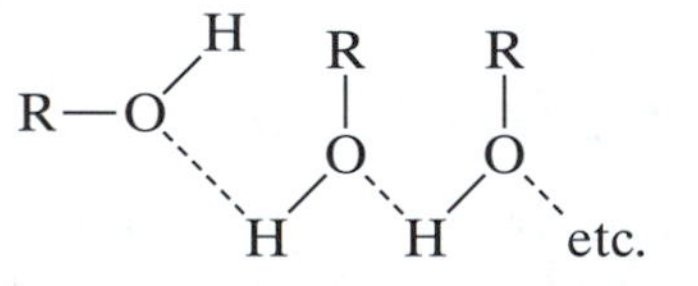

Strong intramolecular hydrogen bonding occurs in *o*-hydroxyacetophenone. The resulting absorption at 3077 cm^{-1} is broad, shallow, and independent of concentration (Fig. 3.18).

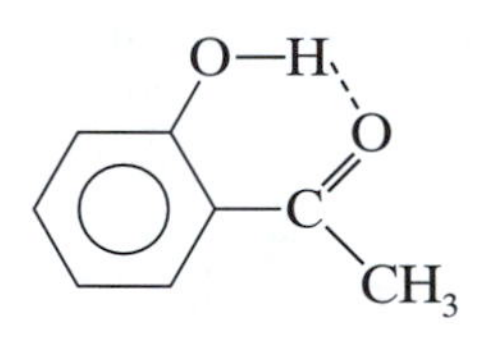

In contrast, *p*-hydroxyacetophenone

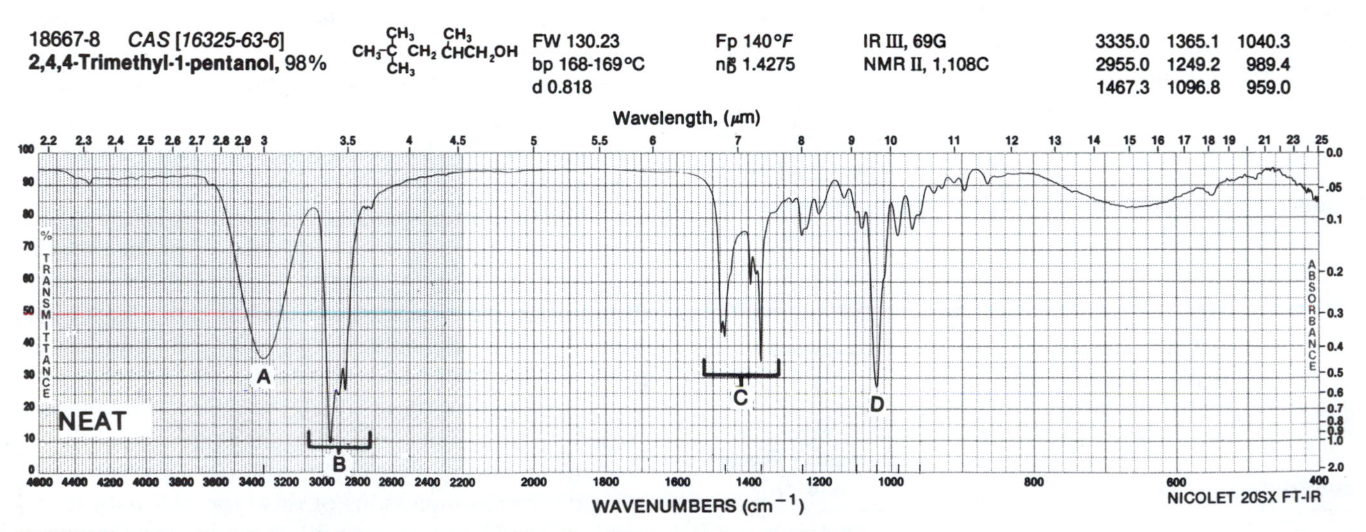

FIGURE 3.15. 24,4-Trimethyl-1-pentanol. A. The O—H stretch, intermolecular hydrogen bonding 3335 cm^{-1}. B. The C—H stretch (see Fig. 3.8, 3000–2800 cm^{-1}). C. The C—H bend (see Fig. 3.8). Note the pair of bands for the *gem*-dimethyl group at 1395 and 1365 cm^{-1}. D. The C—O stretch 1040 cm^{-1}.

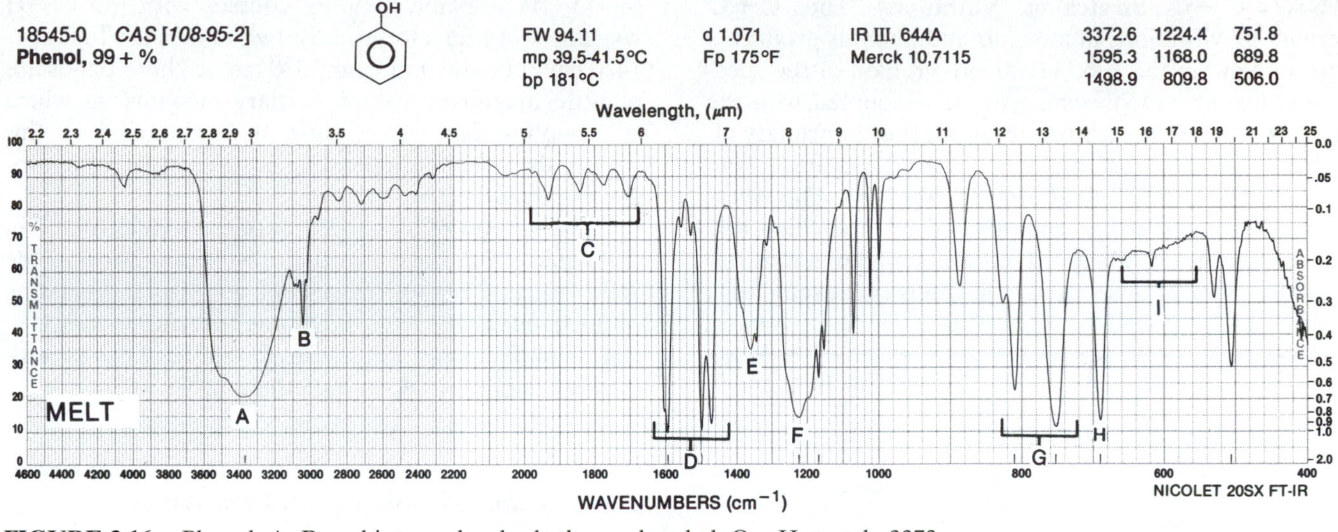

FIGURE 3.16. Phenol. A. Broad intermolecular hydrogen bonded, O—H stretch, 3373 cm^{-1}. B. Aromatic C—H stretch, 3045 cm^{-1}. C. Overtone or combination bands, 2000–1667 cm^{-1}. D. The C⩵C ring stretch, 1595, 1499, 1470 cm^{-1}. E. In-plane O—H bend, 1360 cm^{-1}. F. The C—O stretch, 1224 cm^{-1}. G. Out-of-plane C—H bend, 810, 752 cm^{-1}. H. Out-of-plane ring C⩵C bend, 690 cm^{-1}. I. (Broad) hydrogen-bonded, out-of-plane O—H bend, about 650 cm^{-1}.

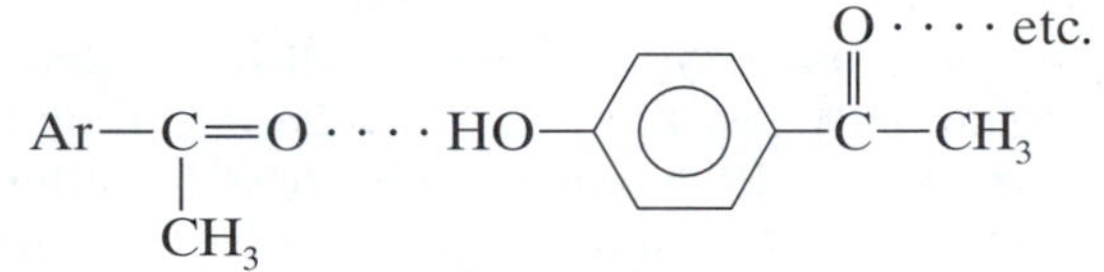

shows a sharp "free" hydroxyl peak at 3600 cm^{-1} in dilute CCl_4 solution as well as a broad, strong intermolecular peak at 3100 cm^{-1} in the spectrum of a neat sample.

In structures such as 2,6-di-*t*-butylphenol, in which steric hindrance prevents hydrogen bonding, no bonded hydroxyl band is observed, not even in spectra of neat samples.

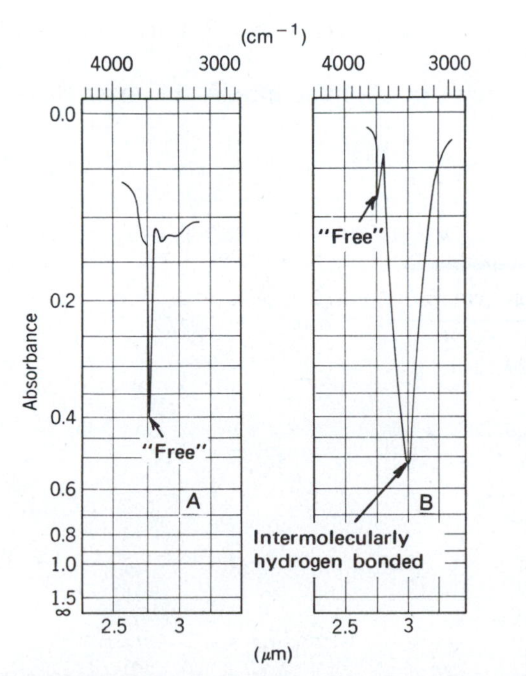

FIGURE 3.17. Infrared spectrum of the O—H stretching region of cyclohexylcarbinol in CCl_4. Peak A at 0.03 *M* (0.406-mm cell); Peak B at 1.00 *M* (0.014-mm cell).

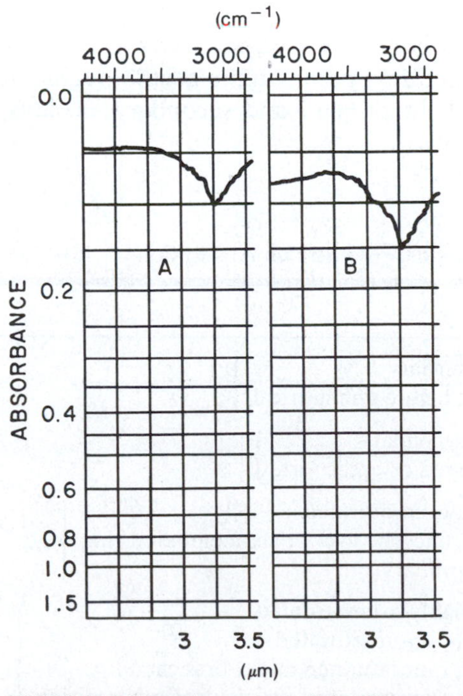

FIGURE 3.18. A portion of the IR spectra of *o*-hydroxyacetophenone. Peak A at 0.03 *M*, cell thickness: 0.41 mm. Peak B at 1.0 *M*, cell thickness: 0.015 mm.

3.6.8.2 C—O Stretching Vibrations The C—O stretching vibrations in alcohols and phenols produce a strong band in the 1260–1000 cm^{-1} region of the spectrum. The C—O stretching mode is coupled with the adjacent C—C stretching vibration; thus in primary alcohols the vibration might better be described as an asymmetric C—C—O stretching vibration. The vibrational mode is further complicated by branching and α,β-unsaturation. These effects are summarized as follows for a series of secondary alcohols (neat samples):

Secondary Alcohol	Absorption (cm^{-1})
2-Butanol	1105
3-Methyl-2-butanol	1091
1-Phenylethanol	1073
3-Buten-2-ol	1058
Diphenylmethanol	1014

The absorption ranges of the various types of alcohols appear in Table 3.2, below. These values are for neat samples of the alcohols.

Mulls, pellets, or melts of phenols absorb at 1390–1330 and 1260–1180 cm^{-1}. These bands apparently result from interaction between O—H bending and C—O stretching. The long-wavelength band is the stronger and both bands appear at longer wavelengths in spectra observed in solution. The spectrum of Figure 3.16 was determined on a melt, to show a high degree of association.

3.6.8.3 O—H Bending Vibrations The O—H in-plane bending vibration occurs in the general region of 1420–1330 cm^{-1}. In primary and secondary alcohols, the O—H in-plane bending couples with the C—H wagging vibrations to produce two bands; the first near 1420 cm^{-1}, the second near 1330 cm^{-1}. These bands are of little diagnostic value. Tertiary alcohols, in which no coupling can occur, show a single band in this region, the position depending on the degree of hydrogen bonding.

The spectra of alcohols and phenols determined in the liquid state, show a broad absorption band in the 769–650 cm^{-1} region because of out-of-plane bending of the bonded O—H group. Some spectra showing typical alcoholic absorptions are shown in Appendix B: ethyl alcohol (No. 16) and methanol (No. 15).

3.6.9 Ethers, Epoxides, and Peroxides

3.6.9.1 C—O Stretching Vibrations The characteristic response of ethers in the IR is associated with the stretching vibration of the C—O—C system. Since the vibrational characteristics of this system would not be expected to differ greatly from the C—C—C system, it is not surprising to find the response to C—O—C stretching in the same general region. However, since vibrations involving oxygen atoms result in greater dipole moment changes than those involving carbon atoms, more intense IR bands are observed for ethers. The C—O—C stretching bands of ethers, as is the case with the C—O stretching bands of alcohols, involve coupling with other vibrations within the molecule. The spectrum of Figure 3.19 is that of a typical aryl alkyl ether. In addition, the spectra of ethyl ether (No. 22) and *p*-dioxane (a cyclic diether, No. 23) are shown in Appendix B.

In the spectra of aliphatic ethers, the most char-

Table 3.2 Alcoholic C—O Stretch Absorptions

Alcohol Type	Absorption Range (cm^{-1})
(1) Saturated tertiary (2) Secondary, highly symmetrical	1205–1124
(1) Saturated secondary (2) α-Unsaturated or cyclic tert.	1124–1087
(1) Secondary, α-unsaturated (2) Secondary, alicyclic five- or six-membered ring (3) Saturated primary	1085–1050
(1) Tertiary, highly α-unsaturated (2) Secondary, di-α-unsaturated (3) Secondary, α-unsaturated and α-branched (4) Secondary, alicyclic seven- or eight-membered ring (5) Primary, α-unsaturated and/or α-branched	<1050

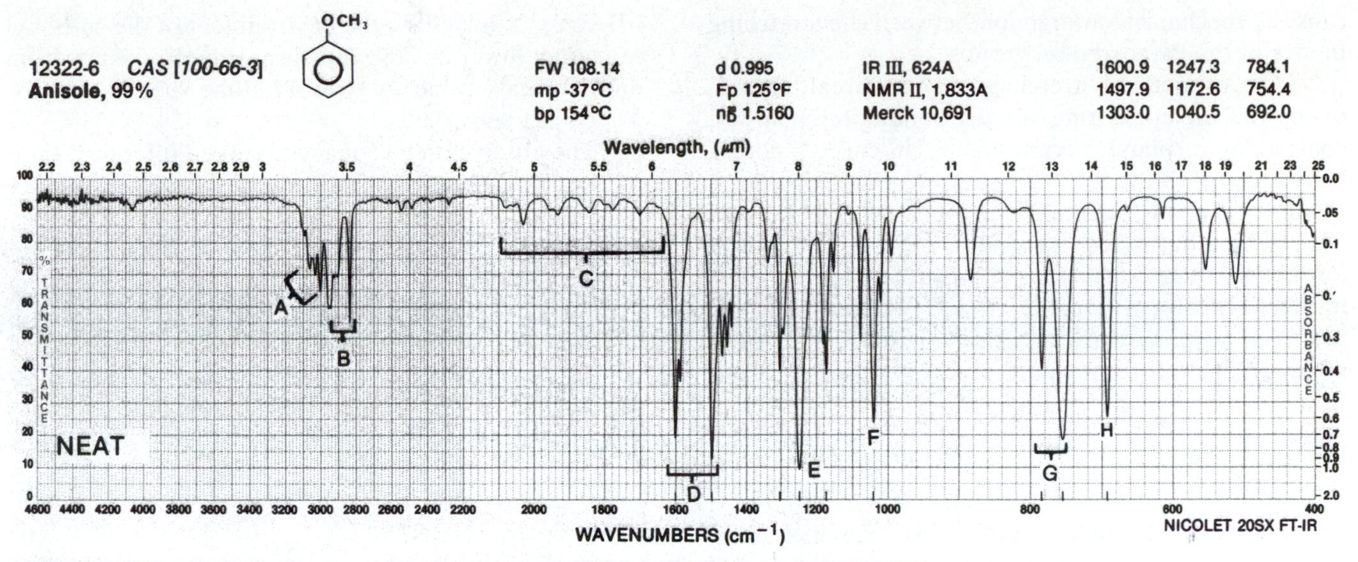

FIGURE 3.19. Anisole. A. Aromatic C—H stretch, 3060, 3030, 3000 cm^{-1}. B. Methyl C—H stretch, 2950, 2835 cm^{-1}. C. Overtone-combination region, 2000–1650 cm^{-1}. D. The C═ ring stretch, 1600, 1498 cm^{-1}. E. Asymmetric C—O—C stretch, 1247 cm^{-1}. F. Symmetric C—O—C stretch, 1040 cm^{-1}. G. Out-of-plane C—H bend, 784, 754 cm^{-1}. H. Out-of-plane ring C═C bend, 692 cm^{-1}.

acteristic absorption is a strong band in the 1150–1085 cm^{-1} region because of asymmetrical C—O—C stretching; this band usually occurs near 1125 cm^{-1}. The symmetrical stretching band is usually weak and is more readily observed in the Raman spectrum.

The C—O—C group in a six-membered ring absorbs at the same frequency as in an acyclic ether. As the ring becomes smaller, the asymmetrical C—O—C stretching vibration moves progressively to lower wavenumbers (longer wavelengths), whereas the symmetrical C—O—C stretching vibration (ring breathing frequency) moves to higher wavenumbers.

Branching on the carbon atoms adjacent to the oxygen usually leads to splitting of the C—O—C band. Isopropyl ether shows a triplet structure in the 1170–1114-cm^{-1} region, the principal band occurring at 1114 cm^{-1}.

Spectra of aryl alkyl ethers display an asymmetrical C—O—C stretching band at 1275–1200 cm^{-1} with symmetrical stretching near 1075–1020 cm^{-1}. Strong absorption caused by asymmetrical C—O—C stretching in vinyl ethers occurs in the 1225–1200 cm^{-1} region with a strong symmetrical band at 1075–1020 cm^{-1}. Resonance, which results in strengthening of the C—O bond, is responsible for the shift in the asymmetric absorption band of arylalkyl and vinyl ethers.

The C═C stretching band of vinyl ethers occurs in the 1660–1610 cm^{-1} region. This alkene band is characterized by its higher intensity compared with the C═C stretching band in alkenes. This band frequently appears as a doublet resulting from absorption of rotational isomers.

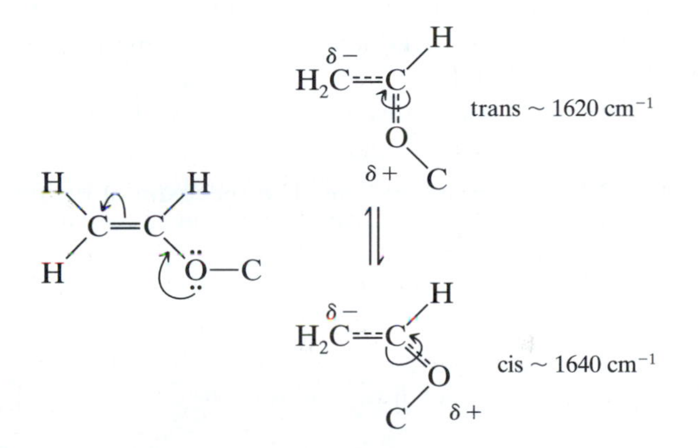

Coplanarity in the trans isomer allows maximum resonance, thus more effectively reducing the double-bond character of the alkene linkage. Steric hindrance reduces resonance in the cis isomer.

The two bands arising from ═C—H wagging in terminal alkenes occur near 1000 and 909 cm^{-1}. In the spectra of vinyl ethers, these bands are shifted to longer wavelengths because of resonance.

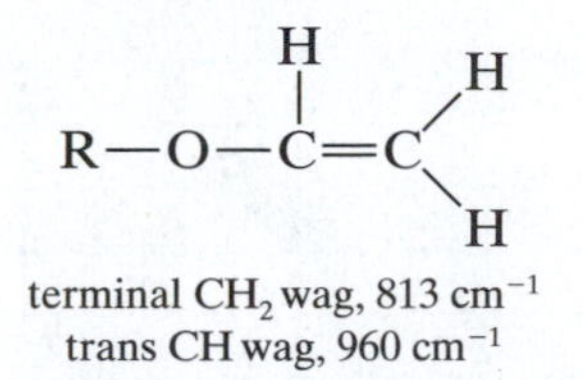

terminal CH_2 wag, 813 cm^{-1}
trans CH wag, 960 cm^{-1}

Alkyl and aryl peroxides display C—C—O absorption in the 1198–1176 cm^{-1} region. Acyl and aroyl peroxides display two carbonyl absorption bands in the 1818–1754 cm^{-1} region. Two bands are observed because of mechanical interaction between the stretching modes of the two carbonyl groups.

The symmetrical stretching, or ring breathing frequency, of the epoxy ring, all ring bonds stretching and contracting in phase, occurs near 1250 cm^{-1}. Another band appears in the 950–810 cm^{-1} region attributed to asymmetrical ring stretching in which the C—C bond is stretching during contraction of the C—O bond. A third band, referred to as the "12 micron band," appears in the 840–750 cm^{-1} region. The C—H stretching vibrations of epoxy rings occur in the 3050–2990 cm^{-1} region of the spectrum.

3.6.10 Ketones

3.6.10.1 C=O Stretching Vibrations Ketones, aldehydes, carboxylic acids, carboxylic esters, lactones, acid halides, anhydrides, amides, and lactams show a strong C=O stretching absorption band in the region of 1870–1540 cm^{-1}. Its relatively constant position, high intensity, and relative freedom from interfering bands make this one of the easiest bands to recognize in IR spectra.

Within its given range, the position of the C=O stretching band is determined by the following factors: (1) the physical state, (2) electronic and mass effects of neighboring substituents, (3) conjugation, (4) hydrogen bonding (intermolecular and intramolecular), and (5) ring strain. Consideration of these factors leads to a considerable amount of information about the environment of the C=O group.

In a discussion of these effects, it is customary to refer to the absorption frequency of a neat sample of a saturated aliphatic ketone, 1715 cm^{-1}, as "normal." For example, acetone and cyclohexanone absorb at 1715 cm^{-1}. Changes in the environment of the carbonyl can either lower or raise the absorption frequency from this "normal" value. A typical ketone spectrum is displayed in Figure 3.20.

The absorption frequency observed for a neat sample is increased when absorption is observed in nonpolar solvents. Polar solvents reduce the frequency of absorption. The overall range of solvent effects does not exceed 25 cm^{-1}.

Replacement of an alkyl group of a saturated aliphatic ketone by a hetero atom (G) shifts the carbonyl absorption. The direction of the shift depends on whether the inductive effect (a) or resonance effect (b) predominates.

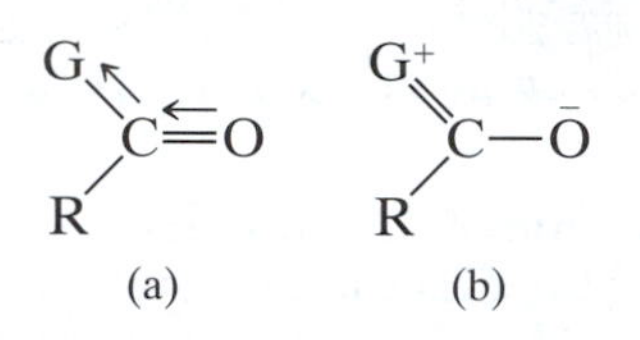

The inductive effect reduces the length of the C=O bond and thus increases its force constant and the frequency of absorption. The resonance effect increases the C=O bond length and reduces the frequency of absorption.

The absorptions of several carbonyl compound classes are summarized in Table 3.3.

Conjugation with a C=C bond results in delocalization of the π electrons of both unsaturated groups. Delocalization of the π electrons of the C=O group reduces the double-bond character of the C—O bond, causing absorption at lower wavenumbers (longer wavelengths). Conjugation with an alkene or phenyl group causes absorption in the 1685–1666 cm^{-1} region. Additional conjugation may cause a slight further re-

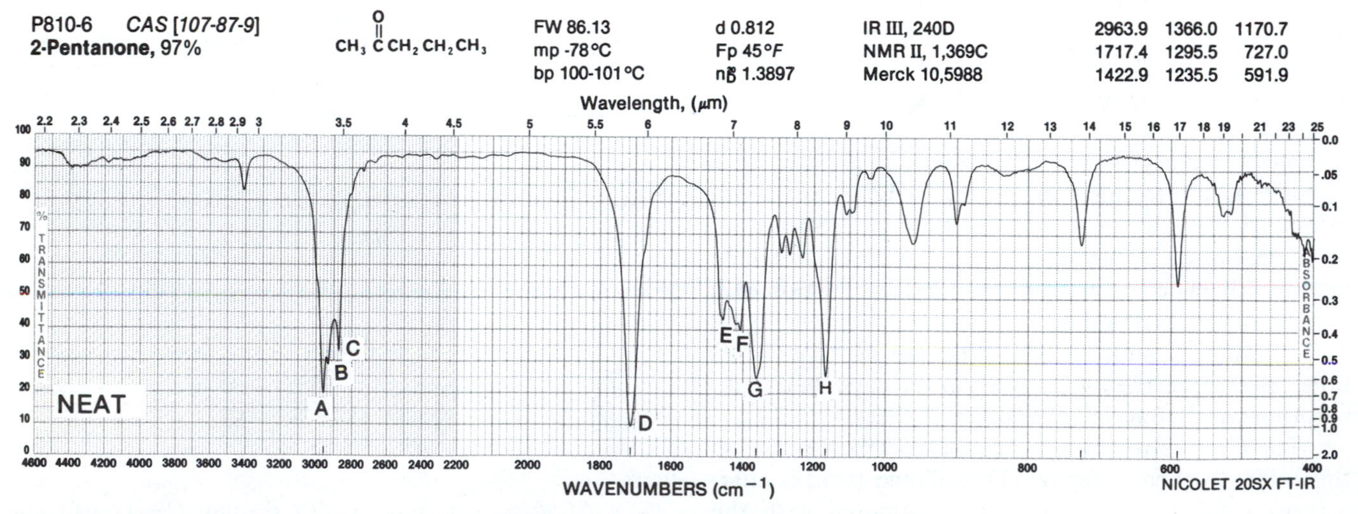

FIGURE 3.20. 2-Pentanone. A. ν_{as}, Methyl, 2964 cm^{-1}. B. ν_{as}, Methylene, 2935 cm^{-1}. C. ν_s, Methyl, 2870 cm^{-1}. D. Normal C=O stretch, 1717 cm^{-1}. E. δ_{as}, CH_3, ~1423 cm^{-1}. F. δ_s, CH_2, ~1410 cm^{-1}. G. δ_s, CH_3 of CH_3CO unit, 1366 cm^{-1}. H. The C—CO—C stretch and bend, 1171 cm^{-1}.

Table 3.3 The Carbonyl Absorption of Various $\mathrm{R\overset{O}{\overset{\|}{C}}G}$ Compounds

G Effect Predominantly Inductive	
G	ν **C=O (cm^{-1})**
Cl	1815–1785
F	~1869
Br	1812
OH (monomer)	1760
OR	1750–1735
G Effect Predominantly Resonance	
G	ν **C=O (cm^{-1})**
NH_2	1695–1650
SR	1720–1690

duction in frequency. This effect of conjugation is illustrated in Figure 3.21.

Steric effects that reduce the coplanarity of the conjugated system reduce the effect of conjugation. In the absence of steric hindrance, a conjugated system will tend toward a planar conformation. Thus, α,β-unsaturated ketones may exist in *s*-cis and *s*-trans conformations. When both forms are present, absorption for each of the forms is observed. The absorption of benzalacetone in CS_2 serves as an example; both the *s*-cis and *s*-trans forms are present at room temperature.

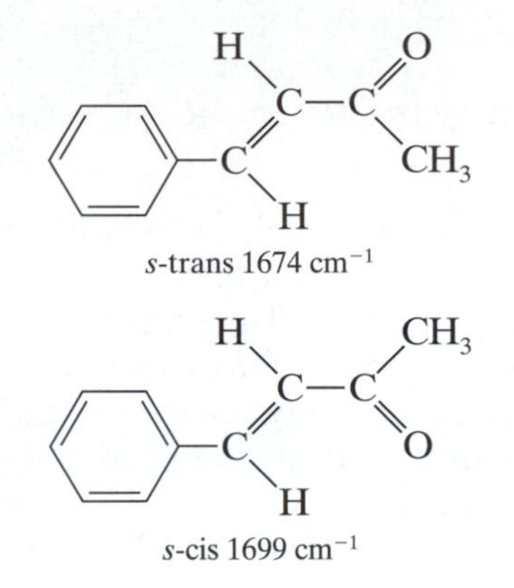

The absorption of the alkene bond in conjugation with the carbonyl group occurs at a lower frequency than that of an isolated C=C bond; the intensity of the conjugated double-bond absorption, when in an *s*-cis system, is greater than that of an isolated double bond.

Intermolecular hydrogen bonding between a ketone and a hydroxylic solvent such as methanol causes a slight decrease in the absorption frequency of the carbonyl group. For example, a neat sample of ethyl methyl ketone absorbs at 1715 cm^{-1}, whereas a 10% solution of the ketone in methanol absorbs at 1706 cm^{-1}.

β-Diketones usually exist as mixtures of tautomeric keto and enol forms. The enolic form does not show the normal absorption of conjugated ketones. Instead, a broad band appears in the 1640–1580 cm^{-1} region, many times more intense than normal carbonyl absorption. The intense and displaced absorption results from intramolecular hydrogen bonding, the bonded structure being stabilized by resonance.

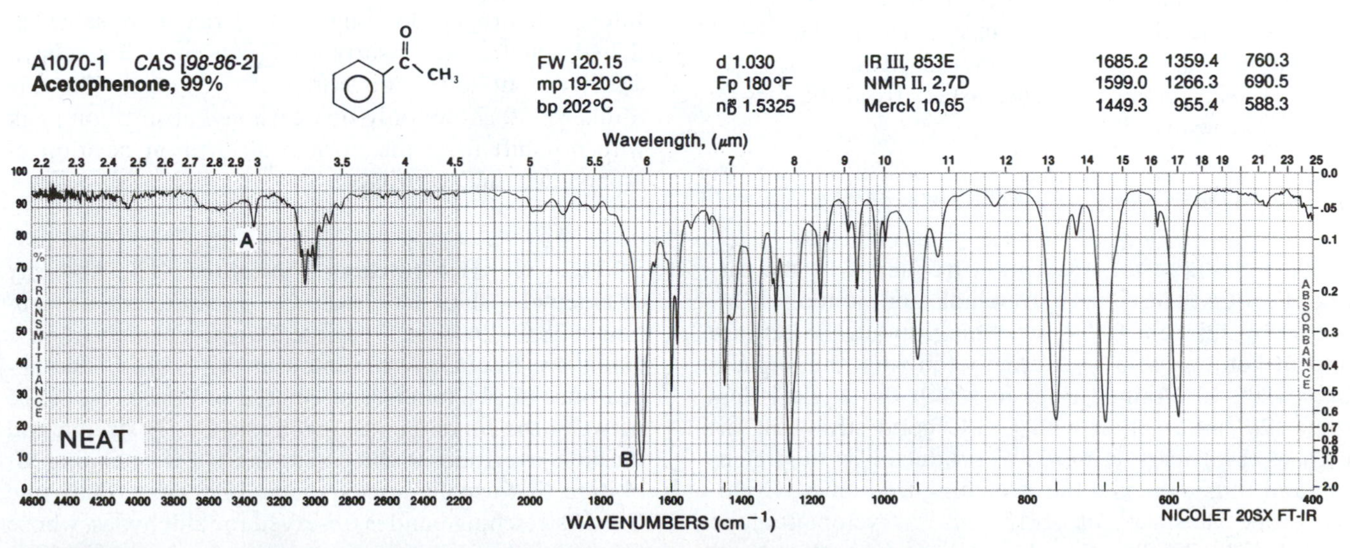

FIGURE 3.21. Acetophenone. A. Overtone of C=O stretch ~3350 cm^{-1}; frequency about twice that of C=O stretch. B. The C=O stretch, 1685 cm^{-1}, lower frequency than observed in Figure 3.20 because of the conjugation with the phenyl group.

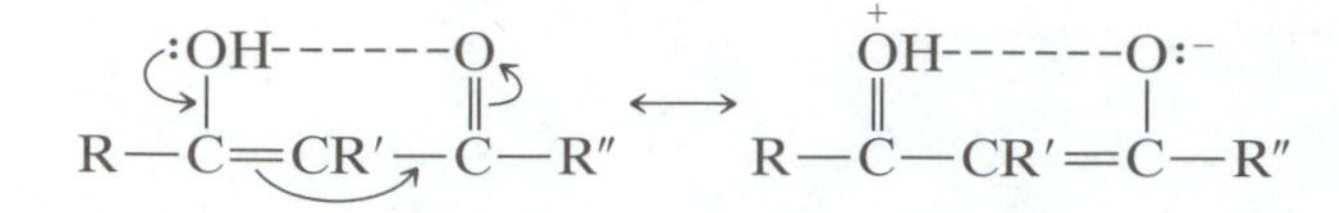

Acetylacetone as a liquid at 40°C exists to the extent of 64% in the enolic form that absorbs at 1613 cm^{-1}. The keto form and a small amount of unbonded enolic form may be responsible for two bands centering near 1725 cm^{-1}. Interaction between the two carbonyl groups in the keto form was suggested as a cause for this doublet. The enolic O—H stretching absorption is seen as a broad shallow band at 3000–2700 cm^{-1}.

α-Diketones, in which carbonyl groups exist in formal conjugation, show a single absorption band near the frequency observed for the corresponding monoketone. Biacetyl absorbs at 1718 cm^{-1}, benzil at 1681 cm^{-1}. Conjugation is ineffective for α-diketones and the C=O groups of these diketones do not couple as do, for example, the corresponding groups in acid anhydrides (see Section 3.6.16).

Quinones, which have both carbonyl groups in the same ring, absorb in the 1690–1655 cm^{-1} region. With extended conjugation, in which the carbonyl groups appear in different rings, the absorption shifts to the 1655–1635 cm^{-1} region.

Acyclic α-chloro ketones absorb at two frequencies because of rotational isomerism. When the chlorine atom is near the oxygen, its negative field repels the nonbonding electrons of the oxygen atom, thus increasing the force constant of the C=O bond. This conformation absorbs at a higher frequency (1745 cm^{-1}) than that in which the carbonyl oxygen and chlorine atom are widely separated (1725 cm^{-1}). In rigid molecules such as the monoketo steroids, α-halogenation results in equatorial or axial substitution. In the equatorial orientation, the halogen atom is near the carbonyl group and the "field effect" causes an increase in the C=O stretching frequency. In the isomer in which the halogen atom is axial to the ring, and distant from the C=O, no shift is observed.

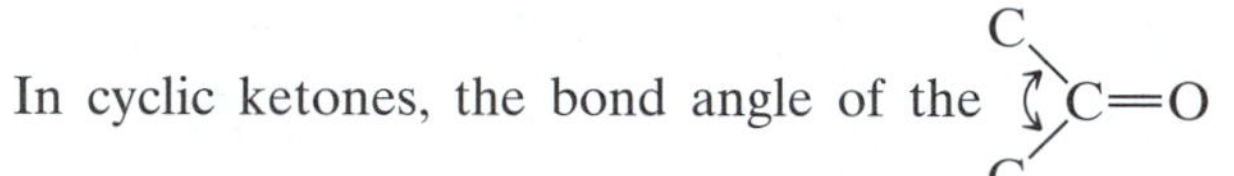

In cyclic ketones, the bond angle of the C(C)C=O group influences the absorption frequency of the carbonyl group. The C=O stretching undoubtedly is affected by adjacent C—C stretching. In acyclic ketones and in ketones with a six-membered ring, the angle is near 120°. In strained rings in which the angle is less than 120°, interaction with C—C bond stretching increases the energy required to produce C=O stretching and thus increases the stretching frequency. Cyclohexanone absorbs at 1715 cm^{-1}, cyclopentanone absorbs at 1751 cm^{-1}, and cyclobutanone absorbs at 1775 cm^{-1}.

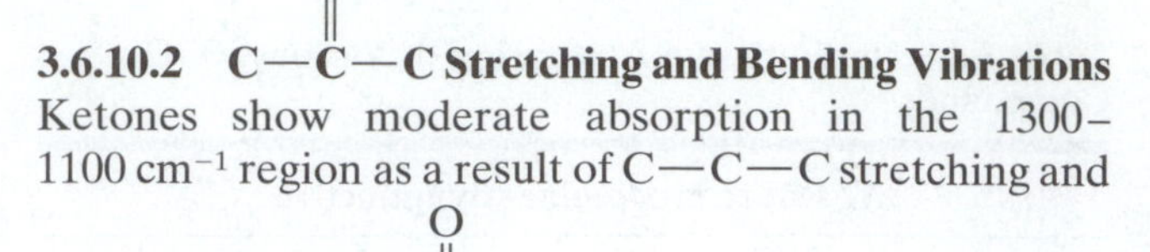

3.6.10.2 C—C(=O)—C Stretching and Bending Vibrations Ketones show moderate absorption in the 1300–1100 cm^{-1} region as a result of C—C—C stretching and bending in the C—C(=O)—C group. The absorption may consist of multiple bands. Aliphatic ketones absorb in the 1230–1100 cm^{-1} region; aromatic ketones absorb at the higher frequency end of the general absorption region.

The spectra of 2-butanone (ethyl methyl ketone, No. 18), acetone (No. 17), and cyclohexanone (No. 19) in Appendix B illustrate ketonic absorptions.

3.6.11 Aldehydes

The spectrum of 2-phenylpropionaldehyde, illustrating typical aldehydic absorption characteristics, is shown in Figure 3.22.

3.6.11.1 C=O Stretching Vibrations The carbonyl groups of aldehydes absorb at slightly higher frequencies than that of the corresponding methyl ketones. Aliphatic aldehydes absorb near 1740–1720 cm^{-1}. Aldehydic carbonyl absorption responds to structural changes in the same manner as ketones. Electronegative substitution on the α carbon increases the frequency of carbonyl absorption. Acetaldehyde absorbs at 1730 cm^{-1}, trichloroacetaldehyde absorbs at 1768 cm^{-1}. Conjugate unsaturation, as in α,β-unsaturated aldehydes and benzaldehydes, reduces the frequency of carbonyl absorption. α,β-Unsaturated aldehydes and benzaldehydes absorb in the region of 1710–1685 cm^{-1}. Internal hydrogen bonding, such as occurs in salicylaldehyde, shifts the absorption (1666 cm^{-1} for salicylaldehyde) to lower wavenumbers. Glyoxal, like the α-diketones, shows only one carbonyl absorption peak with no shift from the normal absorption position of monoaldehydic absorption.

3.6.11.2 C—H Stretching Vibrations The majority of aldehydes show aldehydic C—H stretching absorption in 2830–2695 cm^{-1} region. Two moderately intense bands are frequently observed in this region. The appearance of two bands is attributed to Fermi resonance between the fundamental aldehydic C—H stretch and the first overtone of the aldehydic C—H bending vibration that usually appears near 1390 cm^{-1}. Only one C—H stretching band is observed for aldehydes, whose C—H bending band has been shifted appreciably from 1390 cm^{-1}.

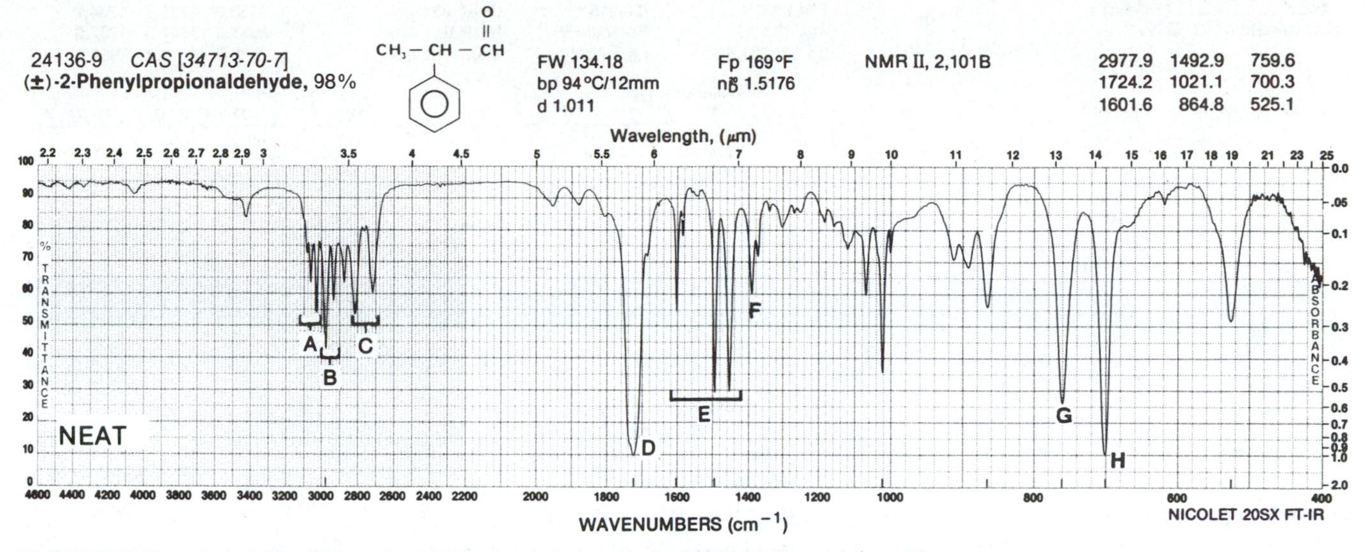

FIGURE 3.22. (±)-2-Phenylpropionaldehyde. A.* Aromatic, 3070, 3040 cm^{-1} (see Fig. 3.13). B.* Aliphatic, 2978, 2940, 2875 cm^{-1} (see Figs. 3.8 and 3.13). C.* Aldehydic, C—H stretch, 2825, 2720 cm^{-1}. Doublet from Fermi resonance with overtone of band at F. D. Normal aldehydic C═O stretch, 1724 cm^{-1}. Conjugated C═O stretch would be about 1700 cm^{-1}, for example, as for C_6H_5CHO. E. Ring C┄C stretch, 1602, 1493, 1455 cm^{-1}. F. Aldehydic C—H bend, 1390 cm^{-1}. G. Out-of-plane C—H bend, 760 cm^{-1}. H. Out-of-plane C┄C bend, 700 cm^{-1}.
*Bands A—C are C—H stretch absorptions
Source: Courtesy of Aldrich Chemical Company.

Some aromatic aldehydes with strongly electronegative groups in the ortho position may absorb as high as 2900 cm^{-1}.

An absorption of medium intensity near 2720 cm^{-1}, accompanied by a carbonyl absorption band is good evidence for the presence of an aldehyde group.

3.6.12 Carboxylic Acids

3.6.12.1 O—H Stretching Vibrations In the liquid or solid state, and in CCl_4 solution at concentrations much over 0.01 *M*, carboxylic acids exist as dimers due to strong hydrogen bonding.

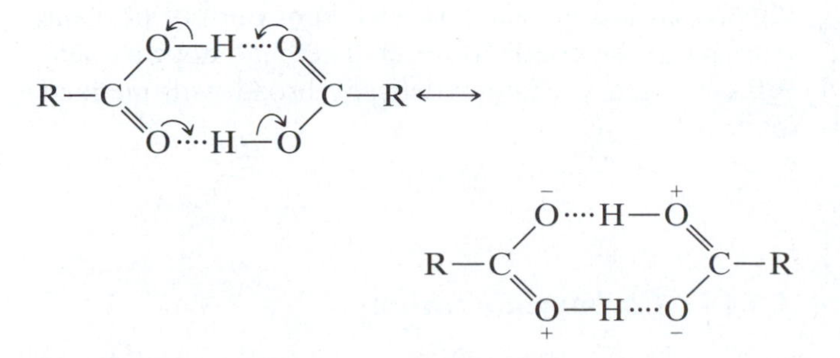

The exceptional strength of the hydrogen bonding is explained on the basis of the large contribution of the ionic resonance structure. Because of the strong bonding, a free hydroxyl stretching vibration (near 3520 cm^{-1}) is observed only in very dilute solution in nonpolar solvents or in the vapor phase.

Carboxylic acid dimers display very broad, intense O—H stretching absorption in the region of 3300–2500 cm^{-1}. The band usually centers near 3000 cm^{-1}. The weaker C—H stretching bands are generally seen superimposed upon the broad O—H band. Fine structure observed on the long-wavelength side of the broad O—H band represents overtones and combination tones of fundamental bands occurring at longer wavelengths. The spectrum of a typical aliphatic carboxylic acid is displayed in Figure 3.23.

Other structures with strong hydrogen bonding, such as β-diketones, also absorb in the 3300–2500-cm^{-1} region, but the absorption is usually less intense. Also, the C═O stretching vibrations of structures such as β-diketones are shifted to lower frequencies than those observed for carboxylic acids.

Carboxylic acids can bond intermolecularly with ethers, such as dioxane and tetrahydrofuran, or with other solvents that can act as proton acceptors. Spectra determined in such solvents show bonded O—H absorption near 3100 cm^{-1}.

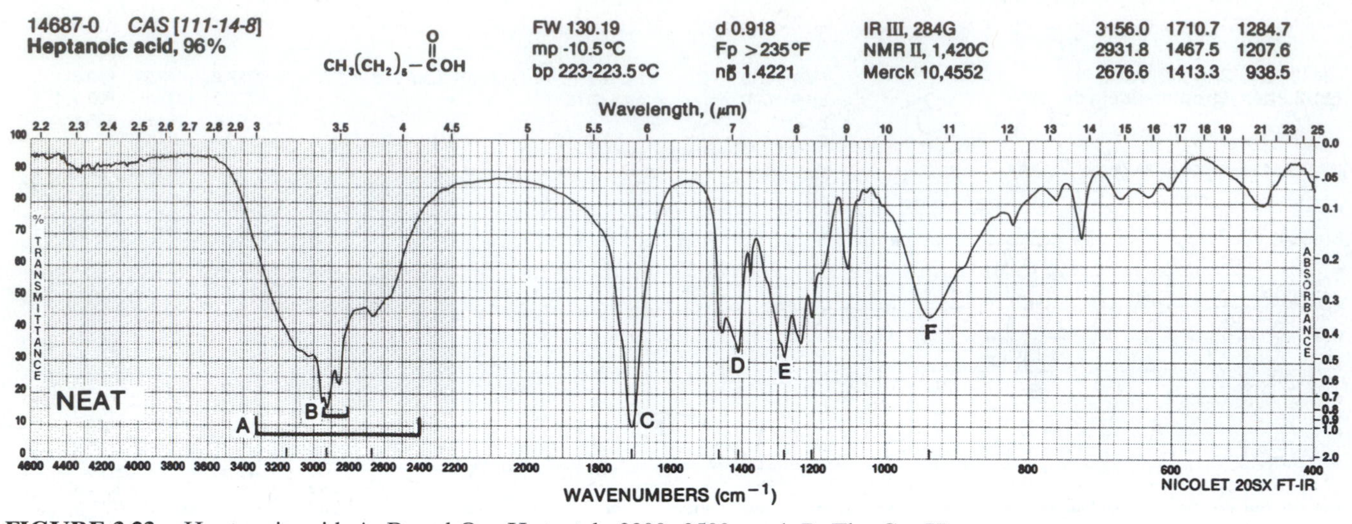

FIGURE 3.23. Heptanoic acid. A. Broad O—H stretch, 3300–2500 cm^{-1}. B. The C—H stretch (see Fig. 3.8), 2950, 2932, 2855 cm^{-1}. Superimposed upon O—H stretch. C. Normal, dimeric carboxylic C=O stretch, 1711 cm^{-1}. D. The C—O—H in-plane bend,* 1413 cm^{-1}. E. The C—O stretch,* dimer, 1285 cm^{-1}. F. The O—H out-of-plane bend, 939 cm^{-1}.
*Bands at D and E involve C—O—H interaction.

3.6.12.2 C=O Stretching Vibrations The C=O stretching bands of acids are considerably more intense than ketonic C=O stretching bands. The monomers of saturated aliphatic acids absorb near 1760 cm^{-1}.

The carboxylic dimer has a center of symmetry; only the asymmetrical C=O stretching mode absorbs in the IR. Hydrogen bonding and resonance weaken the C=O bond, resulting in absorption at a lower frequency than the monomer. The C=O group in dimerized saturated aliphatic acids absorbs in the region of 1720–1706 cm^{-1}.

Internal hydrogen bonding reduces the frequency of the carbonyl stretching absorption to a greater degree than does intermolecular hydrogen bonding. For example, salicylic acid absorbs at 1665 cm^{-1}, whereas *p*-hydroxybenzoic acid absorbs at 1680 cm^{-1}.

Unsaturation in conjugation with the carboxylic carbonyl group decreases the frequency (increases the wavelength) of absorption of both the monomer and dimer forms only slightly. In general, α,β-unsaturated and aryl conjugated acids show absorption for the dimer in the 1710–1680 cm^{-1} region. Extension of conjugation beyond the α,β-position results in very little additional shifting of the C=O absorption.

Substitution in the α-position with electronegative groups, such as the halogens, brings about a slight increase in the C=O absorption frequency (10–20 cm^{-1}). The spectra of acids with halogens in the α-position, determined in the liquid state or in solution, show dual carbonyl bands resulting from rotational isomerism (field effect). The higher frequency band corresponds to the conformation in which the halogen is in proximity to the carbonyl group.

3.6.12.3 C—O Stretching and O—H Bending Vibrations Two bands arising from C—O stretching and O—H bending appear in the spectra of carboxylic acids near 1320–1210 and 1440–1395 cm^{-1}, respectively. Both of these bands involve some interaction between C—O stretching and in-plane C—O—H bending. The more intense band, near 1315–1280 cm^{-1} for dimers, is generally referred to as the C—O stretching band and usually appears as a doublet in the spectra of long-chain fatty acids. The C—O—H bending band near 1440–1395 cm^{-1} is of moderate intensity and occurs in the same region as the CH_2 scissoring vibration of the CH_2 group adjacent to the carbonyl.

One of the characteristic bands in the spectra of dimeric carboxylic acids results from the out-of-plane bending of the bonded O—H. The band appears near 920 cm^{-1} and is characteristically broad with medium intensity.

3.6.13 Carboxylate Anion

The carboxylate anion has two strongly coupled C$\cdots$O bonds with bond strengths intermediate between C=O and C—O.

The carboxylate ion gives rise to two bands: a strong asymmetrical stretching band near 1650–1550 cm^{-1} and a weaker, symmetrical stretching band near 1400 cm^{-1}.

The conversion of a carboxylic acid to a salt can serve as confirmation of the acid structure. This is conveniently done by the addition of a tertiary aliphatic amine, such as triethylamine, to a solution of the carboxylic acid in chloroform (no reaction occurs in CCl_4). The carboxylate ion thus formed shows the two characteristic carbonyl absorption bands in addition to an "ammonium" band in the 2700–2200 cm^{-1} region. The O—H stretching band, of course, disappears. The spectrum of ammonium benzoate, Figure 3.24, demonstrates most of these features.

3.6.14 Esters and Lactones

Esters and lactones have two characteristically strong absorption bands arising from C═O and C—O stretching. The intense C═O stretching vibration occurs at higher frequencies (shorter wavelength) than that of normal ketones. The force constant of the carbonyl bond is increased by the electron-attracting nature of the adjacent oxygen atom (inductive effect). Overlapping occurs between esters in which the carbonyl frequency is lowered, and ketones in which the normal ketone frequency is raised. A distinguishing feature of esters and lactones, however, is the strong C—O stretching band in the region where a weaker band occurs for ketones. There is overlapping in the C═O frequency of esters or lactones and acids, but the OH stretching and bending vibrations and the possibility of salt formation distinguish the acids.

The frequency of the ester carbonyl responds to environmental changes in the vicinity of the carbonyl group in much the same manner as ketones. The spectrum of phenyl acetate, Figure 3.25, illustrates most of the important absorption characteristics for esters.

3.6.14.1 C═O Stretching Vibrations The C═O absorption band of saturated aliphatic esters (except formates) is in the 1750–1735 cm^{-1} region. The C═O absorption bands of formates, α,β-unsaturated, and benzoate esters are in the region of 1730–1715 cm^{-1}. Further conjugation has little or no additional effect upon the frequency of the carbonyl absorption.

In the spectra of vinyl or phenyl esters, with unsaturation adjacent to the C—O— group, a marked rise in the carbonyl frequency is observed along with a lowering of the C—O frequency. Vinyl acetate has a carbonyl band at 1776 cm^{-1}; phenyl acetate absorbs at 1770 cm^{-1}.

α-Halogen substitution results in a rise in the C═O stretching frequency. Ethyl trichloroacetate absorbs at 1770 cm^{-1}.

In oxalates and α-keto esters, as in α-diketones, there appears to be little or no interaction between the two carbonyl groups so that normal absorption occurs

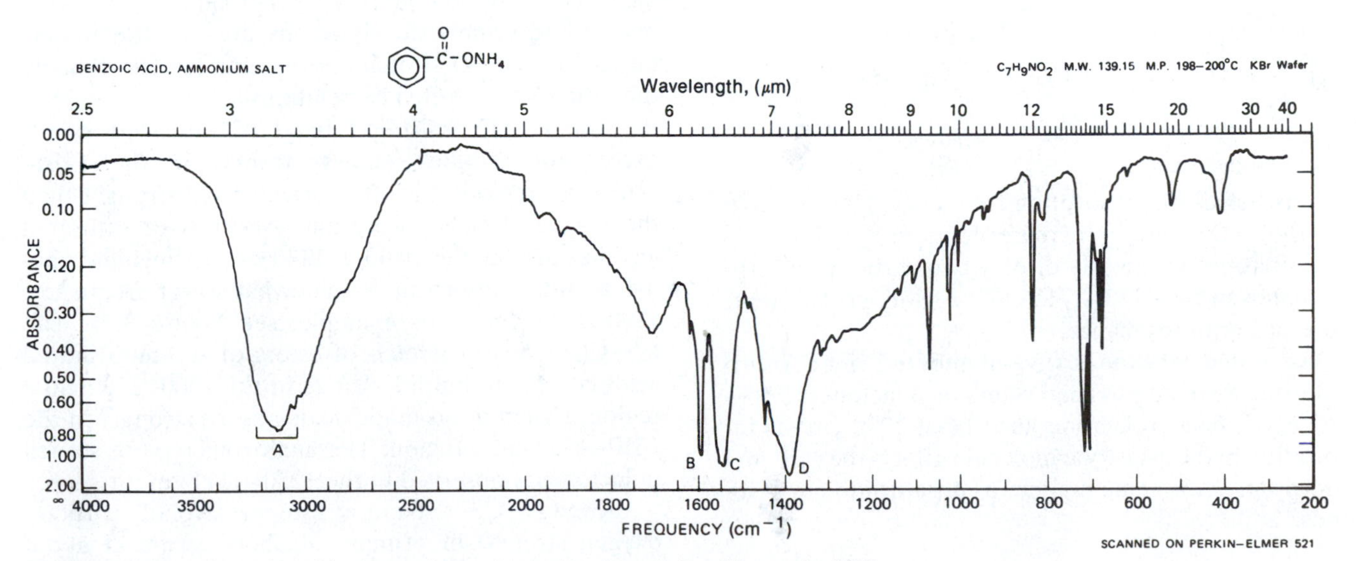

FIGURE 3.24. Benzoic acid, ammonium salt. A. N—H and C—H stretch, 3600–2500 cm^{-1}. B. Ring C═C stretch, 1600 cm^{-1}. C. Asymmetric carboxylate anion $C(\text{=}O)_2^-$ stretch, 1550 cm^{-1}. D. Symmetric carboxylate $C(\text{=}O)_2^-$ stretch, 1385 cm^{-1}.

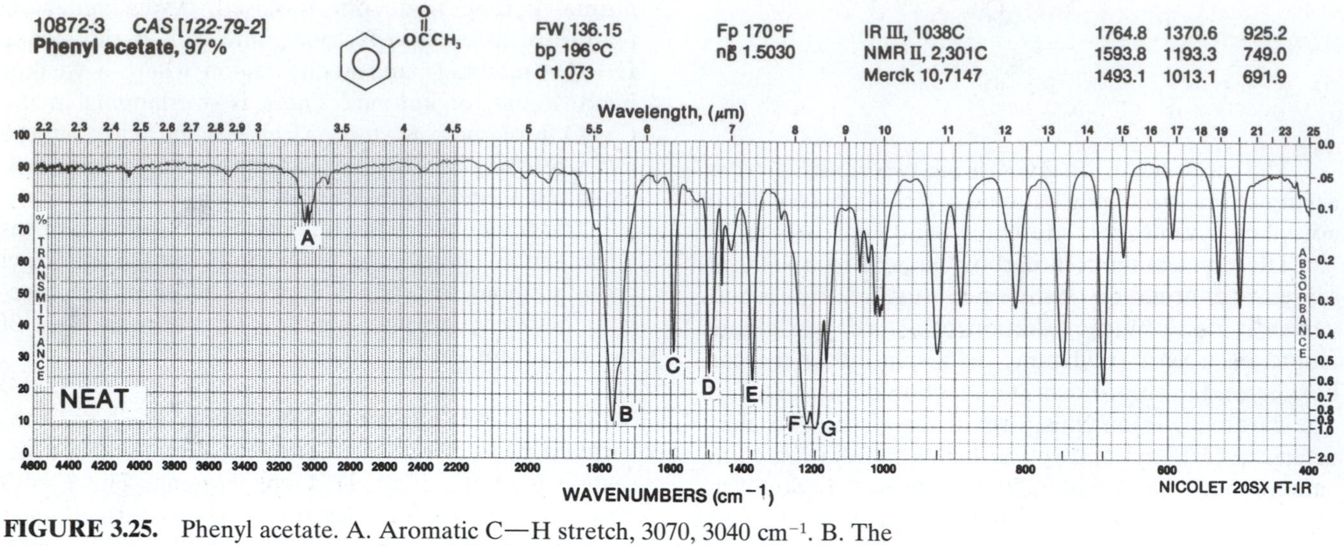

FIGURE 3.25. Phenyl acetate. A. Aromatic C—H stretch, 3070, 3040 cm^{-1}. B. The C=O stretch, 1765 cm^{-1}: this is higher frequency than that from a normal ester C=O stretch (~1740 cm^{-1}: see Table 3.3) because of phenyl conjugation with alcohol oxygen; conjugation of an aryl group or other unsaturation with the carbonyl group causes this C=O stretch to be at lower than normal frequency (e.g., benzoates absorb at about 1724 cm^{-1}). C. Ring C$\overset{...}{=}$C stretch, 1594 cm^{-1}. D. $\delta_{as}CH_3$, 1493 cm^{-1}. E. δ_sCH_3, 1371 cm^{-1}. F. Acetate C(=O)—O stretch, 1215 cm^{-1}. G. The O—C$\overset{...}{=}$C asymmetrical stretch, 1193 cm^{-1}.

in the region of 1755–1740 cm^{-1}. In the spectra of β-keto esters, however, where enolization can occur, a band is observed near 1650 cm^{-1} that results from hydrogen bonding between the ester C=O and the enolic hydroxyl group.

The carbonyl absorption of saturated δ-lactones (six-membered ring) occurs in the same region as straight-chain, unconjugated esters. Unsaturation α to the C=O

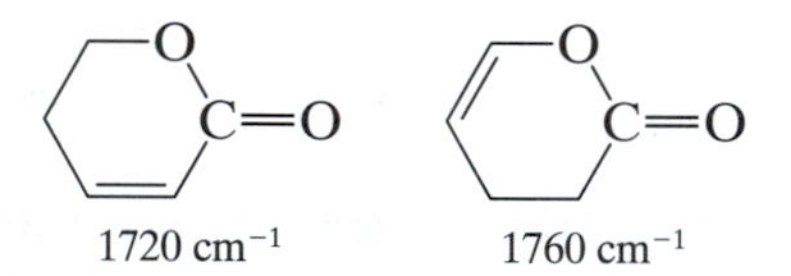

reduces the C=O absorption frequency. Unsaturation α to the —O— group increases it.

α-Pyrones frequently display two carbonyl absorption bands in the 1775–1715 cm^{-1} region, probably because of Fermi resonance.

Saturated γ-lactones (five-membered ring) absorb at shorter wavelengths than esters or δ-lactones: 1795–1760 cm^{-1}; δ-valerolactone absorbs at 1770 cm^{-1}. Unsaturation in the γ-lactone molecule affects the carbonyl absorption in the same manner as unsaturation in δ lactones.

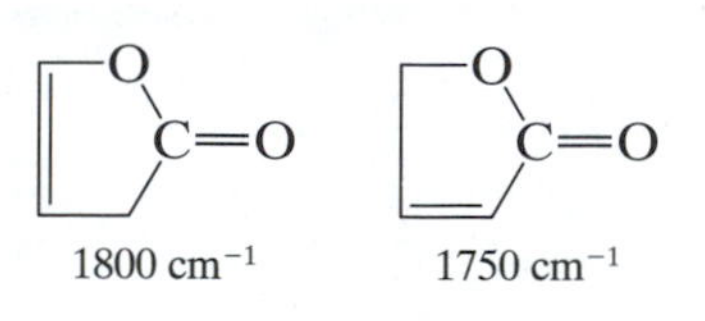

In unsaturated lactones, when the double bond is adjacent to the —O—, a strong C=C absorption is observed in the 1685–1660 cm^{-1} region.

3.6.14.2 C—O Stretching Vibrations The "C—O stretching vibrations" of esters actually consist of two asymmetrical coupled vibrations: C—C(=O)—O and O—C—C, the former being more important. These bands occur in the region of 1300–1000 cm^{-1}. The corresponding symmetric vibrations are of little importance. The C—O stretch correlations are less reliable than the C=O stretch correlations.

The C—C(=O)—O band of saturated esters, except for acetates, shows strongly in the 1210–1163 cm^{-1} region. It is often broader and stronger than the C=O stretch absorption. Acetates of saturated alcohols display this band at 1240 cm^{-1}. Vinyl and phenyl acetates absorb at a somewhat lower frequency, 1190–1140 cm^{-1}; for example, see Figure 3.25. The C—C(=O)—O stretch of esters of α,β-unsaturated acids results in multiple bands in the 1300–1160 cm^{-1} region. Esters of aromatic acids absorb strongly in the 1310–1250 cm^{-1} region. The analogous type of stretch in lactones is observed in the 1250–1111 cm^{-1} region.

The O—C—C band of esters ("alcohol" carbon–oxygen stretch) of primary alcohols occurs at about 1164–1031 cm^{-1} and that of esters of secondary alcohols occurs at about 1100 cm^{-1}. Aromatic esters of primary alcohols show this absorption near 1111 cm^{-1}.

Methyl esters of long-chain fatty acids present

a three-band pattern with bands near 1250, 1205, and 1175 cm^{-1}. The band near 1175 cm^{-1} is the strongest.

Spectra showing typical ester absorptions are shown in Appendix B: ethyl acetate (No. 20) and diethyl phthalate (No. 21).

3.6.15 Acid Halides

3.6.15.1 C=O Stretching Vibrations Acid halides show strong absorption in the C=O stretching region. Unconjugated acid chlorides absorb in the 1815–1785 cm^{-1} region. Acetyl fluoride in the gas phase absorbs near 1869 cm^{-1}. Conjugated acid halides absorb at a slightly lower frequency because resonance reduces the force constant of the C=O bond; aromatic acid chlorides absorb strongly at 1800–1770 cm^{-1}. A weak band near 1750–1735 cm^{-1} appearing in the spectra of aroyl chlorides probably results from Fermi resonance between the C=O band and the overtone of a lower wavenumber band near 875 cm^{-1}. The annotated spectrum of benzoyl chloride is given in Figure 3.26.

3.6.16 Carboxylic Acid Anhydrides

3.6.16.1 C=O Stretching Vibrations Anhydrides display two stretching bands in the carbonyl region. The two bands result from asymmetrical and symmetrical C=O stretching modes. Saturated acyclic anhydrides absorb near 1818 and 1750 cm^{-1}. Conjugated acyclic anhydrides show absorption near 1775 and 1720 cm^{-1}; the decrease in the frequency of absorption is caused by resonance. The higher frequency band is the more intense.

Cyclic anhydrides with five-membered rings show absorption at higher frequencies (lower wavelengths) than acyclic anhydrides because of ring strain; succinic anhydride absorbs at 1865 and 1782 cm^{-1}. The lower frequency (longer wavelength) C=O band is the stronger of the two carbonyl bands in five-membered ring cyclic anhydrides.

3.6.16.2 C—O Stretching Vibrations Other strong bands appear in the spectra of anhydrides as a result of C—C(=O)—O—C(=O)—C stretching vibrations. Unconjugated straight-chain anhydrides absorb near 1047 cm^{-1}. Cyclic anhydrides display bands near 952–909 and 1299–1176 cm^{-1}. The C—O stretching band for acetic anhydride is at 1125 cm^{-1}.

The spectrum of Figure 3.27 is that of a typical aliphatic anhydride.

3.6.17 Amides and Lactams

All amides show a carbonyl absorption band known as the amide I band. Its position depends on the degree of hydrogen bonding and, thus, on the physical state of the compound.

Primary amides show two N—H stretching bands resulting from symmetrical and asymmetrical N—H stretching. Secondary amides and lactams show only one N—H stretching band. As in the case of O—H stretching, the frequency of the N—H stretching is reduced by hydrogen bonding, though to a lesser degree. Overlapping occurs in the observed position of N—H

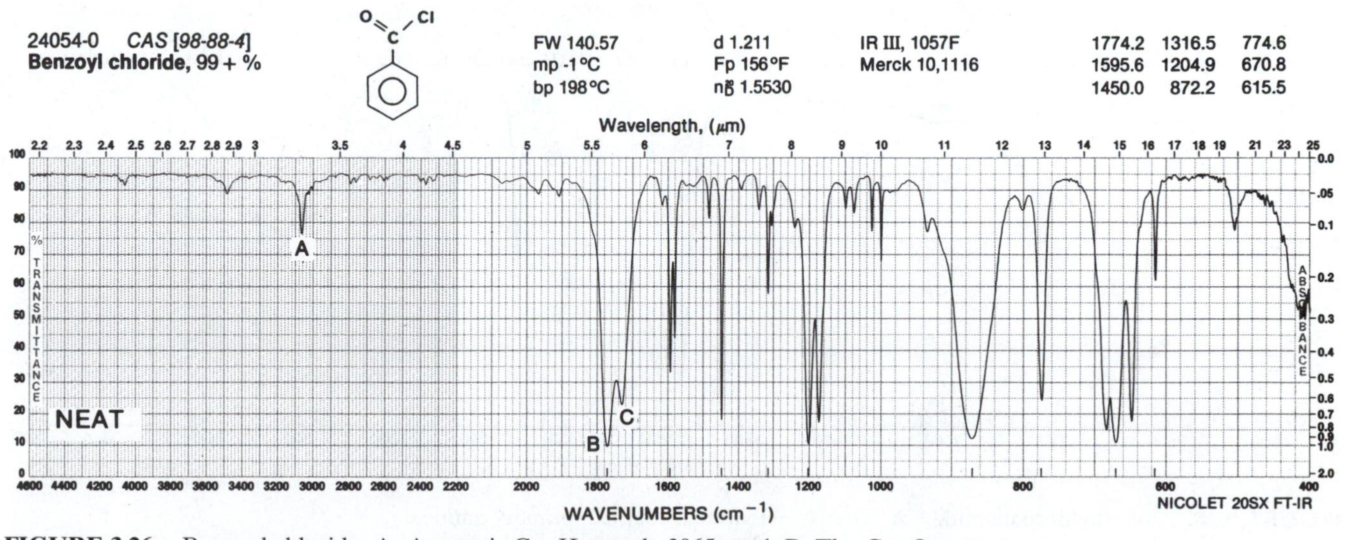

FIGURE 3.26. Benzoyl chloride. A. Aromatic C—H stretch, 3065 cm^{-1}. B. The C=O stretch, 1774 cm^{-1} (see Table 3.3). (Acid chloride C=O stretch position shows very small dependence on conjugation; aroyl chlorides identified by band such as at C.) C. Fermi resonance band (of C=O stretch and overtone of 872 cm^{-1} band), 1730 cm^{-1}.
Source: Courtesy of Aldrich Chemical Co.

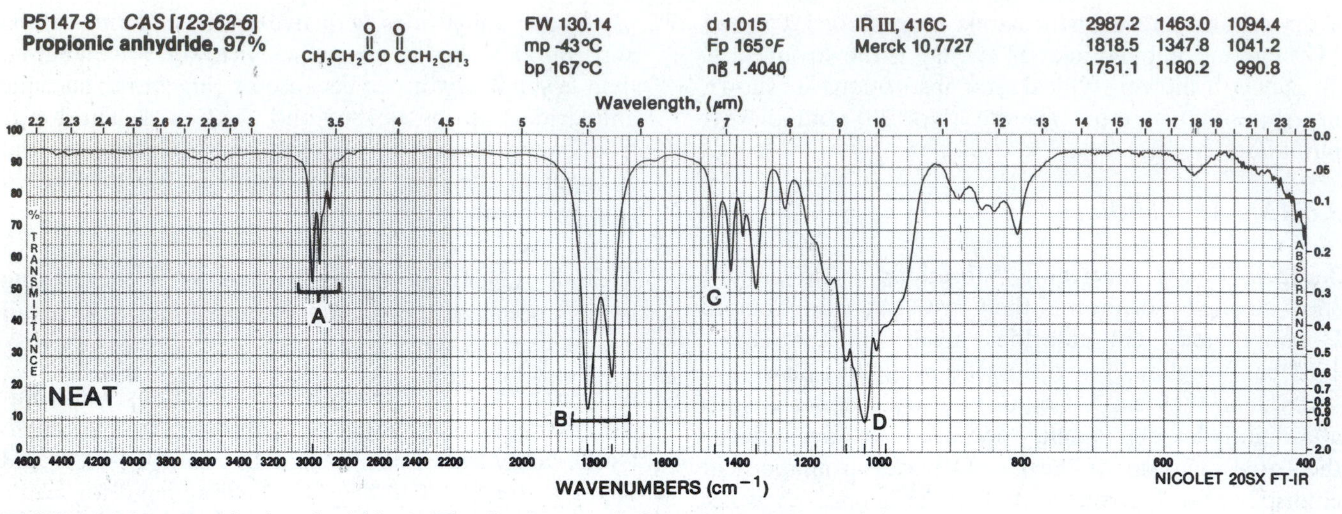

FIGURE 3.27. Propionic anhydride A. The C—H stretch, 2987, 2940, 2880 cm^{-1}. B. Asymetric and symmetric C=O coupled stretching, respectively: 1818, 1751 cm^{-1}. See Table 3.3. C. δ_3CH_2 (scissoring), 1463 cm^{-1}. D. C—CO—O—CO—C stretch, 1041 cm^{-1}.

and O—H stretching frequencies so that an unequivocal differentiation in structure is sometimes impossible.

Primary amides and secondary amides, and a few lactams, display a band or bands in the region of 1650–1515 cm^{-1} caused primarily by NH_2 or NH bending: the amide II band. This absorption involves coupling between N—H bending and other fundamental vibrations and requires a trans geometry.

Out-of-plane NH wagging is responsible for a broad band of medium intensity in the 800–666 cm^{-1} region.

The spectrum of Figure 3.28 is that of a typical primary amide of an aliphatic acid. The spectrum of DMF

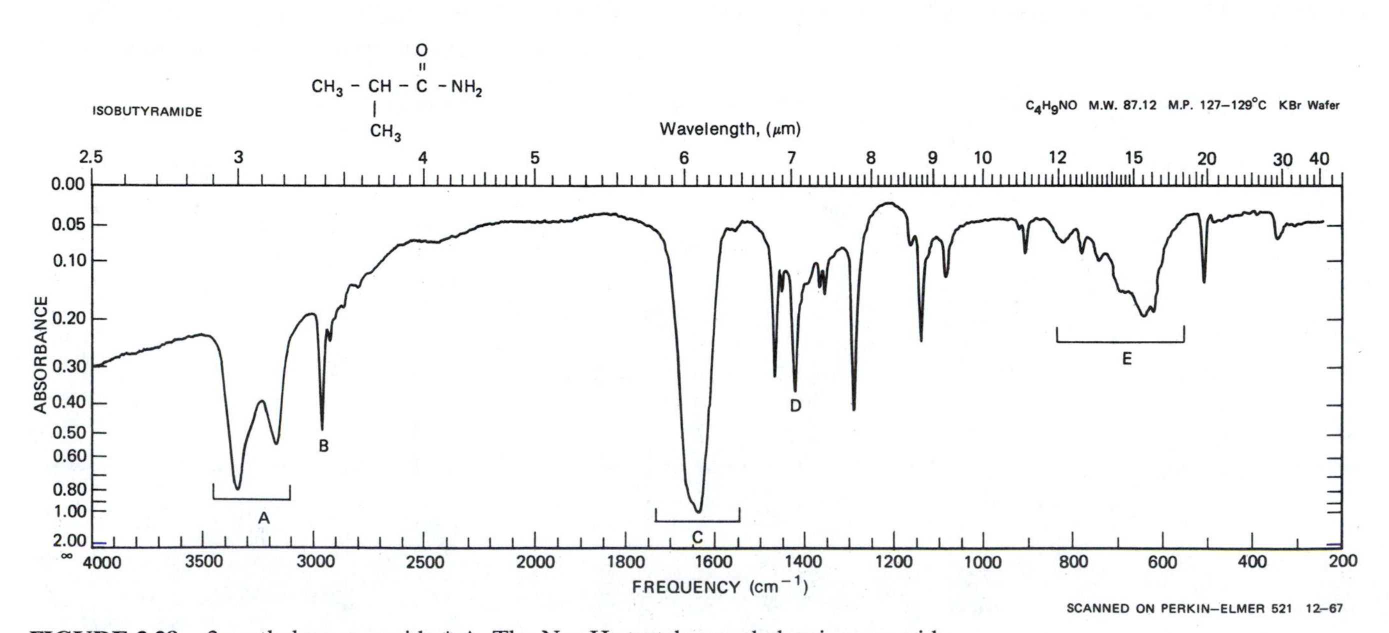

FIGURE 3.28. 2-methylpropanamide.* A. The N—H stretch, coupled, primary amide, hydrogen bonded; asymmetric, 3352 cm^{-1}; symmetric, 3170 cm^{-1}. B. Aliphatic C—H stretch, 2960 cm^{-1}. C. Overlap C=O stretch, amide I band, 1640 cm^{-1}; see Table 3.3. N—H bend, amide II band, 1640 cm^{-1}. D. The C—N stretch, 1425 cm^{-1}. E. Broad N—H out-or-plane bend 700–600 cm^{-1}.

*The CA name for isobutyramide is 2-methylpropanamide.

(*N,N*-dimethylformamide, Appendix B, No. 28) displays typical amide absorptions.

3.6.17.1 N—H Stretching Vibrations In dilute solution in nonpolar solvents, primary amides show two moderately intense NH stretching frequencies corresponding to the asymmetrical and symmetrical NH stretching vibrations. These bands occur near 3520 and 3400 cm^{-1}, respectively. In the spectra of solid samples, these bands are observed near 3350 and 3180 cm^{-1} because of hydrogen bonding.

In IR spectra of secondary amides, which exist mainly in the trans conformation, the free NH stretching vibration observed in dilute solutions occurs near 3500–3400 cm^{-1}. In more concentrated solutions and in solid samples, the free NH band is replaced by multiple bands in the 3330–3060 cm^{-1} region. Multiple bands are observed since the amide group can bond to produce dimers with an *s*-cis conformation and polymers with an *s*-trans conformation.

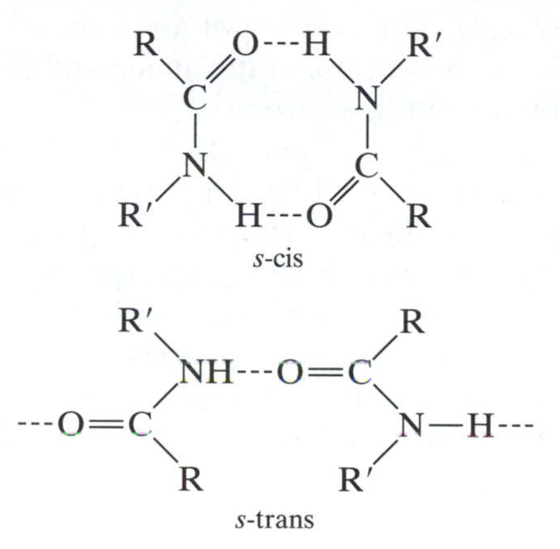

s-cis

s-trans

3.6.17.2 C=O Stretching Vibrations (Amide I Band) The C=O absorption of amides occurs at lower frequencies than "normal" carbonyl absorption due to the resonance effect (see Section 3.6.10.1). The position of absorption depends on the same environmental factors as the carbonyl absorption of other compounds.

Primary amides (except acetamide, whose C=O bond absorbs at 1694 cm^{-1}) have a strong amide I band in the region of 1650 cm^{-1} when examined in the solid phase. When the amide is examined in dilute solution, the absorption is observed at a higher frequency, near 1690 cm^{-1}. In more concentrated solutions, the C=O frequency is observed at some intermediate value, depending on the degree of hydrogen bonding.

Simple, open-chain, secondary amides absorb near 1640 cm^{-1} when examined in the solid state. In dilute solution, the frequency of the amide I band may be raised to 1680 cm^{-1} and even to 1700 cm^{-1} in the case of the anilides. In the anilide structure there is competition between the ring and the C=O for the nonbonded electron pair of the nitrogen.

The carbonyl frequency of tertiary amides is independent of the physical state since hydrogen bonding with another tertiary amide group is impossible. The C=O absorption occurs in the range of 1680–1630 cm^{-1}. The absorption range of tertiary amides in solution is influenced by hydrogen bonding with the solvent: *N,N*-Diethylacetamide absorbs at 1647 cm^{-1} in dioxane and at 1615 cm^{-1} in methanol.

Electron-attracting groups attached to the nitrogen increase the frequency of absorption since they effectively compete with the carbonyl oxygen for the electrons of the nitrogen, thus increasing the force constant of the C=O bond.

3.6.17.3 N—H Bending Vibrations (Amide II Band) All primary amides show a sharp absorption band in dilute solution (amide II band) resulting from NH_2 bending at a somewhat lower frequency than the C=O band. This band has an intensity of one-half to one-third of the C=O absorption band. In mulls and pellets the band occurs near 1655–1620 cm^{-1} and is usually under the envelope of the amide I band. In dilute solutions, the band appears at lower frequency, 1620–1590 cm^{-1}, and normally is separated from the amide I band. Multiple bands may appear in the spectra of concentrated solutions, arising from the free and associated states. The nature of the R group $\left(R{-}\overset{\overset{\large O}{\|}}{C}{-}NH_2\right)$ has little effect upon the amide II band.

Secondary acyclic amides in the solid state display an amide II band in the region of 1570–1515 cm^{-1}. In dilute solution, the band occurs in the 1550–1510 cm^{-1} region. This band results from interaction between the N—H bending and the C—N stretching of the C—N—H group. A second, weaker band near 1250 cm^{-1} also results from interaction between the N—H bending and C—N stretching.

3.6.17.4 Other Vibration Bands The C—N stretching band of primary amides occurs near 1400 cm^{-1}. A broad, medium band in the 800–666 cm^{-1} region in the spectra of primary and secondary amides results from out-of-plane N—H wagging.

In lactams of medium ring size, the amide group is forced into the *s*-cis conformation. Solid lactams absorb strongly near 3200 cm^{-1} because of the N—H stretching vibration. This band does not shift appreciably with dilution since the *s*-cis form remains associated at relatively low concentrations.

3.6.17.5 C=O Stretching Vibrations of Lactams The C=O absorption of lactams with six-membered rings or larger is near 1650 cm^{-1}. Five-membered ring (γ) lactams absorb in the 1750–1700 cm^{-1} region. Four-membered ring (β) lactams, unfused, absorb at 1760–1730 cm^{-1}. Fusion of the lactam ring to another ring generally increases the frequency by 20–50 cm^{-1}.

Most lactams do not show a band near 1550 cm^{-1} that is characteristic of *s*-trans noncyclic secondary amides. The N—H out-of-plane wagging in lactams causes broad absorption in the 800–700-cm^{-1} region.

3.6.18 Amines

The spectrum of a typical primary, aliphatic amine appears in Figure 3.29.

3.6.18.1 N—H Stretching Vibrations Primary amines, examined in dilute solution, display two weak absorption bands: one near 3500 cm^{-1} and the other near 3400 cm^{-1}. These bands represent, respectively, the "free" asymmetrical and symmetrical N—H stretching modes. Secondary amines show a single weak band in the 3350–3310 cm^{-1} region. These bands are shifted to longer wavelengths by hydrogen bonding. The associated N—H bands are weaker and frequently sharper than the corresponding O—H bands. Aliphatic primary amines (neat) absorb at 3400–3300 and 3330–3250 cm^{-1}. Aromatic primary amines absorb at slightly higher frequencies (shorter wavelengths). In the spectra of liquid primary and secondary amines, a shoulder usually appears on the low-frequency side of the N—H stretching band, arising from the overtone of the NH bending band intensified by Fermi resonance. Tertiary amines do not absorb in this region.

3.6.18.2 N—H Bending Vibrations The N—H bending (scissoring) vibration of primary amines is observed in the 1650–1580 cm^{-1} region of the spectrum. The band is medium to strong in intensity and is moved to slightly higher frequencies when the compound is associated. The N—H bending band is seldom detectable in the spectra of aliphatic secondary amines, whereas secondary aromatic amines absorb near 1515 cm^{-1}.

Liquid samples of primary and secondary amines display medium to strong broad absorption in the 909–666 cm^{-1} region of the spectrum arising from NH wagging. The position of this band depends on the degree of hydrogen bonding.

3.6.18.3 C—N Stretching Vibrations Medium to weak absorption bands for the unconjugated C—N linkage in primary, secondary, and tertiary aliphatic amines appear in the region of 1250–1020 cm^{-1}. The vibrations responsible for these bands involve C—N stretching coupled with the stretching of adjacent bonds in the molecule. The position of absorption in this region depends on the class of the amine and the pattern of substitution on the α carbon.

Aromatic amines display strong C—N stretching absorption in the 1342–1266 cm^{-1} region. The absorption appears at higher frequencies (shorter wavelengths) than the corresponding absorption of aliphatic amines because the force constant of the C—N bond is increased by resonance with the ring.

Characteristic strong C—N stretching bands in the spectra of aromatic amines have been assigned as in Table 3.4.

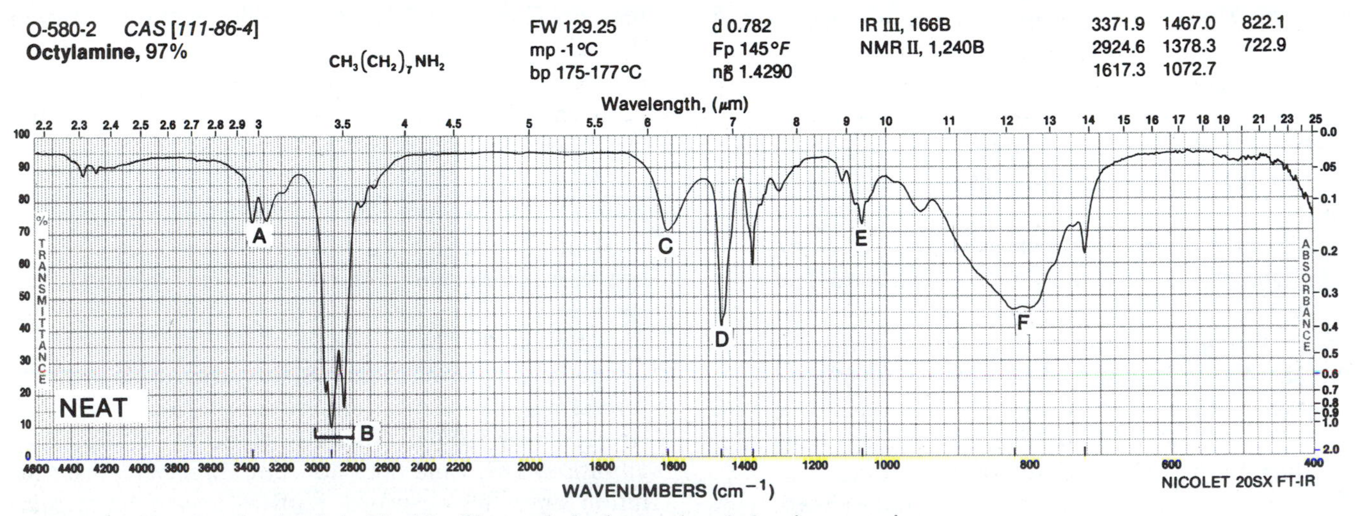

FIGURE 3.29. Octylamine.* A. The N—H stretch, hydrogen-bonded, primary amine coupled doublet: asymmetric, 3372 cm^{-1}. Symmetric, 3290 cm^{-1}. (Shoulder at about 3200 cm^{-1}, Fermi resonance band with overtone of band at C.) B. Aliphatic C—H stretch, 2925, 2850 cm^{-1}; $\nu_s CH_2$, 2817 cm^{-1}. C. The N—H bend (scissoring) 1617 cm^{-1}. D. $\delta_s CH_2$ (scissoring), 1467 cm^{-1}. E. The C—N stretch, 1073 cm^{-1}. F. The N—H wag (neat sample), ~900–700 cm^{-1}.

*The CA name for octylamine is octanamine.

Table 3.4 C—N Stretch of Primary, Secondary, and Tertiary Aromatic Amines

Aromatic Amine	Absorption Region (cm^{-1})
Primary	1340–1250
Secondary	1350–1280
Tertiary	1360–1310

3.6.19 Amine Salts

3.6.19.1 N—H Stretching Vibrations The ammonium ion displays strong, broad absorption in the 3300–3030 cm^{-1} region because of N—H stretching vibrations (see Fig. 3.24). There is also a combination band in the 2000–1709 cm^{-1} region.

Salts of primary amines show strong, broad absorption between 3000 and 2800 cm^{-1} arising from asymmetrical and symmetrical stretching in the NH_3^+ group. In addition, multiple combination bands of medium intensity occur in the 2800–2000 cm^{-1} region, the most prominent being the band near 2000 cm^{-1}. Salts of secondary amines absorb strongly in the 3000–2700 cm^{-1} region with multiple bands extending to 2273 cm^{-1}. A medium band near 2000 cm^{-1} may be observed. Tertiary amine salts absorb at longer wavelengths than the salts of primary and secondary amines (2700–2250 cm^{-1}). Quaternary ammonium salts can have no N—H stretching vibrations.

3.6.19.2 N—H Bending Vibrations The ammonium ion displays a strong, broad NH_4^+ bending band near 1429 cm^{-1}. The NH_3^+ group of the salt of a primary amine absorbs near 1600–1575 and 1550–1504 cm^{-1}. These bands originate in asymmetrical and symmetrical NH_3^+ bending, analogous to the corresponding bands of the CH_3 group. Salts of secondary amines absorb near 1620–1560 cm^{-1}. The N—H bending band of the salts of tertiary amines is weak and of no practical value.

3.6.20 Amino Acids and Salts of Amino Acids

Amino acids are encountered in three forms:

1. The free amino acid (zwitterion).*

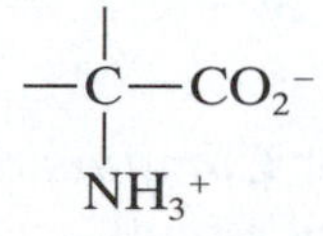

*Aromatic amino acids are not zwitterions. Thus, *p*-aminobenzoic acid is

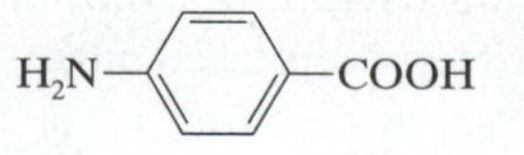

2. The hydrochloride (or other salt).

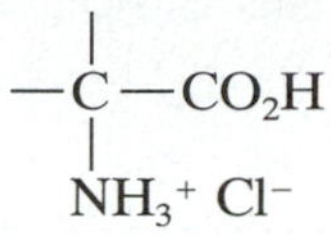

3. The sodium (or other cation) salt.

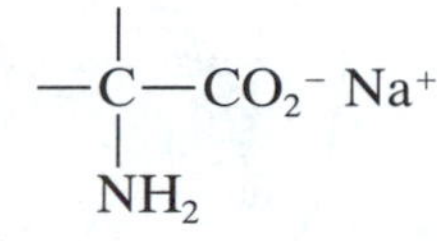

Free primary amino acids are characterized by the following absorptions (most of the work was done with α-amino acids, but the relative positions of the amino and carboxyl groups seem to have little effect):

1. A broad, strong NH_3^+ stretching band in the 3100–2600 cm^{-1} region. Multiple combination and overtone bands extend the absorption to about 2000 cm^{-1}. This overtone region usually contains a prominent band near 2222–2000 cm^{-1} assigned to a combination of the asymmetrical NH_3^+ bending vibration and the torsional oscillation of the NH_3^+ group. The torsional oscillation occurs near 500 cm^{-1}. The 2000 cm^{-1} band is absent if the nitrogen atom of the amino acid is substituted.
2. A weak asymmetrical NH_3^+ bending band near 1660–1610 cm^{-1}, a fairly strong symmetrical bending band near 1550–1485 cm^{-1}.
3. The carboxylate ion group (—C(O)O⁻, shown as —C with two partial C···O bonds) absorbs strongly near 1600–1590 cm^{-1} and more weakly near 1400 cm^{-1}. These bands result, respectively, from asymmetrical and symmetrical $C(\overset{\cdots}{=}O)_2$ stretching.

The spectrum of the amino acid leucine, including assignments corresponding to the preceding three categories, is shown in Figure 3.30.

Hydrochlorides of amino acids present the following patterns:

1. Broad, strong absorption in the 3333–2380-cm^{-1} region resulting from superimposed O—H and NH_3^+ stretching bands. Absorption in this region is characterized by multiple fine structure on the low-wavenumber side of the band.
2. A weak, asymmetrical NH_3^+ bending band near 1610–1590 cm^{-1}; a relatively strong, symmetrical NH_3^+ bending band at 1550–1481 cm^{-1}.

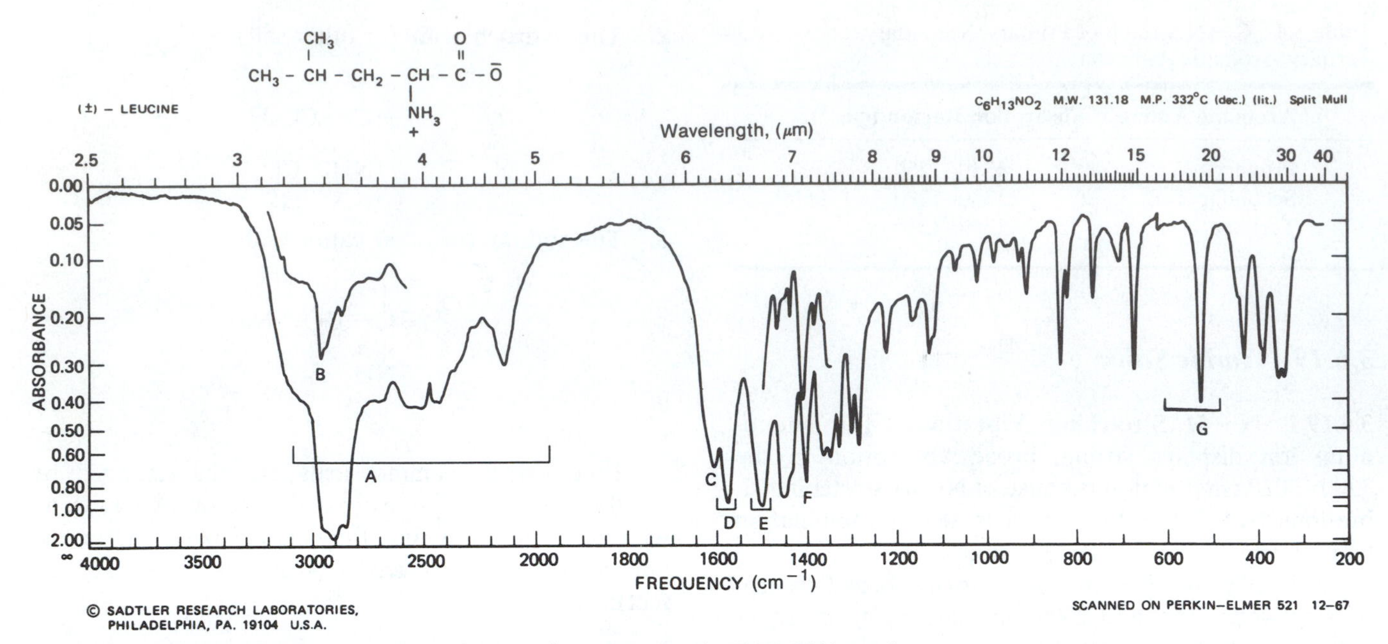

FIGURE 3.30. (±)-Leucine. A. Broad (—NH_3^+) N—H stretch, 3100–2000 cm^{-1}, extended by combination band at 2140 cm^{-1}, and other combination-overtone bands. B. Aliphatic C—H stretch (superimposed on N—H stretch), 2967 cm^{-1}. C. Asymmetric (—NH_3^+) N—H bend, 1610 cm^{-1}. D. Asymmetric carboxylate (C⩦O)$_2$ stretch, 1580 cm^{-1}. E. Symmetric (—NH_3^+) N—H bend, 1505 cm^{-1}. F. Symmetric carboxylate (C⩦O)$_2$ stretch, 1405 cm^{-1}. G. Torsional (—NH_3^+) N—H oscillation, 525 cm^{-1}.

3. A strong band at 1220–1190 cm^{-1} arising from C—C(=O)—O stretching.
4. Strong carbonyl absorption at 1755–1730 cm^{-1} for α-amino acid hydrochlorides, and at 1730–1700 cm^{-1} for other amino acid hydrochlorides.

Sodium salts of amino acids show the normal N—H stretching vibrations at 3400–3200 cm^{-1} common to other amines. The characteristic carboxylate ion bands appear near 1600–1590 cm^{-1} and near 1400 cm^{-1}.

3.6.21 *Nitriles*

The spectra of nitriles (R—C≡N) are characterized by weak to medium absorption in the triple-bond stretching region of the spectrum. Aliphatic nitriles absorb near 2260–2240 cm^{-1}. Electron-attracting atoms, such as oxygen or chlorine, attached to the carbon atom α to the C≡N group reduce the intensity of absorption. Conjugation, such as occurs in aromatic nitriles, reduces the frequency of absorption to 2240–2222 cm^{-1} and enhances the intensity. The spectrum of a typical nitrile, with an aryl group in conjugation with the cyano function, is shown in Figure 3.31.

3.6.22 *Isonitriles* (R—$\overset{+}{N}$≡$\overset{-}{C}$), *Cyanates* (R—O—C≡N), *Isocyanates* (R—N=C=O), *Thiocyanates* (R—S—C≡N), *and Isothiocyanates* (R—N=C=S)

These groups show the triple bond or cumulative bond stretch in the 2280–2000 cm^{-1} region.

3.6.23 *Compounds Containing* —N=N—*Group*

The N=N stretching vibration of a symmetrical *trans*-azo compound is forbidden in the IR but absorbs in the 1576 cm^{-1} region of the Raman spectrum. Unsymmetrical para-substituted azobenzenes in which the substituent is an electron-donating group absorb near 1429 cm^{-1}. The bands are weak because of the nonpolar nature of the bond.

3.6.24 *Covalent Compounds Containing Nitrogen–Oxygen Bonds*

Nitro compounds, nitrates, and nitramines contain an NO_2 group. Each of these classes shows absorption caused by asymmetrical and symmetrical stretching of the NO_2 group. Asymmetrical absorption results in a

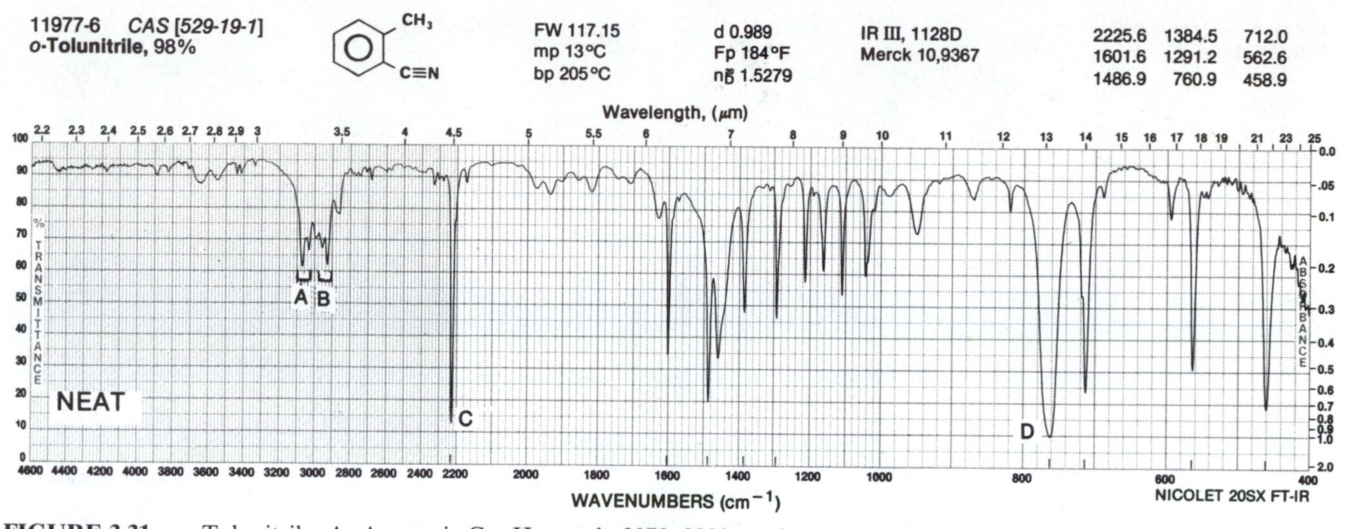

FIGURE 3.31. *o*-Tolunitrile. A. Aromatic C—H stretch, 3070, 3030 cm^{-1}. B. Aliphatic C—H stretch, 2960, 2930 cm^{-1}. C. The C≡N stretch, 2226 cm^{-1} (intensified by aryl conjugation; aliphatic nitriles absorb at higher frequency). D. Out-of-plane C—H bend (aromatic ring) D, 761 cm^{-1}.

strong band in the 1661–1499 cm^{-1} region; symmetrical absorption occurs in the region between 1389–1259 cm^{-1}. The exact position of the bands is dependent on substitution and unsaturation in the vicinity of the NO_2 group.

3.6.24.1 N⋯O Stretching Vibrations ***Nitro Compounds*** In the nitroalkanes, the bands occur near 1550 and 1372 cm^{-1}. Conjugation lowers the frequency of both bands, resulting in absorption near 1550–1500 and 1360–1290 cm^{-1}. Attachment of electronegative groups to the α carbon of a nitro compound causes an increase in the frequency of the asymmetrical NO_2 band and a reduction in the frequency of the symmetrical band; chloropicrin, Cl_3CNO_2, absorbs at 1610 and 1307 cm^{-1}.

Aromatic nitro groups absorb near the same frequencies as observed for conjugated aliphatic nitro compounds. Interaction between the NO_2 out-of-plane bending and ring C—H out-of-plane bending frequencies destroys the reliability of the substitution pattern observed for nitroaromatics in the long-wavelength region of the spectrum. Nitroaromatic compounds show a C—N stretching vibration near 870 cm^{-1}. The spectrum of nitrobenzene, with assignments corresponding to the preceding discussion, is shown in Figure 3.32.

Because of strong resonance in aromatic systems containing NO_2 groups and electron-donating groups such as the amino group, ortho or para to one another, the symmetrical NO_2 vibration is shifted to lower frequencies and increases in intensity. *p*-Nitroaniline absorbs at 1475 and 1310 cm^{-1}.

The positions of asymmetric and symmetric NO_2 stretching bands of nitramines (>N—NO_2) and the NO stretch of nitrosoamines are given in Appendix C.

Nitrates Organic nitrates show absorption for N—O stretching vibrations of the NO_2 group and for the O—N linkage. Asymmetrical stretching in the NO_2 group results in strong absorption in the 1660–1625 cm^{-1} region; the symmetrical vibration absorbs strongly near 1300–1255 cm^{-1}. Stretching of the π bonds of the N—O linkage produces absorption near 870–833 cm^{-1}. Absorption observed at longer wavelengths, near 763–690 cm^{-1}, likely results from NO_2 bending vibrations.

Nitrites Nitrites display two strong N=O stretching bands. The band near 1680–1650 cm^{-1} is attributed to the trans isomer; the cis isomer absorbs in the 1625–1610-cm^{-1} region. The N—O stretching band appears in the region between 850 and 750 cm^{-1}. The nitrite absorption bands are among the strongest observed in IR spectra.

Nitroso Compounds Primary and secondary aliphatic *C*-nitroso compounds are usually unstable and rearrange to oximes or dimerize. Tertiary and aromatic nitroso compounds are reasonably stable, existing as monomers in the gaseous phase or in dilute solution and as dimers in neat samples. Monomeric, tertiary, aliphatic nitroso compounds show N=O absorption in the 1585–1539 cm^{-1} region; aromatic monomers absorb between 1511 and 1495 cm^{-1}.

The N → O stretching absorption of dimeric nitroso

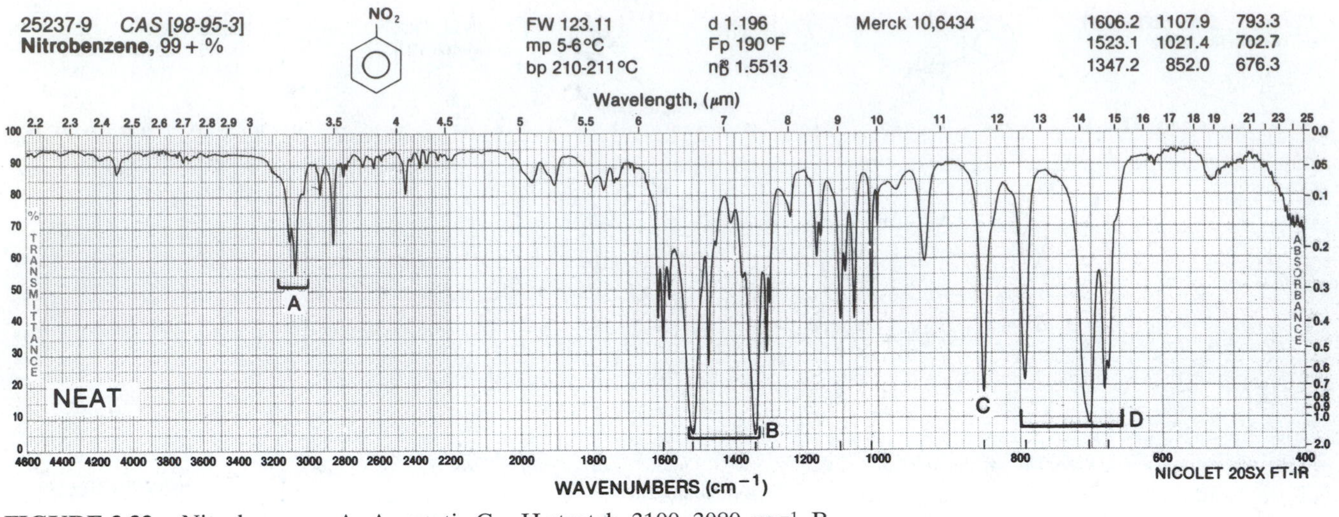

FIGURE 3.32. Nitrobenzene. A. Aromatic C—H stretch, 3100, 3080 cm^{-1}. B. Asymmetric ($ArNO_2$) $(N{=}O)_2$ stretch, 1523 cm^{-1}. Symmetric ($ArNO_2$) $(N{\equiv}O)_2$ stretch 1347 cm^{-1}. C. C—N stretch for $ArNO_2$, 852 cm^{-1}. D. Low-frequency bands are of little use in determining the nature of ring substitution since these absorption patterns result from interaction of NO_2 and C—H out-of-plane bending frequencies. The inability of the "oop" region to reveal structural information is typical of aromatic compounds with highly polar substituents.

compounds are categorized in Appendix C as to cis versus trans and aliphatic versus aromatic. Nitrosoamine absorptions are given in Appendix C.

3.6.25 *Organic Sulfur Compounds*

3.6.25.1 S—H Stretching Vibrations ***Mercaptans*** Aliphatic mercaptans and thiophenols, as liquids or in solution, show S—H stretching absorption in the range of 2600–2550 cm^{-1}. The S—H stretching band is characteristically weak and may go undetected in the spectra of dilute solutions or thin films. However, since few other groups show absorption in this region, it is useful in detecting S—H groups. The spectrum of benzyl mercaptan in Figure 3.33 is that of a mercaptan with a detectable S—H stretch band. The band may be obscured by strong carboxyl absorption in the same region. Hydrogen bonding is much weaker for S—H groups than for O—H and N—H groups.

The S—H group of thiol acids absorbs in the same region as mercaptans and thiophenols.

3.6.25.2 C—S and C═S Stretching Vibrations ***Sulfides*** The stretching vibrations assigned to the C—S linkage occur in the region of 700–600 cm^{-1}. The weakness of absorption and variability of position make this band of little value in structural determination.

Disulfides The S—S stretching vibration is very weak and falls between 500 and 400 cm^{-1}.

Thiocarbonyl Compounds Aliphatic thials or thiones exist as trimeric, cyclic sulfides. Aralkyl thiones may exist either as monomers or trimers, whereas diaryl thiones, such as thiobenzophenone, exist only as monomers. The C═S group is less polar than the C═O group and has a considerably weaker bond. In consequence, the band is not intense, and it falls at lower frequencies, where it is much more susceptible to coupling effects. Identification is therefore difficult and uncertain.

Compounds that contain a thiocarbonyl group show absorption in the 1250–1020 cm^{-1} region. Thiobenzophenone and its derivatives absorb moderately in the 1224–1207 cm^{-1} region. Since the absorption occurs in the same general region as C—O and C—N stretching, considerable interaction can occur between these vibrations within a single molecule.

Spectra of compounds in which the C═S group is attached to a nitrogen atom show an absorption band in the general C═S stretching region. In addition, several other bands in the broad region of 1563–700 cm^{-1} can be attributed to vibrations involving interaction between C═S stretching and C—N stretching.

Thioketo compounds that can undergo enolization exist as thioketo-thioenol tautomeric systems; such systems show S—H stretching absorption. The thioenol tautomer of ethyl thiobenzoylacetate,

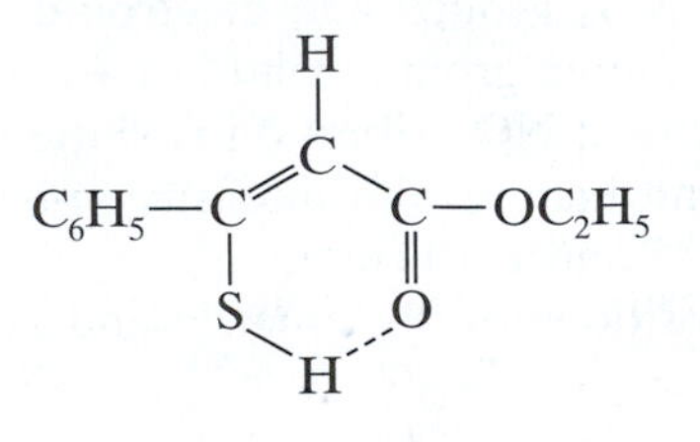

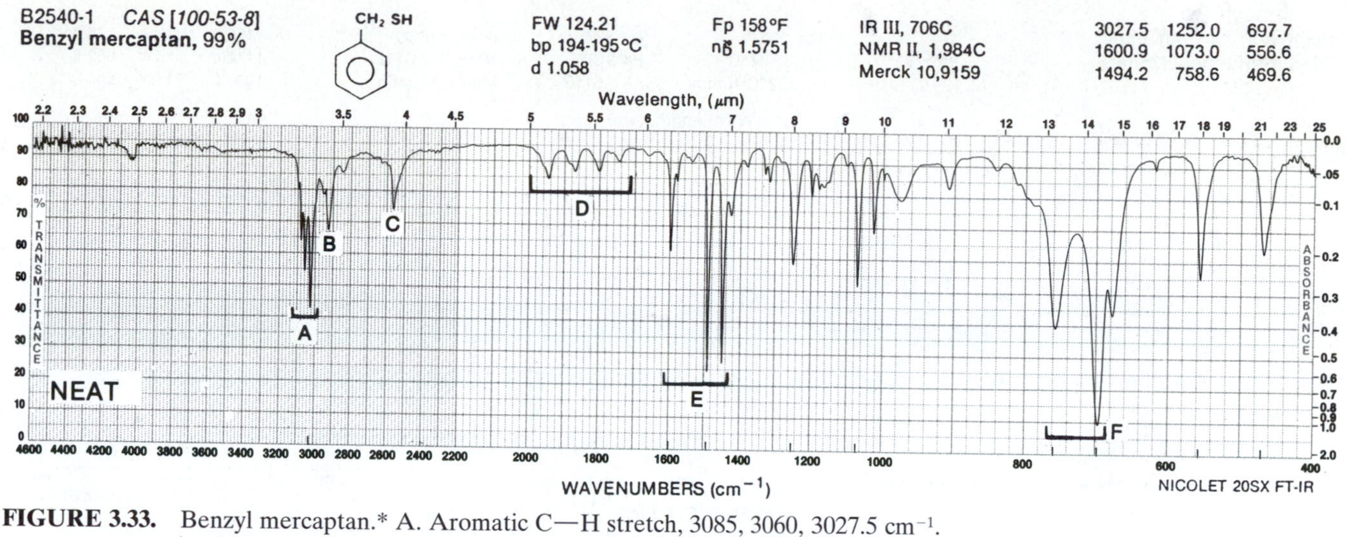

FIGURE 3.33. Benzyl mercaptan.* A. Aromatic C—H stretch, 3085, 3060, 3027.5 cm^{-1}. B. Aliphatic C—H stretch, 2930 cm^{-1}. C. Moderately weak S—H stretch, 2565 cm^{-1}. D. Overtone or combination band pattern indicative of monosubstituted aromatic, 2000–1667 cm^{-1}. E. The C═C ring stretch, 1601, 1495, 1455 cm^{-1}. F. Out-of-plane aromatic C—H bend (monosubstituted benzene ring), 759, 698 cm^{-1}.

*Benzyl mercaptan is another common name for α-toluenethiol.

absorbs broadly at 2415 cm^{-1} because of hydrogen-bonded S—H stretching absorption.

3.6.26 Compounds Containing Sulfur–Oxygen Bonds

3.6.26.1 S═O Stretching Vibrations ***Sulfoxides*** Alkyl and aryl sulfoxides as liquids or in solution show strong absorption in the 1070–1030 cm^{-1} region. This absorption occurs at 1050 cm^{-1} for dimethyl sulfoxide (DMSO, methyl sulfoxide) as may be seen in Appendix B, No. 26. Conjugation brings about a small change in the observed frequency, in contrast to the marked reduction in frequency of the C═O bond accompanying conjugation. Diallyl sulfoxide absorbs at 1047 cm^{-1}. Phenyl methyl sulfoxide and cyclohexyl methyl sulfoxide absorb at 1055 cm^{-1} in dilute solution in carbon tetrachloride. The sulfoxide group is susceptible to hydrogen bonding, the absorption shifting to slightly lower frequencies from dilute solution to the liquid phase. The frequency of S═O absorption is increased by electronegative substitution.

Sulfones Spectra of sulfones show strong absorption bands at 1350–1300 and 1160–1120 cm^{-1}. These bands arise from asymmetric and symmetric SO_2 stretching, respectively. Hydrogen bonding results in absorption near 1300 and 1125 cm^{-1}. Splitting of the high-frequency band often occurs in CCl_4 solution or in the solid state.

Sulfonyl Chlorides Sulfonyl chlorides absorb strongly in the regions of 1410–1380 and 1204–1177 cm^{-1}. This increase in frequency, compared with the sulfones, results from the electronegativity of the chlorine atom.

Sulfonamides Solutions of sulfonamides absorb strongly at 1370–1335 and 1170–1155 cm^{-1}. In the solid phase, these frequencies are lowered by 10–20 cm^{-1}. In solid samples, the high-frequency band is broadened and several submaxima usually appear.

Primary sulfonamides show strong N—H stretching bands at 3390–3330 and 3300–3247 cm^{-1} in the solid state; secondary sulfonamides absorb near 3265 cm^{-1}.

Sulfonates, Sulfates, and Sulfonic Acids The asymmetric (higher frequency, shorter wavelength) and symmetric S═O stretching frequency ranges for these compounds are as follows:

Class	Stretching Frequencies (cm^{-1})
Sulfonates (covalent)	1372–1335, 1195–1168
Sulfates (organic)	1415–1380, 1200–1185
Sulfonic acids	1350–1342, 1165–1150
Sulfonate salts	~1175 ~1055

The spectrum of a typical alkyl arenesulfonate is given in Figure 3.34. In virtually all sulfonates, the asymmetric stretch occurs as a doublet. Alkyl and aryl sul-

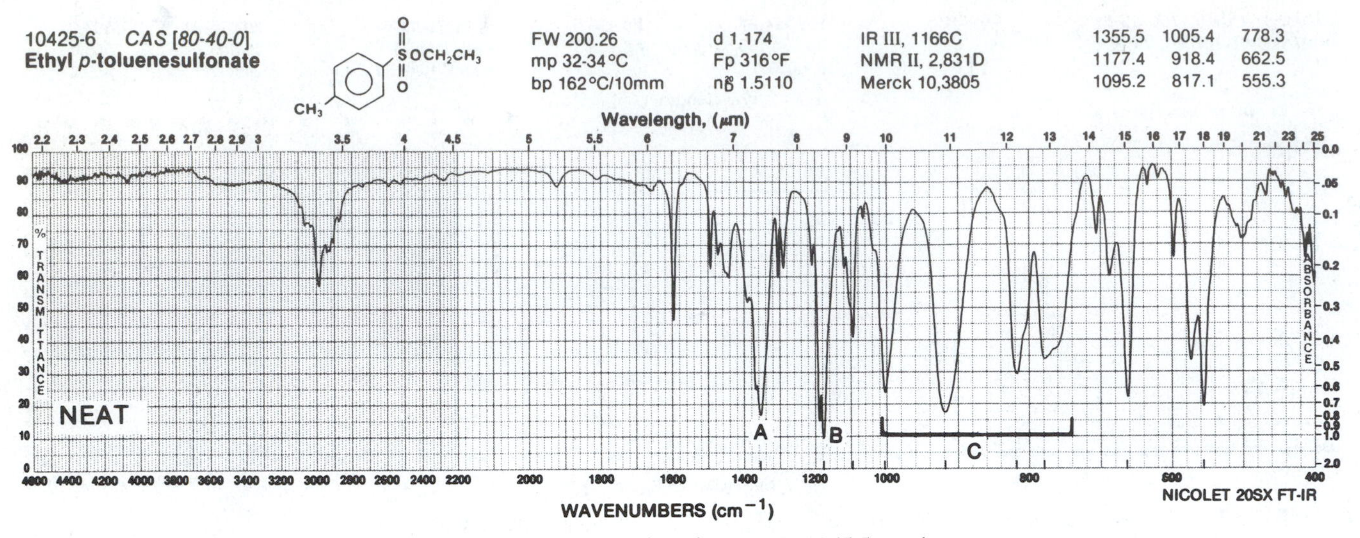

FIGURE 3.34. Ethyl *p*-toluenesulfonate. A. Asymmetric $S(=O)_2$ stretch, 1355.5 cm^{-1}. B. Symmetric $S(=O)_2$ stretch, 1177 cm^{-1}. C. Various strong S—O—C stretching, 1000–769 cm^{-1}.

fonates show negligible differences; electron-donating groups in the para position of arenesulfonates cause higher frequency absorption.

Sulfonic acids are listed in narrow ranges above; these apply only to anhydrous forms. Such acids hydrate readily to give bands that are probably a result of the formation of hydronium sulfonate salts, in the 1230–1120 cm^{-1} range.

3.6.27 *Organic Halogen Compounds*

The strong absorption of halogenated hydrocarbons arises from the stretching vibrations of the carbon-halogen bond.

Aliphatic C—Cl absorption is observed in the broad region between 850 and 550 cm^{-1}. When several chlorine atoms are attached to one carbon atom, the band is usually more intense and at the high-frequency end of the assigned limits. Carbon tetrachloride (see Appendix B, No. 10) shows an intense band at 797 cm^{-1}. The first overtones of the intense fundamental bands are frequently observed. Spectra of typical chlorinated hydrocarbons are shown in Appendix B: Nos. 10–13. Brominated compounds absorb in the 690–515 cm^{-1} region, iodo compounds in the 600–500 cm^{-1} region. A strong CH_2 wagging band is observed for the CH_2X (X = Cl, Br, and I) group in the 1300–1150 cm^{-1} region.

Fluorine-containing compounds absorb strongly over a wide range between 1400 and 1000 cm^{-1} because of C—F stretching modes. A monofluoroalkane shows a strong band in the 1100–1000 cm^{-1} region. As the number of fluorine atoms in an aliphatic molecule increases, the band pattern becomes more complex, with multiple strong bands appearing over the broad region of C—F absorption (see Fluorolube spectrum, Appendix C). The CF_3 and CF_2 groups absorb strongly in the 1350–1120-cm^{-1} region. The spectrum of Fluorolube®, Appendix B, No. 14, illustrates many of the preceding absorption characteristics.

Chlorobenzenes absorb in the 1096–1089 cm^{-1} region. The position within this region depends on the substitution pattern. Aryl fluorides absorb in the 1250–1100 cm^{-1} region of the spectrum. A monofluorinated benzene ring displays a strong, narrow absorption band near 1230 cm^{-1}.

3.6.28 *Silicon Compounds*

3.6.28.1 Si—H Vibrations Vibrations for the Si—H bond include the Si—H stretch (~2200 cm^{-1}) and the Si—H bend (800–950 cm^{-1}). The Si—H stretching frequencies are increased by the attachment of an electronegative group to the silicon.

3.6.28.2 SiO—H and Si—O Vibrations The OH stretching vibrations of the SiOH group absorb in the same region as the alcohols, 3700–3200 cm^{-1}, and strong Si—O bands are at 830–1110 cm^{-1}. As in alcohols, the absorption characteristics depend on the degree of hydrogen bonding.

The spectrum of silicone lubricant, Appendix B (No. 27), illustrates some of the preceding absorptions.

3.6.28.3 Silicon–Halogen Stretching Vibrations Absorption caused by Si—F stretch is in the 800–1000 region.

Bands resulting from Si—Cl stretching occur at frequencies below 666 cm^{-1}.

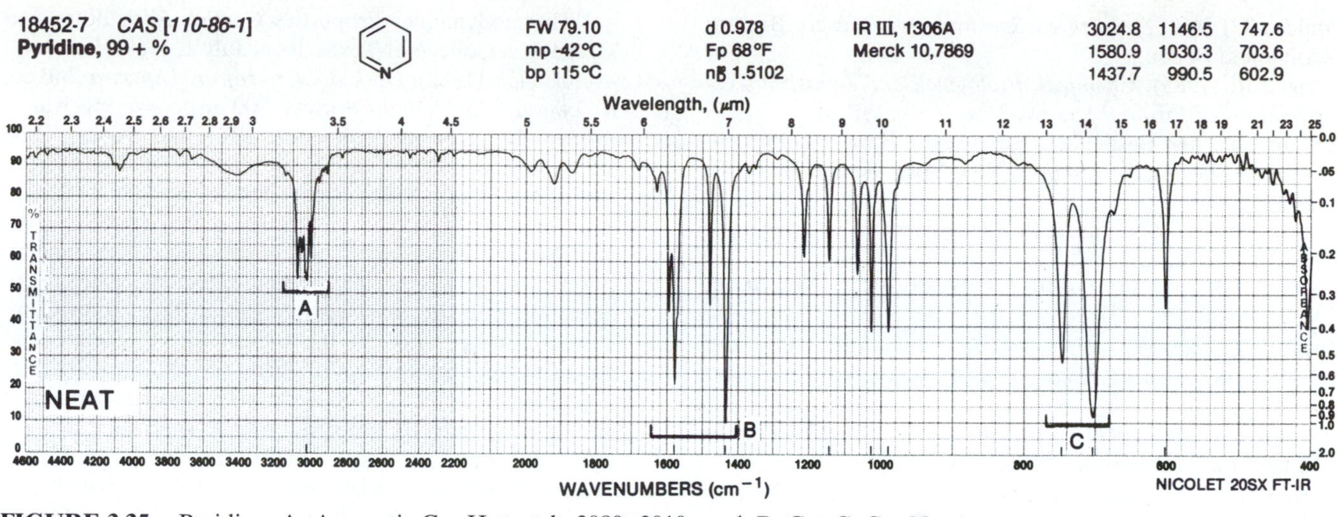

FIGURE 3.35. Pyridine. A. Aromatic C—H stretch, 3080–3010 cm^{-1}. B. C⩵C, C⩵N ring stretching (skeletal bands), 1600–1430 cm^{-1}. C. The C—H out-of-plane bending, 748, 704 cm^{-1}. See Appendix F, Table F-1 for patterns in region C for substituted pyridines.

3.6.29 Phosphorus Compounds

3.6.29.1 P=O and P—O Stretching Vibrations Such absorptions are listed in Appendix E, Table E-1.

3.6.30 Heteroaromatic Compounds

The spectra of heteroaromatic compounds result primarily from the same vibrational modes as observed for the aromatics.

3.6.30.1 C—H Stretching Vibrations Heteroaromatics, such as pyridines, pyrazines, pyrroles, furans, and thiophenes, show C—H stretching bands in the 3077–3003 cm^{-1} region.

3.6.30.2 N—H Stretching Frequencies Heteroaromatics containing an N—H group show N—H stretching absorption in the region of 3500–3220 cm^{-1}. The position of absorption within this general region depends on the degree of hydrogen bonding, and hence upon the physical state of the sample or the polarity of the solvent. Pyrrole and indole in dilute solution in nonpolar solvents show a sharp band near 3495 cm^{-1}; concentrated solutions show a widened band near 3400 cm^{-1}. Both bands may be seen at intermediate concentrations.

3.6.30.3 Ring Stretching Vibrations (Skeletal Bands) Ring stretching vibrations occur in the general region between 1600 and 1300 cm^{-1}. The absorption involves stretching and contraction of all of the bonds in the ring and interaction between these stretching modes. The band pattern and the relative intensities depend on the substitution pattern and the nature of the substituents.

Pyridine (Fig. 3.35) shows four bands in this region and, in this respect, closely resembles a monosubstituted benzene. Furans, pyrroles, and thiophenes display two to four bands in this region.

3.6.30.4 C—H Out-of-Plane Bending The C—H out-of-plane bending (γ-CH) absorption pattern of the heteroaromatics is determined by the number of adjacent hydrogen atoms bending in phase. The C—H out-of-plane and ring bending (β ring) absorption of the alkylpyridines are summarized in Appendix F, Table F-1.

Absorption data for the out-of-phase C—H bending (γ-CH) and ring bending (β ring) modes of three common five-membered heteroaromatic rings are presented in Appendix F, Table F-2. The ranges in Table F-2 include polar as well as nonpolar substituents on the ring.

References

Introductory and Theoretical

Bellamy, L.J. (1975). *The Infrared Spectra of Complex Molecules,* 3rd ed. London: Chapman and Hall: New York: Halsted–Wiley.

Bellamy, L.J. (1968). *Advances in Infrared Group Frequencies.* London: Methuen.

Coleman, P.A. (1991). *Practical Sampling Techniques for Infrared Analysis.* Boca Raton, FL: CRC Press.

Colthup, N.B., Daly, L.H., and Wiberley, S.E. (1990). *Introduction to Infrared and Raman Spectroscopy,* 3rd ed. New York and London: Academic Press.

Conley, R.T. (1972). *Infrared Spectroscopy,* 2nd ed. Boston: Allyn and Bacon.

Durig, J.R. (1985). *Chemical, Biological, and Industrial Applications of Infrared Spectra.* New York: Wiley.

Ferraro, J.R., and Basile, L.J., Eds. (1985). *Fourier Transform Infrared Spectroscopy,* 4 vols. New York: Academic Press. Applications.

George, W.O., and McIntire, P. (1986). *Infrared Spectroscopy.* New York: Wiley.

Griffiths, P.R., and DeHaseth, J.A. (1986). *FTIR,* New York: Wiley, 1986.

Herres, W. (1987). *HRGC-FTIR, Capillary GC, and FTIR Spectroscopy: Theory and Applications.* New York: Huthig.

Mattson, J.S., Ed. (1977). *Infrared, Correlation and Fourier Transform Spectroscopy.* New York: Marcel Dekker.

Nakanishi, K., and Solomon, P.H. (1977). *Infrared Absorption Spectroscopy–Practical,* 2nd ed. San Francisco: Holden–Day.

Roeges, N.P.G. (1984). *Guide to Interpretation of Infrared Spectra of Organic Structures.* New York: Wiley.

Smith, B.C. (1995). *Fundamentals of Fourier Transform Infrared Spectroscopy.* Boca Raton, FL: CRC Press.

Szymanski, H.A. Ed. (1969, 1970). *Raman Spectroscopy: Theory and Practice,* Vols. 1, 2. New York: Plenum Press.

Szymanski, H.A. (1989). *Correlation of Infrared and Raman Spectra of Organic Compounds.* Hertillon Press.

Szymanski, H.A. (1964). *IR Theory and Practice of Infrared Spectroscopy.* New York: Plenum Press.

Szymanski, H.A. (1964–1967). *Interpreted Infrared Spectra,* Vols. I–III. New York: Plenum Press.

Szymanski, H.A. (1963). *Infrared Band Handbook.* New York: Plenum Press.

Compilations of Spectra or Data; Workbooks, Special Topics

Aldrich Library of FTIR Spectra (1985), 3 Vols. Milwaukee, WI: Includes vapor-phase spectrum Vol. 3

ASTM-Wyandotte Index, Molecular Formula List of Compounds, Names and References to Published Infrared Spectra. Philadelphia, PA: ASTM Special Technical Publications 131 (1962) and 131-A (1963). Lists about 57,000 compounds. Covers IR, near-IR, and far-IR spectra.

Bentley, F.F., Smithson, L.D., and Rozek, A.L. (1988). *Infrared Spectra and Characteristic Frequencies ~700 to 300 cm^{-1}, A Collection of Spectra, Interpretation and Bibliography.* New York: Wiley–Interscience.

Brown, C. R., Ayton, M.W., Goodwin, T.C., and Derby, T.J. (1954). *Infrared—A Bibliography.* Washington, DC: Library of Congress, Technical Information Division.

Catalog of Infrared Spectrograms, American Petroleum Institute Research Project 44. Pittsburgh, PA: Carnegie Institute of Technology.

Catalog of Infrared Spectrograms. Philadelphia: Sadtler Research Laboratories, PA. 19104. Spectra are indexed by name and by the Spec-Finder. The latter is an index that tabulates major bands by wavelength intervals. This allows quick identification of unknown compounds.

Catalog of Infrared Spectral Data, Manufacturing Chemists Association Research Project, Chemical and Petroleum Research Laboratories. Pittsburgh, PA: Carnegie Institute of Technology, to June 30, 1960; College Station, TX: Chemical Thermodynamics Properties Center, Agriculture and Mechanical College of Texas, from July 1, 1960.

Craver, C., Ed. (1982). *Desk Book of Infrared Spectra,* 2nd ed. Kirkwood MO: Coblenz Society. 900 dispersive spectra.

Dobriner, K., Katzenellenbogen, E.R., and Jones, R.N. (1953). *Infrared Absorption Spectra of Steriods—An Atlas,* Vol. 1. New York: Wiley–Interscience.

Documentation of Molecular Spectroscopy (DMS), London: Butterworths Scientific Publications, and Weinheim/Bergstrasse. West Germany: Verlag Chemie GMBH, in cooperation with the Infrared Absorption Data Joint Committee, London, and the Institut für Spectrochemie und Angewandte Spectroskopie, Dortmund. Spectra are presented on coded cards. Coded cards containing abstracts of articles relating to IR spectrometry are also issued.

Dolphin, D., and Wick, A.E. (1977). *Tabulation of Infrared Spectral Data.* New York: Wiley.

Finch, A., Gates, P.N., Radcliffe, K., Dickson, F.N., and Bentley, F.F. (1970). *Chemical Applications of Far Infrared Spectroscopy.* New York: Academic Press.

Flett, M. St. C. (1969). *Characteristic Frequencies of Chemical Groups in the Infrared.* New York: American Elsevier.

Hershenson, H.H. (1959, 1964). *Infrared Absorption Spectra Index.* New York and London: Academic Press. Two volumes cover 1945–1962.

Infrared Band Handbook. Supplements 1 and 2. (1964). New York: Plenum Press, 259 pp.

Infrared Band Handbook. Supplements 3 and 4. (1965). New York: Plenum Press.

Infrared Band Handbook. 2nd rev. ed. (1970). New York: IFI/Plenum Press. Two volumes.

IRDC Cards: *Infrared Data Committee of Japan* (S. Mizushima). Haruki-cho, Tokyo: Handled by Nankodo Co.

Katritzky, A.R., Ed. (1963). *Physical Methods in Heterocyclic Chemistry,* Vol. II. New York and London: Academic Press.

Lin-Vien, D., Colthup, N.B. Fately, W.G., and Grasselli, J.G. (1991). *Handbook of Infrared and Raman Characteristic Frequencies of Organic Molecules.* New York: Academic Press.

Loader, E.J. (1970). *Basic Laser Raman Spectroscopy.* London, New York: Heyden–Sadtler.

Maslowsky, E. (1977). *Vibrational Spectra of Organometallic Compounds.* New York: Wiley–Interscience.

Ministry of Aviation Technical Information and Library Services, Ed. (1960). *An Index of Published Infra-Red Spectra,* Vols. 1 and 2. London: Her Majesty's Stationery Office.

NRC-NBS (Creitz) File of Spectrograms. Issued by National Research Council-National Bureau of Standards Committee on Spectral Absorption Data, National Bureau of Standards, Washington 25, DC. Spectra presented on edge-punched cards.

Nyquist, R.A. (1984). *Interpretation of Vapor-Phase Infrared Spectra.* Philadelphia: Sadtler-Heyden.

Porro, T.J., and Pattacini, S.C. (1993). *Spectroscopy* **8**(7), 40–47. Sample handling.

Sadtler Standard Infrared Grating Spectra. (1972). Philadelphia: Sadtler Research Laboratories, Inc. 26,000 spectra in 26 volumes.

Sadtler Standard Infrared Prism Spectra. (1972). Philadelphia: Sadtler Research Laboratories, Inc. 43,000 spectra in 43 volumes.

Sadtler Reference Spectra—Commonly Abused Drugs IR & UV Spectra. (1972). Philadelphia: Sadtler Research Laboratories, Inc. 600 IR Spectra, 300 UV.

Sadtler Reference Spectra—Gases & Vapors High Resolution Infrared. (1972). Philadelphia: Sadtler Research Laboratories, Inc. 150 Spectra.

Sadtler Reference Spectra—Inorganics IR Grating. (1967). Philadelphia: Sadtler Research Laboratories, Inc. 1300 spectra.

Sadtler Reference Spectra—Organometallics IR Grating. (1966). Philadelphia: Sadtler Research Laboratories, Inc. 400 spectra.

Sadtler Digital FTIR Libraries. (1989). Philadelphia: Sadtler Research Labs. 100,000 digitized spectra.

Socrates, G. (1994). *IR Characteristic Group Frequencies.* New York: Wiley.

Stewart, J.E. (1965). "Far Infrared Spectroscopy." In S.K. Freeman (Ed.), *Interpretive Spectroscopy.* New York: Reinhold, p. 131.

Tichy, M. (1964). "The Determination of Intramolecular Hydrogen Bonding by Infrared Spectroscopy and Its Applications in Stereochemistry." In R.R. Raphael (Ed.), *Advances in Organic Chemistry: Methods and Results,* Vol. 5. New York: Wiley–Interscience.

Problems

3.1 Either benzonitrile or phenylacetonitrile shows a band of medium intensity at 2940 cm^{-1}; the other compound shows nothing in the range 3000–2500 cm^{-1}. Explain.

3.2 Select a compound that best fits each of the following sets of IR bands (in cm^{-1}). Each set corresponds to a list of just a few important bands for each compound.

Benzamide	Diphenyl sulfone
Benzoic acid	Formic acid
Benzonitrile	Isobutylamine
Biphenyl	1-Nitropropane
Dioxane	

a. 3080 (w), nothing 3000–2800, 2230 (s), 1450 (s), 760 (s), 688 (s)

b. 3380 (m), 3300 (m), nothing 3200–3000, 2980 (s), 2870 (m), 1610 (m), ~900 to ~700 (b)

c. 3080 (w), nothing 3000–2800, 1315 (s), 1300 (s), 1155 (s)

d. 2955 (s), 2850 (s), 1120 (s)

e. 2946 (s), 2930 (m), 1550 (s), 1386 (m)

f. 2900 (b, s), 1720 (b, s)

g. 3030 (m), 730 (s), 690 (s)

h. 3200–2400 (s), 1685 (b, s), 705 (s)

i. 3350 (s), 3060 (m), 1635 (s)

s = strong, m = medium, w = weak, b = broad

For Problems 3.3–3.6, match the name from each list to the proper IR spectrum. Identify the diagnostic bands in each spectrum.

3.3 SPECTRA A–D
1,3-Cyclohexadiene
Diphenylacetylene
1-Octene
2-Pentene

3.4 SPECTRA E–I
Butyl acetate
Butyramide
Isobutylamine
Lauric acid
Sodium propionate

3.5 SPECTRA J–M
Allyl phenyl ether
Benzaldehyde
o-Cresol
m-Toluic acid

3.6 SPECTRA N–R
Aniline
Azobenzene
Benzophenone oxime
Benzylamine
Dimethylamine hydrochloride

3.7 Deduce the structure of a compound (S) whose formula is C_2H_3NS from the spectrum below.

3.8 Point out evidence for enol formation of 2,4-pentanedione (Compound T). Include explanations of the two bands in the 1700–1750-cm^{-1} range, the 1650 band, and the very broad band with multiple maxima running from 3400 to 2600 cm^{-1} (only the peaks at 3000–2900 result from C—H stretching).

Problem 3.3 Spectrum A

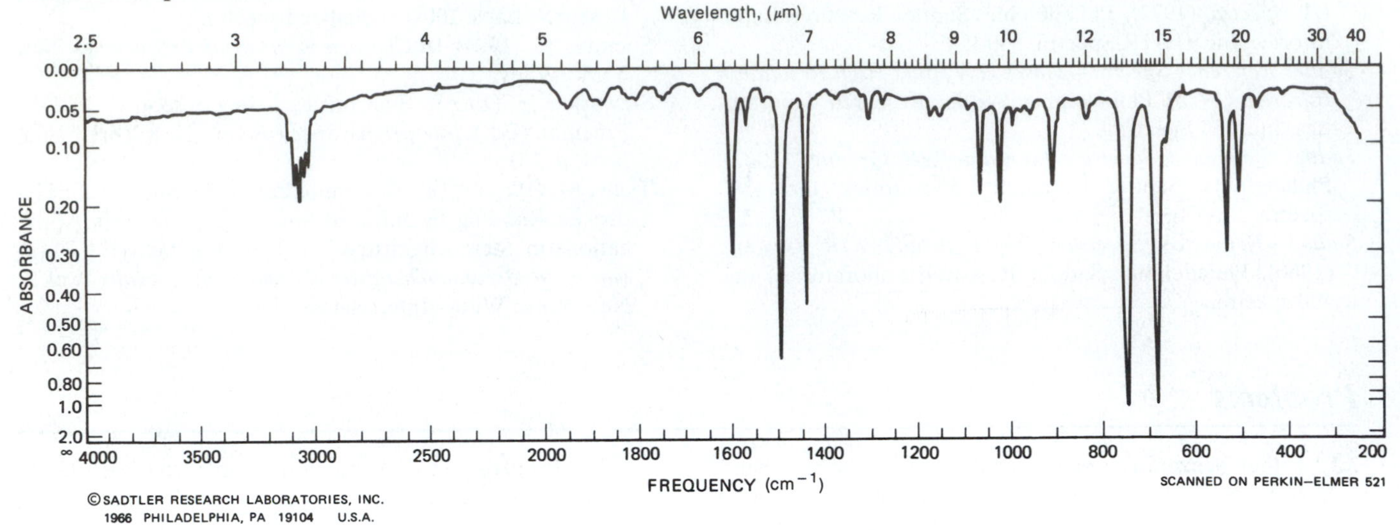

Problem 3.3 Spectrum B

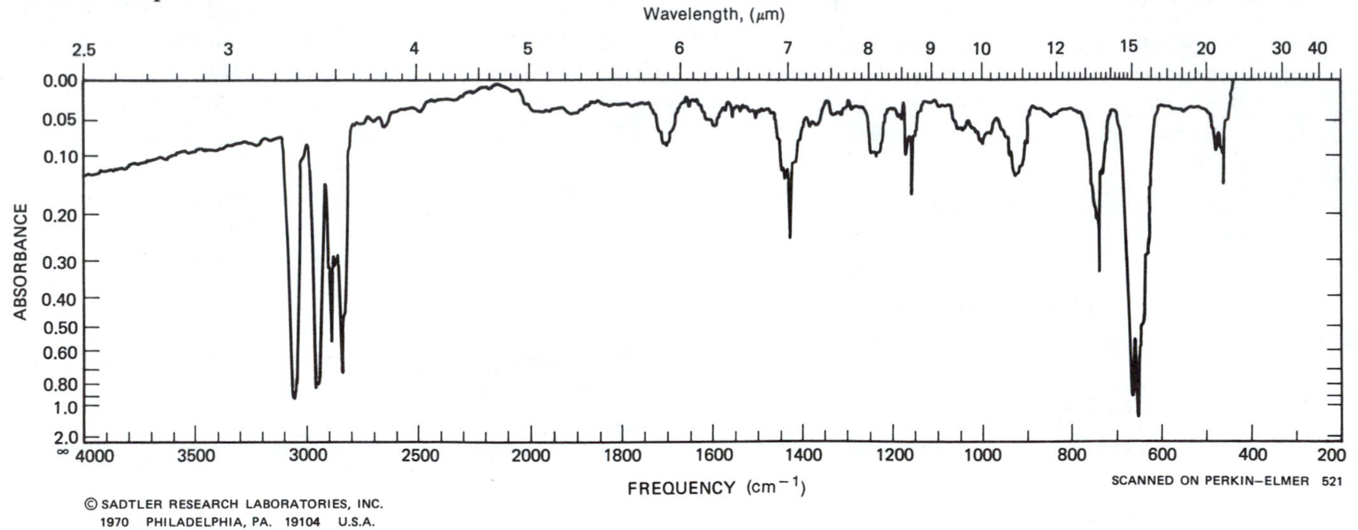

Problem 3.3 Spectrum C

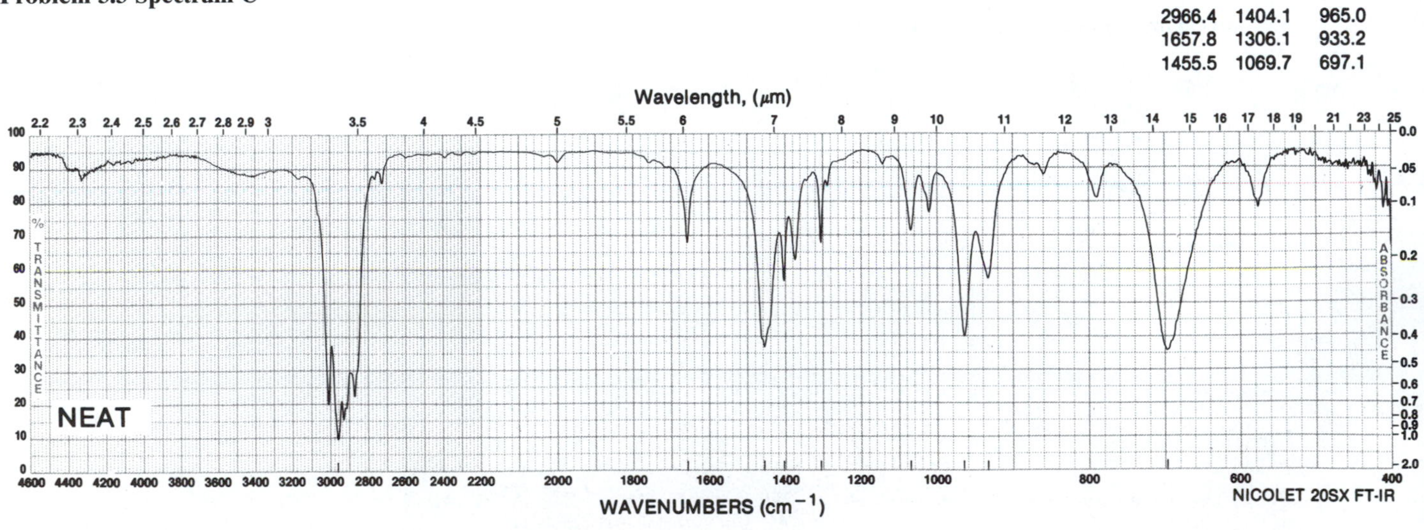

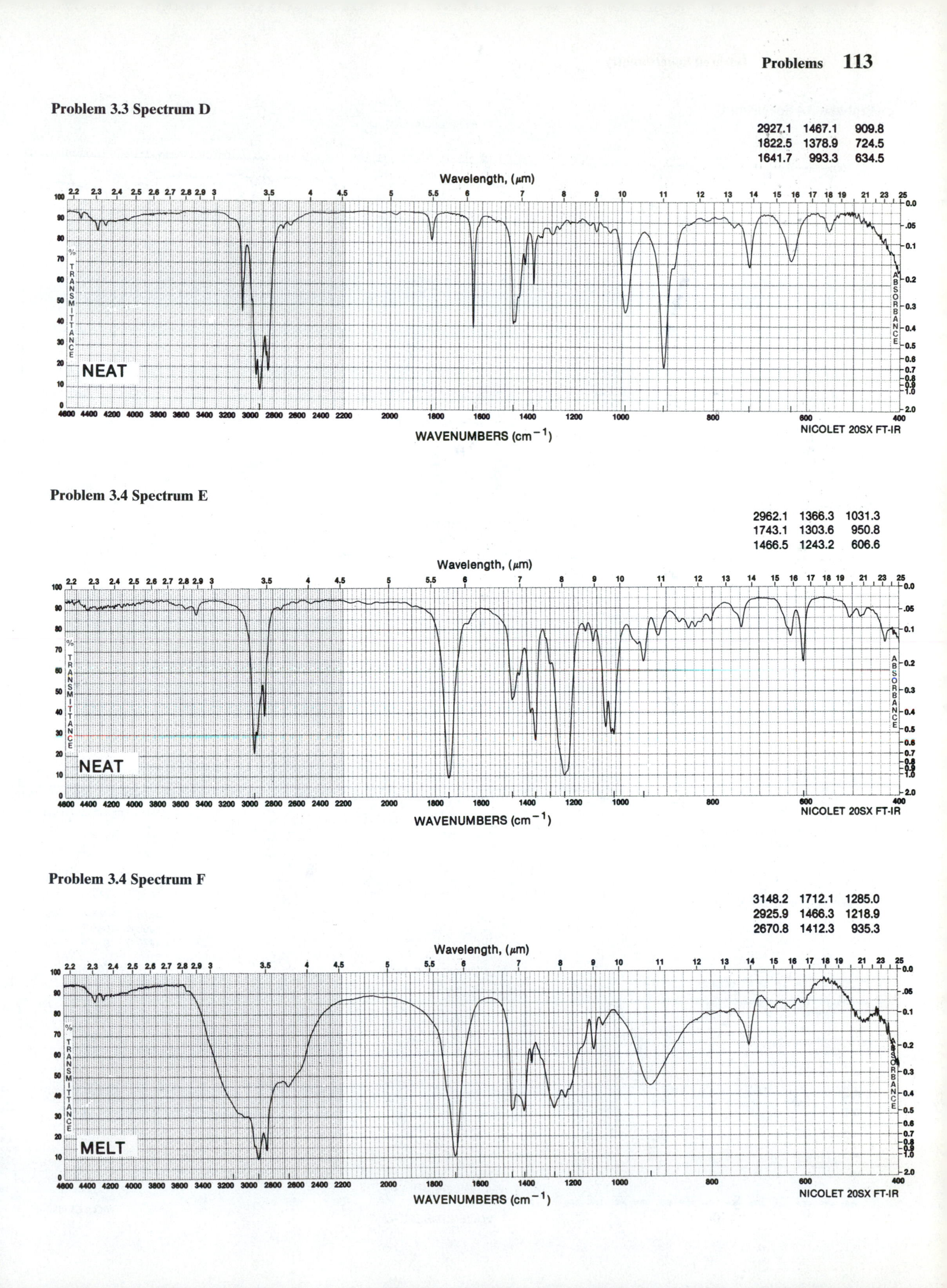
Problem 3.3 Spectrum D
2927.1 1467.1 909.8
1822.5 1378.9 724.5
1641.7 993.3 634.5
Wavelength, (μm)
NEAT
% TRANSMITTANCE
ABSORBANCE
WAVENUMBERS (cm⁻¹)
NICOLET 20SX FT-IR
Problem 3.4 Spectrum E
2962.1 1366.3 1031.3
1743.1 1303.6 950.8
1466.5 1243.2 606.6
Wavelength, (μm)
NEAT
% TRANSMITTANCE
ABSORBANCE
WAVENUMBERS (cm⁻¹)
NICOLET 20SX FT-IR
Problem 3.4 Spectrum F
3148.2 1712.1 1285.0
2925.9 1466.3 1218.9
2670.8 1412.3 935.3
Wavelength, (μm)
MELT
% TRANSMITTANCE
ABSORBANCE
WAVENUMBERS (cm⁻¹)
NICOLET 20SX FT-IR

Problem 3.4 Spectrum G

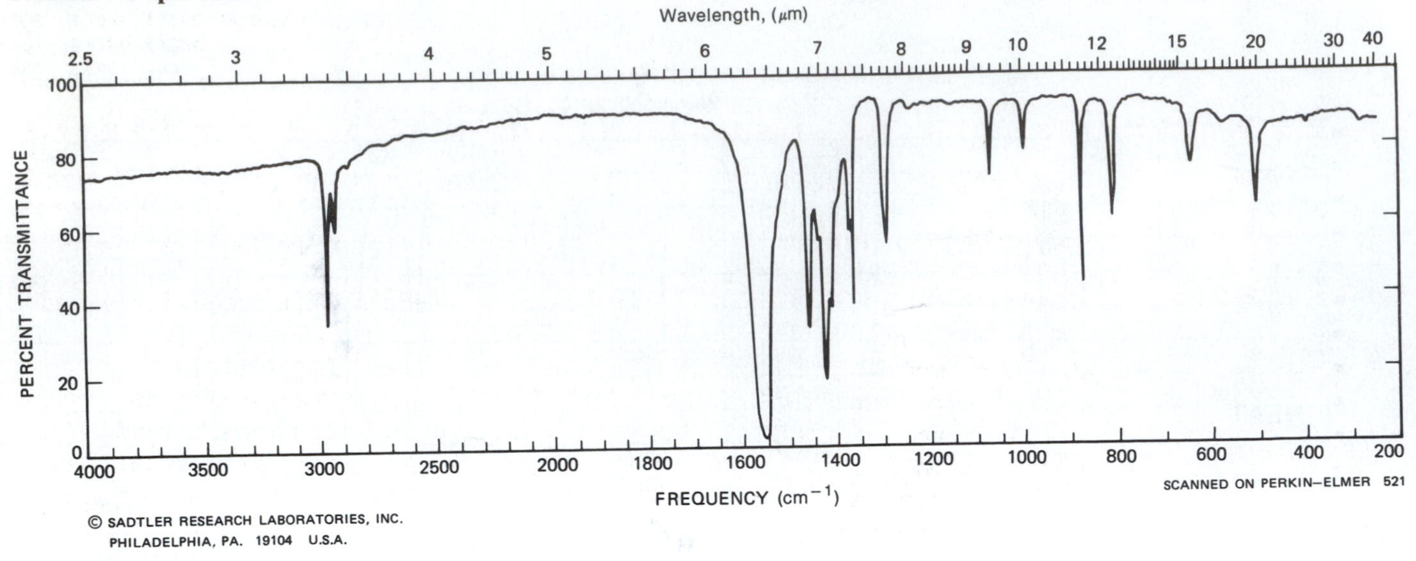

Problem 3.4 Spectrum H

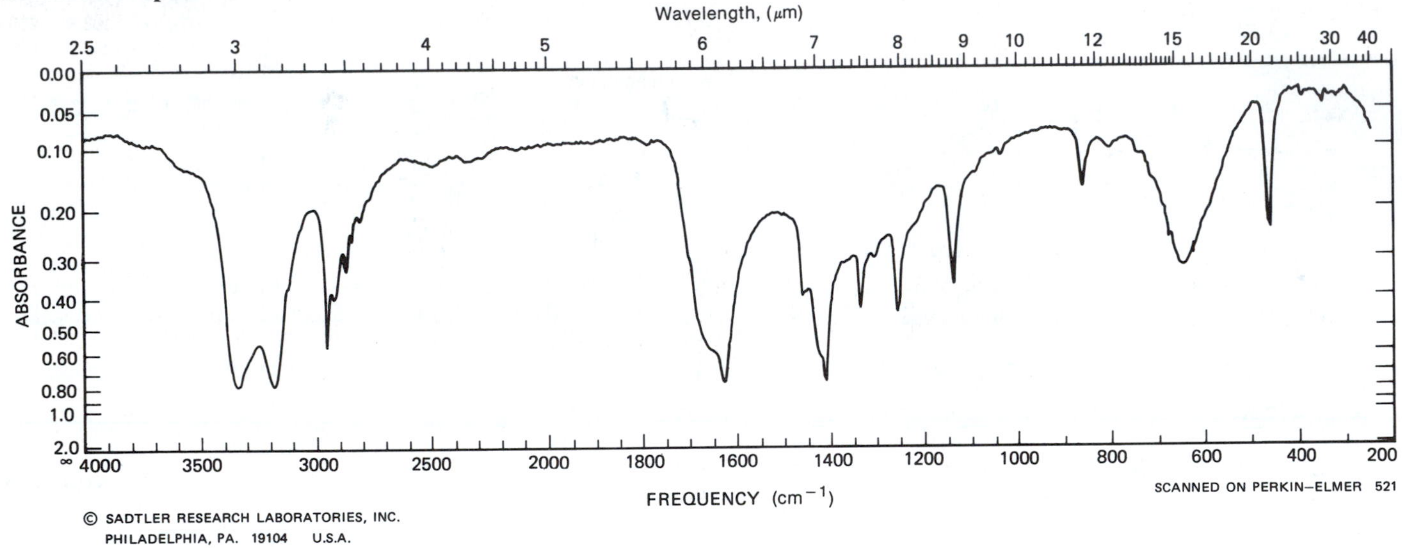

Problem 3.4 Spectrum I

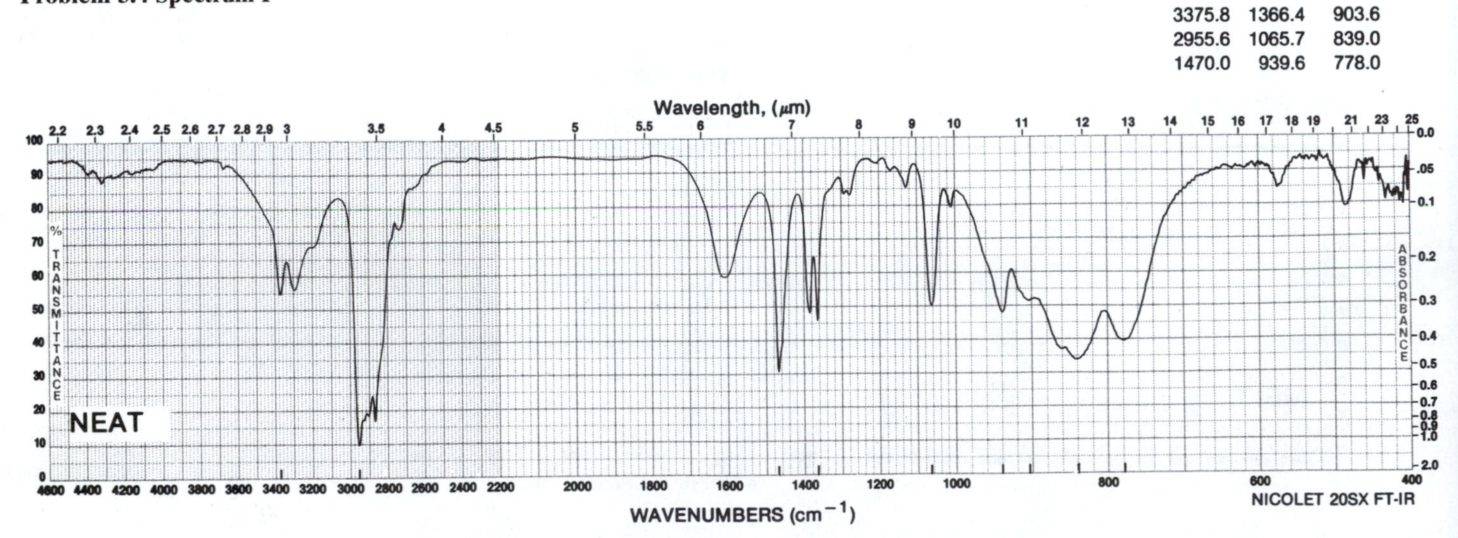

Problem 3.5 Spectrum J

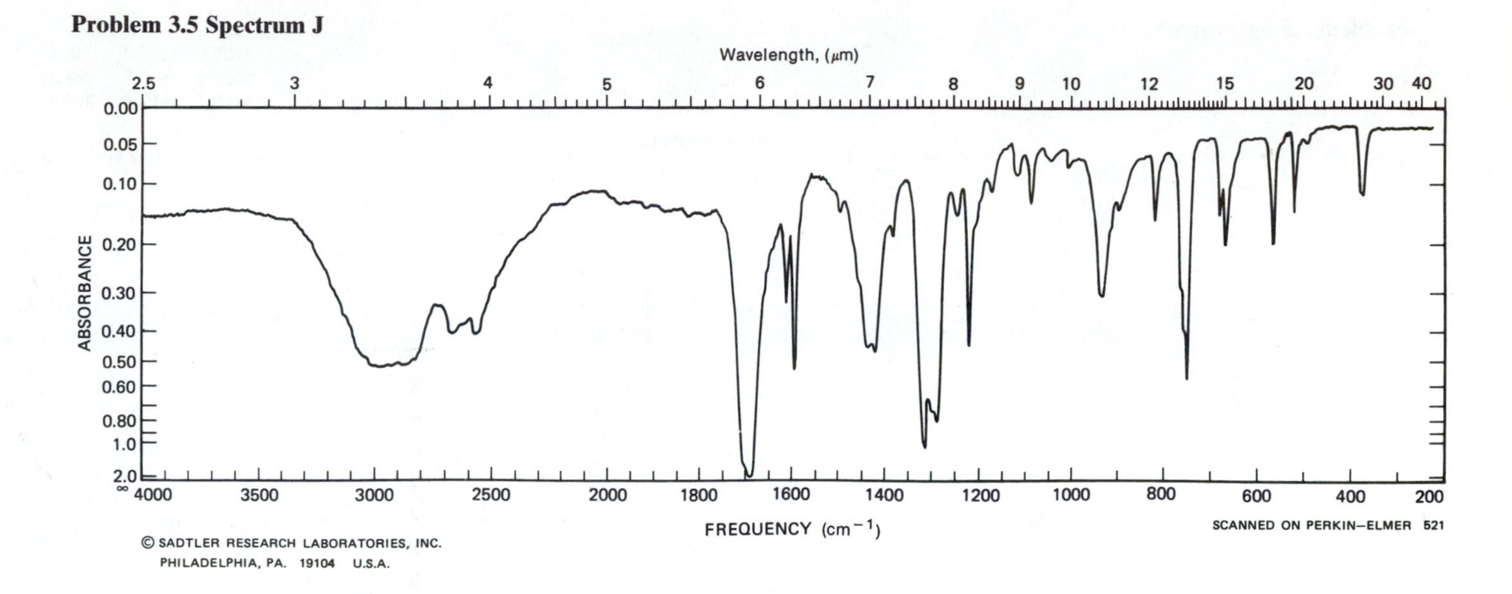

Problem 3.5 Spectrum K

3447.7	1327.7	1106.2
1592.6	1239.7	843.1
1464.2	1170.0	751.3

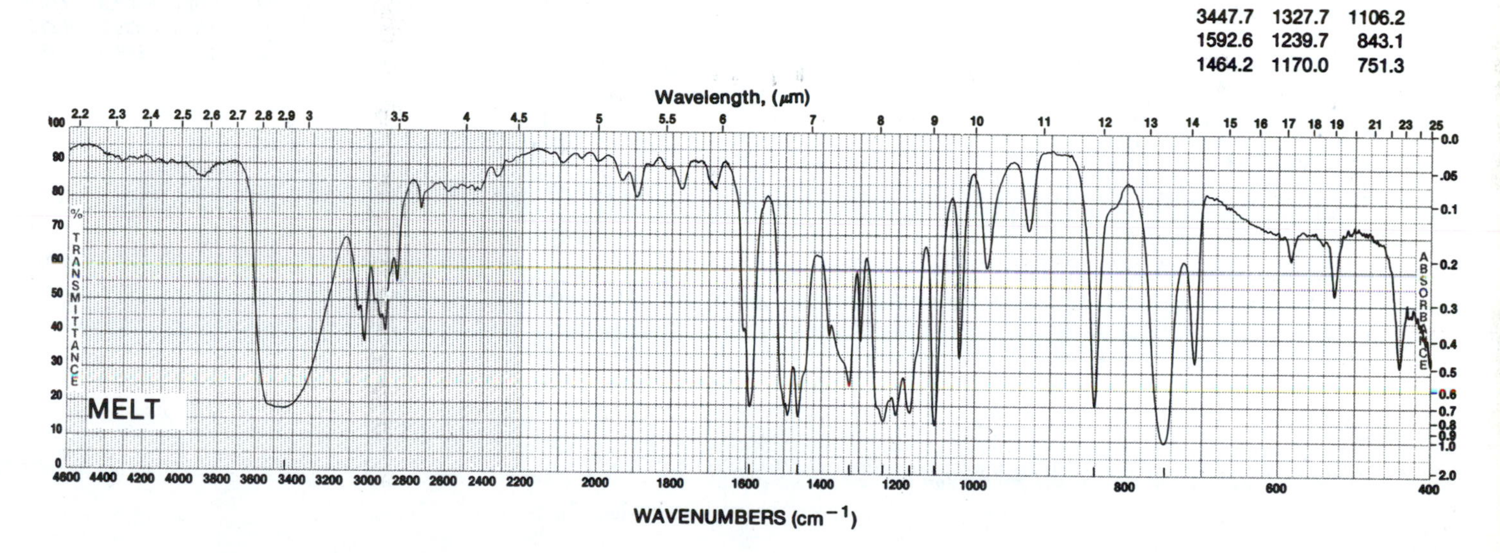

Problem 3.5 Spectrum L

1599.2	1173.0	926.6
1495.8	1033.3	753.8
1242.5	991.7	691.3

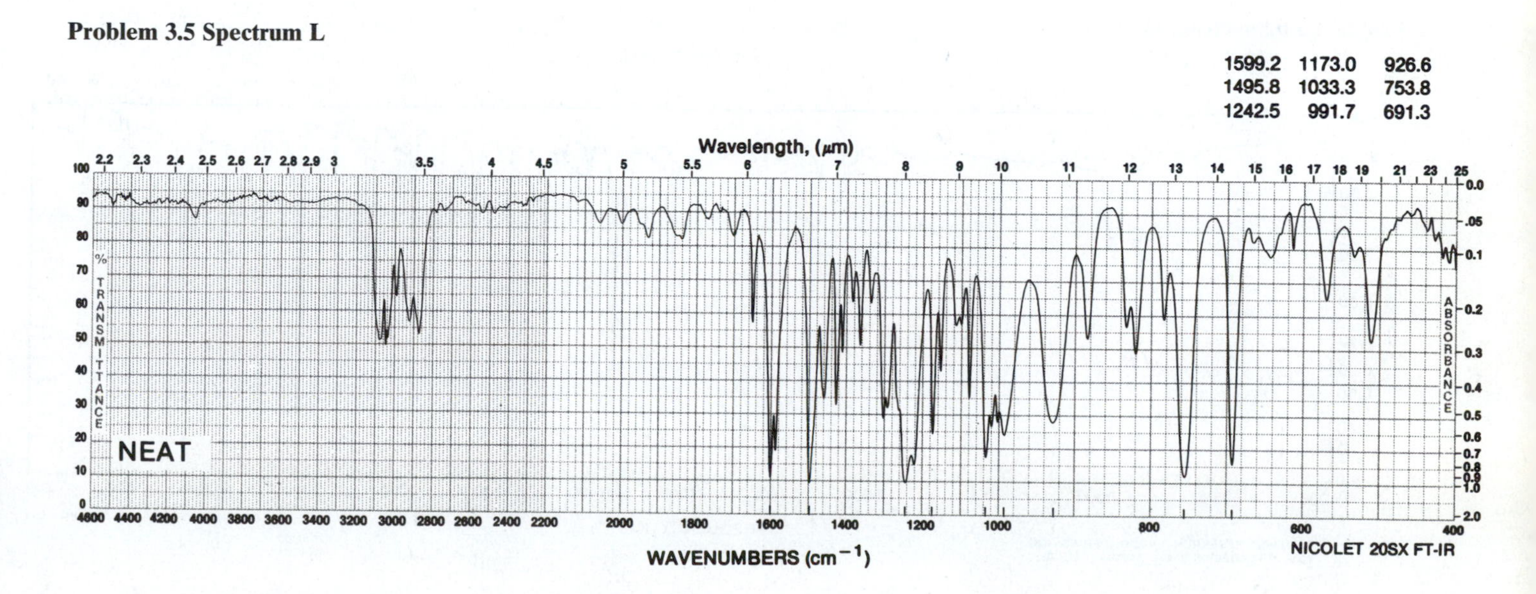

Problem 3.5 Spectrum M

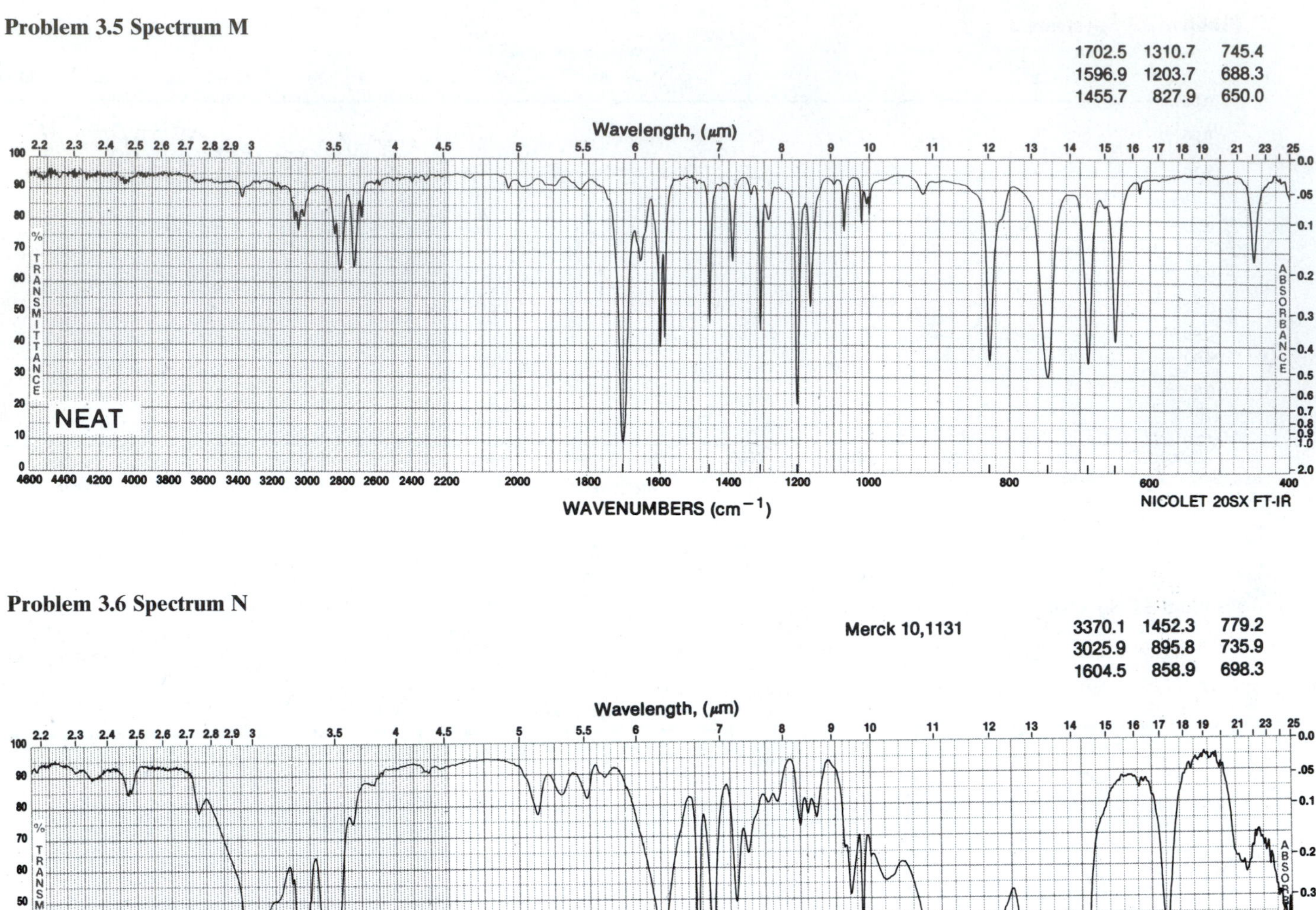

Problem 3.6 Spectrum N

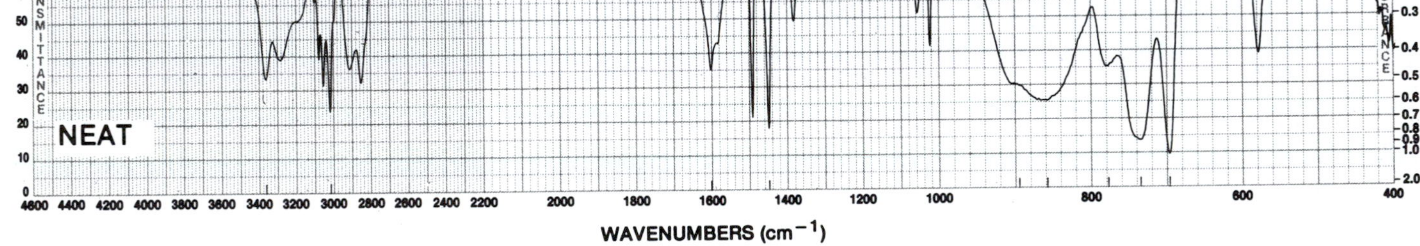

Problem 3.6 Spectrum O

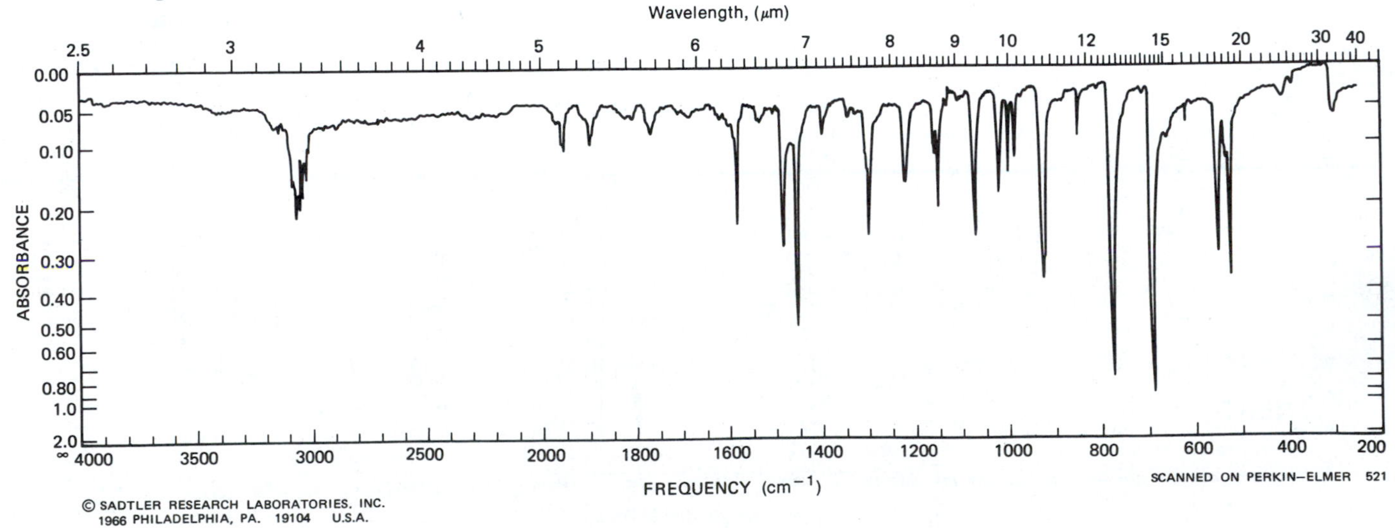

Problem 3.6 Spectrum P

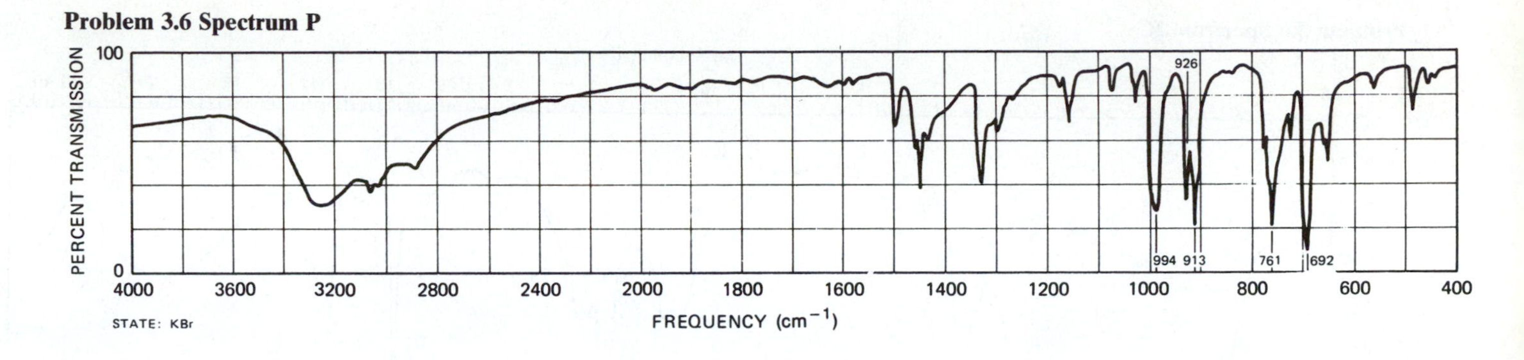

Problem 3.6 Spectrum Q

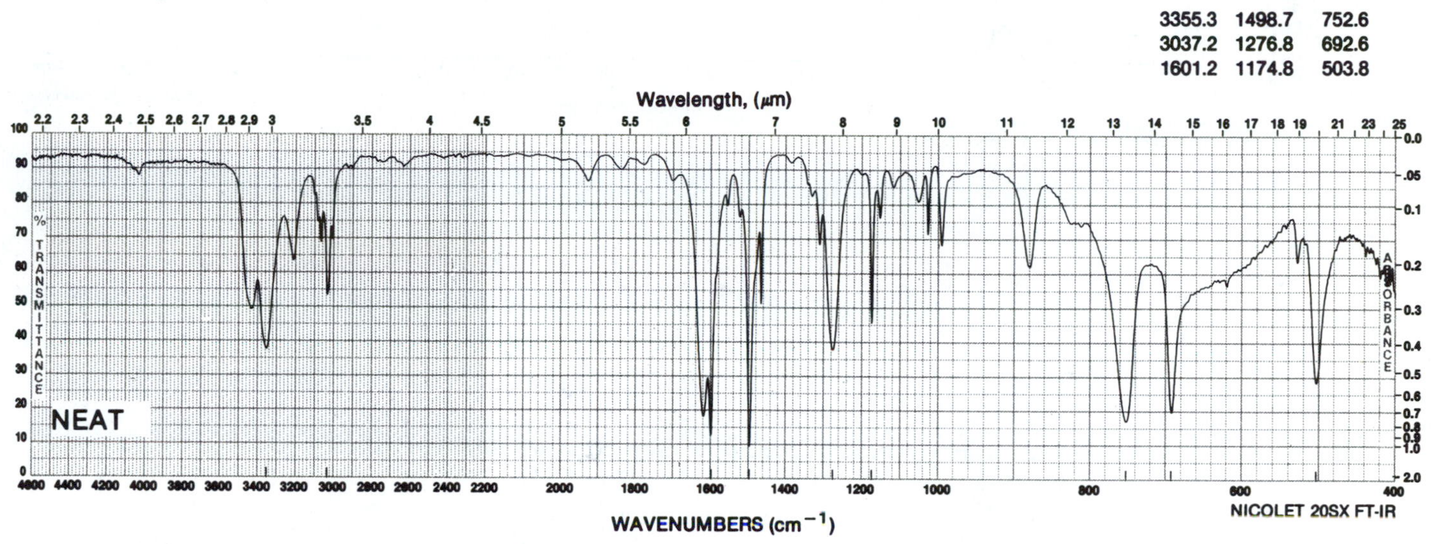

Problem 3.6 Spectrum R

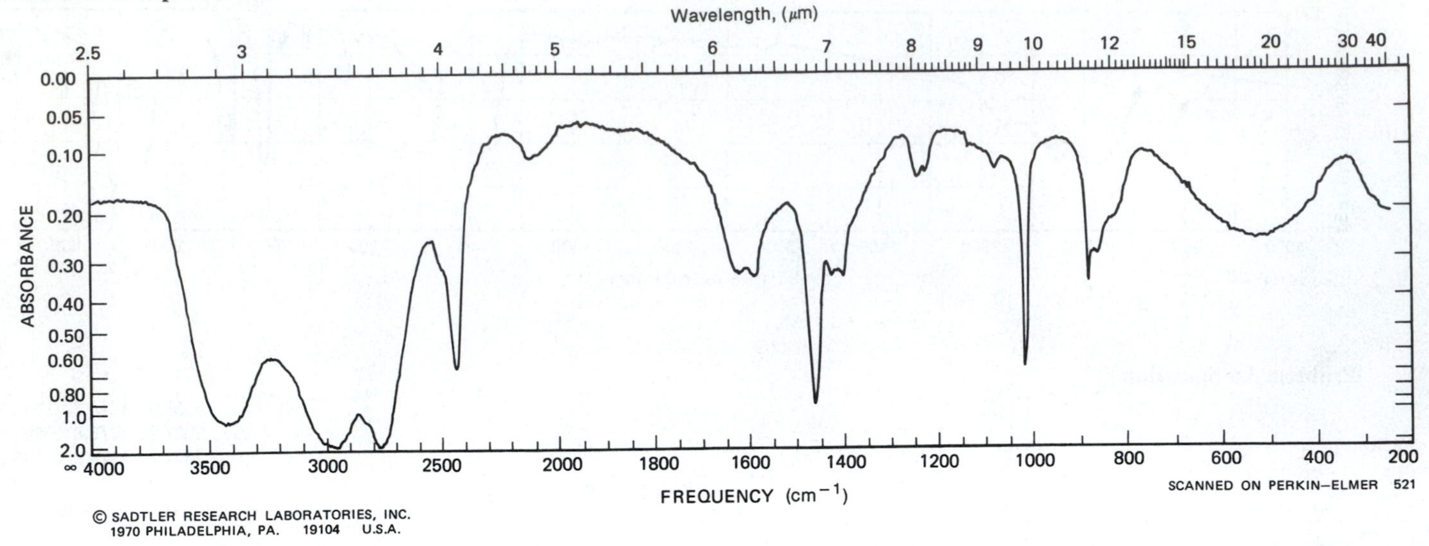

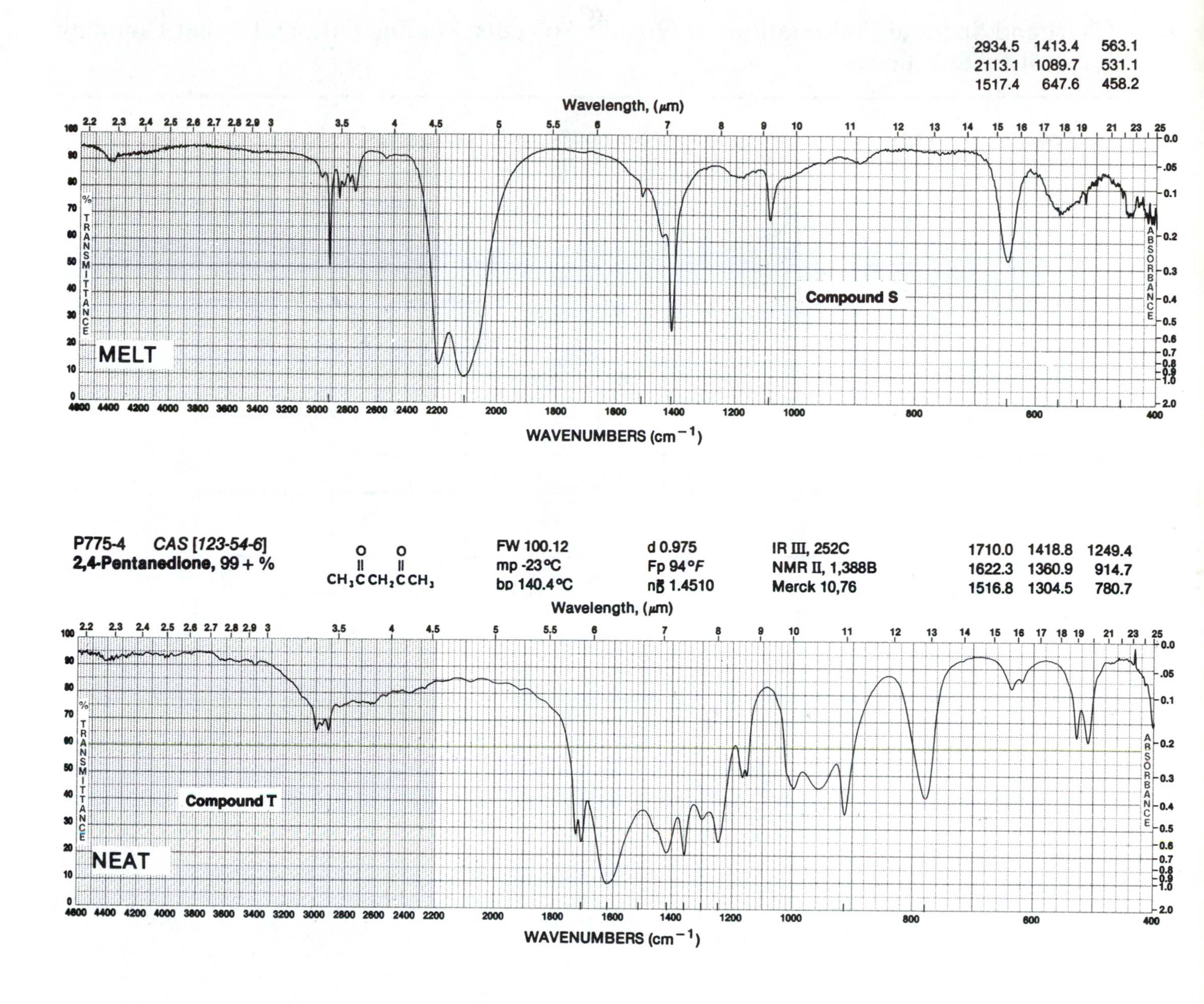
2934.5 1413.4 563.1
2113.1 1089.7 531.1
1517.4 647.6 458.2
Wavelength, (μm)
MELT
Compound S
% TRANSMITTANCE
ABSORBANCE
WAVENUMBERS (cm⁻¹)
P775-4 CAS [123-54-6]
2,4-Pentanedione, 99 + %
CH₃CCH₂CCH₃
FW 100.12
mp -23°C
bp 140.4°C
d 0.975
Fp 94°F
n 1.4510
IR III, 252C
NMR II, 1,388B
Merck 10,76
1710.0 1418.8 1249.4
1622.3 1360.9 914.7
1516.8 1304.5 780.7
Wavelength, (μm)
Compound T
NEAT
% TRANSMITTANCE
ABSORBANCE
WAVENUMBERS (cm⁻¹)

Chart and Spectral Presentations of Organic Solvents, Mulling Oils, and Other Common Laboratory Substances

Appendix A Transparent Regions of Solvents and Mulling Oils

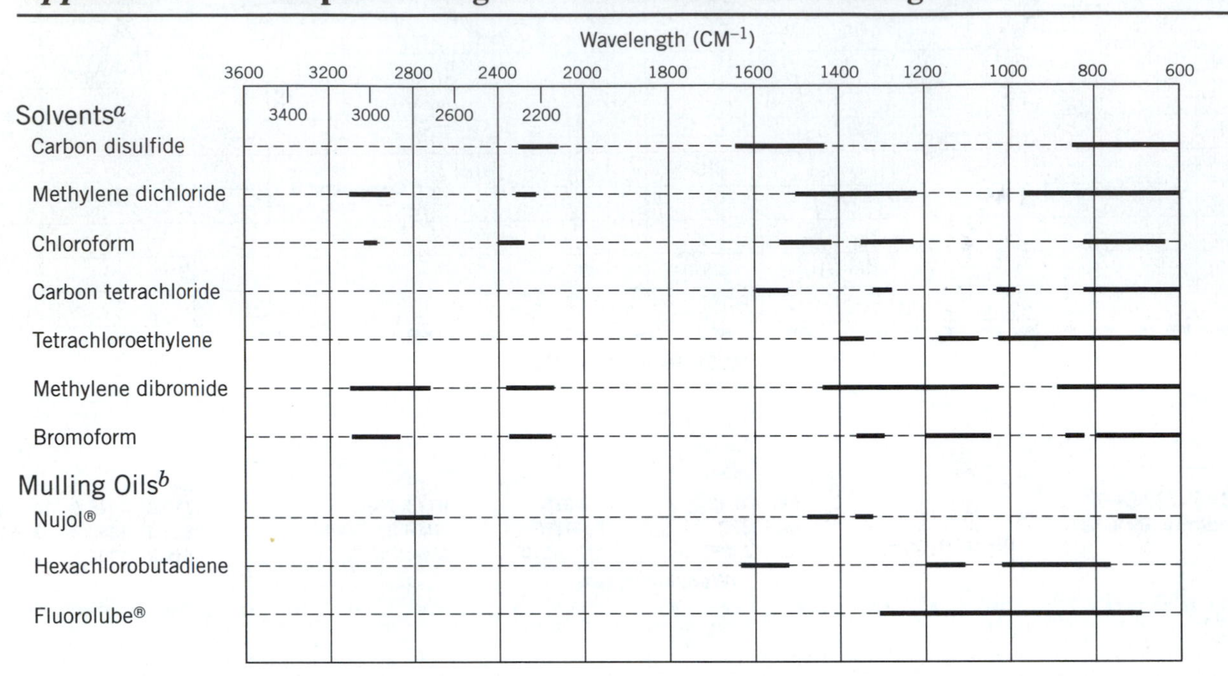

[a] The open regions are those in which the solvent transmits more than 25% of the incident light at 1 mm thickness.
[b] The open regions for mulling oils indicate transparency of thin films.

Appendix B Spectra of Common Laboratory Substances[a]

Alphabetical Listing of Spectra Shown on Succeeding Pages

	Spectrum No.		Spectrum No.
Acetone	17	Ethyl ether	22
Benzene	4	Fluorolube®	14
2-Butanone	18	Hexane	1
Carbon disulfide	25	Indene	8
Carbon tetrachloride	10	Methanol	15
Chloroform	9	Nujol®	2
Cyclohexane	3	Petroleum ether	24
Cyclohexanone	19	Phthalic acid, diethyl ester	21
1,2-Dichloroethane	11	Polystyrene	7
N,N-Dimethylformamide (DMF)	28	Silicone lubricant	27
Dimethyl sulfoxide (DMSO)	26	Tetrachloroethylene	13
p-Dioxane	23	Toluene	5
Ethyl acetate	20	Trichloroethylene	12
Ethanol	16	*m*-Xylene	6

Source: Spectra courtesy of Sadtler Laboratories and Aldrich Chemical Co.

[a] Carbon dioxide (Sadtler Prism Spectrum No. 1924) has bands in the 3700–3550 and 2380–2222 cm^{-1} regions and at ~720 cm^{-1}.

NO. 1

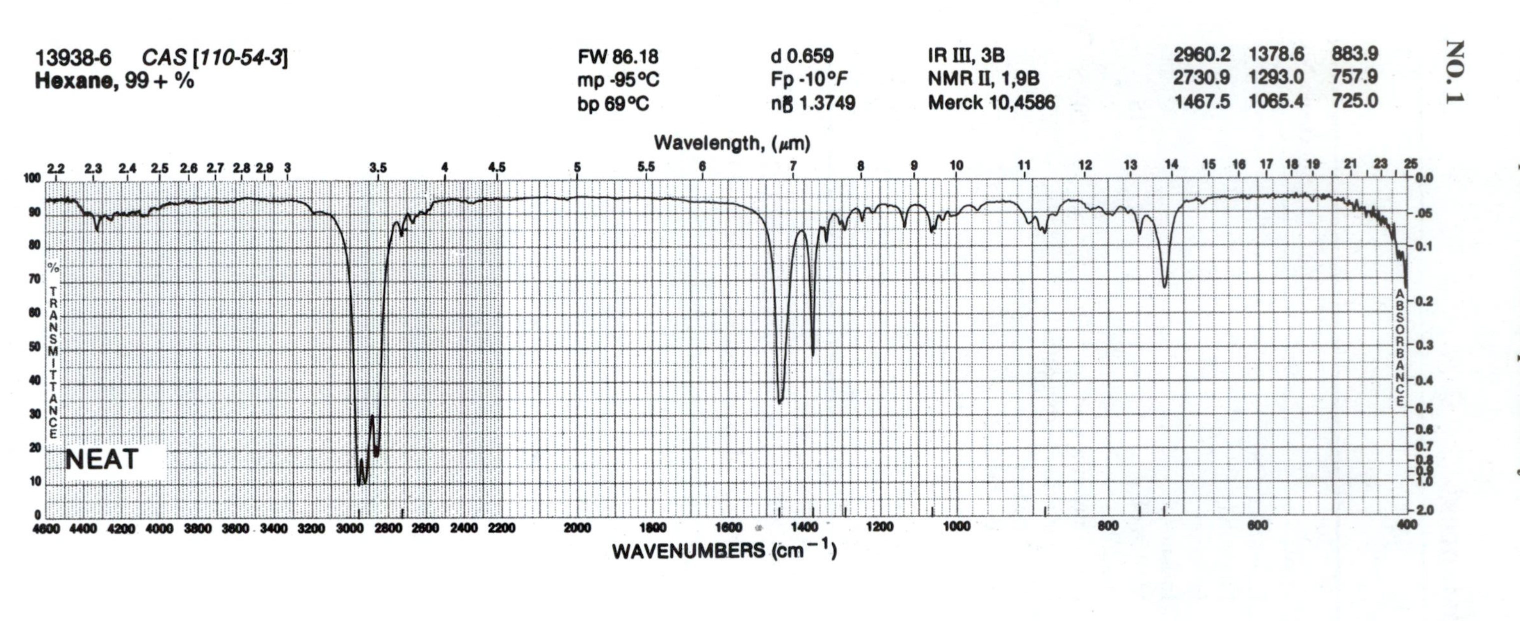

NO. 2

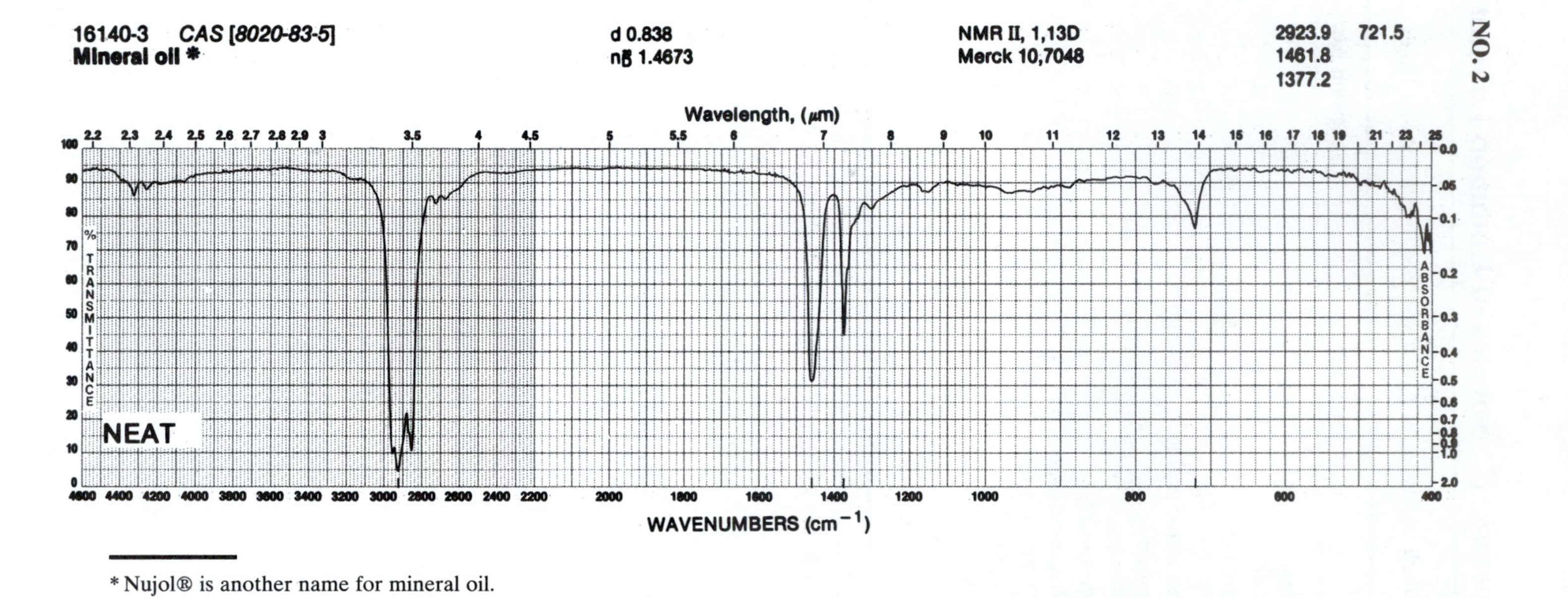

* Nujol® is another name for mineral oil.

NO. 3

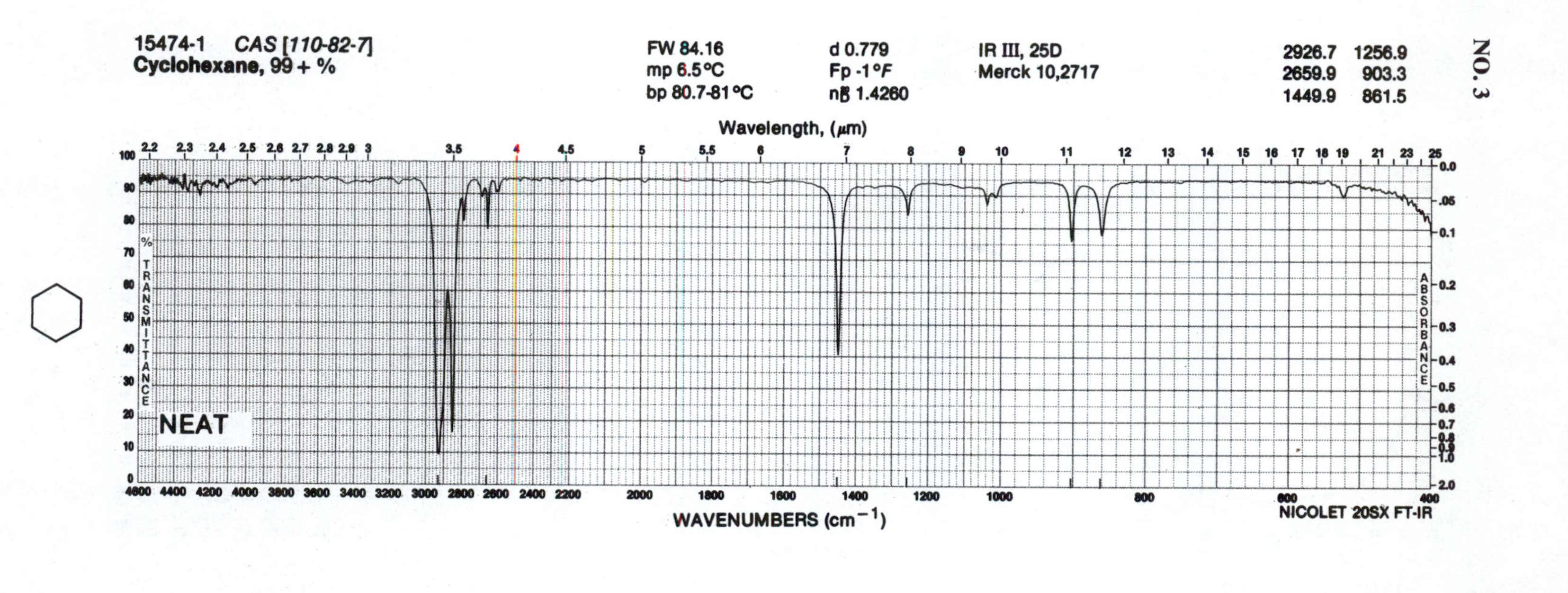
15474-1 CAS [110-82-7]
Cyclohexane, 99 + %
FW 84.16
mp 6.5°C
bp 80.7-81°C
d 0.779
Fp -1°F
n$_D^{20}$ 1.4260
IR III, 25D
Merck 10,2717
2926.7 1256.9
2659.9 903.3
1449.9 861.5
Wavelength, (µm)
% TRANSMITTANCE
ABSORBANCE
NEAT
WAVENUMBERS (cm^{-1})
NICOLET 20SX FT-IR

NO. 4

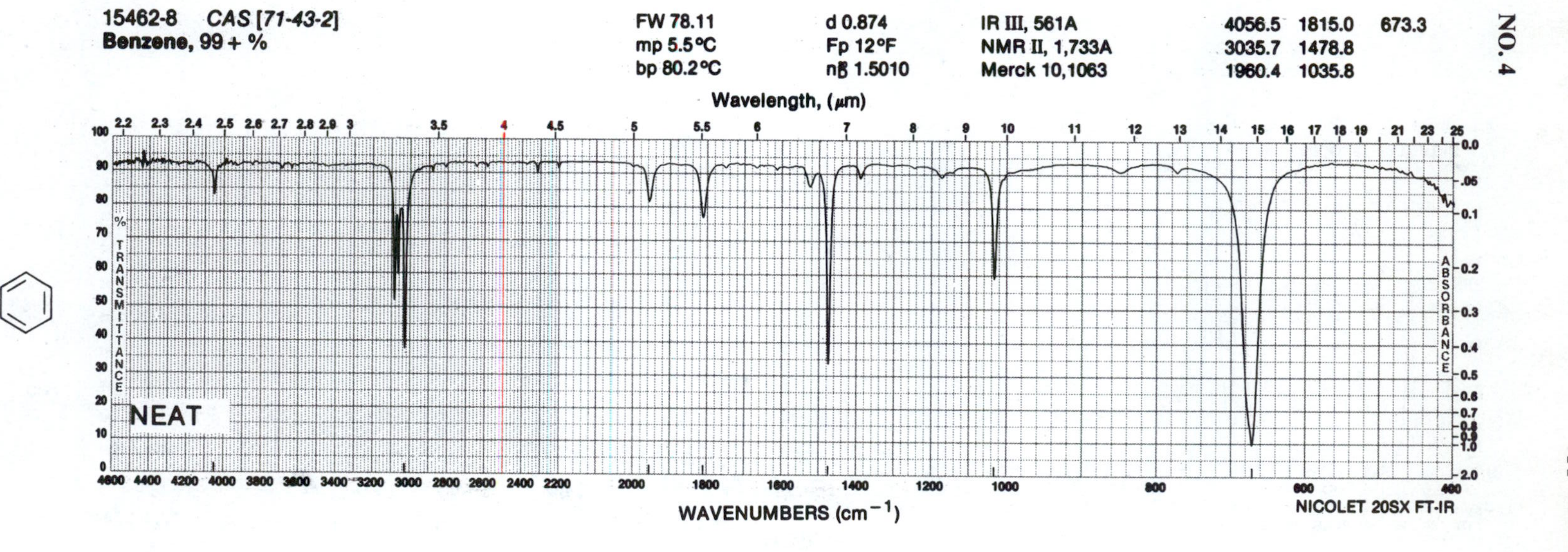
15462-8 CAS [71-43-2]
Benzene, 99 + %
FW 78.11
mp 5.5°C
bp 80.2°C
d 0.874
Fp 12°F
n$_D^{20}$ 1.5010
IR III, 561A
NMR II, 1,733A
Merck 10,1063
4056.5 1815.0 673.3
3035.7 1478.8
1960.4 1035.8
Wavelength, (µm)
% TRANSMITTANCE
ABSORBANCE
NEAT
WAVENUMBERS (cm^{-1})
NICOLET 20SX FT-IR

NO. 5

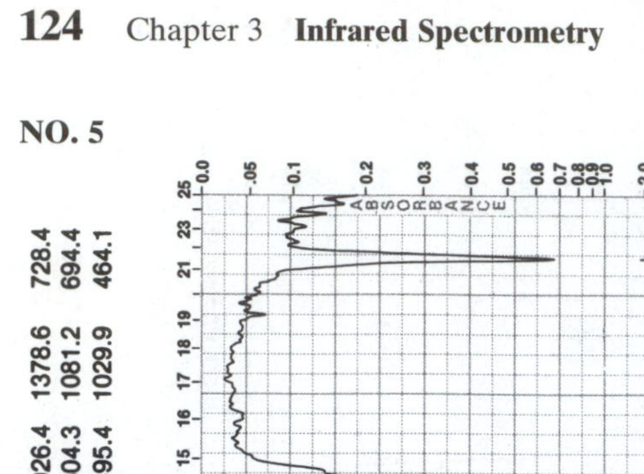
15500-4 CAS [108-88-3]
Toluene, 99+%
FW 92.14
mp -93°C
bp 111°C
d 0.867
Fp 40°F
n20D 1.4968
IR III, 561B
NMR II, 1,733B
Merck 10,9357
3026.4 1378.6 728.4
1604.3 1081.2 694.4
1495.4 1029.9 464.1
NEAT
Wavelength, (μm)
WAVENUMBERS (cm−1)
NICOLET 20SX FT-IR
% TRANSMITTANCE
ABSORBANCE

NO. 6

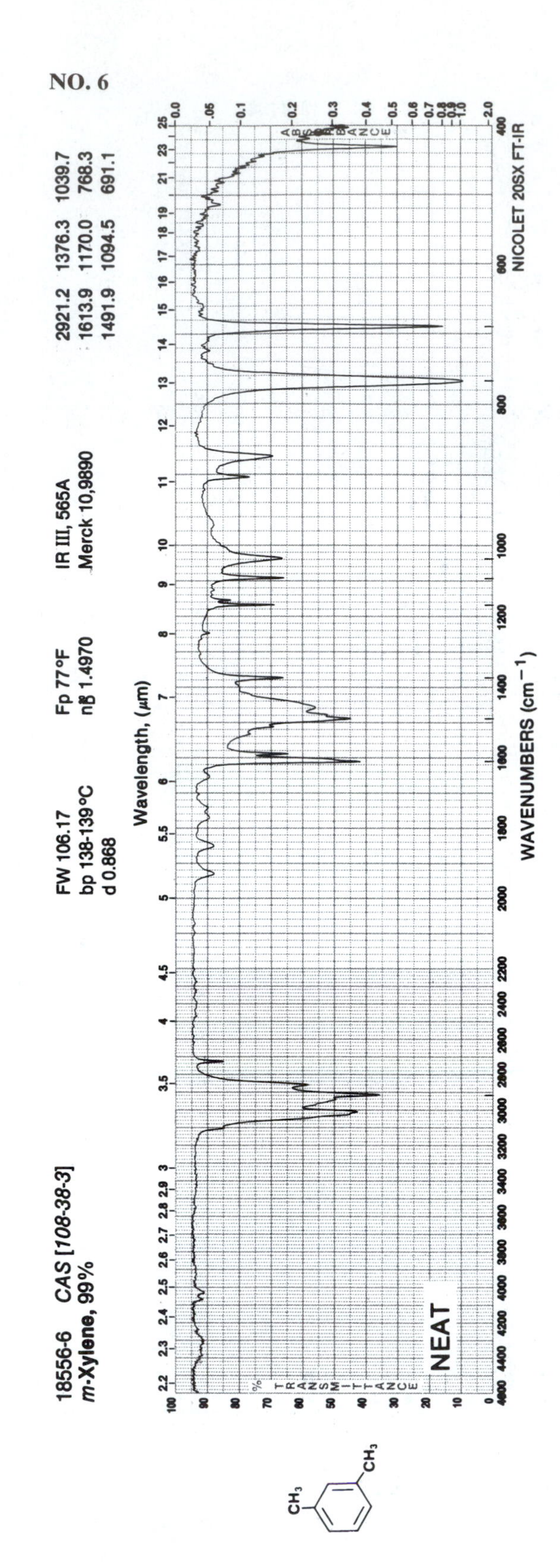
18556-6 CAS [108-38-3]
m-Xylene, 99%
FW 106.17
bp 138-139°C
d 0.868
Fp 77°F
n20D 1.4970
IR III, 565A
Merck 10,9890
2921.2 1376.3 1039.7
1613.9 1170.0 768.3
1491.9 1094.5 691.1
NEAT
Wavelength, (μm)
WAVENUMBERS (cm−1)
NICOLET 20SX FT-IR
% TRANSMITTANCE
ABSORBANCE

NO. 7

NO. 8

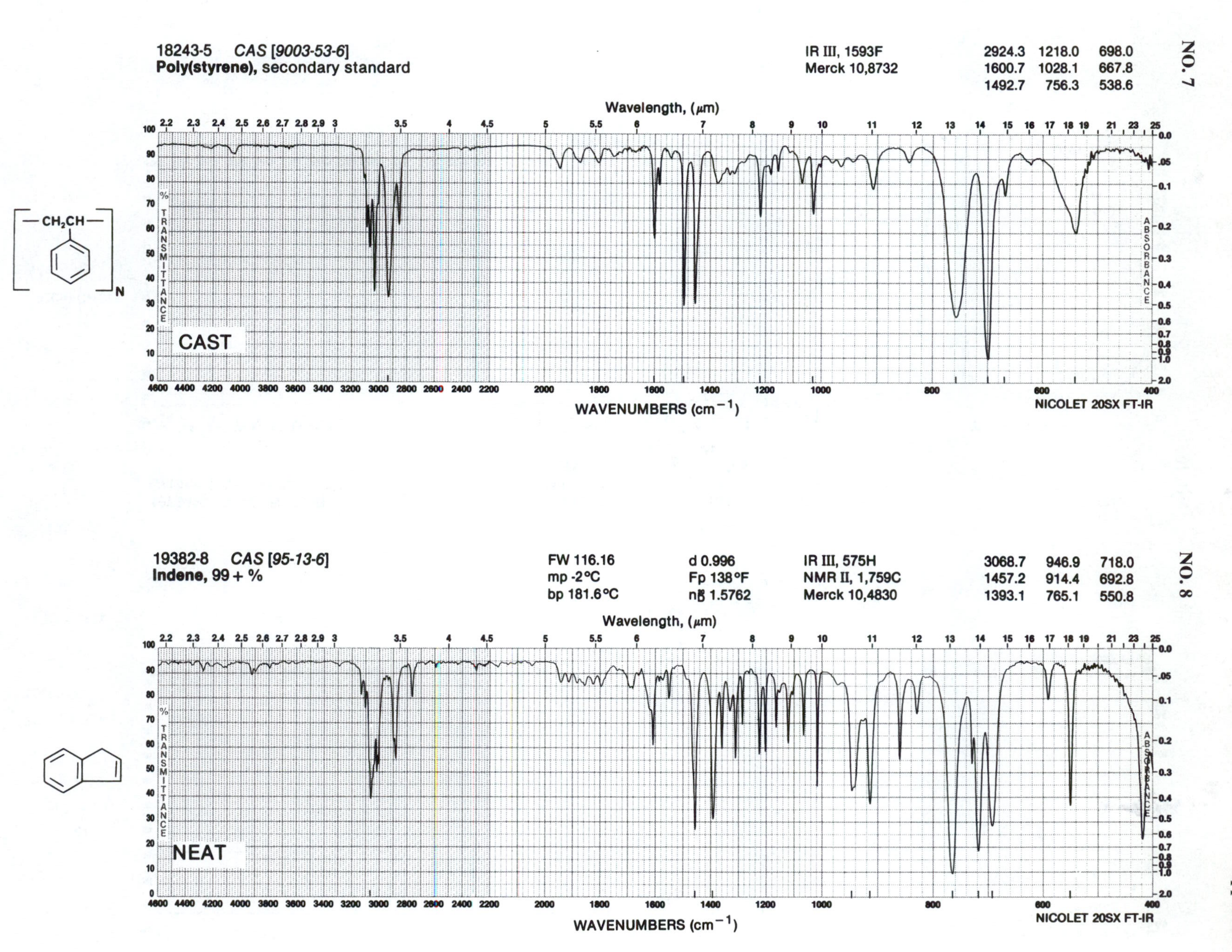
18243-5 CAS [9003-53-6]
Poly(styrene), secondary standard
IR III, 1593F
Merck 10,8732
2924.3 1218.0 698.0
1600.7 1028.1 667.8
1492.7 756.3 538.6
Wavelength, (μm)
CAST
TRANSMITTANCE
ABSORBANCE
WAVENUMBERS (cm−1)
NICOLET 20SX FT-IR
19382-8 CAS [95-13-6]
Indene, 99 + %
FW 116.16
mp -2°C
bp 181.6°C
d 0.996
Fp 138°F
n 1.5762
IR III, 575H
NMR II, 1,759C
Merck 10,4830
3068.7 946.9 718.0
1457.2 914.4 692.8
1393.1 765.1 550.8
Wavelength, (μm)
NEAT
TRANSMITTANCE
ABSORBANCE
WAVENUMBERS (cm−1)
NICOLET 20SX FT-IR

NO. 9

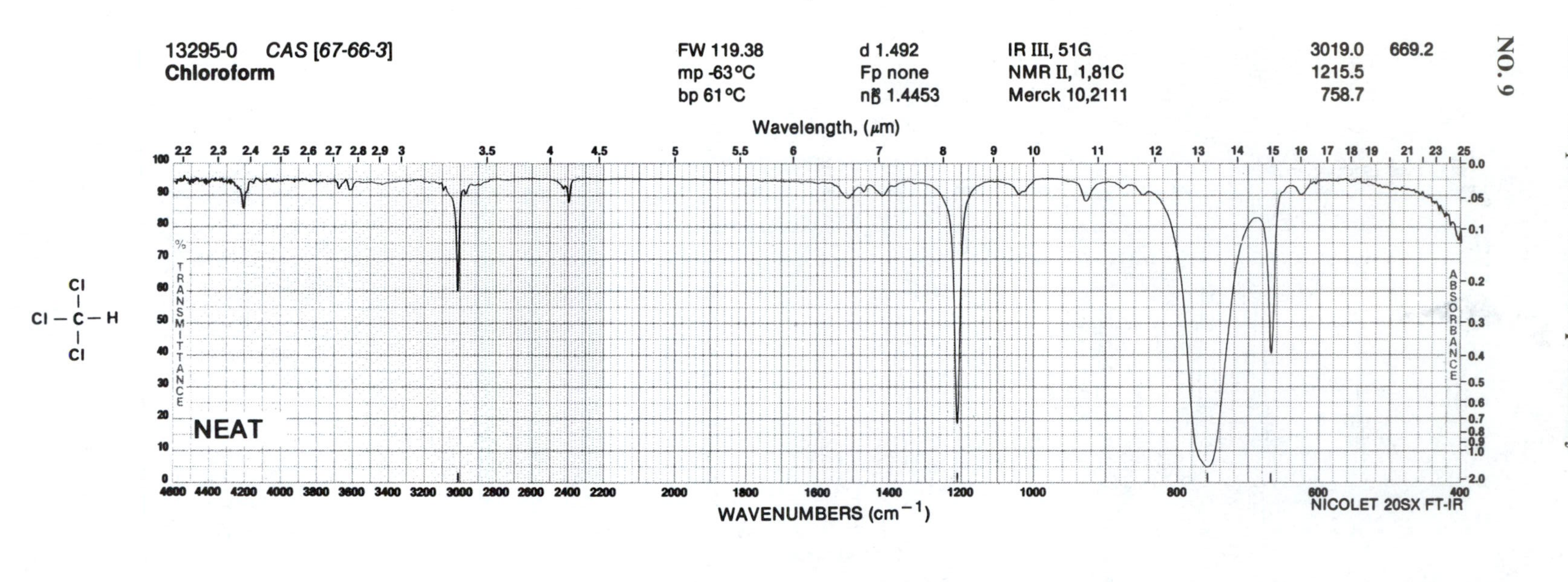
13295-0 CAS [67-66-3]
Chloroform
FW 119.38
mp -63°C
bp 61°C
d 1.492
Fp none
n 1.4453
IR III, 51G
NMR II, 1,81C
Merck 10,2111
3019.0
669.2
1215.5
758.7
Wavelength, (μm)
WAVENUMBERS (cm⁻¹)
NICOLET 20SX FT-IR
NEAT
% TRANSMITTANCE
ABSORBANCE
Cl–C–H

NO. 10

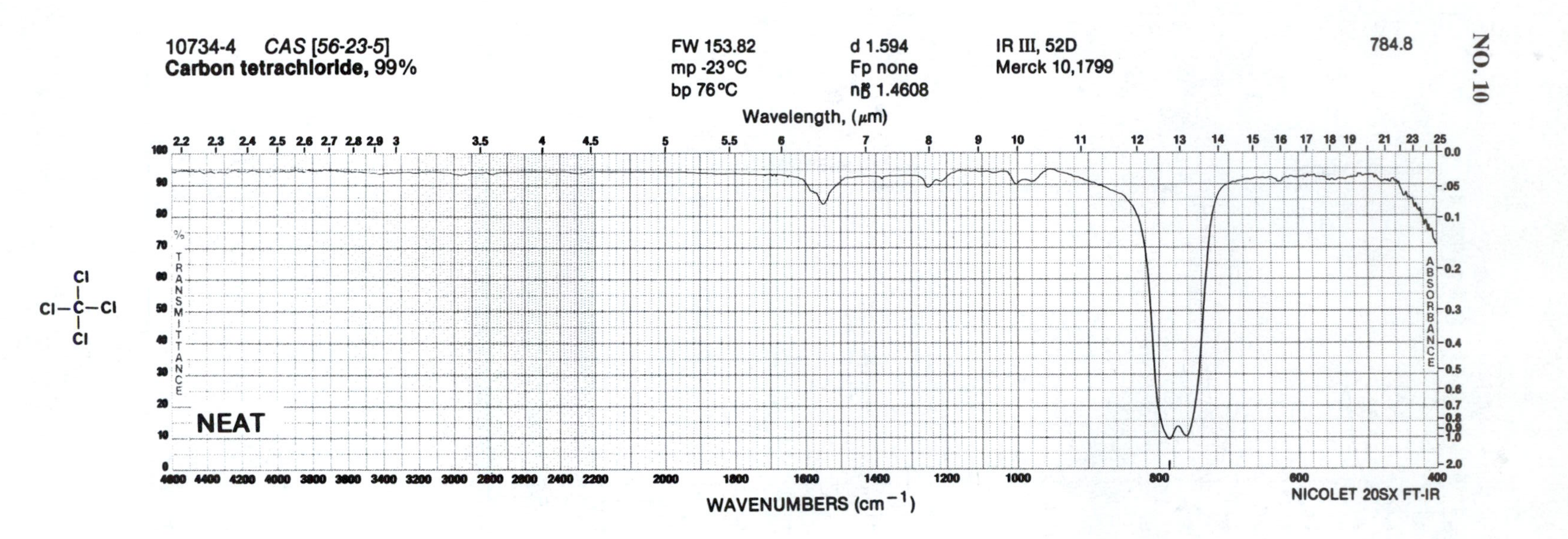
10734-4 CAS [56-23-5]
Carbon tetrachloride, 99%
FW 153.82
mp -23°C
bp 76°C
d 1.594
Fp none
n 1.4608
IR III, 52D
Merck 10,1799
784.8
Wavelength, (μm)
WAVENUMBERS (cm⁻¹)
NICOLET 20SX FT-IR
NEAT
% TRANSMITTANCE
ABSORBANCE
Cl–C–Cl

NO. 11

NO. 12

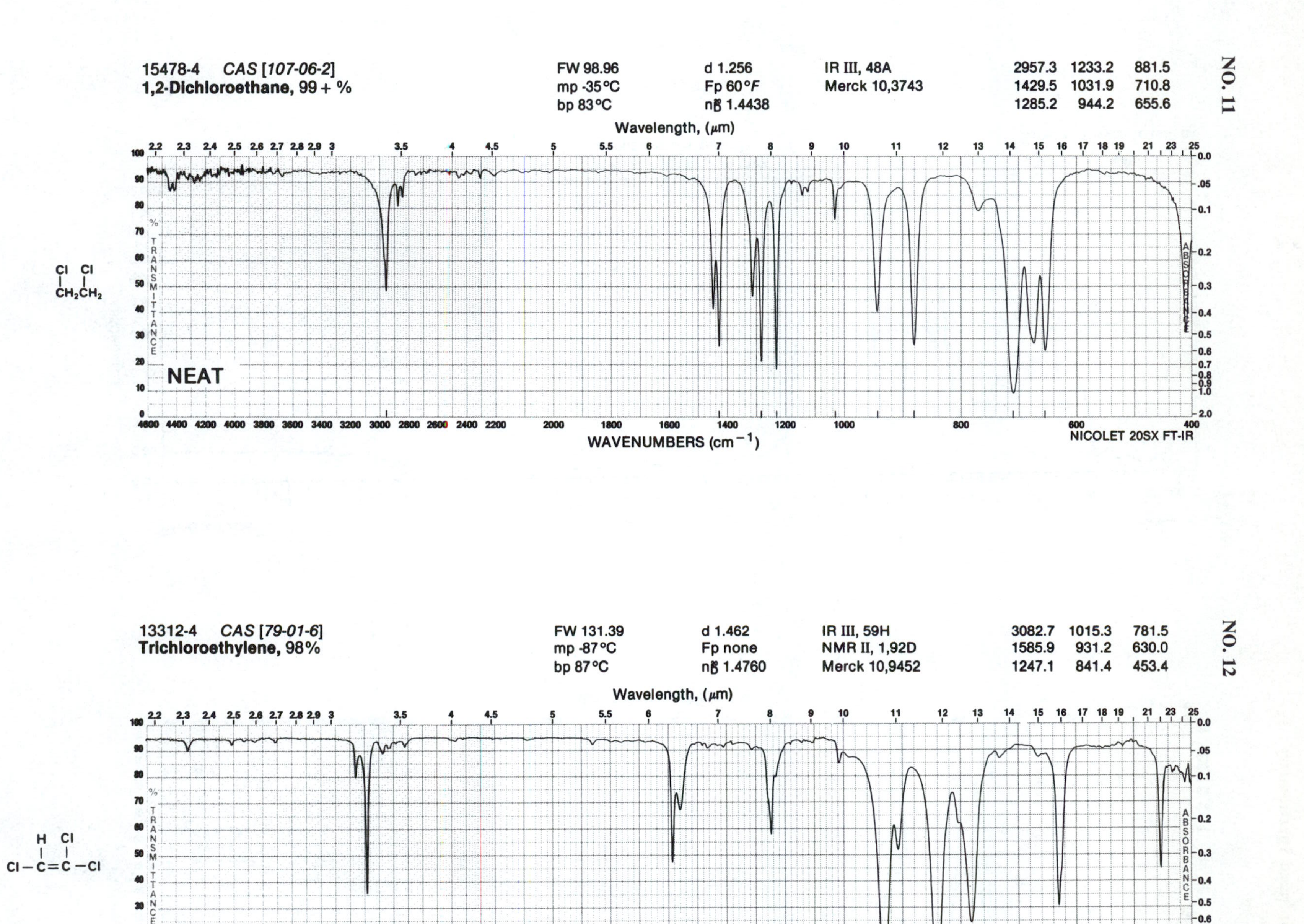
15478-4 CAS [107-06-2]
1,2-Dichloroethane, 99+%
FW 98.96
mp -35°C
bp 83°C
d 1.256
Fp 60°F
n_D^{20} 1.4438
IR III, 48A
Merck 10,3743
2957.3 1233.2 881.5
1429.5 1031.9 710.8
1285.2 944.2 655.6
Wavelength, (μm)
NEAT
WAVENUMBERS (cm^{-1})
NICOLET 20SX FT-IR
%
TRANSMITTANCE
ABSORBANCE
Cl Cl
CH$_2$CH$_2$
13312-4 CAS [79-01-6]
Trichloroethylene, 98%
FW 131.39
mp -87°C
bp 87°C
d 1.462
Fp none
n_D^{20} 1.4760
IR III, 59H
NMR II, 1,92D
Merck 10,9452
3082.7 1015.3 781.5
1585.9 931.2 630.0
1247.1 841.4 453.4
Wavelength, (μm)
NEAT
WAVENUMBERS (cm^{-1})
NICOLET 20SX FT-IR
%
TRANSMITTANCE
ABSORBANCE
H Cl
Cl—C=C—Cl

NO. 13

NO. 14

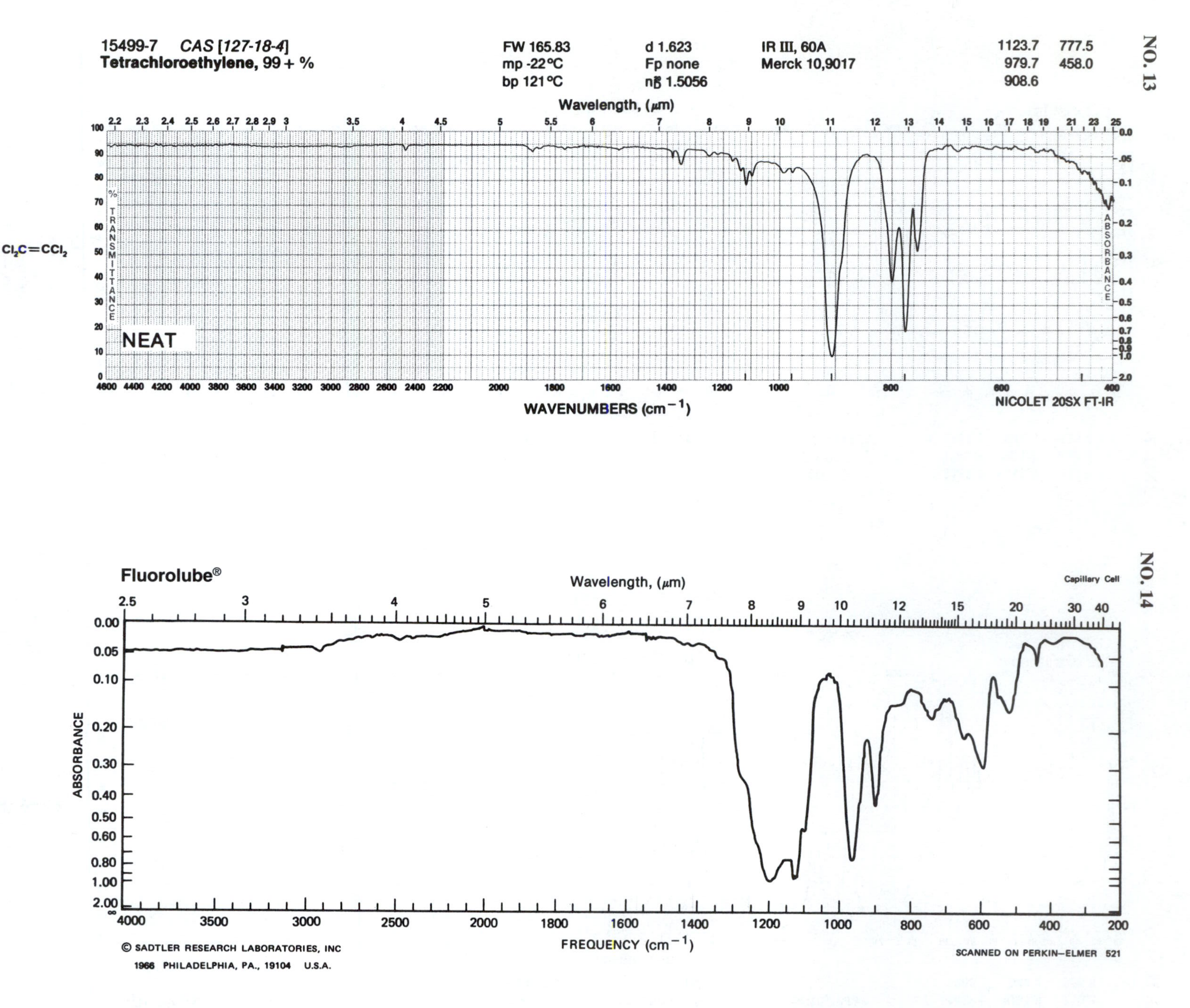
15499-7 CAS [127-18-4]
Tetrachloroethylene, 99 + %
FW 165.83
mp -22 °C
bp 121 °C
d 1.623
Fp none
n_D^{20} 1.5056
IR III, 60A
Merck 10,9017
1123.7 777.5
979.7 458.0
908.6
Wavelength, (μm)
% TRANSMITTANCE
ABSORBANCE
NEAT
WAVENUMBERS (cm⁻¹)
NICOLET 20SX FT-IR
$Cl_2C{=}CCl_2$
Fluorolube®
Wavelength, (μm)
Capillary Cell
ABSORBANCE
FREQUENCY (cm⁻¹)
© SADTLER RESEARCH LABORATORIES, INC
1966 PHILADELPHIA, PA., 19104 U.S.A.
SCANNED ON PERKIN–ELMER 521

NO. 15

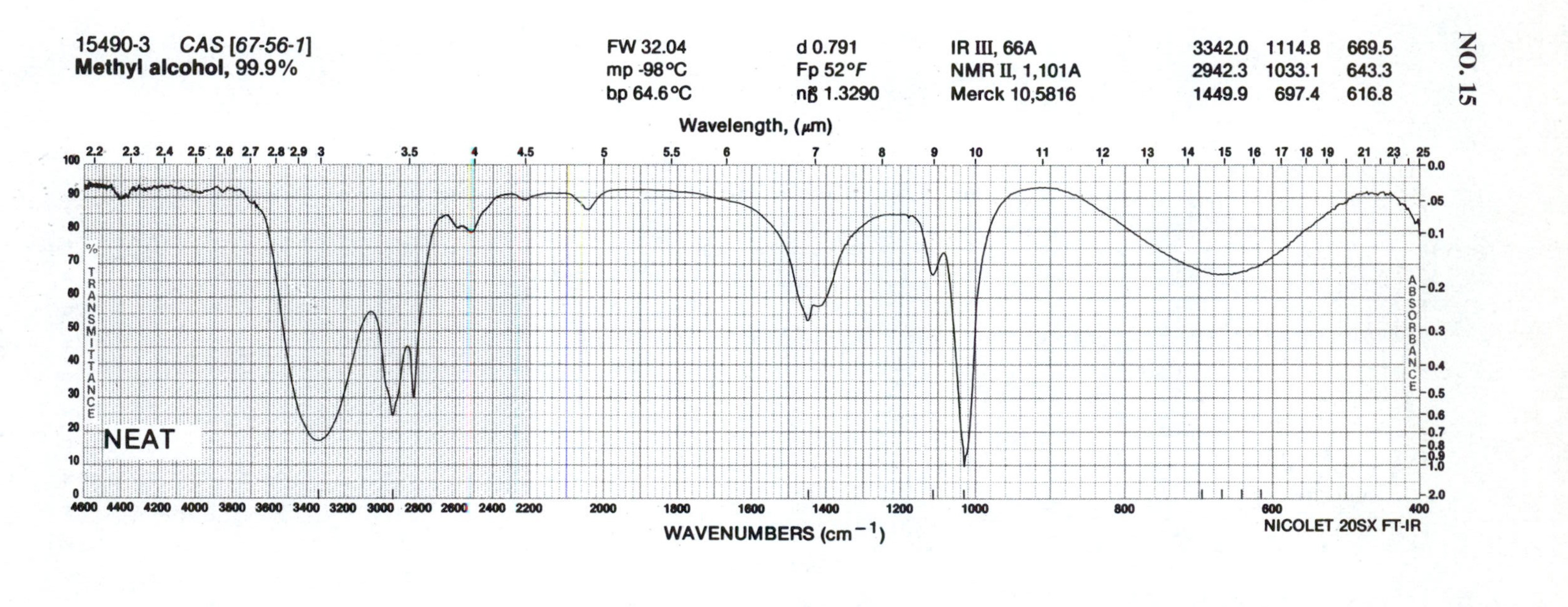

CH_3OH

NO. 16

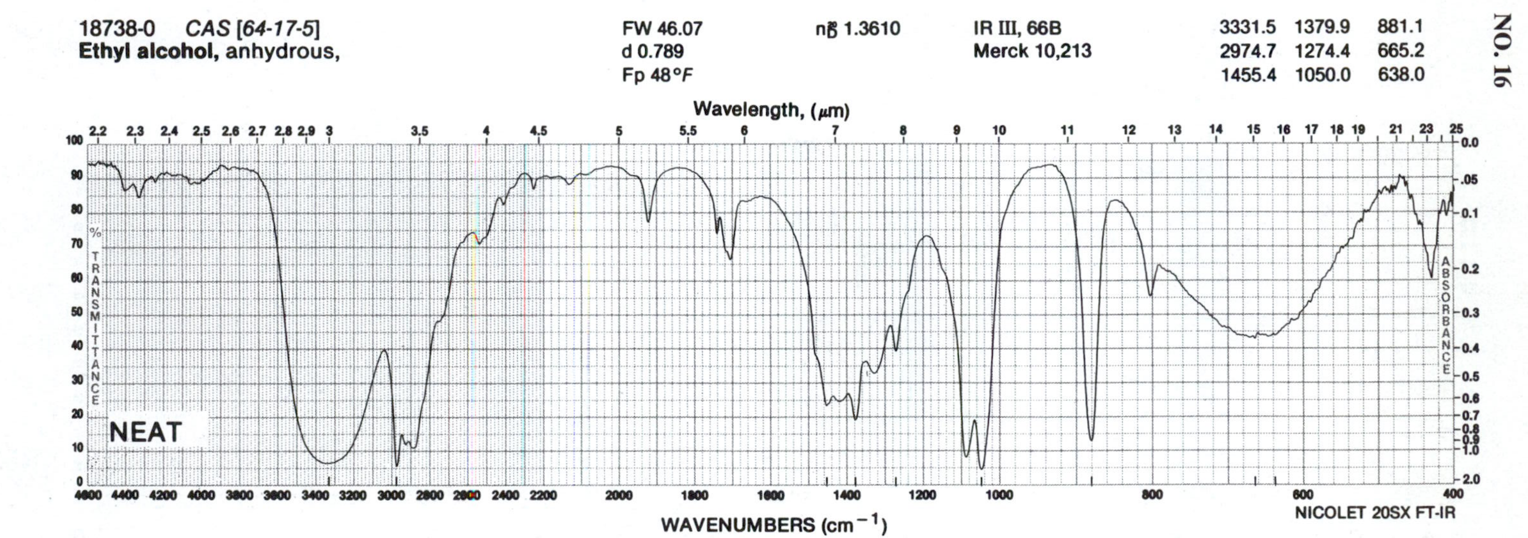

CH_3CH_2OH

NO. 17

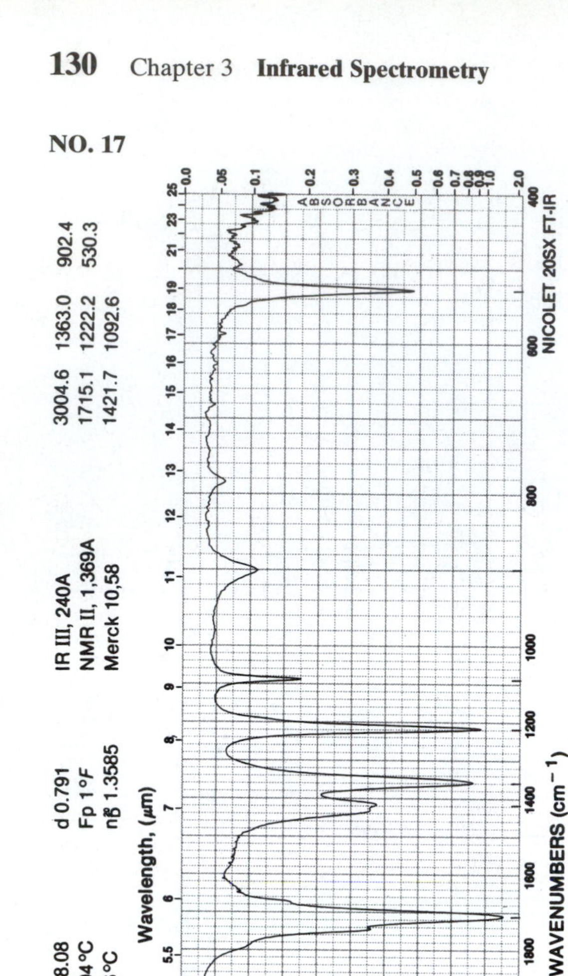

O
‖
CH_3CCH_3

NO. 18

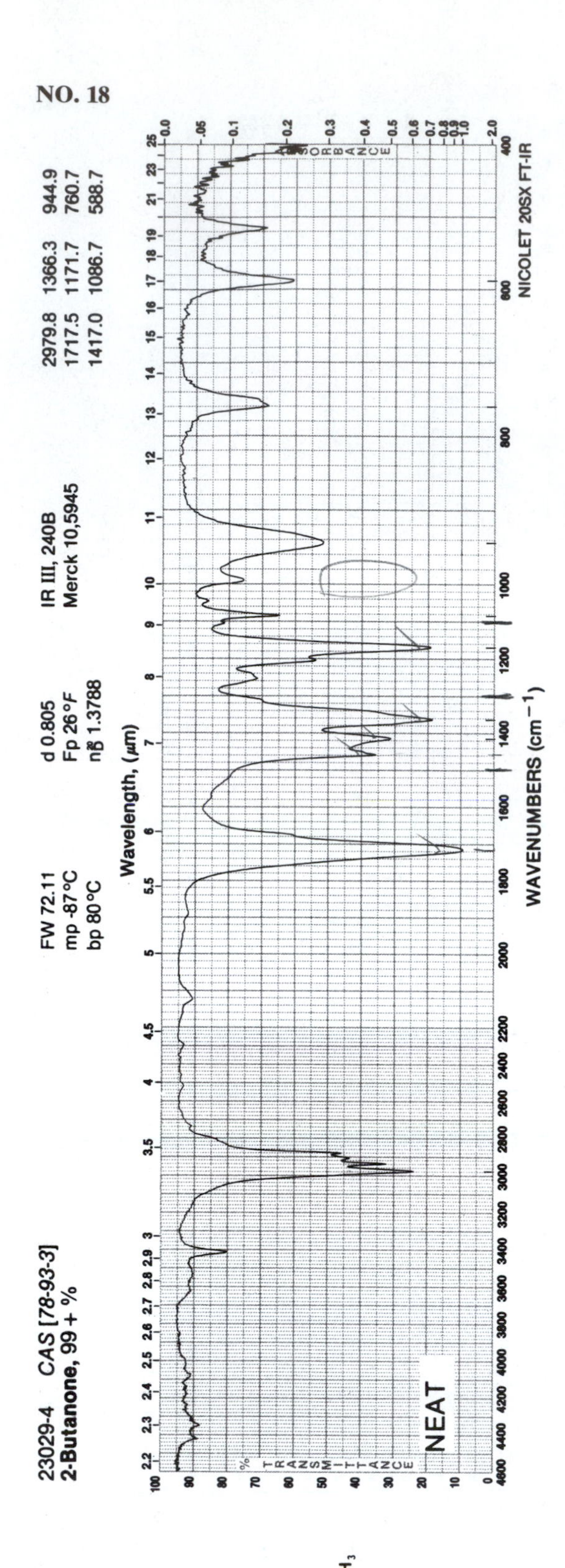

O
‖
$CH_3CH_2-C-CH_3$

NO. 19

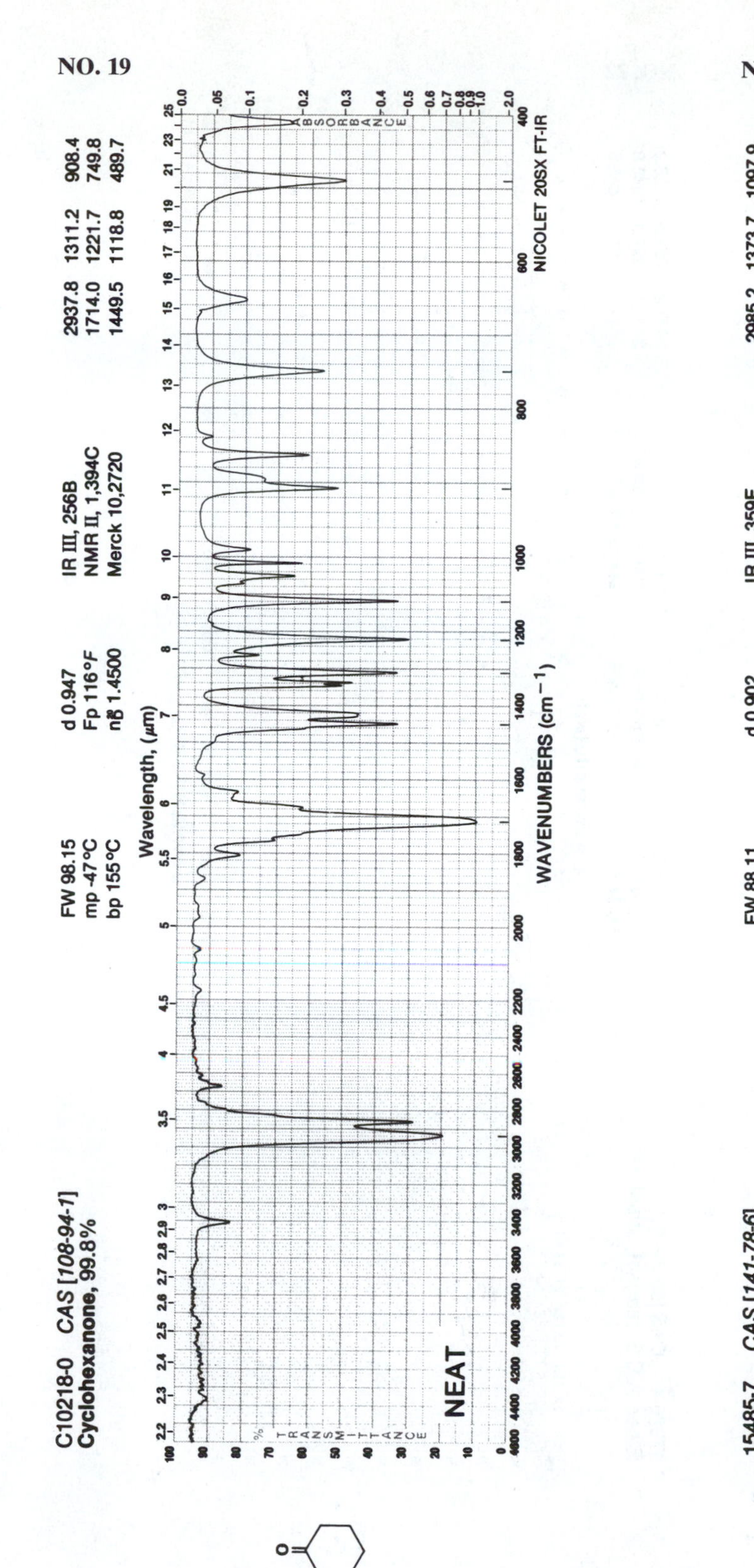

NO. 20

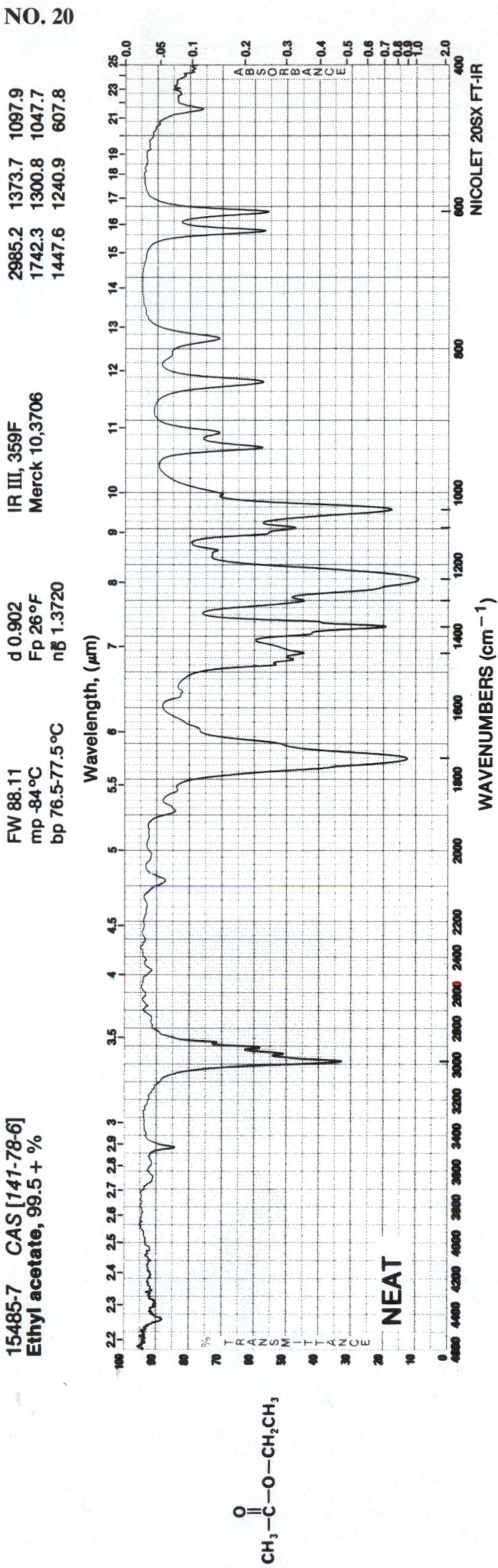

$CH_3-C(=O)-O-CH_2CH_3$

NO. 21

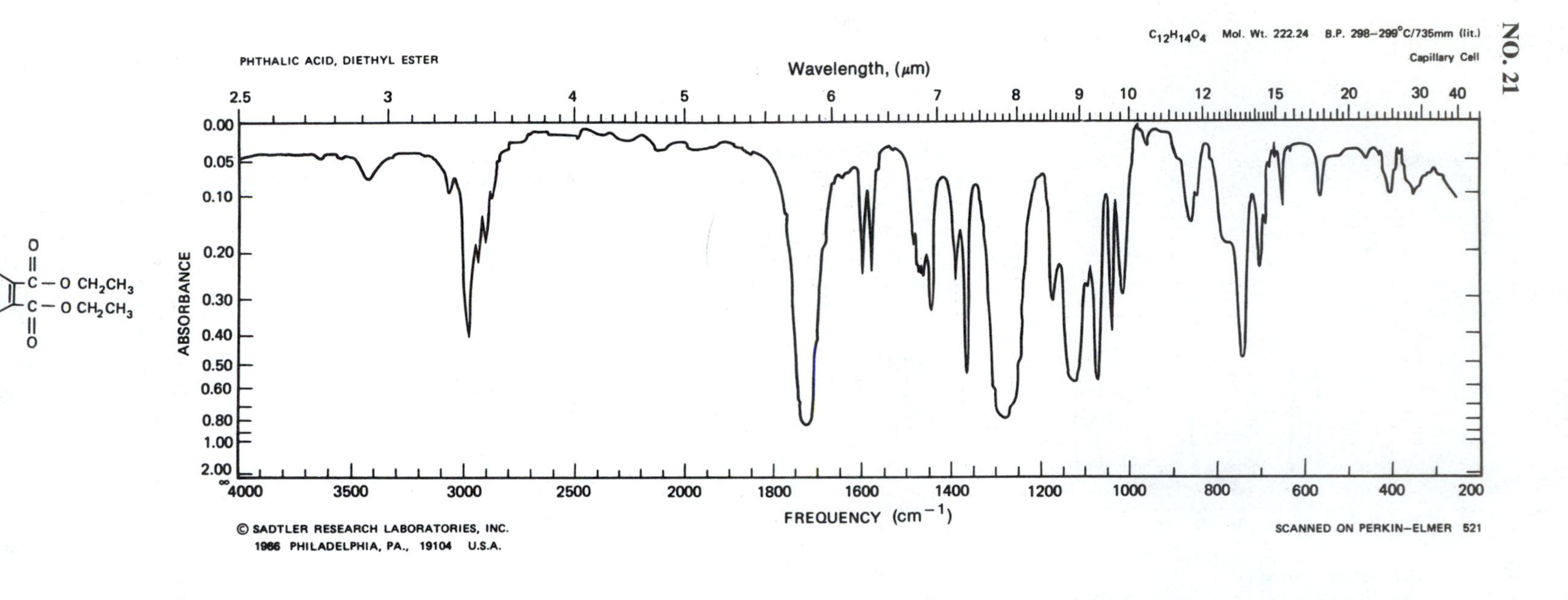
PHTHALIC ACID, DIETHYL ESTER
C12H14O4 Mol. Wt. 222.24 B.P. 298–299°C/735mm (lit.)
Capillary Cell
Wavelength, (μm)
ABSORBANCE
FREQUENCY (cm−1)
© SADTLER RESEARCH LABORATORIES, INC.
1966 PHILADELPHIA, PA., 19104 U.S.A.
SCANNED ON PERKIN-ELMER 521

NO. 22

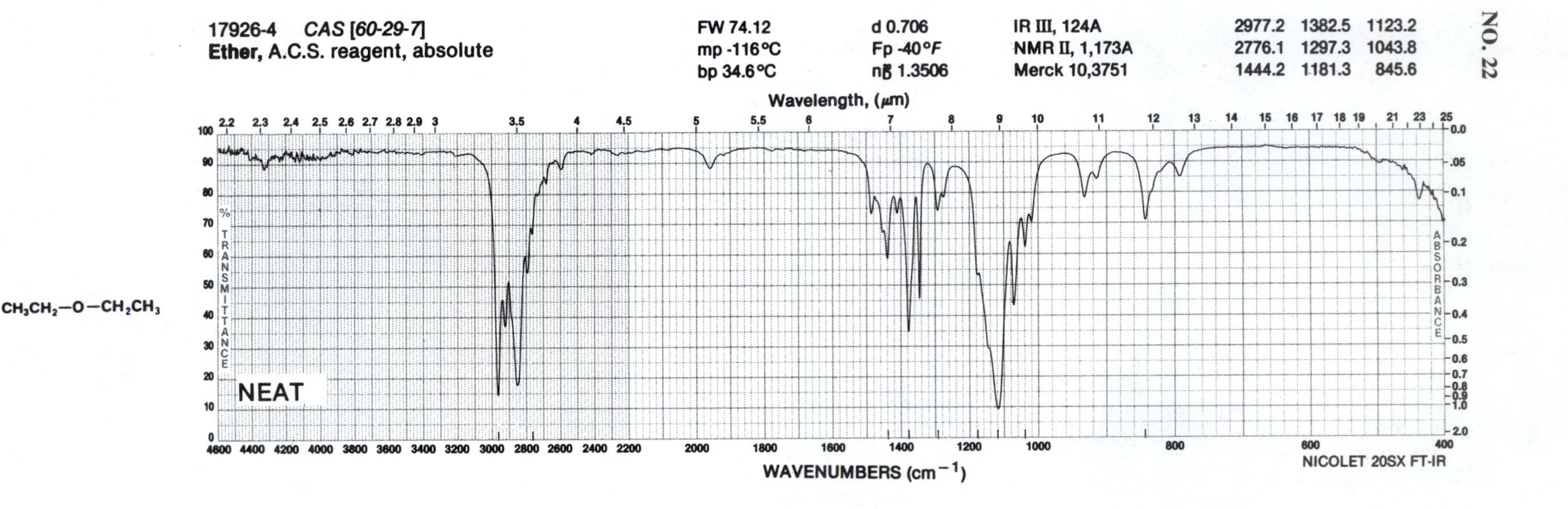
17926-4 CAS [60-29-7]
Ether, A.C.S. reagent, absolute
FW 74.12
mp -116°C
bp 34.6°C
d 0.706
Fp -40°F
IR III, 124A
NMR II, 1,173A
Merck 10,3751
2977.2 1382.5 1123.2
2776.1 1297.3 1043.8
1444.2 1181.3 845.6
Wavelength, (μm)
NEAT
WAVENUMBERS (cm−1)
NICOLET 20SX FT-IR
CH3CH2—O—CH2CH3

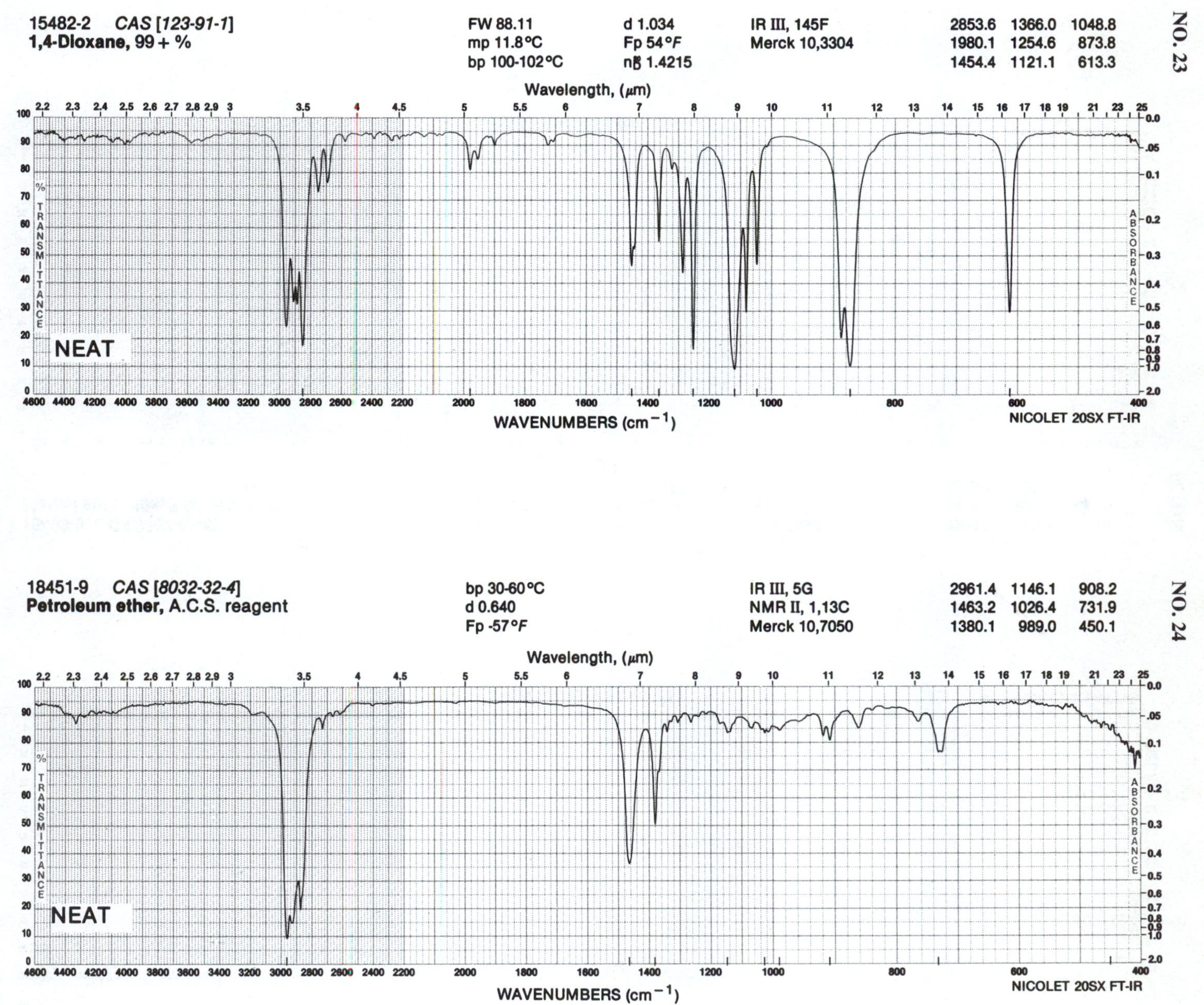

NO. 23
15482-2 CAS [123-91-1]
1,4-Dioxane, 99 + %
FW 88.11
mp 11.8°C
bp 100-102°C
d 1.034
Fp 54°F
n20D 1.4215
IR III, 145F
Merck 10,3304
2853.6 1366.0 1048.8
1980.1 1254.6 873.8
1454.4 1121.1 613.3
Wavelength, (μm)
% TRANSMITTANCE
ABSORBANCE
NEAT
WAVENUMBERS (cm−1)
NICOLET 20SX FT-IR
NO. 24
18451-9 CAS [8032-32-4]
Petroleum ether, A.C.S. reagent
bp 30-60°C
d 0.640
Fp -57°F
IR III, 5G
NMR II, 1,13C
Merck 10,7050
2961.4 1146.1 908.2
1463.2 1026.4 731.9
1380.1 989.0 450.1
Wavelength, (μm)
% TRANSMITTANCE
ABSORBANCE
NEAT
WAVENUMBERS (cm−1)
NICOLET 20SX FT-IR

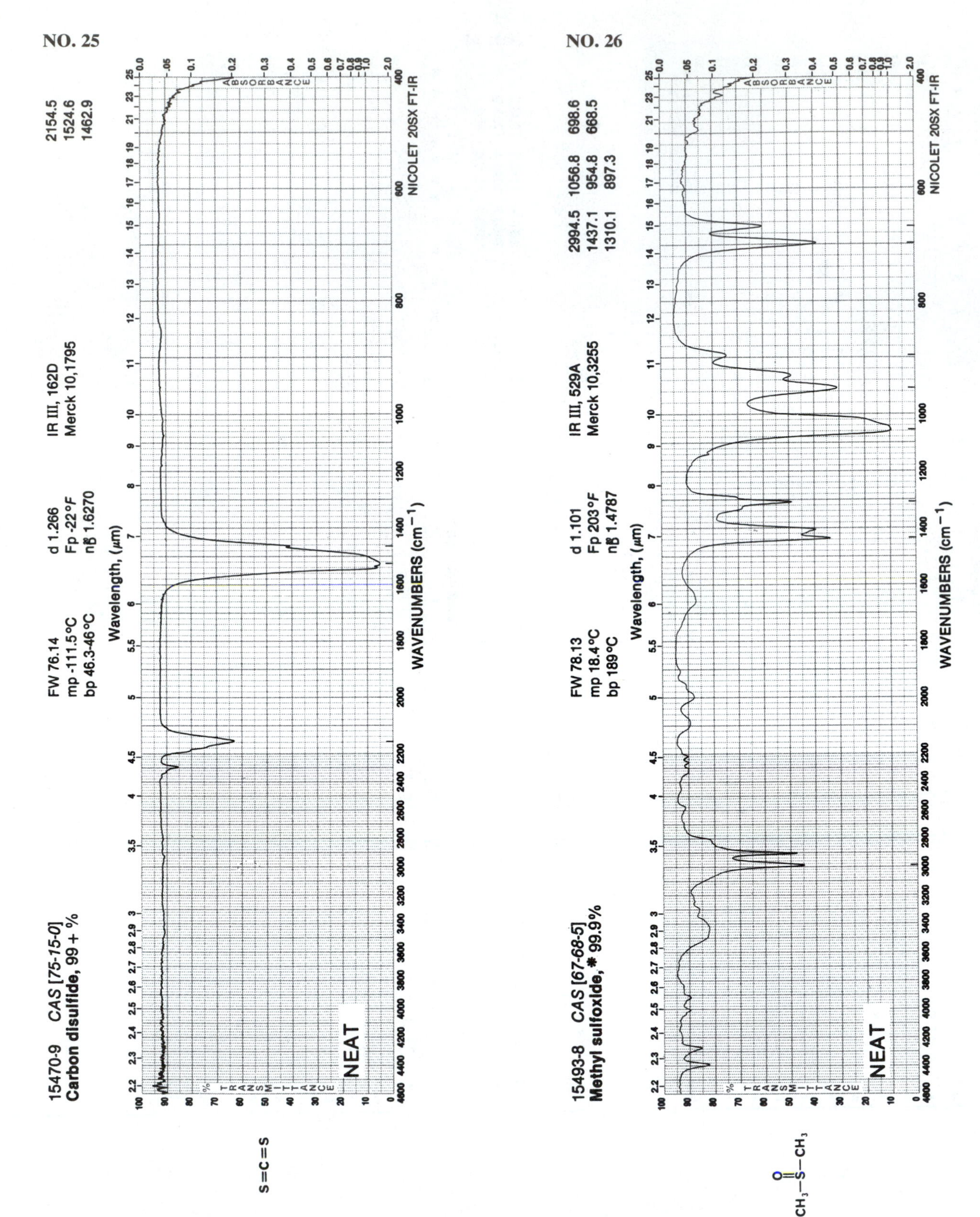
NO. 25
15470-9 CAS [75-15-0]
Carbon disulfide, 99+%
FW 76.14
mp -111.5°C
bp 46.3-46°C
d 1.266
Fp -22°F
n 1.6270
IR III, 162D
Merck 10,1795
2154.5
1524.6
1462.9
NEAT
Wavelength, (μm)
WAVENUMBERS (cm⁻¹)
NICOLET 20SX FT-IR
% TRANSMITTANCE
ABSORBANCE
S=C=S
NO. 26
15493-8 CAS [67-68-5]
Methyl sulfoxide,* 99.9%
FW 78.13
mp 18.4°C
bp 189°C
d 1.101
Fp 203°F
n 1.4787
IR III, 529A
Merck 10,3255
2994.5 1056.8 698.6
1437.1 954.8 668.5
1310.1 897.3
NEAT
Wavelength, (μm)
WAVENUMBERS (cm⁻¹)
NICOLET 20SX FT-IR
% TRANSMITTANCE
ABSORBANCE
CH3—S(=O)—CH3

NO. 27

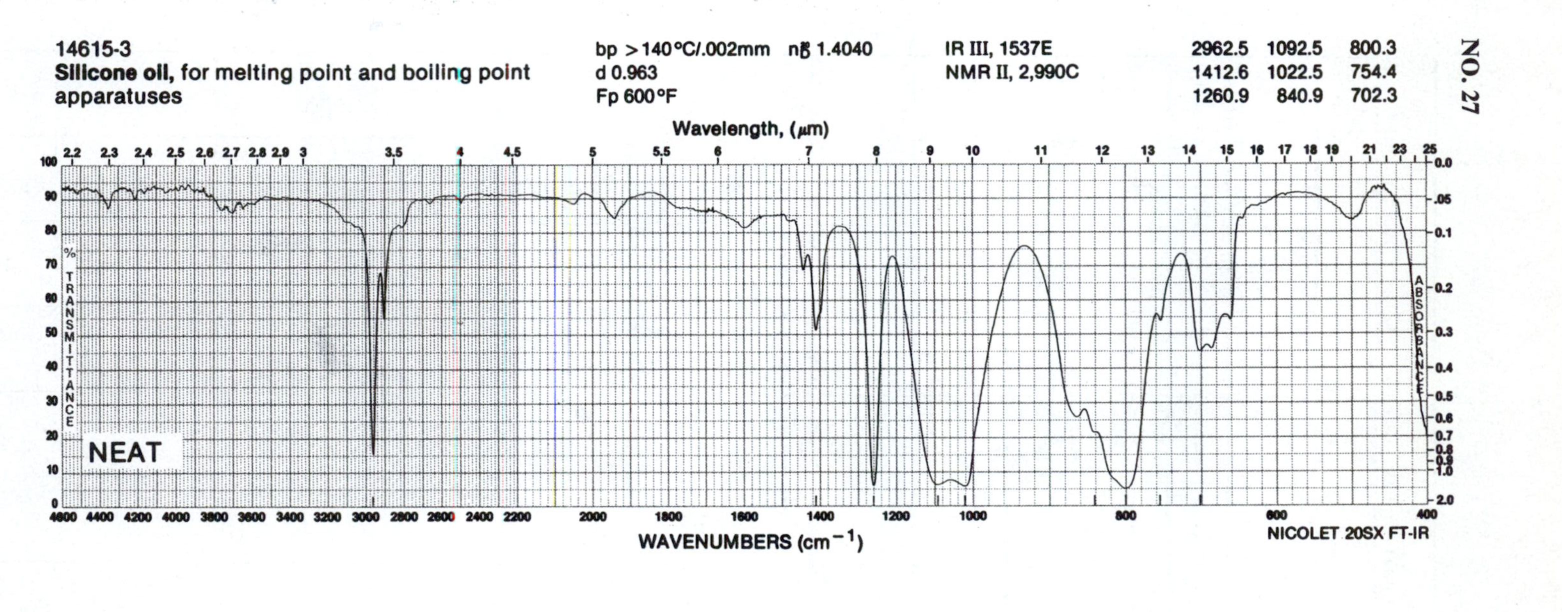
14615-3
Silicone oil, for melting point and boiling point apparatuses
bp >140°C/.002mm n20D 1.4040
d 0.963
Fp 600°F
IR III, 1537E
NMR II, 2,990C
2962.5 1092.5 800.3
1412.6 1022.5 754.4
1260.9 840.9 702.3
Wavelength, (μm)
% TRANSMITTANCE
ABSORBANCE
NEAT
WAVENUMBERS (cm−1)
NICOLET 20SX FT-IR

NO. 28

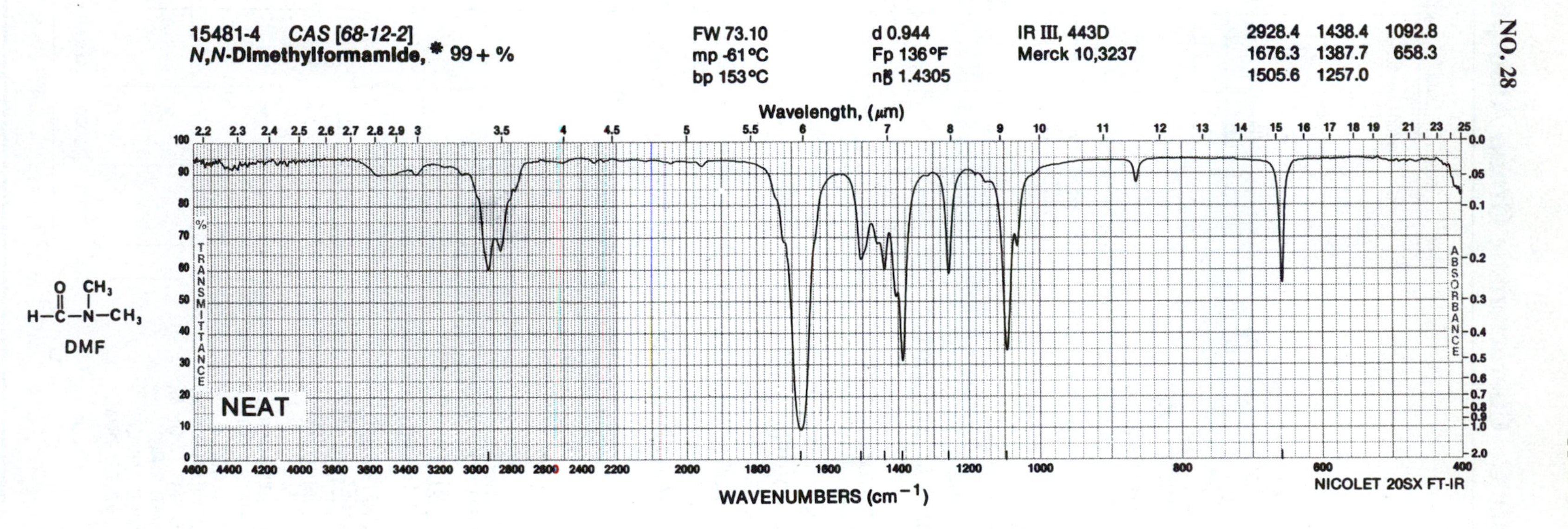
15481-4 CAS [68-12-2]
N,N-Dimethylformamide, * 99+%
FW 73.10
mp -61°C
bp 153°C
d 0.944
Fp 136°F
n20D 1.4305
IR III, 443D
Merck 10,3237
2928.4 1438.4 1092.8
1676.3 1387.7 658.3
1505.6 1257.0
Wavelength, (μm)
% TRANSMITTANCE
ABSORBANCE
NEAT
WAVENUMBERS (cm−1)
NICOLET 20SX FT-IR
DMF

Appendix C Characteristic Group Absorptions[a]

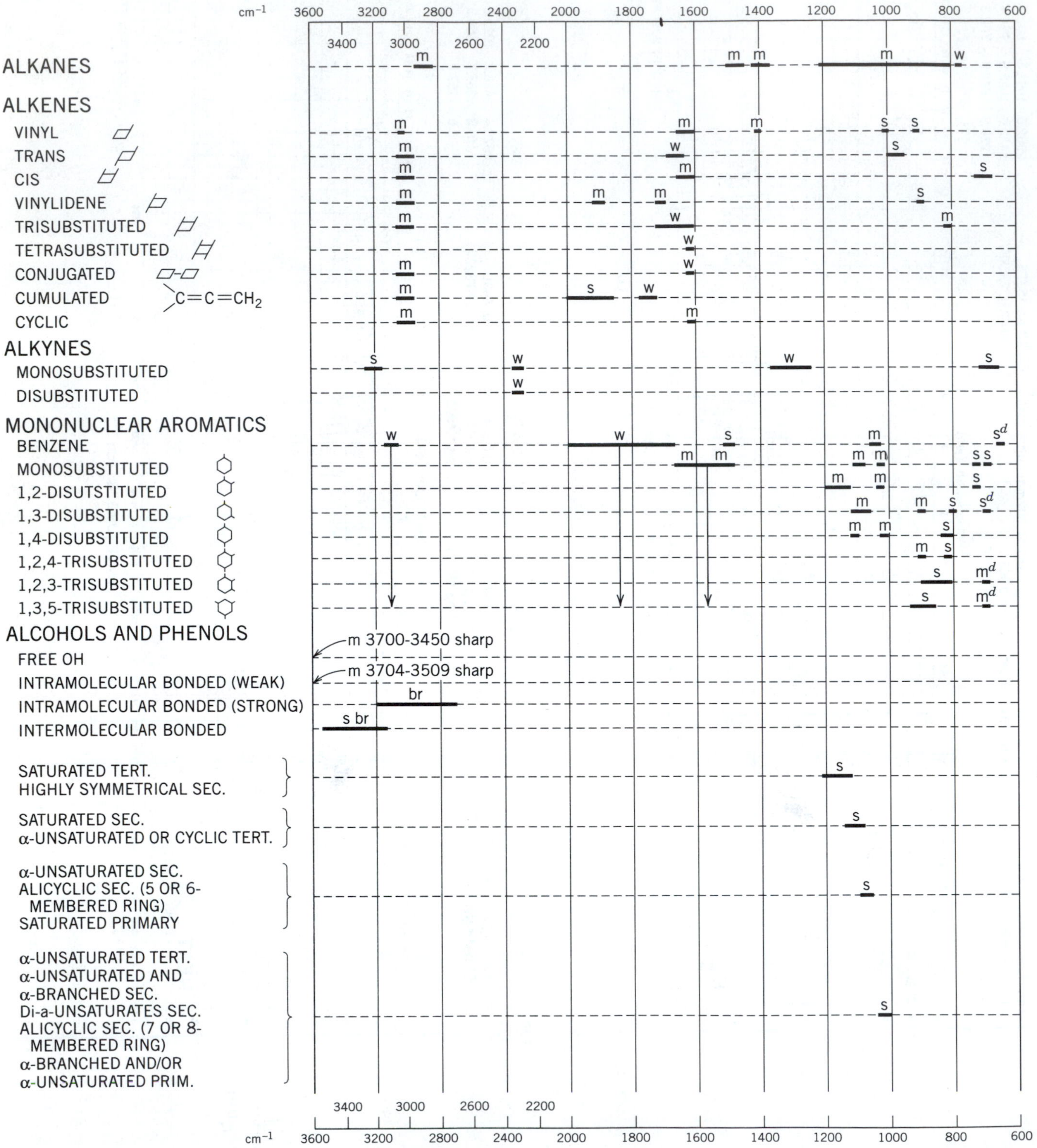

cm⁻¹ 3600 3200 2800 2400 2000 1800 1600 1400 1200 1000 800 600

3400 3000 2600 2200

ALKANES

ALKENES
- VINYL
- TRANS
- CIS
- VINYLIDENE
- TRISUBSTITUTED
- TETRASUBSTITUTED
- CONJUGATED
- CUMULATED $\rangle C{=}C{=}CH_2$
- CYCLIC

ALKYNES
- MONOSUBSTITUTED
- DISUBSTITUTED

MONONUCLEAR AROMATICS
- BENZENE
- MONOSUBSTITUTED
- 1,2-DISUTSTITUTED
- 1,3-DISUBSTITUTED
- 1,4-DISUBSTITUTED
- 1,2,4-TRISUBSTITUTED
- 1,2,3-TRISUBSTITUTED
- 1,3,5-TRISUBSTITUTED

ALCOHOLS AND PHENOLS
- FREE OH — m 3700-3450 sharp
- INTRAMOLECULAR BONDED (WEAK) — m 3704-3509 sharp
- INTRAMOLECULAR BONDED (STRONG) — br
- INTERMOLECULAR BONDED — s br

SATURATED TERT.
HIGHLY SYMMETRICAL SEC.

SATURATED SEC.
α-UNSATURATED OR CYCLIC TERT.

α-UNSATURATED SEC.
ALICYCLIC SEC. (5 OR 6-MEMBERED RING)
SATURATED PRIMARY

α-UNSATURATED TERT.
α-UNSATURATED AND
α-BRANCHED SEC.
Di-a-UNSATURATES SEC.
ALICYCLIC SEC. (7 OR 8-MEMBERED RING)
α-BRANCHED AND/OR
α-UNSATURATED PRIM.

3400 3000 2600 2200

cm⁻¹ 3600 3200 2800 2400 2000 1800 1600 1400 1200 1000 800 600

[a] Absorptions are shown by heavy bars. s = strong, m = medium, w = weak, sh = sharp, br = broad. Two intensity designations over a single bar indicate that two peaks may be present.

[b] May be absent.

[c] Frequently a doublet.

[d] Ring bending bands.

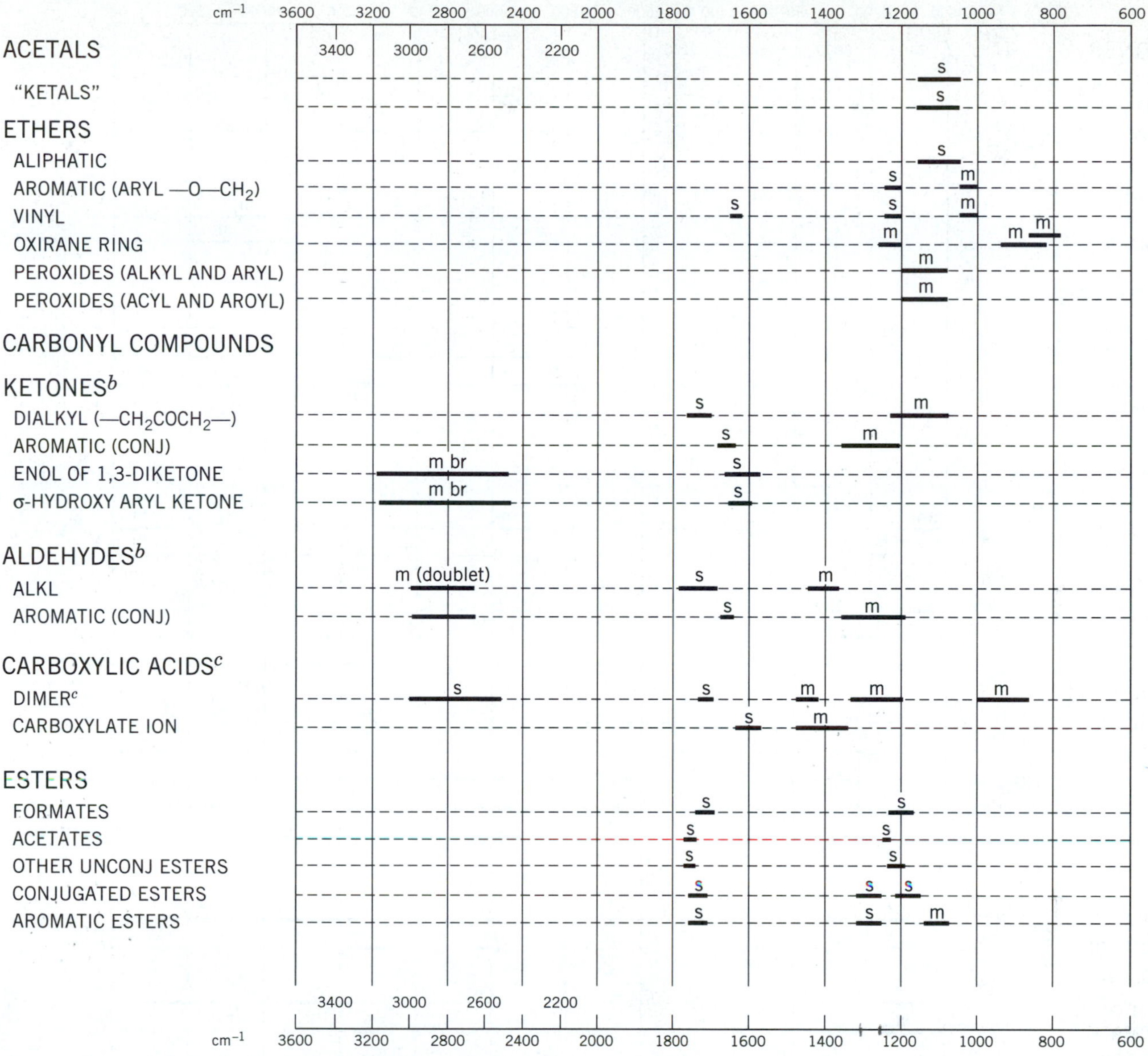

cm⁻¹ 3600 3200 2800 2400 2000 1800 1600 1400 1200 1000 800 600

3400 3000 2600 2200

ACETALS[a]

"KETALS"[a]

ETHERS

ALIPHATIC

AROMATIC (ARYL —O—CH_2)

VINYL

OXIRANE RING

PEROXIDES (ALKYL AND ARYL)

PEROXIDES (ACYL AND AROYL)

CARBONYL COMPOUNDS

KETONES[b]

DIALKYL (—CH_2COCH_2—)

AROMATIC (CONJ)

ENOL OF 1,3-DIKETONE

σ-HYDROXY ARYL KETONE

ALDEHYDES[b]

ALKL

AROMATIC (CONJ)

CARBOXYLIC ACIDS[c]

DIMER[c]

CARBOXYLATE ION

ESTERS

FORMATES

ACETATES

OTHER UNCONJ ESTERS

CONJUGATED ESTERS

AROMATIC ESTERS

3400 3000 2600 2200

cm⁻¹ 3600 3200 2800 2400 2000 1800 1600 1400 1200 1000 800 600

[a] Three bands, sometimes a fourth for ketals, and a fifth band for acetals.

[b] Conjugated aliphatic examples show C=O stretch at virtually the same position as aromatic structures.

[c] Conjugated examples show C=O stretch at lower wavenumbers (1710–1680 cm^{-1}). The O—H stretch (3300–2600 cm^{-1}) is very broad.

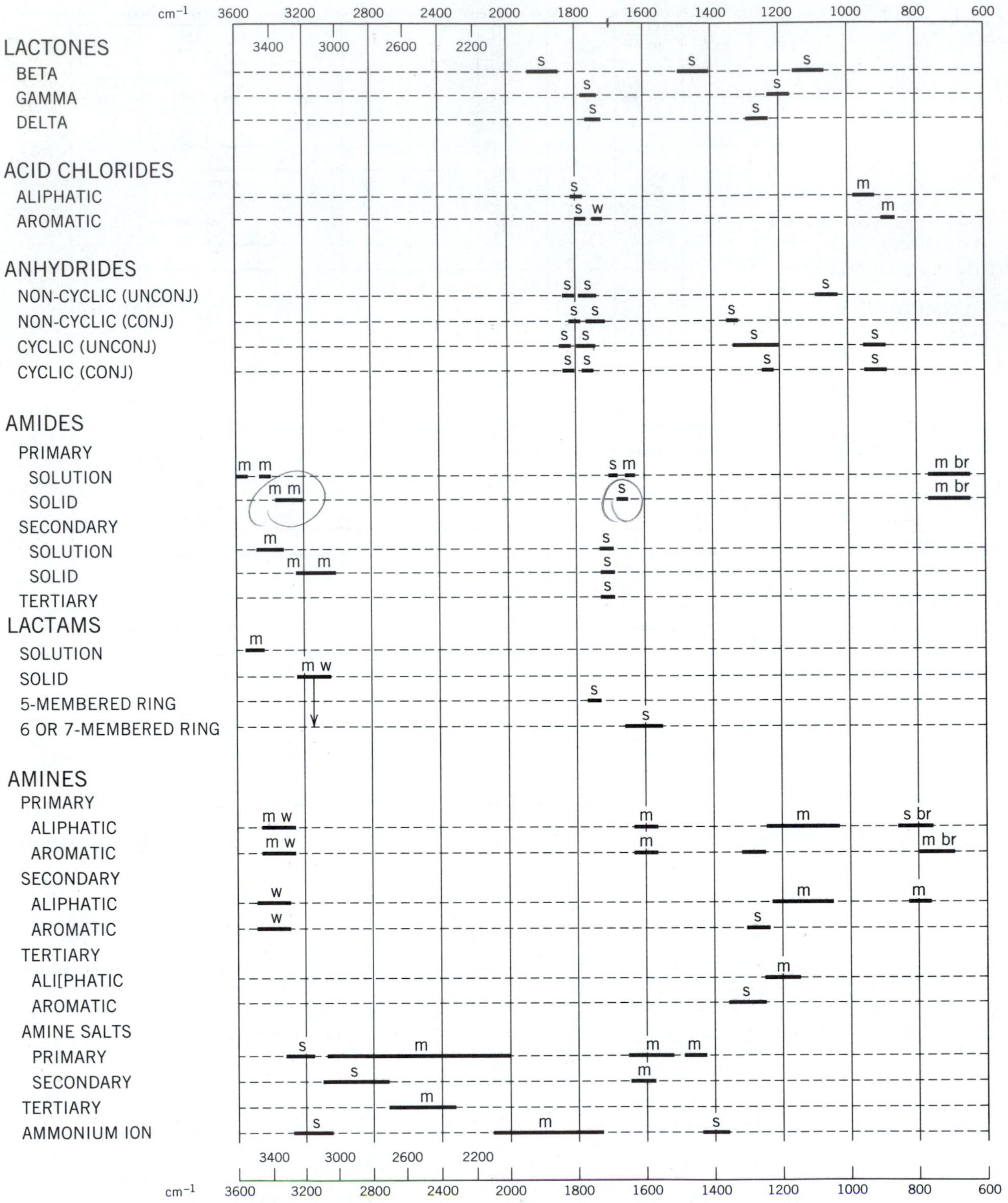

cm^{-1} 3600 3200 2800 2400 2000 1800 1600 1400 1200 1000 800 600

3400 3000 2600 2200

	Band intensities (left to right)
LACTONES	
BETA	s s s
GAMMA	s s
DELTA	s s
ACID CHLORIDES	
ALIPHATIC	s m
AROMATIC	s w m
ANHYDRIDES	
NON-CYCLIC (UNCONJ)	s s s
NON-CYCLIC (CONJ)	s s s
CYCLIC (UNCONJ)	s s s s
CYCLIC (CONJ)	s s s s
AMIDES	
PRIMARY	
SOLUTION	m m s m m br
SOLID	m m s m br
SECONDARY	
SOLUTION	m s
SOLID	m m s
TERTIARY	s
LACTAMS	
SOLUTION	m
SOLID	m w
5-MEMBERED RING	s
6 OR 7-MEMBERED RING	s
AMINES	
PRIMARY	
ALIPHATIC	m w m m s br
AROMATIC	m w m m br
SECONDARY	
ALIPHATIC	w m m
AROMATIC	w s
TERTIARY	
ALI[PHATIC	m
AROMATIC	s
AMINE SALTS	
PRIMARY	s m m m
SECONDARY	s m
TERTIARY	m
AMMONIUM ION	s m s

3400 3000 2600 2200

cm^{-1} 3600 3200 2800 2400 2000 1800 1600 1400 1200 1000 800 600

cm⁻¹ 3600 3200 2800 2400 2000 1800 1600 1400 1200 1000 800 600

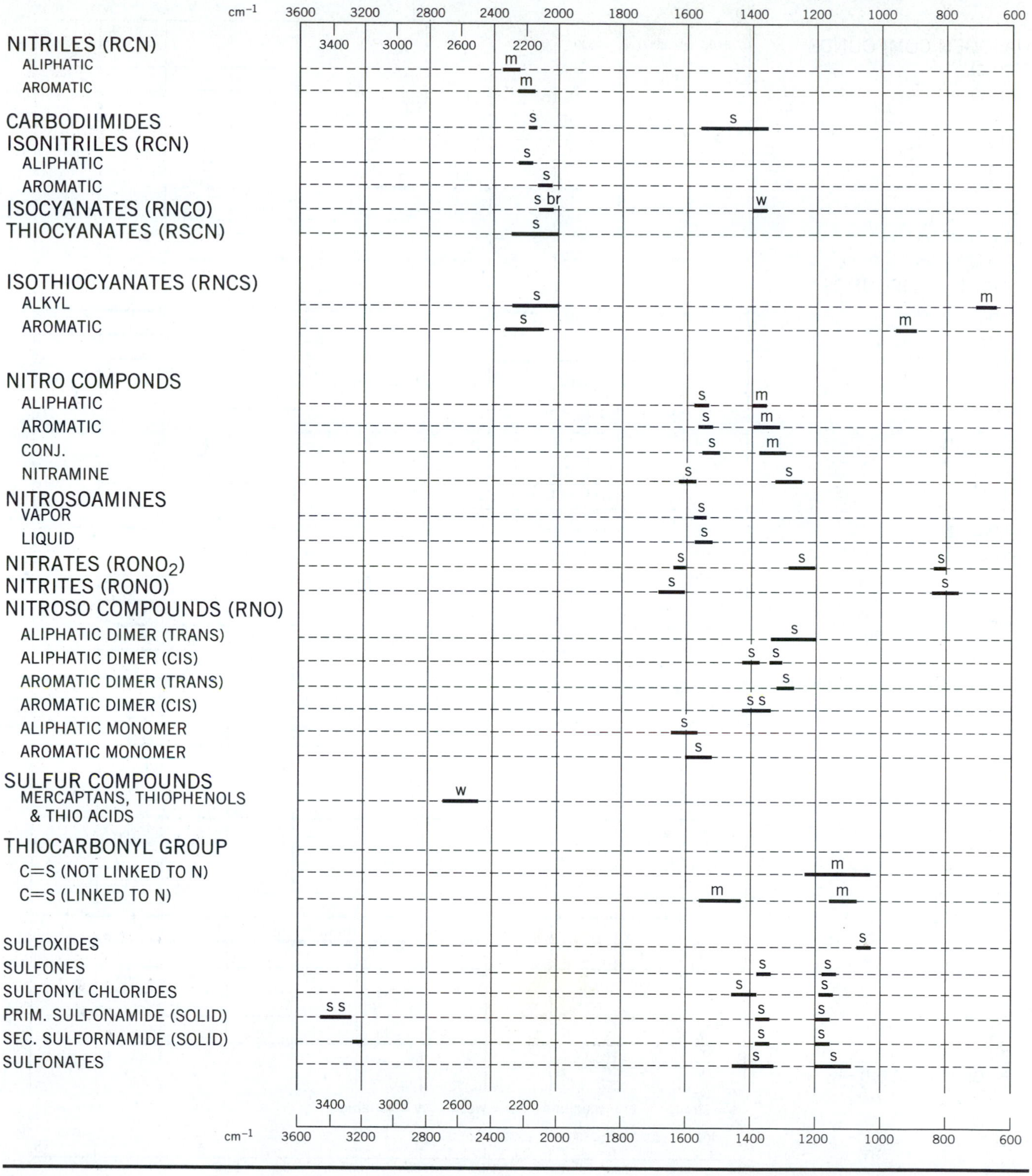

3400 3000 2600 2200

NITRILES (RCN)
- ALIPHATIC — m
- AROMATIC — m

CARBODIIMIDES — s, s

ISONITRILES (RCN)
- ALIPHATIC — s
- AROMATIC — s

ISOCYANATES (RNCO) — s br, w

THIOCYANATES (RSCN) — s

ISOTHIOCYANATES (RNCS)
- ALKYL — s, m
- AROMATIC — s, m

NITRO COMPONDS
- ALIPHATIC — s, m
- AROMATIC — s, m
- CONJ. — s, m
- NITRAMINE — s, s

NITROSOAMINES
- VAPOR — s
- LIQUID — s

NITRATES ($RONO_2$) — s, s, s

NITRITES (RONO) — s, s

NITROSO COMPOUNDS (RNO)
- ALIPHATIC DIMER (TRANS) — s
- ALIPHATIC DIMER (CIS) — s, s
- AROMATIC DIMER (TRANS) — s
- AROMATIC DIMER (CIS) — s s
- ALIPHATIC MONOMER — s
- AROMATIC MONOMER — s

SULFUR COMPOUNDS
- MERCAPTANS, THIOPHENOLS & THIO ACIDS — w

THIOCARBONYL GROUP
- C=S (NOT LINKED TO N) — m
- C=S (LINKED TO N) — m, m

SULFOXIDES — s

SULFONES — s, s

SULFONYL CHLORIDES — s, s

PRIM. SULFONAMIDE (SOLID) — s s, s, s

SEC. SULFORNAMIDE (SOLID) — s, s

SULFONATES — s, s

3400 3000 2600 2200

cm⁻¹ 3600 3200 2800 2400 2000 1800 1600 1400 1200 1000 800 600

cm^{-1} 3600 3400 3200 3000 2800 2600 2400 2200 2000 1800 1600 1400 1200 1000 800 600

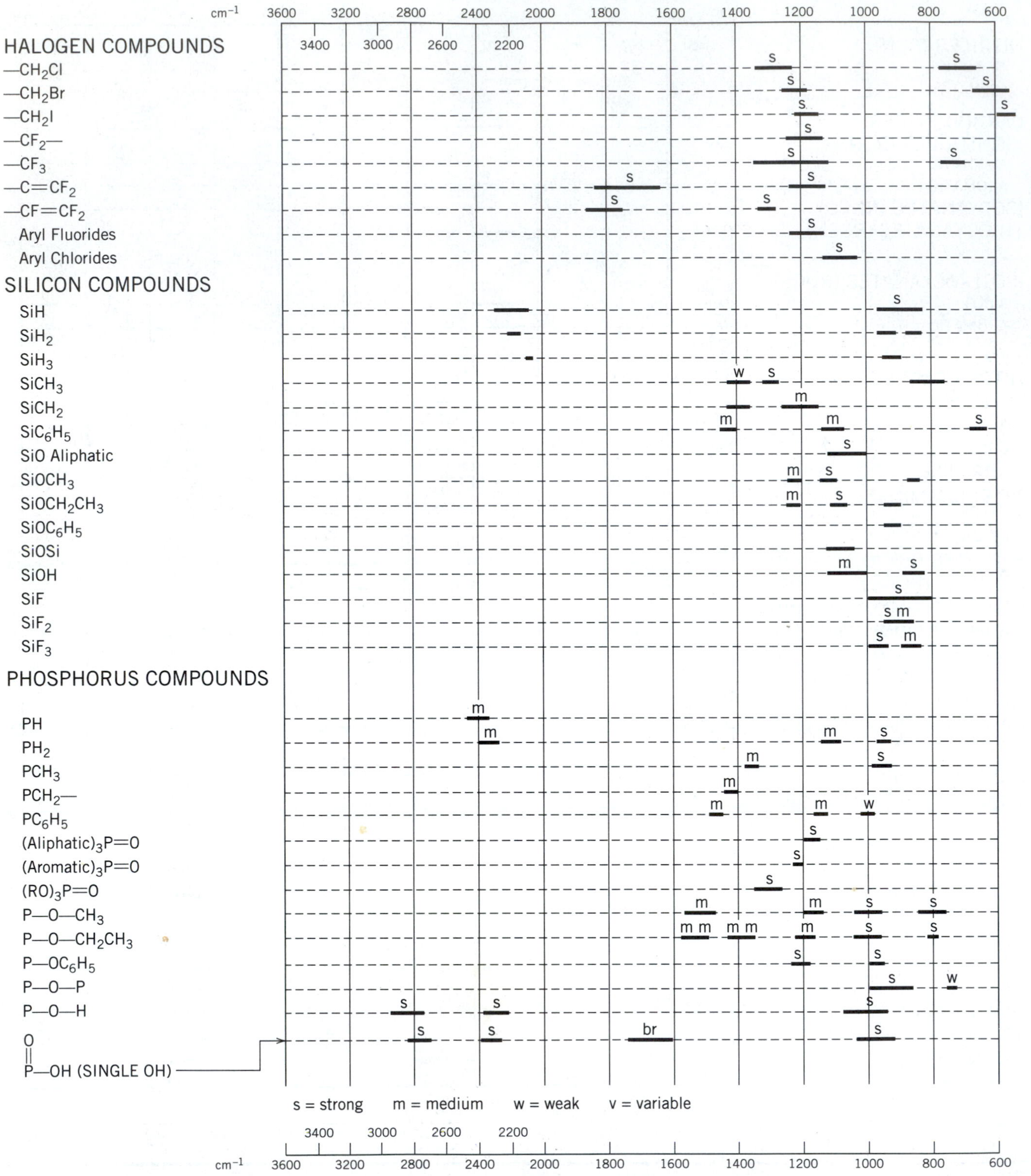

HALOGEN COMPOUNDS

—CH_2Cl
—CH_2Br
—CH_2I
—CF_2—
—CF_3
—$C{=}CF_2$
—$CF{=}CF_2$
Aryl Fluorides
Aryl Chlorides

SILICON COMPOUNDS

SiH
SiH_2
SiH_3
$SiCH_3$
$SiCH_2$
SiC_6H_5
SiO Aliphatic
$SiOCH_3$
$SiOCH_2CH_3$
$SiOC_6H_5$
SiOSi
SiOH
SiF
SiF_2
SiF_3

PHOSPHORUS COMPOUNDS

PH
PH_2
PCH_3
PCH_2—
PC_6H_5
$(Aliphatic)_3P{=}O$
$(Aromatic)_3P{=}O$
$(RO)_3P{=}O$
P—O—CH_3
P—O—CH_2CH_3
P—OC_6H_5
P—O—P
P—O—H
O‖P—OH (SINGLE OH)

s = strong m = medium w = weak v = variable

3400 3000 2600 2200
cm^{-1} 3600 3200 2800 2400 2000 1800 1600 1400 1200 1000 800 600

Appendix D Absorptions for Alkenes

Table D-1 Alkene Absorptions[a]

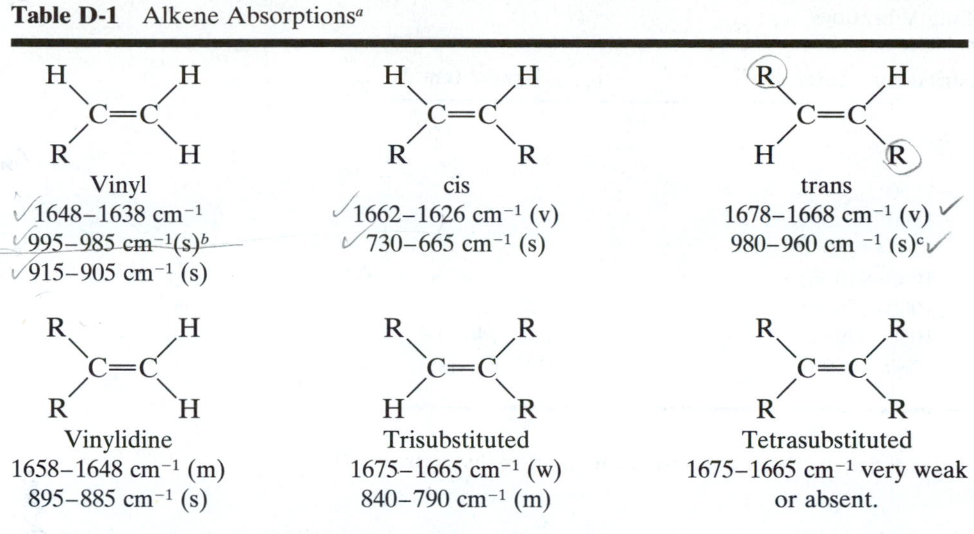

Vinyl	cis	trans
1648–1638 cm^{-1} 995–985 cm^{-1} (s)[b] 915–905 cm^{-1} (s)	1662–1626 cm^{-1} (v) 730–665 cm^{-1} (s)	1678–1668 cm^{-1} (v) 980–960 cm^{-1} (s)[c]
Vinylidine	**Trisubstituted**	**Tetrasubstituted**
1658–1648 cm^{-1} (m) 895–885 cm^{-1} (s)	1675–1665 cm^{-1} (w) 840–790 cm^{-1} (m)	1675–1665 cm^{-1} very weak or absent.

[a] s = strong, m = medium, w = weak, v = variable.
[b] This band also shows a strong overtone band.
[c] This band occurs near 1000 cm^{-1} in conjugated trans–trans systems such as the esters of sorbic acid.

Table D-2 C=C Stretching Frequencies in Cyclic and Acyclic Systems (cm^{-1})

<table>
<tr><th>Ring[a] or Chain</th><th>H(C)C=C(C)H</th><th>H(C)C=C(C)CH_3</th><th>CH_3(C)C=C(C)CH_3</th><th>(C)(C)C=CH_2</th></tr>
<tr><td>Chain cis</td><td>1661</td><td rowspan="2">1681</td><td rowspan="2">1672</td><td rowspan="2">1661</td></tr>
<tr><td>Chain trans</td><td>1676</td></tr>
<tr><td>Three-membered ring</td><td>1641</td><td></td><td>1890</td><td>1780</td></tr>
<tr><td>Four-membered ring</td><td>1566</td><td></td><td>1685</td><td>1678</td></tr>
<tr><td>Five-membered ring</td><td>1611</td><td>1658</td><td>1686</td><td>1657</td></tr>
<tr><td>Six-membered ring</td><td>1649</td><td>1678</td><td>1685</td><td>1651</td></tr>
<tr><td>Seven-membered ring</td><td>1651</td><td>1673</td><td></td><td></td></tr>
<tr><td>Eight-membered ring</td><td>1653</td><td></td><td></td><td></td></tr>
</table>

[a] All rings have cis double bonds.

Appendix E Absorptions for Phosphorus Compounds

Table E-1 P═O and P—O Stretching Vibrations

Group	Position cm^{-1} Intensity[a]	ν_{P-O} Bands[a] (cm^{-1})
P═O stretch		
Phosphine oxides		
Aliphatic	~1150	
Aromatic	~1190	
Phosphate esters[b]	1299–1250	
P—OH	1040–910 (s)	
P—O—P	1000–870 (s)	~700 w
P—O—C (aliph)	1050–970 (s)[c]	830–740 (s)[d]
P—O—C (arom)	1260–1160 (s)	994–855 (s)

[a] s = strong; w = weak
[b] The increase in P═O stretching frequency of the ester, relative to the oxides, results from the electronegativity of the attached alkoxy groups.
[c] May be a doublet.
[d] May be absent.

Appendix F Absorptions for Heteroaromatics

Table F-1 γ-CH and Ring Bending (β-Ring) Bands of Pyridines[a]

Substitution	Number Adjacent H Atoms	γ-CH (cm^{-1})	β-Ring
2-	4	781–740	752–746
3-	3	810–789	715–712
4-	2	820–794	775–709

[a] The γ and β notations are explained in the text (Section 3.6.30.4) and in the book by Katritzky (1963).

Table F-2 Characteristic γ-CH or β-Ring Bands of Furans, Thiophenes, and Pyrroles

Ring	Position of Substitution	Phase	γ-CH or β-Ring Modes[a]			
			cm^{-1}	cm^{-1}	cm^{-1}	cm^{-1}
Furan	2-	$CHCl_3$	~925	~884	835–780	
	2-	Liquid	960–915	890–875		780–725
	2-	Solid	955–906	887–860	821–793	750–723
	3-	Liquid		885–870	741	
Thiophene	2-	$CHCl_3$	~925	~853	843–803	
	3-	Liquid				755
Pyrrole	2-Acyl	Solid			774–740	~755

[a] The γ and β notations are explained in the text (Section 3.6.30.4) and in the book by Katritzky (1963).

CHAPTER 4

Proton Magnetic Resonance Spectrometry

4.1 Introduction

Nuclear magnetic resonance (NMR) spectrometry is basically another form of absorption spectrometry, akin to IR or UV spectrometry. Under appropriate conditions *in a magnetic field,* a sample can absorb electromagnetic radiation in the radio frequency (rf) region at frequencies governed by the characteristics of the sample. Absorption is a function of certain nuclei in the molecule. A plot of the frequencies of the absorption peaks versus peak intensities constitutes an NMR spectrum. This chapter covers proton magnetic resonance (^{1}H NMR) spectrometry.

With some mastery of basic theory, interpretation of NMR spectra merely by inspection is usually feasible in greater detail than is the case for IR or mass spectra. The present account will suffice for the immediate limited objective: identification of organic compounds in conjunction with other spectrometric information. References are given at the end of this chapter.

We begin by describing some magnetic properties of nuclei. All nuclei carry a charge. In some nuclei this charge "spins" on the nuclear axis, and this circulation of nuclear charge generates a magnetic dipole along the axis (Fig. 4.1). The angular momentum of the spinning charge can be described in terms of quantum spin numbers ***I***; these numbers have values of 0, $\frac{1}{2}$, 1, $\frac{3}{2}$, and so on ($I = 0$ denotes no spin). The intrinsic magnitude of the generated dipole is expressed in terms of nuclear magnetic moment, μ.

Relevant properties, including the spin number I, of several nuclei are given in Appendix H. The spin number I can be determined from the atomic mass and the atomic number as shown in the next column.

Spectra of several nuclei can be readily obtained (e.g., $^1_1$H, $^3_1$H, $^{13}_6$C, $^{15}_7$N, $^{19}_9$F, $^{31}_{15}$P) since they have spin numbers I of $\frac{1}{2}$ and a uniform spherical charge distribution (Fig. 4.1). Of these, by far the most widely used in NMR spectrometry are ^{1}H (this chapter) and ^{13}C (Chapter 5).

I	Atomic Mass	Atomic Number	Example (I)
Half-integer	Odd	Odd or even	$^1_1$H($\frac{1}{2}$), $^{17}_8$O($\frac{5}{2}$), $^{15}_7$N($\frac{1}{2}$)
Integer	Even	Odd	$^2_1$H(1), $^{14}_7$N(1), $^{10}_5$B(3)
Zero	Even	Even	$^{12}_6$C(0), $^{16}_8$O(0), $^{34}_{16}$S(0)

Nuclei with a spin number I of 1 or higher have a nonspherical charge distribution. This asymmetry is described by an electrical quadrupole moment which, as we shall see later, affects the relaxation time and, consequently, the linewidth of the signal and the coupling with neighboring nuclei. In quantum mechanical terms, the spin number I determines the number of orientations a nucleus may assume in an external uniform magnetic field in accordance with the formulas $2I + 1$. We are concerned with the proton whose spin number I is $\frac{1}{2}$.

Thus in Figure 4.2, these are two energy levels and a slight excess of proton population in the lower energy state ($N_\alpha > N_\beta$) in accordance with the Boltzmann distribution. The states are labeled α and β or $\frac{1}{2}$ and $-\frac{1}{2}$; ΔE is given by

$$\Delta E = \frac{h\gamma}{2\pi} \boldsymbol{B}_0$$

where h is Planck's constant, which simply states that ΔE is proportional to $\boldsymbol{B}_0$ (as shown in Fig. 4.2) since h, γ, and π are constants. $\boldsymbol{B}_0$ represents the magnetic field strength.*

* The designations ***B*** (magnetic induction or flux density) and ***H*** (magnetic intensity) are often used interchangeably for magnetic field strength in NMR spectrometry. The SI term tesla (T), the unit of measurement for ***B***, supercedes the term gauss (G); 1 T = 10^4 G. The frequency term hertz (Hz) supercedes cycles per second (cps). MHz is megahertz (10^6 Hz).

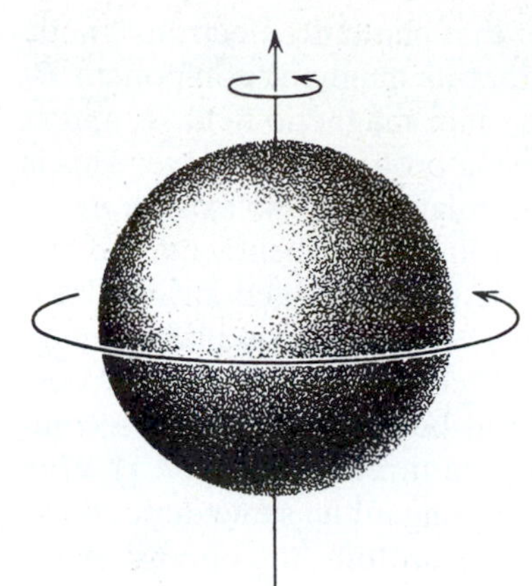

FIGURE 4.1. Spinning charge on proton generates magnetic dipole.

Once two energy levels for the proton have been established, it is possible to introduce energy in the form of radiofrequency radiation (ν_1) to effect a transition between these energy levels in a stationary magnetic field of given strength $\boldsymbol{B}_0$. *The fundamental NMR equation correlating the applied radiofrequency ν_1 with the magnetic field strength is*

$$\nu_1 = \frac{\gamma}{2\pi}\boldsymbol{B}_0$$

since

$$\Delta E = h\nu$$

The introduced radiofrequency ν_1 is given in megahertz (MHz). A frequency of 100 MHz is needed at a magnetic field strength $\boldsymbol{B}_0$ of 2.35 tesla (T) for the proton (or any other desired combination of ν_1 and $\boldsymbol{B}_0$ at the same ratio. See Appendix H). At this ratio, the system is in *resonance;* energy is absorbed by the proton, raising it to the higher energy state, and a spectrum results. Hence the name *nuclear magnetic resonance spectrometry.* The constant γ is called the magnetogyric ratio, a fundamental nuclear constant; it is the proportionality constant between the magnetic moment μ and the spin number I.

$$\gamma = \frac{2\pi\mu}{hI}$$

The radiofrequency ν_1 can be introduced either by continuous-wave (CW) scanning or by a radiofrequency pulse.

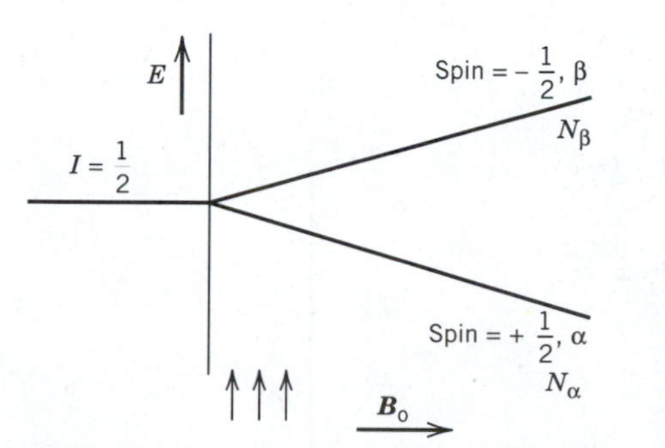

FIGURE 4.2. Two proton energy levels, from quantum mechanics, in a magnetic field of magnitude $\boldsymbol{B}_0$. N is population. The direction of the magnetic field (↑ ↑ ↑) is up, parallel to the ordinate, and $\boldsymbol{B}_0$ increases to the right.

4.2 Continuous-Wave (CW) NMR Spectrometry

The problem is how to apply radiofrequency (rf) electromagnetic energy to protons aligned in a stationary magnetic field and how to measure the energy thus absorbed as the protons are raised to the higher spin state. This can best be explained in classical mechanical terms, wherein we visualize the proton as spinning in an external magnetic field. The magnetic axis of the proton precesses about the z axis of the stationary magnetic field $\boldsymbol{B}_0$ in the same manner in which an off-perpendicular spinning top precesses under the influence of gravity (Fig. 4.3).

An assemblage of equivalent protons precessing in random phase around the z axis (i.e., in the direction of the stationary magnetic field $\boldsymbol{B}_0$) has a net macroscopic magnetization $\boldsymbol{M}_0$ along the z axis, but none in the xy plane (Fig. 4.4).

When an applied rf (ν_1) is equal to the precessional frequency of the equivalent protons (Larmor frequency ν_L in MHz), the state of nuclear magnetic resonance is

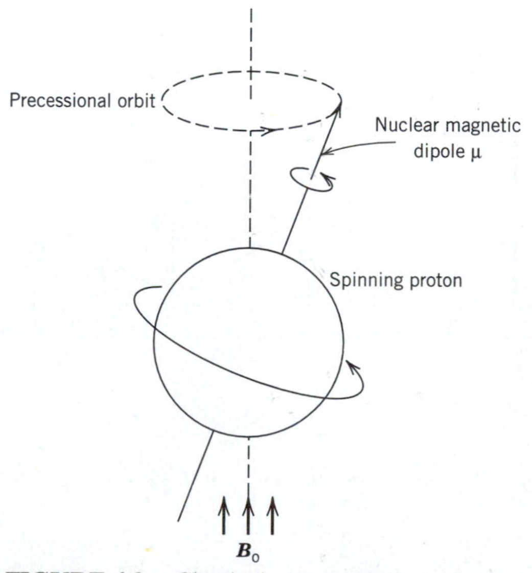

FIGURE 4.3. Classical representation of a proton precessing in a magnetic field of magnitude $\boldsymbol{B}_0$ in analogy with a precessing spinning top.

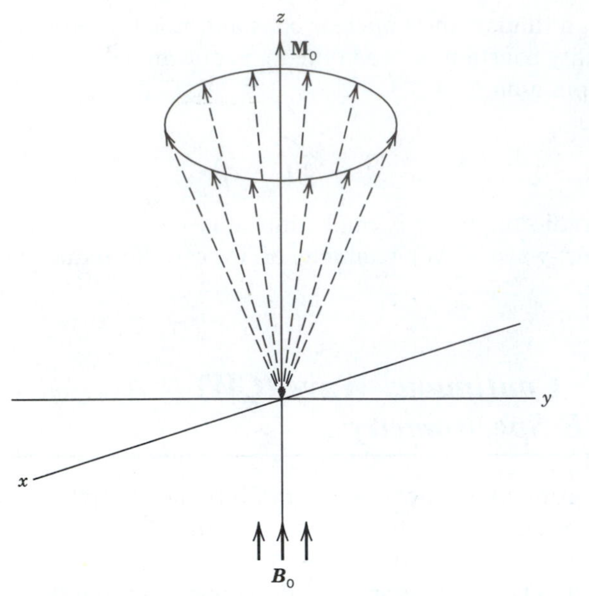

FIGURE 4.4. Assemblage of precessing nuclei with net macroscopic magnetization $\mathbf{M}_0$ in the direction of the stationary magnetic field $\boldsymbol{B}_0$.

attained, and the basic NMR relationship can be written:

$$\nu_L = \nu_1 = \frac{\gamma}{2\pi} \boldsymbol{B}_0$$

This equation applies to an isolated proton (see Sections 4.6 and 4.7).

The aim is to tip the net magnetization $\boldsymbol{M}_0$ toward the xy horizontal plane of the stationary Cartesian frame of reference and measure the resulting component of magnetization in that plane. Rf electromagnetic energy ν_1 is applied so that its magnetic component $\boldsymbol{B}_1$ is at right angles to the main magnetic field $\boldsymbol{B}_0$ and is rotating with the precessing proton assemblage. This is accomplished by an rf oscillator with its axis (conventionally along the x coordinate) perpendicular to the axis of the main magnetic field $\boldsymbol{B}_0$. Such an oscillator will generate a continuous-wave (CW), oscillating, magnetic field $\boldsymbol{B}_1$ along the direction of the x axis. An oscillating magnetic field can be resolved into two components rotating in opposite directions (Fig. 4.5). One of these components is rotating in the same direction as the precessional orbit of the protons; the oppositely rotating component is ineffective. When the oscillator frequency ν_1 is varied (frequency "scan"), the frequency of the rotating magnetic field will come into *resonance* with the precessing Larmor frequencies ν_L of the protons, induce phase coherence, and tip the net magnetization $\boldsymbol{M}_0$ toward the horizontal plane (Fig. 4.6, *a* and *b*). The magnetic component thus generated in the xy plane can be detected by the receiver coil mounted in the xy plane. Thus, the maximum signal intensity is attained with a tip angle of 90°.

4.3 Relaxation

Having in classical mechanical terms tipped the net magnetization ($\boldsymbol{M}_0$) toward the xy plane, we need to discuss how $\boldsymbol{M}_0$ returns to the z axis.

There are two "relaxation" processes. The spin–lattice or longitudinal relaxation process, designated by the time T_1, involves transfer of energy from the "ex-

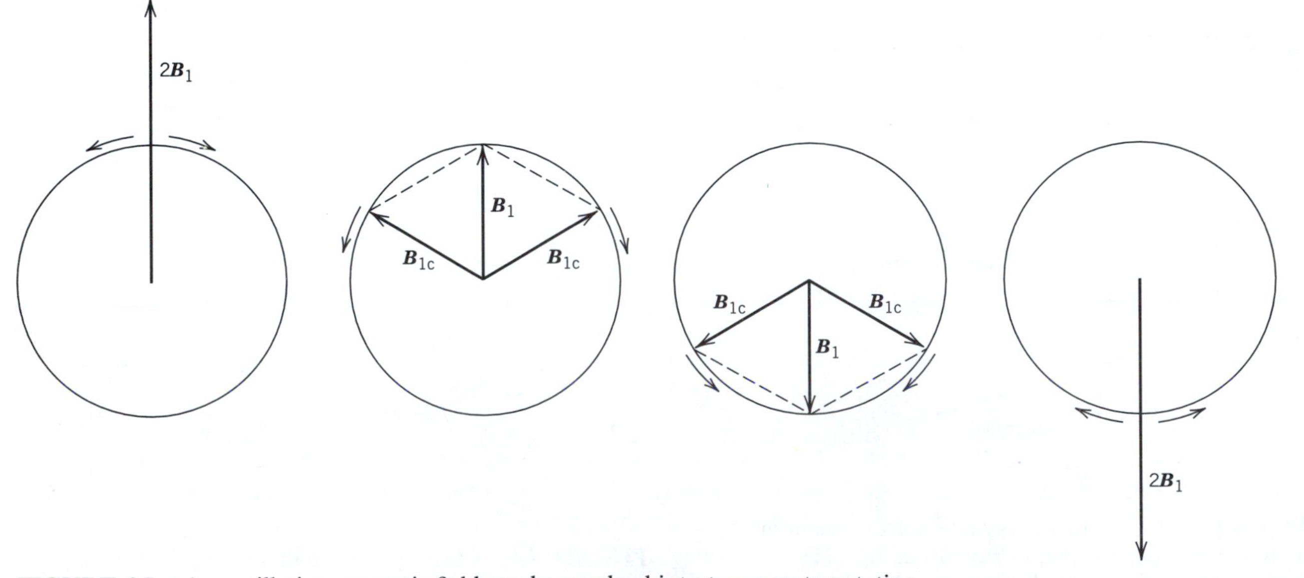

FIGURE 4.5. An oscillating magnetic field can be resolved into two counterrotating components.

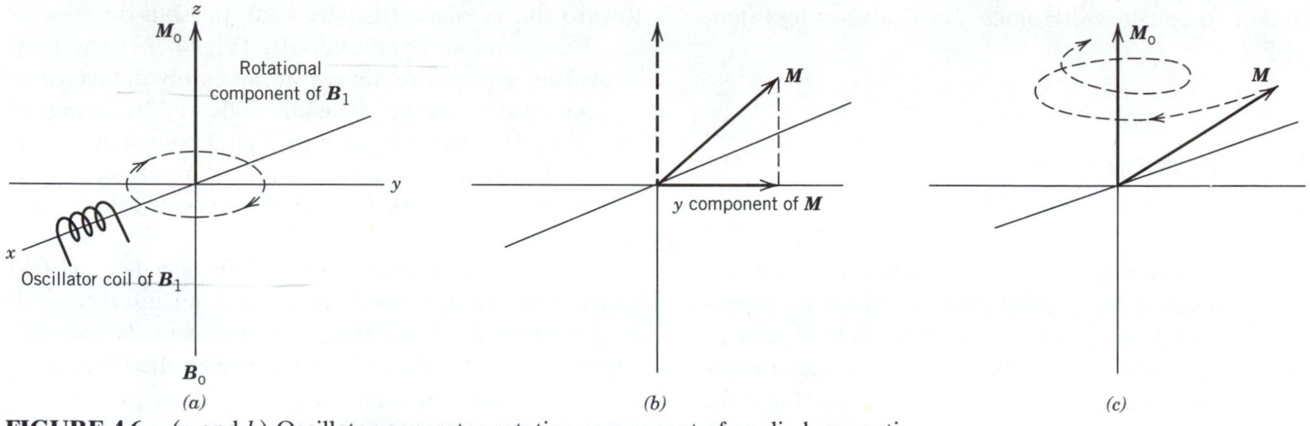

FIGURE 4.6. (*a* and *b*) Oscillator generates rotating component of applied magnetic field B_1. The net magnetization $\mathbf{M}_0$ is tipped to **M**, which precesses about the *z* axis generating a component of magnetization in the horizontal plane. (*c*) Longitudinal relaxation of **M** to $\mathbf{M}_0$ follows a decreasing spiral. Transverse relaxation T_2 (dephasing of **M**) is omitted. The Cartesian frame is stationary.

cited" protons to the surrounding protons that are tumbling at the appropriate frequencies. Fig. 4.6*c* shows the loss of the *xy* component by the T_1 process as the net magnetization returns to the *z* axis in a decreasing spiral.

The spin–spin or transverse relaxation, characterized by the time T_2, involves transfer of energy among the processing protons, which results in dephasing (fanning out), line broadening, and signal loss (Fig. 4.7*a*). The designation T_2* is used to denote the time for all the contributory factors to the transverse signal loss. This term includes both T_2 (the time of the actual spin dynamics) and the effect of magnetic field inhomogeneities, which usually dominates.

For protons in the usual nonviscous solutions, the T_1 and T_2 relaxation times are such that sharp peaks are obtained, and their intensities are proportional to the number of protons involved. Thus, the relative number of different kinds of protons in a spectrum can be determined by measuring the areas under the peaks (see Section 4.6).

However, for ^{13}C and ^{15}N nuclei, the relaxation times must be considered since they are longer than those of protons and vary widely. Several preliminary, general statements follow, and further relevant details are found in Chapters 5 and 7.

^{13}C and ^{15}N nuclei undergo T_1 dipole–dipole interactions with attached protons and, to a lesser extent, with other nearby protons. There are further complications with ^{15}N nuclei. In routine spectra of ^{13}C and ^{15}N, large T_1 values result in only partial recovery of the signal so that a delay interval must be inserted between the individual pulses (see Figure 5.1*c*). Thus, we see that T_1 relaxation is intimately involved with peak intensity.

In contrast, T_2 is involved with peak width in accordance with the Heisenberg uncertainty principle, which states that the product of the uncertainty of the frequency range and the uncertainty of the time interval is a constant:

$$\Delta\nu \cdot \Delta t \geq 1$$

If Δt (i.e., T_2*) is small, then $\Delta\nu$ is large, and the peak is therefore broad. T_2*, whose major component is field inhomogeneity is the principal determin-

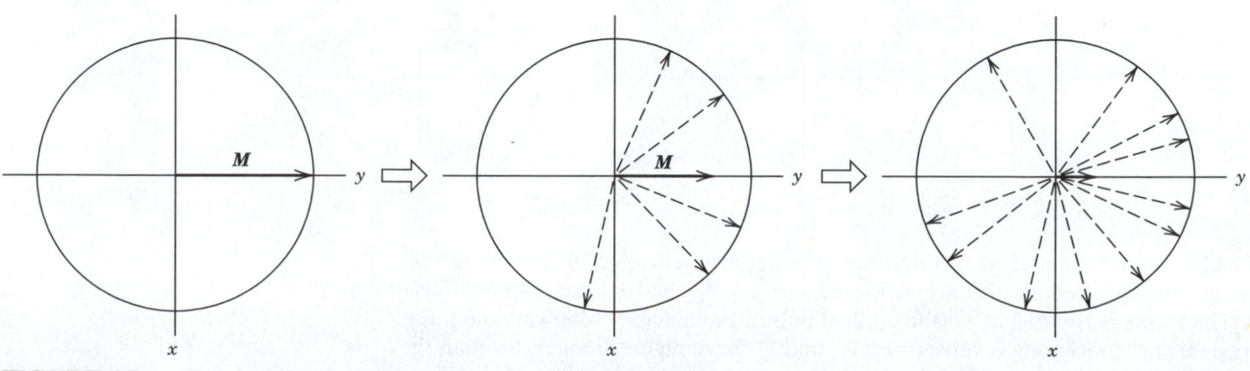

FIGURE 4.7*a*. T_2 relaxation in the *xy* plane of a rotating frame.

ing factor for peak width since T_2^* is always less than T_1 or T_2.

4.4 *Pulsed Fourier Transform Spectrometry*

So far, we have described the interaction between the net magnetization $\boldsymbol{M}_0$ of an assemblage of identical protons in a static, homogeneous, magnetic field $\boldsymbol{B}_0$ and an oscillating rf field ν_1—actually one of two circular components of the rf-generated magnetic field $\boldsymbol{B}_1$ (Fig. 4.5). To obtain a spectrum, the oscillator frequency ν_1 is scanned over the proton frequency range; alternatively, the oscillator frequency may be held constant and the field $\boldsymbol{B}_0$ scanned. Each different kind of proton must be brought into resonance one by one. This mode is called continuous-wave (CW) spectrometry and was employed in the early instruments. CW is still used in some of the lower resolution instruments, but CW has been almost completely superceded by pulsed Fourier transform (FT). However, since the CW mode is grasped more readily, we have discussed it first, using the familiar, stationary *xyz* Cartesian frame of reference. We now present an introduction to pulsed FT spectrometry and the rotating Cartesian frame of reference.

The pulsed technique was developed largely in response to the need for much higher sensitivity in ^{13}C spectrometry (Chapter 5). This higher sensitivity is achieved by exciting all of the nuclei of interest simultaneously (in this chapter, protons), then collecting all of the signals simultaneously. In a sense, a pulse may be described as an instantaneous "scan." A short (microseconds, μs), powerful, rf pulse of center frequency ν_1 applied along the x axis generates the entire, desired frequency range and has essentially the same effect as the scanning oscillator: It tips the net magnetization $\boldsymbol{M}_0$ toward the *xy* plane (usually a 90° tip) but does so for all of the protons simultaneously (Fig. 4.7*b*). The magnetization signals are almost immediately detected, after the pulse, in the *xy* plane and collected by an on-line computer (following analog to digital conversion) over a period of time, called the acquisition period. During this period, the signals from the precessing relaxing nuclei decay.

The result is a so-called *free induction decay* (FID), which may be described as a decaying interferogram (see Section 5.1 for examples). The signals collected represent the *difference* between the applied frequency ν_1 and the Larmor frequency ν_L of each proton. The FIDs are then Fourier transformed by computer into a conventional NMR spectrum. Since relaxation times for protons are usually on the order of a few seconds or fractions of a second, rapid repetitive pulsing with signal accumulation is possible. Some ^{13}C nuclei—those that have no attached protons to provide T_1 relaxation—require much longer intervals between pulses to allow for relaxation; lack of adequate intervals results in weak signals and inaccurate peak areas (see Section 5.1).

Assume that the total net magnetization $\boldsymbol{M}_0$ (a broad vector in Fig. 4.7*b*) aligned with the z axis consists of three individual net magnetizations representing three different kinds of protons in a chemical compound. With a 90° pulse, each net magnetization is in resonance with a different frequency in the pulse, and all are rotated simultaneously onto the y axis (Fig. 4.7*b*). Each of the component net magnetizations (narrow vectors) now begins to precess, each at its own Larmor frequency, while relaxing by the T_1 and T_2 mechanisms. There are two practical problems: First, it is difficult to measure accurately absolute frequencies that differ over the range of, say, 5000 Hz around a pulsed central frequency of, say, 300,000,000 Hz (ν_1), also called a carrier frequency. For example, our three different kinds of protons comprising the net magnetization have Larmor

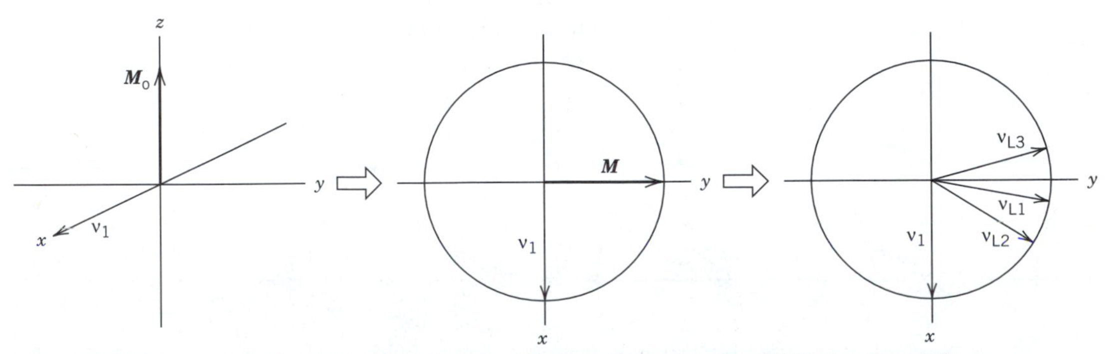

FIGURE 4.7*b*. A rotating frame of reference. The net magnetization **M** (following a 90° pulse) has three components with Larmor frequencies ν_{L1}, δ_{L2}, and ν_{L3} (i.e., three different protons). The frame is rotating at ν_1 (the applied pulse). Immediately following the pulse, the components are precessing *relative* to ν_1: ν_{L1} and ν_{L2} have higher frequencies than the frequency of the applied pulse ν_1, but the frequency of ν_{L3} is lower than that of ν_1.

frequencies (ν_L) of 300,002,000 Hz, 300,000,800 Hz, and 299,999,000 Hz. This problem is solved by measuring the difference between each Larmor frequency and the carrier frequency, which is applied in the middle of the spectral window. The frequency differences are 2000 Hz, 800 Hz, and (−) 1000 Hz; that is, two frequencies are higher than the carrier frequency and one frequency is lower. As mentioned above, these frequency differences comprise the free induction decay (FID) as the signal intensities from the precessing, relaxing vectors decrease (see Section 5.1).

The second problem is pictorial. How can these phenomena be presented in the conventional, static, Cartesian frame of reference? We avoid the complexities by using a rotating frame of reference.

4.5 Rotating Frame of Reference

Let us imagine that the frame of reference is not static but is rotating clockwise around the z axis at the carrier frequency ν_1. In Figure 4.7*b*, we are looking down the z axis toward the xy plane and place the ν_1 vector on the x axis, where it apparently remains even though it is really precessing at the carrier frequency of, say, 300 MHz. The two faster Larmor vectors (ν_{L1} and ν_{L2}) appear to precess clockwise, whereas the slowest vector (ν_{L3}) appears to precess counterclockwise, that is, at their difference frequencies: 2000 Hz, 800 Hz, and (−) 1000 Hz—precisely those used to produce the FID.

The rotating frame is used in Chapter 6 to illustrate multipulse manipulation of the precessing vectors. For other purposes, the frame may be rotating at the Larmor frequency of a particular carbon atom or at the frequency of the midpoint of the peaks of a ^{13}CH group, a $^{13}CH_2$ group, or a $^{13}CH_3$ group.

4.6 Instrumentation and Sample Handling

Beginning in 1953 with the first commercial NMR spectrometer, the early instruments used permanent magnets or electromagnets with fields of 1.41, 1.87, 2.20, or 2.35 T corresponding to 60, 80, 90, or 100 MHz, respectively, for proton resonance (the usual way of describing an instrument).

The "horsepower race," driven by the need for higher resolution and sensitivity, has resulted in wide use of 200–500 MHz instruments and in the production of 800-MHz instruments. All of the instruments above 100 MHz are based on helium-cooled superconducting magnets (solenoids) and operate in the pulsed FT mode. The other basic requirements besides high field are frequency-field stability, field homogeneity, and a computer interface.

The sample (routinely a solution in a deuterated solvent in a 5-mm tube) is placed in the probe, which contains the transmitter and receiver coils and a spinner to spin the tube about its vertical axis in order to average out field inhomogeneities. Figure 4.8 shows the probe elements between the poles of an electromagnet or a permanent magnet, and Figure 4.9 shows the arrangement for a superconducting magnet. Note that, in the electromagnet, the tube spins at right angles to the z axis, which is horizontal, whereas in the superconducting magnet, the tube fits in the bore of the solenoid and spins about the z axis, which is vertical. The transmitter and receiver are coupled through the sample nuclei (protons in this chapter).

The spectrum obtained either by CW scan or pulse FT at constant magnetic field is shown as a series of peaks whose areas are proportional to the number of protons they represent. Peak areas are measured by an electronic integrator that traces a series of steps with heights proportional to the peak areas (see Fig. 4.22).* A proton count from the integration is useful to determine or confirm molecular formulas, detect hidden peaks, determine sample purity, and do quantitative analysis. Peak positions (chemical shifts, Section 4.7) are measured in frequency units from a reference peak.

A routine sample for proton NMR on a 300-MHz instrument consists of about 2 mg of the compound in about 0.4 mL of solvent in a 5-mm o.d. glass tube. Under favorable conditions, it is possible to obtain a spectrum on 1 μg of a compound of modest molecular weight in a microtube (volume 185 μl) in a 300-MHz pulsed instrument. Microprobes that accept a 2.5 mm or 3-mm o.d. tube are convenient and provide high sensitivity.† A capillary microprobe that accepts a few nanograms of material in a few nanoliters of solvent is under development.‡

The ideal solvent should contain no protons and be inert, low boiling, and inexpensive. Since pulsed instruments depend on deuterium in the field-frequency lock, deuterated solvents are necessary.§ Deuterated chlo-

* "Chemically different protons" absorb rf energy at very slightly different frequencies—differences up to around 5000 hertz at a frequency of 300 MHz (see Section 4.7). The utility of NMR spectrometry for the organic chemist dates from the experiment at Varian Associates that obtained three peaks from the chemically different protons in CH_3CH_2OH; the peak areas were in the ratio 3 : 2 : 1. [J.T. Arnold, S.S. Dharmatti, and M.E. Packard, *J. Chem. Phys.* **19,** 507 (1951).]

† Nalorac, 538 Arnold Drive, Suite 600, Martinez, CA 94553.

‡ D.L. Olson et al., *Science* **270,** 1967 (1995).

§ A field-frequency internal lock provides corresponding changes in the irradiating frequency for minor variations in field strength to furnish a constant field/frequency ratio. The frequencies are locked to a master oscillator.

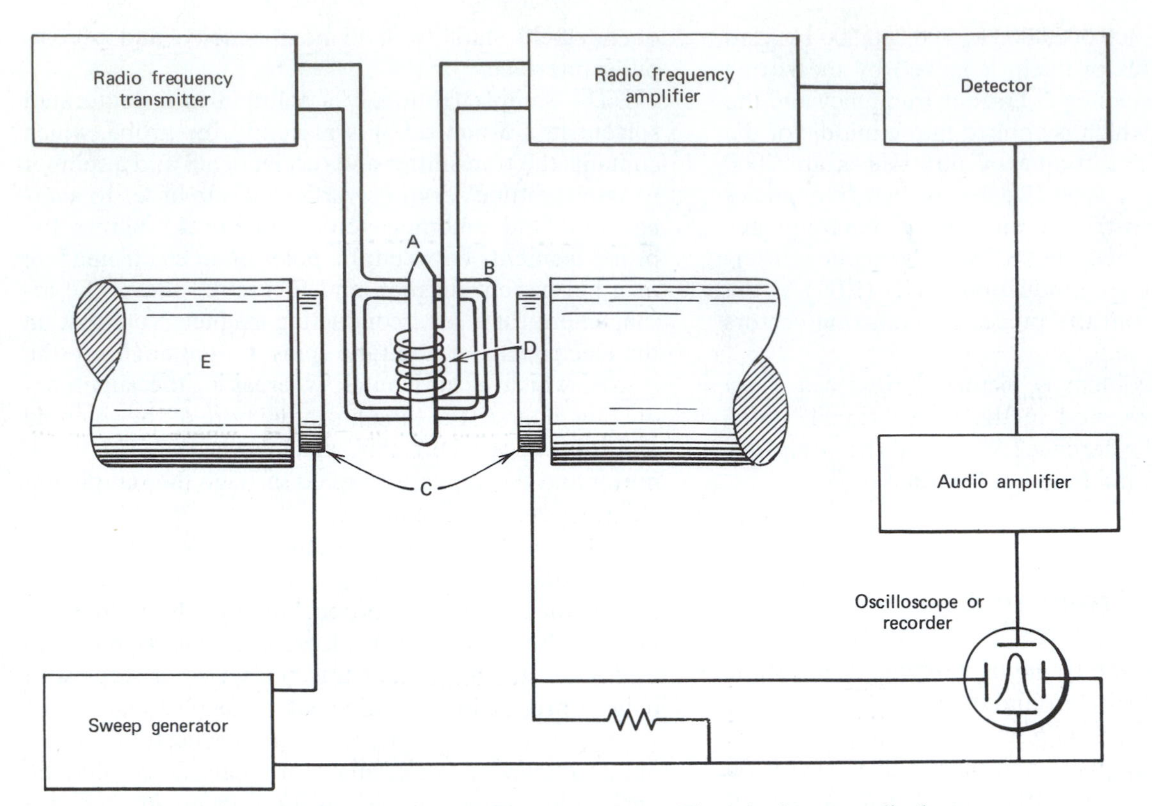

FIGURE 4.8. Schematic diagram of an NMR spectrometer. The tube is perpendicular to the *z* axis of the magnet. A, sample tube; B, transmitter coil; C, sweep coils; D, receiver coil; E, magnet. Courtesy of Varian Associates, Palo Alto, California.

roform ($CDCl_3$) is used whenever circumstances permit—in fact most of the time. The small sharp proton peak from $CHCl_3$ impurity present at δ 7.26 rarely interferes seriously. For very dilute samples, $CDCl_3$ can be obtained in "100% purity". A list of common, commercially available solvents with the positions of proton impurities is given in Appendix G.

Small "spinning side bands" (Fig. 4.10) are some-

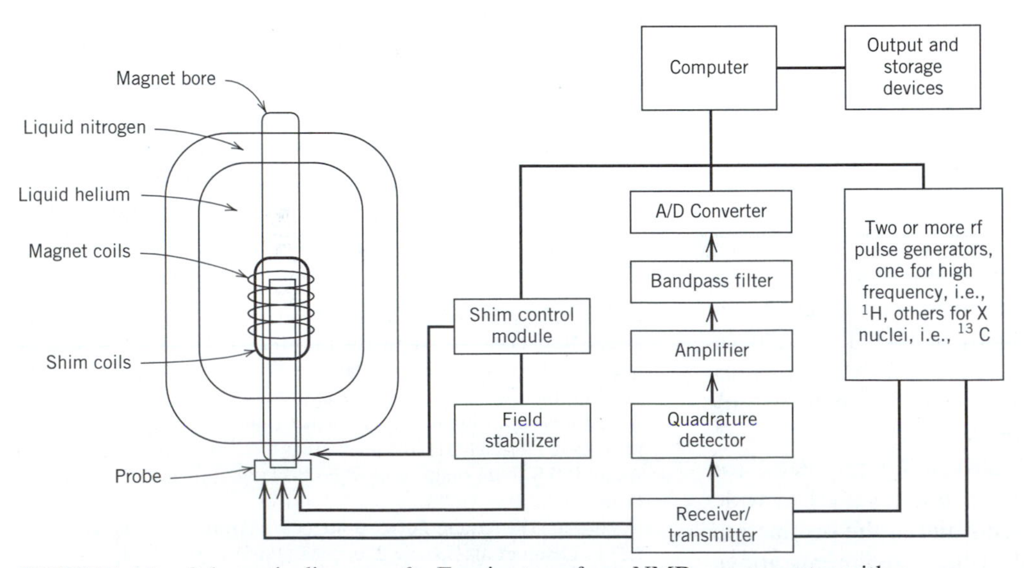

FIGURE 4.9. Schematic diagram of a Fourier transform NMR spectrometer with a superconducting magnet. The probe is parallel with the *z* axis of the magnet, which is cooled with liquid helium surrounded by liquid nitrogen in a large Dewar flask. From Kiemle, D.J., and Winter, W.T. (1995). In *Kirk-Othmer Encyclopedia of Chemical Technology,* 4th ed., Vol. 15. New York: Wiley page 789, with permission.

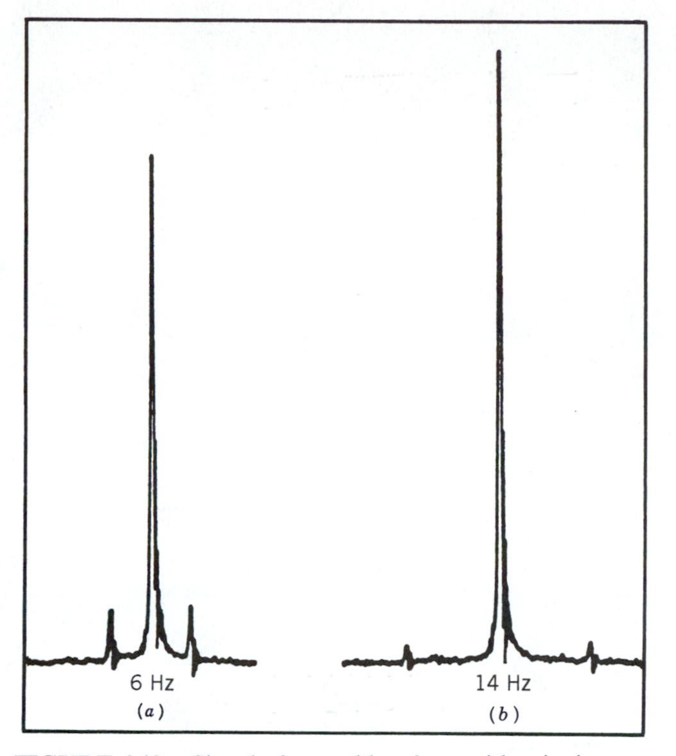

FIGURE 4.10. Signal of neat chloroform with spinning side bands produced by spinning rate of (*a*) 6 Hz and (*b*) 14 Hz. From Bovey, F.A. (1969). *NMR Spectroscopy.* New York: Academic Press, with permission.

times seen symmetrically disposed on both sides of a strong absorption peak; these result from inhomogeneities in the magnetic field and in the spinning tube. They are readily recognized because of their symmetrical appearance and because their separation from the absorption peak is equal to the rate of spinning, typically 10–40 Hz.

The oscillations seen only in scanned (CW) spectra at the low-frequency end of a strong sharp peak are called "ringing" (Fig. 4.11). These are "beat" frequencies resulting from passage through the absorption peak.

Traces of ferromagnetic impurities cause severe broadening of absorption peaks because of reduction of

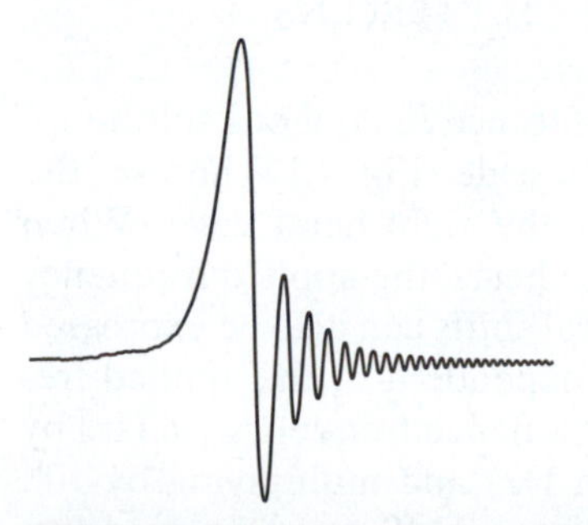

FIGURE 4.11. Ringing (or wiggles) seen after passage through resonance in a scanned spectrum. Direction of scan is from left to right.

T_2 relaxation times. Common sources are tapwater, steel wool, Raney nickel, and particles from metal spatulas or fittings (Fig. 4.12). These impurities can be removed by filtration.

4.7 Chemical Shift

Only a single proton peak should be expected from the interaction of rf energy and a strong magnetic field on an isolated proton in accordance with the basic NMR equation (Section 4.1): $\nu_1 = \frac{\gamma}{2\pi}\boldsymbol{B}_0$, where ν_1 is the applied frequency, $\boldsymbol{B}_0$ is the flux density of the stationary magnetic field, and $\frac{\gamma}{2\pi}$ is a constant. Fortunately, the situation is not quite so simple. A proton in a molecule is *shielded* to a very small extent by its electron cloud, the density of which varies with the chemical environment. This variation gives rise to small differences in absorption positions, usually within the range of about 1000 Hz in a magnetic field, corresponding to 60 MHz, or about 8000 Hz in a field, corresponding to 300 MHz. The ability to discriminate among the individual absorptions describes high-resolution NMR spectrometry.

The basic NMR equation for the isolated proton is now modified for an individual proton in the molecule:

$$\nu_{\text{eff}} = \frac{\gamma}{2\pi}\boldsymbol{B}_0(1 - \sigma)$$

The symbol σ is the "shielding constant" whose value is proportional to the degree of shielding. At a given value of $\boldsymbol{B}_0$, the *effective* frequency at resonance, ν_{eff}, is less than the applied frequency ν_1.

Electrons under the influence of a magnetic field circulate and, in circulating, generate their own magnetic field opposing the applied field; hence, the "shielding" effect (Fig. 4.13). This effect accounts for the diamagnetism exhibited by all organic materials. In the case of materials with an unpaired electron, the paramagnetism associated with the net electron spin far overrides the diamagnetism of the circulating, paired electrons.

The degree of shielding depends on the density of the circulating electrons, and, as a first, very rough approximation, the degree of shielding of a proton on a carbon atom will depend on the inductive effect of other groups attached to the carbon atom. The difference in the absorption position of a particular proton from the absorption position of a *reference* proton is called the *chemical shift* of the particular proton.

We now have the concept that protons in "different" chemical environments have different chemical shifts. Conversely, protons in the "same" chemical en-

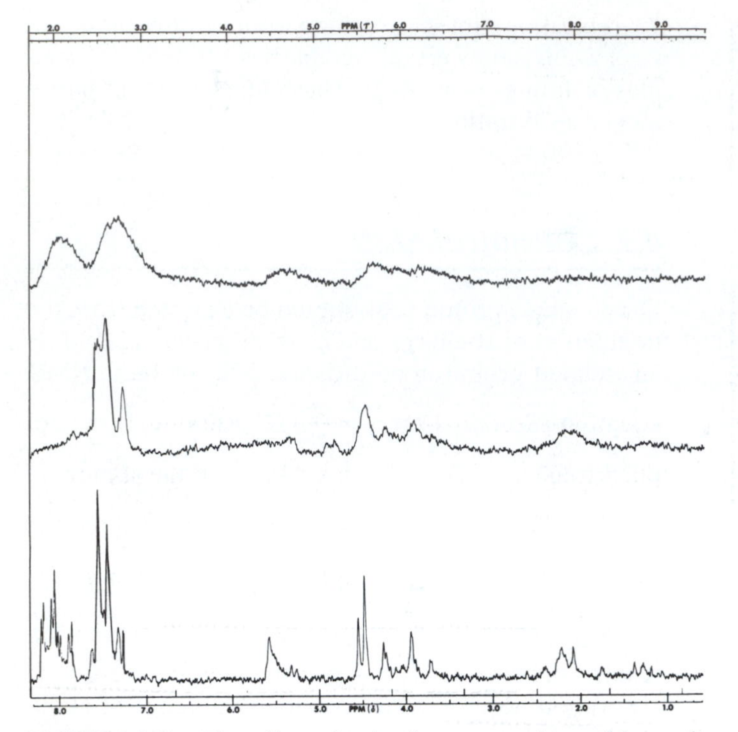

FIGURE 4.12. The effect of a tiny ferromagnetic particle on the proton resonance spectrum of a benzoylated sugar. The top and middle curves are repeated runs with the particle present; the bottom curve is the spectrum with the particle removed. From Becker, E.D. (1980). *High Resolution NMR,* 2nd ed. New York: Academic Press with permission.

vironment have the same chemical shift. But what do we mean by "different" and "same"? It is intuitively obvious that the chemically different methylene groups of $ClCH_2CH_2OH$ have different chemical shifts and that the protons in either one of the methylene groups have the same chemical shift. But it may not be so obvious, for example, that the individual protons of the methylene group of $C_6H_5CH_2CHBrCl$ do not have the same chemical shift. For the present, we shall deal with obvious cases and postpone a more rigorous treatment of chemical shift equivalence to Section 4.12.

The most generally useful reference compound is tetramethylsilane (TMS).

$$\begin{array}{c} CH_3 \\ | \\ H_3C-Si-CH_3 \\ | \\ CH_3 \end{array}$$

This has several advantages: it is chemically inert, symmetrical, volatile (bp 27°C), and soluble in most organic solvents; it gives a single, intense, sharp, absorption peak, and its protons are more "shielded" than almost all organic protons. When water or deuterium oxide is the solvent, TMS can be used as an "external reference" in a concentric capillary. The methyl protons of the water-soluble sodium 2,2-dimethyl-2-silapentane-5-sulfonate (DSS)

$$(CH_3)_3SiCH_2CH_2CH_2SO_3Na$$

are used as an internal reference in aqueous solution.

Let us set up an NMR scale (Fig. 4.14) and set the TMS peak at zero Hz at the right-hand edge. When chemical shifts are given in hertz, the applied frequency must be specified. Chemical shifts can also be expressed in dimensionless units, independent of the applied frequency, by dividing the resonance frequency (in Hz) by the applied frequency (in Hz) and multiplying by 10^6. Thus, a peak at 300 Hz from TMS at an applied frequency of 300 MHz would be at δ 1.00 (δ scale).

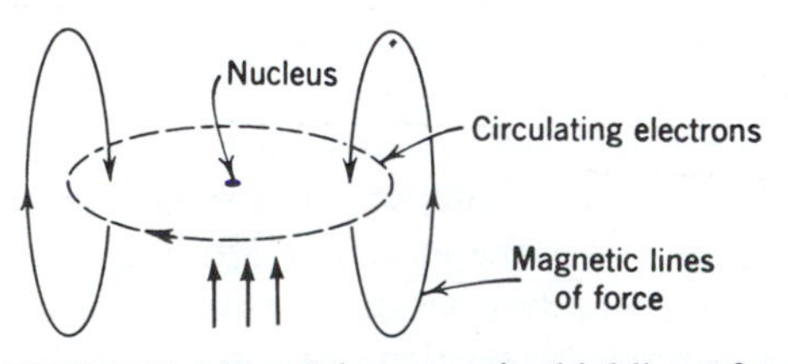

FIGURE 4.13. Diamagnetic shielding of nucleus by circulating electrons. ↑ ↑ ↑ represents the direction of the stationary magnetic field of magnitude $\boldsymbol{B}_0$. The circulating electrons comprise the electrical current, but the current direction is shown conventionally as flow of positive charge.

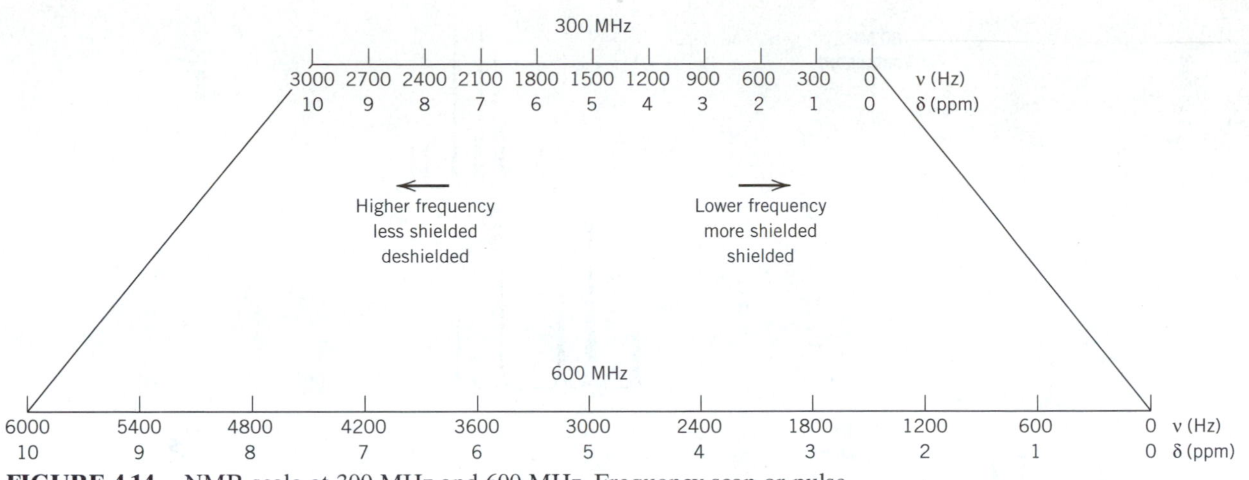

FIGURE 4.14. NMR scale at 300 MHz and 600 MHz. Frequency scan or pulse.

$$\frac{300}{300 \times 10^6} \times 10^6 = \delta\ 1.00, \text{ or } 1.00 \text{ ppm}$$

Since δ units are expressed in parts per million, the expression ppm is often used. The same peak at an applied frequency of 600 MHz would be at 600 Hz but would still be at δ 1.00 or 1.00 ppm.

$$\frac{600}{600 \times 10^6} \times 10^6 = \delta\ 1.00, \text{ or } 1.00 \text{ ppm}$$

The strongest magnetic field necessary and available should be used to spread out the chemical shifts. This is made clear in Figure 4.14 and in Figure 4.15 in which increased applied magnetic field in the NMR spectrum of acrylonitrile means increased separation of signals.

The concept of electronegativity of substituents near the proton in question is a dependable guide, up to a point, to chemical shifts. It tells us that the electron density around the protons of TMS is high (silicon is electropositive relative to carbon), and these protons will therefore be highly shielded.* Since C is more electronegative that H, the sequence of proton absorptions in the alkyl series CH_4, RCH_3, R_2CH_2, and R_3CH is from right to left in the spectrum (Appendix A, Chart A.1). We could make a number of good estimates as to chemical shifts, using concepts of electronegativity and proton acidity. For example, the following values are reasonable on these grounds:

Compound	δ
$(CH_3)_2O$	3.27
CH_3F	4.30
RCO_2H	~10.80

But finding the protons of acetylene at δ 1.80, that is, more shielded than ethylene protons (δ 5.25), is unsettling. Finding the aldehydic proton of acetaldehyde at δ 9.97 definitely calls for some augmentation of the electronegativity concept. We shall use diamagnetic anisotropy to explain these and other apparent anomalies, such as the unexpectedly large deshielding effect of the benzene ring (benzene protons δ 7.27).

Let us begin with acetylene. The molecule is linear, and the triple bond is symmetrical about the axis. If this axis is aligned with the applied magnetic field, the π electrons of the bond can circulate at right angles to the

* By convention, the TMS reference peak is placed at the right-hand edge of the spectrum and designated zero on the δ scale. Positive δ numbers increase to the left of TMS, negative numbers increase to the right; i.e., the *values* increase to the left. The term "shielded" means toward the right; "deshielded" means toward the left.

It follows that the strongly *deshielded* protons of dimethyl ether, for example, are more exposed than those of TMS to the applied field; hence, resonance occurs at higher frequency—i.e., to the left—relative to the TMS proton peak. Thus the δ scale reflects the increase in applied frequency, *at constant field,* toward the left of the TMS resonance frequency, and the decrease in applied frequency toward the right. (Reread the first four paragraphs of this section.)

This convention, however, posed a conflict with the practice, in the earlier days of NMR, of increasing the magnetic field strength (field scan) from left to right at constant frequency and reporting the *magnetic field strength in terms of the resonance frequency units.* This "upfield scan" meant that the resonance frequency units increased from left to right, at odds with the conventional δ scale.

To resolve this conflict, the τ scale was introduced ($\tau = 10 - \delta$) and found its way into the literature from the late 1950s through the 1970s—the British literature in particular. Chemical shifts published during this period should be carefully checked for the convention used. For example, see Bovey (1967).

The τ scale disappeared with the development of frequency-scan instruments and of the pulsed FT mode, which is essentially an instantaneous frequency "scan." The terms "upfield" and "downfield" are now obsolete and have been replaced, respectively, by shielded (lower δ, or to the right) and deshielded (higher δ, or to the left).

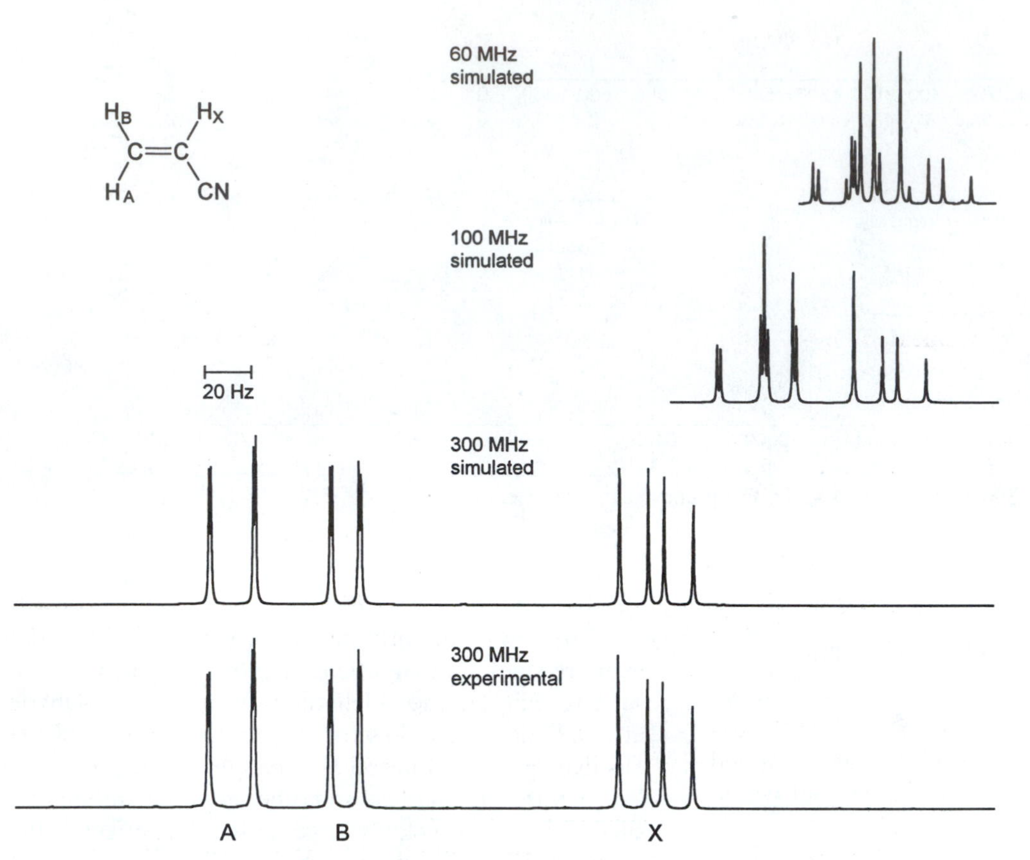

FIGURE 4.15. Simulated 60-, 100-, and 300-MHz spectra of acrylonitrile; 300-MHz experimental spectrum (in $CDCl_3$) for comparison. For reference to simulation of spectra, see footnote reference to Bruker Instruments program in Section 4.8.

applied field, thus inducing their own magnetic field opposing the applied field. Since the protons lie along the magnetic axis, the magnetic lines of force induced by the circulating electrons act to shield the protons (Fig. 4.16), and the NMR peak is found further to the right than electronegativity would predict. Of course, only a small number of the rapidly tumbling molecules are aligned with the magnetic field, but the overall average shift is affected by the aligned molecules.

This effect depends on diamagnetic anisotropy, which means that shielding and deshielding depend on the orientation of the molecule with respect to the applied magnetic field. Similar arguments can be adduced to rationalize the unexpected deshielded position of the aldehydic proton. In this case, the effect of the applied magnetic field is greatest along the transverse axis of the C=O bond (i.e., in the plane of the page in Fig. 4.17). The geometry is such that the aldehydic proton, which lies in front of the page, is in the deshielding portion of the induced magnetic field. The same argument can be used to account for at least part of the rather large deshielding of alkene protons.

The so-called "ring-current effect" is another ex-

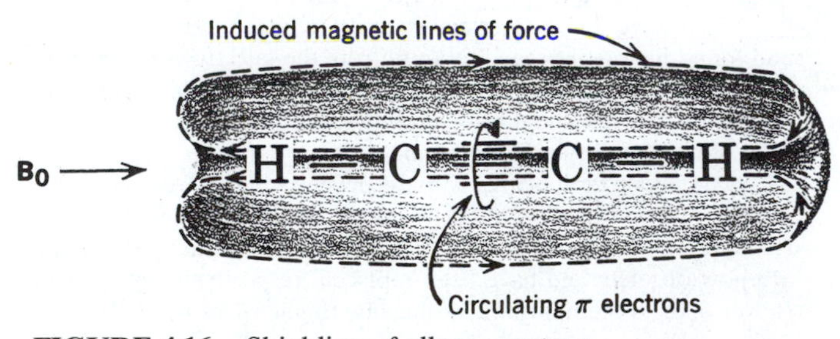

FIGURE 4.16. Shielding of alkyne protons.

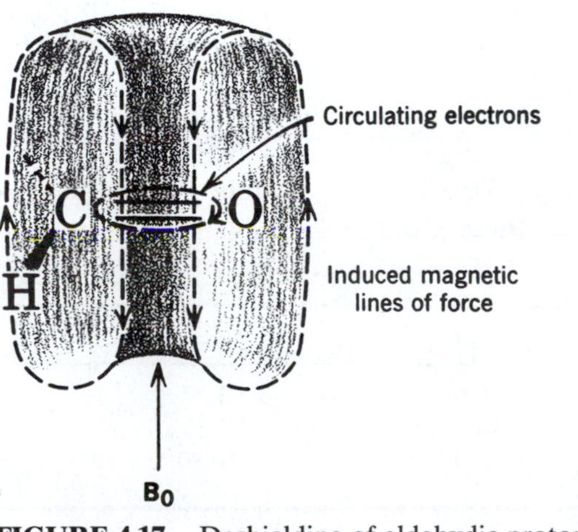

FIGURE 4.17. Deshielding of aldehydic protons.

ample of diamagnetic anisotropy and accounts for the large deshielding of benzene ring protons. Figure 4.18 shows this effect. It also indicates that a proton held directly above or below the aromatic ring should be shielded. This has been found to be the case for some of the methylene protons in 1,4-polymethylenebenzenes.

All the ring protons of acetophenone are deshielded because of the ring current effect. Moreover, the ortho protons are further deshielded (meta, para $\delta \sim 7.40$; ortho $\delta \sim 7.85$) because of the additional deshielding effect of the carbonyl group. In Figure 4.19, the carbonyl bond and the benzene ring are coplanar. If the molecule is oriented so that the applied magnetic field $\boldsymbol{B}_0$ is perpendicular to the plane of the molecule, the circulating π electrons of the C=O bond shield the conical zones above and below them and deshield the lateral zones in which the ortho protons are located. Both ortho protons are equally deshielded since another, equally populated, conformation can be written in which the "left-hand" ortho proton is deshielded by the anisotropy cone. Nitrobenzene shows a similar effect.

A spectacular example of shielding and deshielding by ring currents is furnished by some of the annulenes. At about −60°C, the protons outside the ring of [18]annulene are strongly deshielded (δ 9.3) and those inside are strongly shielded (δ − 3.0, i.e., more shielded than TMS).

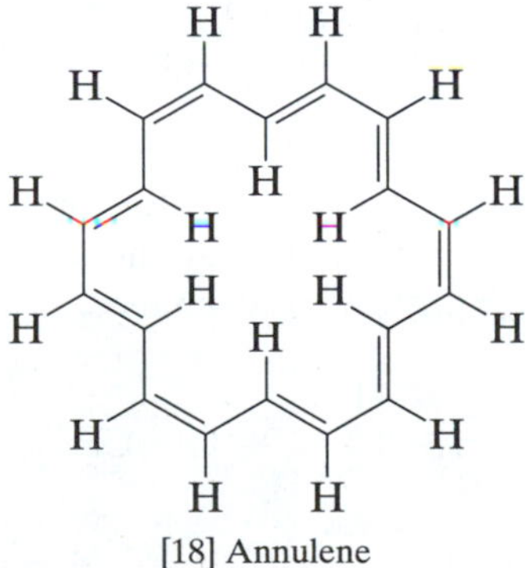

[18] Annulene

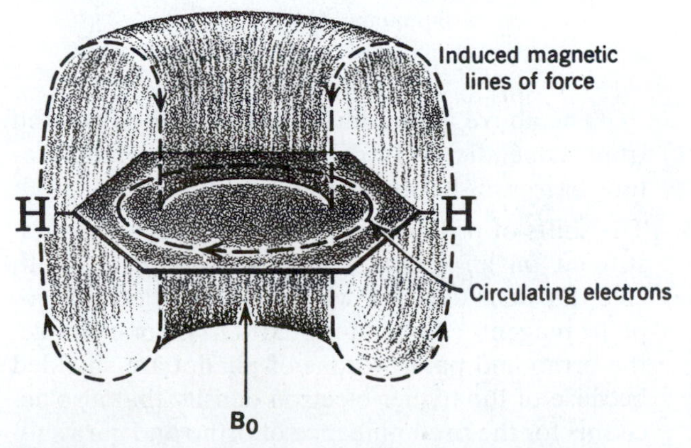

FIGURE 4.18. Ring current effects in benzene.

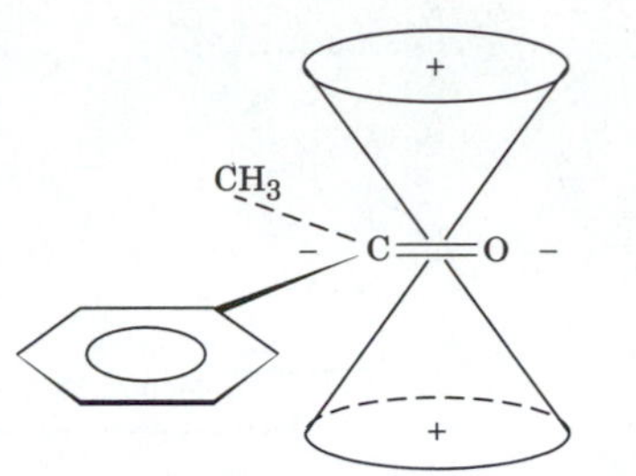

FIGURE 4.19. Shielding (+) and deshielding (−) zones of acetophenone.

Demonstration of such a ring current is good evidence for planarity and aromaticity, at least at low temperature. As the temperature is raised, the signals broaden because of slow interchanges in ring conformations. At about 110°C, a single averaged peak appears at approximately δ 5.3 because of rapid interchanges in ring conformations to give an averaged chemical shift.

In contrast with the striking anisotropic effects of circulating π electrons, the σ electrons of a C—C bond produce a small effect. For example, the axis of the C—C bond in cyclohexane is the axis of the deshielding cone (Fig. 4.20). The observation that an equatorial proton is consistently found further to the left by 0.1–0.7 ppm than the axial proton on the same carbon atom in a rigid six-membered ring can thus be rationalized. The axial and equatorial protons on C_1 are oriented similarly with respect to C_1—C_2 and C_1—C_6, but the equatorial proton is within the deshielding cone of the C_2—C_3 bond (and C_5—C_6).

Extensive tables and charts of chemical shifts in the Appendices give the useful impression that chemical shifts of protons in organic compounds fall roughly into eight regions as shown in Figure 4.21.

To demonstrate the use of some of the material in the Appendices, we predict the chemical shifts of the protons in benzyl acetate (Fig. 4.22).

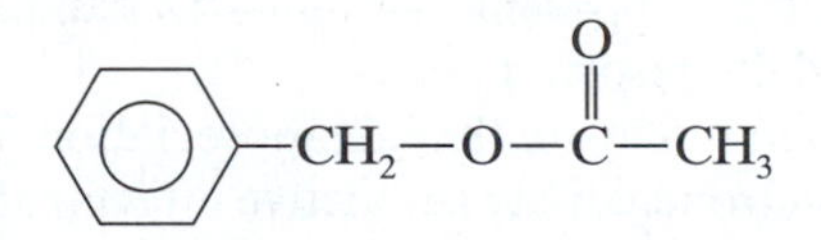

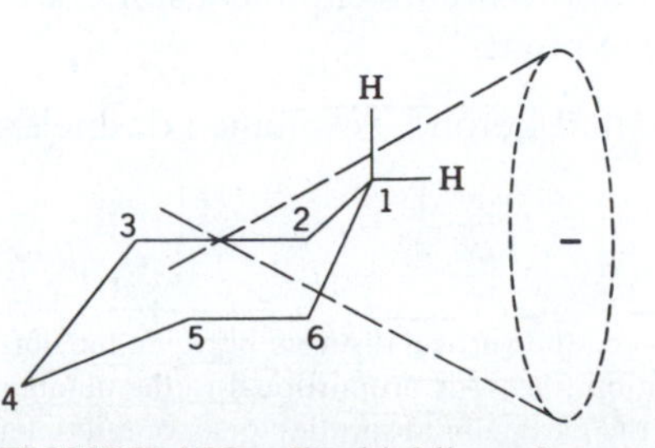

FIGURE 4.20. Deshielding of equatorial proton of a rigid six-membered ring.

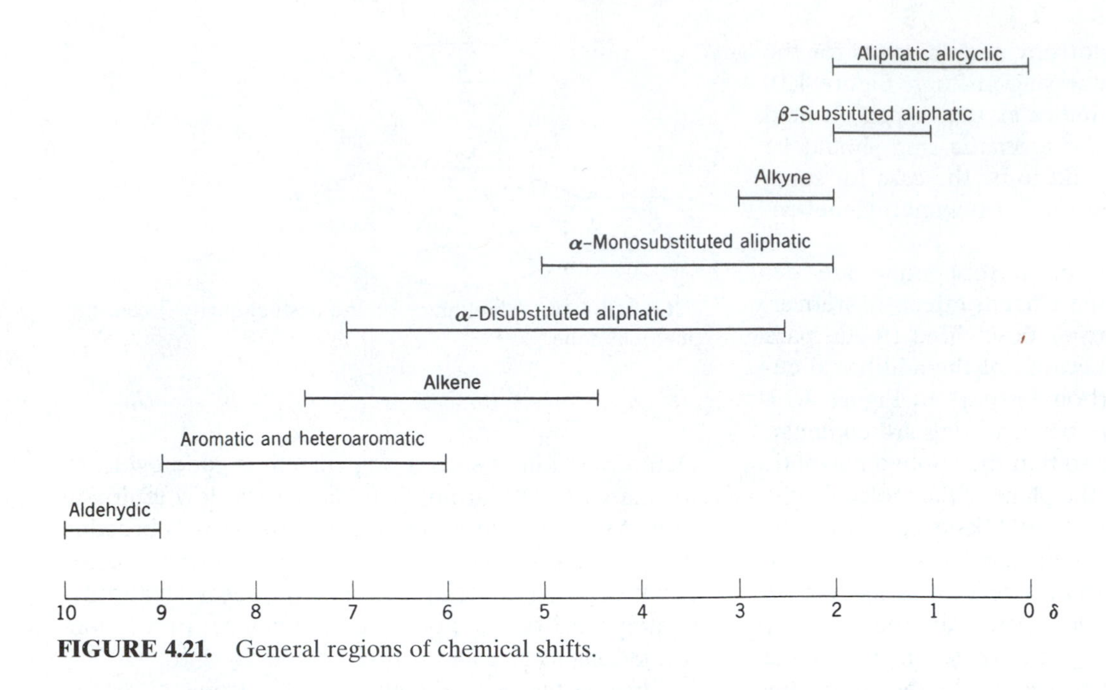

FIGURE 4.21. General regions of chemical shifts.

In Appendix A, Chart A.1, we see that the chemical shift of the CH_3 group is $\sim\delta$ 2.0. From Table B.1, we find that the CH_2 group is at $\sim\delta$ 5.07. In Appendix D, Chart D.1, the aromatic protons are at $\sim\delta$ 7.2. In the spectrum of benzyl acetate (Fig. 4.22), we see three sharp peaks from right to left at δ 1.96, δ 5.00, and δ 7.22; the "integration steps" are in the ratios 3:2:5, corresponding to CH_3, CH_2, and five ring-protons.* The peaks are all singlets. This means that the CH_3 and CH_2 groups are isolated; that is, there are no protons on the adjacent carbon atoms for coupling (see Section 4.8). However, there is a problem with the apparent singlet representing the ring protons, which are not chemical shift equivalent (Section 4.12.1) and do couple with one another. At higher resolution, we would see a multiplet rather than an apparent singlet. The expanded inset shows partially resolved peaks.

We point out again that an appreciation of the concepts of electronegativity (inductive effects) and of electron delocalization—combined with an understanding of diamagnetic anisotropy—permits both rationalization and prediction of approximate chemical shift. Several examples make the point:

1. In an α,β-unsaturated ketone, resonance deshields the β-proton;

$$CH_3-\underset{\beta}{\overset{H}{C}}=\underset{H}{\overset{\alpha}{C}}-\underset{\|}{\underset{O}{C}}-CH_3 \longleftrightarrow CH_3-\underset{+}{\overset{H}{C}}-\underset{H}{C}=\underset{|}{\underset{O^-}{C}}-CH_3$$

α-proton = $\delta\sim$6.2
β-proton = $\delta\sim$6.8

2. In a substituted vinyl ether, the oxygen atom deshields the α-proton by an inductive effect and shields the β-proton by resonance.

$$CH_3-\underset{\beta}{\overset{H}{C}}=\underset{H}{\overset{\alpha}{C}}\cdots O-CH_3$$

$$CH_3-\overset{H}{C}=\underset{H}{C}-\ddot{O}-CH_3 \longleftrightarrow CH_3-\underset{\ominus}{\overset{H}{C}}-\underset{H}{C}=\overset{\oplus}{O}-CH_3$$

α-proton = $\delta\sim$6.2
β-proton = $\delta\sim$4.6

The above approximate values were calculated from Appendix D. In comparison the olefinic protons of *trans*-3-hexene are at δ 5.40.

3. The shifts of protons ortho, meta, or para to a substituent on an aromatic ring are correlated with electron densities and with the effects of electrophilic reagents (Appendix Chart D.1). For example, the ortho and para protons of phenol are shielded because of the higher electron density that also accounts for the predominance of ortho and para substitution by electrophilic reagents. Conversely, the

* The "integration step"—i.e., the vertical distance between the horizontal lines of the integration trace—is proportional to the number of protons represented by the particular absorption peak or multiplet of peaks. These steps give ratios, not absolute numbers of protons. The ratios actually represent areas under the peaks.

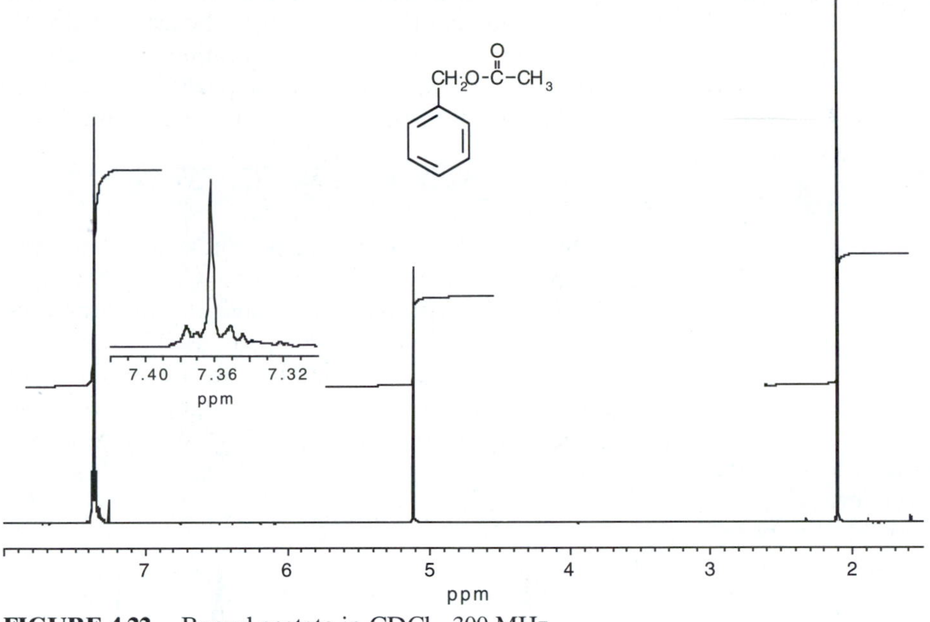

FIGURE 4.22. Benzyl acetate in $CDCl_3$, 300 MHz.

ortho and para protons of nitrobenzene are deshielded.

Since chemical shift increments are approximately additive, it is possible to calculate the ring proton shifts in polysubstituted benzene rings from the monosubstituted values in Appendix Chart D.1. The chemical shift increments for the ring protons of *m*-diacetylbenzene, for example, are calculated as follows.

Chemical shift increments are the shifts from that of the protons of benzene (δ 7.27). Thus for a $CH_3C{=}O$ substituent (line 26, Appendix Chart D.1), the ortho increment is +0.63, and the meta and para increments are both +0.28 (+ being at higher frequency than δ 7.27). The C-2 proton has two ortho substituents; the C-4 and C-6 protons are equivalent and have ortho and para substituents; the C-5 proton has two meta substituents. Thus the calculated increment for C-2 is +1.26, for C-4 and C-6 is +0.91, and for C-5 is +0.56. The spectrum shows increments of +1.13, +0.81, and +0.20, respectively. This agreement is adequate for determining the substitution pattern.* Integration of the spectrum would show the expected ratio of 1:2:1. Furthermore, both the ortho- (with identical substituents) and para-substituted compounds would show the characteristic, symmetrical patterns of Figures 4.41 and 4.42. Finally, the spin coupling pattern for each isomer would be distinctive, as will become evident at the end of Section 4.8 and on study of Appendix F. Obviously, proton NMR spectrometry is a powerful tool for elucidating aromatic substitution patterns—as is carbon-13 NMR (see Chapter 5). Two-dimensional NMR spectrometry offers another powerful tool (see Chapter 6).

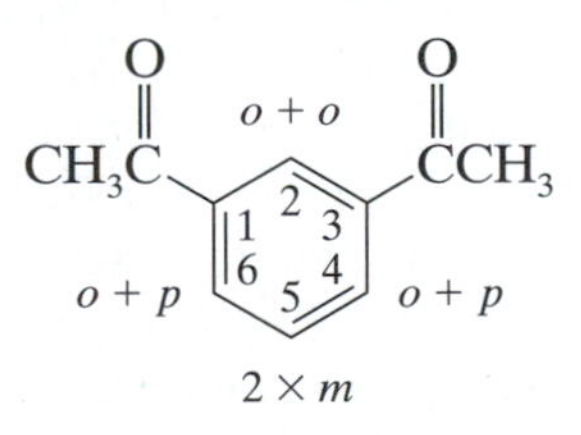

* Calculations for ortho-disubstituted compounds are less satisfactory because of steric or other interactions between the ortho substituents.

4.8 Simple Spin Coupling

We have obtained a series of absorption peaks representing protons in different chemical environments, each absorption area (from integration) being proportional to the number of protons it represents. We have now to consider one further phenomenon, *spin coupling*. This can be described as the indirect coupling of proton spins through the intervening bonding electrons. Very briefly, it occurs because there is some tendency for a bonding electron to pair its spin with the spin of the nearest proton; the spin of a bonding electron hav-

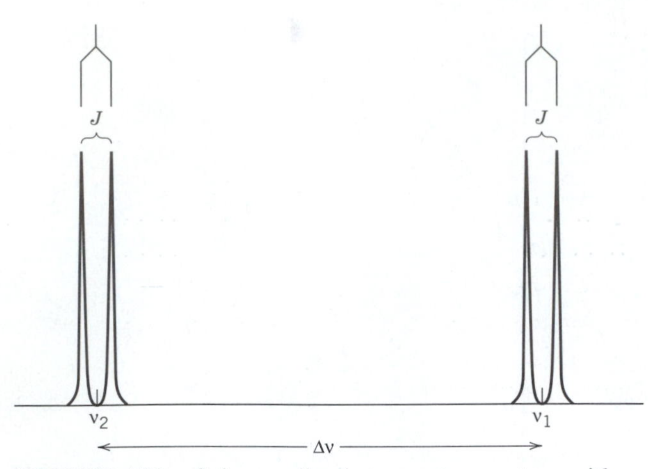

FIGURE 4.23. Spin coupling between two protons with very different chemical shifts.

ing been thus influenced, the electron will affect the spin of the other bonding electron, and so on, through to the next proton. Coupling is ordinarily not important beyond three bonds unless there is ring strain as in small rings or bridged systems, delocalization as in aromatic or unsaturated systems, or four connecting bonds in a W configuration (Section 4.18). Two-bond coupling is termed *geminal;* three-bond coupling, *vicinal:*

H—C—H	H—C—C—H
Geminal coupling	Vicinal coupling
2 bonds (^2J)	3 bonds (^3J)

Suppose that two vicinal protons are in very different chemical environments from one another, as in the compound RO—CH(OR)—CH(CR$_3$)—CR$_3$. Each proton will give rise to an absorption, and the absorptions will be quite widely separated, but the spin of each proton is affected slightly by the two orientations of the other proton through the intervening electrons, so that each absorption appears as a doublet (Fig. 4.23). The frequency difference in Hz between the component peaks of a doublet is proportional to the effectiveness of the coupling and is denoted by a coupling constant, J, which is independent of the applied magnetic field $\boldsymbol{B}_0$.* Whereas chemical shifts usually range over about 1250 Hz at 100 MHz, coupling constants between protons rarely exceed 20 Hz (see Appendix F).

So long as the chemical shift difference in hertz ($\Delta\nu$) is much larger than the coupling constant (arbitrarily $\Delta\nu/J$ is greater than about 8), the simple pattern of two doublet appears. As $\Delta\nu/J$ becomes smaller, the doublets approach one another, the inner two peaks increase in intensity, and the outer two peaks decrease (Fig. 4.24). The shift position of each proton is no longer midway between its two peaks as in Figure 4.23 but is at the "center of gravity" (Fig. 4.25); it can be estimated with fair accuracy by inspection or determined precisely by the following formula, in which the peak positions (1, 2, 3, and 4 from left to right) are given in hertz from TMS.

$$\Delta\nu = \sqrt{(1-4) \times (2-3)}$$

The shift position of each proton is $\Delta\nu/2$ from the midpoint of the pattern. When $\Delta\nu = J\sqrt{3}$, the two pairs could be mistaken for a quartet, which results from splitting by three equivalent vicinal protons (Fig. 4.24, *d*, is almost at this stage). Failure to note the small outer peaks (i.e., 1 and 4) may lead to mistaking the two large inner peaks for a doublet (Fig. 4.23, *e*). When the chem-

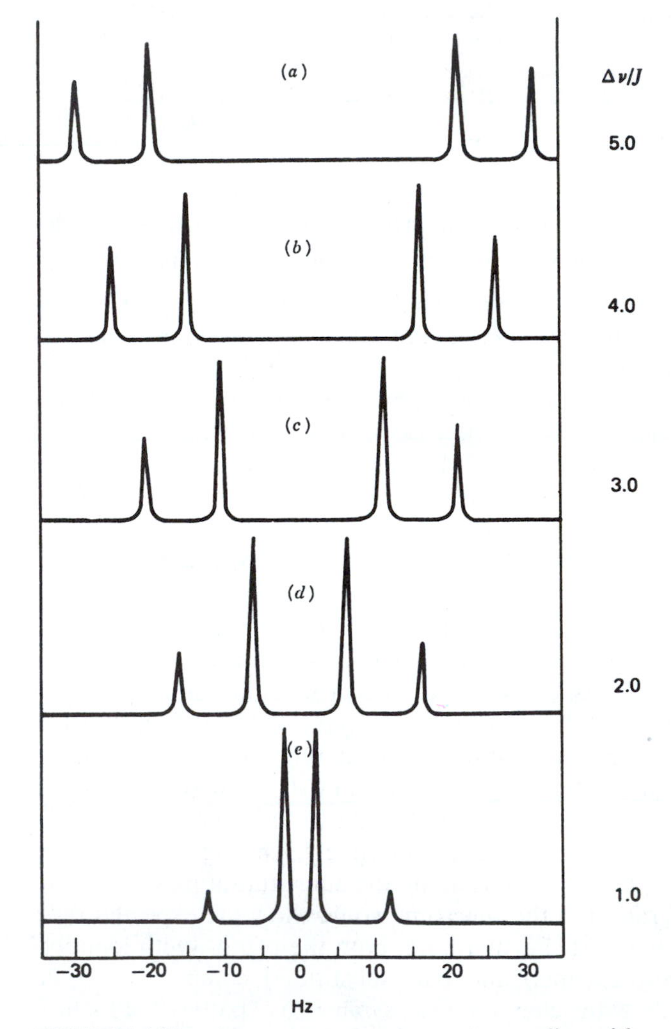

FIGURE 4.24. A two-proton system, spin coupling with a decreasing difference in chemical shifts and a large J value (10 Hz); the difference between AB and AX notation is explained in the text.

* The number of bonds between coupled nuclei (protons in this chapter) is designated by J and a left superscript. For example, H—C—H is ^2J, H—C—C—H is ^3J, H—C=C—C—H is ^4J. Double or triple bonds are counted as single bonds.

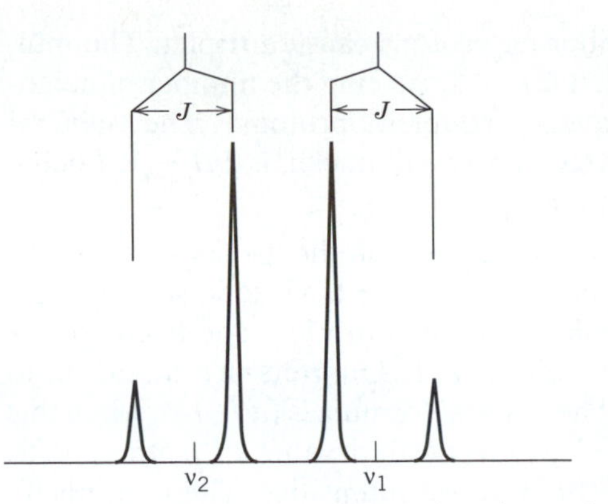

FIGURE 4.25. "Center of gravity," instead of linear midpoints, for shift location (in the case of "low" $\Delta\nu/J$ ratio).

ical shift difference becomes zero, the middle peaks coalesce to give a single peak, and the end peaks vanish; that is, the protons are equivalent.

Note that the inner lines of coupled protons merge but do not cross. A further point to be noted is the obvious one that the spacing between the peaks of two coupled multiplets is the same.

The dependence of chemical shift on the applied magnetic field and the independence of the spin coupling afford a method of distinguishing between them. The spectrum is merely run on two different instruments, for example, at 100 and 300 MHz. Chemical shifts are also solvent dependent, but J values are usually only slightly affected by change of solvent, at least to a far lesser degree than are chemical shifts.

The chemical shifts of the methyl and alkyne protons of methylacetylene are coincident (δ 1.80) when the spectrum is obtained in a $CDCl_3$ solvent, whereas the spectrum of a neat sample of this alkyne shows the alkyne proton at δ 1.80 and the methyl protons at δ 1.76. Figure 4.26 illustrates the chemical shift dependence of the protons of biacetyl (2,3-butanedione) on solvent. The change from a chlorinated solvent (e.g., $CDCl_3$) to an aromatic solvent (e.g., C_6D_6) often drastically influences the position and appearance of NMR signals.

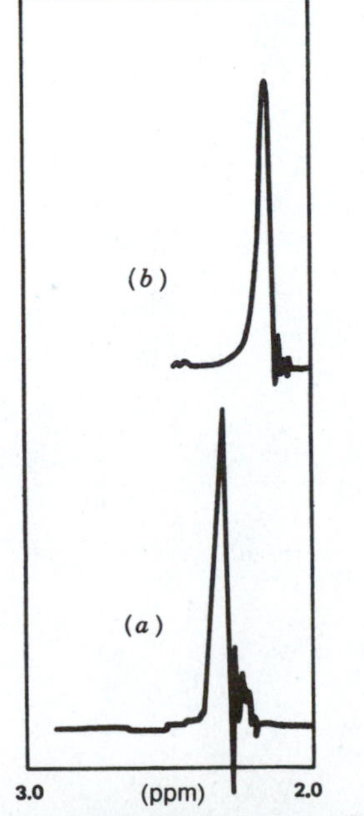

FIGURE 4.26. The 60-MHz spectrum of biacetyl (2,3-butanedione): (*a*) in $CDCl_3$; (*b*) in C_6D_6.

FIGURE 4.27. Spin coupling between CH and CH_2 with very different chemical shifts.

Look at the next stage in complexity of spin coupling (Fig. 4.27). Consider the system —HC—CH_2—

in the compound $RO-\overset{\overset{\displaystyle OR}{|}}{C}H-CH_2-CR_3$ in which the single methine proton is in a very different chemical environment from the two methylene protons. As before, we see two sets of absorptions widely separated, and now the absorption areas are in the ratio of 1 : 2. The methine proton couples with the methylene protons and splits the methylene proton absorption into a symmetrical doublet, as explained above.

The following orientations of the methylene protons (*a* and *b*) exist as shown in Figure 4.28, where the

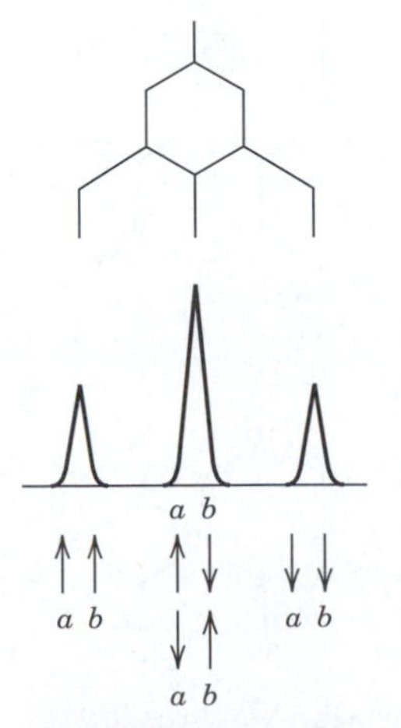

FIGURE 4.28. Energy levels for the three spin states of the methylene group (protons *a* and *b*) that produce the triplet shown in Figure 4.27.

"up" arrows are parallel to the magnetic field and the "down" arrows are opposed:

up, up up, down / down, up down, down

Thus, the neighboring methine proton "sees" three energy levels of the methylene protons in the ratio 1:2:1 and produces a 1:2:1 triplet.

The triplet can be shown as the result of two consecutive splittings of the CH proton by the two equivalent CH_2 protons. Both splittings have the same coupling constant (Fig. 4.28).

When the methine and methylene protons in the system $—CH—CH_2—$ are in rather similar environments (i.e., $\Delta\nu/J$ is small), the simple doublet-triplet pattern degenerates to a complex pattern of from seven to nine lines as a result of higher order splitting; analysis by inspection is no longer possible, since the peak spacings may not correspond to the coupling constants.

Simple splitting patterns that are produced by the coupling of protons that have very different chemical shifts ($\Delta\nu/J$ is greater than about 8 or so) are called *first-order* splitting patterns. These can usually be interpreted by using two rules.

1. Splitting of a proton absorption is done by neighboring protons, and the multiplicity of the split is determined by the number of these protons. Thus, one proton causes a doublet, and two equally coupled neighboring protons cause a triplet. The multiplicity then is $n + 1$, n being the number of neighboring equally coupled protons. The general formula, which covers all nuclei, is $2nI + 1$, I being the spin number.
2. The relative intensities of the peaks of a multiplet also depend on n. We have seen that doublet ($n = 1$) peaks are in the ratio 1:1, and triplet peaks are in the ratio 1:2:1. Quartets are in the ratio 1:3:3:1. The general formula is $(a + b)^n$; when this is expanded to the desired value of n, the coefficients give the relative intensities. The multiplicity and relative intensities may be easily obtained from Pascal's triangle (Fig. 4.29), in which n is the number of equally coupled protons.

In the Pople notation,* protons that have the same chemical shift are placed in the same *set;* each set is designated by a capital letter. The difference *in hertz* between two sets is designated $\Delta\nu$. The coupling constant J is also determined. The Pople notation depends upon the ratio of $\Delta\nu/J$. If the ratio is large (arbitrarily greater than about 8), the sets are weakly coupled; i.e., they are well separated, and they are designated by well-separated letters of the alphabet (e.g., AM or AX). If

* J.A. Pople, W.G. Schneider, H.J. Bernstein. *High Resolution NMR.* New York: McGraw-Hill, 1959.

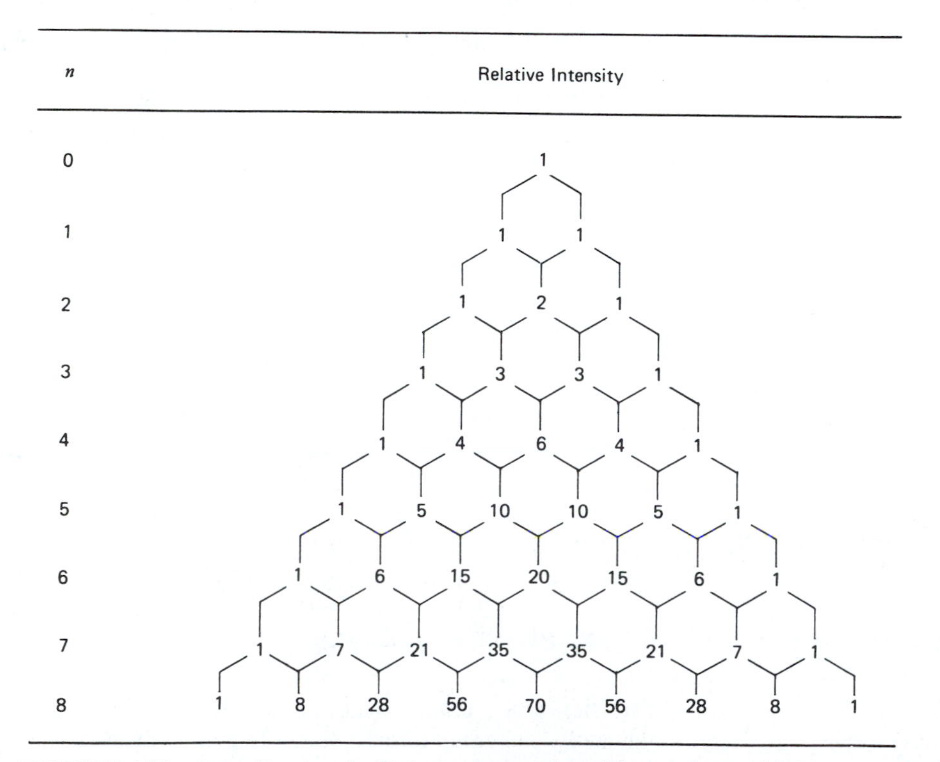

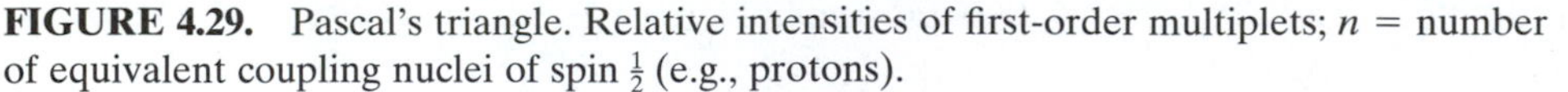

FIGURE 4.29. Pascal's triangle. Relative intensities of first-order multiplets; n = number of equivalent coupling nuclei of spin $\frac{1}{2}$ (e.g., protons).

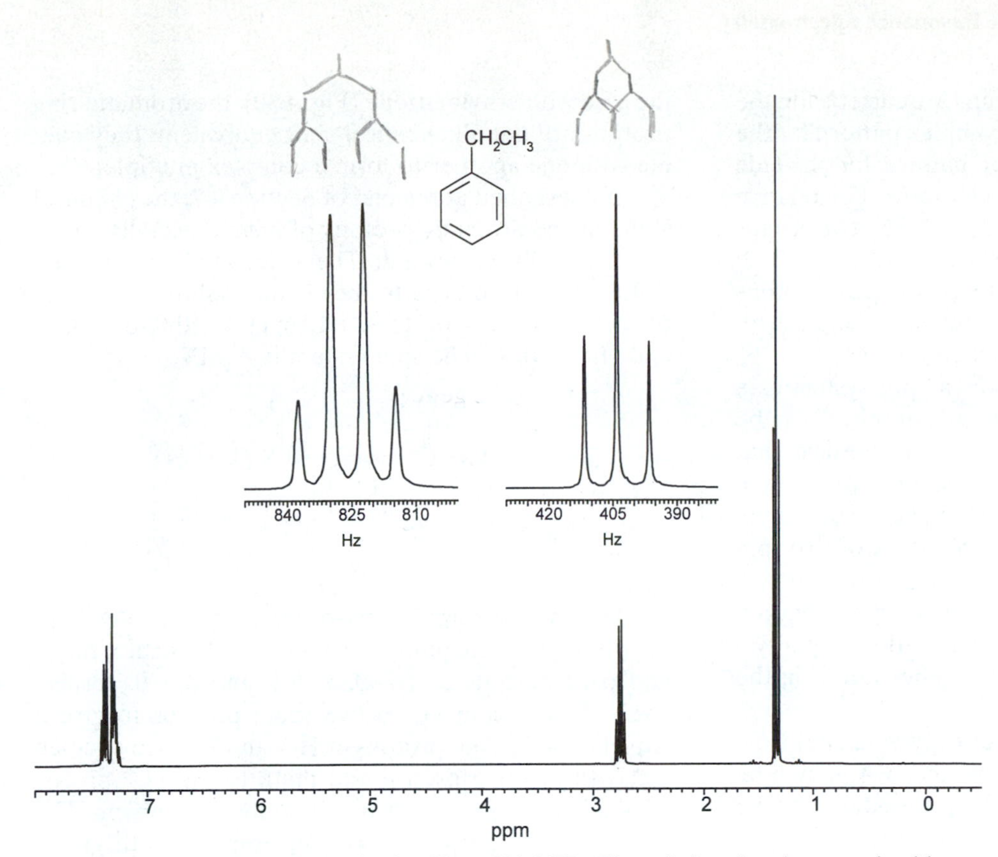

FIGURE 4.30. Ethylbenzene in $CDCl_3$ at 300 MHz. The ethyl moiety is recognized by the CH_3 triplet and the CH_2 quartet.

the ratio is small (arbitrarily less than about 8), letters such as AB are used; these sets are described as strongly coupled. The number of protons in a set is designated by a subscript number; if there is only one proton in a set, no number is used—as above. Such a collection of sets insulated from further coupling form a *spin system.*

Thus the first case examined (Figure 4.23) is an AX *spin system.* The second case (Fig. 4.25) is an AB spin system, and the third case we examined (Fig. 4.27) is an A_2X spin system. As $\Delta\nu/J$ decreases, the A_2X spin system approaches an A_2B spin system, and the simple first-order splitting of the A_2X spin system becomes more complex.

Thus far, we have dealt with two sets of protons; every proton in each set is equally coupled to every proton in the other set; that is, a single coupling constant is involved. Given these conditions and the condition that $\Delta\nu/J$ be large (≥ 8), the two rules in this Section apply, and we obtain a *first-order spin system.* In general, these are the A_aX_x (*a* and *x* are the number of protons in each set); the first-order rules apply only to these systems, but as we have seen, there is a gradual change in the appearances of spectra changing from an AX to an AB system, which is not first-order. In a similar way, it is frequently possible to relate complex patterns back to first-order patterns. *With practice, a fair amount of deviation from first-order may be tolerated.* Wilberg's (1962) and Bovey's (1988) collections of calculated spectra can be used to match fairly complex splitting patterns (see References).*

A system of three sets of protons, each set separated by a large chemical shift, can be designated $A_aM_mX_x$. If two sets are separated from each other by a small chemical shift, and the third set is widely separated from the other two, we use an $A_aB_bX_x$ designation. If all shift positions are close, the system is $A_aB_bC_c$. Both end sets are coupled to the middle set with different coupling constants, whereas the end sets may or may not be coupled to one another. The AMX systems are first-order; ABX systems can be approximated by using first-order rules, but ABC systems cannot be analyzed by inspection. These more complex patterns are treated in Section 4.12.

We can now appreciate the three main features of an NMR spectrum: chemical shifts, peak intensities, and spin splittings that are first order or that approximate first-order patterns. The term "weakly coupled" is used for first-order coupling ($\Delta\nu/J > \sim 8$) and "strongly coupled" is used for couplings whose $\Delta\nu/J$ ratio is less than about eight.

The 300-MHz spectrum of ethylbenzene (Fig. 4.30)

* Alternatively, these sets can be simulated on the computer of a modern NMR spectrometer or on a PC. For example, see the Win-Daisy program, available from Bruker Instruments Incorp.

shows a triplet for the CH_3 group, a quartet* for the CH_2 group (A_3X_2 system) and a complex pattern for the aromatic protons. The first-order pattern for the side chain is rationalized by the high $\Delta\nu/J$ ratio. The relative integrations from right to left are 3:2:5. The two groups of the five aromatic protons show a ratio of 3:2; at this point, we merely note that these aromatic protons are not all chemical-shift equivalent and couple with one another to produce a complex pattern.†

The definition given above for a spin system as a "collection of sets 'insulated' from one another" can be formalized: A spin system consists of sets of nuclei that spin couple with one another but do not spin couple with any nuclei outside the spin system; hence insulated. For example, ethyl isopropyl ether consists of two spin systems: the ethyl protons and the isopropyl protons, which are "insulated" from each other by the oxygen atom. It is not necessary for all nuclei within a spin system to be spin coupled with all the other nuclei in the spin system.

The spectrum of cumene (isopropylbenzene) (Fig. 4.31) shows the isopropyl side chain as an A_6X system: a six-proton doublet for the six CH_3 protons split by the CH proton, and a one-proton septet for the CH proton split by the six methyl protons. (See Figure 4.29 for the relative intensities of the peaks of each multiplet.) As is the case with ethylbenzene (Fig. 4.30), the aromatic-ring protons are not all chemical-shift equivalent; they couple with one another to form a complex multiplet.

As described at the end of Section 4.7, the chemical shifts of the aromatic protons of *meta*-diacetylbenzene are discretely separated: The calculated incremental shifts—with reference to the chemical shift of the protons of benzene—are H-2, +1.26; H-4 and H-6, +0.91; and H-5, +0.56. The spin system is A_2MX.

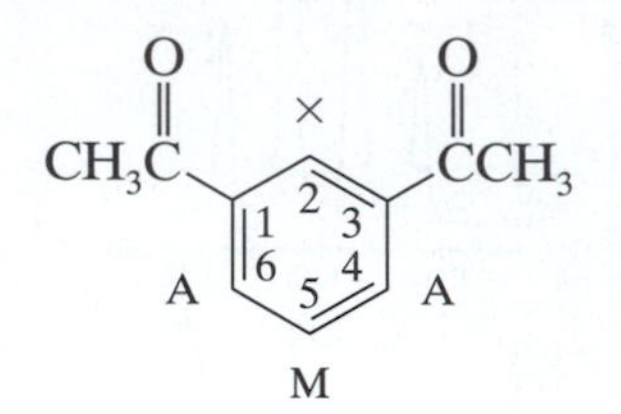

We can now apply first-order coupling rules. Appendix F gives coupling constants for the ortho, meta, and para protons as $J = 9$, $J = 2$, and $J \sim 0$, respectively. H-2 is coupled to two meta protons to give a triplet ($J = 2$). The protons at H-4 and H-6 are coupled ortho to the H-5 proton and meta to the H-2 proton ($J = 9$ and 2); the result is a doublet of doublets. The H-5 proton is coupled to two ortho protons to give a triplet ($J = 9$). The small meta coupling may appear only as peak broadening. The first-order spectrum of *meta*-diacetylbenzene is diagrammed as shown below.

Obviously, the analysis is more complex when two different substituents are present, but it is workable if the absorptions are separated enough so that first-order rules apply.

* Broadening is caused by long-range coupling to ring protons (see Section 4.18).

† Note that the expanded insets are calibrated in Hz. Since this is a 300 MHz spectrum—i.e., one δ unit is equal to 300 Hz—the center of the expanded quartet at 825 Hz is $825/300 = \delta\ 2.75$.

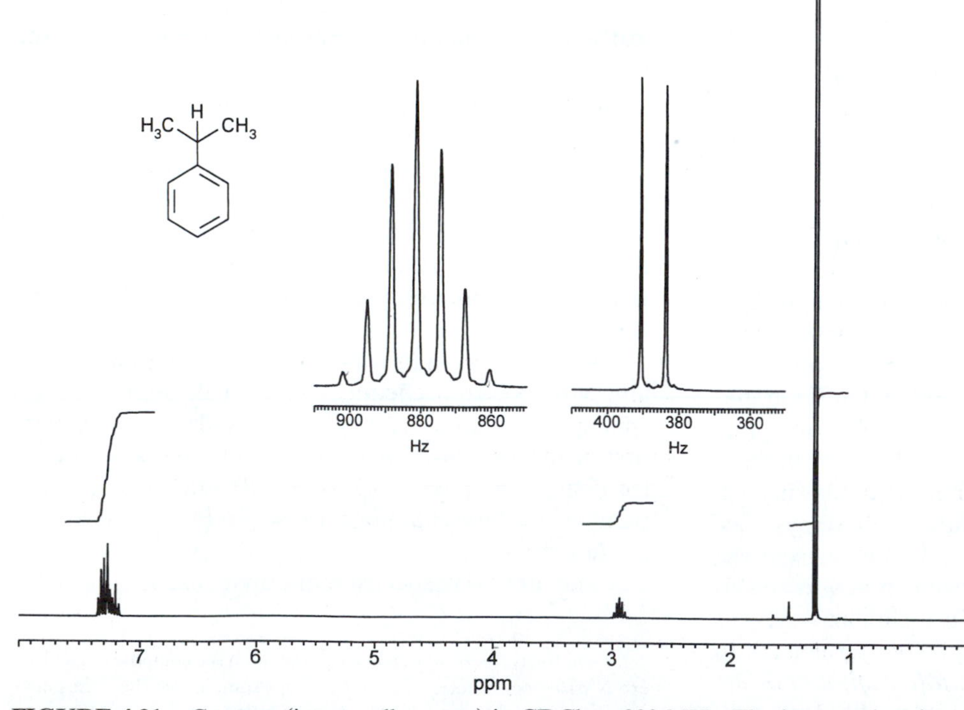

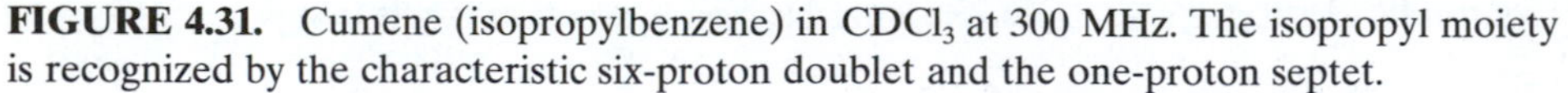

FIGURE 4.31. Cumene (isopropylbenzene) in $CDCl_3$ at 300 MHz. The isopropyl moiety is recognized by the characteristic six-proton doublet and the one-proton septet.

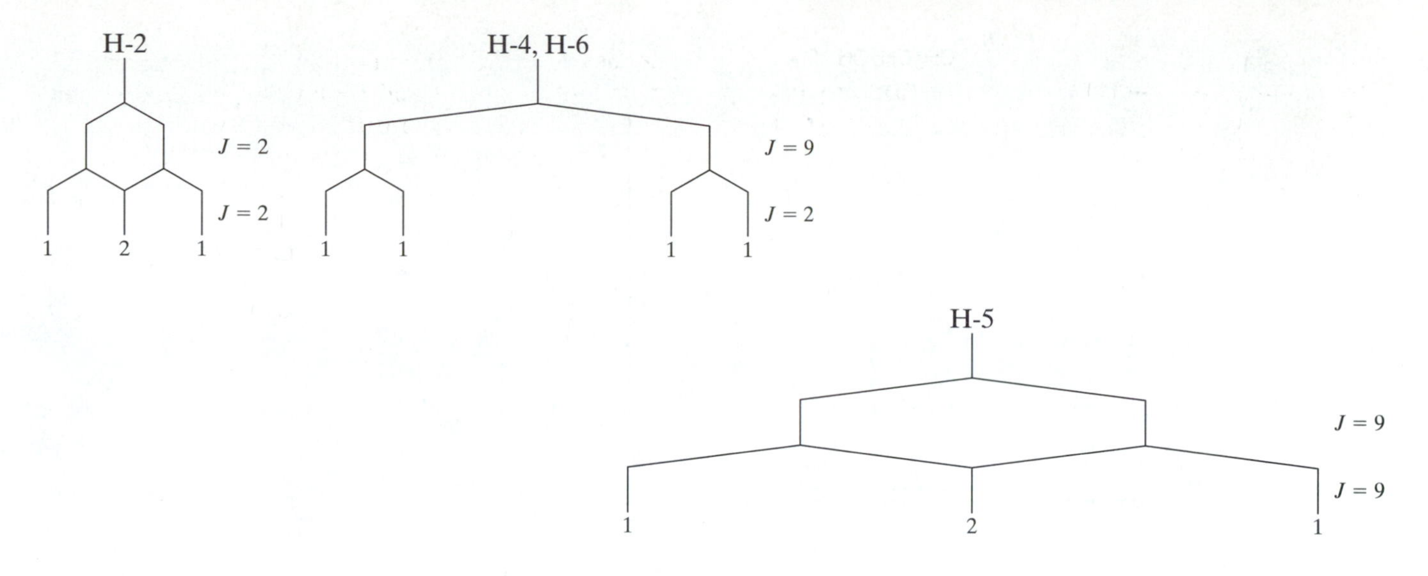

4.9 *Protons on Oxygen, Nitrogen, and Sulfur Atoms*

Protons directly bonded to an oxygen, nitrogen, or sulfur atom differ from protons on a carbon atom in that:

1. They are exchangeable.
2. They are subject to hydrogen bonding.
3. Those on a nitrogen (^{14}N) atom are subject to partial or complete decoupling by the electrical quadrupole moment of the ^{14}N nucleus.

Shift ranges for such protons are given in Appendix E. These variations in shift depend on concentration, temperature, and solvent effects.

4.9.1 *Protons on an Oxygen Atom*

4.9.1.1 Alcohols Depending on concentration, the hydroxylic peak in alcohols is found between ~ δ 0.5 and ~ δ 4.0. A change in temperature or solvent will also shift the peak position.

Intermolecular hydrogen bonding (see Section 3.2.2) explains why the shift depends on concentration, temperature, and polarity of solvent. Hydrogen bonding decreases the electron density around the proton, thus moving the proton peak to higher frequency. Decrease in concentration in a nonpolar solvent disrupts such hydrogen bonding, and the peak appears at lower frequency—i.e., the alcohol molecules become less "polymeric." Increased temperature has a similar effect.

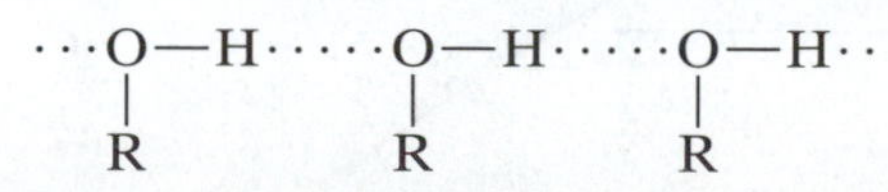

Intramolecular hydrogen bonds are less affected by their environment than are intermolecular hydrogen bonds. In fact, the enolic hydroxylic absorption of β-diketones, for example, is hardly affected by change of concentration or solvent, though it can be shifted upfield somewhat by warming. Nuclear magnetic resonance spectrometry is a powerful tool for studying hydrogen bonding.

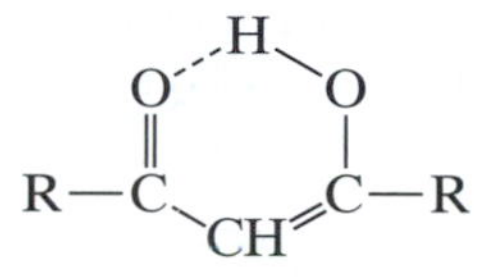

Rapid exchangeability explains why the hydroxylic peak of ethanol is usually seen as a singlet (Fig. 4.32). Under ordinary conditions—exposure to air, light, and water vapor—acidic impurities develop in $CDCl_3$ solution and catalyze rapid exchange of the hydroxylic proton.* The proton is not on the oxygen atom of an individual molecule long enough for it to be affected by the methylene protons; therefore there is no coupling. The OH proton shows a singlet, the CH_2 a quartet, and the CH_3 a triplet.

The rate of exchange can be decreased by lowering the temperature or by treating the solvent with anhydrous sodium carbonate, alumina or grade 3A or 4A molecular sieves, then filtering immediately before obtaining the spectrum. Purified, dry deuterated DMSO or deuterated acetone as solvent, in addition to allowing a lower rate of exchange, shifts the hydroxylic proton peak to the left by hydrogen bonding between solute and solvent. Since the solvent-bonded hydroxylic pro-

* $CDCl_3$ in small vials from Aldrich is pure enough so that a spectrum of CH_3CH_2OH taken within several hours showed the OH peak as a triplet. On standing for about 24 hours exposed to air, the sample gave a spectrum with the OH peak as a singlet (Fig. 4.32). The high dilution used with modern instruments also accounts for the persistence of the vicinal coupling of the OH proton.

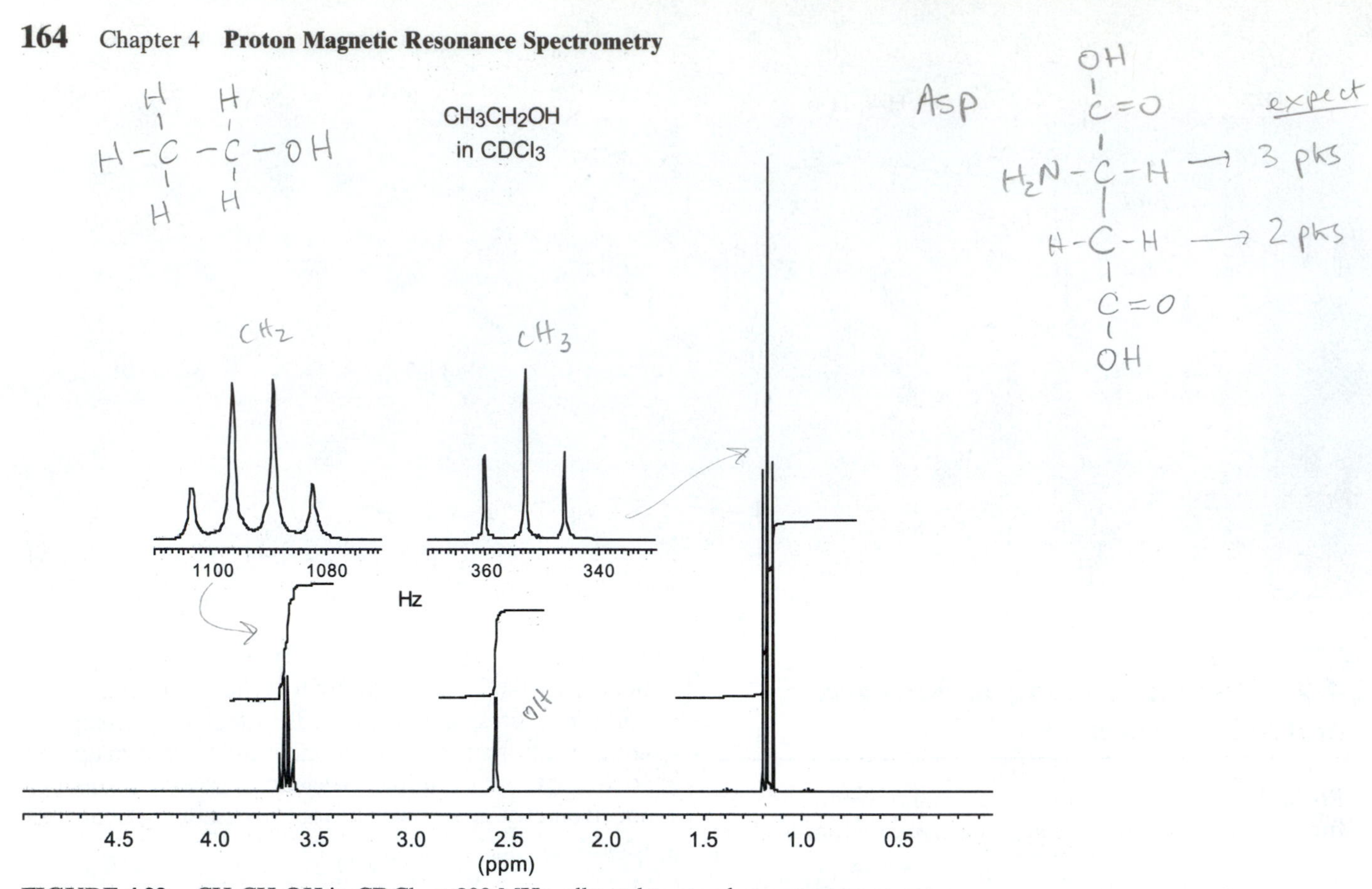

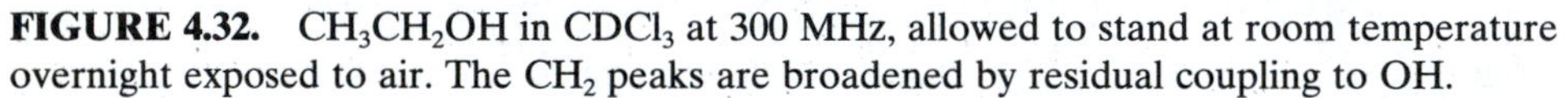

FIGURE 4.32. CH_3CH_2OH in $CDCl_3$ at 300 MHz, allowed to stand at room temperature overnight exposed to air. The CH_2 peaks are broadened by residual coupling to OH.

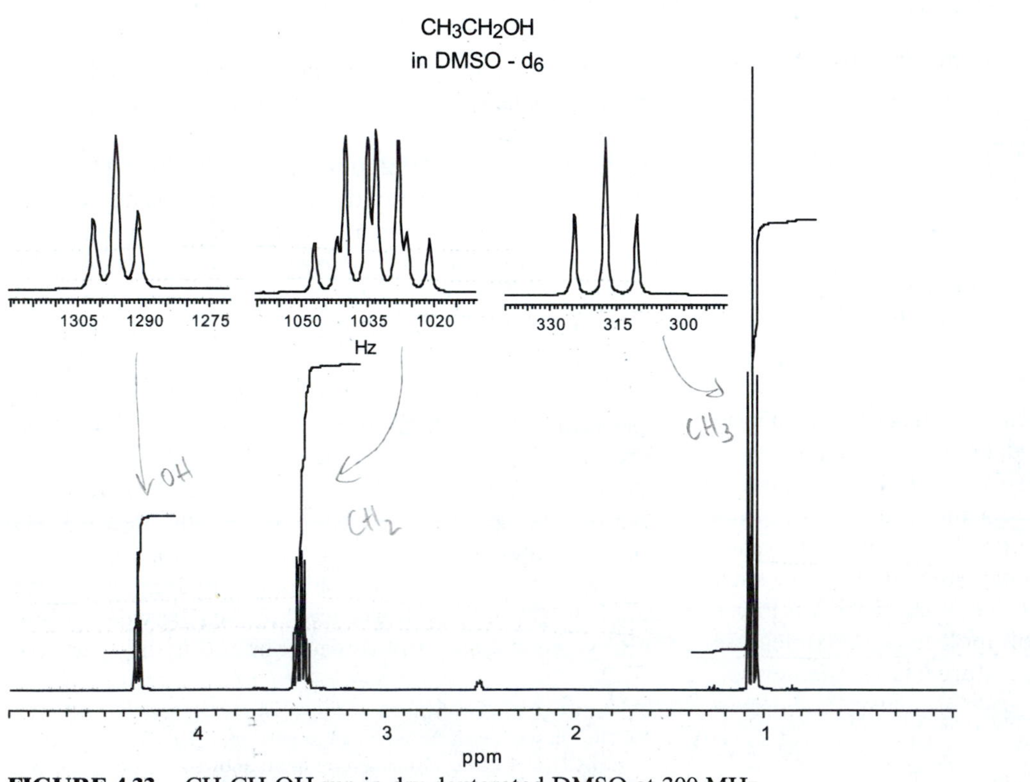

FIGURE 4.33. CH_3CH_2OH run in dry deuterated DMSO at 300 MHz.

ton can now couple with the protons on the α carbon, a primary alcohol will show a hydroxylic triplet, a secondary alcohol, a doublet, and a tertiary alcohol a singlet. The CH_2 protons of ethanol, now coupled to the hydroxylic proton, show a quartet of doublets at high resolution; the $J_{H,OH}$ coupling is 5 Hz, whereas the J_{H,CH_3} coupling is 7 Hz (Fig. 4.33).

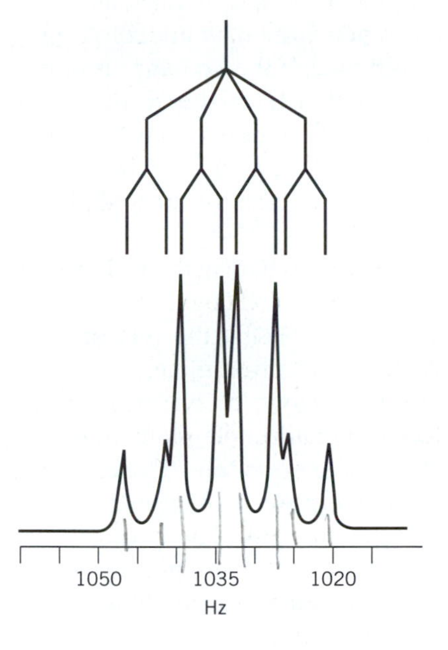

At intermediate rates of exchange, the hydroxylic multiplet merges into a broad band, which progresses to a singlet at higher exchange rates (Fig. 4.32).*

A diol may show separate absorption peaks for each hydroxylic proton; in this case, the rate of exchange in hertz is much less than the difference in hertz between the separate absorptions. As the rate increases (trace of acid catalyst), the two absorption peaks broaden and then merge to form a single broad peak; at this point, the exchange rate (k) in hertz is approximately equal to twice the original signal separation in hertz. As the rate increases, the single peak becomes sharper. The relative position of each peak depends on the extent of hydrogen bonding of each hydroxylic proton; steric hindrance to hydrogen bonding moves the peak to the right.

The spectrum of a compound containing rapidly exchangeable protons can be simplified, and the exchangeable proton absorption removed, simply by shaking the solution with excess deuterium oxide or by obtaining a spectrum in deuterium oxide solution if the compound is soluble. A peak resulting from HOD will appear, generally between δ 5 and δ 4.5 in nonpolar solvents, and near δ 3.3 in DMSO (see Appendix E). A $CDCl_3$ or CCl_4 solution in a stoppered NMR tube may be shaken vigorously for several seconds with 1 or 2 drops of D_2O, and the mixture allowed to stand (or centrifuged) until the layers are clearly separated. The top aqueous layer does not interfere.

Acetylation or benzoylation of a hydroxyl group moves the absorption of the CH_2OH protons of a primary alcohol to the left about 0.5 ppm, and the CHOH proton of a secondary alcohol about 1.0–1.2 ppm. Such shifts provide a confirmation of the presence of a primary or secondary alcohol.

* H_2O as an impurity may exchange protons with other exchangeable protons to form a single peak at an averaged position between the proton peaks involved.

4.9.1.2 Water Aside from the problems of exchangeability, as just discussed, water is an ubiquitous impurity that faithfully obeys Murphy's law by interfering with critically important peaks. "Bulk" water as suspended droplets or wall films gives a peak at $\sim\delta$ 4.7 in $CDCl_3$ (HOD occurs in the D_2O exchange experiment mentioned in Section 4.9.1.1).

Dissolved (monomeric) water absorbs at $\sim\delta$ 1.55 in $CDCl_3$ and can be a serious interference in a critical region of the spectrum in dilute solutions.* Use of C_6D_6 (dissolved H_2O at δ 0.4) avoids this interference. A table of water peaks in the common deuterated solvents appears in Appendix Table E.1.

4.9.1.3 Phenols The behavior of a phenolic proton resembles that of an alcoholic proton. The phenolic proton peak is usually a sharp singlet (rapid exchange, no coupling), and its range, depending on concentration, solvent, and temperature, is generally to the left ($\delta \sim 7.5$ to $\delta \sim 4.0$) compared with the alcoholic proton. A carbonyl group in the ortho position shifts the phenolic proton absorption to the range of about δ 12.0–δ 10.0 because of intramolecular hydrogen bonding. Thus, *o*-hydroxyacetophenone shows a peak at about δ 12.05 almost completely invariant with concentration. The much weaker intramolecular hydrogen bonding in *o*-chlorophenol explains its shift range (δ ~6.3 at 1 *M* concentration to δ ~5.6 at infinite dilution), which is broad compared with that of *o*-hydroxyacetophenone but narrow compared with that of phenol.

4.9.1.4 Enols The familiar tautomeric equilibrium of keto and enol forms of acetylacetone is described in Section 4.12.3.1 (see Fig. 4.39). The enol form predominates over the keto form under the conditions described.

* Webster, F.X., and Silverstein, R.M. (1985). *Aldrichimica Acta* **18** (No. 3), 58.

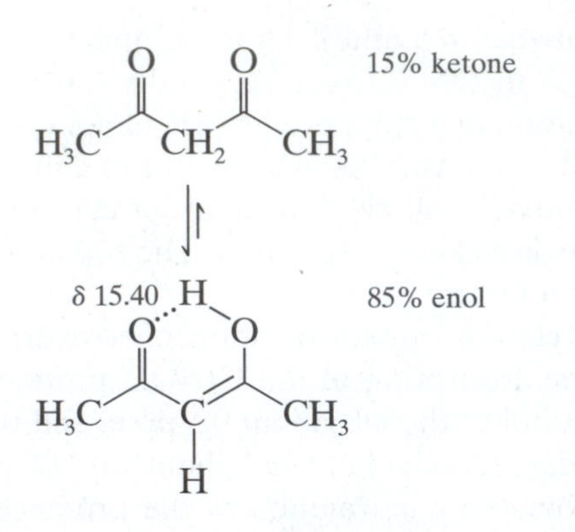

Ordinarily we do not write the enol form of acetone or the keto form of phenol, although minuscule amounts do exist at equilibrium. But both forms of acetylacetone are seen in the NMR spectrum because equilibration is slow enough on the NMR scale and the enol form is stabilized by intramolecular hydrogen bonding (see Sections 3.2.2 and 3.6.10). The enol form of acetone and the keto form of benzene are not thus stabilized; furthermore, the aromatic resonance stabilization of phenol strongly favors the enol form. Note the deshielded chemical shift of the enol proton in Figure 4.39 (see also Appendix Chart E.1).

Ordinarily only the keto form of α-diketones such as 2,3-butanedione is seen in NMR spectra. However, if the enol form of an α-diketone is stabilized by hydrogen bonding—as in the following cyclic α-diketones—only the stabilized enol form appears in the NMR spectra.

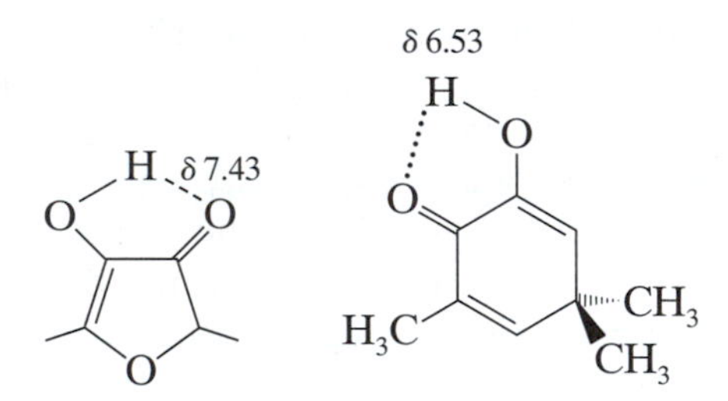

4.9.1.5 Carboxylic Acids Carboxylic acids exist as stable hydrogen-bonded dimers in nonpolar solvents even at high dilution. The carboxylic proton therefore absorbs in a characteristically narrow range $\delta \sim 13.2 - \delta \sim 10.0$ and is affected only slightly by concentration. Polar solvents partially disrupt the dimer and shift the peak accordingly.

The peak width at room temperature ranges from sharp to broad, depending on the exchange rate of the particular acid. The carboxylic proton exchanges quite rapidly with protons of water and alcohols (or hydroxyl groups of hydroxy acids) to give a single peak whose averaged position depends on concentration. Sulfhydryl or enolic protons do not exchange rapidly with carboxylic protons, and individual peaks are observed.

4.9.2 Protons on Nitrogen

The ^{14}N nucleus* has a spin number I of 1 and, in accordance with the formula $2I + 1$, should cause a proton attached to it and a proton on an adjacent carbon atom to show three equally intense peaks. There are two factors, however, that complicate the picture: the rate of exchange of the proton on the nitrogen atom and the electrical quadrupole moment of the ^{14}N nucleus.

The proton on a nitrogen atom may undergo rapid, intermediate, or slow exchange. If the exchange is rapid, the NH proton(s) is decoupled from the N atom and from protons on adjacent carbon atoms. The NH peak is therefore a sharp singlet, and the adjacent CH protons are not split by NH. Such is the case for most aliphatic amines.†

At an intermediate rate of exchange, the NH proton is partially decoupled, and a broad NH peak results. The adjacent CH protons are not split by the NH proton. Such is the case for *N*-methyl-*p*-nitroaniline.

If the NH exchange rate is low, the NH peak is still broad because the electrical quadrupole moment of the nitrogen nucleus induces a moderately efficient spin relaxation and, thus, an intermediate lifetime for the spin states of the nitrogen nucleus. The proton thus sees three spin states of the nitrogen nucleus (spin number = 1), which are changing at a moderate rate, and the proton responds by giving a broad peak. In this case, coupling of the NH proton to the adjacent protons is observed. Such is the case for pyrroles, indoles, secondary and primary amides, and carbamates (Fig. 4.34).

Note that $\underline{H}—N—C—\underline{H}$ coupling takes place through the C—H, C—N, and N—H bonds, but coupling between nitrogen and protons on adjacent carbon atoms is negligible. The proton–proton coupling is observed in the signal caused by hydrogen on carbon; the N—H proton signal is severely broadened by the quadrupolar interaction.

In the spectrum of ethyl *N*-methylcarbamate (Fig. 4.34), $CH_3NH\underset{\underset{O}{\|}}{C}OCH_2CH_3$, the NH proton shows a broad absorption centered about δ 5.16, and the N—CH_3 absorption at δ 2.78 is split into a doublet ($J \sim 5$ Hz) by the NH proton. The ethoxy protons are represented by the triplet at δ 1.23 and the quartet at δ 4.14.

Aliphatic and cyclic amine NH protons absorb from $\sim\delta$ 3.0 to 0.5; aromatic amines absorb from $\sim\delta$ 5.0 to

* ^{15}N spectra are discussed in Chapter 7.

† H—C—N—H coupling in several amines was observed following rigorous removal (with Na–K alloy) of traces of water. This effectively stops proton exchange on the NMR time scale. [K.L. Henold, *Chem. Commun.*, 1340 (1970).]

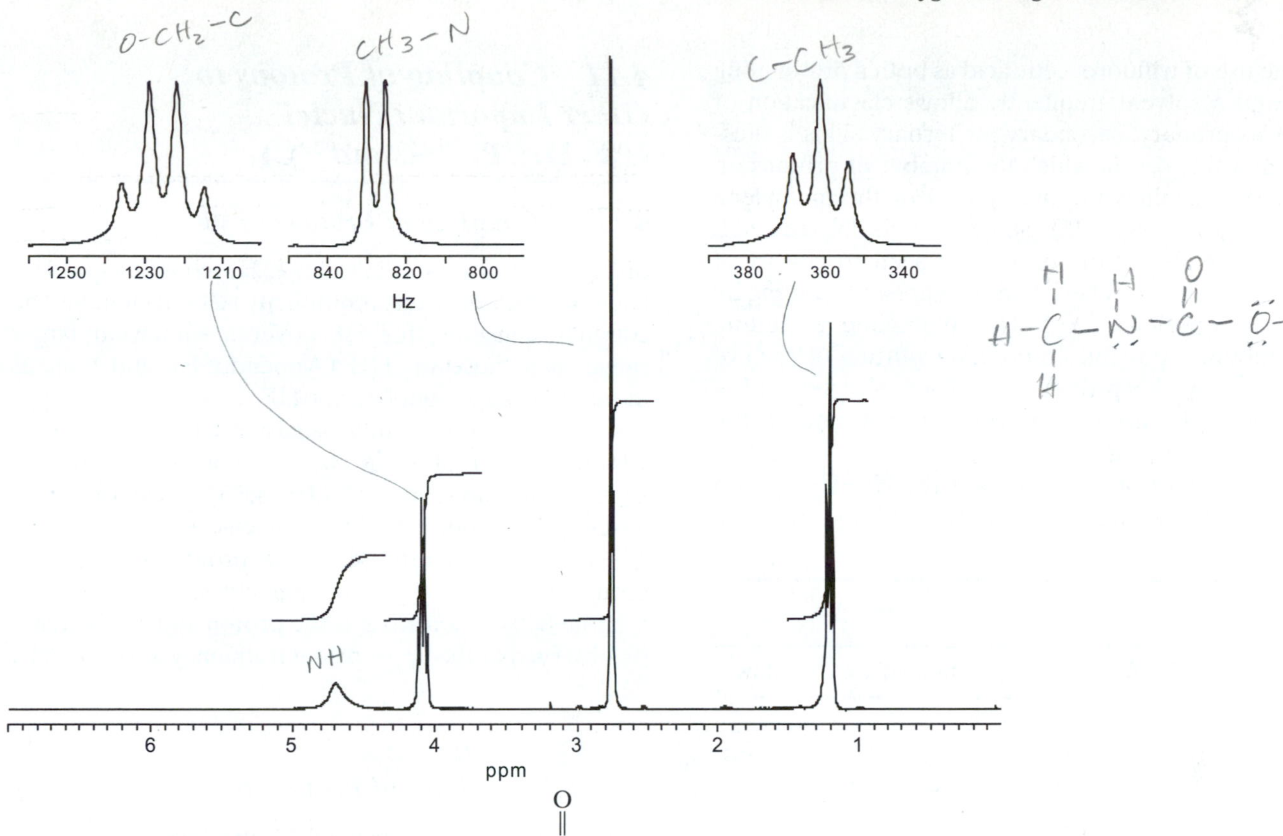

FIGURE 4.34. Ethyl *N*-methylcarbamate, $CH_3NHC(=O)OCH_2CH_3$, at 300 MHz in $CDCl_3$.

3.0. Because amines are subject to hydrogen bonding, the shift depends on concentration, solvent, and temperature. Amide, pyrrole, and indole NH groups absorb from ~ δ 8.5 to 5.0; the effect on the absorption position of concentration, solvent, and temperature is generally smaller than in the case of amines. The nonequivalence of the protons on the nitrogen atom of a primary amide and of the methyl groups of *N,N*-dimethylamides is caused by slow rotation around the C(=O)—N bond because of the contribution of the resonance form $C{=}N^+$ (with O^- on C).

Table 4.1 Classification of Amines by NMR of Their Ammonium Salts in Trifluoroacetic Acid

Amine Precursor Class	Ammonium Salt Structure	Multiplicity of Methylene Unit
Primary	$C_6H_5CH_2NH_3^+$	Quartet (Fig. 4.35)
Secondary	$C_6H_5CH_2NH_2R^+$	Triplet
Tertiary	$C_6H_5CH_2NHR_2^+$	Doublet

Source: Anderson, W.R. Jr., and Silverstein R.M. *Anal. Chem.,* **37,** 1417 (1965).

(Section 4.12.3.2).

Protons on the nitrogen atom of an amine salt exchange at a moderate rate; they are seen as a broad peak, (δ ~8.5 to δ ~6.0), and they are coupled to protons on adjacent carbon atoms (*J* ~ 7 Hz).

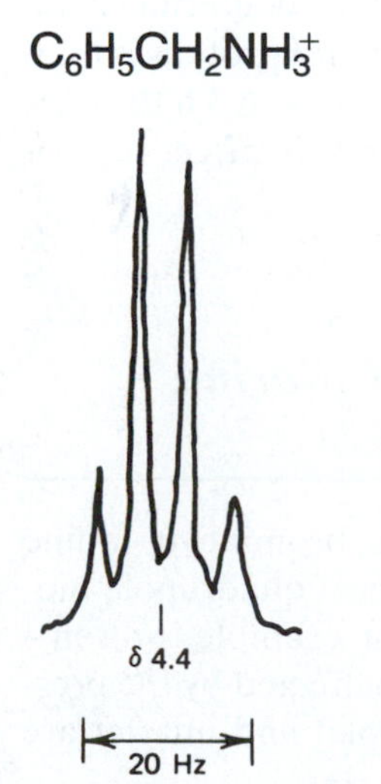

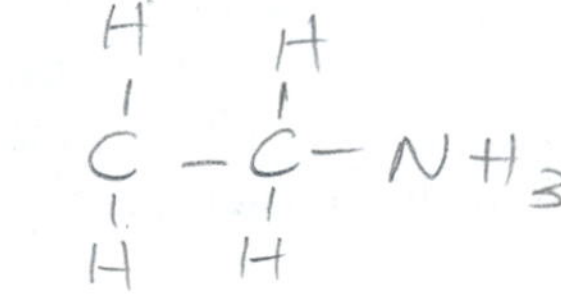

FIGURE 4.35. NMR spectrum of α-methylene unit of a primary amine at 100 MHz in CF_3CO_2H; corresponds to Table 4.1, first line.

The use of trifluoroacetic acid as both a protonating agent and a solvent frequently allows classification of amines as primary, secondary, or tertiary. This is illustrated in Table 4.1, in which the number of protons on nitrogen determines the multiplicity of the methylene unit in the salt (Fig. 4.35). Sometimes the broad ^+NH, $^+NH_2$, or $^+NH_3$ absorption can be seen to consist of three broad humps. These humps represent splitting by the nitrogen nucleus ($J \sim 50$ Hz). With good resolution, it is sometimes possible to observe splitting of each of the humps by the protons on adjacent carbons ($J \sim$ 7 Hz), but it is easier to observe the splitting on the sharper α-CH signals (Table 4.1).

The behavior of the protons in the H—C—N—H sequence may be summarized as follows.*

	Rate of NH Exchange		
	Fast	**Intermediate**	**Slow**
Effect on N—H	Singlet, sharp	Singlet, broad	Singlet, broad
Effect on C—H	No coupling	No coupling	Coupling

4.9.3 Protons on Sulfur

Sulfhydryl protons usually exchange at a low rate so that at room temperature they are coupled to protons on adjacent carbon atoms ($J \sim 8$ Hz). They do not exchange rapidly with hydroxyl, carboxylic, or enolic protons on the same or on other molecules; thus, separate peaks are seen. However, exchange is rapid enough that shaking the solution for a few minutes with deuterium oxide replaces sulfhydryl protons with deuterium. The absorption range for aliphatic sulfhydryl protons is δ 1.6 to δ 1.2; for aromatic sulfhydryl protons, δ 3.6 to δ 2.8. Concentration, solvent, and temperature affect the position within these ranges.

4.10 Protons on or near Chlorine, Bromine, or Iodine Nuclei

Protons are not coupled to chlorine, bromine, or iodine nuclei because of the strong electrical quadrupole moments of these halogen nuclei. For example, proton–proton coupling in CH_3CH_2Cl is unaffected by the presence of a chlorine nucleus; the triplet and quartet are sharp.

* Courtesy of Dr. Donald C. Dittmer (Syracuse University).

4.11 Coupling of Protons to Other Important Nuclei (^{19}F, D, ^{31}P, ^{29}Si, and ^{13}C)

4.11.1 Coupling of Protons to ^{19}F

Since ^{19}F has a spin number of $\frac{1}{2}$, HF coupling and HH coupling obey the same multiplicity rules; in general, the coupling constants for HF cover a somewhat larger range than those for HH (Appendix F), and there is more long-range coupling for HF.

The spectrum of fluoroacetone CH_3—(C=O)—CH_2F, in $CDCl_3$ at 300 MHz (Fig. 4.36) shows the CH_3 group as a doublet at δ 2.2 (J = 4.3 Hz) resulting from long-range coupling by the F nucleus. The doublet at δ 4.75 (J = 48 Hz) represents the protons of the CH_2 group coupled to the geminal F nucleus. The ^{19}F nucleus is about 80% as sensitive as the proton and can be readily observed at the appropriate frequency and magnetic field.

4.11.2 Couplings of Protons to D

Deuterium (D or ^{2}H) usually is introduced into a molecule to detect a particular group or to simplify a spectrum. Deuterium has a spin number of 1, a small coupling constant with protons (see Appendix G), and a small electrical quadrupole moment. The ratio of the J values for HH to those of HD is about 6.5.

Suppose the protons on the α-carbon atom of a ketone

$$X-\overset{\gamma}{C}H_2-\overset{\beta}{C}H_2-\overset{\alpha}{C}H_2-(C{=}O)-Y$$

(X and Y contain no protons)

were replaced by deuterium to give

$$X-\overset{\gamma}{C}H_2-\overset{\beta}{C}H_2-\overset{\alpha}{C}D_2-(C{=}O)-Y$$

The spectrum of the undeuterated compound consists of a triplet for the α protons, a quintet for the β protons—assuming equal coupling for all protons—and a triplet for the γ protons. For the deuterated compound, the α-proton absorption would be absent, the β-proton absorption would appear, at modest resolution, as a slightly broadened triplet, and the γ-proton absorption would be unaffected. Actually, at very high resolution, each peak of the β-proton triplet would appear as a very closely spaced quintet ($J_{H-C-C-D} \sim 1$ Hz) since $2nI + 1 = 2 \times 2 \times 1 + 1 = 5$, where n is the number of D nuclei coupled to the β protons.

Most deuterated solvents have residual proton im-

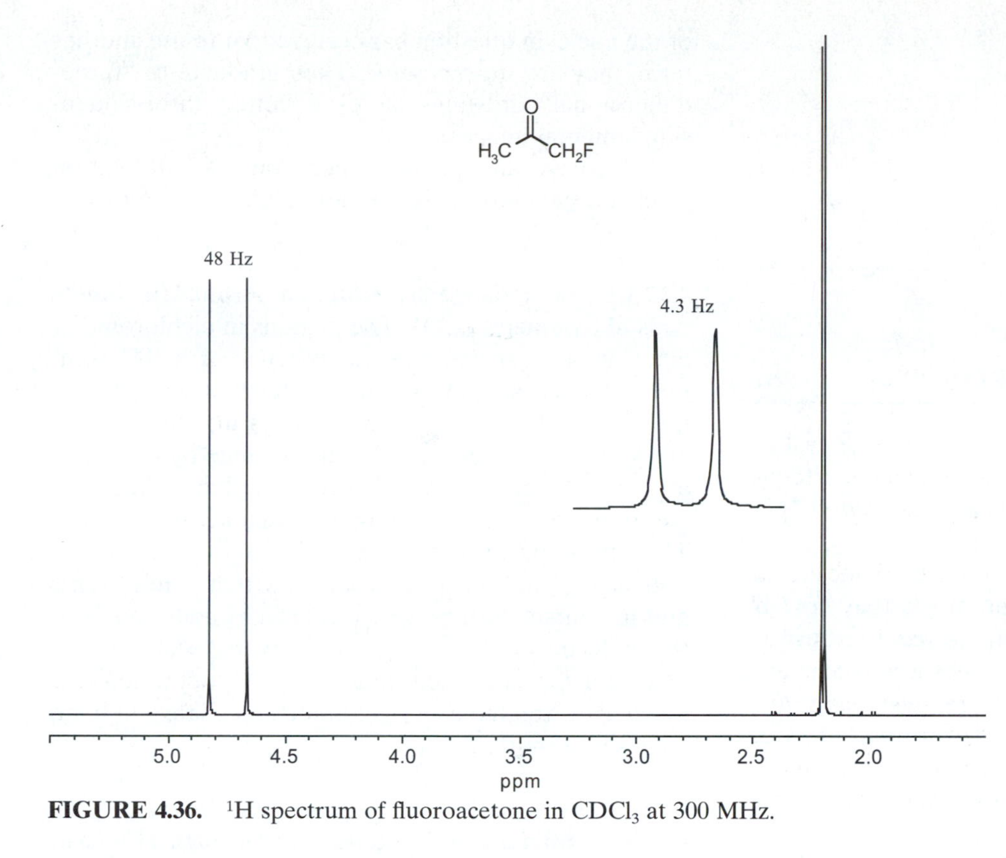

FIGURE 4.36. ^1H spectrum of fluoroacetone in $CDCl_3$ at 300 MHz.

purities in an otherwise completely deuterated molecule; thus, deuterated dimethyl sulfoxide, $(CD_3)_2S{=}O$, contains a few molecules of $(CHD_2)_2S{=}O$, which show a closely spaced quintet ($J \sim 2$ Hz, intensities 1:2:3:2:1) in accordance with $2nI + 1$ (see Appendix G).

Because of the electrical quadrupole moment of D only broad absorption peaks can be obtained from a spectrum of deuterium nuclei.

4.11.3 Coupling of Protons to ^{31}P

The nucleus ^{31}P has a natural abundance of 100% and a spin number of $\frac{1}{2}$ (therefore no electrical quadrupole moment). The multiplicity rules for proton–phosphorus splitting are the same as those for proton–proton splitting. The coupling constants are large ($J_{H-P} \sim$ 200–700 Hz, and J_{HC-P} is 0.5–20 Hz) (Appendix F) and are observable through at least four bonds. The ^{31}P nucleus can be observed at the appropriate frequency and magnetic field (Chapter 7).

4.11.4 Coupling of Protons to ^{29}Si

The ^{29}Si isotope has a natural abundance of 4.70% (^{28}Si = 92.28%) and a spin number of $\frac{1}{2}$. The value of J $^{29}Si{-}CH$ is about 6 Hz. The low-intensity doublet caused by the $^{29}Si{-}CH_3$ coupling can often be seen straddling (± 3 Hz) an amplified peak of TMS; the low-intensity $^{13}C{-}H_3$ "satellite" doublet can also be seen at ± 59 Hz (Section 4.11.5). ^{29}Si spectra can be obtained at the appropriate frequency and magnetic field (Chapter 7).

4.11.5 Coupling of Protons to ^{13}C

The isotope ^{13}C has a natural abundance relative to ^{12}C of 1.1% and a spin number of $\frac{1}{2}$. Protons directly attached to ^{13}C are split into a doublet with a large coupling constant, about 115–270 Hz for $^{13}C{-}H$. The $CH_3{-}CH_2$ group, for example, is predominantly $^{12}CH_3{-}^{12}CH_2$ but contains a small amount of $^{13}CH_3{-}^{12}CH_2$ and of $^{12}CH_3{-}^{13}CH_2$. Thus, the $^{13}CH_3$ protons are split into a doublet by ^{13}C ($J \sim 120$ Hz), and each peak of the doublet is split into a triplet by the $^{12}CH_2$ protons ($J \sim 7$ Hz) as shown below. These "^{13}C satellite" peaks are small because of the small number of molecules containing the $^{13}CH_3$ group and can usually be seen disposed on both sides of a large $^{12}CH_3$ peak (e.g., the large $^{12}CH_3$ triplet shown below). The chemical shift of the $^{12}CH_3$ protons is midway between the satellites. See Chapter 5 for ^{13}C NMR spectrometry.

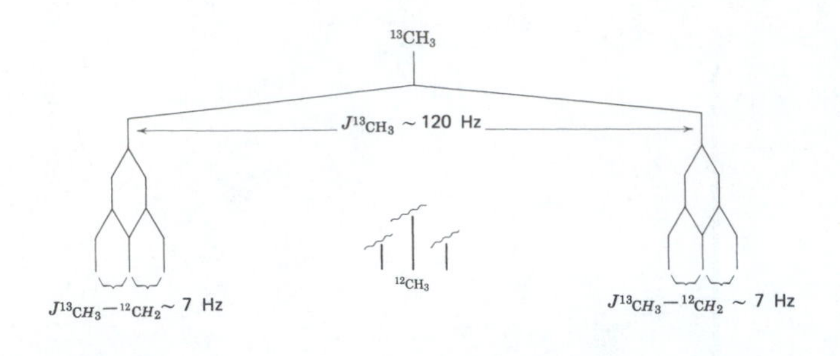

4.12 Chemical Shift Equivalence

The concept of chemical shift equivalence is central to NMR spectrometry. Chemical-shift equivalent (isochronous) nuclei comprise a *set* within a *spin system* (Pople notation, Section 4.8).

The immediate question is: Are selected nuclei in a molecule chemical shift equivalent, or are they not? If they are, they are placed in the *same set.* The answer can be framed as succinctly as the question: Nuclei are chemical shift *equivalent if they are interchangeable through any symmetry operation or by a rapid process.* This broad definition assumes an achiral environment (solvent or reagent) in the NMR experiment; the common solvents are achiral (Section 4.16).

We deal first with symmetry operations and later with rapid processes (Section 4.12.3).

4.12.1 Determination of Chemical Shift Equivalence by Interchange Through Symmetry Operations

There are three symmetry operations: *rotation about a simple axis of symmetry* (C_n), *reflection through a plane of symmetry* (σ), and *inversion through a center of symmetry* (i). More rigorously, symmetry operations may be described under two headings: C_n and S_n. The latter is *rotation around an alternating axis of symmetry.* It turns out that S_1 is the same as σ, S_2 is the same as i, and higher subscripts for S_n are rare. We shall use C_n, σ, and i. The subscripts denote the number of such rotations required to make a 360° rotation. Thus C_1 is a 360° rotation, C_2 is a 180° rotation, etc. The designation S_1 requires a 360° rotation followed by a reflection through the plane at a right angle to the axis. S_2 requires a 180° rotation followed by a reflection, and so forth.

Before we describe the three common symmetry operations to effect interchange of nuclei within a molecule, we raise another question: How do we know whether or not a symmetry operation has indeed resulted in interchange? Simply enough: Build two models of the molecule—one before the operation, one after the operation. If they are *indistinguishable,* interchange of the nuclei in question has occurred; or to use another term, they are *superposable.* Then graduate to "three-dimensional" drawings, or even better, three-dimensional mental images.

Of course, an "identity" operation—a 360° rotation around a symmetry axis—is not valid.

4.12.1.1 Interchange by Rotation Around a Simple Axis of Symmetry (C_n)

The protons in dichloromethane (Fig. 4.37, structure a) interchange by a 180° rotation around a simple axis of symmetry in the plane of the paper. This C_2 interchange shows that the protons are *homotopic;* they can also interchange by a σ operation (next section) through a vertical plane at right angles to the page, but the C_2 operation takes precedence. The homotopic protons are chemical-shift equivalent in both achiral and chiral environments, that is, in solvents and in contact with reagents. Drawings and models of the molecule are indistinguishable by inspection before and after the operation. During the C_n operation, the molecule is treated as a rigid object; no change in bond angles is permitted.

4.12.1.2 Interchange by Reflection Through a Plane of Symmetry (σ)

A plane of symmetry exists if one half of the molecule is the mirror image of the other half. The protons in chlorofluoromethane (Fig. 4.37, structure b) are interchanged only by reflection at a right angle through a vertical plane of symmetry at a right angle to the page; there is no other symmetry element. The protons are mirror images *(enantiotopes)* of each other and are chemical-shift equivalent only in an achiral solvent or in contact with an achiral reagent. Chiral solvents or chiral reagents can distinguish between a pair of enantiotopic protons in an NMR experiment (Sections 4.16 and 4.21).

The methylene protons of propanoic acid (Fig. 4.37, structure c) are exchangeable by reflection through the plane of symmetry in the plane of the paper. There are no other symmetry elements. The protons are enantiotopes of each other and have the same chemical shift only in an achiral environment. Drawings and models of the molecule are indistinguishable by inspection before and after the operation.

4.12.1.3 Interchange by Inversion Through a Center of Symmetry (i)

A center of symmetry exists if a line drawn from each nucleus through the selected center encounters the same nucleus or group at the same distance on the other side of the center (Fig. 4.37, structure d).

The inversion is equivalent to an S_2 operation, which consists of a 180° rotation around an alternating

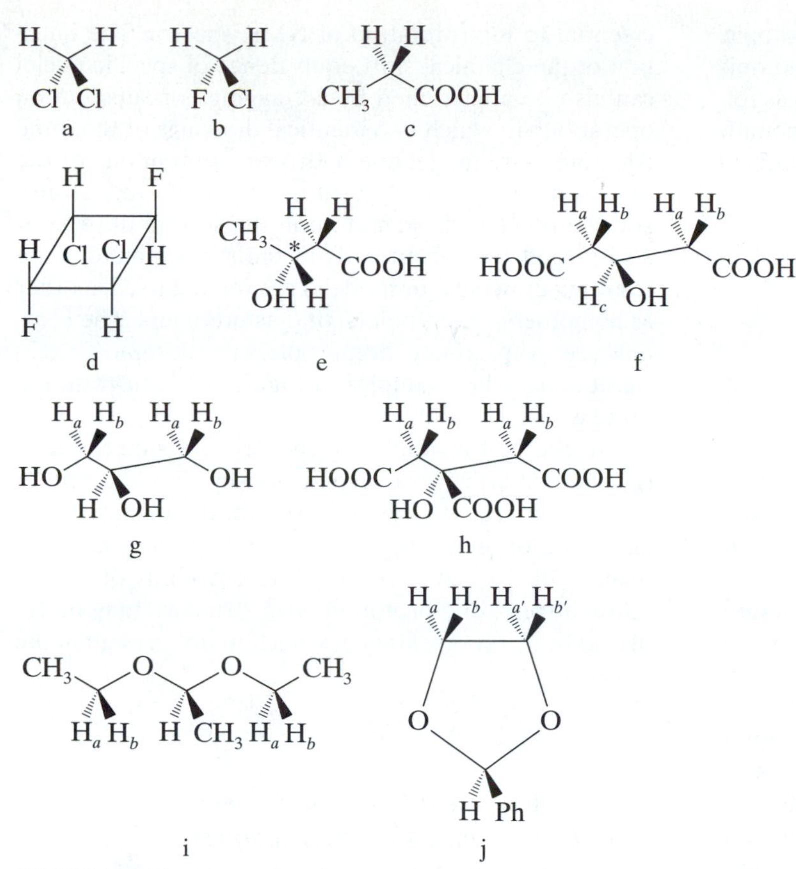

FIGURE 4.37. Examples of proton interchange by symmetry operations:

a. Homotopic protons by C_2
b, c. Enantiotopic protons by σ
d. Enantiotopic protons by i
e. Diastereotopic protons in a chiral molecule (* denotes chiral center)
f, g, h, i. Diastereotopic protons in achiral molecules (f is 3-hydroxyglutaric acid; g is glycerol; h is citric acid; i is diethyl acetal; j is a cyclic acetal of benzaldehyde, 2-phenyl-1,3-dioxolane.)

axis of symmetry followed by reflection through a symmetry plane at a right angle to the axis. Therefore, the inversion interchange involves a reflection, so that, strictly speaking, the interchangeable nuclei or groups must be *mirror images* of one other. This distinction is not necessary for nuclei or for achiral groups, but it is necessary for interchange of chiral groups. Thus, an *R* group and an *S* group are interchangeable, but two *R* groups (or two *S* groups) are not. Of course, the same restrictions hold for interchange through a plane of symmetry (Section 4.12.1.2). In both operations, the interchanged nuclei or groups are enantiotopic to each other and are chemical shift equivalent only in an achiral environment.

4.12.1.4 No Interchangeability by a Symmetry Operation If geminal protons (CH_2) in a molecule cannot be interchanged through a symmetry element, those protons are *diastereotopic* to one another; each has a different chemical shift—except for coincidental overlap. The diastereotopic geminal protons couple with each other (through two bonds). In principle, each geminal proton should show different coupling constants with other neighboring nuclei, although the difference may not always be detectable. A CH_2 group consisting of a pair of diastereotopic protons is shown in Figure 4.37, structure e; the chiral center is shown by an asterisk, but a chiral center is not necessary for the occurrence of diastereotopic protons (see Fig. 4.37, structure f). This achiral molecule, 3-hydroxyglutaric acid, has a plane of symmetry, perpendicular to the page through the middle carbon atom, through which the two H_a protons interchange and the two H_b protons interchange, as enantiotopes. Since there is no plane of symmetry passing between the protons of each CH_2 group, protons *a* and *b* of each CH_2 group are diastereotopic. An idealized first-order spectrum for the diastereotopic protons of 3-hydroxyglutaric acid is diagrammed as follows, but in practice most of these types of compounds even at high resolution, would show partially resolved peaks, because $\delta\nu/J$ ratios are small.

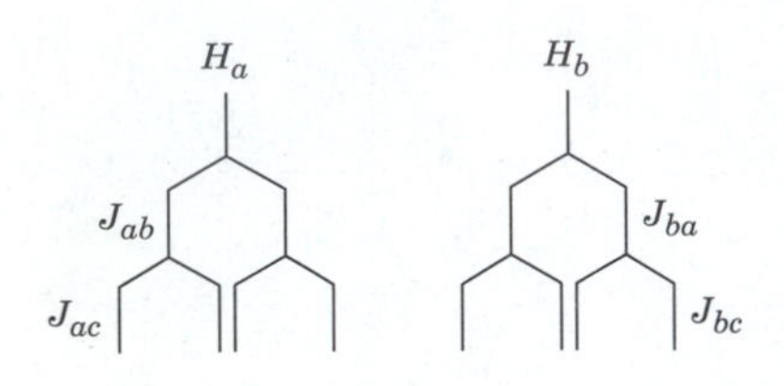

The following similar achiral molecules contain diastereotopic methylene protons: 3-hydroxyglutaric acid, glycerol, citric acid, diethyl acetal, and a cyclic acetal (respectively, Fig. 4.37, structures f, g, h, i, and j); structure j involves the additional concept of magnetic equivalence (Section 4.13).

From the above discussion, diastereotopic protons cannot be placed in the same set since they are not chemical-shift equivalent. However it is not uncommon for diastereotopic protons to *appear* to be chemical-shift equivalent in a given magnetic field in a particular solvent. Such *accidental* chemical shift equivalence can usually be detected by using an instrument with a higher magnetic field or by changing solvents.

Diastereotopic protons (or other ligands) must be in constitutionally equivalent locations; that is, they cannot differ in *connectivity.* For example in structure e of Figure 4.37, the geminal protons have the same connectivity but differ in the sense that they are not interchangeable; thus they are diastereotopic. On the other hand, the proton on C-3 has a different connectivity from those on C-2, and the term, "diastereotopic" does not apply.

Students are familiar with the terms applied to relationships between stereoisomeric molecules: *homomeric* molecules (superposable molecules), *enantiomeric* molecules (nonsuperposable mirror images), and *diastereomeric* molecules (stereoisomers that are not mirror images of one another). These familiar terms are parallel to the terms that we have introduced above: *homotopic, enantiotopic,* and *diastereotopic,* which are applied to nuclei or groups within the molecule.

4.12.2 Determination of Chemical Shift Equivalence by Tagging (or Substitution)

As noted in Section 4.12.1, a clear understanding of the concepts of chemical shift equivalence and its relationship to symmetry elements and symmetry operations is essential to interpretation of NMR spectra. The question of the chemical shift equivalence of specific nuclei can also be approached by a "tagging" or substitution operation* in which two identical drawings of the same compound are made; one hydrogen atom in one of the drawings is tagged (or substituted by a different atom), and the other hydrogen atom in the second drawing is also tagged (or substituted) in the same manner. The resulting drawings (or models) are related to each other as homomers, enantiomers, or diastereomers. The H atoms are, respectively, homotopic, enantiotopic, or diastereotopic. The examples in Figure 4.38 illustrate the process.

In the first example, the models are superposable (i.e., homomers); in the second, nonsuperposable mirror images (i.e., enantiomers); in the third, not mirror images (i.e., diastereomers). Note that the tags are permanent; i.e., H and Ⓗ are different kinds of atoms. Alternatively, one proton in each structure may be replaced by Z, representing any nucleus not present in the molecule.

4.12.3 Chemical Shift Equivalence by Rapid Interconversion of Structures

If chemical structures can interconvert, the result depends on temperature, catalyst, solvent, and concentration. We assume a given concentration and absence of catalyst, and we treat four systems.

4.12.3.1 Keto–Enol Interconversion The tautomeric interconversion of acetylacetone (Fig. 4.39) at room temperature is slow enough that the absorption peaks of both forms can be observed—i.e., there are two spectra. The equilibrium keto/enol ratio can be determined from the relative areas of the keto and enol CH_3 peaks, as shown. At higher temperatures the interconversion rate will be increased so that a single "averaged" spectrum will be obtained. Chemical shift equivalence for all of the interconverting protons has now been achieved. Note that the NMR time scale is of the same order of magnitude as the chemical shift separation of interchanging signals expressed in hertz, i.e., about 10^1–10^{-3} Hz. Processes occurring faster than this will lead to averaged signals. Note also that the enolic OH proton peak is deshielded relative to the OH proton of alcohols because the enolic form is strongly stabilized by intramolecular hydrogen bonding.

* Ault, A. (1974). *J. Chem. Educ.,* **51,** 729.

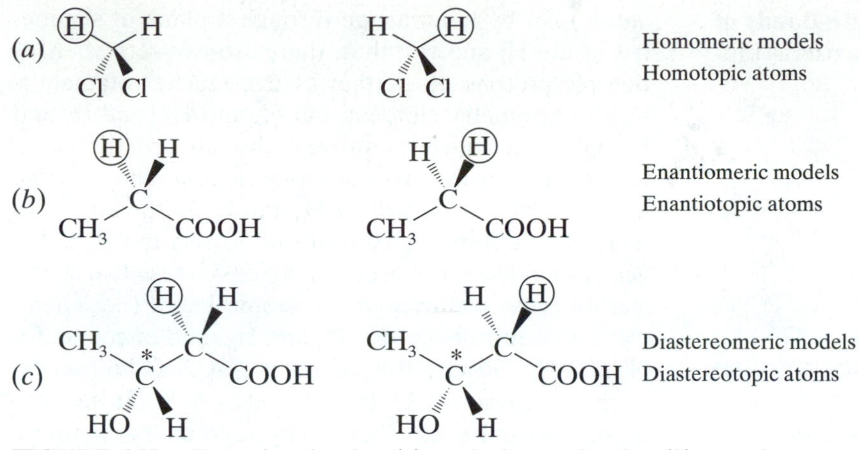

FIGURE 4.38. Tagged molecules: (*a*) equivalent molecules, (*b*) enantiomers, (*c*) diastereomers.

4.12.3.2 Interconversion Around a "Partial Double Bond" (Restricted Rotation) At room temperature, a neat sample of dimethylformamide shows two CH_3 peaks because the rate of rotation around the hindered "partial double bond" is slow. At ~123°C, the rate of exchange of the two CH_3 groups is rapid enough so that the two peaks merge.

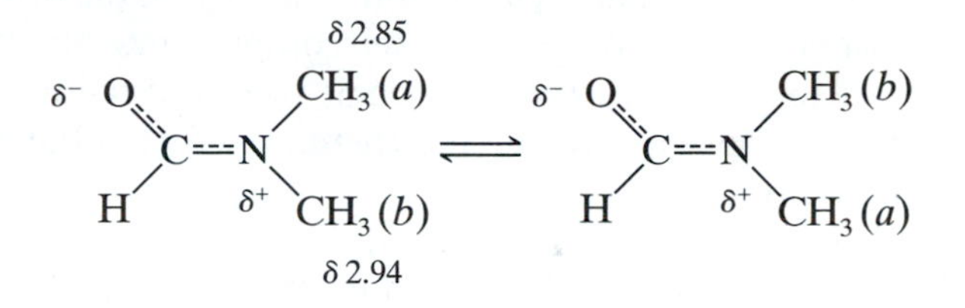

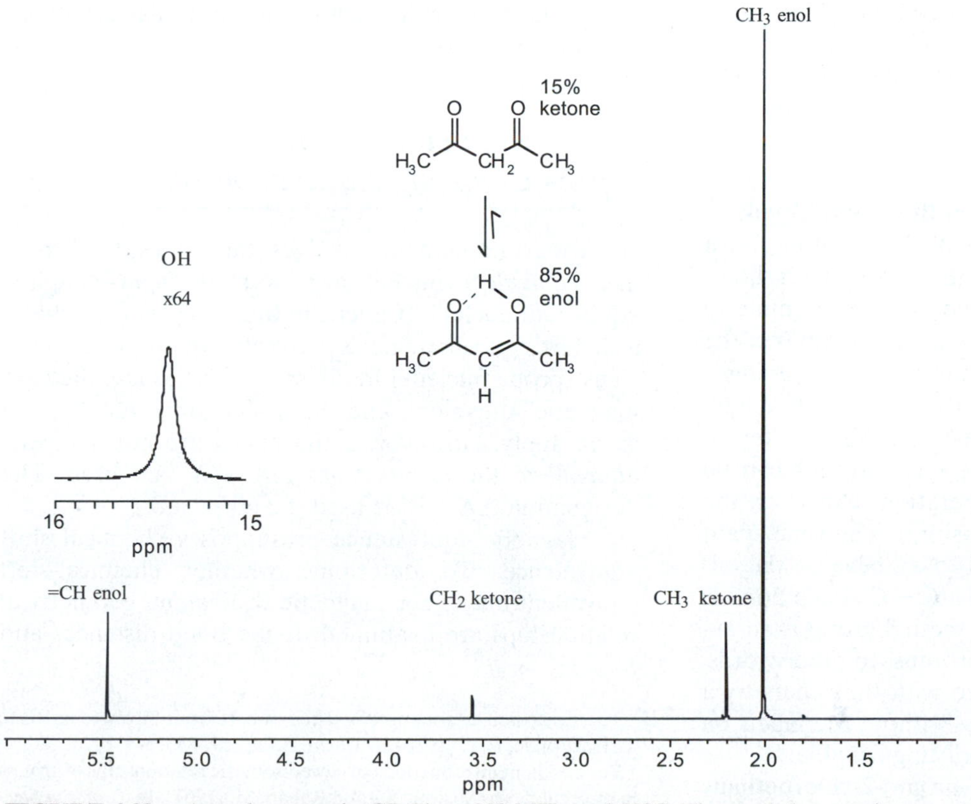

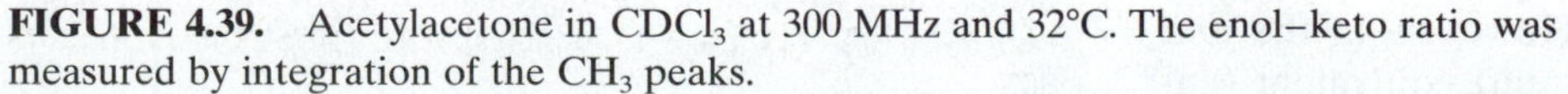

FIGURE 4.39. Acetylacetone in $CDCl_3$ at 300 MHz and 32°C. The enol–keto ratio was measured by integration of the CH_3 peaks.

4.12.3.3 Interconversion Around the Single Bonds of Rings Cyclohexane at room temperature exists as rapidly interconverting chair forms.

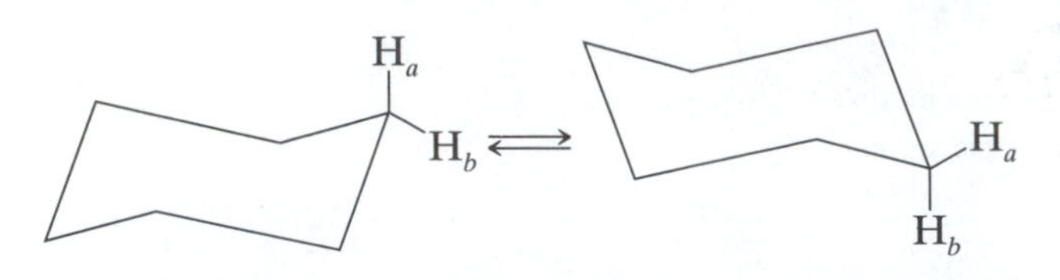

An axial proton becomes an equatorial proton and vice versa in the interconverting structures, and the spectrum consists of a single "averaged" peak. As the temperature is lowered, the peak broadens and at a sufficiently low temperature two peaks appear—one for the axial protons, one for the equatorial protons. In other words, at room temperature, the axial and equatorial protons are chemical-shift equivalent by rapid interchange. At very low temperatures, they are not chemical-shift equivalent; in fact, in each "frozen" chair form, the protons of each CH_2 group are diastereotopic pairs, but at room temperature, the rate of chair interconversion is sufficiently high to average the chemical shifts of these geminal protons.

The methylene protons of methylcyclohexane are diastereotopic pairs at any temperature. The chair forms are not equivalent (i.e., not superposable); they are diastereomers. Substitution on a ring may also create chiral centers (see Section 4.16 and Problem 8.3).

In a fused cyclohexane ring, such as those of steroids, the rings are "frozen" at room temperature and the axial and equatorial protons of each CH_2 group are not chemical-shift equivalent.

4.12.3.4 Interconversion Around the Single Bonds of Chains Chemical shift equivalence of protons on a CH_3 group results from rapid rotation around a carbon–carbon single bond even in the absence of a symmetry element. Figure 4.40*a* shows Newman projections of the three staggered rotamers of a molecule containing a methyl group attached to another sp^3 carbon atom having four different substituents, that is, a chiral center. In any single rotamer, none of the CH_3 protons can be interchanged by a symmetry operation. However, the protons are rapidly changing position. The time spent in any one rotamer is short ($\sim 10^{-6}$ s), because the energy barrier for rotation around a C—C single bond is small. The chemical shift of the methyl group is an average of the shifts of the three protons. In other words, each proton can be interchanged with the others by a rapid rotational operation. Thus, without the labels on the protons, the rotamers are indistinguishable.

The staggered rotamers of 1-bromo-2-chloroethane (Fig. 4.40*b*) are distinguishable. However, in the *anti* rotamer, H_a and H_b are chemical-shift equivalent (enantiotopic) by interchange through a plane of symmetry, as are H_c and H_d; thus, there are two sets of enantiotopic protons. In neither of the gauche rotamers is there a symmetry element, but H_a and H_b, and H_c and H_d, are chemical-shift equivalent by rapid rotational interchange between two enantiomeric rotamers. Now we have one chemical shift for H_a and H_b in the *anti* rotamer, and a different chemical shift for H_a and H_b in the *gauche* rotamers. By rapid averaging of these two chemical shifts, we obtain a single chemical shift (i.e., chemical shift equivalence) for H_a and H_b, and of course for H_c and H_d. Strictly the system is AA′XX′ but, as explained in Section 4.13, it is treated as A_2X_2. In general, if protons can be interchanged by a symmetry operation (through a plane of symmetry) in one of the rotamers, they are also chemical-shift equivalent (enantiotopic) by rapid rotational interchange.*

Consider a methylene group next to a chiral center, as in 1-bromo-1,2-dichloroethane (Fig. 4.40*c*). Protons H_a and H_b are not chemical-shift equivalent since they cannot be interchanged by a symmetry operation in any conformation; the molecule has no simple axis, plane, center, or reflection axis of symmetry. Although there is a rapid rotation around the carbon–carbon single bond, the CH_2 protons are not interchangeable by a rotational operation; the averaged chemical shifts of H_a and H_b are not identical. An observer can detect the difference before and after rotating the methylene group: the protons in each rotamer are diastereotopic. The system is ABX.

4.13 Magnetic Equivalence (Spin–Coupling Equivalence)†

A further refinement involves the concept of *magnetic-equivalent* nuclei, also termed "spin–coupling equivalent nuclei". If nuclei in the same set (i.e., chemical-shift equivalent nuclei) couple equally to any nucleus (probe nucleus) in *the same spin system,* they are magnetic equivalent, and the designations A_2, X_2, and so on apply. However, if the nuclei are not magnetic equivalent, the designations *AA′ XX′* are used. The designation AA′BB′ is used if $\Delta\nu/J < \sim 8$.

Magnetic equivalence presupposes chemical shift equivalence. To determine whether chemical-shift equivalent nuclei are magnetic equivalent, geometrical relationships are examined. If the bond distances and

* The discussion of rotamers is taken in part from Silverstein, R.M., and LaLonde, R.T. (1980). *J. Chem. Educ.,* **57,** 343.

† For excellent introductions to stereoisomeric relationships of groups in molecules, see Mislow, K., and Raban, M. (1967). In *Topics in Stereochemistry,* Vol. I. page 1; Jennings, W.B. (1975) *Chem. Rev.,* **75,** 307.

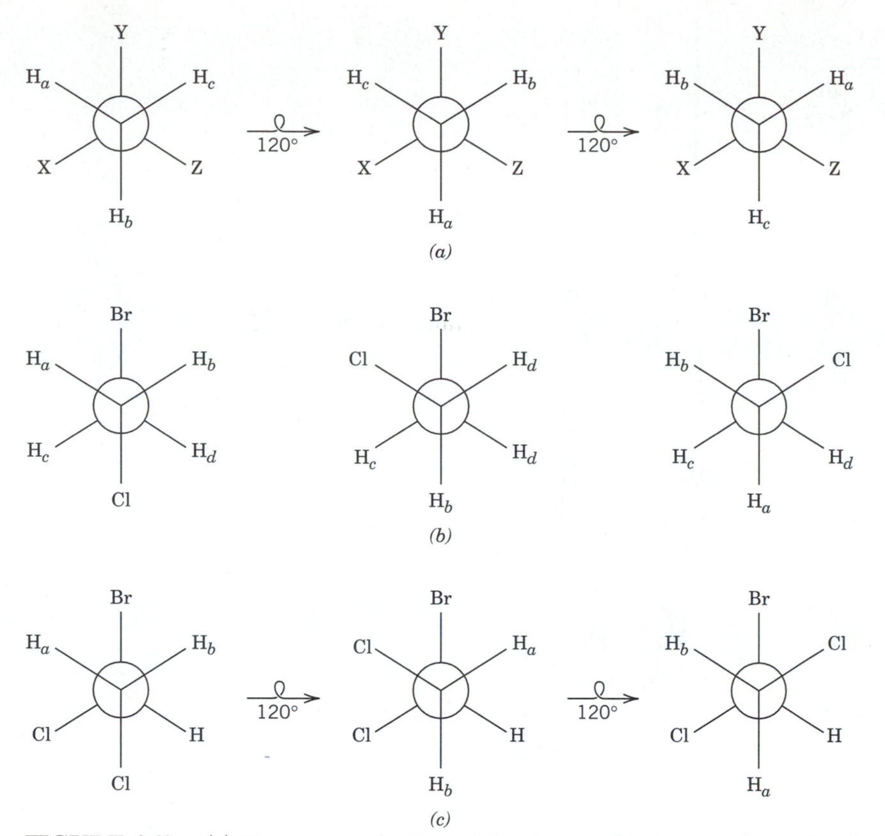

FIGURE 4.40. (*a*) Newman projection of the staggered rotamers of a molecule with a methyl group attached to a chiral sp^3 carbon atom. (*b*) 1-Bromo-1,2-dichloroethane. (*c*) 1-bromo-2-chloroethane.

angles from each nucleus in relation to the probe nucleus are identical, the nuclei in question are magnetic equivalent. In other words, two chemical-shift equivalent nuclei are magnetic equivalent if they are symmetrically disposed with respect to each nucleus (probe) in the same spin system. Magnetic-equivalent nuclei *couple identically* with any *other* nucleus in the same spin system; hence the synonymous term "spin–coupling equivalence".* Note that a test for magnetic equivalence is valid only when the *two nuclei are chemical shift equivalent;* that is, they are in the same set.

These rules are applied readily to conformationally restricted structures. Thus, in *p*-chloronitrobenzene (Fig. 4.41) the protons ortho to the nitro group (H_A and $H_{A'}$) are chemical shift equivalent to each other, and the protons ortho to the chlorine atom (H_X and $H_{X'}$) are chemical-shift equivalent to each other. In general for para-disubstituted benzene rings, J_{AX} and $J_{AX'}$ are the same, approximately 7–10 Hz; $J_{A'X}$ and $J_{A'X'}$ are also the same but much smaller, approximately 1 Hz. Since H_A and $H_{A'}$ couple differently to another specific proton, they are not magnetic equivalent, and first-order rules do not apply. (Similarly, H_X is not magnetic equivalent to $H_{X'}$). To test this conclusion, we draw a line connecting H_A and $H_{A'}$, select $H_{X'}$ (or H_X) as a probe nucleus, then draw a probe line from the probe nucleus, at a right angle to the line connecting H_A and $H_{A'}$. If the probe line were to intersect the connecting line at midpoint, H_A and $H_{A'}$ would be magnetic equivalent. Obviously, they are not.

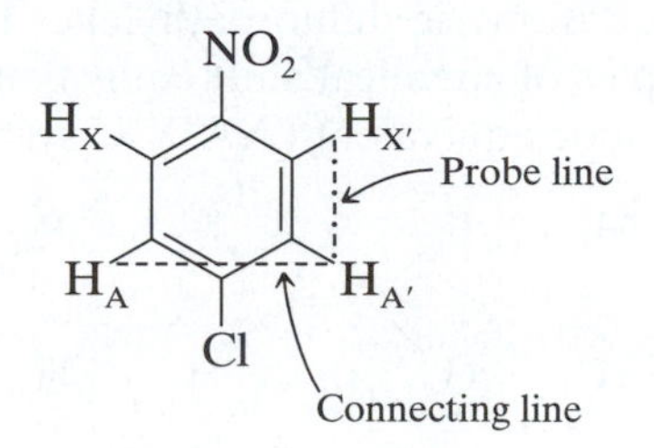

The system is described as AA′XX′, and the spec-

* See references to Mislow and Raban, and to Jennings (this Section). Those authors object to the term "magnetic equivalence" because it has been used ambiguously in the literature; they prefer the term "spin–coupling equivalence." However "magnetic equivalence" is widely used in the literature and will be retained here.

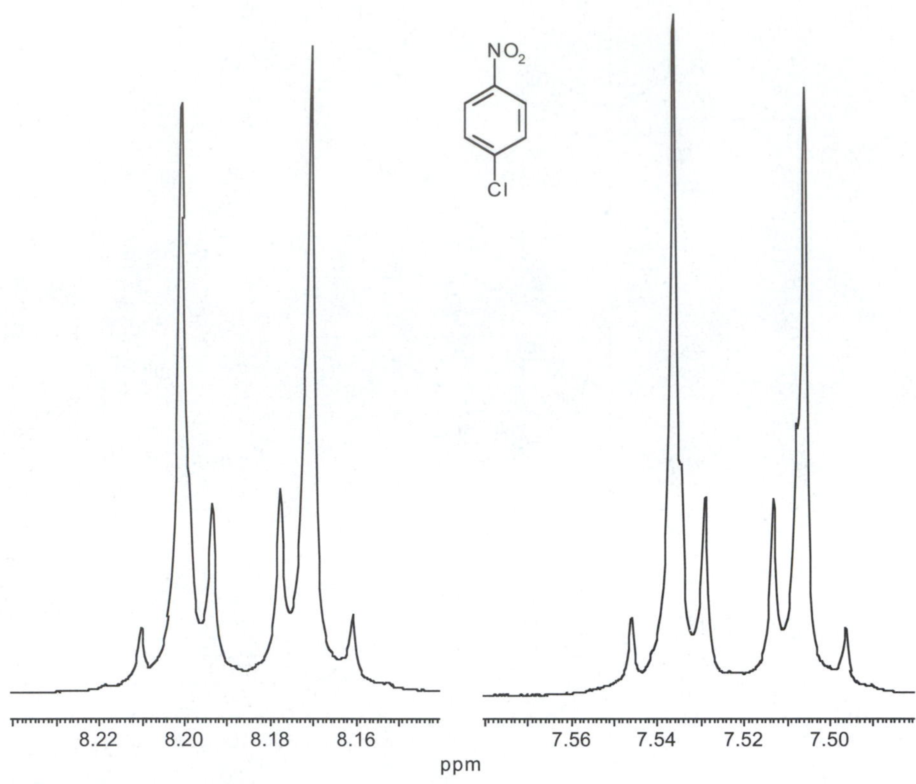

FIGURE 4.41. *p*-Chloronitrobenzene in $CDCl_3$ at 300 MHz.

trum is very complex (Fig. 4.41): Fortunately, the pattern is readily recognized because of its symmetry; the two multiplets are mirror images. In fact, at lower resolution, the pattern is two tight clusters somewhat resembling that of an AX system, which might be naively predicted if magnetic nonequivalence and the small meta coupling are ignored. The pattern is "deceptively simple."

In the present example (*p*-chloronitrobenzene), mental reduction of the AA′XX′ spectrum to an *apparent* AX spectrum (two apparent doublets) would predict the para disubstitution, as would the deliberately naive interpretation in Section 7.3.

The aromatic protons of symmetrically ortho-disubstituted benzenes also give a symmetrical AA′XX′ spectrum. An example is *o*-dichlorobenzene (Fig. 4.42).

The three isomeric difluoroethylenes furnish additional examples of chemical-shift equivalent nuclei that are not magnetic equivalent (AA′XX′ systems).

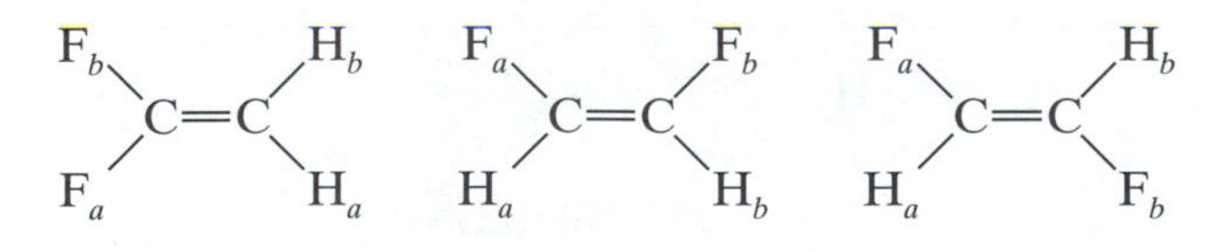

In each case, the protons H_a and H_b comprise a set, and fluorines F_a and F_b comprise a set (of chemical shift equivalent nuclei), but the nuclei in each set are not magnetic equivalent and the spectra are complex.*

With some mastery of the concepts of chemical shift equivalence and magnetic equivalence, we return to structure j in Figure 4.37 (2-phenyl-1,3-dioxolane), whose spectrum is given in Figure 4.43.

Structure j resembles structure f–i in that all of the CH_2 protons are diastereotopic since there is no plane of symmetry in the plane of the paper; in other words, the protons of any of the methylene groups cannot interchange.

However, the CH_2 groups in structure j are in a ring system with limited flexibility and are directly coupled to each other. Thus, there is a significant difference between the coupling constant of protons *a* and *b* compared with the coupling constant of protons *a* and *b*′ (or *a*′ and *b*′ compared with *a*′ and *b*). The spin system, therefore, is AA′BB′ and the spectrum is strikingly similar in complexity and symmetry to the spectrum of *p*-chloronitrobenzene (Fig. 4.41) and the spectrum of

* Note that the protons A and A′ (and X and X′) are coupled to each other even though they are chemical shift equivalent; however, the ratio $\frac{\Delta\nu}{J}$ is zero regardless of the strength of the electromagnetic field. Therefore, *an AA'XX' spectrum cannot be transformed into a first order spectrum by increasing the magnetic field strength*—this in contrast with for example, an ABX spectrum.

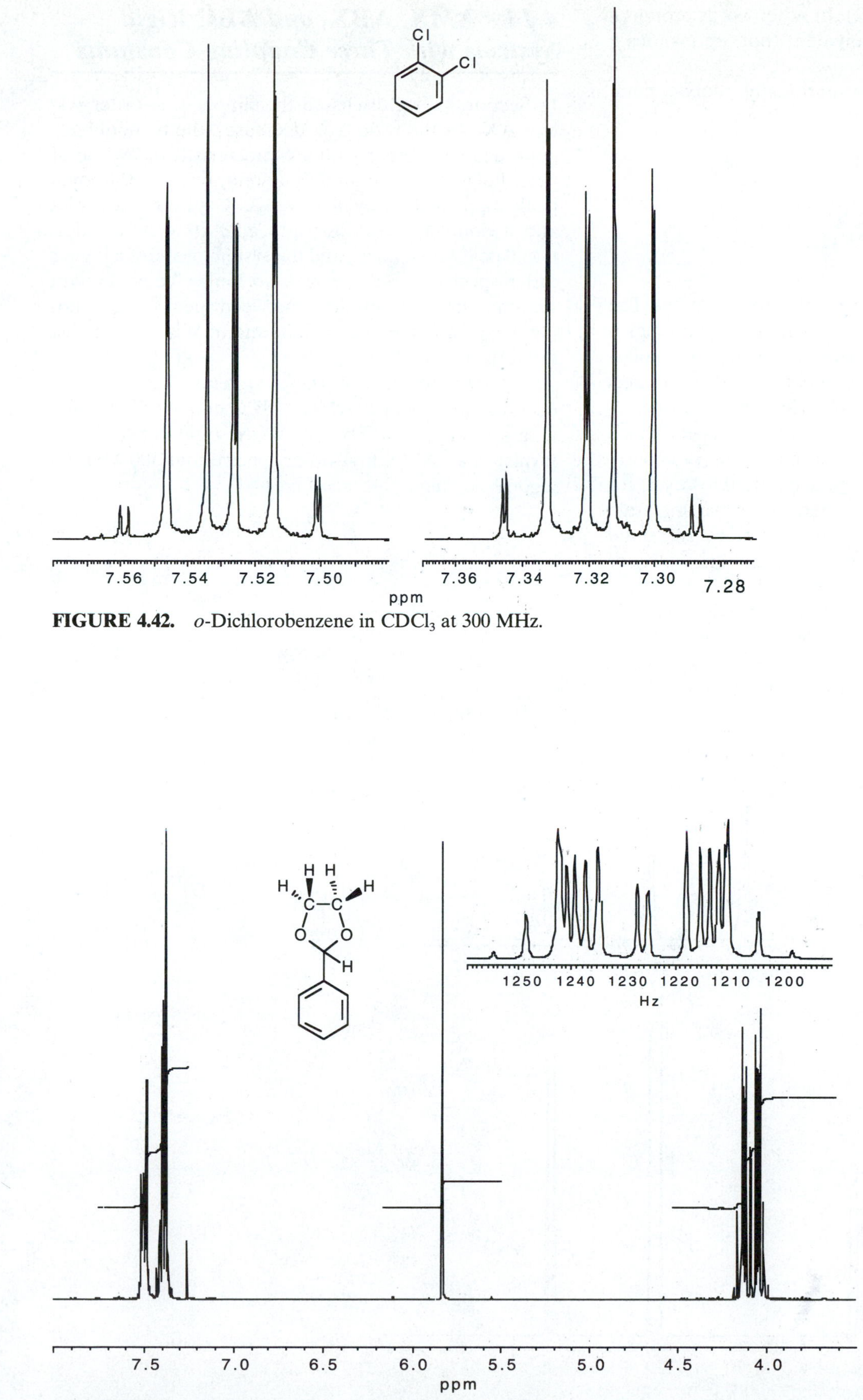

FIGURE 4.42. *o*-Dichlorobenzene in $CDCl_3$ at 300 MHz.

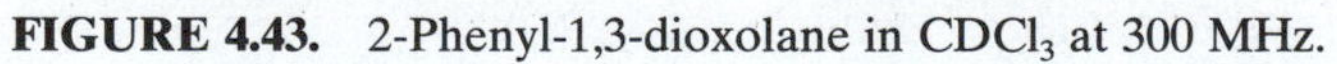

FIGURE 4.43. 2-Phenyl-1,3-dioxolane in $CDCl_3$ at 300 MHz.

o-dichlorobenzene (Fig. 4.42). In other words protons *a* and *a*′ are not magnetic equivalent (nor are protons *b* and *b*′).

The open-chain, conformationally mobile compounds of the type

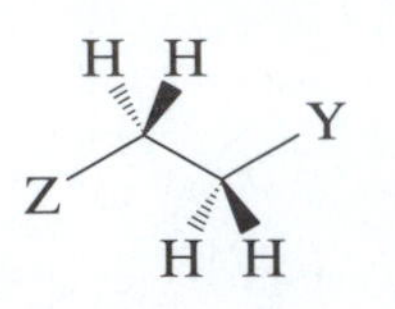

consist of two sets of protons coupled to each other. The groups Z and Y contain no chiral element and no protons that couple to the two sets shown; Z and Y polarities determine the difference between shifts of the sets (i.e., $\Delta\nu$). If this molecule were rigid, it would be regarded as an AA′XX′ or AA′BB′ system depending on the magnitude of $\Delta\nu/J$, and at a low enough temperature, this description would be accurate. However, for such a molecule at room temperature—barring large conformational preferences—the *J* values are quite similar, and, in practice, spectra resembling A_2X_2 or A_2B_2 spectra result. The "weakly coupled" A_2X_2 system would show two triplets, and a "strongly coupled" A_2B_2 system would show a complex, higher order spectrum. We shall treat these conformationally mobile systems as A_2X_2 or A_2B_2 rather than AA′XX′ or AA′BB′.

4.14 AMX, ABX, *and* ABC *Rigid Systems with Three Coupling Constants*

In Section 4.8 we discussed the simple, first-order system AX. As the ratio $\Delta\nu/J$ decreases, the two doublets approach each other with a characteristic distortion of peak heights to give an AB system, but no additional peaks appear. However, as an A_2X system—a triplet and a doublet—develops into an A_2B system, additional peaks do appear, and the system presents a higher order spectrum; the *J* values no longer coincide with the measured differences between peaks [see simulated spectra in Bovey (1988) and in Wiberg and Nist (1962)].

Having considered these systems, we can now examine the systems AMX, ABX, and ABC, starting with a rigid system. Styrene, whose rigid vinylic group furnishes an AMX first-order spectrum at 300 MHz, is a good starting point (see Fig. 4.44).

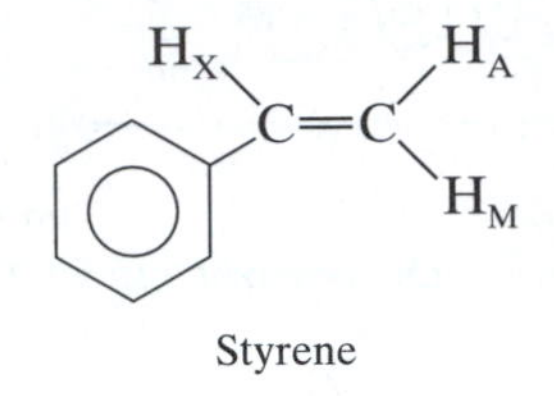

Correlation of the structure of a compound

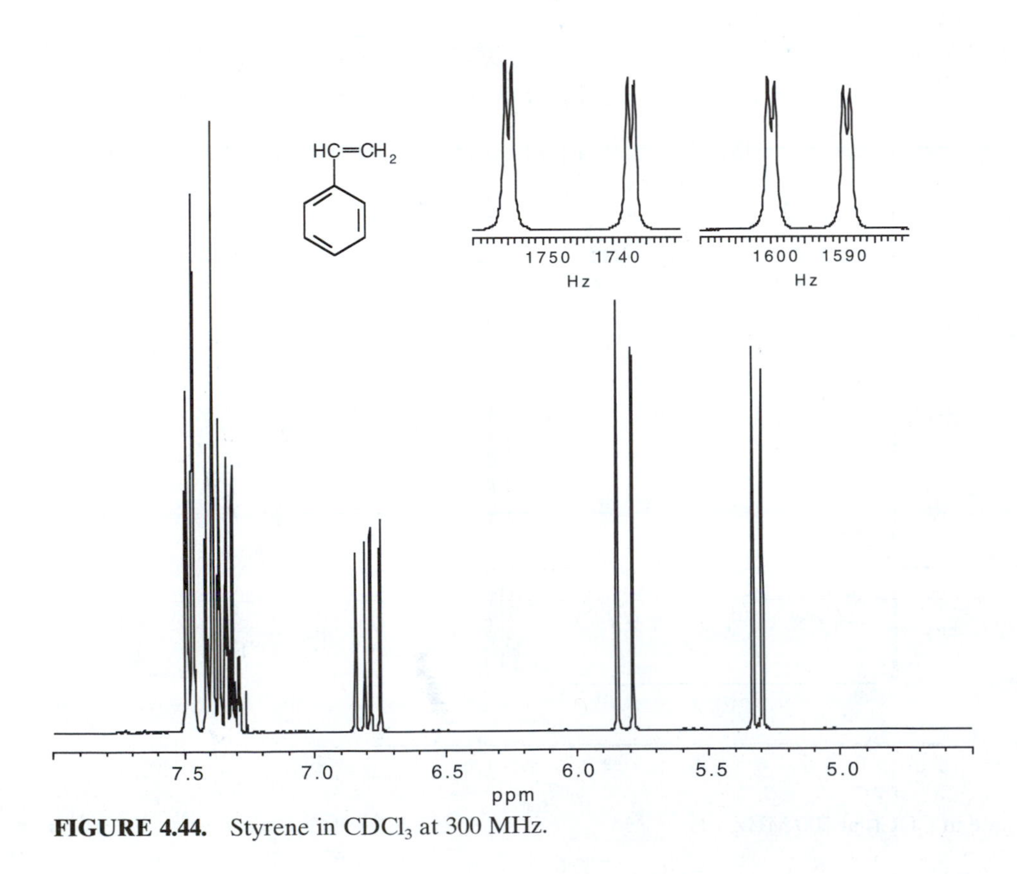

FIGURE 4.44. Styrene in $CDCl_3$ at 300 MHz.

with its NMR spectrum requires the following information:

- Number of spin systems
- Number of sets in each spin system
- Number of protons in each set
- Chemical shift (δ value) of each set
- Spacing of sets ($\Delta\nu$ values in Hz)
- Number of couplings and magnitude of couplings (J values)
- $\Delta\nu/J$ values
- Number and kinds of symmetry elements, and conformational mobility
- Exchange of labile protons
- Effect of substituents (functional groups) on the chemical shift of protons

The following data are obtained from the styrene structure and the spectrum (Fig. 4.44), the vinylic portion being of interest:

- There are two spin systems in the molecule; the vinylic portion constitutes one of these systems.
- There are three sets in the vinylic system AMX.
- Each set contains one proton.
- Chemical shifts are: X = δ 6.80, M = δ 5.82, A = δ 5.31.
- $\Delta\nu_{XM}$ = 599 Hz, $\Delta\nu_{XA}$ = 447 Hz, $\Delta\nu_{MA}$ = 83 Hz.
- There are three coupling constants: J_{XM} = 18 Hz, J_{XA} = 11 Hz, J_{AM} = 1.0 Hz.

$$\Delta\nu_{MA}/J_{MX} = 83/18 = 4.6^*$$

Since there is free rotation around the C—C bond, there are two symmetry planes through the molecule, but the vinylic protons cannot be interchanged since they lie *in* the planes; hence the three sets.

- There are no labile protons.
- The phenyl substituent deshields H_X and to some extent H_M.

The spectrum of the vinylic system of styrene consists of three sets—each a doublet of doublets representing a single proton. The X proton (δ 6.80) is coupled trans across the double bond to the M proton (J_{XM} = 18 Hz) and cis across the double bond to the A proton (J_{XA} = 11 Hz). The M proton (δ 5.82) is coupled trans across the double bond to the X proton (J_{MX} = 18 Hz) and geminally to the A proton (J_{MA} = 1.0 Hz). The A proton (δ 5.31) is coupled cis across the double bond to the X proton (J_{AX} = 11 Hz) and geminally to the M proton (J_{AM} = 1.0 Hz).

At 60 MHz, the spectrum of the vinylic system of styrene becomes a borderline ABX system, but with some imagination the structure can be resolved by recognizing that the ABX spectrum can be regarded as though it were a first-order AMX spectrum—which it is at 300 MHz. Again, it must be emphasized that the order of a spectrum depends partly on the sophistication of available instrumentation. At the extreme, an ABC spectrum—a very low $\Delta\nu/J$ ratio for all of the sets—may not be resolvable with available instrumentation.

* In an AMX spectrum, the smallest shift difference ($\Delta\nu$) should be at least twice as large as the largest coupling constant; a small ratio gives an ABX spectrum. The present ratio is 4.6 [Jackman, L.M., and Sternhell, S. (1969). *Application of Nuclear Magnetic Resonance Spectroscopy in Organic Chemistry,* 2nd ed. Oxford: Pergamon Press, p. 133.]

4.15 Conformationally Mobile, Open-Chain Systems. Virtual Coupling

We shall limit ourselves to some of the more common spin systems and provide a few examples of pitfalls that bedevil students. For more thorough treatments, see the references at the end of the chapter.

4.15.1 Unsymmetrical Chains

4.15.1.1 1-Nitropropane As mentioned in Section 4.13, most open-chain compounds—barring severe steric hindrance—are conformationally mobile at room temperature, and protons in each set average out and become practically magnetic equivalent. Thus a 300-MHz, room-temperature spectrum of 1-nitropropane is described as an $A_3M_2X_2$ system rather than $A_3MM'XX'$, and first-order rules apply (Fig. 4.45).

$$\overset{A_3}{CH_3}—\overset{M_2}{CH_2}—\overset{X_2}{CH_2}NO_2$$

The X_2 protons are strongly deshielded by the NO_2 group, the M_2 protons less so, and the A_3 protons very slightly. There are two coupling constants, J_{AM} and J_{MX}, that are very similar but not exactly equal. In fact, at 300 MHz, the M_2 absorption is a deceptively simple, slightly broadened sextet ($n_A + n_X + 1 = 6$). At sufficient resolution, 12 peaks are possible: $(n_A + 1)(n_X + 1) = 12$.

The A_3 and X_2 absorptions are triplets with slightly different coupling constants. The system is described as *weakly coupled,* and we can justify mentally cleaving the system for analysis.

4.15.1.2 1-Hexanol In contrast, consider the 300 MHz spectrum of 1-hexanol (Figure 4.46).

$$\overset{X_2}{HOCH_2}—\overset{A_2}{CH_2}—\overset{B_2}{CH_2}—\overset{C_2}{CH_2}—\overset{D_2}{CH_2}—\overset{M_3}{CH_3}$$

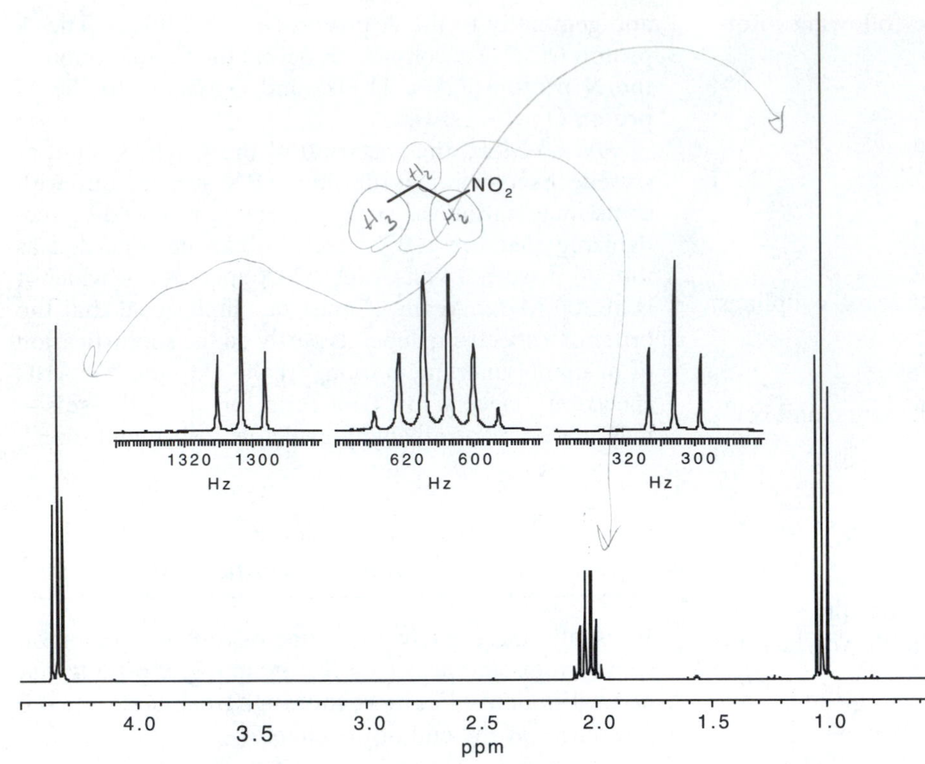

FIGURE 4.45. 1-Nitropropane in $CDCl_3$ at 300 MHz.

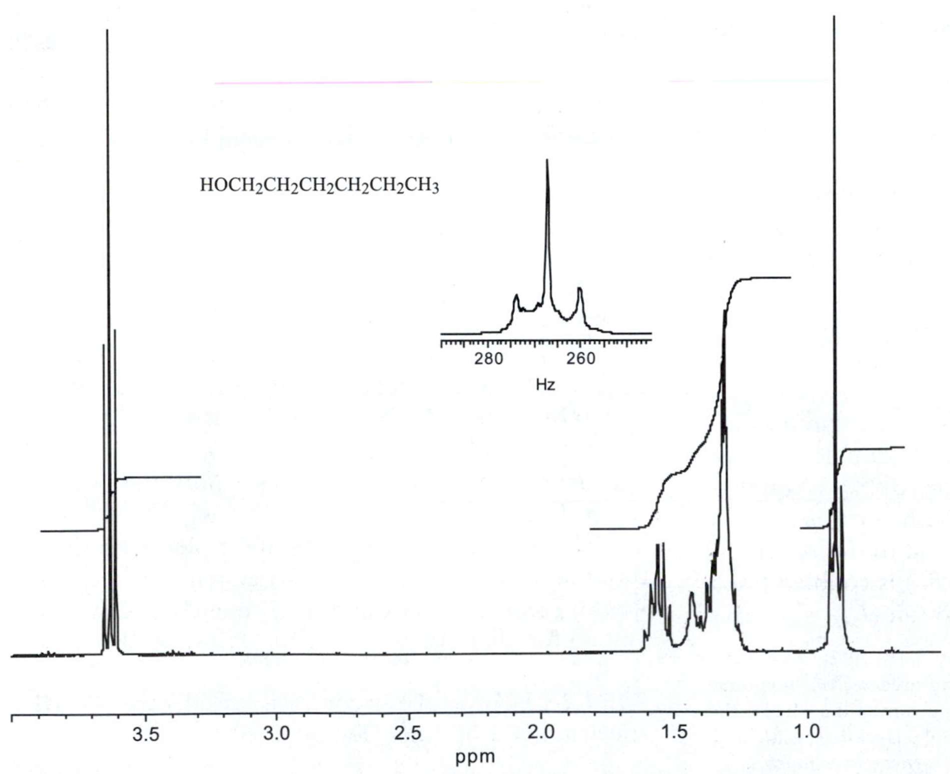

FIGURE 4.46. 1-Hexanol in $CDCl_3$ at 300 MHz. The CH_3 peak is broadened and "filled in."

This very *strongly coupled* system cannot be treated by first-order rules. The CH_3 absorption is described as a broadened, distorted, filled-in "triplet." The strongly deshielded CH_2 group (X_2) is a classical triplet at δ 3.63. The A_2 protons are somewhat deshielded and appear as a distorted quintet. The remaining CH_2 groups, which are very similar in chemical shift, are strongly coupled to one another; they appear as a partially resolved band and act as a "conglomerate" of spins in coupling to the CH_3 group. The severe distortion of the CH_3 group (M_3), which is formally coupled only to the adjoining group, is a result of the "conglomerate" coupling; the effect is termed *virtual coupling* and is characteristic of a strongly coupled hydrocarbon chain.* Note that X_2 protons show a clean triplet because the A_2 protons are somewhat deshielded i.e., separated from the other CH_2 groups; thus there is no virtual coupling. We shall see more examples of virtual coupling. It is a difficult concept.

4.15.2 Symmetrical Chains

4.15.2.1 Dimethyl Succinate The symmetrical, conformationally mobile, open-chain diesters are worth examining. Dimethyl succinate

$$\underset{}{\text{MeOOC}}-\underset{A_2}{CH_2}-\underset{A_2}{CH_2}-\text{COOMe}$$

obviously gives a four-proton singlet.

4.15.2.2 Dimethyl Glutarate

$$\text{MeOOC}-\underset{X_2}{CH_2}-\underset{A_2}{CH_2}-\underset{X_2}{CH_2}-\text{COOMe}$$

Dimethyl glutarate at 300 MHz is an $X_2A_2X_2$ system, which can be written as A_2X_4 and gives a quintet and a triplet. Less electronegative substituents in place of the COOMe groups result in a complex A_2B_4 spectrum.

4.15.2.3 Dimethyl Adipate We move to dimethyl adipate:

$$\overset{6}{\text{MeOOC}}-\underset{X_2}{\overset{5}{CH_2}}-\underset{A_2}{\overset{4}{CH_2}}-\underset{A_2}{\overset{3}{CH_2}}-\underset{X_2}{\overset{2}{CH_2}}-\overset{1}{\text{COOMe}}$$

and confidently cleave the molecule at midpoint to produce two identical A_2X_2 systems; thus we obtain a deshielded triplet and a less deshielded triplet.

The 300-MHz spectrum comes as a shock (see Fig. 4.47). Obviously, this is by no means a first-order spectrum even though $\Delta\nu/J$ for the A_2X_2 coupling is approximately 21, assuming a J value of about 7. The equiva-

* Musher, J.I., and Corey, E.J. (1962). *Tetrahedron,* **18,** 791.

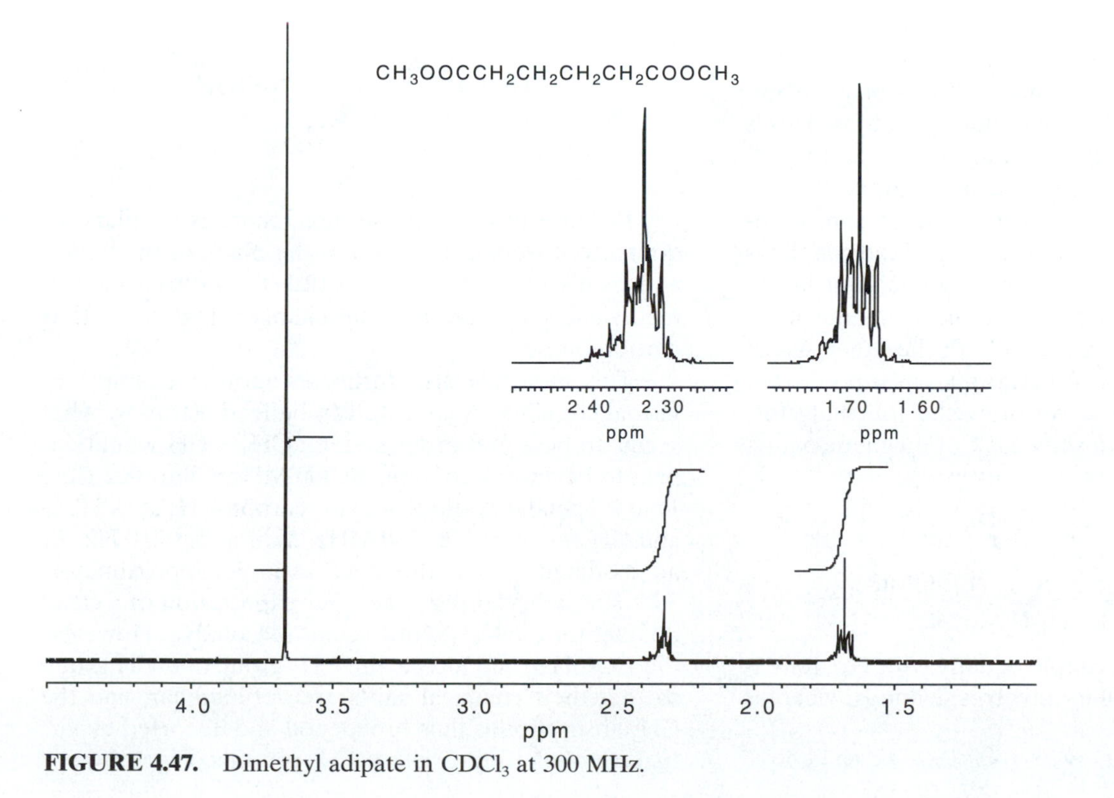

FIGURE 4.47. Dimethyl adipate in $CDCl_3$ at 300 MHz.

lent Δ_2–Δ_2 (inner) methylene groups are strongly coupled. The $\Delta\nu$ value in Hz is zero, and the J value is about 7 Hz. The $\Delta\nu/J$ ratio is 0/7, certainly strongly coupled. And therein lies the problem: We have made the common mistake of attempting to cleave a higher order, very strongly coupled system into two first-order segments.

As mentioned in Section 4.13, we treat open-chain, conformationally mobile compounds as systems in which the protons in each set are magnetic equivalent to each other, at least in practice, because of near averaging of the coupling constants. The same treatment was accorded to the present molecule; the system was thus presented as $X_2A_2A_2X_2$, and we erred in cleaving between two strongly coupled A_2 sets. How then do we treat the system?

One problem lies in the incorrect assumption that "if there is a zero coupling constant between two protons, the spectrum of one of these is not made more complex (or split) by the other."* In other words, we assumed that in dimethyl adipate the 2-CH_2 protons are not split by the 4-CH_2 protons (nor the 5-CH_2 protons by the 3-CH_2 protons) since the coupling constant is formally zero. We have just shown in Section 4.15.1.1 that the assumption of no splitting holds for the C-1 proton and the C-3 protons of 1-nitropropane—a weakly coupled system—but not for the strongly coupled CH_2 protons (a "conglomerate") of 1-hexanol. The term "virtual coupling" was invoked (Section 4.15.1.2) for the interaction between the CH_3 group and the "conglomerate" of CH_2 groups of 1-hexanol.

What we have in the symmetric molecule dimethyl adipate is an extreme case of strong coupling between the sets labeled A_2, for which $\Delta\nu/J$ is zero. That is, the protons of the A_2 sets are chemical-shift equivalent and couple as a "conglomerate" with the X_2 protons. This is another example of virtual coupling.

There is another possible complication. We assumed near magnetic equivalence of the protons of individual CH_2 groups by near averaging of the couplings as is the case for free rotation in a chain. But in the $X_2A_2A_2X_2$ formulation, we also assume averaging of the coupling constants of one A_2 set with the other A_2 set and of one X_2 set with the other. But let us suppose that they do not average, and we draw the following formulation using primes to show lack of magnetic equivalence.

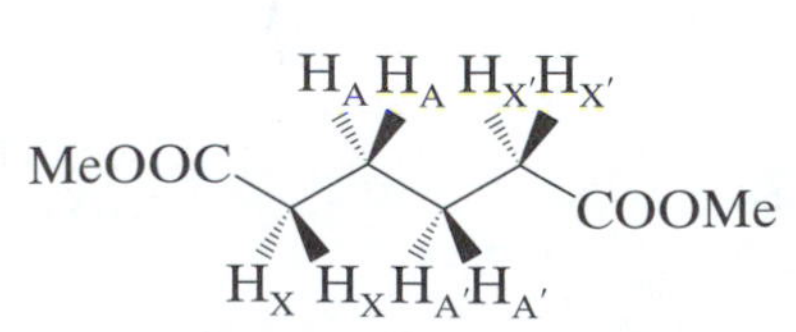

H_A and $H_{A'}$ do not couple equally with H_X, for example; the H_AH_X coupling involves 3 bonds, whereas the $H_{A'}H_X$ coupling involves 4 bonds. These differences are designated by prime marks (Section 4.13) as shown in the above formulation. We write XXAAA′A′X′X′, or more compactly, $X_2A_2A_2'X_2'$. This lack of magnetic equivalence accounts for part of the complexity of the spectrum.

It may be useful at this point to recapitulate briefly the requirements for a first-order system:

- $\Delta\nu/J$ is large for all couplings at a given magnetic field ($\boldsymbol{B}_0$); i.e., the couplings are weak (Section 4.8).
- Protons in a set must be both chemical-shift equivalent and magnetic equivalent (Section 4.13).

If these requirements are met, the sets (at an adequate magnetic field) will be symmetrical. The peak heights and couplings will coincide with first-order theory, and these characteristics will not change with increased magnetic field; nor will additional peaks appear. As the magnetic field is gradually decreased (i.e., $\Delta\nu/J$ is decreased), the first-order characteristics will gradually be lost. (See Fig. 4.29 and Section 4.22.)

4.15.3 Less Symmetrical Chains

4.15.3.1 3-Methylglutaric Acid A somewhat less symmetrical series of open-chain compounds can be obtained by placing a substituent on the center carbon atom of the chain (see Fig. 4.37, structures f, g, h, and i). For example, consider 3-methylglutaric acid:

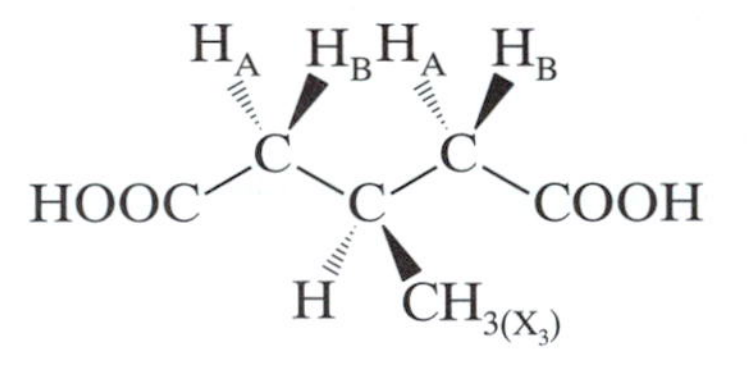

Because of the 3-substituent, there is no plane of symmetry through the chain in the plane of the paper, and, as discussed in Section 4.12.1, the protons in each methylene group are not interchangeable and are thus diastereotopes.

This molecule also furnishes another example of virtual coupling. Again, we can be lead astray by what seems to be a first-order system: CH_3—CH would appear to be a weak coupling at 300 MHz. Chart A.1 (line 1) in Appendix A shows a hydrocarbon CH_3 at $\sim\delta\,0.85$ and CH at $\sim\delta\,1.55$; at 300 MHz, $\Delta\nu$ is about 210 Hz. At an assumed $J = 7$, the $\Delta\nu/J$ ratio is approximately 30—surely first order—and our expectation of a clean doublet for the CH_3 group seems reasonable. However, the COOH group affects the CH_2 group and CH group so that their chemical shifts are very similar, and the CH_3 absorption is thus broadened and distorted by virtual coupling to the distant CH_2 protons (Fig. 4.48).

* Musher, J.I., and Corey, E.J. (1962). *Tetrahedron,* **18,** 791–809.

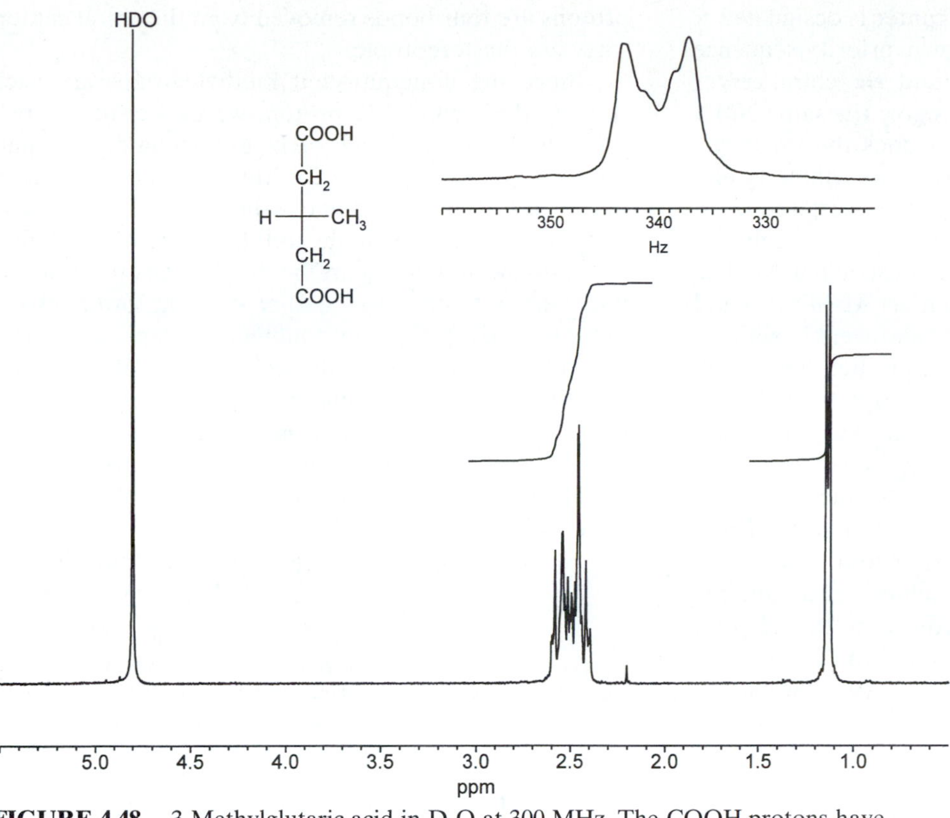

FIGURE 4.48. 3-Methylglutaric acid in D_2O at 300 MHz. The COOH protons have exchanged with D_2O and appear in the HDO peak. The CH_3 peak is broadened and "filled in."

Again the caveat: Look for the pitfall of strongly coupled bonds in the spin system and avoid the temptation to treat this portion of the system in isolation. Strictly, a first-order analysis requires that all sets in the spin system be weakly coupled.

The modus operandi of the organic chemist in interpreting an NMR spectrum is to look initially for first-order systems, or for systems that can be recognized as slight distortions of first-order absorptions. Frequently the emphasis is on portions of *weakly coupled* spin systems, but one must watch for higher-order and virtual-coupling effects as well as for spin–coupling and chemical shift nonequivalence. The above considerations should be of help in setting limits for this convenient, although not rigorous, approach to the interpretation of NMR spectra.

4.16 Chirality

The organic chemist—in particular, the natural products chemist—must always be conscious of *chirality* when interpreting NMR spectra. The topic was mentioned in Section 4.12. A formal definition and a brief explanation will suffice here:

> Chirality expresses the necessary and sufficient condition for the existence of enantiomers.*

Impeccably rigorous but possibly a bit cryptic. The following comments may help. Enantiomers are nonsuperposable mirror images. The ultimate test for a chiral molecule is thus nonsuperposability of its mirror image. If the mirror image is superposable, the molecule is achiral. The most common feature in chiral molecules is a chiral center also called a stereogenic center. A chiral molecule possesses no element of symmetry other than possibly a simple axis or axes. For examples, see Figure 4.37, structure e, and the solved Problem 8.3. For reassurance, consider the human hand, which has no symmetry element. The left and right hands are nonsuperposable mirror images (i.e., enantiomers). The term "chirality" translates from Greek as "handedness."

4.16.1 One Chiral Center

The familiar carbon chiral center has four different substituents as shown in 3-hydroxybutanoic acid (com-

* Cahn, R.S., Ingold, C., and Prelog, V. (1966). Angew: *Chem., Int. Ed. Engl.*, **5**, 385.

pound e in Fig. 4.37). This chiral center is designated *R* in accordance with the well-known priority-sequence rules; in the enantiomeric compound, the chiral center is designated *S*. Both enantiomers give the same NMR spectrum in an achiral solvent, as does the racemate. Because of the chiral center, there is no symmetry element, and the methylene protons are diastereotopes, designated A and B. The C-3 proton is strongly deshielded by the OH group and is designated M. The methyl group is X_3. The spin system, is $ABMX_3$. A and B couple strongly with each other and weakly with M. M couples weakly with X_3. We can predict that X_3 will be a doublet, M will be a complex multiplet as will both A and B. The OH and COOH protons will usually appear as one peak because of rapid interchange.

Chemical shift nonequivalence of the methyl groups of an isopropyl moiety near a chiral center is frequently observed; the effect has been measured through as many as seven bonds between the chiral center and the methyl protons. The methyl groups in the terpene alcohol 2-methyl-6-methylen-7-octen-4-ol (ipsenol) are not chemical shift equivalent (Fig. 4.49) even though the protons are four bonds removed from the chiral center. They are diastereotopic.

Since the nonequivalent methyl groups are each split by the vicinal CH proton, we expect to see two separate doublets. At 300 MHz, unfortunately, the pattern appears to be a classical triplet, usually an indication of a CH_3—CH_2 moiety—impossible to reconcile with the structural formula and the integration. Higher resolution would pull apart the middle peak to show two doublets. Actually, in an earlier study at lower resolution (100 MHz), the two doublets overlapped to show four peaks. To remove the coincidence of the inner peaks that caused the apparent triplet, we used the very useful technique of "titration" with deuterated benzene,* which gave convincing evidence of two doublets at 20% C_6D_6/80% $CDCl_3$ and optimal results at about a 50:50 mixture (Fig. 4.49).

Note also that the chiral center accounts for the

* Sanders, J.K.M., and Hunter, B.K. (1993). *Modern NMR Spectroscopy,* 2nd ed. Oxford: Oxford University Press, p. 289.

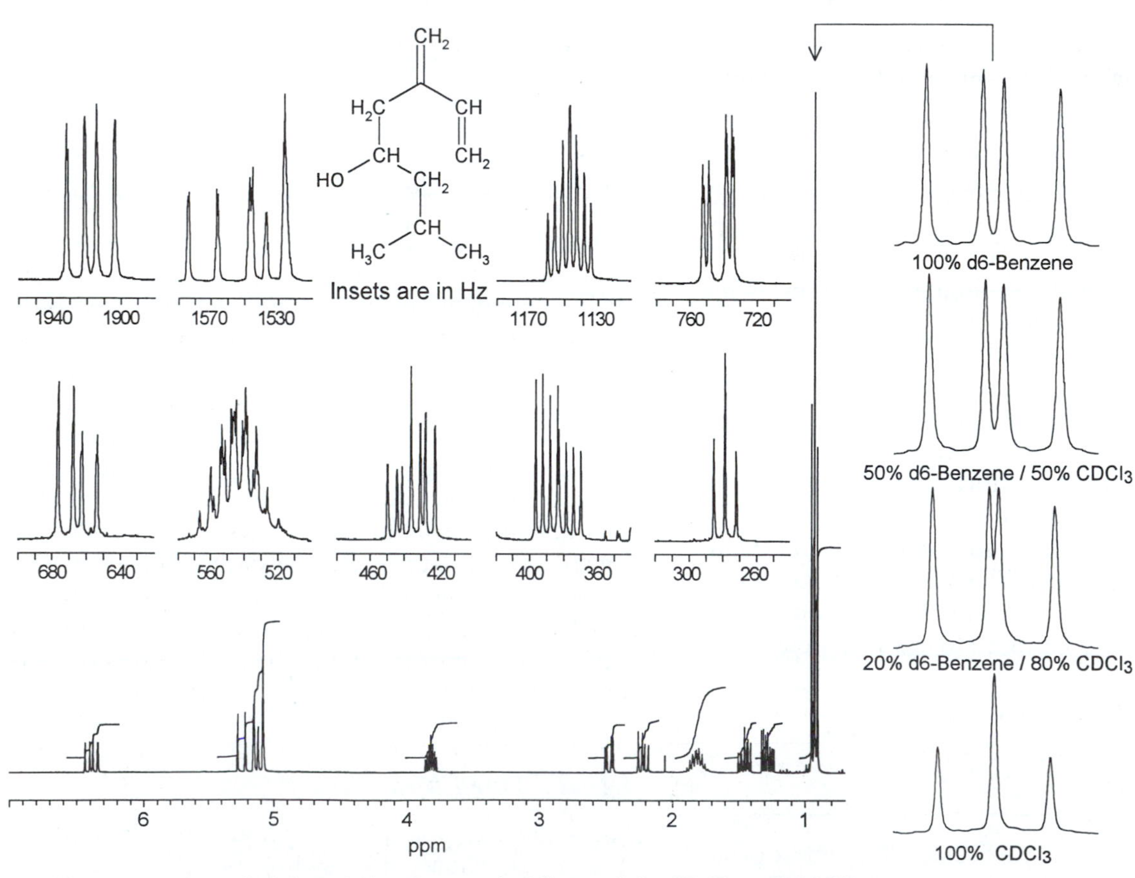

FIGURE 4.49. 2-Methyl-6-methylen-7-octen-4-ol (ipsenol) in $CDCl_3$ at 300 MHz. "Titration" with C_6D_6. The sample was a gift from Phero Tech, Inc., Vancouver, BC, Canada.

fact that the protons of each of the two aliphatic CH_2 are diastereotopic at about δ 1.28, δ 1.42, δ 2.21, and δ 2.48.

Most of the protons can be assigned on the basis of chemical shifts, integration ratios, and coupling patterns as follows.

The entry points for analysis of the spectrum are protons that have distinctive chemical shifts and/or couplings, such as the methyl groups just discussed. The one-proton multiplet at δ 3.82 must be the deshielded *H*COH proton that is coupled to two sets of diastereotopic protons. If all couplings were equal, the multiplicity would be 5; obviously they are not equal.

The conjugated olefinic protons are distinctively at the deshielded end of the spectrum. The isolated $=CH_2$ group is at δ 5.08. The $CH=CH_2$ moiety accounts for the remainder of the patterns with the CH= proton centered at about δ 6.40 and the $=CH_2$ protons centered at about δ 5.25 and δ 5.14. This is a slightly distorted AMX system in which the AM coupling is barely detectable at the expansion shown (see Section 4.14).

At the shielded end of the spectrum, from right to left, we see the identified diastereotopic methyl groups, each of the H-3 diastereotopic protons, a mysterious two-proton multiplet at δ 1.8, and the individual H-5 protons, each of which consists of a doublet of doublets with different coupling constants. The mysterious absorption at δ 1.8 consists of the highly coupled H-2 proton superimposed on the broad OH absorption.

The CH, CH_2, and CH_3 peaks thus identified are confirmed by the ^{13}C/DEPT spectrum (Section 5.5). Finally, the ^{1}H and ^{13}C peaks can be correlated (Chapter 6).

Tetrahedral atoms, in addition to carbon, with four different substituents are also chiral centers. For example, see the phosphorus-containing structure in Section 7.5. Replacement of one of the $-OCH_2CH_3$ substituents by an $-OCH_3$ substituent would produce a chiral center.

4.16.2 Two Chiral Centers

1,3-Dibromo-1,3-diphenylethane has a methylene group between two chiral centers (Fig. 4.50). In the 1R, 3R compound (one of a racemic pair), H_a and H_b are equivalent and so are H_c and H_d, because of a C_2 axis. In the 1S, 3R compound (a meso compound), attempted C_2 rotation gives a distinguishable structure. But H_a and H_b are enantiotopes by interchange through the plane of symmetry shown perpendicular to the plane of the page. H_c and H_d, however, cannot be interchanged since they are *in* the plane of symmetry; they are diastereotopes.*

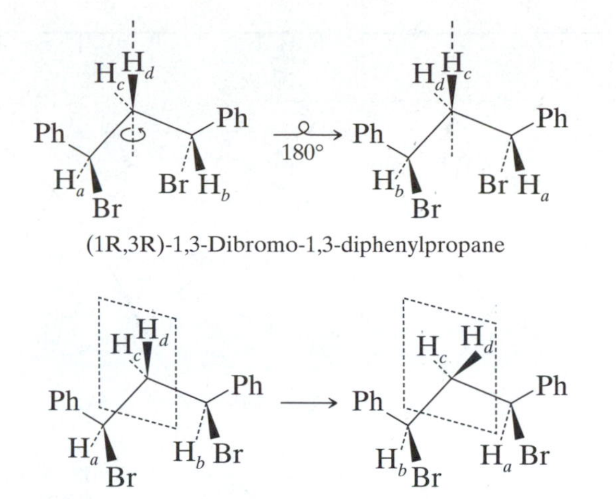

FIGURE 4.50. Two isomers of 1,3-dibromo-1,3-diphenylpropane. In the (1R, 3R) isomer, H_a and H_b are chemical-shift equivalent, as are H_c and H_d. In the (1S, 3R) isomer, H_a and H_b are chemical-shift equivalent, but H_c and H_d are not.

In the (1R, 3R) compound, H_a and H_b are not magnetic equivalent since they do not identically couple to H_c or to H_d; H_c and H_d also are not magnetic equivalent since they do not identically couple to H_a or H_b. But since the J values approximately average out by free rotation, the spin system is treated as A_2X_2 and the spectrum would show two triplets. In the (1S, 3R) compound $J_{ad} = J_{bd}$ and $J_{ac} = J_{bc}$; thus in this molecule, H_a and H_b are magnetic equivalent. The question of magnetic equivalence of H_c and H_d is not relevant since they are not chemical-shift equivalent. The spin system is ABX_2.

4.17 Vicinal and Geminal Coupling

Coupling between protons on vicinal carbon atoms depends primarily on the dihedral angle ϕ between the H—C—C′ and the C—C′—H′ planes. This angle can be visualized by an end-on view of the bond between the vicinal carbon

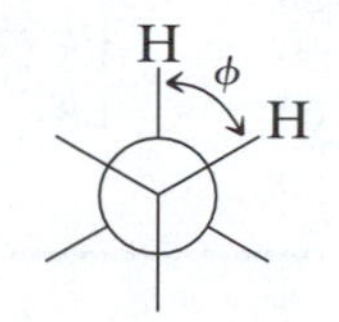

atoms and by the perspective in Figure 4.51 in which the

* For another example containing a CH_2 group flanked by two chiral (stereogenic) centers, see Jennings, W.B. (1975). *Chem. Rev.* **75,** 307.

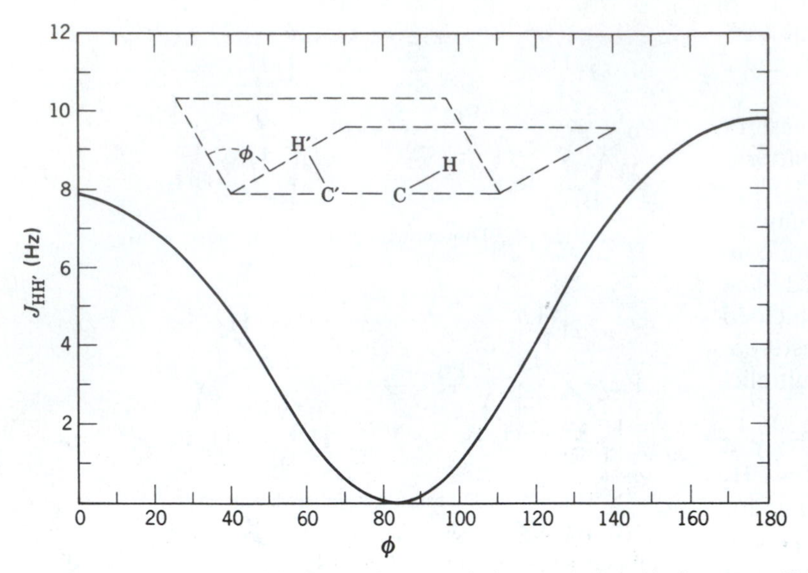

FIGURE 4.51. The vicinal Karplus correlation. Relationship between dihedral angle (ϕ) and coupling constant for vicinal protons.

relationship between dihedral angle and vicinal coupling constant is graphed. Karplus* emphasized that his calculations are approximations and do not take into account such factors as electronegative substituents, the bond angles Θ ($\angle$H—C—C′ and the $\angle$C—C′—H′), and bond lengths. Deductions of dihedral angles from measured coupling constants are safely made only by comparison with closely related compounds. The correlation has been very useful in cyclopentanes, cyclohexanes, carbohydrates, and polycyclic systems. In cyclopentanes, the observed values of about 8 Hz for vicinal cis protons and about 0 Hz for vicinal trans protons are in accord with the corresponding angles of about 0° and about 90°, respectively. In substituted cyclohexane or pyranose rings, the chair is the preferred conformation; the following relations hold and dihedral angles of substituents follow from these ^3J proton couplings.

	Dihedral Angle	Calculated J(Hz)	Observed J(Hz)
Axial–axial	180°	9	8–14 (usually 8–10)
Axial–equatorial	60°	1.8	1–7 (usually 2–3)
Equatorial–equatorial	60°	1.8	1–7 (usually 2–3)

Note the near-zero coupling at the 90° dihedral angle. This has been a source of frustration in attempts at fitting proposed structures to the NMR spectra.

* Karplus, M. (1959). *J. Chem. Phys.*, **30**, 11.

A modified Karplus correlation can be applied to vicinal coupling in alkenes. The prediction of a larger trans coupling (ϕ = 180°) than cis coupling (ϕ = 0°) is borne out. The cis coupling in unsaturated rings decreases with decreasing ring size (increasing bond angle) as follows: cylohexenes ^3J = 8.8–10.5, cyclopentenes ^3J = 5.1–7.0, cyclobutenes ^3J = 2.5–4.0, and cyclopropenes ^3J = 0.5–2.0. Two-bond geminal CH_2 coupling depends on the H—C—H bond angle Θ as shown in Figure 4.52. This relationship is quite susceptible to other influences and should be used with due caution. However, it is useful for characterizing methylene groups in a fused cyclohexane ring (approximately tetrahedral, $^2J \sim$ 12–18), methylene groups of a cyclopropane ring ($^2J \sim$ 5), or a terminal methylene group, i.e., $=CH_2$, ($^2J \sim$ 0–3). Electronegative substituents reduce the geminal coupling constant, whereas sp^2 or sp hybridized carbon atoms increase it.

Geminal coupling constants are usually negative numbers, but this can be ignored except for calculations.* Note that geminal couplings are seen in routine spectra only when the methylene protons are diastereotopic.

In view of the many factors other than angle dependence that influence coupling constants, it is not surprising that there have been abuses of the Karplus correlation. Direct "reading off" of the angle from the magnitude of the ^2J value is risky. The limitations of the Karplus correlations are discussed in Jackman and Sternhell (1969).

* Coupling constants are positive if antiparallel spin states have lower energy than parallel states; the opposite is true for negative coupling constants (see Section 4.1).

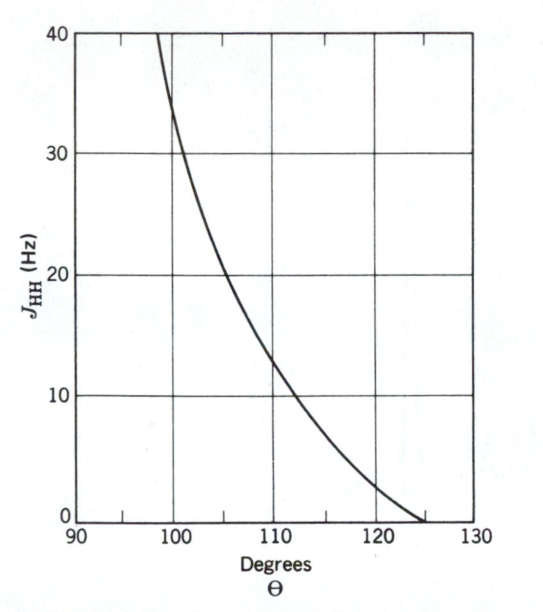

FIGURE 4.52. The geminal Karplus correlation. J_{HH} for CH_2 groups as function of ∠H—C—H.

4.18 *Long-Range Coupling*

Proton–proton coupling beyond three bonds (^3J) is usually less than 1 Hz, but appreciable long-range coupling ($>^3J$) may occur in alkenes, alkynes, aromatics, and heteroaromatics, and in strained ring systems (small or bridged rings). Allylic (H—C—C=C—H) couplings may be as much as 1.6 Hz. Coupling through conjugated polyalkyne chains may occur through as many as nine bonds. Meta coupling in a benzene ring is 1–3 Hz, and para, 0–1 Hz. In five-membered heteroaromatic rings, coupling between the 2 and 4 protons is 0–2 Hz.

$^4J_{AB}$ in the bicyclo[2.1.1]hexane system is about 7 Hz.

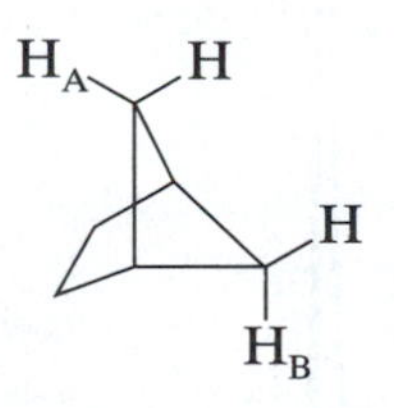

This unusually high long-range coupling constant is attributed to the W conformation" of the four σ bonds between H_A and H_B.

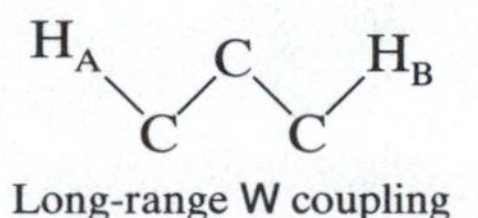

Long-range W coupling

As resolution increases, small couplings beyond three bonds often become noticeable as line broadening of the benzylic CH_2 and CH_3 peaks in Figure 4.30 and Problem 8.2, respectively.

4.19 *Spin Decoupling*

Irradiation of one proton in a spin-coupled system removes its coupling effect on the neighboring protons to which it had been coupled. Thus successive irradiation of the protons of 1-propanol for example, yields the following results:

$$\underset{\text{Triplet}}{CH_3}-\underset{\text{Sextet}}{\overset{\downarrow\text{(irradiate)}}{CH_2}}-\underset{\text{Triplet}}{CH_2OH} \longrightarrow \underset{\text{Triplet}}{CH_3}-CH_2-\underset{\text{Quartet}}{CH_2OH}$$

$$CH_3-\overset{\downarrow\text{(irradiate)}}{CH_2}-CH_2OH \longrightarrow \underset{\text{Singlet}}{CH_3}-CH_2-\underset{\text{Singlet}}{CH_2OH}$$

$$\overset{\downarrow\text{(irradiate)}}{CH_3}-CH_2-CH_2OH \longrightarrow CH_3-\underset{\text{Triplet}}{CH_2}-\underset{\text{Triplet}}{CH_2OH}$$

Thus, we have a powerful tool for determining the connectivity of protons through bonds and assigning proton peaks. Furthermore, overlapping peaks can be simplified by removing one of the couplings.

An extra probe is added to irradiate a particular proton (or set of protons) with a strong continuous-wave frequency (ν_2) at its resonance frequency while observing the other protons with the conventional ν_1 pulse. The intense ν_2 irradiation causes saturation and rapid exchange of the two energy levels of the irradiated proton, thus removing through-bond coupling. Figure 4.53 illustrates a more realistic example.

Protons can be readily decoupled provided they are more than about 100 Hz apart. The utility of proton–proton decoupling is shown in the 100-MHz partial spectrum of methyl 2,3,4-tri-*O*-benzoyl-β-*L*-lyxopyranoside (Fig. 4.53*a*). The integration (not shown) gives the following ratios in Figure 4.53*a*. from low to high δ values 3:1:1:1:1:2. The sharp peak at δ 3.53 represents the OCH_3 group. Decoupling the two-proton multiplet at δ 5.75 causes the multiplet at δ 5.45 to collapse to four peaks, and the doublet at δ 5.00 to a sharp singlet (Fig. 4.53*b*). Decoupling the multiplet at δ 5.45 partially collapses the multiplet at δ 5.75 and collapses the two pairs of doublets (at δ 4.45 and δ 3.77) to two doublets (Fig. 4.53*c*). The H_5 absorptions should be at lower δ values since these two protons are deshielded by an ether oxygen, whereas H_2, H_3, and H_4 are deshielded by benzoyloxy groups. The two H_5 protons are the AM portion of an AMX pattern; the H_4 proton is the X portion (with additional splitting). The pair of doublets at δ 4.45 represents one (H_{5E}) proton at C_5 strongly deshielded by the benzoyloxy group on C_4, and the pair of doublets at δ 3.77 represents the other (H_{5A}) absorption.

Further confirmation is provided by the collapse of

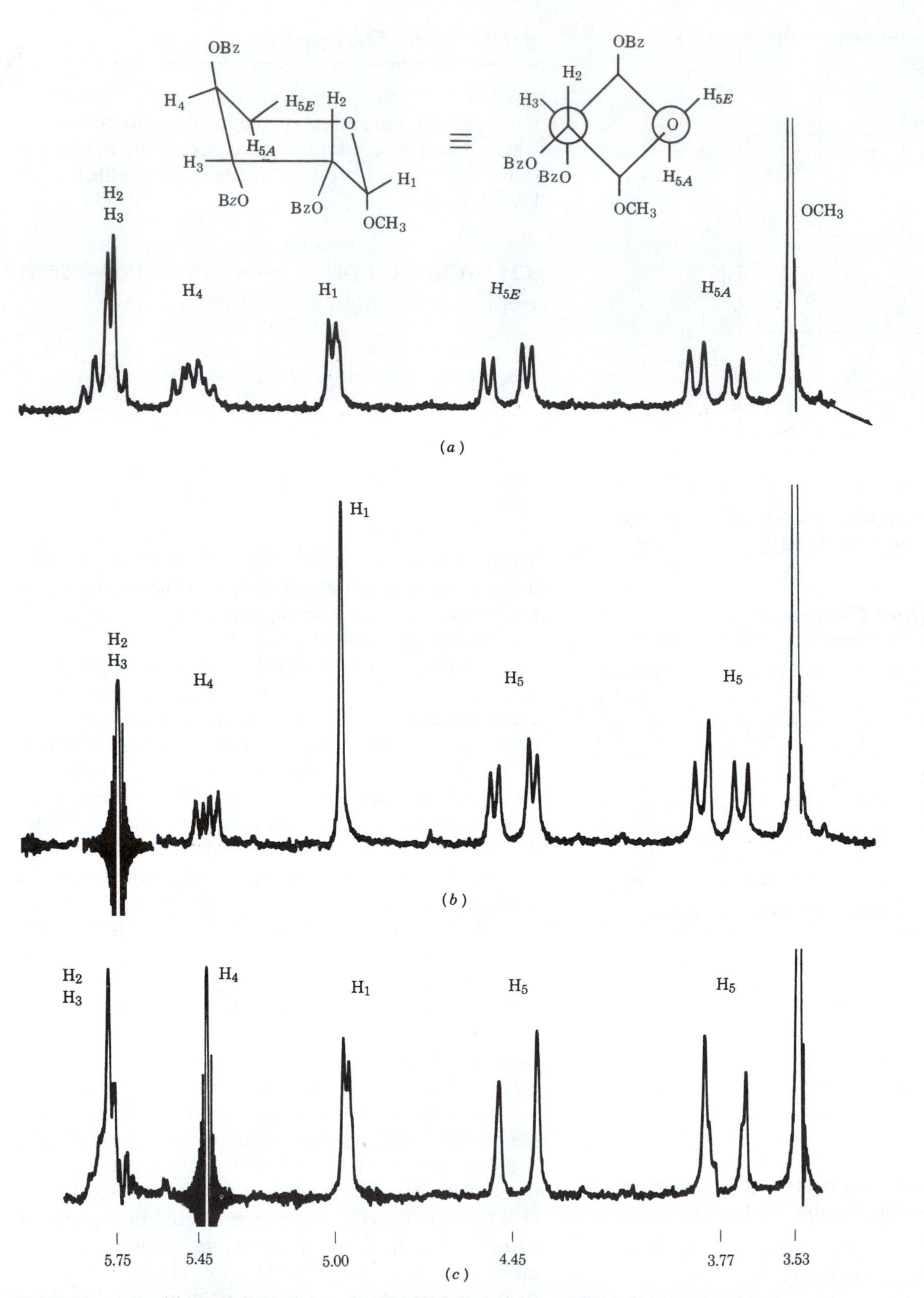

FIGURE 4.53. (*a*) Partial spectrum of methyl 2,3,4-tri-*O*-benzoyl-*β*-L-lyxopyranoside at 100 MHz in $CDCl_3$. (*b*) H_2 and H_3 decoupled. (*c*) H_4 decoupled. Note that there are two diastereotopic H_5 protons. Bz = C_6H_5CO. Irradiation may cause a detectable change in chemical shift of nearby peaks.

each pair of doublets of doublets to simple doublets with the characteristic large geminal coupling (J = 12.5 Hz) on irradiation of the multiplet at δ 5.45, which must therefore be the H_4 absorption (i.e., the X proton that was further split). The multiplet at δ 4.75 must therefore represent H_2 and H_3 since irradiation of this multiplet collapsed the multiplet that we identified as H_4 (this now appears as the X portion of the AMX pattern); and also collapsed the doublet at δ 5.00, which must be the H_1 absorption.

A multiplet caused by coupling to two nonequivalent protons can be completely collapsed to a singlet by irradiating both coupling proton frequencies simultaneously. Other methods for establishing $^1H^1H$, $^1H^{13}C$, or $^{13}C^{13}C$ relationships are described in Chapter 6.

4.20 Nuclear Overhauser Effect Difference Spectrometry, $^1H^1H$ Proximity Through Space

As described under *spin decoupling* (Section 4.19), it is possible to irradiate strongly a particular nucleus at its resonance frequency (ν_2) and detect loss of coupling (decoupling) in all the nuclei that were coupled to the irradiated nucleus. Detection of the decoupled nuclei is carried out with the usual pulsed detector probe (ν_1); the added decoupler probe delivers continuous-wave (CW) irradiation at the resonance frequency (ν_2) of the irradiated nucleus. Section 4.19 describes $^1H—^1H$ spin decoupling through the bonds. Broad-band $^1H—^{13}C$ spin decoupling is briefly mentioned in Section 5.1 (Chapter 5) as is the nuclear Overhauser effect (NOE), which causes enhancement of the ^{13}C signal as a beneficent by-product of the intense decoupling irradiation. The nuclear Overhauser effect also occurs in $^1H—^1H$ spin decoupling, but the effect is small and rarely noticed.

Both spin decoupling and the nuclear Overhauser effect are examples of double resonance involving use of an additional probe to irradiate one nucleus at ν_2 and detect the effect on a different nucleus with a ν_1 pulse. At this point, we are interested in determining $^1H^1H$ proximity within a molecule by means of the through-space nuclear Overhauser effect, which can be brought about by using a much weaker irradiation by ν_2 than that used for decoupling. A proton that is close in space to the irradiated proton is affected by the NOE *whether or not it is coupled to the irradiated proton;* if it is coupled, it remains at least partially coupled because the irradiation is weak in comparison with that used for a decoupling experiment. Polarization—that is, a change in population of the energy levels—by the weak ν_2 irradiation results, through space, in an increase in the population of the higher energy level in the nearby nonirradiated proton. This excess population undergoes T_1 relaxation to a lower energy level, thereby increasing the signal intensity of the nearby proton(s). In very large molecules, other factors intervene, and the results may be a decrease in signal intensity.

The through-space distances involved in $^1H^1H$ proximity determinations are quite small, and the effect decreases as the inverse of the sixth power of the distance, through space, between the protons. The usual observable enhancement is less than 20%. To increase the sensitivity, we use the NOE *difference* experiment, in which a conventional ^1H spectrum* is subtracted from a specific proton-irradiated spectrum; this subtraction leaves *only the enhanced absorptions.* Under these conditions, a measurable effect can be expected between ^1H nuclei over a distance of up to about 4 Å (0.4 nm)†; for example, the distance between 1,3-diaxial protons in the cyclohexane chair form is ~2.6 Å. This procedure is a powerful tool in distinguishing among isomers, all of which should be available, if feasible, for NOE difference spectra (Fig. 4.54).

A number of precautions must be taken in carrying out these $^1H^1H$ proximity experiments [see Derome (1987) or Sanders and Hunter (1993) for example]. Thus, the irradiating transmitter is "gated" off (turned off) during signal acquisition to avoid possible interference from decoupling effects. This precaution is effective since the NOE builds up slowly during irradiation and persists during acquisition, whereas decoupling effects are instantaneous. Furthermore, since relaxation rates are very sensitive to the presence of paramagnetic substances, dissolved oxygen (a diradical) should be removed by several cycles of freeze, pump, thaw, or by bubbling nitrogen or argon through the solution. The tube should then be sealed.

NOE difference spectrometry determined the substitution pattern of a natural product, whose structure was either I or II. The readily available dimethyl homologue III was submitted to NOE difference spectrometry (Fig. 4.54).

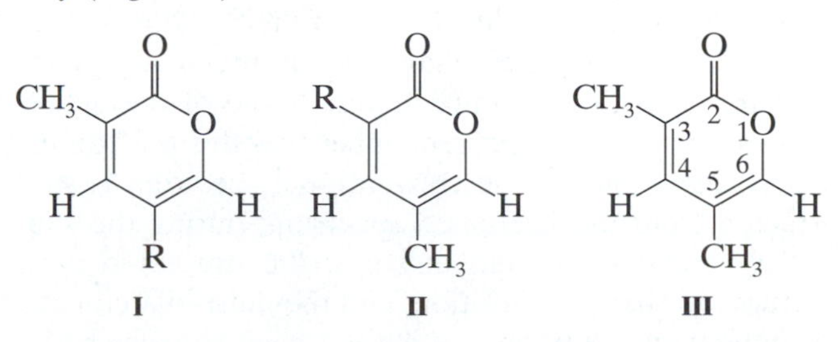

Irradiation of the 5-methyl group resulted in enhancement of both H-4 and H-6, whereas irradiation of the 3-methyl group enhanced only H-4; the assignments of

* Irradiation is applied to a blank region of the spectrum.
† In the familiar Dreiding models, 1 cm = 0.4 Å (0.04 nm).

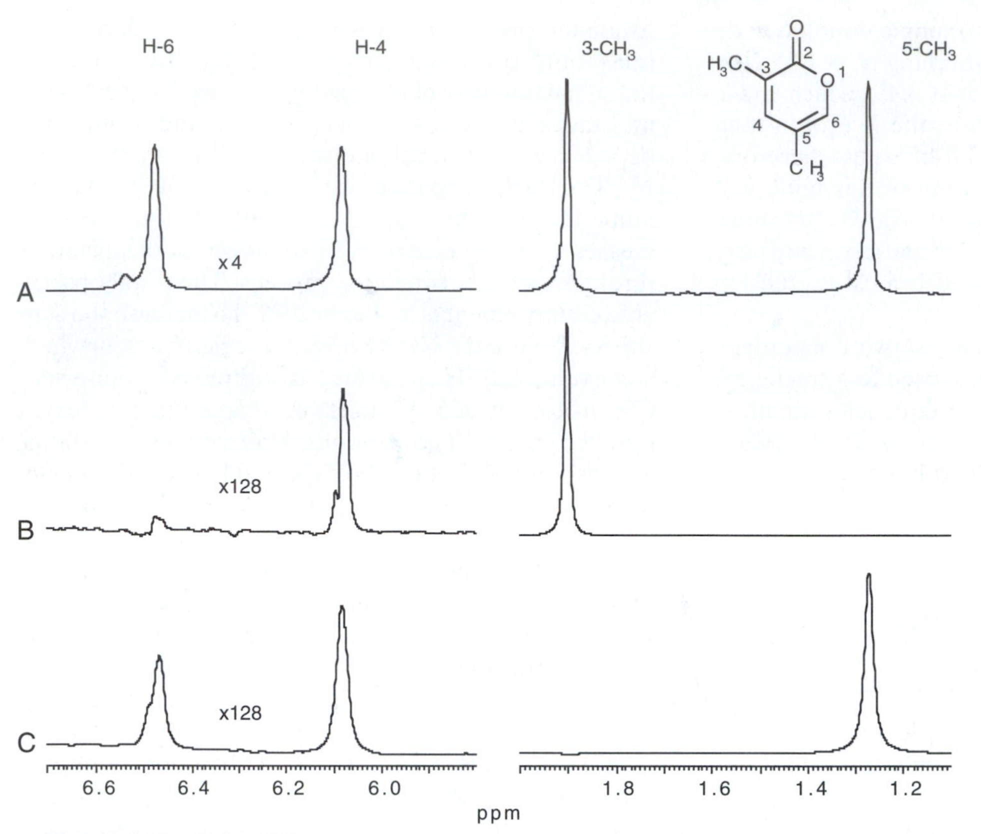

FIGURE 4.54. NOE difference spectrometry for compound **III.** A shows the ^{1}H NMR spectrum. B shows enhancement of the H-4 proton on irradiation of the 3-methyl group. C shows enhancement of both protons on irradiation of the 5-methyl group. 300 MHz in $CDCl_3$.

these entities to the absorption peaks is now clear. Since the chemical shift of the methyl absorption peak of the natural product was almost identical with that of the 3-methyl group of structure III, the natural product is structure **I**.

In Figure 4.54, the conventional proton spectrum is labeled A. Spectrum B shows the result of irradiation of the 3-methyl group: only H-4 is enhanced. Spectrum C shows the result of irradiation of the 5-methyl group: both H-4 and H-6 are enhanced—all predictable since enhancement is proximity driven. Some of the details may seem puzzling, but remember that this is a "difference" experiment: The conventional spectrum is subtracted from the irradiated spectrum; during the subtraction and enhancement steps, the irradiated peak returns to the "up" position; and the unirradiated peak is not detectable because it has not been enhanced. The sensitivity numbers—x4, x128, and x128—refer to the increased instrumental sensitivity of the left side of each split spectrum relative to the right side, which remains X1. These sensitivity ratios (left side : right side) are 4 : 1 for spectrum A, and 128 : 1 for spectra B and C. The purpose is merely to keep all peaks at convenient heights.

Irradiation of an OH group results in nuclear Overhauser enhancement of nearby protons. However, the usual rapid exchange rate of hydroxylic protons must be retarded either by using dry deuterated acetone or deuterated dimethyl sulfoxide as a solvent, or by simply cooling a $CDCl_3$ solution. Either process also moves the OH peak to the left; this movement can be controlled to prevent overlap of absorptions, and, of course, it serves to identify the OH peak (Section 4.9.1.1).

Distinguishing between a trisubstituted (E) and a (Z) double bond is not a trivial assignment. The use of NOE difference spectrometry for this purpose is illustrated in Problem 8.4.

This examination of Problem 9.50 is offered here to demonstrate the difference between an enhanced absorption signal and a nonenhanced dispersion signal. This will all make sense on solution of the stereoisomerism of Problem 9.50.

In Problem 9.50, the results of irradiating each of the three singlet methyl peaks are shown. Irradiation of the peak at the right-hand edge of the spectrum is indicated by the intense negative peak at the same position. None of the other signals shows enhancement; they all show dispersion signals—i.e., positive and negative peaks.

Although the remaining two methyl singlets are closely spaced, each peak was irradiated individually. In each case, an intense negative peak resulted, with a small contribution of a much narrower (less intense) negative peak from the other methyl singlet. Irradiation of the more shielded absorption gave the following enhancement results (from left to right): no, yes, no, no, yes, no, no. The enhancement results from the less shielded irradiated peak were: no, yes, yes, yes, yes, no, no.

4.21 NMR Shift Reagents

Addition of a "shift reagent" to a compound

$$\overset{G}{CH_3}-\overset{F}{CH_2}-\overset{E}{CH_2}-\overset{D}{CH_2}-\overset{C}{CH_2}-\overset{B}{CH_2}-\overset{A}{CH_2OH}$$

containing an appropriate functional group spreads out the NMR absorption pattern, as shown in Figure 4.55, with obvious advantages. Most of the commercially available shift reagents are paramagnetic chelates of europium.

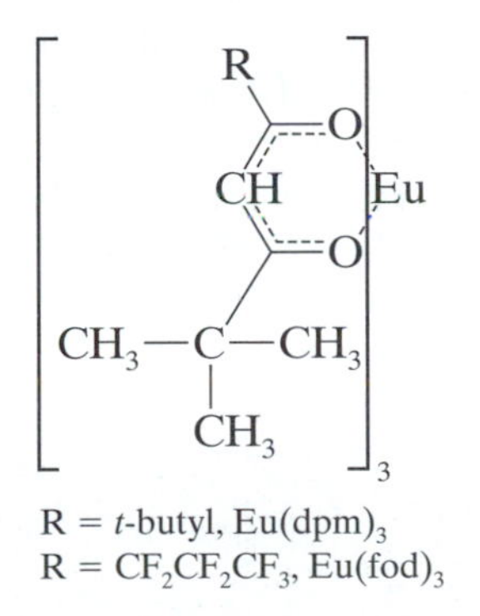

R = *t*-butyl, $Eu(dpm)_3$
R = $CF_2CF_2CF_3$, $Eu(fod)_3$

The chelate forms a complex with the functional group of the substrates.*

As mentioned in Section 4.16, a chiral shift reagent may be used to resolve mixtures of enantiomers.

During the 1970s and 1980s, shift reagents were used quite extensively at the commonly available frequencies of 60 to 100 MHz, at which peak broadening associated with shift was not a problem. However, at higher frequencies, peak broadening may pose limitations.† There are other drawbacks: The hygroscopicity of the reagents and the susceptibility of the complexes to traces of moisture demand dry-box techniques, especially with small samples; several experiments at various reagent–substrate ratios are usually necessary.

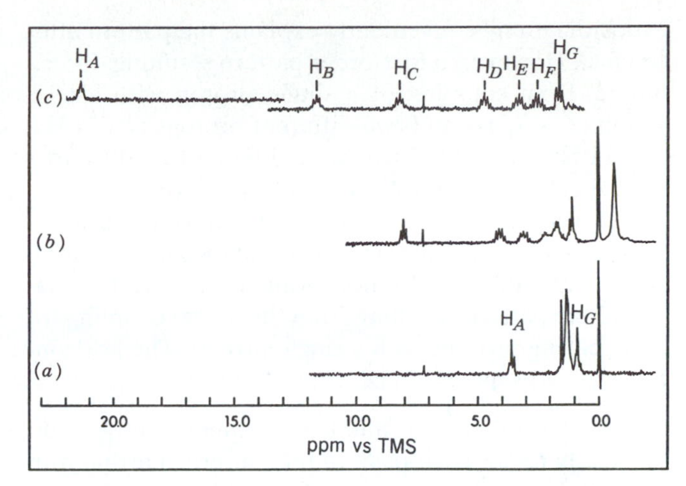

FIGURE 4.55. The 60-MHz proton NMR spectra of 0.40 mL of $CDCl_3$ solution containing 0.300 *M* 1-heptanol at various mole ratios [moles of $Eu(dpm)_3$ per mole of 1-heptanol]: (*a*) 0.00, (*b*) 0.19, (*c*) 0.78. Temperature, 30°C. From Rabenstein, D.L. *Anal. Chem.*, Vol. 43, p. 1599, Copyright © 1971 by the American Chemical Society. Reprinted by permission of the copyright owner.

Thus, in recent years, the approach has been to utilize the much higher frequency instruments now available in preference to shift reagents.

Use of a chiral shift reagent requires a functional group that is a strong enough base to form a stable complex, sufficient resolution to separate the corresponding peaks in each diastereomeric complex, and freedom from interfering peaks. The ideal protons are those of an uncoupled CH_3 group near the functional group; a similarly situated, uncoupled CH group, although less intense, is also useful. Protons of a methylene group can be troublesome; they may be enantiotopic as in 1-nitropropane, for example, and thus chemical shift equivalent in an achiral environment but not so in the presence of a chiral shift reagent. Protons of a methylene group may also be diastereotopic and thus not chemical shift equivalent in any environment.

For mixtures of enantiomers, methods other than the use of chiral shift reagents, such as gas chromatography or high-performance liquid chromatography with chiral columns, are now used. Alternatively, the substrate is derivatized with a pure enantiomer of a chiral reagent, and the resulting derivatives are separated by standard gas or liquid chromatography.

Information on NMR shift reagents is available in Morrill (ed.) (1986).

4.22 Addendum—Analysis of First-Order Patterns

It has recently been pointed out that students are not taught how to analyze a first-order pattern. The usual

* The chelate is a Lewis acid, capable also of catalyzing reactions.

† Line broadening has been reported to be 25 times more severe on a 500 MHz instrument than on a 100 MHz instrument. Parker, D. (1991). *Chem. Rev.*, **91**, 1441.

"stick diagram" conveniently explains the pattern after the fact. However, a first-order pattern resulting, for example, from coupling of a single proton with a CH_3 group (J = 7 Hz) and two different protons (J = 4 Hz, J = 12 Hz) (Fig. 4.56) would be difficult to reduce to a stick diagram without prior coupling information. B.E. Mann has proposed a routine procedure for constructing such a stick diagram.* The following is a brief, slightly modified description of his procedure for analyzing the pattern resulting from the above couplings of neighboring protons with a single proton. The horizontal scale is 10 mm = 5 Hz.

1. Assign an integer intensity number centered directly under each peak in the pattern, starting with intensity 1 for the outermost peaks. A first-order pattern must have a centrosymmetric distribution of intensities (level a). If it does not, the pattern is higher order.
2. The sum of the assigned peak intensities in level a must be 2^n, n being the number of coupling protons. In this example, $n = 5$, which corresponds to coupling of three equivalent neighboring protons and two nonequivalent protons as specified above.
3. Draw a vertical line (stick) centered directly under each peak of the pattern, i.e., beneath each intensity number. All lines are drawn to equal height (level b).

* Mann, B.E. (1995). *J. Chem. Educ.*, **72,** 614.

4. The intensity numbers of the two outer lines at either end determine the multiplicity at level b. Thus, in this example, an intensity of 1 : 1 ordains a series of doublets. An intensity of 1 : 2 would ordain a series of 1 : 2 : 1 triplets. An intensity of 1 : 3 would ordain a series of 1 : 3 : 3 : 1 quartets, and so forth through the Pascal triangle (Fig. 4.29). Note that some of the doublets have intensities of 3 : 3, but the ratio is still 1 : 1.
5. The distance on the Hz scale between the first two lines in level b gives the coupling constant for the doublets in level b. All doublets, of course, have the same coupling constant as that of the end pairs. Although there is overlapping of the doublets, there is no difficulty in sorting out the doublets on the basis of equal coupling and the 1 : 1 ratio of paired lines. The Hz scale shows that these are the 4-Hz doublets.
6. At level c, the "stem" of each pair of lines at level b is assigned an intensity number, which is the intensity number of the first line of each pair at level b. These "stems" now become the lines of the multiplets at level c—in this example, two overlapping quartets (J = 7 Hz) in accordance with the 1 : 3 ratio of the first two lines. In turn, the stems of these quartets are assigned the intensity number of the first line of each quartet. The stems now become the lines of the multiplet of level d—in this example a doublet with J = 12 Hz. Note that at each level

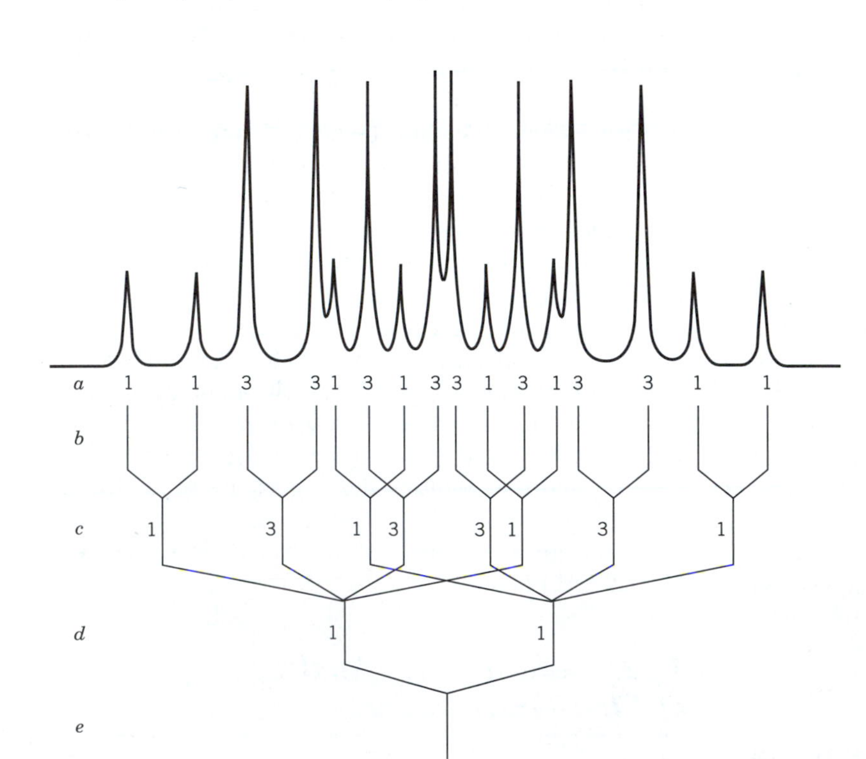

FIGURE 4.56. The coupling constants for this first-order multiplet are doublets of 4 and 12 Hz, and a quartet of 7 Hz. The horizontal scale is 10 mm = 5 Hz.

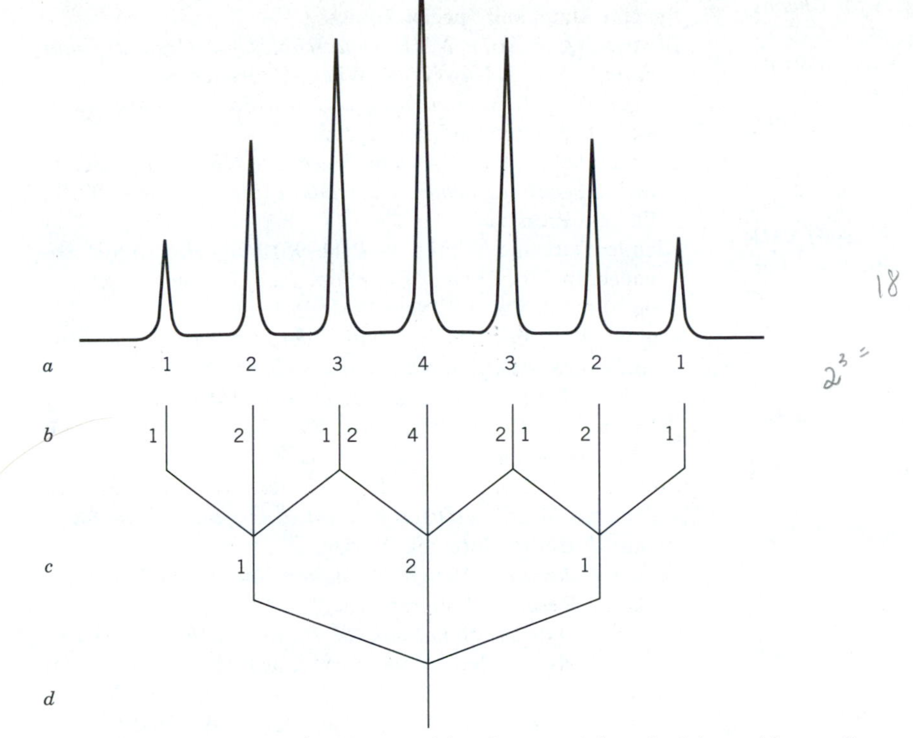

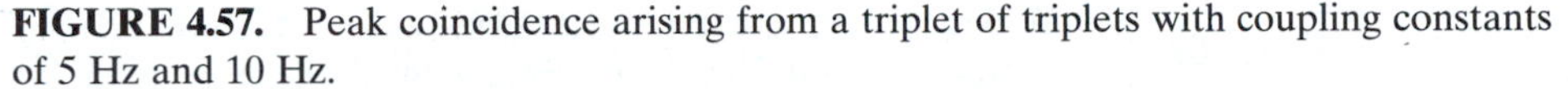

FIGURE 4.57. Peak coincidence arising from a triplet of triplets with coupling constants of 5 Hz and 10 Hz.

the lines should be checked for centrosymmetry of intensity and coupling.

7. In summary, we confirm, starting at level d, that there is coupling to a CH proton to give a doublet (J = 12 Hz). Each line of this doublet is split into a quartet (J = 7 Hz) by vicinal CH_3 protons at level c. Each line of the quartets is split by another CH proton to give a series of doublets (J = 4 Hz) at level b. The entire scheme rests on the intensity number of the first two lines and the distance in Hz between these lines. These intensity numbers ordain the type of multiplet at each level of lines; the separation in Hz gives the coupling constant at each level.

It should be noted that accidental degeneracy of peaks—total or partial coincidence—can cause problems in assigning intensity numbers. Several examples are given in the footnote reference.* See also Problem 4.5. Note the distinction between overlapping and coincident peaks. In Figure 4.56, the doublets overlap but cause no difficulty. In Figure 4.57, the end peaks of three triplets are coincident. Thus the apparent intensities of level a are incorrect and are corrected in level b. The erroneous assignment is detectable because the outer two peaks on either side are in the ratio 1:2 and must be part of a 1:2:1 triplet. The middle triplet now has the intensities 2:4:2.

References

General

Abraham, R.J., Fisher, J., and Loftus, P. (1989). *Introduction to NMR Spectroscopy,* 2nd ed. London-New York: Wiley.

Akitt, J.W. (1992). *NMR and Chemistry: An Introduction to Modern NMR Spectroscopy,* 3rd ed. London: Chapman and Hall.

Atta-ur-Rahman. (1986). *Nuclear Magnetic Resonance.* New York: Springer–Verlag.

Ault, A., and Dudek, G.O. (1976). *NMR, An Introduction to Proton Nuclear Magnetic Resonance Spectrometry.* San Francisco: Holden–Day.

Becker, E.D. (1980). *High Resolution NMR,* 2nd ed. New York: Academic Press.

Bovey, F.A. (1988). *Nuclear Magnetic Resonance Spectroscopy.* 2nd ed. New York: Academic Press.

* Mann, B.E. (1995). *J. Chem. Educ.,* **72,** 614.

Breitmaier, E. (1993). *Structure Elucidation by NMR in Organic Chemistry. A Practical Guide.* New York: Wiley.

Derome, A.E. (1987). *Modern NMR Techniques for Chemistry Research.* Oxford: Pergamon.

Farrar, T.C. (1987). *An Introduction to Pulse NMR Spectroscopy.* Chicago: Farragut Press.

Friebolin, H. (1993). *Basic One- and Two-Dimensional Spectroscopy,* 2nd ed. New York: VCH.

Guñther, H. (1995). *NMR Spectroscopy,* 2nd ed. New York: Wiley.

Jackman, L.M., and Sternhell, S. (1969). *Applications of NMR Spectroscopy in Organic Chemistry,* 2nd ed. New York: Pergamon Press.

Macomber, R.S. (1988). *NMR Spectroscopy—Essential Theory and Practice.* New York: Harcourt.

Neuhaus, D.N., and Williamson, M.P. (1989). *The Nuclear Overhauser Effect in Structural and Conformational Analysis.* New York: VCH.

Paudler, W.W. (1987). *Nuclear Magnetic Resonance.* New York: Wiley.

Sanders, J.K., and Hunter, B.K. (1993). *Modern NMR Spectroscopy,* 2nd ed. Oxford: Oxford University Press.

Schwartz, L.J. (1988). A step-by-step picture of pulsed (time-domain) NMR. *J. Chem. Ed.* **65,** 752–756.

Shaw, D. (1987). *Fourier Transform NMR Spectroscopy.* Amsterdam: Elsevier.

Shoolery, J.N. (1972). *A Basic Guide to NMR.* Palo Alto, CA: Varian Associates.

Williams, K.R., and King, R.W. (1990). *J. Chem. Educ.,* **67,** A125. See References therein for previous papers.

Yoder, C.H., and Schaeffer, C.D., Jr. (1987). *Introduction to Multinuclear NMR.* Menlo Park, CA: Benjamin–Cummings.

Spectra, Data, and Special Topics

Bovey, F.A. (1967). *NMR Data Tables for Organic Compounds,* Vol. 1. New York: Wiley–Interscience.

Brugel, W. (1979). *Handbook of NMR Spectral Parameters,* Vols. 1–3. Philadelphia: Heyden.

Chamberlain, N.F. (1974). *The Practice of NMR Spectroscopy with Spectra–Structure Correlation for* ^1H. New York: Plenum Press.

Kiemle, D.J., and Winter, W.T. (1995) Magnetic Spin Resonance. In *Kirk-Othmer Encyclopedia of Chemical Technology,* 4th ed. Vol. 15., New York, Wiley.

Morrill, T.C., Ed. (1988) *Lanthanide Shift Reagents in Stereochemical Analysis.* New York: VCH.

Pouchert, C.J., and Behnke, J. (1993). *Aldrich Library of* ^{13}C *and* ^1H *FT-NMR Spectra, 300 MHz.* Milwaukee, WI: Aldrich Chemical Co.

Pretsch, E., Clerc, T., Seibl, J., and Simon, W. (1989). *Spectra Data for Structure Determination of Organic Compounds,* 2nd ed. Berlin: Springer–Verlag.

Sadtler Collection of High Resolution Spectra. Philadelphia: Sadtler Research Laboratories.

Sasaki, S. (1985). *Handbook of Proton-NMR Spectra and Data,* Vols. 1–5. New York: Academic Press. (4000 spectra)

Varian Associates. (1962, 1963). *High Resolution NMR Spectra Catalogue,* Vol. 1, Vol. 2. Palo Alto, CA: Varian.

Wiberg, K.B., and Nist, B.J. (1962). *The Interpretation of NMR Spectra.* New York: Benjamin.

Problems

4.1 Match the underlined proton in each compound with the correct chemical shift (δ). Describe all sets, spin systems, and symmetry elements (Section 4.12).

a. $CH_3—CH{=}C\underline{H}—CH_3$ ( ) 1.2

b. $C_6H_5—C(=O)—C\underline{H}_3$ ( ) 2.4

c. $C\underline{H}_3—CH_2—O—CH_3$ ( ) 5.3

d. $(\underline{H})(CH_3)C{=}C(COOCH_3)(CH_3)$ ( ) 8.2

e. $C_6H_4(NO_2)\underline{H}$ (H ortho to NO_2) ( ) 5.0

f. $C_6H_5—C\underline{H}_2OC(=O)CH_3$ ( ) 6.7

g. $C_6H_5—C(\underline{H})(CH_3)—O—C\underline{H}_2—CH_3$ ( ) 4.6

h. $C\underline{H}_2{=}C(CH_3)—CH(CH_3)_2$ ( ) 3.3

4.2 Draw an ^1H NMR spectrum for each of the following compounds in $CDCl_3$. Assume sufficient resolution to provide a first-order spectrum except for benzene ring protons. Systematically spin decouple each set of nonring protons and show the results of decoupling (Section 4.19).

a.

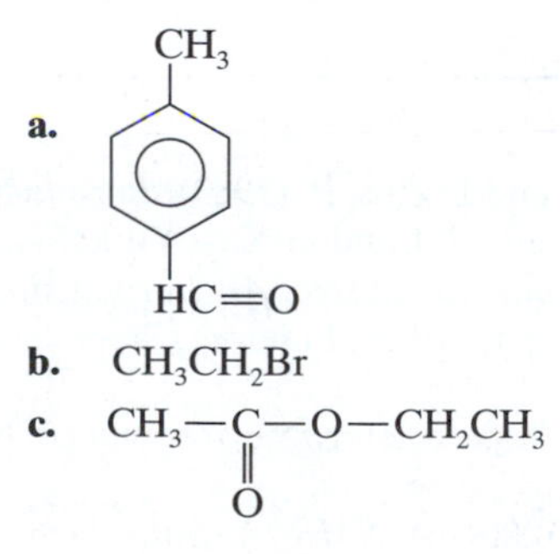

b. CH_3CH_2Br

c. $CH_3—C(=O)—O—CH_2CH_3$

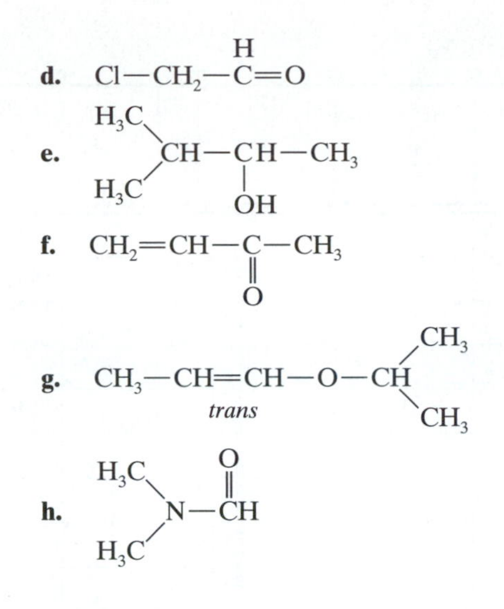

4.3 Deduce the structures of Compounds A-I from the 300-MHz proton spectra and assign all ^1H signals. The top spectrum in each case is a ^{13}C spectrum; these will be assigned at the end of Chapter 5. All spectra were taken in $CDCl_3$ except for 4.3F, which was taken in DMSO-d_6. The peaks are designated as follows: s = singlet, d = doublet, t = triplet, q = quartet, quint = quintet, sext = sextet, sept = septet, m = multiplet.

Problem 4.3 A $C_6H_{12}O$ ketone

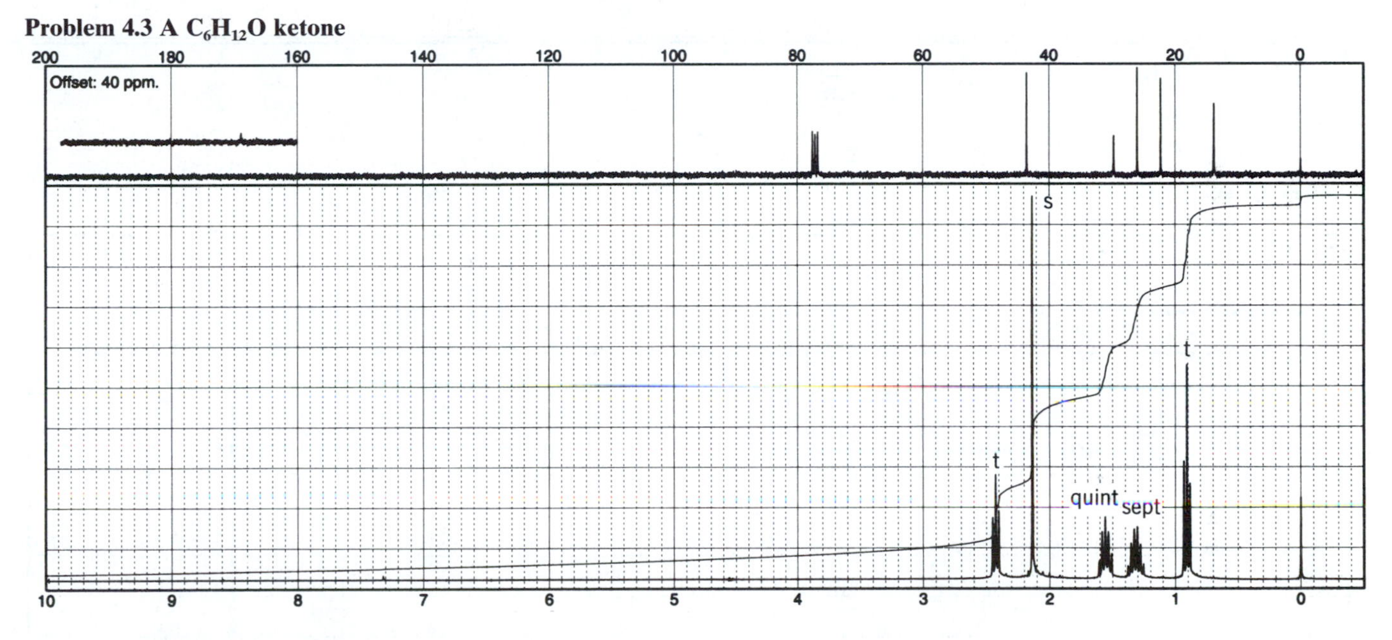

Problem 4.3 B $C_8H_{10}O_2$ alcohol ether

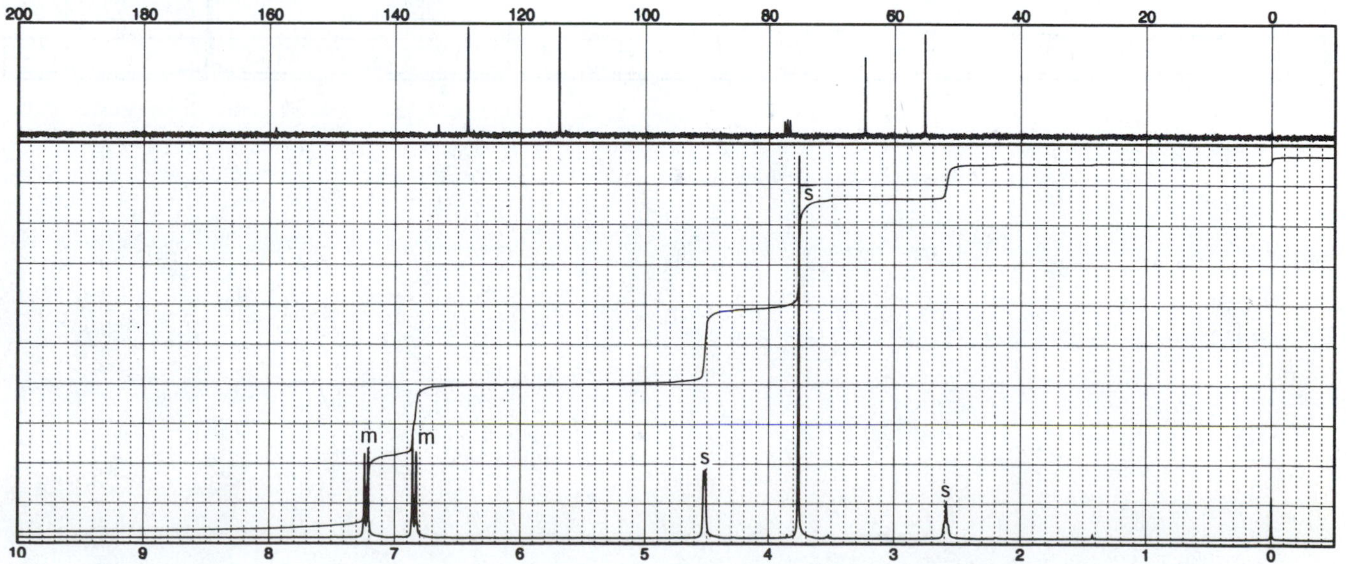

Problem 4.3 C $C_5H_8O_2$ ester

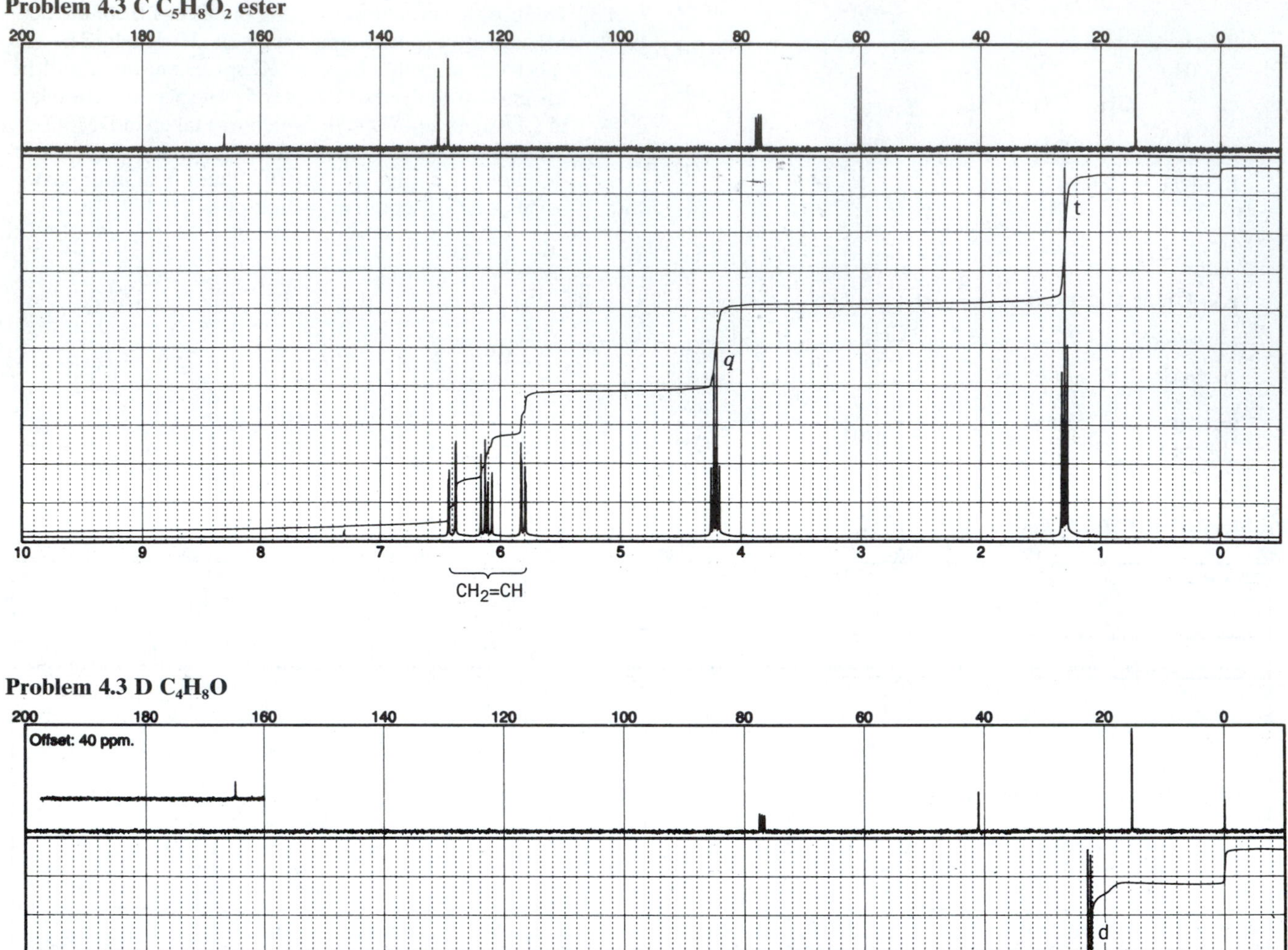

Problem 4.3 D C_4H_8O

Problem 4.3 E $C_7H_7NO_2$

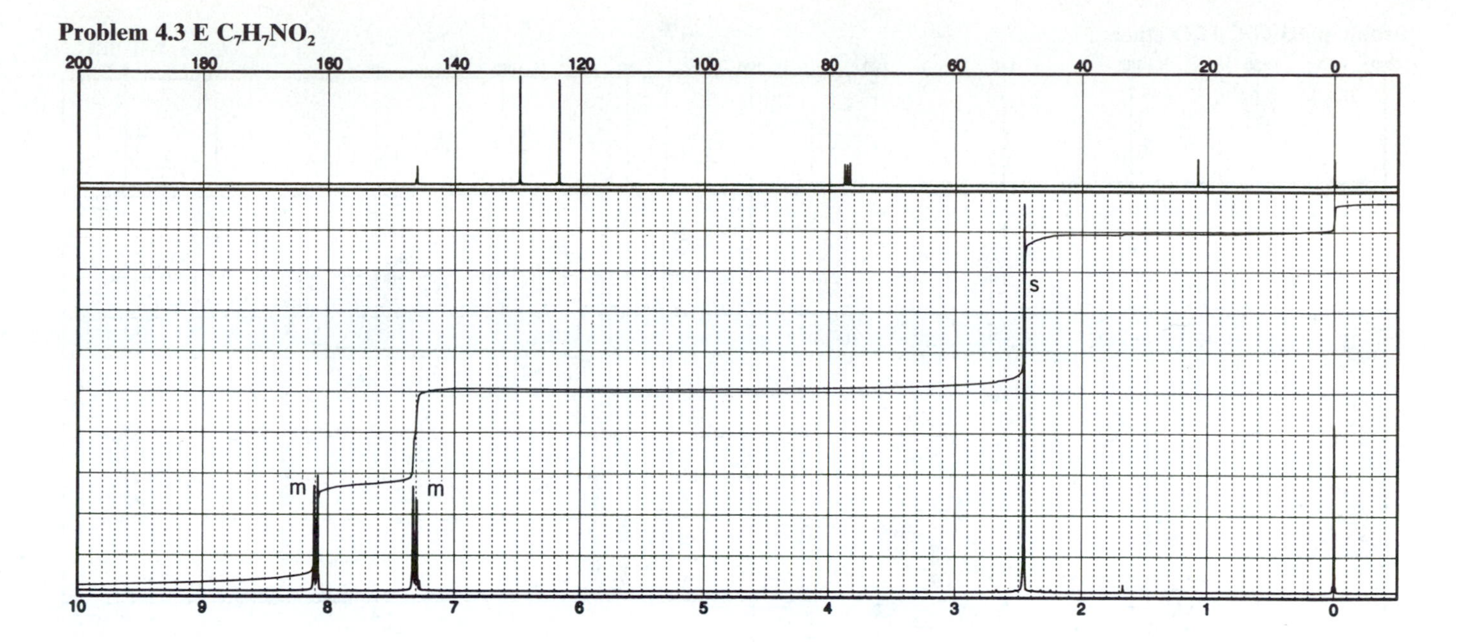

Problem 4.3 F C_8H_9NO amide

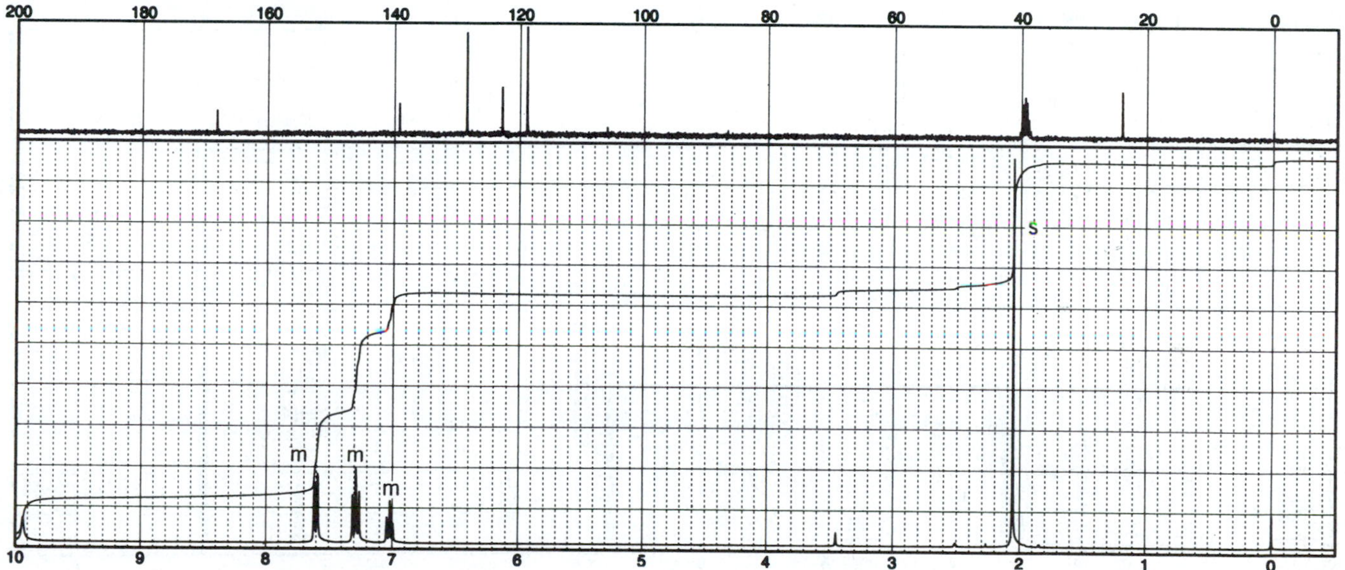

Problem 4.3 G C_4H_8O ether

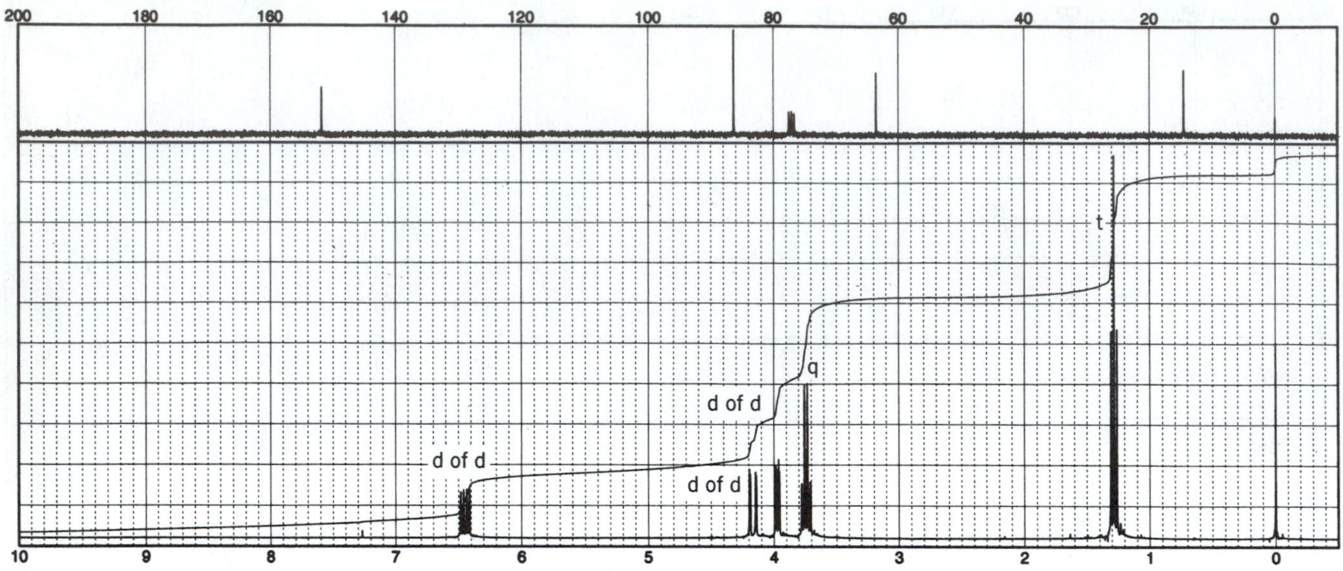

Problem 4.3 H $C_{10}H_{12}O_2$ ester

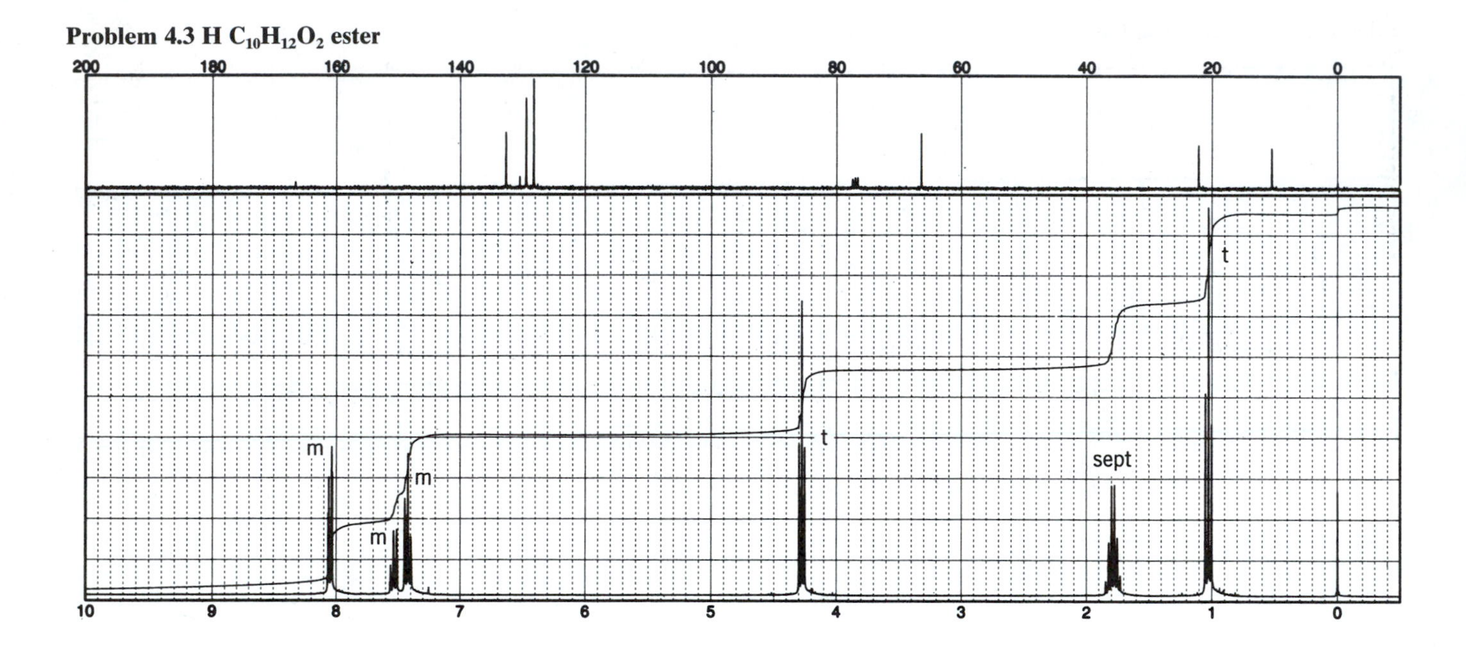

Problem 4.3 I C_3H_5FO ketone

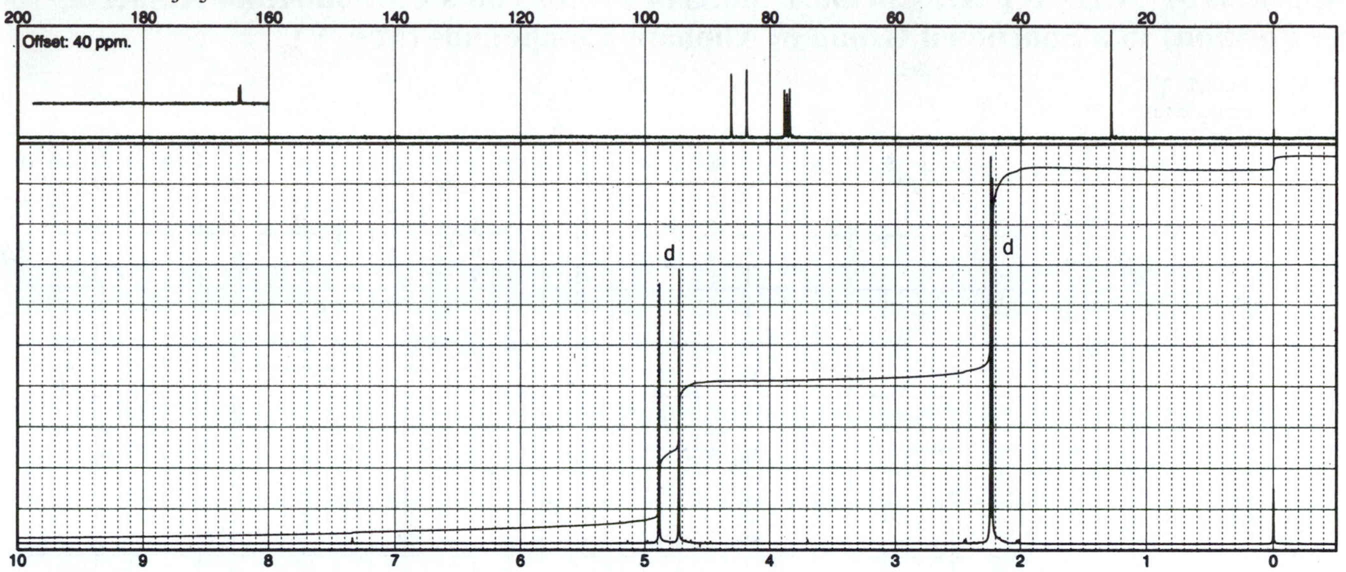

4.4 In Figure 4.46, the spectrum of hexanol

$$HOCH_2CH_2CH_2CH_2CH_2CH_3$$

shows the CH_3 group as a "filled-in" triplet because of virtual coupling to three CH_2 groups that have almost identical chemical shifts. However, the $HOC\underline{H}_{-2}$ protons show a first-order triplet. Explain.

4.5 Draw a stick diagram for the following first-order multiplet (10 mm = 5 Hz). Hint: There are three pairs of totally coincident peaks. Thus, the apparent intensities must be adjusted.

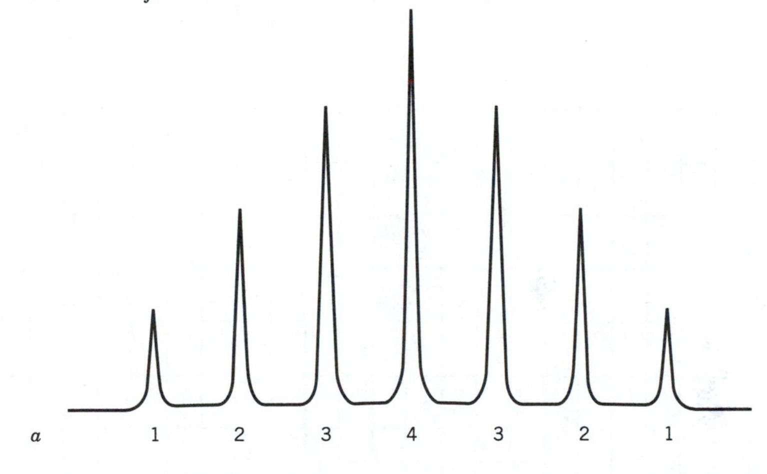

Appendix A CHART A.1 Chemical Shifts of Protons on a Carbon Atom Adjacent (α Position) to a Functional Group in Aliphatic Compounds (M—Y)

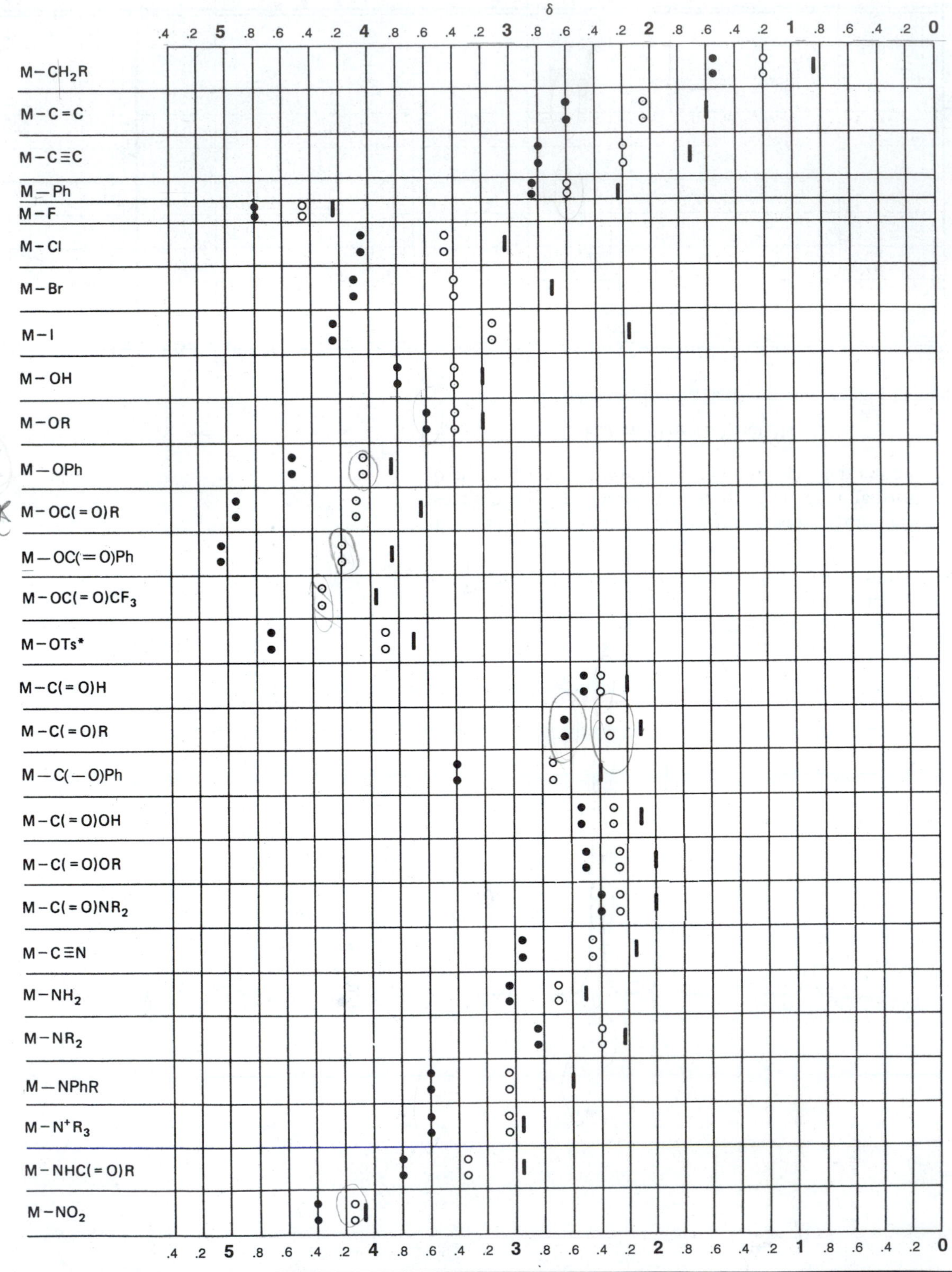

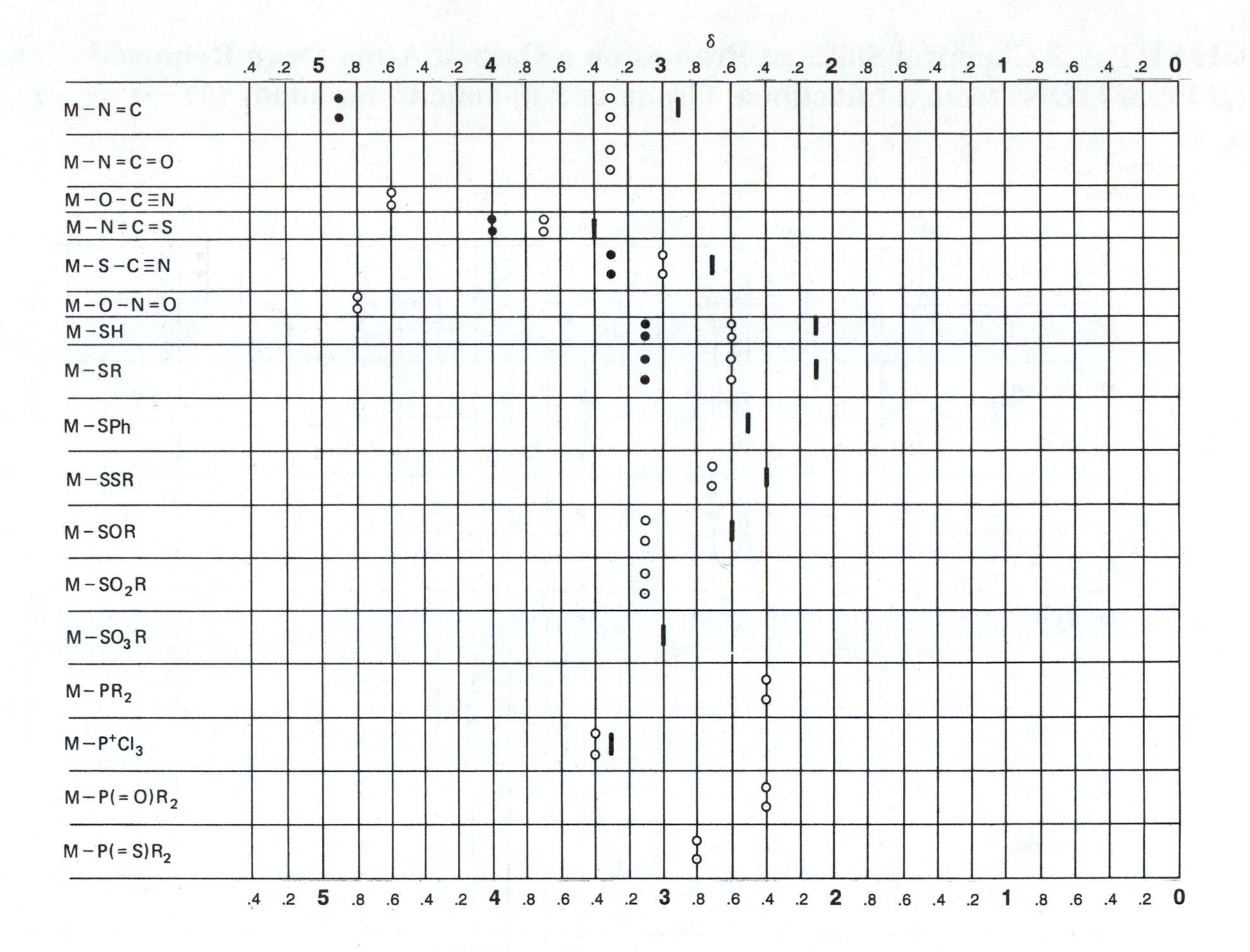
δ
M–N=C
M–N=C=O
M–O–C≡N
M–N=C=S
M–S–C≡N
M–O–N=O
M–SH
M–SR
M–SPh
M–SSR
M–SOR
M–SO$_2$R
M–SO$_3$R
M–PR$_2$
M–P$^+$Cl$_3$
M–P(=O)R$_2$
M–P(=S)R$_2$
.4 .2 5 .8 .6 .4 .2 4 .8 .6 .4 .2 3 .8 .6 .4 .2 2 .8 .6 .4 .2 1 .8 .6 .4 .2 0

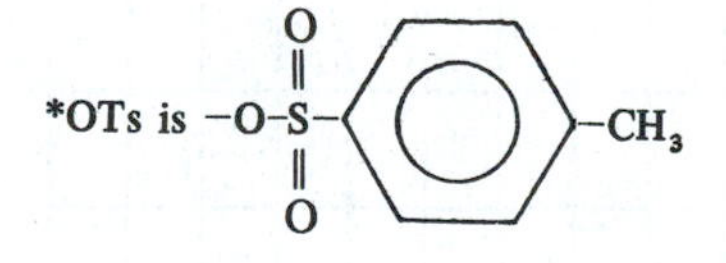
*OTs is
O
O
CH$_3$

CHART A.2 Chemical Shifts of Protons on a Carbon Atom Once Removed (β POSITION) from a Functional Group in Aliphatic Compounds (M—C—Y)

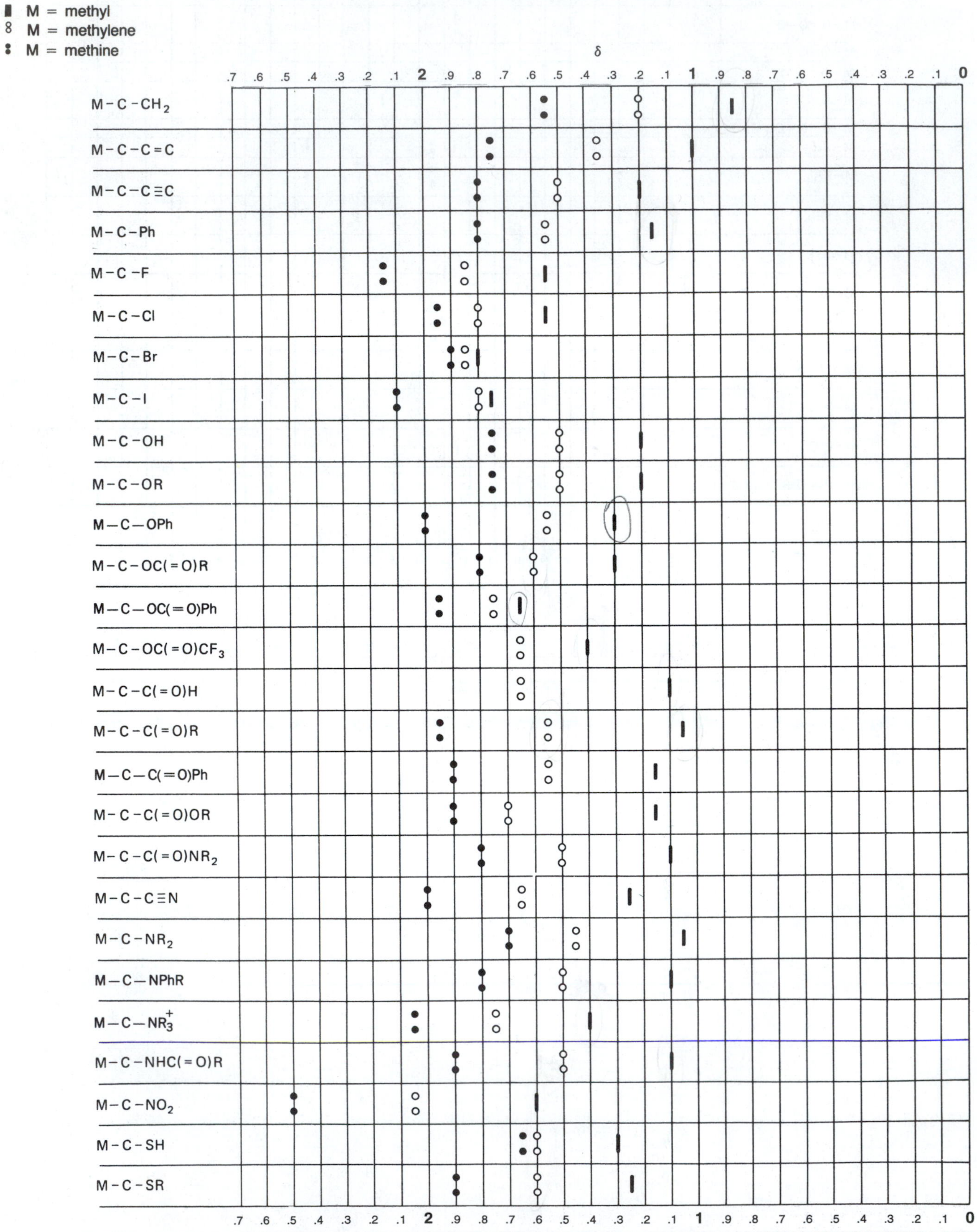

- ▮ M = methyl
- 8 M = methylene
- ⁚ M = methine

δ

.7 .6 .5 .4 .3 .2 .1 **2** .9 .8 .7 .6 .5 .4 .3 .2 .1 **1** .9 .8 .7 .6 .5 .4 .3 .2 .1 **0**

M-C-CH_2
M-C-C=C
M-C-C≡C
M-C-Ph
M-C-F
M-C-Cl
M-C-Br
M-C-I
M-C-OH
M-C-OR
M-C-OPh
M-C-OC(=O)R
M-C-OC(=O)Ph
M-C-OC(=O)CF_3
M-C-C(=O)H
M-C-C(=O)R
M-C-C(=O)Ph
M-C-C(=O)OR
M-C-C(=O)NR_2
M-C-C≡N
M-C-NR_2
M-C-NPhR
M-C-$\overset{+}{N}R_3$
M-C-NHC(=O)R
M-C-NO_2
M-C-SH
M-C-SR

.7 .6 .5 .4 .3 .2 .1 **2** .9 .8 .7 .6 .5 .4 .3 .2 .1 **1** .9 .8 .7 .6 .5 .4 .3 .2 .1 **0**

Appendix B Effect on Chemical Shifts by Two or Three Directly Attached Functional Groups

$$Y{-}CH_2{-}Z \text{ and } Y{-}\underset{\displaystyle W}{\underset{|}{CH}}{-}Z$$

The chemical shift of a methylene group attached to two functional groups can be calculated by means of the substituent constants (σ values) in Table B.1. Shoolery's rule* states that the sum of the constants for the attached functional groups is added to δ 0.23, the chemical shift for CH_4:

$$\delta Y{-}CH_2{-}Z = 0.23 + \sigma_Y + \sigma_Z$$

The chemical shift for the methylene protons of $C_6H_5CH_2Br$, for example, is calculated from the σ values in Table B.1.

$$\begin{aligned} & 0.23 \\ \sigma_{Ph} &= 1.85 \\ \sigma_{Br} &= 2.33 \\ \hline \delta &= 4.41 \quad \text{Found, } \delta\ 4.43 \end{aligned}$$

Shoolery's original constants have been revised and extended in Table B.1. The observed and calculated chemical shifts for 62% of the samples tested were within ±0.2 ppm, 92% within ±0.3 ppm, 96% within 0.4 ppm, and 99% within ±0.5 ppm.† Table B.1 contains substituent constants (Friedrich and Runkle, 1984) for the more common functional groups. Note that chemical shifts of methyl protons can be calculated by using the constant for H (0.34). For example $H{-}CH_2{-}Br$ is equivalent to CH_3Br.

* Shoolery, J.N. (1959). *Varian Technical Information Bulletin,* Vol. 2, No. 3. Palo Alto, CA: Varian Associates.

† Data from Friedrich, E.C., and Runkle, K.G. (1984). *J. Chem. Educ.* **61,** 830; (1986)**63,** 127.

Table B.1 Substituent Constants for Alkyl Methylene (and Methyl) Protons

Y or Z	Substituent Constants (σ)	Y or Z	Substituent Constants (σ)
—H	0.34	—OC(=O)R	3.01
$-CH_3$	0.68	—OC(=O)Ph	3.27
—C=C	1.32	—C(=O)R	1.50
—C≡C	1.44	—C(=O)Ph	1.90
—Ph	1.83	—C(=O)OR	1.46
$-CF_2$	1.12	$-C(=O)NR_2(H_2)$	1.47
$-CF_3$	1.14	—C≡N	1.59
—F	3.30	$-NR_2(H_2)$	1.57
—Cl	2.53	—NHPh	2.04
—Br	2.33	—NHC(=O)R	2.27
—I	2.19	$-N_3$	1.97
—OH	2.56	$-NO_2$	3.36
—OR	2.36	—SR(H)	1.64
—OPh	2.94	$-OSO_2R$	3.13

Tables B.2a, B.2b, and B.2c: Chemical Shift Correlations for Methine Protons

Table B.2a gives the substituent constants* to be used with the formulation

$$\delta\ CHXYZ = 2.50 + \sigma_X + \sigma_Y + \sigma_Z$$

which is satisfactory if at least two of the substituents are electron-withdrawing groups. In other words, only a single substituent may be an alkyl group (R). Within these limits, the standard error of estimate is 0.20 ppm. For example, the chemical shift of the methine proton in

$$CH_3{-}\overset{\displaystyle OEt}{\overset{|}{CH}}{-}OEt$$

is calculated from Table B.2a as follows:

* Bell, H.M., Berry, L.K., and Madigan, E.A. (1984). *Org. Magn. Reson.,* **22,** 693. With permission.

Table B.2a Substituent Constants for Methine Protons

Group	(σ)
—F	1.59
—Cl	1.56
—Br	1.53
$-NO_2$	1.84
$-NH_2$	0.64
$-NH_3^+$	1.34
—NHCOR	1.80
—OH, —OR	1.14
—OAr	1.79
—OCOR	2.07
—Ar	0.99
—C=C	0.46
—C≡C	0.79
—C≡N	0.66
—COR, —COOR, —COOH	0.47
$-CONH_2$	0.60
—COAr	1.22
—SH, —SR	0.61
$-SO_2R$	0.94
—R	0

$$\delta = 2.50 + 1.14 + 1.14 + 0.00 = 4.78$$

The found value is 4.72.

Tables B.2b and B.2c are used jointly for methine protons that are substituted by at least two alkyl groups (or other groups of low polarity). Friedrich and Runkle proposed the relationship

$$\delta_{CHXYZ} = \delta_{(CH_3)_2CHZ} + \Delta xy$$

in which the X and Y substituents are alkyl groups or other groups of low polarity. The Z substituent covers a range of polarities. Δxy is a correction factor. The relationship states that the chemical shift of a methine proton with at least two low-polarity groups is equivalent to the chemical shift of an isopropyl methine proton plus correction factor.

The substituent constants for a Z substituent on an isopropyl methine proton are given in Table B.2b. The Δxy correction factors are given in Table B.2c.

The following example illustrates the joint use of Tables B.2b and B.2c, with CH_3, $CH{=}CH_2$, and C_6H_5 as substituents. The most polar substituent is always designated Z.

Table B.2b Observed Methine Proton Chemical Shifts of Isopropyl Derivatives

$(CH_3)_2CHZ$ Z	δ (ppm) obs	$(CH_3)_2CHZ$ Z	δ (ppm) obs
H	1.33	HO	3.94
H_3C	1.56	RO	3.55
R	1.50	C_6H_5O	4.51
XCH_2	1.85	R(H)C(=O)O	4.94
R(H)C(=O)	2.54	$C_6H_5C(=O)O$	5.22
$C_6H_5C(=O)$	3.58	$F_3CC(=O)O$	5.20
R(H)OC(=O)	2.52	$ArSO_2O$	4.70
$R_2(H_2)NC(=O)$	2.44		
C_6H_5	2.89	R(H)S	3.16
$R_2(H_2)C{=}CR(H)$	2.62	RSS	2.63
R(H)C≡C	2.59		
N≡C	2.67	F	4.50
		Cl	4.14
$R_2(H_2)N$	3.07	Br	4.21
R(H)C(=O)NH	4.01	I	4.24
O_2N	4.67		

$$\delta\, X{-}\underline{CH}(Z){-}Y = \delta\, CH_3{-}\underline{CH}(Z){-}CH{=}CH_2 = \delta\, CH_3{-}\underline{CH}(C_6H_5){-}CH_3 + \Delta xy$$

From Table B.2b, $\delta = 2.89$ for $CH_3{-}\underline{CH}(C_6H_5){-}CH_3$.

From Table B.2c, $\Delta xy = 0.00$ for CH_3. $\Delta xy = 0.40$ for $CH{=}CH_2$.

Therefore, $\delta\, CH_3{-}\underline{CH}(C_6H_5){-}CH{=}CH_2 = 2.89 + 0.00 + 0.40 = 3.29$ (Found: $\delta = 3.44$).

Table B.2c Correction Factors for Methine Substituents of Low Polarity

Open-Chain Methine Proton Systems	Δxy	Cyclic Methine Proton Systems	Δxy
$CH_3{-}\underline{CH}(Z){-}CH_3$	0.00	Z, H	−1.0
$CH_3{-}\underline{CH}(Z){-}R$	−0.20	Z, H	+0.40
$R{-}\underline{CH}(Z){-}R$	−0.40	Z, H	+0.20
$CH_3{-}\underline{CH}(Z){-}CH_2X$	+0.20	Z, H	monosub. −0.20
			axial H −0.45
$CH_3{-}\underline{CH}(Z){-}CH{=}CH_2$	+0.40		equat. H +0.25
$CH_3{-}\underline{CH}(Z){-}C_6H_5$	+1.15	Z, H	0.00
$R{-}\underline{CH}(Z){-}C_6H_5$	+0.90	Z, H	0.00

Appendix C Chemical Shifts in Alicyclic and Heterocyclic Rings

Table C.1 Chemical Shifts in Alicyclic Rings

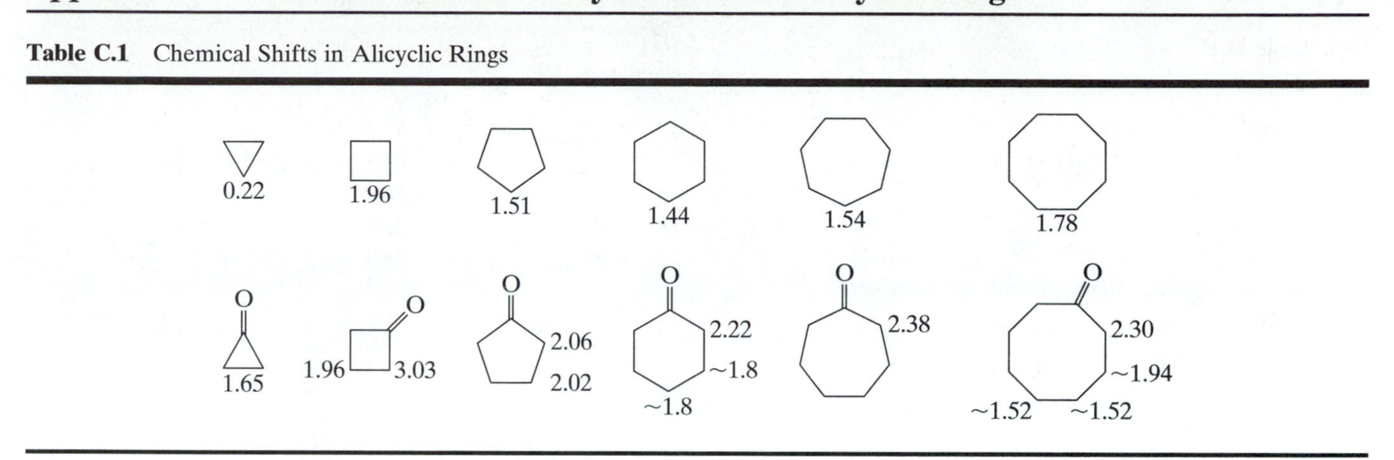

Table C.2 Chemical Shifts in Heterocyclic Rings

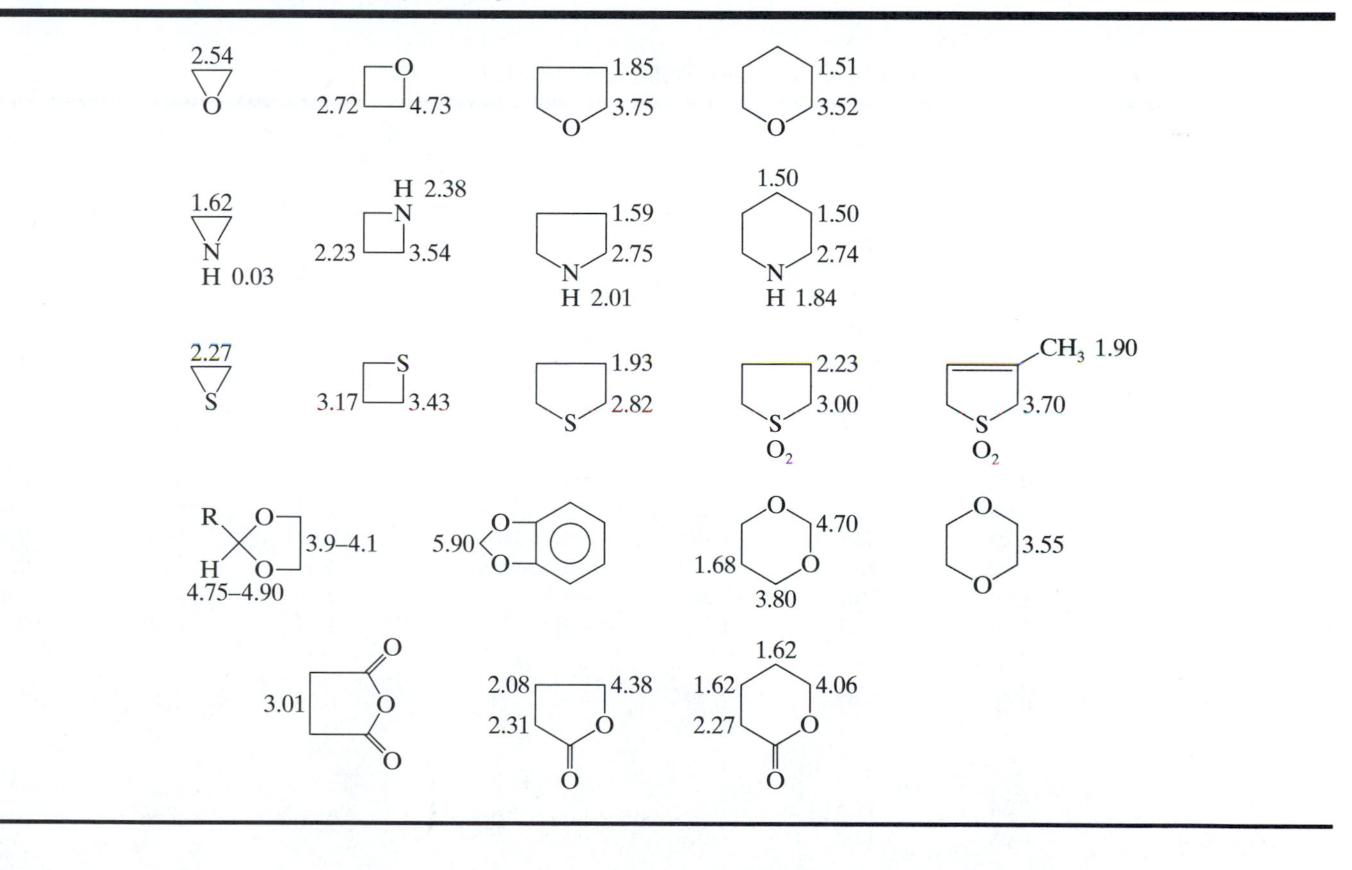

Appendix D Chemical Shifts in Unsaturated and Aromatic Systems

(See Table D.1)

R_{cis}(H)C=C(R_{trans})(R_{gem})

$$\delta_H = 5.25 + Z_{gem} + Z_{cis} + Z_{trans}$$

For example, the chemical shifts of the alkene protons in

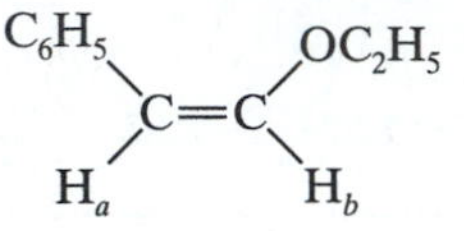

are calculated:

H_a	$C_6H_{5\,gem}$	1.35	5.25
	OR_{trans}	−1.28	0.07
		0.07	δ 5.32
H_b	OR_{gem}	1.18	5.25
	$C_6H_{5\,trans}$	−0.10	1.08
		1.08	δ 6.33

Table D.1 Substituent Constants *(Z)* for Chemical Shifts of Substituted Ethylenes

Substituent R	Z gem	Z cis	Z trans	Substituent R	Z gem	Z cis	Z trans
—H	0	0	0				
—Alkyl	0.44	−0.26	−0.29	—C(H)=O	1.03	0.97	1.21
—Alkyl-ring[a]	0.71	−0.33	−0.30				
$—CH_2O$, $—CH_2I$	0.67	−0.02	−0.07				
$—CH_2S$	0.53	−0.15	−0.15	—C(N)=O	1.37	0.93	0.35
$—CH_2Cl$, $—CH_2Br$	0.72	0.12	0.07				
$—CH_2N$	0.66	−0.05	−0.23	—C(Cl)=O	1.10	1.41	0.99
—C≡C	0.50	0.35	0.10				
—C≡N	0.23	0.78	0.58	—OR, R: aliph	1.18	−1.06	−1.28
—C=C	0.98	−0.04	−0.21	—OR, R: conj[b]	1.14	−0.65	−1.05
—C=C conj[b]	1.26	0.08	−0.01	—OCOR	2.09	−0.40	−0.67
—C=O	1.10	1.13	0.81	—Aromatic	1.35	0.37	−0.10
—C=O conj[b]	1.06	1.01	0.95	—Cl	1.00	0.19	0.03
—COOH	1.00	1.35	0.74	—Br	1.04	0.40	0.55
—COOH conj[b]	0.69	0.97	0.39	$—NR_2$ R: aliph	0.69	−1.19	−1.31
—COOR	0.84	1.15	0.56	$—NR_2$ R: conj[b]	2.30	−0.73	−0.81
—COOR conj[b]	0.68	1.02	0.33				
				—SR	1.00	−0.24	−0.04
				$—SO_2$	1.58	1.15	0.95

[a] Alkyl ring indicates that the double bond is part of the ring R(C=C).

[b] The *Z* factor for the conjugated substituent is used when either the substituent or the double bond is further conjugated with other groups.

Source: Pascual C., Meier, J., and Simon, W. (1966) *Helv. Chim. Acta,* **49,** 164.

Table D.2 Chemical Shifts of Miscellaneous Alkenes

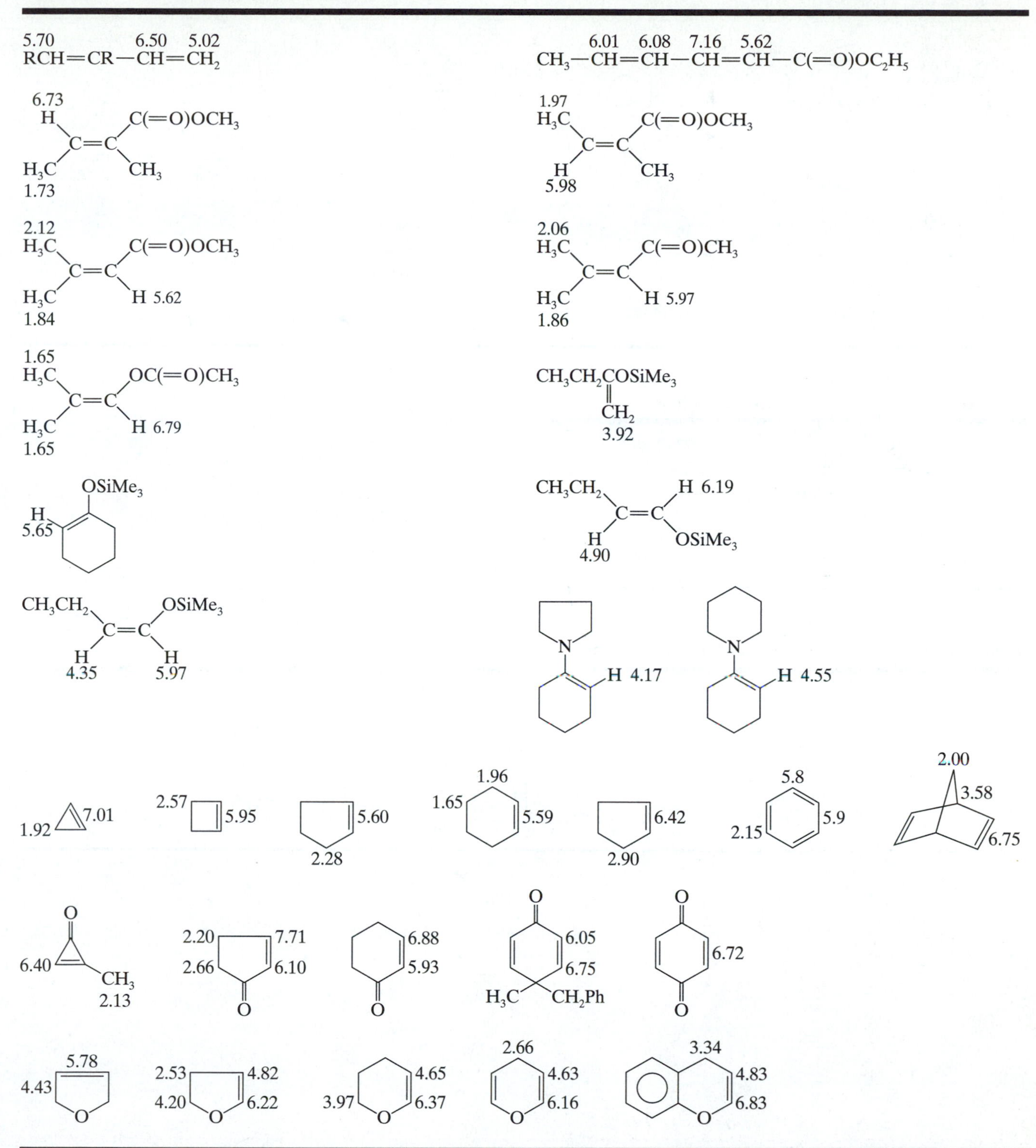

Table D.2 *Continued*

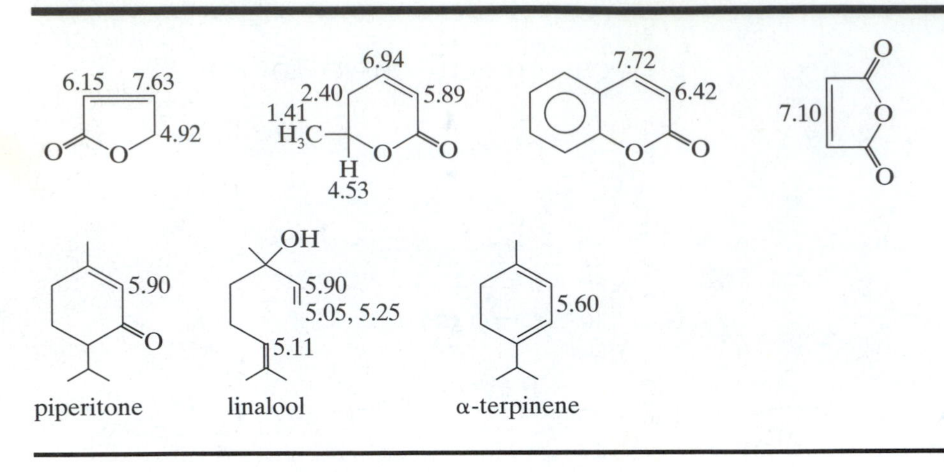

Table D.3 Chemical Shifts of Alkyne Protons

HC≡CR	1.73–1.88
HC≡C—COH	2.23
HC≡C—C≡CR	1.95
HC≡CH	1.80
HC≡CAr	2.71–3.37
HC≡C—C=CR	2.60–3.10

Table D.4 Chemical Shifts of Protons on Fused Aromatic Rings

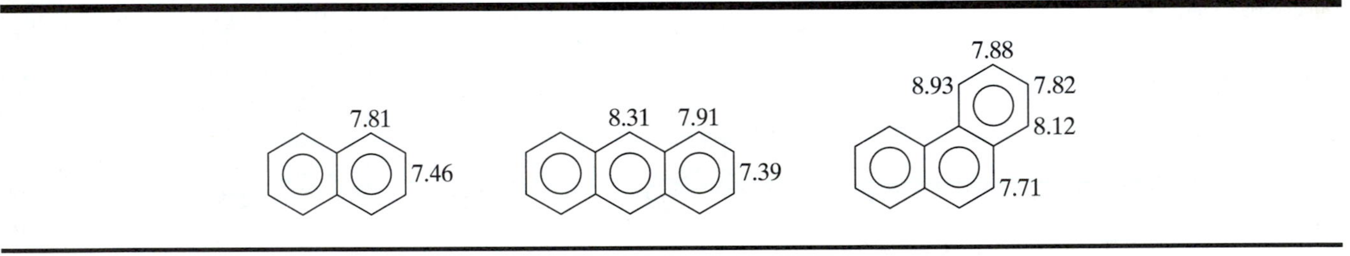

Chart D.1. Chemical Shifts of Protons on Monosubstituted Benzene Rings

	9	.8	.6	.4	.2	8	.8	.6	.4	.2	7	.8	.6	.4	.2	6				δ
Benzene[a]									:											
CH_3 (omp)										:										
CH_3CH_2 (omp)										:										
$(CH_3)_2CH$ (omp)										:										
$(CH_3)_3C$ o,m,p									:	: :										
$C{=}CH_2$ (omp)									:											
$C{\equiv}CH$ o, (mp)								:	:											
Phenyl o, m, p								:	:	:										
CF_3 (omp)								:												
CH_2Cl (omp)									:											
$CHCl_2$ (omp)									:											
CCl_3 o, (mp)					:			:												
CH_2OH (omp)										:										
CH_2OR (omp)									:											
$CH_2OC({=}O)CH_3$ (omp)									:											
CH_2NH_2 (omp)									:											
F m,p.o									:	:	:									
Cl (omp)									:											
Br o, (pm)								:	:											
I o,p,m							:		:	:										
OH m,p,o										:	:	:								
OR m, (op)									:	:										
$OC({=}O)CH_3$ (mp), o									:	:										
OTs[b] (mp), o									:	:										
$CH({=}O)$o,p,m						:	:	:												
$C({=}O)CH_3$ o, (mp)						:		:												
$C({=}O)OH$ o, p, m						:		: :	:											
$C({=}O)OR$ o, p, m					:			: :												
$C({=}O)Cl$ o, p, m					:			: :												
$C{\equiv}N$ (omp)								:												
NH_2 m,p,o										:	:		:							
$N(CH_3)_2$ m(op)										:		:								
$NHC({=}O)R$ o,m,p								:	:	:										
NH_3^+ o (mp)						:	:													
NO_2 o,p,m				:			:	:												
SR (omp)									:											
$N{=}C{=}O$ (omp)										:										

[a] The benzene ring proton is at δ 7.27, from which the shift increments are calculated as shown at the end of Section 4.7.
[b] OTS = *p*-toluenesulfonyloxy group.

Table D.5 Chemical Shifts of Protons on Heteroaromatic Rings

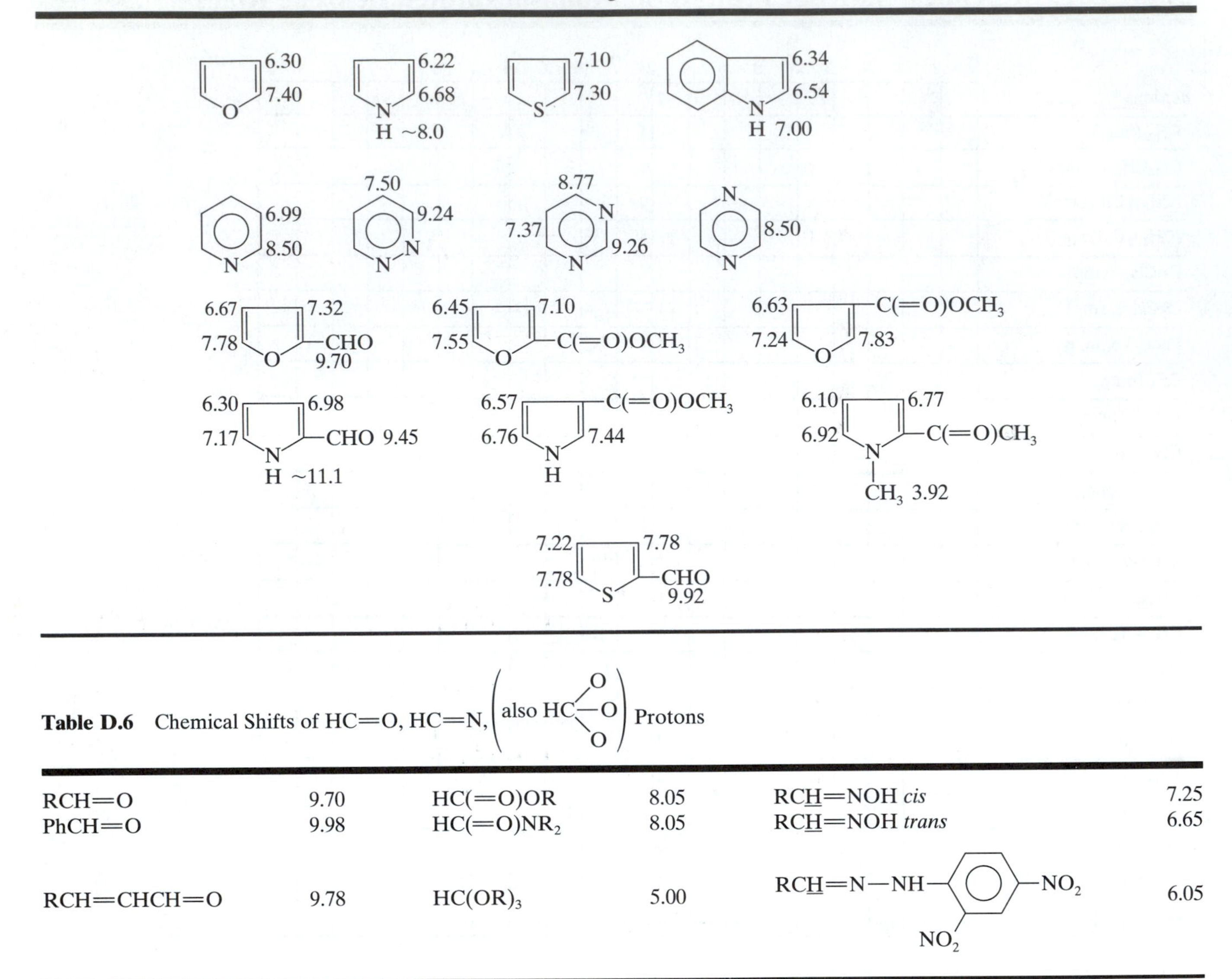

Table D.6 Chemical Shifts of HC=O, HC=N, (also HC(O)(O)O) Protons

RCH=O	9.70	HC(=O)OR	8.05	RC<u>H</u>=NOH *cis*	7.25
PhCH=O	9.98	HC(=O)NR_2	8.05	RC<u>H</u>=NOH *trans*	6.65
RCH=CHCH=O	9.78	$HC(OR)_3$	5.00	RC<u>H</u>=N—NH—C_6H_3(NO_2)—NO_2 (2,4-dinitrophenyl)	6.05

Appendix E Protons on Heteroatoms

Chart E.1. Protons Subject to Hydrogen-Bonding Effects (Protons on Heteroatoms)[a]

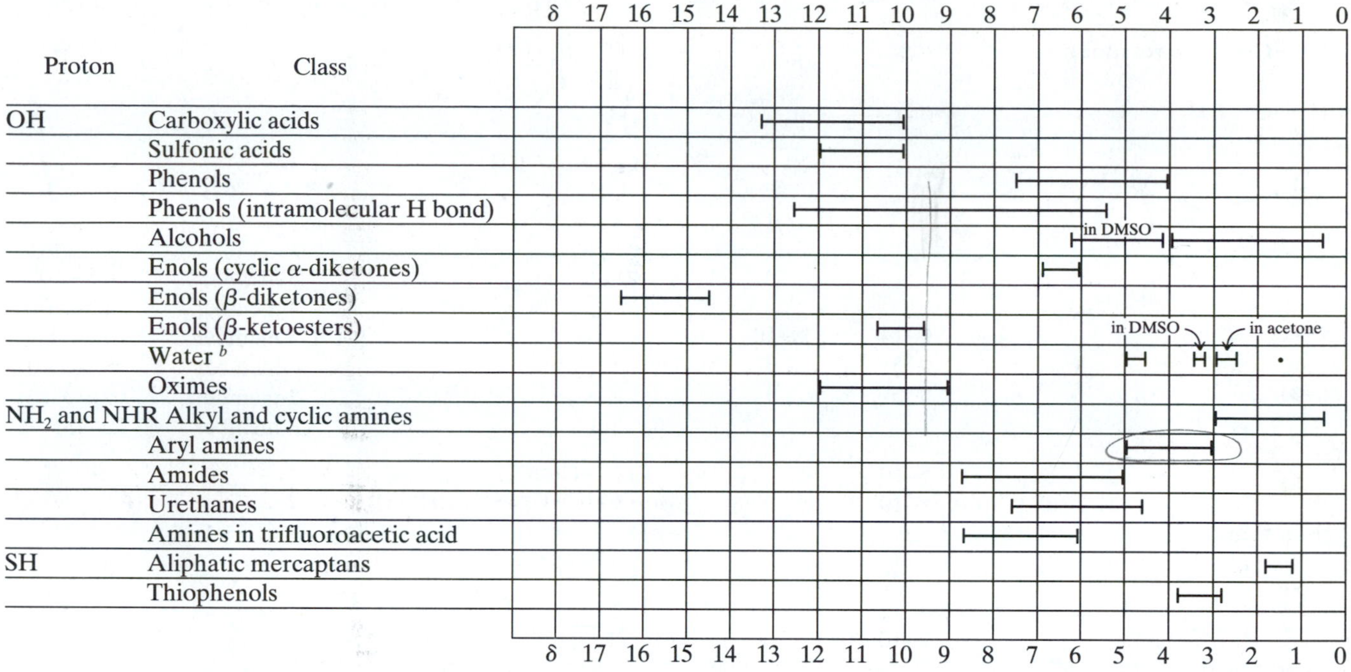

δ 17 16 15 14 13 12 11 10 9 8 7 6 5 4 3 2 1 0

Proton	Class
OH	Carboxylic acids
	Sulfonic acids
	Phenols
	Phenols (intramolecular H bond)
	Alcohols (in DMSO)
	Enols (cyclic α-diketones)
	Enols (β-diketones)
	Enols (β-ketoesters)
	Water[b] (in DMSO; in acetone)
	Oximes
NH_2 and NHR	Alkyl and cyclic amines
	Aryl amines
	Amides
	Urethanes
	Amines in trifluoroacetic acid
SH	Aliphatic mercaptans
	Thiophenols

δ 17 16 15 14 13 12 11 10 9 8 7 6 5 4 3 2 1 0

[a] Solvent $CDCl_3$. Chemical shifts within a range are a function of concentration.
[b] See Section 4.9.1.2.

Table E.1 Chemical Shift of Dissolved Water in Deuterated Solvents

Solvent	δ
Chloroform-d_3	1.5
Benzene-d_6	0.4
Acetone-d_6	2.75
Methylene chloride-d_2	1.55
Dimethylformamide-d_7	3.0
Pyridine-d_5	5.0
Toluene-d_8	0.1–0.2
Methanol-d_4	4.9
Acetonitrile-d_3	2.1
Dimethyl sulfoxide-d_6	3.35
Water-d_2	~4.75 (HDO)

Appendix F Proton Spin-Coupling Constants

Type	J_{ab} (Hz)	J_{ab} Typical
H_a—C—H_b (geminal, $>C(H_a)(H_b)$)	0–30	12–15
$CH_a—CH_b$ (free rotation)	6–8	7
$CH_a—C—CH_b$	0–1	0
Cyclohexane, H_a / H_b:		
ax–ax	6–14	8–10
ax–eq	0–5	2–3
eq–eq	0–5	2–3
Cyclopentane, H_a / H_b (cis or trans)	*cis* 5–10 *trans* 5–10	
Cyclobutane, H_a / H_b (cis or trans)	*cis* 4–12 *trans* 2–10	
Cyclopropane, H_a / H_b (cis or trans)	*cis* 7–13 *trans* 4–9	
$CH_a—OH_b$ (no exchange)	4–10	5
$>CH_a—CH_b$(=O)	1–3	2–3
$C=CH_a—CH_b$(=O)	5–8	6
H_a—C=C—H_b (trans)	12–18	17
$>C=C(H_a)(H_b)$	0–3	0–2
H_a—C=C—H_b (cis)	6–12	10
CH_a—C=C—CH_b (cis)	0–3	1–2

Type	J_{ab} (Hz)	J_{ab} Typical
$>C=C(CH_a)(H_b)$	4–10	7
H_a—C=C—CH_b (trans)	0–3	1.5
H_a—C=C—CH_b (cis)	0–3	2
$C=CH_a—CH_b=C$	9–13	10
H_a—C=C—H_b (ring)	3 member	0.5–2.0
	4 member	2.5–4.0
	5 member	5.1–7.0
	6 member	8.8–11.0
	7 member	9–13
	8 member	10–13
$CH_a—C≡CH_b$	2–3	
$—CH_a—C≡C—CH_b—$	2–3	
Epoxide, H_a, H_b on same carbon		6
Epoxide, H_a, H_b cis		4
Epoxide, H_a, H_b trans		2.5
Benzene, H_a / H_b: *J* (ortho)	6–10	9
J (meta)	1–3	3
J (para)	0–1	~0
Pyridine: *J* (2–3)	5–6	5
J (3–4)	7–9	8
J (2–4)	1–2	1.5
J (3–5)	1–2	1.5
J (2–5)	0–1	1
J (2–6)	0–1	~0
Furan: *J* (2–3)	1.3–2.0	1.8
J (3–4)	3.1–3.8	3.6
J (2–4)	0–1	~0
J (2–5)	1–2	1.5
Thiophene: *J* (2–3)	4.9–6.2	5.4
J (3–4)	3.4–5.0	4.0
J (2–4)	1.2–1.7	1.5
J (2–5)	3.2–3.7	3.4

Source: Compiled by Varian Associates. Absolute values. Reproduced with permission.

Appendix F Continued

Type		J_{ab} (Hz)	J_{ab} Typical
Pyrrole (N–H, positions 1–5)	J (1–2)	2–3	
	J (1–3)	2–3	
	J (2–3)	2–3	
	J (3–4)	3–4	
	J (2–4)	1–2	
	J (2–5)	1.5–2.5	
Pyrimidine (positions 1–6)	J (4–5)	4–6	
	J (2–5)	1–2	
	J (2–4)	0–1	
	J (4–6)	2–3	
Thiazole (positions 2–5)	J (4–5)	3–4	
	J (2–5)	1–2	
	J (2–4)	~0	

Proton–Carbon-13
(See Tables 5.17, 5.18)

Proton–Fluorine

Type	J_{ab} (Hz)	J_{ab} Typical
$>C(H_a)(F_b)$		44–81
$>CH_a-CF_b$		3–25
$>CH_a-C-CF_b$		0–4
H_a and F_b on the two carbons of C=C, same side		1–8
H_a and F_b on the two carbons of C=C, opposite sides		12–40
Fluorobenzene with ring H_a		*o* 6–10 *m* 5–6 *p* 2
$\alpha\ H_3C-C(=O)-CH_2F\ \gamma$ (CH_2 = β)		$\alpha\gamma$ 4.3 $\beta\gamma$ 48

Proton–Phosphorus

Type	J_{ab} (Hz)	
$>P(=O)H$	630–707	
$(CH_3)_3P$	2.7	
$(CH_3)_3P{=}O$	13.4	
$(CH_3CH_2)_3P$	0.5 (HCCP)	13.7 (HCP)
$(CH_3CH_2)_3P{=}O$	11.9 (HCCP)	16.3 (HCP)
$CH_3P(=O)(OR)_2$	10–13	
$CH_3C{-}P(=O)(OR)_2$	15–20	
$CH_3OP(OR)_2$	10.5–12	
$P[N(CH_3)_2]_3$	8.8	
$O{=}P[N(CH_3)_2]_3$	9.5	

Source: Compiled by Varian Associates. Absolute values. Reproduced with permission.

3.144 25.5 22.5 .72387 .49343 3.11

Appendix G Chemical Shifts and Multiplicities of Residual Protons in Commercially Available Deuterated Solvents (Merck & Co., Inc.)

Compound,[a] Molecular Weight	δ_H (mult)
Acetic acid-d_4	11.53 (1)
64.078	2.03 (5)
Acetone-d_6	2.04 (5)
64.117	
Acetonitrile-d_3	1.93 (5)
44.071	
Benzene-d_6	7.15 (br)
84.152	
Chloroform-d	7.26 (1)
120.384	
Cyclohexane-d_{12}	1.38 (br)
96.236	
Deuterium oxide	4.63 (ref. DSS)[c]
20.028	4.67 (ref. TSP)[c]
1,2-Dichloroethane-d_4	3.72 (br)
102.985	
Diethyl-d_{10} ether	3.34 (m)
84.185	1.07 (m)
Diglyme-d_{14}	3.49 (br)
148.263	3.40 (br)
	3.22 (5)
N,N-Dimethylformamide-d_7	8.01 (br)
80.138	2.91 (5)
	2.74 (5)
Dimethyl-d_6 sulphoxide	2.49 (5)
84.170	
p-Dioxane-d_8	3.53 (m)
96.156	
Ethyl alcohol-d_6 (anh)	5.19 (1)
52.106	3.55 (br)
	1.11 (m)
Glyme-d_{10}	3.40 (m)
100.184	3.22 (5)
Hexafluoroacetone deuterate	5.26 (1)
198.067	
HMPT-d_{18}	2.53 (2 × 5)
197.314	
Methyl alcohol-d_4	4.78 (1)
36.067	3.30 (5)
Methylene chloride-d_2	5.32 (3)
86.945	
Nitrobenzene-d_5	8.11 (br)
128.143	7.67 (br)
	7.50 (br)

[a] Purity (Atom % D) up to 99.96% ("100%") for several solvents.

[b] The residual proton consists of one proton of each kind in an otherwise completely deuterated molecule. For example, deuterated acetic acid has two different kinds of residual protons: CD_2H—COOD and CD_3—COOH. The CD_2H proton, coupled to two D nuclei is at δ 2.03 with a multiplicity of 5 (i.e., $2nI + 1 = 2 \times 2 \times 1 + 1 = 5$). The carboxylic proton is a singlet at δ 11.53.

[c] DSS is 3-trimethylsilyl)-1-propane sulfonic acid, sodium salt. TSP is sodium-3-trimethylpropionate-2,2,3,3-d_4. Both are reference standards used in aqueous solutions.

Appendix G Continued

Compound,[a] Molecular Weight	δ_H (mult)
Nitromethane-d_3	4.33 (5)
64.059	
Isopropyl alcohol-d_8	5.12 (1)
68.146	3.89 (br)
	1.10 (br)
Pyridine-d_5	8.71 (br)
84.133	7.55 (br)
	7.19 (br)
Tetrahydrofuran-d_8	3.58 (br)
80.157	1.73 (br)
Toluene-d_8	7.09 (m)
100.191	7.00 (br)
	6.98 (m)
	2.09 (5)
Trifluoroacetic acid-d	11.50 (1)
115.030	
2,2,2-Trifluoroethyl alcohol-d_3	5.02 (1)
103.059	3.88 (4 × 3)

[a] Purity (Atom % D) up to 99.96% ("100%") for several solvents.

[b] The residual proton consists of one proton of each kind in an otherwise completely deuterated molecule. For example, deuterated acetic acid has two different kinds of residual protons: CD_2H—COOD and CD_3—COOH. The CD_2H proton, coupled to two D nuclei is at δ 2.03 with a multiplicity of 5 (i.e., $2nI + 1 = 2 \times 2 \times 1 + 1 = 5$). The carboxylic proton is a singlet at δ 11.53.

[c] DSS is 3-trimethylsilyl)-1-propane sulfonic acid, sodium salt. TSP is sodium-3-trimethylpropionate-2,2,3,3-d_4. Both are reference standards used in aqueous solutions.

Appendix H Properties of Several Nuclei

Isotope	NMR Frequency MHz for a 1-T Field	Natural Abundance %	Relative Sensitivity at Constant Field	Magnetic Moment (μ)	Spin Number (I)	Electrical Quadrupole Moment ($e \times 10^{-24}$ cm^2)	Magnetogyric Ratio[a] γ (10^7 rad $T^{-1}s^{-1}$)
^{1}H	42.576	99.9844	1.000	2.79268	$\frac{1}{2}$		26,753
^{2}H	6.5357	1.56×10^{-2}	9.64×10^{-3}	0.85739	1	2.77×10^{-3}	4,107
^{3}H	45.414		1.21	2.9788	$\frac{1}{2}$		
^{10}B	4.575	18.83	1.99×10^{-2}	1.8005	3	7.4×10^{-2}	
^{11}B	13.660	81.17	0.165	2.6880	$\frac{3}{2}$	3.55×10^{-2}	
^{12}C		98.9			0		
^{13}C	10.705	1.108	1.59×10^{-2}	0.70220	$\frac{1}{2}$		6,728
^{14}N	3.076	99.635	1.01×10^{-3}	0.40358	1	7.1×10^{-2}	
^{15}N	4.315	0.365	1.04×10^{-3}	-0.28304	$\frac{1}{2}$		$-2,712$
^{16}O		99.76			0		
^{17}O	5.772	3.7×10^{-2}	2.91×10^{-2}	-1.8930	$\frac{5}{2}$	-4.0×10^{-3}	$-3,628$
^{19}F	40.055	100	0.834	2.6273	$\frac{1}{2}$		25,179
^{28}Si		92.28			0		
^{29}Si	8.458	4.70	7.85×10^{-3}	-0.55548	$\frac{1}{2}$		$-5,319$
^{30}Si		3.02			0		
^{31}P	17.236	100	6.64×10^{-2}	1.1305	$\frac{1}{2}$		10,840
^{32}S		95.06			0		
^{33}S	3.266	0.74	2.26×10^{-3}	0.64274	$\frac{3}{2}$	-0.053	2,054
^{34}S		4.2			0		
^{35}Cl	4.172	75.4	4.71×10^{-3}	0.82091	$\frac{3}{2}$	-7.9×10^{-2}	2,624
^{37}Cl	3.472	24.6	2.72×10^{-3}	0.68330	$\frac{3}{2}$	-6.21×10^{-2}	2,184
^{79}Br	10.667	50.57	7.86×10^{-2}	2.0991	$\frac{3}{2}$	0.34	
^{81}Br	11.499	49.43	9.84×10^{-2}	2.2626	$\frac{3}{2}$	0.28	
^{127}I	8.519	100	9.35×10^{-2}	2.7937	$\frac{5}{2}$	-0.75	

[a] T = Tesla.

Source: Varian Associates NMR Table, 4th ed., 1964, with permission.

CHAPTER 5

^{13}C *NMR Spectrometry*

*5.1 Introduction**

Faced with a choice during the early development of nuclear magnetic resonance spectrometry, most organic chemists would certainly have selected the carbon nucleus over the hydrogen nucleus for immediate investigation. After all, the carbon skeletons of rings and chains are central to organic chemistry. The problem, of course, is that the carbon skeleton consists almost completely of the ^{12}C nucleus, which is not accessible to NMR spectrometry. The spectrometrist is left to cope with the very small amount of the ^{13}C nucleus.

There are enough differences between ^{13}C and ^{1}H NMR spectrometry to justify separate chapters on pedagogical grounds. With an understanding of the basic concepts of NMR spectrometry in Chapter 4, mastery of ^{13}C spectrometry will be rapid.

The ^{12}C nucleus is not magnetically "active" (spin number, I, is zero), but the ^{13}C nucleus, like the ^{1}H nucleus, has a spin number of $\frac{1}{2}$. However, since the natural abundance of ^{13}C is only 1.1% that of ^{12}C, and its sensitivity is only about 1.6% that of ^{1}H, the overall sensitivity of ^{13}C compared with ^{1}H is about 1/5700.

The earlier, continuous-wave, slow-scan procedure requires a large sample and a prohibitively long time to obtain a ^{13}C spectrum, but the availability of pulsed Fourier transform (FT) instrumentation, which permits simultaneous irradiation of all ^{13}C nuclei, has resulted in an increased activity in ^{13}C spectrometry, beginning in the early 1970s, comparable to the burst of activity in ^{1}H spectrometry that began in the late 1950s.

An important development was the use of broadband decoupling (i.e., irradiation) of protons. Because of the large J values for $^{13}C—H$ ($\sim$110–320 Hz) and appreciable values for $^{13}C—C—H$ and $^{13}C—C—C—H$, proton-coupled ^{13}C spectra usually show complex overlapping multiplets that are difficult to interpret; but some proton-coupled spectra such as that of diethyl phthalate (Fig. 5.1*a*) are quite simple. Irradiation (Fig. 5.1*b*) of the protons over a broad frequency range by means of a broadband generator removes these couplings. Figure 5.1*c* shows the effect of a delay between pulses (see Section 5.2).

The result, in the absence of other coupling nuclei, such as ^{31}P or ^{19}F, is a single sharp peak† for each chemically nonequivalent ^{13}C atom except for the infrequent coincidence of ^{13}C chemical shifts.‡ Furthermore, an increase in signal (up to 200%) accrues from the nuclear Overhauser effect (NOE) (see Section 4.20). This enhancement results from an increase in population of the lower energy level of the ^{13}C nuclei concomitant with the increase in population of the high-energy level of the ^{1}H nuclei on irradiation of the ^{1}H nuclei. The net effect is a very large reduction in the time needed to obtain a broadband-decoupled spectrum (Fig. 5.1*b*) as compared with a coupled spectrum (Fig. 5.1*a*).

As described in Section 4.4 for pulsed Fourier transform spectrometry of protons, a short, powerful, rf pulse (on the order of a few microseconds) excites all of the ^{13}C nuclei simultaneously. At the same time, the broadband decoupler is turned on in order to remove the $^{13}C—^{1}H$ coupling. Since the pulse frequencies are slightly off resonance for all of the nuclei, each nucleus shows a free induction decay (FID), which is an exponentially decaying sine wave with a frequency equal to the difference between the applied frequency and the resonance frequency for that nucleus. Figure 5.2*a* shows the result for a single-carbon compound.

The FID display for a compound containing more than one ^{13}C nucleus consists of superimposed sine waves, each with its characteristic frequency, and an in-

* Familiarity with Chapter 4 is assumed.

† Because of the low natural abundance of ^{13}C, the occurrence of adjacent ^{13}C atoms has a low probability; thus we are free of the complication of $^{13}C—^{13}C$ coupling.

‡ But note the coincidence of these ^{13}C peaks in Figure 6.12.

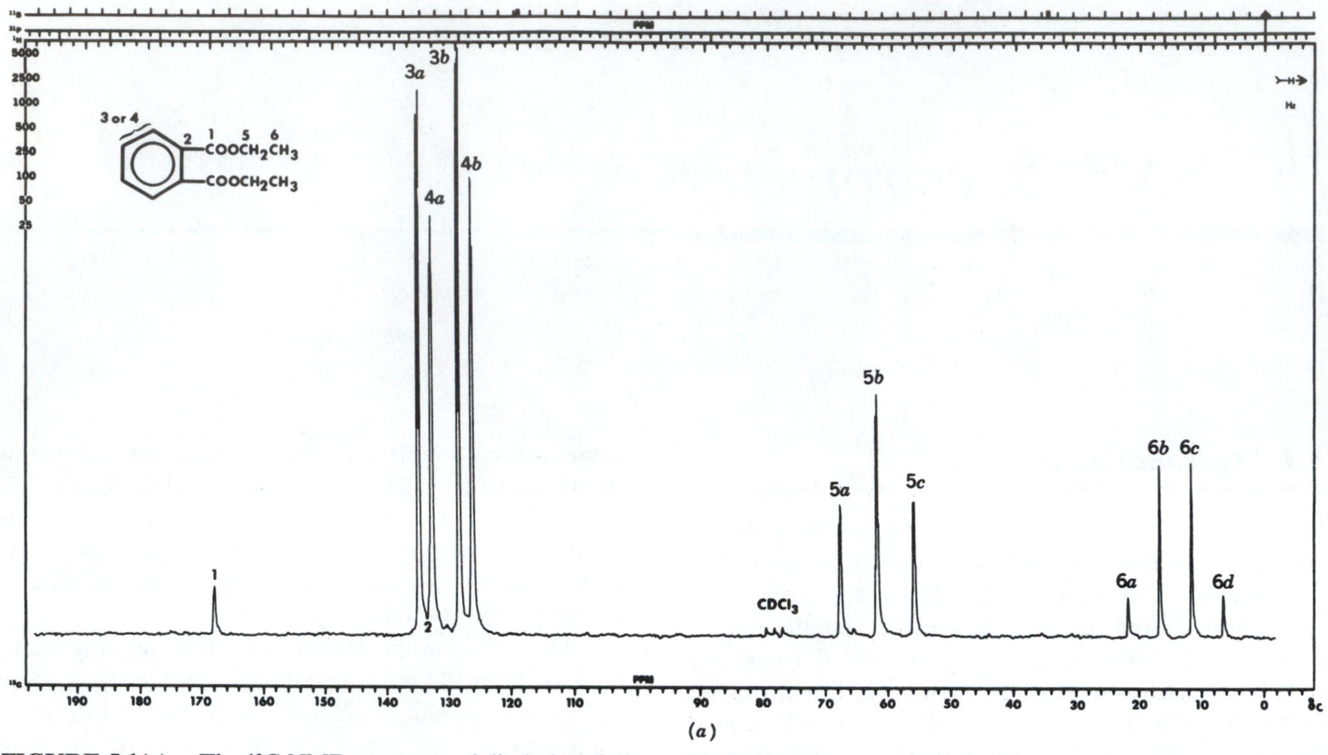

(*a*)

FIGURE 5.1(*a*). The ^{13}C-NMR spectrum of diethyl phthalate with the protons completely coupled. The solvent used was $CDCl_3$ at 25.2 MHz.

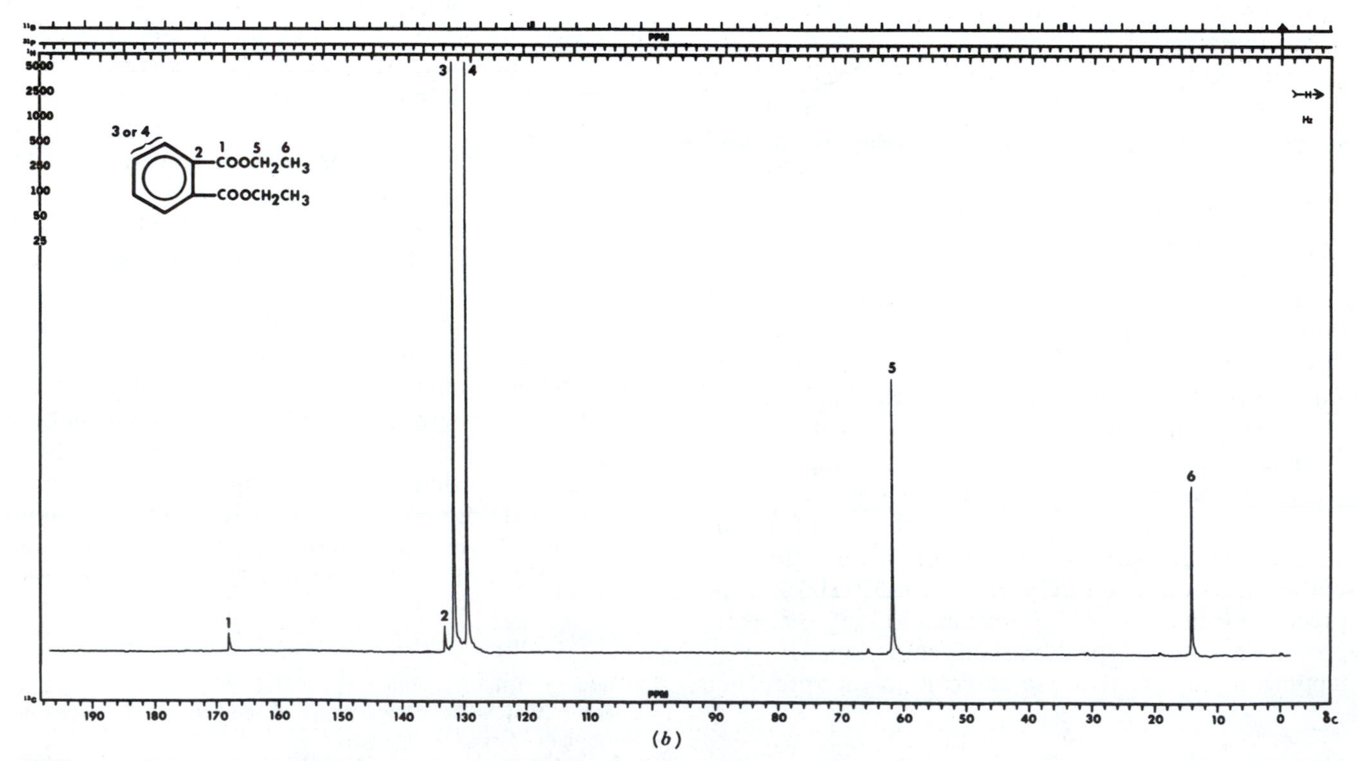

(*b*)

FIGURE 5.1(*b*). The ^{13}C-NMR spectrum of diethyl phthalate with the protons completely decoupled by the broadband decoupler. The solvent used was $CDCl_3$ at 25.2 MHz.

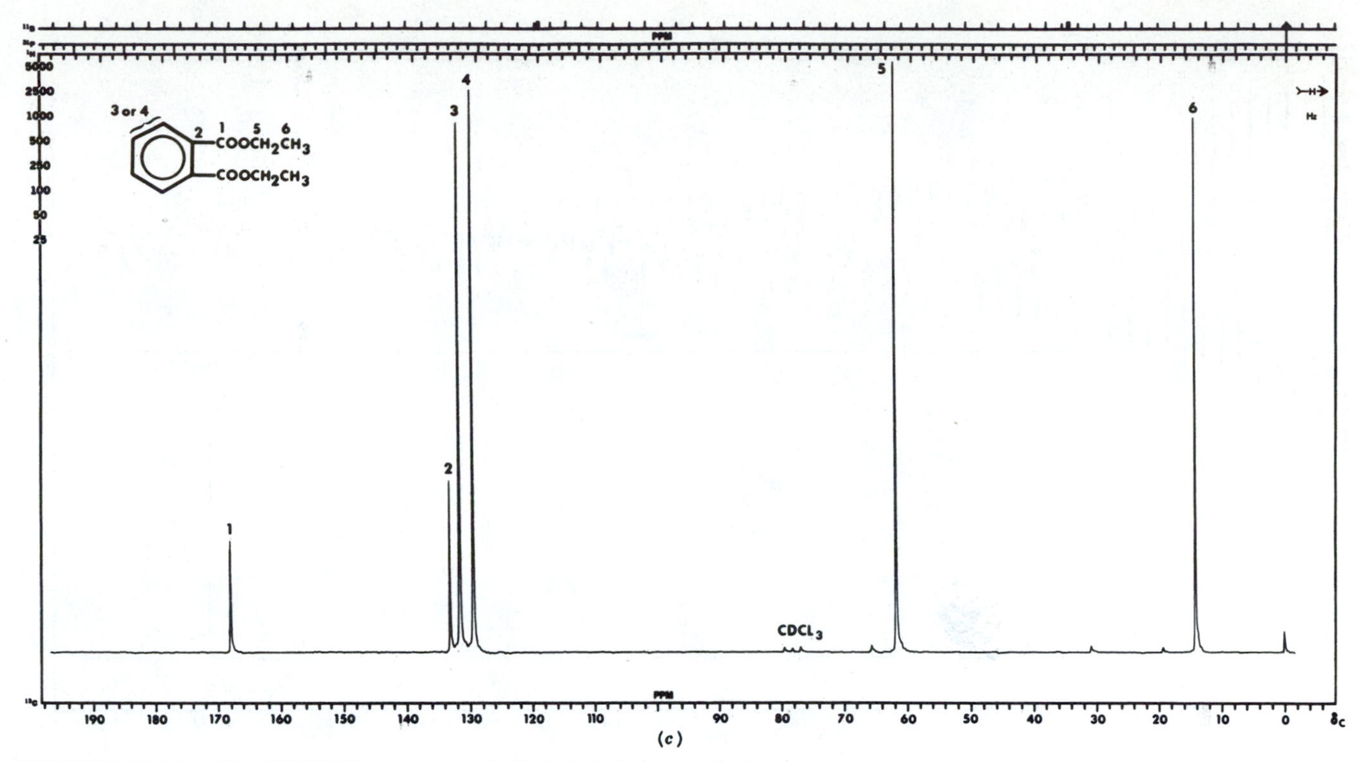

FIGURE 5.1(*c*). The ^{13}C-NMR spectrum of diethyl phthalate with the protons completely decoupled and a 10-s delay between pulses. The solvent used was $CDCl_3$ at 25.2 MHz.

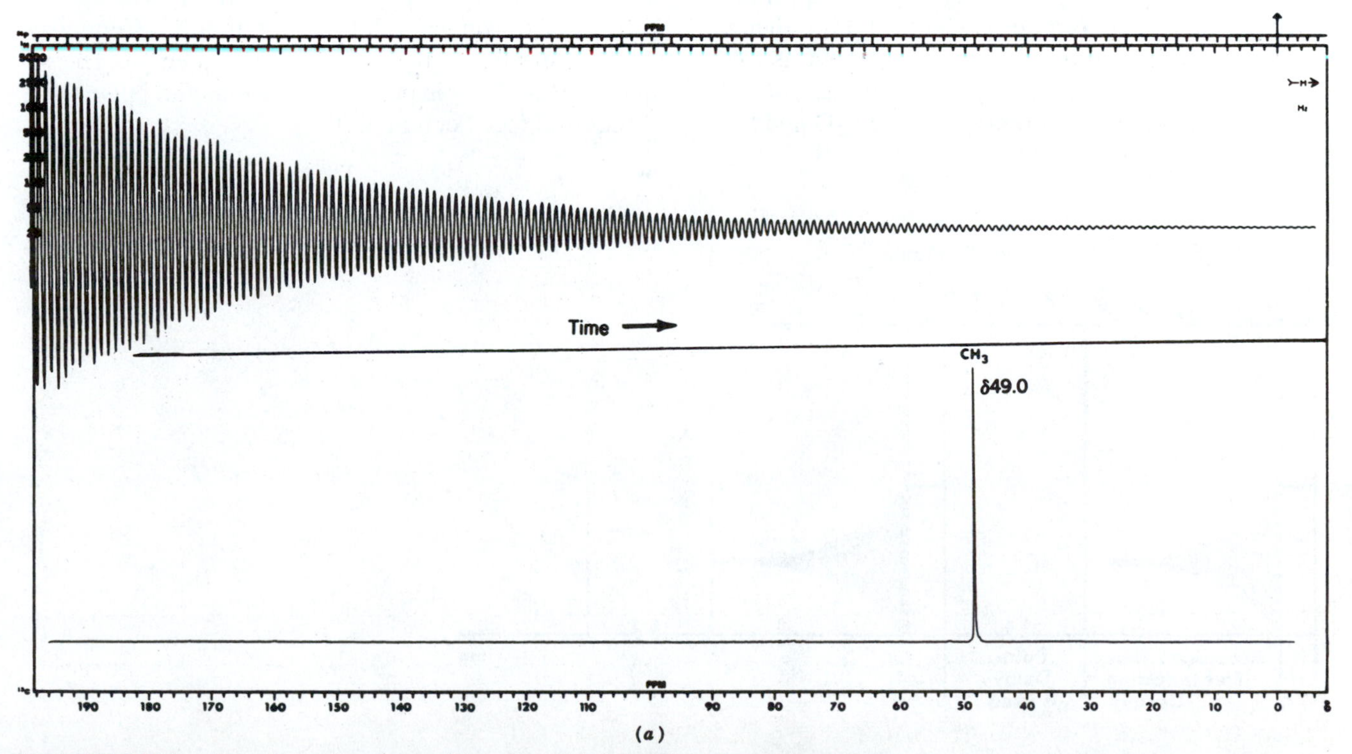

FIGURE 5.2(*a*). Free induction decay (time domain) (above) and transformed ^{13}C-spectrum (frequency domain) of methanol (60% in $CDCl_3$ at 25.2 MHz), decoupled.

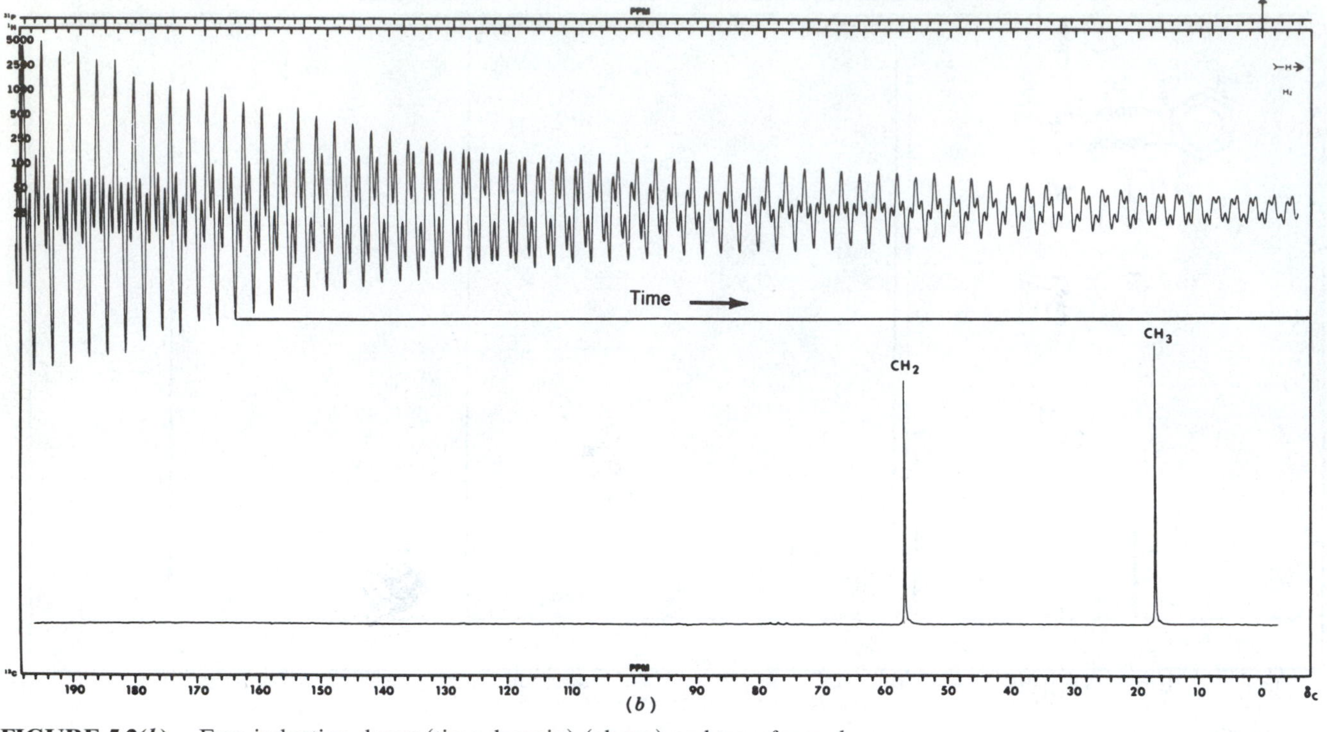

FIGURE 5.2(*b*). Free induction decay (time domain) (above) and transformed ^{13}C-spectrum (frequency domain) of ethanol (60% in $CDCl_3$ at 25.2 MHz), decoupled.

terference (beat) pattern results (Fig. 5.2*b*). These data are automatically digitized and stored in a computer, and a series of repetitive pulses, with signal acquisition and accumulation between pulses, builds up the signal. Fourier transformation by the computer converts this information to the conventional presentation of a ^{13}C NMR spectrum. Figure 5.2*a* represents the FID and the conventional ^{13}C spectrum for CH_3OH. Figure 5.2*b* shows the same spectra for C_2H_5OH. The FID is a time-domain spectrum (the abscissa is time), whereas the transformed, conventional presentation is a frequency-domain spectrum (the abscissa is frequency). The sequence (Fig. 5.2*c*) is pulse, acquisition, and pulse delay if required (see Section 5.2).

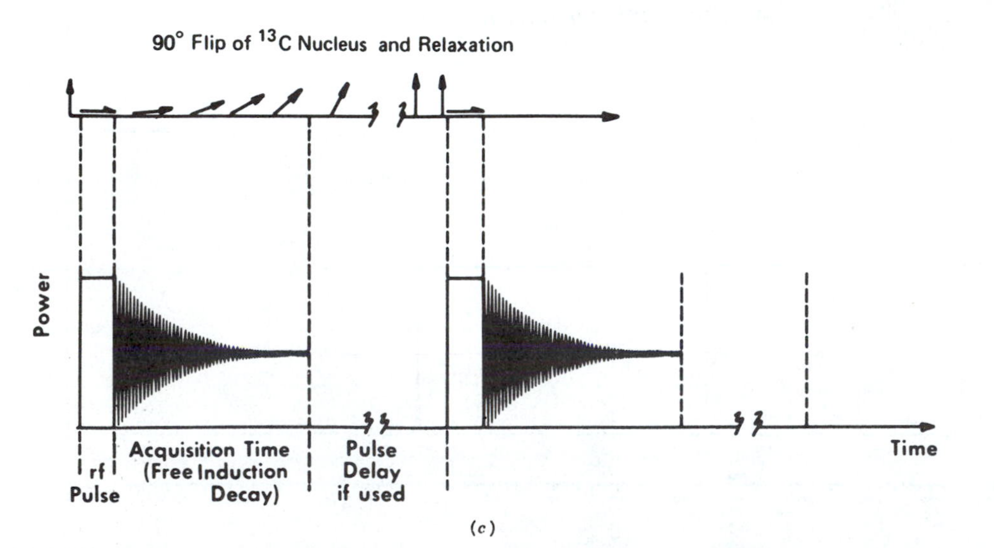

FIGURE 5.2(*c*). Schematic representation of the rf pulse followed by the FID and pulse delay.

Samples for ^{13}C spectrometry are usually dissolved in $CDCl_3$, and the ^{13}C peak of tetramethylsilane (TMS) is used as the internal reference.* A list of the common deuterated solvents is given in Appendix A. The scale is in δ units (ppm). The shifts in routine ^{13}C spectra range over about 240 ppm from TMS—about 20 times that of routine ^{1}H spectra (~12 ppm). As a result of the large range and the sharpness of the decoupled peaks, impurities are readily detected and mixtures may be readily analyzed. Even stereoisomers that are difficult to analyze by means of ^{1}H spectrometry usually show discrete ^{13}C peaks.

An NMR instrument is often described in terms of the resonance frequency for a particular nucleus at a given magnetic field intensity. An instrument with a 7.05-T magnet is thus described at 300-MHz instrument for protons; an 11.7-T magnet corresponds to a 500-MHz instrument for protons. The resonance frequencies for ^{13}C are about $\frac{1}{4}$ those of ^{1}H (the ratio of the γ values for these nuclides). Thus for a 7.05-T magnet, the ^{13}C resonance frequency is about 75.5 MHz, and for an 11.7-T magnet, the resonance frequency is about 126.0 MHz.

A routine ^{13}C spectrum at 75.5 MHz would require about 10 mg of sample in 0.4 mL of solvent in a 5-mm o.d. tube. Samples on the order of 100 μg can be handled in a concentric microcell with an inner tube of 40–100 μL capacity; the annular volume is left empty or is filled with solvent. A probe that accepts a 3-mm i.d. tube gives high sensitivity (See Section 4.6). Reference to further details, collections of spectra, and spectra–structure correlations are appended at the end of the chapter.

5.2 Peak Assignments

5.2.1 Peak Intensity

In pulsed Fourier transform ^{1}H spectra, there is usually a satisfactory relationship between integrated peak areas and the number of protons under the areas because there is sufficient time between pulses for relaxation to occur. In routine pulsed ^{13}C spectra, however, the ^{13}C nuclei, whose relaxation times (T_1) vary over a wide range, are not equally relaxed between pulses, and the peak areas do not integrate for the correct number of nuclei. Long delays between pulses can be used to obtain quantitative results, but the time needed may be prohibitive for routine work. Furthermore, the NOE response is not the same for all ^{13}C nuclei, resulting in further loss of quantitation (see Section 5.6).

However, one advantage does result. It is usually possible by inspection of a ^{13}C spectrum to recognize the nuclei that do not bear protons by their low intensity (peaks 1 and 2 in Fig. 5.1*b*). The common spin-lattice relaxation mechanism for ^{13}C results from dipole-dipole interaction with directly attached protons. Thus, nonprotonated carbon atoms have longer T_1 relaxation times, which together with little or no NOE, results in small peaks. It is therefore often possible to recognize carbonyl groups (except formyl), nitriles, nonprotonated alkene and alkyne carbon atoms, and other quaternary† carbon atoms readily.

However, care must be taken to allow a sufficient number of pulses or a long enough interval between pulses (to compensate for the long T_1) so that these weak signals are not completely lost in the baseline noise. In Figure 5.1*c*, a 10-s interval (pulse delay) was used to increase the relative intensities of peaks 1 and 2).

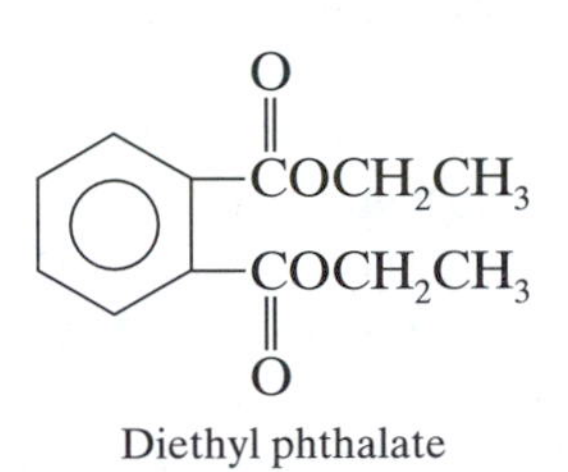

Diethyl phthalate

In the broadband-decoupled spectrum of diethyl phthalate (Fig. 5.1*b*), we can assign the small peak at 167.75 ppm to the two equivalent C=O groups, the small peak at 132.85 ppm to the equivalent substituted aromatic carbon atoms, the large peaks at 131.33 and 129.19 ppm to the remaining aromatic carbon atoms, the medium peak at 61.63 ppm to the two equivalent CH_2 groups, and the medium peak at 14.15 ppm to the two equivalent CH_3 groups. These assignments can be made on the basis of Appendices B and C and on the assumption that the quaternary carbon atoms are responsible for the weak peaks; their relative intensity can be increased by inserting a pulse delay (an interval between the acquisition period and the next pulse) as in Figure 5.1*c*. Note that the 10-s pulse delay nearly equalizes the intensities of all the peaks except for those representing the quaternary carbon atoms.

The most effective procedure for peak assignments is DEPT (Distortionless Enhancement by Polarization Transfer), whereby peaks can be classified as representing CH_3, CH_2, CH (or C). This discussion is reserved for Section 5.5.

* With modern instrumentation, TMS is not actually added; instead, the ^{13}C peak of the deuterated solvent is used as a reference. The spectrum, however, is presented with the ^{13}C peak of TMS at δ 0.00 at the right-hand edge of the scale.

† For want of a better general term, "quaternary" is used to describe any carbon atom without an attached hydrogen atom.

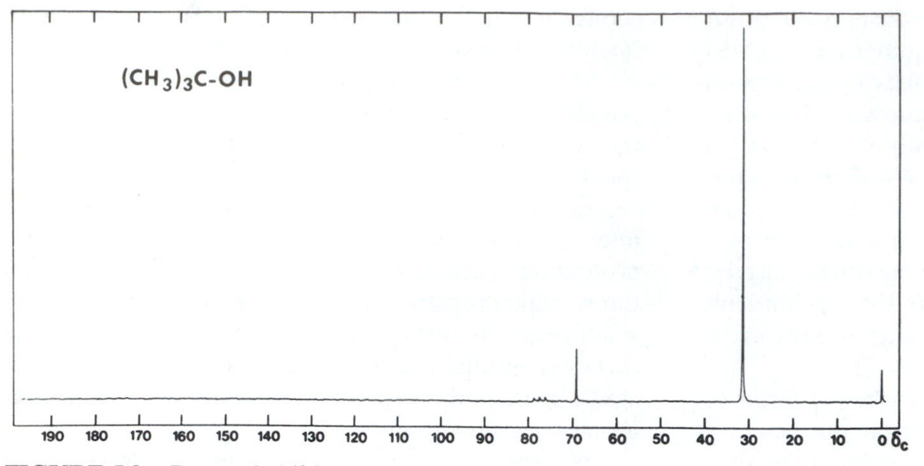

FIGURE 5.3. Decoupled ^{13}C-spectrum of *t*-butyl alcohol. Solvent, $CDCl_3$ at 25.2 MHz; 5000-Hz sweepwidth. From Johnson and Jankowski (1972) spectrum No. 88, with permission.

5.2.2 *Deuterium Substitution*

Substitution of D for H on a carbon results in a dramatic diminution of the height of the ^{13}C signal in a broad-band-decoupled spectrum for the following reasons. Since deuterium has a spin number of 1 and a magnetic moment 15% that of ^{1}H, it will split the ^{13}C absorption into three lines (ratio 1:1:1) with a J value equal to $0.15 \times J_{CH}$. Furthermore, T_1 for ^{13}C—D is longer than that for ^{13}C—H because of decreased dipole-dipole relaxation. Finally, the NOE is lost, since there is no irradiation of deuterium.* A separate peak may also be seen for any residual ^{13}C—H, since the isotope effect usually results in a slight upfield shift of the ^{13}C—D absorption (~0.2 ppm per D atom). The isotope effect may also slightly shift the absorption of the carbon atoms once removed from the deuterated carbon.

5.2.3 *Chemical Shift Equivalence*

The definition of chemical shift equivalence given for protons also applies to carbon atoms: interchangeability by a symmetry operation or by a rapid mechanism. The presence of equivalent carbon atoms (or coincidence of shift) in a molecule results in a discrepancy between the apparent number of peaks and the actual number of carbon atoms in the molecule.

* The same explanation also accounts for the relatively weak signal shown by deuterated solvents. In addition, small solvent molecules tumble rapidly; this rapid movement makes for a longer T_1, hence for smaller peaks. Deuterated chloroform, $CDCl_3$, shows a 1:1:1 *triplet*, deuterated *p*-dioxane a 1:2:3:2:1 *quintet*, and deuterated DMSO $(CD_3)_2SO$, a 1:3:6:7:6:3:1 *septet* in accordance with the $2nI + 1$ rule (Chapter 4). The chemical shifts, coupling constants, and multiplicities of the ^{13}C atoms of common NMR solvents are given in Appendix A.

Thus, ^{13}C atoms of the methyl groups in *t*-butyl alcohol (Fig. 5.3) are equivalent by rapid rotation in the same sense in which the protons of a methyl group are equivalent. The ^{13}C spectrum of *t*-butyl alcohol shows two peaks, one much larger than the other, but not necessarily exactly three times as large; the carbinyl carbon peak (quaternary) is much less than $\frac{1}{3}$ the intensity of the peak representing the carbon atoms of the methyl groups.

In the chiral molecule 2,2,4-trimethyl-1,3-pentanediol (Fig. 5.4), we note that CH_3a and CH_3b are not equivalent, and two peaks are seen. Even though the two methyl groups labeled *c* are not equivalent, they coincidently show only one peak. Two peaks may be seen at higher resolution.

In Section 4.12.3 we noted that the CH_3 protons of $(CH_3)_2NCH{=}O$ gave separate peaks at room temperature, but became chemical shift equivalent at about 123°. Of course, the ^{13}C peaks show similar behavior.

5.3 *Chemical Classes and Chemical Shifts*

In this section, chemical shifts will be discussed under the headings of the common chemical classes of organic compounds. As noted earlier, the range of shifts generally encountered in routine ^{13}C studies is about 240 ppm.

As a first reassuring statement, we can say that trends in chemical shifts of ^{13}C are somewhat parallel to those of ^{1}H, so that some of the "feeling" for ^{1}H spectra may carry over to ^{13}C spectra. Furthermore, the concept of *additivity of substituent effects* (see Sections 4.7 and 5.3.6) is useful for both spectra. The ^{13}C shifts are related mainly to hybridization and substituent electro-

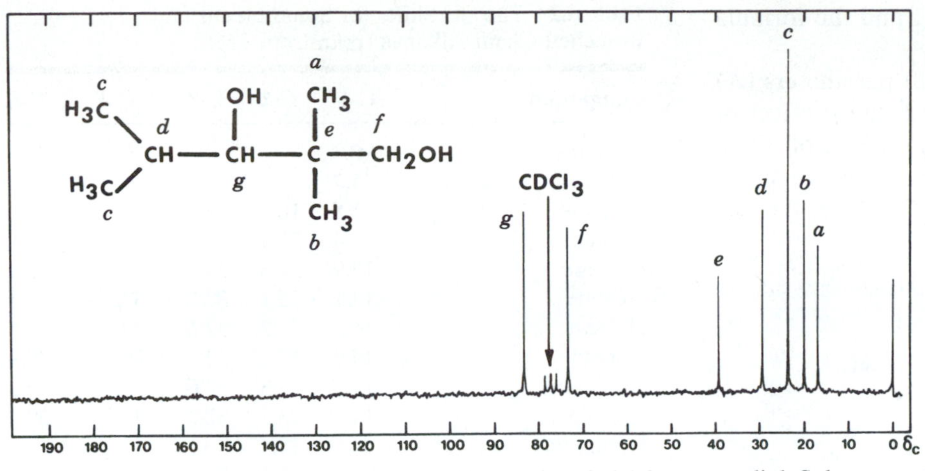

FIGURE 5.4. Decoupled ^{13}C-spectrum of 2,2,4-trimethyl-1,3-pentanediol. Solvent used was $CDCl_3$ at 25.2 MHz; 5000-Hz sweepwidth. From Johnson and Jankowski (1972), spectrum No. 324, with permission.

negativity; solvent effects are important in both spectra. Chemical shifts for ^{13}C are affected by substitutions as far removed as the δ position; in the benzene ring, pronounced shifts for ^{13}C are caused by substituents at the ortho, meta, and para positions. The ^{13}C chemical shifts are also moved significantly to the right by the γ-gauche effect (see Section 3.5.1.1). Shifts to the right as much as several parts per million may occur on dilution. Hydrogen-bonding effects with polar solvents may cause shifts to the left.

Appendix B gives credence to the statement that ^{13}C chemical shifts somewhat parallel those of ^{1}H, but we note some divergences that are not readily explainable and require development of another set of interpretive skills. In general, in comparison with ^{1}H spectra, it seems more difficult to correlate ^{13}C shifts with substituent electronegativity.

As in other types of spectrometry, peak assignments are made on the basis of reference compounds. Reference material for many classes of compounds has accumulated in the literature. The starting point is a general correlation chart for chemical shift regions of ^{13}C atoms in the major chemical classes (see Appendices B and C); then, minor changes within these regions are correlated with structure variations in the particular chemical class. The chemical shift values in the following tables must not be taken too literally because of the use of various solvents and concentration. (Furthermore, much of the early work used various reference compounds, and the values were corrected to give parts per million from TMS.) For example, the C=O absorption of acetophenone in $CDCl_3$ appears at 2.4 ppm further to the left than in CCl_4; the effect on the other carbon atoms of acetophenone ranges from 0.0 to 1.1 ppm.

A ^{13}C spectrum will often distinguish substitution patterns on an aromatic ring. If, for example, there are two identical (achiral) substituents, the symmetry elements alone will distinguish among the para, ortho, and meta isomers if the chemical shifts of the ring carbon atoms are sufficiently different. The para isomer has two simple axes and two planes. The ortho and meta isomers have one simple axis and one plane, but in the meta isomer, the elements pass through two carbon atoms. There is also a symmetry plane in the plane of the ring in each compound, which does not affect the ring carbon atoms.

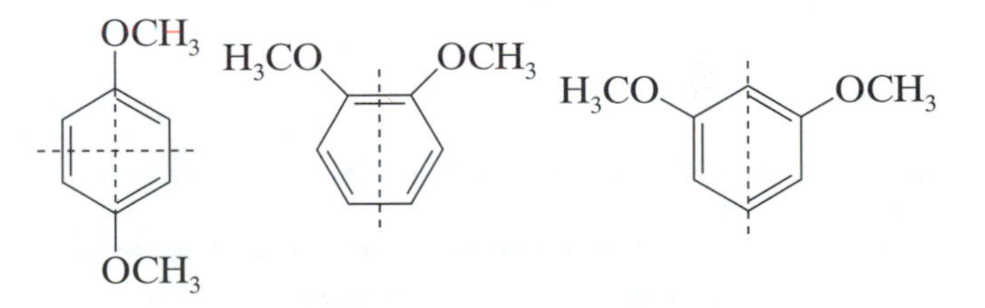

The aromatic region of the ^{13}C spectrum for the para isomer shows two peaks; for the ortho isomer, three peaks; and for the meta isomer, four peaks. The quaternary carbon peaks are much less intense than the unsubstituted carbon peaks.

The additivity of shift increments is demonstrated in Section 5.3.6.

5.3.1 Alkanes

5.3.1.1 Linear and Branched Alkanes We know from the general correlation chart (Appendix C) that alkane groups unsubstituted by heteroatoms absorb to about 60 ppm. (Methane absorbs at -2.5 ppm). Within this range, we can predict the chemical shifts of individual ^{13}C atoms in a straight-chain or branched-chain hydrocarbon from the data in Table 5.1 and the formula given below.

This table shows the additive shift parameters (A) in hydrocarbons: the α effect of $+9.1$, the β effect of $+9.4$ ppm, the γ effect of -2.5, the δ effect of $+0.3$, the ϵ effect of $+0.1$, and the corrections for branching effects. The calculated (and observed) shifts for the carbon atoms of 3-methylpentane are

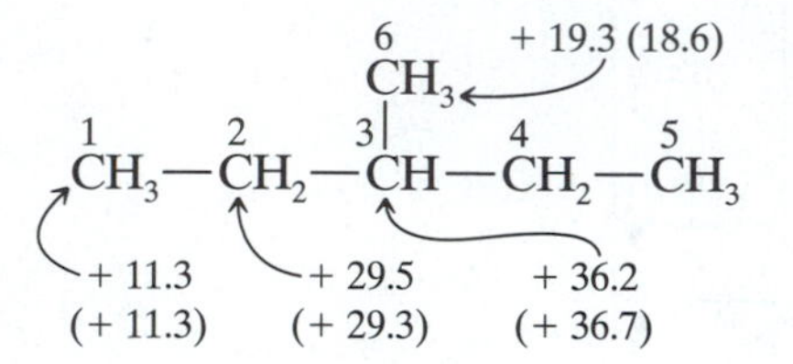

Calculations of shift are made from the formula: $\delta = -2.5 + \Sigma nA$, where δ is the predicted shift for a carbon atom; A is the additive shift parameter; and n is the number of carbon atoms for each shift parameter (-2.5 is the shift of the ¹³C of methane). Thus, for carbon atom 1, we have 1 α-, 1 β-, 2 γ-, and 1 δ-carbon atoms.

$$\delta_1 = -2.5 + (9.1 \times 1) + (9.4 \times 1) + (-2.5 \times 2) + (0.3 \times 1) = +11.3$$

Carbon atom 2 has 2 α-, 2 β-, and 1 γ-carbon atoms. Carbon atom 2 is a 2° carbon with a 3° carbon attached [2°(3°) = −2.5].

$$\delta_2 = -2.5 + (9.1 \times 2) + (9.4 \times 2) + (-2.5 \times 1) + (-2.5 \times 1) = 29.5$$

Table 5.1 The ¹³C Shift Parameters in Some Linear and Branched Hydrocarbons

¹³C Atoms	Shift (ppm) (A)
α	+9.1
β	+9.4
γ	−2.5
δ	+0.3
ϵ	+0.1
1° (3°)[a]	−1.1
1° (4°)[a]	−3.4
2° (3°)[a]	−2.5
2° (4°)	−7.2
3° (2°)	−3.7
3° (3°)	−9.5
4° (1°)	−1.5
4° (2°)	−8.4

[a] The notations 1° (3°) and 1° (4°) denote a CH_3 group bound to a R_2CH group and to a R_3C group, respectively. The notation 2° (3°) denotes a RCH_2 group bound to a R_2CH group, and so on.

Table 5.2 The ¹³C Shifts for Some Linear and Branched-Chain Alkanes (ppm from TMS)

Compound	C-1	C-2	C-3	C-4	C-5
Methane	−2.3				
Ethane	5.7				
Propane	15.8	16.3	15.8		
Butane	13.4	25.2	25.2		
Pentane	13.9	22.8	34.7	22.8	13.9
Hexane	14.1	23.1	32.2	32.2	23.1
Heptane	14.1	23.2	32.6	29.7	32.6
Octane	14.2	23.2	32.6	29.9	29.9
Nonane	14.2	23.3	32.6	30.0	30.3
Decane	14.2	23.2	32.6	31.1	30.5
Isobutane	24.5	25.4			
Isopentane	22.2	31.1	32.0	11.7	
Isohexane	22.7	28.0	42.0	20.9	14.3
Neopentane	31.7	28.1			
2,2-Dimethylbutane	29.1	30.6	36.9	8.9	
3-Methylpentane	11.5	29.5	36.9	(18.8, 3-CH_3)	
2,3-Dimethylbutane	19.5	34.3			
2,2,3-Trimethylbutane	27.4	33.1	38.3	16.1	
2,3-Dimethylpentane	7.0	25.3	36.3	(14.6, 3-CH_3)	

Carbon atom 3 has 3 α- and 2 β-carbon atoms, and it is a 3° atom with two 2° atoms attached [3°(2°) = −3.7]. Thus,

$$\delta_3 = -2.5 + (9.1 \times 3) + (9.4 \times 2) + (-3.7 \times 2) = +36.2$$

Carbon atom 6 has 1 α-, 2 β-, and 2 γ-carbon atoms, and it is a 1° atom with a 3° atom attached [1°(3°) = −1.1]. Thus,

$$\delta_6 = -2.5 + (9.1 \times 1) + (9.4 \times 2) + (-2.5 \times 2) + (-1.1 \times 1) = +19.3$$

The agreement with the determined values for such calculations is very good. Another useful calculation has been given.* The ¹³C γ shift to lower frequency resulting from the γ carbon has been attributed to the steric compression of a *gauche* interaction but has no counterpart in ¹H spectra. It accounts, for example, for the shift to the right of an axial methyl substituent on a conformationally rigid cyclohexane ring, relative to an equatorial methyl, and for the shift to the right of the γ carbon atoms of the ring.

* Lindeman, L.P., and Adams, J.Q. (1971). *Anal. Chem.*, **43,** 1245.

Table 5.2 lists the shifts in some linear and branched alkanes.

5.3.2 *Effect of Substituents on Alkanes*

Table 5.3 shows the effects of a substituent on linear and branched alkanes. The effect on the α carbon parallels the electronegativity of the substituent except for bromine and iodine. The effect at the β carbon seems fairly constant for all the substituents except for the carbonyl, cyano, and nitro groups. The shift to the right at the γ carbon results (as above) from steric compression of a *gauche* interaction. For Y = N, O, and F, there is also a shift to the right with Y in the *anti* conformation, attributed to hyperconjugation.

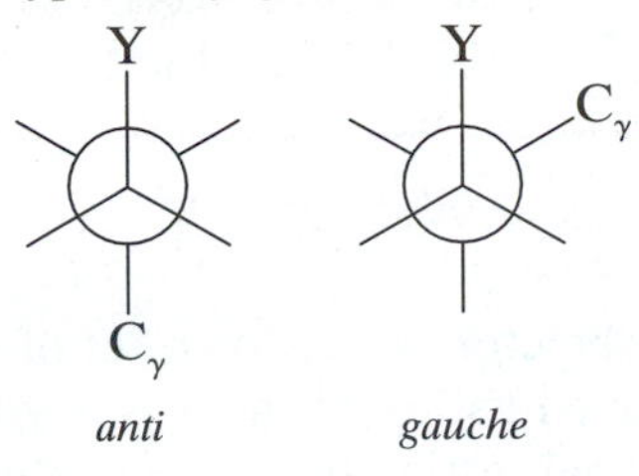

Table 5.3 Incremental Substituent Effects (ppm) on Replacement of H by Y in Alkanes. Y is Terminal or Internal[a] (+ left, − right)

γ α β Y (Terminal) — γ β Y α β γ (Internal)

	α		β		γ
Y	**Terminal**	**Internal**	**Terminal**	**Internal**	
CH_3	+9	+6	+10	+8	−2
$CH{=}CH_2$	+20		+6		−0.5
$C{\equiv}CH$	+4.5		+5.5		−3.5
COOH	+21	+16	+3	+2	−2
COO^-	+25	+20	+5	+3	−2
COOR	+20	+17	+3	+2	−2
COCl	+33	+28		+2	
$CONH_2$	+22		+2.5		−0.5
COR	+30	+24	+1	+1	−2
CHO	+31		0		−2
Phenyl	+23	+17	+9	+7	−2
OH	+48	+41	+10	+8	−5
OR	+58	+51	+8	+5	−4
OCOR	+51	+45	+6	+5	−3
NH_2	+29	+24	+11	+10	−5
NH_3^+	+26	+24	+8	+6	−5
NHR	+37	+31	+8	+6	−4
NR_2	+42		+6		−3
NR_3^+	+31		+5		−7
NO_2	+63	+57	+4	+4	
CN	+4	+1	+3	+3	−3
SH	+11	+11	+12	+11	−4
SR	+20		+7		−3
F	+68	+63	+9	+6	−4
Cl	+31	+32	+11	+10	−4
Br	+20	+25	+11	+10	−3
I	−6	+4	+11	+12	−1

[a] Add these increments to the shift values of the appropriate carbon atom in Table 5.2 or to the shift value calculated from Table 5.1.

Source: Wehrli, F.W., Marchand, A.P., and Wehrli, S. (1983). *Interpretation of Carbon-13 NMR Spectra,* 2nd ed. London: Heyden.

Table 5.3 provides the functional-group increments that must be added to the appropriate shift values for alkanes given in Table 5.2. For example, we can calculate the ^{13}C shifts for 3-pentanol.

$$\overset{\gamma}{CH_3}-\overset{\beta}{CH_2}-\overset{\alpha}{\underset{\displaystyle OH}{\underset{|}{CH}}}-\overset{\beta}{CH_2}-\overset{\gamma}{CH_3}$$

The OH substituent is attached "internally" (rather than "terminally") to the linear alkane chain of pentane; the point of attachment is labeled α, which corresponds to C-3 of pentane, for which the shift value of 34.7 is given in Table 5.3. To this value is added the increment +41, this being the increment for an OH group attached internally to the α carbon of 3-pentanol (see line 12, 2nd column of numbers). The shift, therefore, for the point of attachment (the α carbon) is calculated as 75.8. The β and γ shifts are calculated as follows. All of the calculated shifts are in reasonable agreement with the determined values (Table 5.11).

	Calculated	Determined (See Table 5.11)
C_α	34.7 + 41 = 75.8	73.8
C_β	22.8 + 8 = 30.8	29.7
C_γ	13.9 − 5 = 8.9	9.8

1-Pentanol would be treated similarly but as a "terminal" alcohol, and the carbon atom to which the OH group is attached would again be labeled α.

5.3.3 *Cycloalkanes and Saturated Heterocyclics*

The chemical shifts of the CH_2 groups in monocyclic alkanes are given in Table 5.4.

The striking feature here is the strong shift to the right of cyclopropane, analogous to the shift of its proton absorptions.

Table 5.4 Chemical Shifts of Cycloalkanes (ppm from TMS)

C_3H_6	−2.9	C_7H_{14}	28.4
C_4H_8	22.4	C_8H_{16}	26.9
C_5H_{10}	25.6	C_9H_{18}	26.1
C_6H_{12}	26.9	$C_{10}H_{20}$	25.3

Each ring skeleton has its own set of shift parameters, but a detailed listing of these is beyond the scope of this text. Rough estimates for substituted rings can be made with the substitution increments in Table 5.3.

One of the striking effects in rigid cyclohexane rings is the shift to the right caused by the *γ-gauche* steric compression. Thus an axial methyl group at C-1 causes a shift to the right of several parts per million at C-3 and C-5.

Table 5.5 presents chemical shifts for several saturated heterocyclics.

5.3.4 *Alkenes*

The sp^2 carbon atoms of alkenes substituted only by alkyl groups absorb in the range of about 110–150 ppm.

Table 5.5 Chemical Shifts for Saturated Heterocyclics (ppm from TMS, neat)

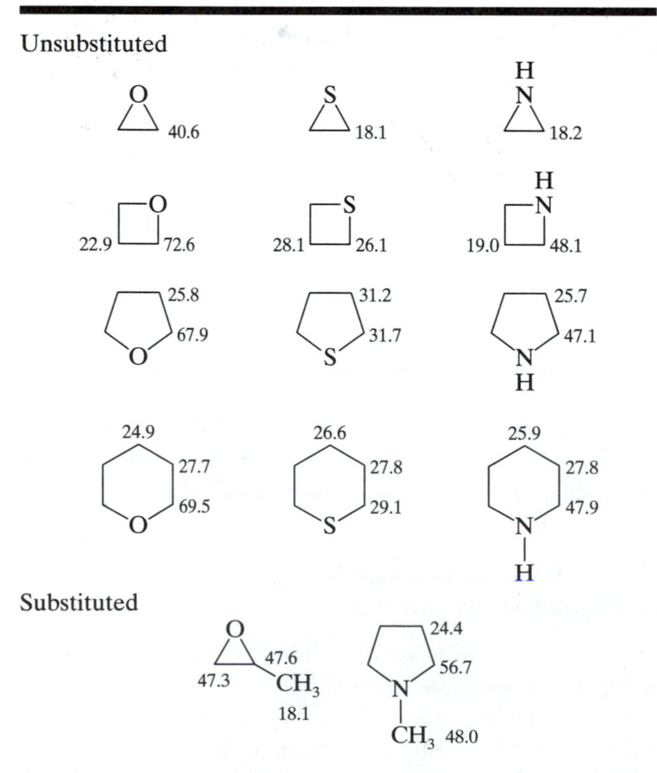

The double bond has a rather small effect on the shifts of the sp^3 carbons in the molecule as the following comparisons demonstrate.

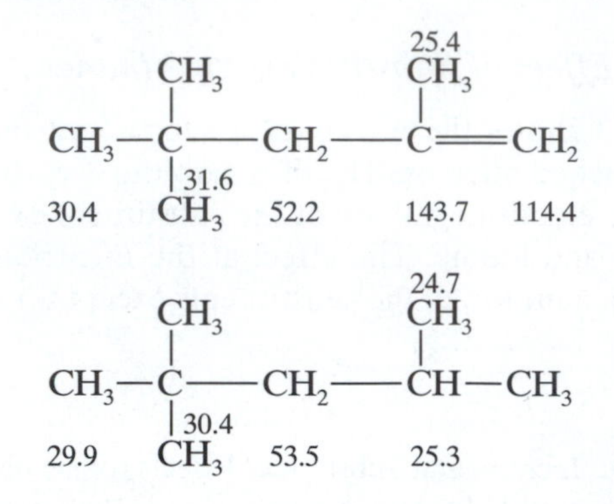

The methyl signal of propene is at 18.7 ppm, and of propane at 15.8 ppm. In (*Z*)-2-butene, the methyl signals are at 12.1 ppm, compared with 17.6 ppm in (*E*)-2-butene, because of the γ effect. (For comparison, the methyl signals of butane are at 13.4 ppm). Note the γ effect on one of the geminal methyl groups in 2-methyl-2-butene (Table 5.6).

In general, the terminal $=CH_2$ group absorbs to the right relative to an internal =CH— group, and *Z* —CH=CH— signals are to the right from those of corresponding *E* groups. Calculations of approximate shifts can be made from the following parameters where α, β, and γ represent substitutents on the same end of the double bond as the alkene carbon of interest, and α', β', and γ' represent substituents on the far side.

α	+10.6
β	+7.2
γ	−1.5
α'	−7.9
β'	−1.8
γ'	−1.5
Z(cis) correction	−1.1

These parameters are added to 123.3 ppm, the shift for ethylene. We can calculate the values of C-3 and C-2 for (*Z*)-3-methyl-2-pentene as follows:

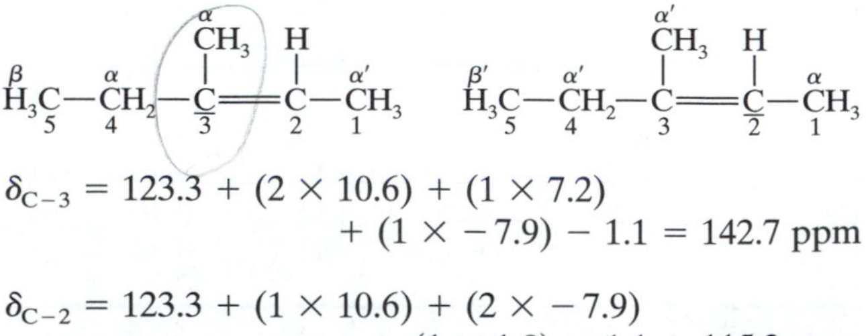

$$\delta_{C-3} = 123.3 + (2 \times 10.6) + (1 \times 7.2) + (1 \times -7.9) - 1.1 = 142.7 \text{ ppm}$$

$$\delta_{C-2} = 123.3 + (1 \times 10.6) + (2 \times -7.9) + (1 \times 1.8) - 1.1 = 115.2 \text{ ppm}$$

The measured values are C-3 = 137.2 and C-2 = 116.8. The agreement is fair.

Table 5.6 Alkene and Cycloalkene Chemical Shifts (ppm from TMS)

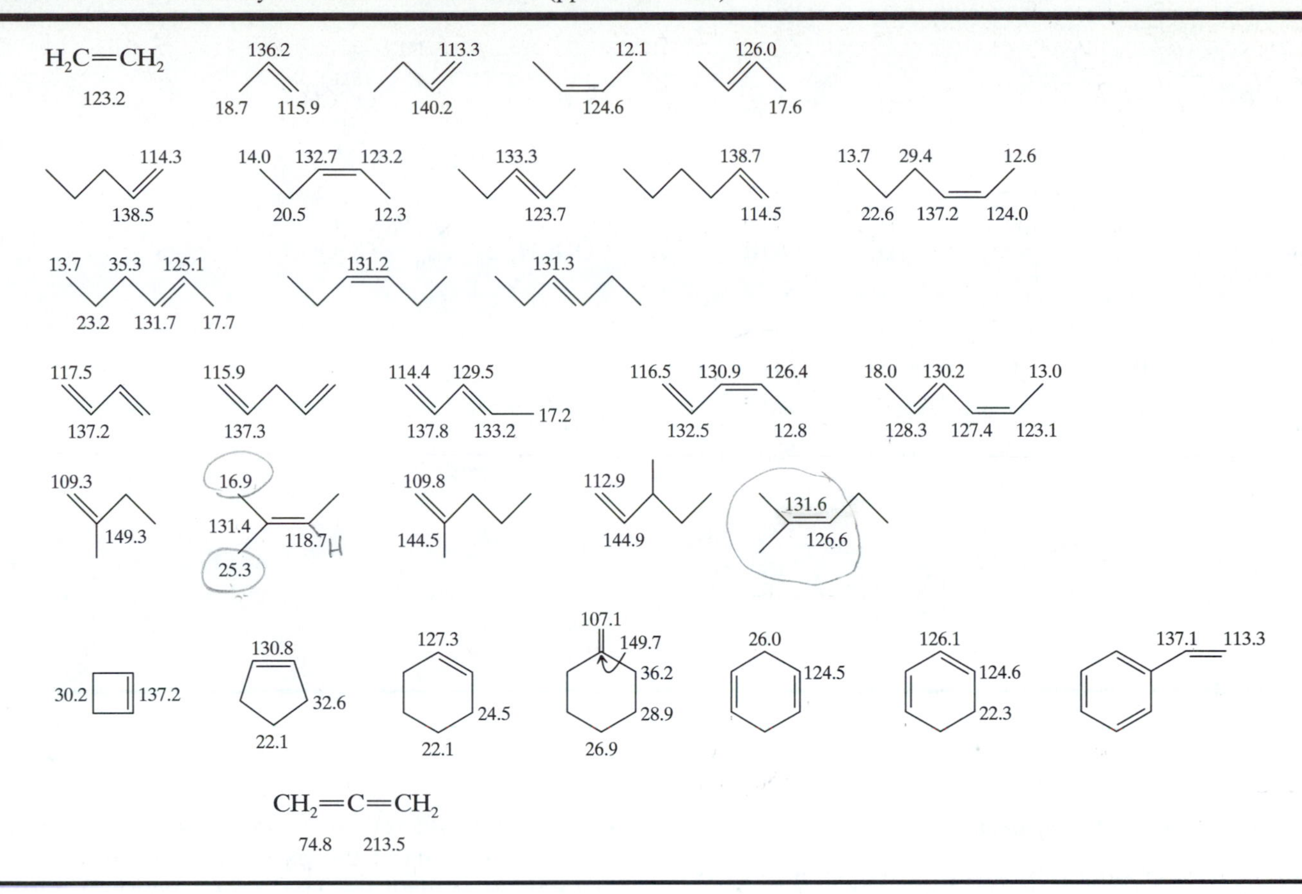

Carbon atoms directly attached to a (*Z*) C=C group are more shielded than those attached to the stereoisomeric (*E*) group by 4–6 ppm (Table 5.6). Alkene carbon atoms in polyenes are treated as though they were alkane carbon substituents on one of the double bonds. Thus, in calculating the shift of C-2 in 1,4-pentadiene, C-4 is treated like a β-sp^3 carbon atom.

Representative alkenes are presented in Table 5.6.

There are no simple rules to handle polar substituents on an alkene carbon. The shifts for vinyl ethers can be rationalized on the basis of electron density of the contributor structures

$$\underset{84.2}{CH_2}=\underset{153.2}{CH}-\ddot{\underset{..}{O}}-CH_3 \longleftrightarrow {}^{-}\ddot{C}H_2-CH=\underset{..}{O}^{+}-CH_3$$

as can the shifts for α,β-unsaturated ketones.

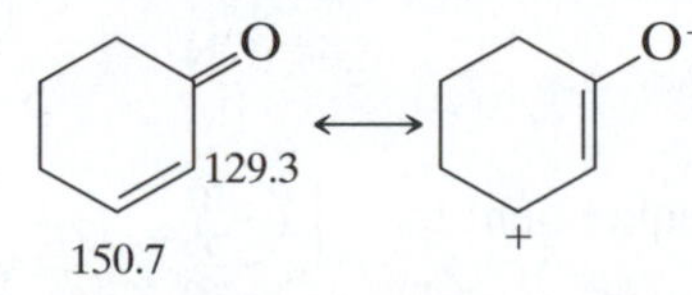

The same rationalization applies to the proton shifts in these compounds. Shifts for several substituted alkenes are presented in Table 5.7.

The central carbon atom (=C=) of alkyl-substituted allenes absorbs in the range of about 200–215 ppm, whereas the terminal atoms (*C*=C=*C*) absorb in the range of about 75–97 ppm.

5.3.5 *Alkynes*

The *sp* carbon atoms of alkynes substituted only by alkyl groups absorb in the range of approximately 65–90 ppm (Table 5.8). The triple bond shifts the sp^3 carbon atoms, directly attached, about 5–15 ppm to the right relative to the corresponding alkane. The terminal ≡CH absorbs further to the right than the internal ≡CR. Alkyne carbon atoms with a polar group directly attached absorb from about 20 to 95 ppm.

Polar resonance structures explain these shifts for alkynyl ethers, which are analogous to the shifts for vinyl ethers (Section 4.34).

$$\overset{23.2}{HC}\equiv\overset{89.4}{C}-OCH_2CH_3 \qquad CH_3-\overset{28.0}{C}\equiv\overset{88.4}{C}-O-CH_3$$

Table 5.7 Chemical Shifts of Substituted Alkenes (ppm from TMS)

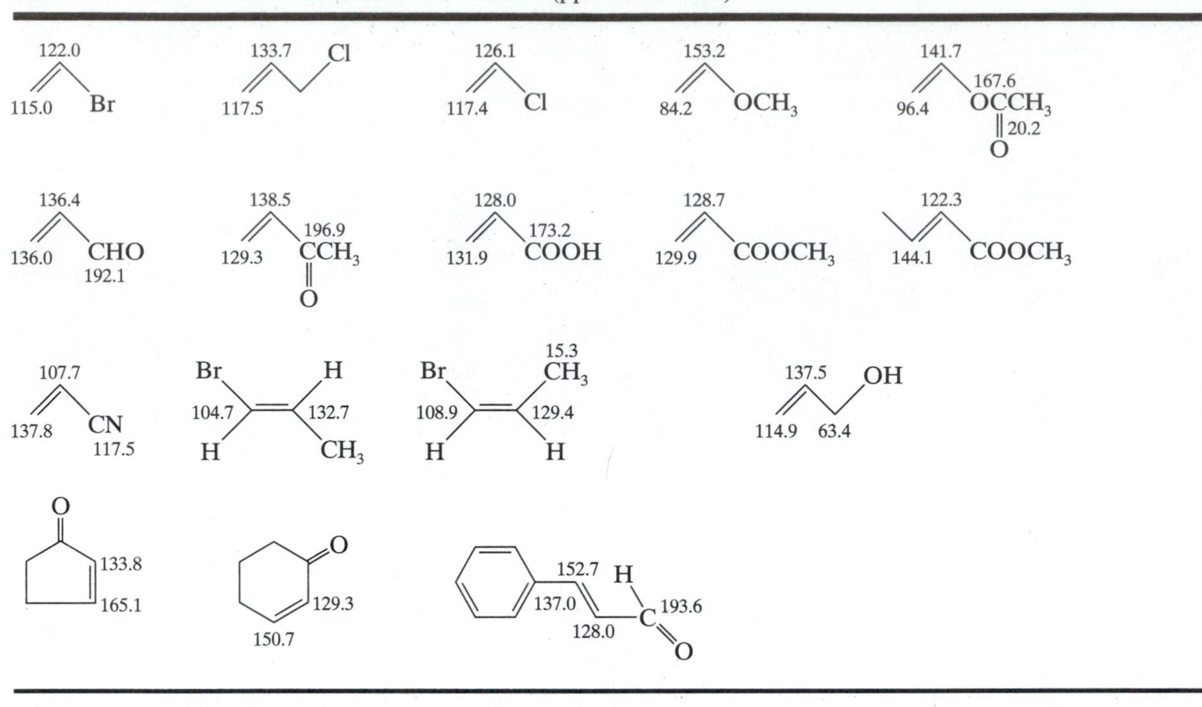

5.3.6 *Aromatic Compounds*

Benzene carbon atoms absorb at 128.5 ppm, neat or as a solution in $CDCl_3$. Substituents shift the attached aromatic carbon atom as much as ± 35 ppm. Fused-ring absorptions are as follows:

Naphthalene: C-1, 128.1; C-2, 125.9; C-4a, 133.7.

Anthracene: C-1, 130.1; C-2, 125.4; C-4a, 132.2; C-9, 132.6.

Phenanthrene: C-1, 128.3; C-2, 126.3; C-3, 126.3; C-4, 122.2; C-4a, 131.9*; C-9, 126.6; C-10a, 130.1.*

Shifts of the aromatic carbon atom directly attached to the substituent have been correlated with substituent electronegativity after correcting for magnetic anisotropy effects; shifts at the para aromatic carbon have been correlated with the Hammett σ constant. Ortho shifts are not readily predictable and range over about 15 ppm. Meta shifts are generally small—up to several parts per million for a single substituent.

The substituted aromatic carbon atoms can be distinguished from the unsubstituted aromatic carbon atom by its decreased peak height; that is, it lacks a proton and thus suffers from a longer T_1 and a diminished NOE.

Incremental shifts from the carbon atoms of benzene for the aromatic carbon atoms of representative monosubstituted benzene rings (and shifts from TMS of carbon-containing substituents) are given in Table 5.9. Shifts from benzene for polysubstituted benzene ring carbon atoms can be approximated by applying the principle of increment additivity. For example, the shift from benzene for C-2 of the disubstituted compound 4-chlorobenzonitrile is calculated by adding the effect for an ortho CN group (+3.6) to that for a meta Cl group (+1.0):

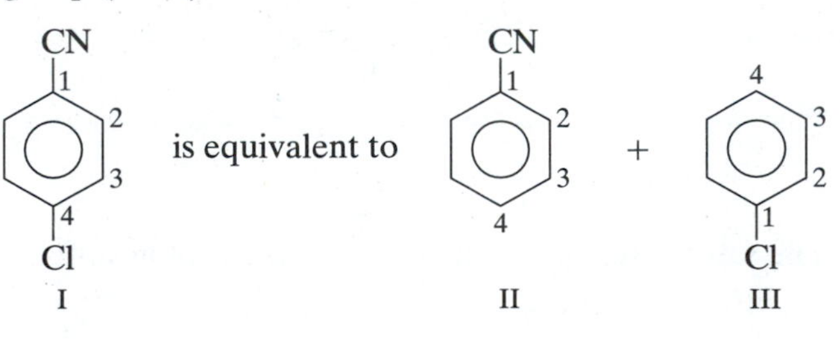

* Assignment uncertain.

Table 5.8 Alkyne Chemical Shifts (ppm)

Compound	C-1	C-2	C-3	C-4	C-5	C-6
1-Butyne	67.0	84.7				
2-Butyne		73.6				
1-Hexyne	68.1	84.5	18.1	30.7	21.9	13.5
2-Hexyne	2.7	73.7	76.9	19.6	21.6	12.1
3-Hexyne	15.4	13.0	80.9			

Table 5.9 Incremental Shifts of the Aromatic Carbon Atoms of Monosubstituted Benzenes (ppm from Benzene at 128.5 ppm, + to the left, − to the right). Carbon Atom of Substituents in parts per million from TMS[a]

Substituent	C-1 (Attachment)	C-2	C-3	C-4	C of Substituent (ppm from TMS)
H	0.0	0.0	0.0	0.0	
CH_3	+9.3	+0.7	−0.1	−2.9	21.3
CH_2CH_3	+15.6	−0.5	0.0	−2.6	29.2 (CH_2), 15.8 (CH_3)
$CH(CH_3)_2$	+20.1	−2.0	0.0	−2.5	34.4 (CH), 24.1 (CH_3)
$C(CH_3)_3$	+22.2	−3.4	−0.4	−3.1	34.5 (C), 31.4 (CH_3)
$CH{=}CH_2$	+9.1	−2.4	+0.2	−0.5	137.1 (CH), 113.3 (CH_2)
$C{\equiv}CH$	−5.8	+6.9	+0.1	+0.4	84.0 (C), 77.8 (CH)
C_6H_5	+12.1	−1.8	−0.1	−1.6	
CH_2OH	+13.3	−0.8	−0.6	−0.4	64.5
$CH_2OC(=O)CH_3$	+7.7	~0.0	~0.0	~0.0	20.7 (CH_3), 66.1 (CH_2), 170.5 (C=O)
OH	+26.6	−12.7	+1.6	−7.3	
OCH_3	+31.4	−14.4	+1.0	−7.7	54.1
OC_6H_5	+29.0	−9.4	+1.6	−5.3	
$OC(=O)CH_3$	+22.4	−7.1	−0.4	−3.2	23.9 (CH_3), 169.7 (C=O)
$C(=O)H$	+8.2	+1.2	+0.6	+5.8	192.0
$C(=O)CH_3$	+7.8	−0.4	−0.4	+2.8	24.6 (CH_3), 195.7 (C=O)
$C(=O)C_6H_5$	+9.1	+1.5	−0.2	+3.8	196.4 (C=O)
$C(=O)CF_3$	−5.6	+1.8	+0.7	+6.7	
$C(=O)OH$	+2.9	+1.3	+0.4	+4.3	168.0
$C(=O)OCH_3$	+2.0	+1.2	−0.1	+4.8	51.0 (CH_3), 166.8 (C=O) 168.5
$C(=O)Cl$	+4.6	+2.9	+0.6	+7.0	
$C(=O)NH_2$	+5.0	−1.2	0.0	+3.4	
$C{\equiv}N$	−16.0	+3.6	+0.6	+4.3	119.5
NH_2	+19.2	−12.4	+1.3	−9.5	
$N(CH_3)_2$	+22.4	−15.7	+0.8	−11.8	40.3
$NHC(=O)CH_3$	+11.1	−9.9	+0.2	−5.6	
NO_2	+19.6	−5.3	+0.9	+6.0	
$N{=}C{=}O$	+5.7	−3.6	+1.2	−2.8	129.5
F	+35.1	−14.3	+0.9	−4.5	
Cl	+6.4	+0.2	+1.0	−2.0	
Br	−5.4	+3.4	+2.2	−1.0	
I	−32.2	+9.9	+2.6	−7.3	
CF_3	+2.6	−3.1	+0.4	+3.4	
SH	+2.3	+0.6	+0.2	−3.3	
SCH_3	+10.2	−1.8	+0.4	−3.6	15.9
SO_2NH_2	+15.3	−2.9	+0.4	+3.3	
$Si(CH_3)_3$	+13.4	+4.4	−1.1	−1.1	

[a] See Ewing, D.E., (1979). *Org. Magn. Reson.*, **12,** 499, for 709 chemical shifts of monosubstituted benzenes.

I			II		III	
C Atom	Calculated	Observed	C Atom	Observed	C Atom	Observed
1	−18.0	−16.6	1	−16.0	4	−2.0
2	+ 4.6	+ 5.1	2	+ 3.6	3	+1.0
3	+ 0.8	+ 1.3	3	+ 0.6	2	+0.2
4	+10.7	+10.8	4	+ 4.3	1	+6.4

5.3.7 *Heteroaromatic Compounds*

Complex rationalizations have been offered for the shifts of carbon atoms in heteroaromatic compounds. As a general rule, C-2 of oxygen- and nitrogen-containing rings is further to the left than C-3. Large solvent and pH effects have been recorded. Table 5.10 gives values for neat samples of several five- and six-membered heterocyclic compounds.

5.3.8 *Alcohols*

Substitution of H in an alkane by an OH group moves the signal to the left by 35–52 ppm for C-1, 5–12 ppm for C-2, and to the right by about 0–6 ppm for C-3. Shifts for several acyclic and alicyclic alcohols are given in Table 5.11. Acetylation provides a useful diagnostic test for an alcohol: The C-1 absorption moves to the left by about 2.5–4.5 ppm, and the C-2 absorption moves to the right by a similar amount; a 1,3-diaxial interaction may cause a slight (~1 ppm) shift to the left of C-3. Table 5.3 may be used to calculate shifts for alcohols as described earlier.

5.3.9 *Ethers, Acetals, and Epoxides*

An alkoxy substituent causes a somewhat larger shift to the left at C-1 (~11 ppm larger) than that of a hydroxy substituent. This is attributed to the C-1′ of the alkoxy group having the same effect as a β-C relative to C-1. The O atom is regarded here as an "α-C" to C-1.

$$\underset{17.6}{CH_3}-\underset{57.0}{CH_2}-OH \qquad \underset{14.7}{\overset{2}{CH_3}}-\underset{67.9}{\overset{1}{CH_2}}-O-\underset{57.6}{\overset{1'}{CH_3}}$$

Note also that the "γ effect" (shift to the right) on C-2 is explainable by similar reasoning. Conversely, the ethoxy group affects the OCH_3 group (compare CH_3OH). Table 5.12 gives shifts of several ethers.

The dioxygenated carbon of acetals absorbs in the

Table 5.10 Shifts for Carbon Atoms of Heteroaromatics (neat, ppm from TMS)

Compound	C-2	C-3	C-4	C-5	C-6	Substituent
Furan	142.7	109.6				
2-Methylfuran	152.2	106.2	110.9	141.2		13.4
Furan-2-carboxaldehyde	153.3	121.7	112.9	148.5		178.2
Methyl 2-furoate	144.8	117.9	111.9	146.4		159.1 (C=O), 51.8 (CH_3)
Pyrrole	118.4	108.0				
2-Methylpyrrole	127.2	105.9	108.1	116.7		12.4
Pyrrole-2-carboxaldehyde	134.0	123.0	112.0	129.0		178.9
Thiophrene	124.4	126.2				
2-Methylthiophene	139.0	124.7	126.4	122.6		14.8
Thiophene-2-carboxaldehyde	143.3	136.4	128.1	134.6		182.8
Thiazole	152.2		142.4	118.5		
Imidazole	136.2		122.3	122.3		
Pyridine	150.2	123.9	135.9			
Pyrimidine	159.5		157.4	122.1	157.4	
Pyrazine	145.6					
2-Methylpyrazine	154.0	141.8[a]	143.8[a]	144.7[a]		21.6

[a] Assignment not certain

Table 5.11 Chemical Shifts of Alcohols (neat, ppm from TMS)

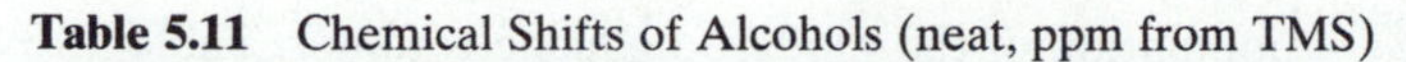

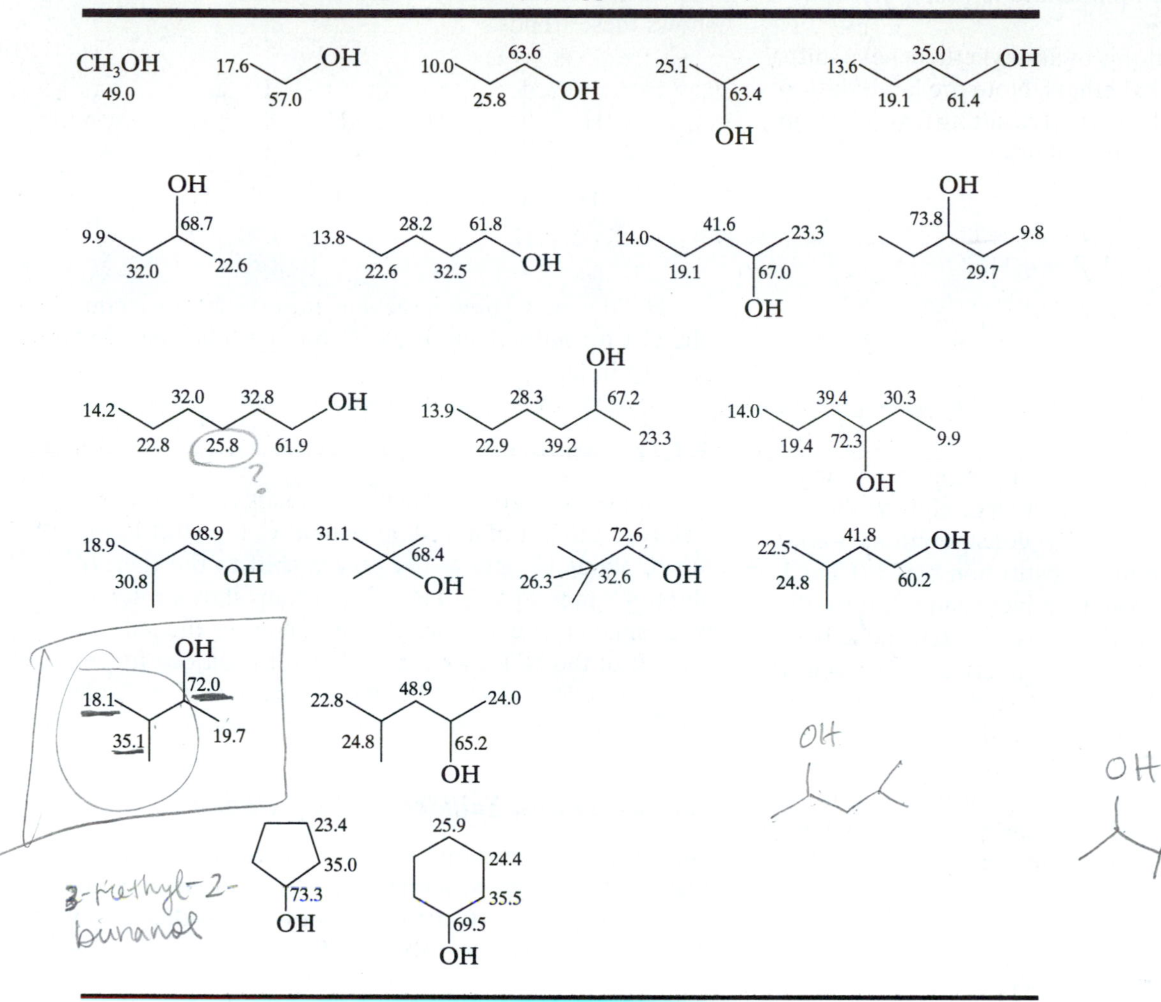

Table 5.12 Chemical Shifts of Ethers, Acetals, and Epoxides (ppm from TMS)

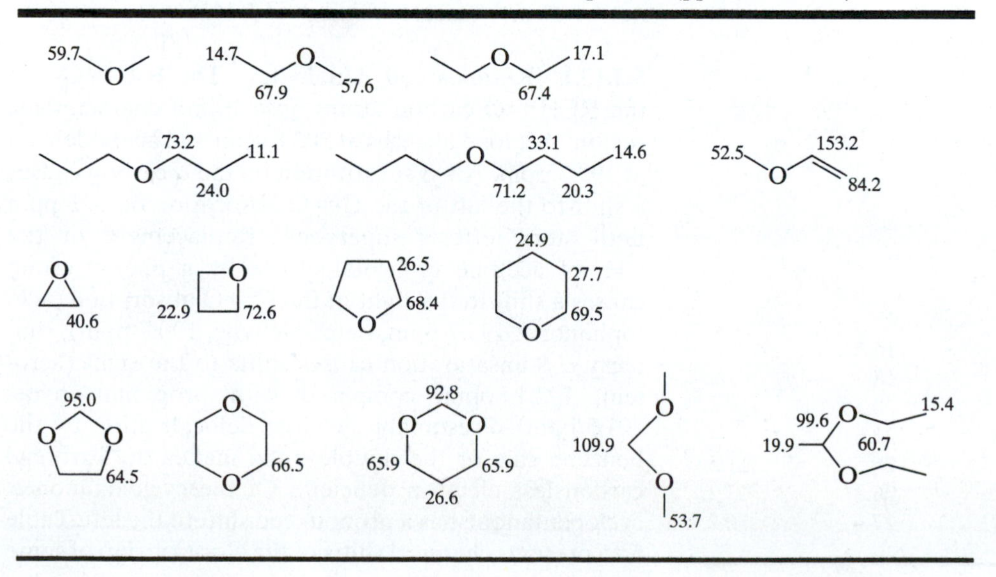

range of about 88–112 ppm. Oxirane (an epoxide) absorbs at 40.6 ppm.

The alkyl carbon atoms of aralkyl ethers have shifts similar to those of dialkyl ethers. Note the large shift to the right of the ring ortho carbon resulting from electron delocalization as in the vinyl ethers.

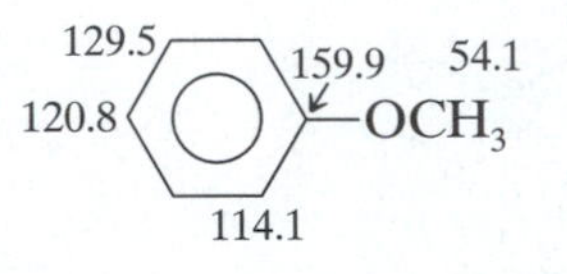

5.3.10 *Halides*

The effect of halide substitution is complex. A single fluorine atom (in CH_3F) causes a large shift to the left from CH_4 as electronegativity considerations would suggest. Successive geminal substitution by Cl (CH_3Cl, CH_2Cl_2, $CHCl_3$, CCl_4) results in increasing shifts to the left—again expected on the basis of electronegativity. But with Br and I, the "heavy atom effect" supervenes. The carbon shifts of CH_3Br and CH_2Br_2 move progressively to the right. A strong progression to the right for I commences with CH_3I, which is to the right of CH_4. There is a progressive shift to the left at C-2 in the order I > Br > F. Chlorine and Br show *γ-gauche* shielding at C-3, but I does not, presumably because of the low population of the hindered gauche rotamer. Table 5.13 shows these trends.

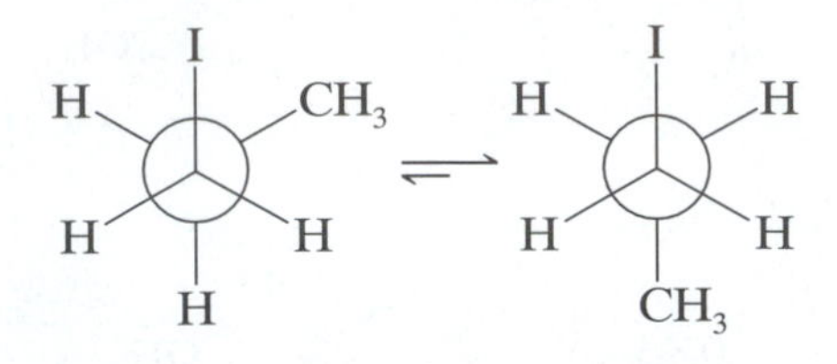

Halides may show large solvent effects; for example, C-1 for iodoethane is at −6.6 in cyclohexane, and at −0.4 in DMF.

Table 5.13 Shift Positions for Alkyl Halides (neat, ppm from TMS)

Compound	C-1	C-2	C-3
CH_4	−2.3		
CH_3F	75.4		
CH_3Cl	24.9		
CH_2Cl_2	54.0		
$CHCl_3$	77.5		
CCl_4	96.5		
CH_3Br	10.0		
CH_2Br_2	21.4		
$CHBr_3$	12.1		
CBr_4	−28.5		
CH_3I	−20.7		
CH_2I_2	−54.0		
CHI_3	−139.9		
CI_4	−292.5		
CH_3CH_2F	79.3	14.6	
CH_3CH_2Cl	39.9	18.7	
CH_3CH_2Br	28.3	20.3	
CH_3CH_2I	−0.2	21.6	
$CH_3CH_2CH_2Cl$	46.7	26.5	11.5
$CH_3CH_2CH_2Br$	35.7	26.8	13.2
$CH_2CH_2CH_2I$	10.0	27.6	16.2

5.3.11 *Amines*

A terminal NH_2 group attached to an alkyl chain causes a shift to the left of about 30 ppm at C-1, a shift to the left of about 11 ppm at C-2, and a shift to the right of about 4.0 ppm at C-3. The NH_3^+ group shows a somewhat smaller effect. *N*-alkylation increases the shift to the left of the NH_2 group at C-1. Shift positions for selected acyclic and alicyclic amines are given in Table 5.14A (see Table 5.5 for heterocyclic amines).

5.3.12 *Thiols, Sulfides, and Disulfides*

Since the electronegativity of sulfur is considerably less than that of oxygen, sulfur causes a correspondingly smaller chemical shift. Examples of thiols, sulfides, and disulfides are given in Table 5.14B.

5.3.13 *Functional Groups Containing Carbon*

Carbon-13 NMR spectrometry permits direct observation of carbon-containing functional groups; the shift ranges for these are given in Appendix A. With the exception of CH=O, the presence of these groups could not be directly ascertained by ¹H NMR.

5.3.13.1 Ketones and Aldehydes The $R_2C{=}O$ and the RCH=O carbon atoms absorb in a characteristic region. Acetone absorbs at 203.8 ppm and acetaldehyde at 199.3 ppm. Alkyl substitution on the *α* carbon causes a shift to the left of the C=O absorption of 2–3 ppm until steric effects supervene. Replacement of the CH_3 of acetone or acetaldehyde by a phenyl group causes a shift to the right of the C=O absorption (acetophenone, 195.7 ppm; benzaldehyde, 190.7 ppm); similarly *α*,*β*-unsaturation causes shifts to the right (acrolein, 192.1 ppm, compared with propionaldehyde, 201.5 ppm). Presumably, charge delocalization by the benzene ring or the double bond makes the carbonyl carbon less electron deficient. Of the cycloalkanones, cyclopentanone has a pronounced shift to the left. Table 5.15 presents chemical shifts of the C=O group of some

Table 5.14A Shift Positions of Acyclic and Alicyclic Amines (neat, ppm from TMS)

Compound	C-1	C-2	C-3	C-4
CH_3NH_2	26.9			
$CH_3CH_2NH_2$	35.9	17.7		
$CH_3CH_2CH_2NH_2$	44.9	27.3	11.2	
$CH_3CH_2CH_2CH_2NH_2$	42.3	36.7	20.4	14.0
$(CH_3)_3N$	47.5			
$CH_3CH_2N(CH_3)_2$	58.2	13.8		
Cyclohexylamine	50.4	36.7	25.7	25.1
N-Methylcyclohexylamine	58.6	33.3	25.1	26.3 (N—CH_3 33.5)

ketones and aldehydes. Because of rather large solvent effects, there are differences of several parts per million from different literature sources. Replacement of CH_2 of alkanes by C=O causes a shift to the left at the α carbon (~10–14 ppm) and a shift to the right at the β carbon (several ppm in acyclic compounds).

5.3.13.2 Carboxylic Acids, Esters, Chlorides, Anhydrides, Amides, and Nitriles The C=O groups of carboxylic acids and derivatives are in the range of 150–185 ppm. Dilution and solvent effects are marked for carboxylic acids; anions appear further to the left. The effects of substituents and electron delocalization are generally similar to those for ketones. Nitriles absorb in the range of 115–125 ppm. Alkyl substituents on the nitrogen of amides cause a small (up to several ppm) shift to the right of the C=O group (see Table 5.16).

5.3.13.3 Oximes The quaternary carbon atom of simple oximes absorb in the range of 145–165 ppm. It is possible to distinguish between *E* and *Z* isomers since the C=N shift is to the right in the sterically more compressed form, and the shift of the more hindered substituent (syn to the OH) is farther to the right than the less hindered.

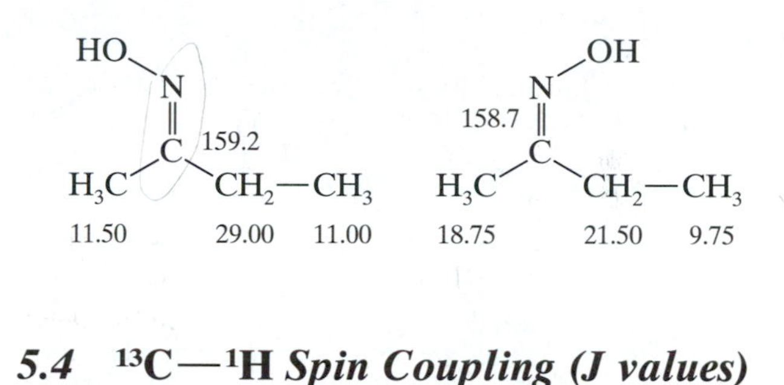

Table 5.14B Shift Positions of Thiols, Sulfides, and Disulfides (ppm from TMS)

Compound	C-1	C-2	C-3
CH_3SH	6.5		
CH_3CH_2SH	19.8	17.3	
$CH_3CH_2CH_2SH$	26.4	27.6	12.6
$CH_3CH_2CH_2CH_2SH$	23.7	35.7	21.0
$(CH_3)_2S$	19.3		
$(CH_3CH_2)_2S$	25.5	14.8	
$(CH_3CH_2CH_2)_2S$	34.3	23.2	13.7
$(CH_3CH_2CH_2CH_2)_2S$	34.1	31.4	22.0
CH_3SSCH_3	22.0		
$CH_3CH_2SSCH_2CH_3$	32.8	14.5	

5.4 $^{13}C—^{1}H$ *Spin Coupling (J values)*

Spin-coupling *J* values—at least as an initial consideration—are less important in ^{13}C NMR than in ^{1}H NMR. Since routine ^{13}C spectra are usually decoupled, $^{13}C—^{1}H$ coupling values are discarded in the interest of obtaining a spectrum in a short time or on small samples—a spectrum, furthermore, free of complex, overlapping absorptions. $^{13}C—^{1}H$ *J* values are given in tables 5.17–5.19.*

One-bond $^{13}C—^{1}H$ coupling ($^{1}J_{CH}$) ranges from about 110 to 320 Hz, increasing with increased *s* character of the $^{13}C—^{1}H$ bond, with substitution on the carbon atom of electron-withdrawing groups, and with angular distortion. Appreciable $^{13}C—^{1}H$ coupling also extends over two or more (*n*) bonds ($^{n}J_{CH}$). Table 5.17 gives some representative $^{1}J_{CH}$ values. Table 5.18 gives some representative $^{2}J_{CH}$ values, which range from about −5 to 60 Hz.

The $^{3}J_{CH}$ values are roughly comparable to $^{2}J_{CH}$ values for sp^3 carbon atoms. In aromatic rings, however, the $^{3}J_{CH}$ values are characteristically larger than $^{2}J_{CH}$ values. In benzene itself, $^{3}J_{CH}$ = 7.4 Hz and $^{2}J_{CH}$ = 1.0 Hz.

Coupling of ^{13}C to several other nuclei, the most

* $^{13}C—^{13}C$ coupling is not observed in a routine ^{13}C spectrum, except in compounds that have been deliberately enriched with ^{13}C, because of the low probability of two adjacent ^{13}C atoms in a molecule.

Table 5.15 Shift Positions of the C=O Group and Other Carbon Atoms of Ketones and Aldehydes (ppm from TMS)

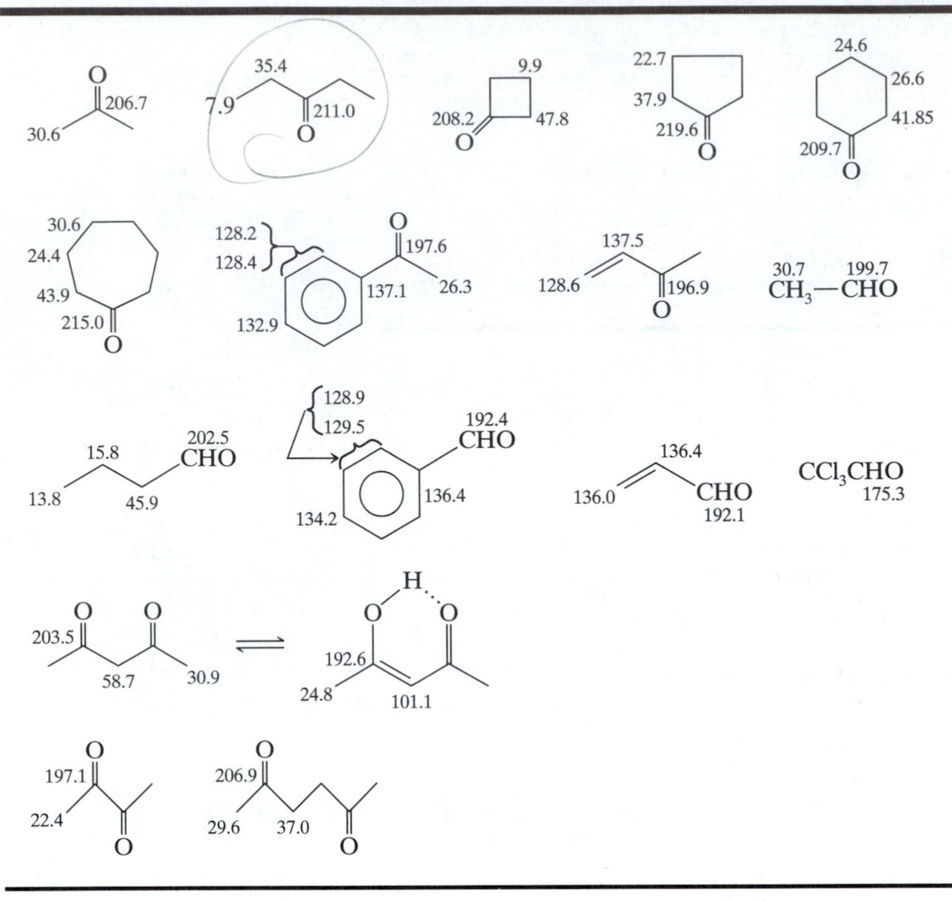

important of which are ^{31}P, ^{19}F and D, may be observed in proton-decoupled spectra. Representative coupling constants are given in Table 5.19.

To retain at least part of the NOE and still maintain ^{13}C—^{1}H coupling, the "gated decoupling" technique may be employed (Figure 5.5). Briefly, the broadband proton decoupler is kept on during the delay period before the ^{13}C pulse. Simultaneously with the ^{13}C pulse, the broadband-proton decoupler is "gated" (switched) off throughout the ^{13}C pulse and acquisition period, then gated on throughout the pulse delay. Thus, the NOE (a slow process), which has built up prior to the ^{13}C pulse, decays only partially during the pulse and acquisition period. However, coupling (a fast process) is established immediately after the ^{13}C pulse and remains throughout the acquisition period. The result is a coupled spectrum in which at least part of the NOE has been retained.

A more useful kind of ^{13}C—^{1}H coupling information is the *number of protons attached to a ^{13}C atom.* In other words, are we dealing with a CH_3 group, a CH_2 group, a CH group, or a quaternary carbon atom? (Refer to Section 5.5 for a discussion of the DEPT spec-

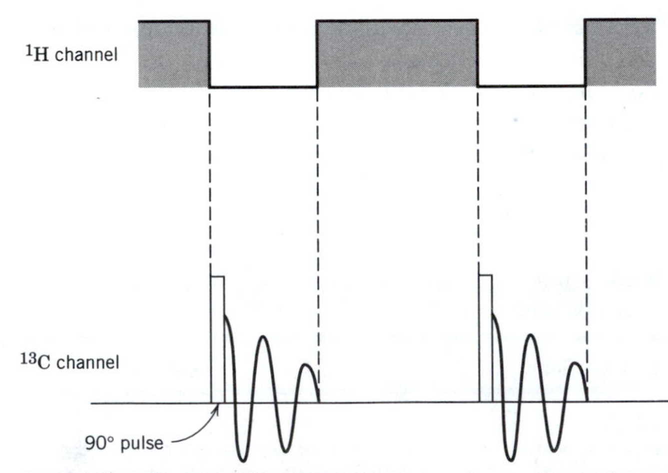

FIGURE 5.5. Gated broadband proton decoupling pulse sequence. The ^{13}C pulse is on the order of several microseconds. The acquisition time is from about 0.1 s to 1.0 s; the delay period is usually slightly longer. The stippled areas represent the periods during which the broadband proton decoupler is gated on. The acquisition period is represented by the free induction decay (FID). The sequence is pulse, acquisition, and delay; periods are not shown in proportion.

Table 5.16 Shift Positions for the C=O group and other Carbon Atoms of Carboxylic Acids, Esters, Lactones, Chlorides Anhydrides, Amides, Carbamates, and Nitriles (ppm from TMS)

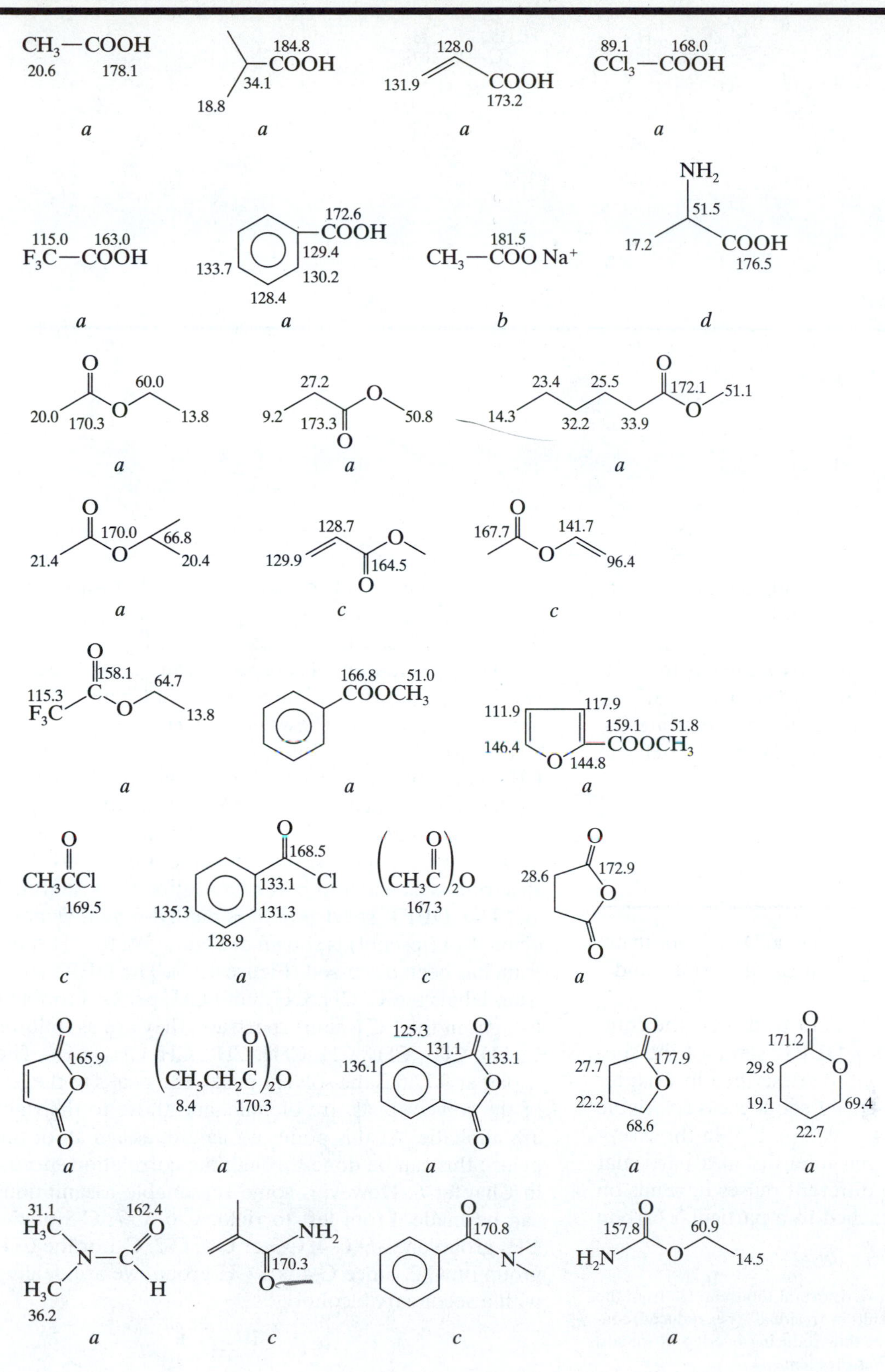

(continued)

Table 5.16 *(Continued)*

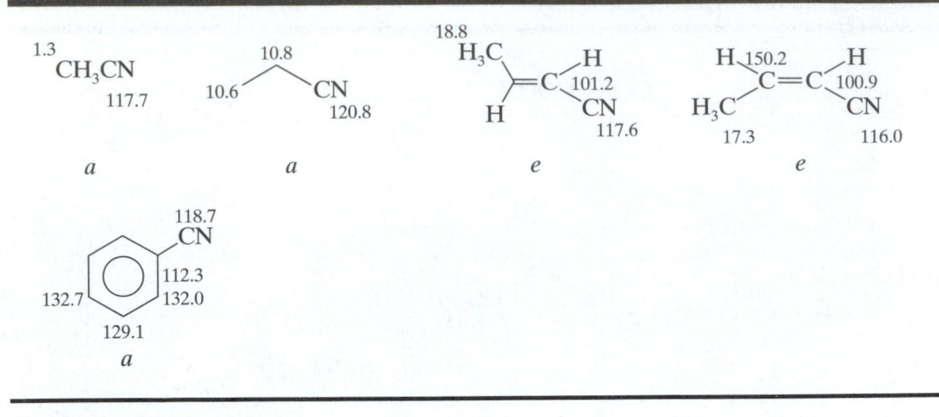

[a] In $CHCl_3$ (~50%).
[b] Saturated aqueous solution of CH_3COONa.
[c] Neat or saturated solution.
[d] In D_2O.
[e] In DMSO.
[f] In dioxane (~50%).

trum). So-called off-resonance decoupling* is obsolete for this purpose and has been completely replaced by DEPT (Section 5.5).

Beyond the number of protons attached to a ^{13}C atom, it would be nice to know *which* protons are attached to each ^{13}C atom. In other words, what are the correlations? Chapter 6 deals with both $^1H—^1H$ and $^1H—^{13}C$ correlations.

5.5 *DEPT*†

We mentioned in Section 5.4 that a DEPT spectrum distinguishes between a CH_3 group, a CH_2 group, and a CH group.

No attempt will be made here to discuss the complex multipulse sequence for DEPT; some of the simpler multipulse sequences will be described in Chapter 6. The novel feature in the DEPT sequence is a variable proton pulse that is set at 45°, 90°, or 135° in three separate experiments. The signal intensity at a particular time for each of the three different pulses depends on the number of protons attached to a particular carbon atom. A separate subspectrum is recorded for each of the CH_3, CH_2, and CH groups. The broadband-decoupled ^{13}C spectrum is also acquired.

The presentation of subspectra can be condensed to two lines as shown in Figure 5.6 and in Chapter 6. One line (B subspectrum) shows peaks CH_3 and CH up, and CH_2 down. The other line (A subspectrum) shows the CH peaks up. Quaternary carbon atoms are not recorded in a subspectrum since there is no attached proton, but of course the main (conventional ^{13}C) spectrum does show these peaks. In many laboratories, a DEPT spectrum is considered part of a routine ^{13}C spectrum.

The DEPT spectrum of 2-methyl-6-methylene-7-octen-4-ol (ipsenol) is shown in Figure 5.6; the ^1H spectrum has been discussed (Figure 4.49). The DEPT spectrum labels the C, CH, CH_2, and CH_3 peaks. From left to right in the ^{13}C (main) spectrum, they are as follows: C, CH, CH_2, CH_2, CH, CH_2, CH_2, CH, CH_3, CH_3. The triplet at δ 77.0 is the solvent peak. The peaks to the left of the solvent peak are olefinic, and those to the right are aliphatic. At this point, we cannot assign all of the peaks; this can be done through the correlation spectra in Chapter 6. However, some reasonable assumptions can be made. From left to right: C-6, C-7, C-8 or the CH_2 group on C-6; C-4, C-3 or C-5; C-2, C-1 or the CH_3 group on C-2. Since C-4 is a CH group, we are dealing with a secondary alcohol.

5.6 *Quantitative Analysis*

Quantitative ^{13}C NMR is desirable in two situations. First, in structural determinations, it is clearly useful to

* The broadband decoupler is moved several thousand Hz from the proton frequency of TMS; this results in "residual" (i.e., reduced) coupling by directly attached proton, thus reducing overlap of signals. However, overlap remains a serious problem.

† Distortionless Enhancement by Polarization Transfer. As the name implies, sensitivity is increased by polarization transfer from the more sensitive coupled proton(s) to the less sensitive ^{13}C atom. The APT (Attached Proton Test; see Sanders (1993)) gives similar information but lacks the sensitivity of DEPT.

Table 5.17 Some $^1J_{CH}$ Values

Compound	J (Hz)
*sp*³	
CH_3CH_3	124.9
$CH_3\underline{C}H_2CH_3$	119.2
$(CH_3)_3CH$	114.2
CH_3NH_2	133.0
CH_3OH	141.0
CH_3Cl	150.0
CH_2Cl_2	178.0
$CHCl_3$	209.0
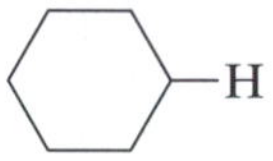 (cyclohexane)—H	123.0
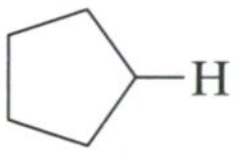 (cyclopentane)—H	128.0
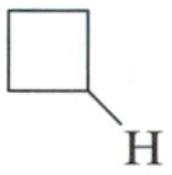 (cyclobutane)—H	134.0
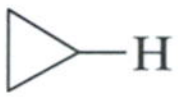 (cyclopropane)—H	161.0
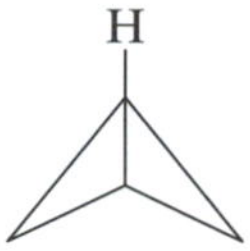 (bicyclobutane)—H	205.0
*sp*²	
$CH_2{=}CH_2$	156.2
$CH_3\underline{C}H{=}C(CH_3)_2$	148.4
$CH_3\underline{C}H{=}O$	172.4
$NH_2\underline{C}H{=}O$	188.3
C_6H_6	159.0
sp	
$CH{\equiv}CH$	249.0
$C_6H_5C{\equiv}CH$	251.0
$HC{\equiv}N$	269.0

Table 5.18 Some $^2J_{CH}$ Values

Compound	J (Hz)
*sp*³	
$C\underline{H}_3\underline{C}H_3$	−4.5
$C\underline{H}_3\underline{C}Cl_3$	5.9
$\underline{C}H_3C\underline{H}{=}O$	26.7
*sp*²	
$C\underline{H}_2{=}\underline{C}H_2$	−2.4
$(C\underline{H}_3)_2\underline{C}{=}O$	5.5
$CH_2{=}\underline{C}HC\underline{H}{=}O$	26.9
$^*C_6H_6$	1.0
sp	
$C\underline{H}{\equiv}\underline{C}H$	49.3
$C_6H_5O\underline{C}{\equiv}C\underline{H}$	61.0

* ^2J = 7.6 (>^2J)

know whether a signal results from more than one shift-equivalent carbon. Second, quantitative analysis of a mixture of two or more components requires that the area of a signal be proportional to the number of carbon atoms causing that signal.

There are four reasons that broadband-decoupled ^{13}C spectra are usually not susceptible to quantitative analysis.

1. ^{13}C nuclei with long T_1 relaxation times may not return to the equilibrium Boltzmann distribution between pulses. Thus the signals do not achieve full amplitude.
2. The nuclear Overhauser enhancement (NOE, see Section 5.1) varies among the ^{13}C nuclei, and the signal intensities vary accordingly.
3. The number of data points used to record the peak may not be sufficient to record the proper shape and area of the peak.
4. The pulse consists of a central frequency (ν_1) of maximum amplitude with frequencies of decreasing amplitude on both sides. Peaks resulting from these different pulse amplitudes vary in amplitude.

The first problem—a long T_1—can often be resolved by inserting a pulse delay after the acquisition period to reestablish equilibrium. However, the length of time required may be prohibitive, especially for quaternary ^{13}C nuclei.

To deal with the NOE variations, "inverse gated decoupling" (Fig. 5.7) is useful (compare "gated decoupling," Section 5.4). Briefly, the ^1H broadband decoupler is "gated" (switched) on only during the ^{13}C pulse and the acquisition period; it is gated off during the pulse delay period. The NOE, a slow process, builds up only slightly during the pulse and acquisition period. Decoupling, a fast process, is established almost immediately on irradiating with the ^1H broadband decoupler, so that the end result is a number of singlets whose intensities are proportional to the protons they represent.

Of course, loss of part of the NOE means many repetitions to build up the signal intensity. The time required can be shortened by addition of a paramagnetic relaxation reagent, such as the metal–organic complex $Cr(acac)_3$ to reduce all of the T_1 and T_2 relaxations.

The remaining problems, which are instrumental in nature, are more difficult to deal with but are usually less important.

Table 5.19 Coupling Constants for ^{19}F, ^{31}P, and D Coupled to ^{13}C

Compound	^1J (Hz)	^2J (Hz)	^3J (Hz)	^4J (Hz)
CH_3CF_3	271			
CF_2H_2	235			
CF_3CO_2H	284	43.7		
C_6H_5F	245	21.0	7.7	3.3
$(CH_3CH_2)_3P$	5.4	10.0		
$(CH_3CH_2)_4P^+ Br^-$	49	4.3		
$(C_6H_5)_3P^+CH_3 I^-$	88 (CH_3 52)	10.9		
$CH_3CH_2\overset{O}{\overset{\|}{P}}(OCH_2CH_3)_2$	143	7.1 (J_{COP} 6.9)	J_{CCOP} 6.2	
$(C_6H_5)_3P$	12.4	19.6	6.7	0.0
$CDCl_3$	31.5			
$CD_3\overset{O}{\overset{\|}{C}}CD_3$	19.5			
$(CD_3)_2SO$	22.0			
C_6D_6	25.5			

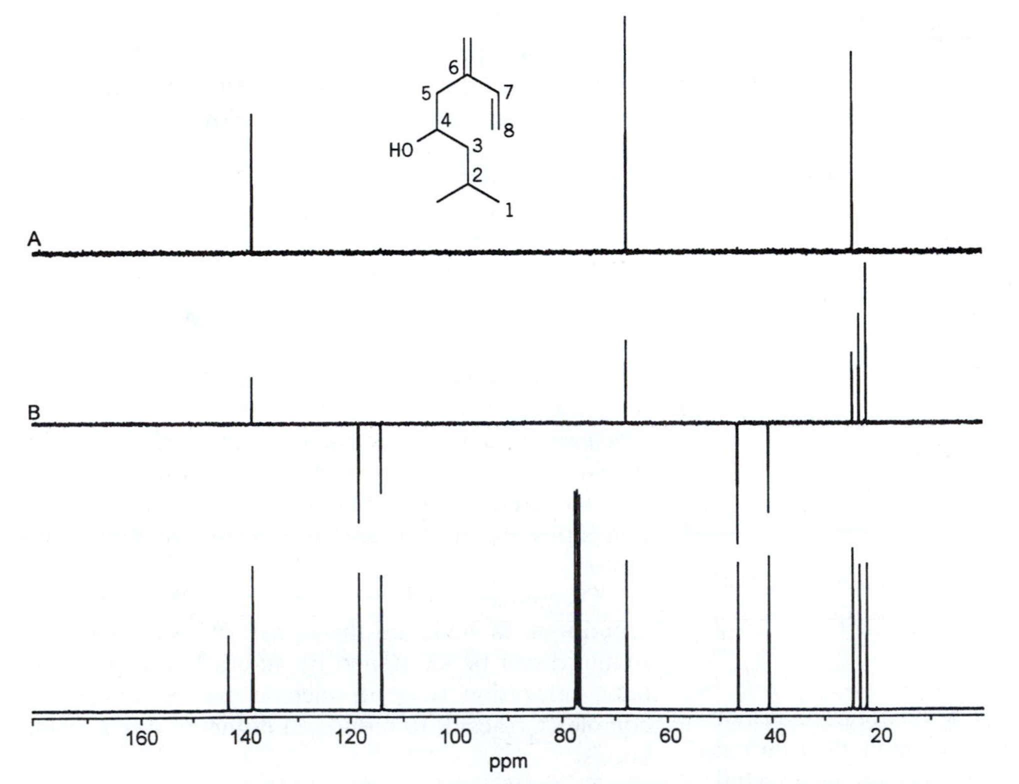

FIGURE 5.6. DEPT spectrum of ipsenol in $CDCl_3$ at 75.6 MHz. Subspectrum A, CH up. Subspectrum B, CH_3 and CH up, CH_2 down. The conventional ^{13}C spectrum is at the bottom of the figure.

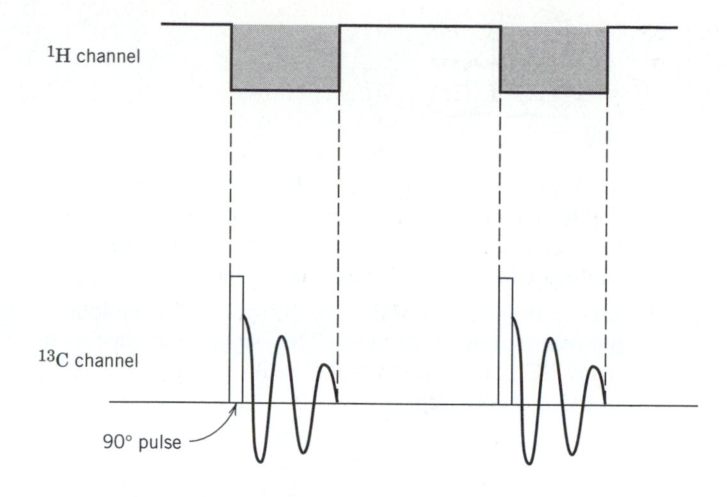

FIGURE 5.7. Inverse gated broadband proton decoupling pulse sequence. See Figure 5.5 for comparison.

References

General

Abraham, R.J., Fisher, J., and Loftus, P. (1988). *Introduction to NMR Spectroscopy.* New York: Wiley.

Atta-Ur-Rahman. (1986). *Nuclear Magnetic Resonance.* New York: Springer-Verlag.

Breitmaier, E. (1993). *Structure Elucidation by NMR in Organic Chemistry. A Practical Guide.* New York: Wiley.

Breitmaier, E., and Voelter, W. (1987). *Carbon-13 NMR Spectroscopy.* 3rd ed. New York: VCH Publishers.

Farrar, T.C. (1987). *An Introduction to Pulse NMR Spectroscopy.* Chicago: Farragut Press.

Kalinowski, H.O., Berger, S., and Braun, S. (1988). *Carbon-13 NMR Spectroscopy.* New York: Wiley.

Levy, G.C., Lichter, R.L., and Nelson, G.L. (1980). *Carbon-13 Nuclear Magnetic Resonance for Organic Chemists,* 2nd ed. New York: Wiley.

Muller, K., and Pregosin, P.S. (1976). *Fourier Transform NMR: A Practical Approach.* New York: Academic Press.

Paudler, W.W. (1987). *Nuclear Magnetic Resonance.* New York, Wiley.

Sanders, J.K.M. and Hunter, B.K. 1993. *Modern NMR Spectroscopy,* 2nd ed. Oxford: Wiley.

Shaw, D. (1984). *Fourier Transform NMR Spectroscopy,* 2nd ed. Amsterdam: Elsevier.

Wehrli, F.W., Marchand, A.P., and Wehrli, S. (1988). *Interpretation of Carbon-13 NMR Spectra,* 2nd ed. New York: Wiley.

Spectra, Data, and Workbooks

Bates, R.B., and Beavers, W.A. (1981). *C-13 NMR Problems.* Clifton, NJ: Humana Press.

Breitmaier, E., Haas, G., and Voelter, W. (1979). *Atlas of C-13 NMR Data,* Vols. 1–3. Philadelphia: Heyden (3017 compounds).

Bremser, W., Ernst, L., Franke, B., Gerhards, R., and Hardt, A. (1987). *Carbon-13 NMR Spectral Data (microfiche),* 4th ed. New York: VCH Publishers (58,108 spectra of 48,357 compounds, tabular).

Fuchs, P.L., and Bunnell, C.A. (1979). *Carbon-13 NMR-Based Organic Spectral Problems.* New York: Wiley.

Johnson, L. F., and Jankowski, W. C. (1972). *Carbon-13 NMR Spectra, a Collection of Assigned, Coded, and Indexed Spectra.* New York: Wiley.

Pouchert, C.J., and Behnke, J. (1993). *Aldrich Library of ^{13}C and ^{1}H FT-NMR Spectra, 300 MHz.* Milwaukee, WI: Aldrich Chemical Co.

Pretsch, E., Clerc, T., Seibl, J., and Simon, W. (1981). *Spectra Data for Structure Determination of Organic Compounds.* Berlin: Springer-Verlag.

Sadtler Research Lab. *^{13}C NMR Spectra.* Philadelphia: Sadtler Research Laboratories.

Problems

5.1 Draw the ^{1}H NMR spectrum and the broadband-decoupled ^{13}C NMR spectrum (designate multiplicities) for the following compounds (solvent $CDCl_3$). Show peaks in proper proportions.

a.

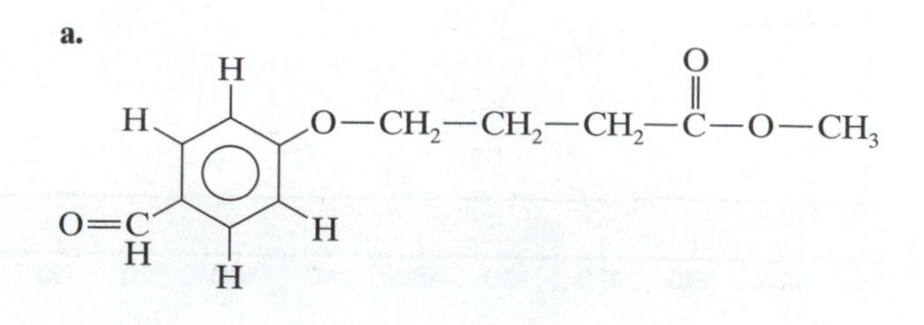

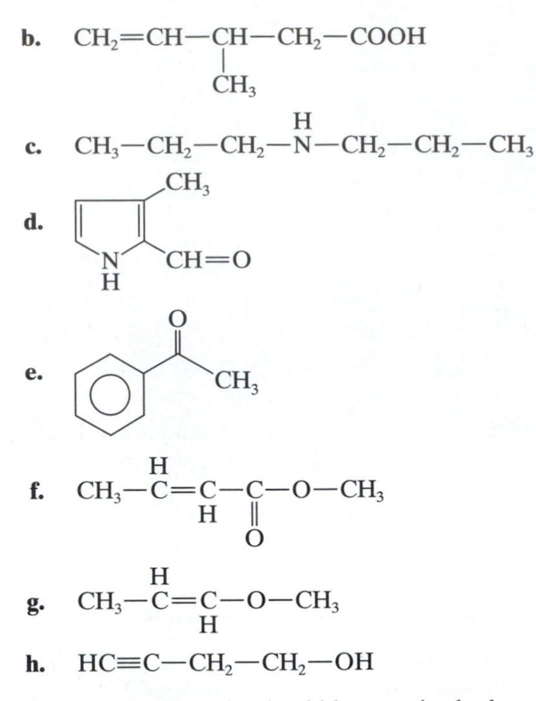

5.2 How many ^{13}C peaks should be seen in the broad-band decoupled spectrum of each of the following compounds? Assign the multiplicity expected for each signal in the off-resonance decoupled spectrum.

a. Benzene

b. Toluene

c. Naphthalene

d. Dodecane

e.

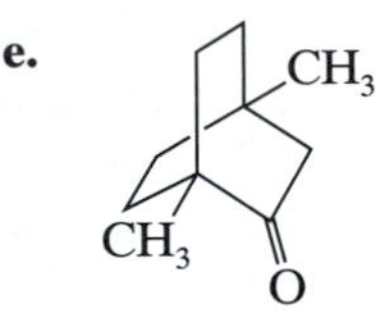

f.

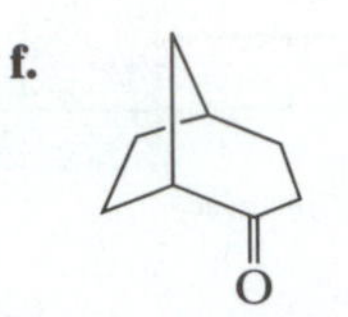

5.3 Because a compound of molecular formula C_6H_8 is highly symmetrical, it shows just two peaks in a broad-band-decoupled ^{13}C spectrum. The DEPT spectrum shows a CH and a CH_2 peak. Draw the structure.

5.4 Predict the number of lines in ^{13}C spectra for the following compounds. Recall the $CDCl_3$ signal that appears in many ^{13}C spectra and adapt Pascal's triangle (Fig. 4.29) to ^{13}C—D coupling.

$Cl—{}^{13}CCl_2—D$, $Cl—{}^{13}CCl(D)—D$, $Cl—{}^{13}CD_2—D$

5.5 What are the symmetry elements in *o*-, *m*-, and *p*-diethyl phthalate, and how many nonequivalent carbon atoms and protons are there in each compound? Draw the broadband-decoupled spectrum of each compound (see Fig. 5.1).

5.6 Assign peaks in the ^{13}C spectra in Problem 4.3 (Chapter 4).

5.7 Deduce the structures of Compounds A—H and assign the ^{13}C signals. The multiplicities are abbreviated: s = singlet, d = doublet, t = triplet, q = quartet. The following spectra were run in $CDCl_3$ at 25.2 MHz.

Problem 5.7 Compound A, $C_5H_{10}O$. The C=O group is offset at 211.8 ppm.

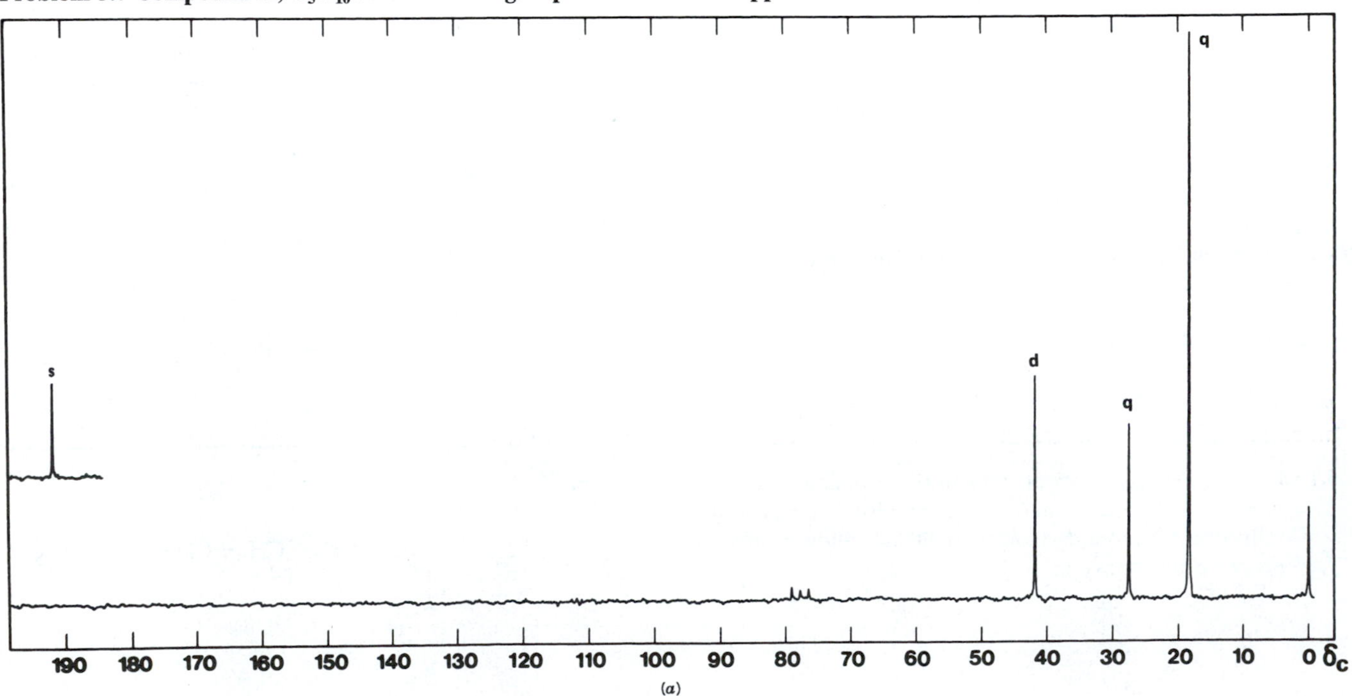

Problem 5.7 Compound B, $C_4H_{10}O$.

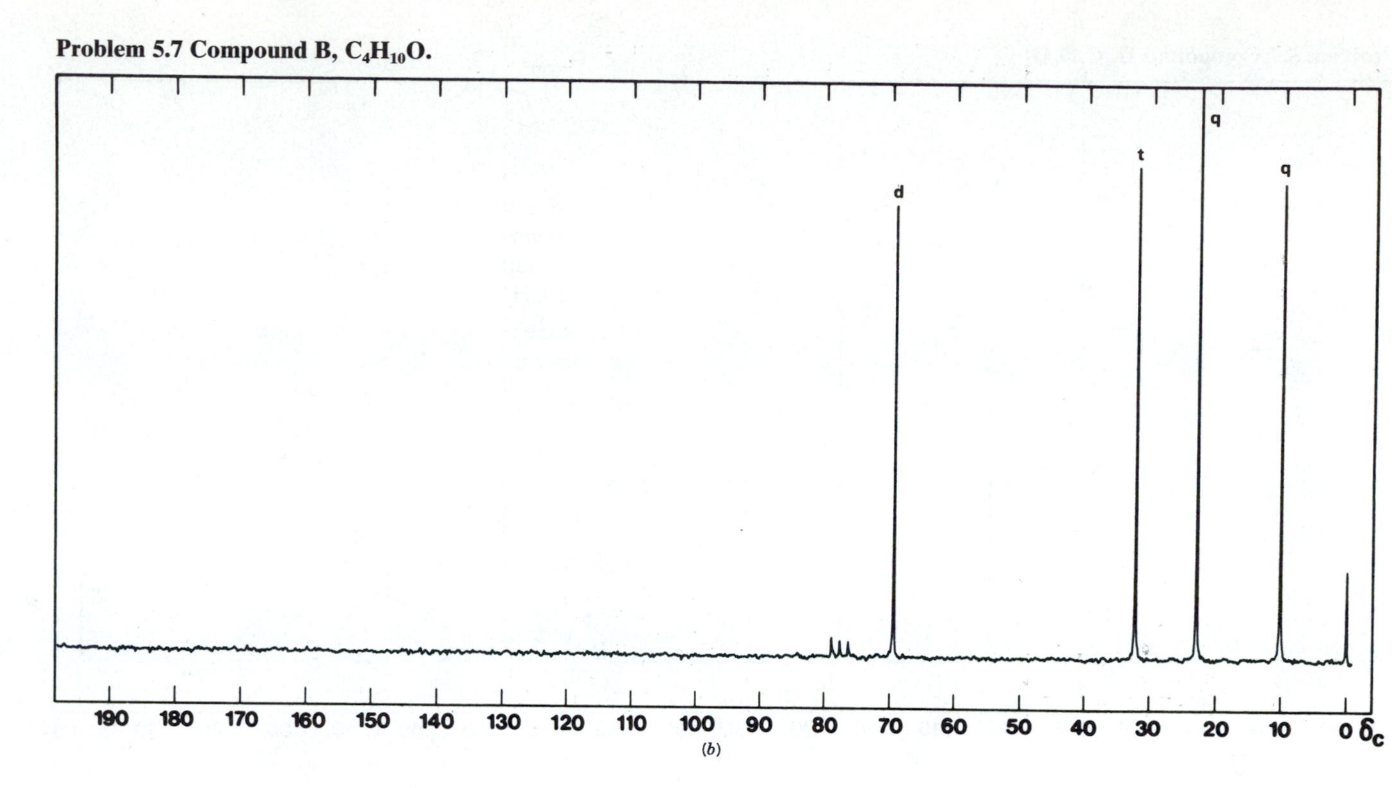

(*b*)

Problem 5.7 Compound C, $C_{11}H_{14}O_2$.

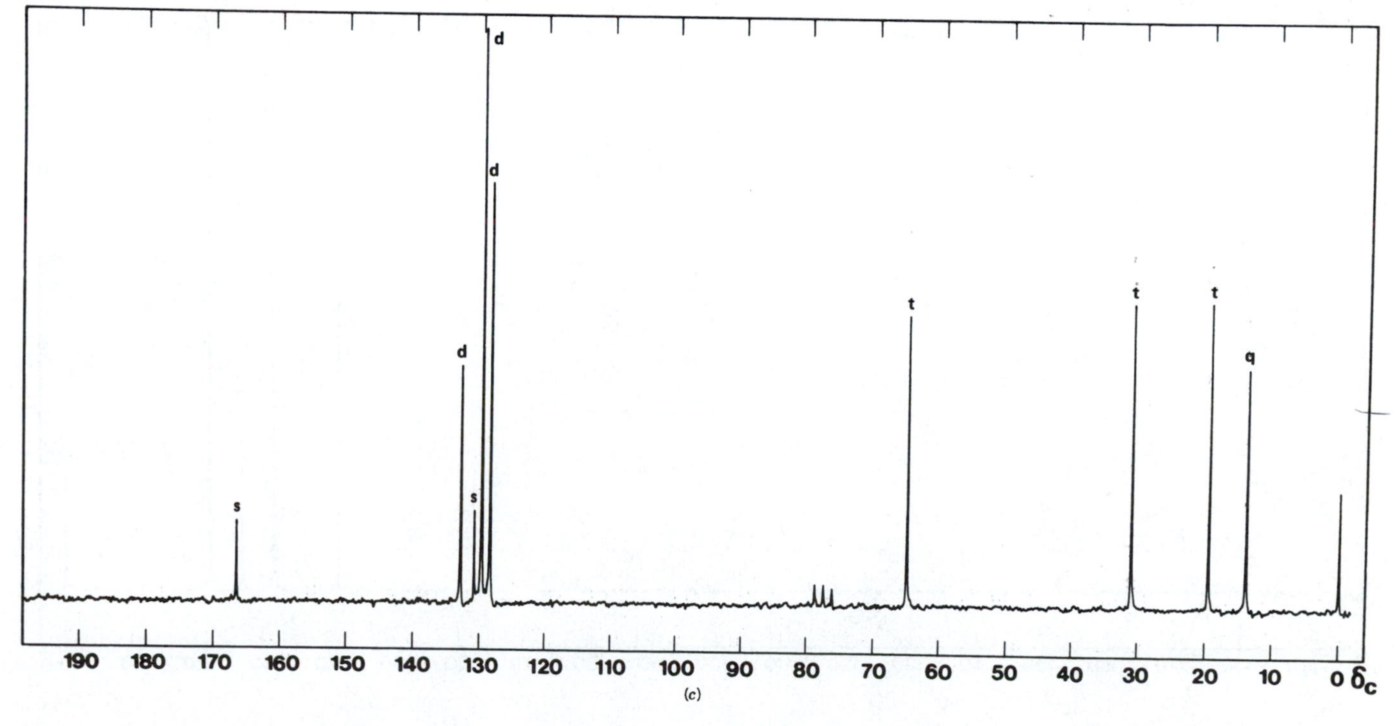

(*c*)

Problem 5.7 Compound D, C_3H_6O.

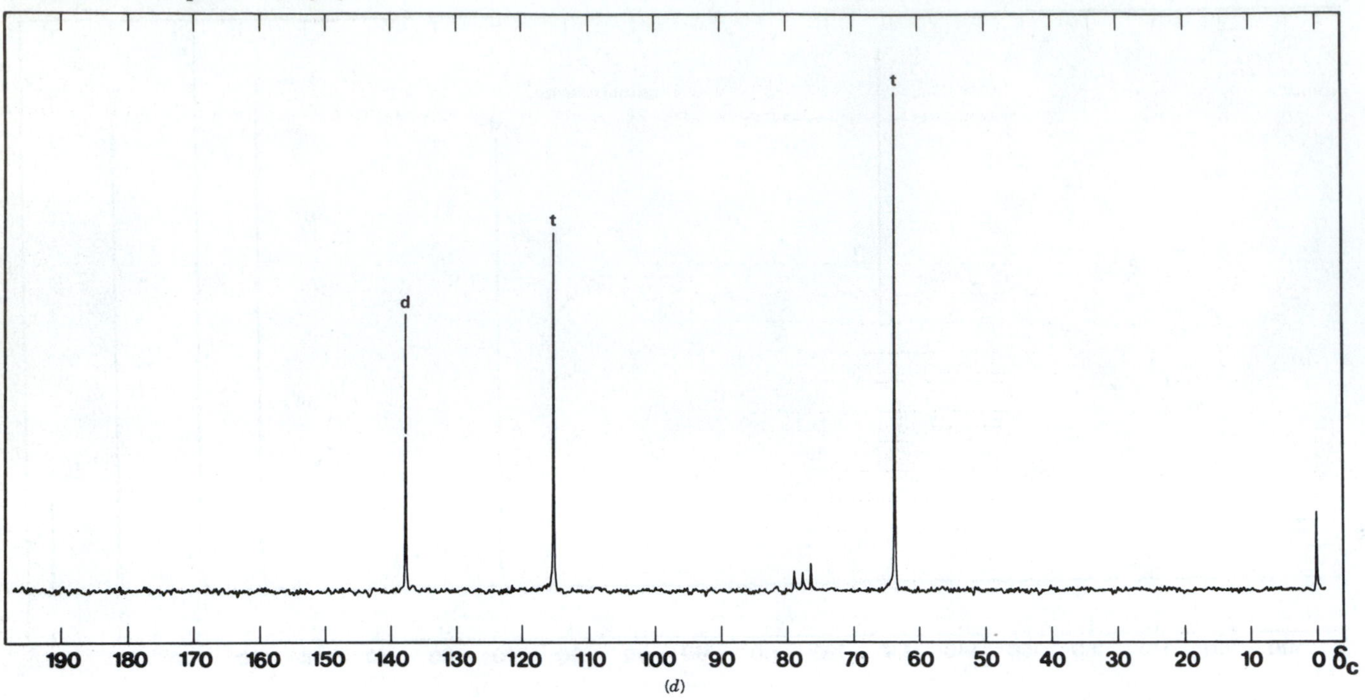

(d)

Problem 5.7 Compound E, $C_5H_{11}Br$.

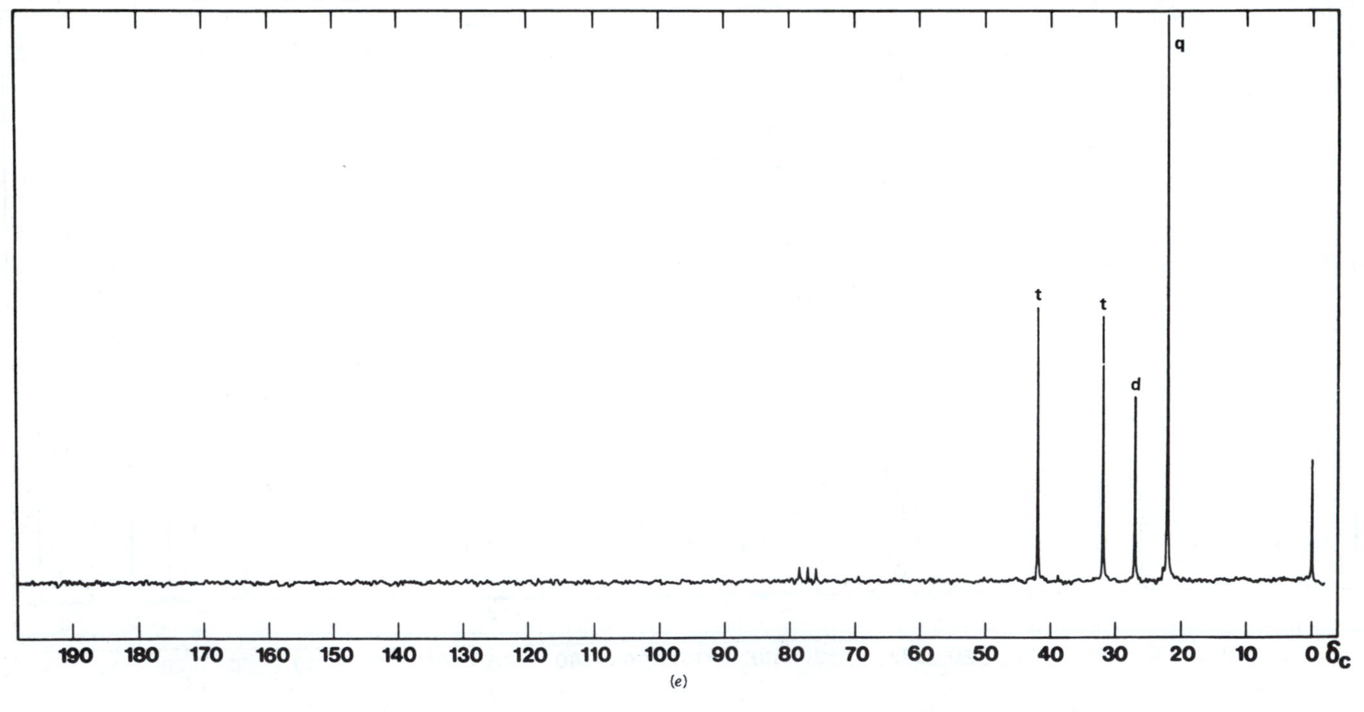

(e)

Problem 5.7 Compound F, C_7H_7ClO.

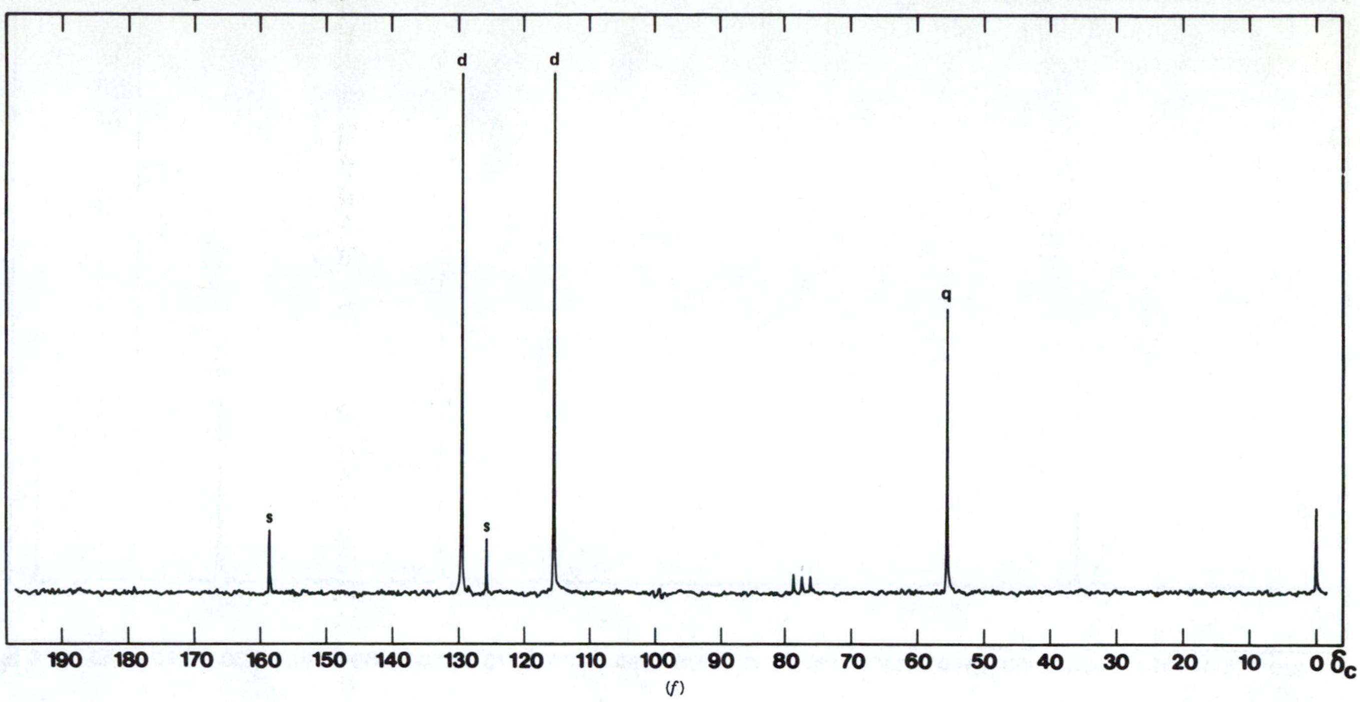

(f)

Problem 5.7 Compound G, $C_8H_{19}N$.

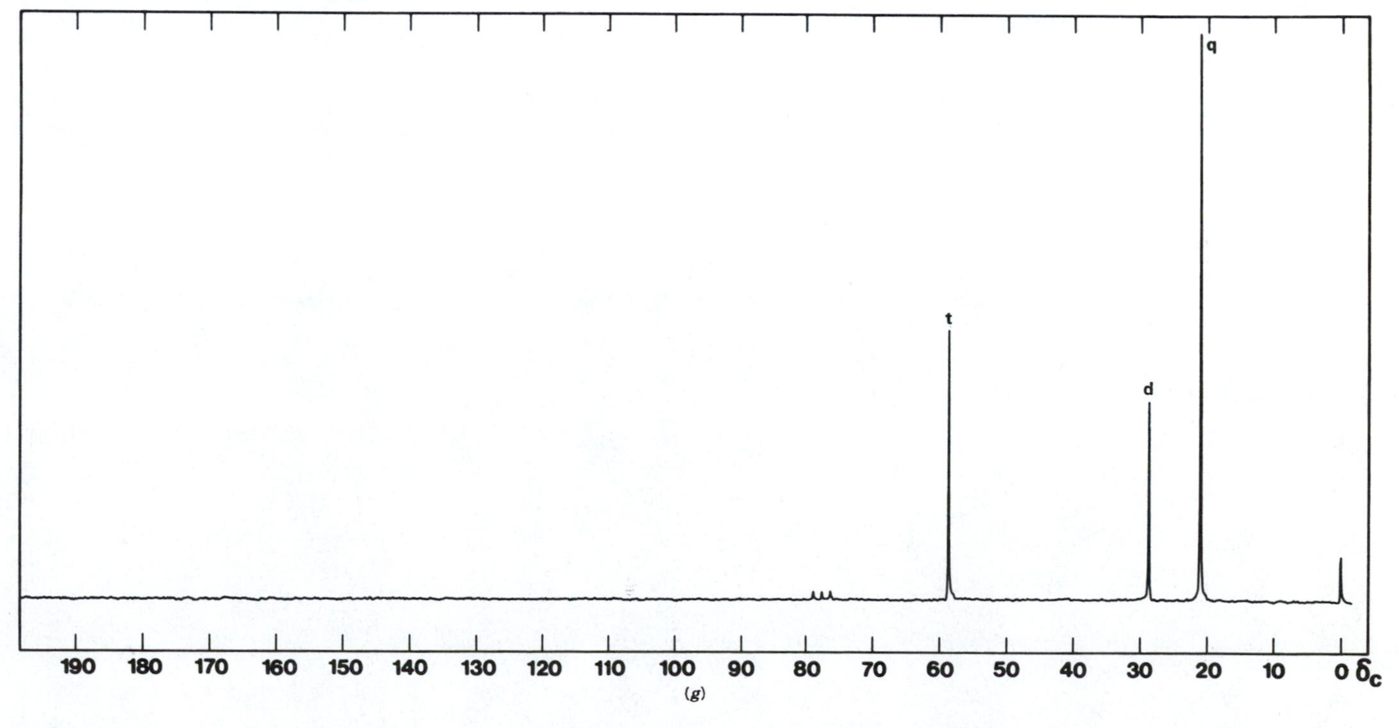

(g)

Problem 5.7 Compound H, $C_4H_9NO_2$.

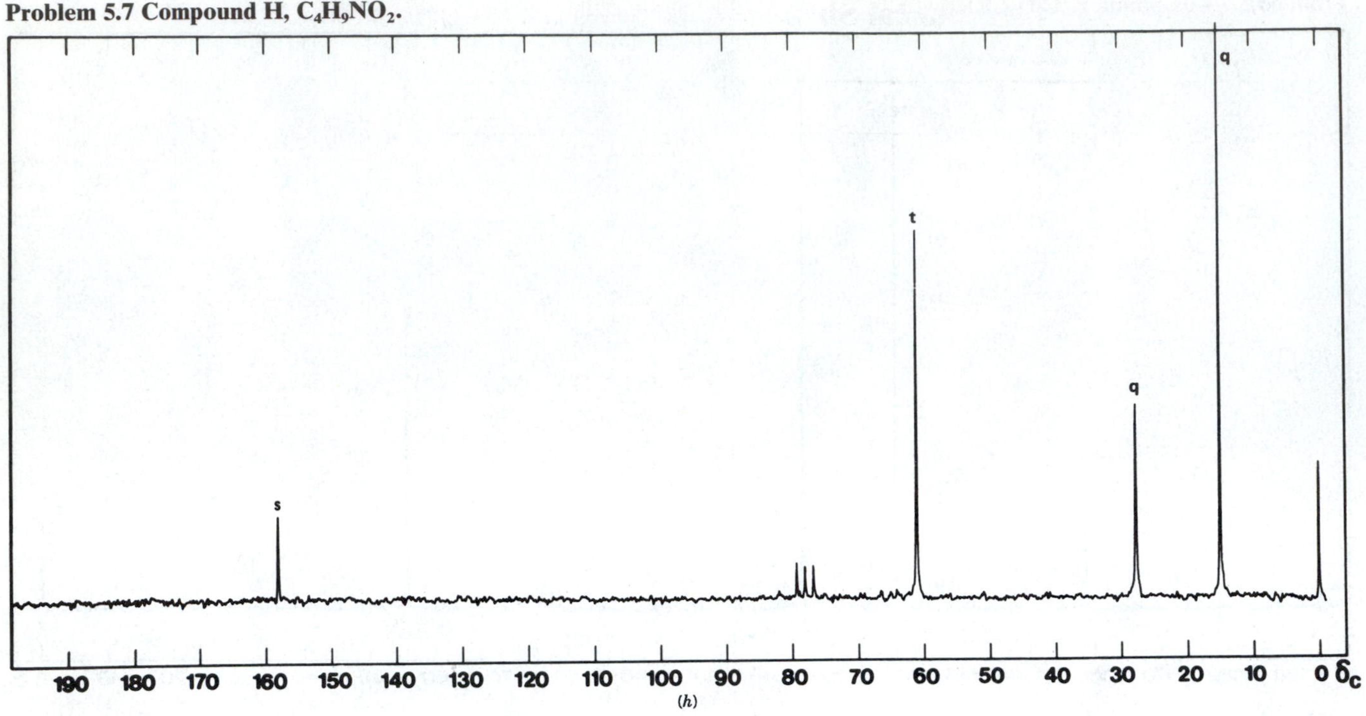

(*h*)

Appendix A The ^{13}C Chemical Shifts, Couplings, and Multiplicities of Common NMR Solvents

Structure	Name	δ(ppm)	J_{C-D} (Hz)	Multiplicity[a]
$CDCl_3$	Chloroform-d_1	77.0	32	Triplet
CD_3OD	Methanol-d_4	49.0	21.5	Septet
CD_3SOCD_3	DMSO-d_6	39.7	21	Septet
$DC(=O)N(CD_3)_2$	DMF-d_7	30.1	21	Septet
		35.2	21	Septet
		167.7	30	Triplet
C_6D_6	Benzene-d_6	128.0	24	Triplet
D_2C—CD_2 / D_2C, CD_2, O (ring)	THF-d_8	25.2	20.5	Quintet
		67.4	22	Quintet
O, D_2C, CD_2, D_2C, CD_2, O (ring)	Dioxane-d_8	66.5	22	Quintet
Pyridine ring with D at each carbon, N	Pyridine-d_5	123.5(C-3,5)	25	Triplet
		135.5(C-4)	24.5	Triplet
		149.2(C-2,6)	27.5	Triplet
$CD_3C(=O)CD_3$	Acetone-d_6	29.8 (methyl)	20	Septet
		206.5 (carbonyl)	<1	Septet[b]
CD_3CN	Acetonitrile-d_3	1.3 (methyl)	32	Septet
		118.2 (CN)	<1	Septet[b]
CD_3NO_2	Nitromethane-d_3	60.5	23.5	Septet
CD_3CD_2OD	Ethanol-d_6	15.8 (C-2)	19.5	Septet
		55.4 (C-1)	22	Quintet
$(CD_3CD_2)_2O$	Ether-d_{10}	13.4 (C-2)	19	Septet
		64.3 (C-1)	21	Quintet
$[(CD_3)_2N]_3P{=}O$	HMPA-d_{18}	35.8	21	Septet
CD_3CO_2D	Acetic acid-d_4	20.0 (C-2)	20	Septet
		178.4 (C-1)	<1	Septet[b]
CD_2Cl_2	Dichloromethane-d_2 (Methylene chloride-d_2)	53.1	29	Quintet

[a] Triplet intensities = 1:1:1, quintet = 1:2:3:2:1, septet = 1:3:6:7:6:3:1.
[b] Unresolved, long-range coupling.
Source: Breitmaier, E., and Voelter, W. (1987). *Carbon-13 NMR Spectroscopy,* 3rd ed. New York: VCH, p. 109; with permission. Also Merck & Co., Inc.

Appendix B Comparison of ¹H and ¹³C Chemical Shifts

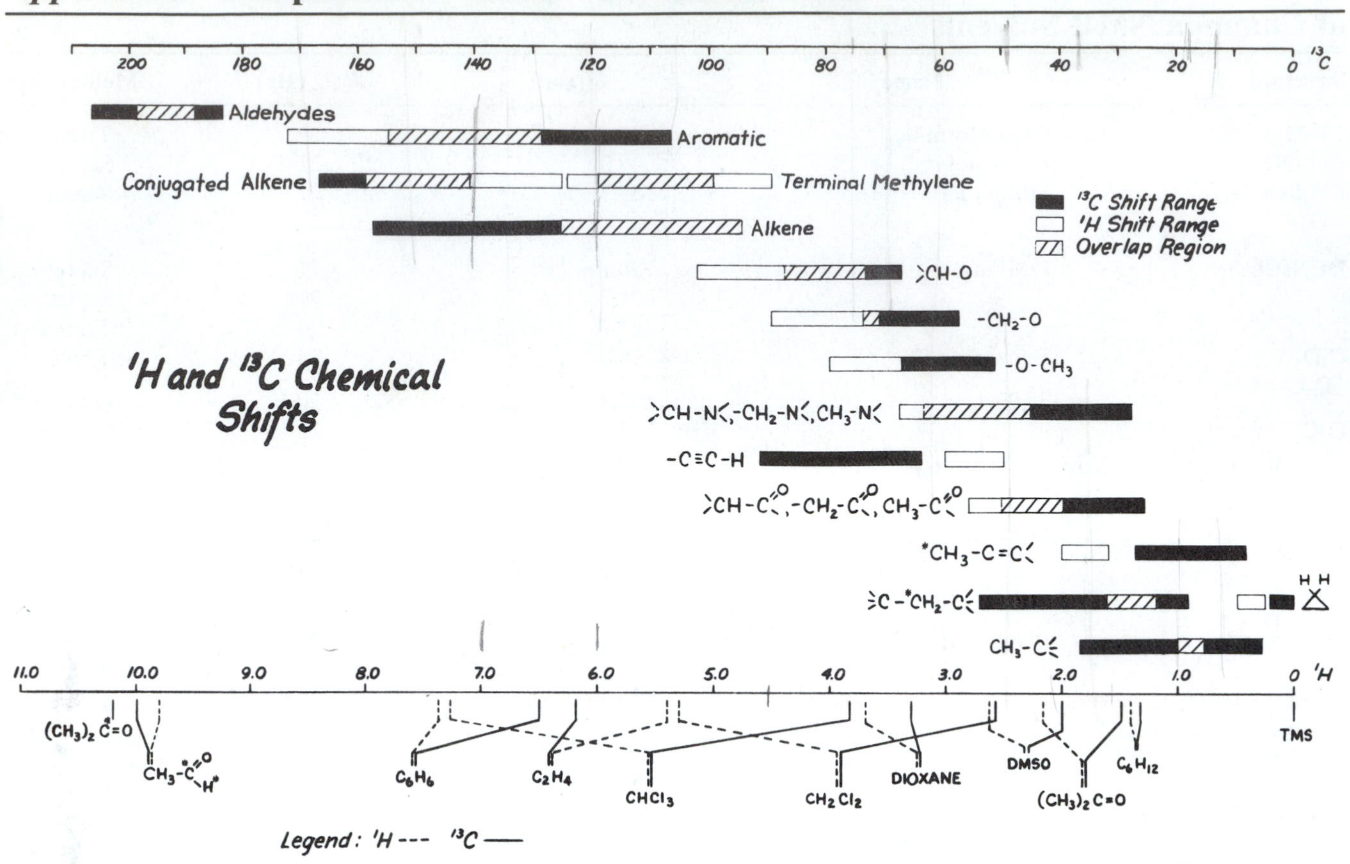

Appendix C The ^{13}C Correlation Chart for Chemical Classes

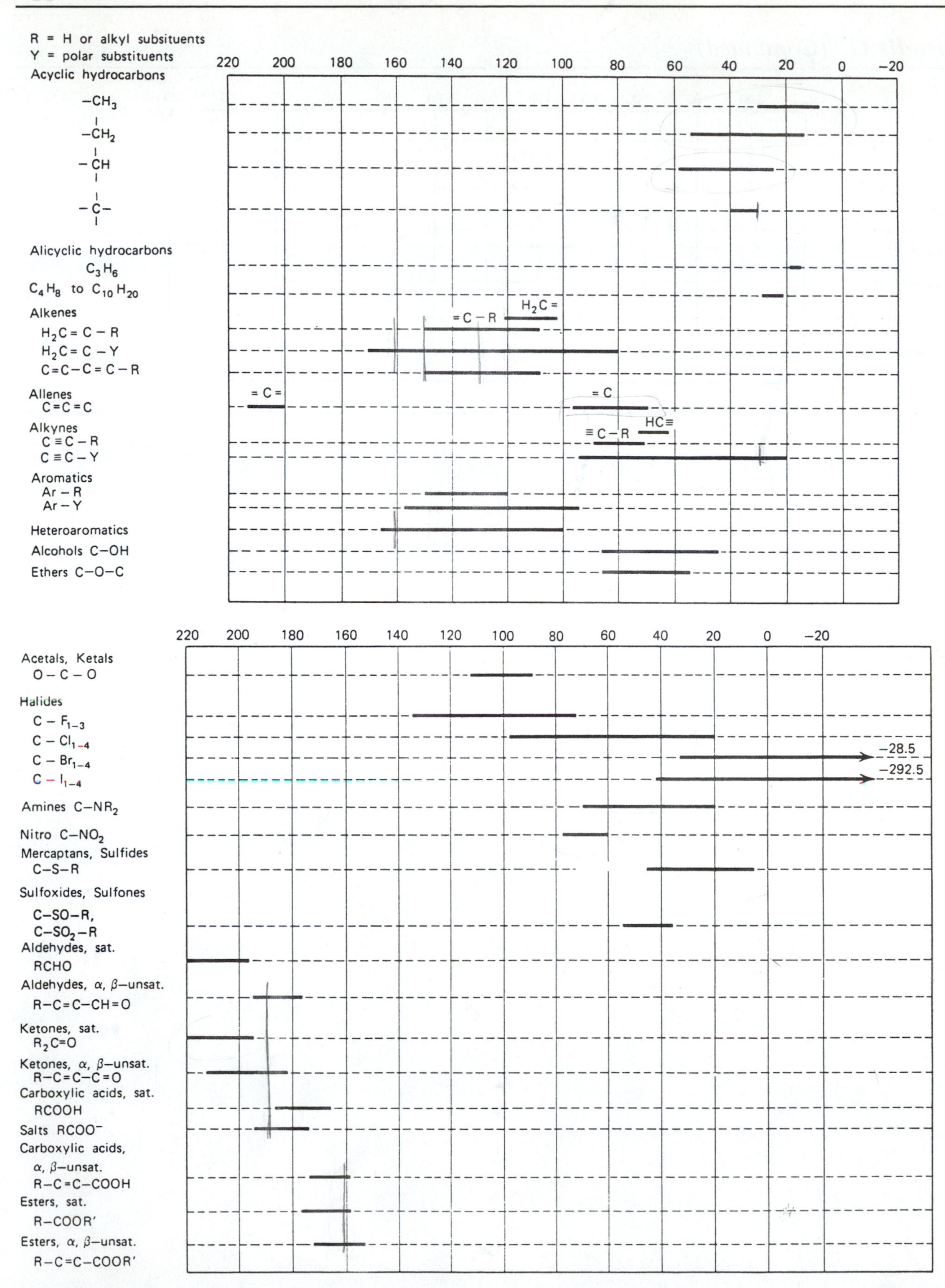

(continued)

Appendix C *(Continued)*

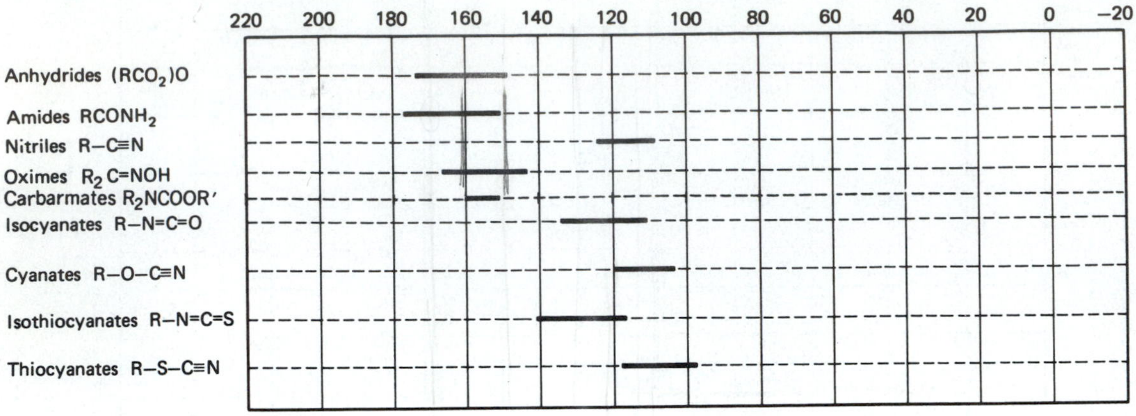

220 200 180 160 140 120 100 80 60 40 20 0 −20

Anhydrides $(RCO_2)O$
Amides $RCONH_2$
Nitriles R—C≡N
Oximes $R_2C{=}NOH$
Carbarmates R_2NCOOR'
Isocyanates R—N=C=O
Cyanates R—O—C≡N
Isothiocyanates R—N=C=S
Thiocyanates R—S—C≡N

Appendix D ^{13}C NMR Data for Several Natural Products (δ)

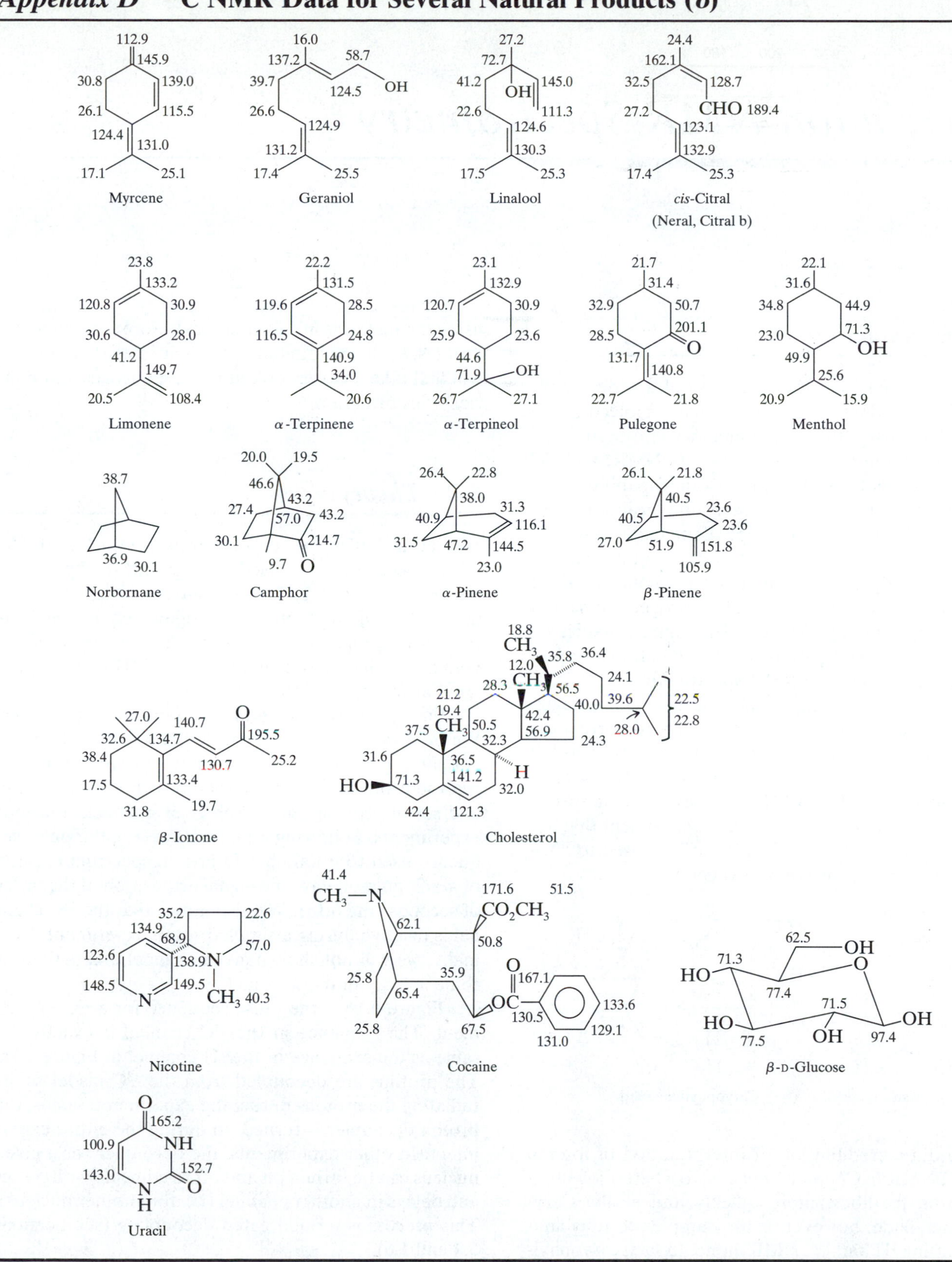

CHAPTER 6

Correlation NMR Spectrometry

6.1 Introduction

By now it is clear that the structures of a wide variety of simple to moderately complex organic molecules can be elucidated using the complementary information described in Chapters 2–5: mass, IR, ^{1}H NMR, and ^{13}C NMR spectrometries. To extend our capabilities, we turn once more to NMR. Let us again consider ipsenol (see Section 4.16.1), and caryophyllene oxide, a sesquiterpene epoxide. These structures strike a nice balance between a modest and a rather severe challenge. Each compound contains diastereotopic hydrogen atoms, and diastereotopic methyl groups. The signals associated with these atoms and groups can be difficult to interpret using simple ^{1}H and ^{13}C NMR spectra.

In this chapter, using ipsenol and caryophyllene oxide as examples, we turn our attention to correlation NMR spectroscopy; most of the useful experiments fall into the category of two-dimensional (2-D) NMR. (The exception is the "HOHAHA" experiment, a special type of one-dimensional, or "1-D," experiment that has a 2-D analogue.) Other examples will be used to illustrate specific points or types of spectra.

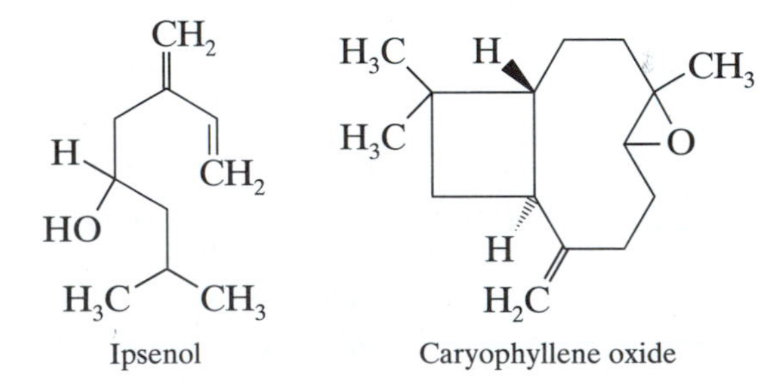

We did a credible job of interpretation of ipsenol using ^{1}H NMR in Chapter 4 but can do a better job using correlation methods, more quickly and easily. Caryophyllene oxide, however, is too complex to fully analyze by using ^{1}H and ^{13}C NMR alone. In fact, caryophyllene oxide is a worthy challenge to fully characterize using the methods in this chapter! Before turning to describe specific experiments and their interpretation, we will first take a closer look at pulse sequences and Fourier transformation.

6.2 Theory

We recall that in order to obtain a routine ^{1}H or ^{13}C NMR spectrum in a pulsed experiment, the "pulse sequence" (Figure 6.1) involves an equilibration in the magnetic field, an rf pulse, and signal acquisition. This sequence is repeated until a satisfactory signal is obtained; Fourier transformation of the FID results in the familiar frequency-domain spectrum.

Figure 6.1 reveals a number of interesting features. We note that there is a separate line for the ^{1}H "channel" and one for the ^{13}C "channel". These "channels" represent the hardware associated with the irradiation and signal acquisition of each relevant nucleus in our experiments. Following equilibration (t_d), the pulse sequence used to obtain a 1-D proton spectrum consists of a $\pi/2_x$ pulse, delay, and signal acquisition of the order of seconds (Fig 6.1a). We also notice that the ^{13}C channel is inactive during a simple proton experiment. Normally, we will not show a given channel unless there is some activity in that channel.

Figure 6.1b is the pulse sequence for a ^{13}C experiment. The sequence in the ^{13}C channel is exactly the same as the sequence in the ^{1}H channel in Figure 6.1a. The protons are decoupled from the ^{13}C nuclei by irradiating the protons during the experiment; that is, the proton decoupler is turned on during the entire experiment. In other experiments, the decoupler for a given nucleus can be turned on and off to coincide with pulses and delays in another channel (i.e. for another nucleus). This process is termed gated decoupling (see Sections 5.4 and 5.6).

It is worthwhile to review here (see Chapters 4 and

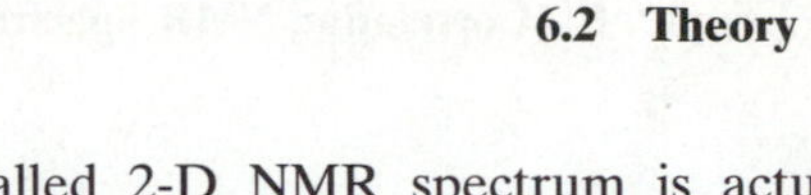

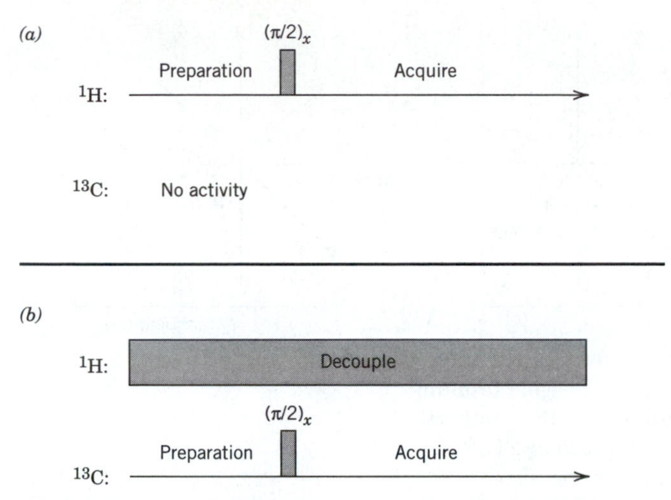

FIGURE 6.1. *(a)* Pulse sequence for standard 1-D ^1H spectrum. *(b)* Pulse sequence for a standard (decoupled) ^{13}C spectrum. $(\pi/2)_x$ is a 90° pulse along the x axis. An equilibration period in the magnetic field before the pulse is assumed (t_d); there is a very brief delay (t_d) to recover from the pulse.

5) what is happening to the net magnetization vector, $\boldsymbol{M_o}$, for a single spin during this pulse sequence when viewed in a rotating frame of reference. In a frame of reference rotating at the Larmor frequency, $\boldsymbol{M_o}$ is stationary on the z axis (equilibration period in Figure 6.1). A $\pi/2$ (90°) pulse brings $\boldsymbol{M_o}$ onto the y axis; when viewed in the rotating frame, the magnetization vector appears to remain stationary although the magnitude of the vector is decreasing with time (T_1 and T_2 relaxation). Returning for a moment to the static laboratory frame, we see that the net magnetization vector is actually not static; it is rotating in the xy plane about the z axis at the Larmor frequency. This rotating vector generates an rf signal that is detected as an FID in an NMR experiment. The net magnetization vector soon returns (in the rotating frame once again) to the z axis, relaxation is complete, and the sequence can be repeated. In a simple one-pulse experiment, a $\pi/2$ pulse is used because it produces the strongest signal. A pulse less than (or greater than) $\pi/2$ leaves some of the possible signal on the z (or $-z$) axis; only the component of the vector on the y axis generates a signal.

We now consider multiple-pulse experiments and two-dimensional NMR. Exactly what is meant by a "dimension" in NMR? The familiar proton spectrum is a plot of frequency (in δ units) versus intensity (arbitrary units)—obviously "two dimensional" but called a "1-D" NMR experiment, the one dimension referring to the frequency axis. It is important to remember that the frequency axis, with which we are comfortable, is derived from the time axis (the acquisition time) of the FID through the mathematical process of Fourier transformation. Thus, *experimentally*, the variable of the abscissa of a 1-D experiment is in time units.

The so-called 2-D NMR spectrum is actually a three-dimensional (3-D) plot; the omitted dimension in all NMR experiments (1-D, 2-D, 3-D, etc.) is always the intensity in arbitrary units. The two dimensions referred to in a 2-D NMR experiment are both frequency axes. It requires two Fourier transformations at right angles to each other on two independent time axes to arrive at two orthogonal frequency axes.

When the simple one-pulse experiment is again considered, there is only one time factor (or variable) that affects the spectrum, namely the acquisition time, t_2. We now consider a multiple-pulse sequence in which the equilibration period is followed by two pulses with an intervening time interval, the final pulse being the $\pi/2$ acquisition pulse. Thus, we have inserted an "evolution" period between the pulses. If we now vary this evolution time interval (t_1) over many different "experiments" and collect the resulting FIDs into one overall experiment, we have the basis of a 2-D experiment. Sequential Fourier transformation of these FIDs yields a set of "spectra" whose peak intensities vary sinusoidally. This first series of Fourier transformations result in the "second" frequency axis, ν_2, derived from the acquisition time, t_2, of each FID. The data are now turned by 90° and a second Fourier transformation is carried out at right angles to the first series of transformations. This second series of Fourier transformations result in the "first" frequency axis, ν_1, a function of the evolution time, t_1, which you recall was changed (i.e. incremented) in the pulse sequence for each successive FID.

A simple prototype of a 2-D experiment should clarify some of these ideas while serving as a "template" for other more useful 2-D experiments. In this simple case, the pulse sequence (Fig. 6.2) consists of a $\pi/2$ pulse, a time interval (t_1, the evolution period), a second $\pi/2$ acquisition pulse, and acquisition (t_2). This pulse sequence (individual experiment) is repeated a number of times (each time resulting in a *separate* FID) with an increased t_1 interval.

We choose for this experiment a simple compound, chloroform ($CHCl_3$), to avoid the complication (for now) of spin coupling. In Figure 6.3, we see that after the first $\pi/2$ pulse along the x axis $(\pi/2)_x$, the magnetization $\boldsymbol{M_o}$ has rotated onto the y axis to $\boldsymbol{M}$. The evolution during t_1 for the spin of the proton of chloroform is shown in a rotating frame in Figure 6.3. In this treat-

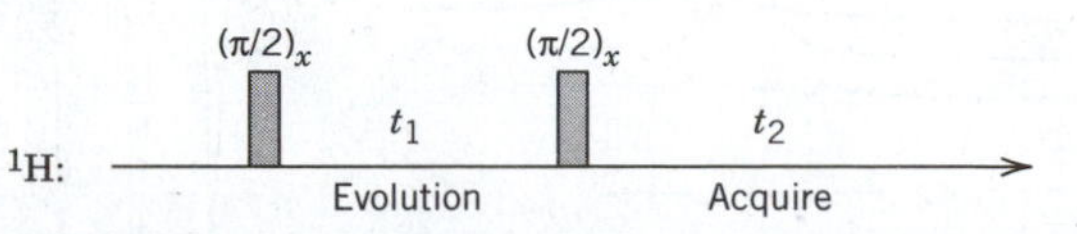

FIGURE 6.2. Prototype pulse sequence for a 2-D NMR experiment. The incremental delay, t_1, and the acquisition time, t_2, are Fourier transformed into frequencies, ν_2 and ν_1, respectively. $(\pi/2)_x$ represents a 90° pulse along the x axis. The interval t_1 is of the order of microseconds; t_2 is of the order of seconds.

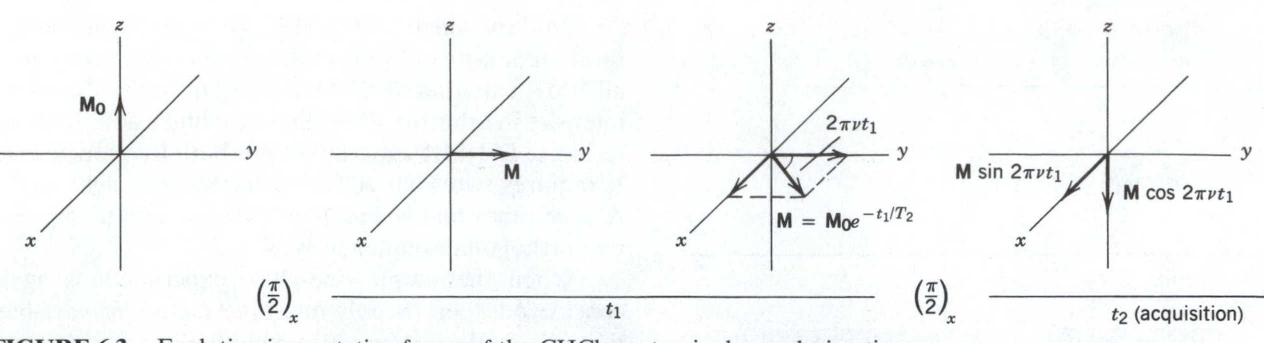

FIGURE 6.3. Evolution in a rotating frame of the $CHCl_3$ proton is shown during time interval t_1 following the first pulse. The second pulse and acquisition give a signal resulting only from the x component of $\boldsymbol{M}$; this signal amplitude varies sinusoidally with t_1. Interval t_1 is on the order of microseconds and milliseconds; t_2 is on the order of seconds. The precessional frequency of the proton is higher than that of the rotating frame. From Derome (1987) with permission.

ment, we ignore spin–lattice relaxation but include transverse relaxation with time constant T_2 (Section 4.3). If the Larmor frequency (ν_2) is at higher frequency than that of the rotating frame, $\boldsymbol{M}$ precesses in the xy plane during the time interval t_1 through the angle $2\pi\nu t_1$. From trigonometry, the y component of $\boldsymbol{M}$ is $\boldsymbol{M}$ $\cos(2\pi\nu t_1)$, and the x component is $\boldsymbol{M}\sin(2\pi\nu t_1)$.

After time t_1, the acquisition pulse $(\pi/2)_x$ rotates the y component downward onto the z axis; this component therefore contributes no signal to the FID. The x component, on the other hand, remains unchanged (in the xy plane) and its "signal" is recorded as the FID. When this FID is Fourier transformed, it gives a peak with frequency ν_2 and amplitude $\boldsymbol{M}\sin(2\pi\nu t_1)$. If we repeat this "experiment" many times (e.g., 1024 or 2^{10}), each time increasing t_1 in a regular way, we obtain 2^{10} FIDs. Successive Fourier transformation of each of these FIDs gives a series of "spectra" each with a single peak of frequency ν_2 and amplitude $\boldsymbol{M}\sin(2\pi\nu t_1)$. In Figure 6.4, 19 of the 1024 "spectra" are plotted in a stacked column;

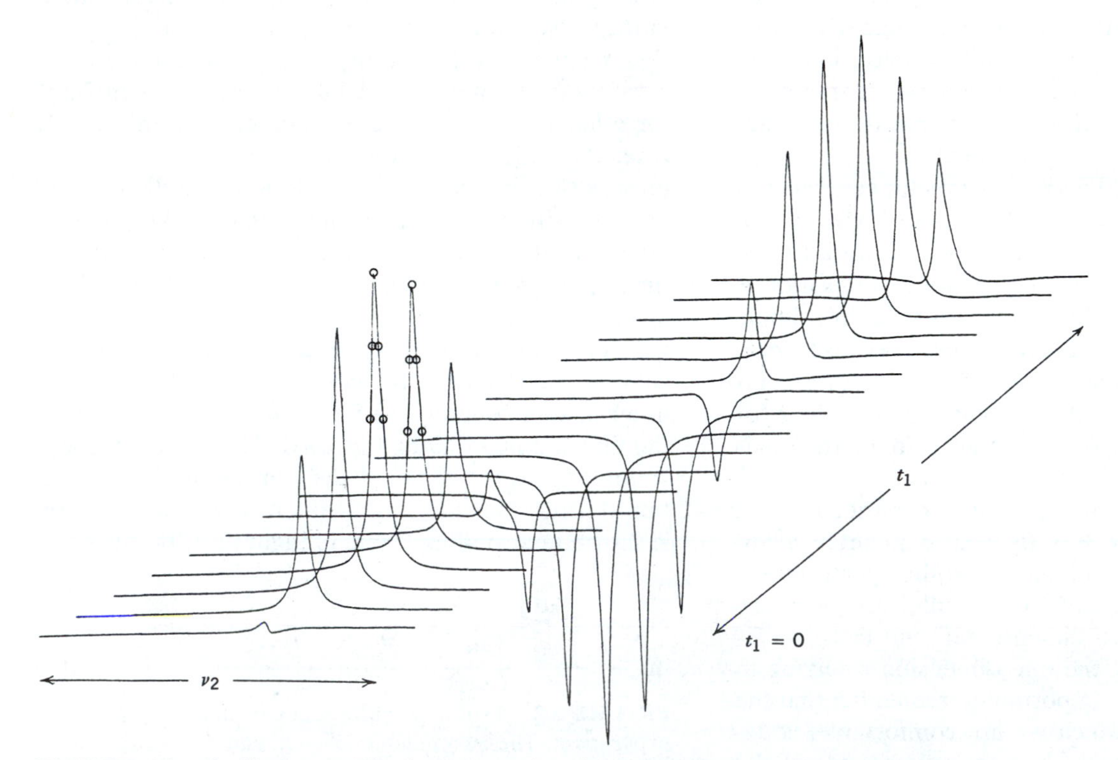

FIGURE 6.4. Oscillating amplitude of the series of acquired peaks with increasing t_1. Only the first few peaks of the series are shown. Data points are shown on two of the rows. Reprinted with permission from A. E. Derome, *Modern NMR Techniques,* Copyright 1987, Pergamon Press PLC.

we see that the amplitude of the chloroform peak varies sinusoidally as a function of t_1. We have now established one of the frequency axes (ν_2) for our prototype 2-D spectrum.

Before we establish our second axis, let us do a little bookkeeping. Remember that each of these "spectra" that we now have is actually a digitized collection of points. Let us assume that each of these 1024 "spectra" is composed of 1024 data points. Thus, we have a square matrix of data. If we mentally rotate this collection of "spectra," we can perform a second series of Fourier transformations on the data orthogonal to the first. Before we perform these transformations though, let us take a closer look at the data in Figure 6.4. If we replot one row of data from Figure 6.4 [let us choose the row that corresponds to the maximum (or minimum) for the chloroform peak], the data we obtain (Fig. 6.5) look like an FID or a time-domain spectrum. In fact, it is a time domain spectrum, now a function of the interval t_1 from our pulse sequence. To distinguish, we refer to data obtained in real time (a function of t_2) as an FID, and to data constructed point by point as a function of t_1 as an *interferogram.*

We now perform our second series of Fourier transformations on each of the 1024 interferograms to produce literally a transform of a transform. This result is the end product: a 2-D spectrum. We are now faced, however, with the challenge of visualizing our results. One way to plot the data is as a "stacked plot" similar to the plot that we have already seen in Figure 6.4. This type of plot, shown in the first part of Figure 6.6, is completely analogous to a topographic map. For this spectrum, this type of plot is satisfactory; there are no peaks being blocked so one perspective is sufficient. For more complex spectra, the data are usually presented as a series of contours just as hills and valleys are represented on a topological map. We see this representation of the data in the second part of Figure 6.6. As long as there are no negative peaks (e.g., phase-sensitive COSY (Section 6.4), not covered in this book), we use this method without comment.

6.3 *Correlation Spectrometry*

The observant reader will have by now realized that the above experiment, which Derome (1987) calls "frequency labeling," provides no additional information beyond the simple ^{1}H spectrum of chloroform. Actually,

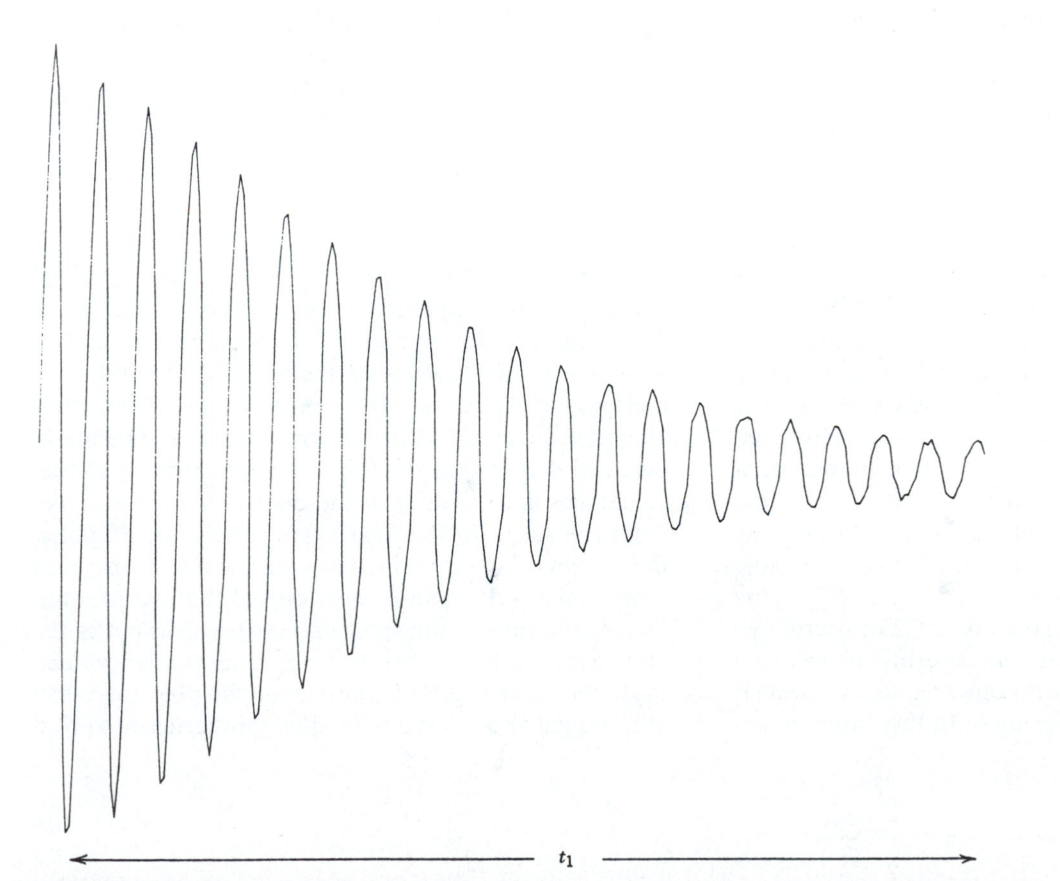

FIGURE 6.5. Transposition of Figure 6.4. Slice parallel with t_1 through the tops of peaks in Figure 6.4. Reprinted with permission from A. E. Derome, *Modern NMR Techniques,* Copyright 1987, Pergamon Press PLC.

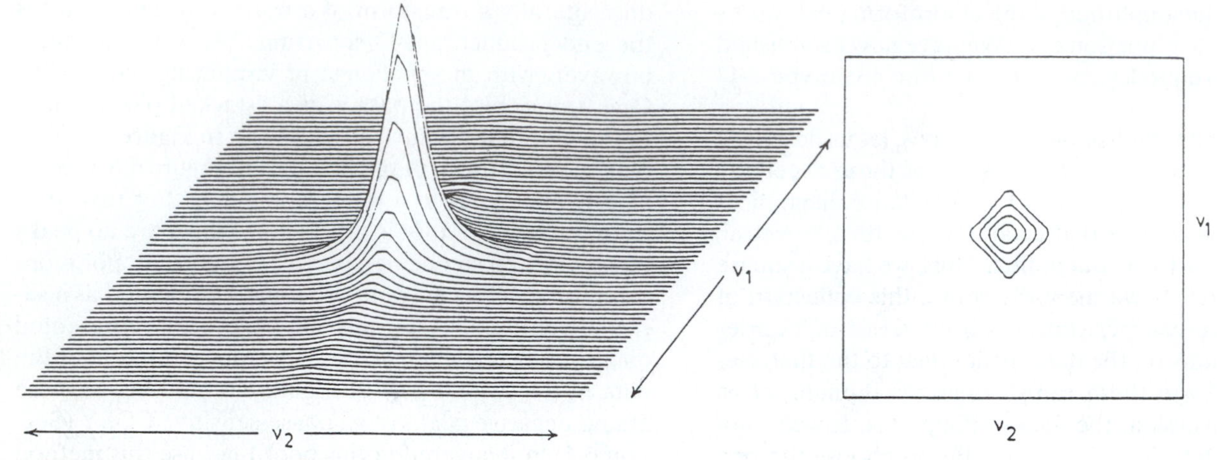

FIGURE 6.6. Fourier transformation of a series of FIDs like the one in Figure 6.5 to give the frequency-domain spectrum as both a peak and a contour. Reprinted with permission from A. E. Derome, *Modern NMR Techniques,* Copyright 1987, Pergamon Press PLC.

that is the beauty of that experiment; it has all of the elements of a 2-D correlation experiment *and* we can completely follow the fate of the net magnetization vector for chloroform using simple vectorial models. Let us turn this prototype pulse sequence into a general format for all 2-D experiments. If we replace the first $\pi/2$ pulse with a generalized "pulse" that contains one or more pulses and concomitant delays, and replace the second acquisition $\pi/2$ pulse with a generalized "acquisition pulse" that also contains one or more pulses and their concomitant delays, we arrive at the general pulse sequence for 2-D correlation experiments, shown in Figure 6.7. We can usually describe, using simple vector models and trigonometry, the result of what is happening inside the boxes and oftentimes we will. For the time being, let us ignore "the goings on" inside the boxes and concentrate on what is happening during t_1 and t_2.

In all 2-D experiments, we detect a signal (during acquisition) as a function of t_2; this signal, however, has been *modulated* as a function of t_1. The chloroform experiment is simple because the magnetization experiences identical modulation during t_1 and t_2. In an experiment in which the magnetization is identically modulated during t_1 and t_2, the resulting peaks will be such that ν_1 equals ν_2; in the parlance of 2-D NMR, the experiment gives rise to diagonal peaks. For useful 2-D information, we are interested in experiments in which the magnetization evolves with one frequency during t_1 and a different frequency during t_2. In this case, our experiment will give rise to peaks in which ν_1 and ν_2 are different; this time we call the peaks off-diagonal or cross peaks. In order to interpret a 2-D NMR spectrum, there are two things that we need to know. First, what frequencies do the axes represent? One axis (ν_2) always represents the nucleus detected during acquisition (t_2). The other axis (ν_1, which obviously depends on t_1) can represent the same nucleus (e.g., $^1H-^1H$ COSY), a different nucleus (e.g., $^1H-^{13}C$ COSY, also labeled HETCOR, Section 6.6), or a coupling constant (e.g. *J*-resolved spectroscopy, not covered in this chapter). Second, we need to know how the magnetizations are related during t_1 and t_2; in this way we can account for and interpret the cross peaks.

If we return to our prototype 2-D experiment and apply this pulse sequence to an AX system, we will be in a better position to appreciate the incremental time, t_1. While we can describe mathematically precisely how the spins evolve during this time, we cannot show this evolution pictorially with vector diagrams. (The mathematical description for this system requires quantum mechanics and solution of the density matrix, well beyond the scope of this text.) After the first $\pi/2$ pulse, the system can be described as a sum of two terms; each term contains the spin of only one of the two protons. During the time t_1, the spins precess (evolve) under the influences of both chemical shifts *and their mutual spin–spin coupling*. The mutual coupling has the effect of changing some of the individual spin terms into prod-

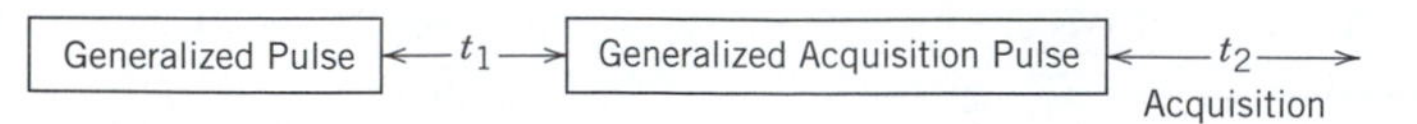

FIGURE 6.7. Generalized pulse sequence for 2-D NMR. The signal detected during acquisition, t_2, is modulated during the incremental time, t_1, thus giving rise to cross peaks in the 2-D spectrum.

ucts containing magnetization components of both nuclei. Next, the second $\pi/2$ pulse causes the spins that have been precessing under both chemical shift and coupling influences to redistribute magnetization among all spins (only one other spin in this case) with which it is associated (coupled). This redistribution of magnetization is detected in t_2; thus, a frequency detected in t_2 has its amplitude modulated as a function of other spins (only one here), with which it is coupled during t_1, leading to cross peaks connecting the coupled nuclei. Because the magnetization is redistributed equally in both directions (i.e., from A to X and from X to A), the cross peaks (at least for this experiment) will be symmetrically disposed about the diagonal. This description of spins precessing and mixing during t_1 and redistribution during the acquisition pulse (and detection during t_2) is admittedly difficult to follow without pictures. As we introduce new pulse sequences, and hence new experiments, we shall provide a simple (i.e. nonmathematical) description of the evolution period, t_1, because it is, after all, this mixing period that defines a 2-D experiment.

6.4 $^1H—^1H$ *COSY*

Our simple 2-D experiment is actually a very important experiment sometimes simply called COSY (COrrelation SpectroscopY), and which we will call $^1H—^1H$ COSY in order to clearly indicate what is being correlated.* The pulse sequence for $^1H—^1H$ COSY is none other than the one we have already described above: two $\pi/2$ proton pulses separated by the required evolution period, t_1, which is systematically incremented, and the acquisition period, t_2. Most modern spectrometers employ a technique known as "phase cycling" in which the phase of the rf pulse is changed in a regular manner (through a "cycle") for each t_1 increment. These phase cycles are extremely important *experimental factors* that help remove artifacts and other peculiarities of quadrature detection. We will ignore phase cyclings in our pulse sequences and discussions because they do not affect our understanding and interpretation of these experiments. The interested reader is referred to Derome (1987) for these and other experimental parameters important for 2-D experiments. The pulse sequence for a simple $^1H—^1H$ COSY is shown in Figure 6.8 (without phase cycling).

In the description of the 2-D experiment above for an AX spin system, we found that during t_1 spins, which

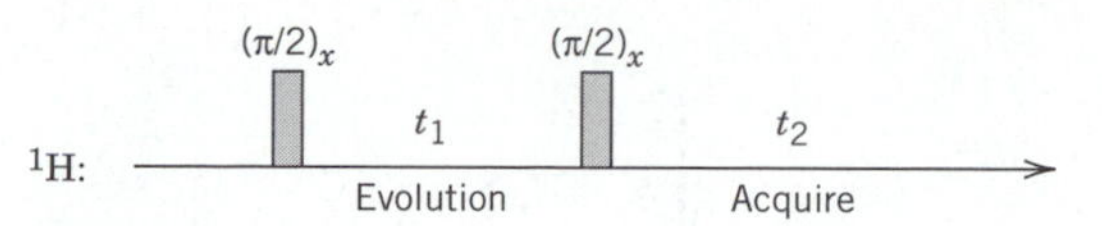

FIGURE 6.8. Pulse sequence for simple $^1H—^1H$ COSY. Note that this sequence is identical to that of Figure 6.2 for the prototypical $CHCl_3$ experiment.

are mutually coupled, precess under the influence of both nuclei's chemical shifts and thus give rise to peaks in which ν_1 does not equal ν_2. In the general case, $^1H—^1H$ COSY spectra are interpreted as giving rise to off-diagonal or cross peaks for all protons that have significant (i.e., measurable) *J–J* coupling; put simply, the cross peaks correlate coupled protons. In a sense, the experiment can be thought of as simultaneously running all pertinent decoupling experiments to see which protons are coupled to which other protons. Of course, no protons are being decoupled and an $^1H—^1H$ COSY should not be thought of as replacing homonuclear decoupling experiments (see Section 4.19).

Let us orient ourselves by considering the $^1H—^1H$ COSY spectrum of ipsenol, the monoterpene alcohol considered in some detail in Section 4.16.1. The contour display of this spectrum is shown in Figure 6.9. The presentation shown here is typical; F2 is found on the bottom, with the proton scale as usual. F1 is displayed on the right, with the proton scale running from top to bottom. A proton spectrum is displayed opposite the F1 and F2 scales and are given for convenience; these 1-D spectra are not part of the $^1H—^1H$ COSY spectrum proper. From the upper right to the lower left runs the "diagonal", a series of absorptions in which ν_1 equals ν_2; these diagonal peaks provide nothing in the way of useful information beyond the simple 1-D ^1H spectrum. On either side of the diagonal and symmetrically disposed (at least theoretically) are the cross peaks. The symmetry in this type of spectrum is oftentimes imperfect.

Before undertaking detailed discussions of $^1H—^1H$ COSY and the structure of ipsenol, there is one further refinement that decreases the "clutter" along the diagonal. Although we can interpret this spectrum without this refinement, there are instances (e.g., caryophyllene oxide) when this improvement makes a great deal of difference.

6.5 *Double-Quantum Filtered* $^1H—^1H$ *COSY*

By simply adding a third $\pi/2$ pulse immediately following the second $\pi/2$ pulse in our simple COSY pulse sequence and changing nothing else, we have the pulse

* Many readers will already be aware that acronyms for 2-D NMR experiments have proliferated along with available experiments. This chapter does not attempt an encyclopedic approach to describing these acronyms or their experimental counterparts. This chapter does, however, cover enough important experiments to enable the reader to interpret nearly any 2-D experiment that one is likely to encounter. Acronyms are listed in the index.

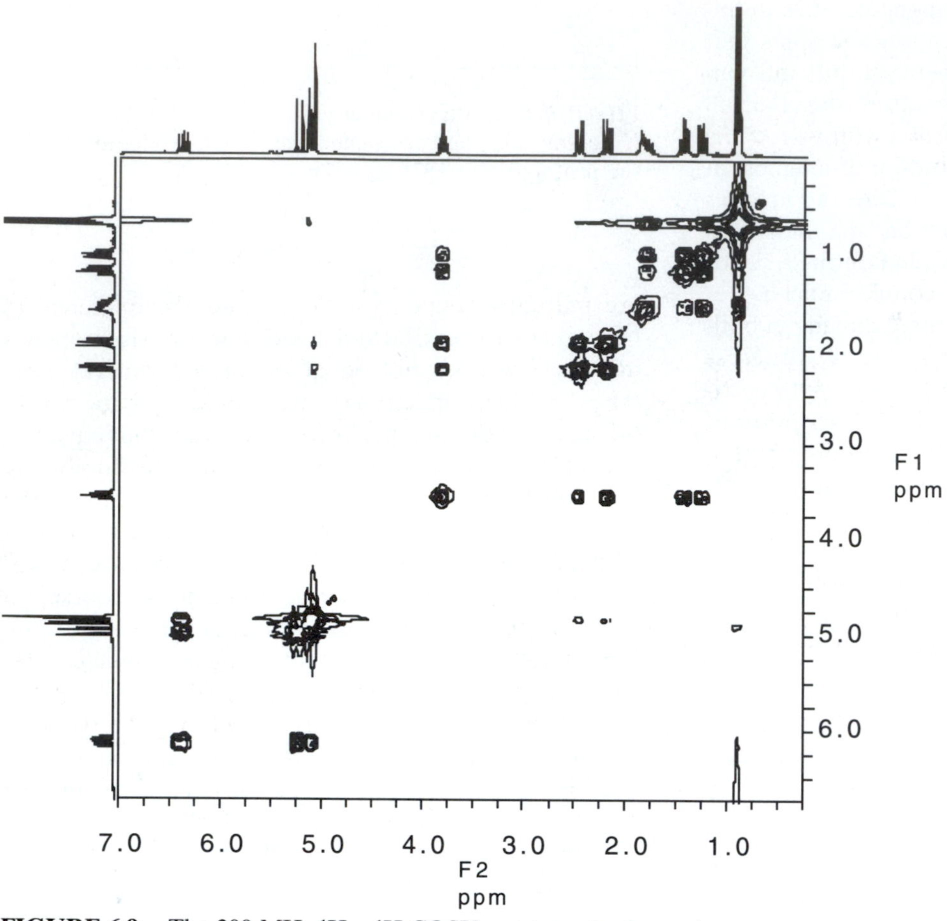

FIGURE 6.9. The 300-MHz $^1H—^1H$ COSY spectrum for ipsenol.

sequence for the very popular double-quantum filtered $^1H—^1H$ COSY (DQF-COSY) experiment (Figure 6.10). The purpose of the third $\pi/2$ pulse is to remove or "filter" single quantum transitions so that only double quantum or higher transitions remain. In practical terms, the double quantum filter will select for systems with at least two spins (minimum AB or AX); thus, methyl singlets (noncoupled) will be greatly reduced. Higher quantum filtering is possible but is generally not used. For instance, in a triple-quantum filtered COSY, only systems with three spins or more are selected so that AB and AX spin systems as well as noncoupled systems will be eliminated.

6.5.1 *Ipsenol*

The DQF $^1H—^1H$ COSY spectrum of ipsenol can be found in Figure 6.11. Note that the spectrum

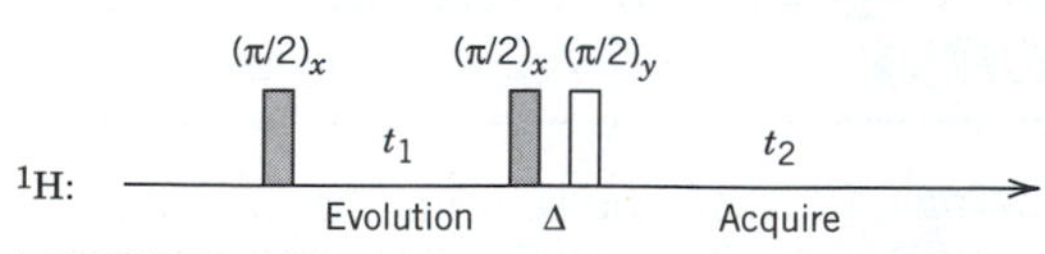

FIGURE 6.10. Pulse sequence for double-quantum filtered $^1H—^1H$ COSY (DQF-COSY).

seems "cleaner" especially along the diagonal making the task of interpretation significantly easier. Because of the greatly improved appearance of DQF-COSY, all COSYs in this book are double quantum filtered.

As we begin our interpretation of the COSY spectrum in Figure 6.11, let us recall that this spectrum shows correlation between coupled protons. The reader may at this point want to return to Chapter 4 and again consider the proton spectrum of ipsenol before continuing. A point of entry (i.e., a distinctive absorption) into a COSY spectrum (and other types of correlation spectrometry as well) is one of the keys to gleaning information from it successfully. The structure of ipsenol allows for more than one useful points of entry, so let us select the carbinol methine at 3.83 ppm. If we begin at the diagonal and trace either directly to the right or directly up (we obtain the same result because the spectrum is symmetrical), we intersect four off-diagonal or cross peaks. By drawing lines through these cross peaks at right angles to the one we just traced, we find the chemical shifts of the four coupled resonances. A quick check of the structure of ipsenol finds the carbinol methine adjacent to two pairs of diastereotopic methylene

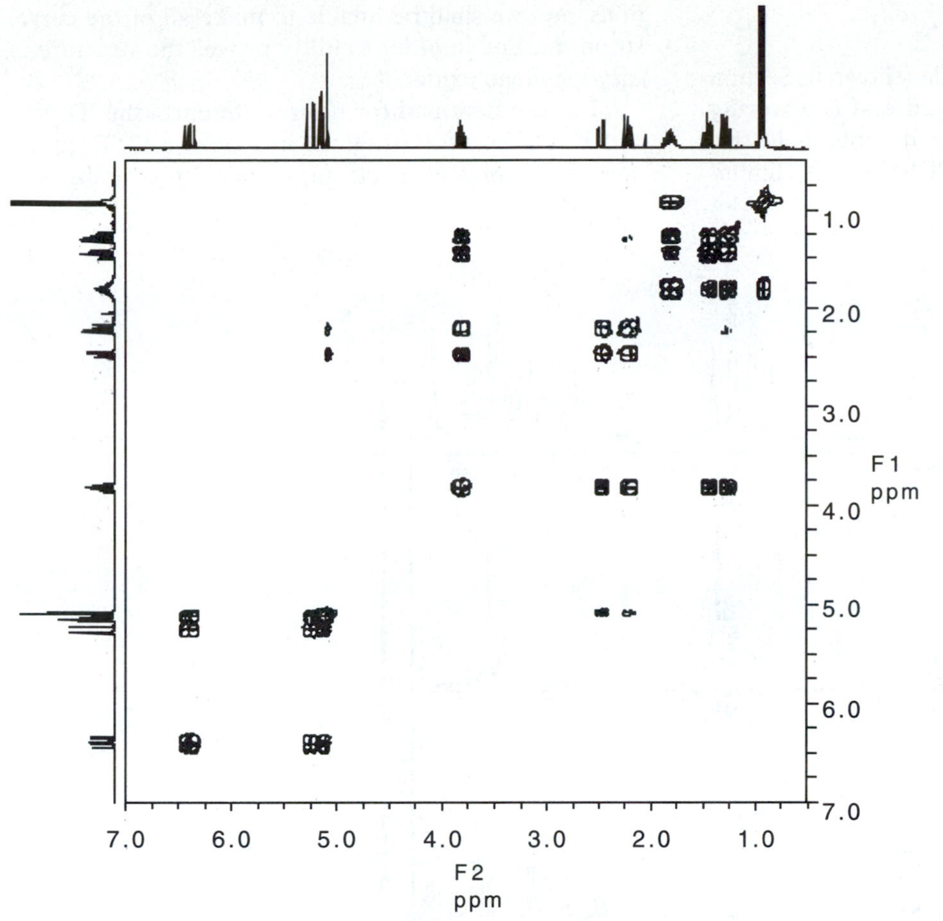

FIGURE 6.11. The 300-MHz DQF-COSY of ipsenol. Compare the diagonal of the DQF-COSY with the diagonal of the simple COSY in Figure 6.9.

protons*; in other words, the proton at 3.83 ppm is coupled to four protons, and the four protons correspond to two adjacent methylene groups.

We could continue to trace correlation paths from these four protons and the reader is invited to do so at the end of this section. Let us instead select another equally useful entry point: the isopropyl methine at 1.82 ppm. We again begin at the diagonal and this time we find that the isopropyl methine is correlated with three distinct resonances. Two of the correlations correspond to the two protons of one of the diastereotopic methylenes that also correlated with the carbinol methine above. In addition, we find a correlation to the two overlapping methyl doublets at 0.93 ppm. These correlations, of course, make perfect sense with the structure; in fact, by only considering these two protons (i.e., at 3.83 and 1.82 ppm) we have established correlations (also called "connectivities") through three-fifths of the molecule.

Next we consider the two protons on the C-5 methylene at 2.48 and 2.22 ppm. We have already seen that they are coupled to the carbinol methine (you can and should verify this from the methylene protons' perspective) and we see that they are also coupled to each other.† In addition, we see weaker cross peaks from both methylene protons correlating to an olefinic proton at 5.08 ppm. This correlation is due to long range coupling ($^4J_{HH}$ or four-bond coupling) of the methylene protons to the cis proton of the adjacent double bond. This is a nice correlation to find because it provides H—H connectivity to the otherwise isolated diene spin system. In the absence of these long range correlations, such isolated spin systems can be "connected" by either HMBC (Section 6.8) or INADEQUATE (Section 6.9) sequences, which are described below. At this point, the reader is invited to complete the correlations for ipsenol in this COSY spectrum. Correlations can be found for all protons except the hydroxylic proton, which is buried at δ1.8 and rapidly exchanging.

* We will reserve further discussion of diastereotopic methylene groups until the next section on $^1H—^{13}C$ COSY or HETCOR.

† Geminal methylene protons (sp^3 hybridized) are always coupled to each other and their coupling constant ($^2J_{HH}$) is always rather large (see Appendices, Chapter 4).

6.5.2 *Caryophyllene Oxide*

The structure of caryophyllene oxide (shown in Section 6.1) is significantly more complicated and is a worthy challenge for the methods we are describing in this chapter. We shall soon see that COSY has its limitations, and we shall be unable to make all of the correlations needed in order to fully "prove" the structure of caryophyllene oxide.

For use here and for future reference, the ^{1}H, ^{13}C, and DEPT spectra are given in Figure 6.12. Focusing now on the proton spectrum, we see three methyl sin-

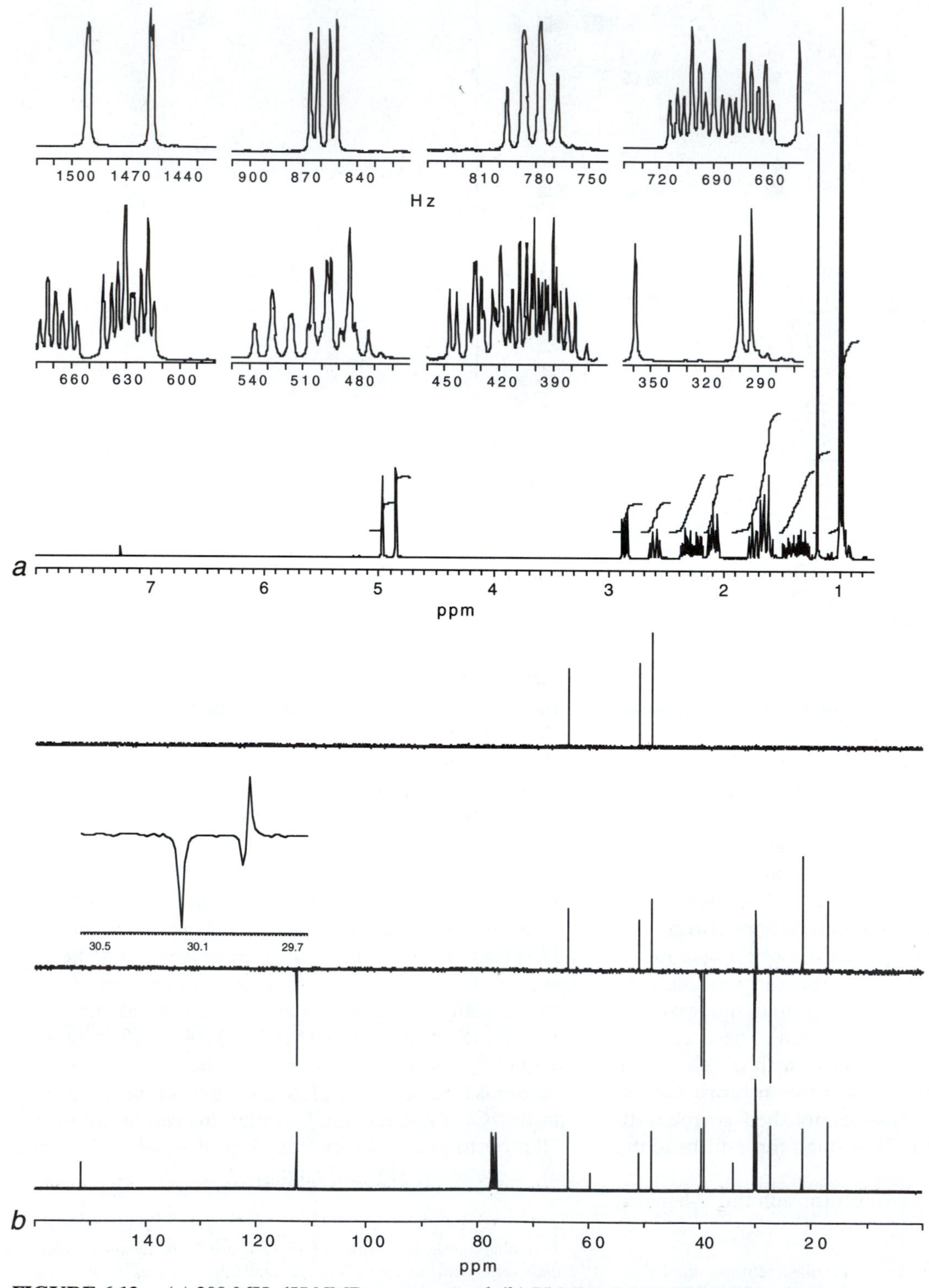

FIGURE 6.12. *(a)* 300-MHz ^{1}H NMR spectrum and *(b)* 75-MHz ^{13}C NMR spectrum and DEPT spectra for caryophyllene oxide.

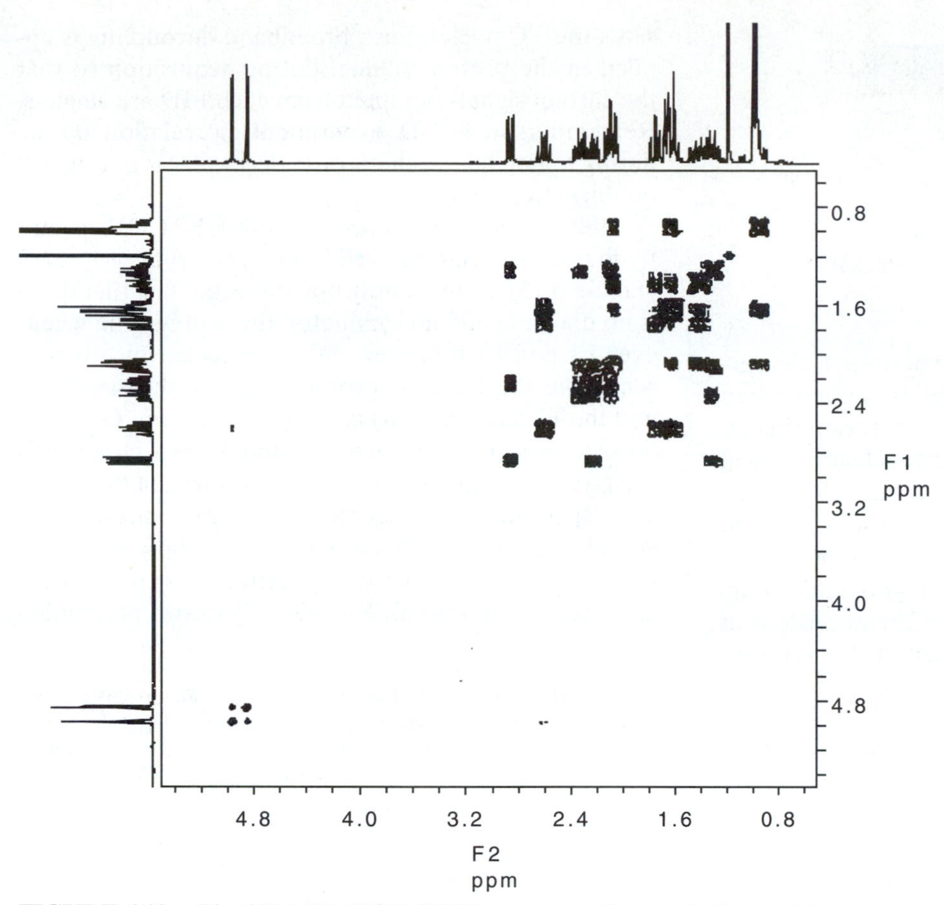

FIGURE 6.13. The 300-MHz DQF-COSY spectrum of caryophyllene oxide.

glets, 0.98, 1.01, 1.19 ppm, two olefinic "doublets" (small geminal olefinic coupling), 4.86 and 4.97 ppm, and resonances from 13 other protons giving multiplets between 0.90 and 3.90 ppm. Even though we know the structure, it is impossible to assign any of these protons unless we make one or more unreasonable assumptions.*

The DQF-COSY spectrum of caryophyllene oxide can be found in Figure 6.13. The problem is that *there is no good entry point.* The previous statement is not trivial. Without an entry point, it is impossible to relate the many obvious correlations that we see to a structural formula. Our approach therefore will be to record some of the correlations that we do see and wait until we have other information (i.e., HETCOR) before we try to translate these correlations into a structure.

The exocyclic olefinic methylene protons show obvious COSY correlations to one another. In addition, we note a very weak cross peak between the deshielded olefinic proton and an apparent quartet at 2.60 ppm. This interaction is reminiscent of the long-range allylic coupling that we saw in ipsenol; we still must ask whether the coupling is to the bridgehead methine or to one of the two diastereotopic methylene protons on the other side of the double bond. Again, we cannot be sure and return to this point later in the chapter.

A look at the extreme low-frequency portion of this COSY spectrum reveals an unexpected interaction. It seems that either one or both of the methyl singlets shows coupling to resonances at 1.65 and at 2.09 ppm. This apparent conflict can be resolved by a close examination of the methyl singlet at 0.98 ppm. There is an unusually low-frequency multiplet, partially buried by the methyl singlets, that we had initially overlooked. This type of unexpected dividend is common in correlation spectra; both partially and completely obscured resonances usually reveal themselves in 2-D spectra (see HETCOR, Section 6.6). Before continuing our discussion of caryophyllene oxide, let us consider $^{1}H-^{13}C$ correlations and how $^{1}H-^{1}H$ correlations interplay with $^{1}H-^{13}C$ correlations.

* If pressed, we might assume that the allylic bridgehead methine would be the furthest downfield and assign the doublet of doublets at 2.86 ppm to this proton (wrong). The methods in this chapter will allow us to make these assignments without making unsubstantiated assumptions.

6.6 $^{1}H-^{13}C$ COSY: HETCOR

The HETCOR experiment correlates ^{13}C nuclei with directly attached (i.e., coupled) protons; these are one-

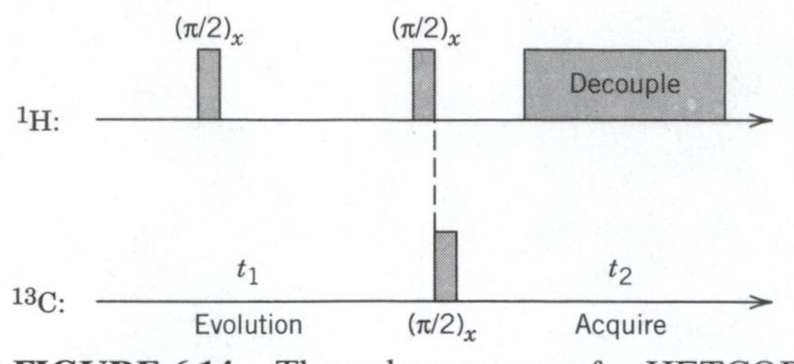

FIGURE 6.14. The pulse sequence for HETCOR.

bond ($^1J_{CH}$) couplings. In an ideal experiment, the carbons resonances should be singlets rather than multiplets.* The pulse sequence for this experiment, commonly called HETCOR (HETeronuclear chemical shift CORrelation), is recorded in Figure 6.14. Three details in this pulse sequence are worth discussing. The F1 axis (ν_1), which is derived from our incremental delay, t_1, is the proton axis. The F2 axis (ν_2) obtained during t_2 is the carbon axis. Thus, our $\pi/2$ read pulse is in the ^{13}C channel, and the FID acquired during t_2 represents the ^{13}C nuclei. Last, broadband decoupling is applied in the proton channel during acquisition so that the carbon signals obtained from each FID are singlets. Remember, in a 2-D experiment, correlation occurs during t_1 and, hence, the proton decoupler is not turned on during this period.

Let us familiarize ourselves with HETCOR spectra by first considering the HETCOR spectrum of ipsenol (Figure 6.15). Immediately obvious is the fact that there is no diagonal and no symmetry; this will be true whenever F1 and F2 represent different nuclei. In this presentation, the F1 axis (proton) is along the right side and the F2 axis (carbon) is along the bottom. Opposite these axes we find the corresponding 1-D spectra, which are given as a convenience and are not part of the actual 2-D spectrum. Interpretation of this spectrum is straightforward. We begin with any carbon atom and mentally drop a line vertically until a cross peak is encountered.† Another line is mentally drawn perpendic-

* In the presence of one-bond couplings, a methyl carbon will appear as a quartet, a methylene as a triplet, etc.

† We could just as well start on the proton axis, and in this case we would obtain exactly the same result. In cases of overlap in the proton spectrum, we will not always be able to find all of the proper starting points. Overlap is usually not a problem on the carbon axis.

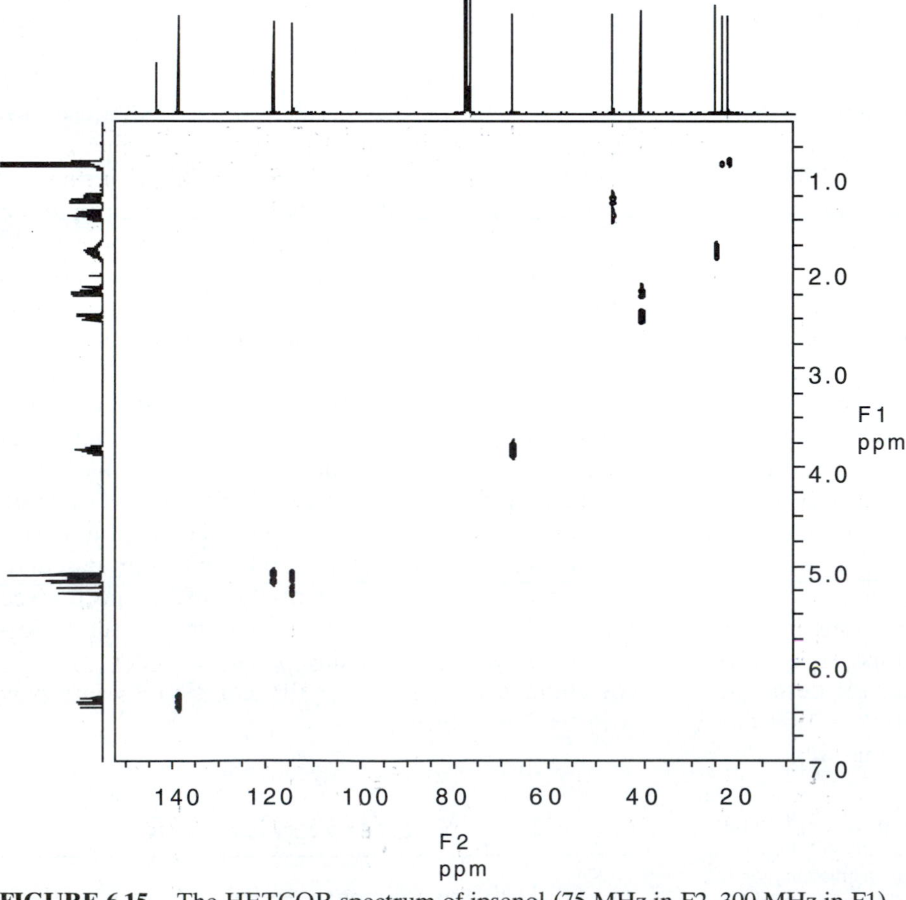

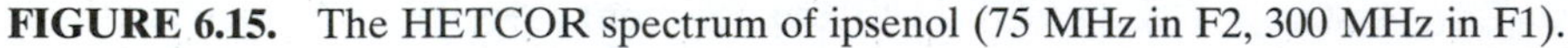
FIGURE 6.15. The HETCOR spectrum of ipsenol (75 MHz in F2, 300 MHz in F1).

ular to the first to find the proton or protons with which it correlates.

There are only three cases possible for each carbon atom. If a line drawn down encounters no cross peaks, then the carbon has no attached hydrogens. If the drawn line encounters only one cross peak, then the carbon may have either 1, 2, or 3 protons attached; if 2 protons are attached, then they are either chemical-shift equivalent or they fortuitously overlap. If the dropline encounters two cross peaks then we have the special case of diastereotopic protons attached to a methylene group. Much of this information will already be available to us from DEPT spectra (see Section 5.5); indeed, the HETCOR should, whenever possible, be considered along with the DEPT.

In ipsenol, there are four methylene groups, all of which possess diastereotopic pairs of protons. Resonance for two of these methylene groups occurs in the carbon spectrum at 41 and 47 ppm. Note with which protons these carbon atoms are correlated and compare these results with what we have found with COSY. As we expect, the results here confirm our assignments from COSY and help build an ever-strengthening basis for our assignments. The other two methylene carbon atoms are found at higher frequency in the olefinic region, and the HETCOR cross peaks for these carbon resonances help clarify the overlapping proton resonances that we find in the proton spectrum. We leave ipsenol for the time being with a question: Can the olefinic methylene carbon resonances be assigned on the basis of combined information from COSY and HETCOR?

We left off with caryophyllene oxide in the COSY section (Section 6.5.2), having made very few assignments. Let us take up the problem again, this time equipped with a HETCOR spectrum (Figure 6.16). From the DEPT spectrum, we already know that caryophyllene oxide has three quaternary carbon resonances (at 34.0, 59.7, and 153.0 ppm), six methylene carbon resonances (27.2, 29.9, 30.1, 39.2, 39.8, and 113.0 ppm), three methine carbon resonances (48.7, 50.9, and 63.6 ppm), and three methyl carbon resonances (16.9, 22.6, and 30.0 ppm).

The olefinic methylene (protons and carbon) and the three methyl groups (protons and carbons) are trivial assignments, and they correspond with our previous discussion. Of more interest and of greater utility, we assign the three methine protons: the doublet of dou-

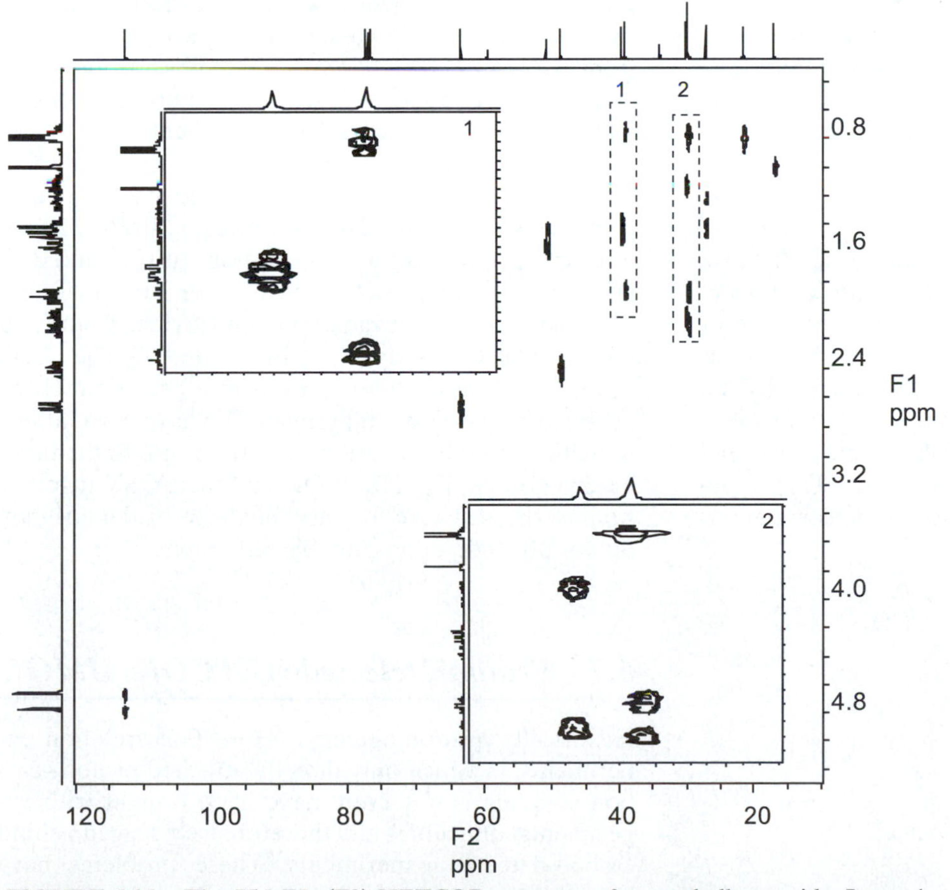

FIGURE 6.16. The-75 MHz (F2) HETCOR spectrum of caryophyllene oxide. Insets 1 and 2 give much better digital resolution for their respective areas.

blets at 2.86 ppm (correlates with the carbon resonance at 64 ppm), the apparent quartet at 2.60 ppm (correlates with the carbon resonance at 49 ppm), and a multiplet (overlapping with at least one other multiplet) at 1.76 ppm (correlates with the carbon resonance at 51 ppm). From the COSY and from the known structure, we assign all three methine resonances and "feed" this information back into the COSY to establish other correlations.

From the long-range, allylic correlation that we found in the COSY, we now know that it correlates not to the methylene group but to the bridgehead methine. The doublet of doublets at 2.86 ppm is assigned to the methine proton of the epoxide ring, and its chemical shift is rationalized on the basis of the deshielding effect of the epoxide oxygen. The other bridgehead methine (adjacent to the *gem*-dimethyl group) is assigned to the multiplet at 1.76 ppm. With these assignments in hand, we could "jump right back" into the COSY spectrum, but instead we will restrain our enthusiasm for now and assign the methylene protons first. Knowing these assignments first will help speed our way through the COSY.

Beginning from the low-frequency end of the ^{13}C spectrum, the following assignments can be made: The methylene carbon at 27.2 ppm correlates with proton resonances at 1.45 and 1.63 ppm, the methylene carbon at 29.9 ppm correlates with proton resonances at 2.11 and 2.37 ppm, the methylene carbon at 30.1 ppm correlates with proton resonances at 1.28 and 2.23 ppm, the methylene carbon at 39.2 ppm correlates with proton resonances at 0.95 and 2.06 ppm, the methylene carbon at 39.8 ppm correlates with proton resonances at 1.43 and 1.47 ppm,* and we have already assigned the olefinic methylene group above. Thus, with little effort we have assigned a chemical shift for all of the protons in caryophyllene oxide and correlated them with a resonance from the ^{13}C spectrum; we have grouped the diastereotopic protons together for each of the methylene groups; and we have obtained three separate entry points for the COSY spectrum when before we had none. We are now ready to return to the COSY spectrum of caryophyllene oxide and assign the correlations in light of the structure.

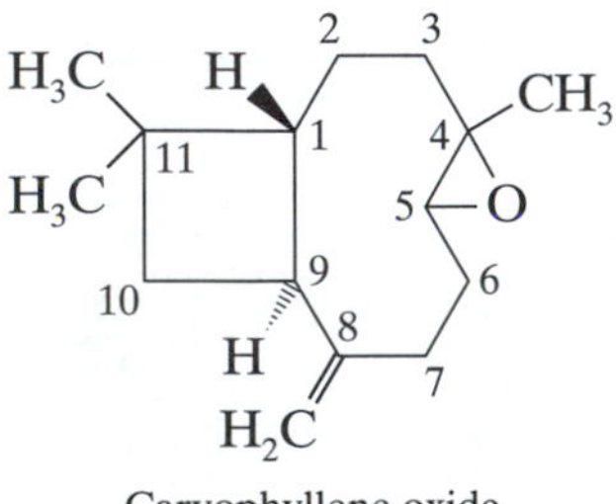

Caryophyllene oxide

* These chemical shift positions are estimated because they badly overlap with each other and with another proton signal at about 1.45 ppm.

An expanded section from 0.8 to 3.0 ppm of the DQF-COSY of caryophyllene oxide is given in Figure 6.17. Included with this figure are lines connecting proton–proton correlations to aid our discussion. The COSY "connectivities" allow us to construct structure fragments or in this case confirm structural segments. To correlate C-5, C-6, and C-7, we start with H-5 at 2.86 ppm. This proton shows cross peaks with two resonances at 1.28 and 2.23 ppm. From the HETCOR we know that these are diastereotopic and assign them as H-6 and H-6′. The protons attached to C-6 give correlations with protons at 2.11 and 2.37 ppm; we assign these protons, which also are diastereotopic, to C-7 at 29.9 ppm. The C-7 protons are coupled as certainly are the C-6 protons. The overlap between 2.1 and 2.4 ppm is severe (both C-7 and one C-6 protons are found here), making our interpretation difficult. Knowing the structure in this case makes the task somewhat easier, but ambiguity remains.

Other correlations are more straightforward. The C-5, C-6, C-7 spin system is isolated, so we must select another entry point. We can start again with the allylic bridgehead methine (H-9) at 2.60 ppm. We have already noted the long-range allylic interaction. In addition, we find three other interactions that the HETCOR helps us to assign. One of the correlations is to a methine proton at 1.76 ppm, which we assign to H-1. The other two correlations find two diastereotopic protons (again, from HETCOR) at 1.43 and 1.47 ppm; we assign them as H-10 and H-10′. The C-10 protons are a dead end and we find no other correlations to them.

H-1 is apparently coupled to only one of the C-2 protons at 1.45 ppm; there is little evidence of a cross peak between H-1 and H-2′ at 1.63 ppm. Both C-2 protons are coupled to both C-3 protons at 0.95 and 2.06 ppm and the appropriate cross peaks can be found. Thus, we have shown indirect connectivities from C-10 through C-9, C-1, and C-2 all the way to C-3. The HETCOR has been invaluable in our interpretation. However, many questions still remain. We have correlations to neither the three quaternary carbons nor to the three methyl groups. The HETCOR and the COSY together *support* the structure for caryophyllene oxide, but they do not preclude other possible structures.

6.7 *Proton-Detected HETCOR: HMQC*

Historically, proton-detected ^{1}H—^{13}C correlation experiments, in which only directly attached proton–carbon coupling is observed, have been fraught with experimental difficulties and therefore have lagged behind carbon-detected experiments. These problems have been overcome and the proton-detected version (also called "inverse detected") is now routine; the name commonly associated with this experiment is HMQC

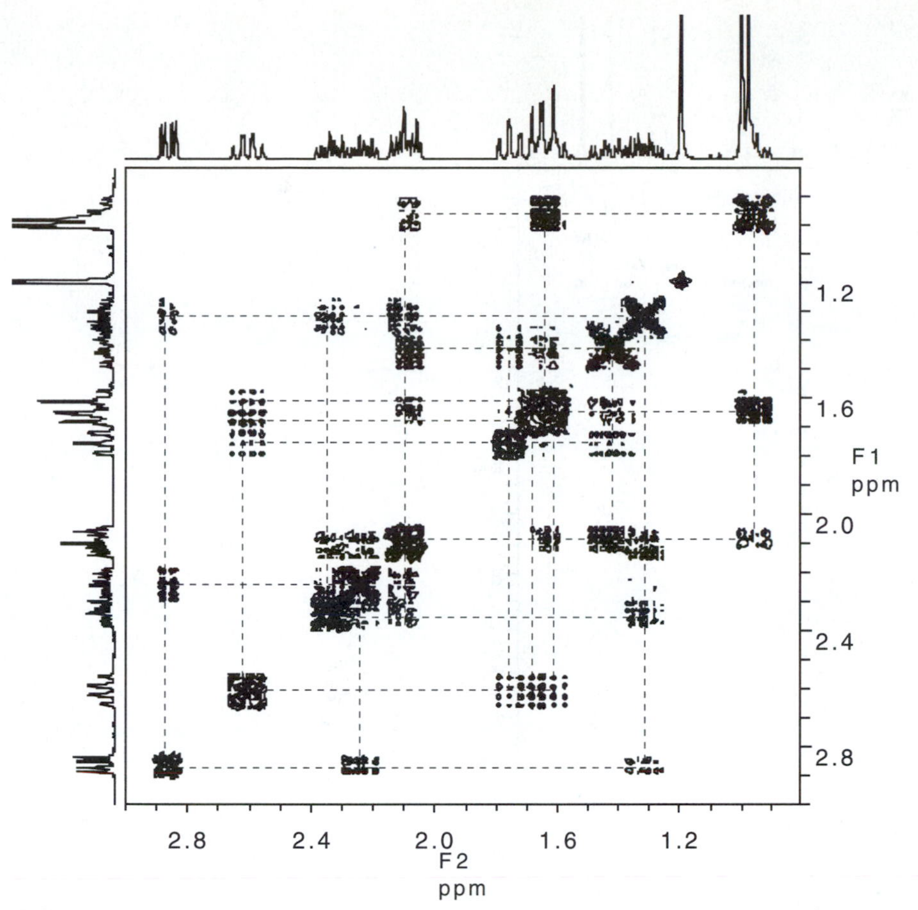

FIGURE 6.17. Expanded view of the DQF-COSY of caryophyllene oxide (see text for discussion of the correlations shown).

(heteronuclear multiple quantum coherence). The F2 axis is assigned the nucleus detected (^1H).

The HMQC spectrum for caryophyllene oxide is given in Figure 6.18. The results here are identical with those of Figure 6.16; the appearance of the spectrum is, however, slightly different. Note that in Figure 6.16 the proton axis is F1 and the carbon axis is F2, whereas in Figure 6.18 the carbon axis is F1 and the proton axis is F2. More importantly, because the carbon axis in the HMQC is derived from t_1, we usually have less digital resolution for the carbon axis in the HMQC than we have in the traditional HETCOR. Experimentally, this is a very important consideration especially when we have very closely spaced ^{13}C resonances. For instance, there are three very closely spaced carbon resonances around 30 ppm, and without good digital resolution, we would be unable to differentiate the five different proton resonances associated with these three carbon atoms.

Are there other theoretical or experimental considerations to distinguish these two heteronuclear correlation experiments? More bluntly, should we prefer one or the other experiments? First, there are different hardware requirements for the two experiments; some instruments are incapable of running the inverse detected experiment. Most new instruments, however, are able to run both. If both experiments are available to us, the main point to consider is sensitivity. For obvious reasons, the proton-detected experiment (HMQC) is at least eightfold more sensitive which translates into a time savings of 8^2. In fact, the historical interest in developing HMQC was sensitivity.

6.8 *Proton-Detected, Long-Range $^1H—^{13}C$ Heteronuclear Correlation: HMBC*

For the HMQC described above, we wanted an experiment that eliminated long-range (i.e., two and three bond) proton–carbon couplings while preserving the directly attached (i.e., one-bond) couplings, which we correlated in a 2-D experiment. The HMBC (Heteronuclear Multiple Bond Coherence) experiment, on the other hand, which is also proton detected,* capitalizes

*There is a carbon detected analogue of the HMBC experiment called COLOC (COrrelated spectroscopy for LOng-range Couplings) that predated the experiment treated here. The COLOC is not used much any more and we will not give any examples.

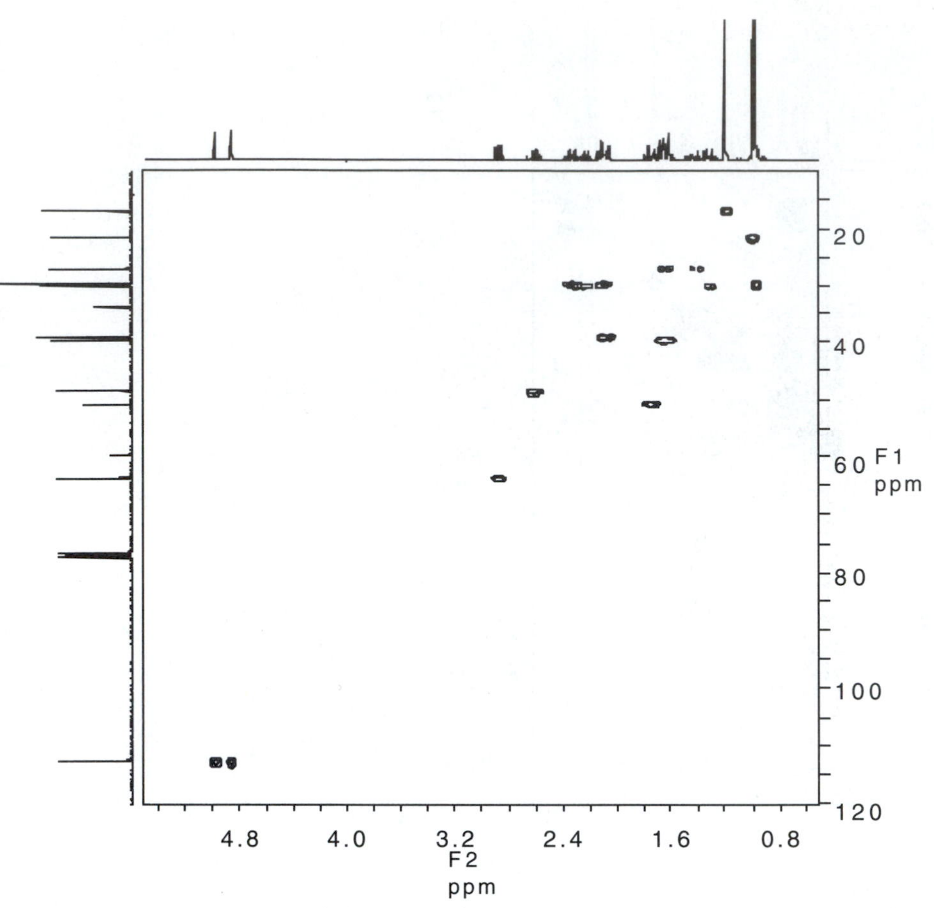

FIGURE 6.18. The 300-MHz (F2) HMQC (also called "inverse detected HETCOR") spectrum of caryophyllene oxide.

on these two- and three-bond couplings providing us with an extremely powerful (although sometimes cluttered) spectrum. In essence, we indirectly obtain carbon–carbon (although not ^{13}C—^{13}C) correlations, and, in addition, we are able to "see" or correlate quaternary carbons with nearby protons. Since both ^2J and ^3J couplings are present, interpretation can be tedious; we must be methodical in our approach and keep in mind the HMQC (or HETCOR) correlations.

Before we tackle the HMBC for caryophyllene oxide, let us practice on ipsenol. The HMBC for ipsenol (Figure 6.19) looks like the HETCOR (with its axes switched) for ipsenol with two obvious differences: There are considerably more correlations and the one-bond correlations (HMQC) are gone. Interpretation of HMBCs requires a degree of flexibility because we do not always find what we expect to find. In particular, whereas two-bond correlations ($^2J_{CH}$) are almost always found, the three-bond ($^3J_{CH}$) correlations are occasionally absent. The variations in correlations that we find result from the variations in the magnitude of $^2J_{CH}$ and $^3J_{CH}$ coupling constants.

Interpretation for ipsenol is straightforward. But first, let us note a common artifact: ^{13}C satellites of intense proton peaks especially methyl groups. If we trace parallel to the proton axis (F2) at about 23 ppm on the carbon axis (F1), we find cross peaks at about 1.0 ppm (proton), which are real. On either side, we find two "cross peaks" that do not line up (correlate) with any protons in F2. These are ^{13}C satellites and should be ignored.

We can begin with either a carbon or a proton resonance and obtain equivalent results. We will use the carbon axis as our starting point because we usually have less overlap there. For example, a line drawn parallel to the proton axis at about 68 ppm on the carbon axis (the carbinol carbon) intersects five cross peaks; none of the five correlations corresponds to the attached proton ($^1J_{CH}$) at 3.8 ppm. Four of the cross peaks correspond to the two pairs of diastereotopic methylene groups (2.48, 2.22, 1.45, and 1.28 ppm) and these represent $^2J_{CH}$, or two-bond couplings. The fifth interaction ($^3J_{CH}$) correlates this carbon atom (68 ppm) to the isopropyl methine proton (1.82 ppm), which is bonded to a carbon atom in the β-position. The other carbon atom in a β-position has no attached protons so we do not have a correlation to it from the carbinol carbon atom. Thus, we have indirect carbon connectivities to two α carbons and to one of two β carbons.

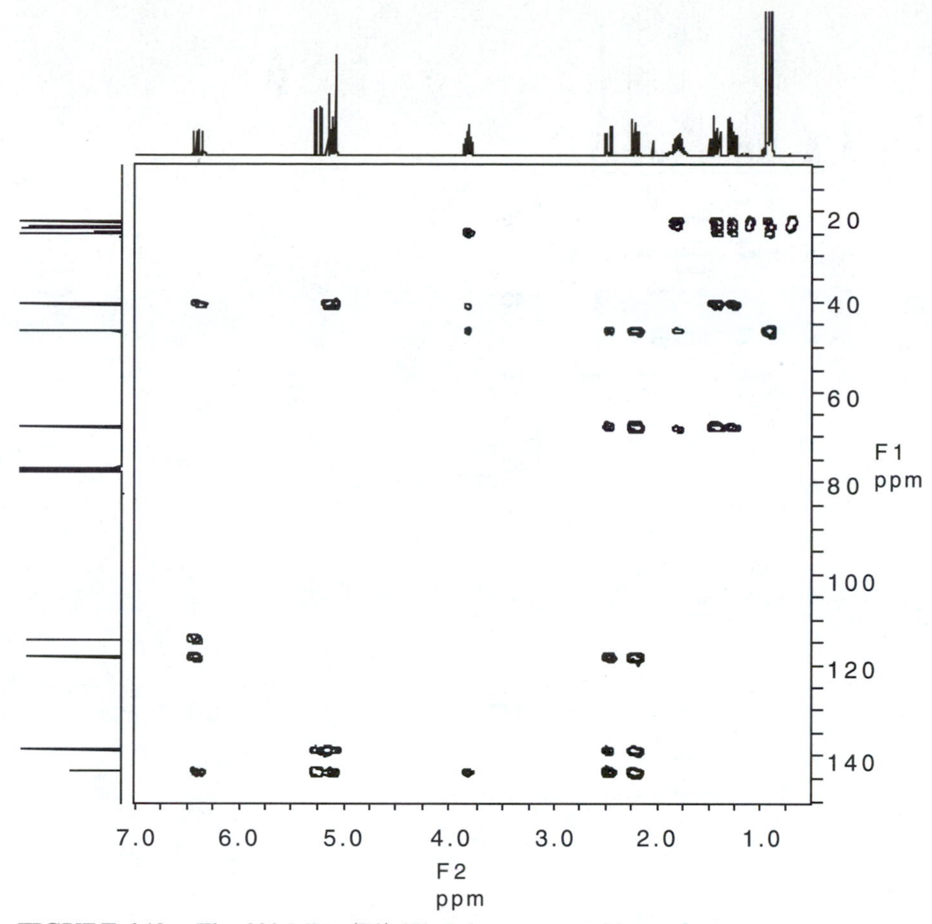

FIGURE 6.19. The 300-MHz (F2) HMBC spectrum of ipsenol.

Another useful example can be found by drawing a line from the carbon resonance at 41 ppm. This carbon is the C-5 methylene and we first note that correlations to the attached protons at 2.48 and 2.22 ppm are absent. There is only one α carbon that has one or more attached protons; its corresponding correlation is found to the C-4 carbinol methine proton at 3.83 ppm.* There are three β carbons and they all have attached protons. The C-3 methylene carbon shows indirect correlation through both of its diastereotopic protons at 1.45 and 1.28 ppm. The C-7 olefinic methine proton gives a cross peak at 6.39 ppm, as do the protons of the olefinic methylene group attached to C-6 at 5.16 and 5.09 ppm. Other assignments are left to the reader as an exercise.

The HMBC for caryophyllene oxide (Figure 6.20) allows us to completely confirm the structure of caryophyllene oxide by giving us the required indirect carbon–carbon connectivities. An analysis of the structure of caryophyllene oxide reveals that there should be 87 cross peaks; this number is derived from considering each of the 15 ^{13}C atoms and counting the number of chemical shift–distinct protons at the α-positions and the number of chemical shift–distinct protons at the β-positions. In order to keep track of all of those interactions, one must be methodical indeed.

One way to keep track of these data is to construct a table listing the carbon resonances in one direction and the proton resonances in the other. In Table 6.1, the carbons are given across the top and protons along the side. The numbering for caryophyllene oxide is shown above.

Our approach for this spectrum is no different from any other spectrum. In this case, it is easier to start on the carbon axis and look for the required cross peaks to the protons as listed in Table 6.1. If we wished to start on the proton axis, we would, of course, obtain the same results, but, because there is severe overlap in the proton spectrum, the interpretation is more difficult.

If we begin at the top left of the table with H-1, we see first that H-1 is bonded to C-1, a result that we already have determined in the HETCOR. From there, we find a total of eight interactions is expected. In the table, each interaction is labeled either α or β depending

* The other α carbon at C-6, which was a β carbon in our first example, also shows no correlation in the HETCOR (HMQC). The reader should show that there are useful correlations to this carbon atom in the current (Fig. 6.19) figure.

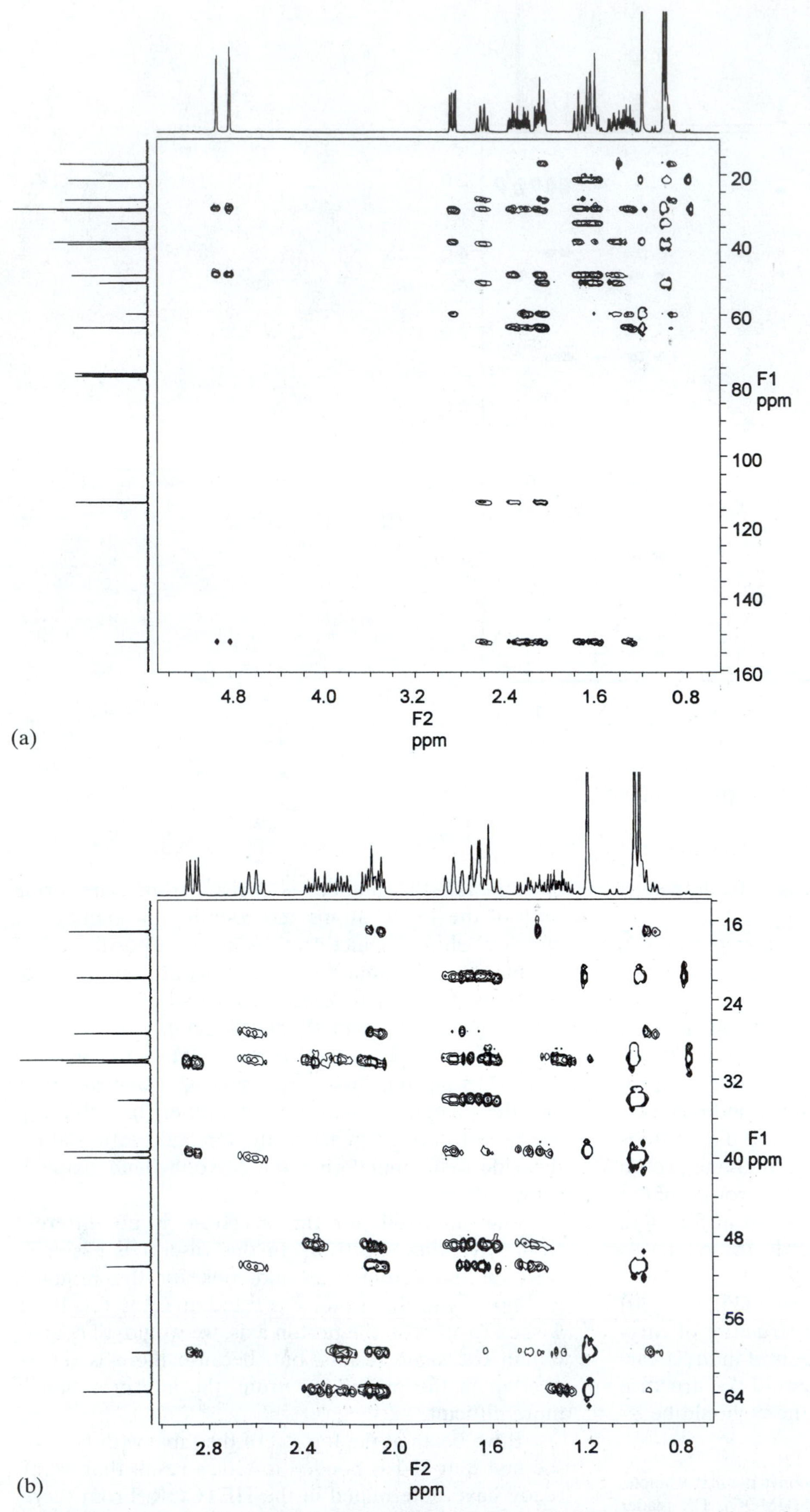

FIGURE 6.20. *(a)* The 300-MHz (F2) HMBC spectrum of caryophyllene oxide, and *(b)* an expanded view showing greater digital resolution.

Table 6.1 HMBC Correlations for Caryophyllene Oxide

	Carbon	C-1	C-2	C-3	C-4	C-12	C-5	C-6	C-7	C-8	C-13	C-9	C-10	C-11	C-14	C-15
Proton	(ppm)	50.9	27.2	39.2	59.7	16.9	63.6	30.1	29.9	153	113	48.7	39.8	34	22.6	30
H-1	1.76	Bonded	α	β						β		α	β	α	β	β
H-2	1.45	α	Bonded	α	β	β						β		β		
H-2′	1.63	α	Bonded	α	β	β						β		β		
H-3	0.95	β	α	Bonded	α	β	β									
H-3′	2.06	β	α	Bonded	α	β	β									
H-12	1.19			β	α	Bonded	β									
H-5	2.86			β	α	β	Bonded	α	β							
H-6	1.28				β		α	Bonded	α	β						
H-6′	2.23				β		α	Bonded	α	β						
H-7	2.11						β	α	Bonded	α	β	β				
H-7′	2.37						β	α	Bonded	α	β	β				
H-13	4.86								β	α	Bonded	β				
H-13′	4.97								β	α	Bonded	β				
H-9	2.60	α	β						β	α	β	Bonded	α	β		
H-10	1.43	β								β		α	Bonded	α	β	β
H-10′	1.47	β								β		α	Bonded	α	β	β
H-14	0.98	β											β	α	Bonded	β
H-15	1.01	β											β	α	β	Bonded

upon whether it results from a two-bond coupling ($^2J_{CH}$) or a three-bond coupling ($^3J_{CH}$). Of course, in the spectrum itself, there is no differentiation of the two types of interactions; we label them that way for our own bookkeeping efforts. Each of the interactions for H-1 designated in the table is found in the spectrum.

There are two protons on C-2, which are labeled H-2 and H-2′; these protons have different chemical shifts, yet we expect them to act much the same way in the HMBC. Thus, we have a useful independent check of our HMBC assignments for each pair of diastereotopic protons in caryophyllene oxide. For H-2 at 1.45 ppm, we have the same five correlations that we have for H-2′ at 1.63 ppm. As we study the spectrum and the table more closely, we find that we have exquisitely detailed structural information that can be deciphered with a methodical approach.

Before we leave HMBC, an important point about quaternary carbons requires comment. Until now, we have had no direct correlations for carbons without protons, nor have we been able "to see through" heteroatoms such as oxygen, sulfur, etc. Both the two- and three-bond coupling correlations of HMBC provide us with both types of critical information. For example, C-4 of caryophyllene oxide has no attached protons and, so far, it has only appeared in the ^{13}C spectrum of the compound, and we know that it is quaternary from DEPT. If we look in Table 6.1 at the C-4 column, we find four two-bond correlations and four three-bond correlations. The HMBC spectrum bears out these expectations and gives us direct evidence of the C-4 position in the molecule.

6.9 ^{13}C—^{13}C Correlations: INADEQUATE

The HMBC experiment allows us to trace the skeleton of organic compounds by way of indirect carbon–carbon connectivities, but the process is tedious because we do not know whether the correlations result from two- or three-bond couplings. The 2-D experiment INADEQUATE (Incredible Natural Abundance DoublE QUAntum Transfer Experiment) completes our set of "basic" through-bond correlations; we have COSY for proton–proton coupling, HETCOR (HMQC) for one-bond and HMBC for two- and three-bond proton–carbon coupling, and now INADEQUATE for directly attached (one-bond) carbon–carbon couplings. For elucidation of structure of organic compounds, this experiment is, without question and without exception, the most powerful and the least ambiguous available, and, to top it off, the experiment is easy to interpret. After reading that last statement, the naturally pessimistic among us inevitably will ask: What's the "catch"? Indeed, the "catch" is plain and simple: sensitivity. Recall from Chapter 5 that the probability of any one carbon atom being a ^{13}C atom is about 0.01. Thus, the probability that any two adjacent carbon atoms will both be ^{13}C atoms (independent events) is 0.01×0.01 or 0.0001; in rounded whole numbers, that is about 1 in 10,000 molecules!

This seemingly impossible task is accomplished with the aid of double-quantum filtering. We recall from our DQF-COSY experiment that double-quantum filtering removes all single-spin transitions, which in this case

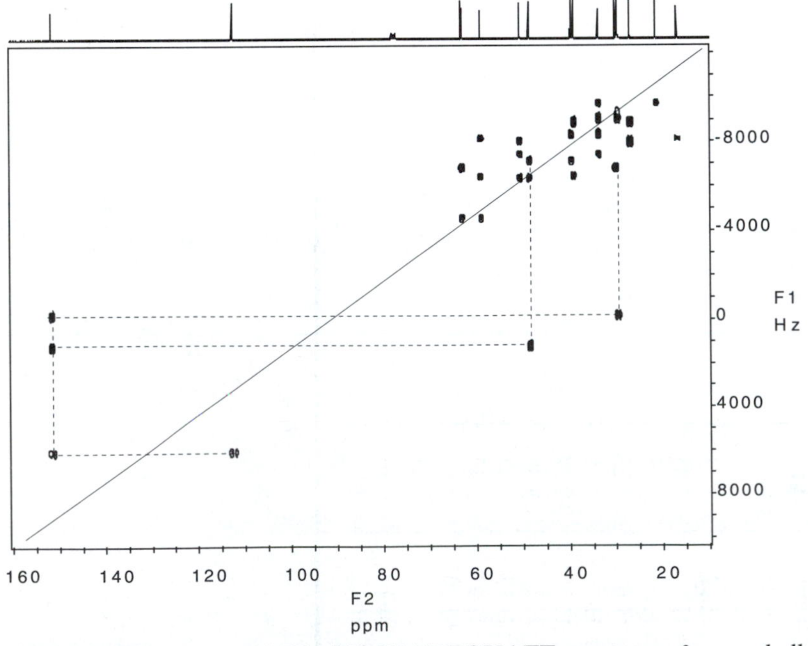

FIGURE 6.21. The 75-MHz (F2) INADEQUATE spectrum of caryophyllene oxide.

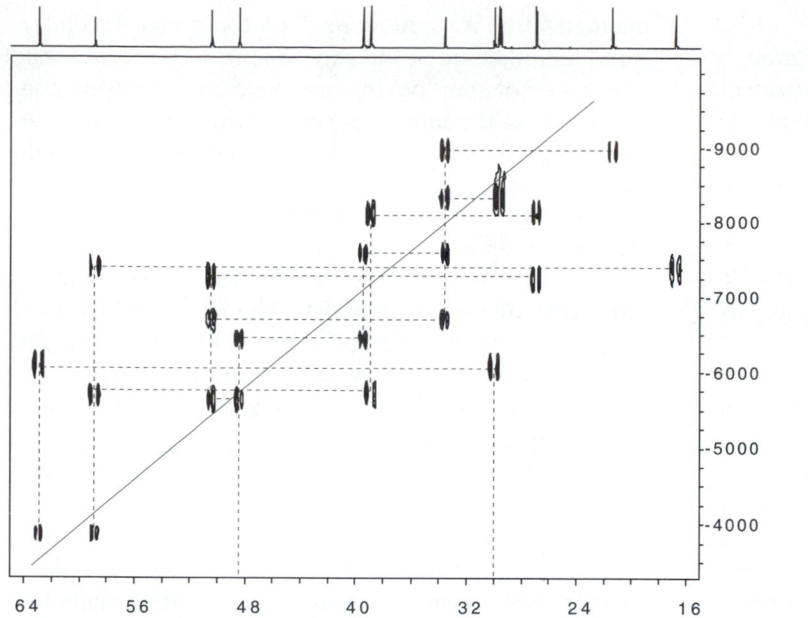

FIGURE 6.22. Expanded view of the 75-MHz (F2) INADEQUATE spectrum of caryophyllene oxide. See Figure 6.21 for the entire spectrum.

corresponds to isolated ^{13}C atoms; only those transitions from systems with two spins (AB and AX systems) and higher* are detected during acquisition (Figure 6.21). The main problem facing us experimentally is sample size, assuming that the compound has the required solubility in an appropriate lock solvent. For low molecular weight compounds (atomic weight < 500) run on a modern high-field spectrometer, 200–300 mg dissolved in 0.5 mL of a deuterated solvent is appropriate.

One way to imagine this experiment is as a carbon analogue of DQF-COSY in which both F1 and F2 would be carbon axes, and theoretically this experiment is possible. For practical considerations related to obtaining complete double-quantum filtration, the INADEQUATE experiment is run slightly differently. In the display of the INADEQUATE spectrum of caryophyllene oxide, we find that the F2 axis is the familiar carbon axis, which we can, of course, relate to t_2 acquisition. The F1 axis looks unfamiliar and requires further explanation.

During t_1, the frequencies that evolve are not the chemical shifts of the coupled nuclei as they are in a typical COSY. Instead, it is the *sum* of the chemical shifts of the coupled nuclei which evolve during t_1, and, because it is double quantum filtered, it is only the two-spin AB and AX systems that contribute significantly to the intensity in the INADEQUATE spectrum. Proper selection of the delays (τ) in the pulse sequence allows us to select the larger one-bond couplings ($^1J_{CC}$) thus ensuring that we are only looking at directly bonded carbon–carbon correlations. The F1 axis is usually given in Hz and it is two times the range in F2.

The 2-D INADEQUATE spectrum of caryophyllene oxide is presented in Figure 6.21. Cross peaks or correlations are found at ($\nu_A + \nu_X$, ν_A) and at ($\nu_A + \nu_X$, ν_X) in the (F1, F2) coordinate system for a given AX system. The actual cross peaks themselves are doublets (see expanded spectrum, Figure 6.22) with a spacing equal to the ($^1J_{CC}$) coupling constant. The midpoint of the line connecting the two sets of doublets is ($\nu_A + \nu_X$), ($\nu_A + \nu_X$)/2; thus, the collection of midpoints for all of the pairs of doublets lies on a line running along the diagonal. This is an important observation because it can be used to distinguish genuine cross peaks from spurious peaks and other artifacts.

With a better understanding of the F1 axis and the "diagonal," we can proceed with interpretation of the spectrum. Table 6.1 lists carbon chemical shifts and carbon numbers based on the structure given earlier; we refer to these numbers in the present discussion. From Figure 6.21, we can make the high-frequency connections quite easily. The carbon at highest frequency is C-8 at 153.0 ppm; by tracing vertically down from this peak on the F2 axis, we intersect three cross peak doublets. These cross peaks "connect" horizontally with C-7 at 29.9 ppm, C-9 at 48.7 ppm, and C-13 at 113.0 ppm. Toward lower frequencies, C-13 at 113.0 ppm comes next, and it has only one cross peak, namely the reciprocal connection to C-8 at 153.0 ppm.

In order to present the low-frequency section more clearly, Figure 6.22 shows an expanded view of that area. The higher resolution of this figure enables us to see the doublet fine structure more readily. Let us trace

* Following the same reasoning as above, the probability of a three-spin system in an unenriched sample is 1 in 1,000,000.

one carbon's connectivities from Figure 6.22. C-11, at 34.0 ppm, is a quaternary carbon, and it accordingly shows four cross peaks. We have connectivities from C-11 to C-14, at 22.6 ppm, C-15 at 30.0 ppm, C-10 at 39.8 ppm, and C-1 at 50.9 ppm.

Before we conclude our discussion, we note that the INADEQUATE spectrum of caryophyllene oxide contains an uncommon phenomenon worth exploring. Carbons 6 and 7 of caryophyllene oxide overlap in the ^{13}C spectrum with each other and with one of the methyls; we list their chemical shifts from Table 6.1: 30.1 and 29.9 ppm. Because they are bonded to one another in caryophyllene oxide, they should show correlation in the INADEQUATE spectrum, but, instead of an AX system, we have an AB system with $\Delta\nu/J$ being much less than 10. For this special case, we no longer expect two doublets whose midpoint lies on the diagonal; instead, we predict that an AB multiplet (see Chapter 4) should fall on the diagonal line itself. This prediction is borne out in Figure 6.22, where we find a cross peak directly below C-6 and C-7, and this cross peak intersects the diagonal line.

The other connectivities found in Figure 6.22 are left to the reader as an exercise. We summarize this section with two points:

- 2-D INADEQUATE provides direct carbon connectivities enabling us to sketch the carbon skeleton unambiguously.
- 2-D INADEQUATE has very limited applicability because of its extremely low sensitivity.

6.10 *Relayed Coherence Transfer: TOCSY*

The common theme so far in our correlation experiments has been to allow spins to evolve during t_1 under the influence of directly coupled nuclear spins. We have seen the power of COSY, HETCOR (HMQC), HMBC, and INADEQUATE to provide us with detailed structural information for ipsenol and caryophyllene oxide. In this section, we will develop another method for showing correlations and apply it to molecules with distinct spin systems, such as carbohydrates, peptides, and nucleic acids.

Our goal is to "relay" or to transfer magnetization beyond directly coupled spins, thus enabling us to see correlations among nuclei that are not directly coupled but within the same spin system. The experiment is called TOCSY (TOtally Correlated SpectroscopY) and we will consider both the 2-D and 1-D versions. The pulse sequence for a 2-D TOCSY resembles our prototype 2-D experiment but, instead of a second $\pi/2$ pulse, we insert a "mixing period" during which the magnetization is "spin locked" on the *y* axis. To understand the outcome of the experiment, we can ignore the particulars of spin locking and concentrate on the consequences of the mixing period. During this mixing period, magnetization is relayed from one spin to its neighbor, and then to its next neighbor, and so on. The longer the mixing period, the further the relay, in theory, throughout an entire spin system.

The appearance of a 2-D TOCSY experiment resembles in all aspects a COSY. The F1 and F2 axes are for proton; the diagonal contains 1-D information; and even the cross peaks have the same appearance. The difference here is that the cross peaks in a COSY result from coupled spins, whereas the cross peaks in the TOCSY spectrum arise from relayed coherence transfer. For long mixing times in a TOCSY spectrum, all spins within a spin system appear to be coupled. To appreciate the advantages of TOCSY, we select the disaccharide lactose which has two distinct (i.e., separate) spin systems.

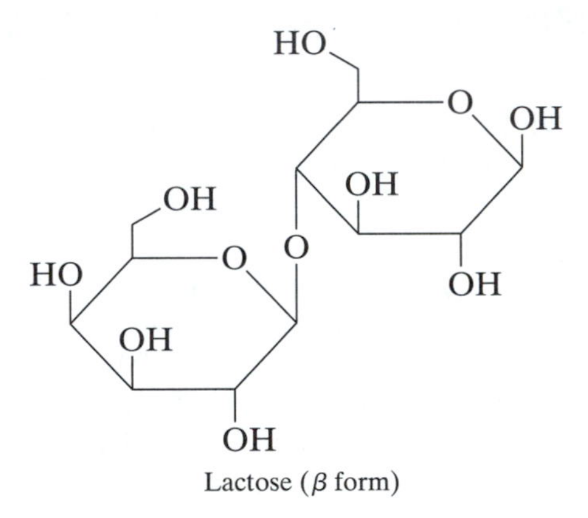

Lactose (β form)

Recall that lactose is a disaccharide in which galactose has formed a β-glycosidic bond to the 4-hydroxyl of glucose. The glucose residue is a reducing moiety, and, in aqueous solution, it exists as a mixture of α- and β-anomers; the β-anomer is drawn above. The proton spectrum of lactose is given in Figure 6.23. The spectrum shows the proton resonances of three sugar residues: one β-D-galactoside ring, one β-D-glucose ring, and one α-D-glucose ring (the two glucose rings both obviously have the galactoside substituent at the 4 position). Resonances for the three anomeric protons are evident and can be assigned; the galactose H-1 is found at 4.43 ppm whereas the β- and α-anomeric protons for glucose are found at 4.64 and 5.21 ppm, respectively. The remaining nonhydroxylic protons overlap badly between 3.5 and 4.0 ppm except for the H-2 proton of β-D-glucose (see COSY spectrum, Fig. 6.24).

The DQF-COSY spectrum of lactose helps a little (Figure 6.24). For instance, we can find correlations from the various anomeric protons to their respective H-2 resonances. If we attempt to trace these H-2 protons to the H-3 protons and further, however, we find

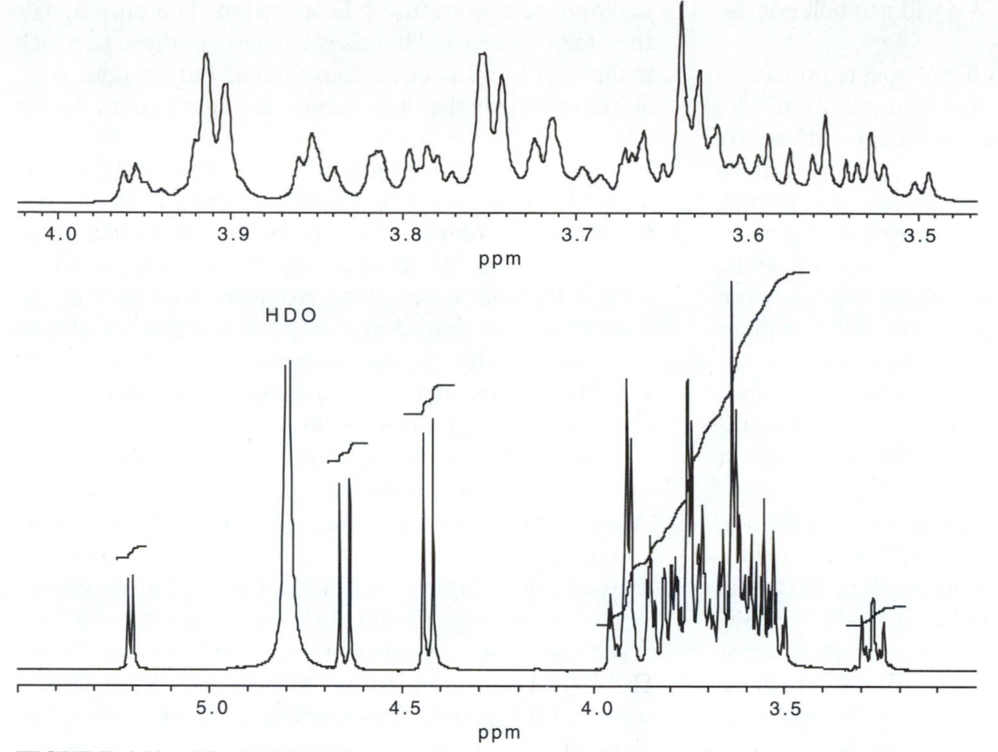

FIGURE 6.23. The 300 MHz ^{1}H NMR spectrum of β-lactose in D_2O. The inset shows severe overlap of the ring protons from both the glucose and the galactose rings.

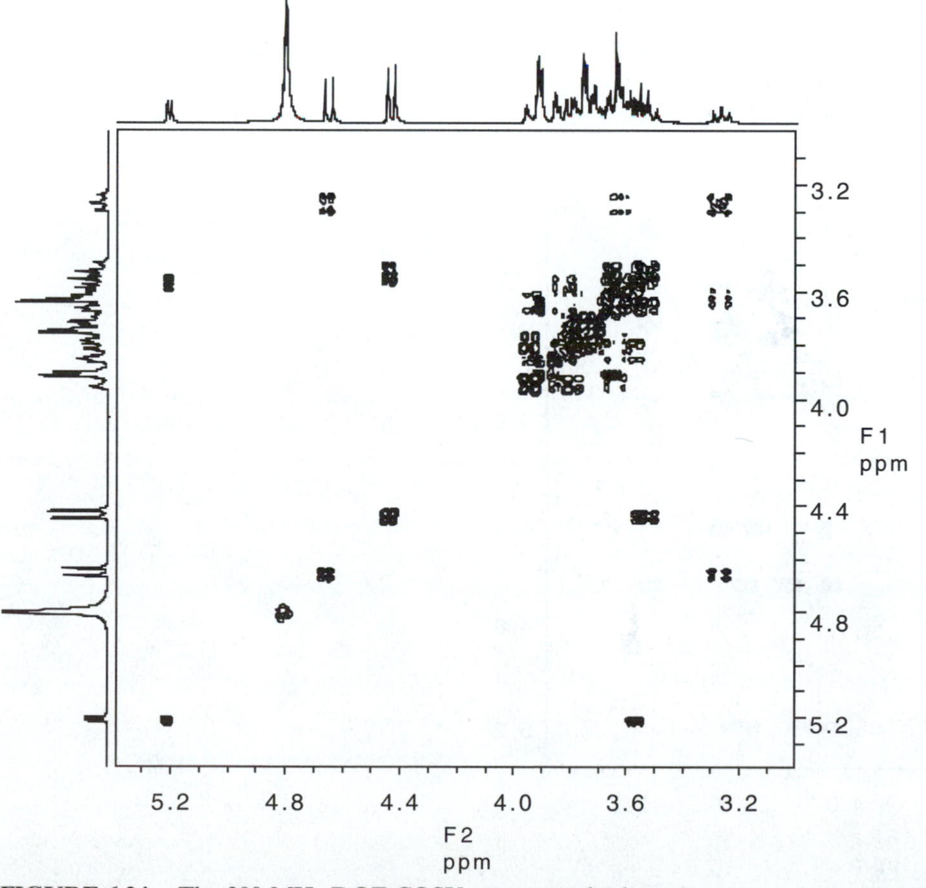

FIGURE 6.24. The 300-MHz DQF-COSY spectrum of β-lactose.

that overlap becomes hopeless. We will not belabor the point.

Turning now to the 2-D TOCSY spectrum of lactose (Figure 6.25), we find an improved situation. The mixing time for this 2-D spectrum has been sufficiently long that magnetic coherence has been transferred more or less throughout each sugar residue's spin system. Resolution is poor and we still have overlap, but most of the individual proton resonances are now discernible. We could try to make assignments from this spectrum (and the ambitious reader is invited to do so), but there is another way to think about this experiment.

Every 2-D experiment has a 1-D analogue, and we tend to think that these 1-D experiments are less efficient, which they usually are. If we think again about our COSY experiment, we have said that homonuclear decoupling would give us the same type of information. We select a proton resonance, irradiate it, and compare the result with the original 1-D proton spectrum (again, see Chapter 4). In similar fashion for our 1-D TOCSY, often called HOHAHA (homonuclear Hartmann-Hann), we select a proton resonance and irradiate it; we allow for an appropriate mixing time for the magnetization to be relayed, during which we apply spin locking; and we acquire the 1-D spectrum. The only signals that will be recorded in this spectrum are those to which magnetization has been transferred. Put another way, all other signals that are outside the spin system do not appear.

An even better scenario is to run a series of 1-D TOCSY experiments in which the mixing time is systematically increased while the proton being irradiated is kept constant. To illustrate these experiments, we irradiate the anomeric proton from the β-anomer of the glucose ring in lactose at 4.43 ppm and run a series of experiments with mixing times ranging from 19 to 232 ms. The results of these experiments are shown in a series of stacked plots in Figure 6.26.

At a mixing time of 19 ms, we find only the H-2 resonance, which is seen clearly as a doublet of doublets. After 33 ms of mixing time, transfer to H-3 is readily apparent (another doublet of doublets, which is almost merged into a triplet) and the H-4 is just barely visible. A plot of the experiment with a mixing time of 62 ms reveals the H-4 resonance strongly and the signal from H-5 is just sprouting from the baseline. After 90 ms, transfer throughout the entire spin system is evident; the H-5 signal is robust and we begin to see the C-6 meth-

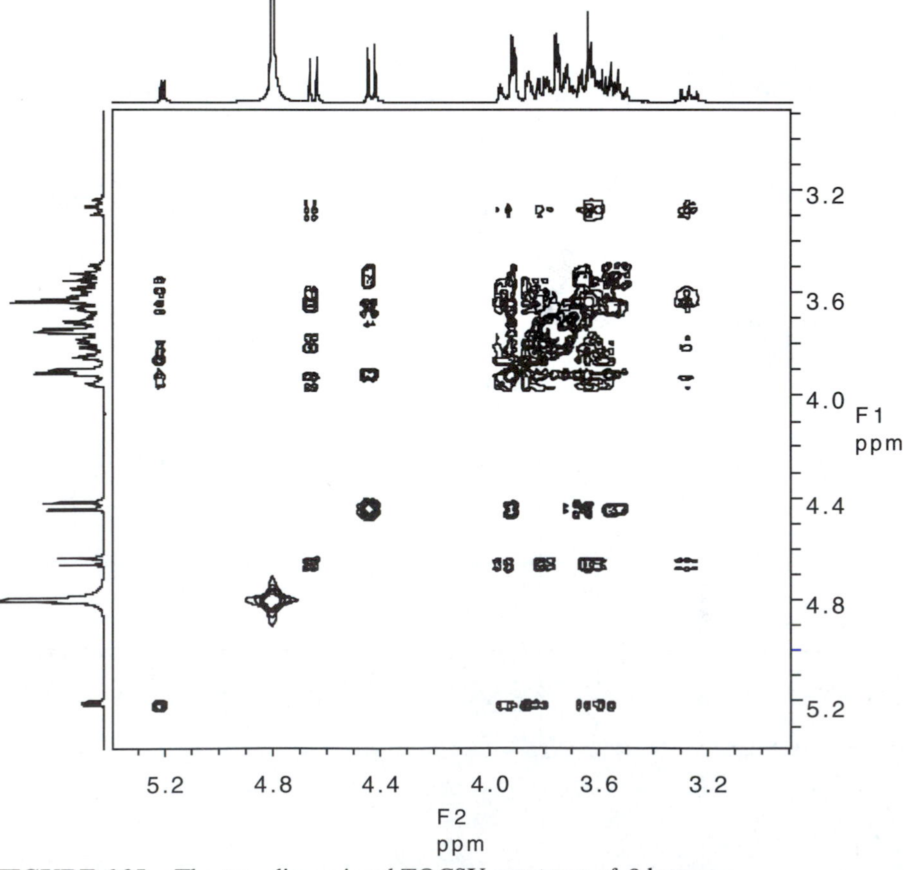

FIGURE 6.25. The two-dimensional TOCSY spectrum of β-lactose.

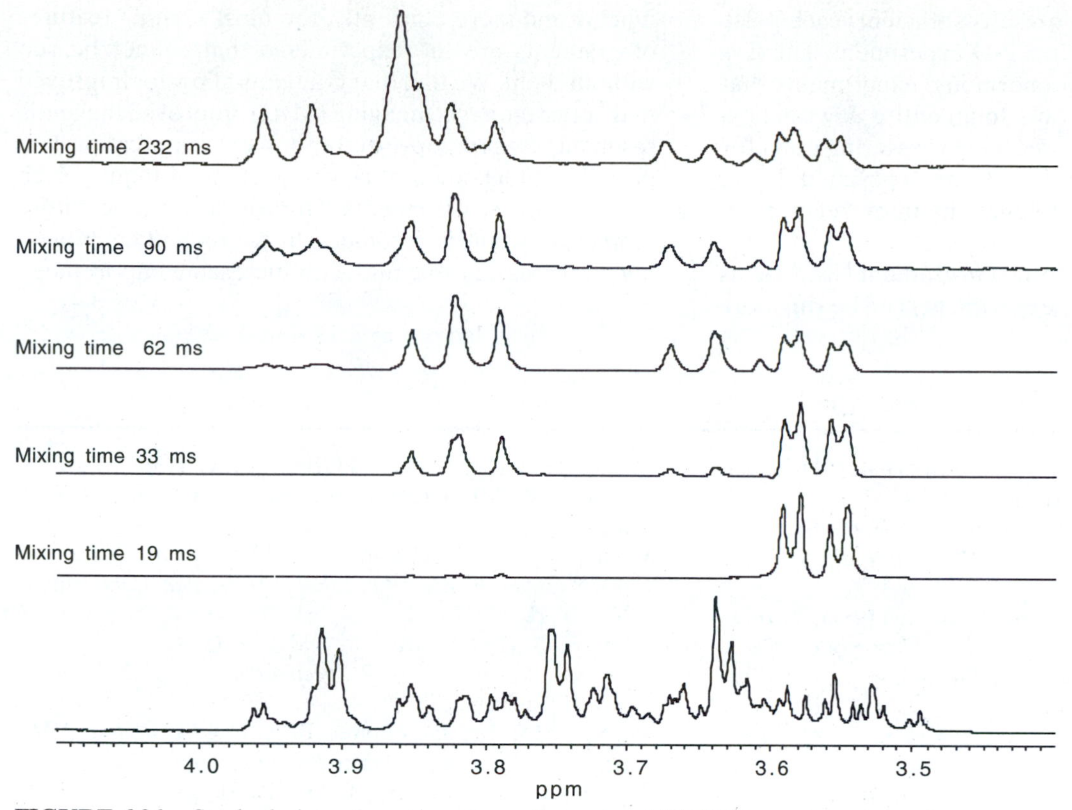

FIGURE 6.26. Stacked plots of a series of one-dimensional TOCSY experiments on β-lactose with increasing "mixing times." Irradiation of the anomeric proton from the β-anomer of the glucose ring at 4.43 ppm. A portion of the ^{1}H NMR spectrum is reproduced for reference in the bottom plot.

ylene group. Finally, the 232-ms mixing time experiment exhibits complete transfer of coherence from the anomeric proton all the way to the two C-6 protons. One negative aspect of long mixing times is that both resolution and signal are lost. Notice for instance how the H-2 doublet of doublets broadens from 19 to 232 ms. Not as readily evident is the loss of signal with a concomitant increase in noise. We can usually offset signal loss by acquiring and summing more FIDs. To obtain the same information for the other two sugar residues, we would simply repeat the same series of experiments using the other anomeric protons as our irradiation resonances.

Both the 1-D and 2-D versions of TOCSY find wide application in deciphering overlapping signals which originate from different spin systems. The 1-D version is particularly exciting as it enables us to "walk" through a spin system as we systematically increase the mixing time.

6.11 *Gradient Field NMR*

One of the fastest growing areas in NMR over the past decade has been the use of "pulsed field gradients," or PFG NMR. It is ironic to consider that so much effort has been expended over so many years to avoid "gradients" in the field or to maintain the field as homogeneous as possible. Today, most modern high-field NMR spectrometers are routinely equipped with hardware (coils) that rapidly ramps the field along one of three mutually orthogonal axes. These magnetic field gradients are incorporated into the pulse sequence in a large range of applications. We include a brief discussion of gradients here because there are many applications to correlation experiments. We treat gradients in a general manner; for a more technical treatment or for a myriad of other applications consult the review by Price (1996).

In Section 6.4 we mention phase cycling as an important part of any pulse sequence. The details of phase cycling are still beyond our treatment (see Derome, 1987) but we can easily appreciate one of their negative aspects: time. In correlation experiments, anywhere from 4 to 64 phase cycles must be summed in order to produce one FID. If signal to noise is poor, then the identical cycle is repeated until sufficient signal is acquired. These phase cycles are wasteful of spectrometer time and account for at least one reason that 2-D experiments take so long.

In a PFG experiment, the pulse sequence can be rewritten so that phase cycling can be eliminated altogether. Thus, if signal to noise is sufficient, each pulse can be saved as one FID in a 2-D experiment. The saving in instrument time is enormous; experiments that previously took several hours to an entire day can now be run in a matter of minutes. One reassuring point for us to realize is that, even though the experiment is run differently, the results and hence the interpretation remain the same.

Although our initial interest in gradient NMR stems from the fact that familiar experiments can be run more quickly and more efficiently, the most exciting features of gradients are the experiments that cannot be run without them. We find that gradients allow for improved magnetic resonance imaging (MRI), improved magnetic resonance microscopy, better solvent suppression (especially water), and entirely new areas of inquiry such as diffusion measurements. Our vision into the future of nuclear magnetic resonance finds gradient fields playing an increasing role and with increasing importance.

References

Brey, W., (Ed.), (1988). *Pulse Methods in 1D and 2D Liquid-Phase NMR.* New York: Academic Press.

Chandrakumar, N., and Subramanian S. (1987). *Modern Techniques in High Resolution FT NMR.* New York: Springer-Verlag.

Croasmum, W. R., and Carlson, R. M. K. (1987). *Two-Dimensional NMR Spectroscopy,* 2nd ed. New York: VCH.

Derome, A. (1987). *Modern NMR Techniques for Chemistry Research.* Oxford: Pergamon Press.

Ernst, R. R., Bodenhausen, G., and Wokaum A. (1987). *Principles of Nuclear Magnetic Resonance in One and Two Dimensions.* Oxford: Clarendon Press.

Farrar, T. C. (1987). *An Introduction to Pulse NMR Spectroscopy.* Chicago: Farragut.

Friebolin, H. (1993). *Basic One- and Two-Dimensional Spectroscopy,* 2nd ed. New York: VCH.

Gunther, H. (1995). *NMR Spectroscopy,* 2nd ed., New York: Wiley.

Kessler, H., Gehrke, M., and Griesinger, C. (1988). Two-Dimensional NMR Spectroscopy. *Angew. Chem. Int. Ed. Engl.* **27,** 490–536.

Martin, G. E., and Zektzer, A. S. (1988). *Two-Dimensional NMR Methods for Establishing Molecular Connectivity.* New York: VCH.

Price, W. S. (1996). Gradient NMR, in G. A. Webb (Ed.), *Annual Reports on NMR Spectroscopy,* Vol. 32, London: Academic Press.

Sanders, J. K. M., and Hunter, B. K. (1993). *Modern NMR Spectroscopy,* 2nd ed., Oxford: Oxford University Press.

Schraml, J., and Bellama, J. M. (1988). *Two-Dimensional NMR Spectrometry.* New York: Wiley.

Williams, K. R., and King, R. W. (1990). The Fourier transform in chemistry—NMR, *J. Chem. Ed.,* **67,** 125–138.

Problems

6.1 For the compounds of Problem 4.3 A–I, draw the following spectra: COSY, HETCOR or HMQC, and INADEQUATE. Be sure to label the F1 and F2 axes. Assume experimental conditions are the same as in Problem 4.3.

6.2 Assign all of the correlations for ipsenol in the DQF-COSY found in Figure 6.11. Indicate each type of coupling as geminal, vicinal, or long range.

6.3 Identify the compound $C_7H_{14}O$ from its COSY spectrum and show all of the correlations. The diagonal runs from upper left to lower right. A 1% solution in $CDCl_3$ at 300 MHz. With permission from Varian Associates.

6.4 Complete the assignments for ipsenol for the HMBC found in Figure 6.19. To aid in bookkeeping, you may want to construct a table similar to Table 6.1.

6.5 Identify the compound $C_5H_{10}O$ from its COSY and HETCOR spectra and show all correlations.

6.6 Identify an isomer of Problem 6.5 from its COSY and HETCOR spectra and again show all correlations.

6.7 Assign all of the carbon connectivities for caryophyllene oxide using the INADEQUATE spectrum found in Figures 6.21 and 6.22.

6.8 Identify compound $C_9H_{12}O_2$ from its ^1H, ^{13}C, and INADEQUATE spectra.

6.9 Make as many correlations as possible for lactose using the 2-D TOCSY found in Figure 6.25. Compare your results for the glucose residue to the results that were found in the 1-D HOHAHA in Figure 6.26.

PROBLEM 6.3. COSY for $C_7H_{14}O$.

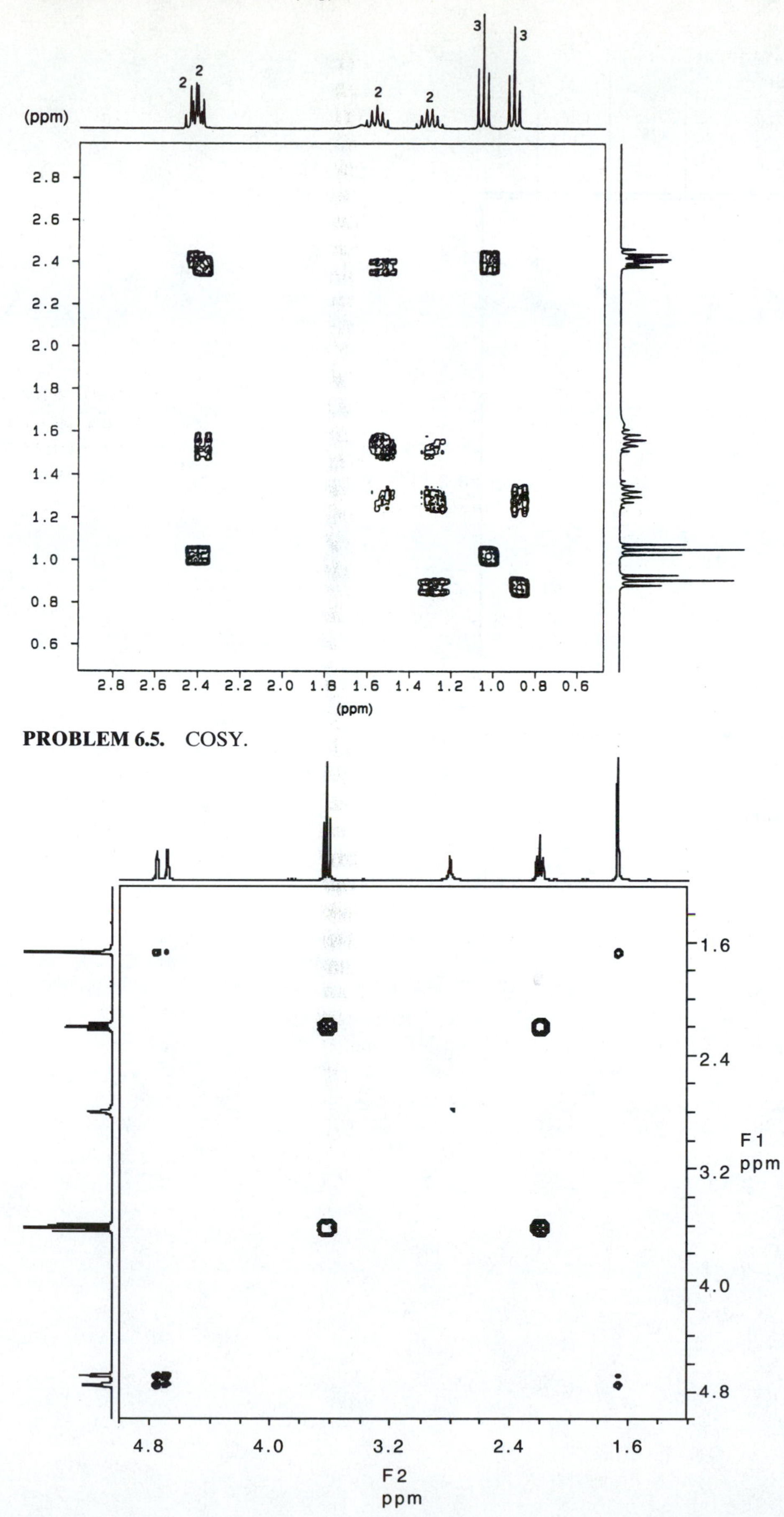

PROBLEM 6.5. COSY.

PROBLEM 6.5. HETCOR.

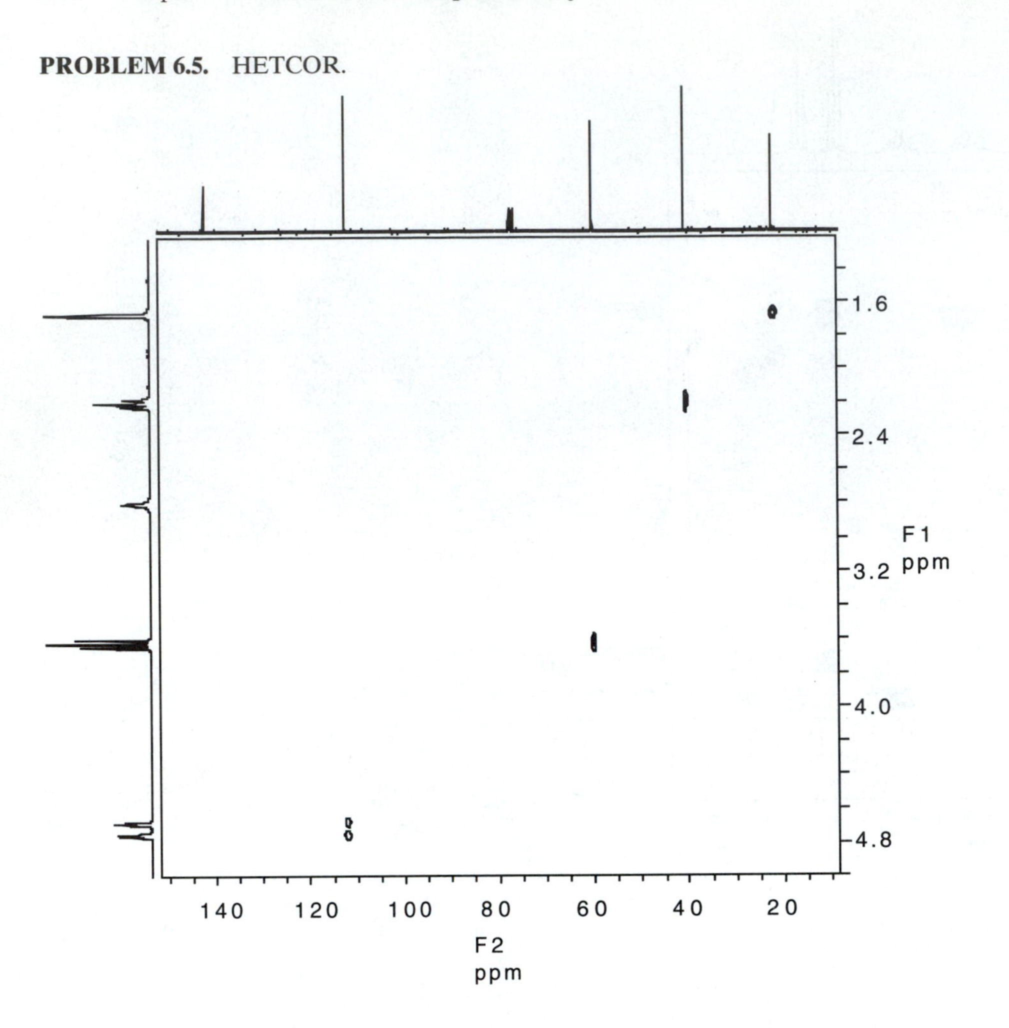

PROBLEM 6.6. COSY.

PROBLEM 6.6. HETCOR.

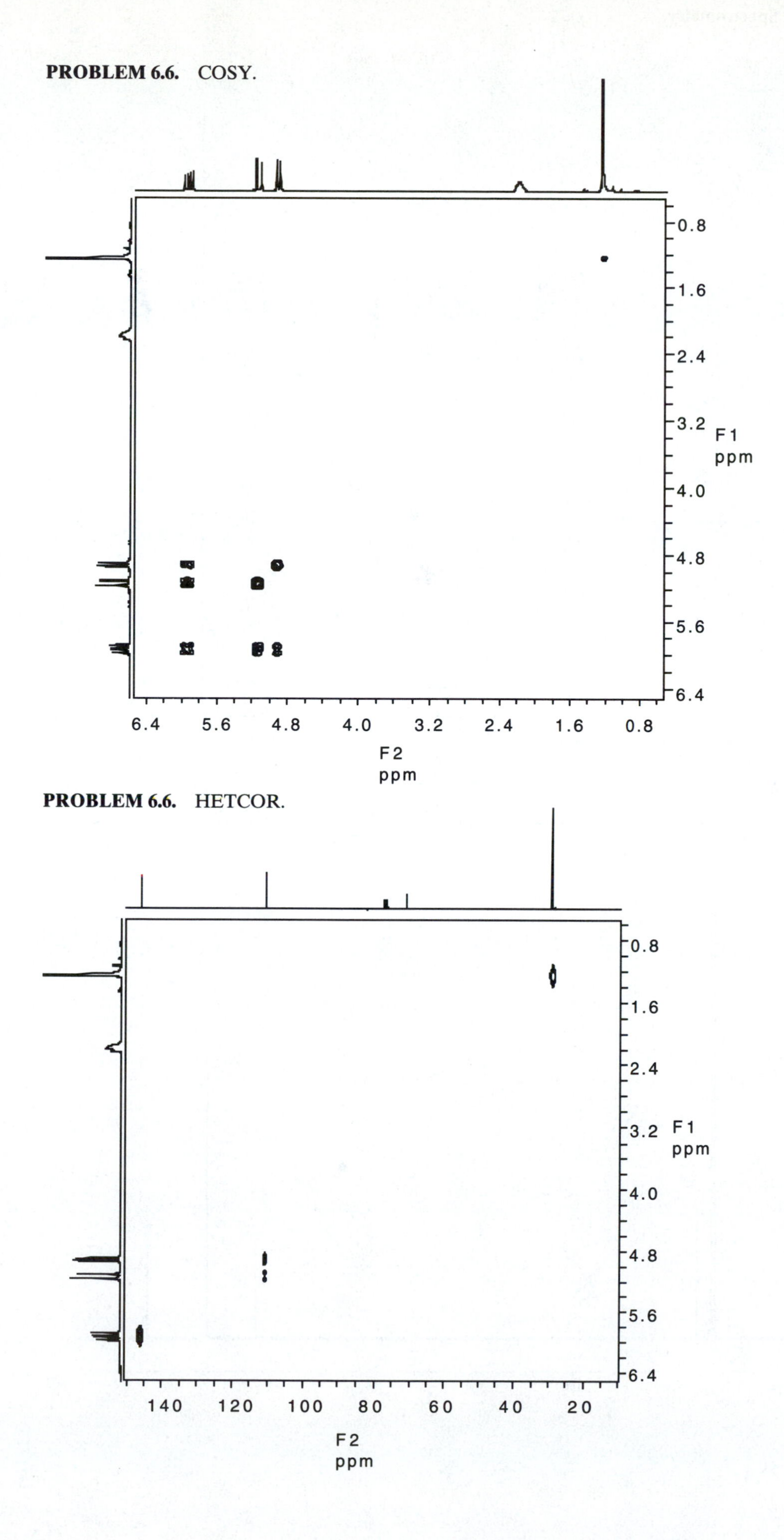

PROBLEM 6.8. ^{1}H NMR spectrum.

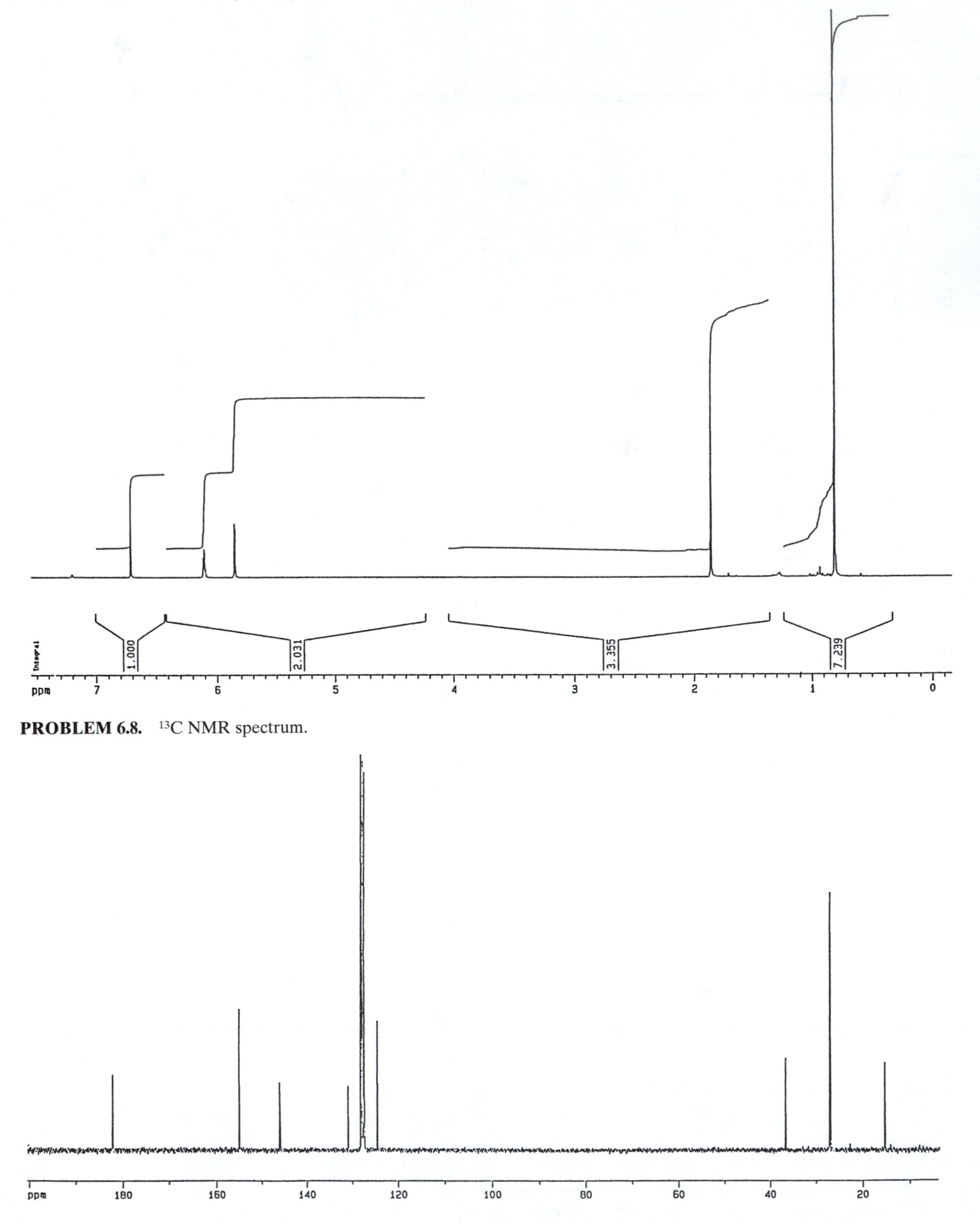

PROBLEM 6.8. ^{13}C NMR spectrum.

PROBLEM 6.8. INADEQUATE spectrum.

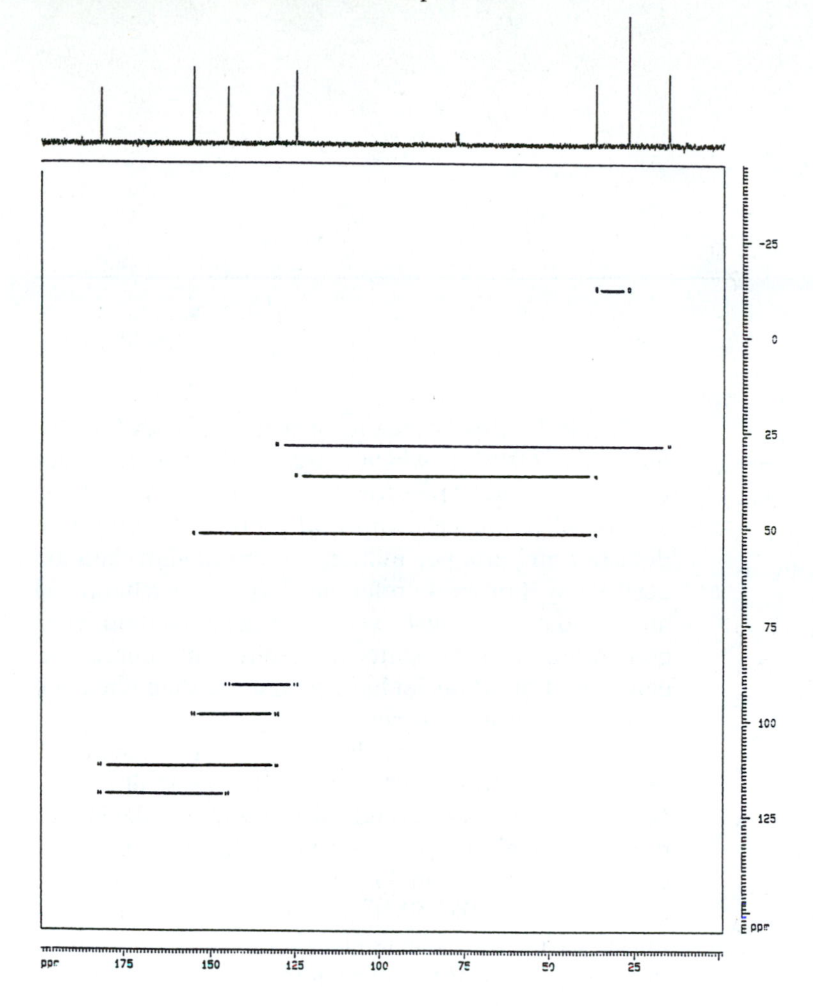

CHAPTER 7

Spectrometry of Other Important Nuclei

7.1 *Introduction*

The previous three chapters have shown that nuclear magnetic resonance experiments with ^{1}H and ^{13}C nuclei are enormously useful to the chemist working with organic compounds. There is no need, however, to limit ourselves to these two important nuclei. Indeed, there are 118 different nuclei whose spin number, I, is greater than zero and, therefore, theoretically observable in an NMR experiment. Of these 118 nuclei, 30 of them are dipolar, which means that their spin number is one-half ($I = \frac{1}{2}$).

Table 7.1 lists the first 23 magnetically active nuclei along with some associated useful data. It is worthwhile to spend some time exploring Table 7.1 and to compare some of the nuclei listed there with ^{1}H and ^{13}C (also listed). First, we find that many elements have more than one magnetically active isotope. Let us consider the element hydrogen and its possibilities as an example. Hydrogen has three isotopes: ^{1}H (protium), ^{2}H (deuterium), and ^{3}H (tritium); each isotope has been used in NMR studies, and each has different advantages to offer. ^{1}H and ^{3}H both have spin numbers of one-half and they both have high relative sensitivities, but ^{3}H has essentially zero natural abundance and it is radioactive. Tritium can therefore only be observed if we intentionally put it into a molecule synthetically, whereas protium is ubiquitous and we are already familiar with its uses.

Deuterium has a spin of one* and a very low natural abundance; it is not radioactive. It is a useful isotope for mechanistic studies in organic chemistry and biochemistry. Its low natural abundance allows us to ignore it when we observe other nuclei (such as ^{1}H and ^{13}C), yet with modern instruments we are still able to observe natural abundance ^{2}H NMR spectra.

Table 7.1 also lists the resonance frequency for each nucleus at 7.0463 T, which corresponds to a resonance frequency of 300 MHz for ^{1}H. Also listed in the table are useful reference compounds and useful ranges of detection in parts per million. Chemical shift data are useful only if properly referenced. In fact, excluding ^{1}H and ^{13}C data, we must exercise extreme caution when comparing chemical shifts from different sources because most other nuclei have seen more than one compound used as a reference.

We start by acknowledging that our goals are modest as we confront such a vast field as multinuclear NMR. In Sections 4.5.2 and 4.6 we have seen the impact of other nuclei that possess a magnetic moment (especially those with spin $\frac{1}{2}$) on proton spectra. We will briefly examine the NMR spectrometry of four spin $\frac{1}{2}$ nuclei, which were selected for their historic importance in organic chemistry (and related natural products and pharmaceutical fields), biochemistry, and polymer chemistry. These four nuclei, ^{15}N, ^{19}F, ^{29}Si, and ^{31}P, are presented with a few simple examples and a brief consideration of important experimental factors and limitations.

The theoretical background for these four nuclei has already been presented in Chapters 4 and 5. Our treatment of spin, coupling, NOE, Fourier transformation, etc., can be applied to these nuclei without modification. The concept of chemical shift we also use without modification, but we must avoid applying the predictive skills that we have developed for ^{1}H and ^{13}C chemical shifts to these nuclei (with some exceptions as noted).

The use of NMR spectrometry of nuclei other than ^{1}H and ^{13}C to characterize and identify organic compounds is now commonplace. The use of other nuclei in NMR experiments ranges from such diverse areas as simply determining whether an unknown compound contains nitrogen to more complex questions of stereochemistry and reaction mechanisms. Although our discussions will be limited to four "other" nuclei, we

* Nuclei with spin greater than one-half are not covered in this treatment. We note that with proper consideration of experimental details, we can record a spectrum from a deuterium sample.

Table 7.1 Useful Magnetic Resonance Data for the First 23 Magnetically Active Nuclei

Isotope	Spin	Natural Abundance (%)	Sensitivity: Relative[a]	Sensitivity: Absolute[b]	MHz at T of 7.0463	Reference Compound	Determined Range (ppm)
^{1}H	$\frac{1}{2}$	99.98	1.00	1.00	300.000	$Si(CH_3)_4$	10 to 0
^{2}H	1	1.5×10^{-2}	9.65×10^{-3}	1.45×10^{-6}	46.051	$Si(CD_3)_4$	10 to 0
^{3}H	$\frac{1}{2}$	0	1.21	0	319.990	$Si(CT_3)_4$	10 to 0
^{3}He	$\frac{1}{2}$	1.3×10^{-4}	0.44	5.75×10^{-7}	228.533		
^{6}Li	1	7.42	8.50×10^{-3}	6.31×10^{-4}	44.146	$^{6}LiCl_2/D_2O$	2 to −10
^{7}Li	$\frac{3}{2}$	92.58	0.29	0.27	116.590	$^{7}LiCl_2/D_2O$	2 to −10
^{9}Be	$\frac{3}{2}$	100	1.39×10^{-2}	1.39×10^{-2}	42.160		
^{10}B	3	19.58	1.99×10^{-2}	3.90×10^{-3}	32.239	$^{10}BF_3/(C_2H_5)_2O$	65 to −130
^{11}B	$\frac{3}{2}$	80.42	0.17	0.13	96.251	$^{11}BF_3/(C_2H_5)_2O$	65 to −130
^{13}C	$\frac{1}{2}$	1.108	1.59×10^{-2}	1.76×10^{-4}	75.432	$Si(CH_3)_4$	0 to 220
^{14}N	1	99.63	1.01×10^{-3}	1.01×10^{-3}	21.671	$^{14}NH_3$ (l)[c]	900 to 0
^{15}N	$\frac{1}{2}$	0.37	1.04×10^{-3}	3.85×10^{-6}	30.398	$^{15}NH_3$ (l)[c]	900 to 0
^{17}O	$\frac{5}{2}$	3.7×10^{-2}	2.91×10^{-2}	1.08×10^{-5}	40.670	H_2O	1700 to −50
^{19}F	$\frac{1}{2}$	100	0.83	0.83	282.231	$CFCl_3$	276 to −280
^{21}Ne	$\frac{3}{2}$	0.257	2.50×10^{-3}	6.43×10^{-6}	23.683		
^{23}Na	$\frac{3}{2}$	100	9.25×10^{-2}	9.25×10^{-2}	79.353	1M $NaCl/H_2O$	10 to −65
^{25}Mg	$\frac{5}{2}$	10.13	2.67×10^{-3}	2.71×10^{-4}	18.358	$MgCl_2/H_2O$	50 to −25
^{27}Al	$\frac{5}{2}$	100	0.21	0.21	78.172	$Al(NO_3)_3$	240 to −240
^{29}Si	$\frac{1}{2}$	4.7	7.84×10^{-3}	3.69×10^{-4}	59.595	$Si(CH_3)_4$	80 to −380
^{31}P	$\frac{1}{2}$	100	6.63×10^{-2}	6.63×10^{-2}	121.442	85% H_3PO_4	270 to −480
^{33}S	$\frac{3}{2}$	0.76	2.26×10^{-3}	1.72×10^{-5}	23.009		
^{35}Cl	$\frac{3}{2}$	75.53	4.70×10^{-3}	3.55×10^{-3}	29.395	^{35}NaCl in H_2O	1200 to −100
^{37}Cl	$\frac{3}{2}$	24.47	2.71×10^{-3}	6.63×10^{-4}	24.467	^{37}NaCl in H_2O	1200 to −100

[a] At constant field for equal number of nuclei.
[b] Product of relative sensitivity and natural abundance.
[c] At 25°C.

should not limit our outlook with respect to the possibilities of other nuclei or other experiments. In fact, our intention here is to broaden our outlooks to the nearly limitless possibilities with NMR and the periodic table.

7.2 ^{15}N *Nuclear Magnetic Resonance*

After carbon and hydrogen, oxygen and nitrogen are the next two most important elements in organic compounds. In the mind of the organic chemist, the presence of either of these elements represents the presence of one or more "functional groups" and the use of IR spectrometry is invoked. Without detracting in any way from IR, this line of reasoning is nonetheless too restrictive, especially with respect to nitrogen. Inspection of Table 7.1 reveals that in the case of oxygen we have but a single choice of a nucleus on which NMR is possible; ^{17}O has a spin of $\frac{5}{2}$ and it is not used much in NMR studies, whereas ^{16}O has a spin of 0 (not listed in Table 7.1).

Nitrogen, on the other hand, has two magnetically active isotopes, ^{14}N and ^{15}N. Because nitrogen compounds are so important* in organic chemistry and its applied fields of natural products, pharmacology, and biochemistry, both of these isotopes have been the subject of intensive NMR investigation.

If we again refer to Table 7.1, we find that neither of the two isotopes of nitrogen is ideal for NMR. The most abundant isotope of nitrogen, ^{14}N, which represents greater than 99% of nitrogen's natural abundance, possesses a spin of 1 and hence an electric quadrupole moment. This nucleus has an inherent low sensitivity and a very broad line because of quadrupolar relaxation. We shall not consider it any further.

The other isotope of nitrogen, ^{15}N, also has an inherent low sensitivity, which, when multiplied by a very low natural abundance, leads to an extremely low absolute sensitivity. Modern instrumentation has largely

* Aside from the various classes of nitrogen-containing functional groups with which we are familiar, entire fields of study have developed based on nitrogen-containing compounds. These fields include alkaloids, peptides and/or proteins, and nucleic acids. For purposes of study by NMR, the nucleic acids are a favorite subject because not only is nitrogen ubiquitous in these compounds but so is phosphorus (see Section 7.5).

overcome the problem of sensitivity and we focus our attention on ^{15}N largely because its spin number is $\frac{1}{2}$ and its line-widths are quite narrow.

There are two important experimental factors that must be accounted for if we are to be successful in running ^{15}N experiments. The ^{15}N nucleus tends to relax very slowly; T_1's of greater than 80 s have been measured. Thus, either long pulse delays must be incorporated into our pulse sequence or, alternatively, we could provide another route for spin relaxation. A common procedure is to add a "catalytic" amount of chromium(III) acetylacetonate, a paramagnetic substance, whose unpaired electrons efficiently stimulate transfer of spin. In cases where T_1's are not known (and not intended to be measured), pulse delays and pulse angles must be considered carefully because the signal from one (or more) ^{15}N resonance can accrue too slowly or be missed altogether.

The other experimental factor, the nuclear Overhauser effect, which has already been discussed in both Chapters 4 and 5, we consider now in more detail. Recall that we routinely run ^{13}C experiments with irradiation of the protons (i.e., proton decoupled) which, aside from producing the desired effect of singlets for all carbon signals, also enhances the signal for carbons with attached protons. This enhancement is caused by NOE; the changes in signal intensity arise from polarization of spin populations away from the predicted Boltzmann distribution. The amount of enhancement depends on two factors: One-half the ratio of the nuclei's magnetogyric ratios (γ) is the maximum possible enhancement, whereas the actual enhancement is proportional to the extent of dipolar relaxation. For a proton-decoupled ^{13}C experiment, the maximum NOE enhancement is $\gamma_H/2\gamma_C$ or 26,753/(2)6728, which equals 1.98. The total sensitivity increase is therefore nearly *threefold* because the NOE enhancement is added to the original intensity.

The actual enhancement for the ^{13}C—^{1}H system can be anywhere from 0 to 1.98 depending on the mechanism of relaxation for each individual nucleus. In practice, for carbons with no attached protons, the enhancement is essentially zero since there is practically no ^{13}C—^{1}H dipolar relaxation. For small to medium-sized organic molecules, ^{13}C—^{1}H dipolar relaxation for carbons with attached protons is very efficient, yielding close to the full 200% increase in signal.

If we apply the same reasoning to the ^{15}N nucleus, we arrive at a very different situation. The magnetogyric ratio for ^{15}N is small and *negative* ($\gamma = -2712$). A quick calculation shows that the maximum NOE enhancement for ^{15}N is $\gamma_H/2\gamma_N$ (26,753/(2) (−2712)) which is equal to −4.93. For the general case, a spin $\frac{1}{2}$ nucleus with a positive magnetogyric ratio gives positive NOE enhancement with proton decoupling, while a spin-one-half nucleus with a negative magnetogyric ratio gives negative NOE enhancement.

For the ^{15}N nucleus, the maximum enhancement is −4.93 + 1, or −3.93. In the case where ^{15}N—^{1}H dipolar spin–lattice relaxation overwhelms, the signal is inverted (negative) and its intensity is nearly four times what it would be in the absence of ^{1}H irradiation. However, since ^{15}N dipolar relaxation is only one of many relaxation mechanisms for ^{15}N, proton decoupling can lead to NOEs ranging from 0 to −4.93 or a signal ranging from +1 to −3.93. The experimental downside to this is that any NOE between 0 and −2.0 lowers the absolute intensity of the observed signal. In fact, an NOE of exactly −1.0 produces no signal at all! In general, as we saw for carbon, and now see for nitrogen, proton decoupling is commonly practiced for routine heteronuclear NMR experiments. In so doing, we must always bear in mind the practical outcome of NOE enhancement.

Let us turn our attention to ^{15}N spectra. As we have already mentioned, natural-abundance ^{15}N spectra can routinely be obtained on modern instruments even though ^{15}N is about an order of magnitude less sensitive than ^{13}C. Today, there is general agreement that liquid ammonia* is the standard reference compound for ^{15}N (used externally) although in the past, many compounds such as ammonium nitrate, nitric acid, nitromethane, and others have been used. When the literature is consulted, reliable chemical shifts can usually be obtained after correcting for their reported standard.

We are by now familiar with the construction of a ppm scale and need not consider the details here. Nitrogen, like carbon, is a second-row element and in many ways experiences similar electronic influences. To a first approximation, the chemical shifts of nitrogen-containing organic compounds closely parallel carbon chemical shifts. The chemical-shift range for nitrogen in common organic compounds is about 500 ppm, which is about twice that for carbon chemical shifts.† Table 7.2 shows the chemical shift ranges for many types of nitrogen-containing compounds. The relatively large chemical-shift range, taken together with the very narrow lines for ^{15}N resonances, means that the chances of fortuitous overlap in an ^{15}N spectrum are even smaller than

* Nitromethane was used occasionally as an internal reference and set to 0 ppm, but the resulting ^{15}N chemical shifts for nitrogen-containing organic compounds were generally negative. The use of liquid ammonia as an external reference precludes the need for negative numbers because virtually all ^{15}N atoms are deshielded by comparison, but handling liquid ammonia is awkward. The usual procedure is to add 380 ppm to the shift obtained by reference to nitromethane in order to report the shift relative to liquid ammonia.

† Both carbon and nitrogen chemical-shift ranges are larger than we are considering here. Chemical shifts outside of these ranges are unusual and not of interest in our discussion (see Table 7.2).

Table 7.2 Chemical Shift Ranges for Various Nitrogen-containing Compounds. Adapted from Levy and Lichter (1979).

Compound class	Type	Range labels
ALIPHATIC AMINES	Primary	
	Secondary	
	Tertiary	
ALIPHATIC AMMONIUM IONS	Primary	
	Secondary	
	Tertiary	
DIMETHYL ENAMINES		
ANILINIUM IONS		
ANILINES		
PIPERIDINES, DECAHYDROQUINOLINES		
CYCLIC ENAMINES		
AMINOPHOSPHINES		
GUANIDINES		Imino; Amino
UREAS, CARBAMATES, LACTAMS		
AMIDES	Primary	
	Secondary	
	Tertiary	
THIOUREAS		
THIOAMIDES		
NITRAMINES		NO_2; –N<
INDOLES, PYRROLES		
HYDRAZONES, TRIAZENES		–H=N–N; –N=N–N; Imino; Amino
IMIDES		
NITRILES, ISONITRILES		Iso
PYRROLE-LIKE NITROGENS		
DIAZO		Terminal; Internal
DIAZONIUM		Terminal; Internal
PYRIDINES		Pyridine
PYRIDINE-LIKE NITROGENS		Benzofurazan; Furazan; Pyrazine; Pyrimidine; Imidazole
IMINES		
OXIMES		
AZOXY		
NITRO		
AZO		
NITROSO		Ar-NO; S-NO; N–N=O

δ ppm: 900 800 700 600 500 400 300 200 100 0

in a ^{13}C spectrum. Our analogy with carbon can be furthered by pointing out that, within a given class of compounds, we can usually derive highly predictive substituent parameters as we did for carbon and that the magnitudes of these parameters are similar to carbon.

Lest we take this analogy with carbon too far, we must remember that nitrogen is unique and has chemical shift features peculiar only to nitrogen. The two most important ones result largely from the unshared pair of electrons found on nitrogen.

Just as this electron pair has a large impact on the chemistry of these compounds, it also has a great influence on the chemical shifts of the nitrogen in certain environments. We find quite often that chemical shift positions are more sensitive to the solvent than are structurally similar carbon resonances. The other way that the lone pair on nitrogen influences chemical shift is through protonation. Surprisingly, we cannot say in which direction the signal will shift upon protonation; we can say, however, that both the direction and the magnitude of the shift are characteristic of the specific type of nitrogen.

The proton-decoupled ^{15}N spectrum of formamide is shown in Figure 7.1. This spectrum looks remarkably like a ^{13}C spectrum with a single resonance. This ^{13}C-like appearance as opposed to an ^{1}H-like appearance will be the norm throughout this chapter as long as we have proton decoupling. The other feature worth noting is the direction of the peak. With proton decoupling, formamide experiences negative NOE enhancement and should therefore be negative. In an FT experiment, however, the initial phase of the FID is random and we

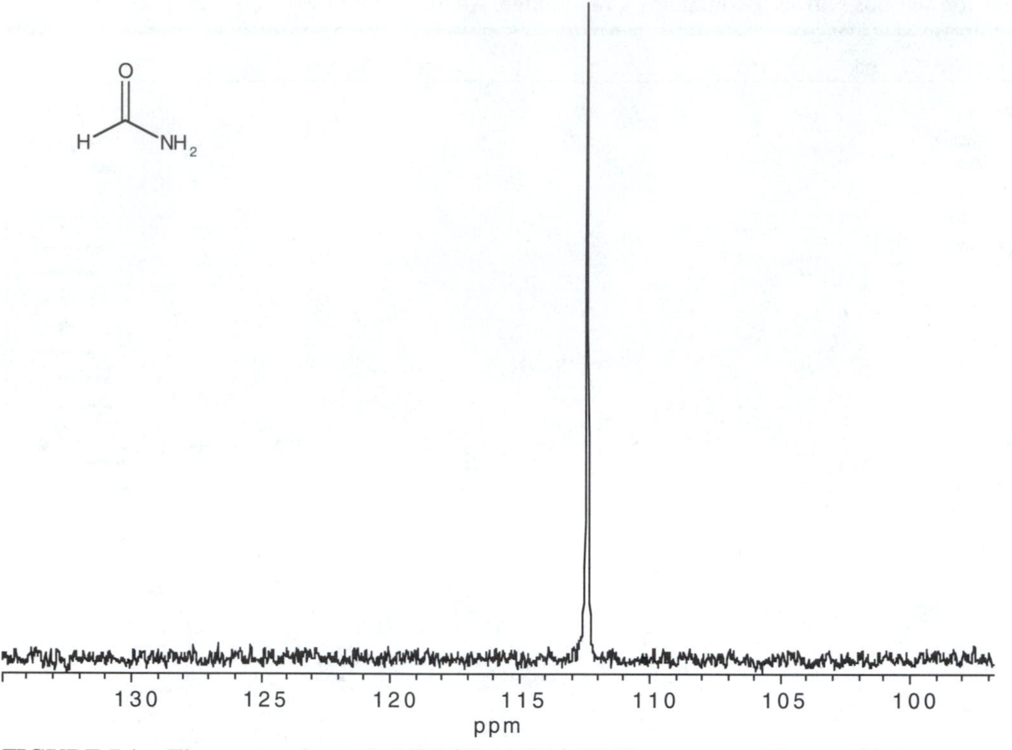

FIGURE 7.1. The proton-decoupled ^{15}N (30.4-MHz) NMR spectrum of formamide in $CDCl_3$ referenced to external NH_3.

could have phased the peak either up or down; the direction of peaks in a spectrum is arbitrary and only has meaning if both types are found in the same spectrum (see, for instance, DEPT). The nitrogen in formamide has partial double-bond character and is shifted accordingly (deshielding effect).

The proton-coupled and proton-decoupled ^{15}N spectra of ethylenediamine are shown in Figure 7.2. First, we note the relatively shielded position of the nitrogen here compared to formamide. The other noteworthy feature in this figure is the inset showing proton coupling. Before we discuss this part of the figure, a few general comments are in order. One-, two-, and three-bond $^{15}N—^{1}H$ couplings are common, whereas long-range couplings usually require intervening π bonds. The magnitude and sign of the coupling constants have been compiled but detailed consideration here is beyond our goal. We note that $^{1}J_{NH}$ varies from about 75 to 135 Hz, $^{2}J_{NH}$ between 0 and 20 Hz, and $^{3}J_{NH}$ between 0 and 10 Hz. If we again consider the inset in Figure 7.2, we note an apparent quintet with a relatively small coupling constant of about 2–3 Hz. Our interpretation is that we see no $^{1}J_{NH}$ coupling because of rapid exchange and that the two- and three-bond coupling constants are about the same. If we have equal coupling constants (careful, not always a good assumption) in a heteronuclear system (i.e., $^{n}J_{XH}$), first-order rules always apply because $\Delta\nu$ is of the order of millions of Hz.

The proton-decoupled ^{15}N spectrum of diisopropylamine can be found in Figure 7.3. We clearly see a trend of deshielding with alkyl substituents and steric crowding. The proton-decoupled ^{15}N spectrum of pyridine, which has an aromatic nitrogen, is recorded in Figure 7.4. A comfortable pattern is now emerging and, in fact, we can safely say that there is nothing extraordinary about this or any of the spectra that we have seen so far. Since our goal in this chapter is not to catalog the literally thousands of chemical shifts that have been reported but to "open the door" to the NMR of other nuclei besides carbon and hydrogen, let us finish the ^{15}N section by briefly considering the proton-decoupled ^{15}N spectrum of quinine (Figure 7.5), a well-studied, naturally occurring alkaloid.

Both nitrogen atoms of quinine are evident and, without hesitation, we can make assignments. If, instead, we had isolated quinine as a new unknown, natural compound we could envision a procedure in which we extend the concept of identifying compounds from a *combination of spectra* to include heteronuclear NMR; ^{15}N NMR would assume a natural place alongside mass, infrared, and other NMR spectrometries. Furthermore, as we made the transition from simple ^{1}H and ^{13}C spectra to correlation spectroscopy in Chapter 6, we asked: Is there more to ^{15}N NMR than proton-decoupled (and proton-coupled) spectra? The question is rhetorical; it is obvious that correlation experiments are possible and

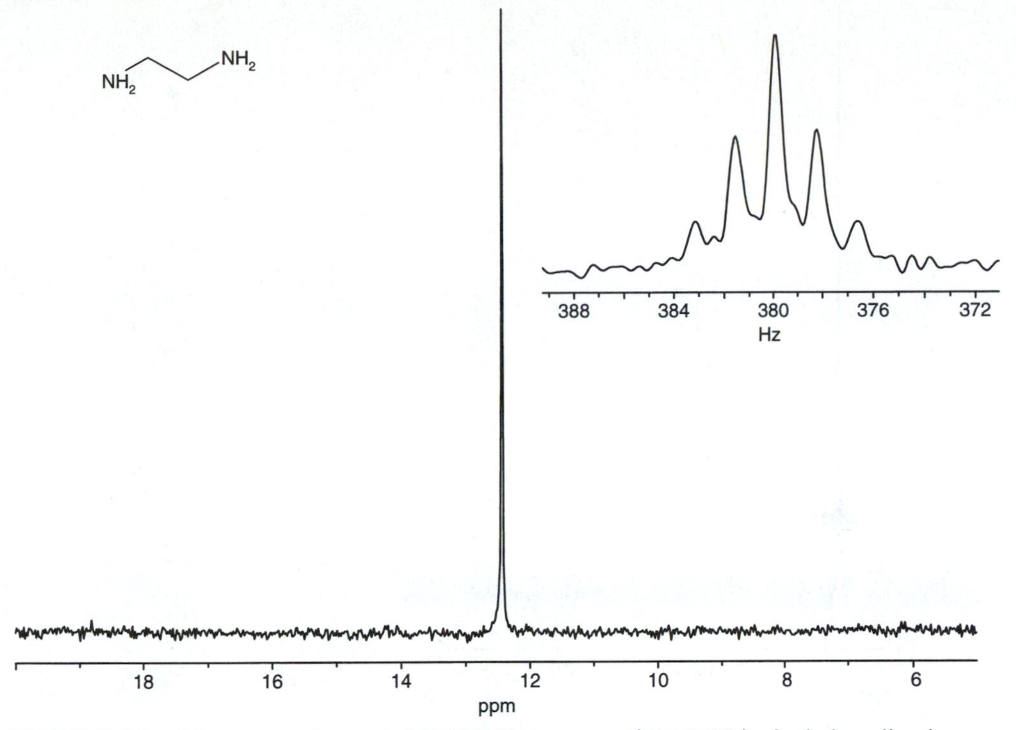

FIGURE 7.2. The proton-decoupled ^{15}N NMR spectrum (30.4 MHz) of ethylenediamine in $CDCl_3$. The proton-coupled spectrum is shown in the inset.

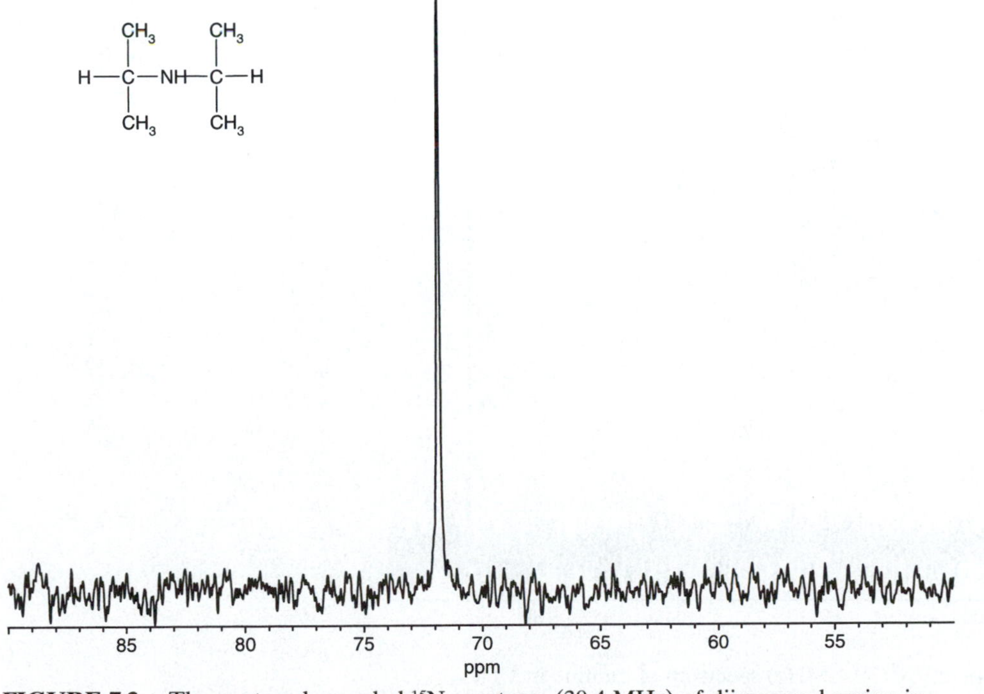

FIGURE 7.3. The proton-decoupled ^{15}N spectrum (30.4 MHz) of diisopropyl amine in $CDCl_3$.

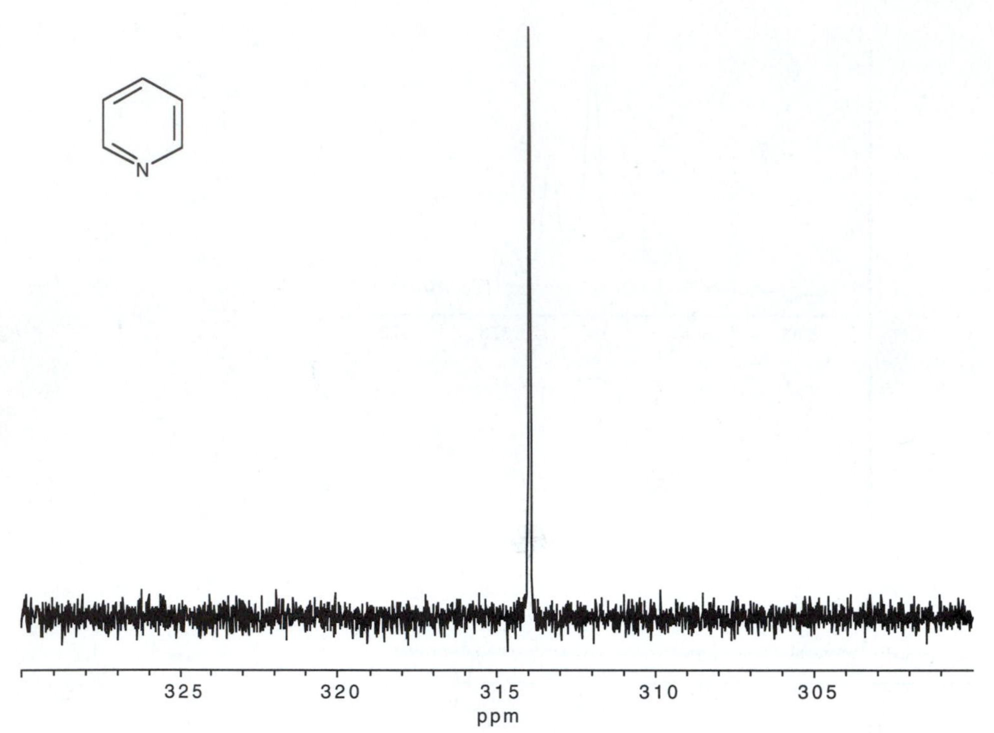

FIGURE 7.4. The proton-decoupled ^{15}N spectrum (30.4 MHz) of pyridine in $CDCl_3$.

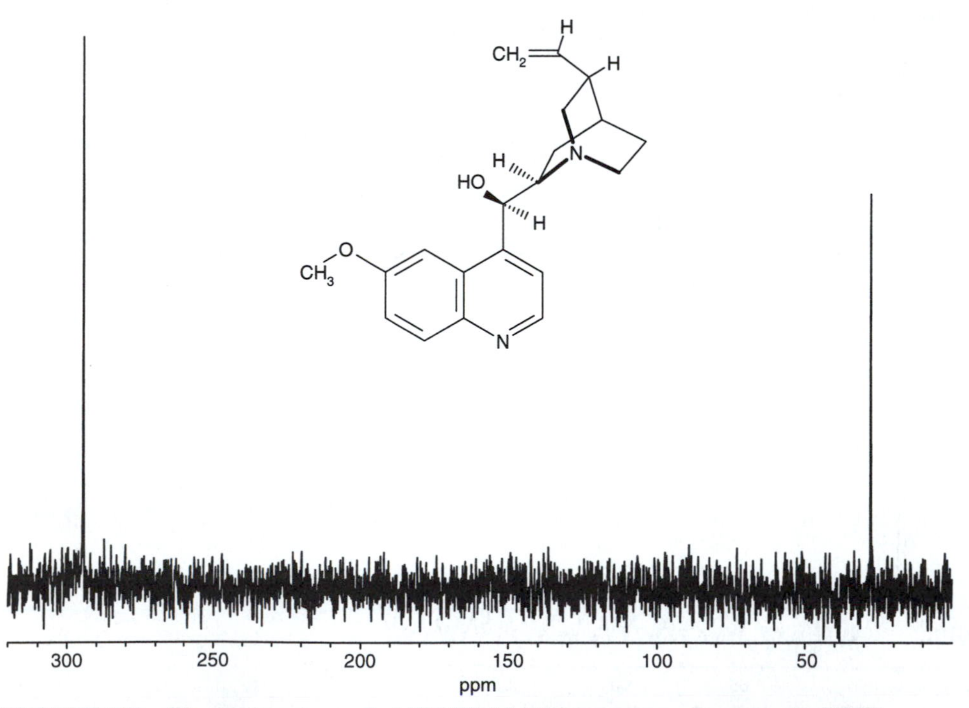

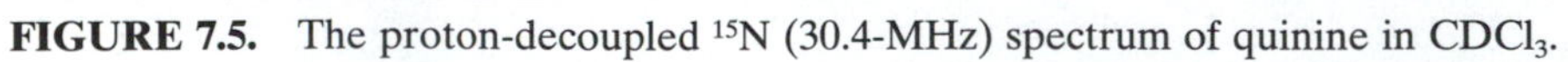

FIGURE 7.5. The proton-decoupled ^{15}N (30.4-MHz) spectrum of quinine in $CDCl_3$.

indeed many have already been developed. Much of the current interest in ^{15}N NMR involves correlation experiments.

7.3 ^{19}F *Nuclear Magnetic Resonance*

The NMR of ^{19}F has great historic importance. Fluorine has only one naturally occurring isotope, ^{19}F, and Table 7.1 reveals that it is an ideal nucleus for study by NMR. The sensitivity of ^{19}F is about 0.83 that of ^{1}H; this fact is the main reason that ^{19}F NMR developed contemporaneously with ^{1}H. The literature on chemical shifts and coupling constants is now old, and, again, some caution is needed when comparing older chemical shifts with newer literature because much of it is referenced to external CF_3COOH. Today, trichlorofluoromethane, $CFCl_3$, is the standard ^{19}F reference compound; it is generally inert, volatile, and gives rise to a single ^{19}F resonance.

Our approach to ^{19}F NMR is quite different than our approach to ^{15}N NMR has been, and this section is brief. Fluorine is monovalent and can be thought of as a substitute for hydrogen in organic compounds. Fluorine is virtually unknown in naturally occurring organic compounds; our interest in fluorine NMR involves synthetic compounds.* Proton decoupling of ^{19}F NMR experiments does not involve NOE enhancement in a critical way, and there are no other experimental factors out of the ordinary. In fact, ^{19}F NMR spectra were routinely recorded using continuous-wave (CW) instruments in exactly the same way as were ^{1}H spectra. Many of the chemical shift and coupling constant data for ^{19}F were published contemporaneously with those of ^{1}H.

The ^{1}H NMR spectrum of fluoroacetone is presented in Figure 4.36 to show the effect of fluorine on ^{1}H spectra. Recall that both the methylene group and the methyl group were split by the ^{19}F atom, each into doublets with different coupling constants. In Figure 7.6, we have an opportunity to see the reciprocal effect of the protons on the fluorine atom in fluoroacetone. In the proton-decoupled spectrum, of course, we see only a singlet for the fluorine atom, reminding us again of a ^{13}C spectrum. In the proton-coupled spectrum, however, we see that the fluorine is coupled to two sets of hydrogens, producing a triplet with a large coupling constant; the triplet is further split into three quartets by a four-bond coupling with the methyl group. We emphasize again that the combination of ^{1}H and ^{19}F spectra is more convincing and more informative than either one is by itself.

* NMR studies on synthetic compounds are initiated for different reasons than are NMR studies on naturally occurring compounds.

An example of an aromatic fluorine-containing compound can be found in Figure 7.7, where we have recorded the ^{19}F spectra (both proton coupled and decoupled) of *p*-fluoroacetophenone. Once again we find a singlet for the fluorine atom in the proton-decoupled spectrum and a complex multiplet for the fluorine atom in the proton-coupled spectrum. The fluorine atom couples differently to the ortho and meta protons in this para-substituted compound. Coupling constants for proton-fluorine can be found in Appendix F of Chapter 4.

The spin system for *p*-fluoroacetophenone is AA′MM′X because of magnetic non-equivalence (Section 4.13); it is not first order.

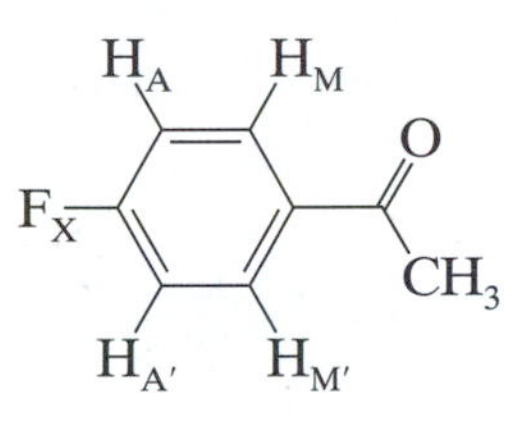

At 300 MHz (for ^{1}H), the broadened peaks overlap; they are not baseline peaks, and there are indications of further splitting. However, it is tempting to think in terms of an A_2M_2X system, look for a triplet of triplets for the ^{19}F nucleus with coupling constants of ~9 Hz and ~5 Hz (Chapter 4, Appendix F).

What we have, however, is a higher-order system, AA′MM′X, where X is a fluorine atom. The fluorine absorption is deceptively simple in being symmetrical; the halves are mirror images. The A and A′ protons are not magnetic equivalent since they do not couple equally with the M proton or the M′ proton. Nor are the M and M′ protons magnetic equivalent since they do not couple equally with the A proton or the A′ proton.

Note the resemblance of the fluorine atom absorption to the symmetry of the absorptions of the magnetic nonequivalent aromatic protons described in Section 4.13. The complexity in all of these spectra is not a result of an inadequate magnetic field. In fact, a more powerful magnetic field would result in greater complexity. This is quite different from the response of a first-order system, which may not be resolved in a weak magnetic field, but would show a first-order spectrum at a sufficiently high field.

Chemical shifts for ^{19}F are difficult to predict, and we will make no attempt here but to refer the interested reader to the literature. One reason that the chemical shifts for ^{19}F are difficult to predict and rationalize is that less than 1% of the shielding of the ^{19}F nucleus in organic compounds is caused by diamagnetic shielding. Rather, paramagnetic shielding is the predominant factor and poorly understood.

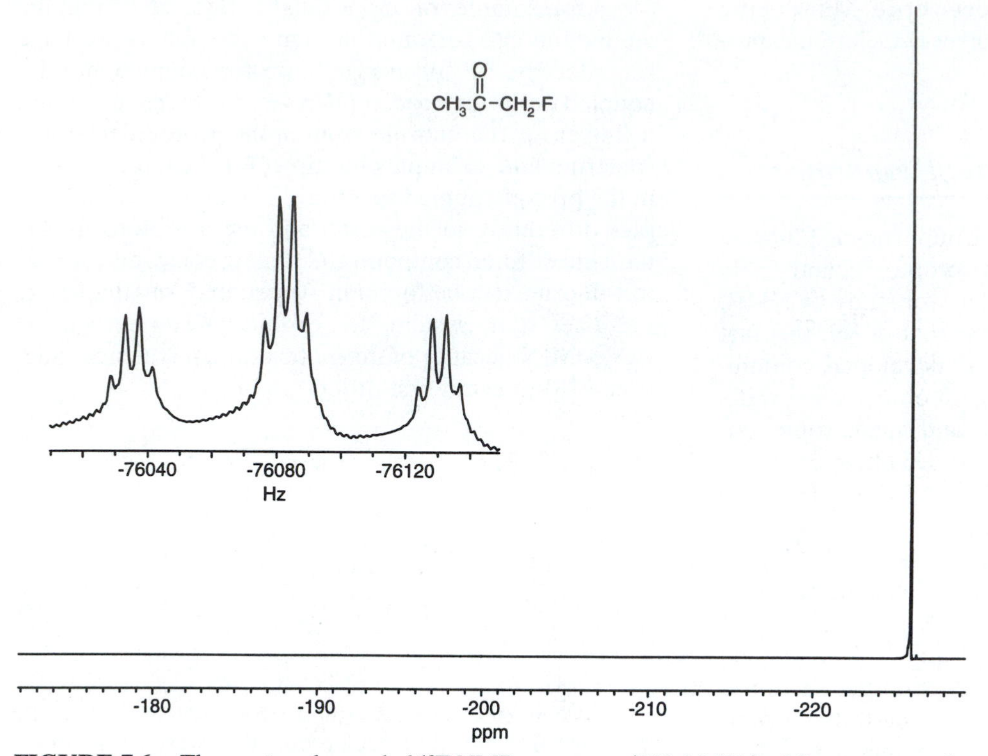

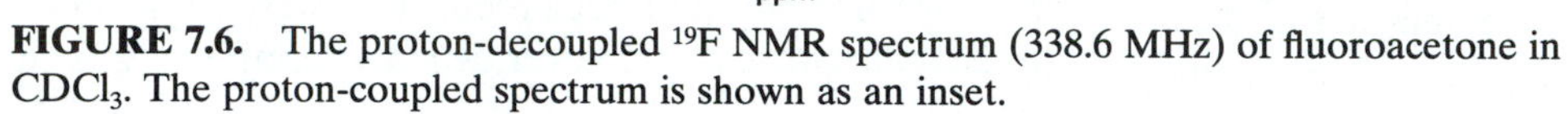

FIGURE 7.6. The proton-decoupled ^{19}F NMR spectrum (338.6 MHz) of fluoroacetone in $CDCl_3$. The proton-coupled spectrum is shown as an inset.

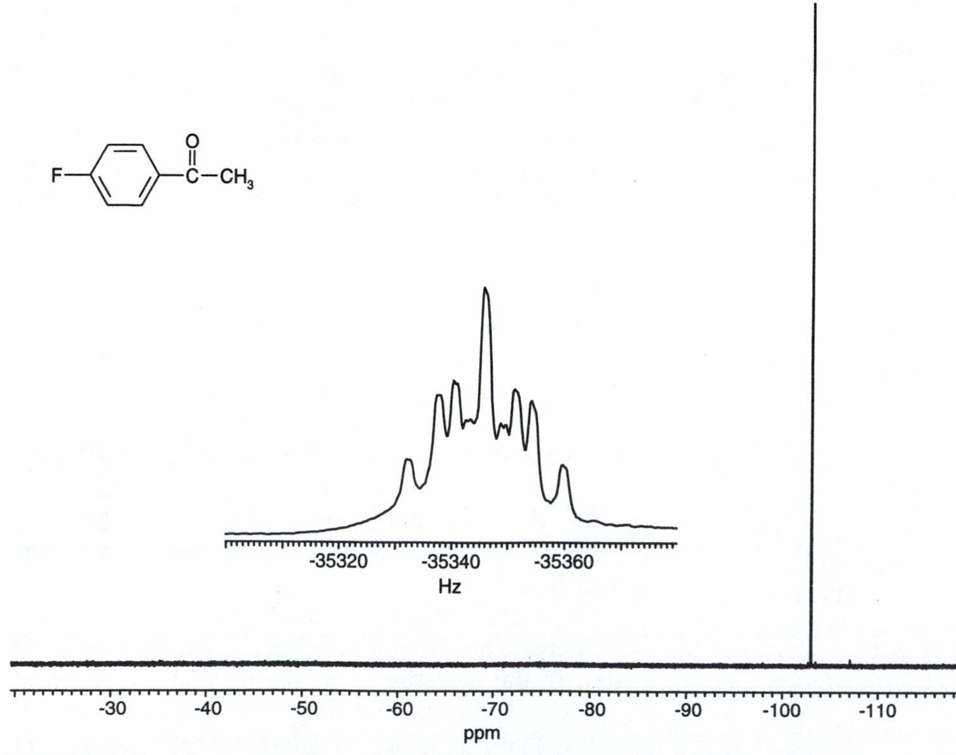

FIGURE 7.7. The proton-decoupled ^{19}F NMR spectrum (338.6 MHz) of *p*-fluoroacetophenone in $CDCl_3$. The proton-coupled spectrum is shown as an inset.

7.4 ^{29}Si *Nuclear Magnetic Resonance*

Silicon, like fluorine, does not occur naturally in organic compounds. Silicon-containing organic compounds, however, are increasingly used by synthetic organic chemists and by polymer chemists. The ^{29}Si nucleus is the only isotope of silicon with a magnetic moment and has a natural abundance of 4.7%. We have already come across the ^{29}Si nucleus in the proton NMR spectrum of TMS; a small doublet, with a coupling constant ($^2J_{SiH}$) of about 6 Hz, with an intensity of 2–3% straddles the sharp, intense TMS singlet (Section 4.11.4). This small doublet represents the 4.7% of spin-$\frac{1}{2}$ ^{29}Si, which naturally occurs in all silicon compounds.

Table 7.1 reveals that the sensitivity of the ^{29}Si nucleus is about two times that of the ^{13}C nucleus when both are recorded at natural abundance. The magnetogyric ratio for ^{29}Si (γ_{Si}) is negative (-5319) so that for routine proton-decoupled spectra we again have the possibility of negative ^{29}Si NOE enhancement depending, of course, on the relative importance of dipolar spin relaxation. In this case, the maximum NOE is -2.51. This situation is much worse than with ^{15}N NOEs because only NOEs between -2.01 and the maximum, -2.51, actually result in an "enhancement." All other values result in a net decrease in signal intensity compared to no proton decoupling. Thus, experimental conditions must be carefully controlled if we are to realize the maximum signal.

The chemical shifts for ^{29}Si in common organic compounds are much smaller than for ^{13}C shifts in common compounds. This smaller shift range probably results from the lack of multiple bonds to silicon, because we find the most deshielding for ^{13}C in sp^2-hybridized atoms. The proton decoupled ^{29}Si spectrum of tetramethylsilane (TMS) is shown in Figure 7.8 with the proton coupled spectrum for comparison as an inset. TMS is the obvious choice for a ^{29}Si reference compound and we set it at 0 ppm. The proton-coupled spectrum is quite interesting because the ^{29}Si nucleus is coupled to 12 equivalent protons in TMS. First-order rules predict a multiplet with 13 peaks. There are 9 peaks clearly visible and 11 with a little imagination; we do not see the full 13 peaks because the outer ones are too weak and are lost in the noise.

The ^{29}Si NMR spectrum of triethylsilane in both proton-decoupled and proton-coupled modes is presented in Figure 7.9. The proton-decoupled spectrum gives a singlet that is only slightly shifted from TMS. In triethylsilane, there is a proton directly attached to silicon resulting in a large one-bond coupling ($^1J_{SiH}$) of about 175 Hz. There are smaller two- and three-bond couplings that leads to identical complex multiplets.

Our last example of ^{29}Si NMR is found in Figure 7.10 where we find the proton-decoupled and proton-

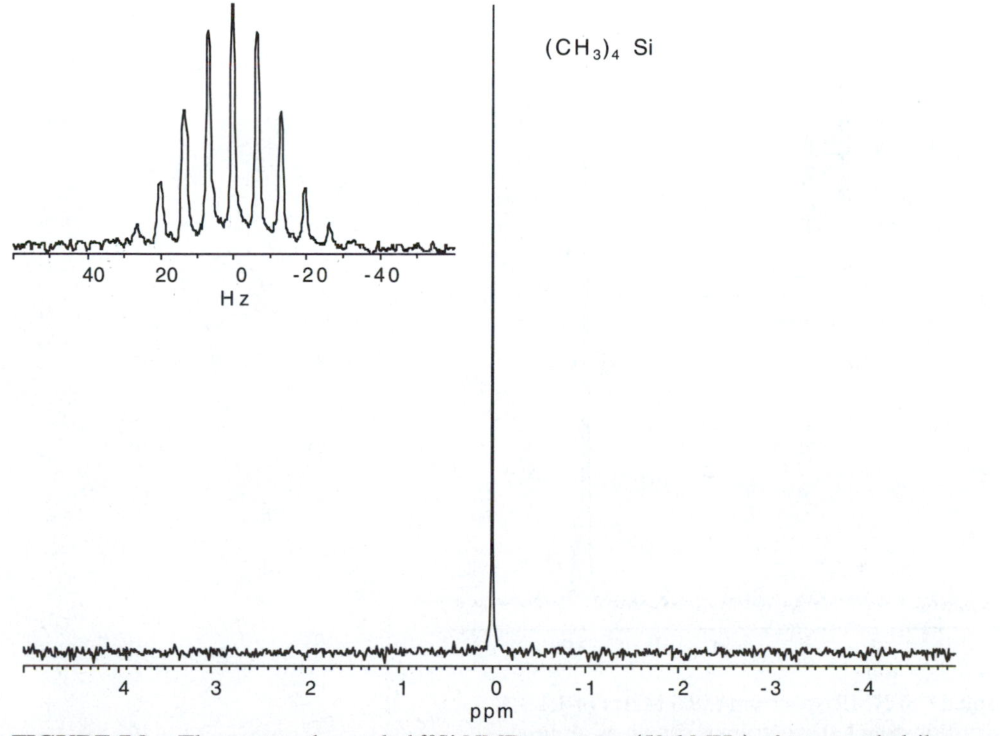

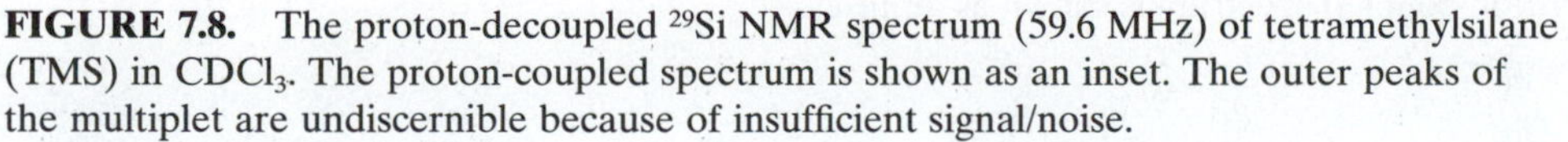

FIGURE 7.8. The proton-decoupled ^{29}Si NMR spectrum (59.6 MHz) of tetramethylsilane (TMS) in $CDCl_3$. The proton-coupled spectrum is shown as an inset. The outer peaks of the multiplet are undiscernible because of insufficient signal/noise.

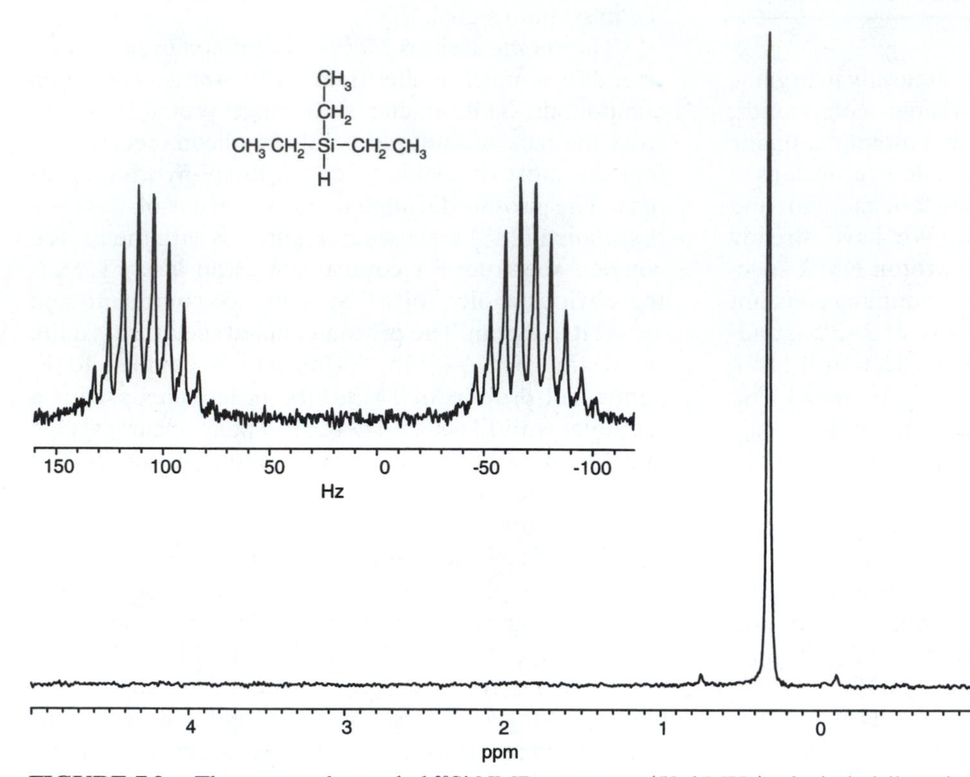

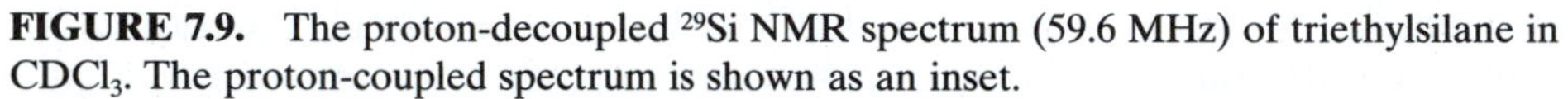

FIGURE 7.9. The proton-decoupled ^{29}Si NMR spectrum (59.6 MHz) of triethylsilane in $CDCl_3$. The proton-coupled spectrum is shown as an inset.

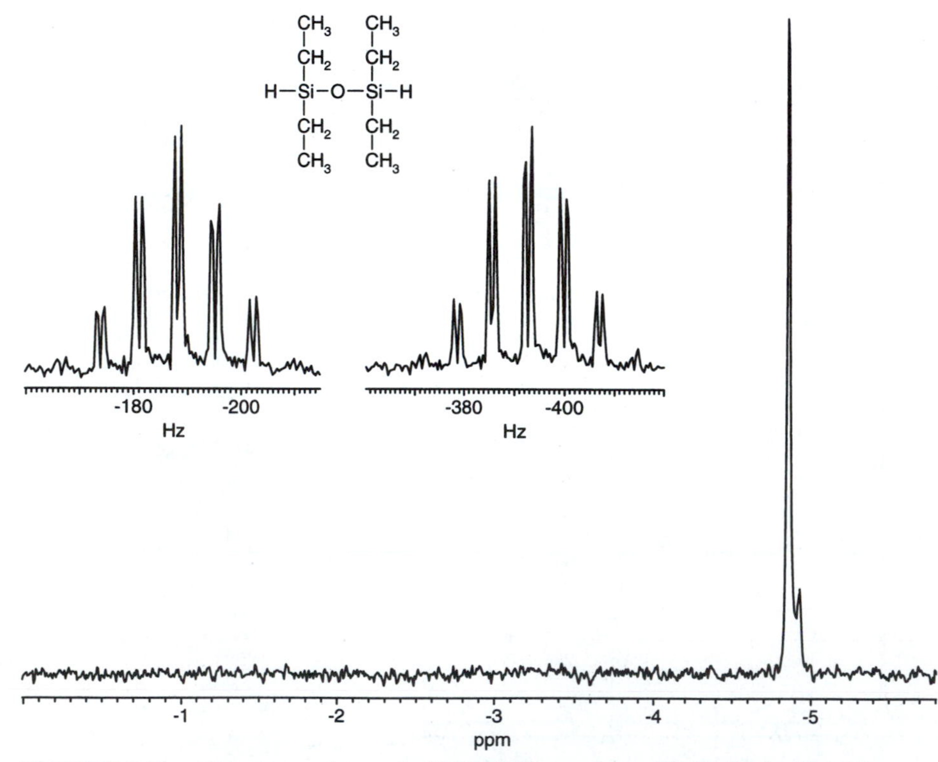

FIGURE 7.10. The proton-decoupled ^{29}Si NMR spectrum (59.6 MHz) of 1,1,3,3-tetraethyldisiloxane in $CDCl_3$. The proton-coupled spectrum is shown as an inset.

coupled ^{29}Si spectra of 1,1,3,3-tetraethyldisiloxane. This compound is commercially available and is widely used to make various silicon-containing polymers. Before using the sample, a conscientious chemist might analyze it using various methods, which we now imagine might include ^{29}Si NMR. In this case, two types of silicon would be found in the sample because we find a peak of about 5–10% on the shoulder of the main peak in the proton-decoupled spectrum. The chemical shift is negative, a common feature of silicon bonded to oxygen.

The proton-coupled spectrum reveals a larger one-bond coupling constant ($^1J_{SiH}$) of about 215 Hz. The coupling pattern derived from the two- and three-bond coupling is complex but the pattern may serve as a starting point in the interpretation of ^{29}Si spectra of reaction products.

7.5 ^{31}P *Nuclear Magnetic Resonance*

The last of the four nuclei that we treat briefly in this chapter is ^{31}P, the only naturally occurring isotope of phosphorus. Phosphorus is of great interest to the organic chemist because reagents containing phosphorus, which range from various inorganic forms of phosphorus to the organic phosphines, phosphites, phosphonium salts, phosphorus ylides, etc., have long been used by organic chemists; the nucleus is of great interest to the biochemists primarily because of the nucleic acids, which contain phosphate esters, and also smaller molecules such as ADP, ATP, etc.

NMR experiments with ^{31}P are rather straightforward; ^{31}P is a spin $\frac{1}{2}$ nucleus with a positive magnetogyric ratio (10,840). ^{31}P NOE enhancement from proton decoupling is positive with a maximum of 1.23. There has been a long history and therefore a rich literature of ^{31}P NMR. ^{31}P chemical shift data are reliable because 85% H_3PO_4 is virtually the only ^{31}P reference compound reported (external) and it remains the preferred reference today (Figure 7.11). The chemical shift range for ^{31}P is rather large and generalizations are dangerous. In fact, even the different valence states of phosphorus do not fall into predictable patterns. All is not lost however, since there many reliable published studies of ^{31}P chemical shifts. Representative proton–phosphorus coupling constants can be found in Appendix F of Chapter 4. More complete lists of ^{31}P coupling constants of phosphorus to protons and other nuclei are available.

The proton-decoupled ^{31}P NMR spectrum of diethyl chlorophosphate is found in Figure 7.12 along with the proton-coupled spectrum. Also included with this figure are the ^1H and ^{13}C NMR spectra for comparison. The details of the proton and carbon spectra are left to the reader to work out. It is important to note, however,

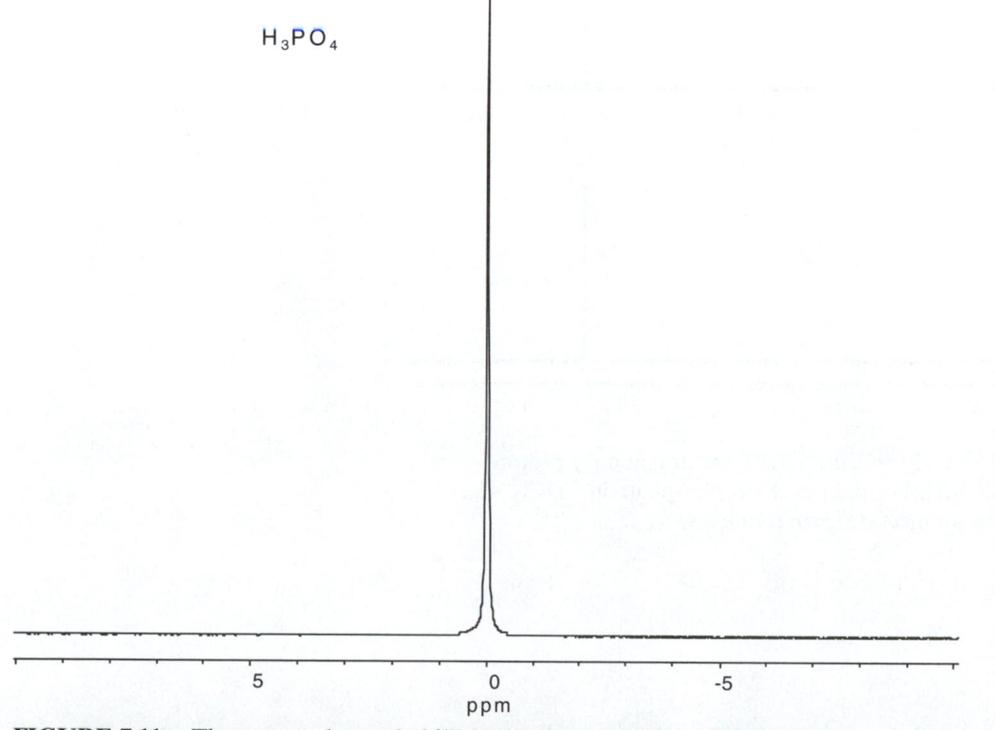

FIGURE 7.11. The proton-decoupled ^{31}P NMR spectrum (121.4 MHz) of 85% phosphoric acid with a small amount of D_2O.

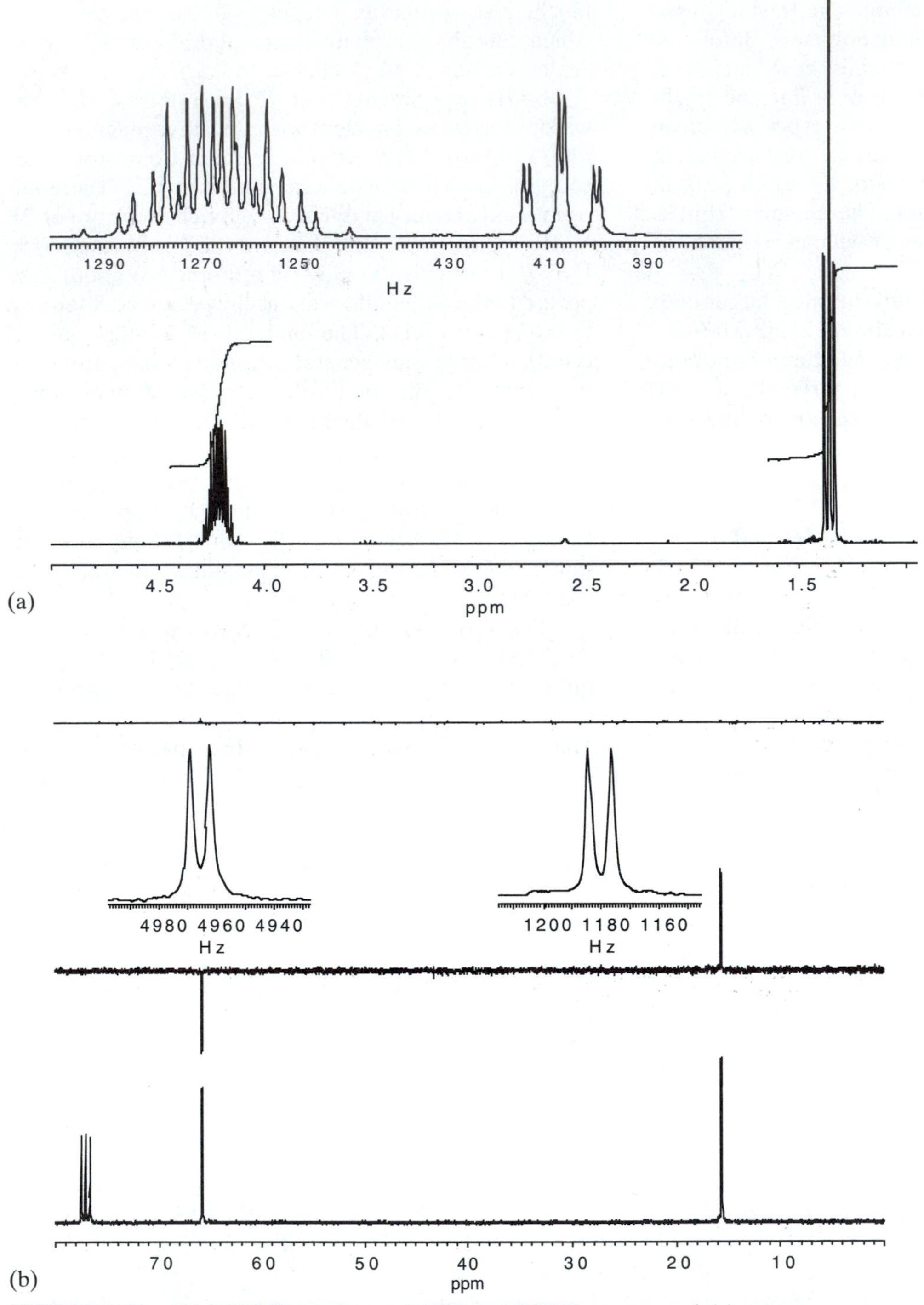

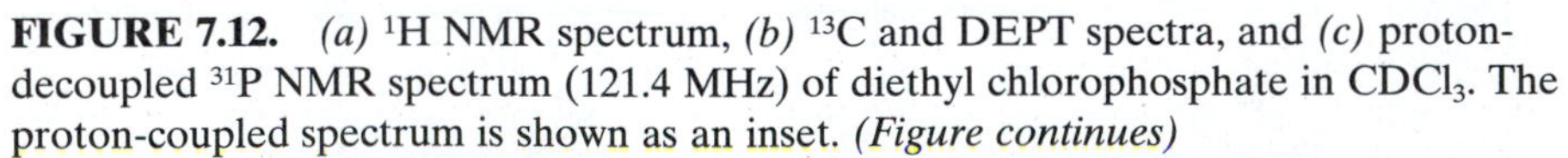

FIGURE 7.12. *(a)* ^{1}H NMR spectrum, *(b)* ^{13}C and DEPT spectra, and *(c)* proton-decoupled ^{31}P NMR spectrum (121.4 MHz) of diethyl chlorophosphate in $CDCl_3$. The proton-coupled spectrum is shown as an inset. *(Figure continues)*

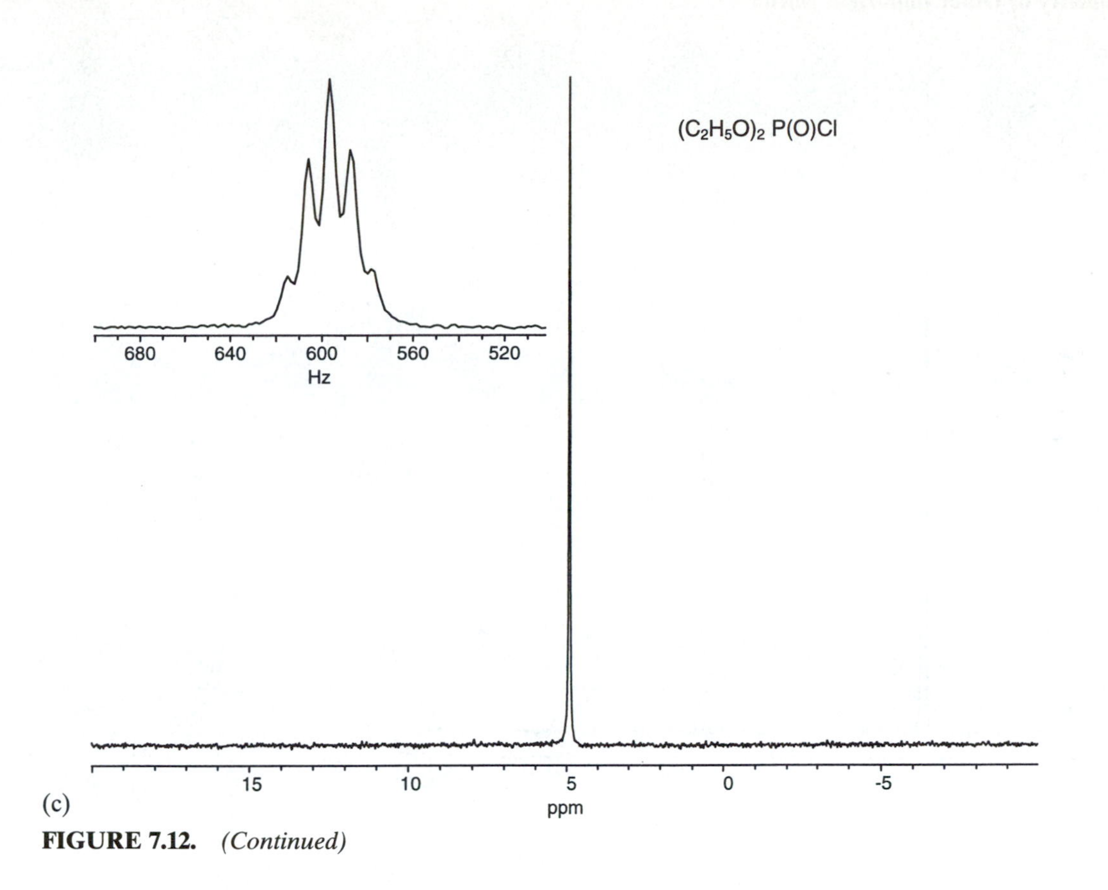

(c)

FIGURE 7.12. *(Continued)*

the coupling of the ^{31}P nucleus with the ^{1}H and ^{13}C nuclei; the appearance of both of these spectra strongly reflects the added coupling. The proton-coupled ^{31}P spectrum shows an apparent quintet, suggesting that there is no appreciable coupling of the phosphorus atom to the methyl protons of the methyl groups. Inspection of the corresponding proton spectrum reveals however, that the methyl groups are actually a triplet of doublets with the four-bond $^{31}P—^{1}H$ coupling constant of ca. 1 Hz. Thus, we conclude that the 1 Hz coupling is not resolved in the ^{31}P spectrum but reveals itself by broadening the lines. The four protons of the two methylene groups couple equally (approximately) to the phosphorus atom to give the observed quintet.

The methylene protons in the ^{1}H spectrum may at first seem "deceptively complex." The matter is resolved by noting that the methylene protons in diethyl chlorophosphate are diastereotopic (see Fig. 4.37), and therefore they are not chemical shift equivalent and show strong coupling. First-order analysis is impossible.

Figure 7.13 displays the proton-decoupled ^{31}P spectrum of triphenylphosphine, a common reagent in organic synthesis. The spectrum itself is unremarkable in that we see a single resonance for the single phosphorus atom in triphenylphosphine. What is remarkable is the fact that there is a chemical shift range of less than 10 ppm for the three compounds shown in Figures 7.11, 7.12, and 7.13 and yet the phosphorus atoms differ widely with respect to valence state, attached groups, and so on. We would have a very difficult job trying to rationalize these figures and it would have been foolish to have tried to predict these ^{31}P chemical shifts.

In Figures 7.14 and 7.15, the proton-decoupled ^{31}P NMR spectra of two phosphites are presented; they are triphenylphosphite and triethylphosphite, respectively. Both compounds give a sharp single peak as expected. The point here is to illustrate that even without a set of predictive rules for chemical shift, we can nonetheless expect useful information with little chance of overlap. By themselves these spectra may not provide much information, but when combined with other spectra, they give one more perspective on composition, structure, and stereochemistry.

7.6 Conclusions

We have attempted in this brief chapter to introduce a few other useful nuclei and some examples of their spectra. The emphasis here is to utilize these spectra *in combination* with other spectra, especially other forms of NMR. We must concede that it is neither possible nor desirable to become "experts" on chemical shifts of these (or other) nuclei and their coupling constants. In making this concession, we can be "comfortable" with these four nuclei, and furthermore, we can easily

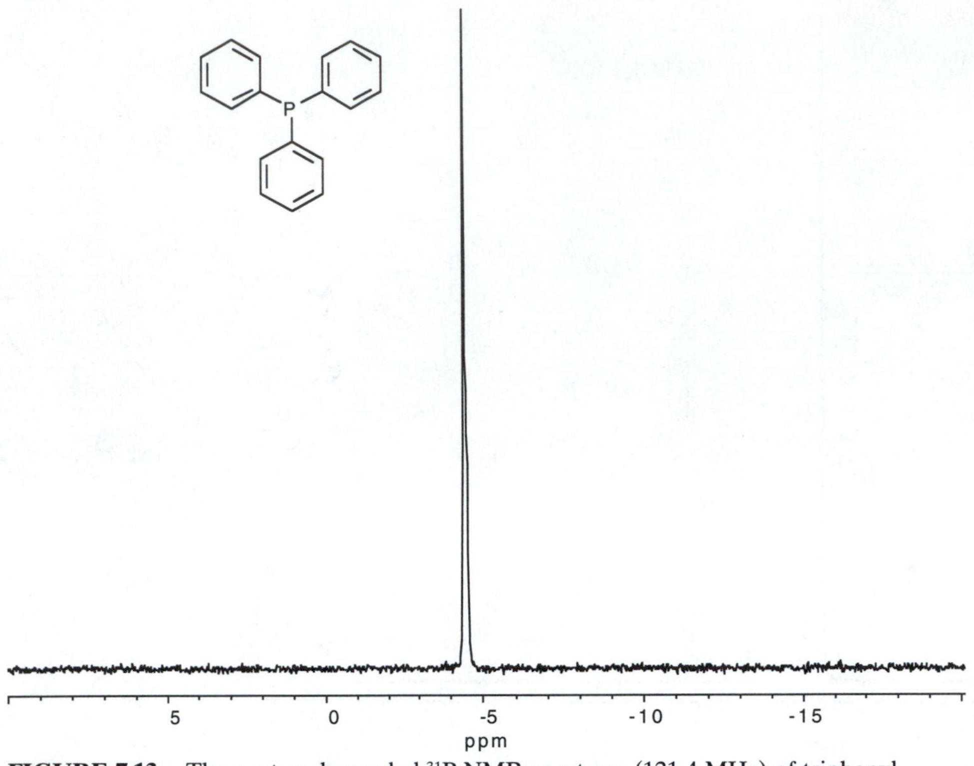

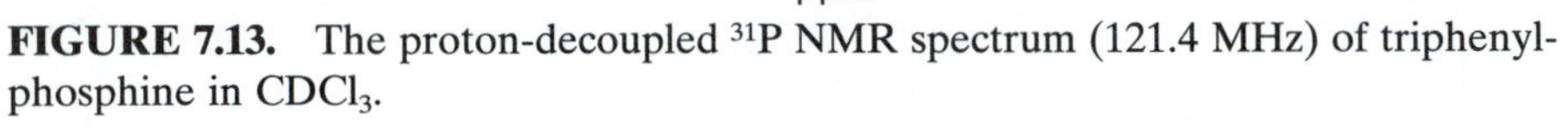

FIGURE 7.13. The proton-decoupled ^{31}P NMR spectrum (121.4 MHz) of triphenylphosphine in $CDCl_3$.

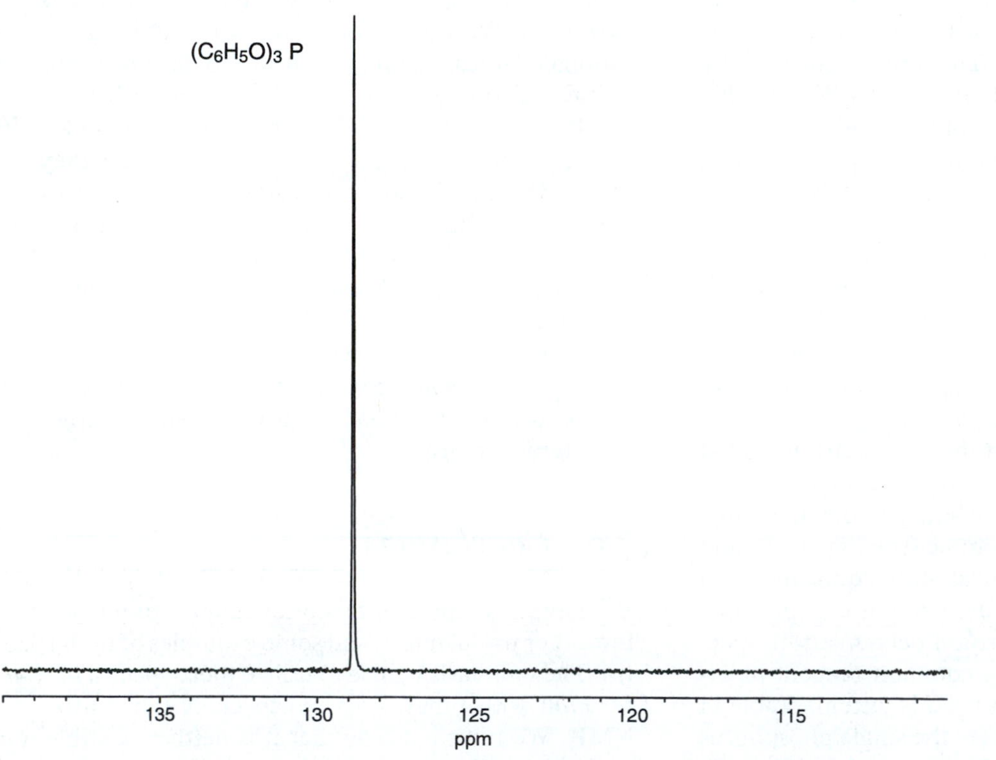

FIGURE 7.14. The proton-decoupled ^{31}P NMR spectrum (121.4 MHz) of triphenylphosphite in $CDCl_3$.

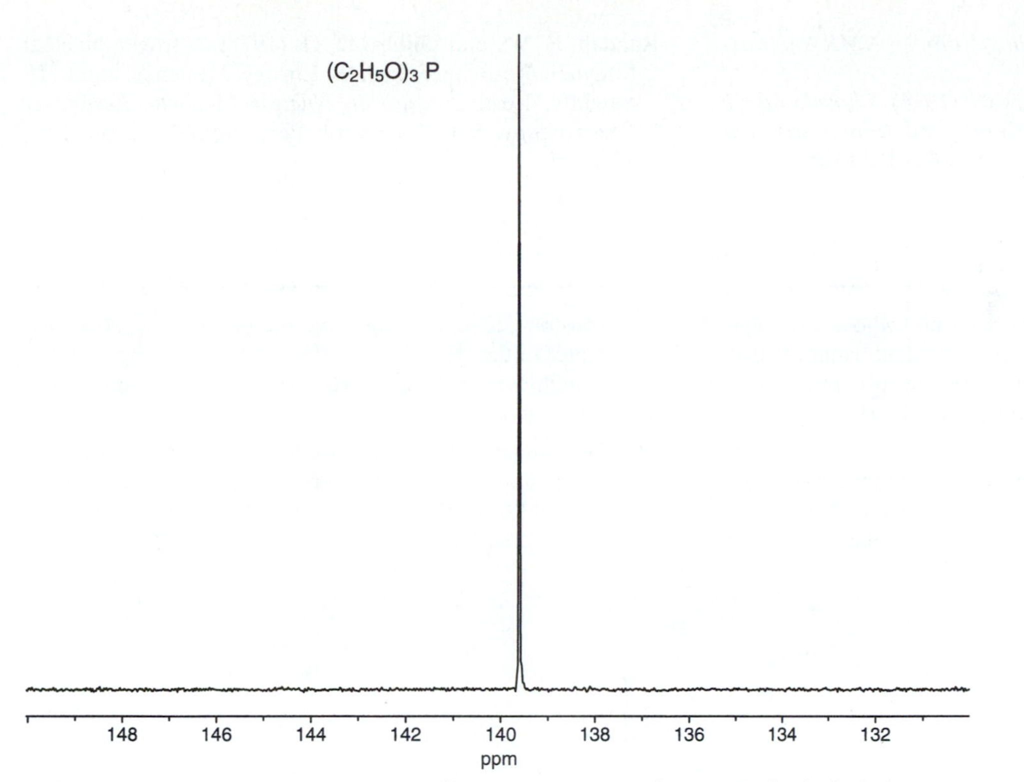

FIGURE 7.15. The proton-decoupled ^{31}P NMR spectrum (121.4 MHz) of triethylphosphite in $CDCl_3$.

broaden our outlook to other elements throughout the periodic table.

Throughout this chapter, we have seen proton-decoupled spectra, which should be considered routine, and we have seen proton-coupled spectra for comparison. As useful as these spectra can be, correlation experiments with these "other" nuclei may be even more useful and exciting. Pulse sequences for these experiments may be beyond our capabilities at this point, but surely interpretation of them is not. The field of NMR is as dynamic today as it was 30 to 40 years ago, and organic and biochemists remain one of the main reasons for many of the advances.

References

Berger, S., Braun, S., and Kalinowski, H. O., (1992). *NMR-Spektroskopie von Nichtmetallen.* Stuttgart: Georg Thieme Verlag.

Band 1: Grundlagen, ^{17}O-, ^{33}S- and ^{129}Xe-NMR Spektroskopie.

Band 2: ^{15}N-NMR Spektroskopie.

Band 3: ^{31}P-NMR Spektroskopie.

Band 4: ^{19}F-NMR Spektroskopie.

Brey, W., Ed. (1988). *Pulse Methods in 1D and 2D Liquid-Phase NMR.* New York: Academic Press.

Chandrakumar, N., and Subramanian, S. (1987). *Modern Techniques in High Resolution FT NMR.* New York: Springer-Verlag.

Dungan, C. H., and Van Wazer, J. R. (1970). *Compilation of Reported F-19 NMR Chemical Shifts.* New York: Wiley-Interscience.

Emsley, J. W., and Phillips, L. (1971). Fluorine Chemical Shifts. In J. W. Emsley, J. Feeney, and L. H. Sutcliffe, (Eds.), *Progress in Nuclear Magnetic Resonance Spectroscopy,* Vol. 7. Oxford: Pergamon Press.

Everett, T. S. (1988). The correlation of multinuclear spectral data for selectively fluorinated organic compounds. *J. Chem. Educ.,* **65,** 422–425.

Gorenstein, D.G. (1984.) Phosphorus-31 NMR. London: Academic.

Laszlo, P., Ed. (1983). *NMR of Newly Accessible Nuclei,* Vol. 1. New York: Academic Press.

Levy, G. C., and Lichter, R. L. (1979). *Nitrogen-15 Nuclear Magnetic Resonance Spectroscopy.* New York: John Wiley & Sons.

Martin, G. J., Martin, M. L., and Gouesnard, J. P. (1981). *^{15}N-NMR Spectroscopy,* Vol. 18. Berlin: Springer-Verlag.

Mason, J. (ed.). (1987). *Multinuclear NMR.* New York: Plenum Press.

Mooney, E. F. (1970). *An Introduction to ^{19}F NMR Spectroscopy.* London: Heyden.

Quin, L. D., and Verkade, J. G., Eds. (1994). *Phosphorus-31 NMR Spectral Properties in Compound Characterization and Structural Analysis.* New York: VCH Publishers.

Randall, E. W., and Gillies, D. G. (1971). Nitrogen Nuclear Magnetic Resonance. In J. W. Emsley, J. Feeney, and L. H. Sutcliffe, Eds., *Progress in Nuclear Magnetic Resonance Spectroscopy,* Vol. 6. Oxford: Pergamon Press, pp. 119–174.

Problems

7.1. Deduce the structure of the compound whose molecular formula is $C_5H_{10}N_2$ from the combined information from the ^{1}H, ^{13}C, DEPT, and ^{15}N NMR spectra. There is no proton-coupled ^{15}N spectrum provided because it provides no useful information.

7.2. The compound for this problem is a reagent commonly used in organic synthesis; its molecular formula is $C_6H_{15}SiCl$. Provided are the ^{1}H, ^{13}C, DEPT, and ^{29}Si (proton-decoupled) spectra. Determine the structure. Note: In the ^{1}H spectrum, the small, straddling doublets (satellites) result from ^{1}H, ^{29}Si couplings (see Sections 4.11.4 and 4.11.5).

7.3. Determine the structure of the phosphorus-containing compound whose molecular formula is $C_{19}H_{18}PBr$ from the ^{1}H, ^{13}C, DEPT, and ^{31}P. The ^{13}C/DEPT spectra show ^{31}P-^{13}C coupling.

PROBLEM 7.1. ^{1}H NMR.

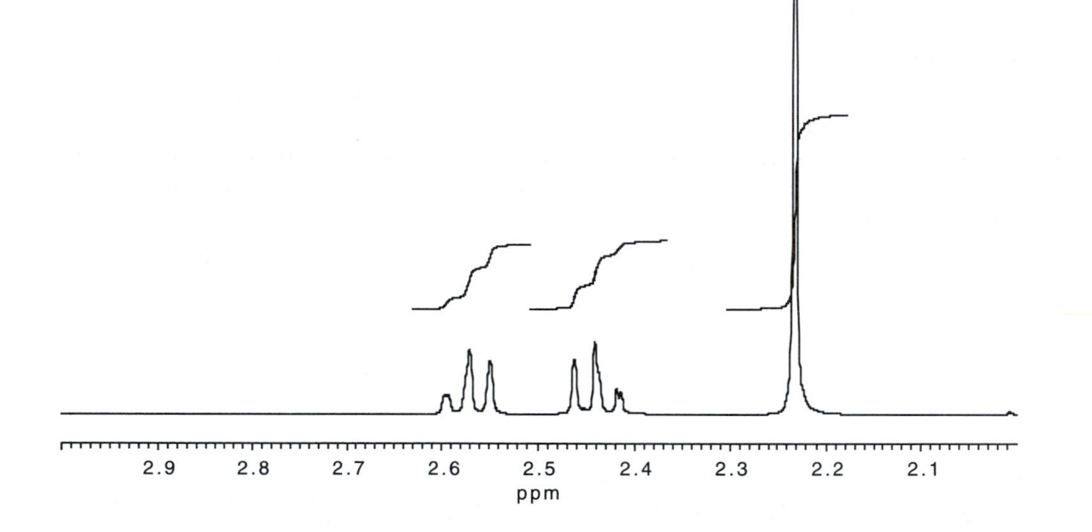

PROBLEM 7.1. ^{13}C/DEPT.

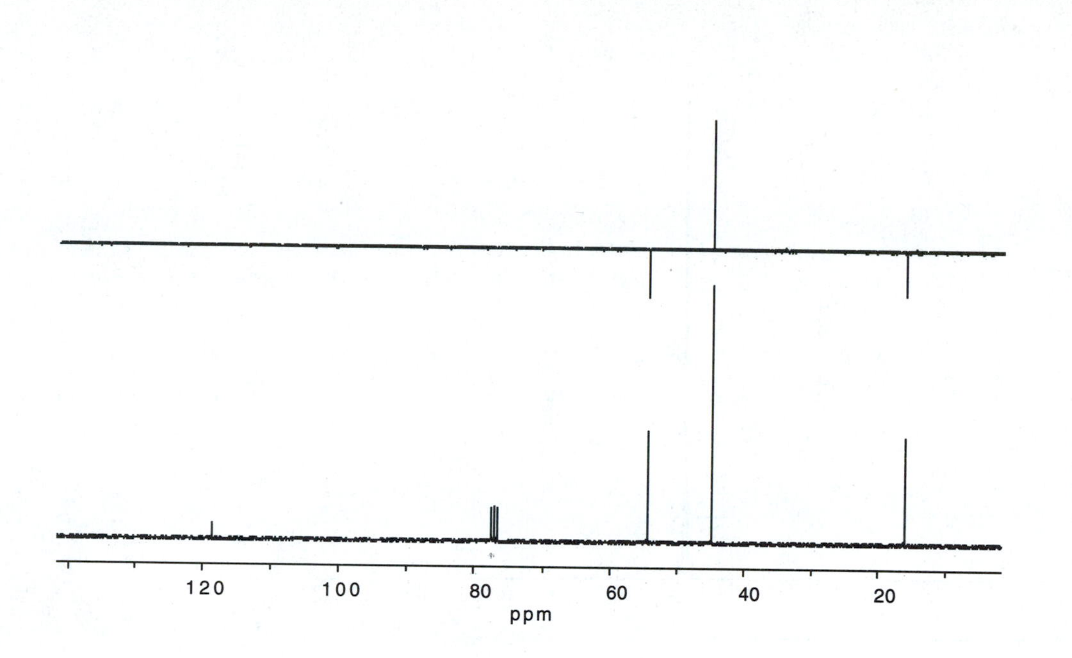

PROBLEM 7.1. ^{15}N NMR (proton decoupled).

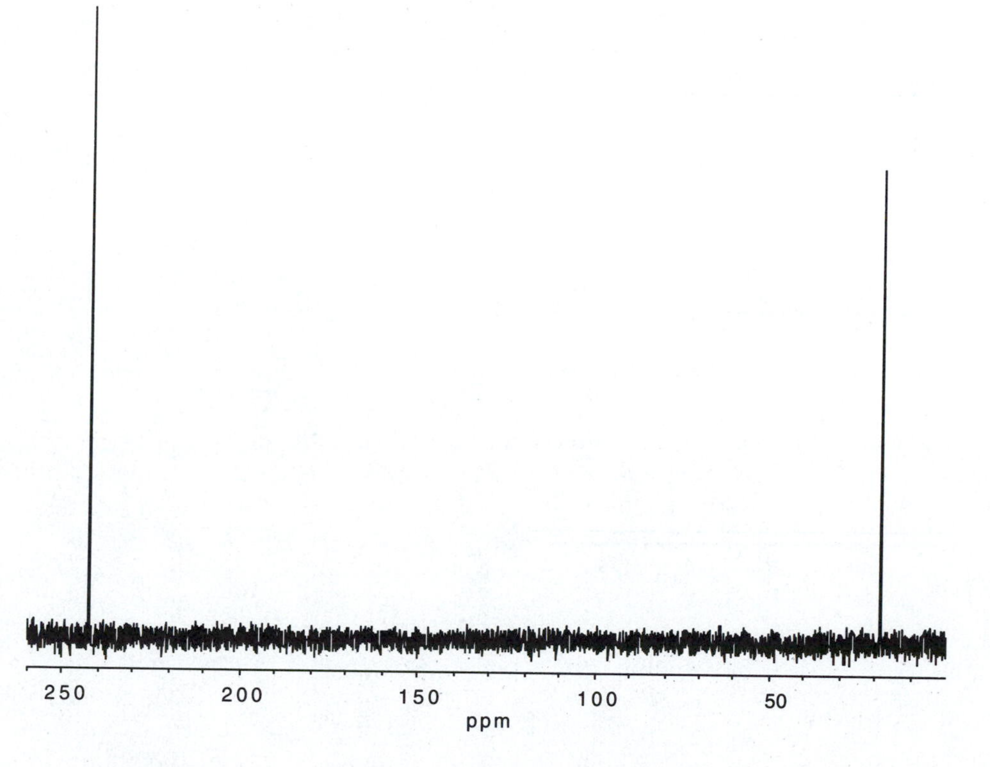

PROBLEM 7.2. ^{1}H NMR.

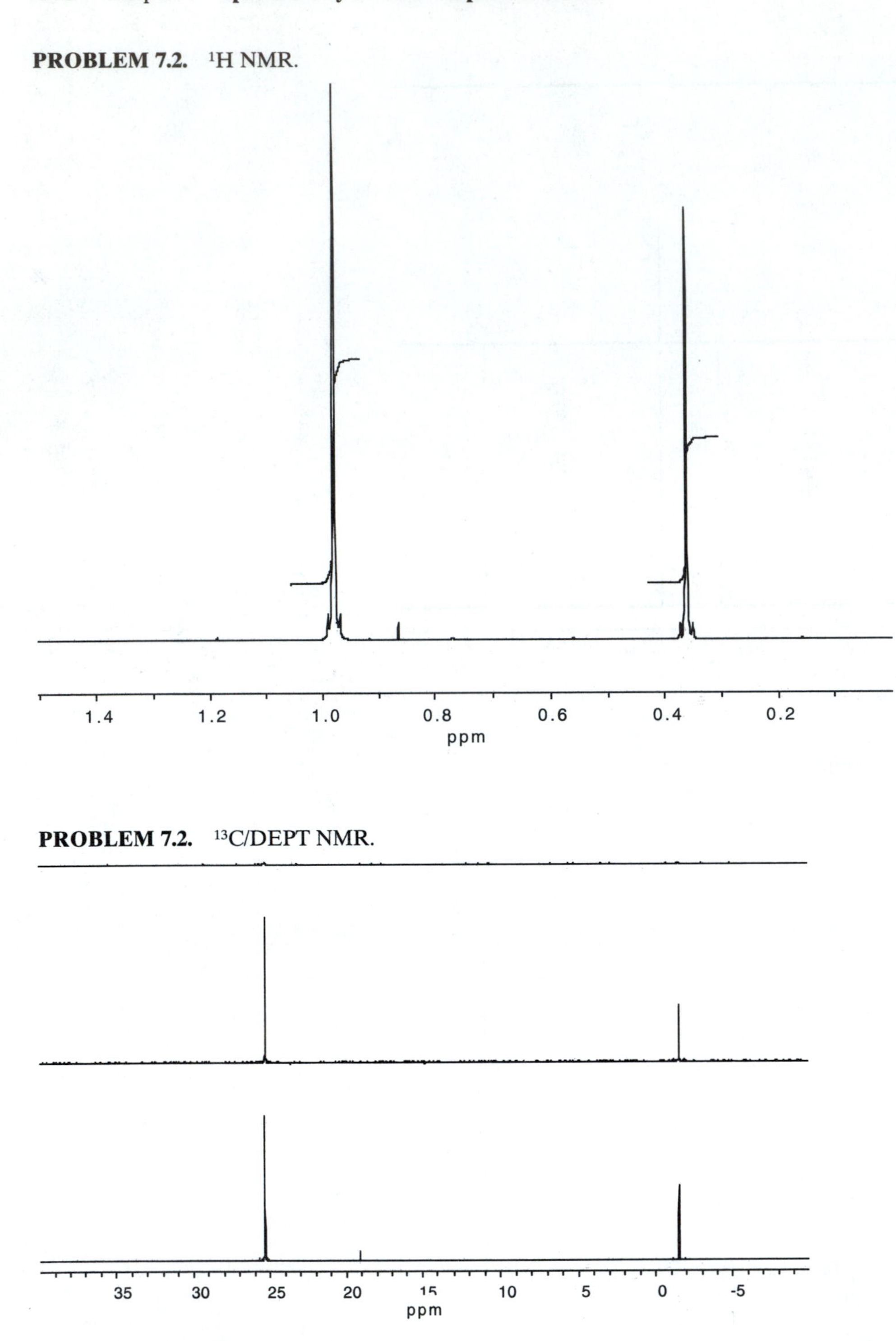

PROBLEM 7.2. ^{13}C/DEPT NMR.

PROBLEM 7.2. ^{27}Si NMR (proton decoupled).

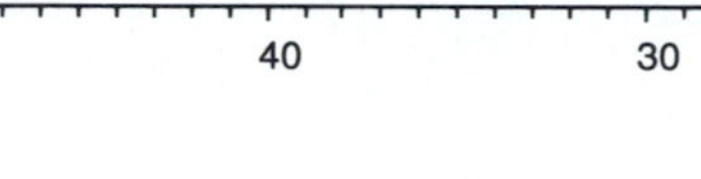

PROBLEM 7.3. ^{1}H NMR.

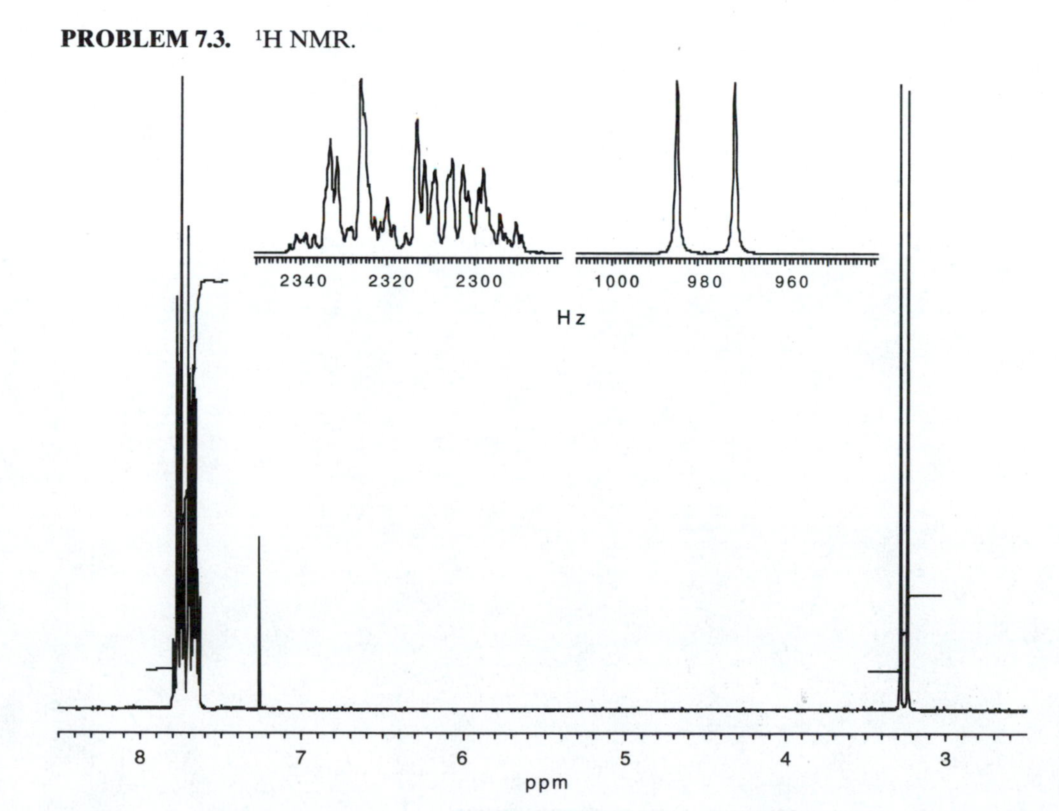

PROBLEM 7.3. ^{13}C/DEPT NMR.

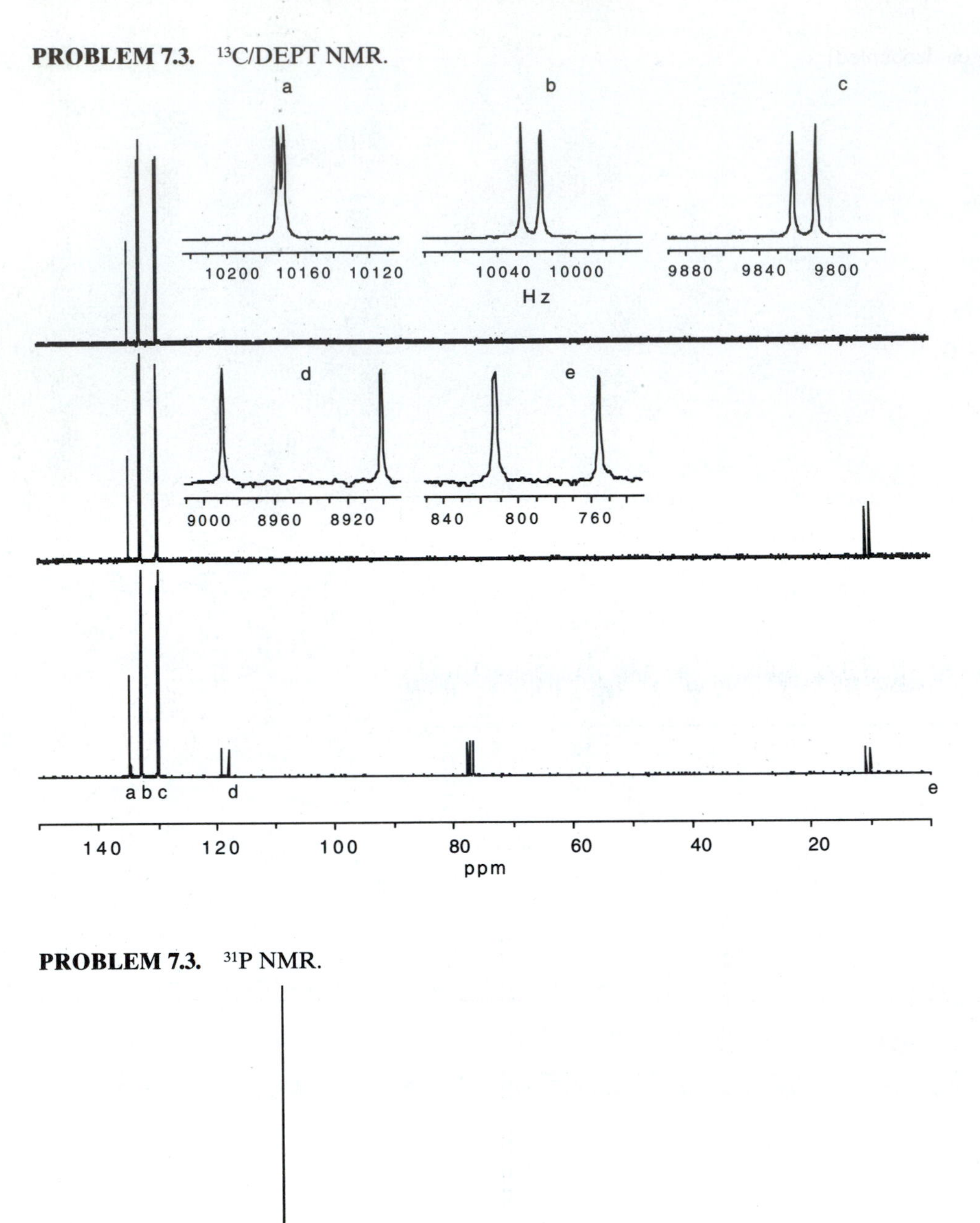

PROBLEM 7.3. ^{31}P NMR.

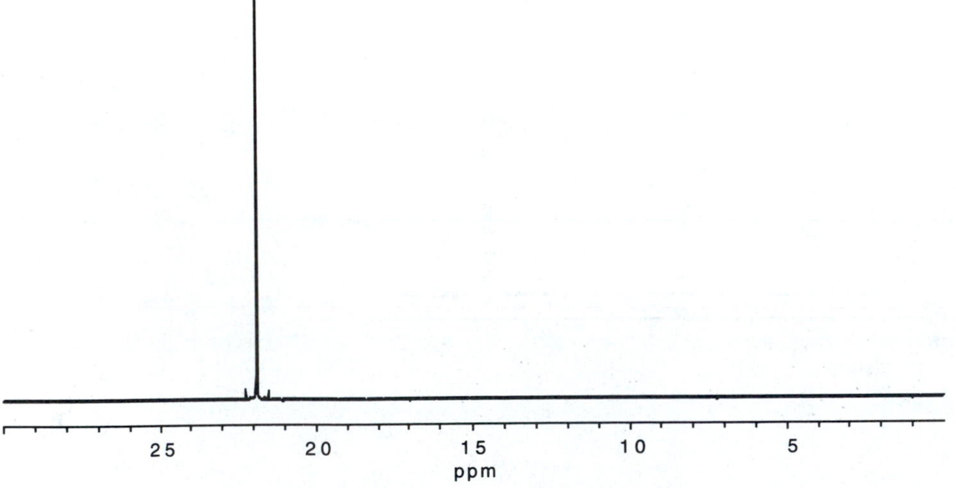

CHAPTER 8

A. *Introduction*

The perennial student question: Where do we start? The instructor will be sympathetic but not rigidly prescriptive. There are, however, guidelines that do start with the prescriptive statement: *Go for the molecular formula.* Why? Simply because it is the single most useful bit of information available to the chemist and is worth the effort sometimes necessary. It provides an overall impression of the molecule (i.e., the number and kinds of atoms), and it provides the *index of hydrogen deficiency*—in other words, the sum of the number of rings and of double and triple bonds (Section 2.6).

Development of the molecular formula starts with recognition of the molecular ion peak (Section 2.5). We assume the usual situation: High-resolution MS instrumentation is not readily available. Let us also assume for now that the peak of highest *m/z* (except for its isotope peaks) is the molecular ion peak and is intense enough so that the isotope peak intensities can be determined accurately, and the presence and number of S, Br, and Cl atoms can be ascertained. If the molecular ion peak is an odd number, an odd number of N atoms is present.

A search of the infrared spectrum for the familiar characteristic groups is now in order. Note in particular the unsaturated functional groups. Look also at the fragmentation pattern of the mass spectrum for recognizable fragments.

With this information in hand, search the proton NMR spectrum for confirmation and further leads. Select an entry point; perhaps a $CH_3C{=}O$ or an OCH_2CH_3 moiety as in Problem 8.1. If the spectrum is first order, or at least resembling one, determine the total proton count and ratios of groups of chemical shift–equivalent protons from the integration.

Look at the ^{13}C/DEPT spectra; determine the carbon and proton counts and the numbers of CH_3, CH_2, and CH groups. Overlap of proton absorptions is common, but absolute coincidence of nonequivalent ^{13}C peaks is quite rare with a high-resolution instrument.

Now, select the most likely molecular formula from Appendix A of Chapter 2, and determine the index of hydrogen deficiency. In addition to difficulties caused by unresolved or overlapping peaks, discrepancies may appear between the selected molecular formula and the ^1H and ^{13}C counts because of the presence of elements of symmetry. But this information also contributes to an understanding of the molecular structure (Section 5.2.3).

Difficulty often starts with uncertainty in the choice of a molecular ion peak. Many laboratories use chemical ionization as a routine supplement to electron impact, and of course, access to a high-resolution instrument is desirable for more difficult problems.

Students are urged to develop their own approaches. To provide practice in the use of the newer techniques, we have sometimes presented more information than needed, but other Problems should provide compensatory frustration to simulate the real world. Remember the overall strategy: Play the spectra against one another, focusing on the more obvious features. Develop a hypothesis from one spectrum; look to the other spectra for confirmation or denial; modify the hypothesis if necessary. *The effect is synergistic,* the total information being greater than the sum of the individual parts.

With the high resolution now available, many NMR spectra are first order, or nearly so, and can be interpreted by inspection with the leads furnished by the mass and infrared spectra. Nevertheless, a rereading of Sections 4.12 through 4.16 may engender caution.

As an example, consider two similar compounds:

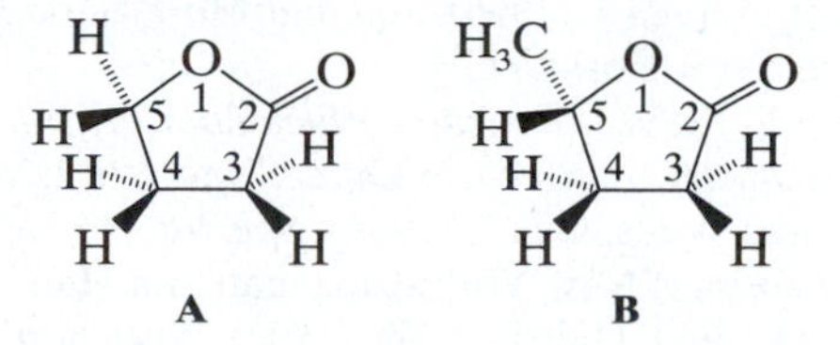

Both rings exist as rapidly flexing ring conformations, but only in Compound **A** do the protons of each

CH_2 group interchange to become chemical-shift equivalent (enantiotopes) (see Section 4.12.3.3). Only Compound **A** has a plane of symmetry in the plane of page through which the protons interchange.

From left to right in the spectrum, we predict for Compound **A**: H-5, a two-proton triplet; H-3, a two-proton triplet; H-4, a two-proton quintet (assuming nearly equal coupling constants). Given modest resolution, the spectrum is first order.

Compound **B** has no symmetry element in the planar conformation. C-5 is a chiral center, and the protons of each CH_2 group are diastereotopic pairs. Each proton of the pair has its own chemical shift. The H-4 protons adjacent to the chiral center are distinctly separated, but the H-3 protons are not, at 300 MHz. Each proton of a diastereotopic pair couples geminally with the other and independently (different coupling constants) with the vicinal protons to give complex multiplets.

The possibility of a chiral center should always be kept in mind (see Problem 8.3); *toujours la stéréochimie.* Compounds **A** and **B** are assigned Problems in Chapter 9.

The power of 2-D spectra will become more evident as we work through the problems in Chapters 8 and 9. It is often not necessary to examine all of the spectra in detail before proposing—tentatively—possible structures or fragments. Spectral features predicted for the postulated structures or fragments are compared with the observed spectra, and structural modifications are made to accommodate discrepancies.

These suggestions are illustrated by the following solved Problems presented in increasing order of difficulty. The assigned Problems of Chapter 9, again in increasing order of difficulty, will provide the essential practice.

Most students enjoy problem solving and rise to the challenge. They also begin to appreciate the elegance of chemical structure as they interpret spectra. Good sleuthing! Be wary of chirality, diastereotopes, virtual coupling, dihedral angles of about 90°, and magnetic nonequivalence.

Finally, what are the requirements for proof of structure? Ultimately, it is congruence of all available spectra with those of a pure, authentic sample obtained under the same conditions and on the same instruments.

Obviously, some compromises are acceptable. Congruence with published spectra or spectral data is considered acceptable for publication, but this cannot apply to a new compound, which must then be synthesized.

Computer programs for simulation of proton NMR spectra are available. If accurate measurements of chemical shifts and coupling constants for all of the protons can be obtained, the simulated spectrum will be congruent with the actual spectrum. In many cases, at least some of the spin systems will be first order. If not, reasonable estimates of shifts and coupling constants may be made, and the iterative computer program will adjust the values until the simulation matches the actual spectrum—assuming, of course, that the identification is valid.*

Unless otherwise labeled here and in Chapter 9, the NMR spectra were obtained at 300 MHz for protons and 75.5 MHz for ^{13}C; $CDCl_3$ was the solvent unless otherwise labeled. The IR spectra were obtained neat (i.e., no solvent) unless otherwise labeled. The mass spectra were obtained by GC/MS.

The COSY spectra are DQF-COSY spectra. The labeled frequency for all 2-D NMR spectra is that of the acquired signal (F2).

The following Problem sets are available for further practice.

References

Atta-ur-Rahman, and Choudhary, M.I. (1996). *Solving Problems with NMR.* New York: Academic Press.

Bates, R.B., and Beavers, W.A. (1981). *Carbon-13 NMR Spectral Problems.* Clifton, NJ: Humana Press.

Braun S., et al. (eds.). (1996). *100 and More Basic NMR Experiments.* New York: VCH

Breitmaier, E. (1993). *Structure Elucidation by NMR in Organic Chemistry. A Practical Guide.* New York: Wiley.

Davis, R., and Wells, C.H.J. (1984). *Spectral Problems in Organic Chemistry.* New York: Chapman and Hall.

Duddeck, H., and Dietrich, W. (1992). *Structural Elucidation by Modern NMR. A Workbook,* 2nd ed. New York: Springer-Verlag.

Field, L.D., Sternhell, S., and Kalman, J.R. (1995). *Organic Structures from Spectra.* New York: Wiley.

Fuchs, P.L., and Bunnell, C.A. (1979). *Carbon-13 NMR Based Organic Spectral Problems.* New York: Wiley.

Sanders, J.K.M., Constable, E.C., and Hunter, B.K. (1989). *Modern NMR Spectroscopy; A Workbook of Chemical Problems.* Oxford: Oxford University Press.

* Spectra can be simulated on the computer of a modern NMR spectrometer or on a PC. For example, see the Win-Daisy program, available from Bruker Instruments Incorp., Billerica, Mass.

CHAPTER 8

B. Solved Problems

Compound 8.1

We start by gathering information in order to establish a molecular formula. We assume that the weak peak at *m/z* 144 is the molecular ion peak. It is so small that the intensities of its isotope peaks cannot be accurately measured. Since *m/z* 144 is an even number, there are 0, 2, 4 . . . N atoms present. To begin, we tentatively assume that are no N, S, or halogen atoms present; this posture, of course, is quite shaky.

From left to right, the proton integrator in the ^1H NMR spectrum reads: 2, 2, 2, 3, 3—calibrated against the presumed methyl singlet at δ 2.17. From high to low frequency the ^{13}C and DEPT spectra read: C, C, CH_2, CH_2, CH_3, CH_2, CH_3. Thus, there are 12 protons and 7 carbon atoms in the molecular formula. Note that the DEPT CH subspectrum is omitted since there are no CH groups. The most likely molecular formula under unit mass 144 is $C_7H_{12}O_3$ (Chapter 2, Appendix A). The index of hydrogen deficiency is 2, and this should be immediately explored.

The IR spectrum shows a strong, broad C=O peak at about ~1725 cm^{-1}, which accounts for one unsaturated site and for one O atom. The ^{13}C spectrum shows a ketone C=O group at ~δ 208, and an ester C=O group at ~δ 172.5; the latter assignment is reinforced by typical ethyl ester peaks in the IR spectrum at ~1160 cm^{-1} and ~1030 cm^{-1}. The broad peak at ~1725 cm^{-1} must represent both C=O groups. The three O atoms in the molecular formula are accounted for.

With this information in hand, interpretation of the ^1H spectrum is straightforward. The methyl singlet mentioned above must be attached to the ketone C=O group to give us one end of the molecule, $CH_3-\underset{\overset{\|}{O}}{C}-$, which also accounts for the base peak in the mass spectrum at *m/z* 43. The three-proton triplet and the strongly deshielded, two-proton quartet account for the $-\underset{\overset{\|}{O}}{C}-O-CH_2-CH_3$ moiety at the other end of the molecule. Confirmation is provided by the strong peak at *m/z* 99 (characteristic loss of $O-CH_2-CH_3$). The NMR spin systems are A_3, A_2X_2, and A_3X_2.

Filling in between the two ends of the molecule requires little imagination. All that remain in the ^1H spectrum are two two-proton triplets—surely two adjacent CH_2 groups. Hence:

$$CH_3-\underset{\overset{\|}{O}}{C}-CH_2-CH_2-\underset{\overset{\|}{O}}{C}-O-CH_2-CH_3$$

Ethyl levulinate,
Ethyl 4-oxopentanoate

Let us return for a moment to the mass spectrum: Note that the loss of 15 units (loss of CH_3) to give a moderate peak of *m/z* 129 provides confirmation that the weak peak at *m/z* 144 is indeed the molecular ion peak. Loss of 45 units to give the strong peak at *m/z* 99 provides further confirmation.

Assignment of the shifts of the CH_2 groups adjacent to the C=O groups is ambiguous. Assignment can be achieved by obtaining an HMBC spectrum (Chapter 6) which would show correlation (long-range coupling) between the groups adjacent to the ketone C=O group (Chapter 6).

For further discussion, consider and reject the following isomers of ethyl levulinate:

$$CH_3-CH_2-\underset{\overset{\|}{O}}{C}-CH_2-\underset{\overset{\|}{O}}{C}-O-CH_2-CH_3$$

$$H\underset{\overset{\|}{O}}{C}-CH_2-CH_2-CH_2-\underset{\overset{\|}{O}}{C}-O-CH_2-CH_3$$

$$CH_3-CH_2-CH_2-\underset{\overset{\|}{O}}{C}-\underset{\overset{\|}{O}}{C}-O-CH_2-CH_3$$

$$CH_3-CH_2-\underset{\overset{\|}{O}}{C}-CH_2-CH_2-\underset{\overset{\|}{O}}{C}-O-CH_3$$

$$CH_3-\underset{\overset{\|}{O}}{C}-CH_2-CH_2-O-\underset{\overset{\|}{O}}{C}-CH_2-CH_3$$

H, CH_2-CH_2
$CH_3-CH_2-O-H_2C$—C (ring: C–CH_2–CH_2–C(=O)–O)

Problem 8.1

INFRARED

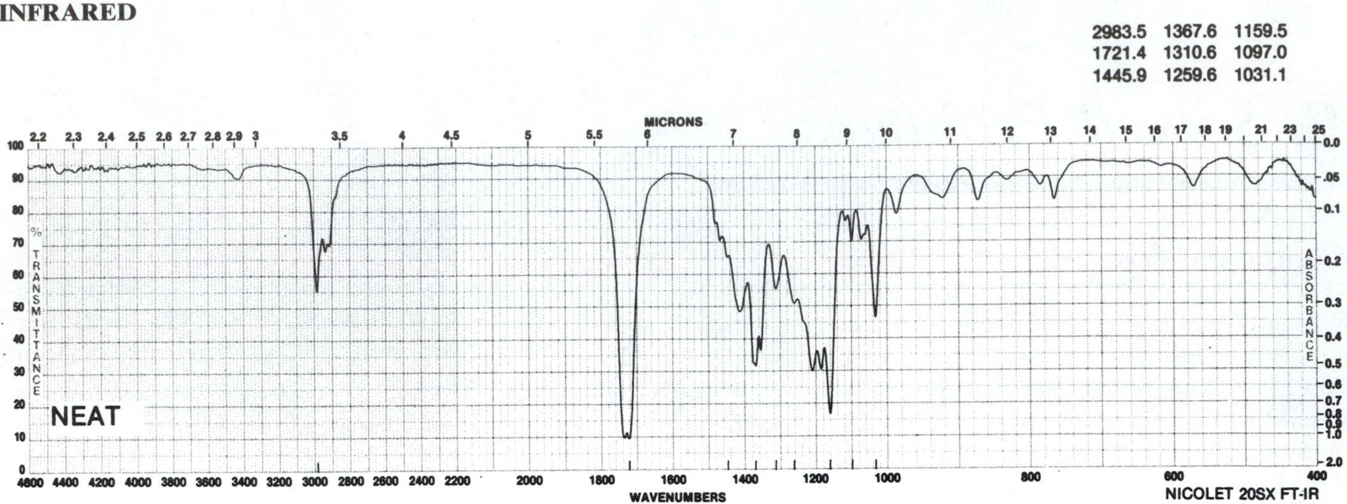

MASS

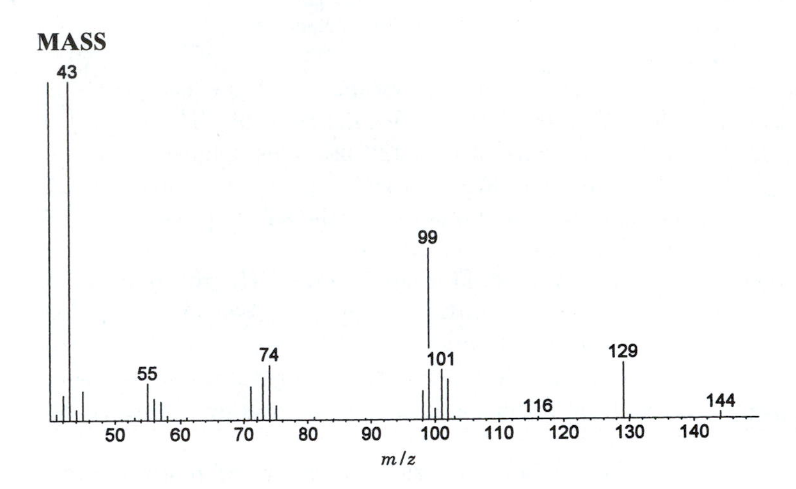

^{1}H NMR

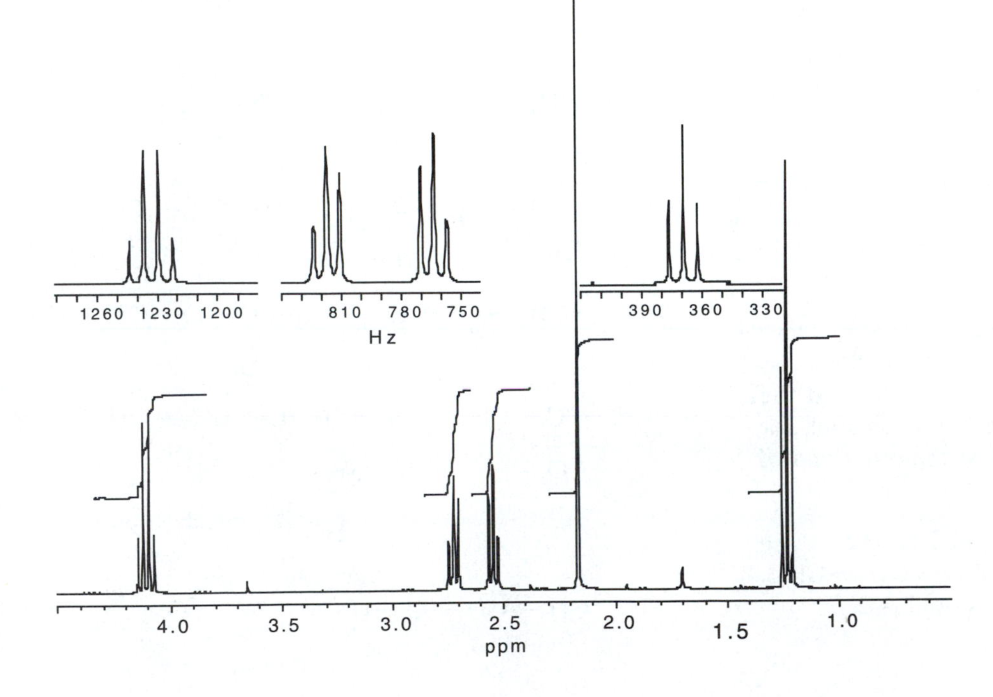

(Continued)

(Continued)
^{13}C/DEPT

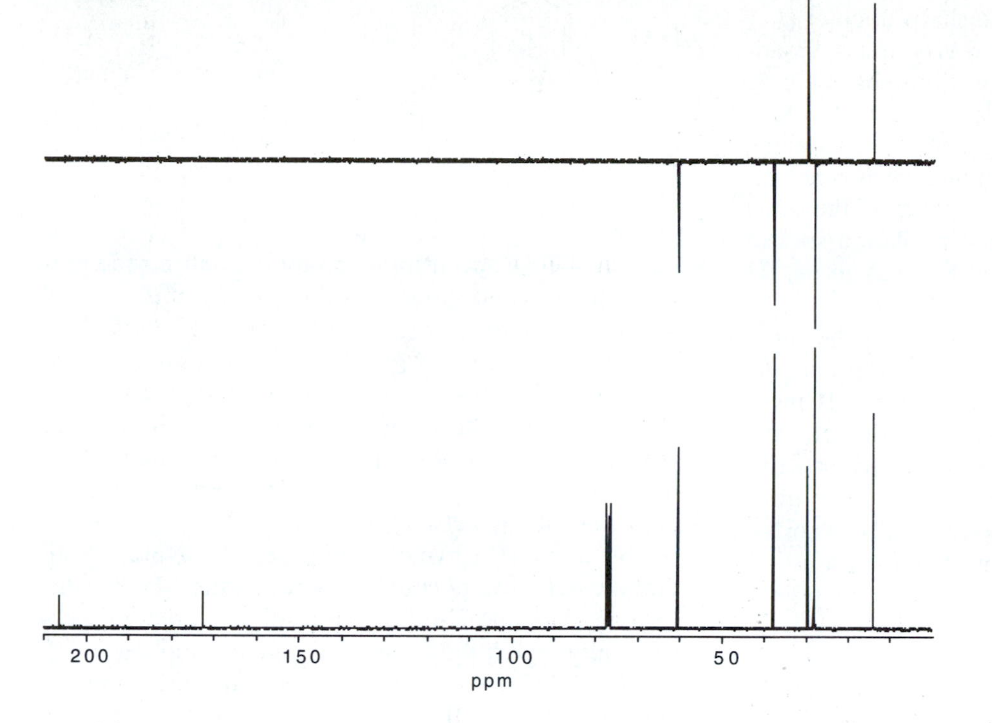

Problem 8.2

The molecular ion is certainly the medium-intensity peak in the mass spectrum at m/z 150; there is a rational loss of a CH_3 group to give the base peak at m/z 135. The isotope peaks for the molecular ion do not permit the presence of S, Cl, or Br. Let us assume, tentatively, that the even-numbered molecular ion peak indicates the absence of N.

The IR spectrum is notable for the intense OH peak at 3400 cm^{-1}. The immediate question is the presence or absence of aromaticity. If an aromatic ring is present, is it attached directly to the OH group to give a phenol? The ^{1}H and ^{13}C spectra provide answers—the answers being yes to both questions. But first, the molecular formula.

There are seven different kinds of protons in the ^{1}H spectrum in the ratios, from left to right, of 1, 1, 1, 1, 1, 3, 6. Hence a total of 14 protons. The six-proton doublet must represent two equivalent CH_3 groups of an isopropyl moiety. The ^{13}C spectrum at 75.5 MHz apparently shows nine peaks, but one of them is suspiciously intense and surely must be correlated with the six-proton doublet; these two superposed CH_3 groups make the total count of 10 carbon atoms. The ^{13}C/DEPT spectra from left to right, specify C, C, C, CH, CH, CH, CH, CH_3 (2), CH_3. Under unit mass 150, the most reasonable molecular formula is $C_{10}H_{14}O$ (Chapter 2, Appendix A) with an index of hydrogen deficiency of four, which fully accounts for a benzene ring: i.e., three double bonds and one ring.

Back to the question of attachment of the OH group: Is it directly attached to the ring—i.e., a phenolic group? Very likely, judging from the shift of the OH singlet at δ 4.70 in the ^{1}H spectrum. Chart E.1 at the end of Chapter 4 gives the absorption ranges as ~δ 7.5 – 4.0 for phenols and ~δ 4.0–δ 0.5 for alcohols. At the high dilutions used for spectra obtained on a modern instrument, these absorption peaks would be at the right-hand end of each range. Thus a phenolic OH assignment is reasonable. Further evidence comes from the most deshielded ^{13}C peak, which can only be explained by direct attachment of the OH group (see Table 5.9).

What else is attached to the ring? Well, we are looking at three substituted aromatic carbon atoms (weak deshielded peaks) in the ^{13}C spectrum of the ring, to which three substituents, including the OH group, are attached. We are left with a CH group and three CH_3 groups to arrange as two of the substituents.

Obviously the deshielded CH_3 singlet at δ 2.27 represents one of the substituents. The six-proton doublet at lowest frequency in the ^{1}H spectrum and the deshielded one-proton septet spell out an isopropyl group. Thus we have the following pieces on an aromatic ring: a hydroxylic proton, a CH_3 group, and an isopropyl group.

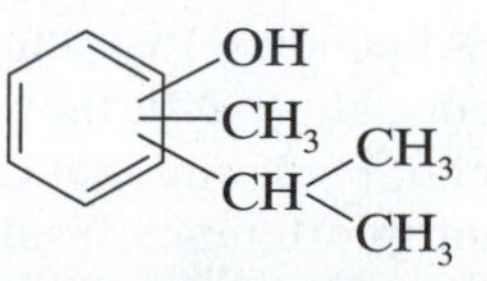

And now we face the question of distribution, which

entails justifying the chemical shifts and coupling constants of the three aromatic protons. In the proton spectrum from left to right, there are a sharp doublet (J = 8 Hz), a broadened doublet (J = 8 Hz) and a broadened singlet—each of these absorptions representing one proton. Interpretation is simple: respectively, ortho coupling, ortho and meta coupling, and meta coupling. The meta couplings cause broadening, which is not resolvable because of long-range coupling to the alkyl substituents, as will be seen in the correlation spectra and in broadening of the benzylic CH_3 peak in the ^1H spectrum.

The distinctly spaced chemical shifts of the three aromatic protons are a result of the OH substituent, the two aliphatic substituent having little effect. If there were two protons ortho to the OH substituent, they would have very similar chemical shifts. Obviously this is not so. As a starting point, therefore, we can draw a structure that has one of the alkyl substituents ortho to the OH substituent (A and B are the alkyl substituents):

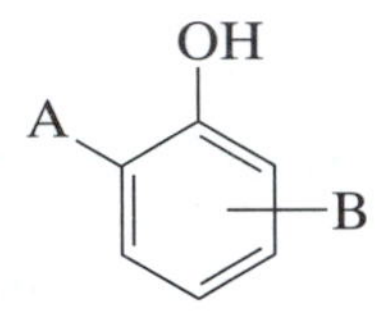

Two of the aromatic protons must be adjacent to each other (ortho coupling); the remaining proton must lie between two substituents (no ortho coupling): There are two possibilities: The alkyl groups either are para to each other or meta to each other.

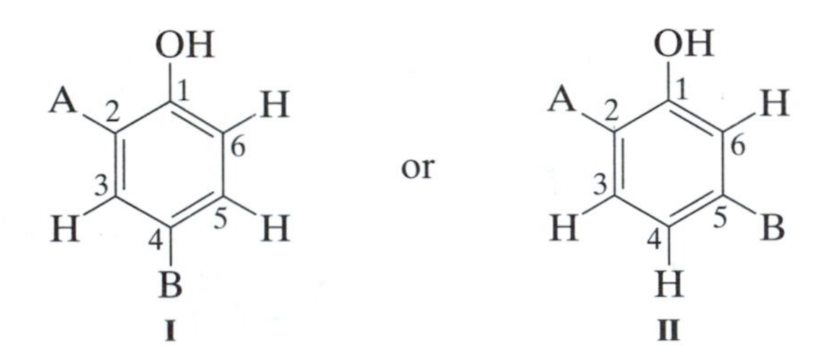

Since we cannot distinguish between the attachment points for A and B, they can be interchanged to give two more possibilities.

Before turning to the 2-D spectra, we can use Chart D.1 (Chapter 4 Appendices) to tentatively assign the ring protons: From right to left, they are H-6, H-4, and H-3—that is, ortho, para, and meta to the OH group. Barring significant interference from the alkyl groups, the para disposition (**II**) of the alkyl groups is favored. The possibilities narrow to **III** and **IV**.

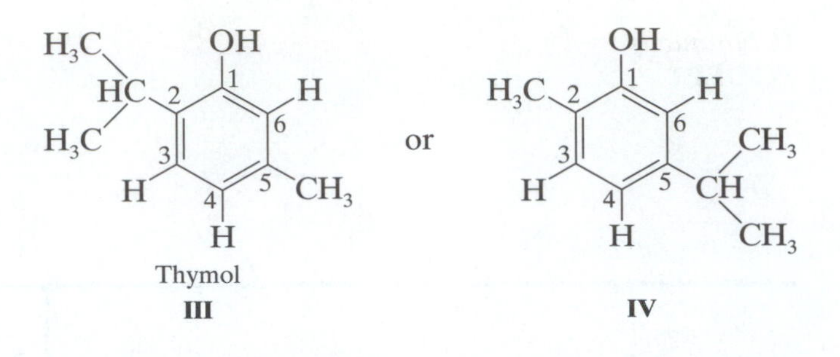

The COSY spectrum (Chapter 6) confirms the previous findings and shows that the *protons of the methyl substituent are long-range coupled* (${}^4J$) *to H-4 and H-6.* Interestingly, the isopropyl CH proton does not show ${}^4J$ coupling to H-3 because of the high multiplicity of the CH absorption, which would produce a very diffuse (not visible) cross peak. As expected, the aromatic protons show meta coupling (${}^4J$) between H-6 and H-4, and ortho (${}^3J$) coupling between H-4 and H-3.

Structure **III** (thymol) is now heavily favored. Note that the definitive ${}^4J$ coupling between the CH_3 substituent and H-4 and H-6 was not resolved in the ^1H spectrum, although it did broaden the meta coupling.

The HMQC spectrum is a proton-detected (inverse detection) HETCOR (Chapter 6); it shows ${}^1J$ CH coupling. Table 5.9 in Chapter 5 allows us to arrange the aromatic unsubstituted carbon atoms as C-6, C-4, C-3 from left to right. The HMQC spectrum confirms the same sequence for H-6, H-4, H-3. The aromatic, unsubstituted carbon atoms can now be correlated with the firmly assigned aliphatic protons. The substituted aromatic carbon atoms cannot yet be assigned.

The HMBC spectrum (Chapter 6) is in effect a long-range (${}^2J_{CH}$ and ${}^3J_{CH}$) HETCOR, but with proton detection for increased sensitivity (Chapter 6). Thus, it affords correlation for quaternary carbon atoms—which of course have no ${}^1J$ correlation:

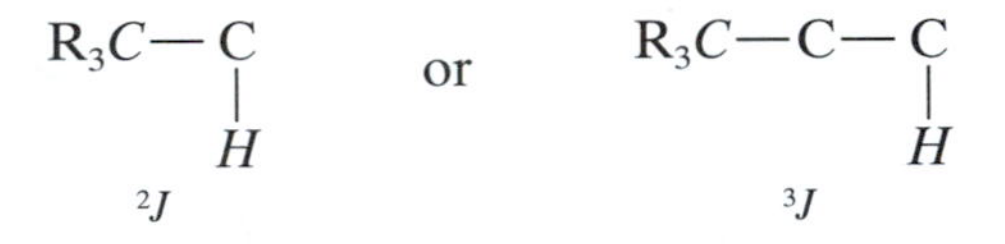

The HMBC spectrum also permits correlation between coupling systems—i.e., bridging such "insulating" atoms as O, S, N, and quaternary carbon atoms.

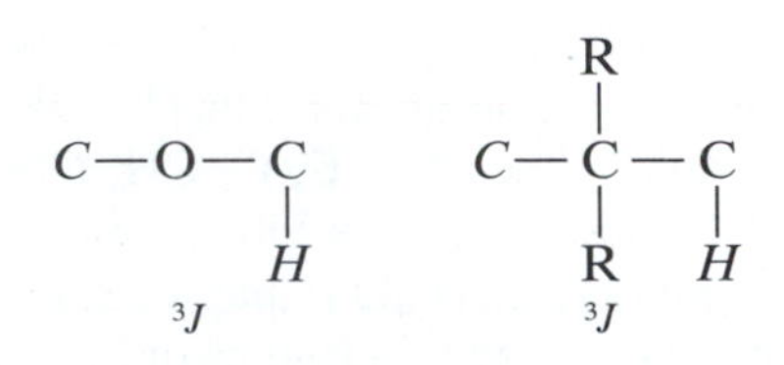

Even in a molecule of modest size, the number of ${}^2J$ and ${}^3J$ couplings can be daunting. Where to start?

Well, simply pose an important question: How do we fully confirm the positions of alkyl substituents? The COSY spectrum did detect the $^4J_{HH}$ coupling for the methyl substituent but not for the isopropyl substituent. But look down from the CH isopropyl septet in the HMBC spectrum and observe three cross peaks that correlate this CH proton with C-2 (^2J), C-3 (^3J) and C-1 (^3J) in the thymol structure. Certainly convincing. As overkill, note that in the HMBC spectrum, the protons of the methyl substituent correlate with C-6 (^3J), C-4 (^3J) and C-5 (^2J). Further note that the six methyl protons of the isopropyl group correlate with C-7 (^2J) and with C-2 (^3J).

The utility of HMBC in correlating quaternary carbon atoms with assigned protons, can be shown by working out the correlations of C-1, C-5, and C-2. The assignment earlier of C-1 on the basis of its shift is sound, but the assignment of C-5 and C-2 on the basis of shift alone should be affirmed by correlations. This exercise is left to the student.

Bridging across quaternary carbon atoms has been demonstrated in the course of the above correlations.

Two final points: (1) There are four contours, designated by arrows, that represent ^1J couplings (large) that have not been completely suppressed. These CH doublets are obvious since they straddle the proton peaks. They can be ignored. (2) The correlations of the OH proton with C-6, C-2, and C-1 should be noted.

Problem 8.2

INFRARED

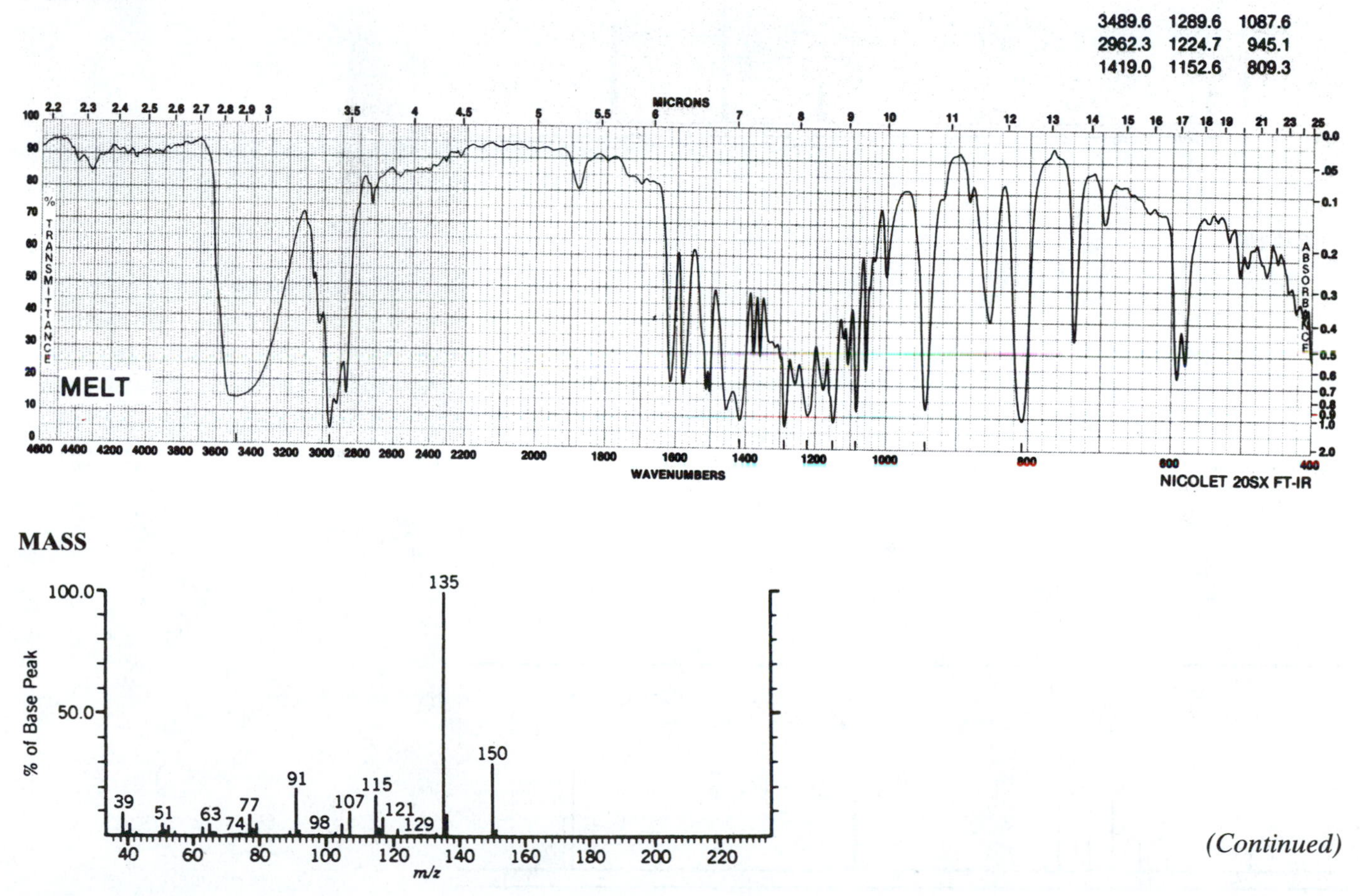

(Continued)

(Continued)
¹H NMR

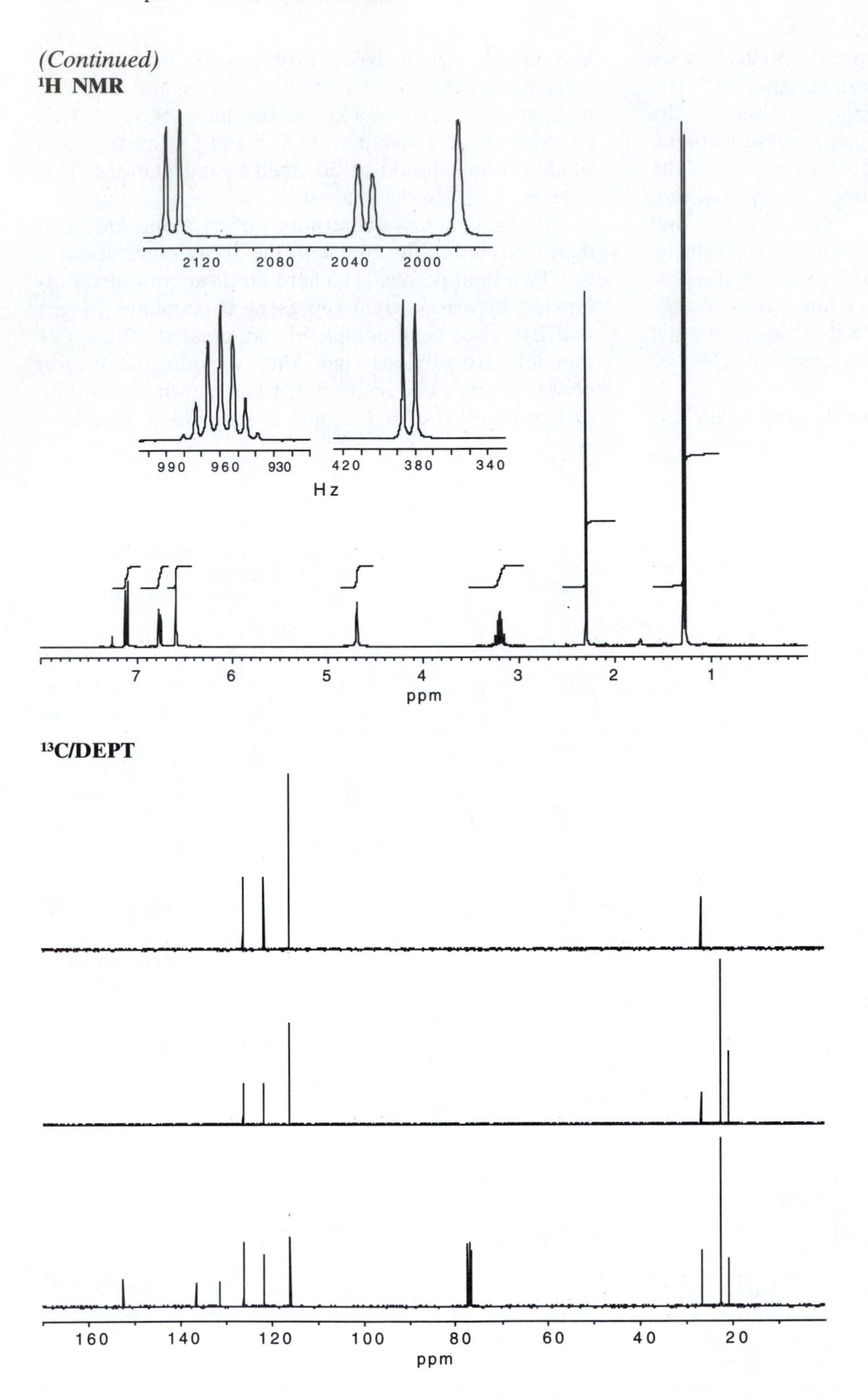

(Continued)

(Continued)

COSY

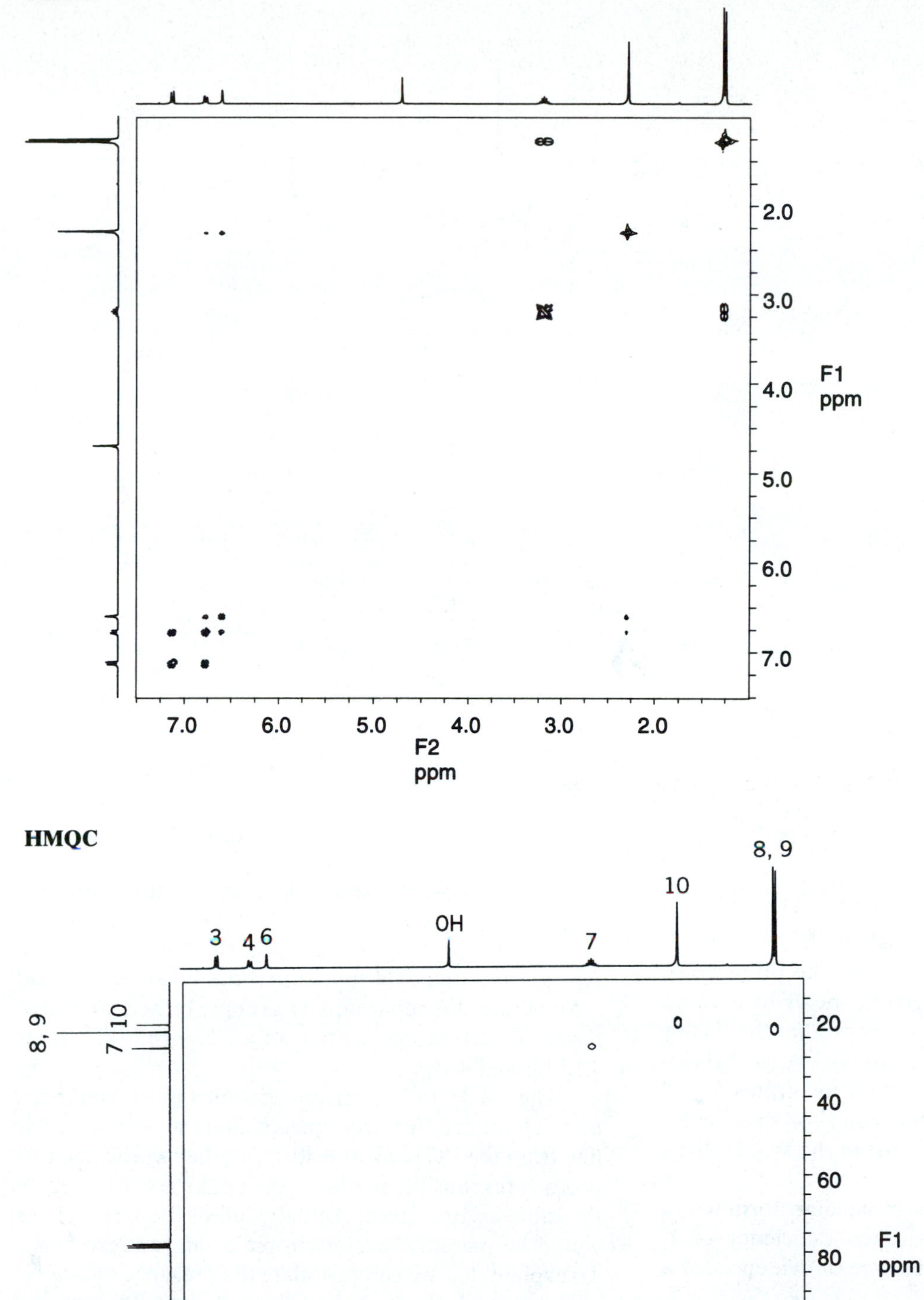

HMQC

(Continued)

(Continued)
HMBC

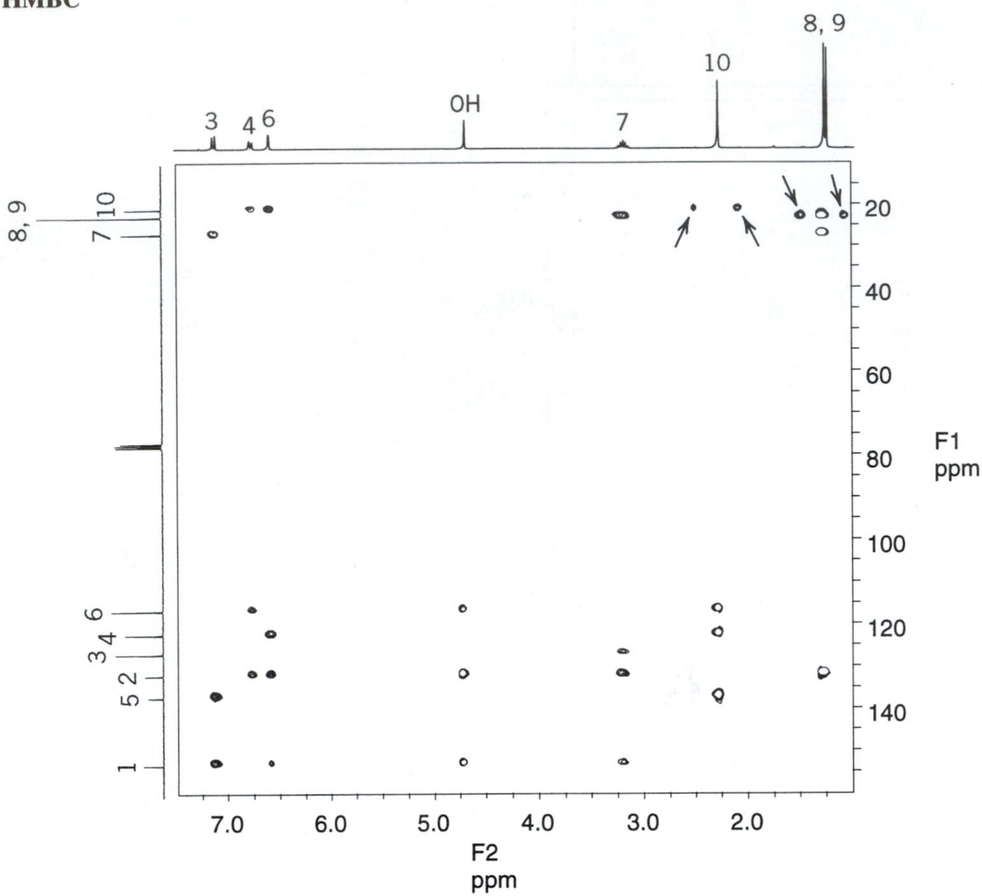

Problem 8.3

The molecular ion peak is 148, and the M + 2 peak is 9.7% of the intensity of M, suggesting the presence of two sulfur atoms. The IR spectrum shows little more than saturated hydrocarbon features; neither the ^{1}H nor the ^{13}C spectrum shows evidence for *sp*- or *sp*^{2}-hybridized carbon atoms. The ^{1}H spectrum integrates for 12 hydrogen atoms and the ^{13}C spectrum shows six peaks. From left to right, the DEPT spectrum shows CH, CH_2, CH_2, CH_2, CH_2, CH_3 groups.

We confidently suggest a molecular formula of $C_6H_{12}S_2$ with an index of hydrogen deficiency of 1, which demands a ring in the absence of evidence for a double bond.

What are the entry points to interpretation of the ^{1}H spectrum? Certainly the classical three-proton triplet at the right-hand end of the spectrum spells out $—CH_2—CH_3$. The CH group at δ 3.57 may be deshielded by one (or both) of the S atoms: hence $—S—\overset{|}{C}H—$.

However, we must cope with an apparent discrepancy: The DEPT spectrum shows one CH, four CH_2, and one CH_3 groups; whereas the integrated proton spectrum apparently shows three CH, three CH_2, and one CH_3 groups. But the proton spectrum is a bit misleading; there are really one CH, four CH_2 groups, and one CH_3 group, in agreement with the DEPT spectrum. Absorptions in the ^{1}H spectrum centered at $\sim\delta$ 2.45 and $\sim\delta$ 1.90 are not separate CH groups; rather, they represent diastereotopic protons of a CH_2 group (Sections 4.12.1.4 and 4.16).

The HETCOR spectrum provides quick confirmation: There are two cross peaks along a perpendicular line from the ^{13}C peak at δ 40.0. In other words, the two protons responsible for the cross peaks are attached to the same carbon atom. Actually, all of the CH_2 groups would be patently diastereotopic at higher resolution. To explain this, we can postulate the presence of a chiral center for which there is an excellent candidate: the strongly deshielded CH group mentioned above. To establish all of the connectivities in the molecule and to show that the CH group has four different ligands, we turn to the COSY spectrum, and summarize the results in the following connectivity diagram.

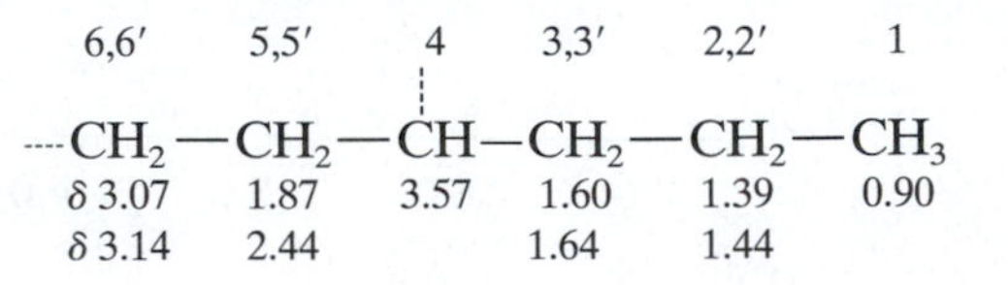

The numbers at the top of the diagram represent the proton assignments in the molecular formula; the bottom numbers are the chemical shifts, which were obtained by simulation; these shifts can be roughly estimated by inspection. Note the individual chemical shift for each proton of the CH_2 groups.

With the proton connectivities and chemical shifts in hand, the HETCOR spectrum establishes the $^1H—^{13}C$ correlations. This is left as an exercise. It remains only to insert the two sulfur atoms in the above diagram to give 3-propyl-1,2-dithiolane.

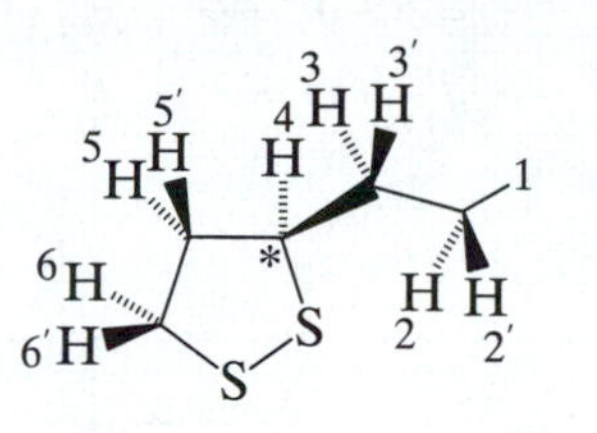

Note that the ring CH_2 protons adjacent to the chiral center show the largest difference in chemical shifts. Each of the other CH_2 groups consists of diastereotopic protons that overlap at the available resolution.

Although the fragmentation pattern of the mass spectrum was not initially very helpful, it is possible, with hindsight, to rationalize the major peaks. The molecular ion fragments by two pathways: loss of SH to give *m/z* 115 and by cleavage at the ring junction with loss of the $CH_2CH_2CH_3$ radical to give *m/z* 105. The base peak *m/z* 55 represents a hydrocarbon fragment C_4H_7.

In review, there were several critical observations. First was the detection of two S atoms by the intensity of the M + 2 peak in the mass spectrum. The ring structure was then based on the index of hydrogen deficiency. Next was the apparent discrepancy between the ^{13}C/DEPT and the ^1H spectrum, which was explained by the HETCOR. Finally was the use of the COSY spectrum to establish all of the connectivities.

Problem 8.3

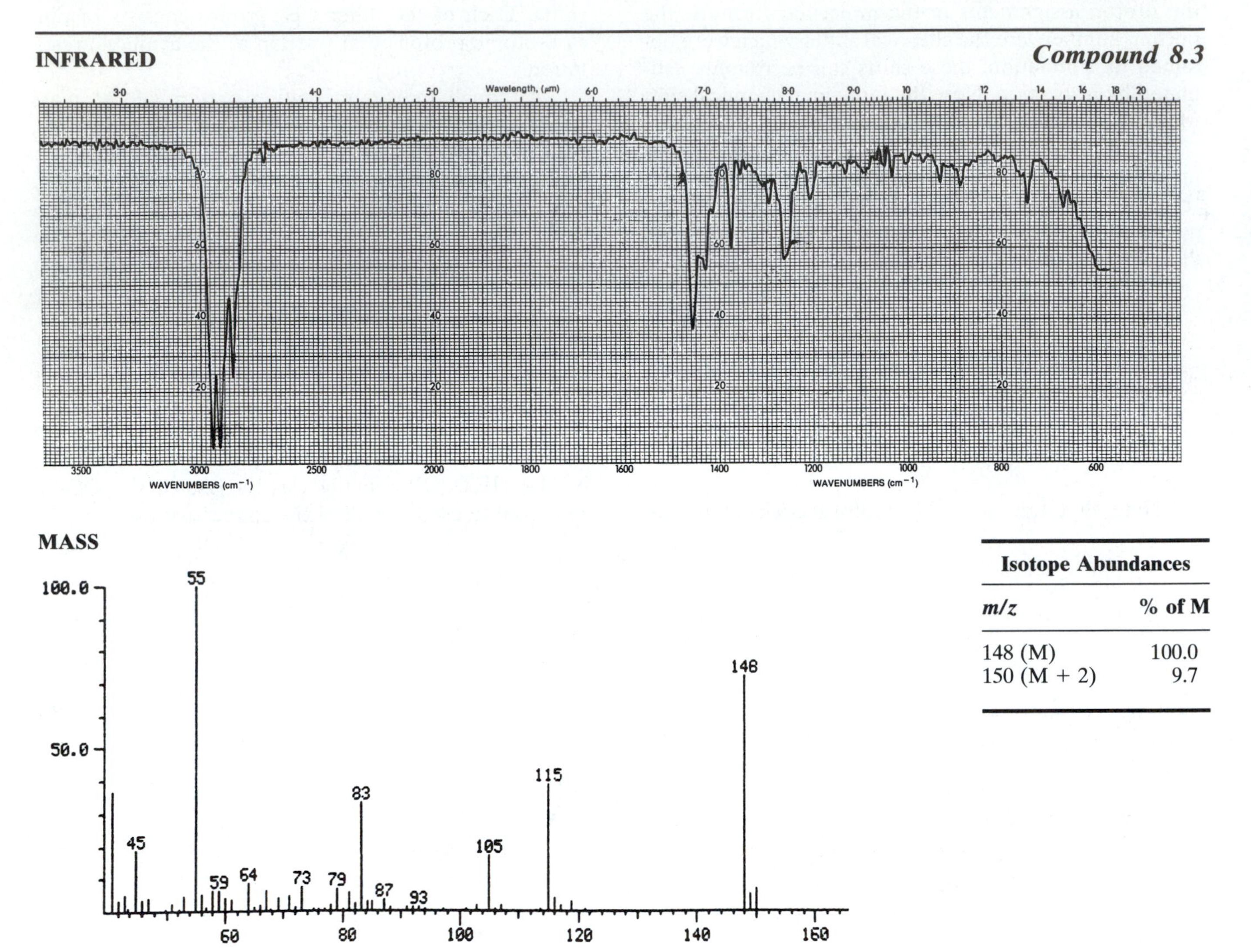

Isotope Abundances	
m/z	% of M
148 (M)	100.0
150 (M + 2)	9.7

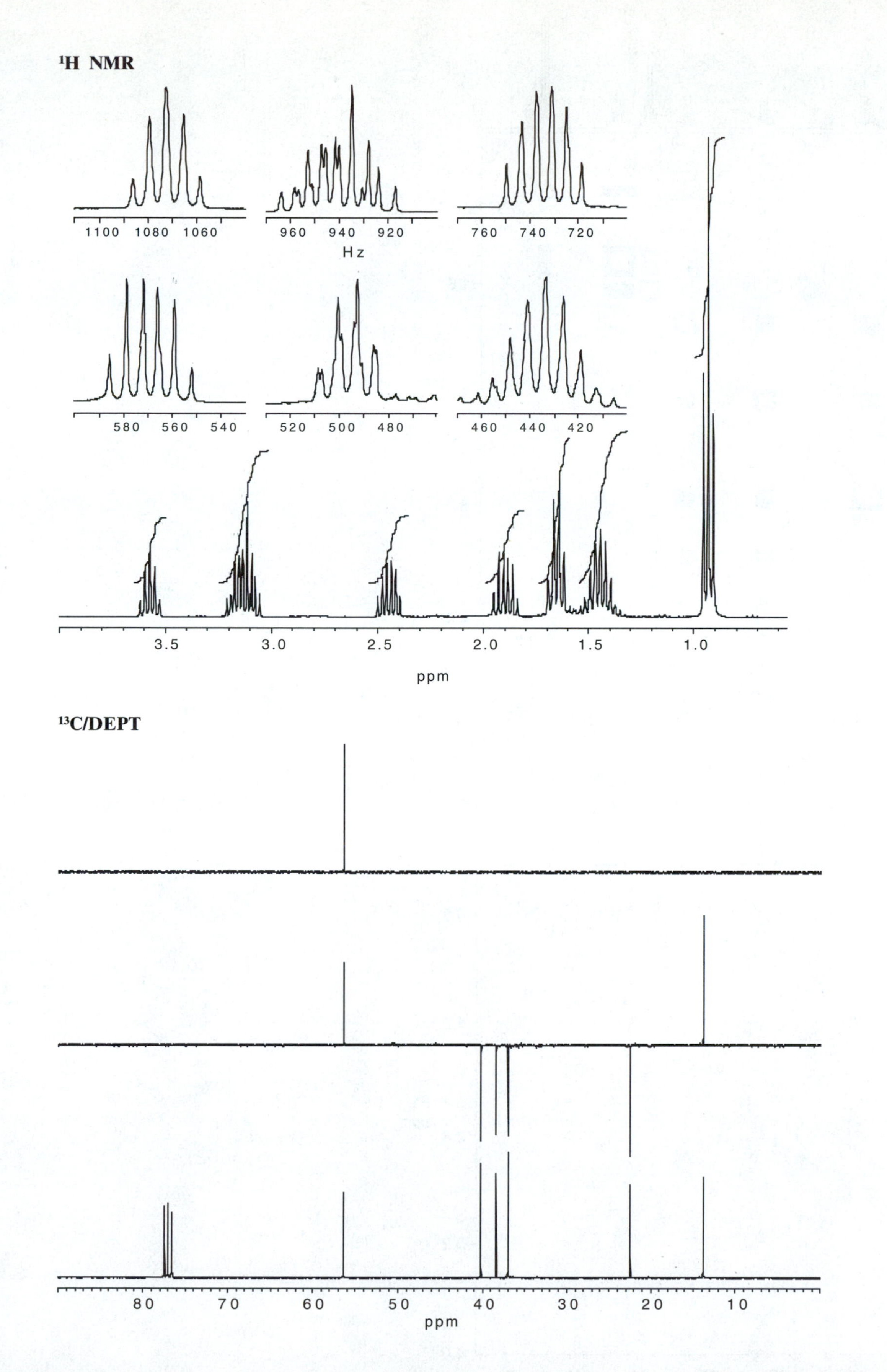

^{1}H NMR
1100 1080 1060
960 940 920
760 740 720
Hz
580 560 540
520 500 480
460 440 420
3.5 3.0 2.5 2.0 1.5 1.0
ppm
^{13}C/DEPT
80 70 60 50 40 30 20 10
ppm

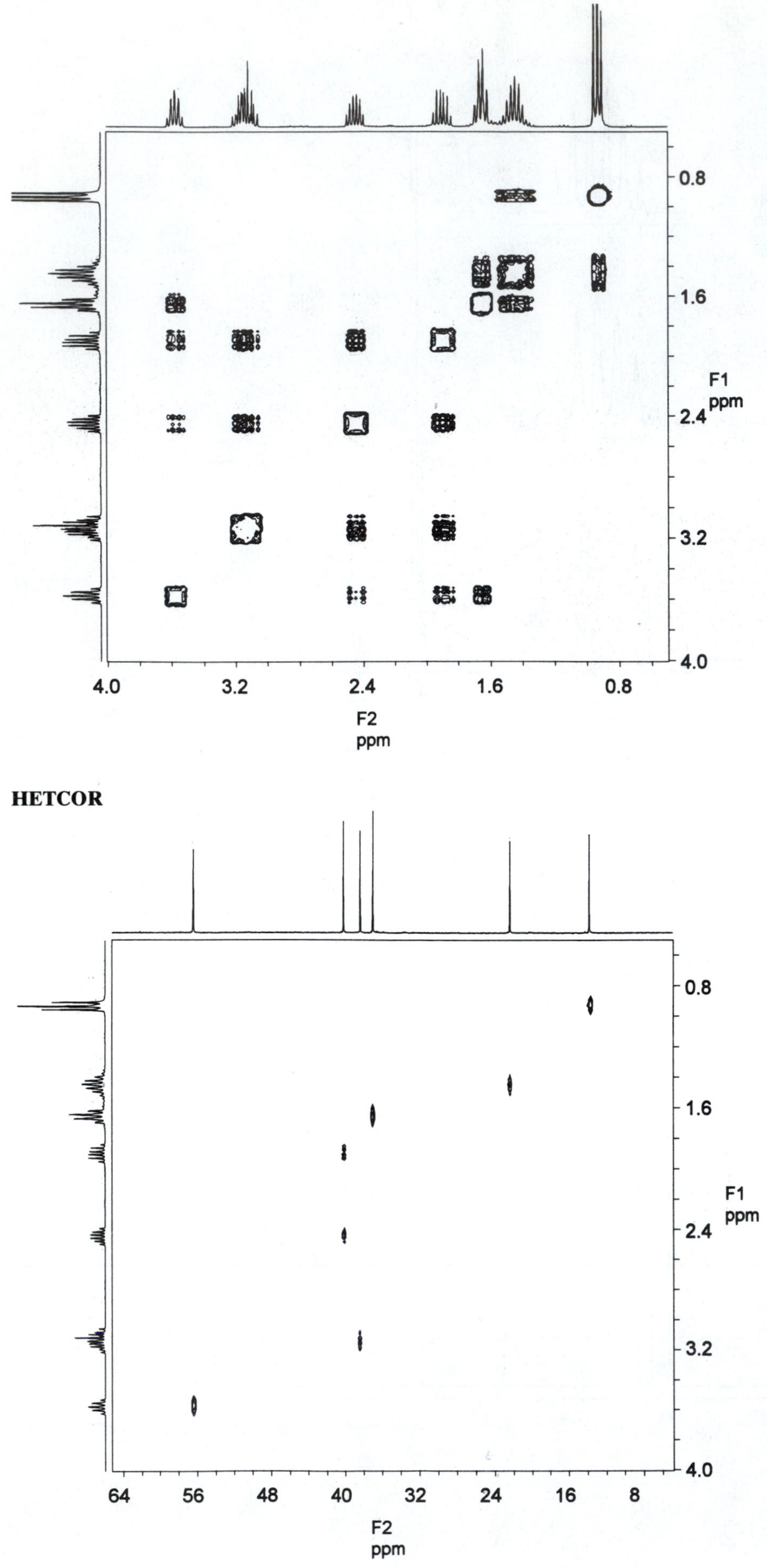

COSY
F1
ppm
0.8
1.6
2.4
3.2
4.0
4.0
3.2
2.4
1.6
0.8
F2
ppm
HETCOR
F1
ppm
0.8
1.6
2.4
3.2
4.0
64
56
48
40
32
24
16
8
F2
ppm

Problem 8.4

It is quite likely that the m/z 154 peak, though small, is the M^+ peak. The m/z 139 peak, also small, results from rational loss of a methyl group. The alert interpreter also notes the M-18 peak at m/z 136 and promptly finds the intense, broad OH peak in the "neat" IR spectrum for confirmation; the intense band at about 1000 cm^{-1} is probably C—O stretching. Again, as in Problem 8.2, we ask: alcohol or phenol; aromatic or not?

The very weak M^+ peak in the present problem, together with loss of H_2O, suggests, but does not prove, an alcohol rather than a phenol. It may be worthwhile at this point to entertain the possibility that the base peak represents the fragment C_5H_9 and results directly from the M^+ peak by a strongly favored mechanism. If so, the intact molecule contains at least one double bond.

At this point, not much can be done with the rather difficult ^1H spectrum, but the ^{13}C and the DEPT spectra are more amenable. There are 10 carbon atoms arranged thus from left to right: C, C, CH, CH, CH_2, CH_2, CH_2, CH_3, CH_3, CH_3. Note that the intensity of the quaternary carbon atom peaks have been increased by pulse delay.

Since we also know that there is at least one OH group and probably one or more double bonds, we tentatively select the molecular formula $C_{10}H_{18}O$—with an index of hydrogen deficiency of two. Since there is no evidence for a C=O, and there are four olefinic carbon atoms, there must be two double bonds.

It is now possible to extract the wealth of detail in the ^1H spectrum. The integration steps agree with the molecular formula and with the ^{13}C/DEPT spectrum: 1:1:2:5:6:3. As mentioned above, the groups at high frequency in the ^{13}C/DEPT spectrum—C, C, CH, CH—represent the olefinic parts of the molecule. There are three deshielded methyl singlets in the ^1H spectrum; these groups must be attached to the two quaternary carbon atoms. This means that two methyl groups must be attached to one of the carbon atoms. The following possibility emerges with no commitment to stereochemistry:

$$H_3C-\overset{H_3C}{\overset{|}{C}}=\overset{H}{\overset{|}{C}}- \quad \text{and} \quad -\overset{H_3C}{\overset{|}{C}}=\overset{H}{\overset{|}{C}}-$$

Note further that both olefinic protons are triplets, albeit with additional, partially resolved, long-range coupling. Hence:

$$H_3C-\overset{H_3C}{\overset{|}{C}}=\overset{H}{\overset{|}{C}}-CH_2- \quad \text{and} \quad -\overset{H_3C}{\overset{|}{C}}=\overset{H}{\overset{|}{C}}-CH_2-$$

Another familiar absorption in the ^1H spectrum is the strongly deshielded two-proton doublet at δ 4.11. This —CH_2OH moiety accounts for the CH triplet in either of the above structures. We now add an OH group in parentheses to both structures. The double bond in either structure accounts for the extra deshielding of the —CH_2OH protons.

$$H_3C-\overset{H_3C}{\overset{|}{C}}=\overset{H}{\overset{|}{C}}-CH_2(OH) \quad \text{or} \quad -\overset{H_3C}{\overset{|}{C}}=\overset{H}{\overset{|}{C}}-CH_2(OH)$$

The structure on the left, being a complete molecule in itself, is eliminated, and the two fragments are now

$$H_3C-\overset{H_3C}{\overset{|}{C}}=\overset{H}{\overset{|}{C}}-CH_2 \quad \text{and} \quad -\overset{H_3C}{\overset{|}{C}}=\overset{H}{\overset{|}{C}}-CH_2-OH$$

In fact, these must be the terminal fragments of a linear molecular. Insertion of the remaining CH_2 group gives

$$H_3C-\overset{H_3C}{\overset{|}{C}}=\overset{H}{\overset{|}{C}}-CH_2-CH_2-\overset{H_3C}{\overset{|}{C}}=\overset{H}{\overset{|}{C}}-CH_2-OH$$

which is a doubly unsaturated terpene alcohol.

There are still several features in the ^1H spectrum that need further analysis. For example, where is the OH peak and what are the multiplicities of the two connected CH_2 groups? Actually, the two questions are interrelated. With a plausible structure in hand, rationalizing a spectrum is easier than analyzing it de novo.

The molecule has no chiral centers. The only symmetry element is a plane of symmetry in the plane of the page. The stereochemistry is more accessible if the structure is redrawn as a conventional terpene.

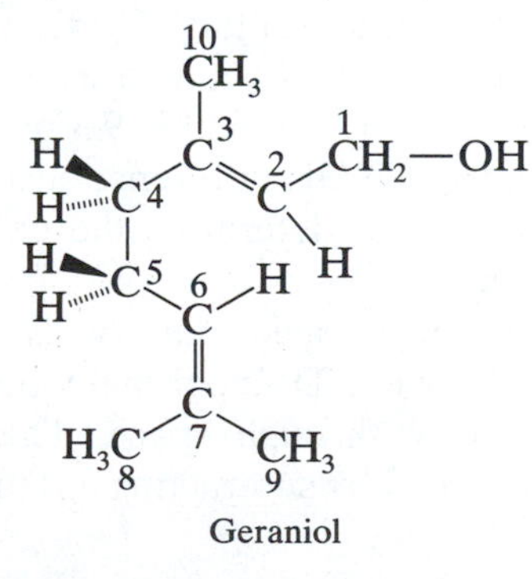

Geraniol

Unfortunately this commits us prematurely to a specific stereoisomer (E). The two methyl groups, CH_3-8 and CH_3-9, are *in* the plane of symmetry, thus not interchangeable.

To respond to the above question of the OH group and the two CH_2 groups, we look at the five-proton absorption centered at about δ 2.08. It is obvious from the expanded insets (which have been obtained at different concentrations) that one peak is missing from the upper inset—obviously the OH peak. The H-4 protons now show a distorted triplet (i.e., coupling to the adjacent H-5 protons) at lower frequency; the H-5 protons show a distorted quartet by coupling to the H-4 and H-6 protons, the couplings being almost identical.

Certainly the story is convincing, but the evidence is based only on chemical shifts and on coupling. The connectivity data of the COSY spectrum is most reassuring. This is a DQF-COSY in which the intensities of the CH_3 singlet peaks are greatly reduced to alleviate overlapping (Chapter 6). The peaks along the diagonal are numbered, but the cross peak assignments are left as an exercise. Note the long-range allylic and homoallylic coupling through the double bonds in this olefinic model (four and five bonds, respectively). Especially noteworthy is the diagnostic coupling between H-8 and H-9 through four single bonds. The OH proton, of course, shows no cross peak because of rapid exchange. At this point, we have assigned all of the protons but still cannot differentiate between H-8 and H-9.

We can now use the proton assignments to assign the ^{13}C peaks through the HETCOR spectrum (Chapter 6). This is left as an exercise. Note, however, that quaternary carbon peaks do not appear on the standard HETCOR. In this case, the ^{13}C spectrum was manually superimposed on the F_2 axis. Since quaternary carbon atoms 3 and 7 are not correlated with protons, they cannot be assigned. Since protons H-8 and H-9 have not yet been specifically assigned, we still cannot differentiate between CH_3-8 and CH_3-9.

The INADEQUATE spectrum (Chapter 6) delineates the connectivities between adjacent ^{13}C atoms. This will serve as a useful student exercise; the ^{13}C peaks are numbered. This spectrum was run in deuterated acetone (septet at δ 30.6); the chemical shifts therefore differ slightly from those of the ^{13}C spectrum in $CDCl_3$. All of the ^{13}C connectivities can be seen, but the differentiation between CH_3-8 and CH_3-9 is not resolved. We therefore turn to an NOE difference spectrum to make this resolution and to determine the geometry of the C-2 double bond.

NOE difference spectrometry is described in Section 4.20. It is a 1-D experiment that determines ^1H—^1H proximity through space. The "difference" spectrum is obtained by subtracting a standard ^1H spectrum from the NOE spectrum; this leaves only the "enhanced" peaks.

The task we face with the present molecule—distinguishing between a trisubstituted (*E*) double bond and the corresponding (*Z*) double bond—is not a trivial assignment. Nor is the task of distinguishing between CH_3-8 and CH_3-9. For conclusive results, we examine both the (*E*) isomer (geraniol) and the (*Z*) isomer (nerol). The NOE difference spectrum (*a*) shows the ^1H spectrum for geraniol and the results of irradiating H-9 and H-10 simultaneously (and unavoidably) and H-8. H-2 is not detectably enhanced—equal positive and negative peaks—whereas H-6 is definitely enhanced.

These results indicate that CH_3-10 and H-2 are on opposite sides of the 2,3-double bond, which therefore has the (*E*) configuration as drawn. They also indicate that CH_3-9 and H-6 are on the same side of the 6,7-double bond as numbered. The latter interpretation is confirmed by irradiation of H-8 with only slight effect on H-6.

In spectrum (*b*) (nerol), irradiation of H-10 strongly enhances H-2, thus confirming the (*Z*) configuration of the 2,3-double bond in nerol. Irradiation of H-9 in nerol gives the same result as for geraniol: enhancement of H-6. Irradiation of H-8 in nerol also gives the same results as for geraniol: no pronounced enhancement of either H-2 or H-6, the latter result showing that CH_3-8 and H-6 are on opposite sides of the 6,7-double bond in both geraniol and nerol. Irradiation of CH_3-10 of geraniol also enhances CH_2-1, but this is not shown in the present spectrum. Note that the small allylic coupling is retained on NOE irradiation.

Geraniol is now completely identified as the (*E*) isomer, and nerol as the (*Z*) isomer. With ineffable hindsight, we can now recognize the fragmentation peak at *m/z* 69 (the base peak) as the result of allylic cleavage of an olefin. Ordinarily, reliance on this cleavage for location of a double bond is dubious, but in geraniol, cleavage of the bis-allylic bond between C-4 and C-5 results in the stabilized fragment *m/z* 69 (see Section 2.1.10.2 for the analogous allylic cleavage of β-myrcene into fragments *m/z* 69 and 67).

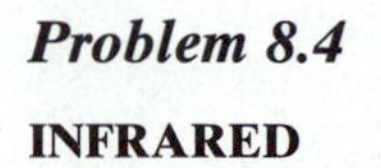

Problem 8.4

INFRARED

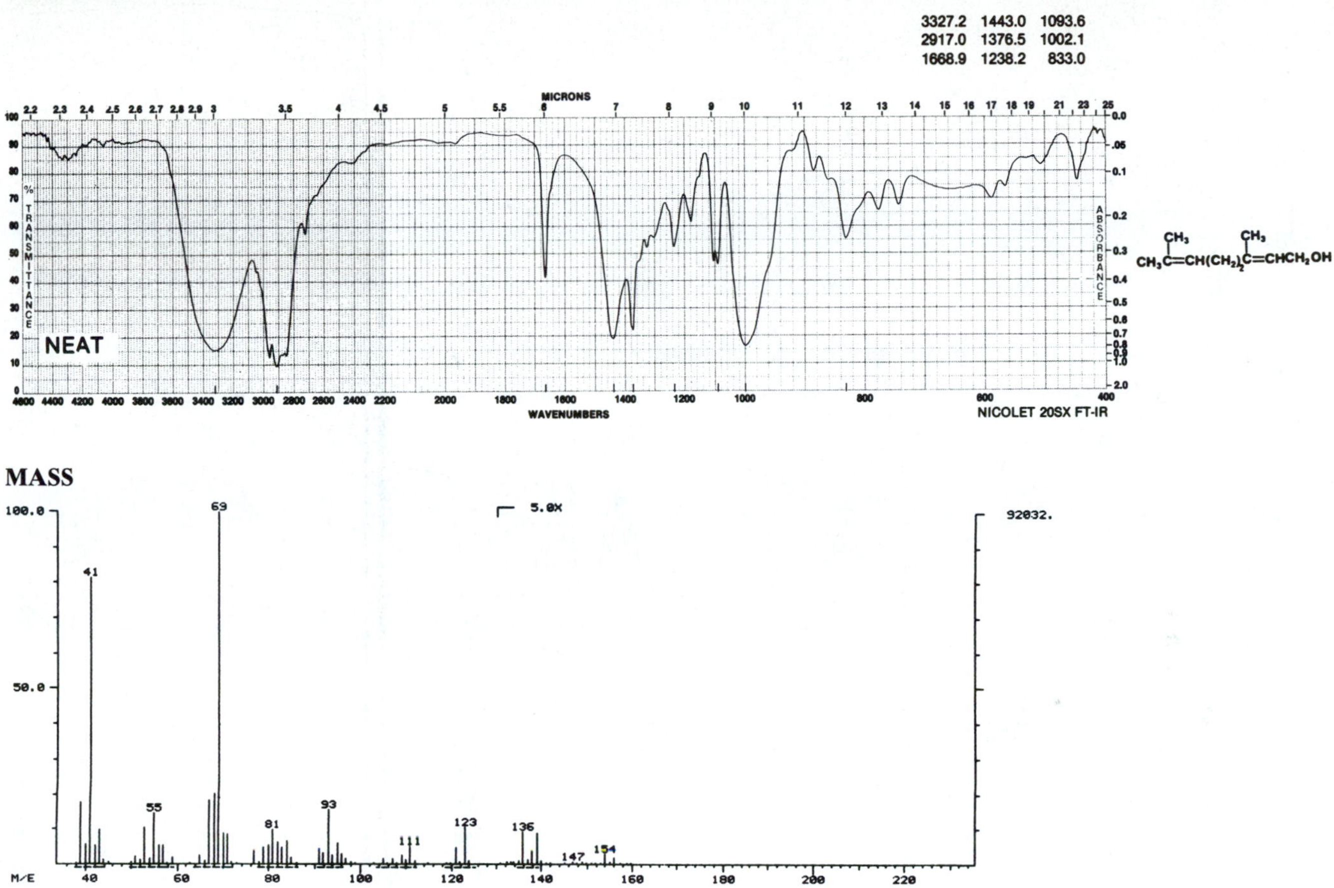

(Continued)

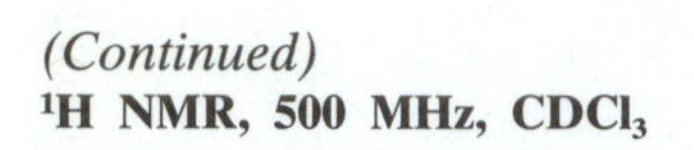

(Continued)
^{1}H NMR, 500 MHz, $CDCl_3$

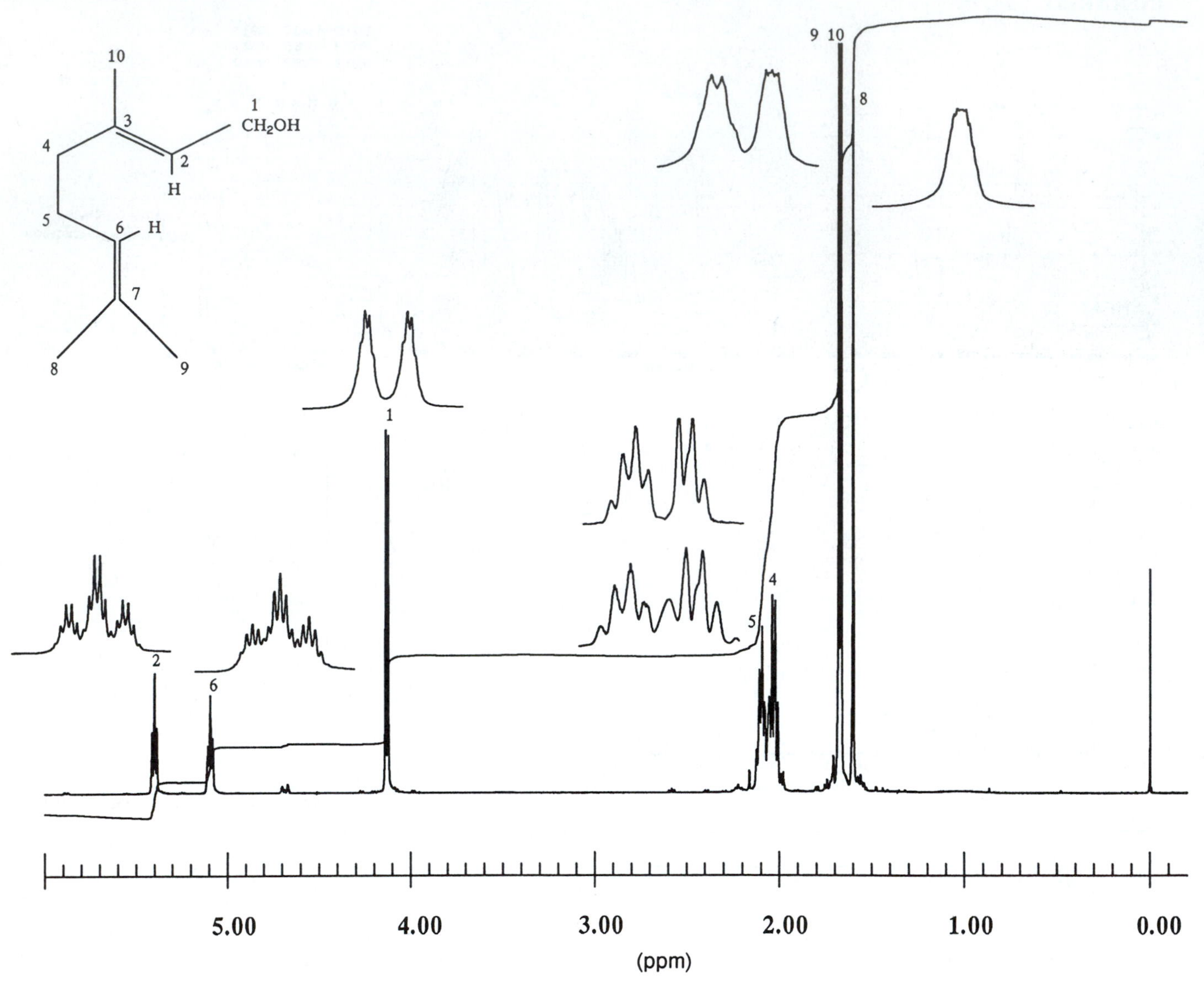

(Continued)

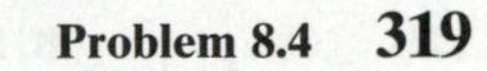

(Continued)

^{13}C NMR, 126 MHz, $CDCl_3$, PULSE DELAY

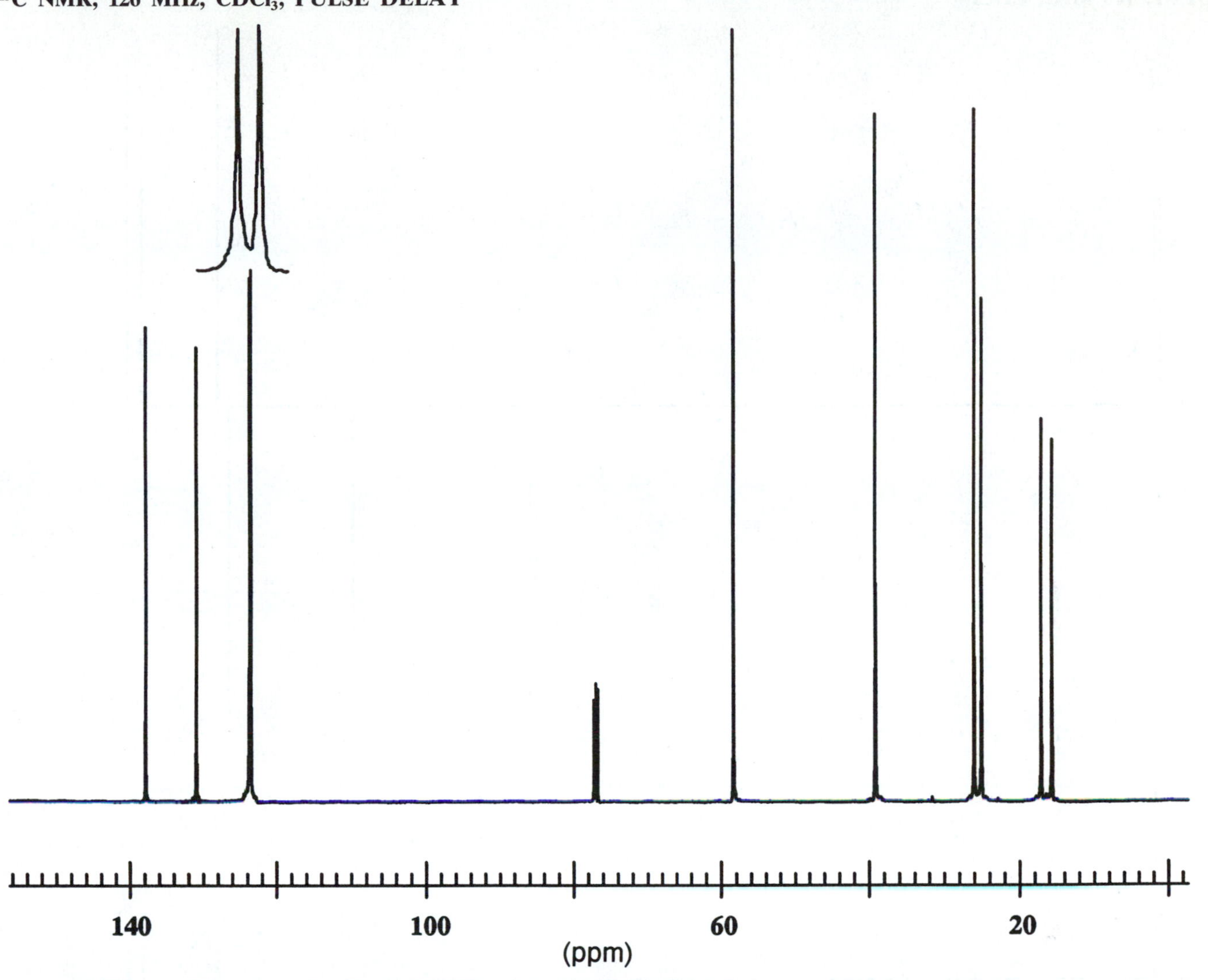

(Continued)

(Continued)
DEPT, 126 MHz, $CDCl_3$*

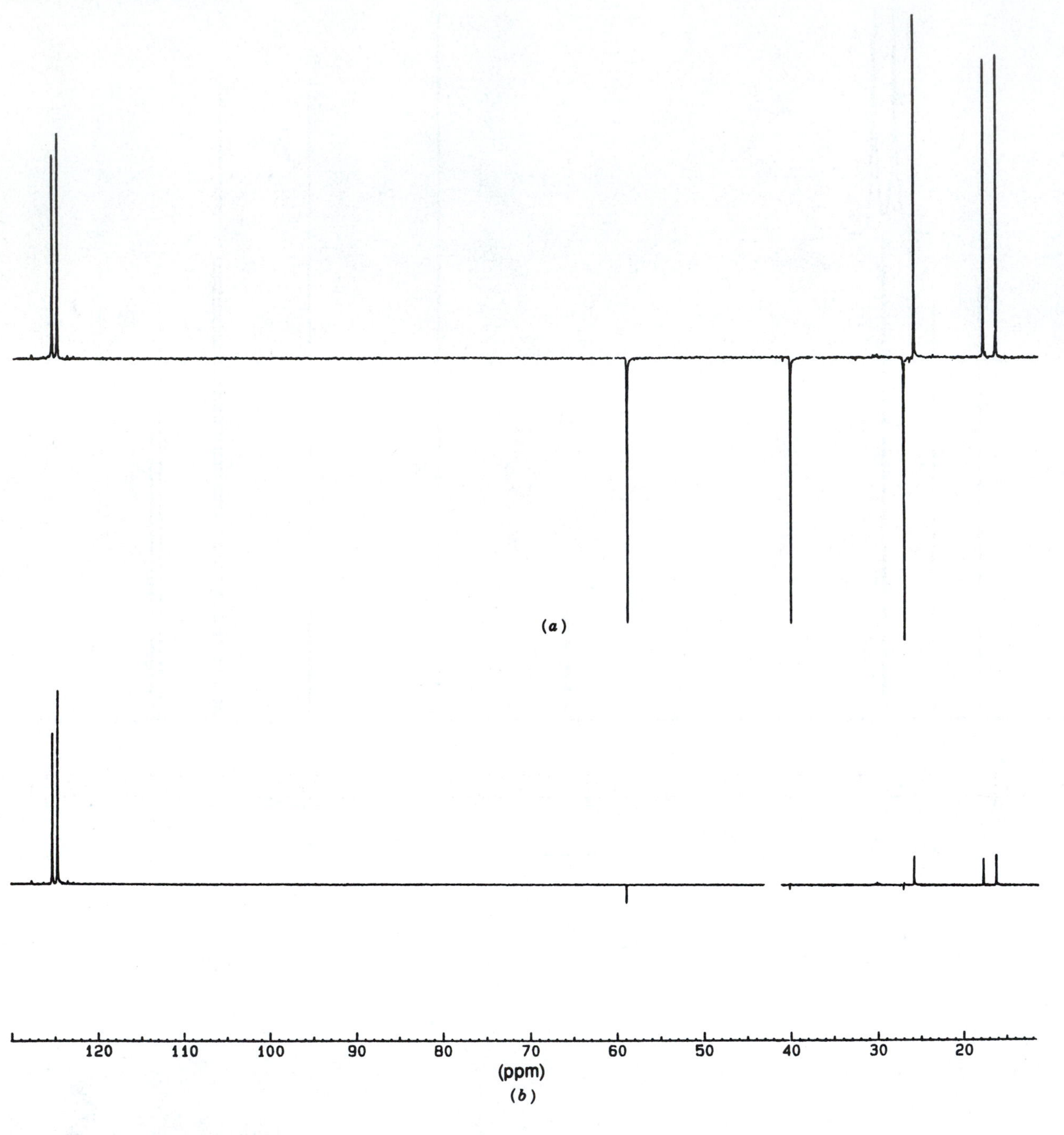

* **The DEPT spectrum of geraniol.** (*a*) Peaks CH_3 and CH up, CH_2 down. (*b*) Peaks CH up (residual CH_3 and CH_2 peaks). In $CDCl_3$ at 75.6 MHz for ^{13}C.

(Continued)

(Continued)
COSY, 500 MHz, $CDCl_3$

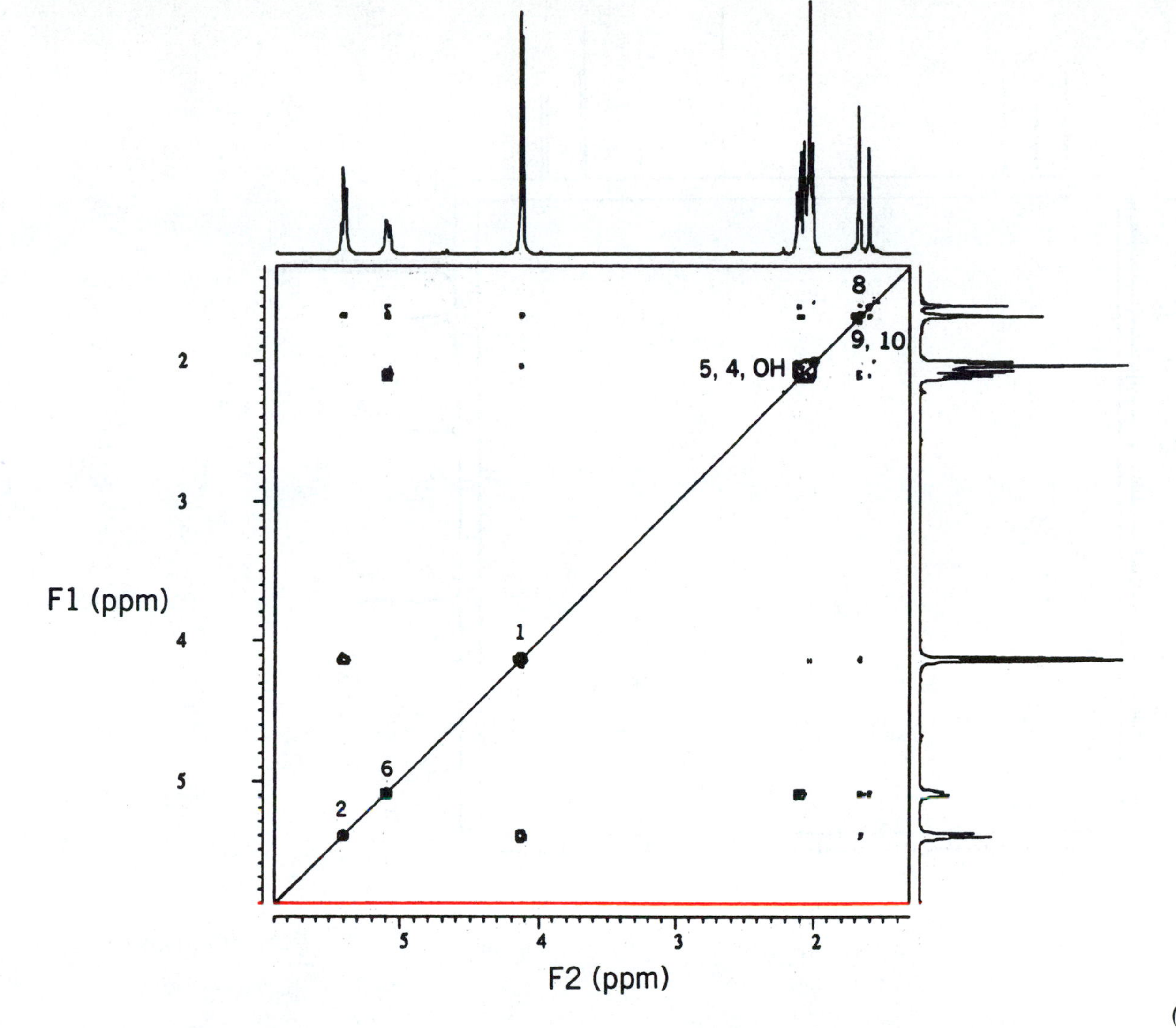

(Continued)

(Continued)
HETCOR, 126 MHz, $CDCl_3$

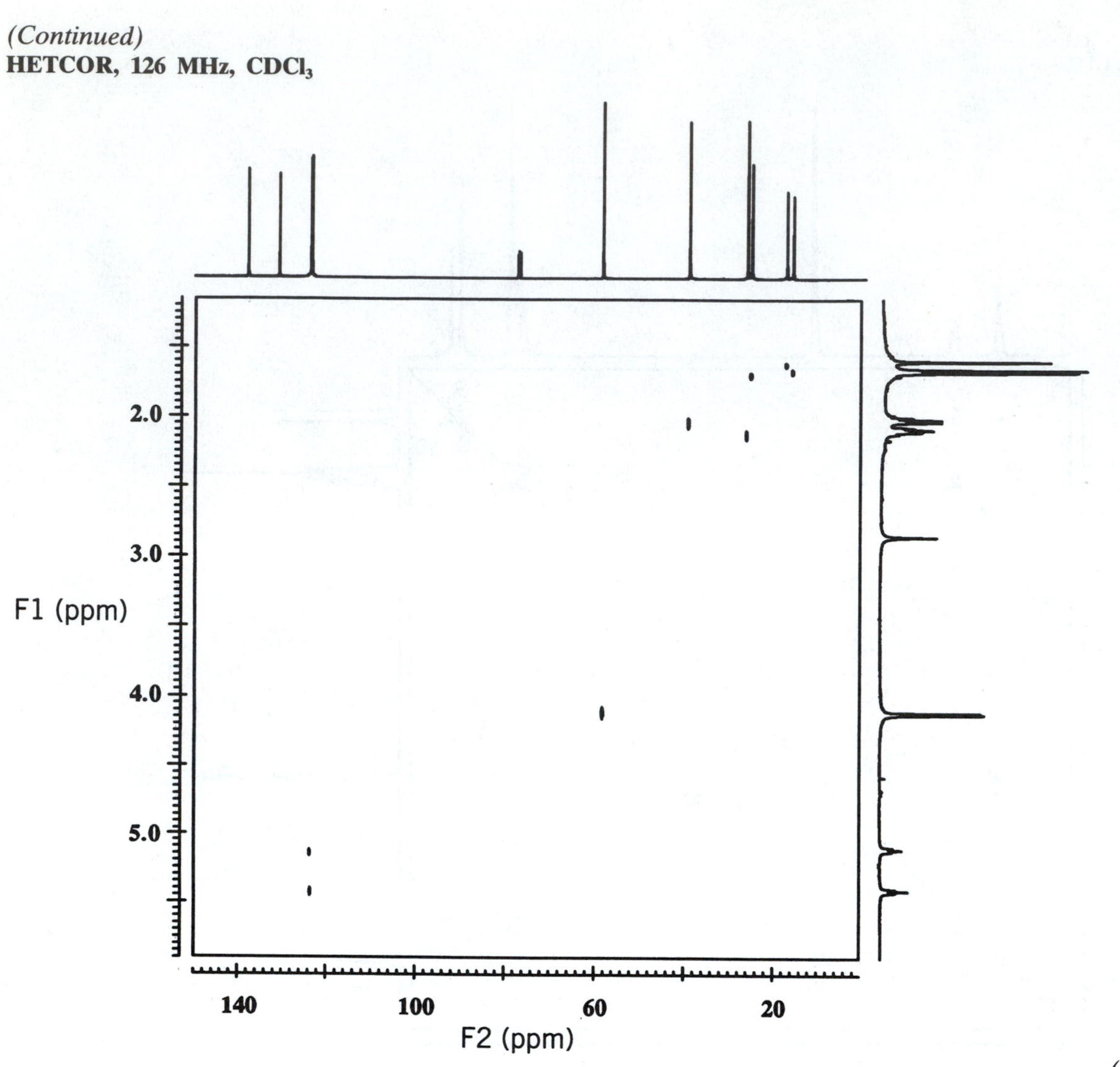

(Continued)

(Continued)
INADEQUATE, 126 MHz, CD_3COCD_3

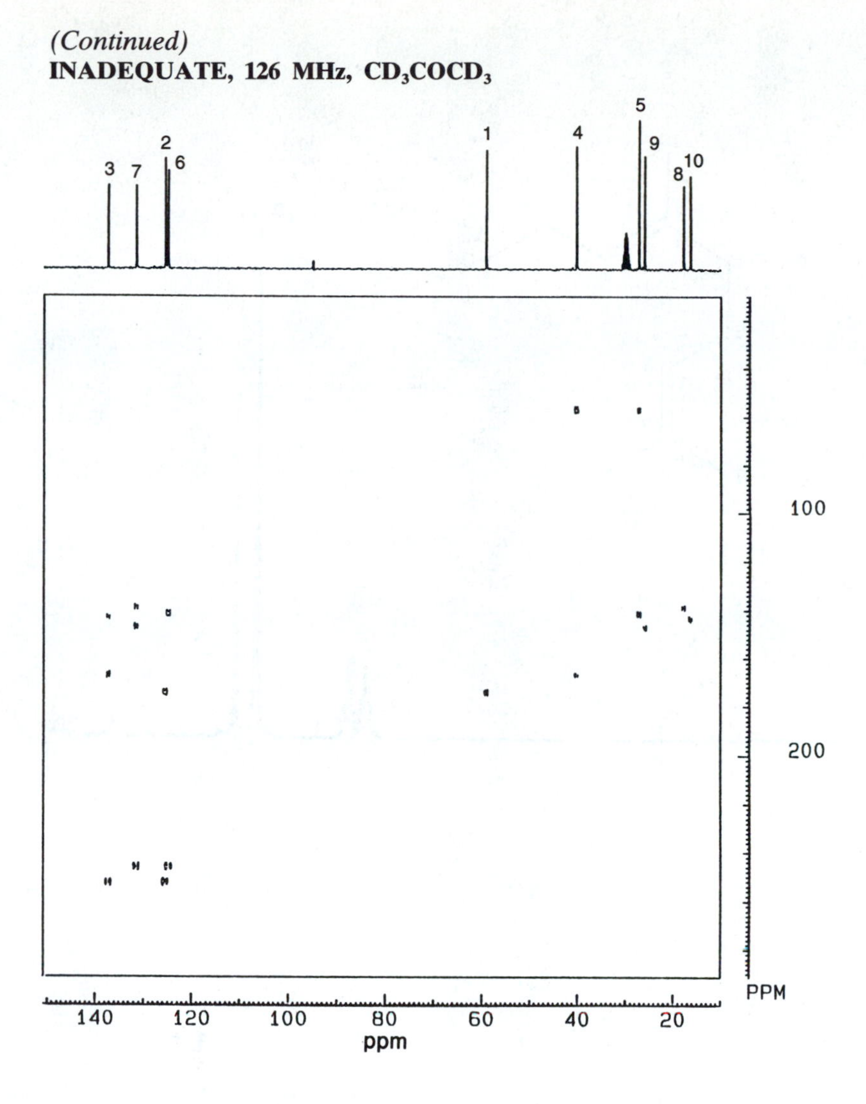

(Continued)

(Continued)
NOE, 500 MHz, $CDCl_3$

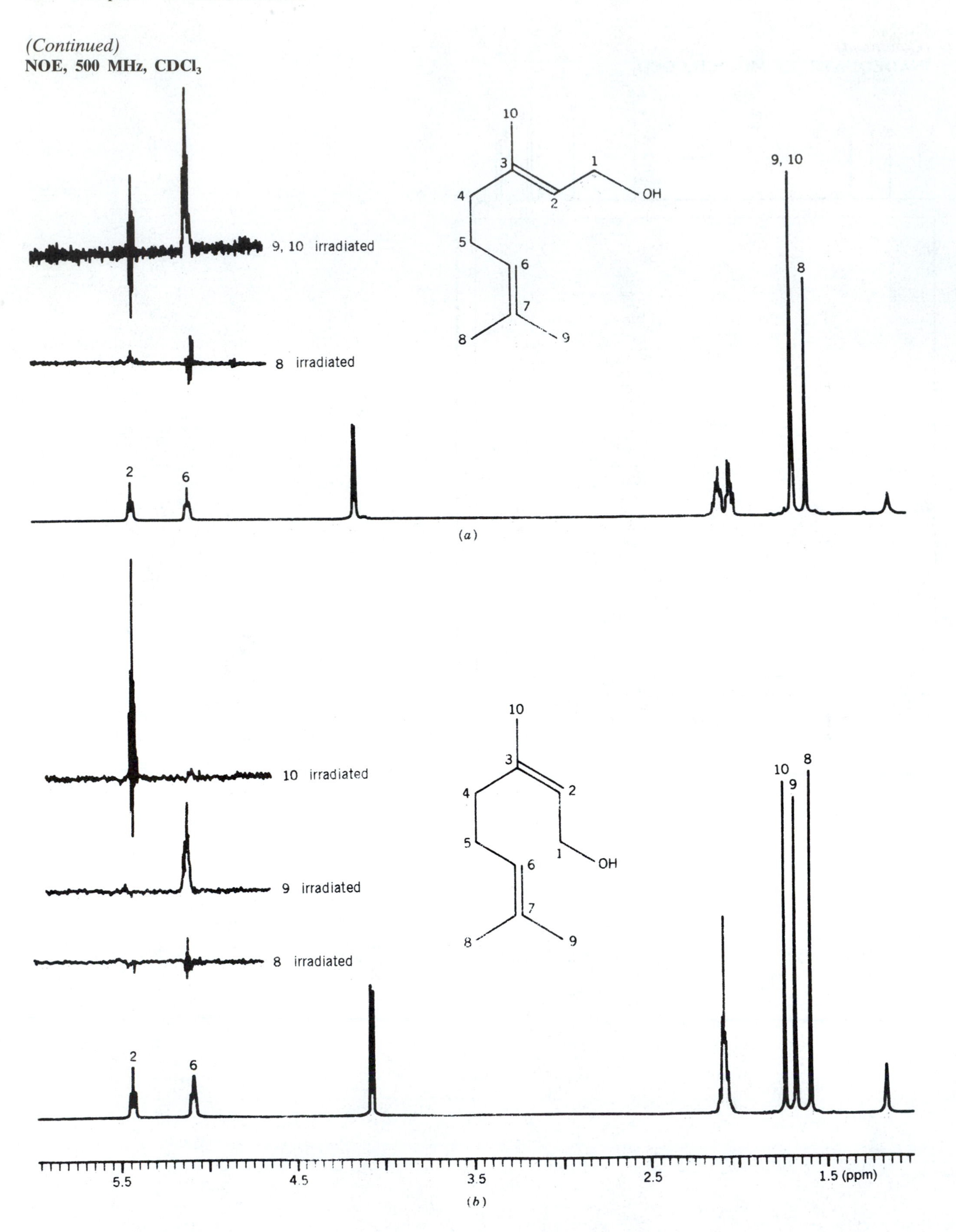

CHAPTER 9

A. *Introduction*

Chapter 8 has been propounded as though unknown Problems were being resolved by the authors. Not so. It was inevitable that the authors knew the structures. The students inevitably consulted the structures as they worked through the explications. The experience, therefore, has been more rationalization of the given structures than analysis of the spectra. Nonetheless, the experience was worthwhile.

Chapter 9 will be an experience in foresight rather than hindsight, but the first few modest Problems should provide encouragement, as did the Problems at the end of each chapter. Admittedly, difficulties will be encountered further along, but satisfactions also.*

*Unless otherwise labeled, the NMR spectra were obtained at 300 MHz for protons, 75.5 MHz for ^{13}C; $CDCl_3$ was the solvent unless otherwise labeled. The IR spectra were obtained neat (i.e., no solvent) unless otherwise labeled. The mass spectra were obtained by GC/MS. CH_4 was used to obtain the chemical ionization mass spectra. The COSY spectra are DQF-COSY spectra. The labeled frequency axis for all 2-D NMR spectra is that of the acquired signal (F2). The F1 axis may also be labeled.

At this point, we may all enjoy a more literary statement: "It is easy to be wise in retrospect, uncommonly difficult in the event."†

†Stegner, W. (1954). *Beyond the Hundredth Meridian.* Boston: Houghton Mifflin. Reprint, Penguin Books, 1992, Chapter 3.

CHAPTER 9

B. Assigned Problems

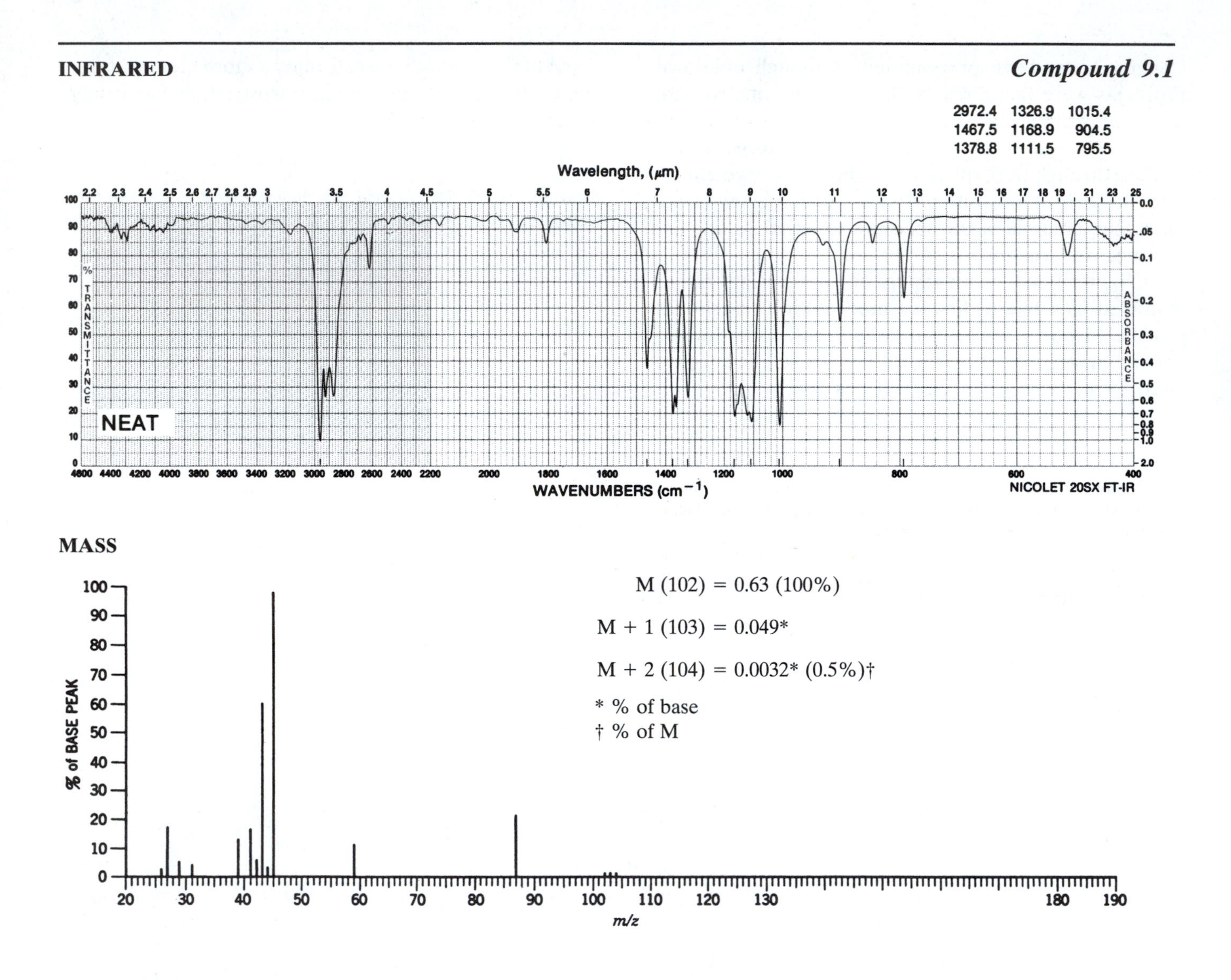

^{1}H NMR (SOLVENT $CDCl_3$, 60 MHz)

^{13}C NMR

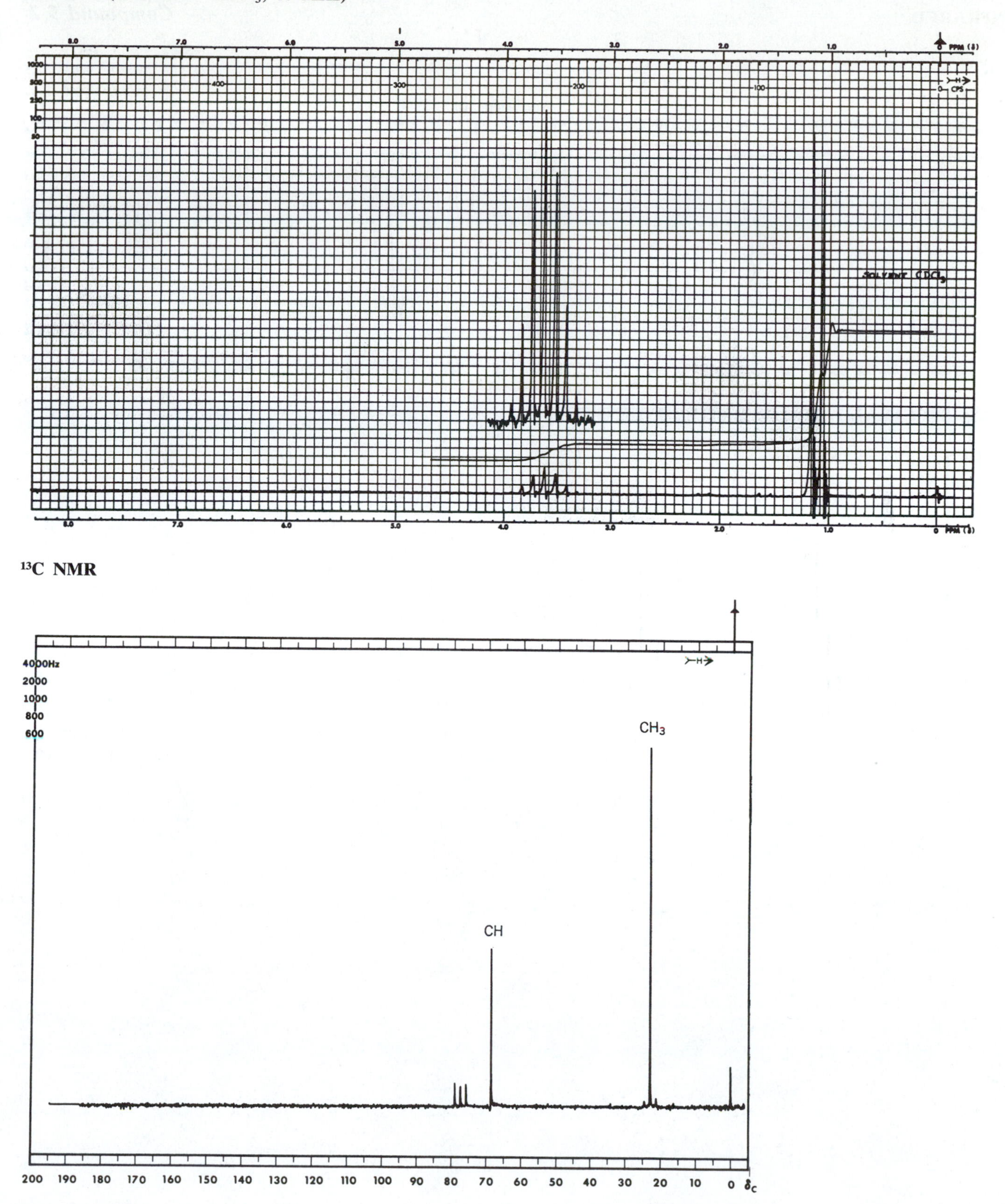

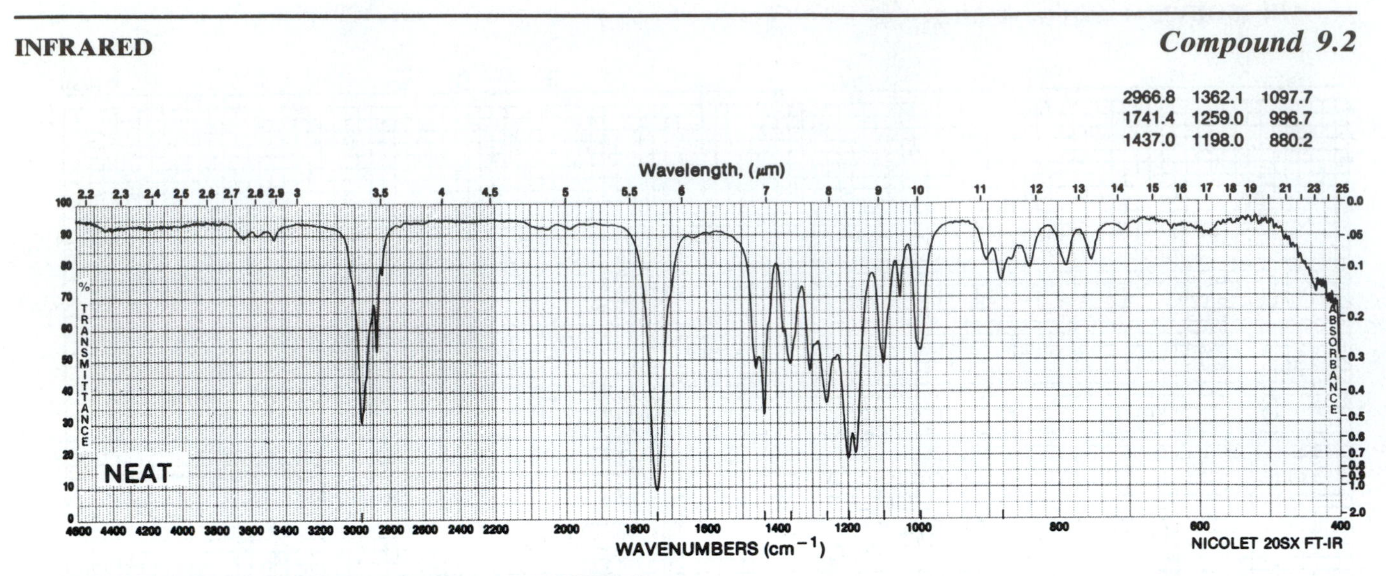

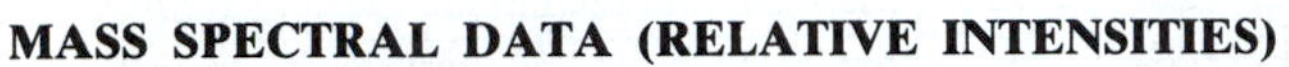
MASS SPECTRAL DATA (RELATIVE INTENSITIES)

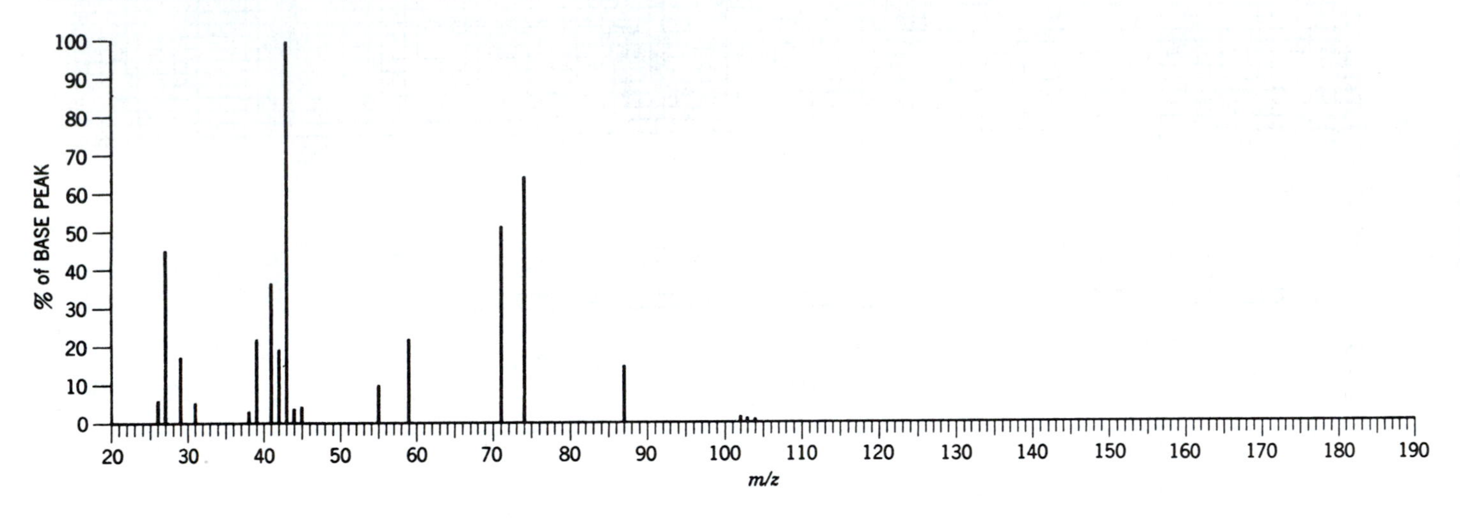

MASS, CHEMICAL IONIZATION

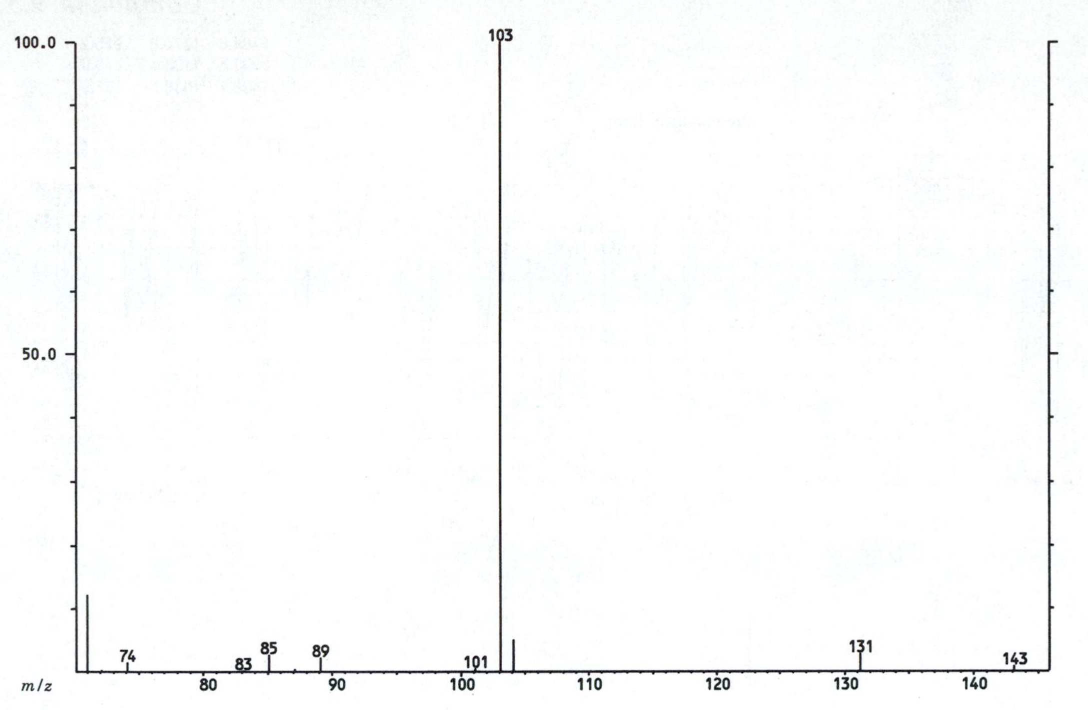

¹H NMR

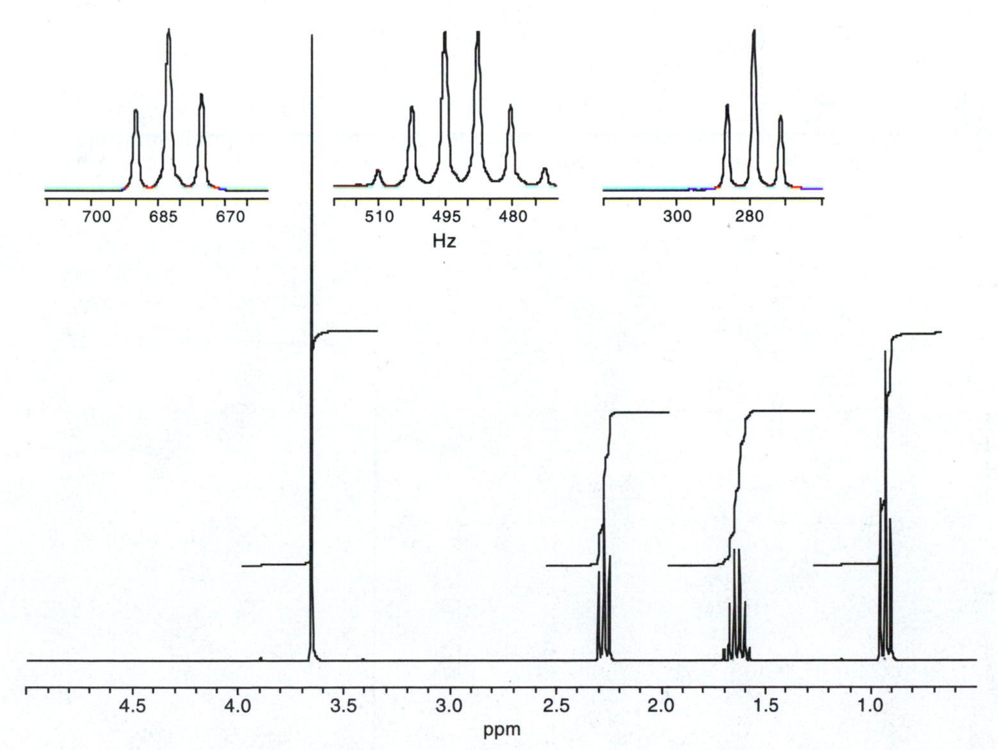

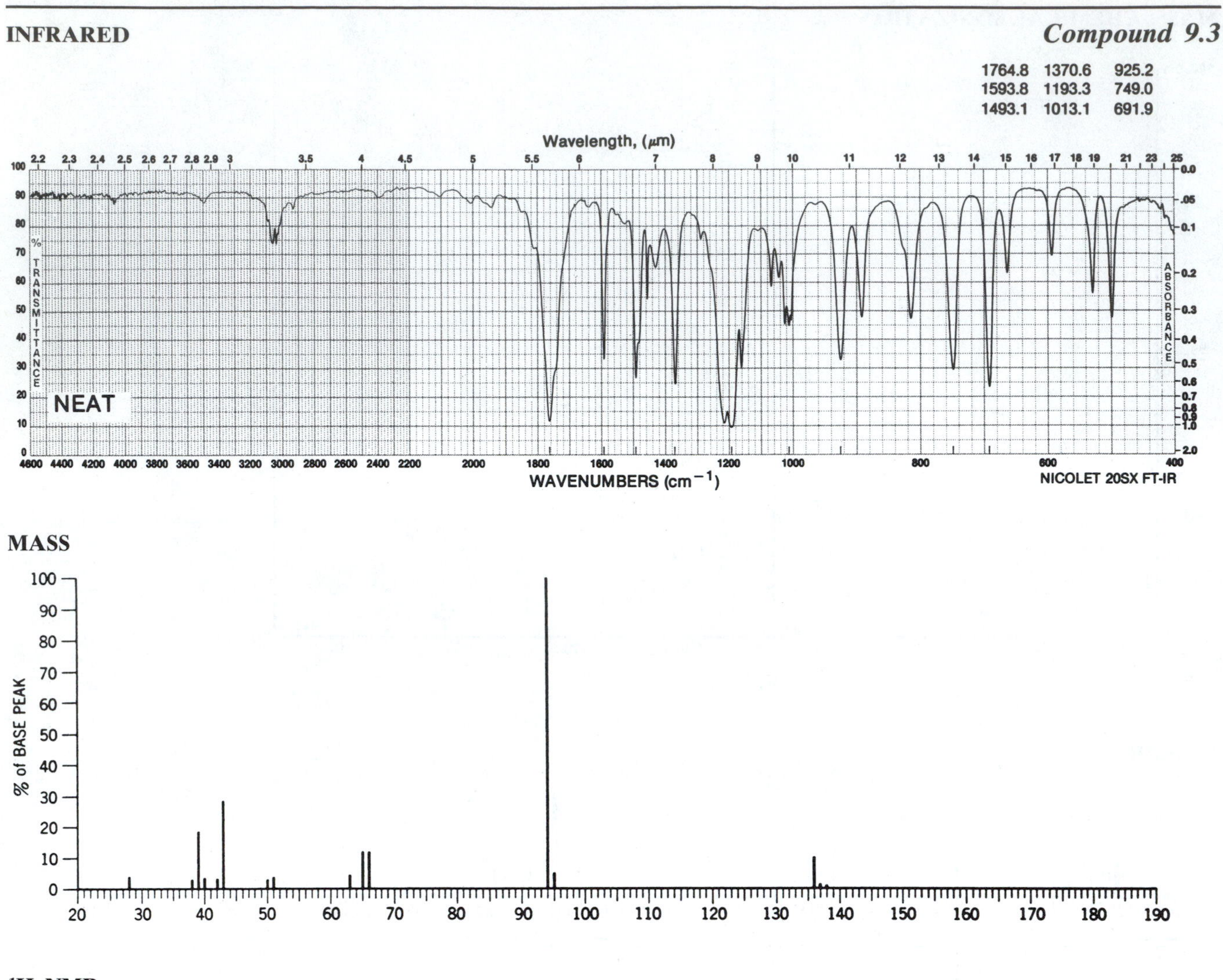

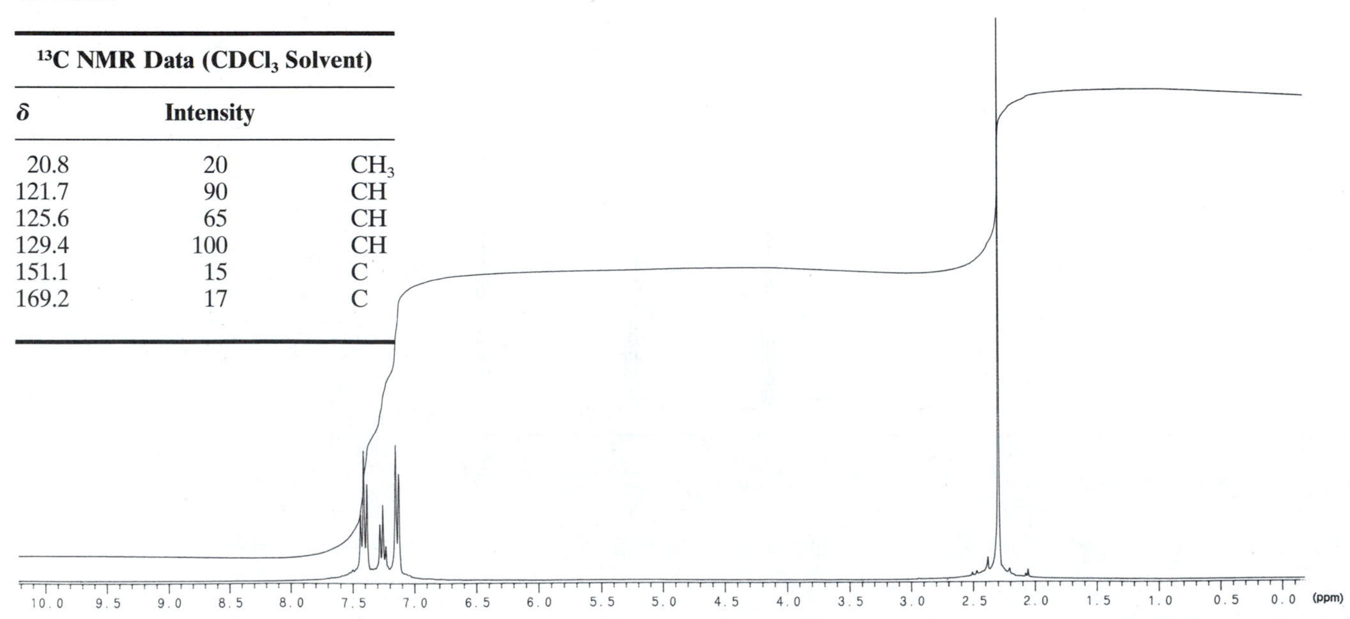

^{13}C NMR Data ($CDCl_3$ Solvent)

δ	Intensity	
20.8	20	CH_3
121.7	90	CH
125.6	65	CH
129.4	100	CH
151.1	15	C
169.2	17	C

INFRARED

Compound 9.4

2981.9	1367.2	1108.5
1718.5	1275.8	1028.5
1451.4	1175.2	710.3

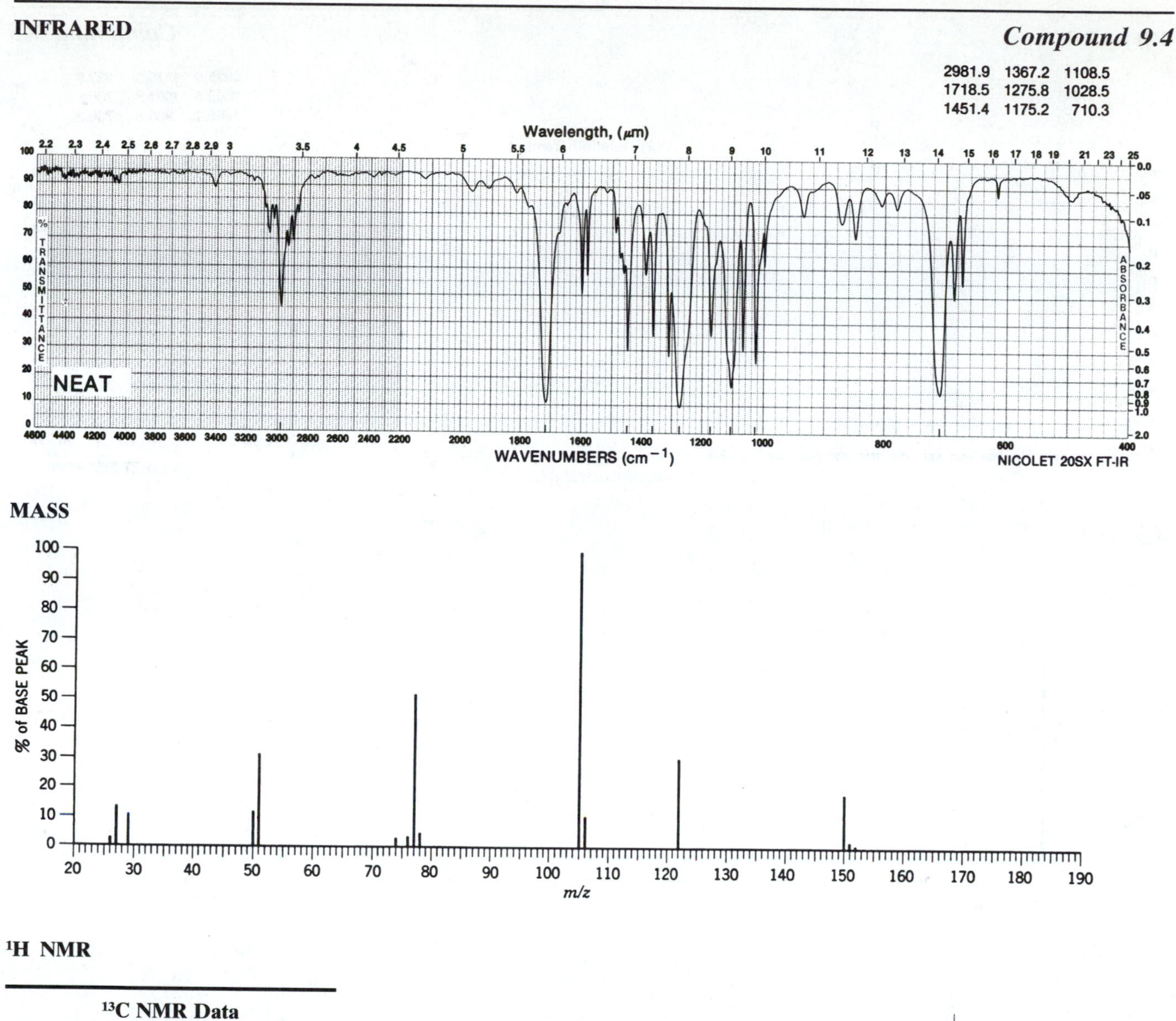

MASS

^{1}H NMR

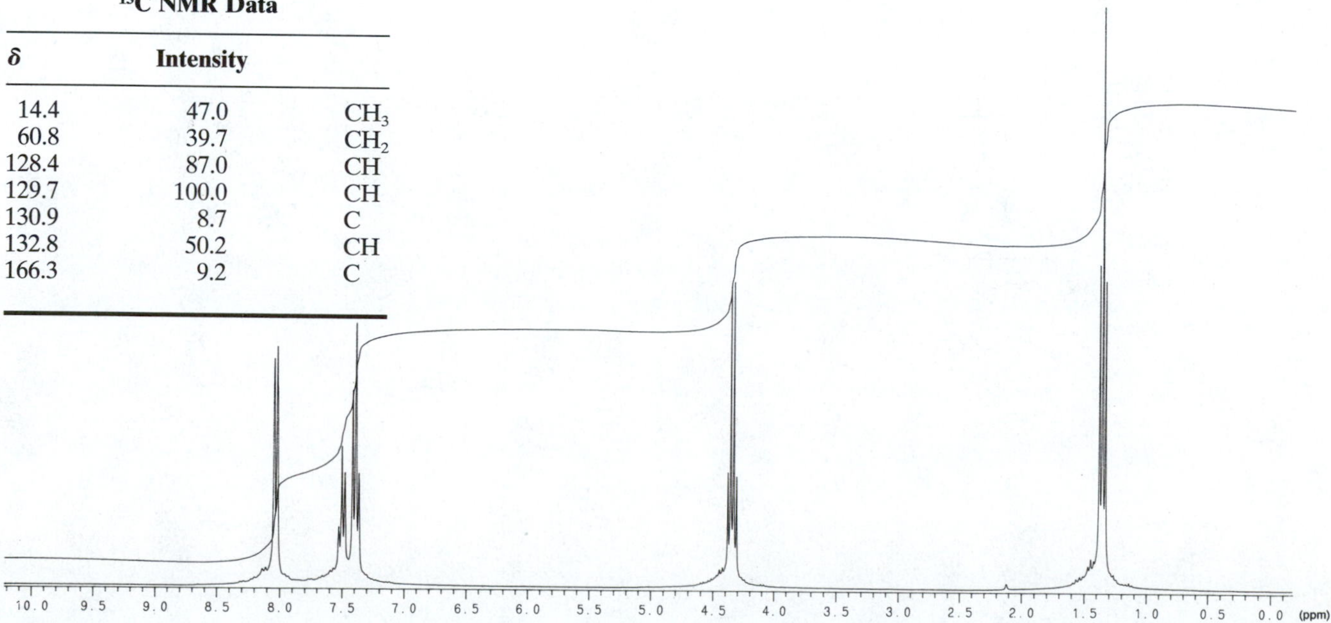

^{13}C NMR Data		
δ	Intensity	
14.4	47.0	CH_3
60.8	39.7	CH_2
128.4	87.0	CH
129.7	100.0	CH
130.9	8.7	C
132.8	50.2	CH
166.3	9.2	C

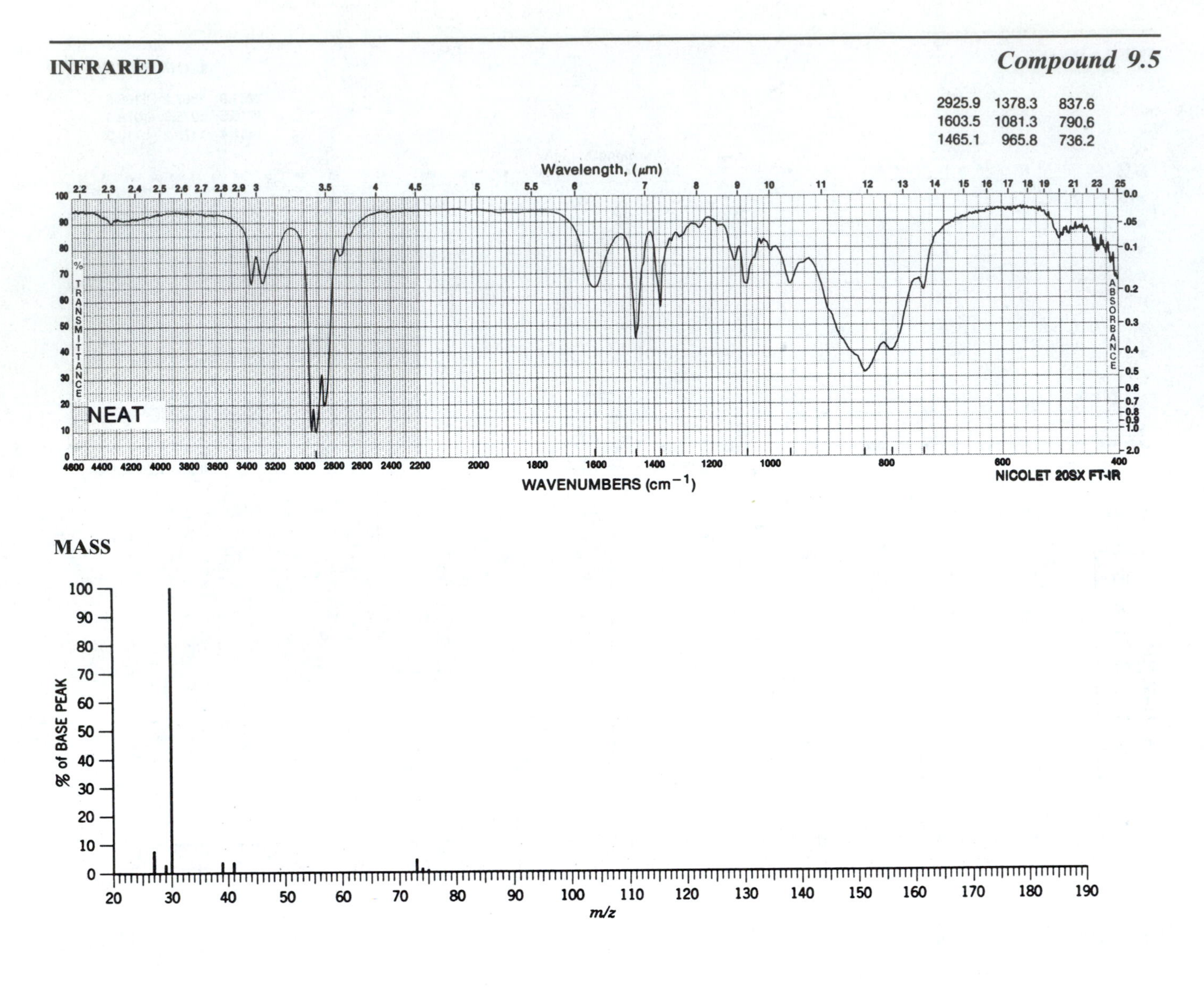

INFRARED
Compound 9.5
2925.9 1378.3 837.6
1603.5 1081.3 790.6
1465.1 965.8 736.2
Wavelength, (μm)
% TRANSMITTANCE
ABSORBANCE
NEAT
WAVENUMBERS (cm⁻¹)
NICOLET 20SX FT-IR
MASS
% of BASE PEAK
m/z

^{1}H NMR SPECTRUM

^{13}C NMR Data ($CDCl_3$ solvent)		
δ	Intensity	
13.9	95.3	CH_3
20.2	100.0	CH_2
36.3	85.3	CH_2
42.0	93.0	CH_2

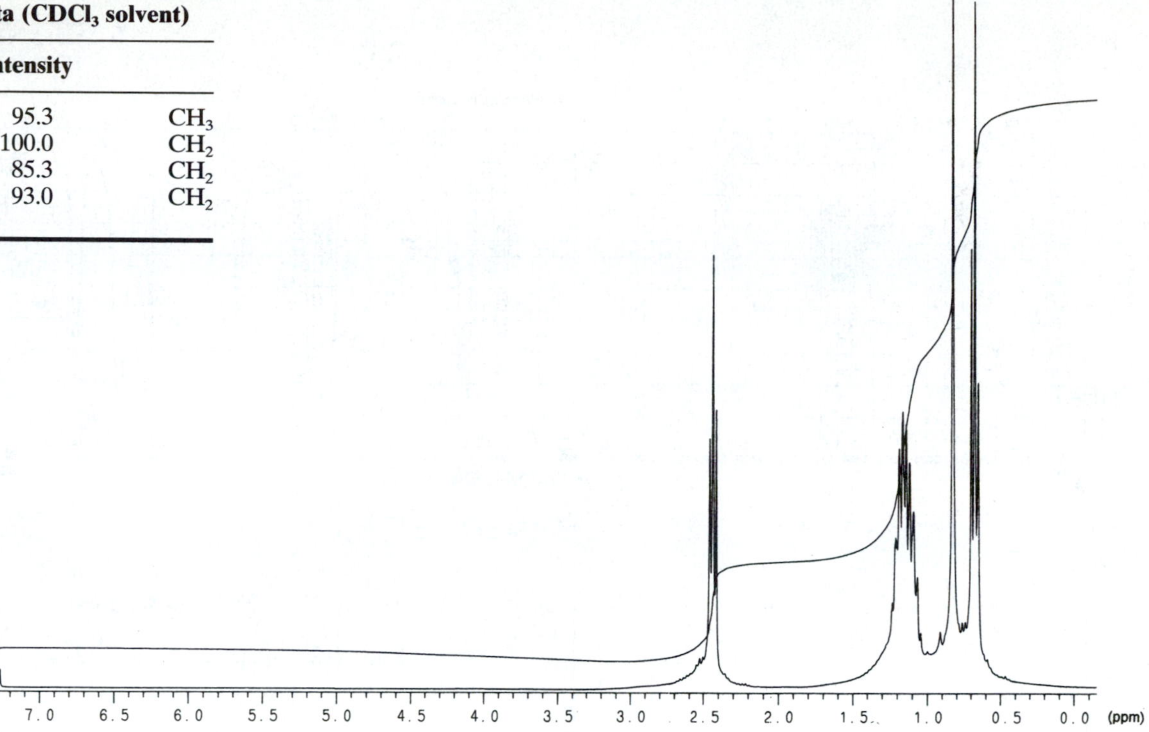

INFRARED

Compound 9.6

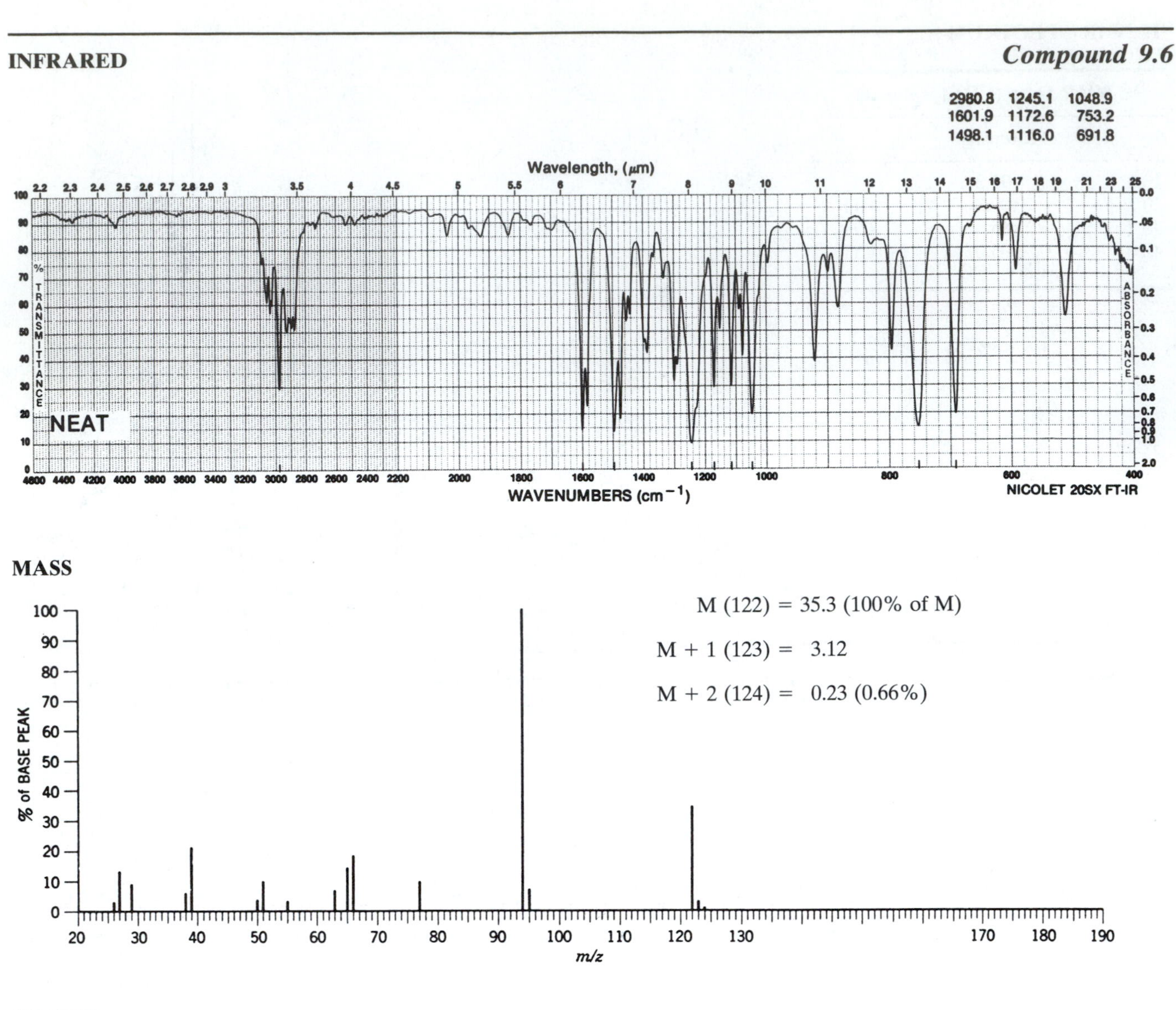

¹H NMR

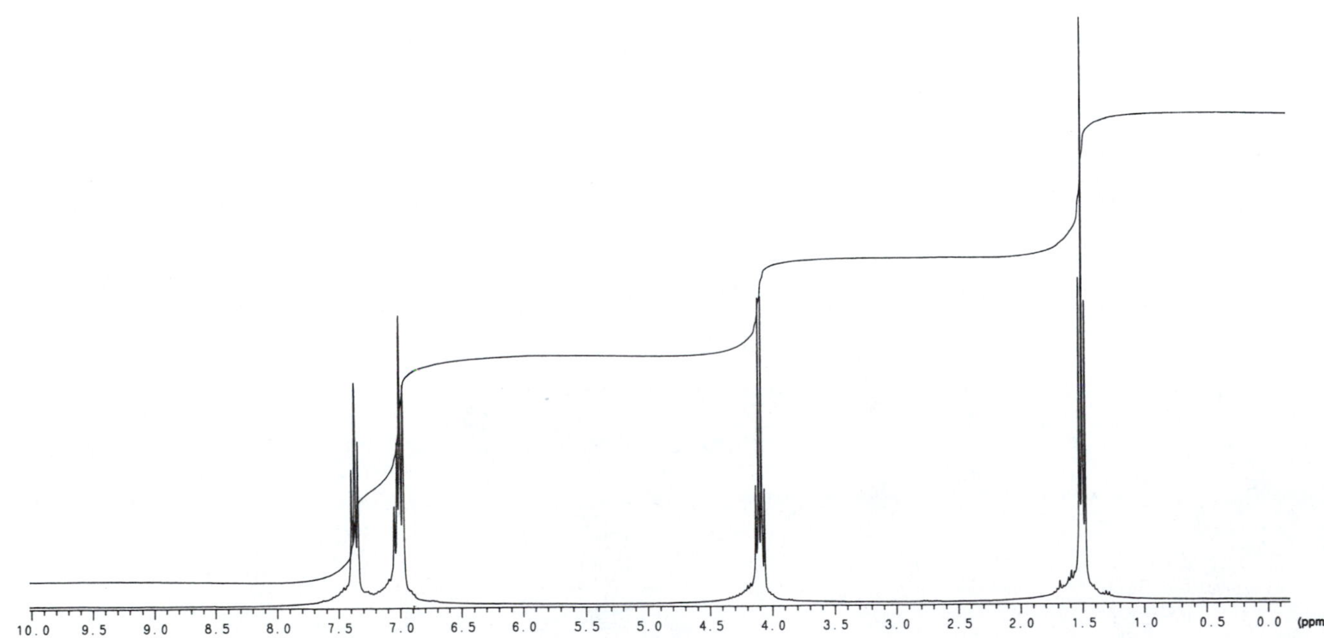

INFRARED

Compound 9.7

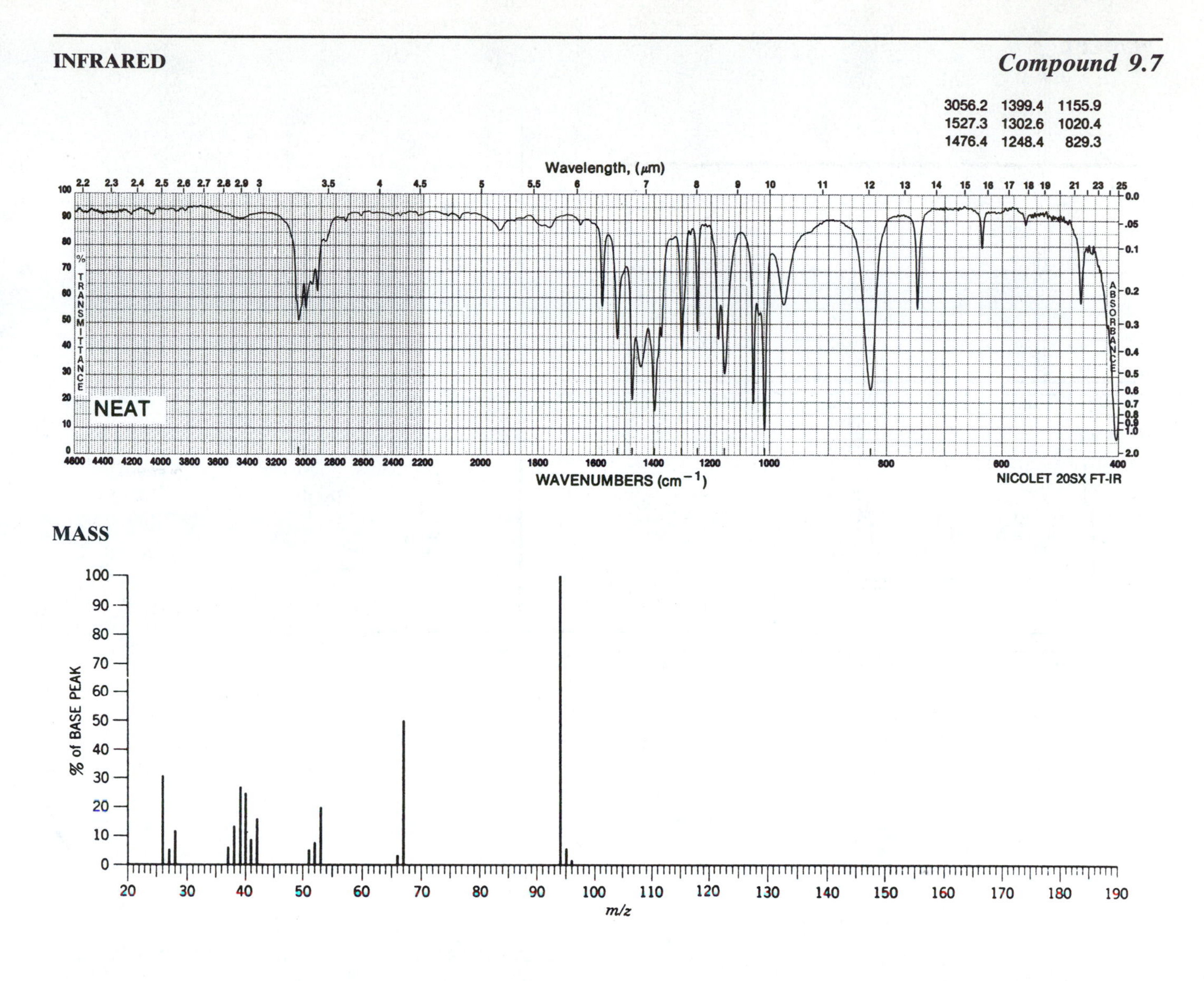

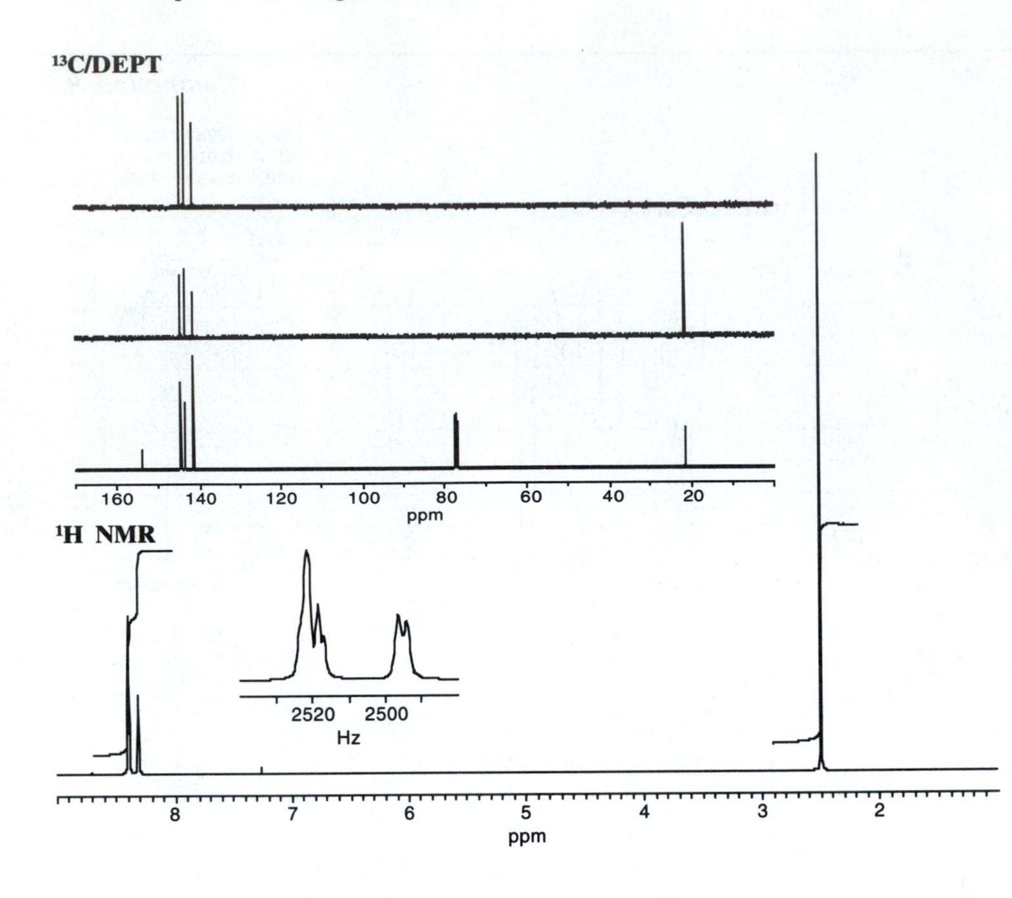

¹³C/DEPT
160
140
120
100
80
60
40
20
ppm
¹H NMR
2520
2500
Hz
8
7
6
5
4
3
2
ppm

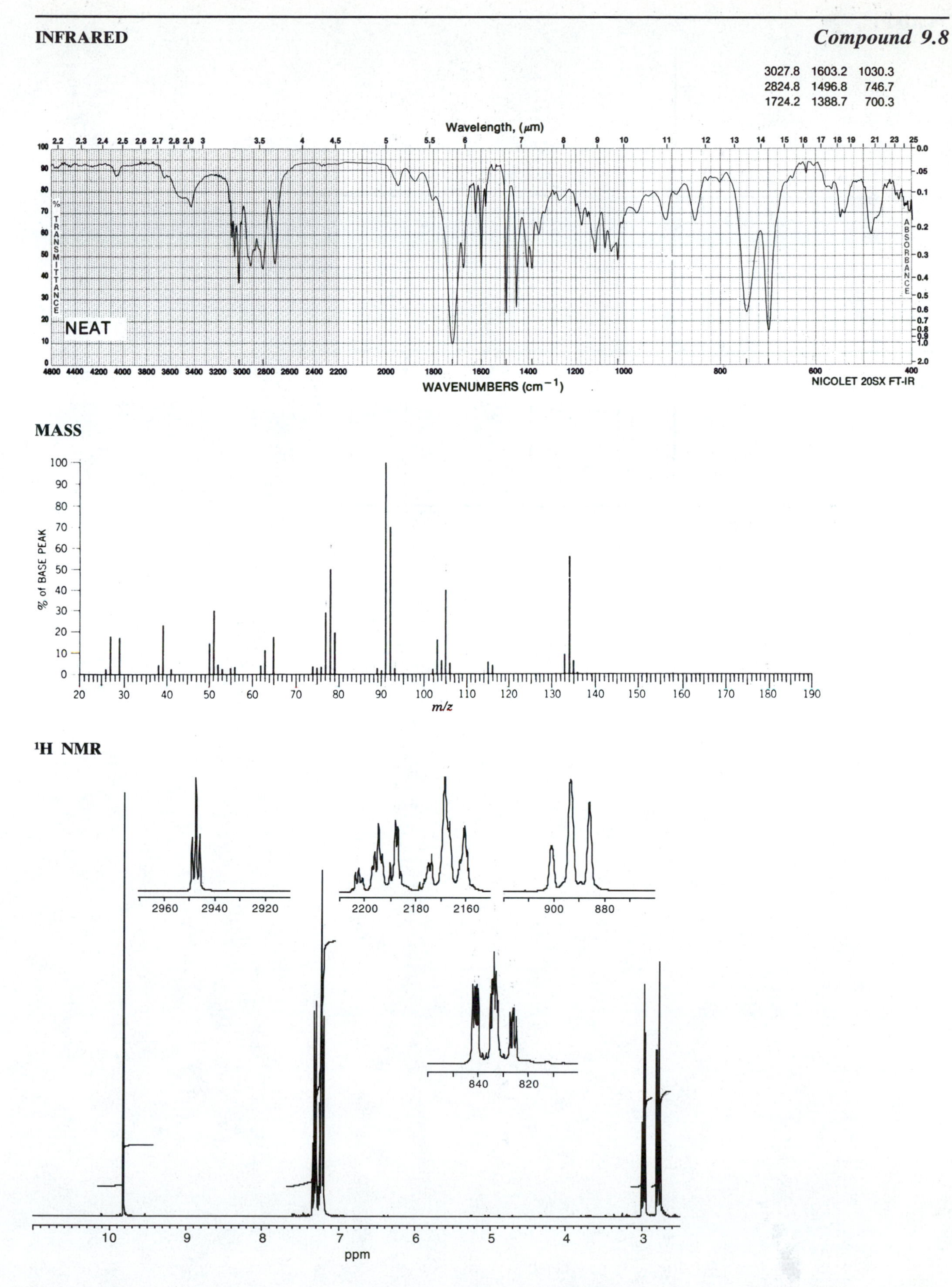

INFRARED
Compound 9.8
3027.8 1603.2 1030.3
2824.8 1496.8 746.7
1724.2 1388.7 700.3
Wavelength, (μm)
% TRANSMITTANCE
NEAT
ABSORBANCE
WAVENUMBERS (cm⁻¹)
NICOLET 20SX FT-IR
MASS
% of BASE PEAK
m/z
¹H NMR
ppm

^{13}C/DEPT

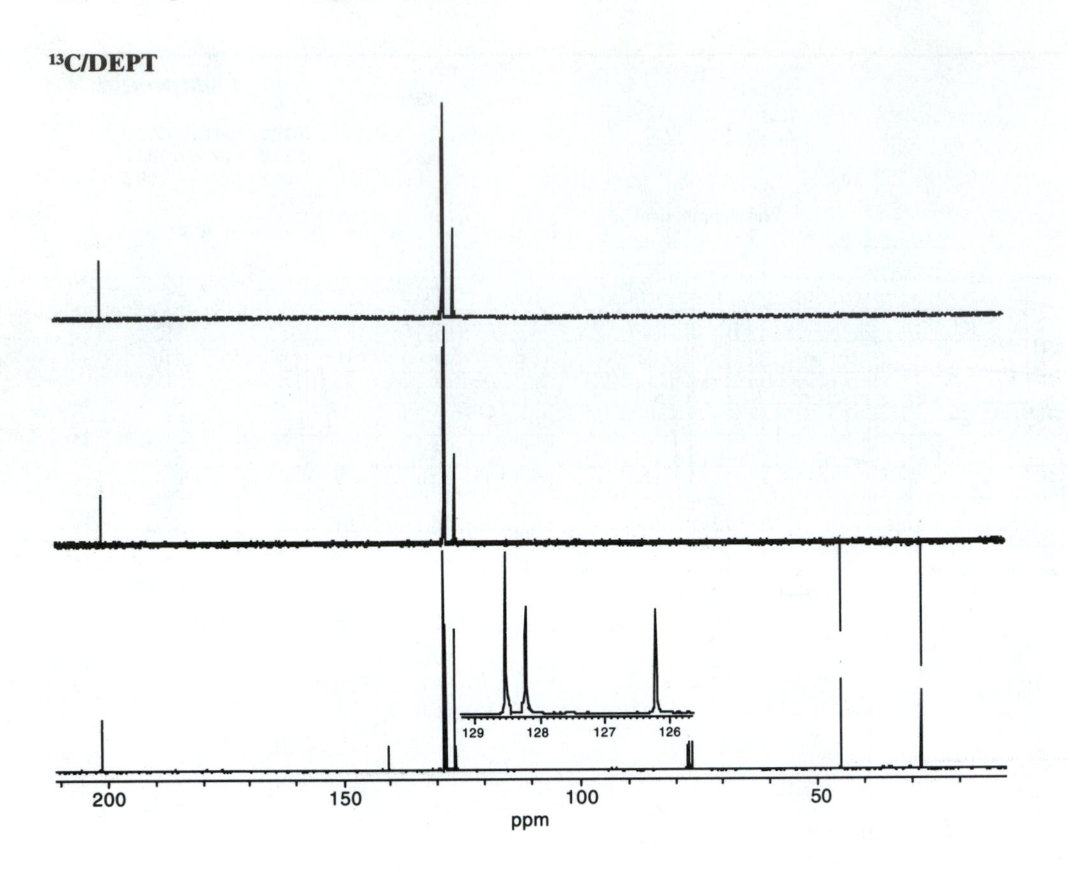

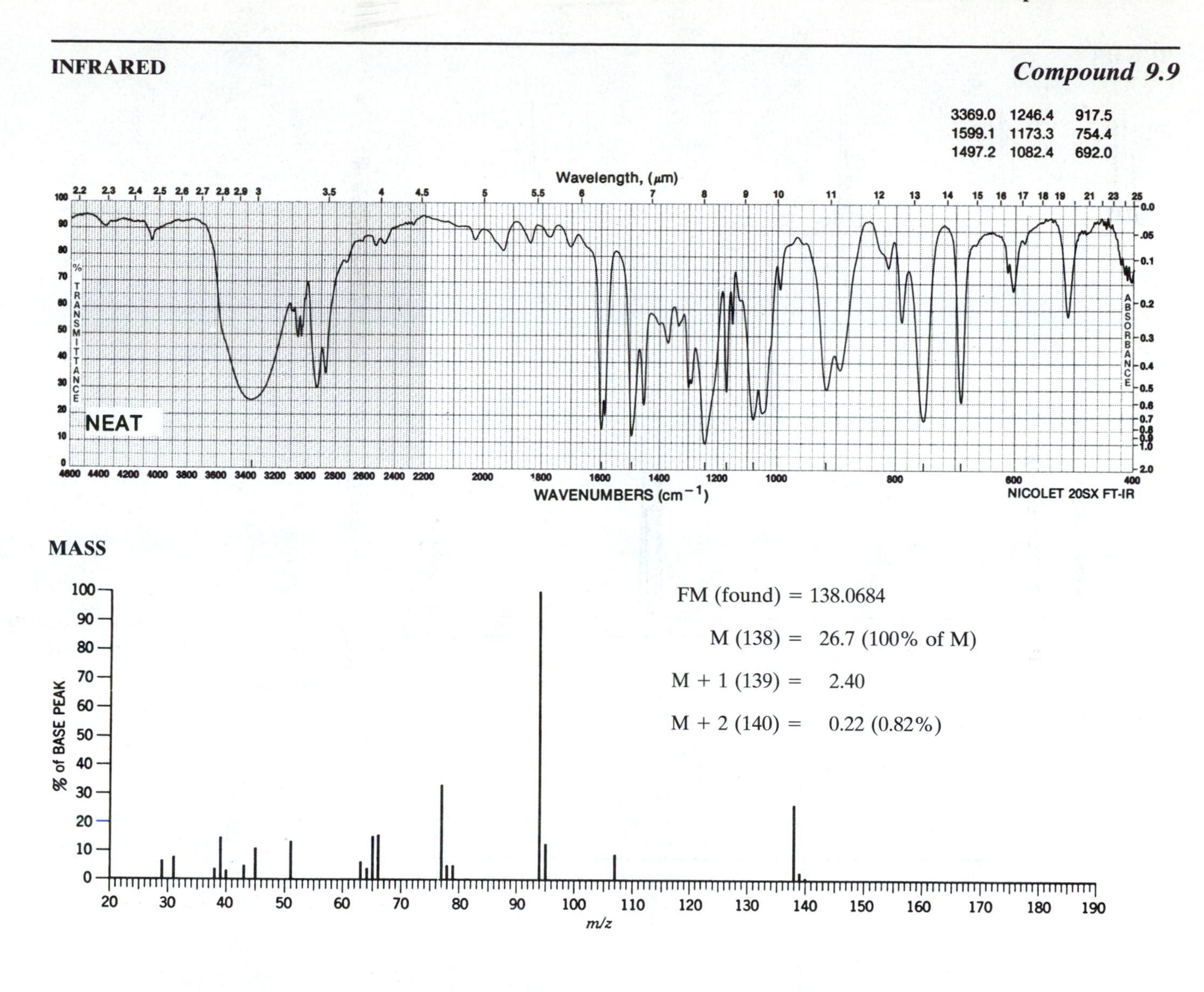
INFRARED
Compound 9.9
3369.0 1246.4 917.5
1599.1 1173.3 754.4
1497.2 1082.4 692.0
Wavelength, (μm)
% TRANSMITTANCE
ABSORBANCE
NEAT
WAVENUMBERS (cm⁻¹)
NICOLET 20SX FT-IR
MASS
% of BASE PEAK
m/z
FM (found) = 138.0684
M (138) = 26.7 (100% of M)
M + 1 (139) = 2.40
M + 2 (140) = 0.22 (0.82%)

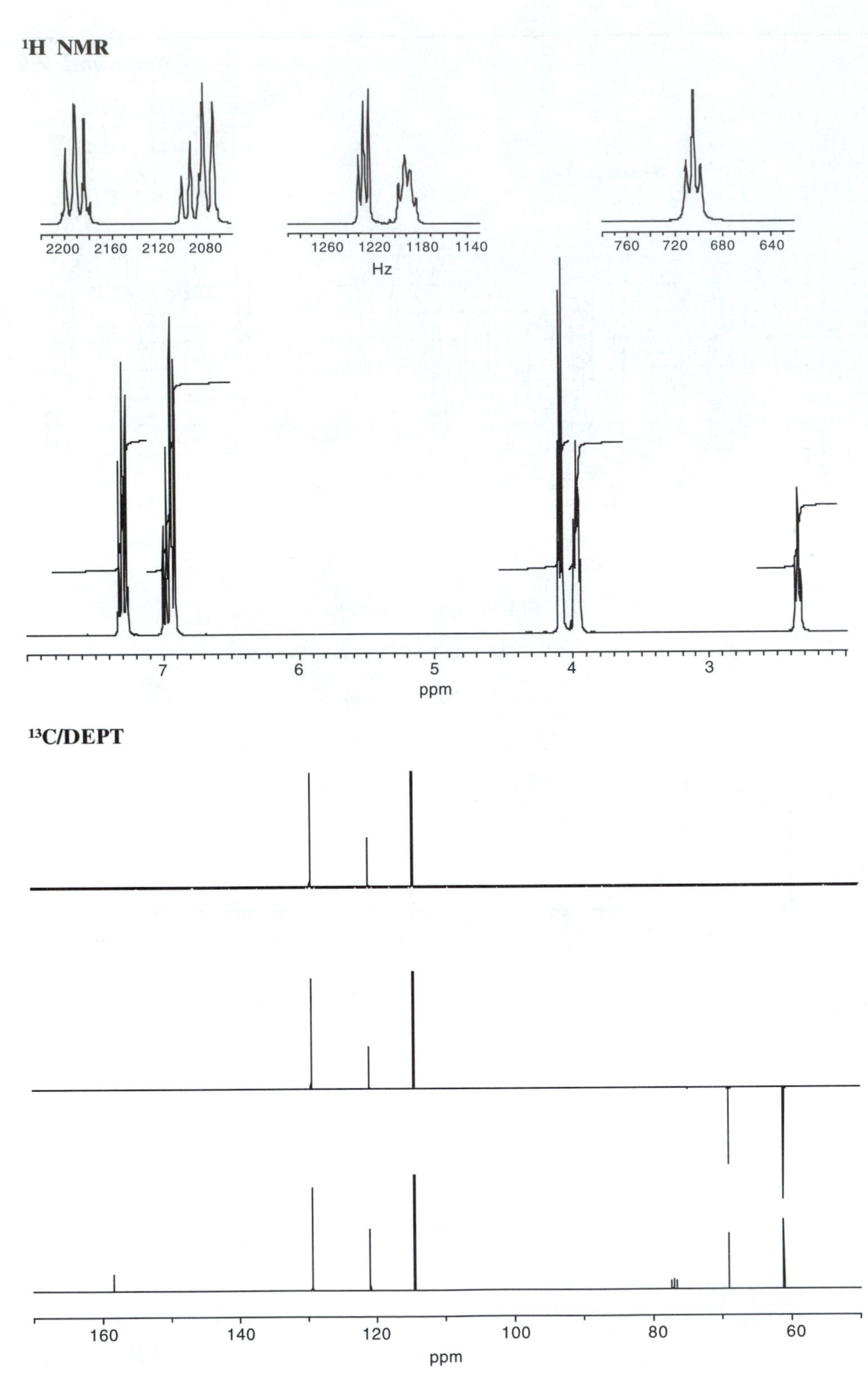

¹H NMR
2200
2160
2120
2080
1260
1220
1180
1140
Hz
760
720
680
640
7
6
5
4
3
ppm
¹³C/DEPT
160
140
120
100
80
60
ppm

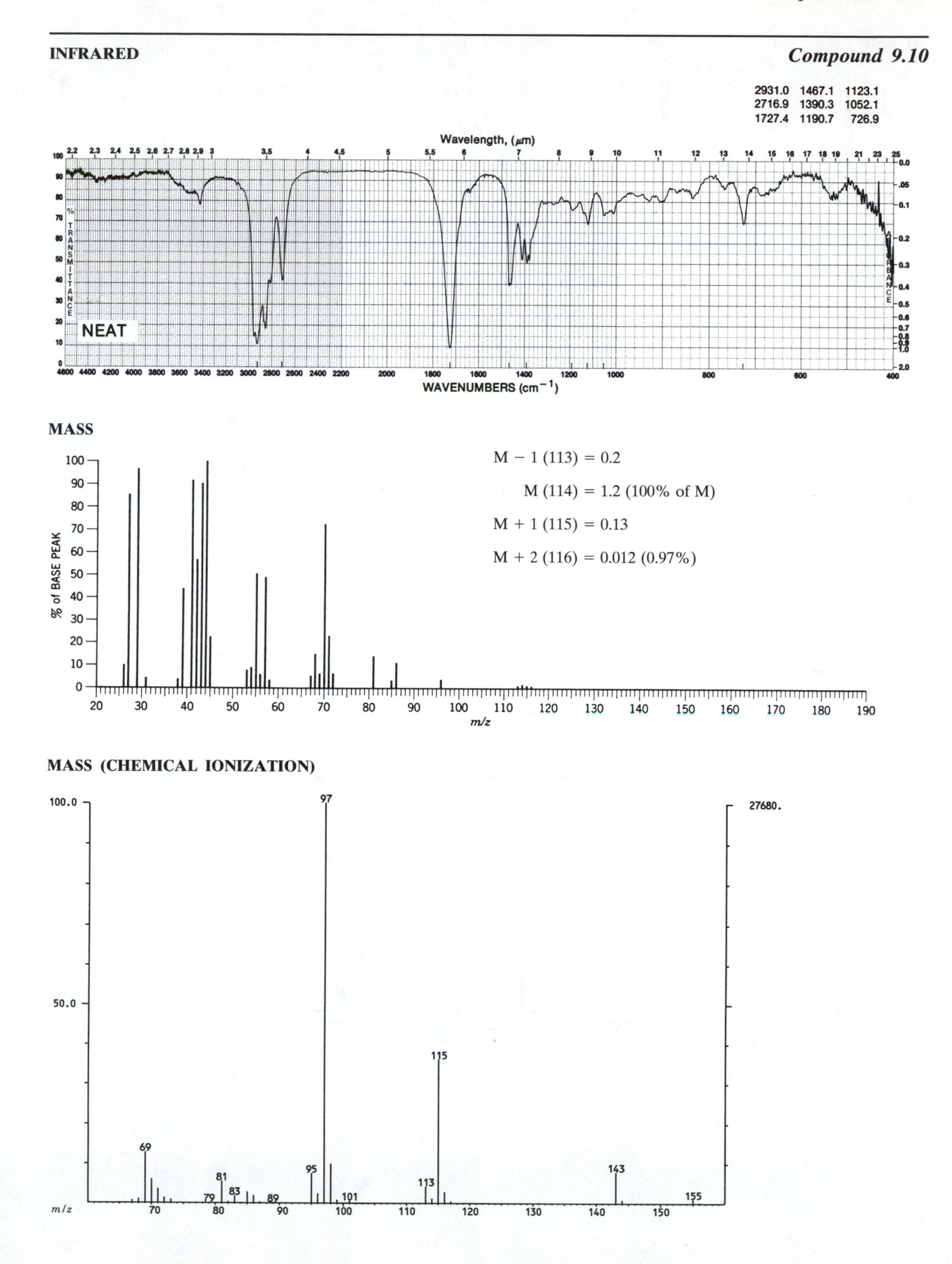

INFRARED
Compound 9.10
2931.0 1467.1 1123.1
2716.9 1390.3 1052.1
1727.4 1190.7 726.9
Wavelength, (μm)
NEAT
% TRANSMITTANCE
ABSORBANCE
WAVENUMBERS (cm⁻¹)
MASS
M − 1 (113) = 0.2
M (114) = 1.2 (100% of M)
M + 1 (115) = 0.13
M + 2 (116) = 0.012 (0.97%)
% of BASE PEAK
m/z
MASS (CHEMICAL IONIZATION)
100.0
50.0
27680.
97
115
69
95
81
143
113
83
79
89
101
155
m/z

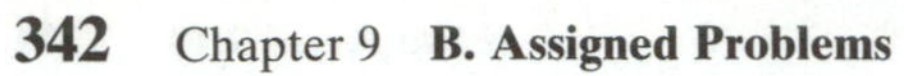

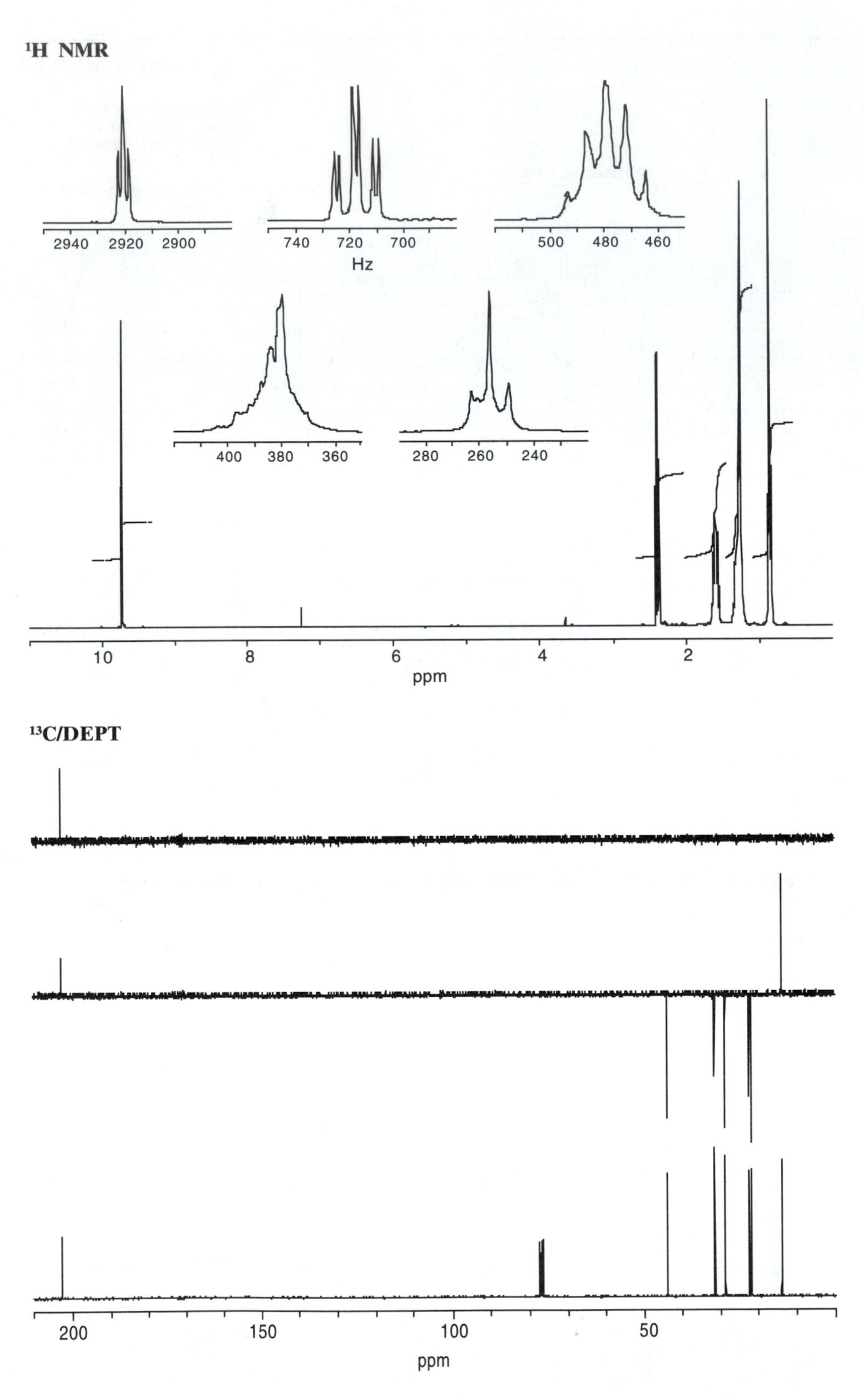
¹H NMR
2940 2920 2900
740 720 700
Hz
500 480 460
400 380 360
280 260 240
10 8 6 4 2
ppm
¹³C/DEPT
200 150 100 50
ppm

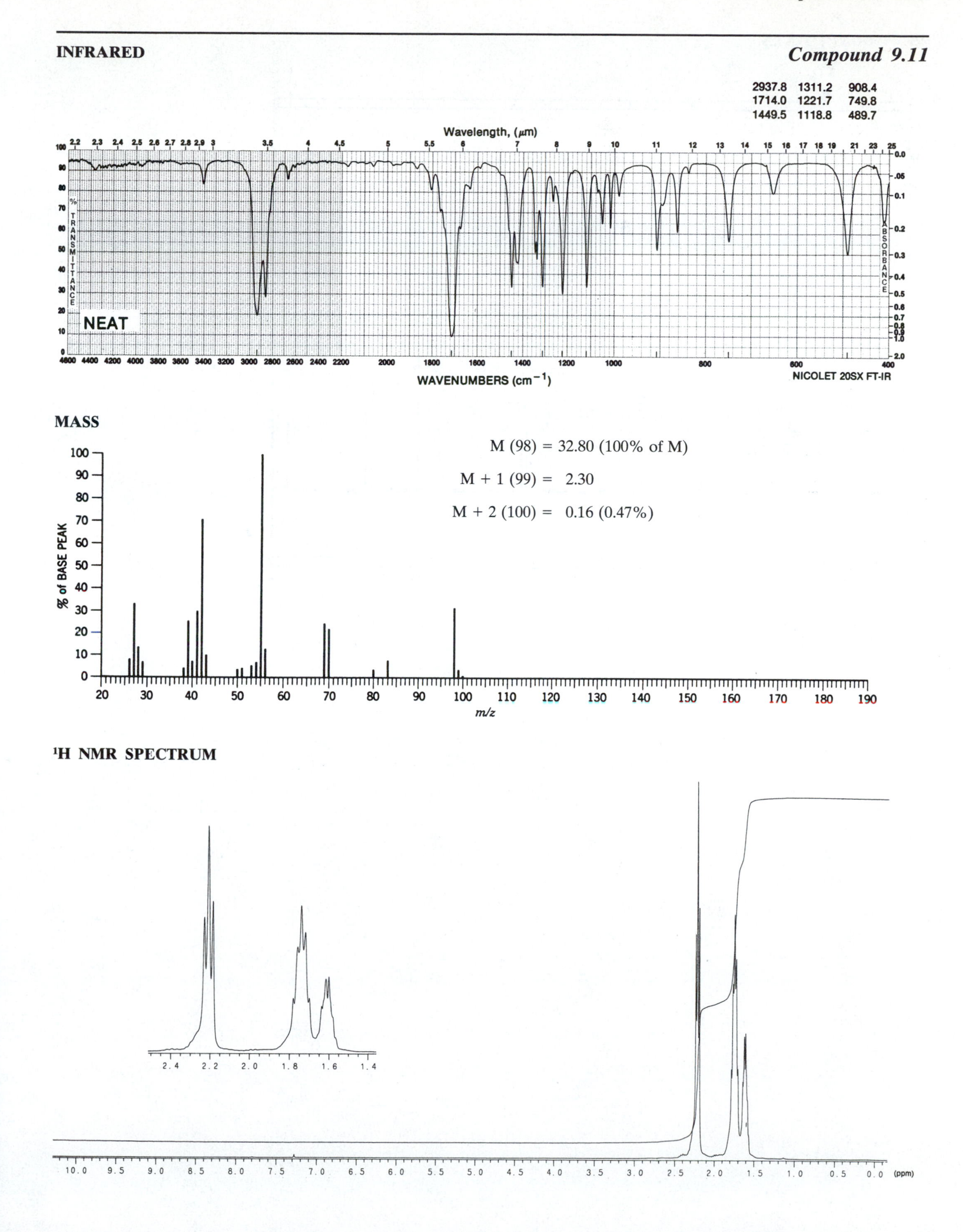
INFRARED
Compound 9.11
2937.8 1311.2 908.4
1714.0 1221.7 749.8
1449.5 1118.8 489.7
Wavelength, (μm)
% TRANSMITTANCE
ABSORBANCE
NEAT
WAVENUMBERS (cm⁻¹)
NICOLET 20SX FT-IR
MASS
M (98) = 32.80 (100% of M)
M + 1 (99) = 2.30
M + 2 (100) = 0.16 (0.47%)
% of BASE PEAK
m/z
¹H NMR SPECTRUM
(ppm)

^{13}C NMR/APT DATA

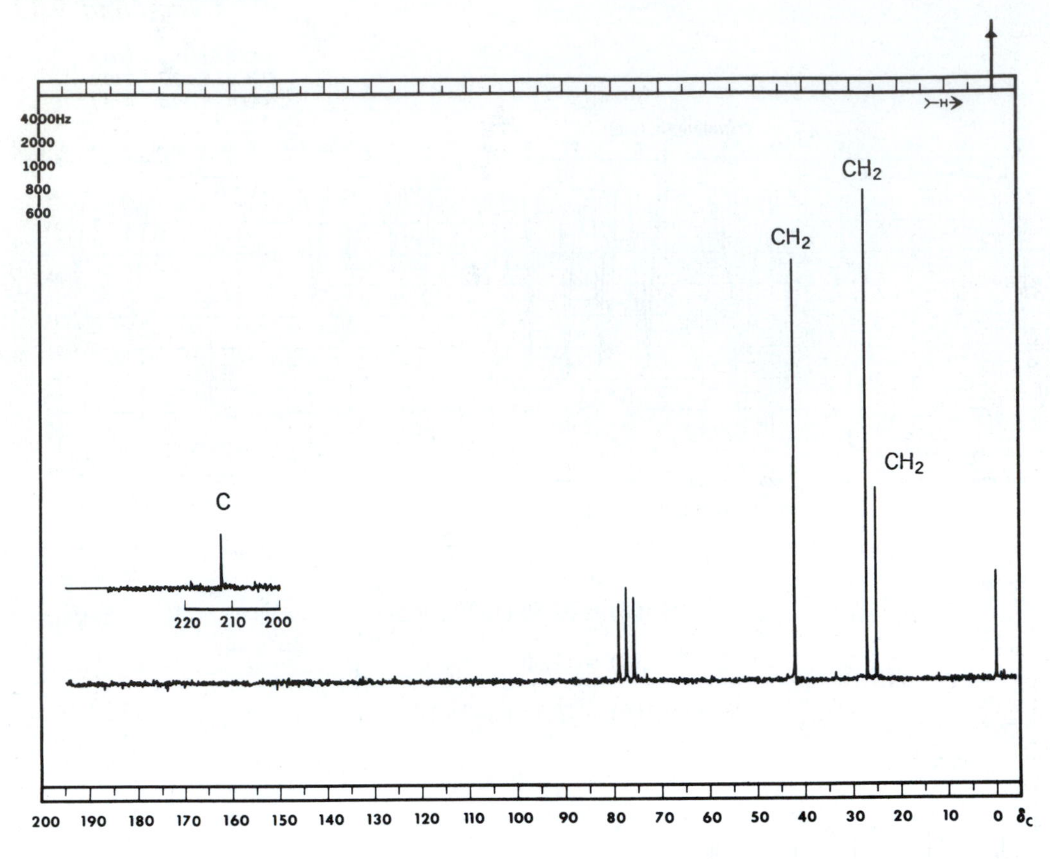

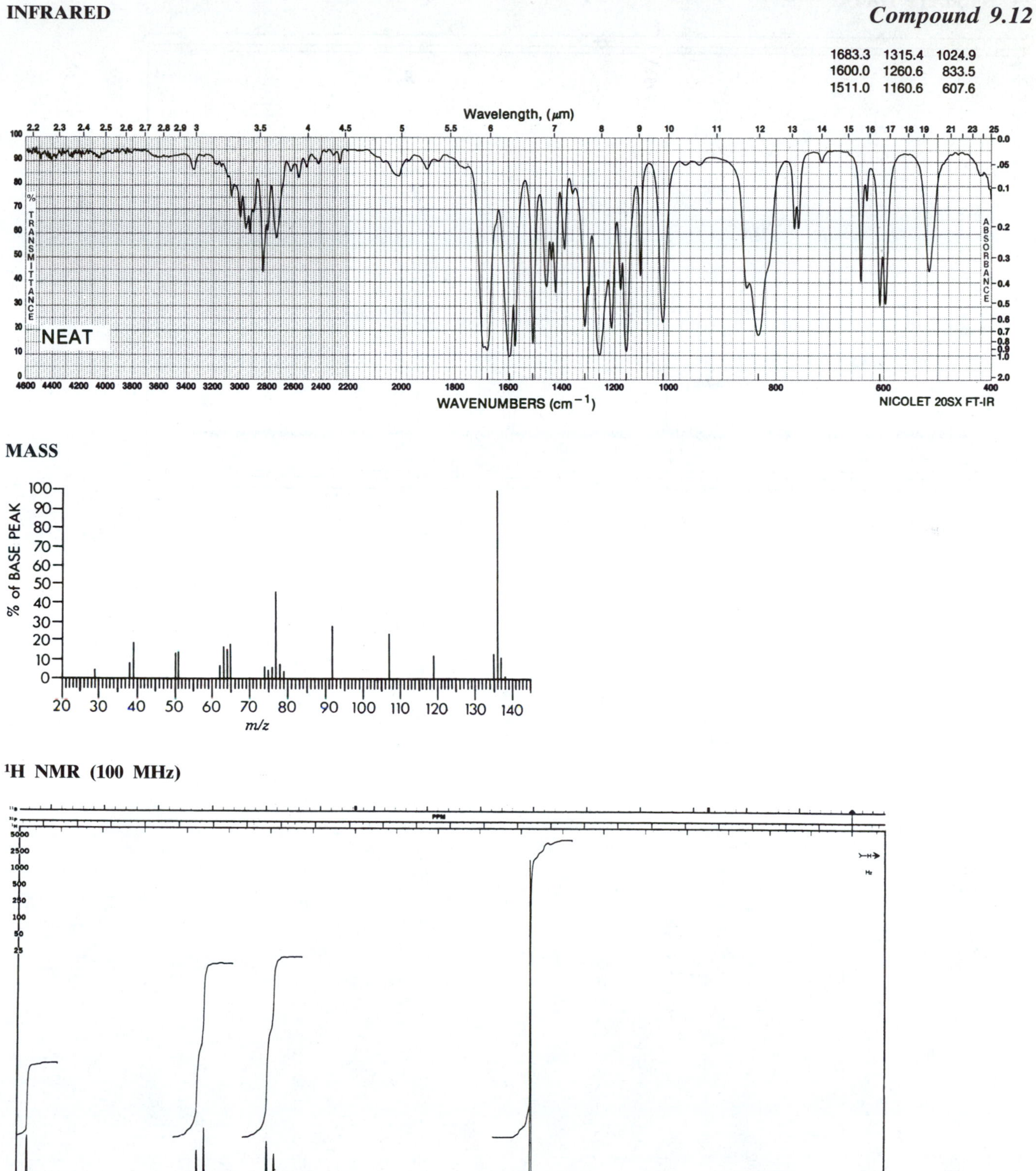
INFRARED
Compound 9.12
1683.3 1315.4 1024.9
1600.0 1260.6 833.5
1511.0 1160.6 607.6
Wavelength, (μm)
% TRANSMITTANCE
ABSORBANCE
NEAT
WAVENUMBERS (cm⁻¹)
NICOLET 20SX FT-IR
MASS
% of BASE PEAK
m/z
¹H NMR (100 MHz)
PPM
δ

^{13}C NMR/ATP DATA

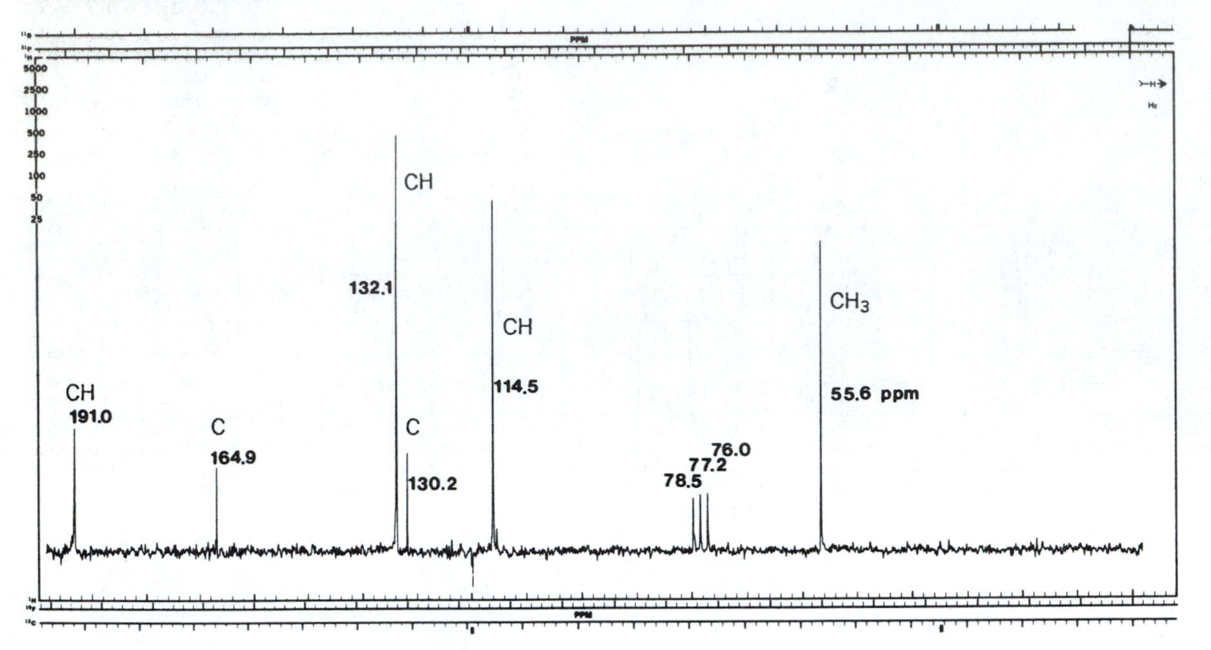

INFRARED

Compound 9.13

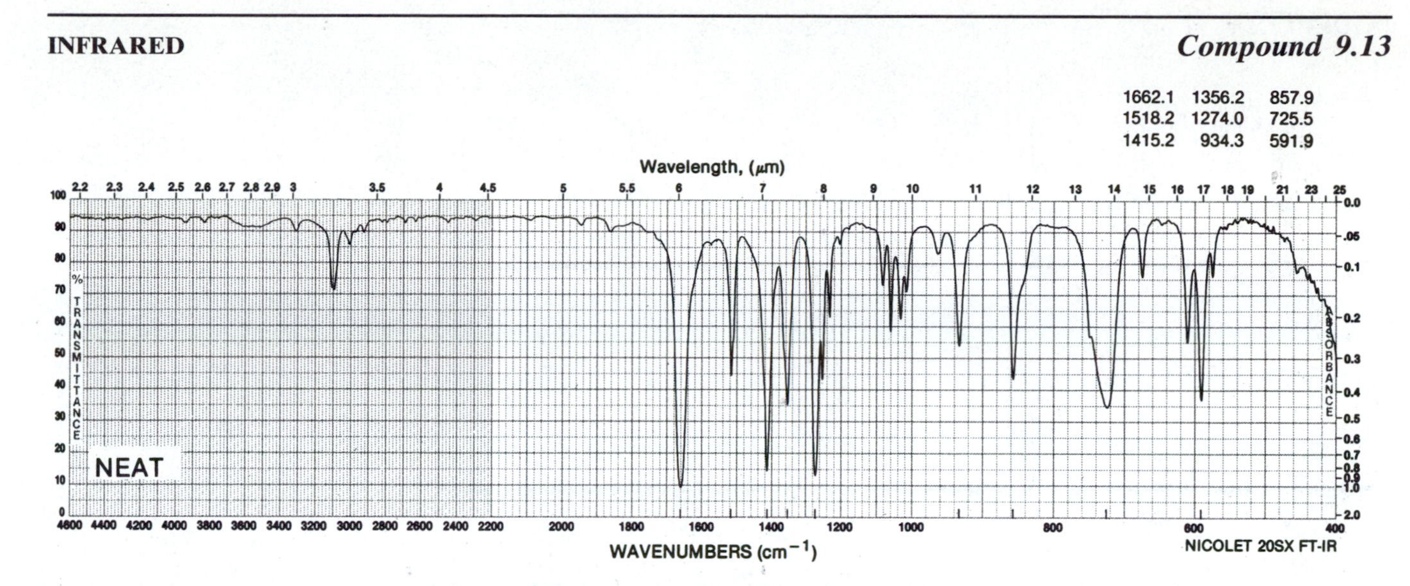

MASS

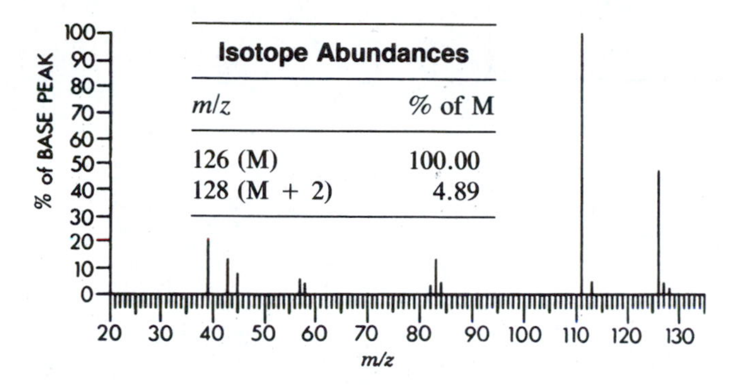

Isotope Abundances	
m/z	% of M
126 (M)	100.00
128 (M + 2)	4.89

¹H NMR

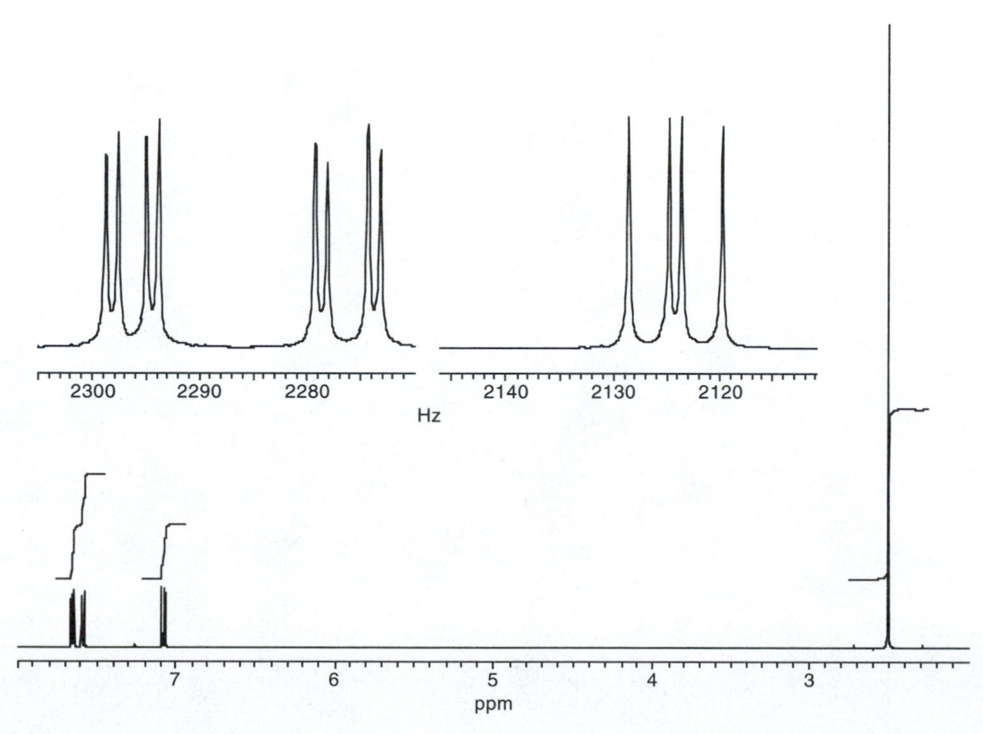

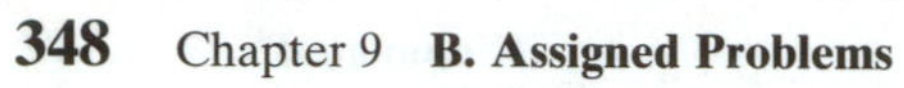

^{13}C/DEPT

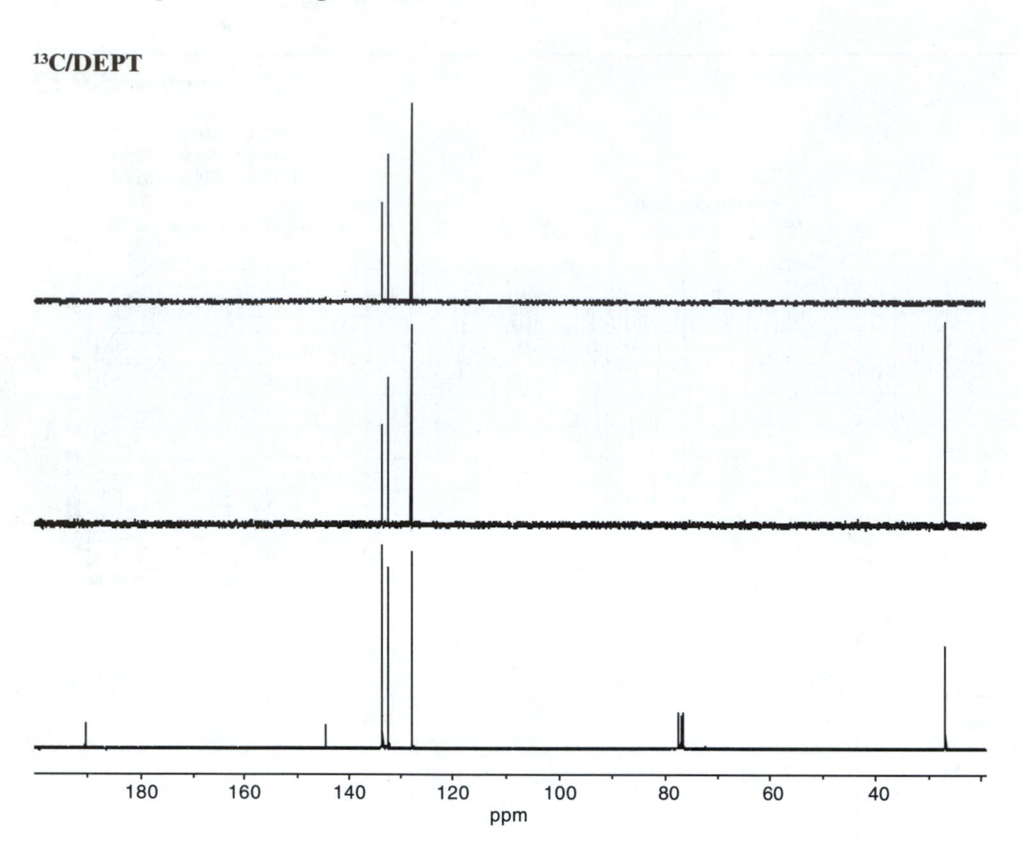

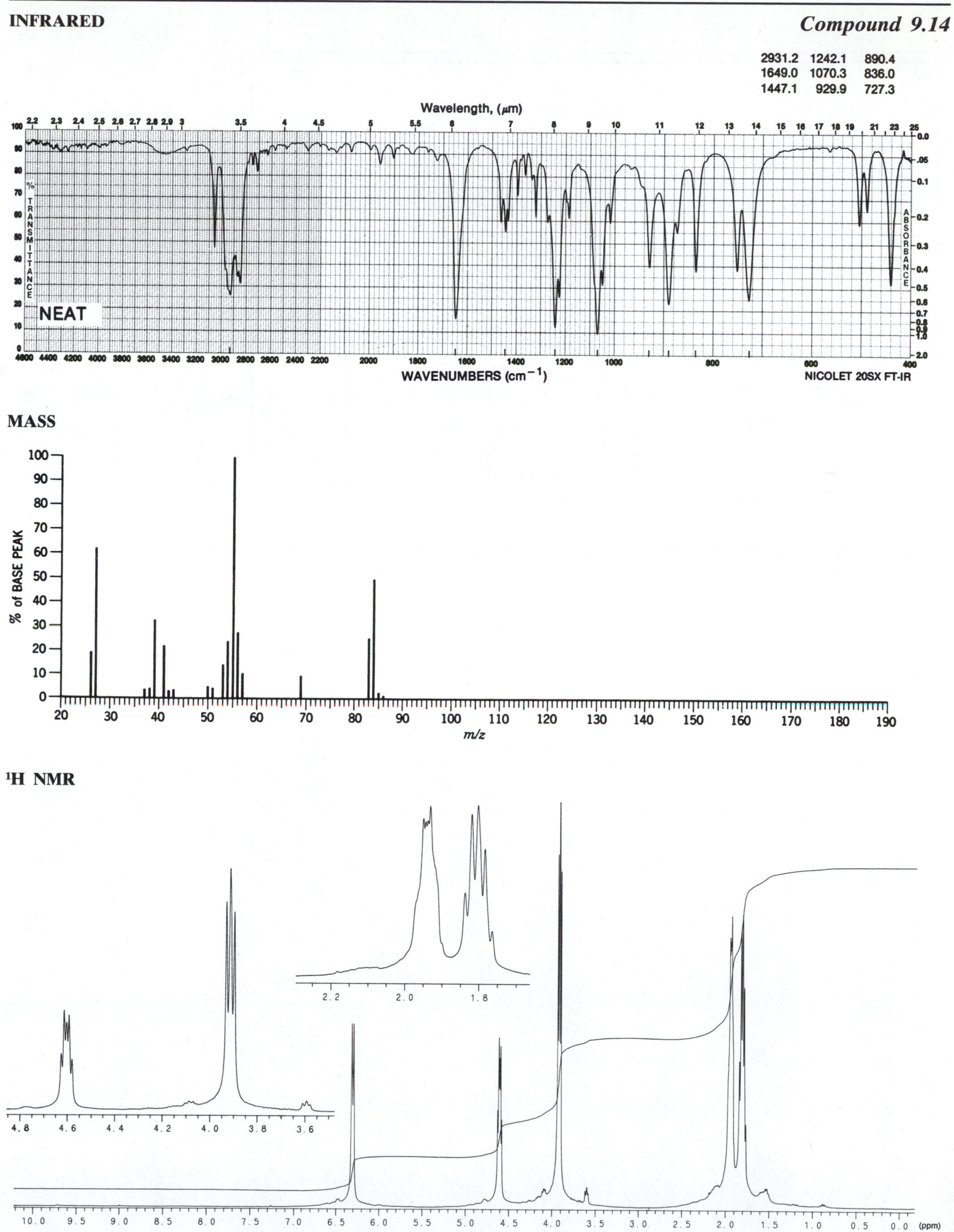

INFRARED
Compound 9.14
2931.2 1242.1 890.4
1649.0 1070.3 836.0
1447.1 929.9 727.3
Wavelength, (μm)
% TRANSMITTANCE
ABSORBANCE
NEAT
WAVENUMBERS (cm⁻¹)
NICOLET 20SX FT-IR
MASS
% of BASE PEAK
m/z
¹H NMR
(ppm)

^{13}C NMR/ATP DATA

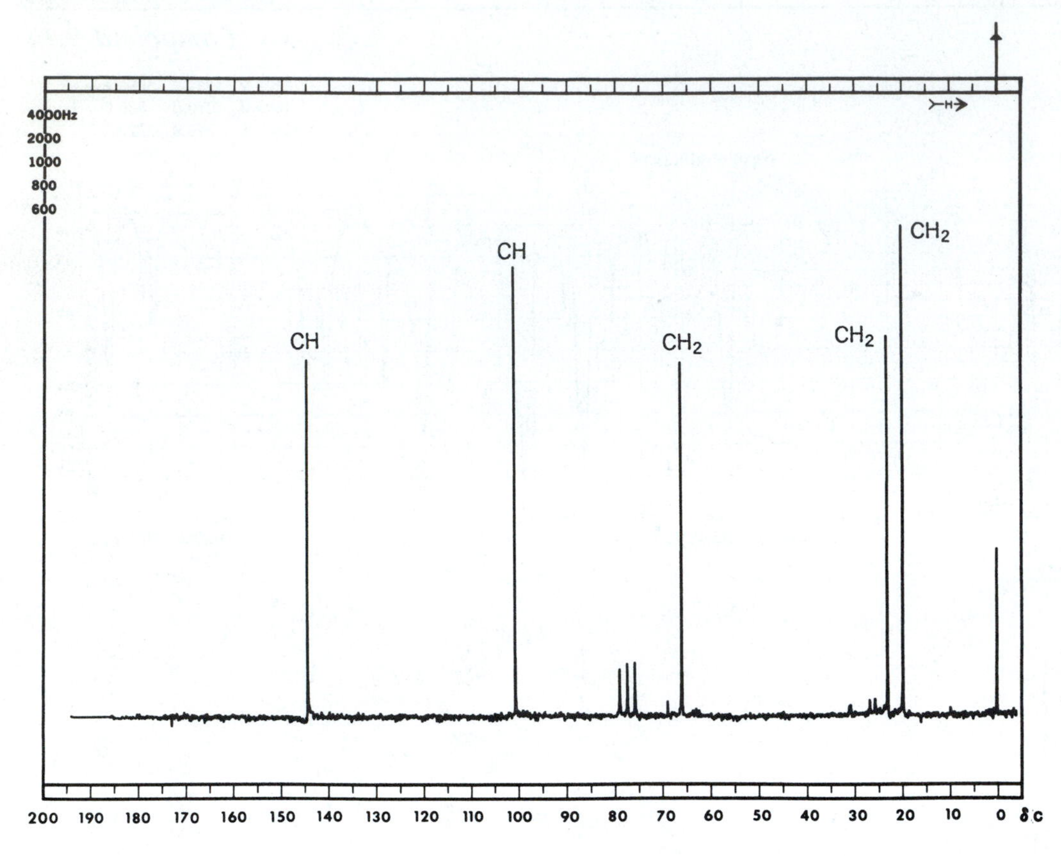

INFRARED

Compound 9.15

2980.7	1267.2	920.7
1641.1	1211.9	678.8
1428.3	993.7	563.7

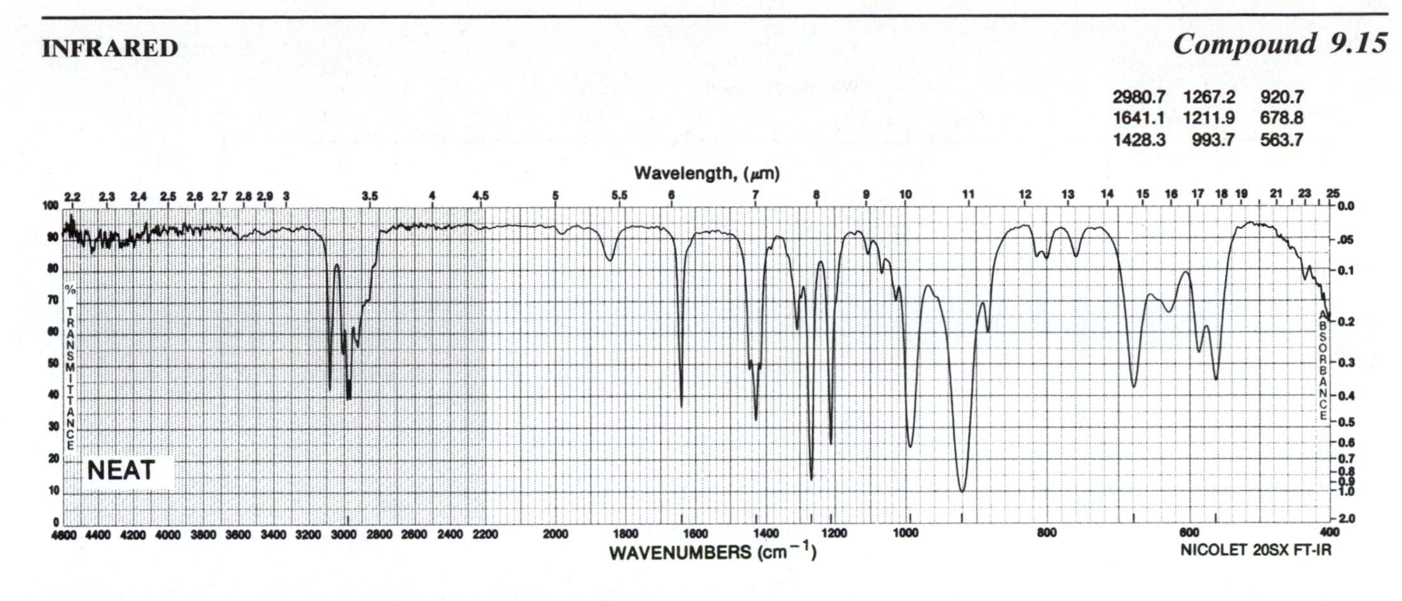

MASS SPECTRAL DATA (RELATIVE INTENSITIES)

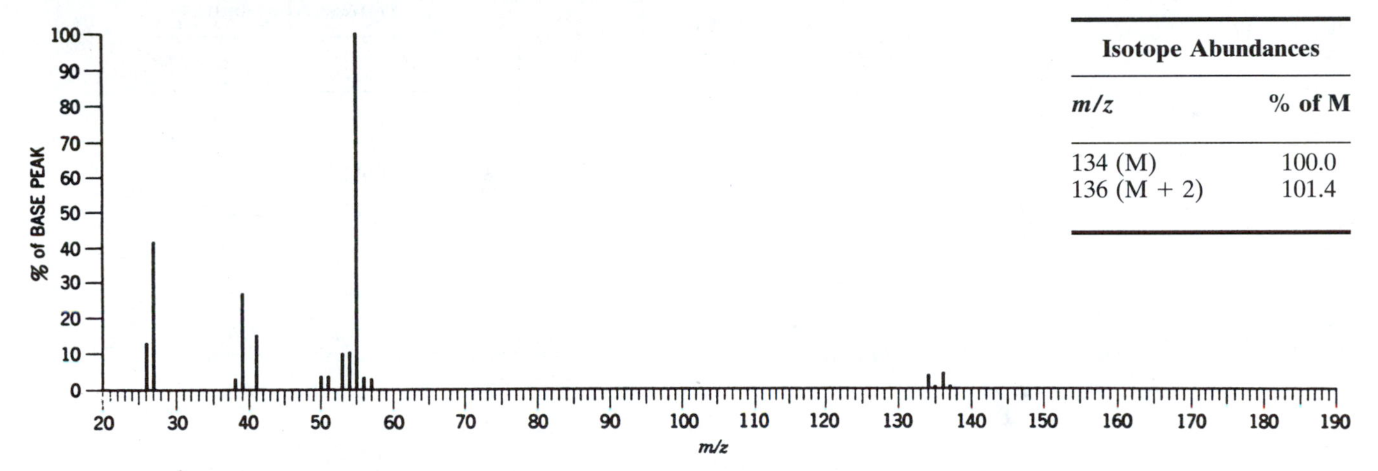

Isotope Abundances	
m/z	% of M
134 (M)	100.0
136 (M + 2)	101.4

¹H NMR SPECTRUM

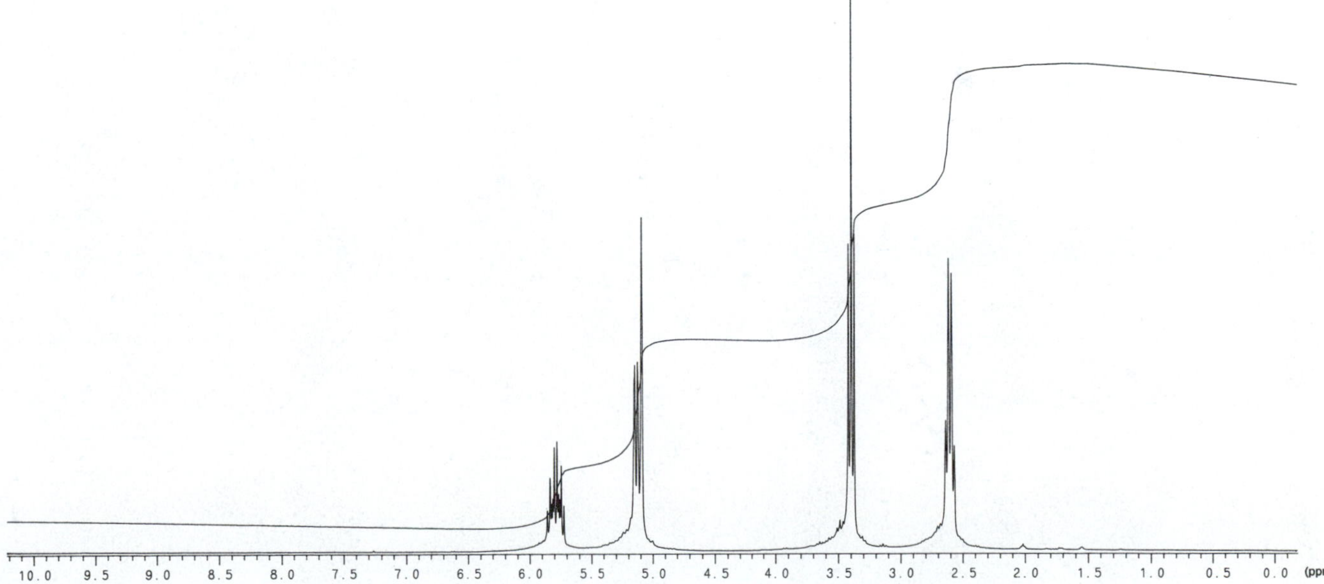

INFRARED *Compound 9.16*

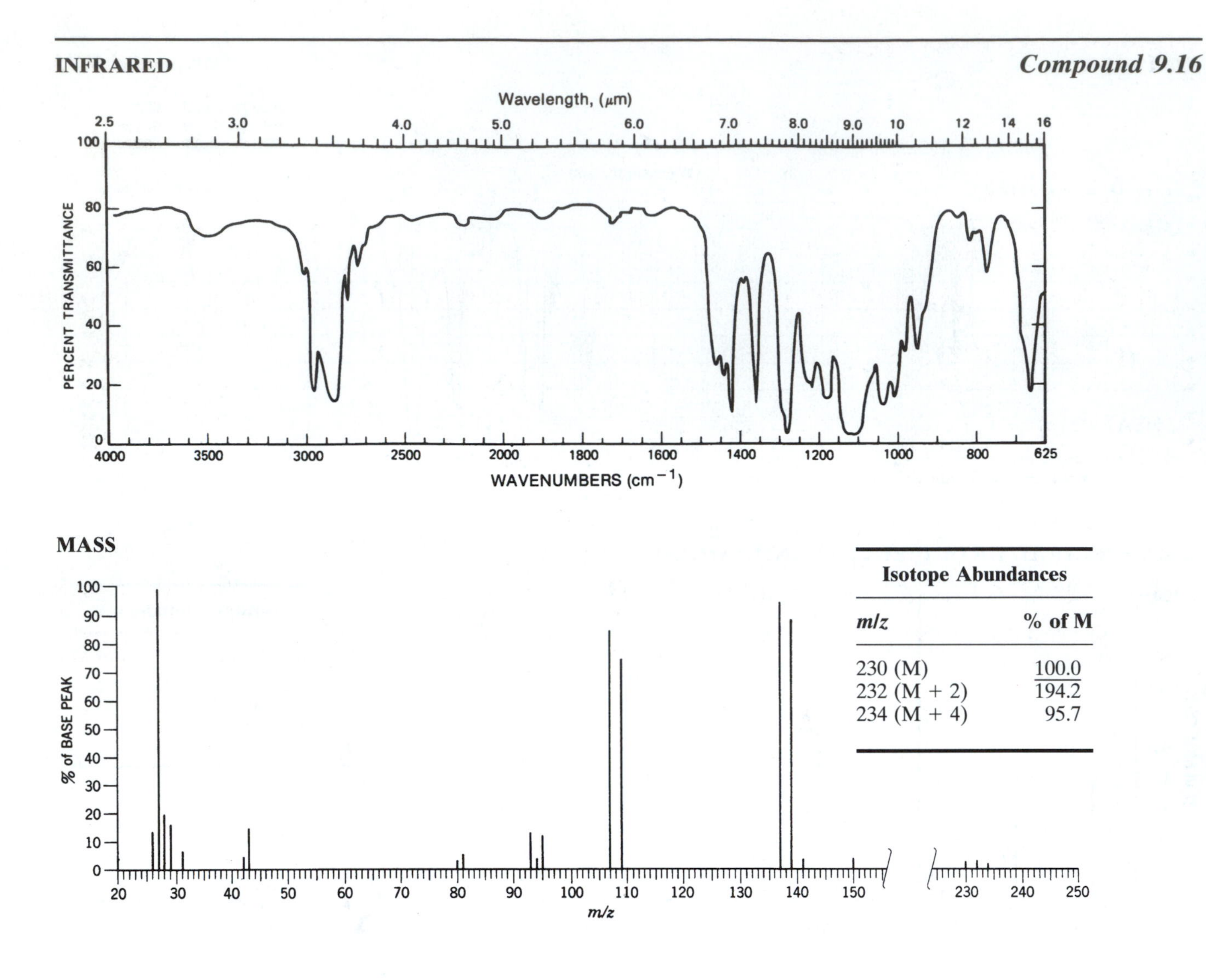

Isotope Abundances	
m/z	% of M
230 (M)	100.0
232 (M + 2)	194.2
234 (M + 4)	95.7

MASS (CHEMICAL IONIZATION)

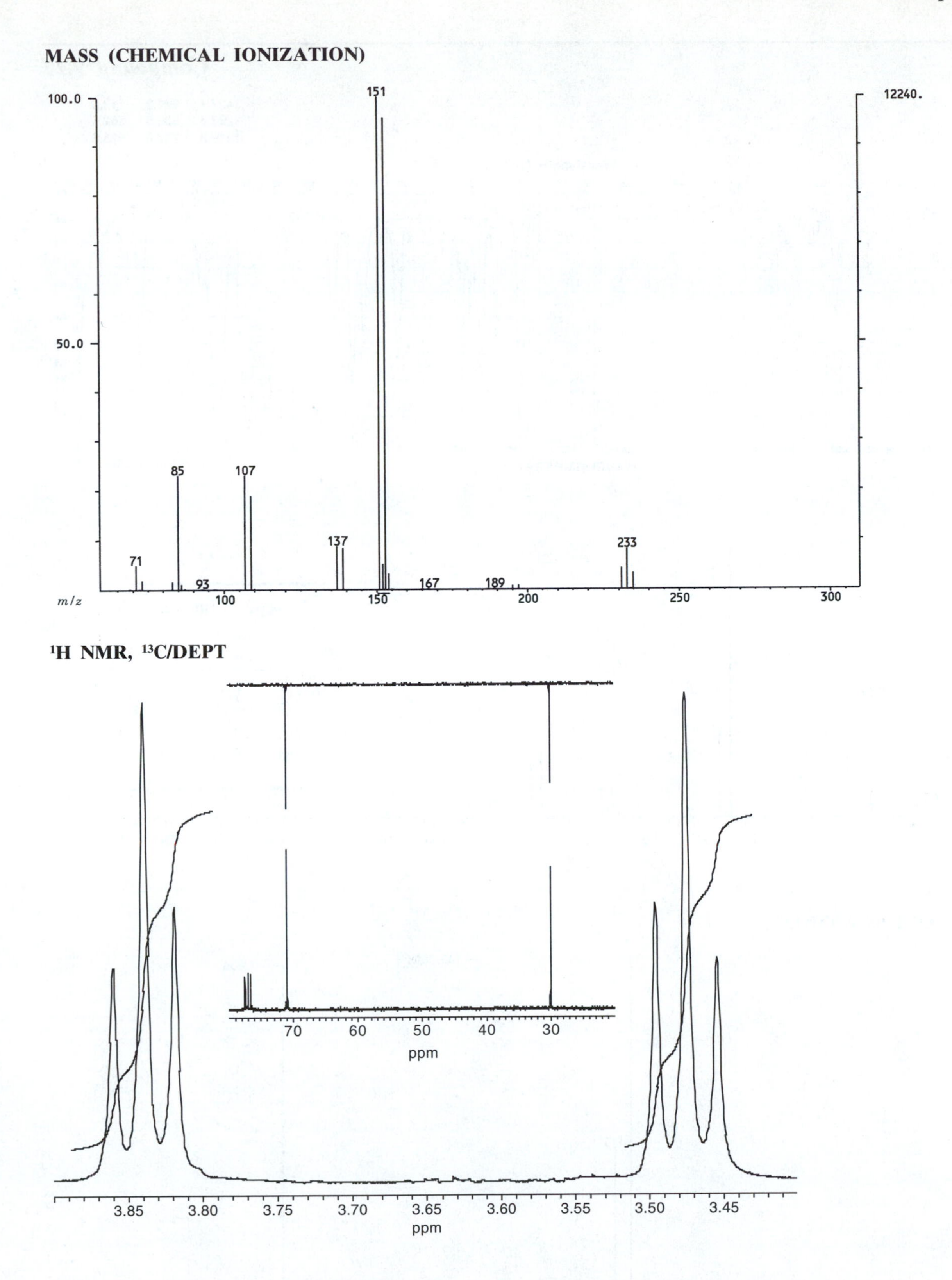

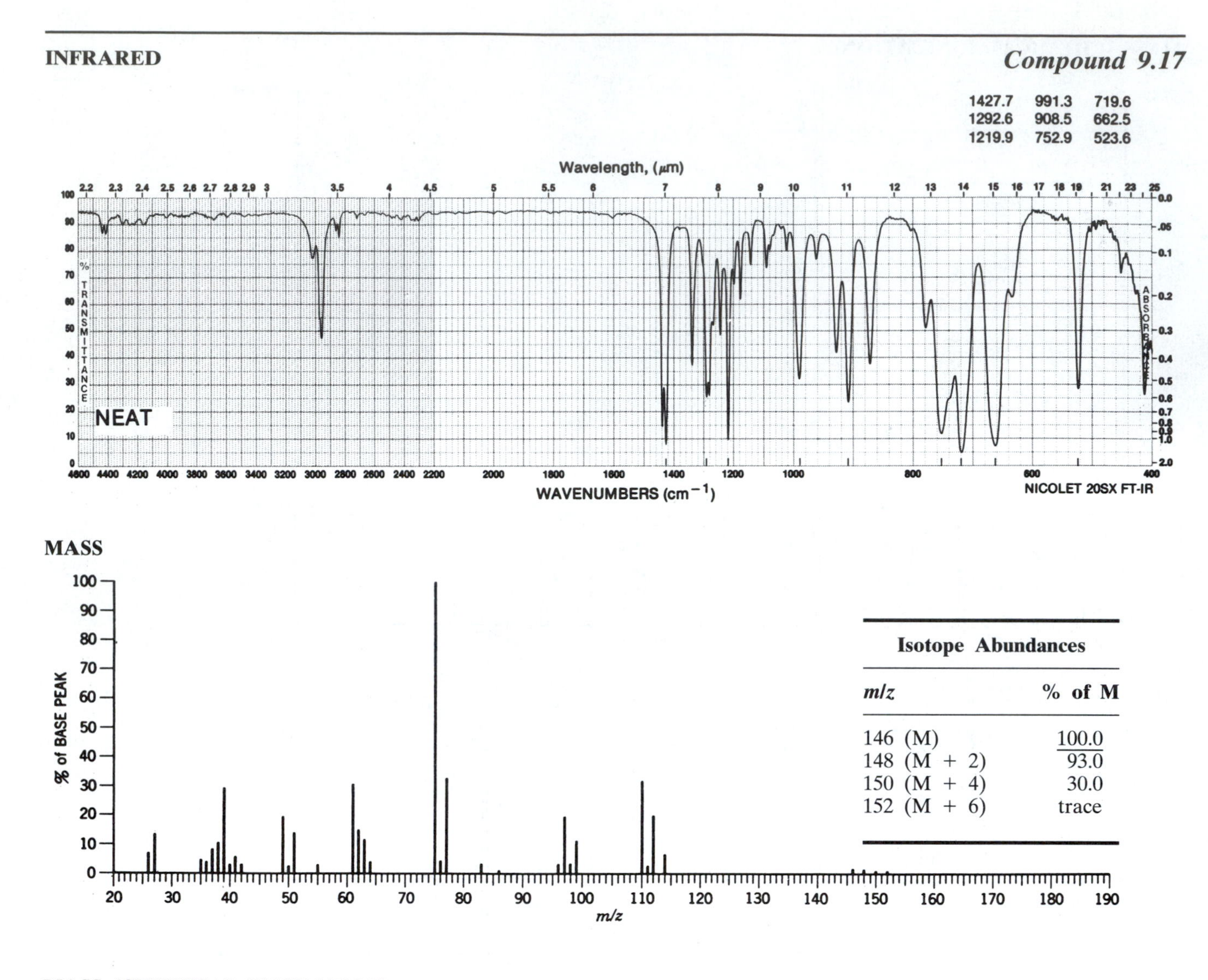

Isotope Abundances

m/z	% of M
146 (M)	100.0
148 (M + 2)	93.0
150 (M + 4)	30.0
152 (M + 6)	trace

MASS (CHEMICAL IONIZATION)

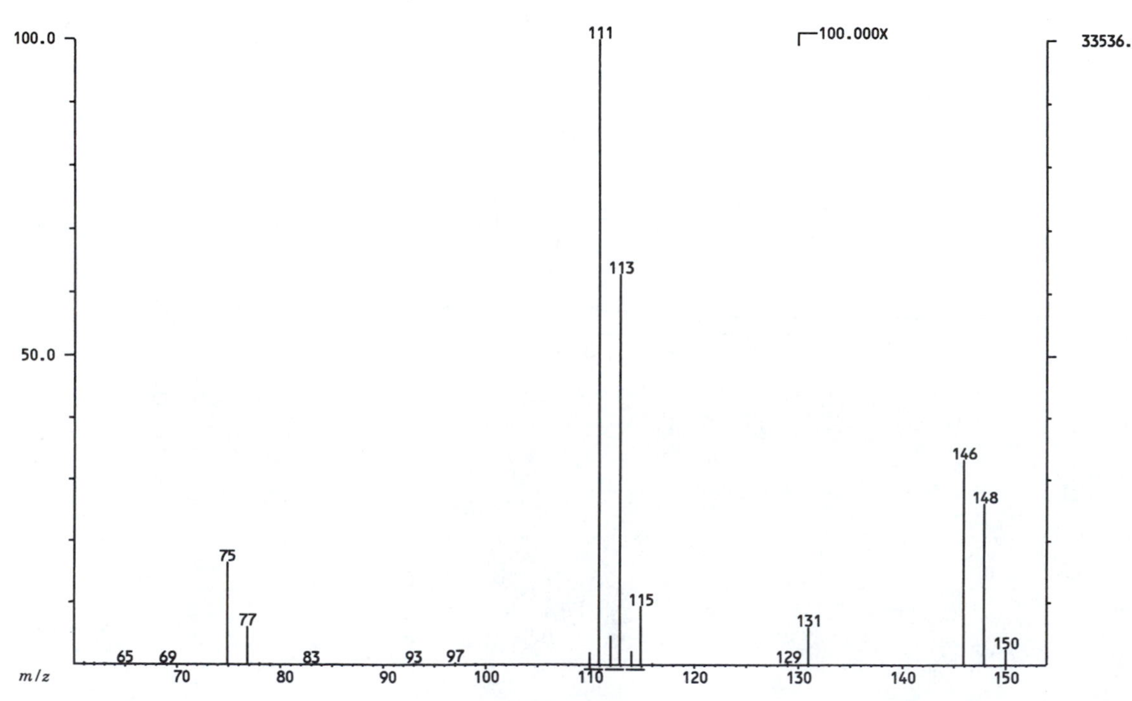

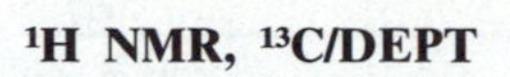

^{1}H NMR, ^{13}C/DEPT

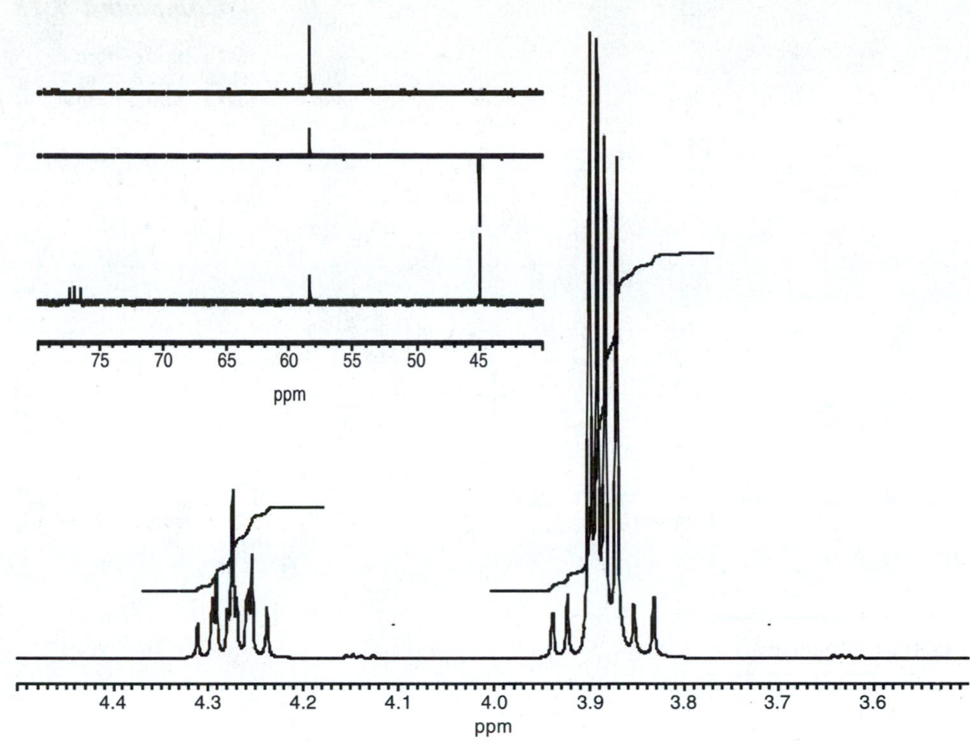

INFRARED ***Compound 9.18***

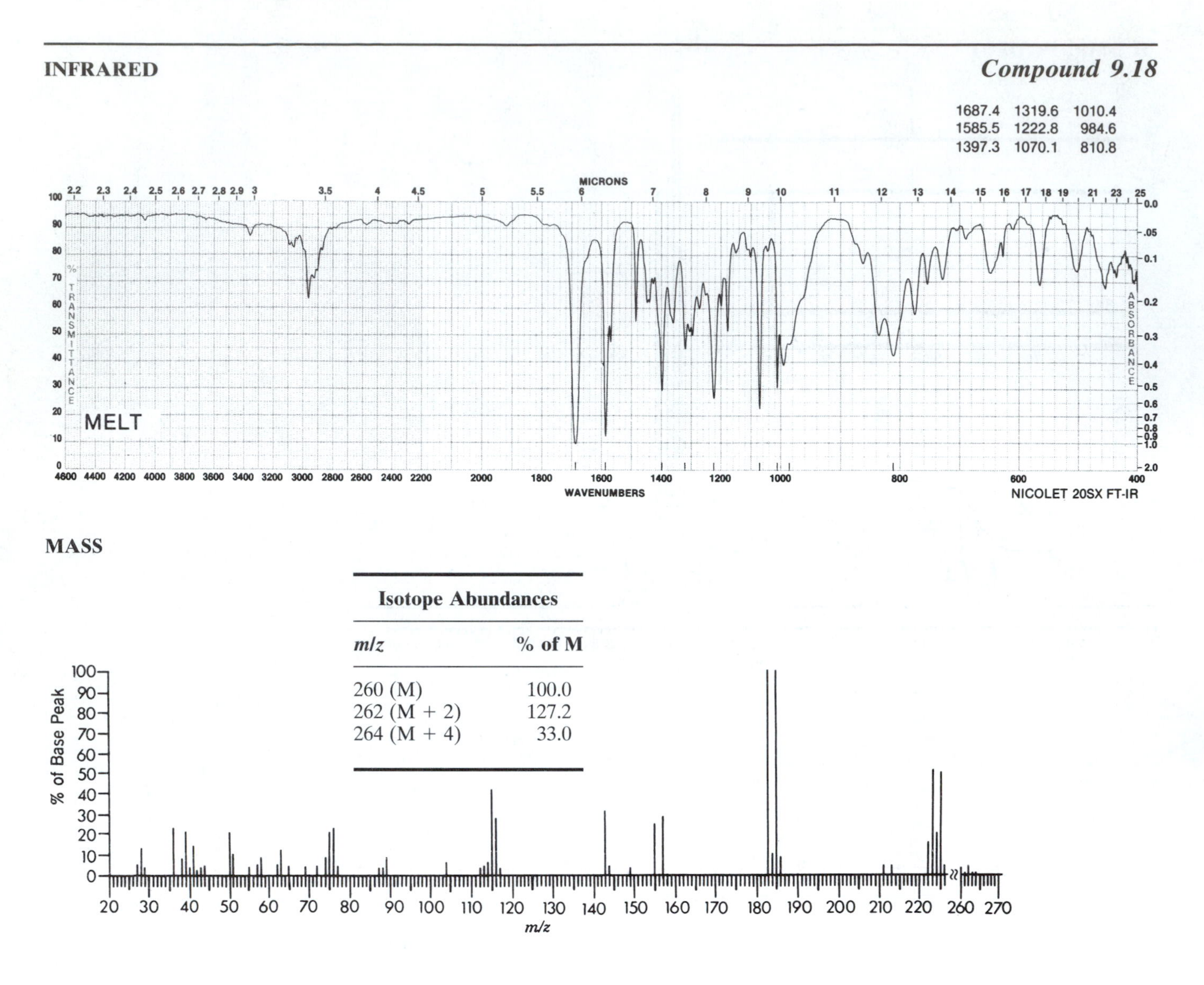

Isotope Abundances

m/z	% of M
260 (M)	100.0
262 (M + 2)	127.2
264 (M + 4)	33.0

MASS (CHEMICAL IONIZATION)

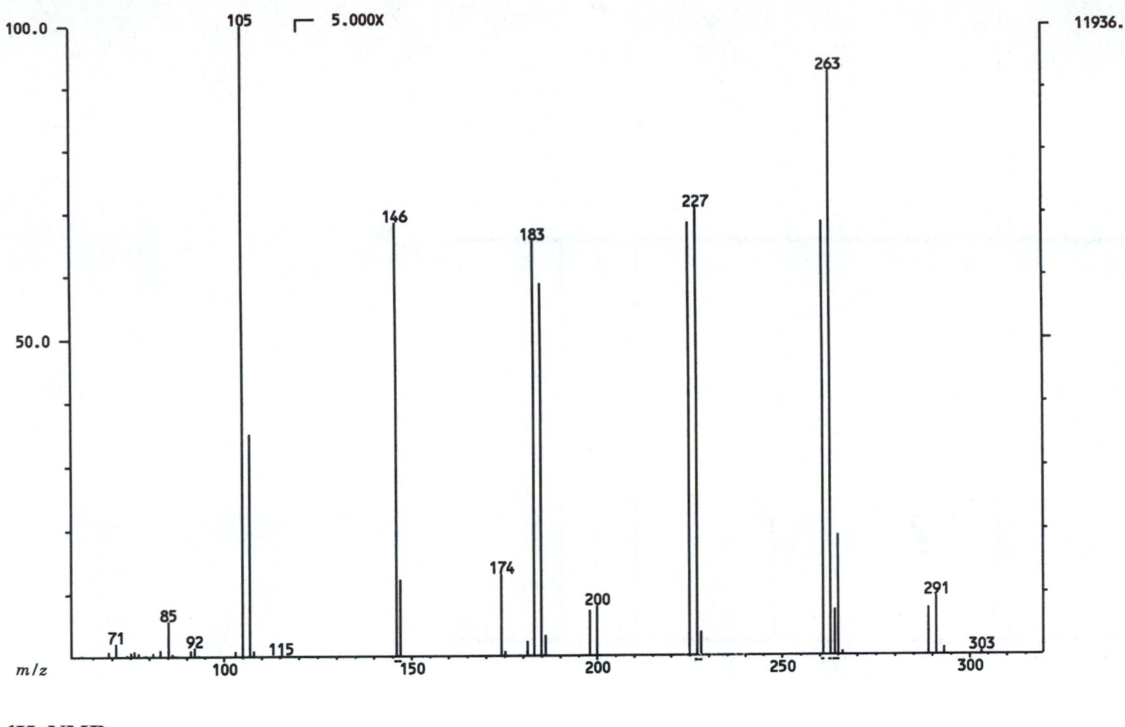

¹H NMR

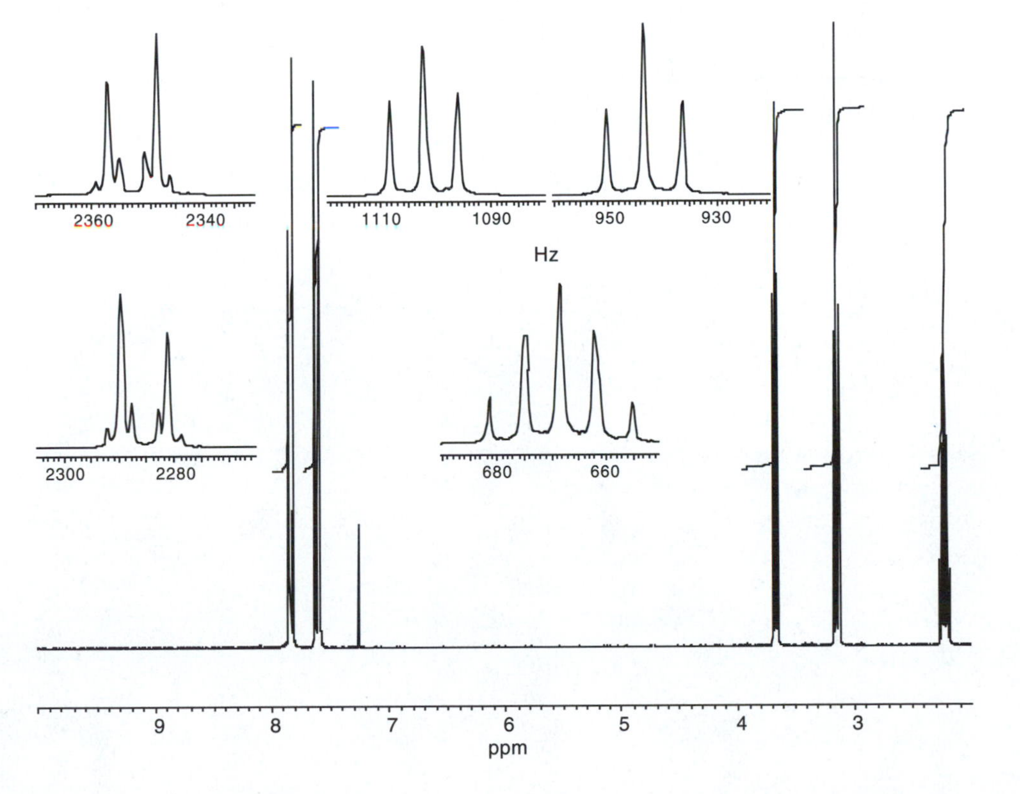

^{13}C/DEPT

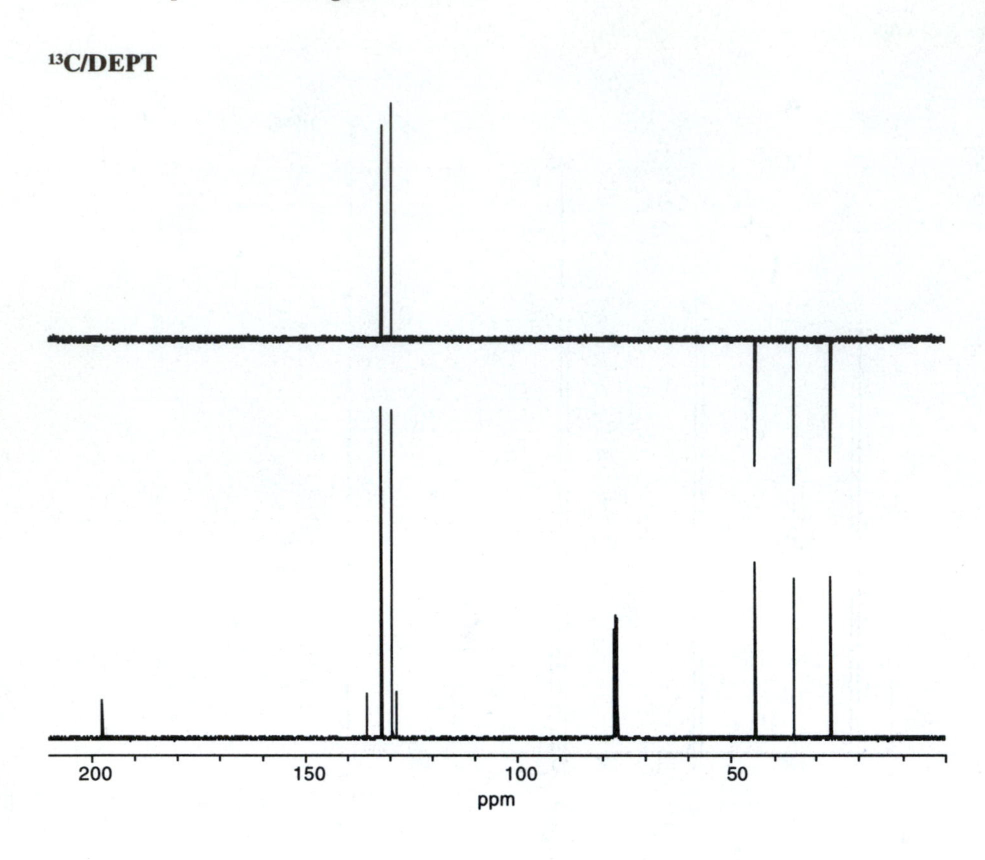

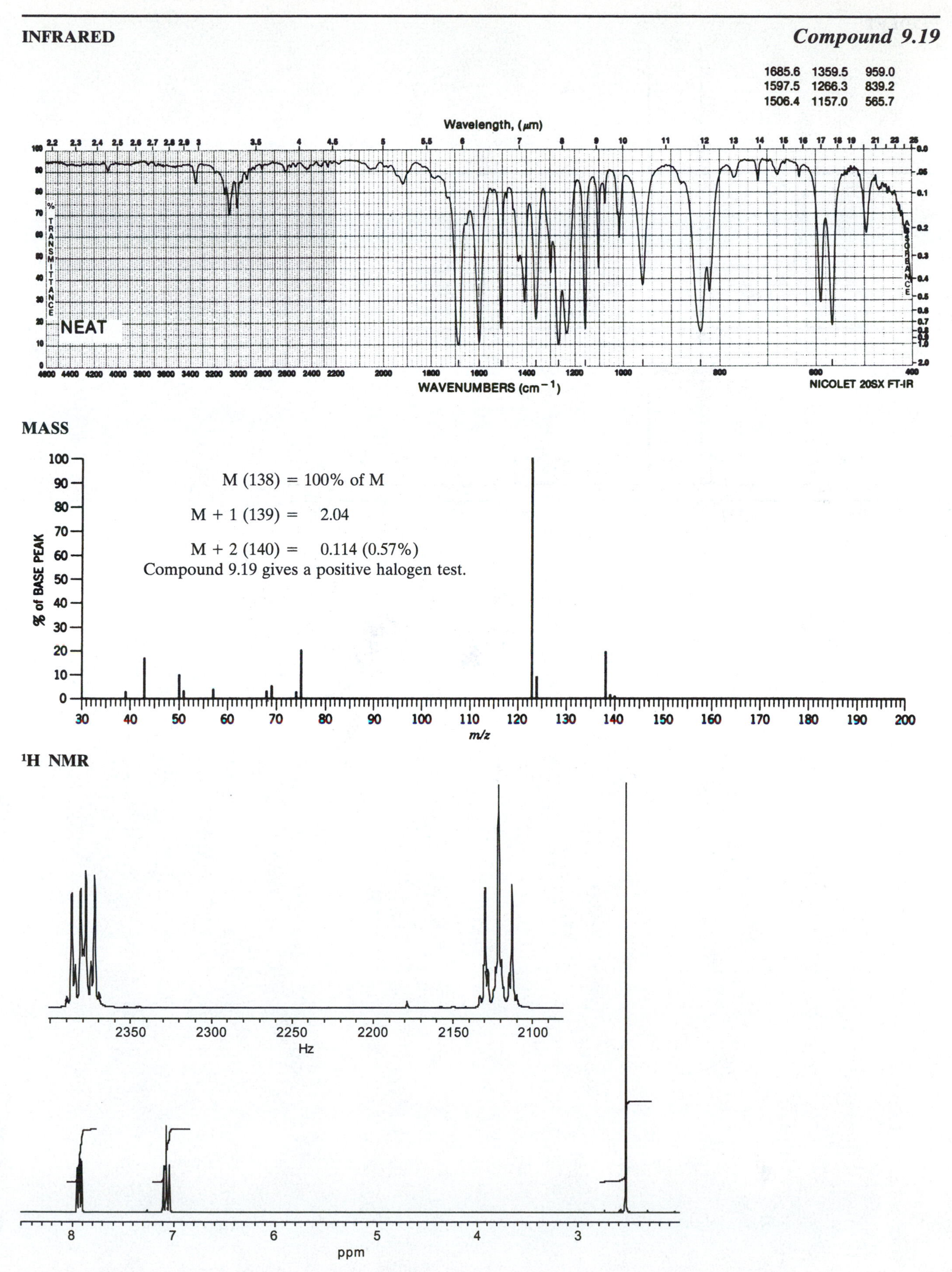
INFRARED
Compound 9.19
1685.6 1359.5 959.0
1597.5 1266.3 839.2
1506.4 1157.0 565.7
Wavelength, (μm)
% TRANSMITTANCE
ABSORBANCE
NEAT
WAVENUMBERS (cm⁻¹)
NICOLET 20SX FT-IR
MASS
M (138) = 100% of M
M + 1 (139) = 2.04
M + 2 (140) = 0.114 (0.57%)
Compound 9.19 gives a positive halogen test.
% of BASE PEAK
m/z
¹H NMR
Hz
ppm

^{13}C/DEPT

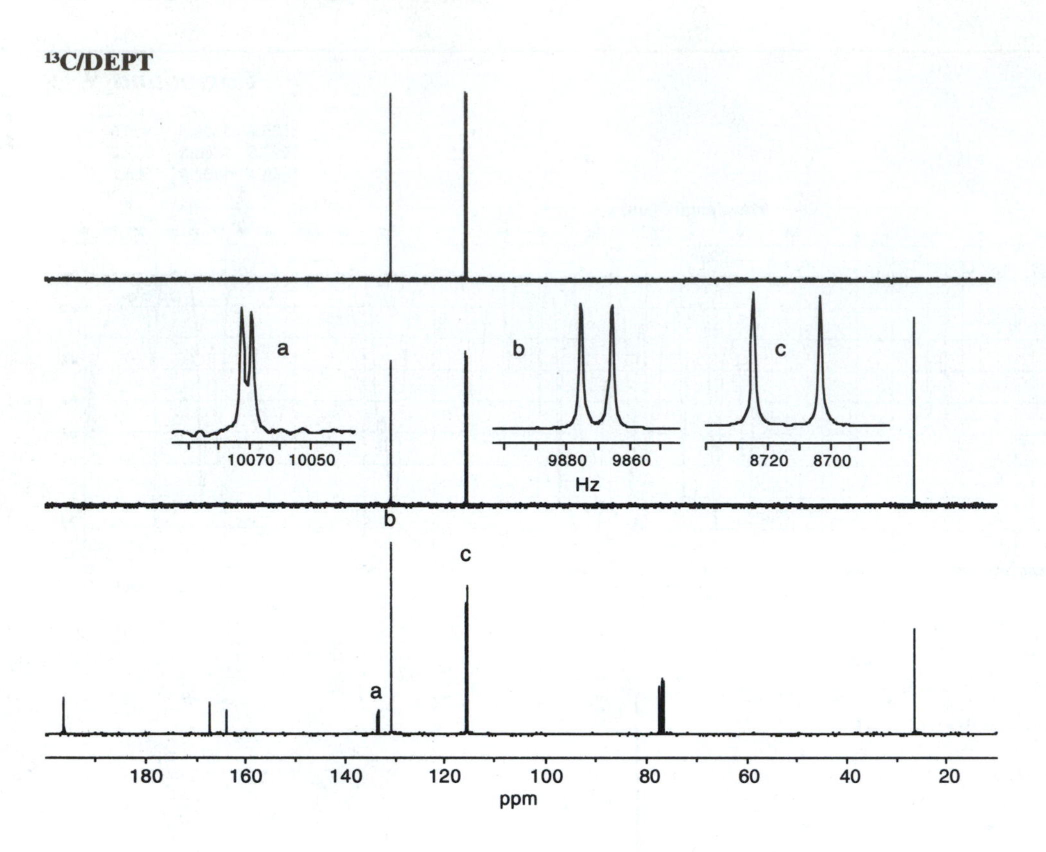
a
b
c
10070
10050
9880
9860
8720
8700
Hz
b
c
a
180
160
140
120
100
80
60
40
20
ppm

INFRARED

Compound 9.20

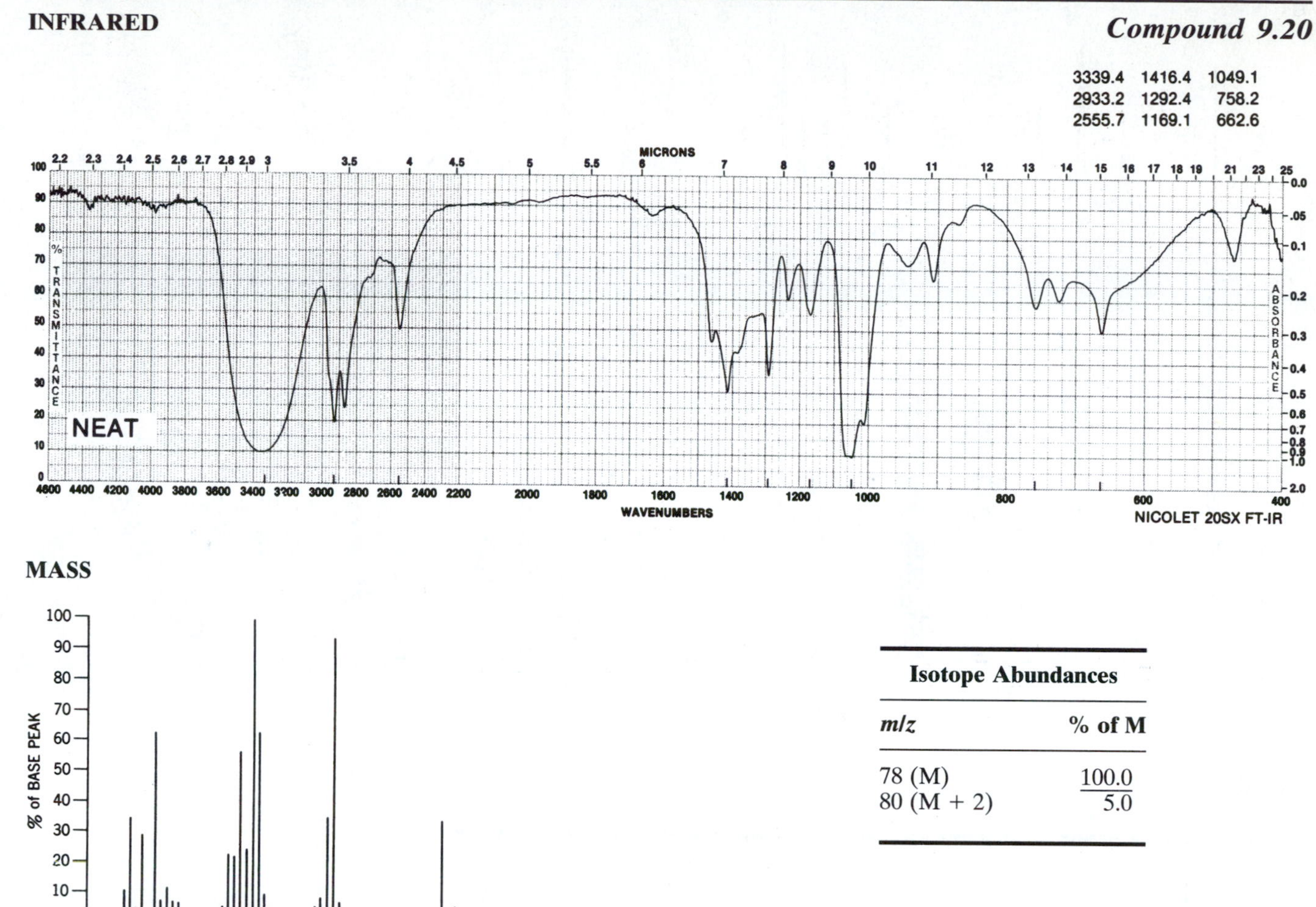

MASS

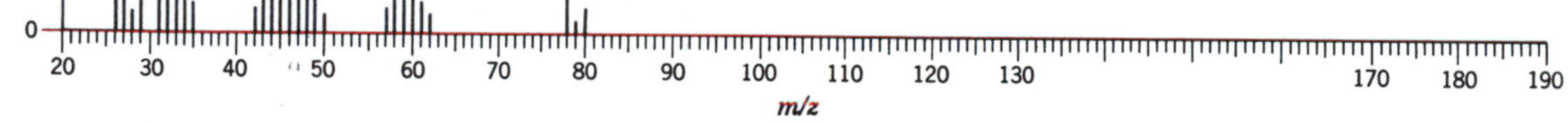

Isotope Abundances	
m/z	% of M
78 (M)	100.0
80 (M + 2)	5.0

^1H NMR

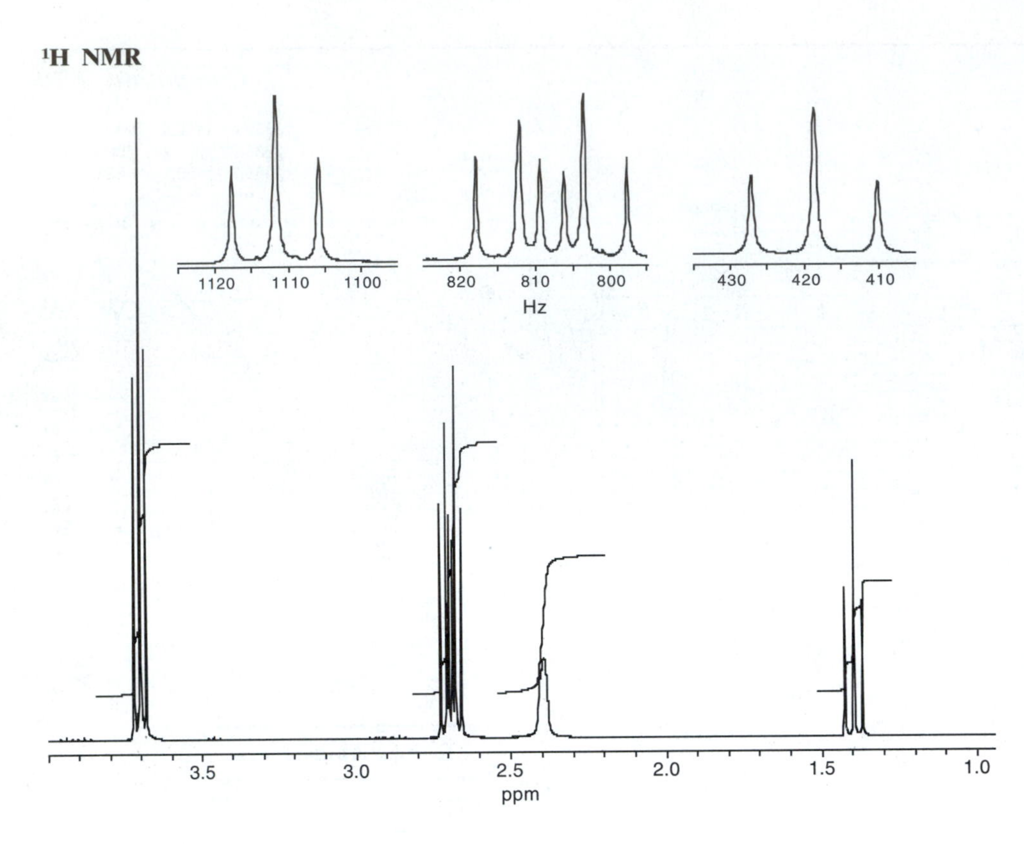

^{13}C/DEPT

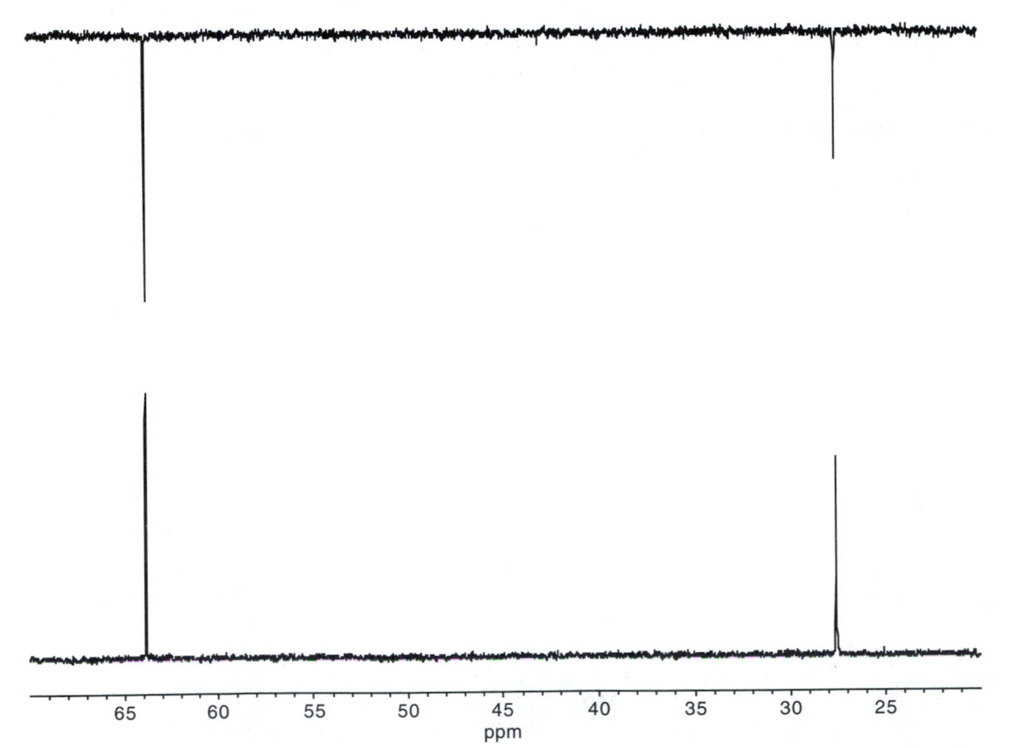

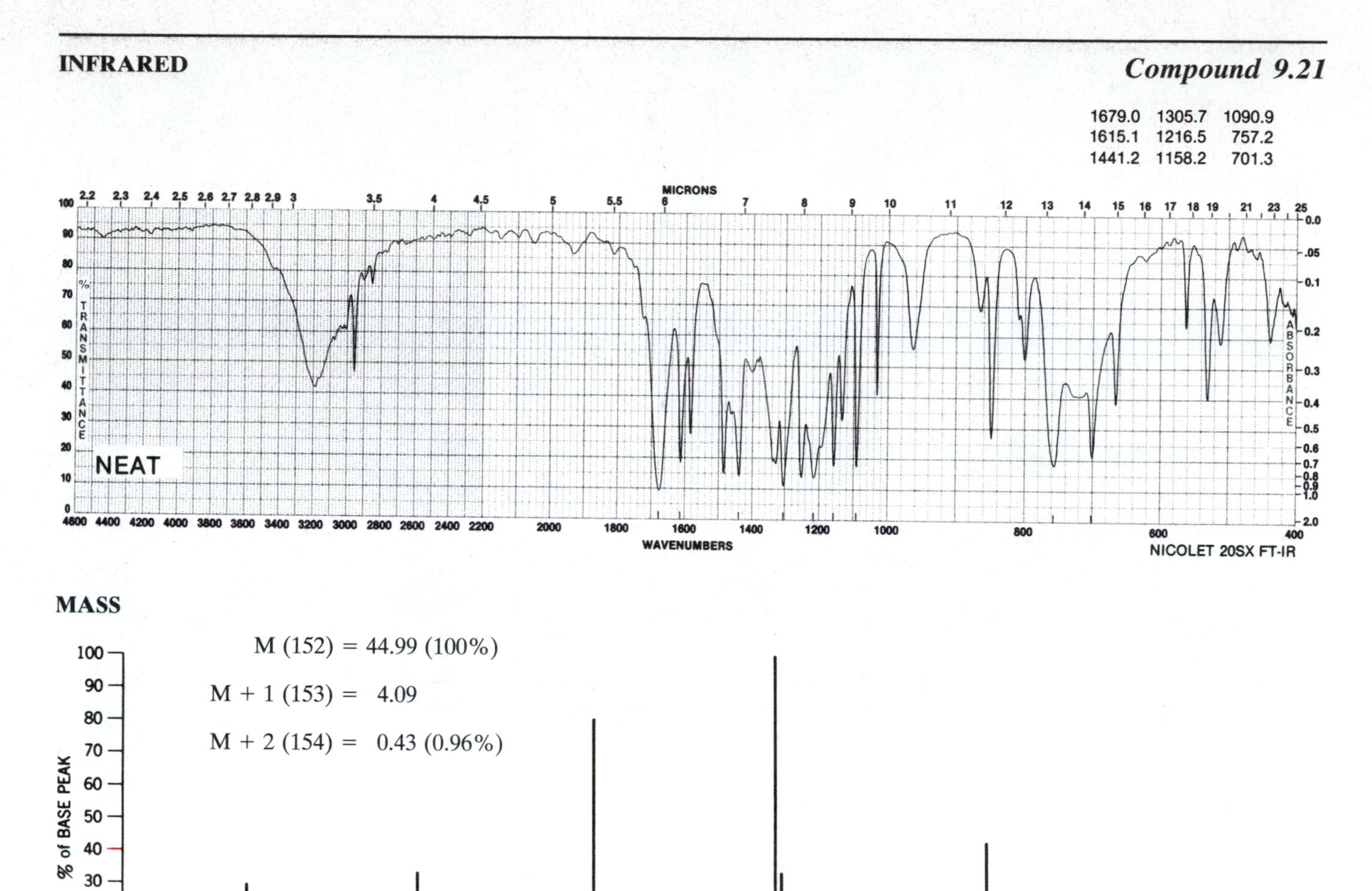
INFRARED
Compound 9.21
1679.0 1305.7 1090.9
1615.1 1216.5 757.2
1441.2 1158.2 701.3
MICRONS
TRANSMITTANCE
ABSORBANCE
NEAT
WAVENUMBERS
NICOLET 20SX FT-IR
MASS
M (152) = 44.99 (100%)
M + 1 (153) = 4.09
M + 2 (154) = 0.43 (0.96%)
% of BASE PEAK
m/z

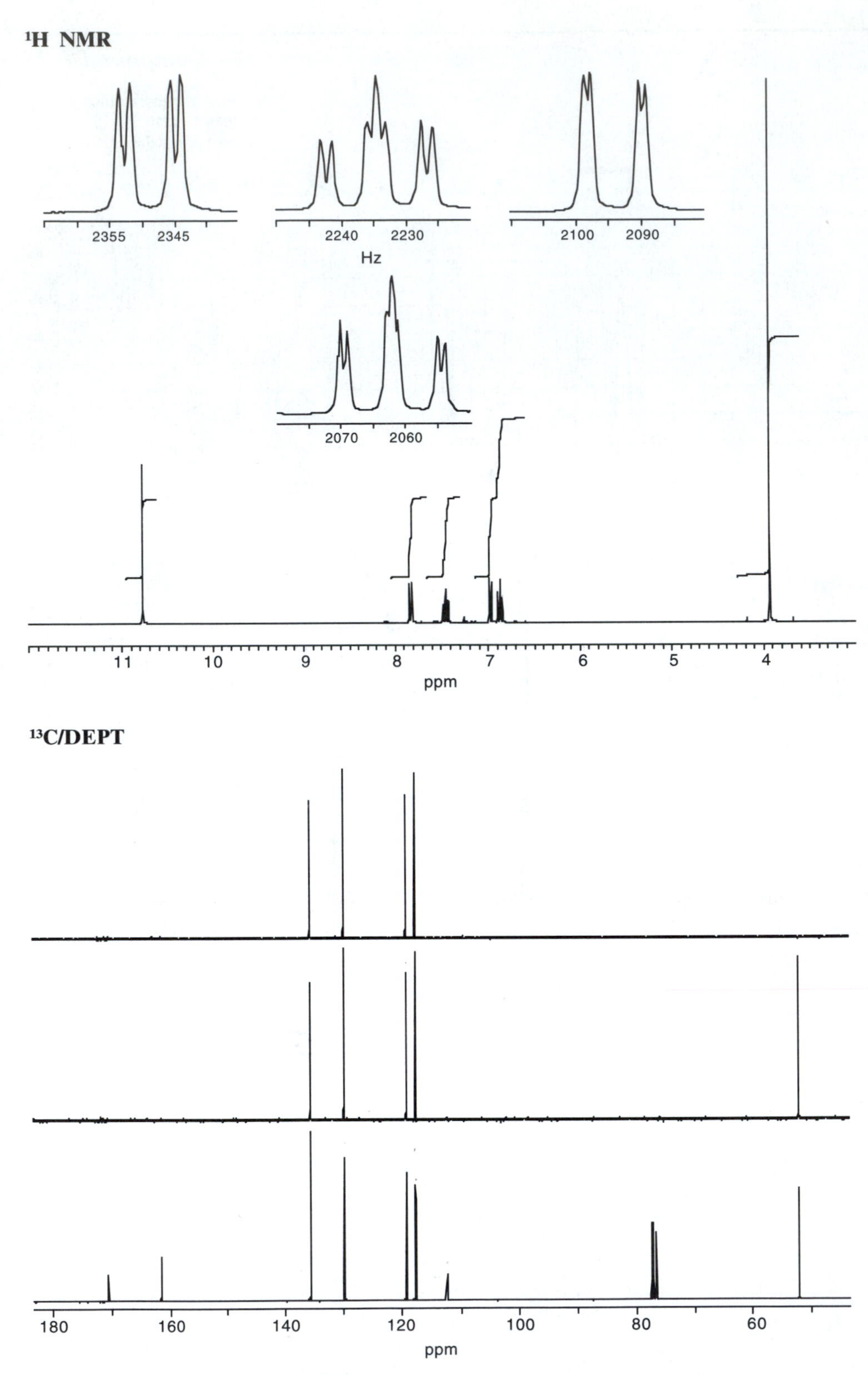

¹H NMR
2355
2345
2240
2230
Hz
2100
2090
2070
2060
11
10
9
8
7
6
5
4
ppm
¹³C/DEPT
180
160
140
120
100
80
60
ppm

COSY

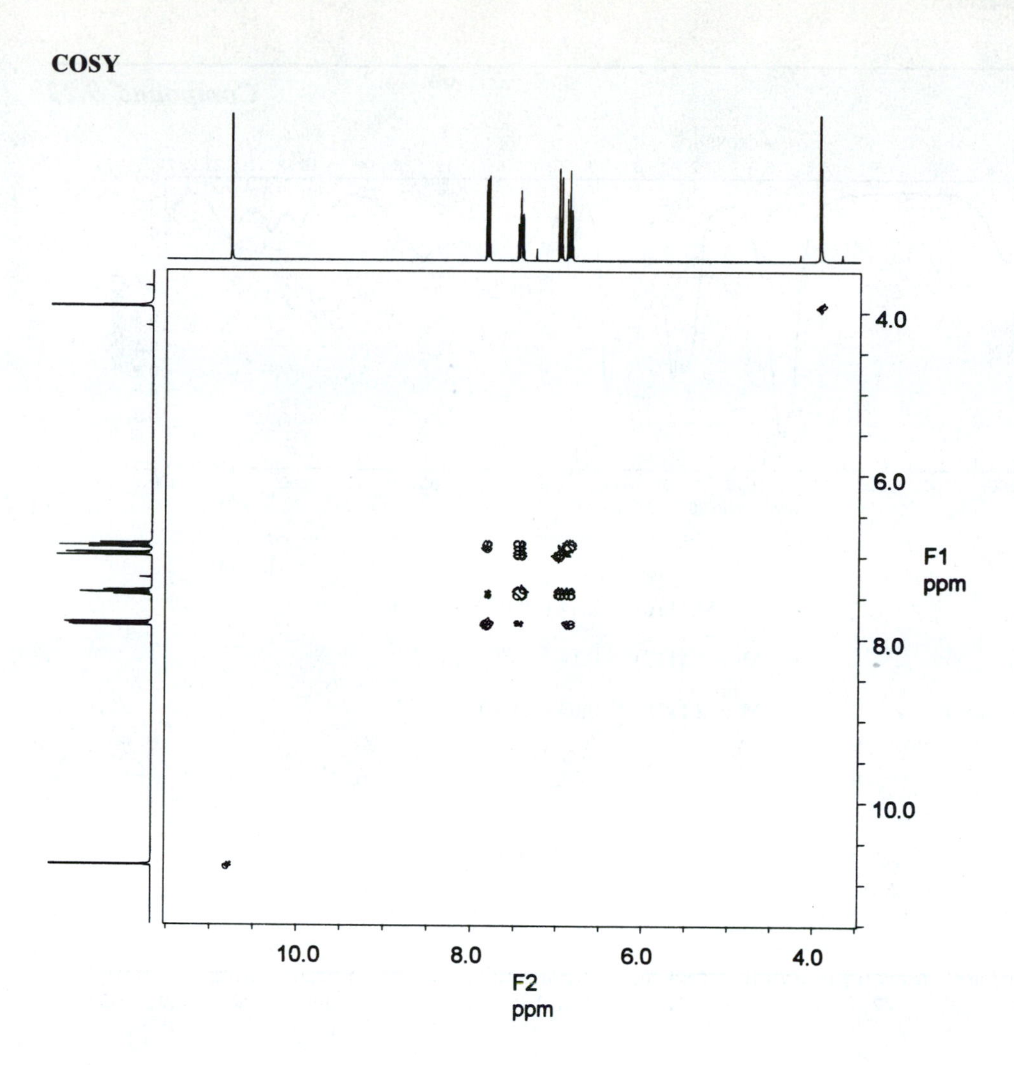

INFRARED ***Compound 9.22***

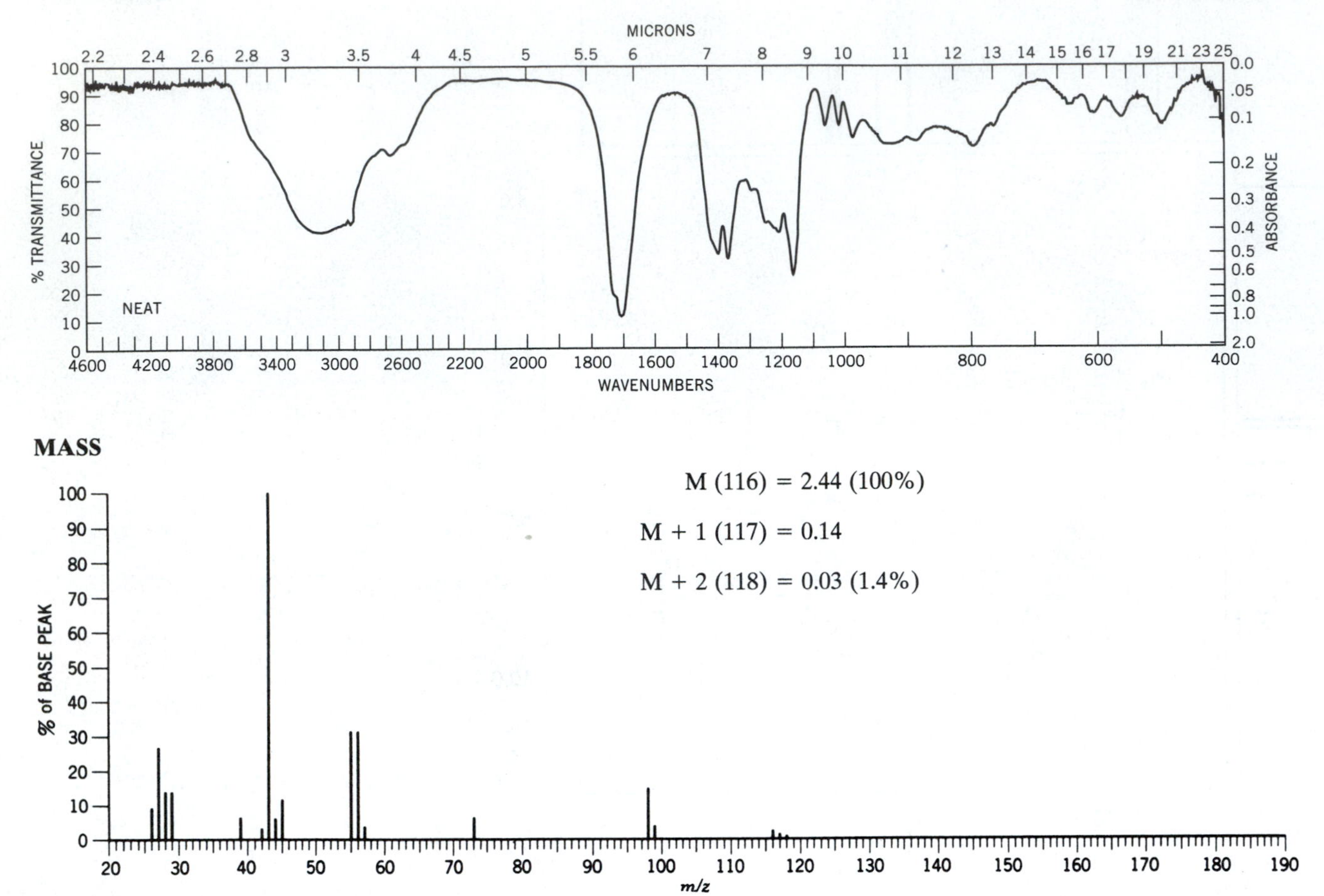

¹H NMR, ¹³C/DEPT

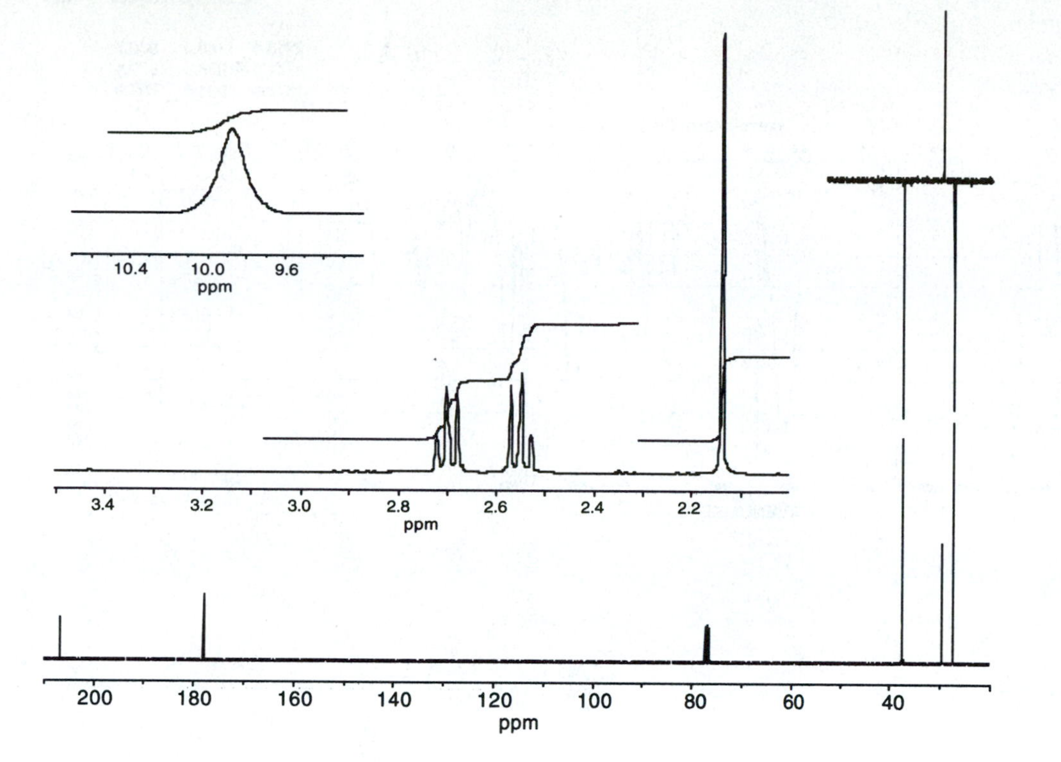

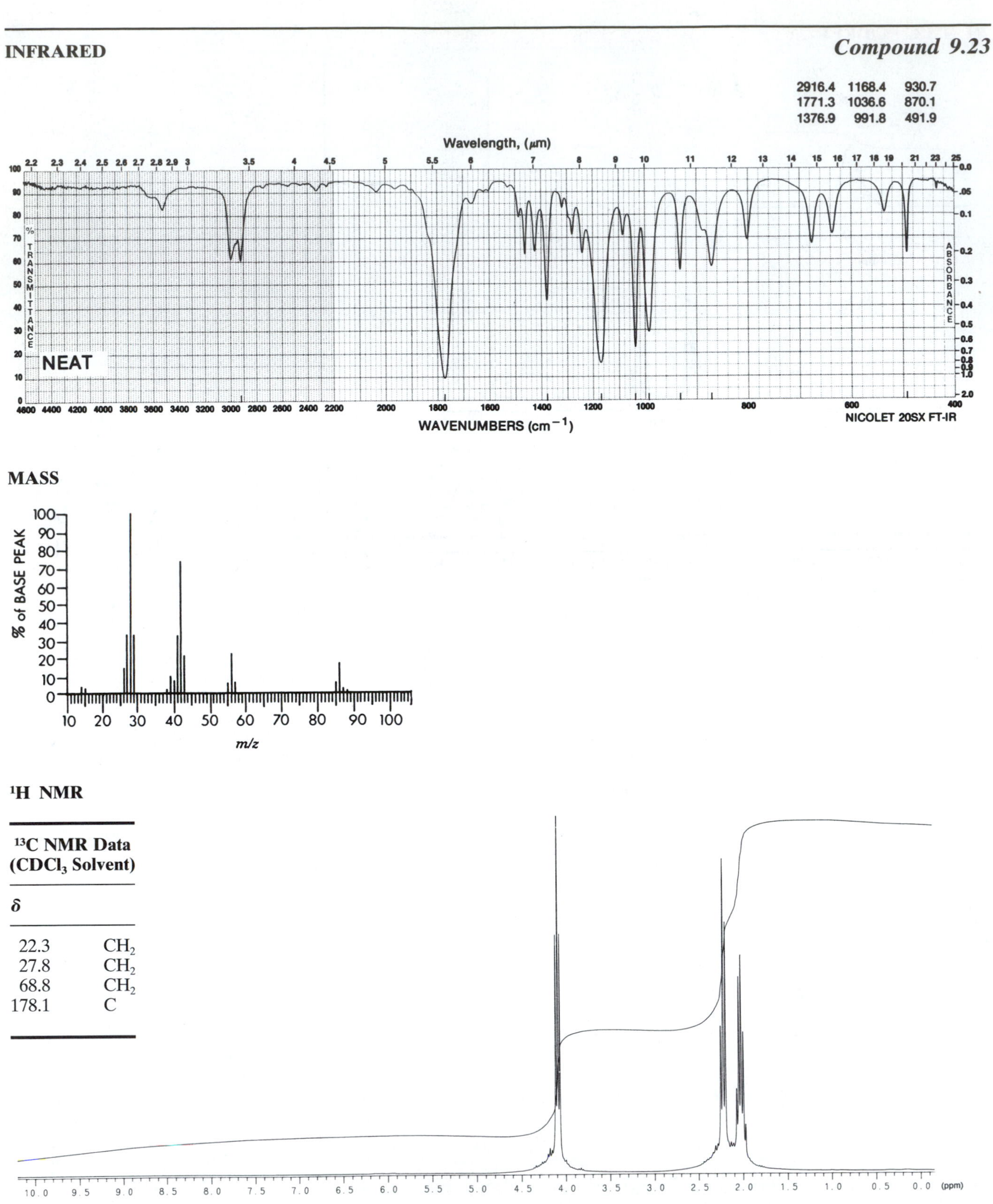

^{13}C NMR Data ($CDCl_3$ Solvent)

δ	
22.3	CH_2
27.8	CH_2
68.8	CH_2
178.1	C

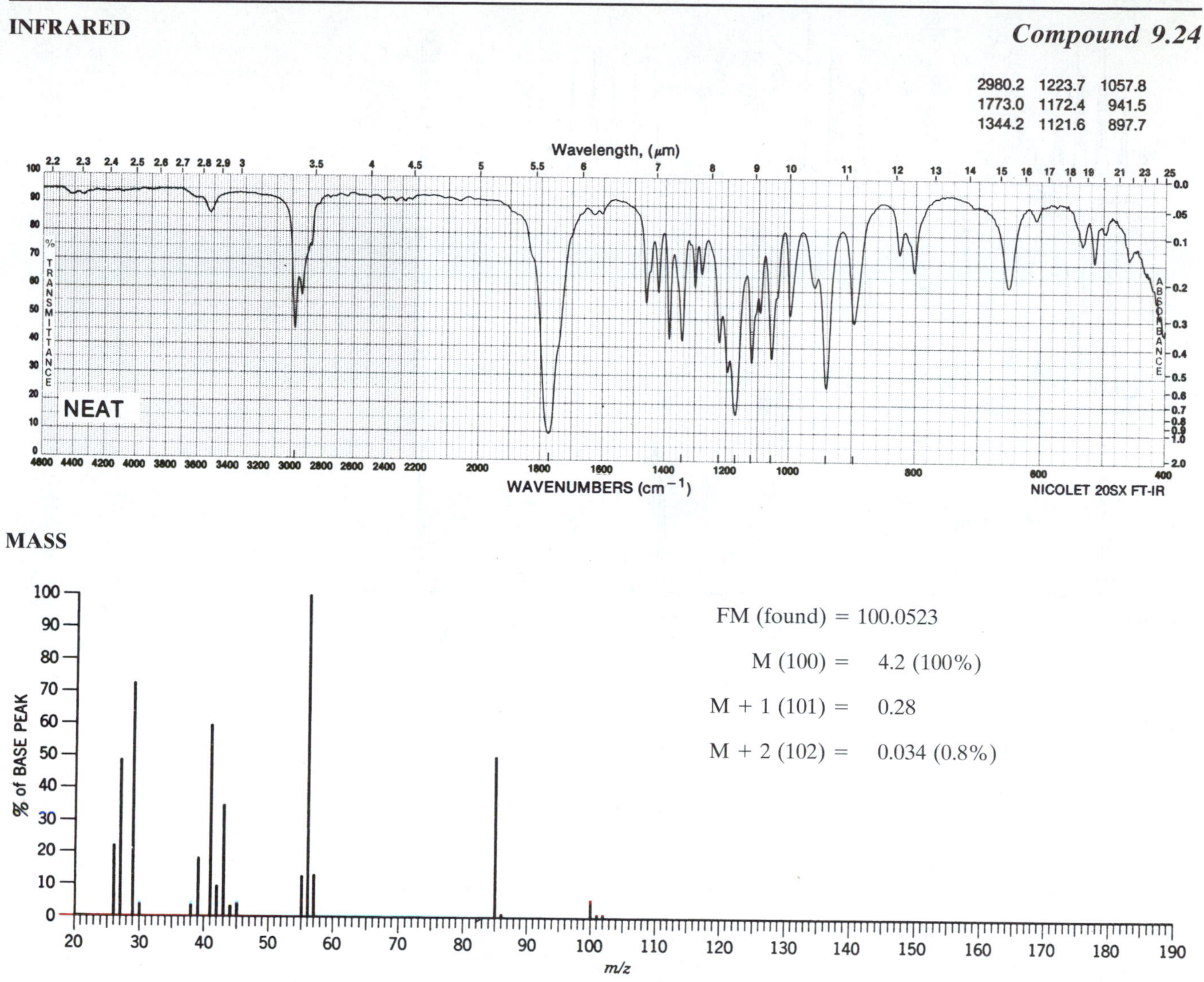
INFRARED
Compound 9.24
2980.2 1223.7 1057.8
1773.0 1172.4 941.5
1344.2 1121.6 897.7
Wavelength, (μm)
NEAT
% TRANSMITTANCE
ABSORBANCE
WAVENUMBERS (cm⁻¹)
NICOLET 20SX FT-IR
MASS
FM (found) = 100.0523
M (100) = 4.2 (100%)
M + 1 (101) = 0.28
M + 2 (102) = 0.034 (0.8%)
% of BASE PEAK
m/z

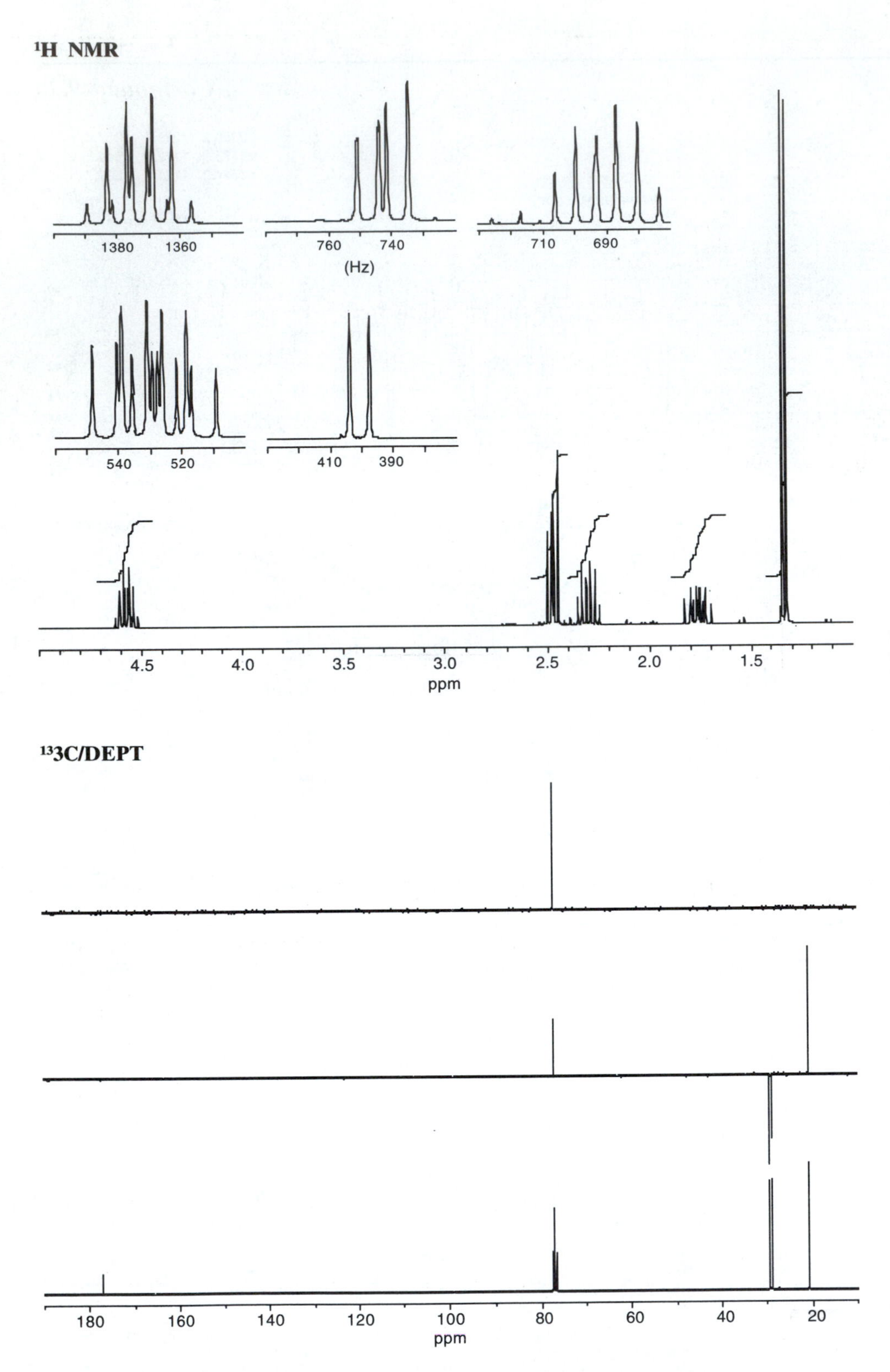
¹H NMR
1380
1360
760
740
(Hz)
710
690
540
520
410
390
4.5
4.0
3.5
3.0
ppm
2.5
2.0
1.5
¹³C/DEPT
180
160
140
120
100
ppm
80
60
40
20

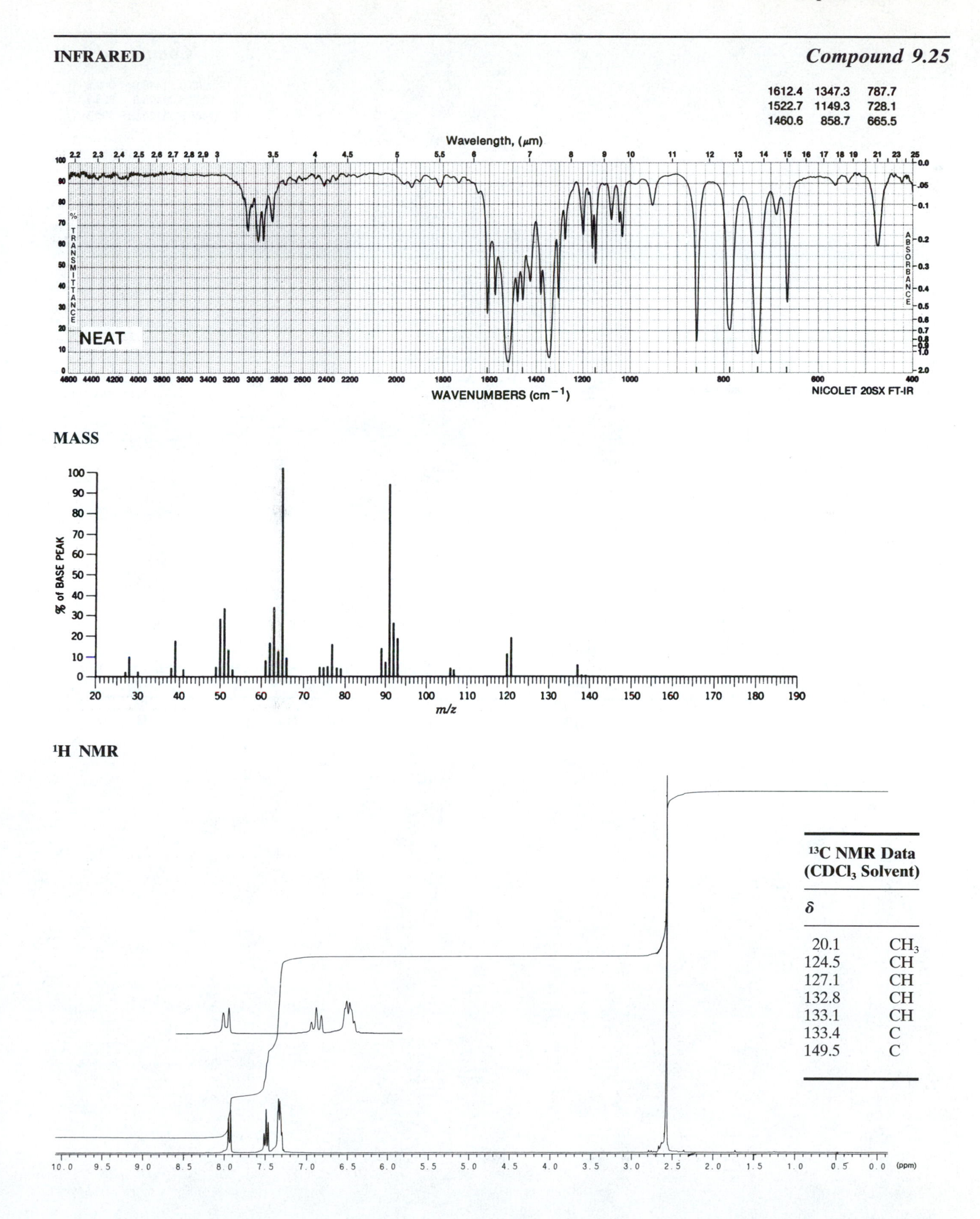

^{13}C NMR Data ($CDCl_3$ Solvent)	
δ	
20.1	CH_3
124.5	CH
127.1	CH
132.8	CH
133.1	CH
133.4	C
149.5	C

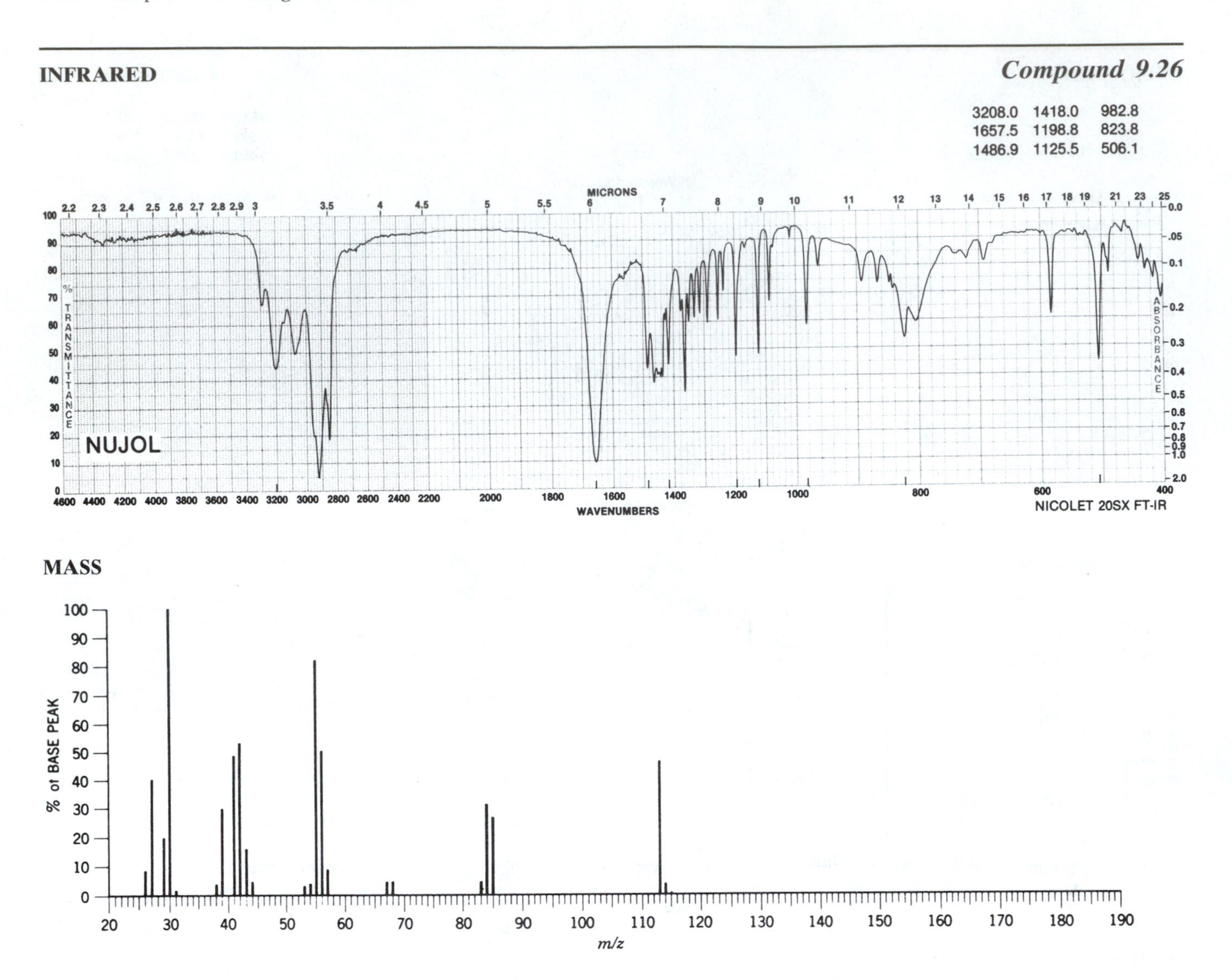
INFRARED
Compound 9.26
3208.0 1418.0 982.8
1657.5 1198.8 823.8
1486.9 1125.5 506.1
MICRONS
% TRANSMITTANCE
ABSORBANCE
NUJOL
WAVENUMBERS
NICOLET 20SX FT-IR
MASS
% of BASE PEAK
m/z

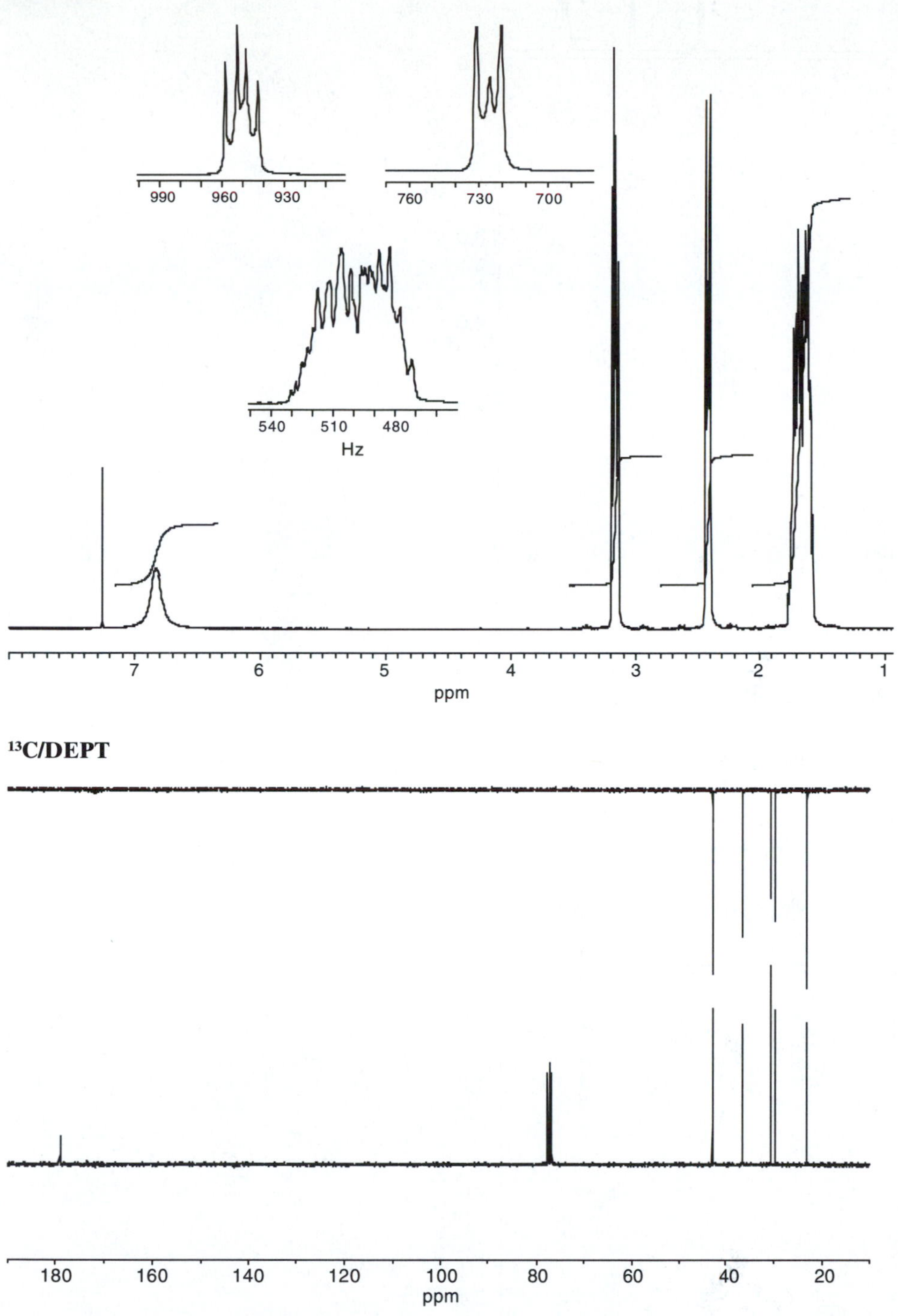

¹H NMR
990
960
930
760
730
700
540
510
480
Hz
7
6
5
4
3
2
1
ppm
¹³C/DEPT
180
160
140
120
100
80
60
40
20
ppm

COSY

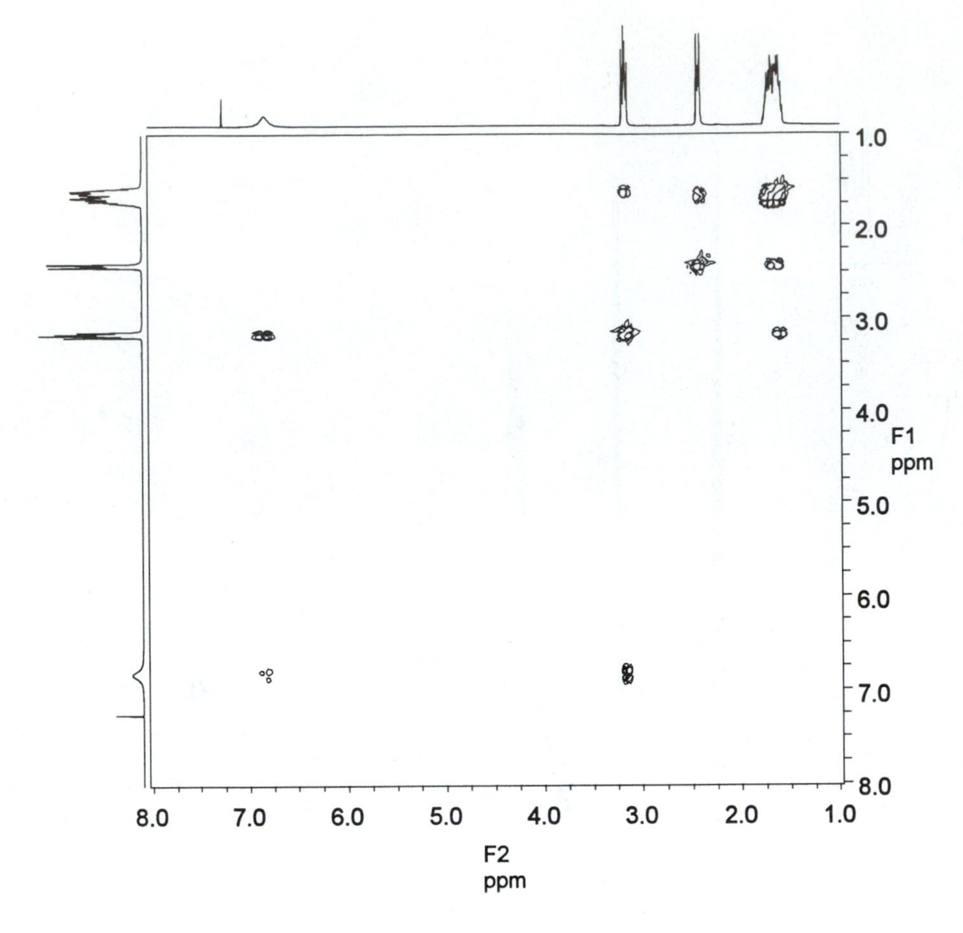
F1
ppm
F2
ppm
1.0
2.0
3.0
4.0
5.0
6.0
7.0
8.0

HETCOR

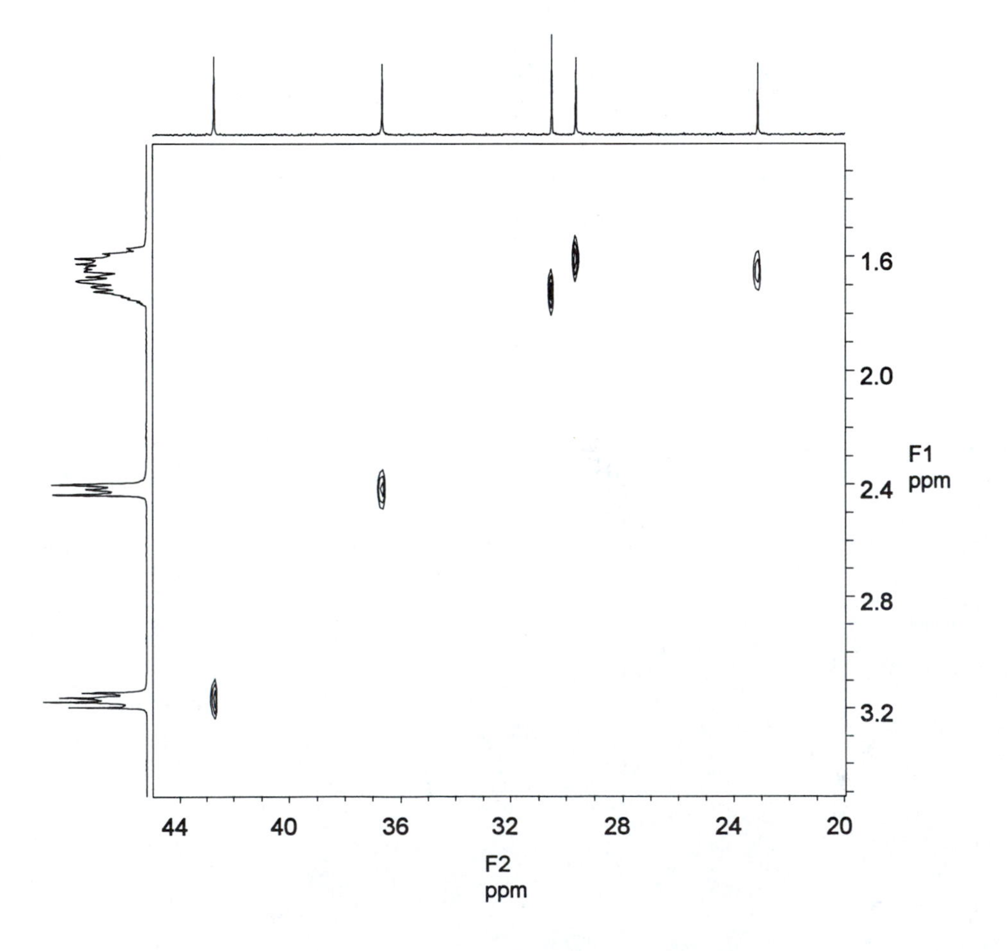
F1
ppm
F2
ppm
1.6
2.0
2.4
2.8
3.2
44
40
36
32
28
24
20

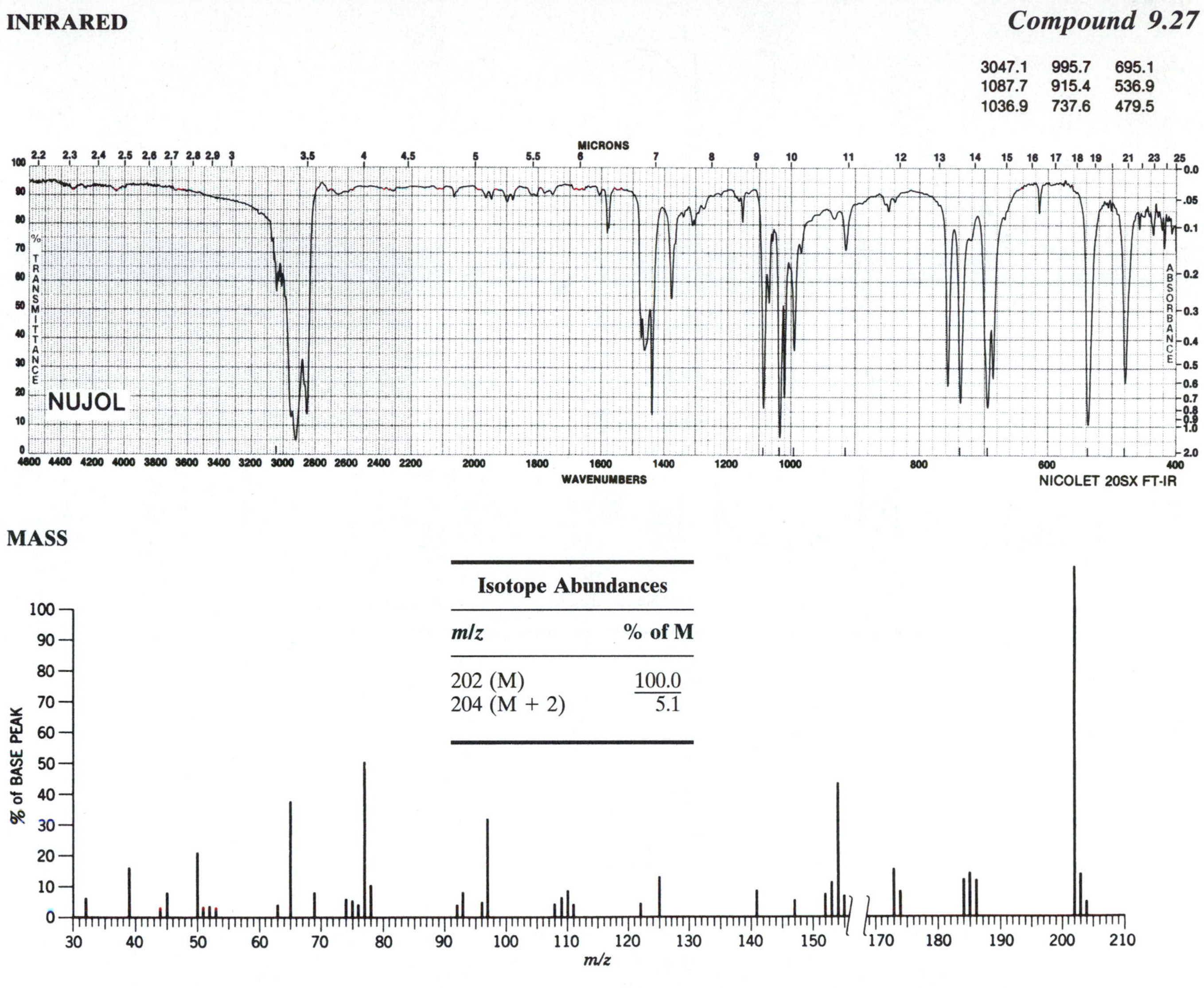

Isotope Abundances

m/z	% of M
202 (M)	100.0
204 (M + 2)	5.1

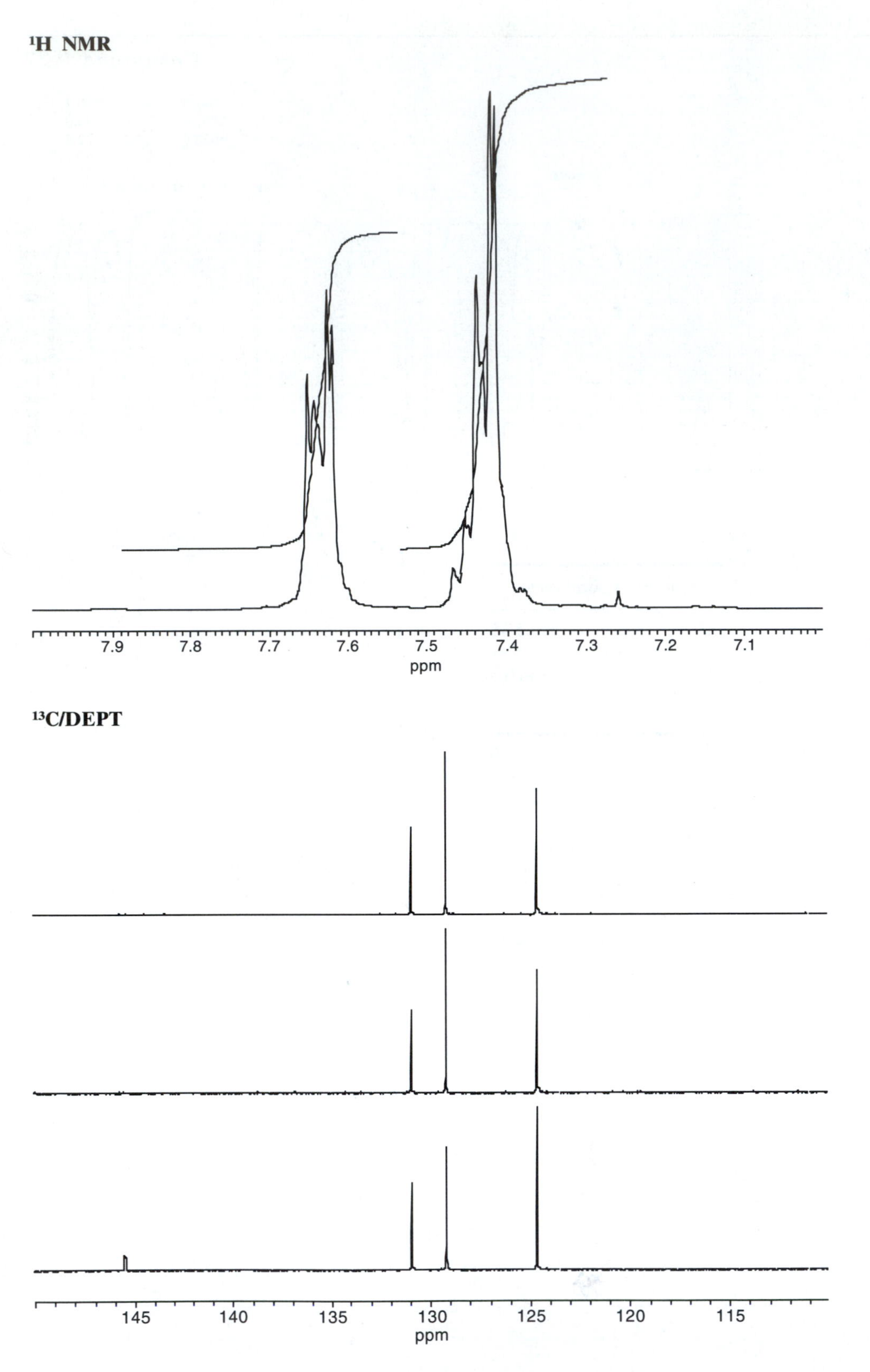
¹H NMR
7.9
7.8
7.7
7.6
7.5
7.4
7.3
7.2
7.1
ppm
¹³C/DEPT
145
140
135
130
125
120
115
ppm

INFRARED

Compound 9.28

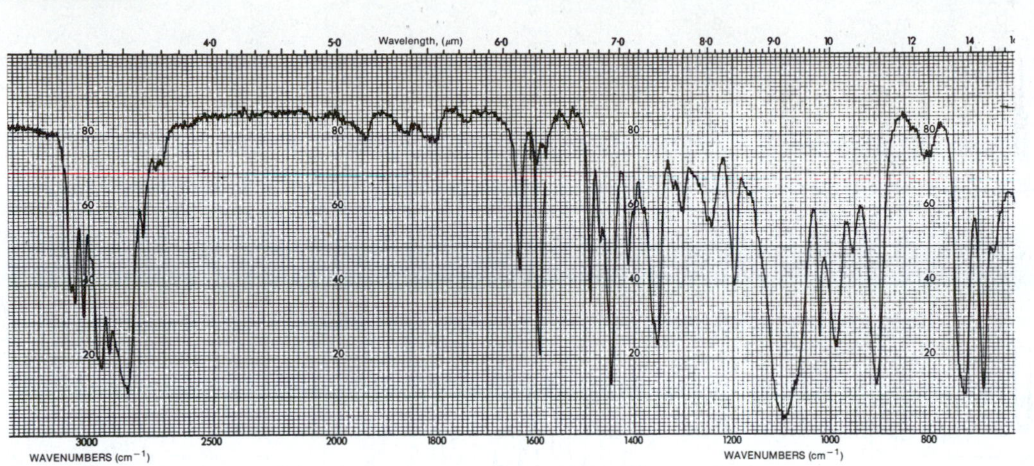

MASS

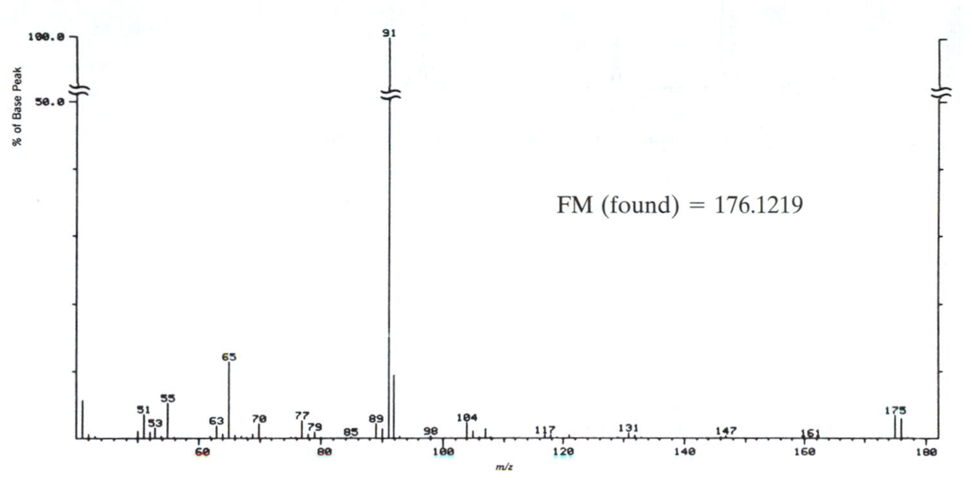

^{1}H NMR SPECTRUM (SOLVENT $CDCl_3$, 500 MHz)

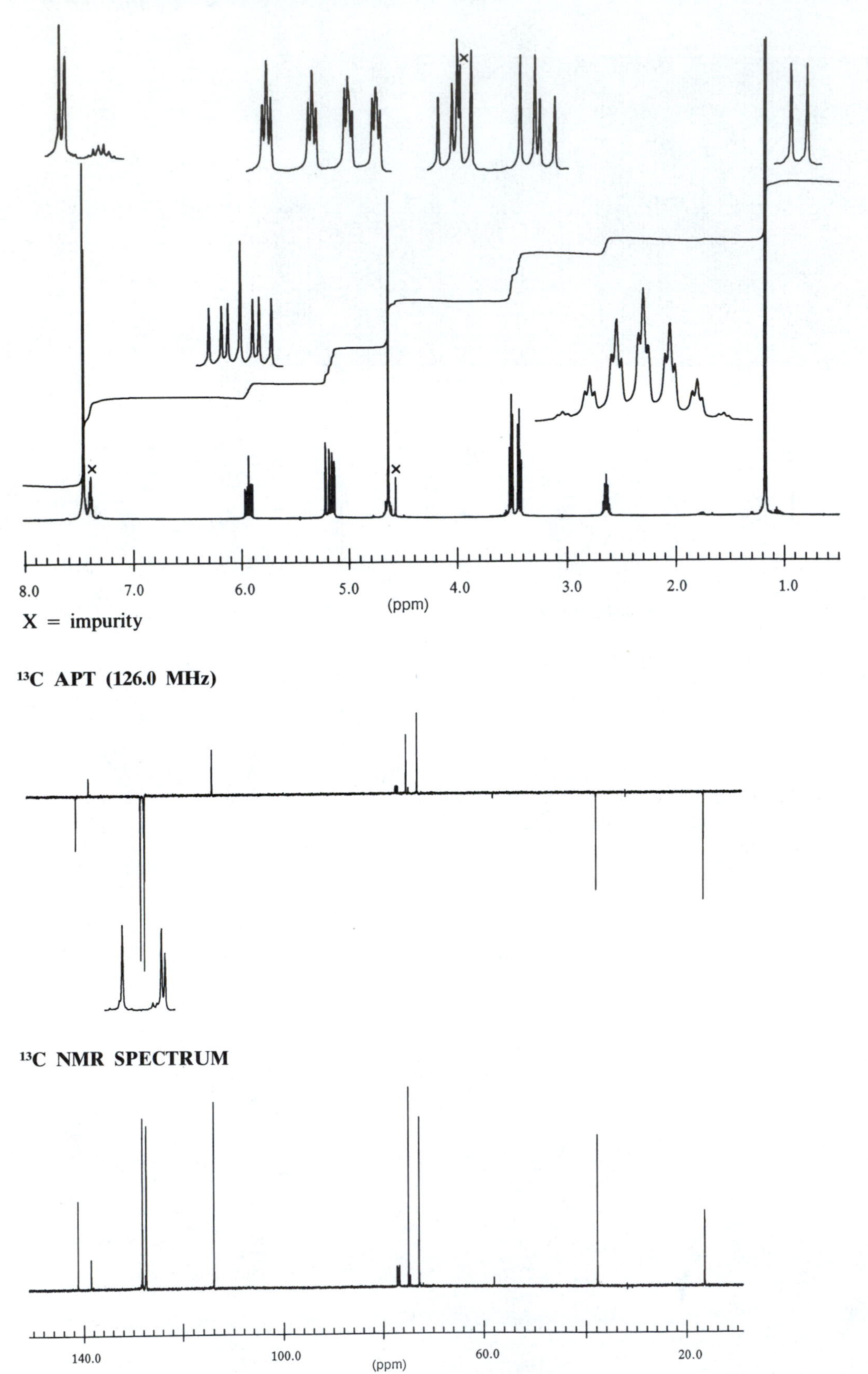

C/CH_2 ↑
CH/CH_3 ↓

H C COSY (HETCOR) (^{13}C 126.0 MHz)

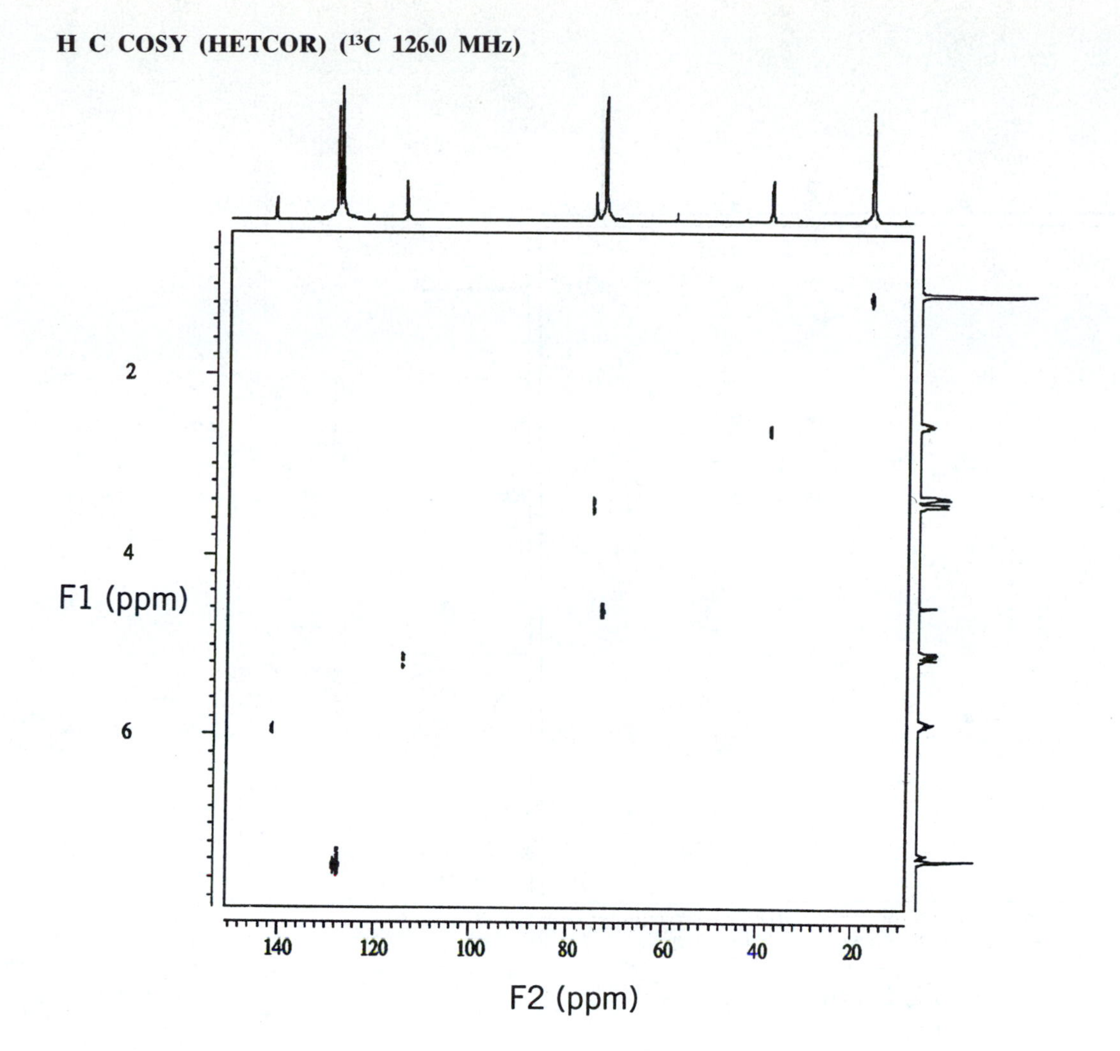

H H COSY (500 MHz)

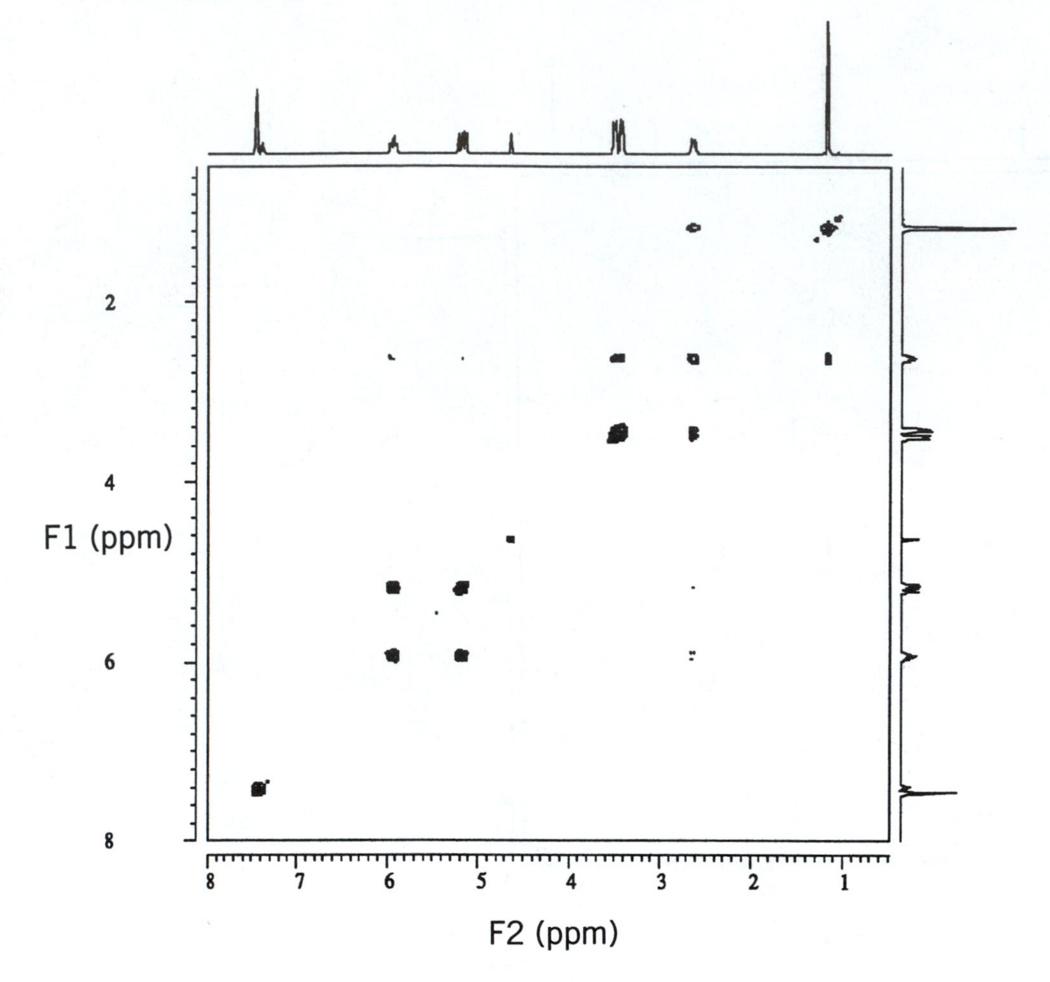

INFRARED SPECTRUM (NEAT)

Compound 9.29

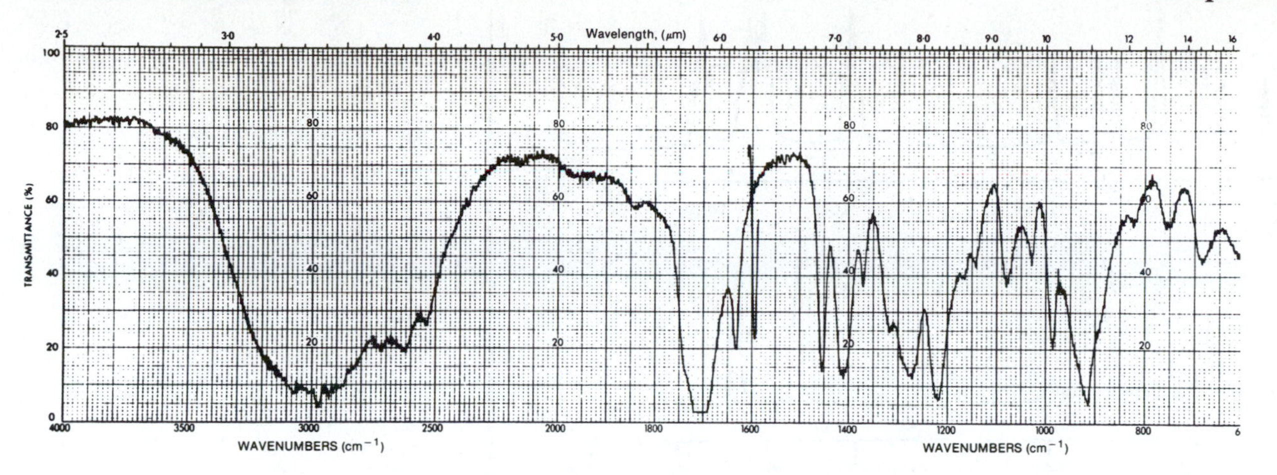

MASS

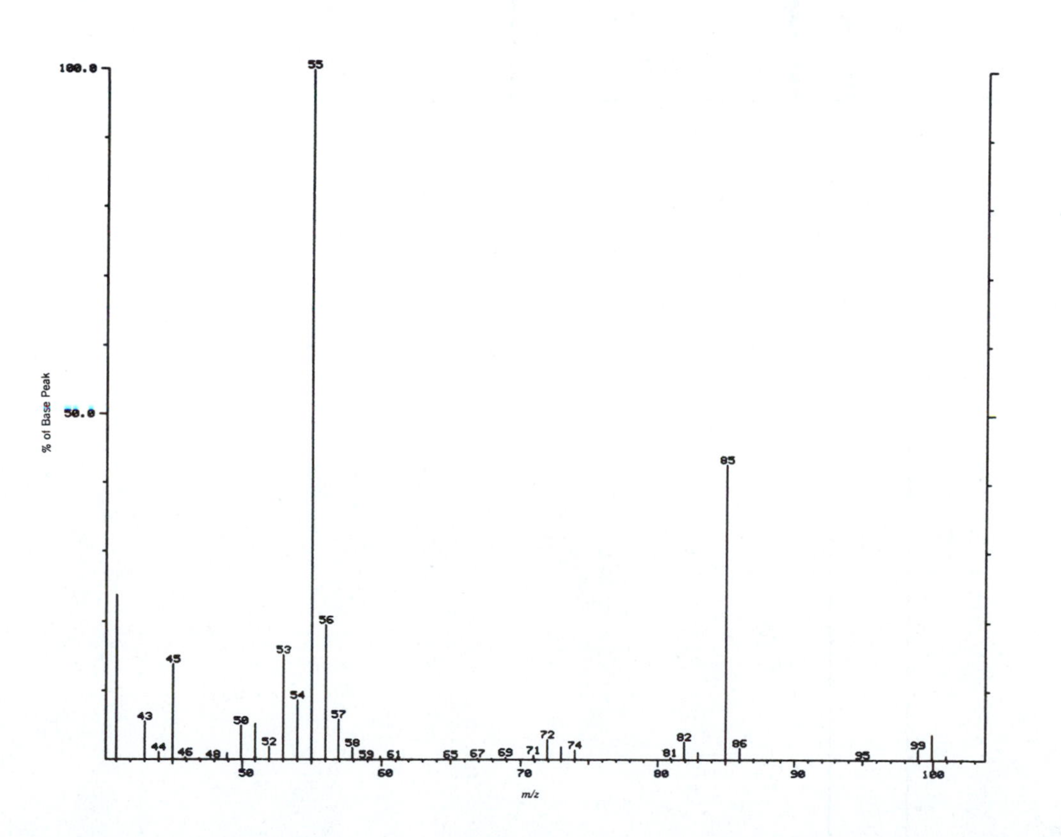

^{1}H NMR SPECTRUM

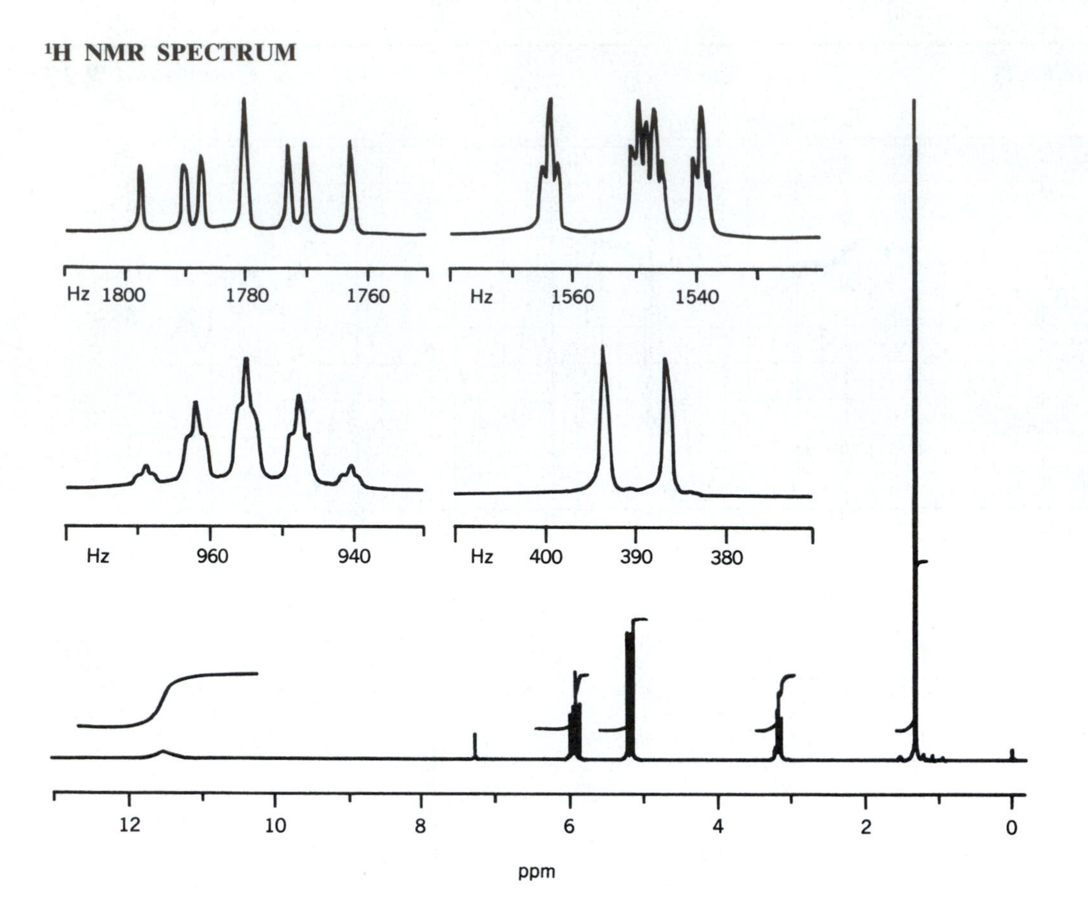

^{13}C NMR

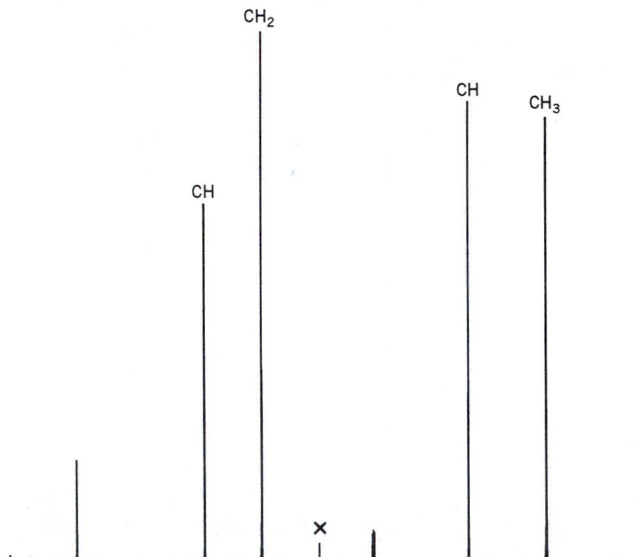

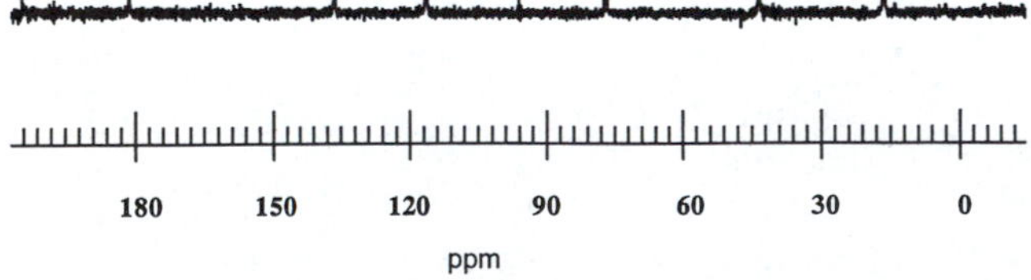

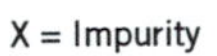
X = Impurity

INFRARED ***Compound 9.30***

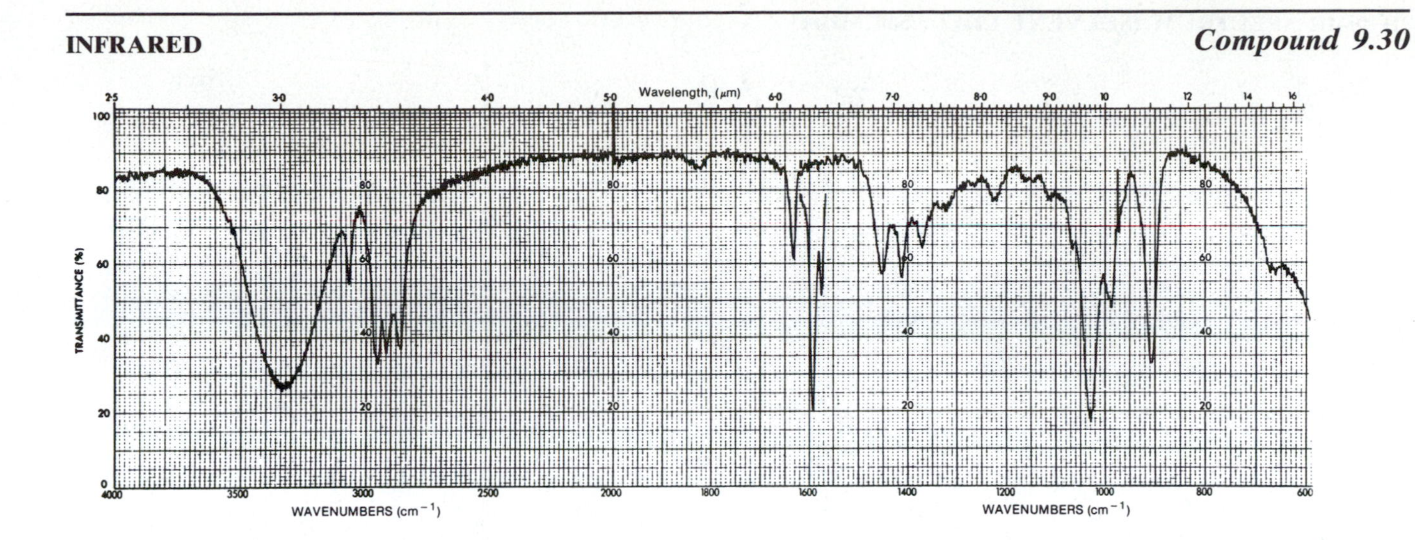

MASS

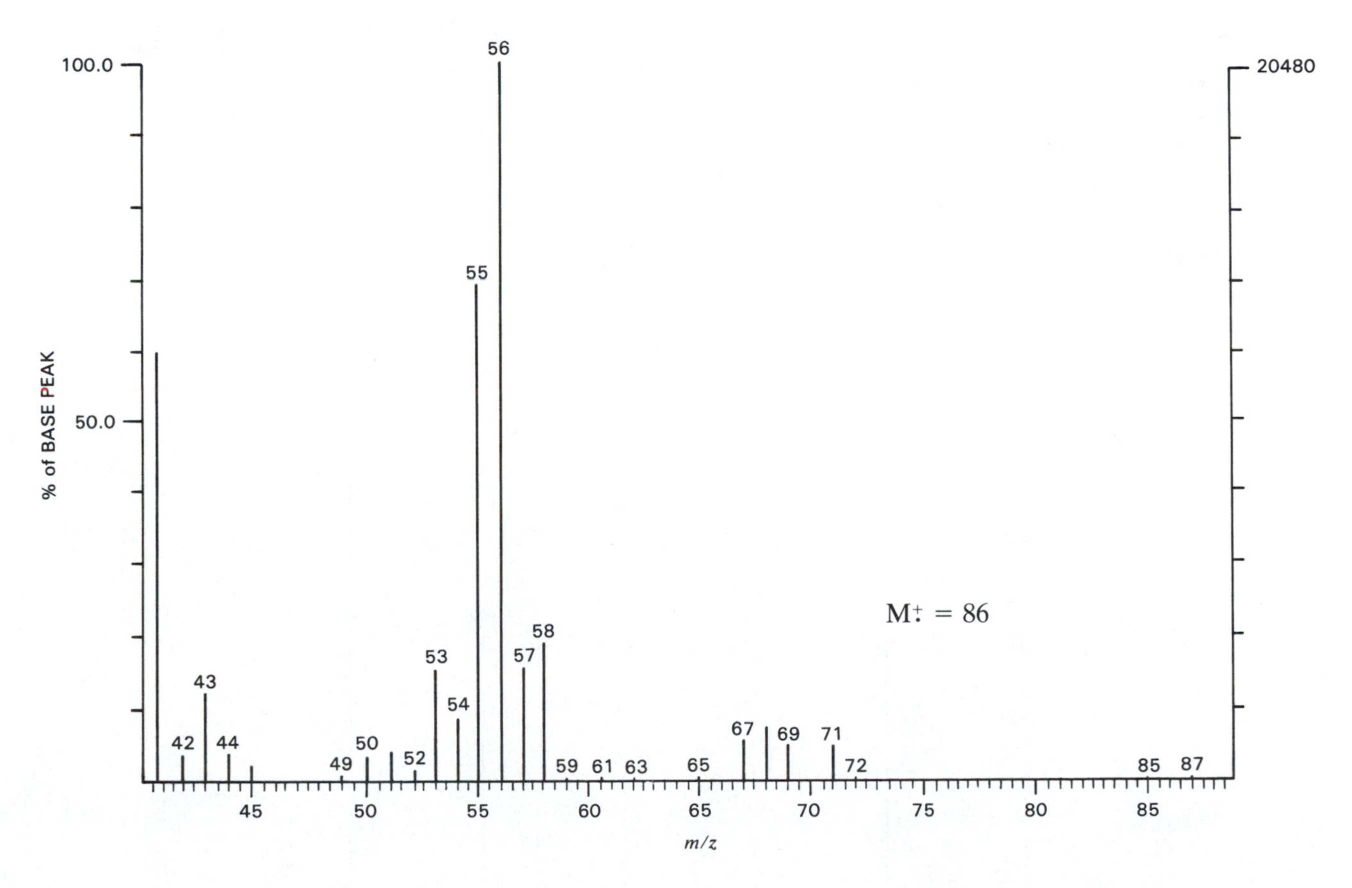

^{1}H NMR SPECTRUM (SOLVENT $CDCl_3$, 500 MHz)

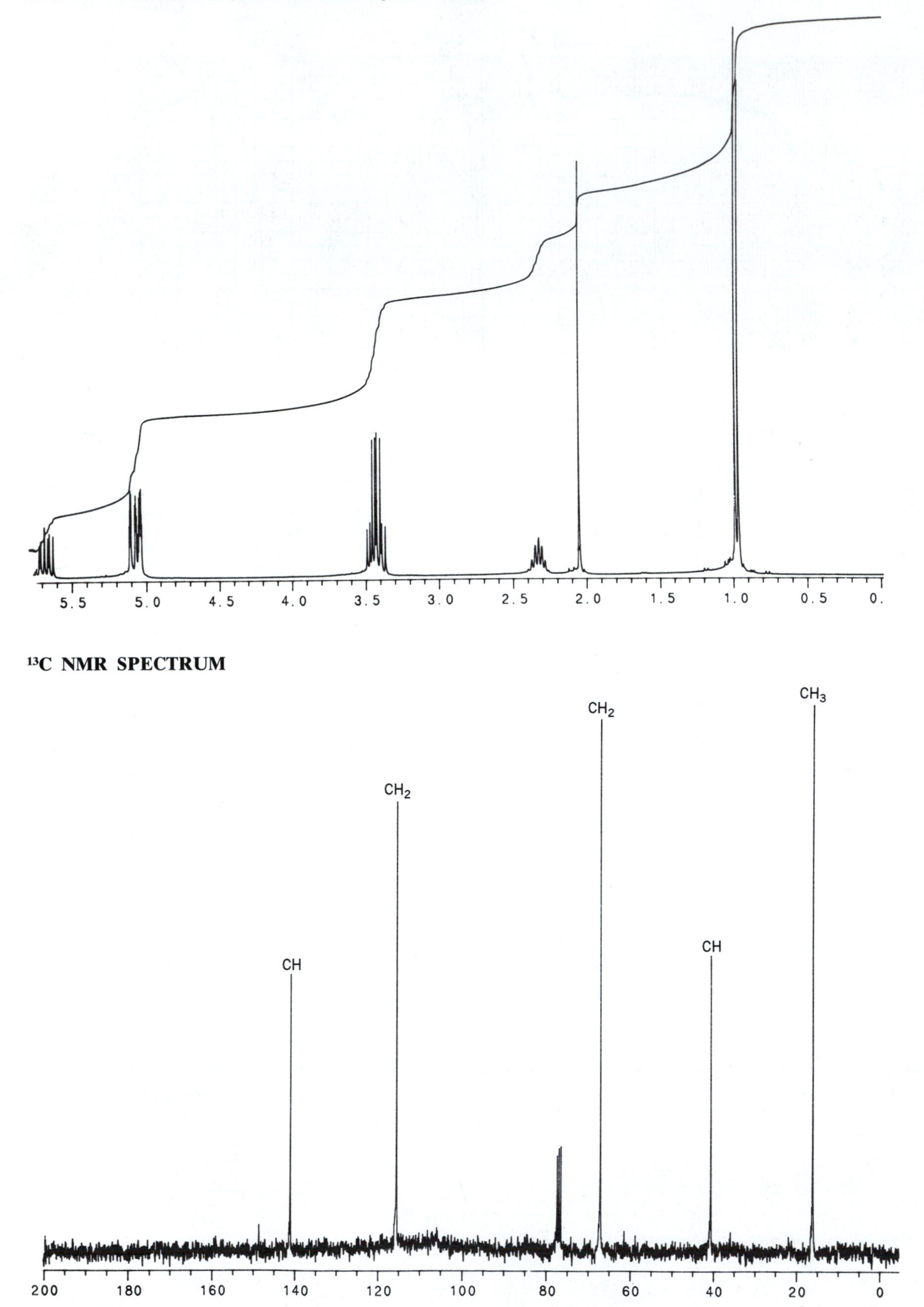

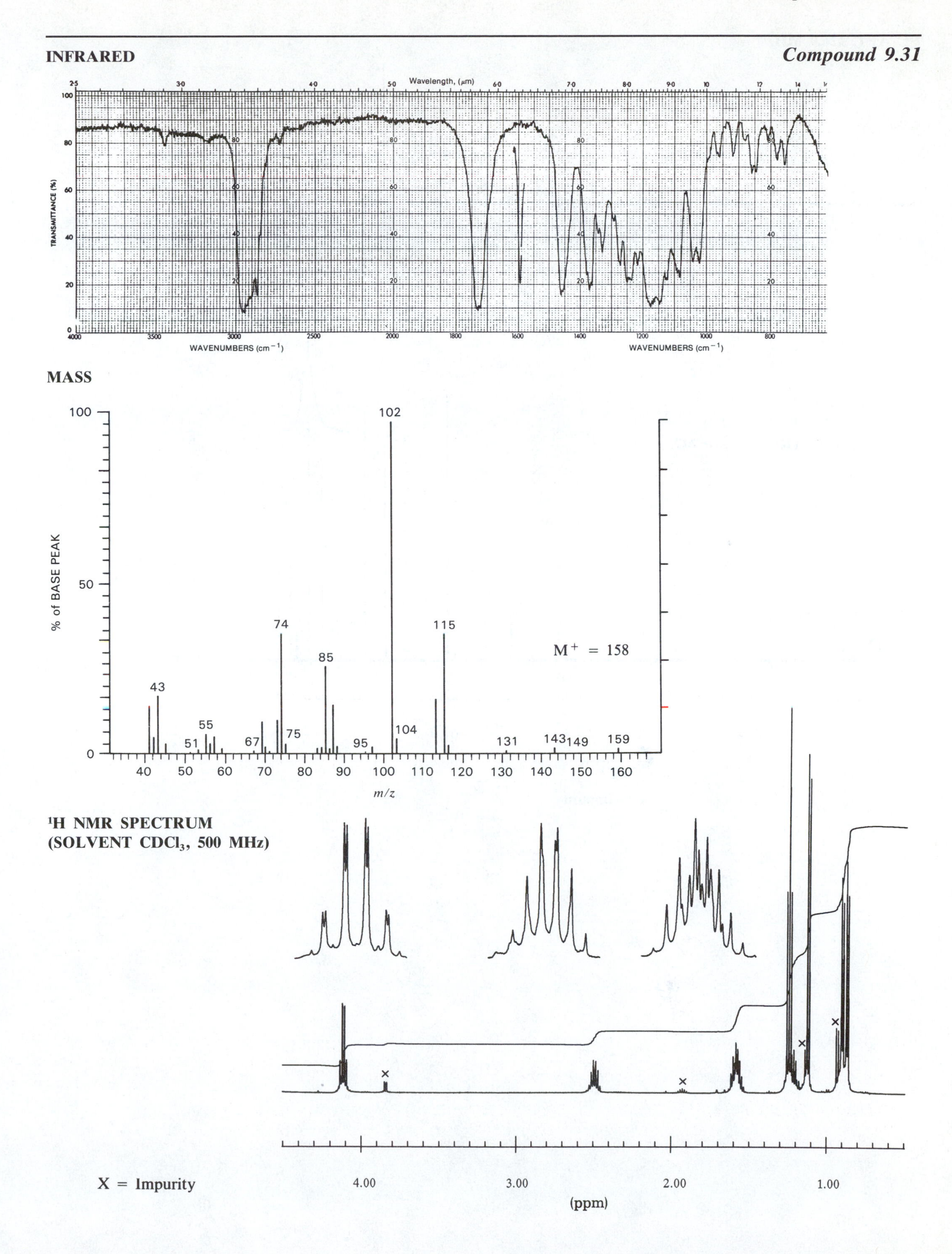
INFRARED
Compound 9.31
Wavelength, (μm)
TRANSMITTANCE (%)
WAVENUMBERS (cm⁻¹)
MASS
% of BASE PEAK
102
74
115
85
43
55
51
67
75
95
104
131
143
149
159
M⁺ = 158
m/z
¹H NMR SPECTRUM
(SOLVENT CDCl₃, 500 MHz)
X = Impurity
4.00
3.00
2.00
1.00
(ppm)

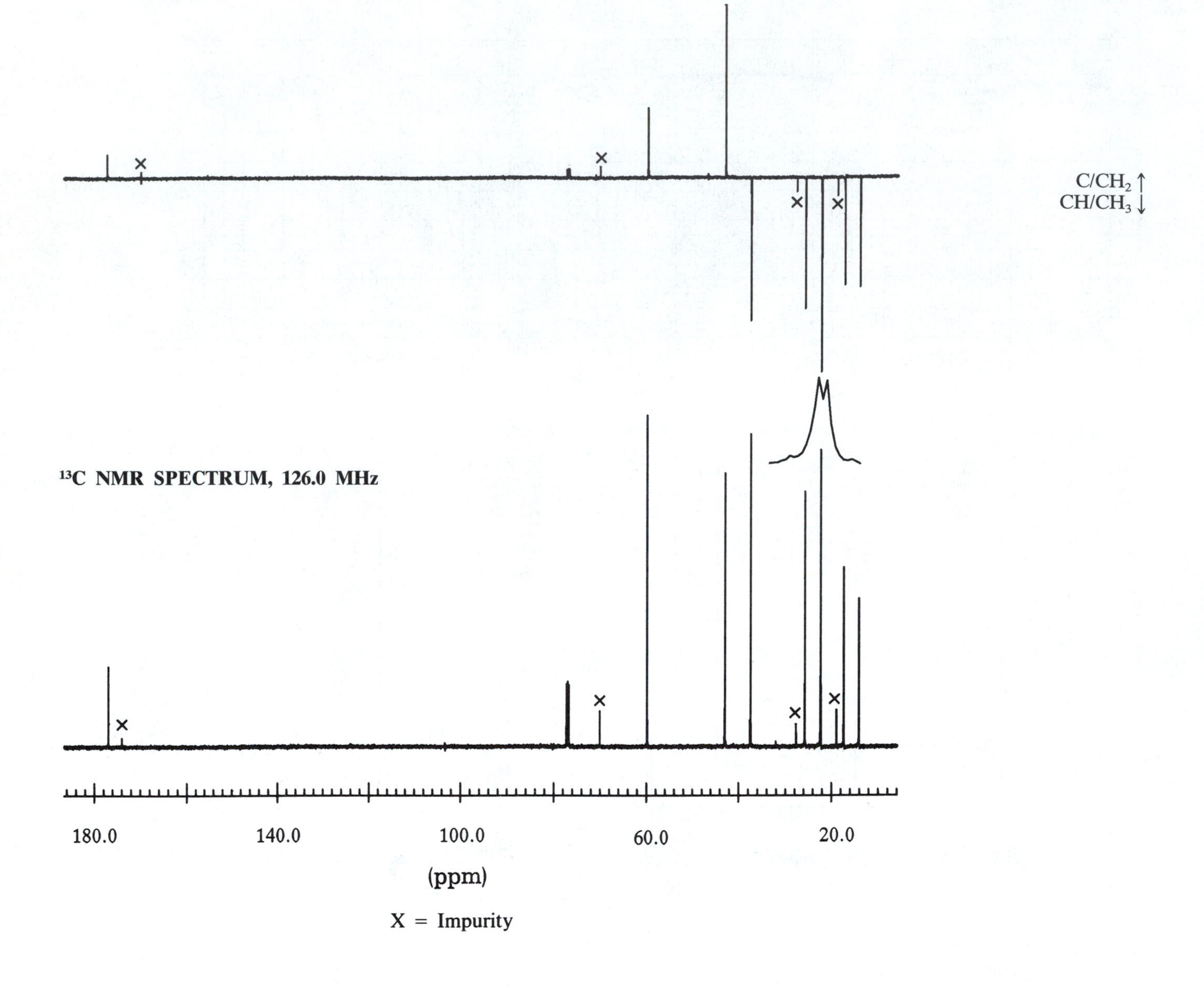

¹³C APT, 126.0 MHz
C/CH₂ ↑
CH/CH₃ ↓
¹³C NMR SPECTRUM, 126.0 MHz
180.0
140.0
100.0
60.0
20.0
(ppm)
X = Impurity

H H COSY, 500 MHz

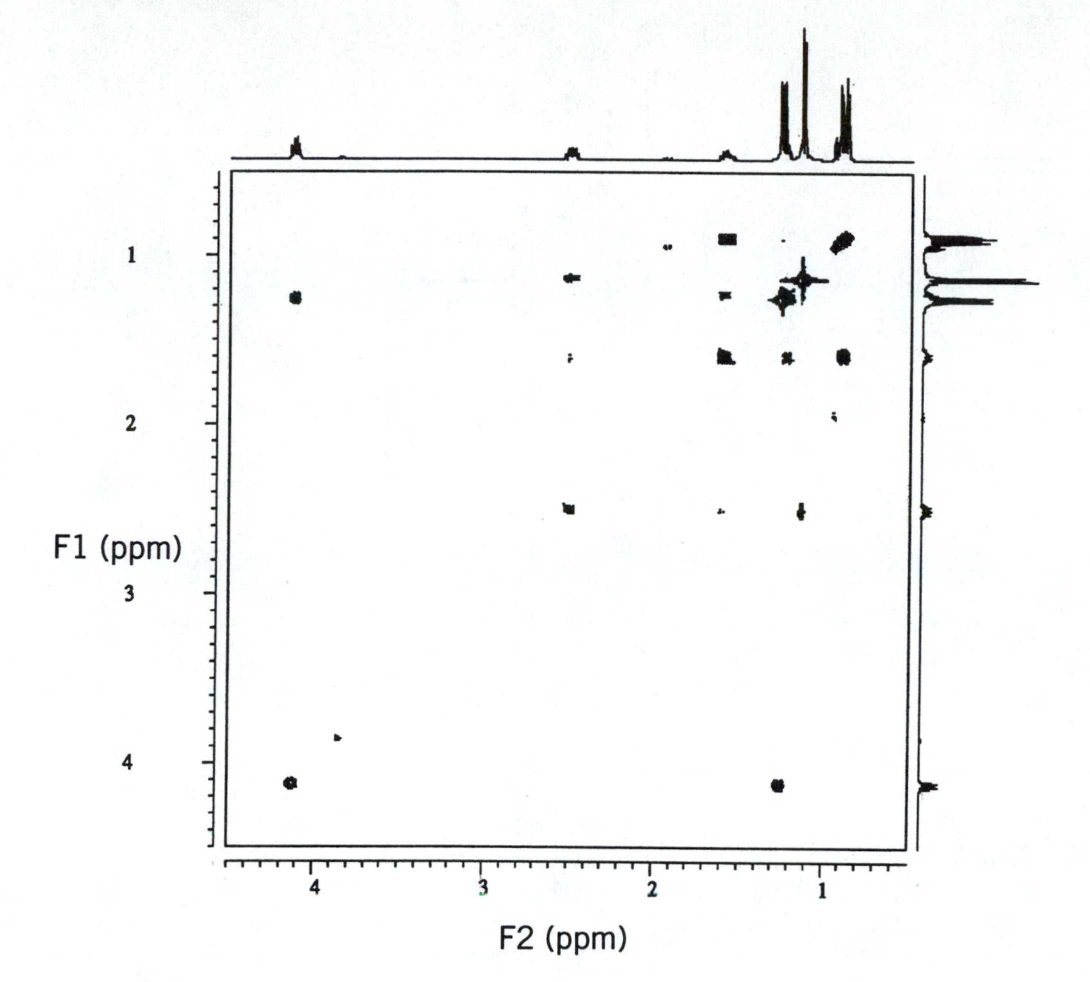

H C COSY (HETCOR), ^{13}C 126 MHz

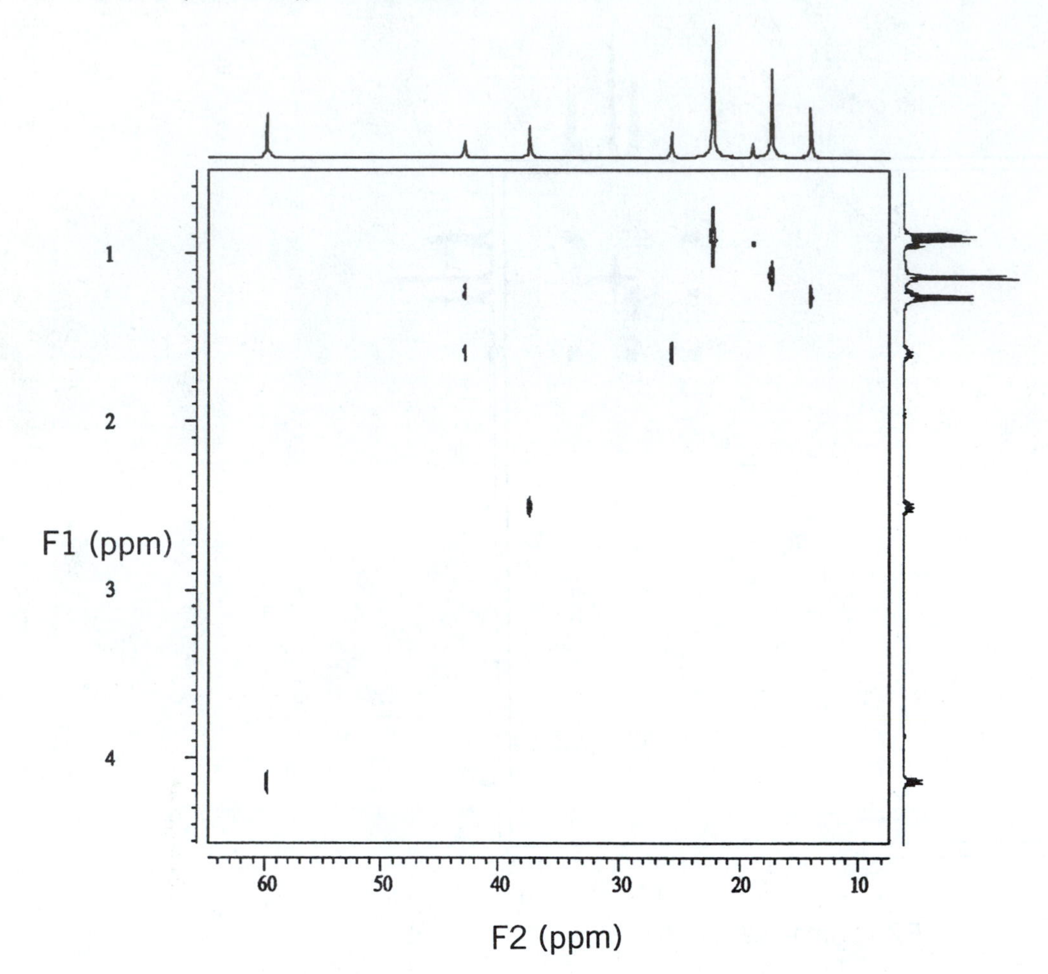

INFRARED SPECTRUM

Compound 9.32

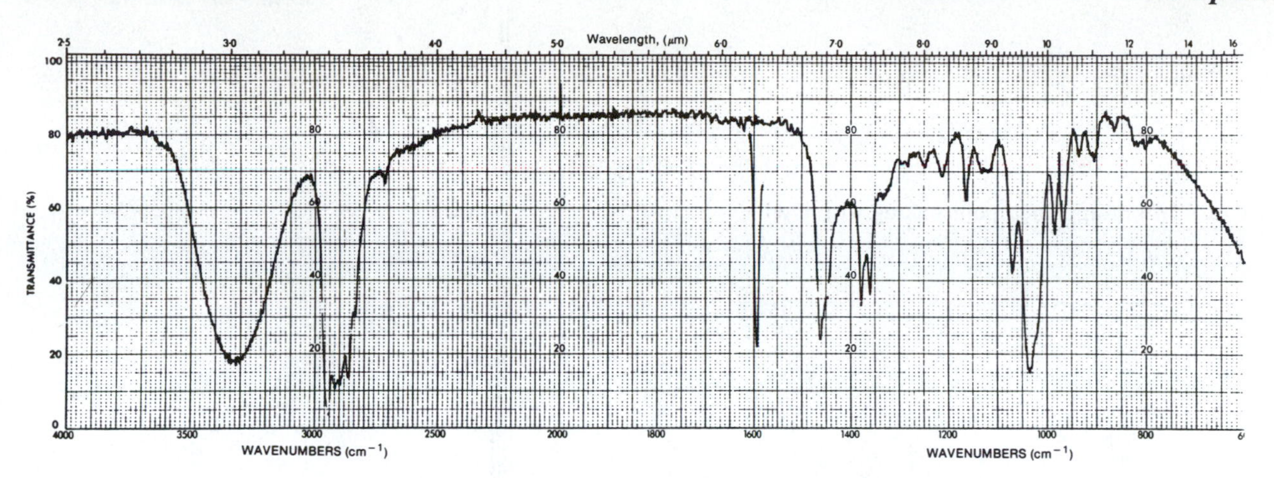

MASS

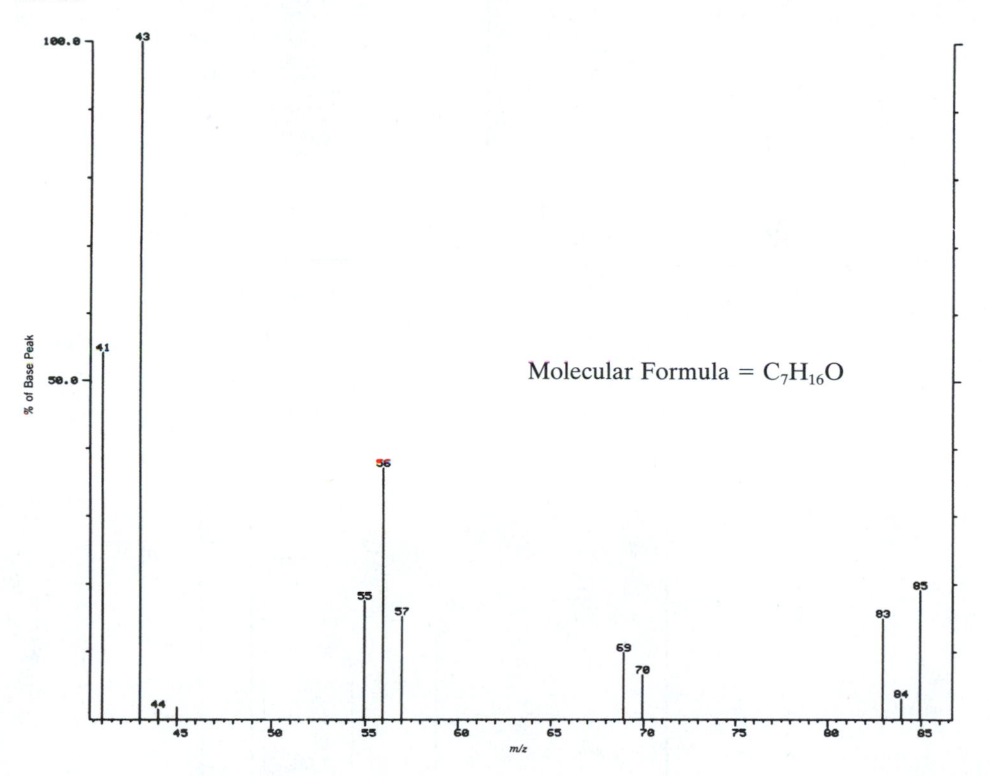

^{1}H NMR SPECTRA (SOLVENT $CDCl_3$, 500 MHz)

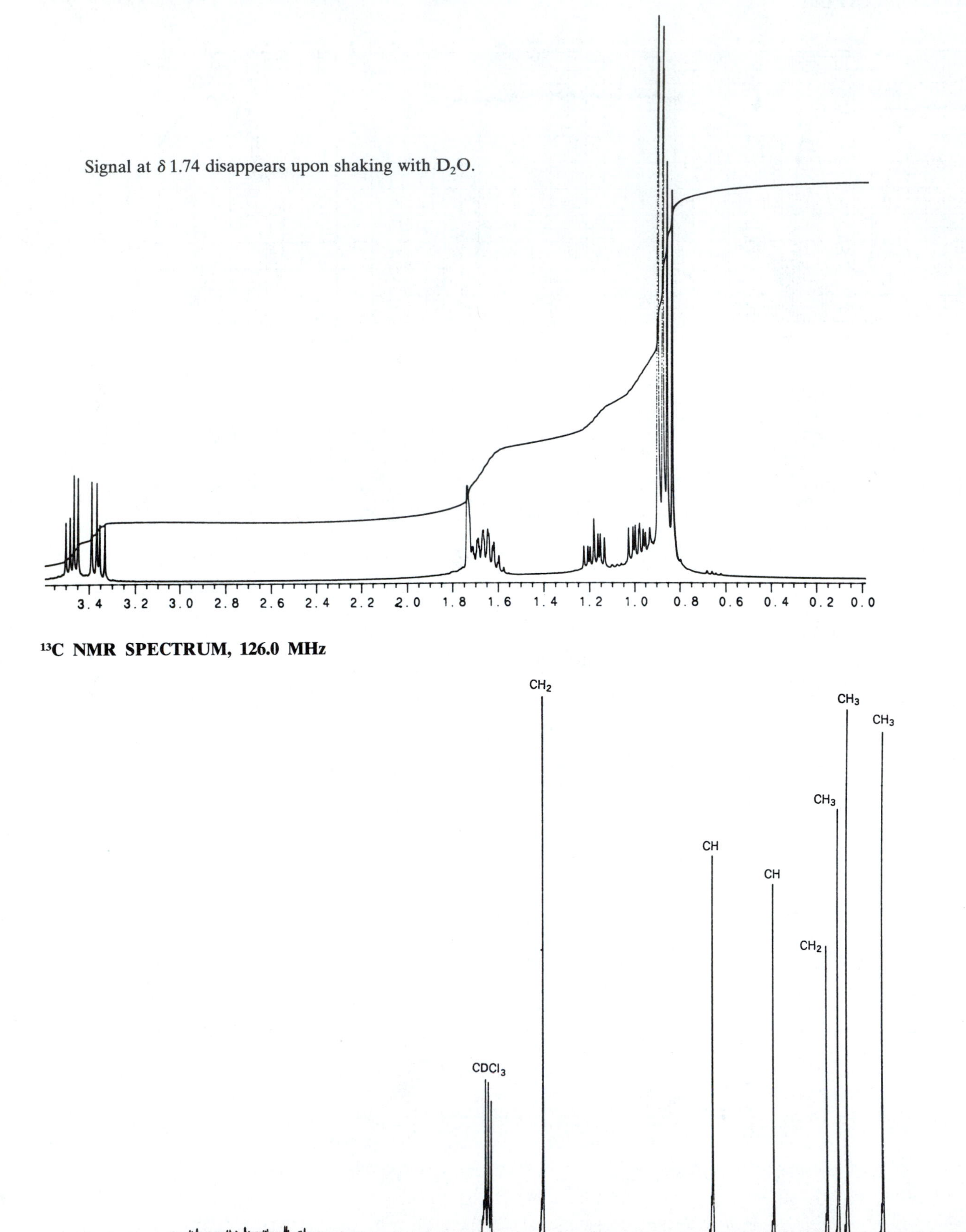

INFRARED

Compound 9.33

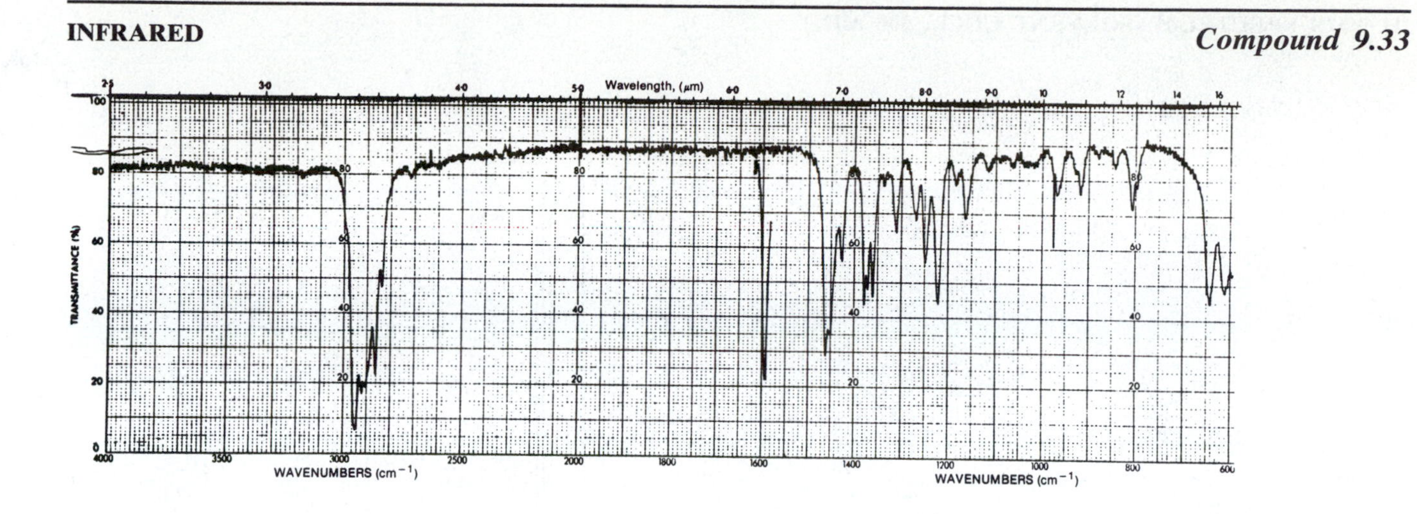

MASS

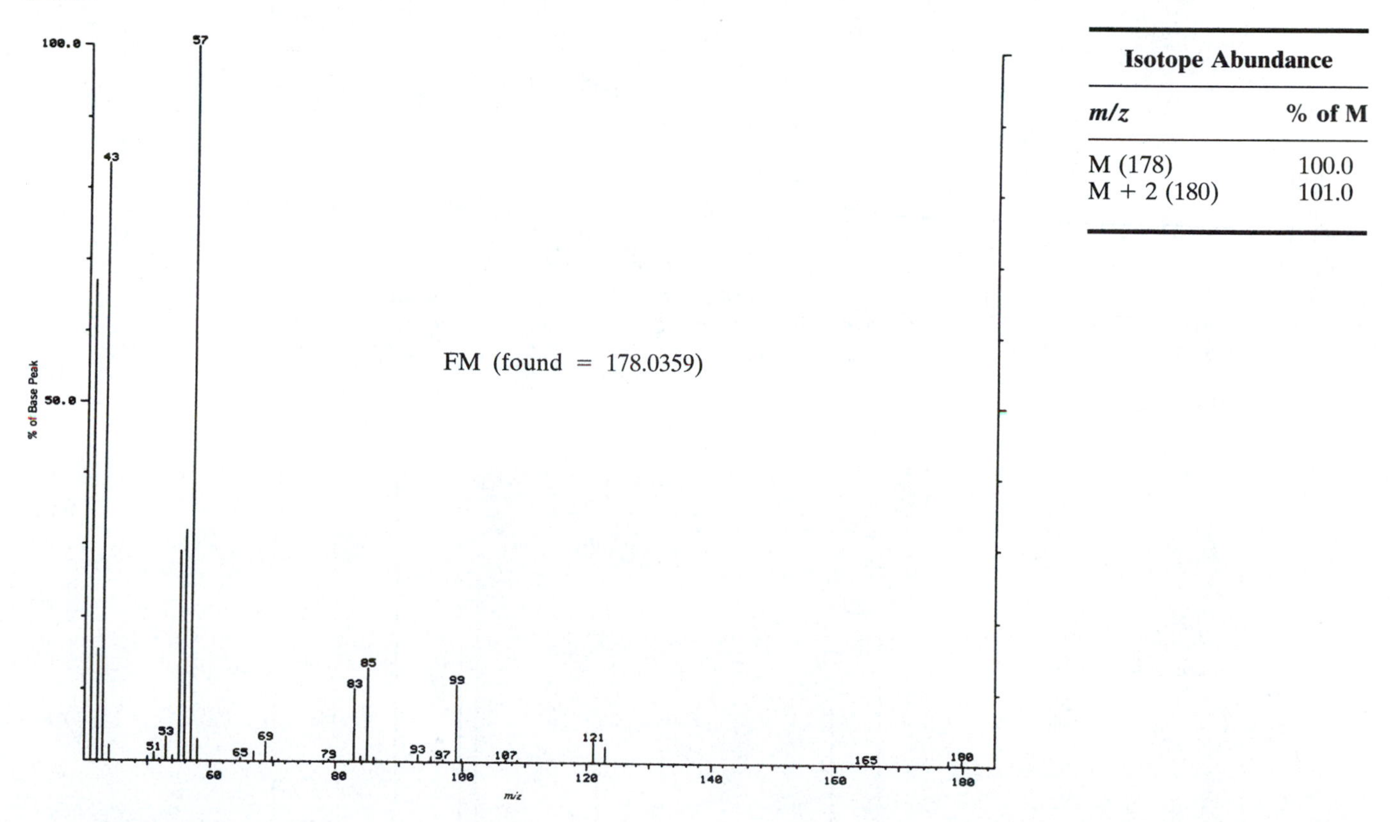

Isotope Abundance	
m/z	% of M
M (178)	100.0
M + 2 (180)	101.0

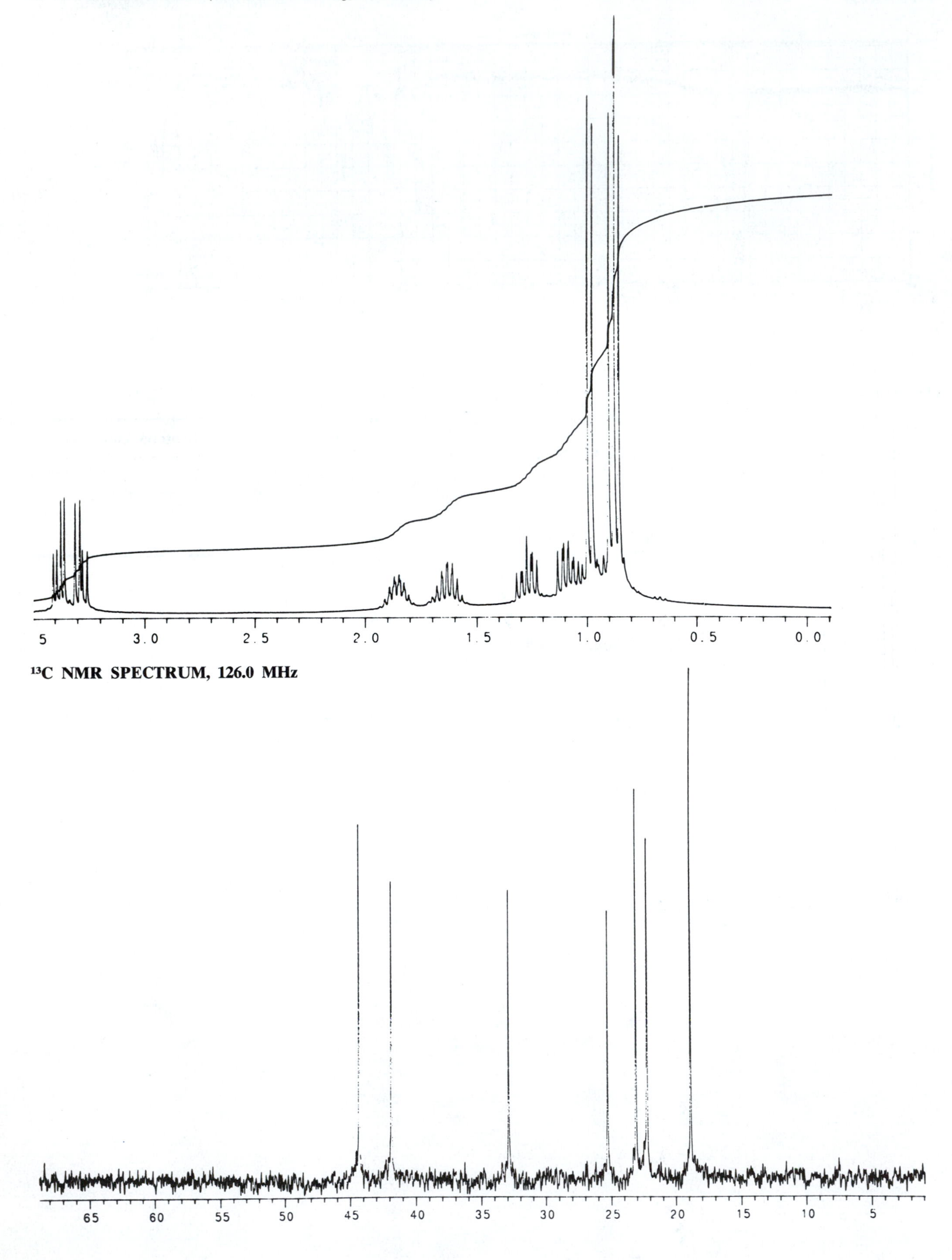

¹H NMR SPECTRUM (SOLVENT CDCl₃, 500 MHz)
5
3.0
2.5
2.0
1.5
1.0
0.5
0.0
¹³C NMR SPECTRUM, 126.0 MHz
65
60
55
50
45
40
35
30
25
20
15
10
5

INFRARED

Compound 9.34

3513.3	1426.2	1154.6
1597.5	1264.9	1033.5
1512.5	1205.4	962.5

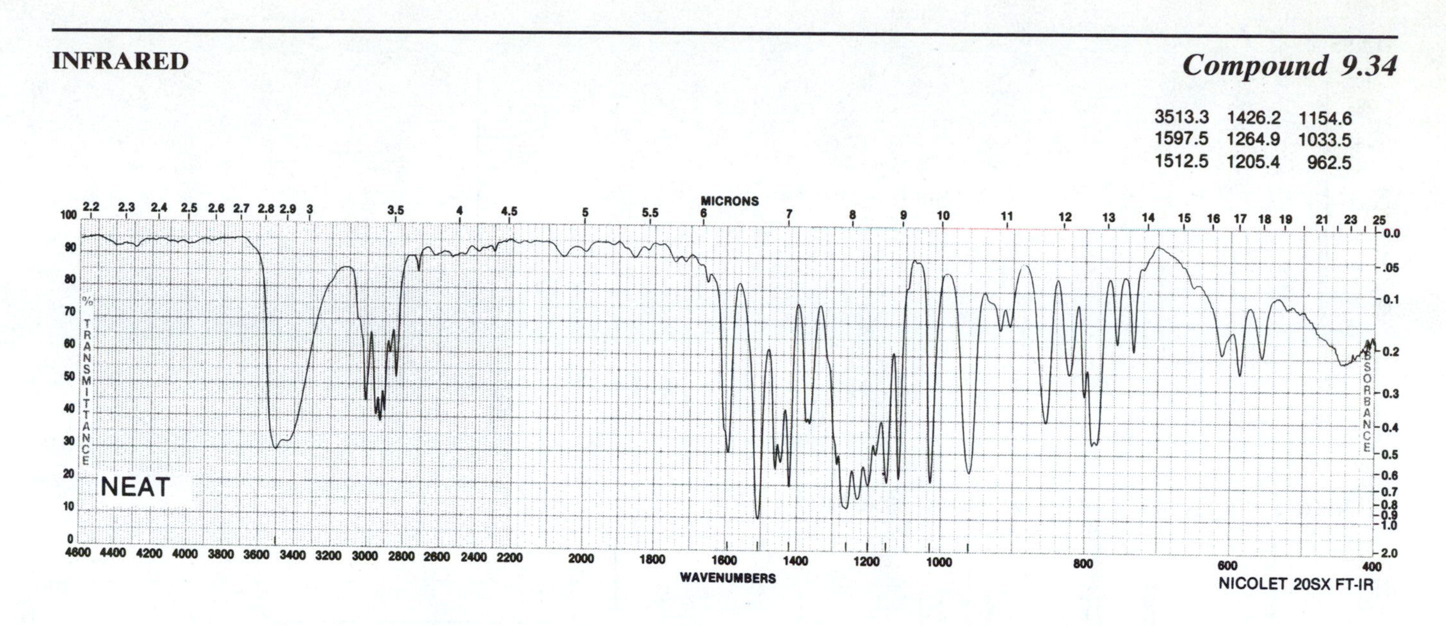

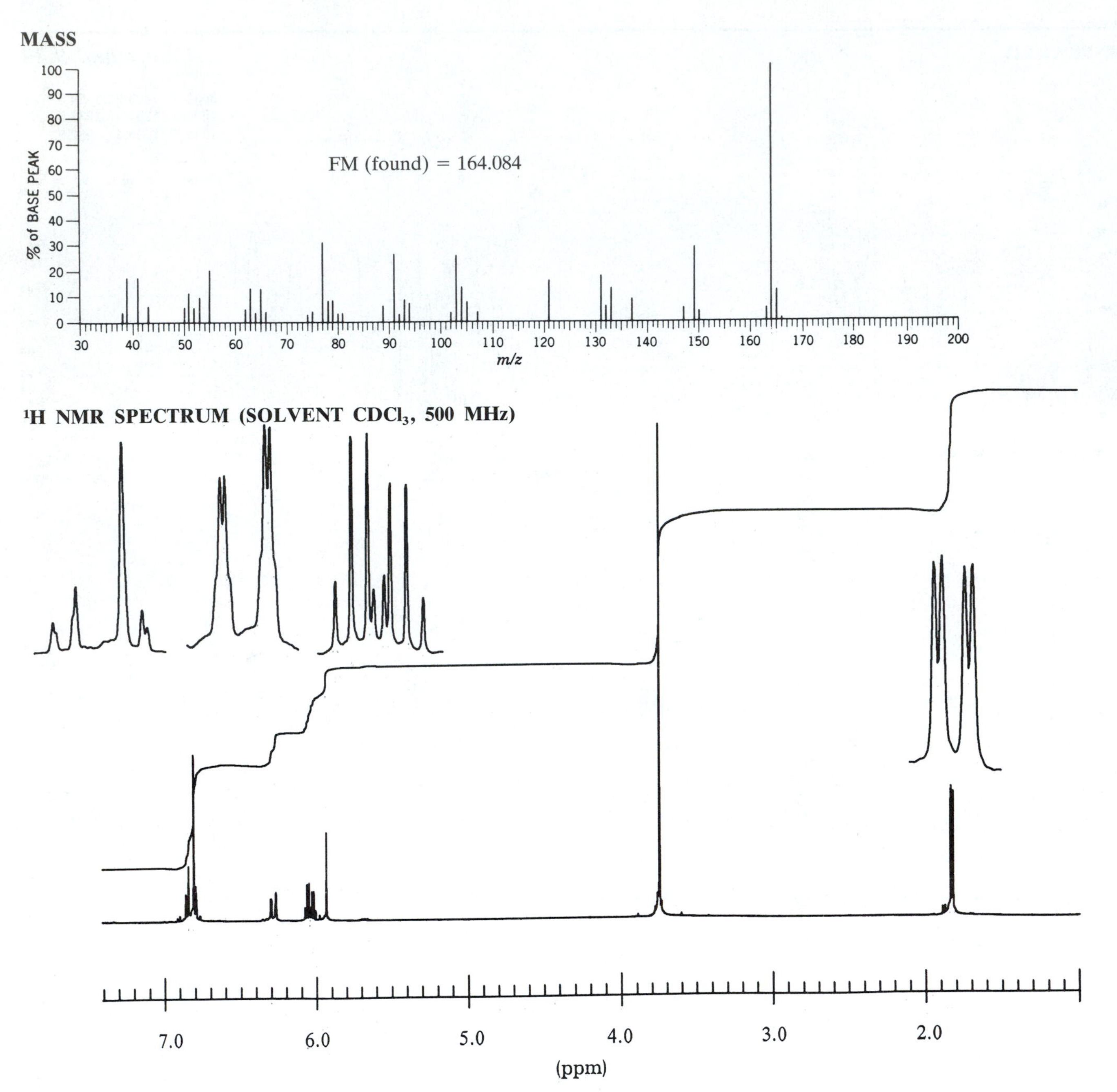
MASS
% of BASE PEAK
FM (found) = 164.084
m/z
¹H NMR SPECTRUM (SOLVENT CDCl₃, 500 MHz)
7.0
6.0
5.0
4.0
3.0
2.0
(ppm)

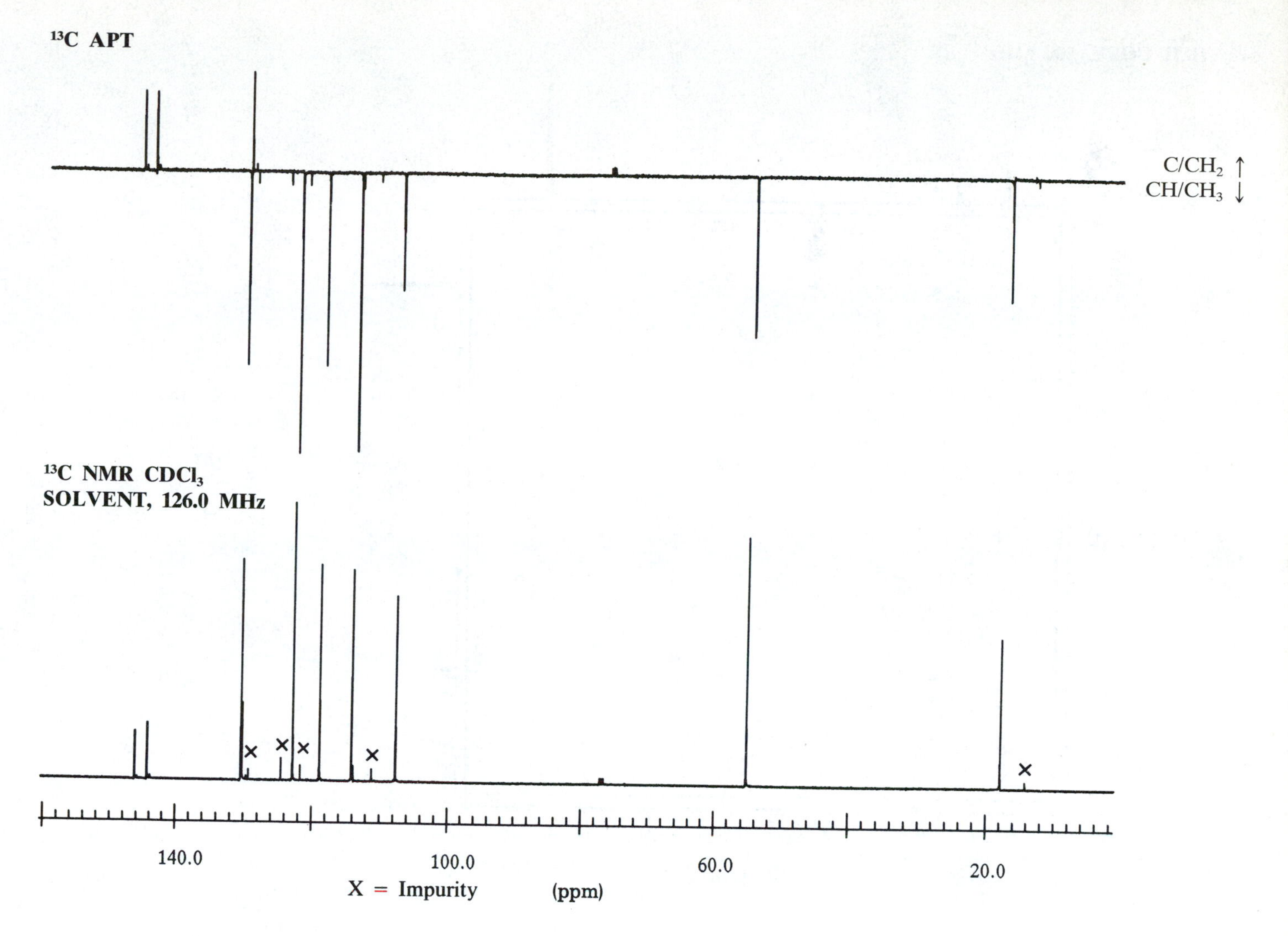

^{13}C APT
C/CH$_2$ ↑
CH/CH$_3$ ↓
^{13}C NMR CDCl$_3$
SOLVENT, 126.0 MHz
140.0
100.0
60.0
20.0
X = Impurity
(ppm)

H H COSY, 500 MHz

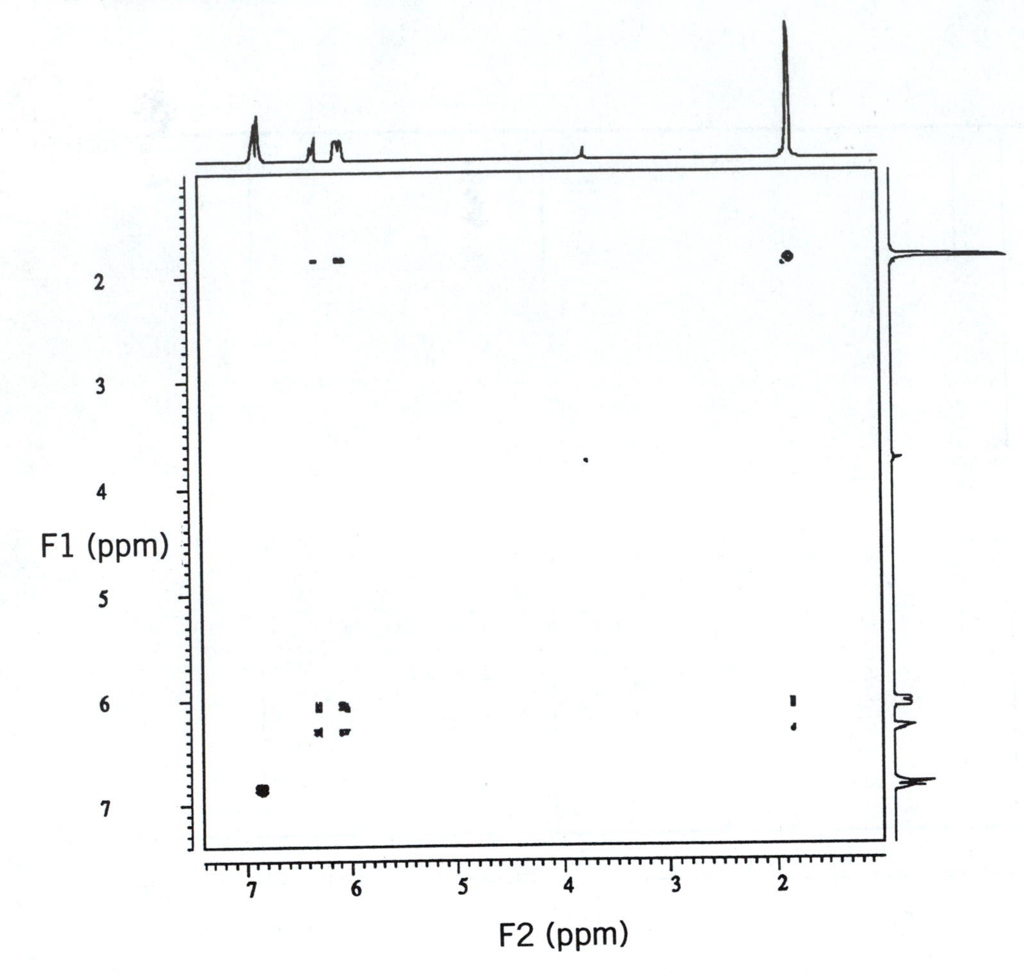

H C COSY (HETCOR), ^{13}C 126.0

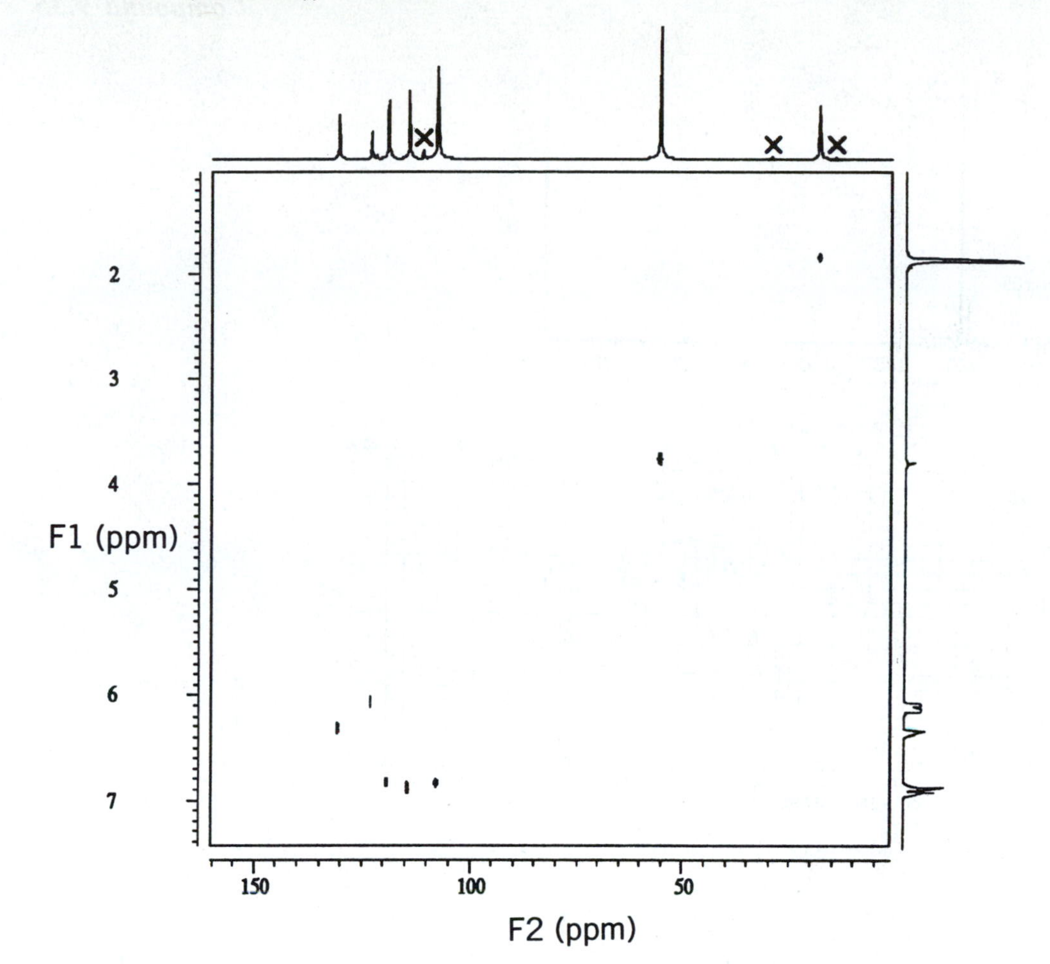

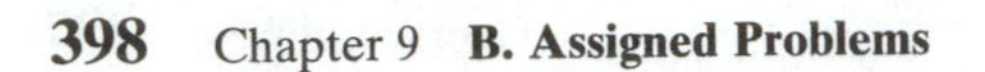

Compound 9.35

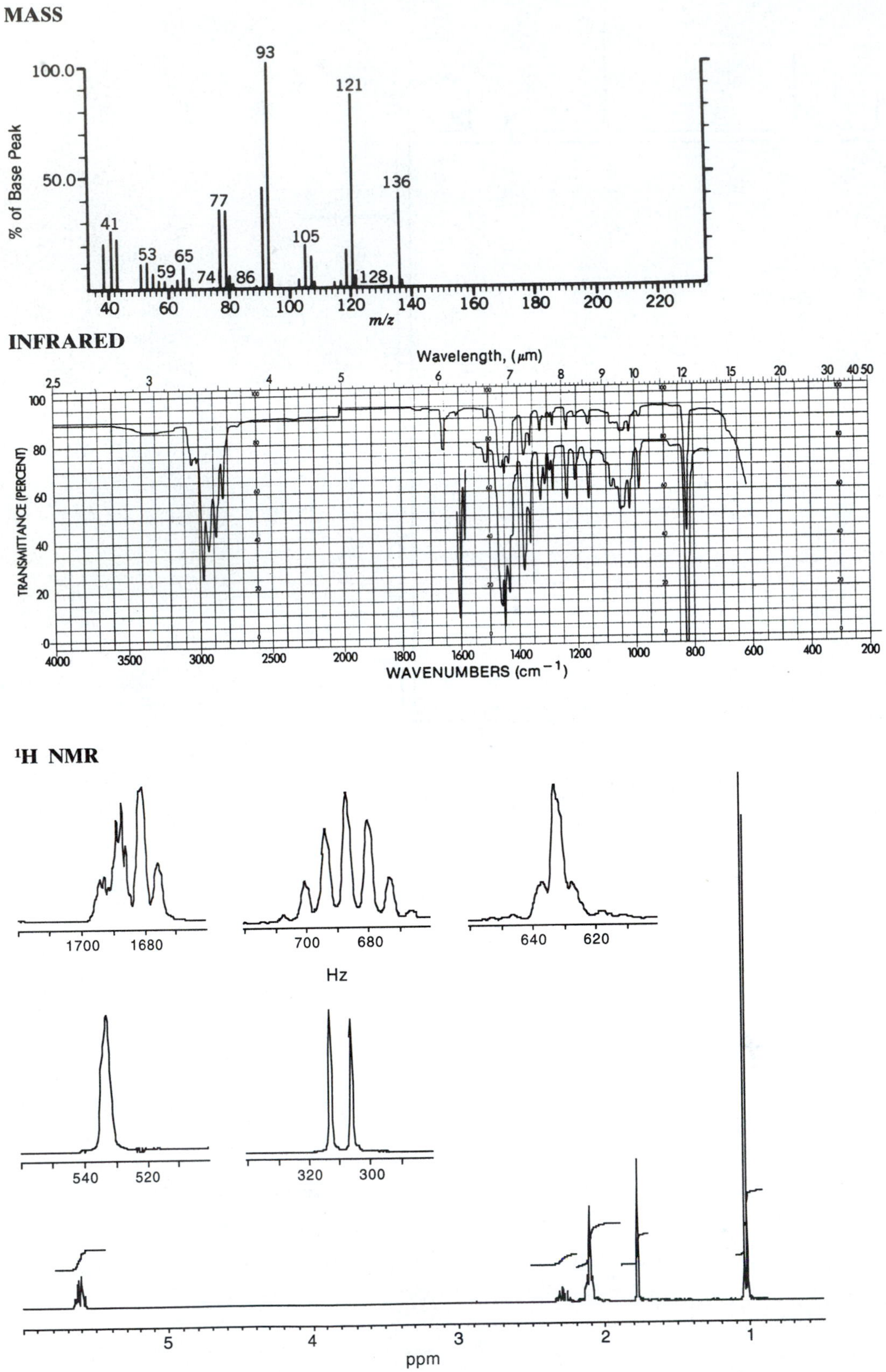

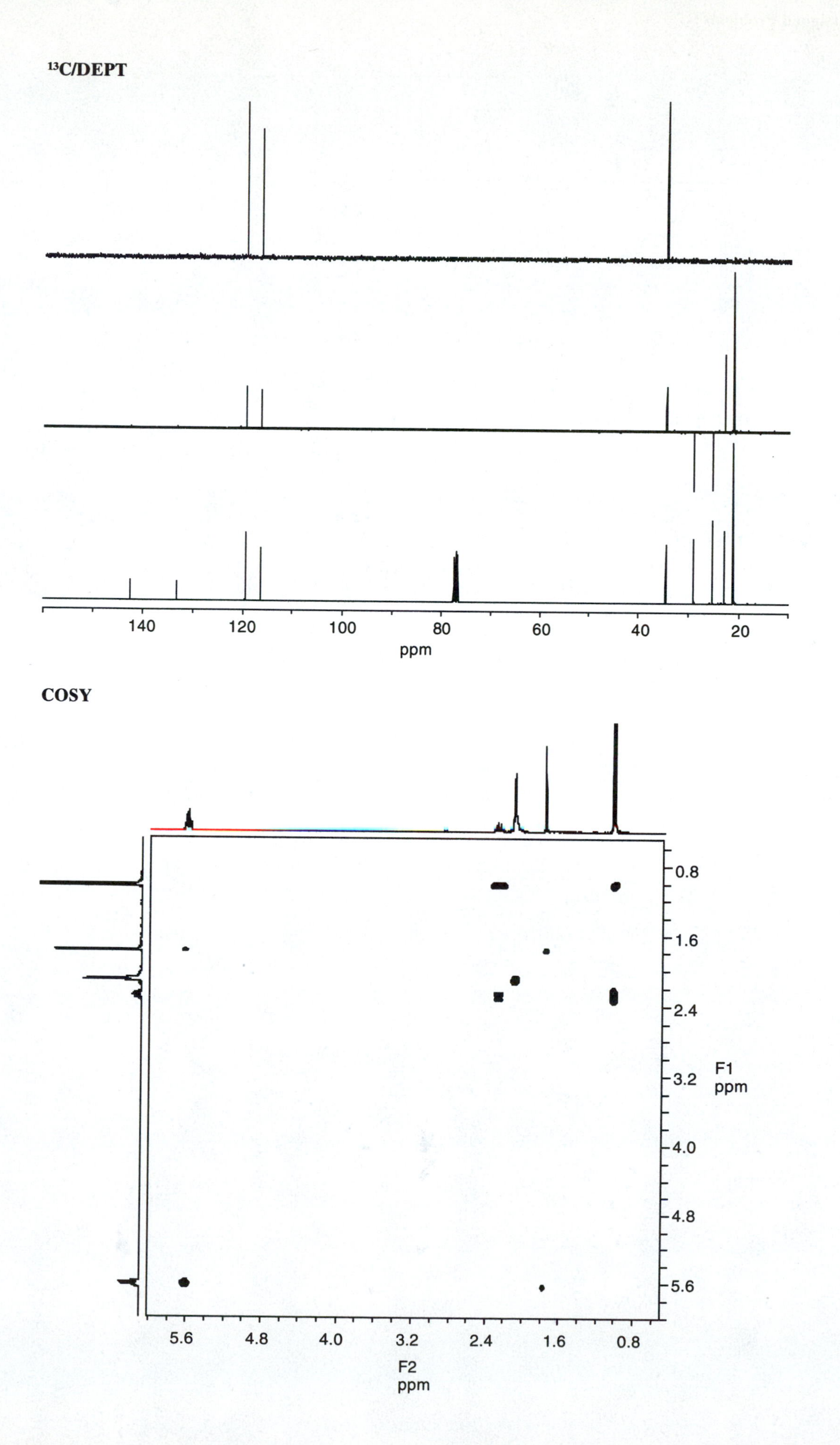

13C/DEPT
140
120
100
80
60
40
20
ppm
COSY
0.8
1.6
2.4
3.2
4.0
4.8
5.6
F1
ppm
5.6
4.8
4.0
3.2
2.4
1.6
0.8
F2
ppm

HETCOR

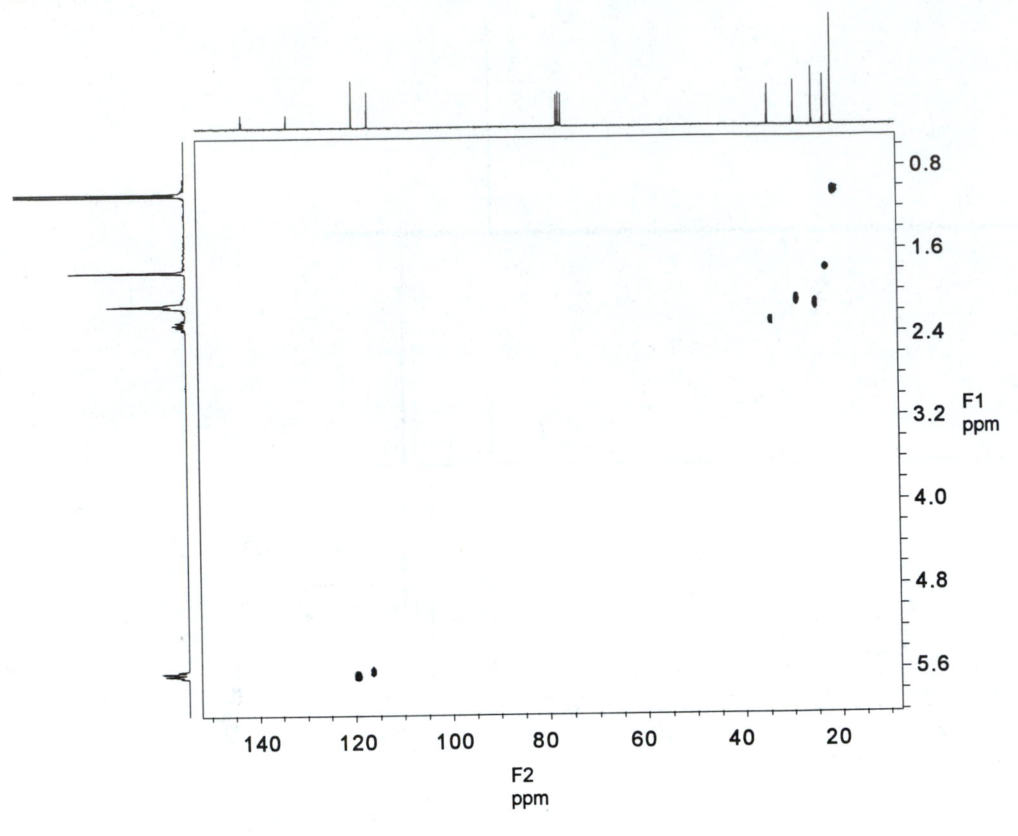

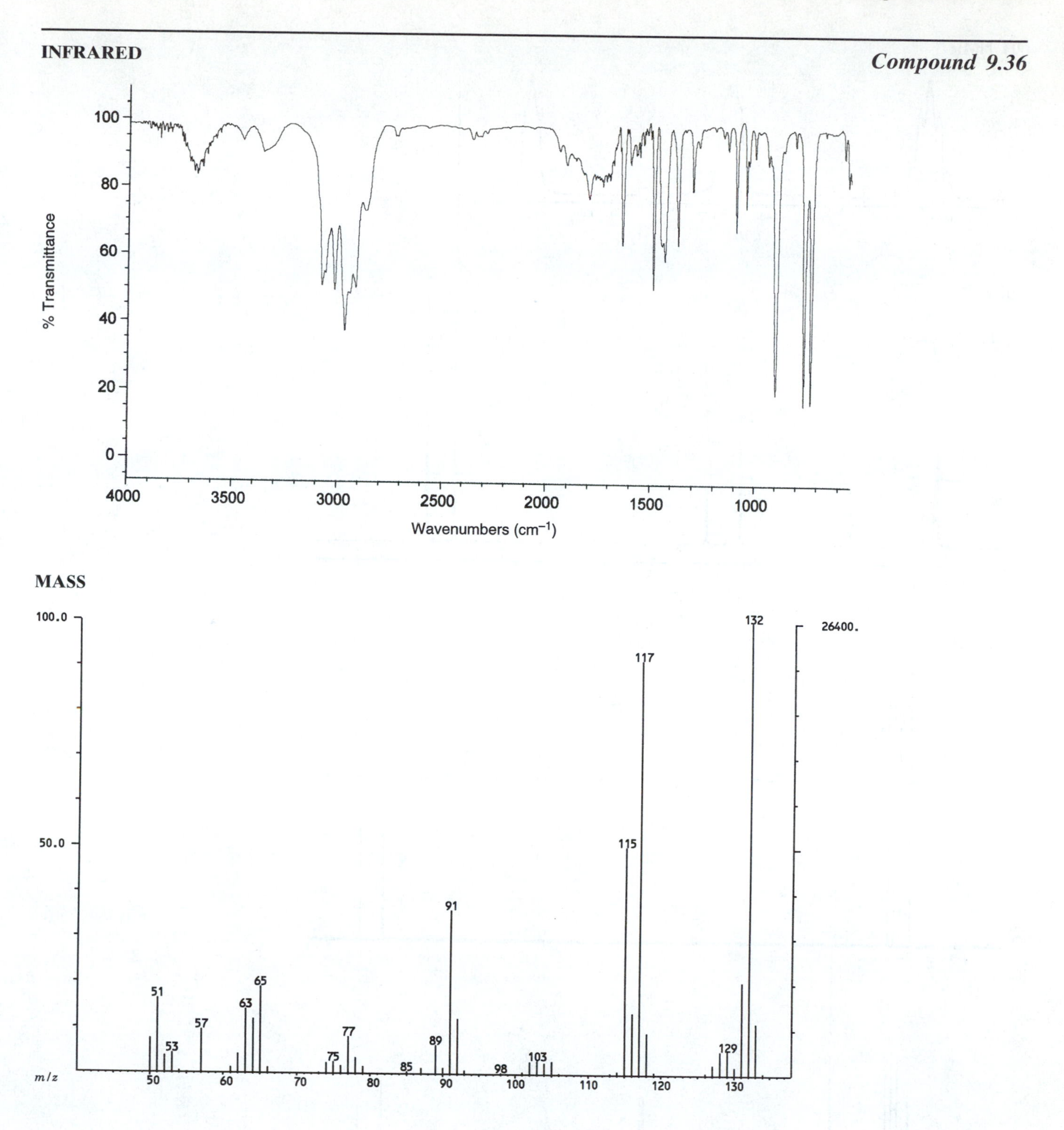

INFRARED
Compound 9.36
% Transmittance
Wavenumbers (cm⁻¹)
MASS
100.0
50.0
26400.
m/z
132
117
115
91
65
63
51
57
53
77
75
89
85
98
103
129

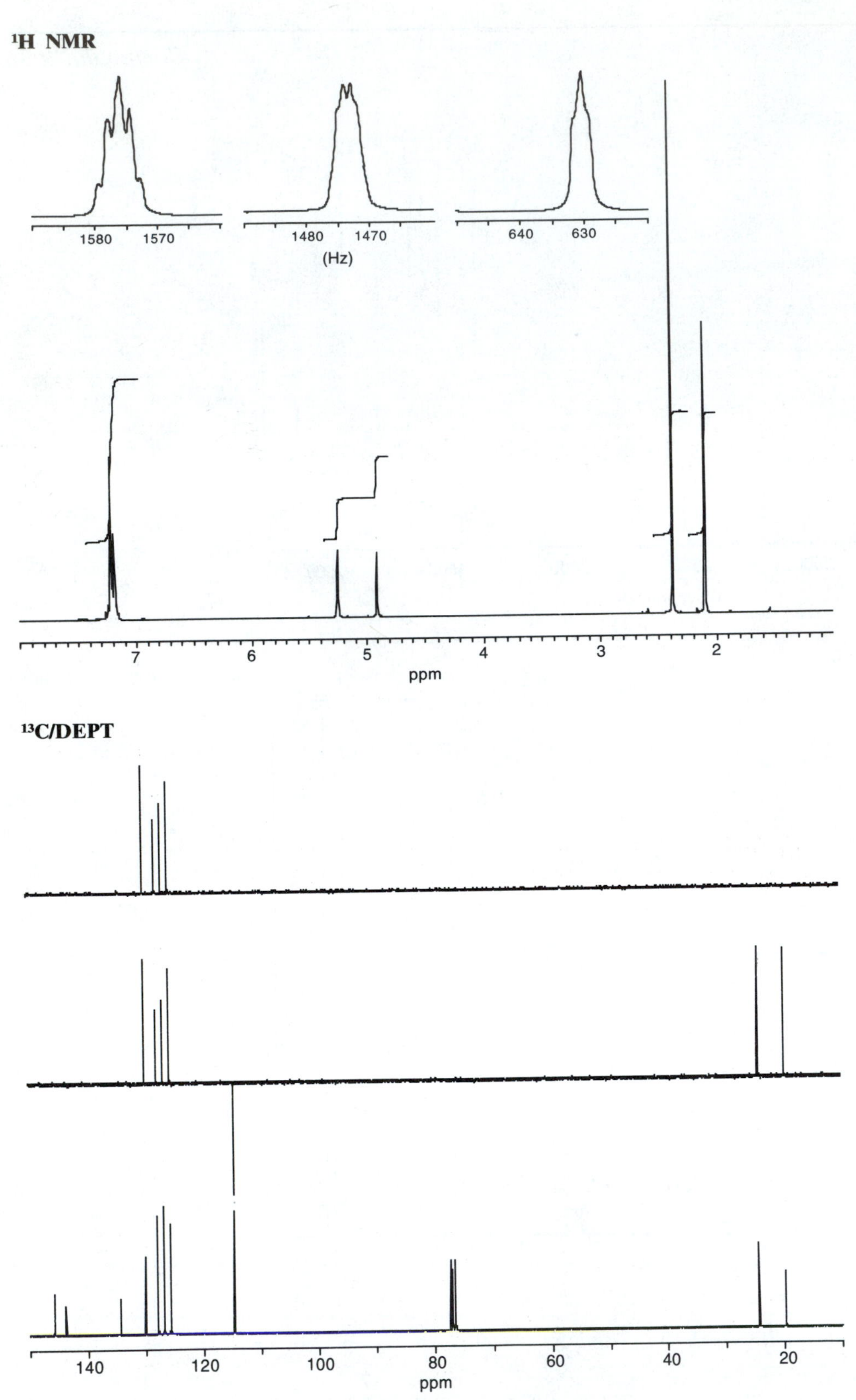

¹H NMR
1580
1570
1480
1470
(Hz)
640
630
7
6
5
4
3
2
ppm
¹³C/DEPT
140
120
100
80
60
40
20
ppm

COSY

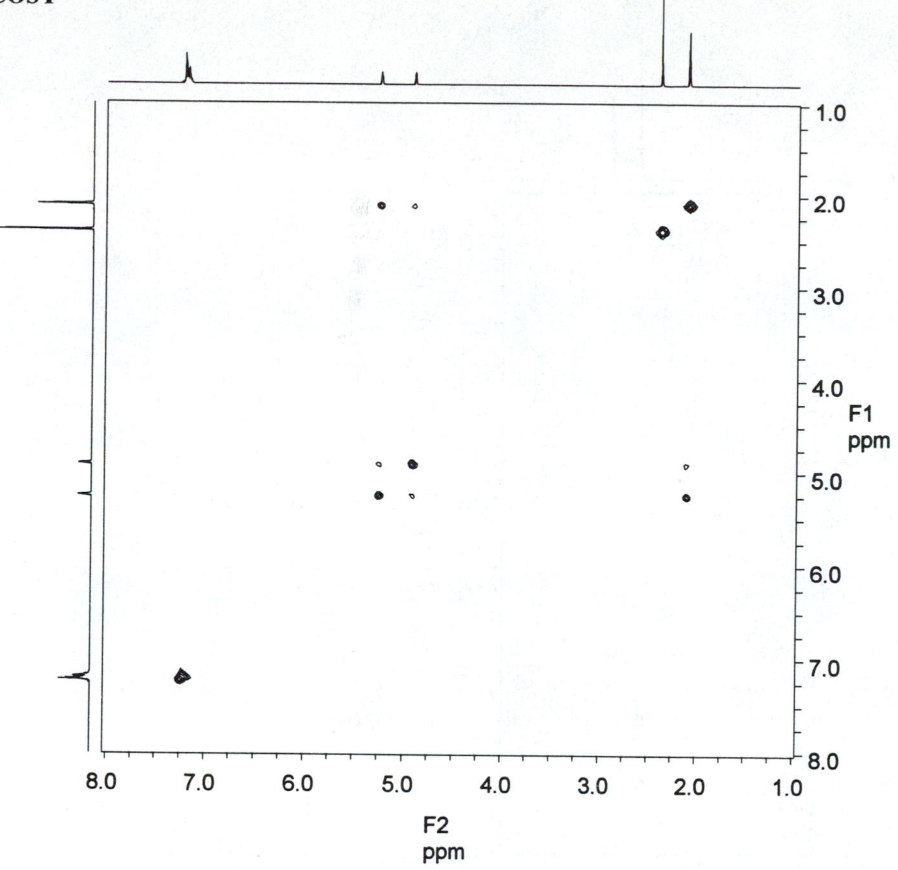

1.0
2.0
3.0
4.0
F1
ppm
5.0
6.0
7.0
8.0
8.0
7.0
6.0
5.0
4.0
3.0
2.0
1.0
F2
ppm

HETCOR

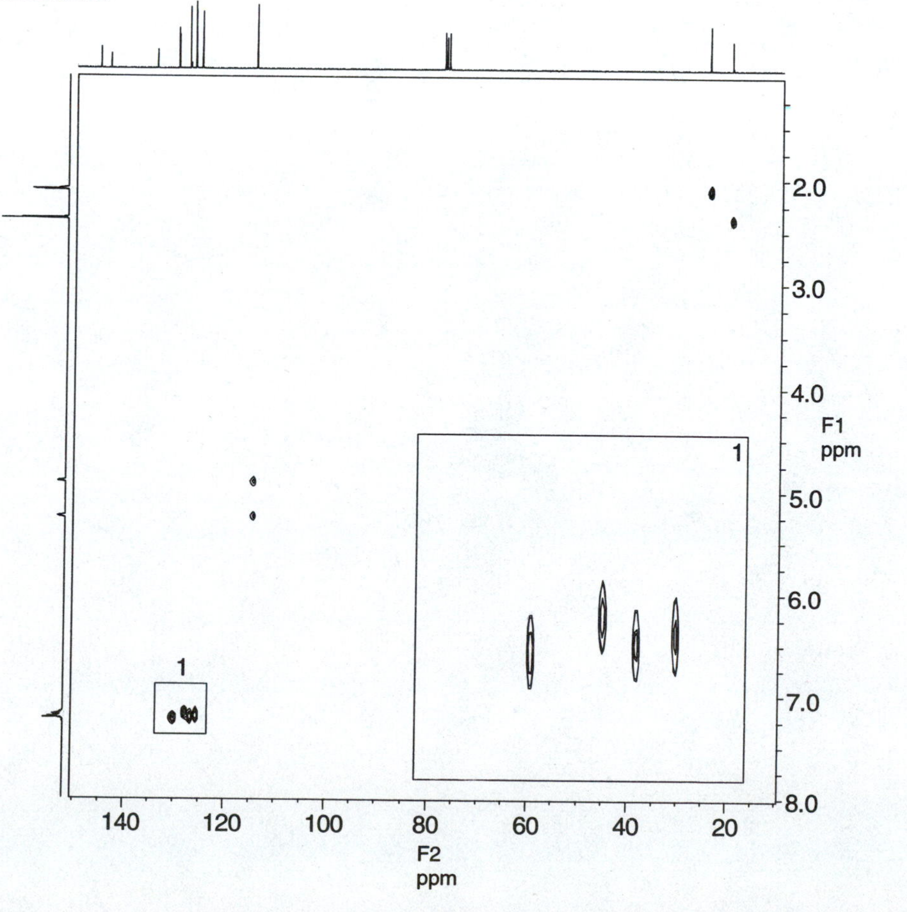

2.0
3.0
4.0
F1
ppm
5.0
6.0
7.0
8.0
1
1
140
120
100
80
60
40
20
F2
ppm

HMBC

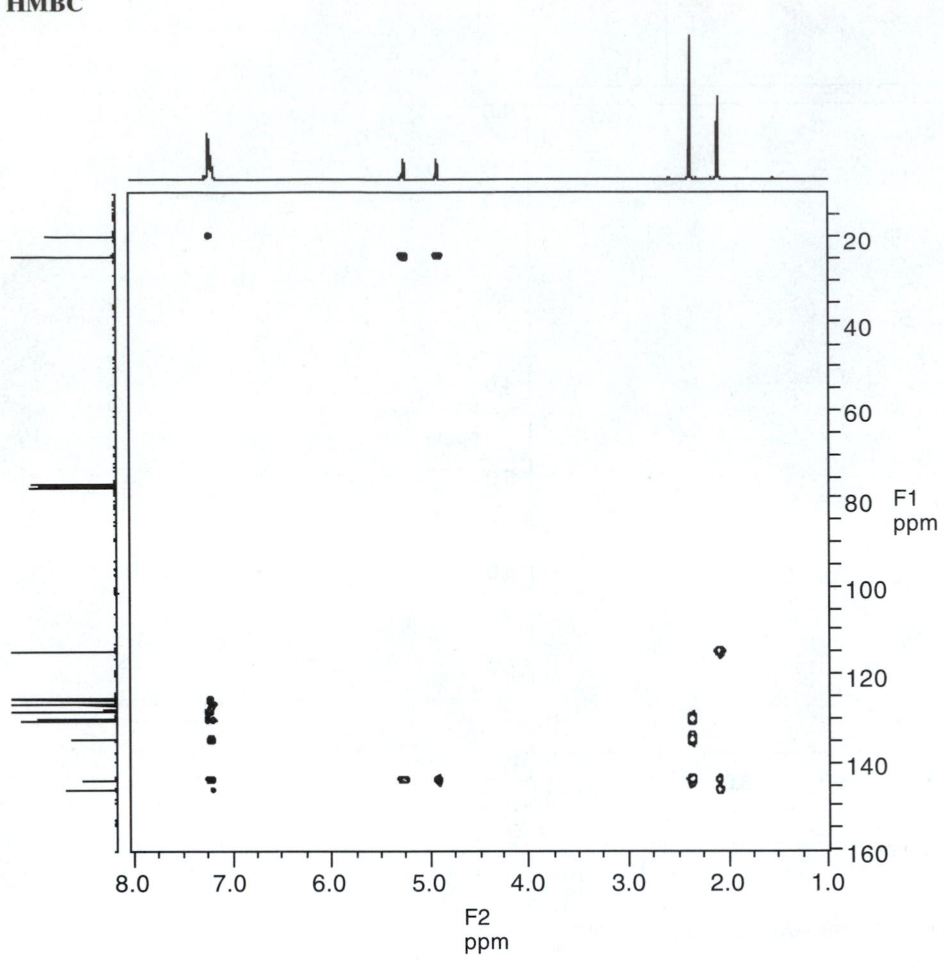

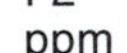

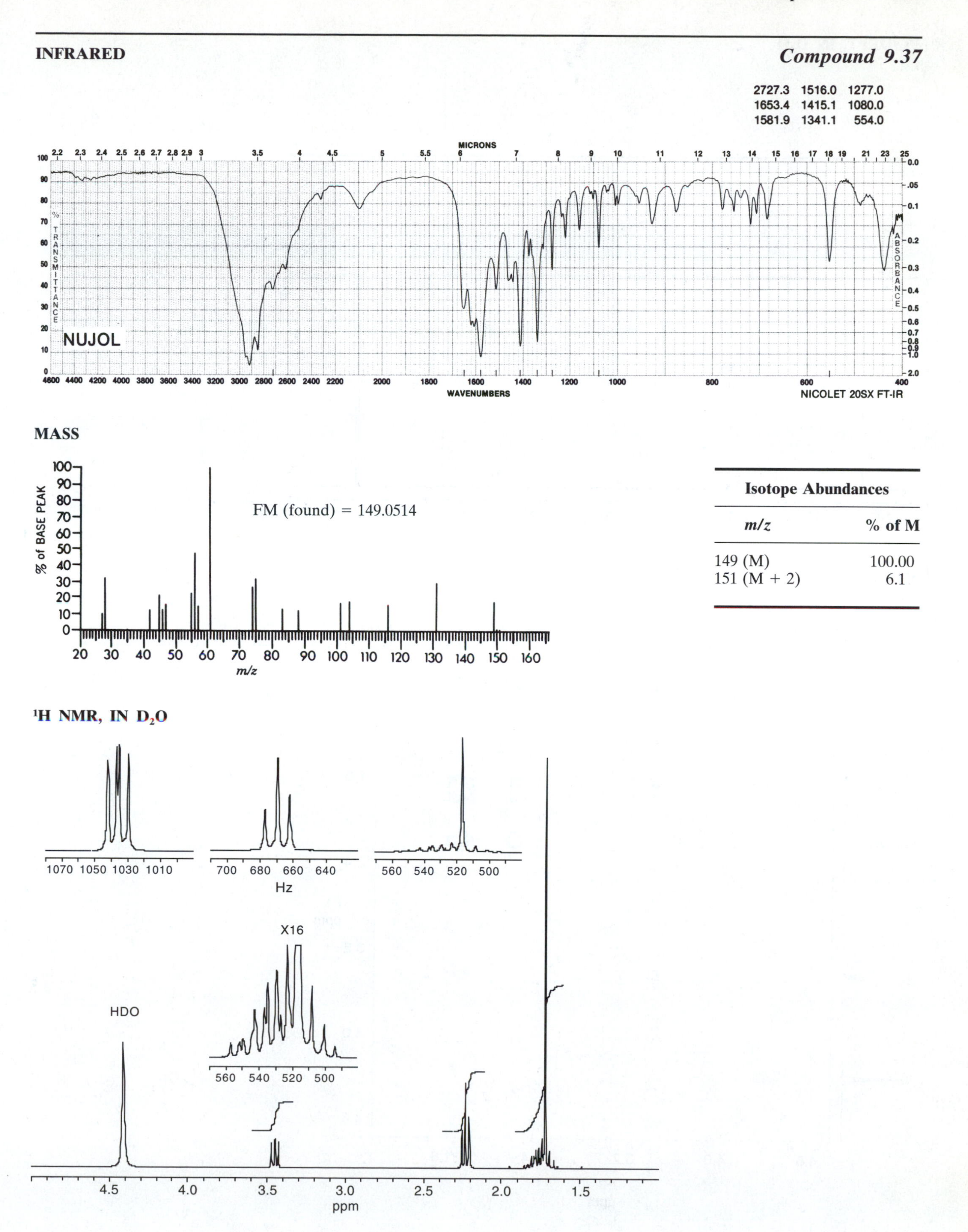

Isotope Abundances	
m/z	% of M
149 (M)	100.00
151 (M + 2)	6.1

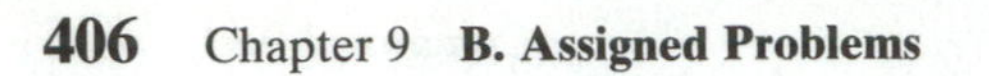

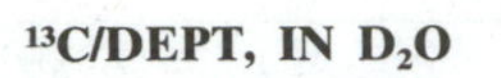

^{13}C/DEPT, IN D_2O

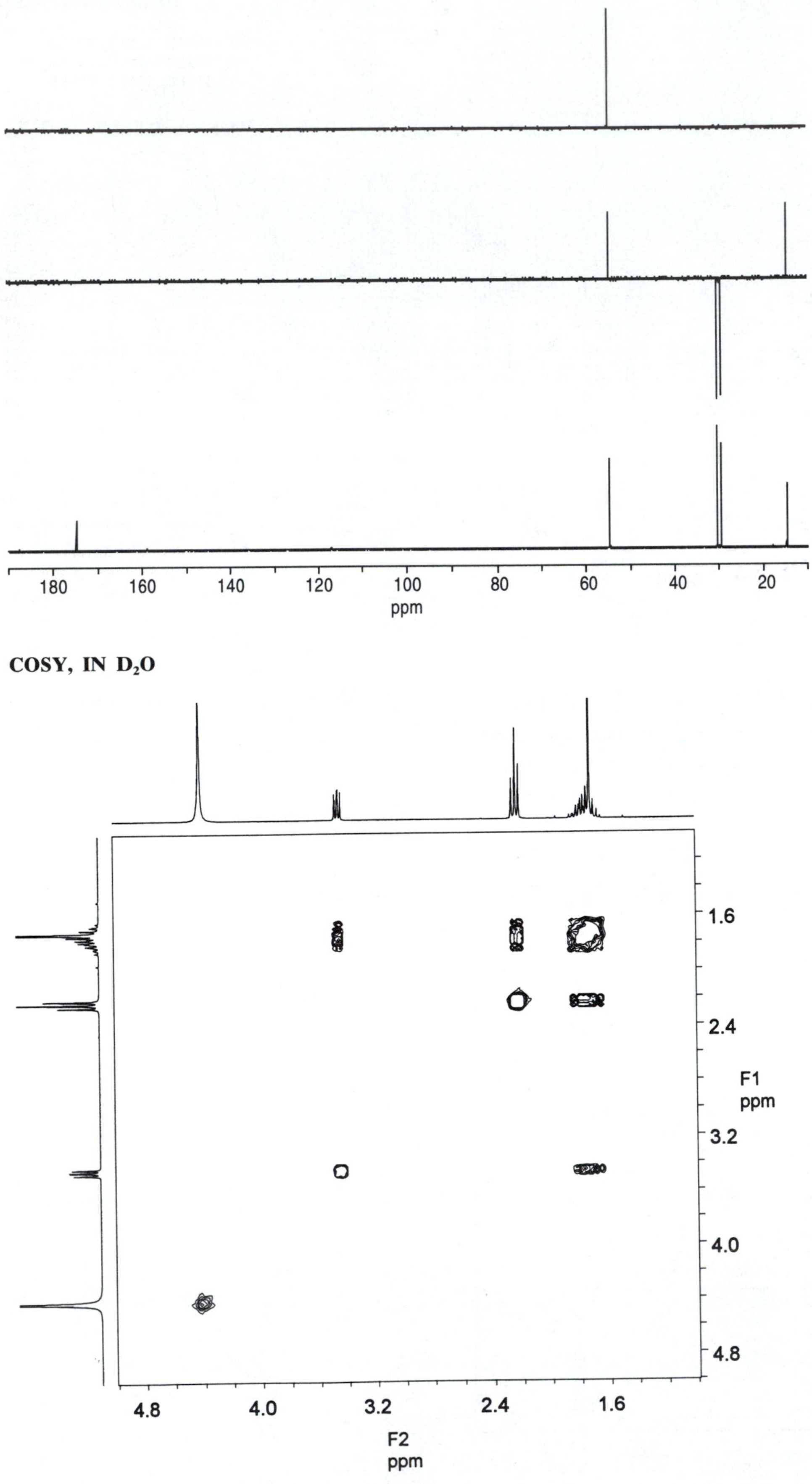

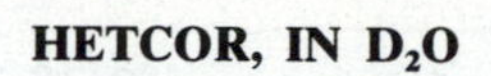

HETCOR, IN D_2O

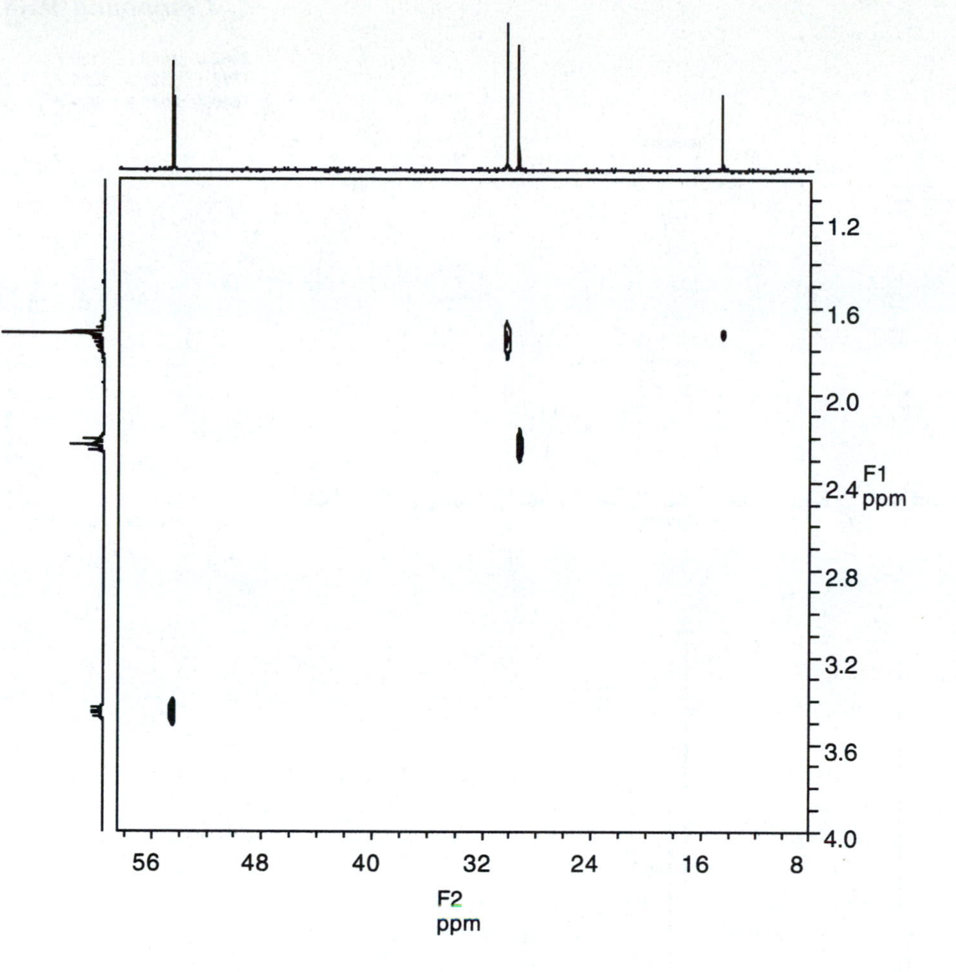

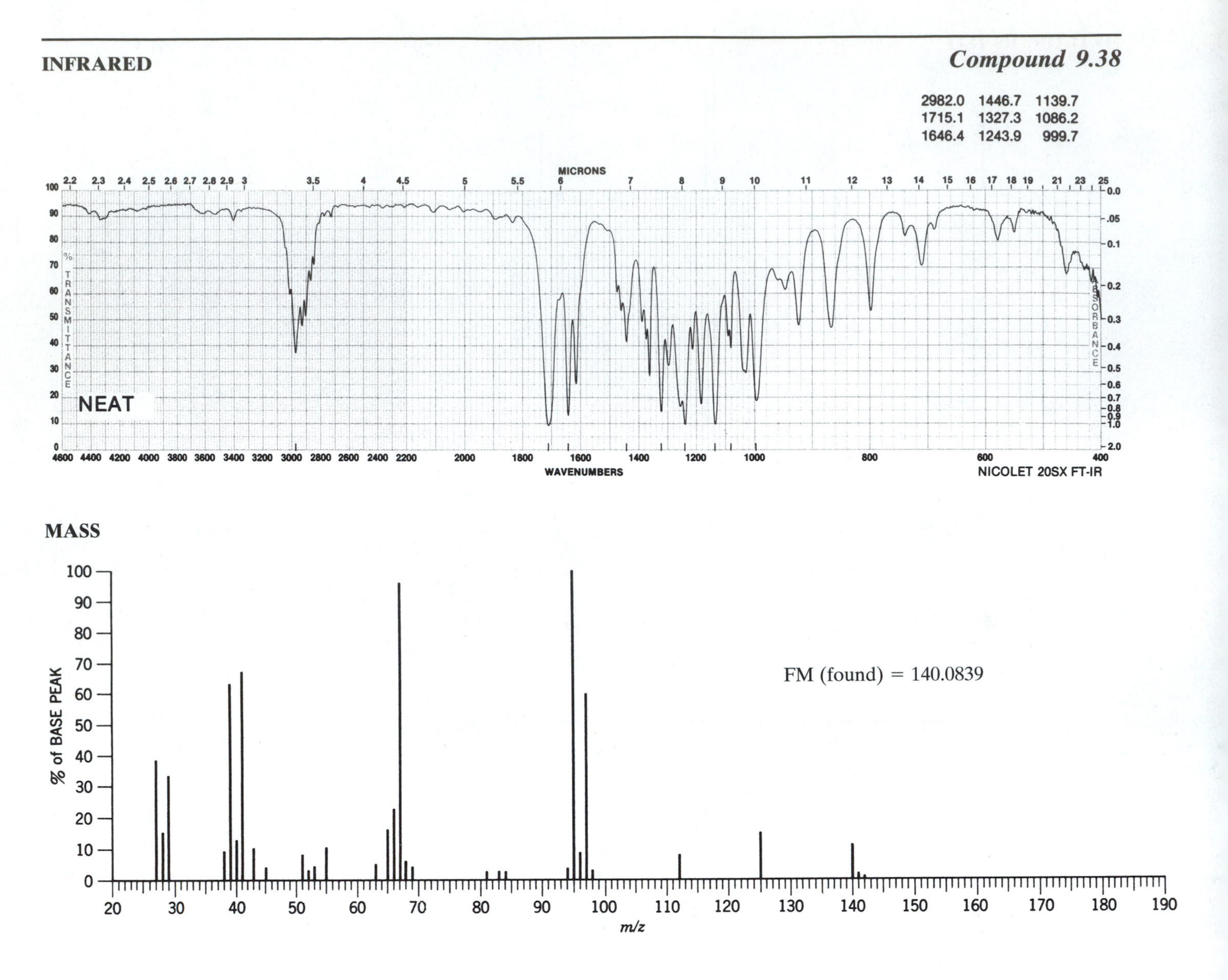
INFRARED
Compound 9.38
2982.0 1446.7 1139.7
1715.1 1327.3 1086.2
1646.4 1243.9 999.7
MICRONS
% TRANSMITTANCE
ABSORBANCE
NEAT
WAVENUMBERS
NICOLET 20SX FT-IR
MASS
% of BASE PEAK
FM (found) = 140.0839
m/z

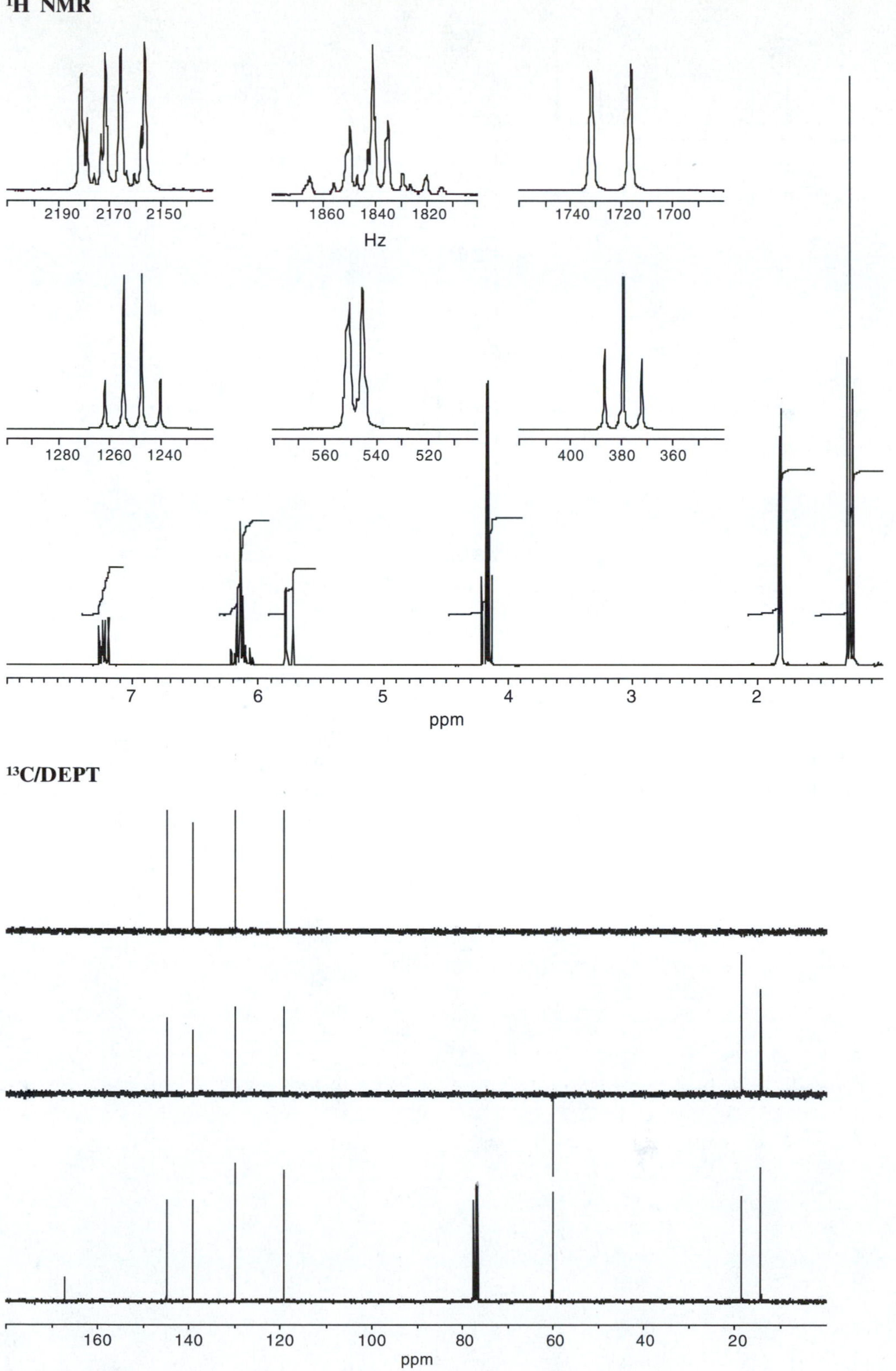

¹H NMR
2190
2170
2150
1860
1840
1820
Hz
1740
1720
1700
1280
1260
1240
560
540
520
400
380
360
7
6
5
4
3
2
ppm
¹³C/DEPT
160
140
120
100
80
60
40
20
ppm

COSY

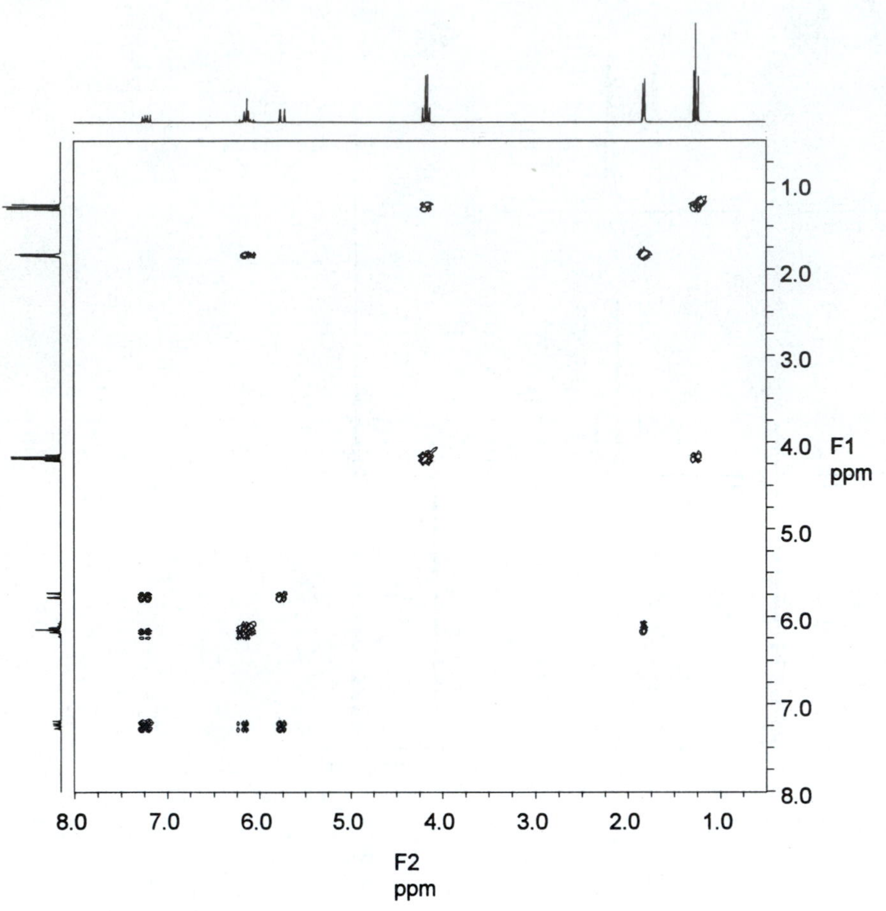

INFRARED

Compound 9.39

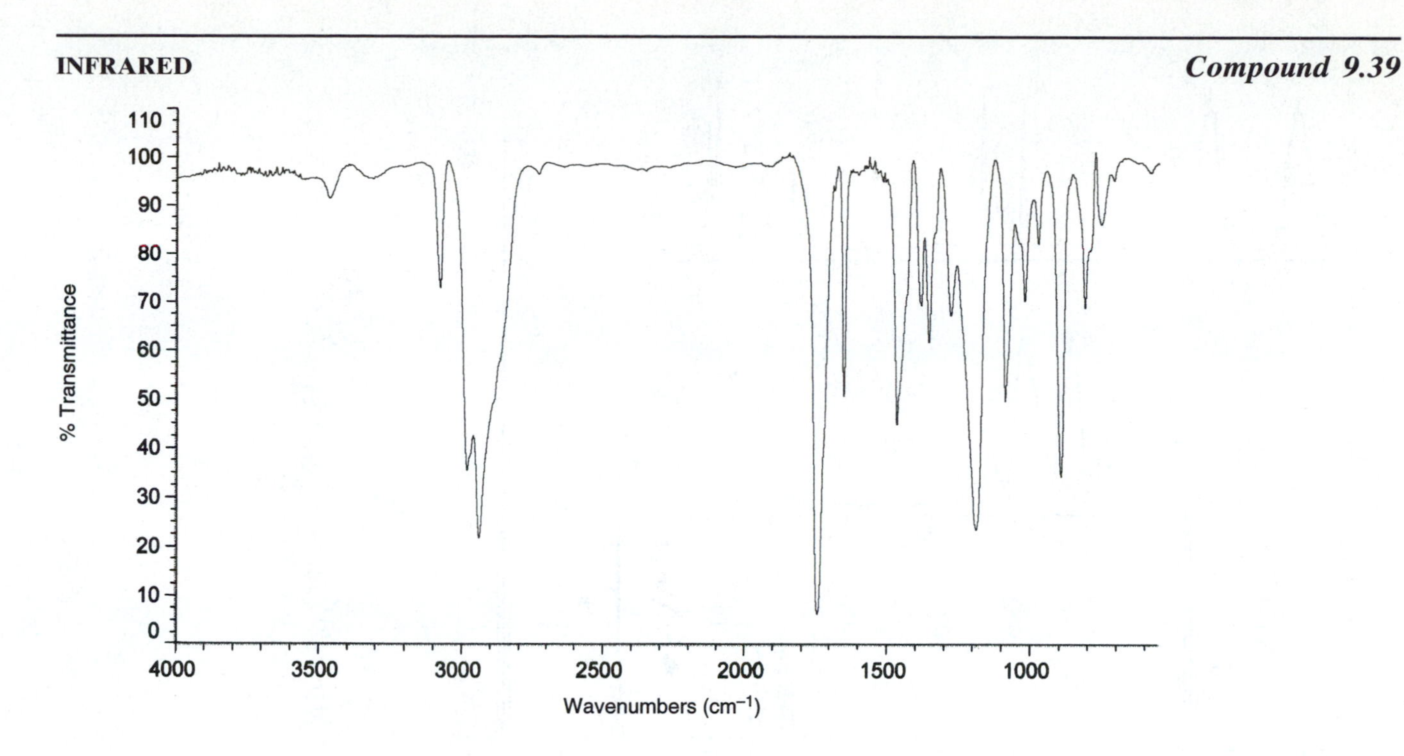

MASS

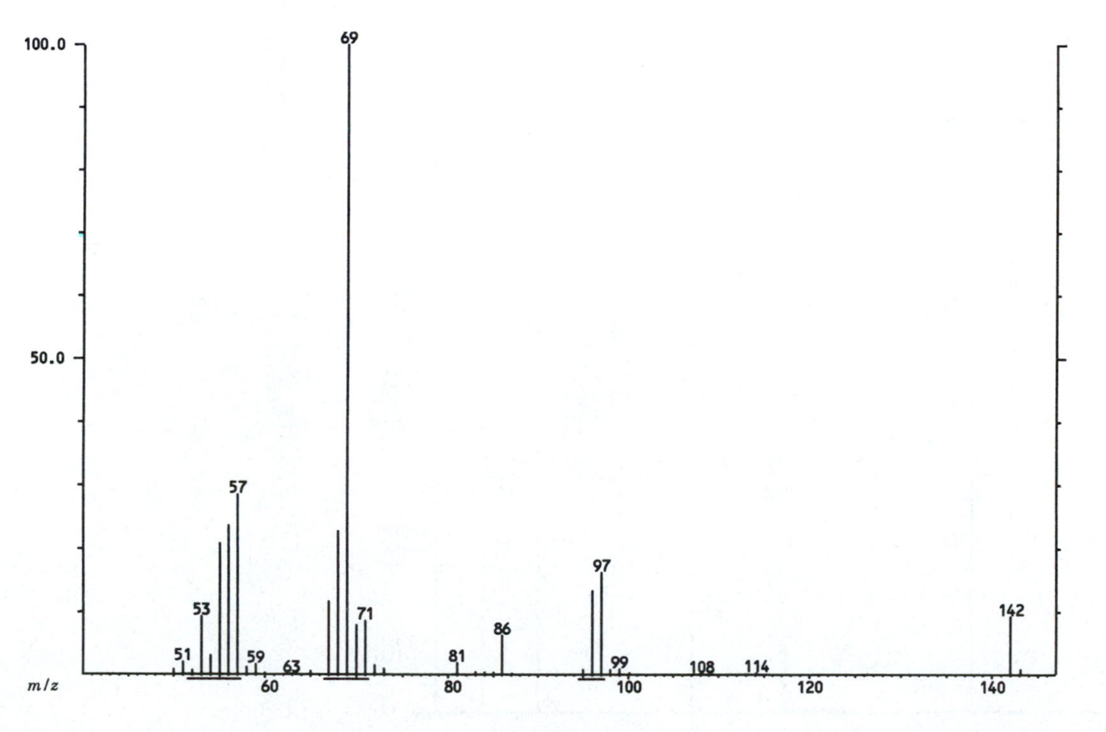

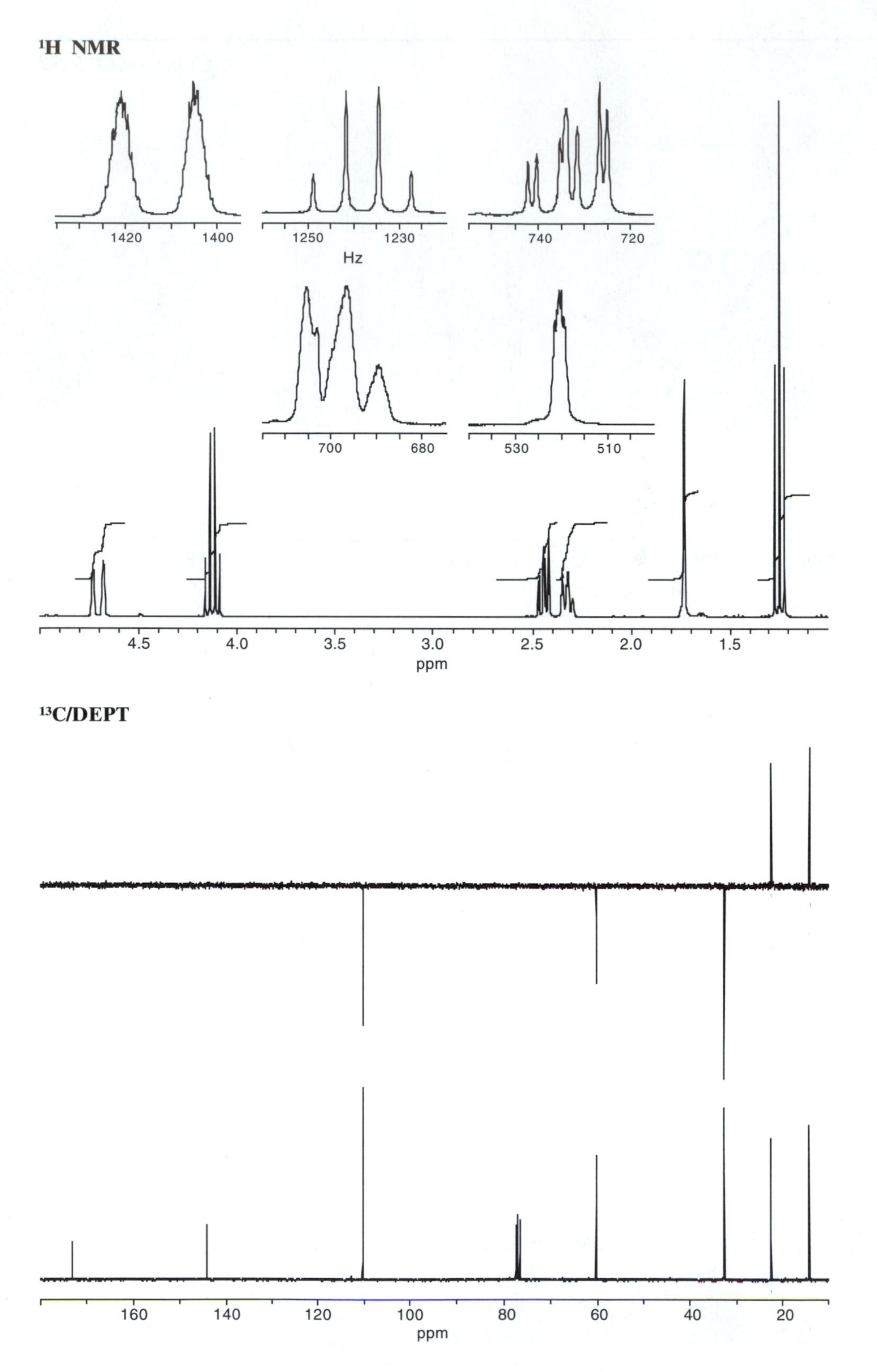
¹H NMR
1420
1400
1250
1230
Hz
740
720
700
680
530
510
4.5
4.0
3.5
3.0
ppm
2.5
2.0
1.5
¹³C/DEPT
160
140
120
100
80
60
40
20
ppm

MASS SPECTRAL DATA ***Compound 9.40***

FM (found) = 128.1203
peaks at *m/z* 128, 113, 110, and 95 (base peak)

INFRARED DATUM

broad, s at ≈ 3300 cm^{-1}

^{1}H NMR

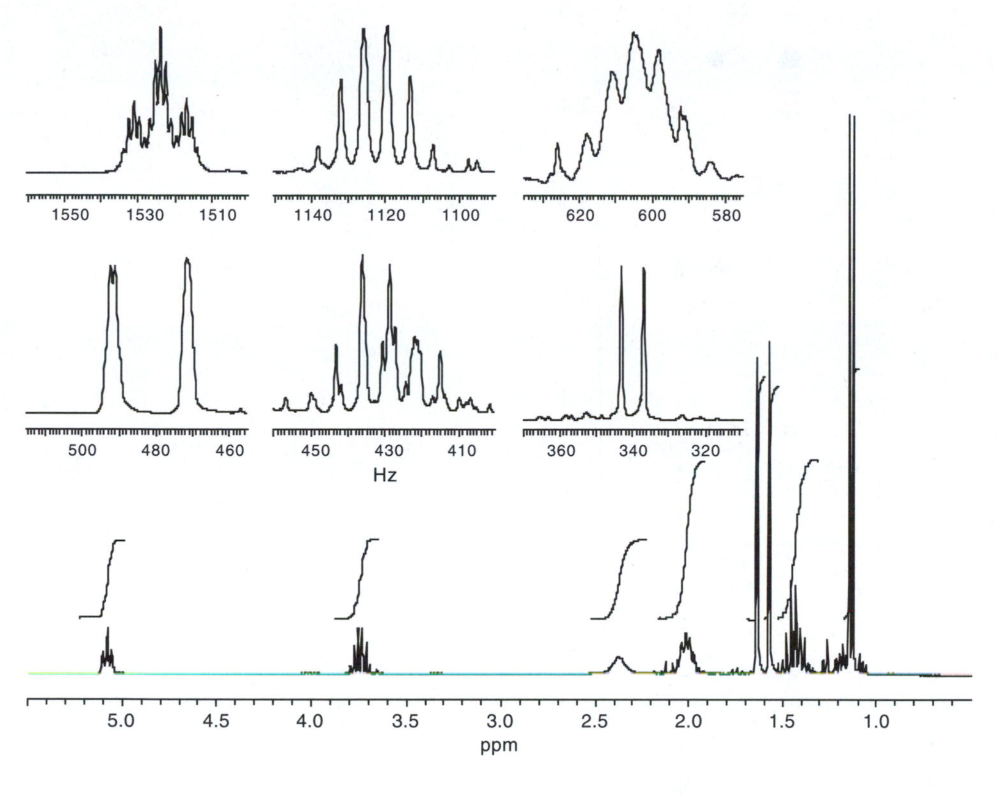

^{13}C/DEPT

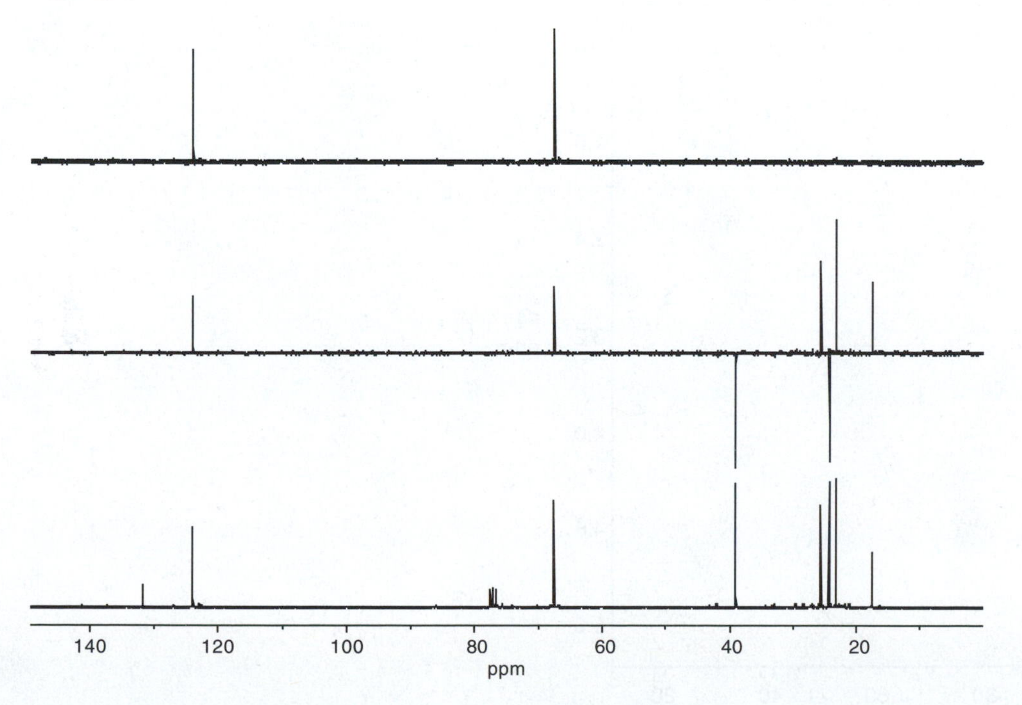

COSY

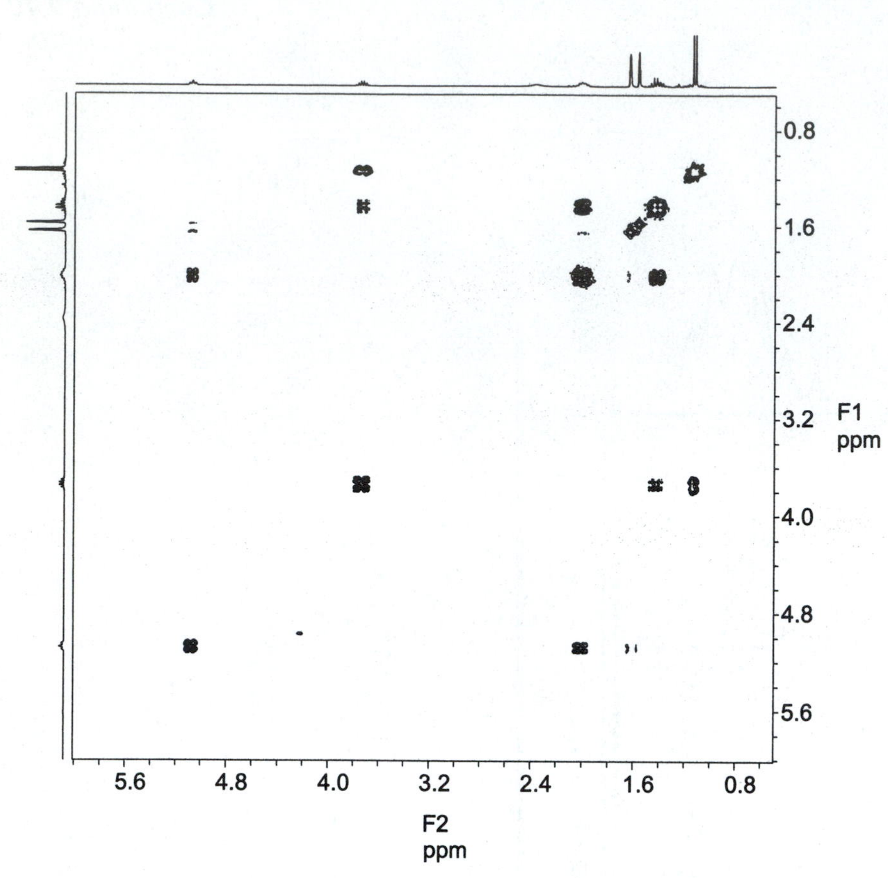
0.8
1.6
2.4
3.2
F1
ppm
4.0
4.8
5.6
5.6
4.8
4.0
3.2
2.4
1.6
0.8
F2
ppm

HETCOR

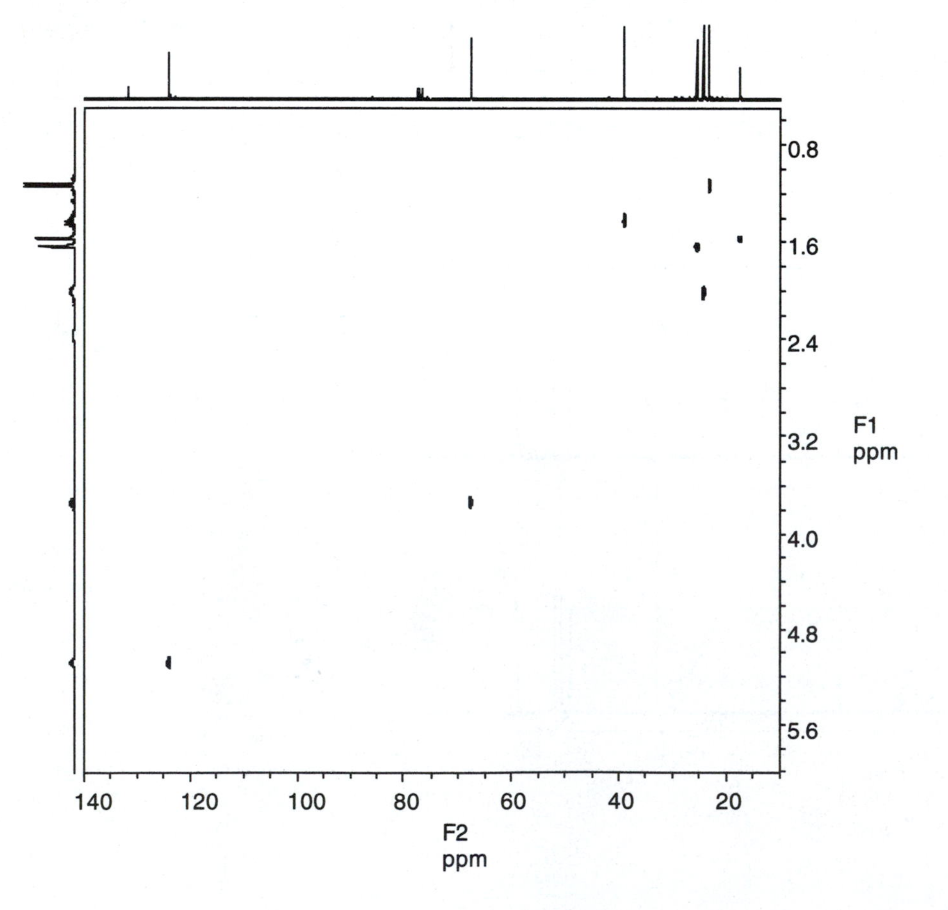
0.8
1.6
2.4
3.2
F1
ppm
4.0
4.8
5.6
140
120
100
80
60
40
20
F2
ppm

NOE, DIFFERENCE IRRADIATED at δ 5.08

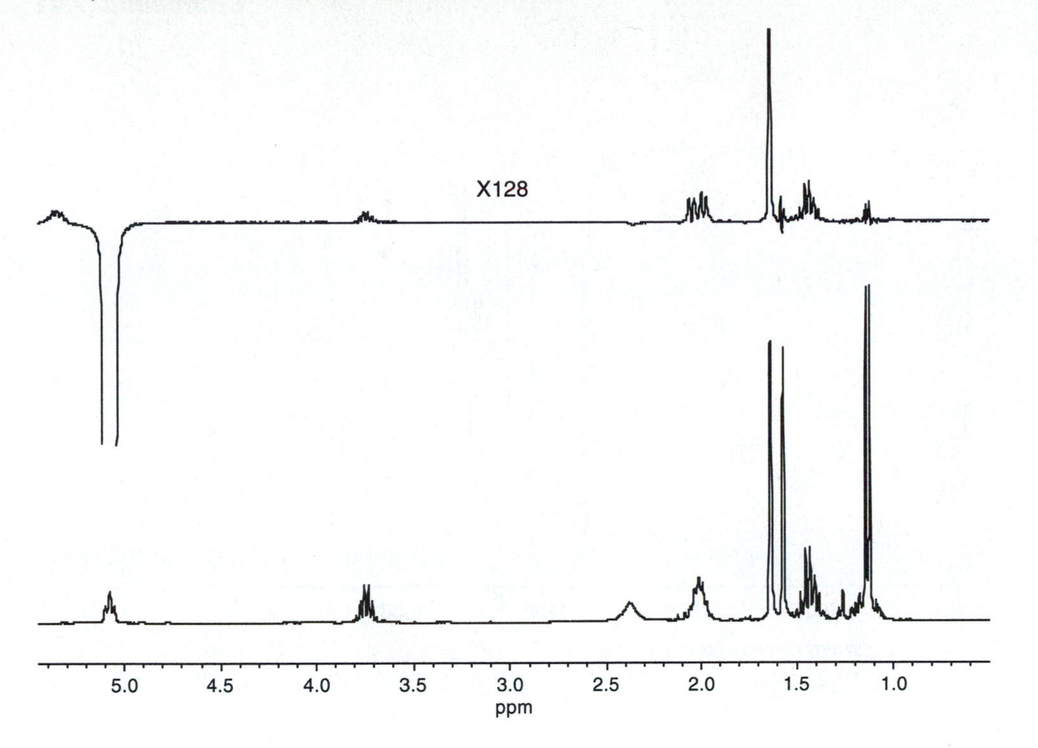

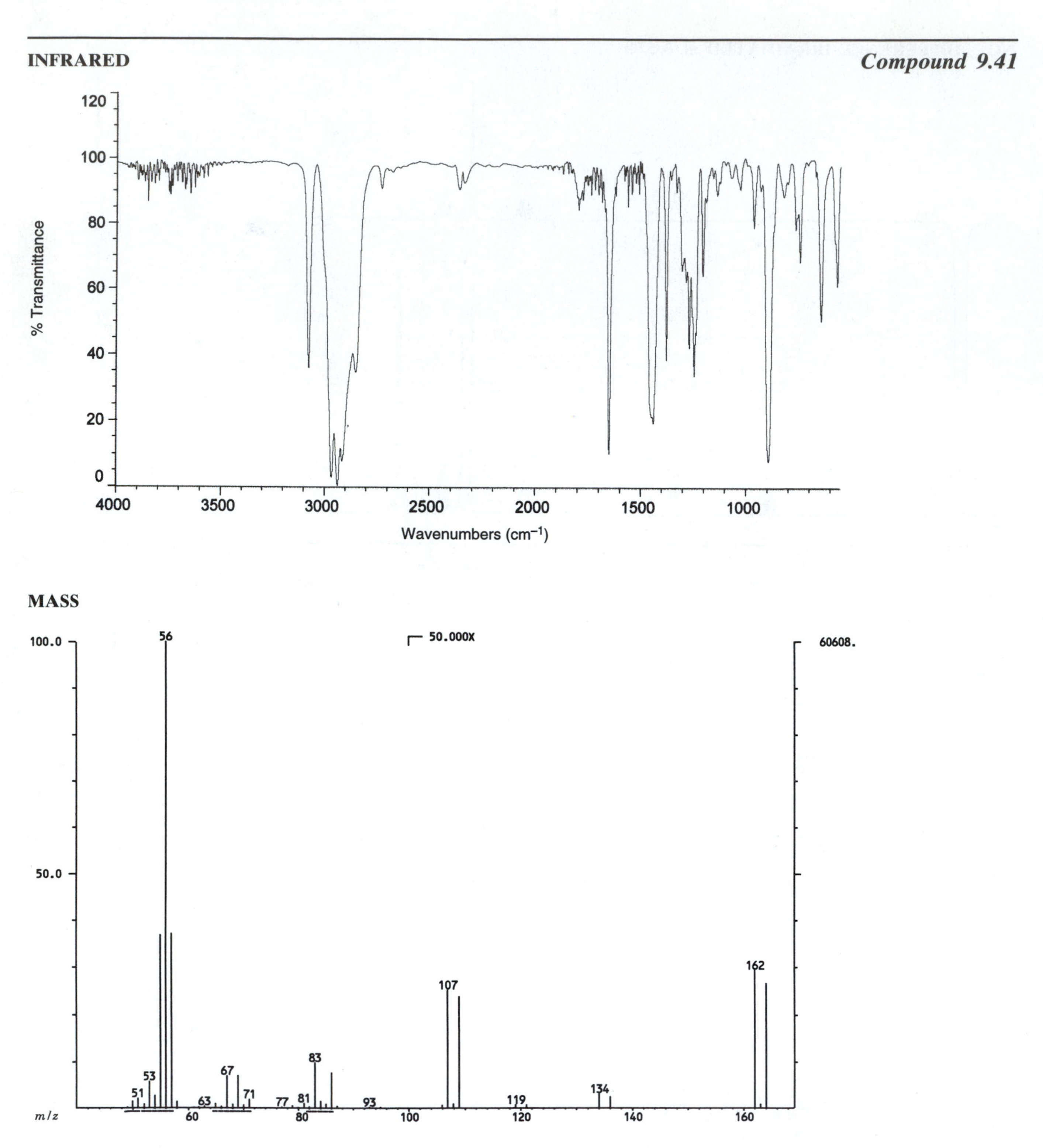
INFRARED
Compound 9.41
% Transmittance
Wavenumbers (cm⁻¹)
4000
3500
3000
2500
2000
1500
1000
MASS
100.0
50.0
50.000X
60608.
56
53
51
63
67
71
77
81
83
93
107
119
134
162
m/z
60
80
100
120
140
160

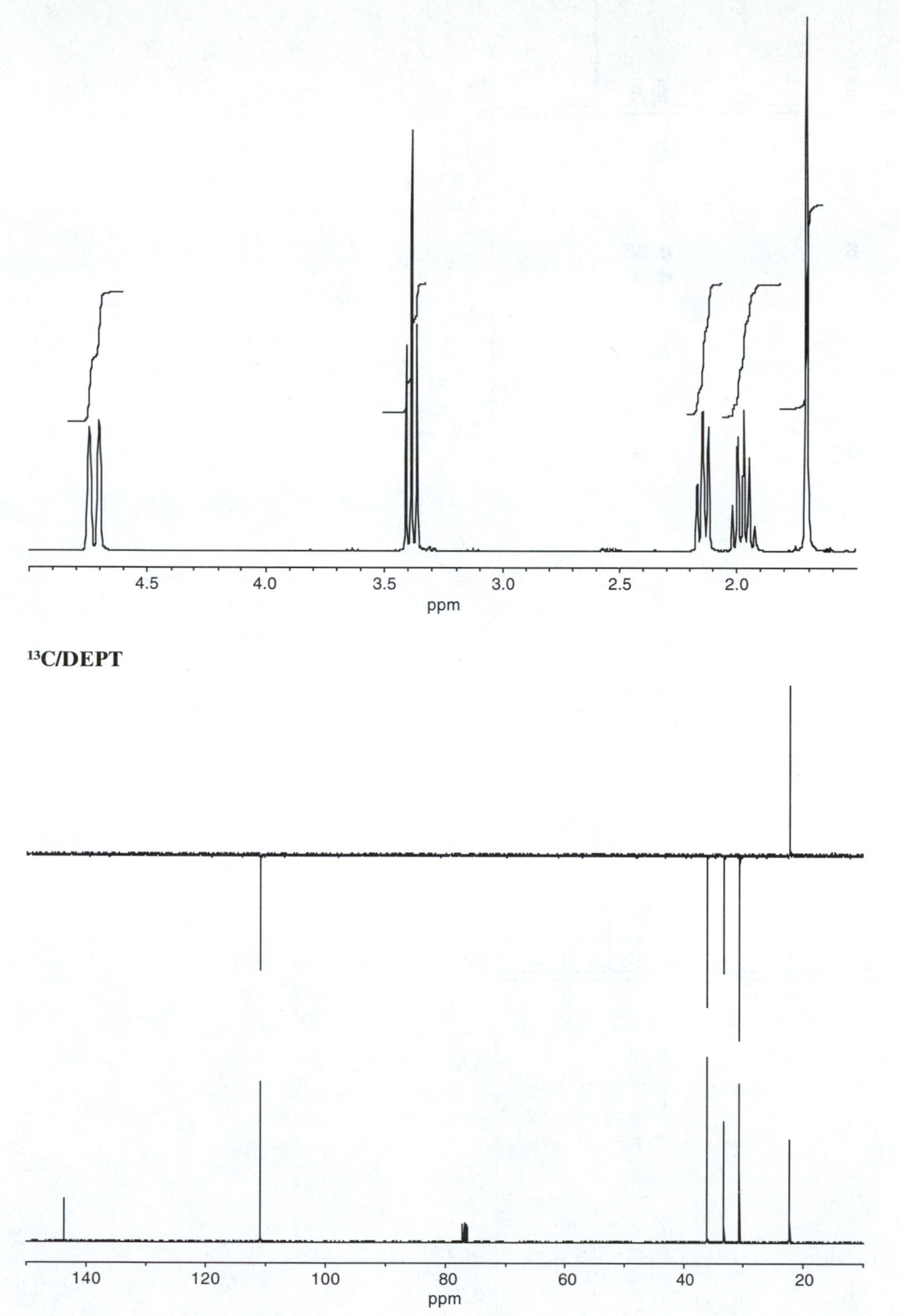
¹H NMR
4.5
4.0
3.5
3.0
2.5
2.0
ppm
¹³C/DEPT
140
120
100
80
60
40
20
ppm

COSY

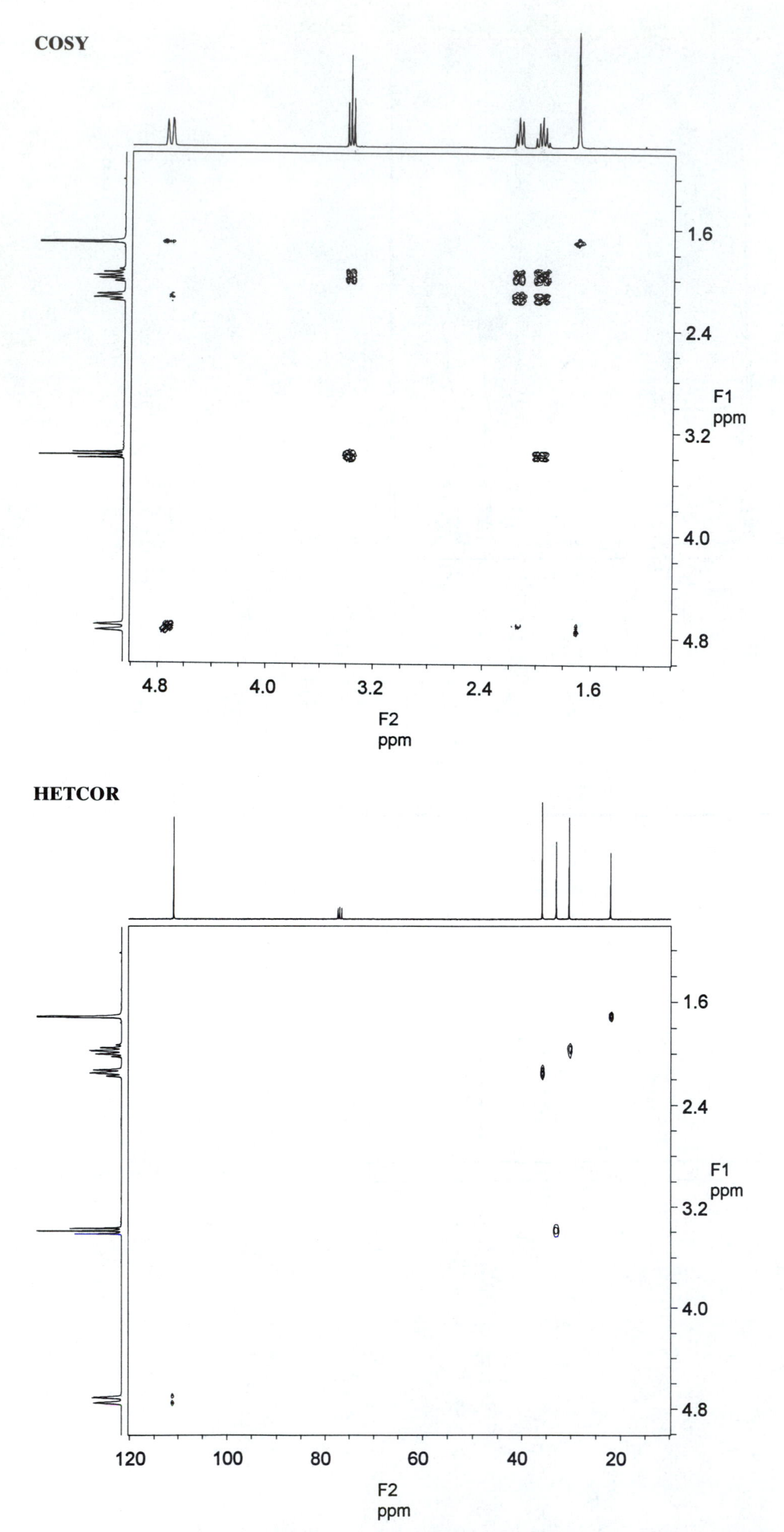
1.6
2.4
F1
ppm
3.2
4.0
4.8
4.8
4.0
3.2
2.4
1.6
F2
ppm

HETCOR

1.6
2.4
F1
ppm
3.2
4.0
4.8
120
100
80
60
40
20
F2
ppm

INFRARED SPECTRUM

Compound 9.42

3402.9	1375.9	995.4
2969.6	1205.3	920.1
1451.7	1113.6	835.6

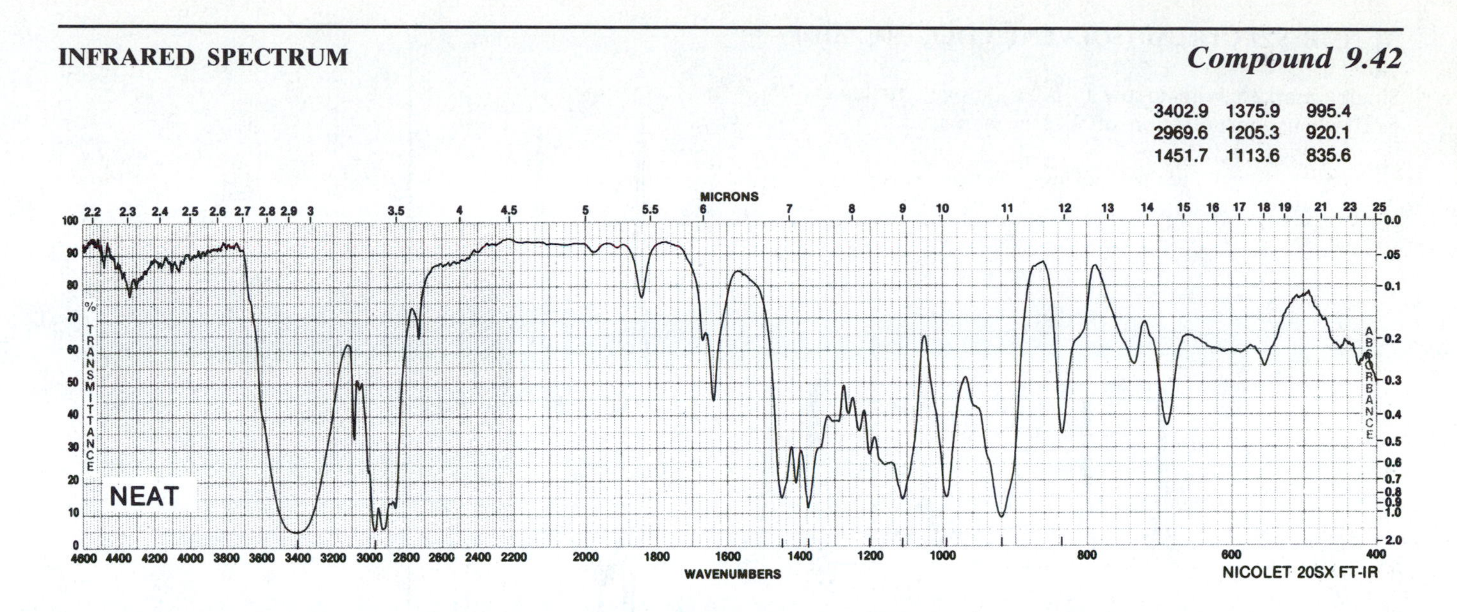

MASS

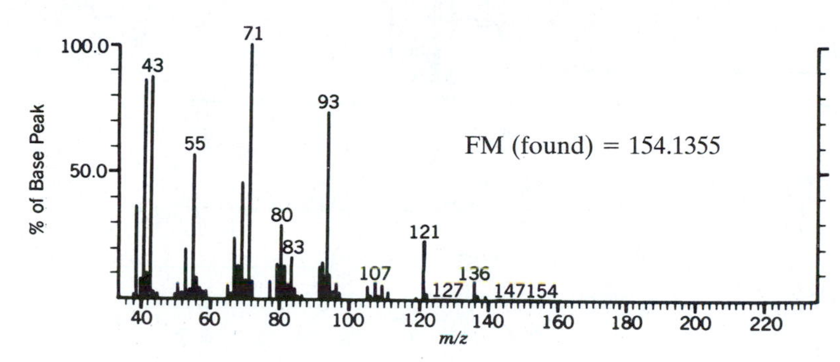

^{1}H NMR SPECTRUM (SOLVENT $CDCl_3$, 500 MHz)

Singlet marked with arrow (~δ 1.6) moved to lower field when concentration of 9.42 is increased

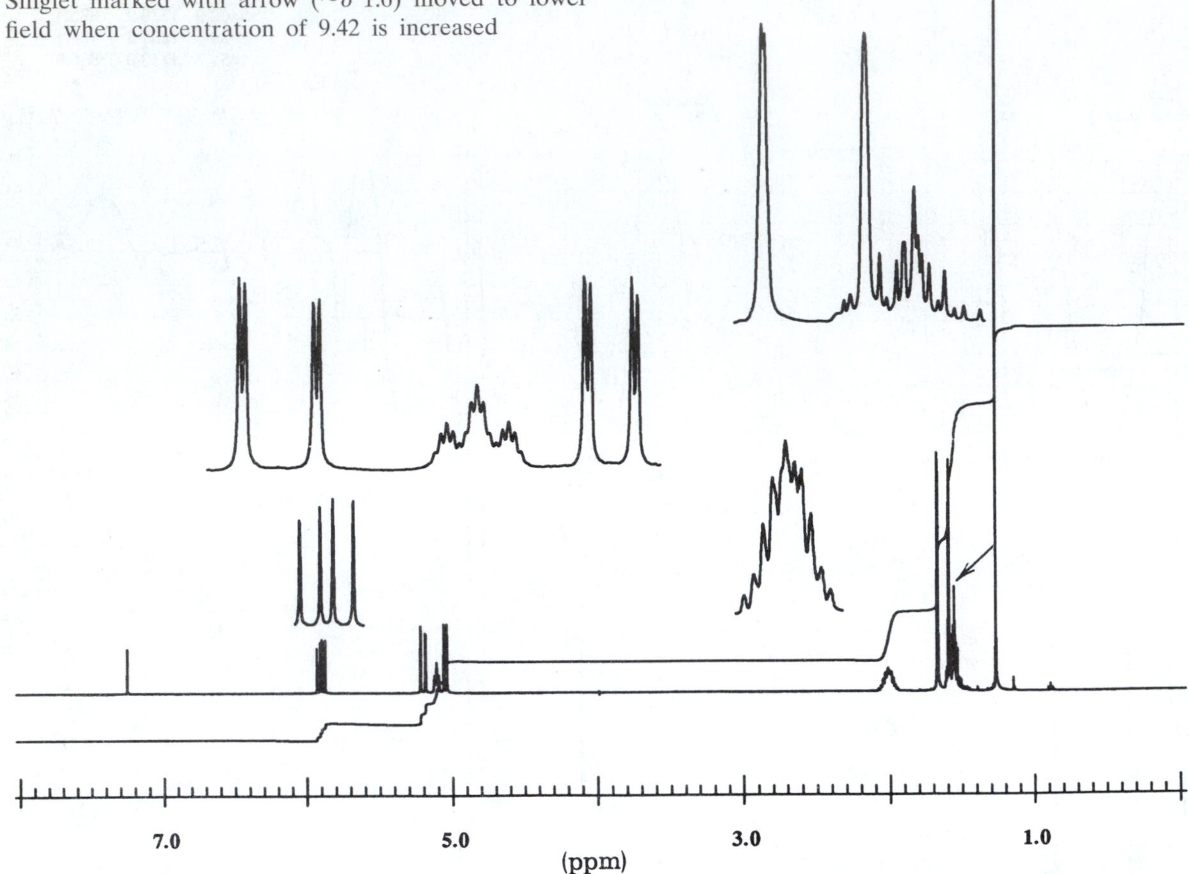

^{13}C NMR SPECTRUM, SOLVENT $CDCl_3$

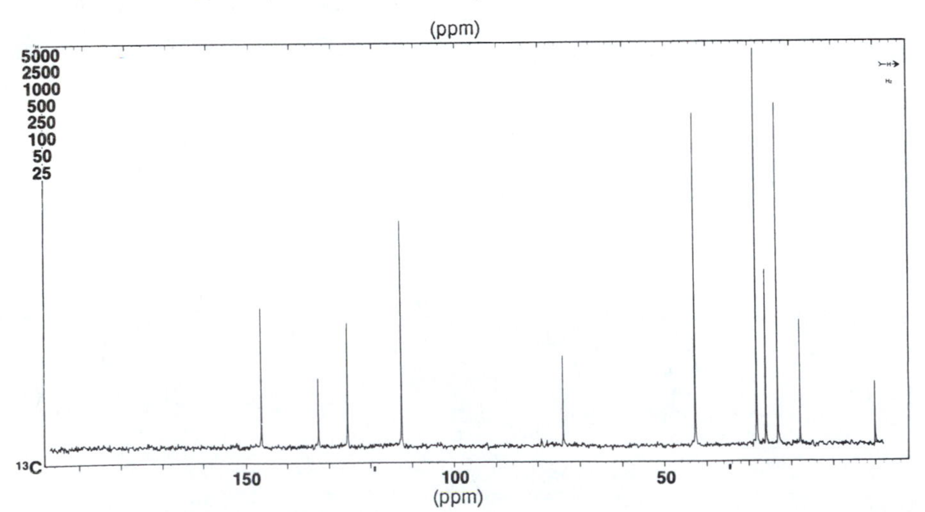

^{13}C NMR Data	
δ	
17.5	CH_3
22.6	CH_2
25.3	CH_3
27.2	CH_3
41.2	CH_2
72.7	C
111.3	CH_2
124.6	CH
130.3	C
145.0	CH

H H COSY

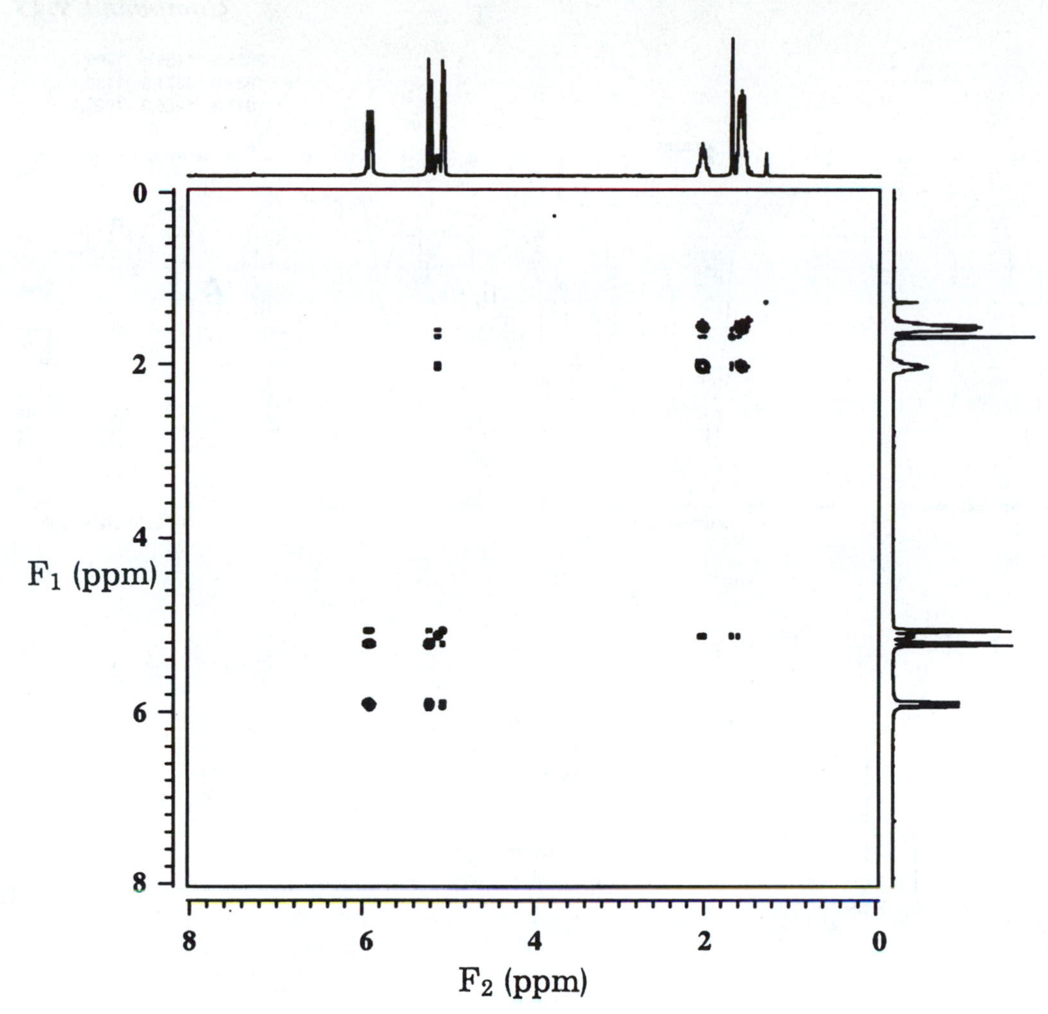

INFRARED

Compound 9.43

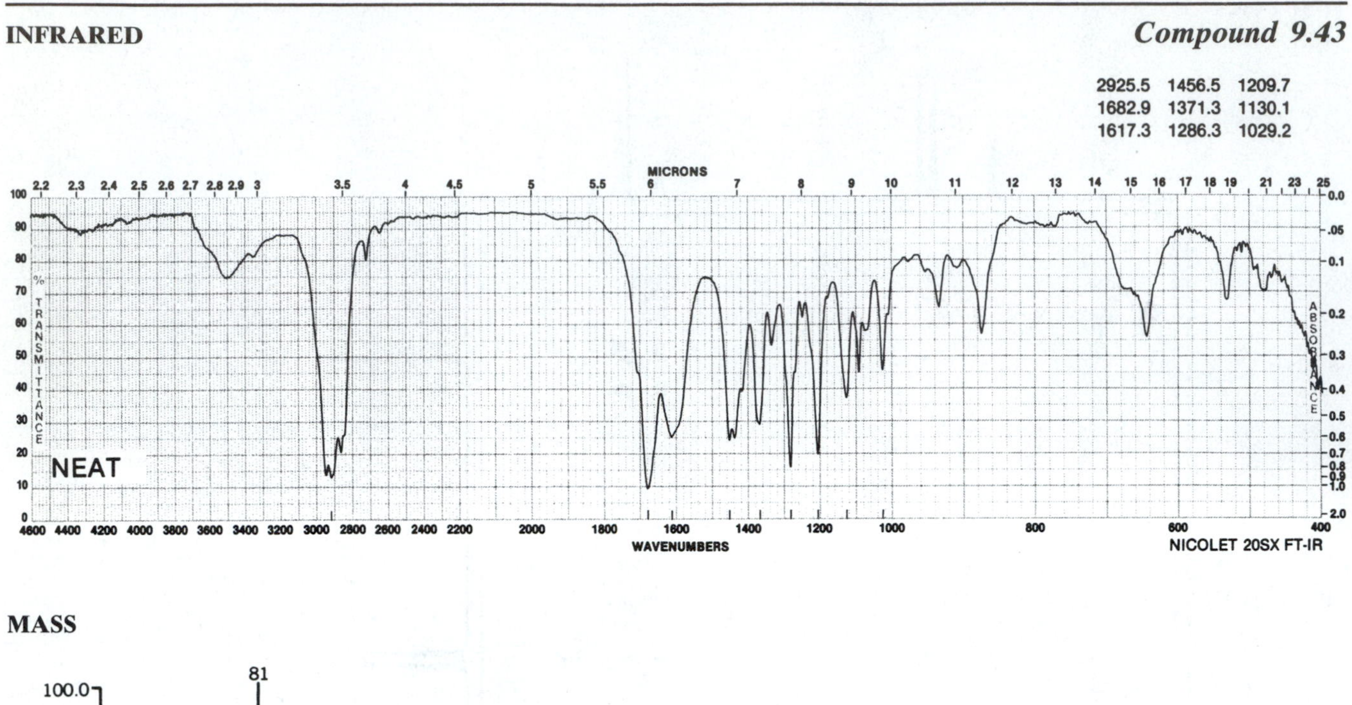

MASS

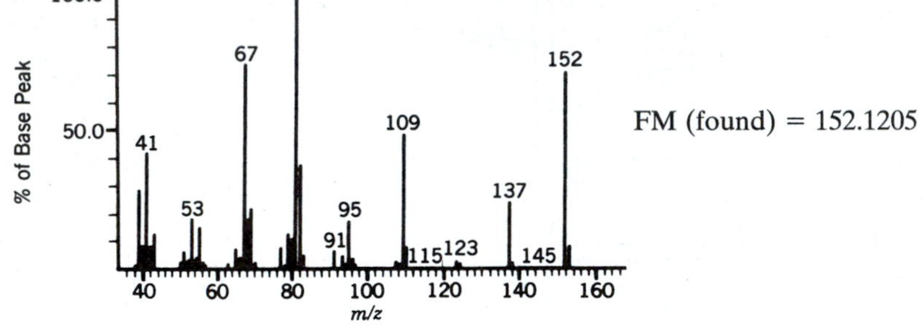

FM (found) = 152.1205

^1H NMR SPECTRUM (SOLVENT $CDCl_3$, 500 MHz)

^{13}C NMR (SOLVENT $CDCl_3$, 126.0 MHz)

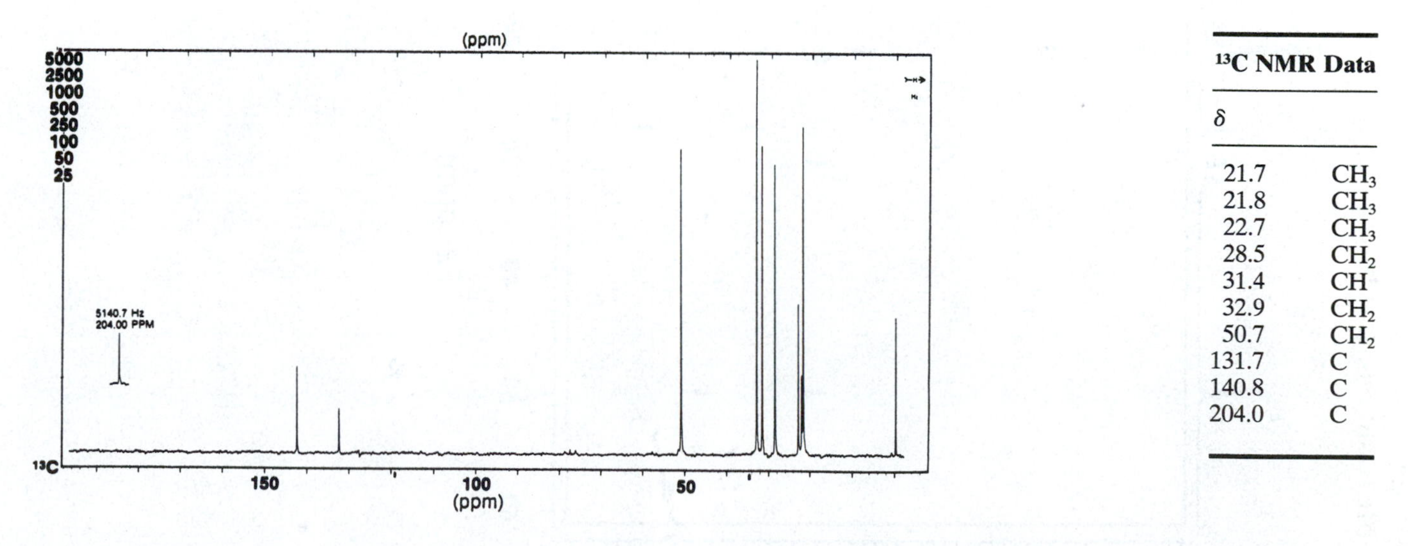

^{13}C NMR Data	
δ	
21.7	CH_3
21.8	CH_3
22.7	CH_3
28.5	CH_2
31.4	CH
32.9	CH_2
50.7	CH_2
131.7	C
140.8	C
204.0	C

H H COSY, ^{1}H, 500 MHz

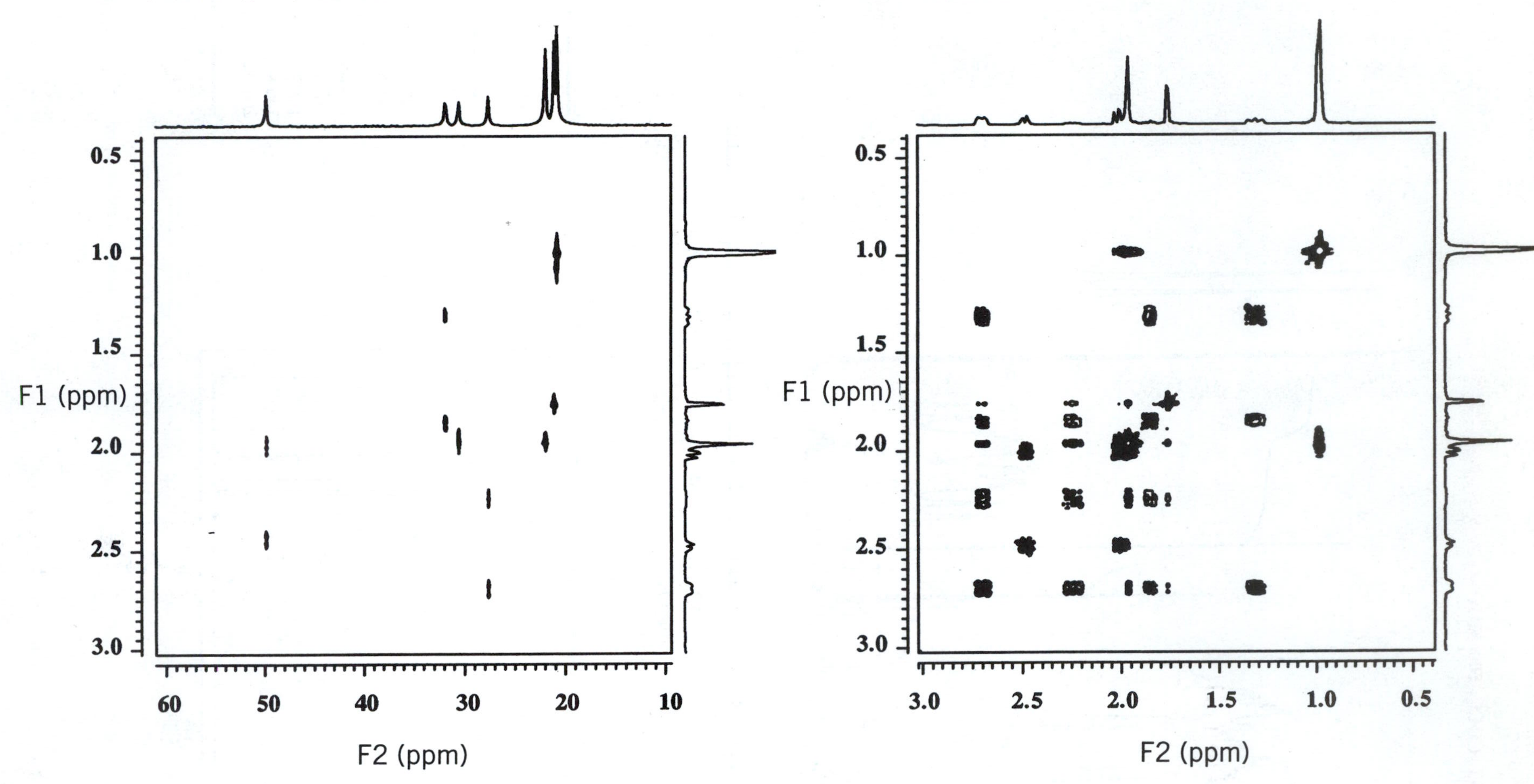

MASS

Compound 9.44

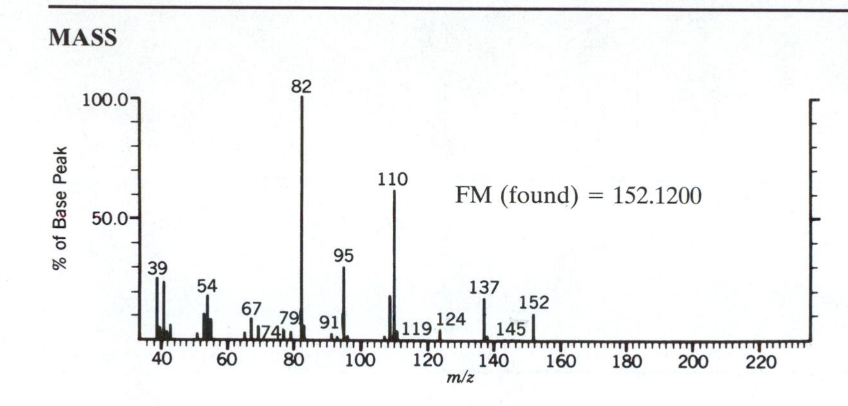

INFRARED SPECTRUM

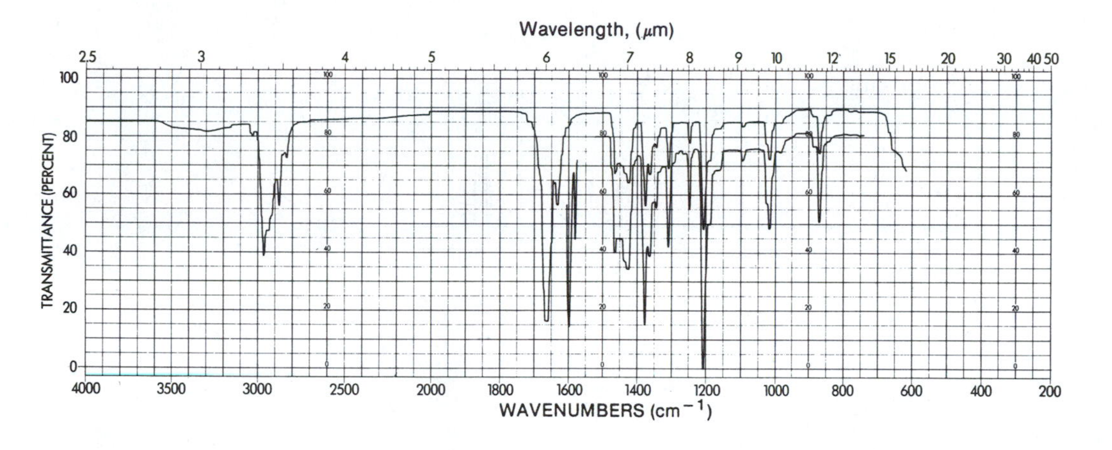

¹H NMR

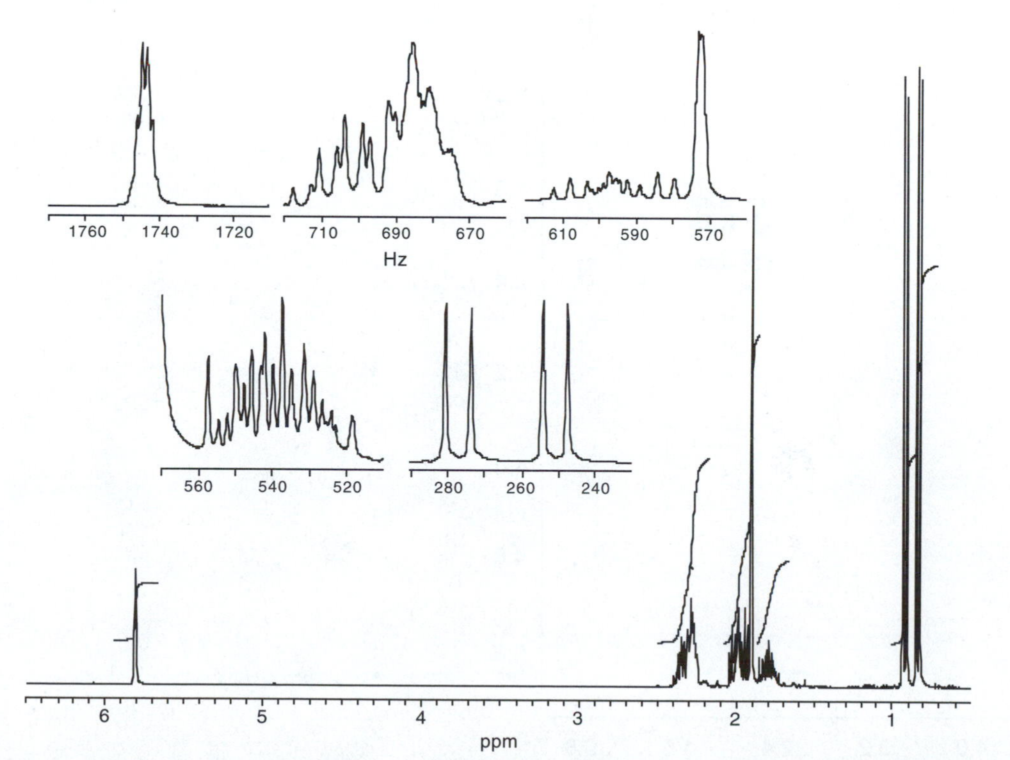

^{13}C/DEPT

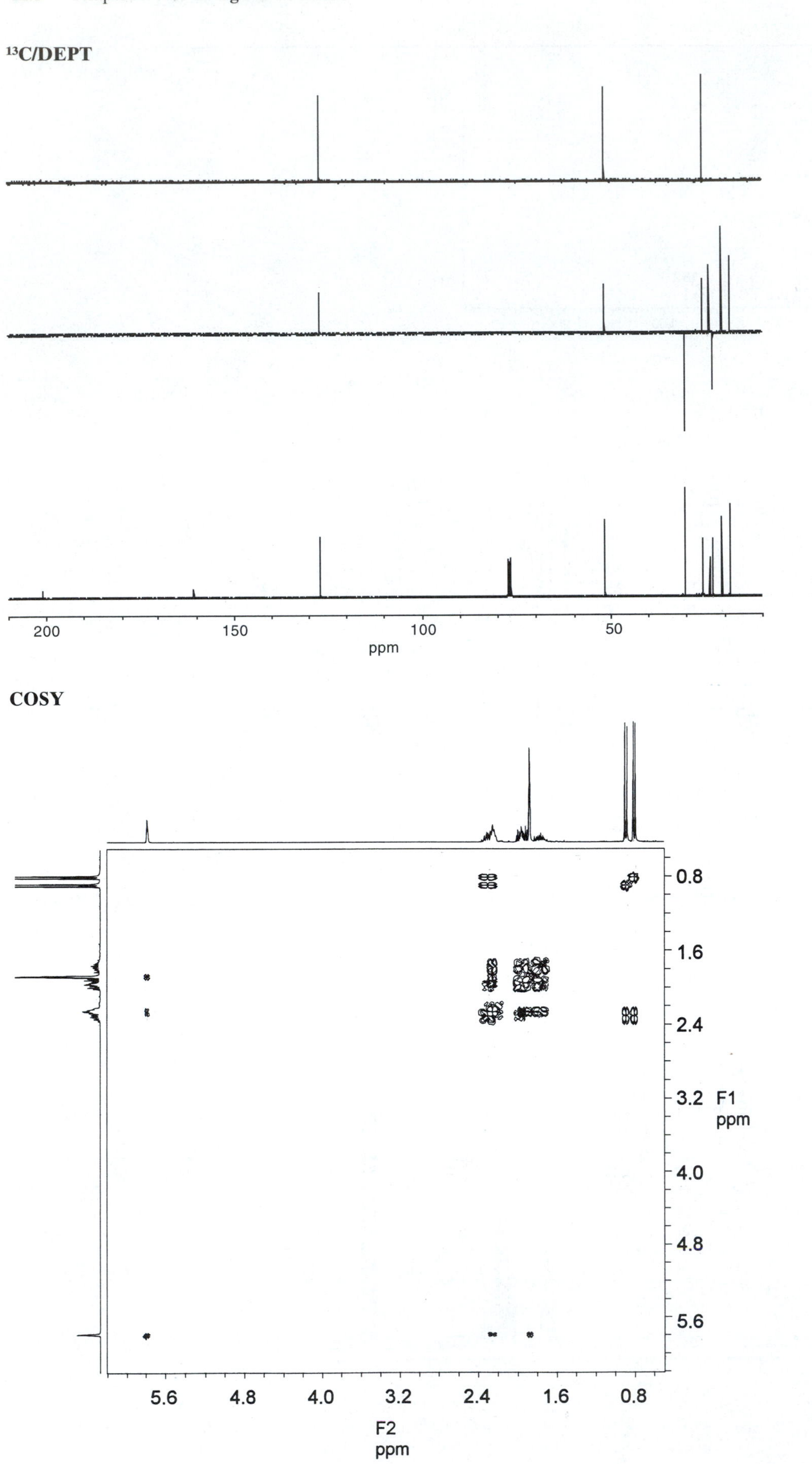
200
150
100
50
ppm
COSY
0.8
1.6
2.4
3.2
4.0
4.8
5.6
F1
ppm
5.6
4.8
4.0
3.2
2.4
1.6
0.8
F2
ppm

HETCOR

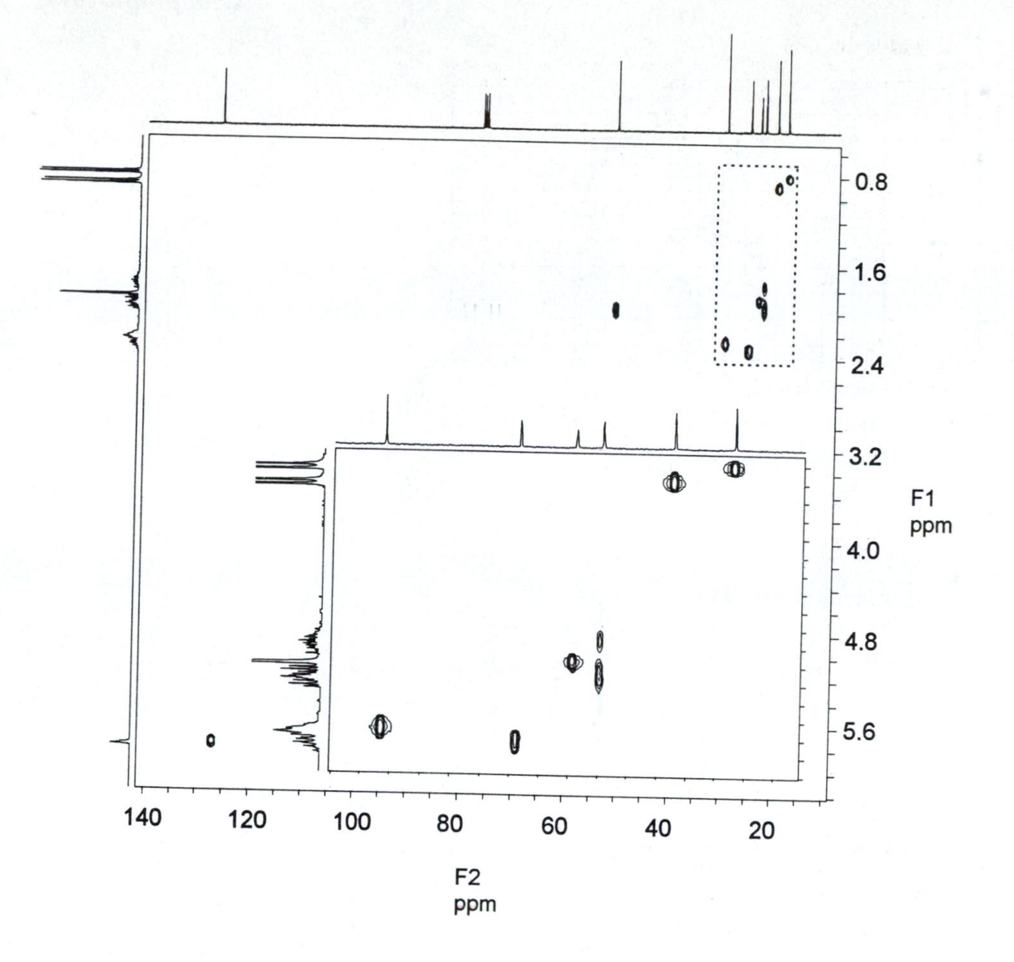

Compound 9.45

INFRARED

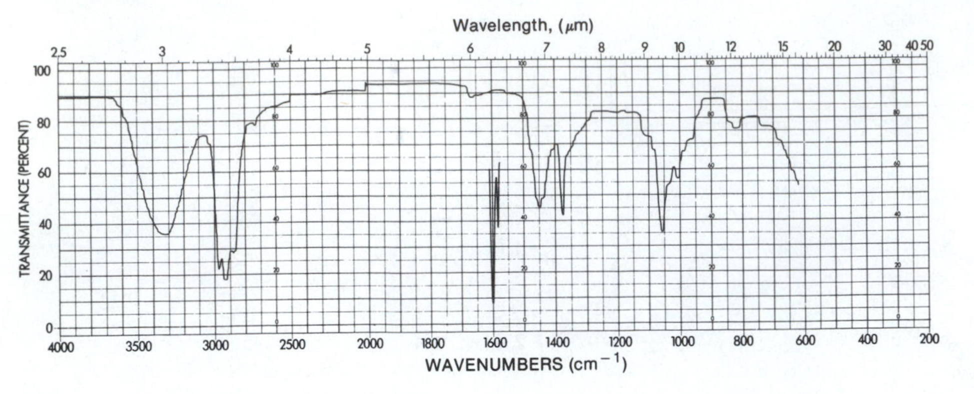

MASS

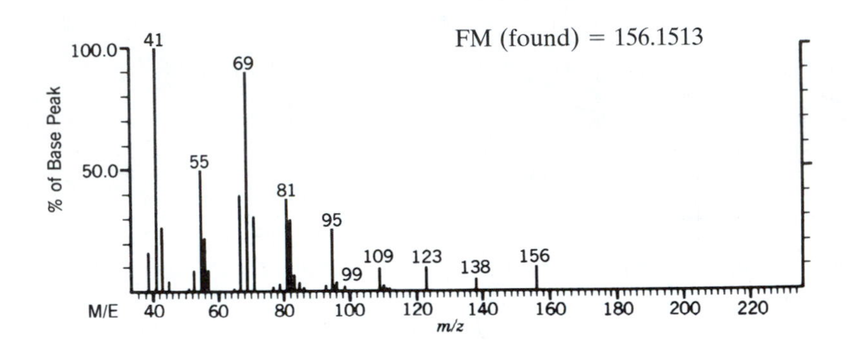

^{1}H NMR SPECTRUM (SOLVENT $CDCl_3$, 500 MHz)

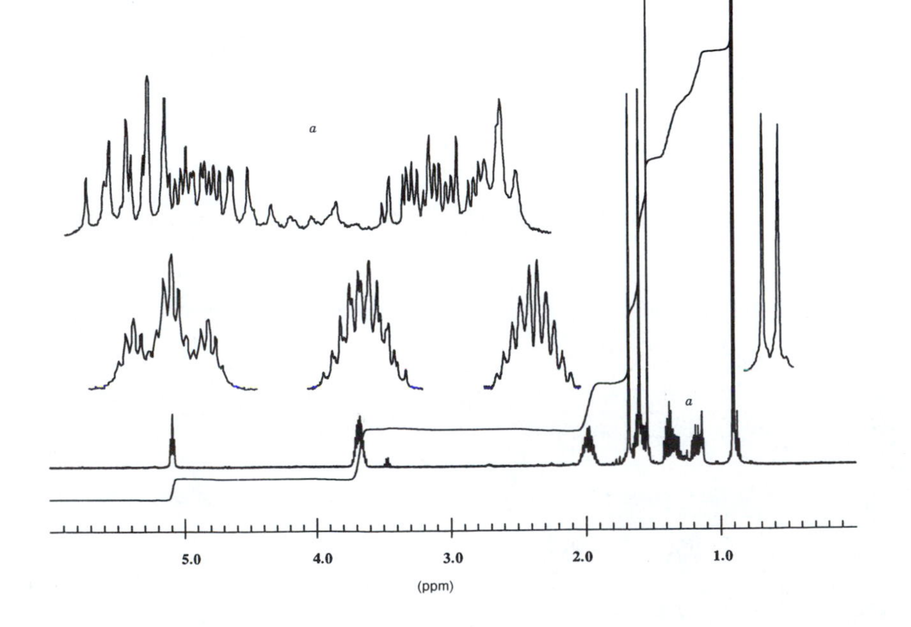

^{13}C NMR SPECTRUM

^{13}C NMR Data

δ	
17.6	CH_3
19.6	CH_3
25.6	CH_2
25.7	CH_3
29.4	CH
37.4	CH_2
39.8	CH_2
60.2	CH_2
125.0	CH
130.7	C

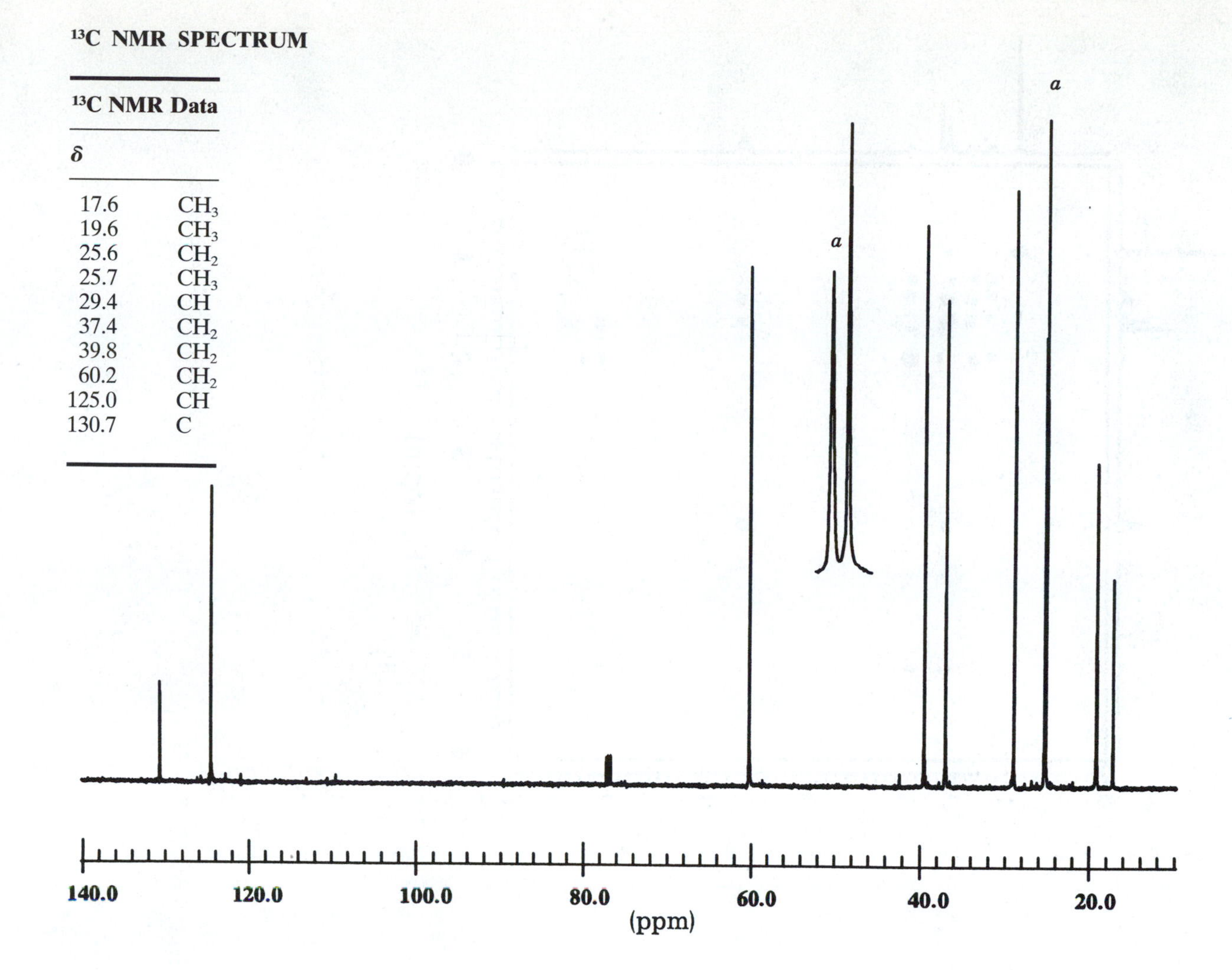

H C COSY (HETCOR)

H H COSY

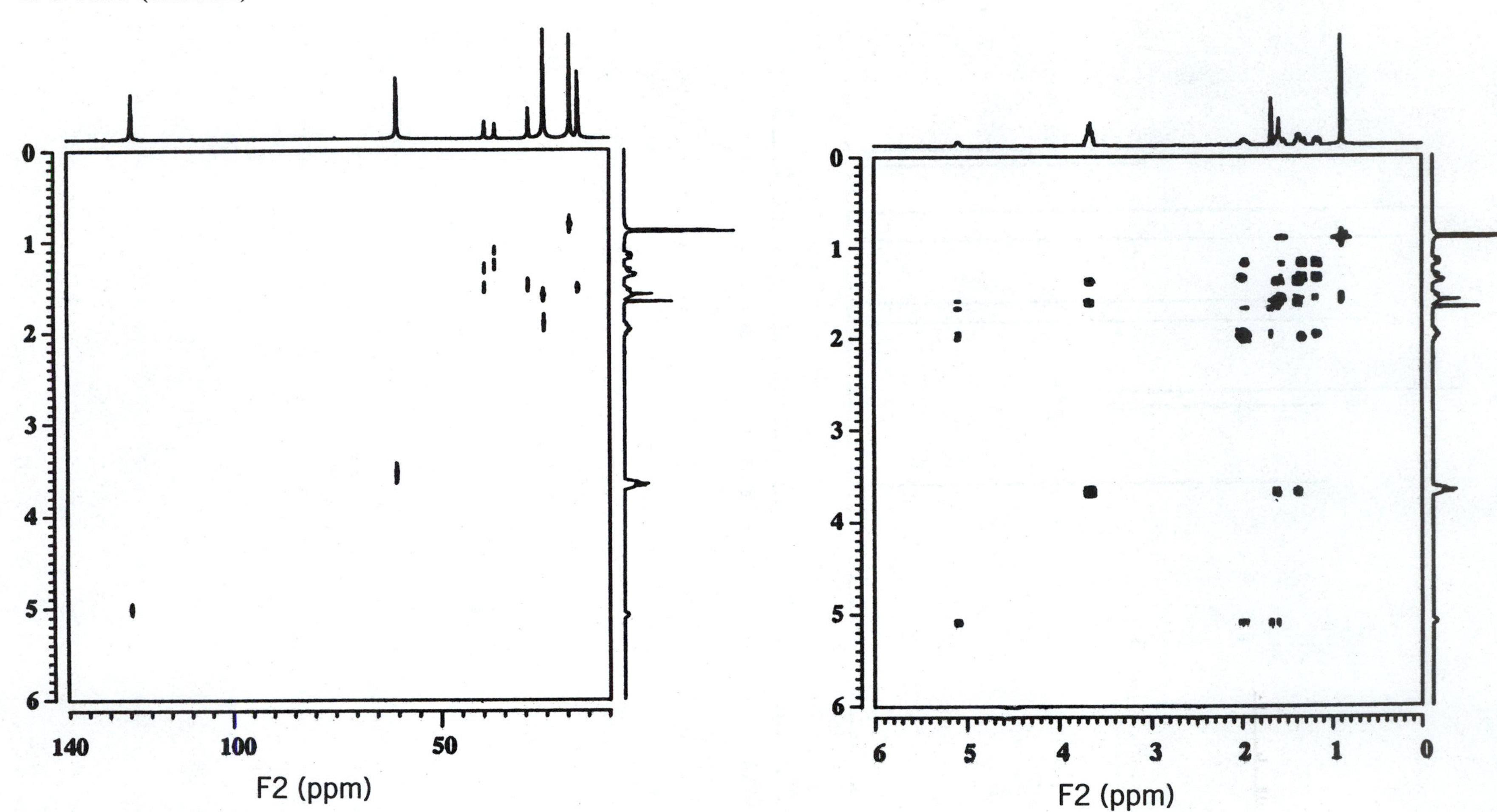

MASS

Compound 9.46

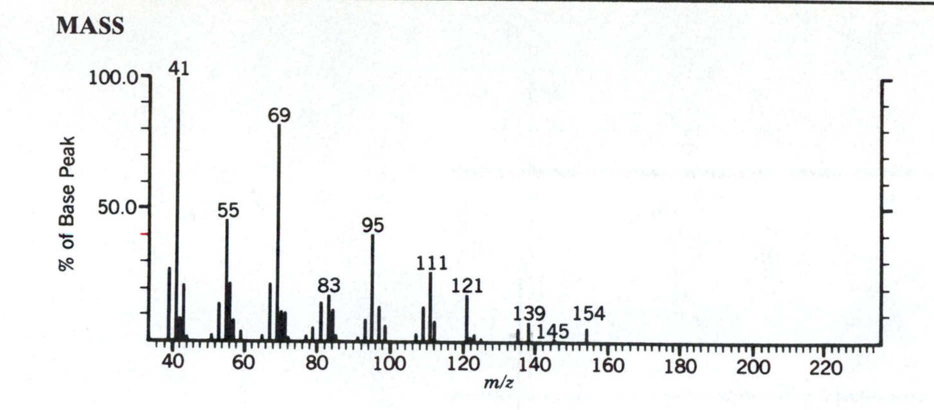

INFRARED

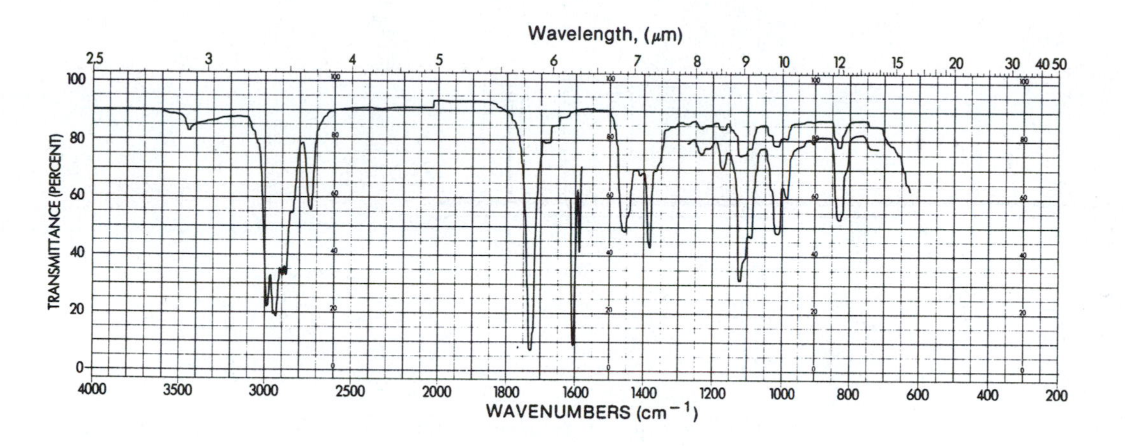

^{1}H NMR

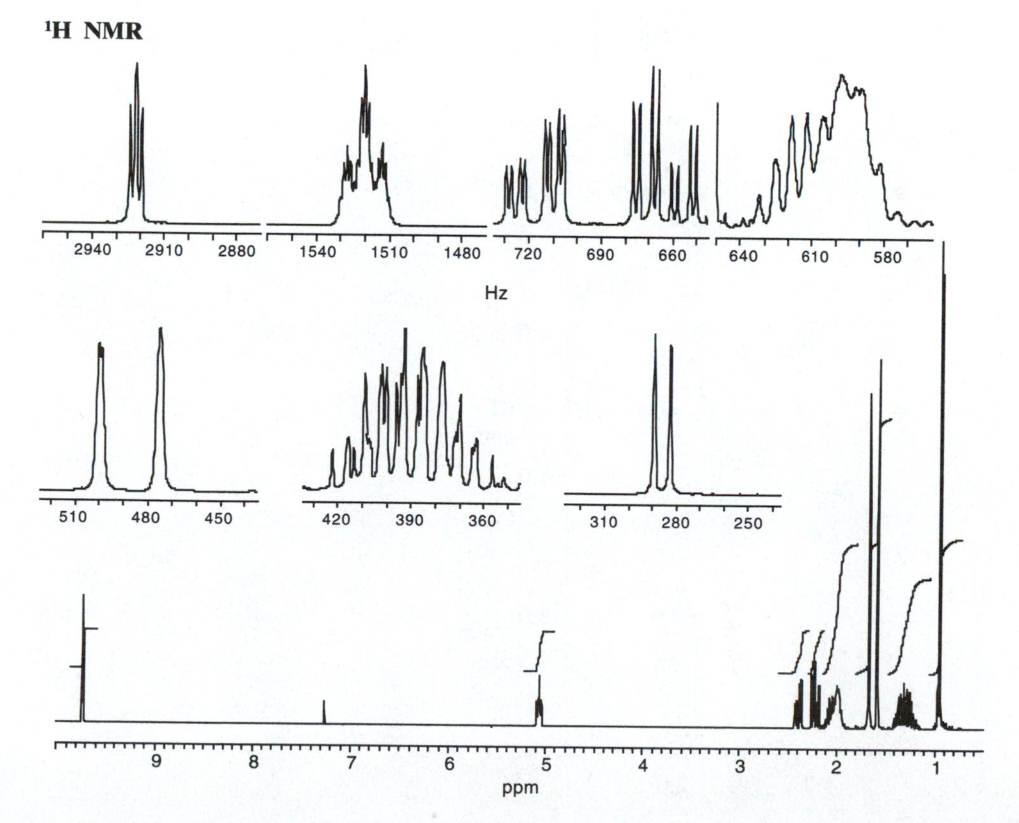

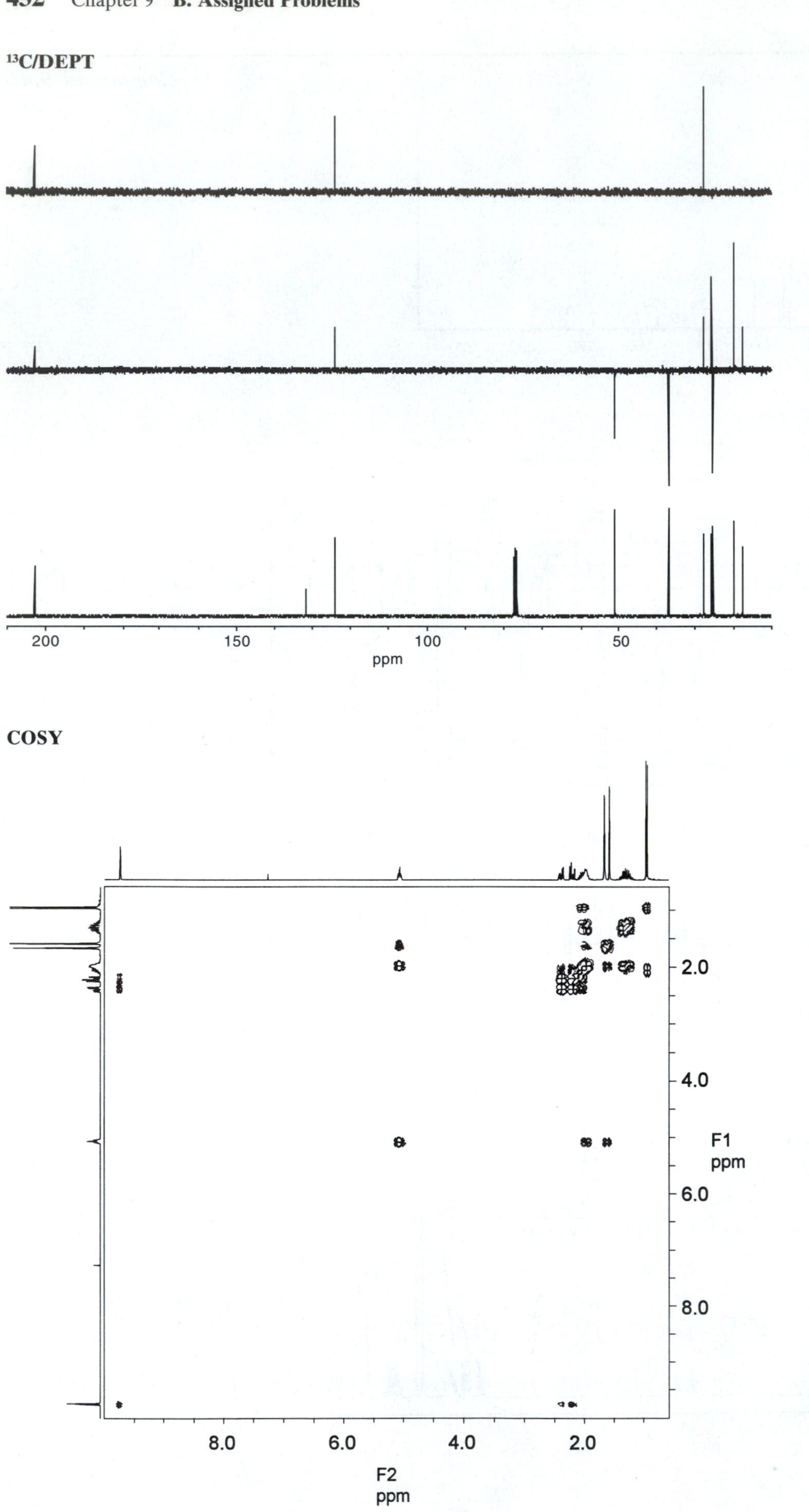

13C/DEPT
200
150
100
50
ppm
COSY
2.0
4.0
F1
ppm
6.0
8.0
8.0
6.0
4.0
2.0
F2
ppm

HMQC

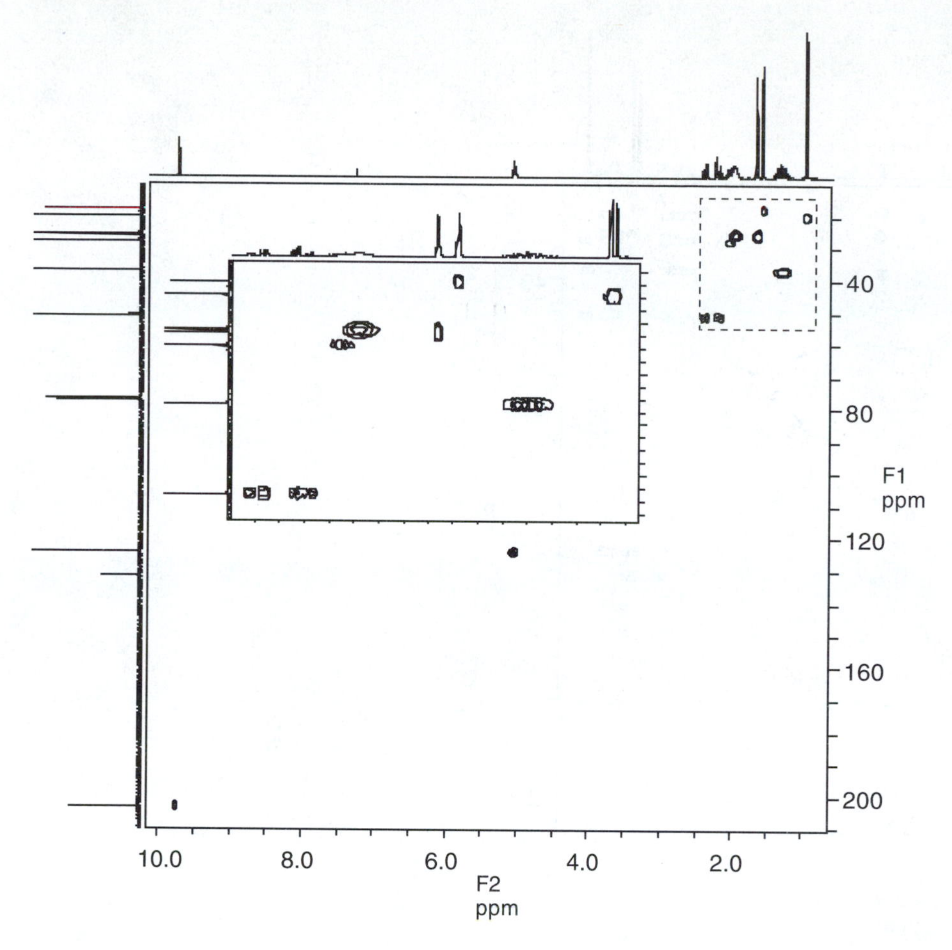

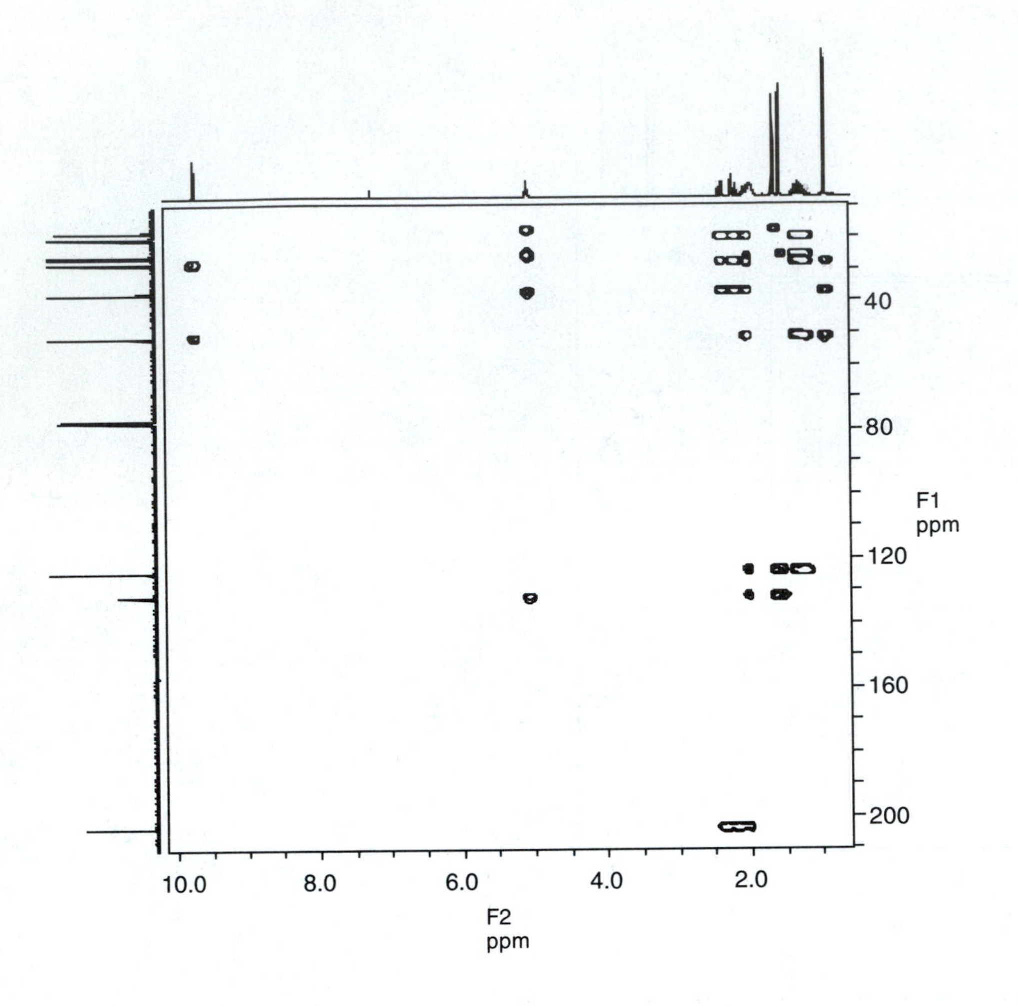

40
80
120
160
200
F1
ppm
10.0
8.0
6.0
4.0
2.0
F2
ppm

INFRARED

Compound 9.47

3238.0	1402.0	1142.9
1732.7	1285.8	794.5
1682.0	1201.6	650.4

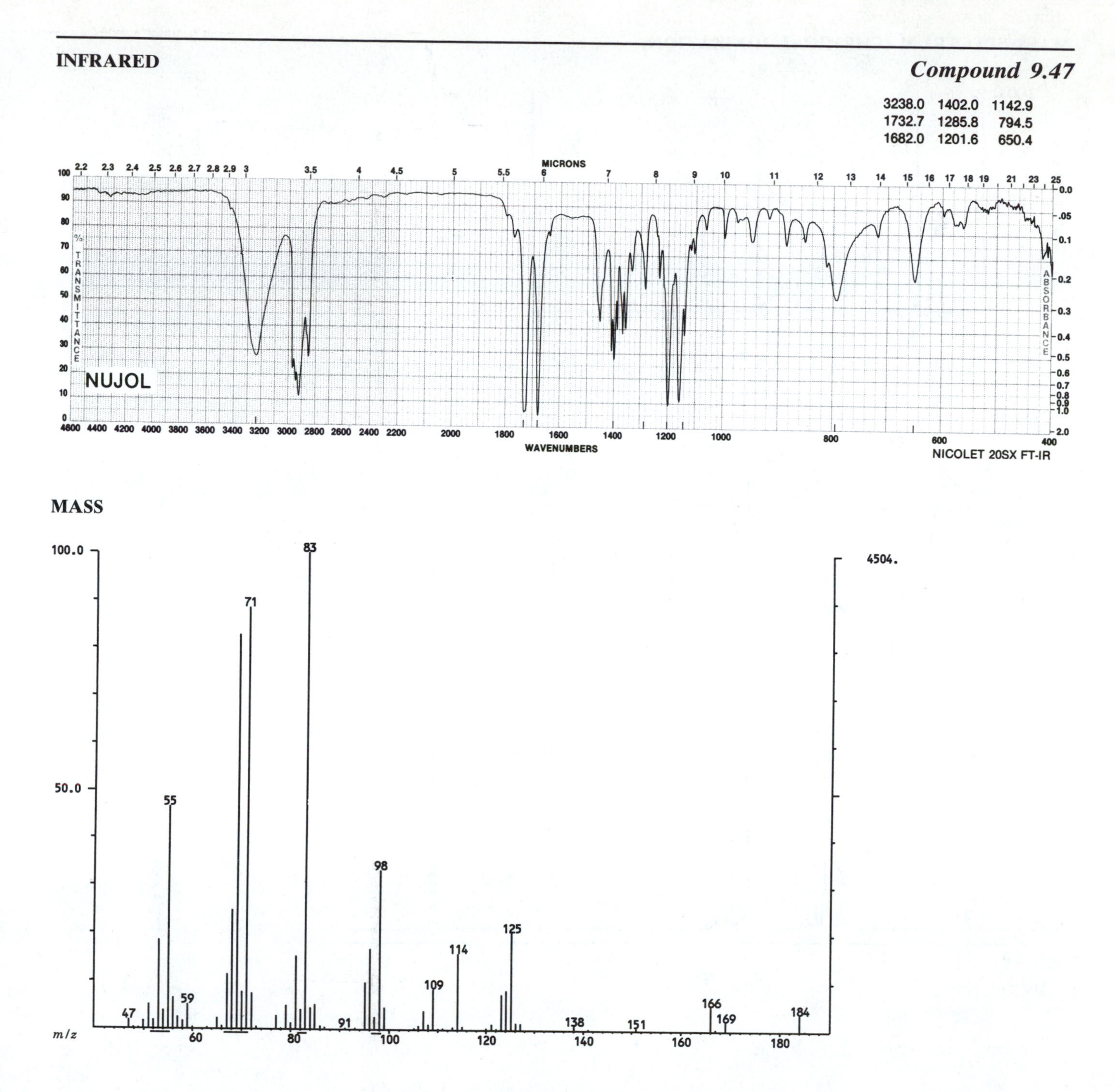

MASS SPECTRUM (CHEMICAL IONIZATION)

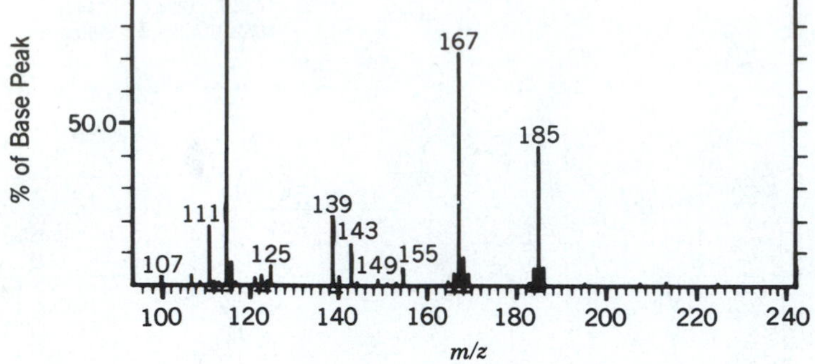

^{1}H NMR

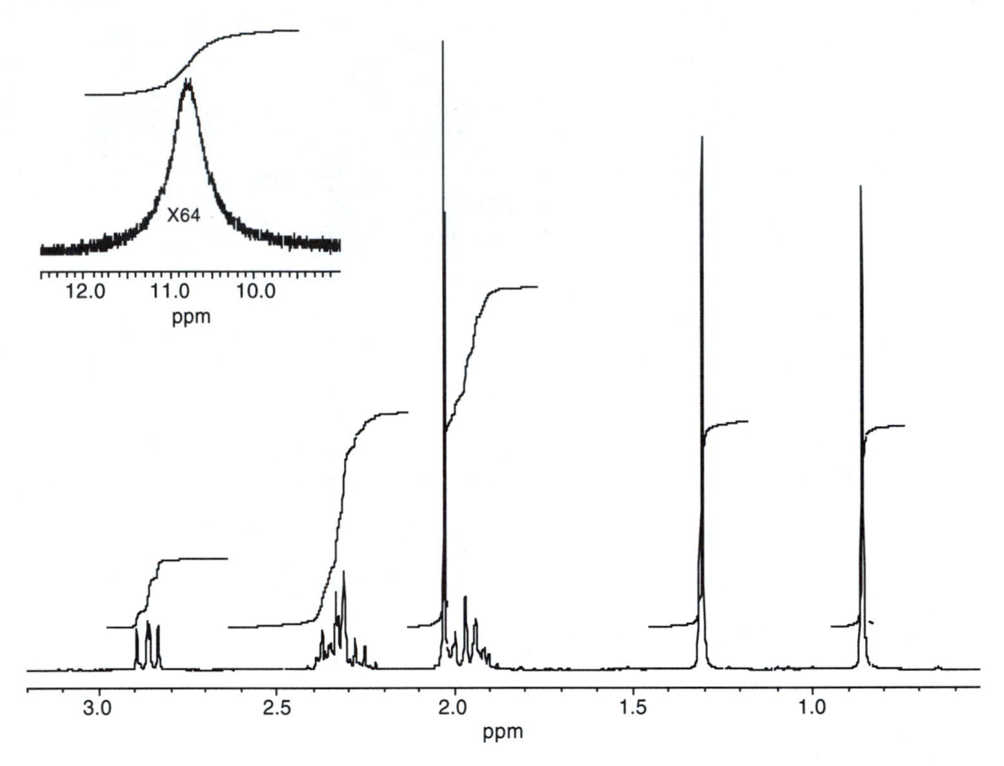

¹³C/DEPT

COSY

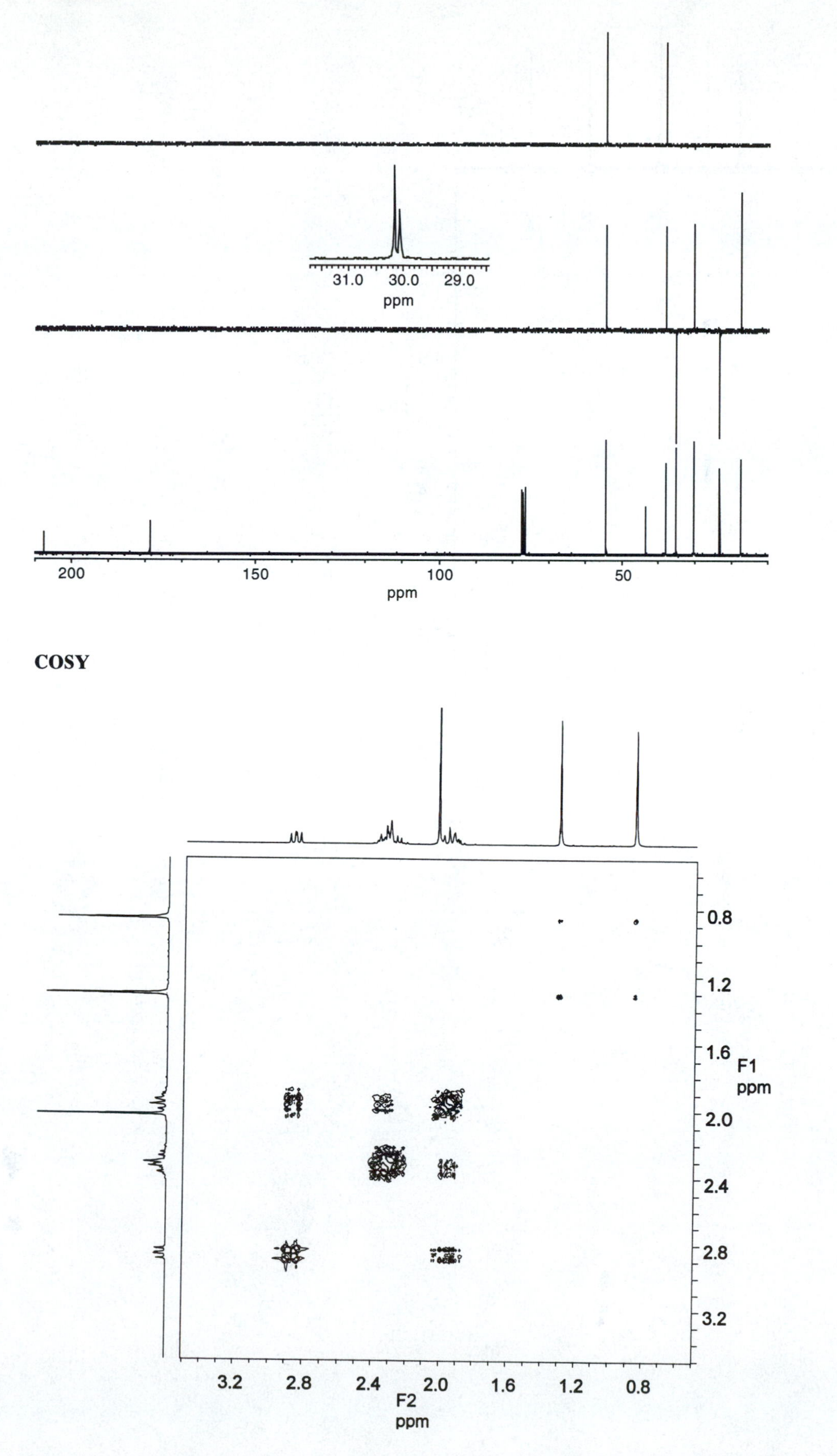

HETCOR

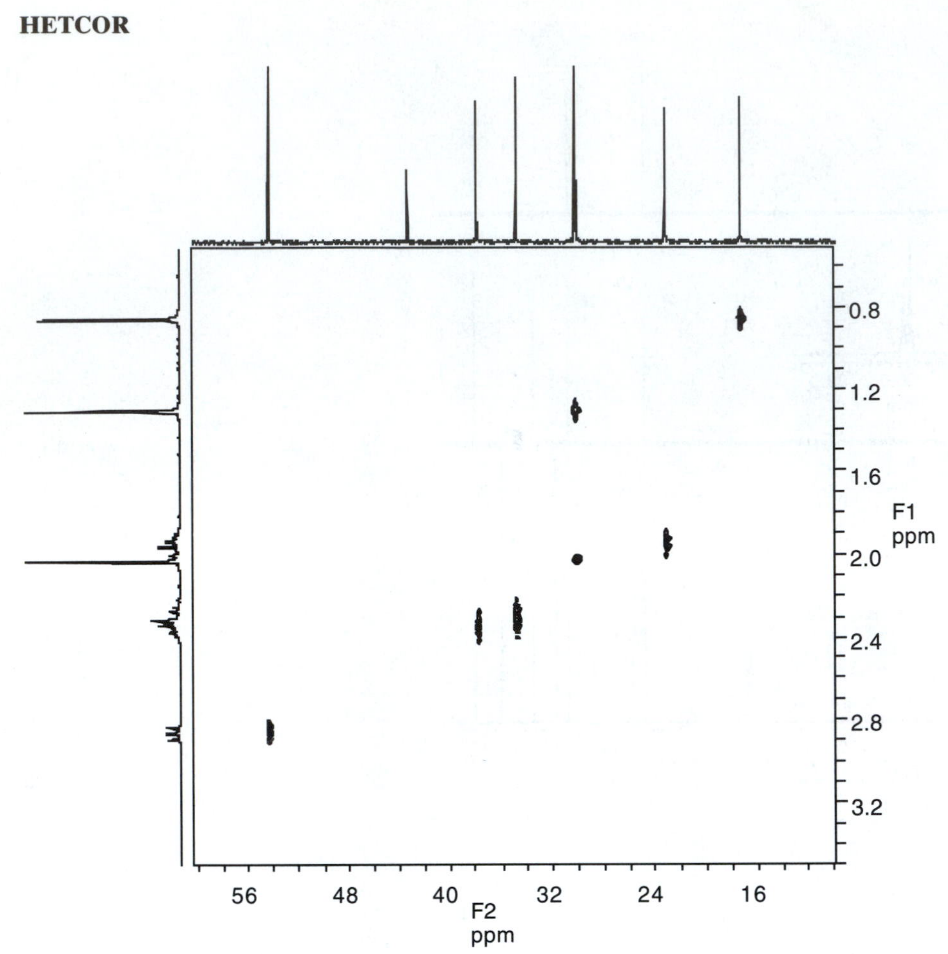

HMBC

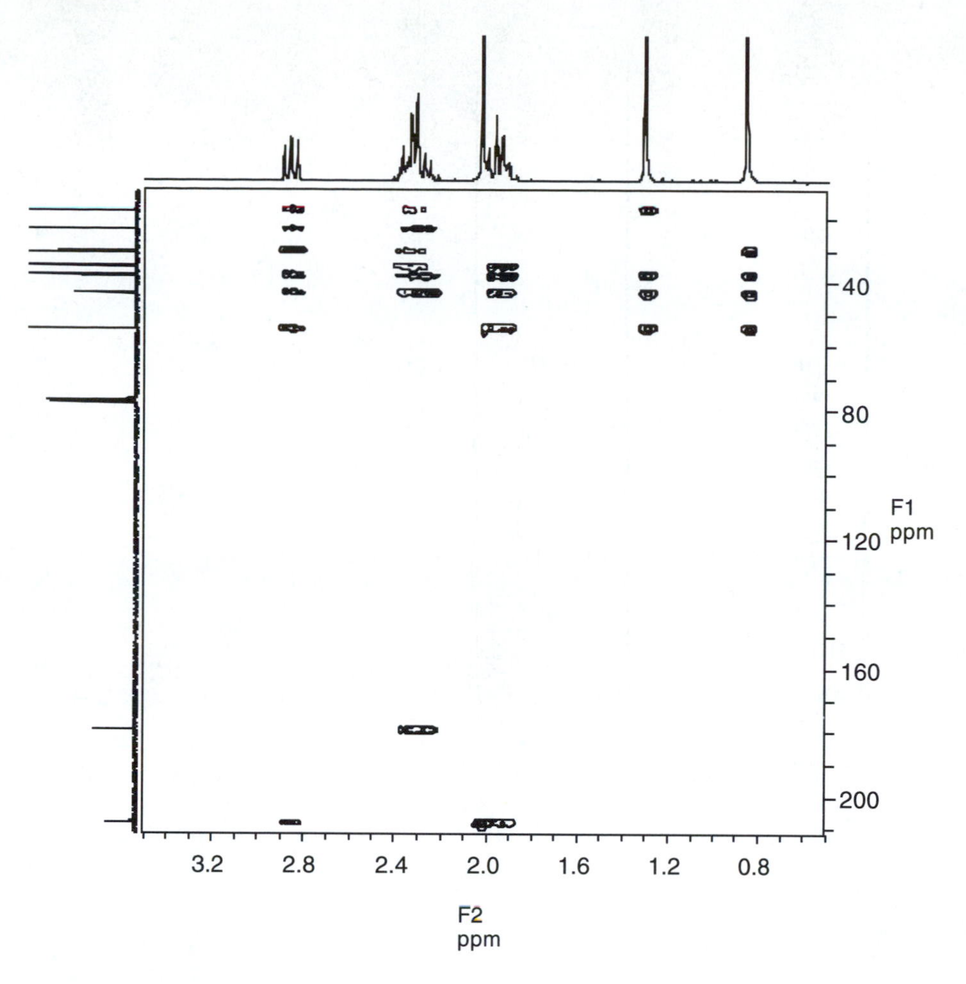

INADEQUATE

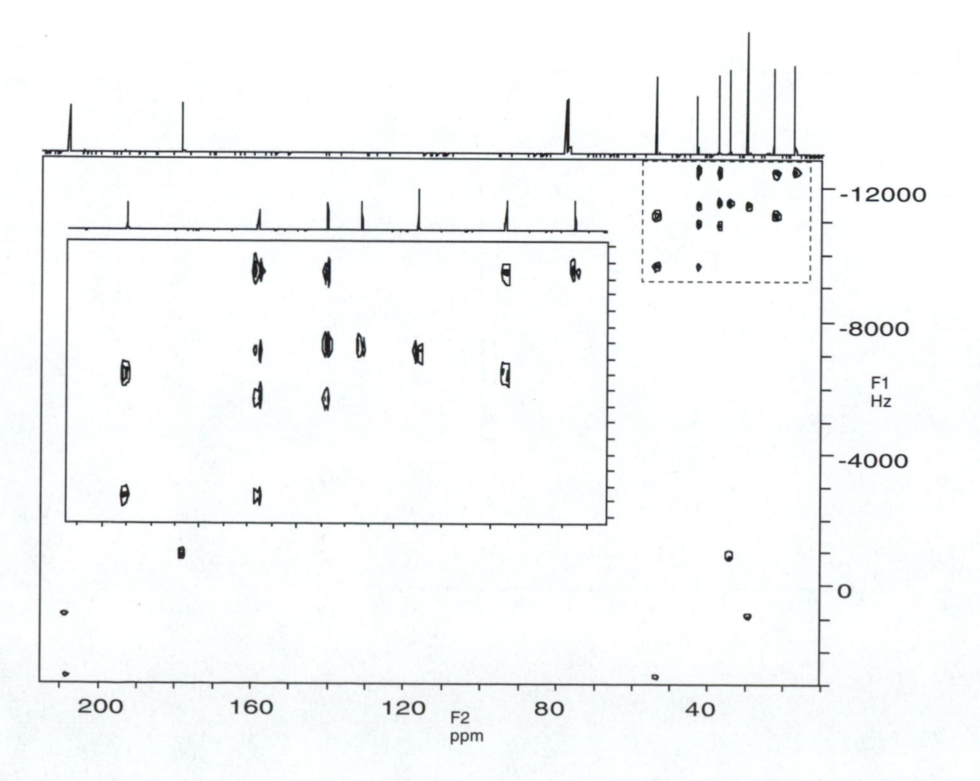

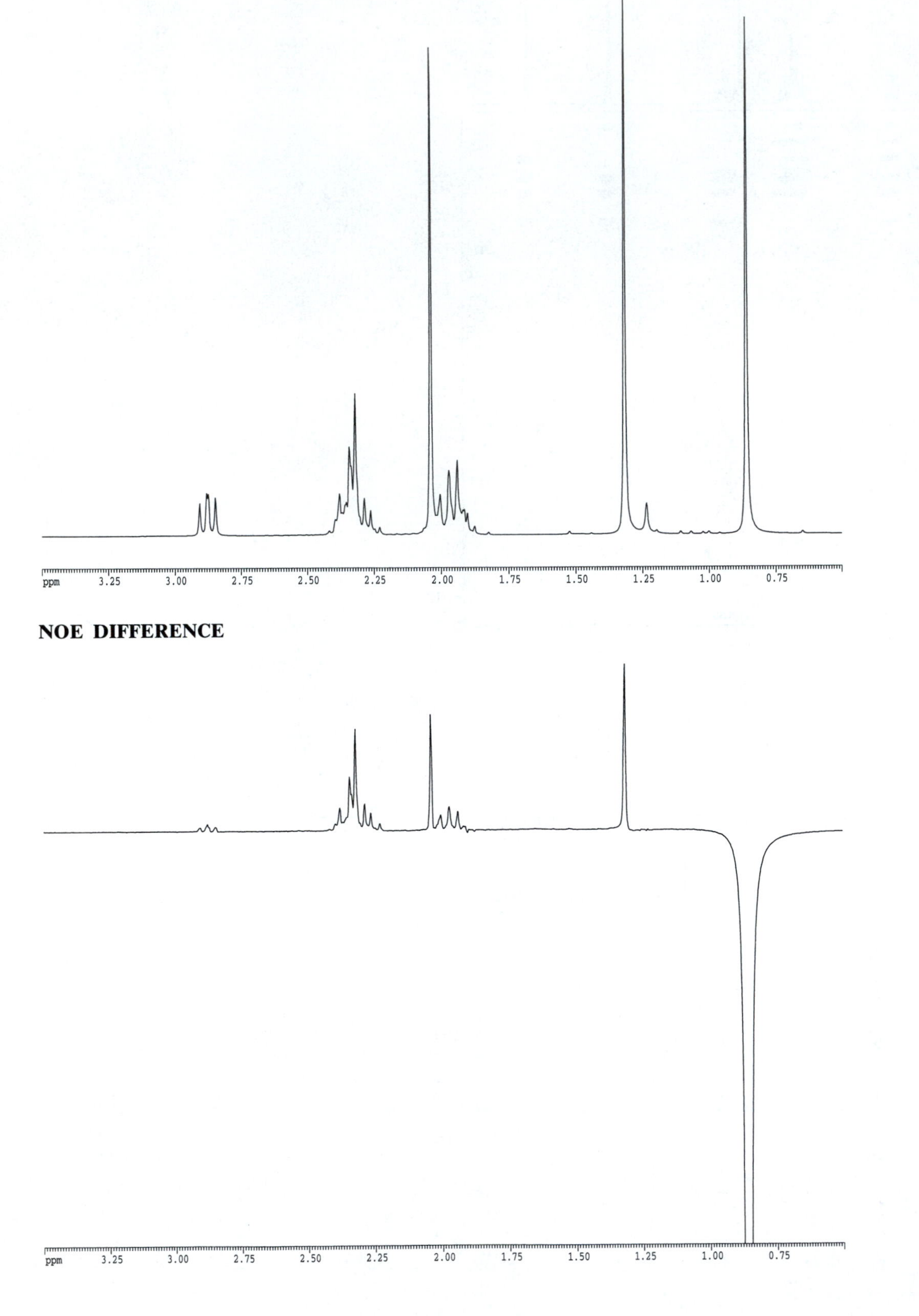
¹H SPECTRUM FOR NOE
ppm 3.25 3.00 2.75 2.50 2.25 2.00 1.75 1.50 1.25 1.00 0.75
NOE DIFFERENCE
ppm 3.25 3.00 2.75 2.50 2.25 2.00 1.75 1.50 1.25 1.00 0.75

NOE DIFFERENCE

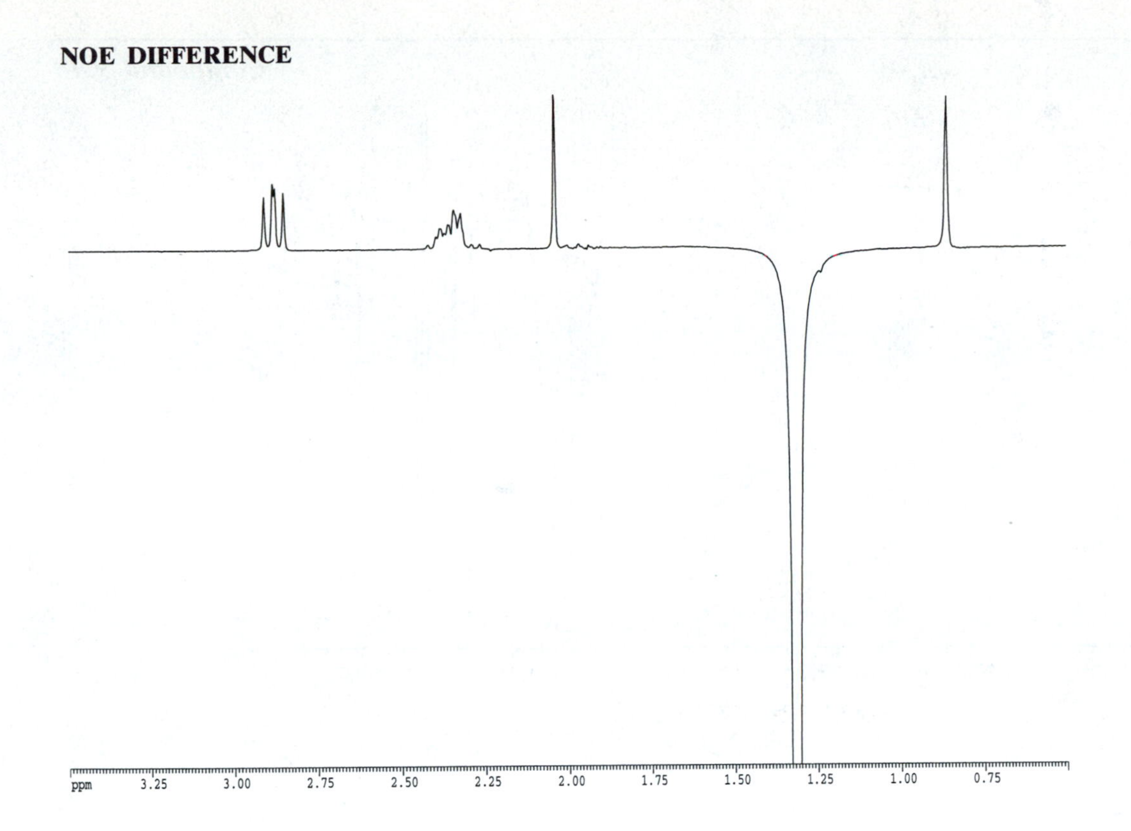

INFRARED IN $CHCl_3$ ***Compound 9.48***

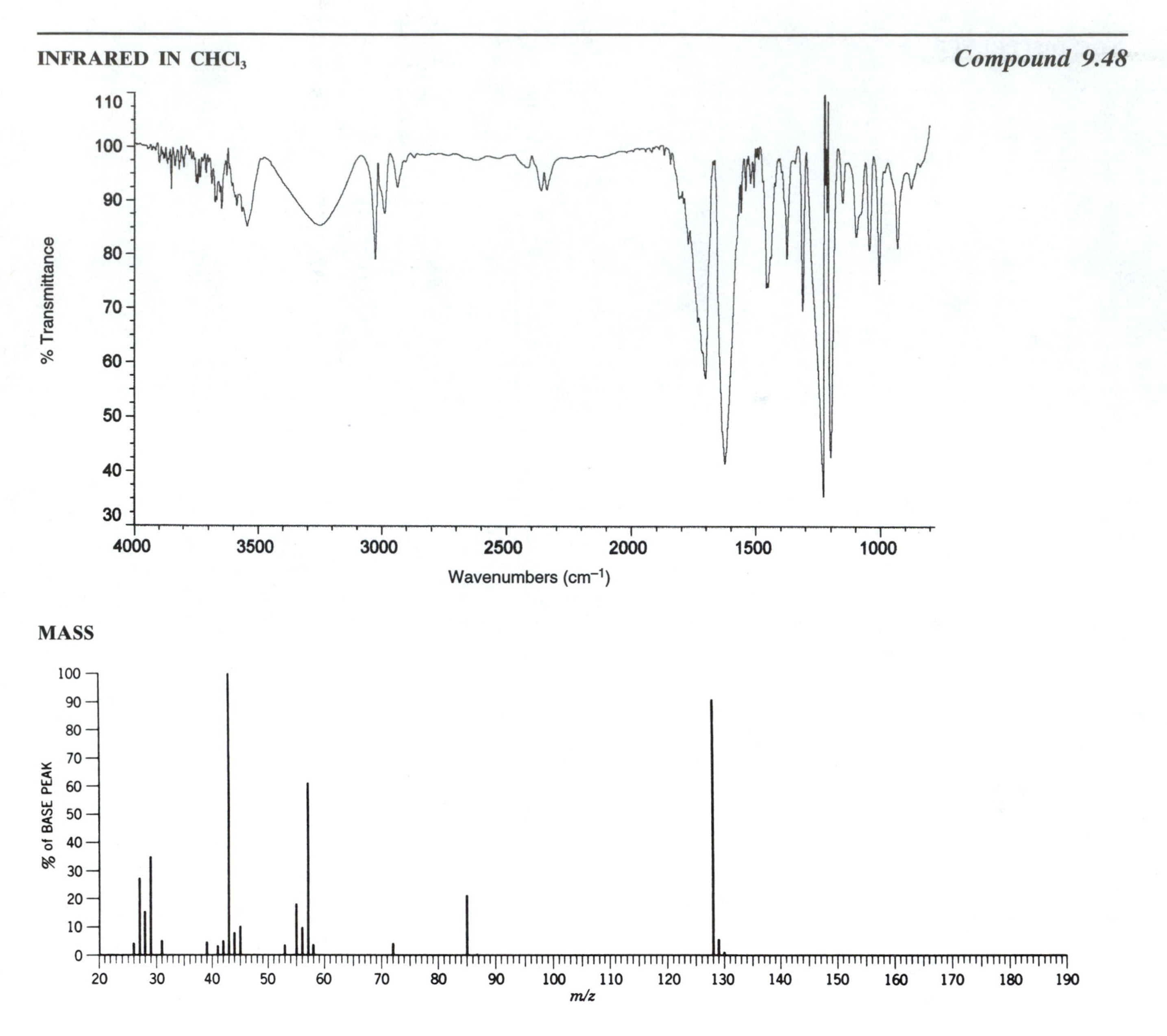

^{1}H NMR

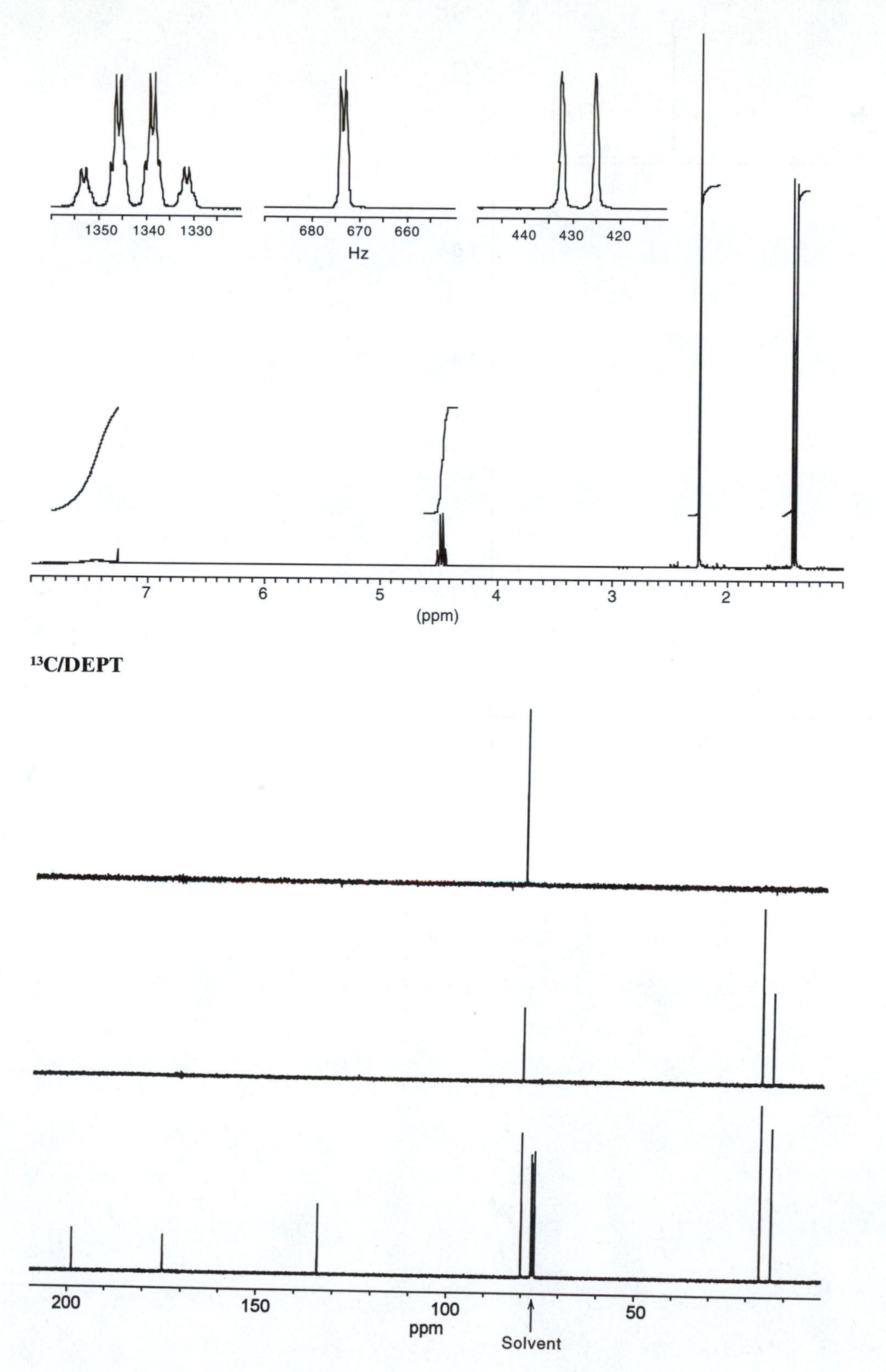
1350 1340 1330
680 670 660
Hz
440 430 420
7 6 5 4 3 2
(ppm)
^{13}C/DEPT
200 150 100 50
ppm
Solvent

COSY

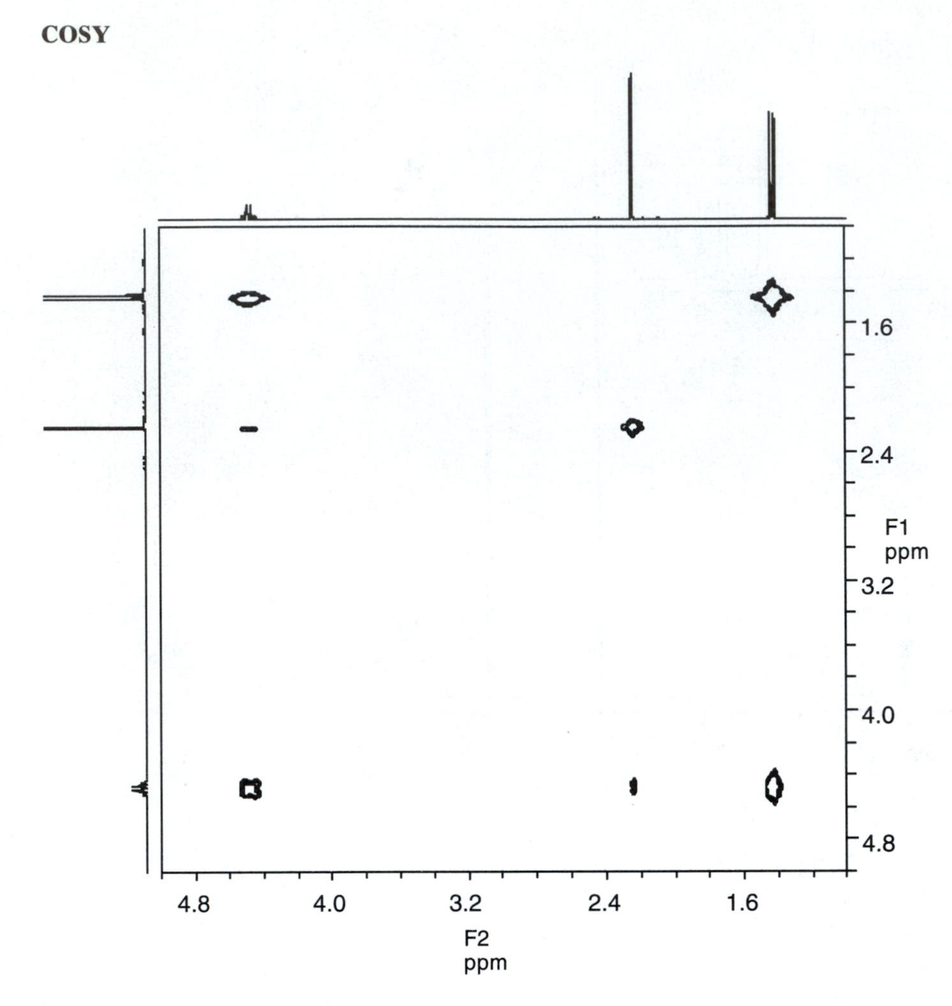

HETCOR

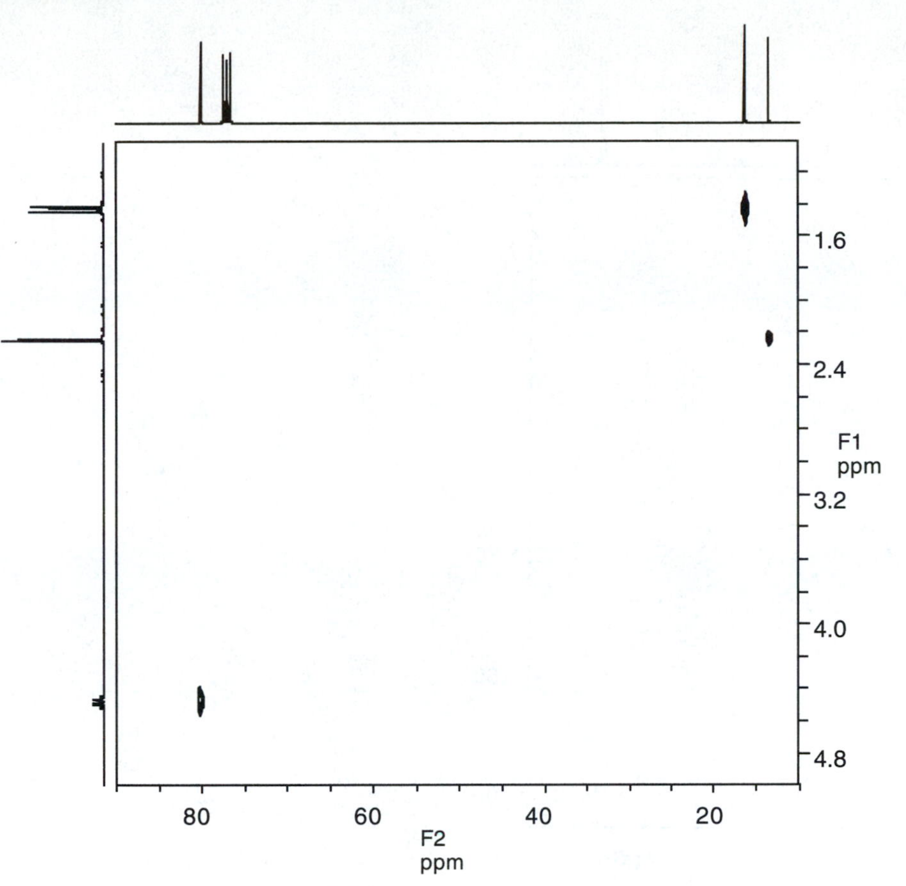

HMBC

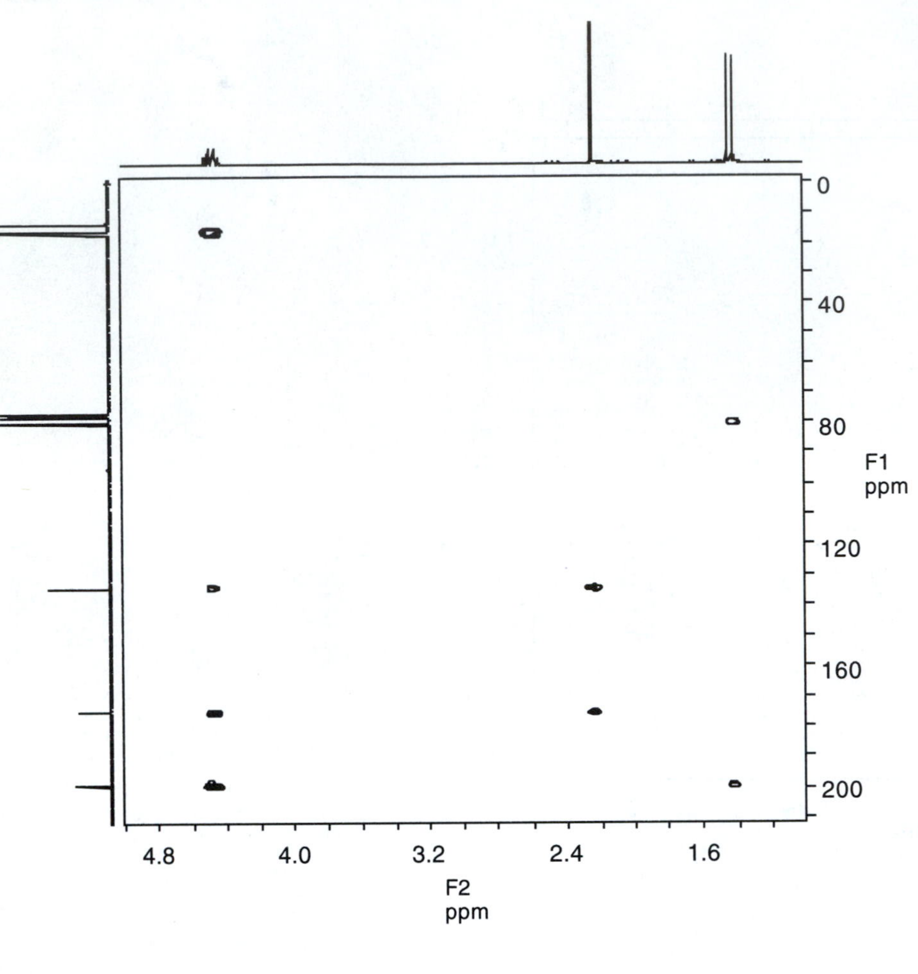

INFRARED *Compound 9.49*

MASS

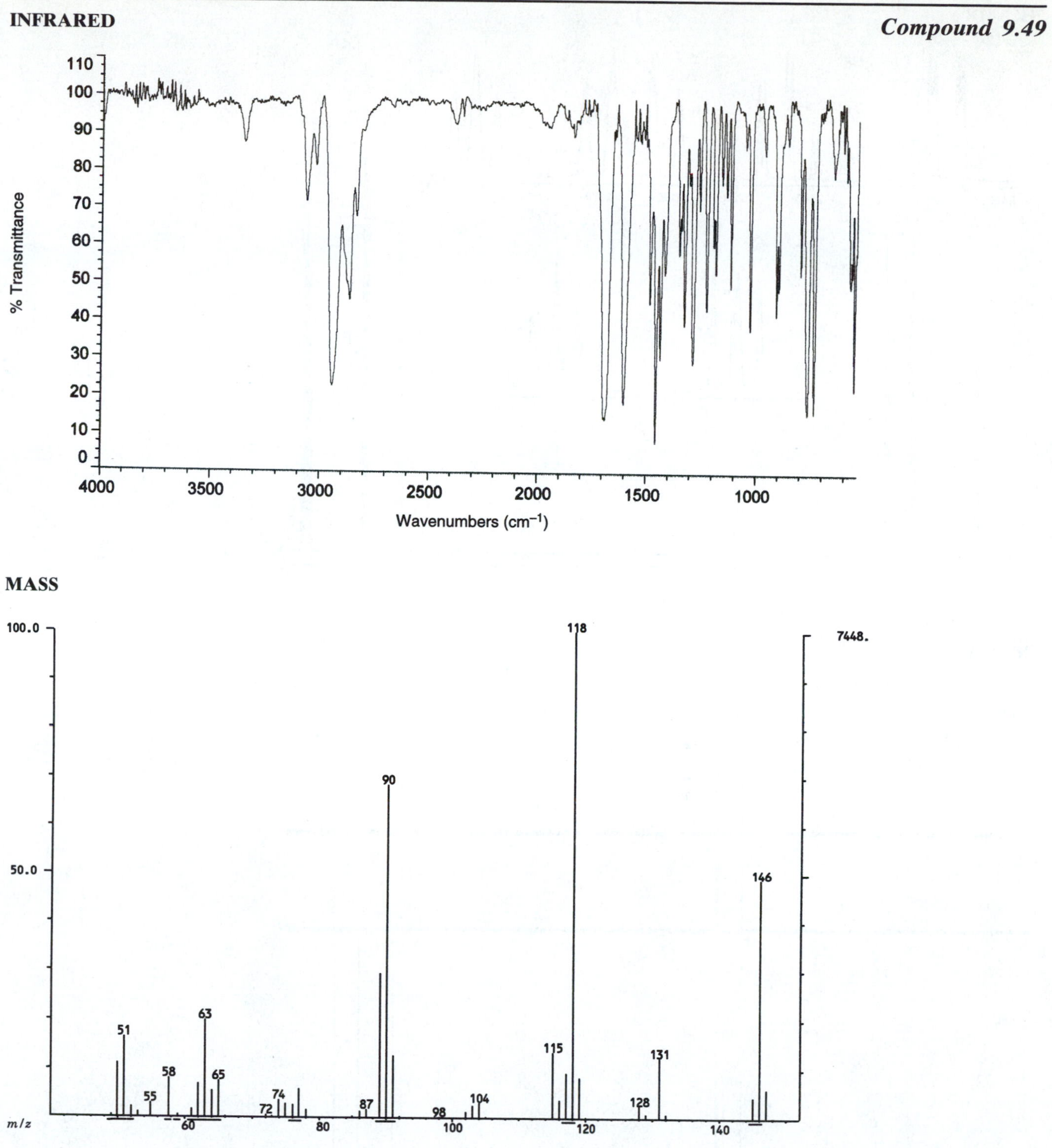

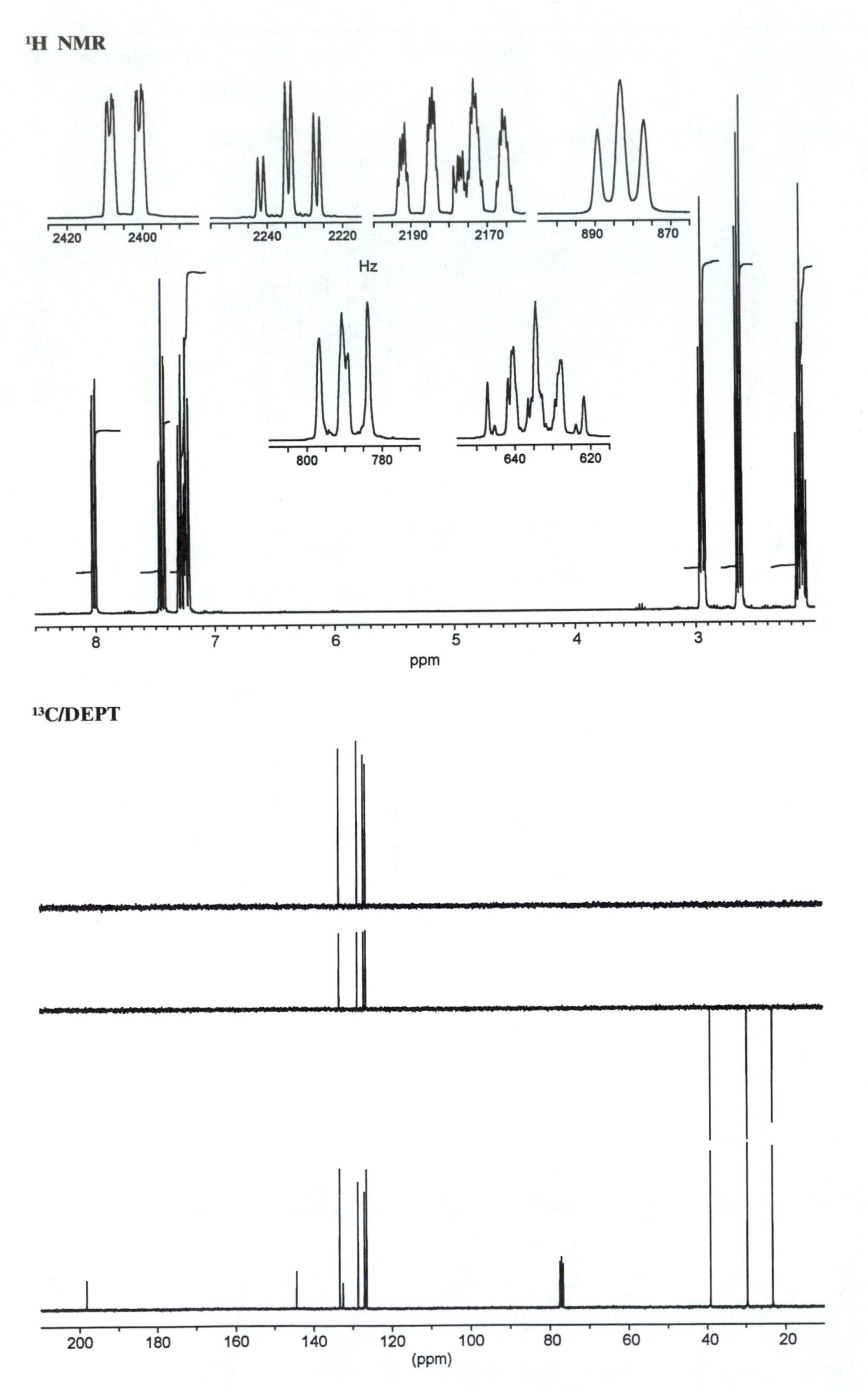

¹H NMR
2420
2400
2240
2220
2190
2170
890
870
Hz
800
780
640
620
8
7
6
5
4
3
ppm
¹³C/DEPT
200
180
160
140
120
100
80
60
40
20
(ppm)

COSY

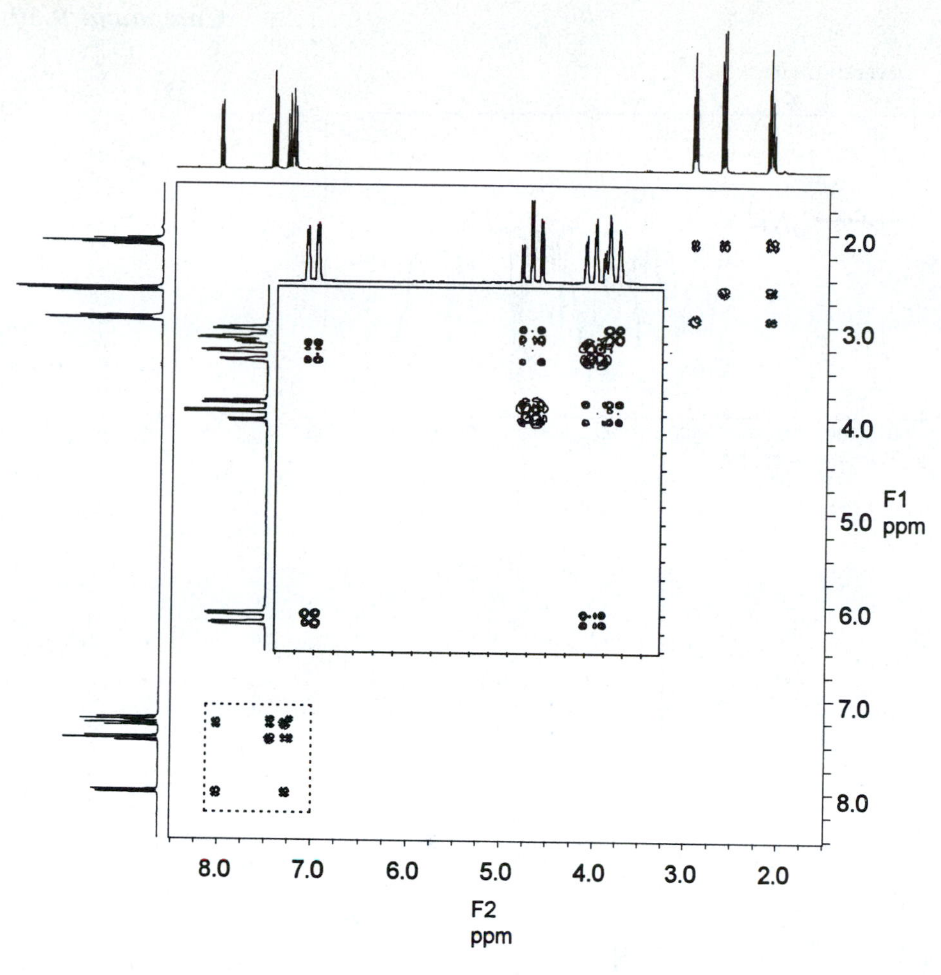

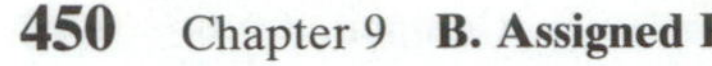

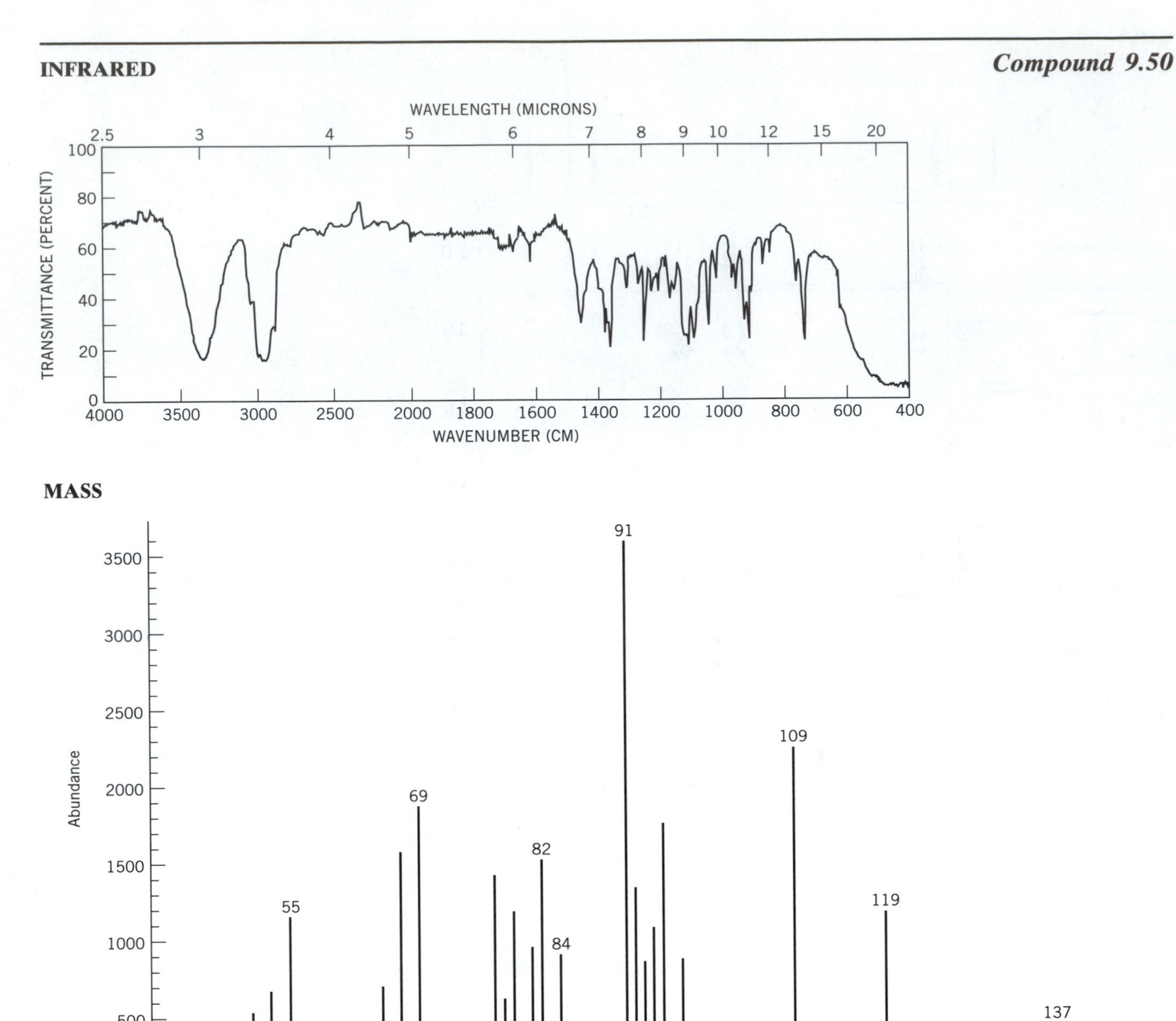

INFRARED
Compound 9.50
WAVELENGTH (MICRONS)
2.5
3
4
5
6
7
8
9
10
12
15
20
TRANSMITTANCE (PERCENT)
100
80
60
40
20
0
4000
3500
3000
2500
2000
1800
1600
1400
1200
1000
800
600
400
WAVENUMBER (CM)
MASS
Abundance
3500
3000
2500
2000
1500
1000
500
0
91
109
69
82
55
84
56
71
103
119
120
132
137
m/z
50
60
70
80
90
100
110
120
130
140
High Res. Molecular Ion = 152.1199

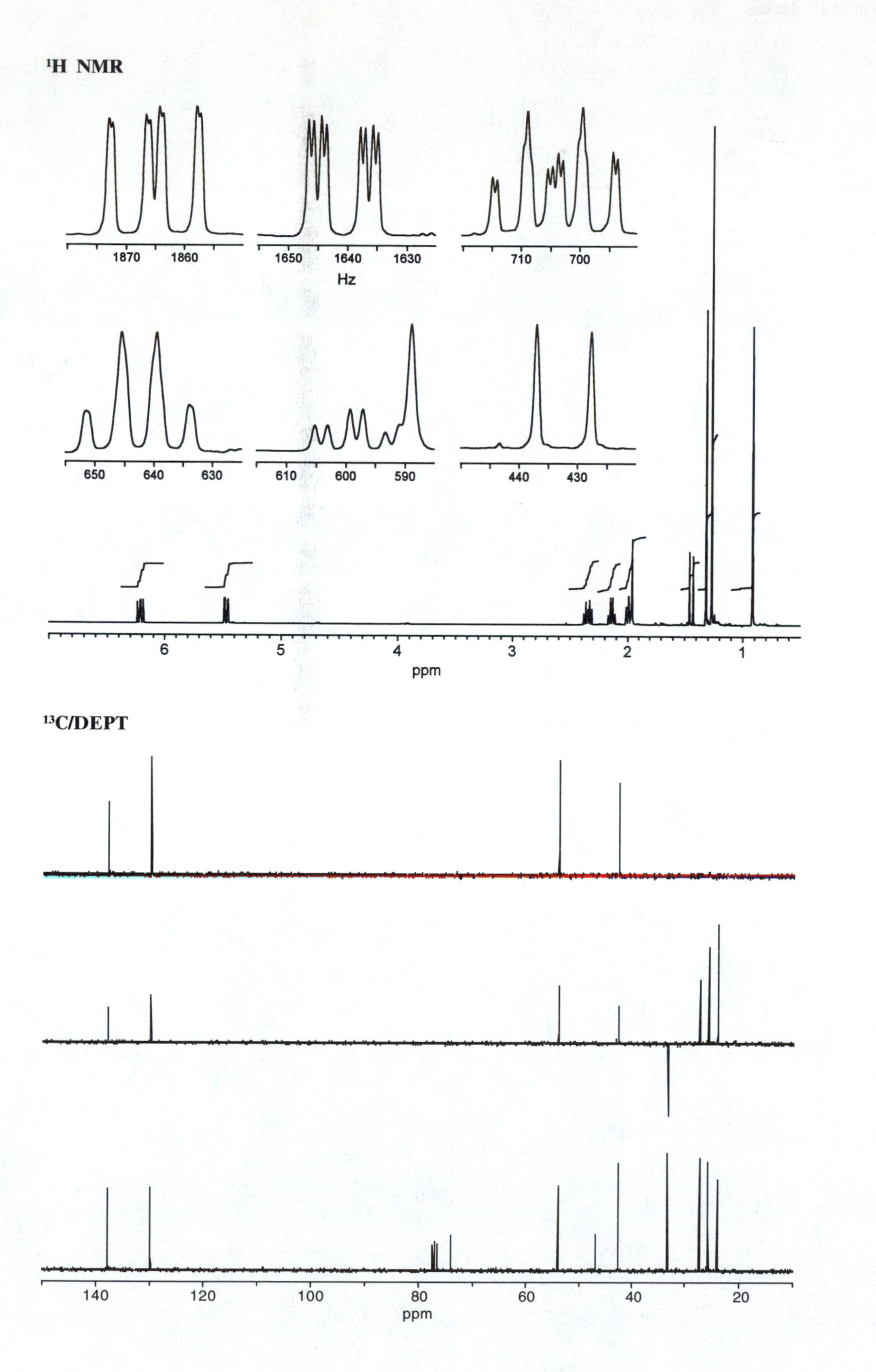

¹H NMR
1870
1860
1650
1640
1630
Hz
710
700
650
640
630
610
600
590
440
430
6
5
4
3
2
1
ppm
¹³C/DEPT
140
120
100
80
60
40
20
ppm

COSY

HETCOR

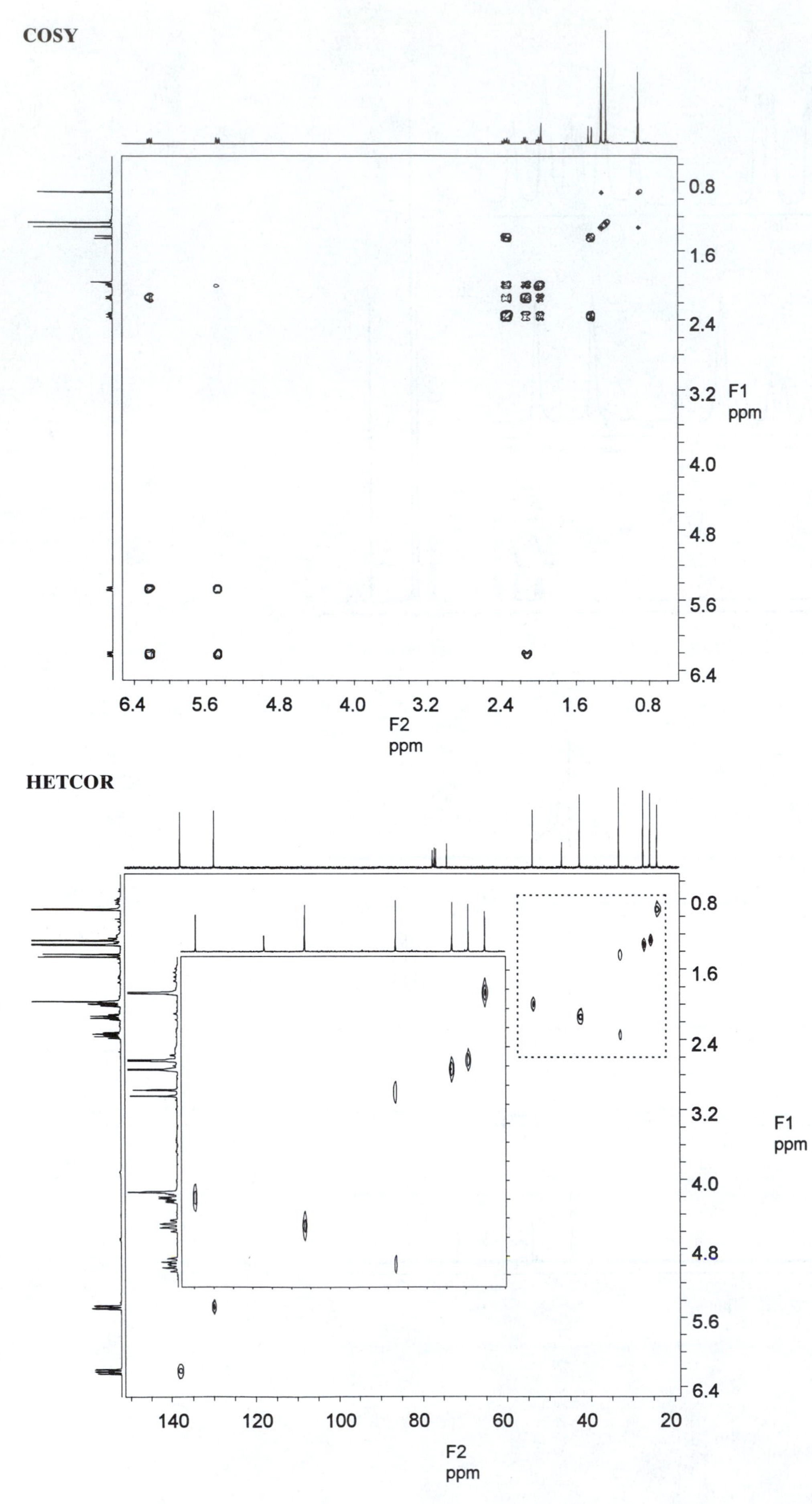

COSY
0.8
1.6
2.4
3.2
4.0
4.8
5.6
6.4
F1
ppm
6.4
5.6
4.8
4.0
3.2
2.4
1.6
0.8
F2
ppm
HETCOR
0.8
1.6
2.4
3.2
4.0
4.8
5.6
6.4
F1
ppm
140
120
100
80
60
40
20
F2
ppm

HMBC

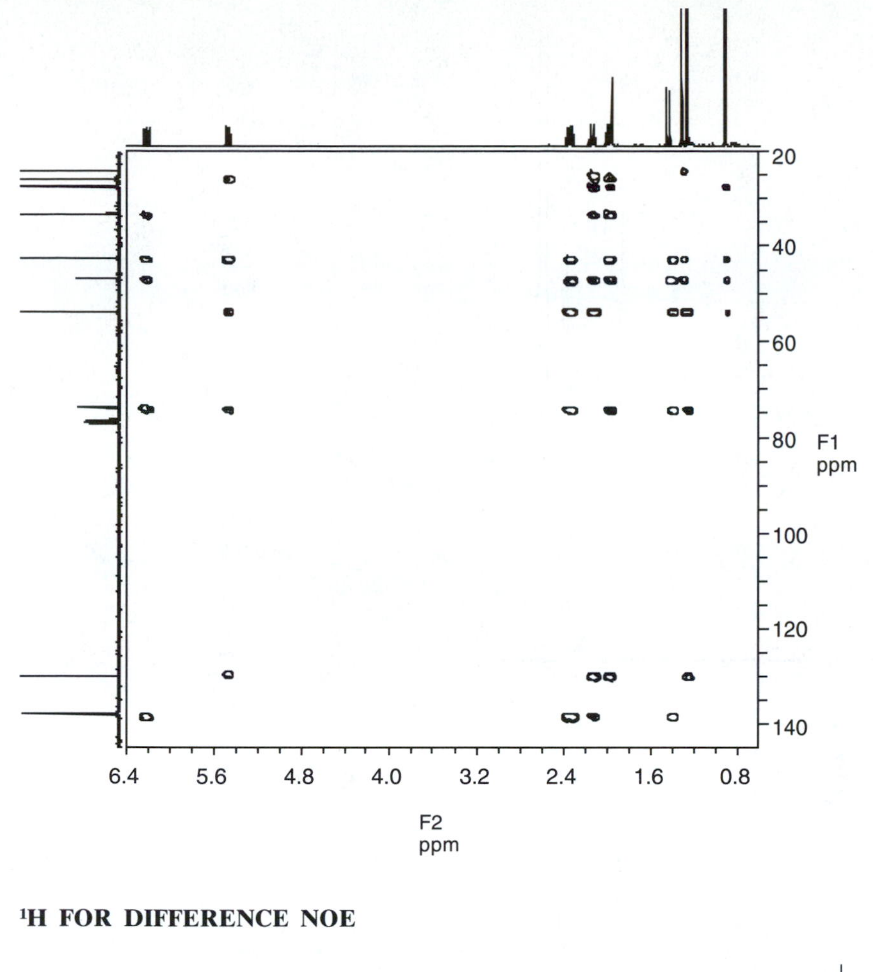

[1]H FOR DIFFERENCE NOE

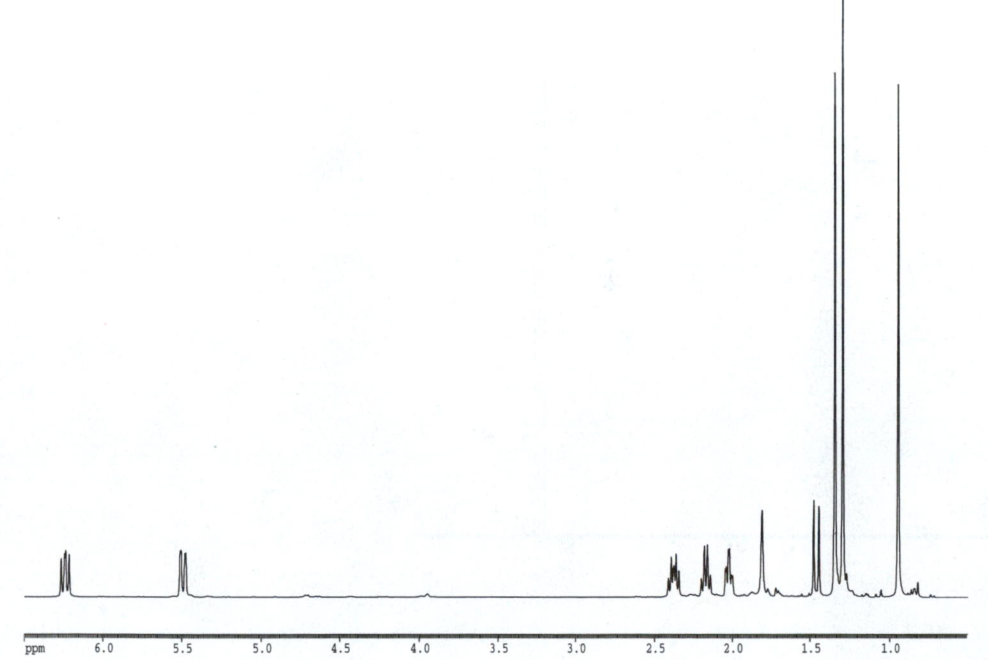

DIFFERENCE NOE (SEE SECTION 4.20)

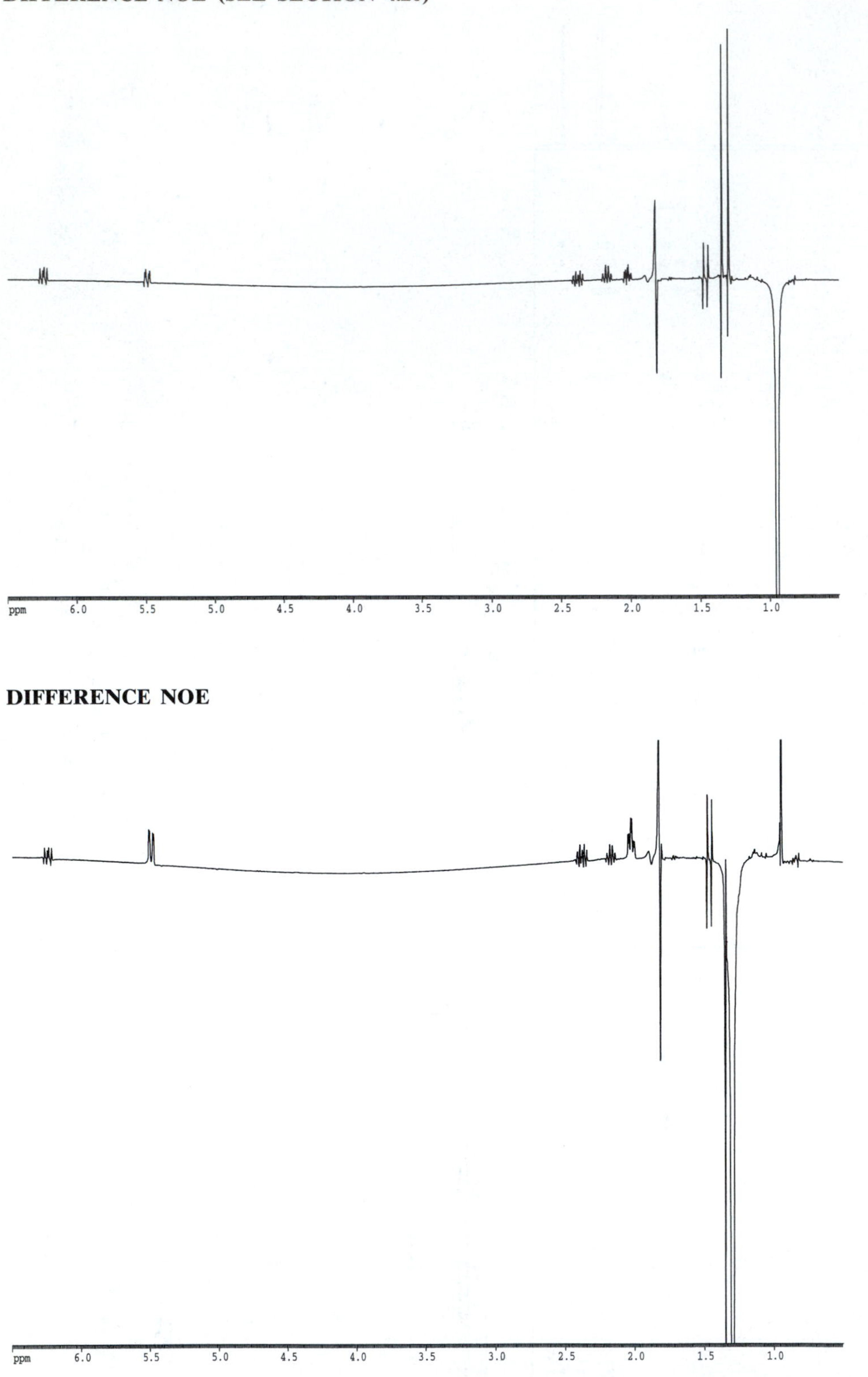

DIFFERENCE NOE

DIFFERENCE NOE

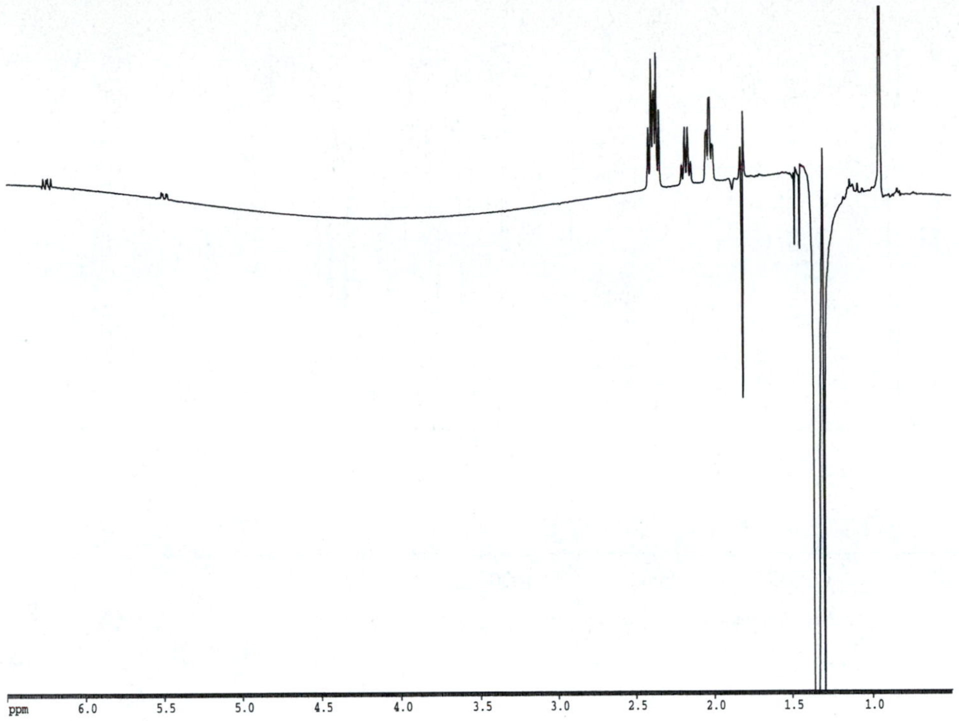

INFRARED ***Compound 9.51***

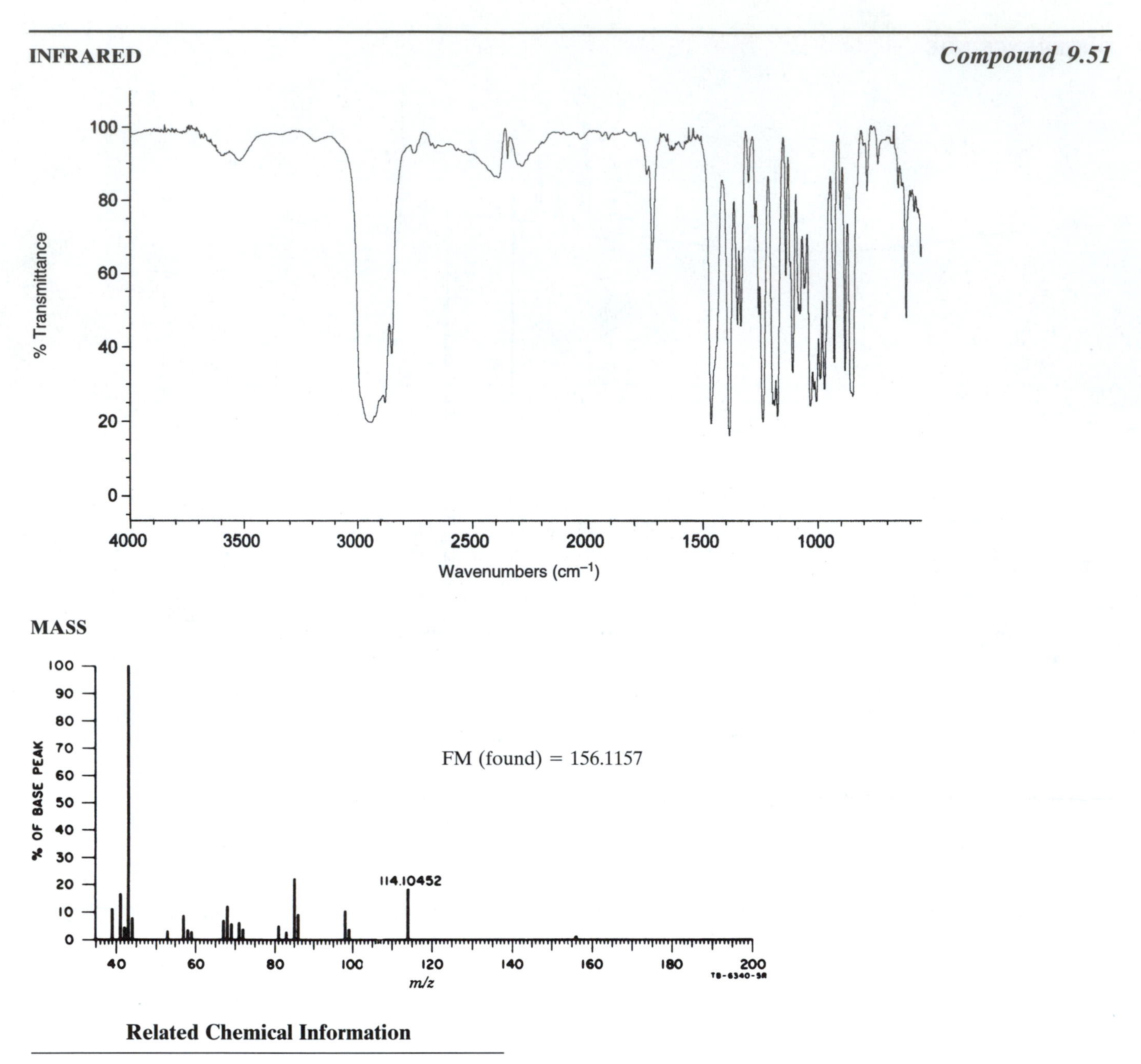

Related Chemical Information

1. Recovered unchanged from treatment with $LiAlH_4$.
2. Catalytic hydrogenolysis (250°C) gave nonane.

^{1}H NMR

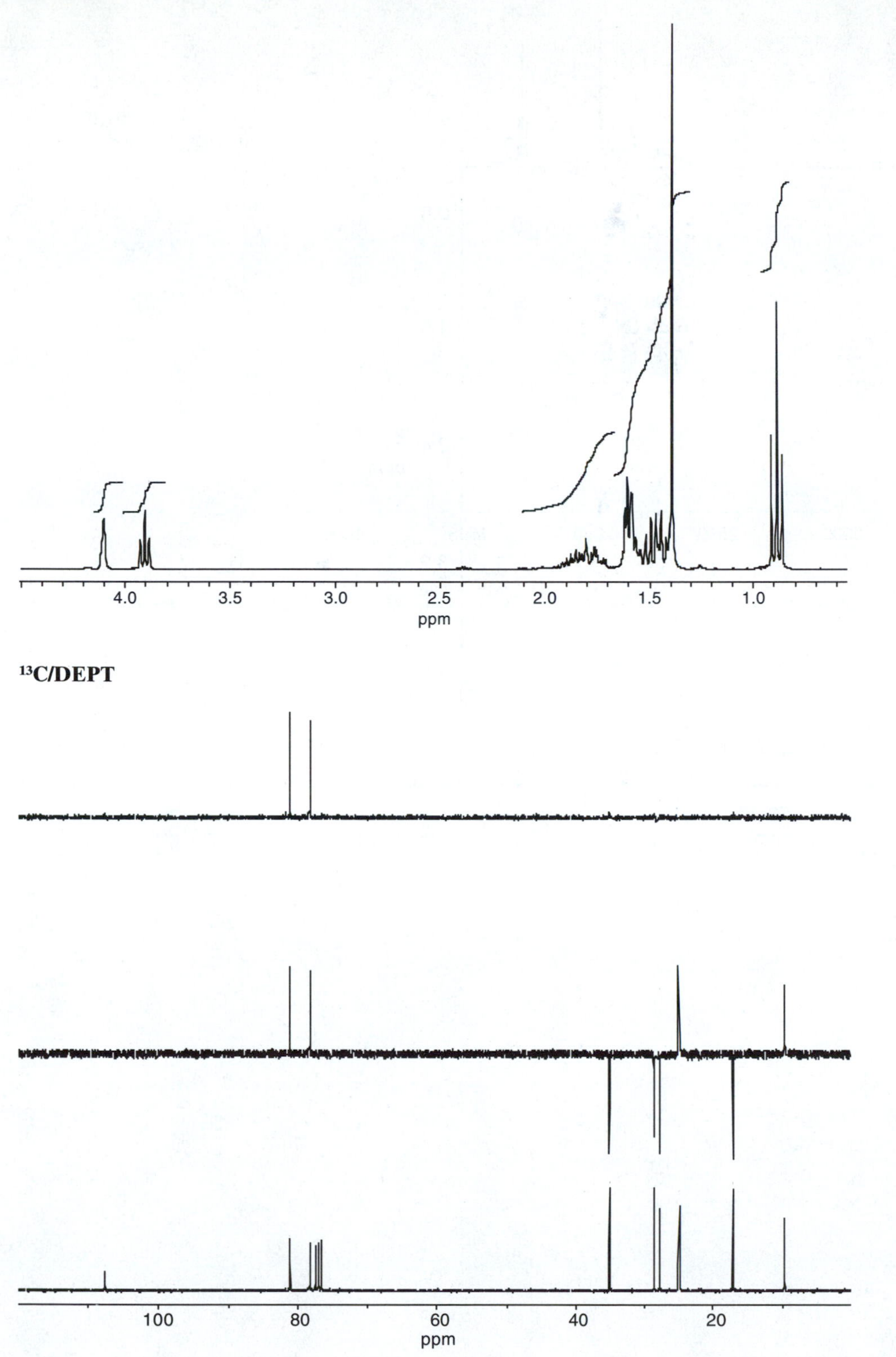

^{13}C/DEPT

COSY

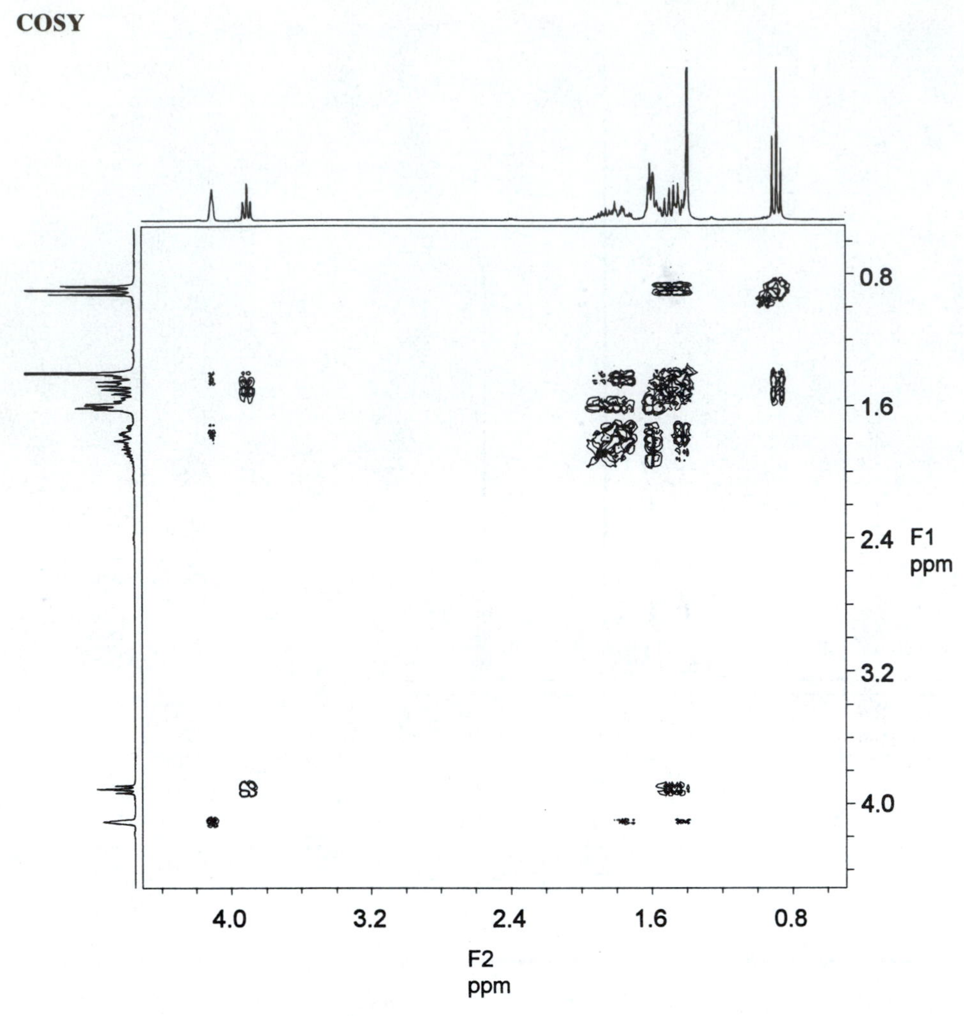

HETCOR

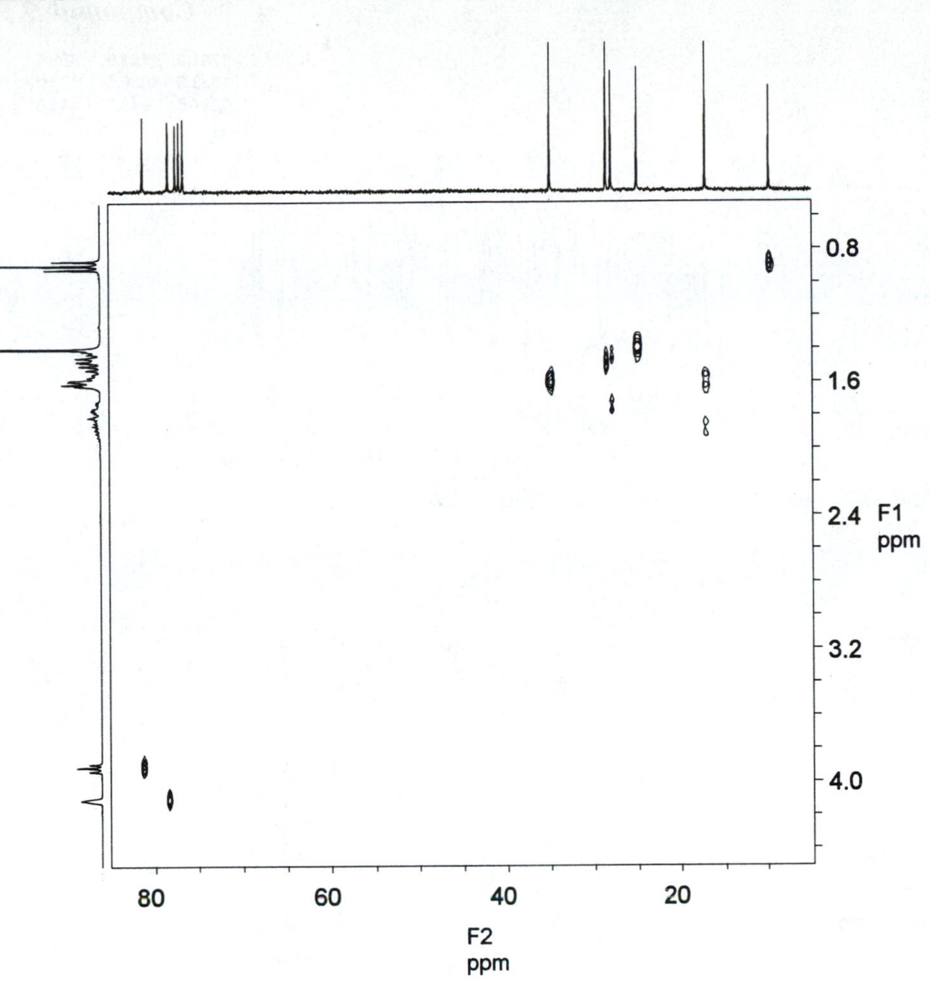

INFRARED

Compound 9.52

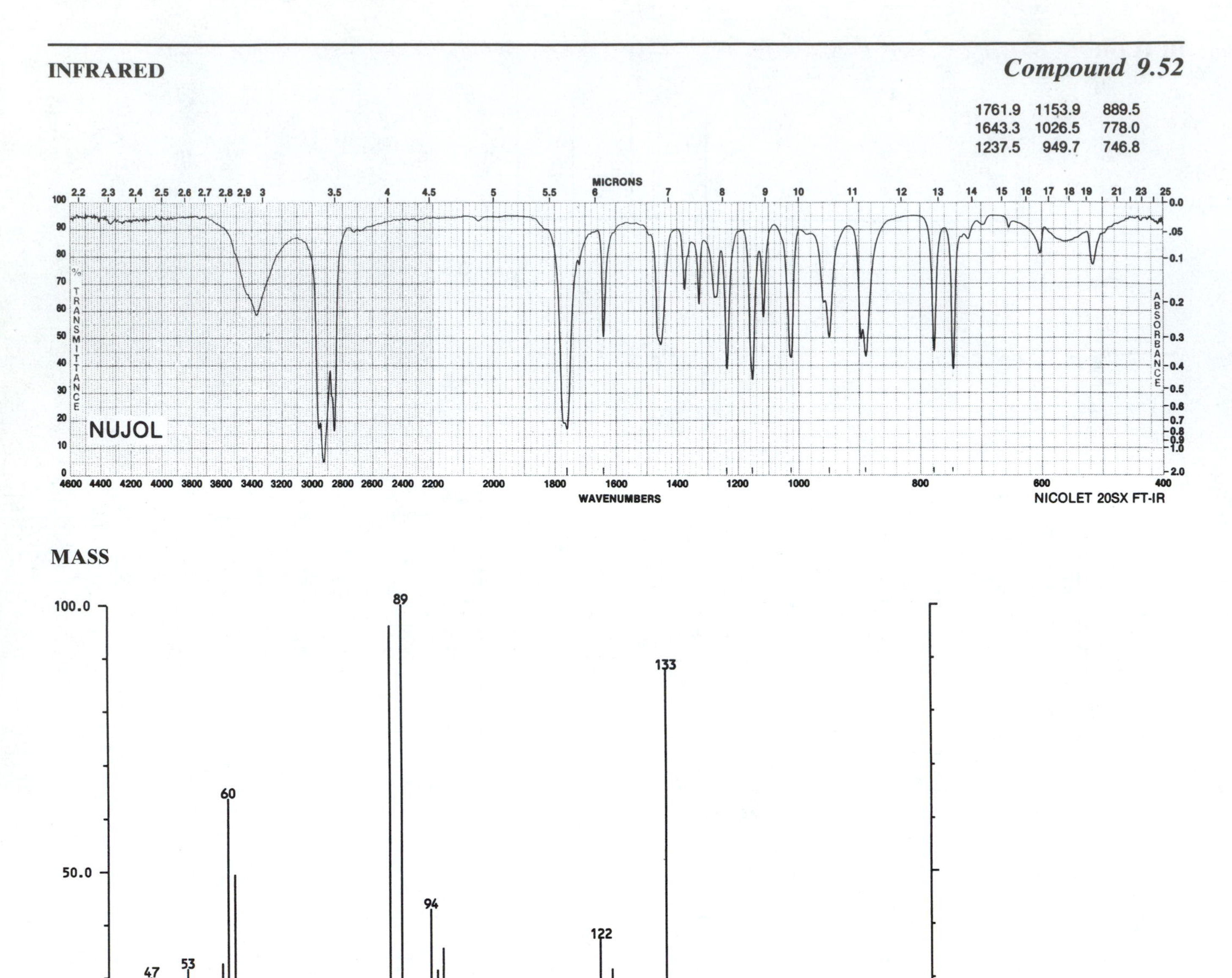

MASS

^{1}H IN ACETONE-d_6

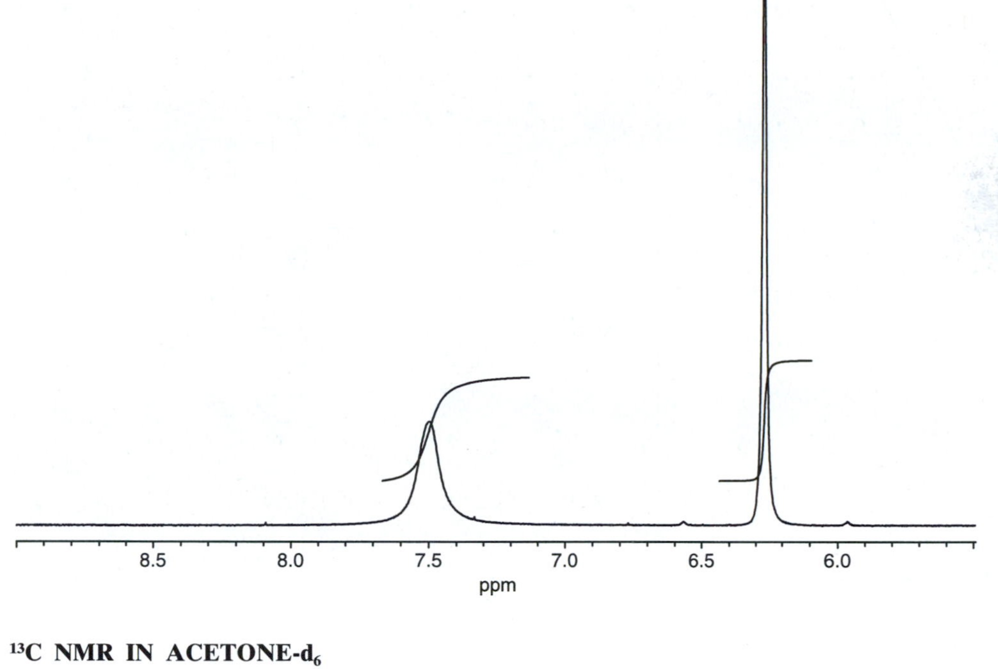

^{13}C NMR IN ACETONE-d_6

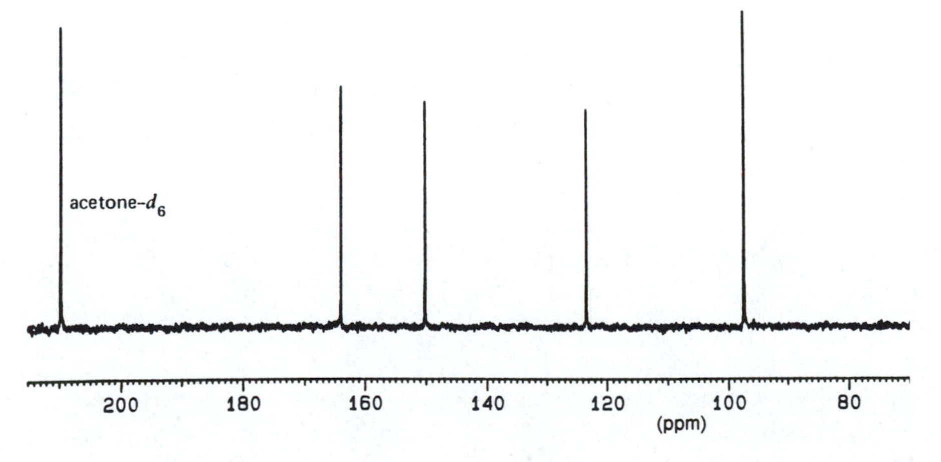

2-D INADEQUATE SPECTRUM IN ACETONE-d_6

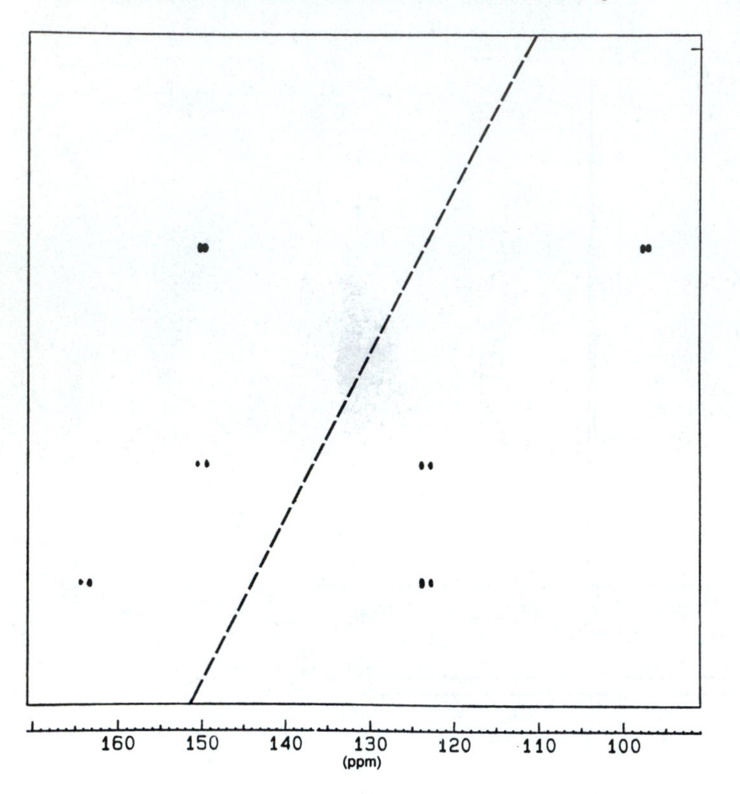

INFRARED

Compound 9.53

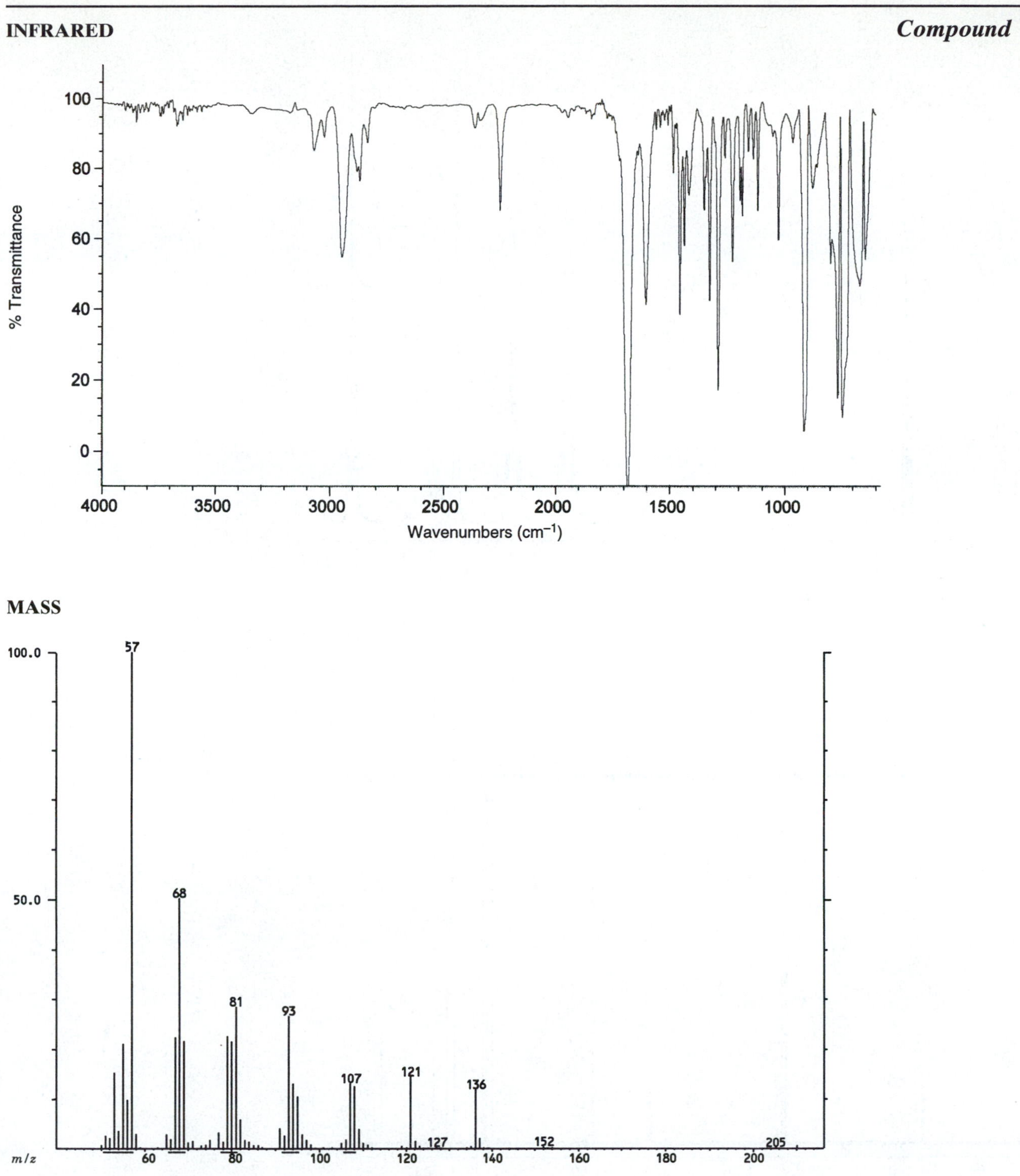

¹H NMR

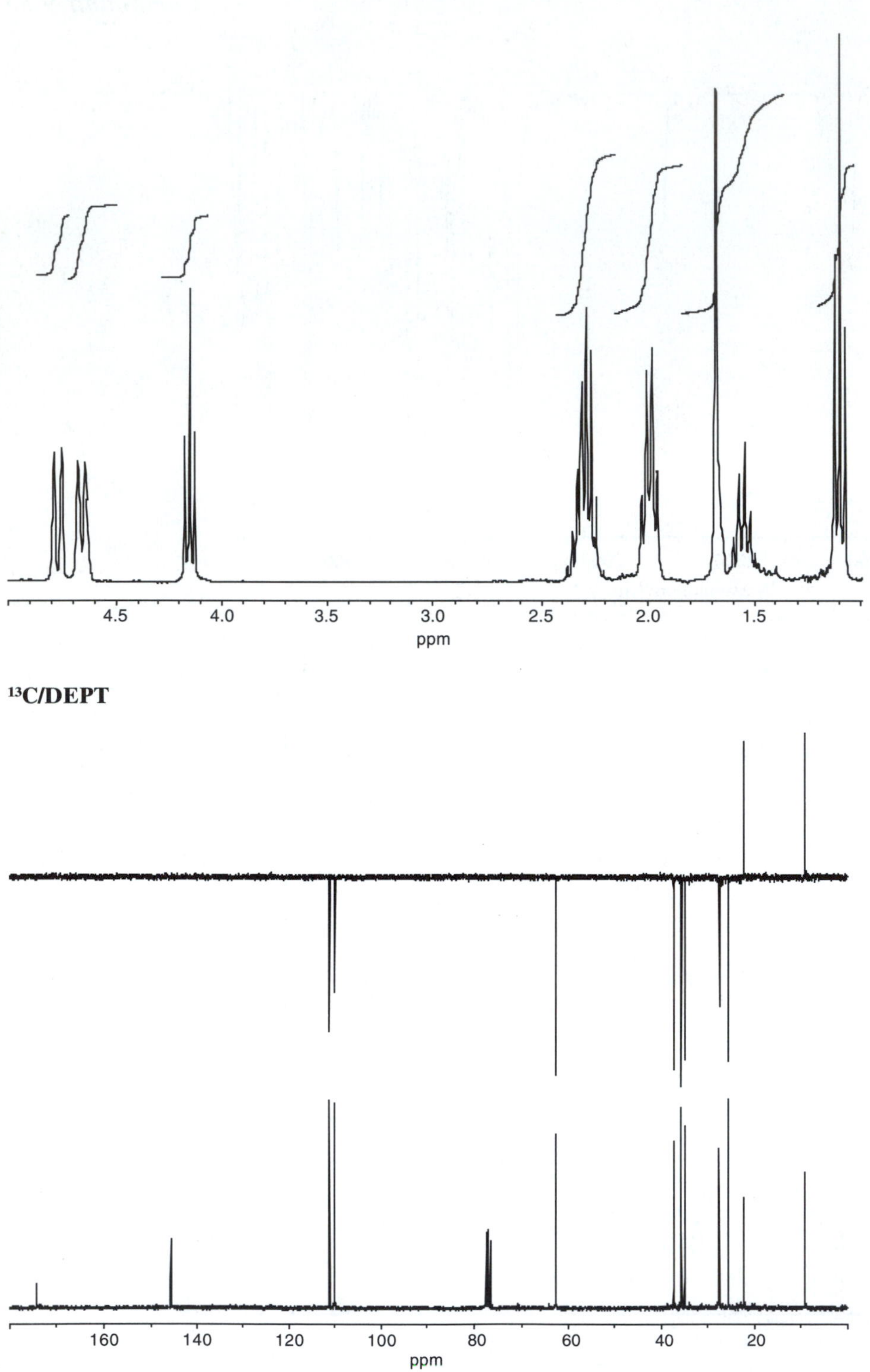
4.5
4.0
3.5
3.0
2.5
2.0
1.5
ppm
¹³C/DEPT
160
140
120
100
80
60
40
20
ppm

COSY

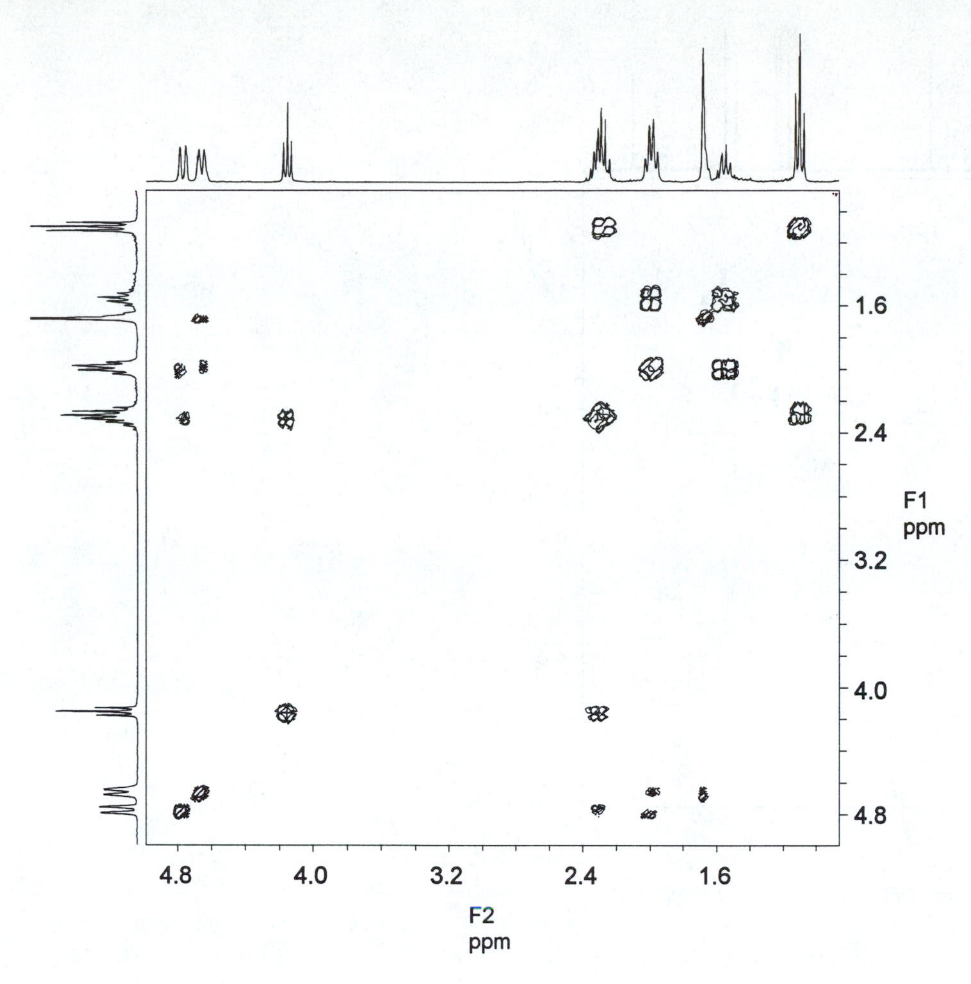
1.6
2.4
F1
ppm
3.2
4.0
4.8
4.8
4.0
3.2
2.4
1.6
F2
ppm

HETCOR

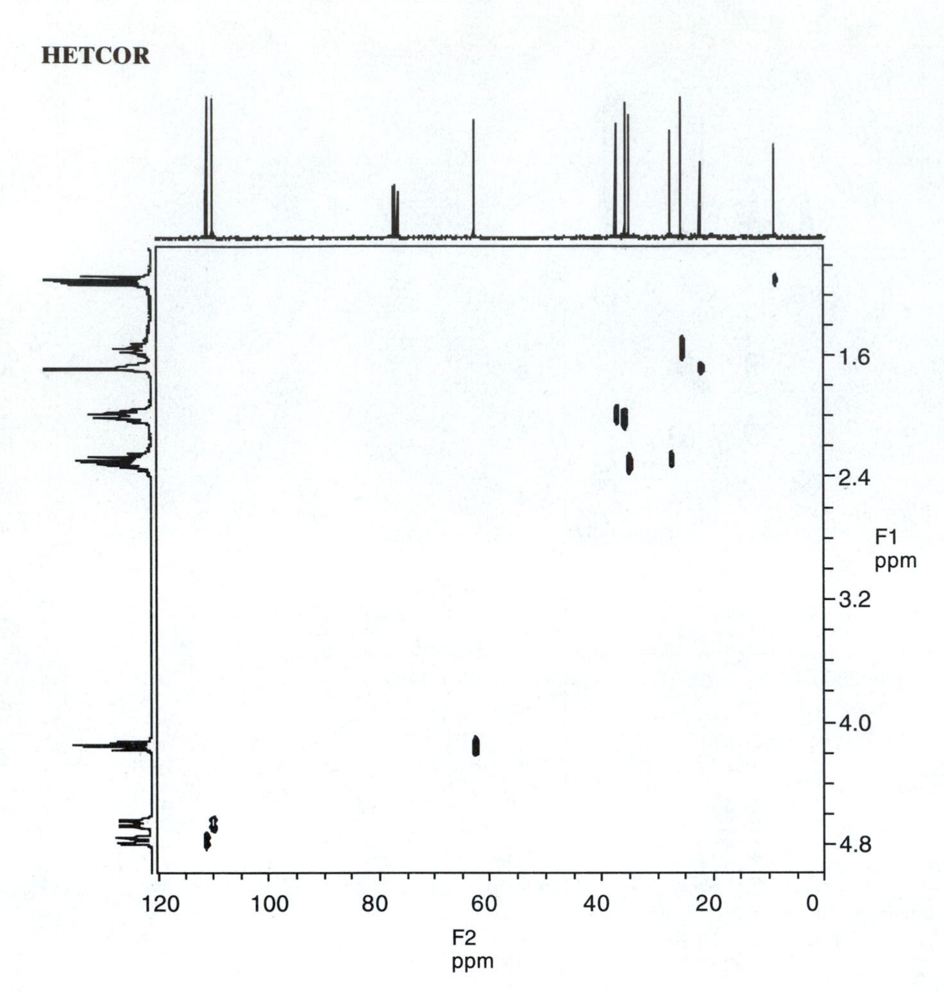

INFRARED

Compound 9.54

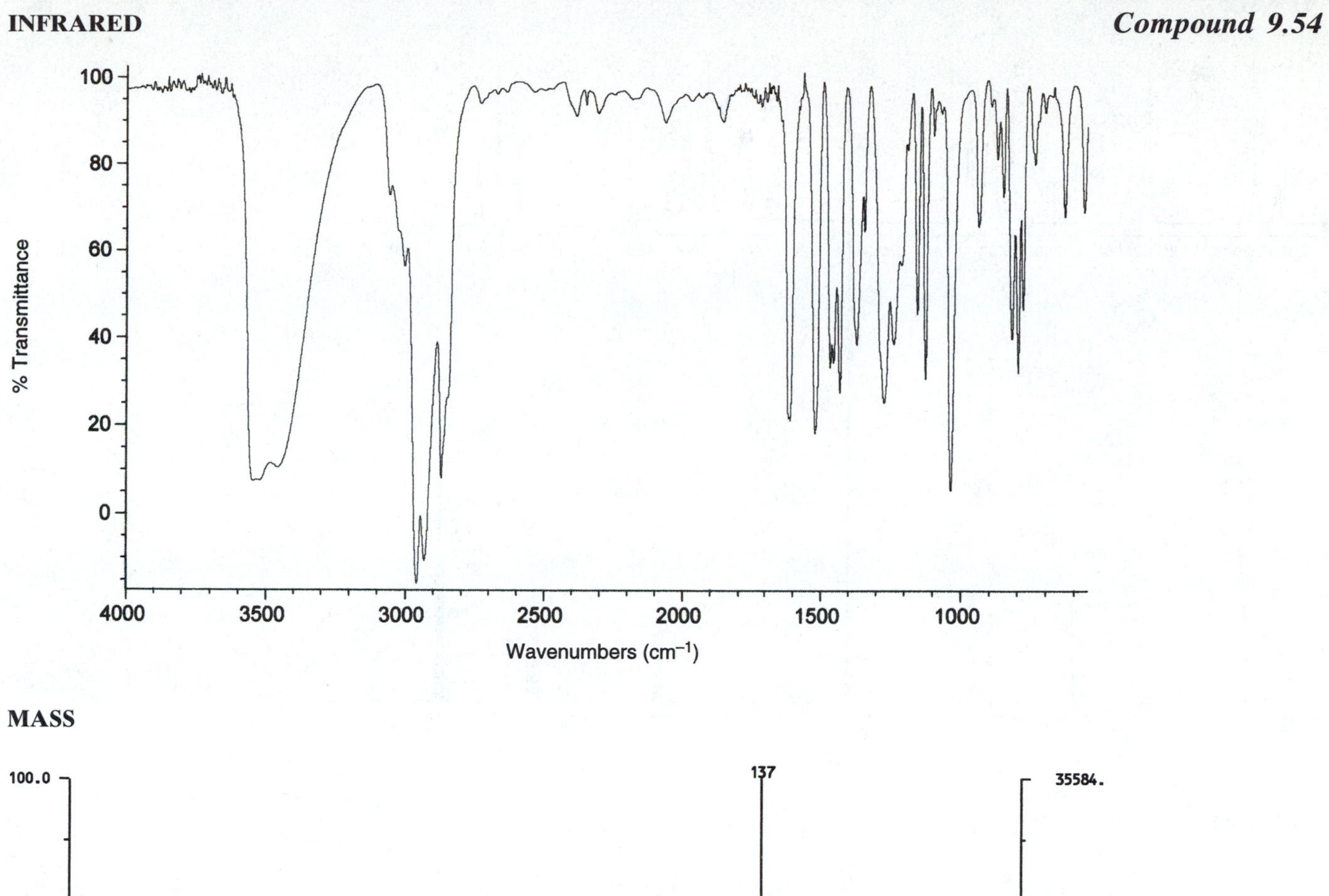

MASS

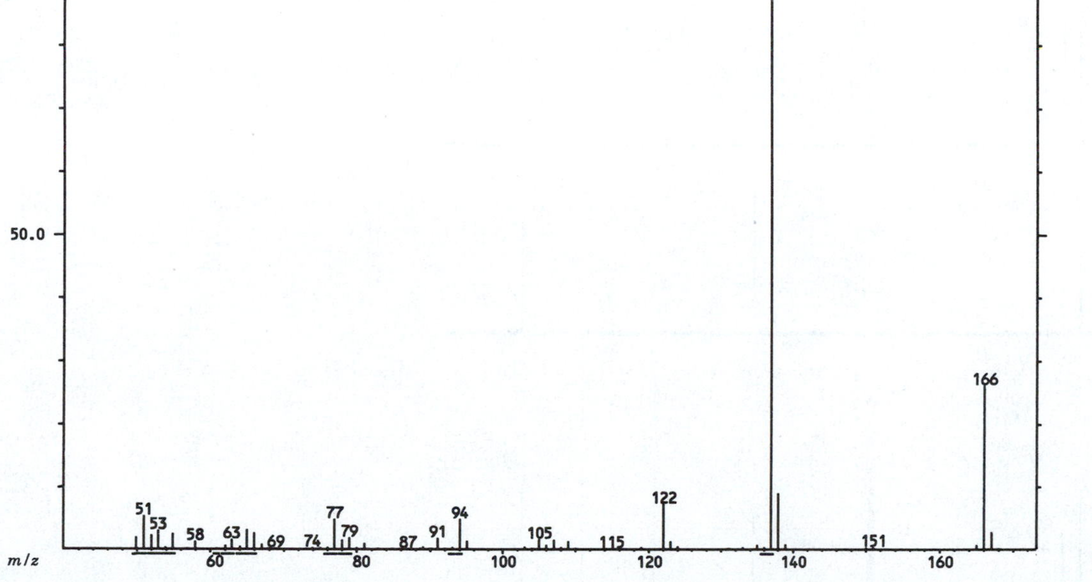

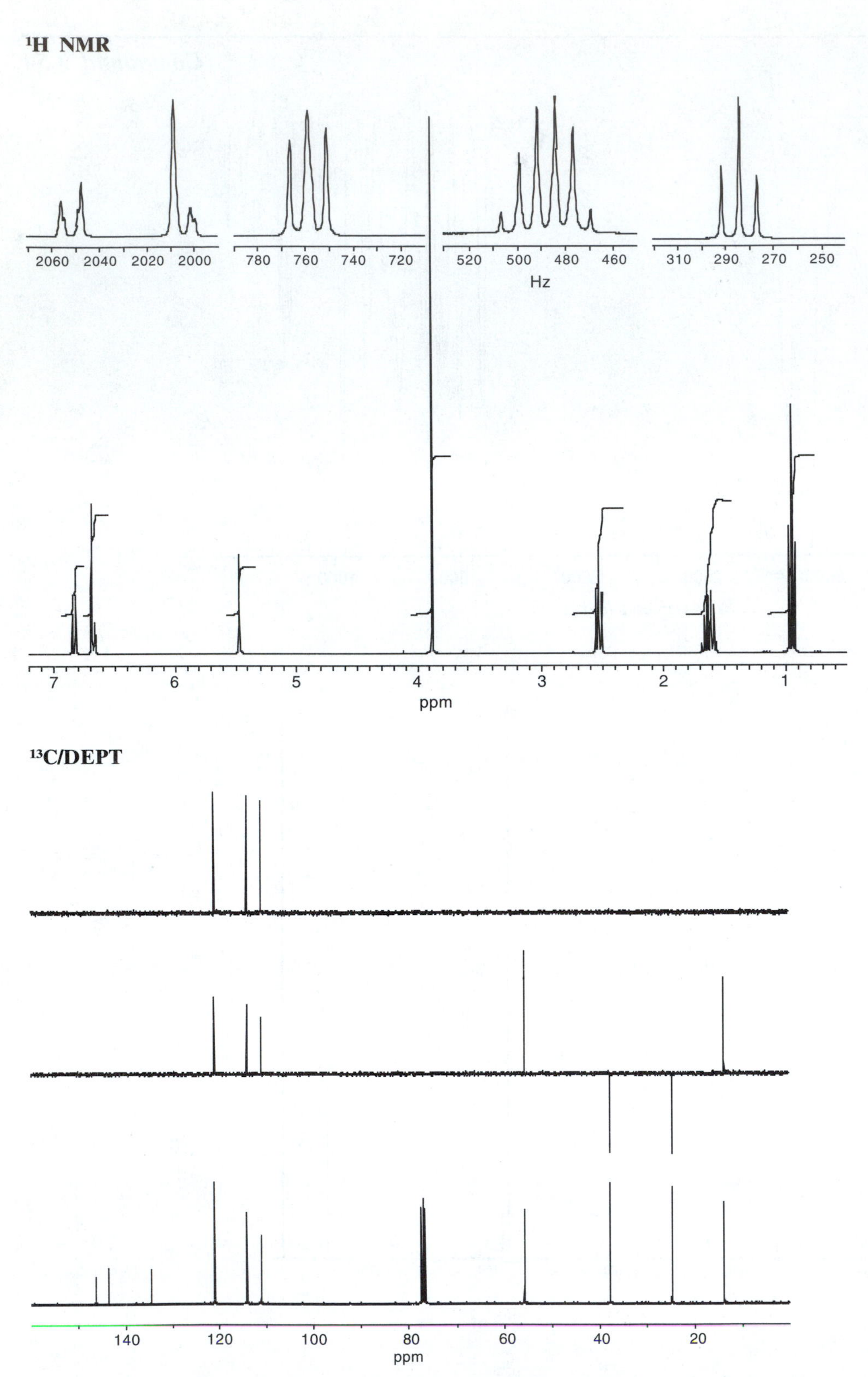

¹H NMR
2060
2040
2020
2000
780
760
740
720
520
500
480
460
Hz
310
290
270
250
7
6
5
4
3
2
1
ppm
¹³C/DEPT
140
120
100
80
60
40
20
ppm

COSY

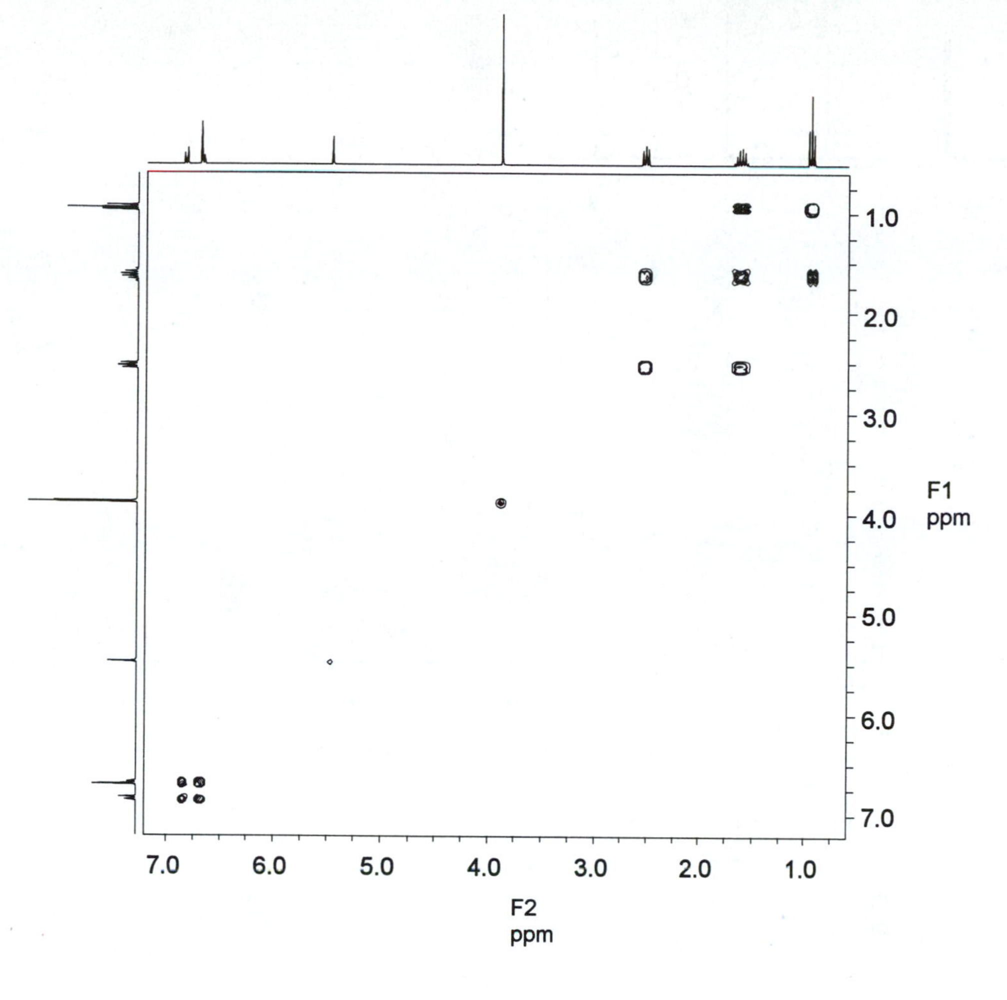

HETCOR

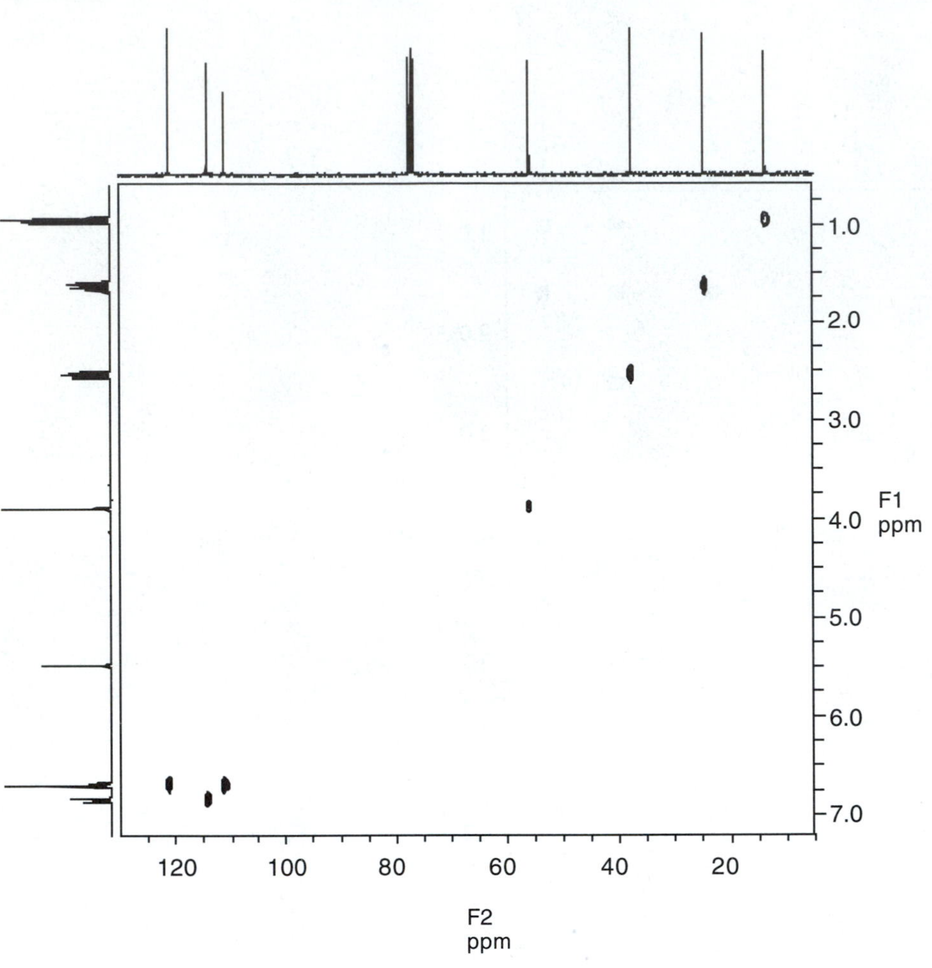

INFRARED

Compound 9.55

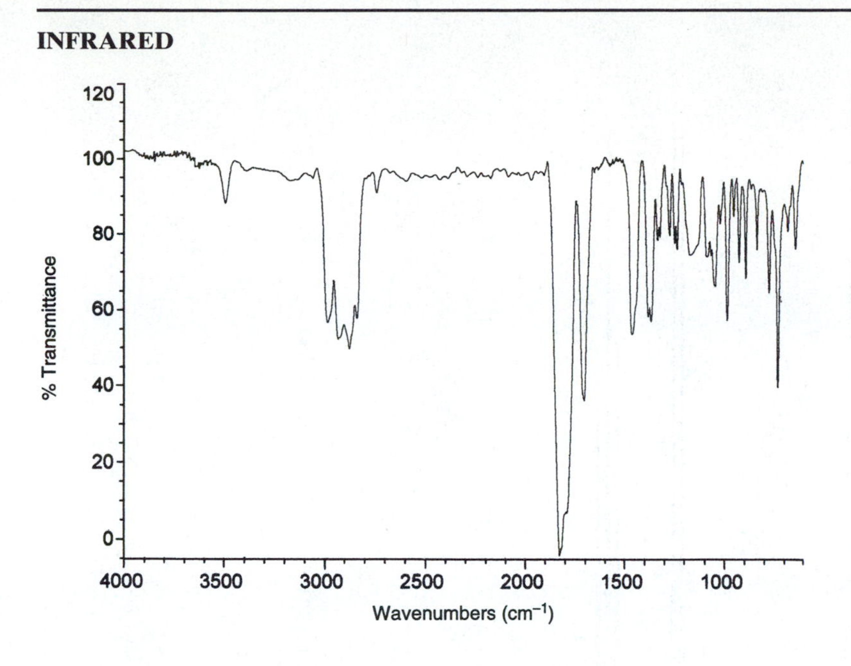

MASS

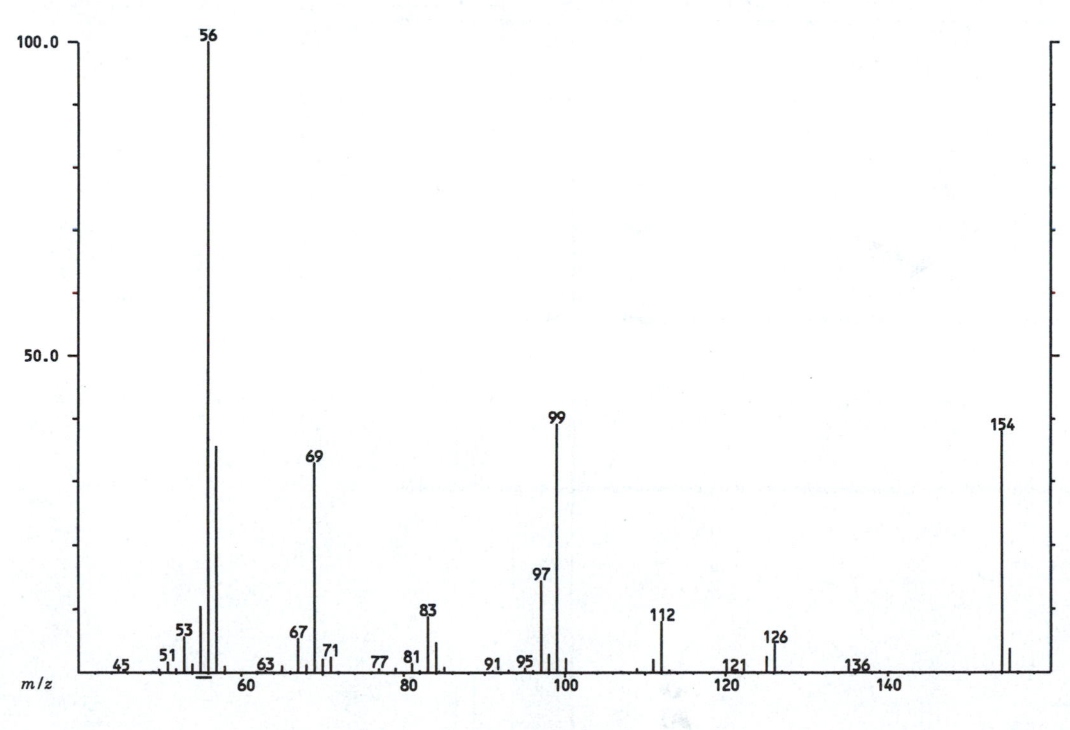

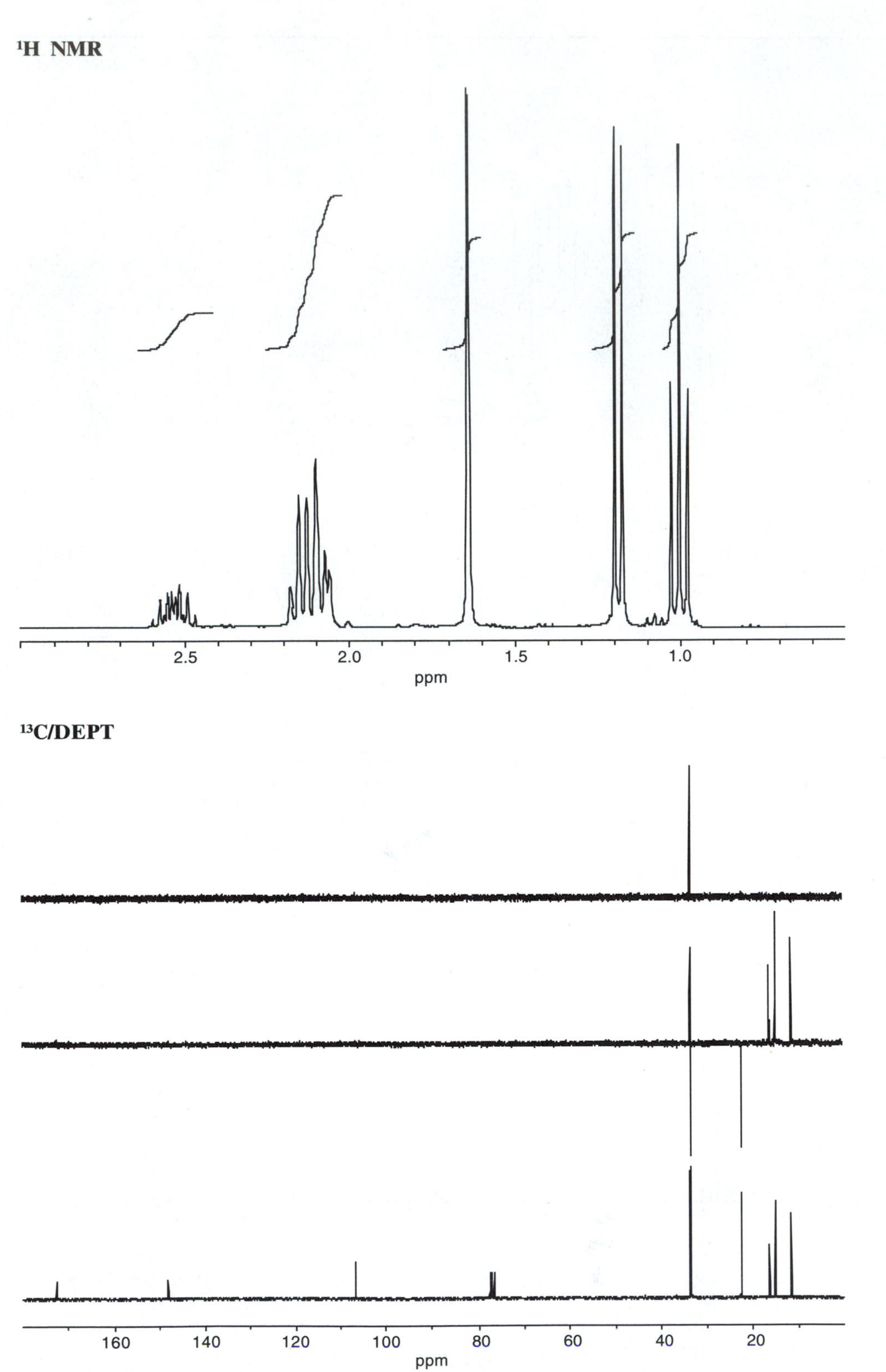
¹H NMR
2.5
2.0
1.5
1.0
ppm
¹³C/DEPT
160
140
120
100
80
60
40
20
ppm

COSY

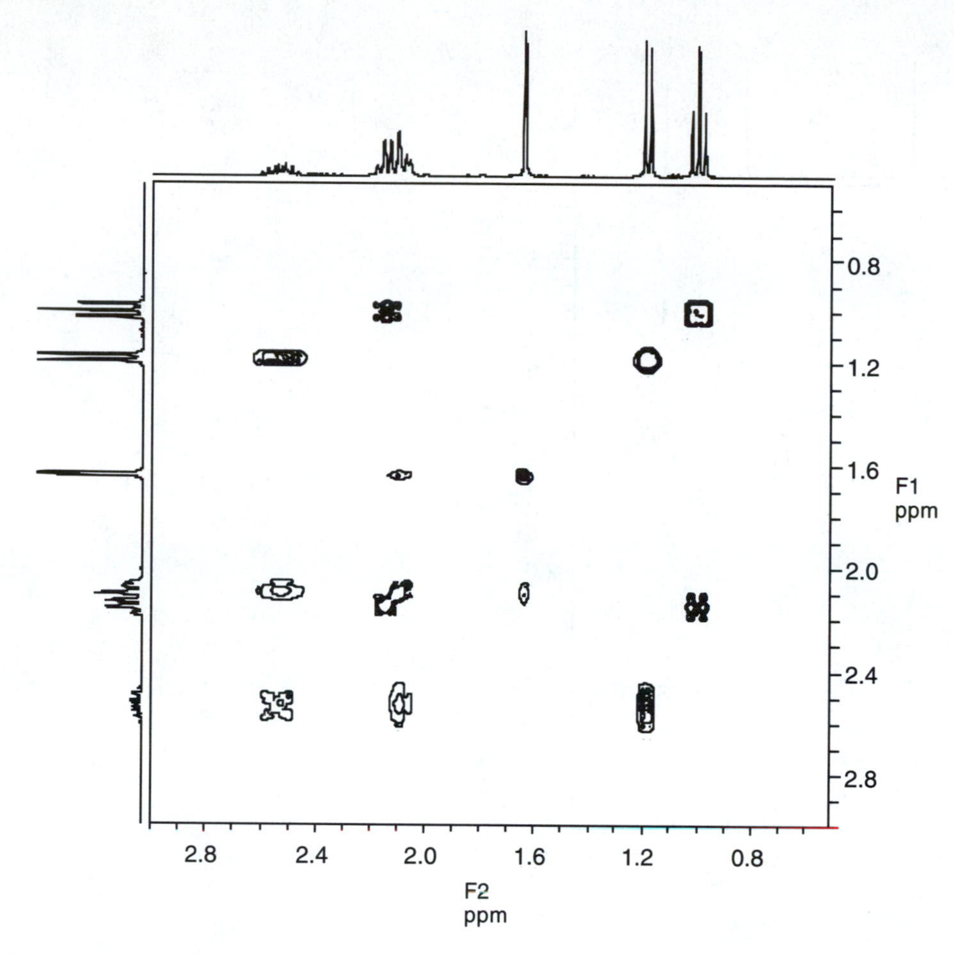

HETCOR

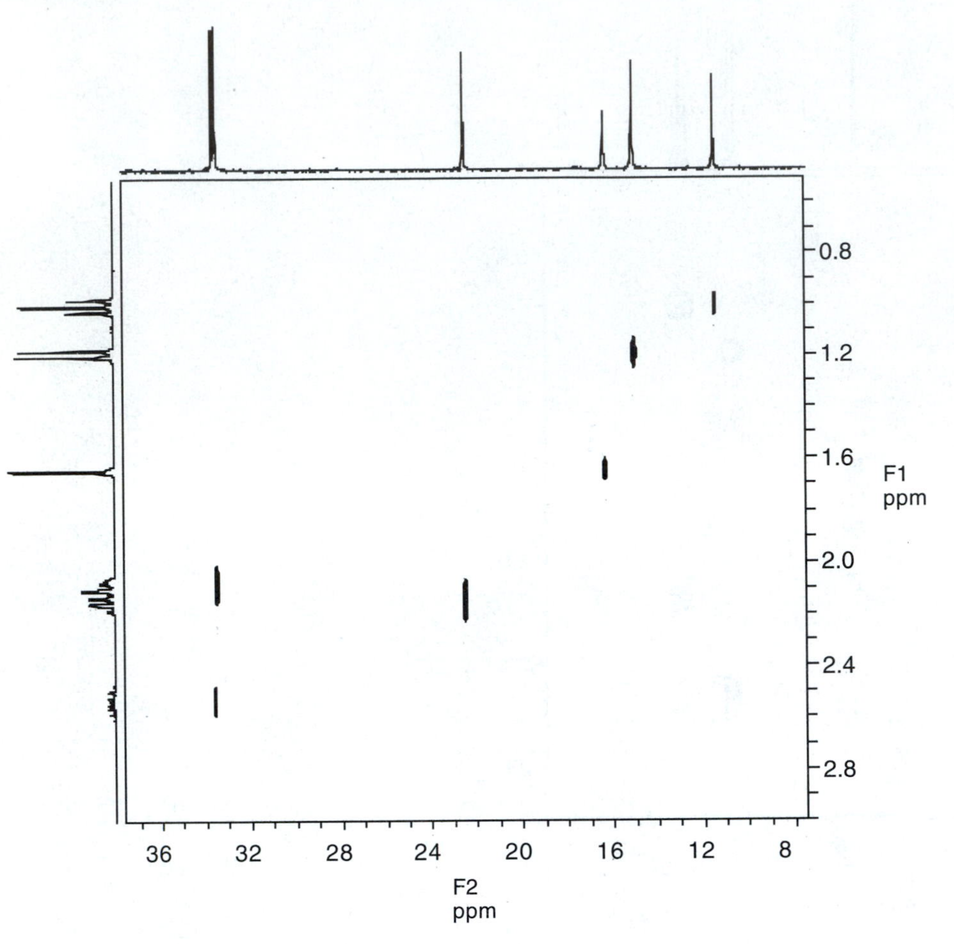
0.8
1.2
1.6
2.0
2.4
2.8
F1
ppm
36
32
28
24
20
16
12
8
F2
ppm

HMBC

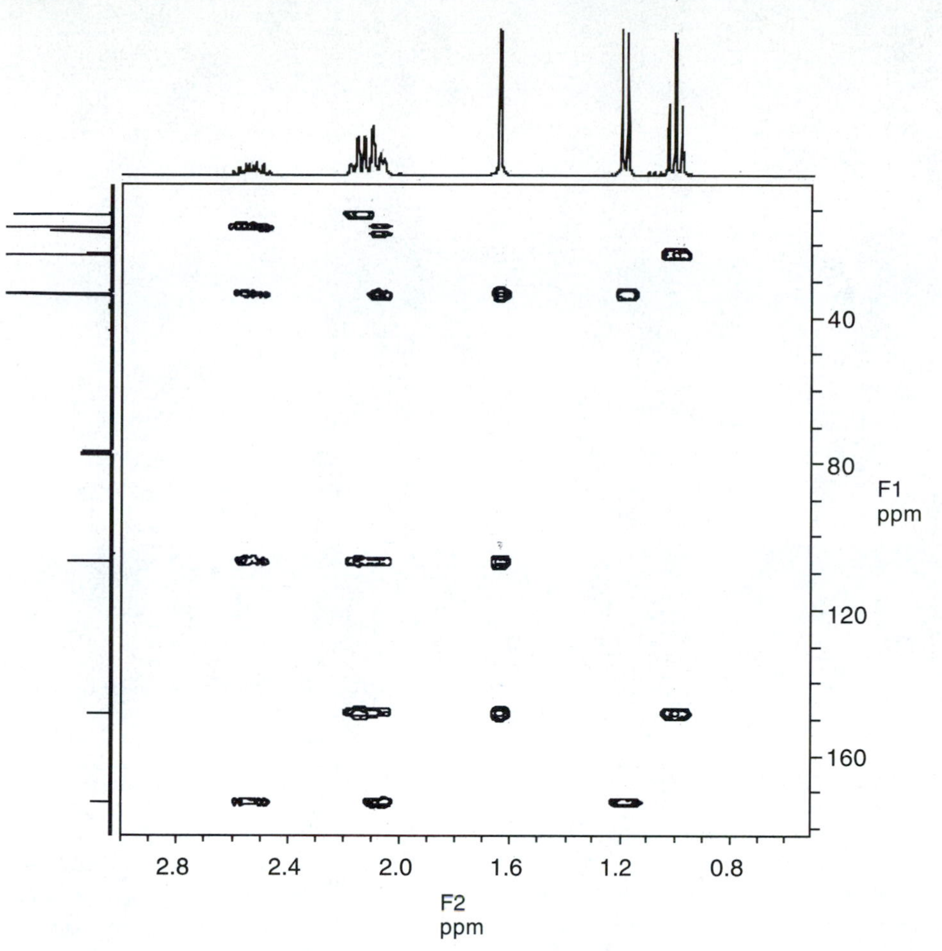

Index